SALINADW23D

JAGUAR

Parts Catalogue

XJ-S 3.6 & 5.3 Litre

RTC9900CA
January 1987 on

Whilst every effort is made to ensure the accuracy of the particulars contained in this Manual the Manufacturing Companies will not, in any circumstances, be held liable for any inaccuracies or the consequences thereof.

Published by
Jaguar Cars Limited

JAGUAR

Replacement parts that are not of genuine Jaguar supply cannot be relied upon to be of correct specification, material, and workmanship and therefore Jaguar cannot be expected to extend Warranty to vehicles which have been fitted with parts not supplied by them.

Jaguar reserves the right to make changes or improvements in the construction or specification of their products at any time.

IN ORDER TO ENSURE PROMPT AND CORRECT EXECUTION OF YOUR ORDERS IT IS ESSENTIAL TO QUOTE:-

1. The relative Unit Number, together with prefixes, suffixes and specification reference where applicable.

   eg. Axle, Body, Vehicle Identification, Engine, Gearbox Numbers etc.

2. The Part Number and description (prefix and suffix letters to Part Numbers must always be quoted).

3. Always state the colour of carpets, painted items, trim items etc., using the colour code from the Vehicle Identification number (VIN) plate and the colour combination key to identify the colour name.

4. The quantity required bearing in mind that the quantities shown are those used in the application actually illustrated. Sometimes only minimum packaged quantities are available. These packs may well contain more parts than are required for a single job; for details of packaged quantities, see the Service Parts List.

THE LISTING OR ILLUSTRATING OF AN ITEM DOES NOT GUARANTEE ITS AVAILABILITY

GROUP A INTRODUCTION

ABBREVIATIONS: Standard abbreviations are employed. The complete range of abbreviations covering all lists is given below.

A/R	As required
(B)	Body number
C	Chassis number
cm	Centimetres
(CN)	Commission number
D/E	Double-ended
dia	Diameter
(E)	Engine number.
E.C.E.	Economic Commission of Europe
(G)	Gearbox number
G.V.W.	Gross Vehicle Weight
HC	High compression
ID	Inside diameter
KPH	Kilometers per hour
LC	Low compression
LH	Left hand
LHD	Left-hand drive
(LP)	Low pressure
LWB	Long wheelbase
m	metres
mm	millimetres
MPH	Miles per hour
N L A	No longer available
No.	Number
NSP	Non-serviceable part
OD	Outside diameter
O/S	Oversize
pr	Pair
psi	Pounds square/inch
RH	Right-hand
RHD	Right-hand drive
STD	Standard
SWB	Short wheelbase
UK	Great Britain and Northern Ireland
U/S	Undersize
U.S.A.	United States of America
(v)	Vehicle identification Number (VIN)
WB	Wheelbase
W.S.E.	When stock exhausted
(+)	Change point not available
s	This item is subject to safety regulations

PART NUMBER SUFFIX LETTERS

E	Exchange scheme item
M	Part supplied in multiples of metres (39.37")
N	New condition (ie not exchange)
R	Reconditioned unit

GROUP A

V.I.N Plate Location

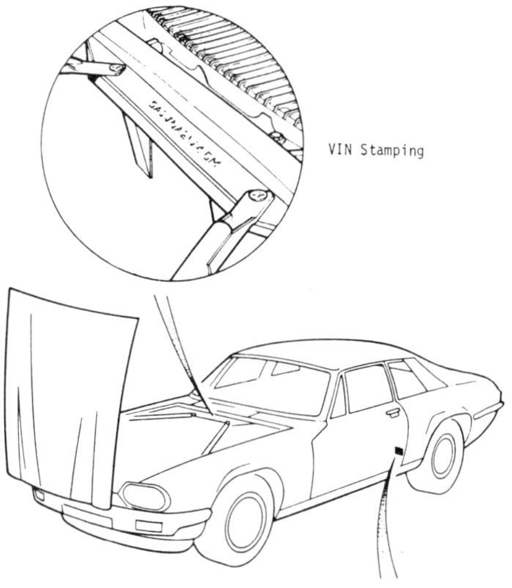

VIN Stamping

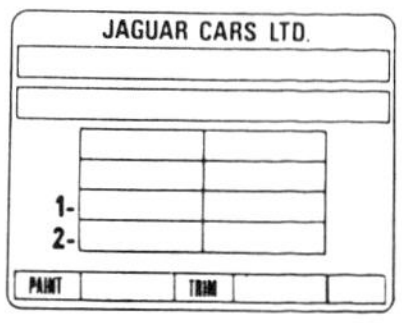

VIN Plate

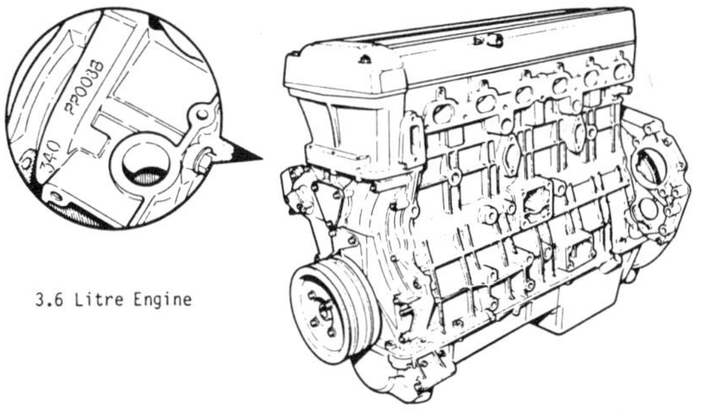

3.6 Litre Engine

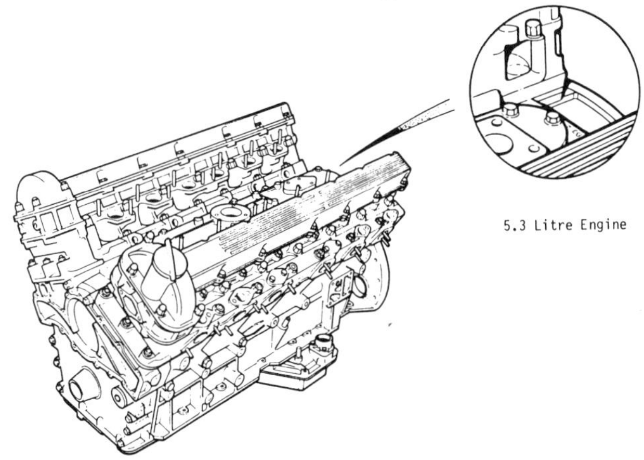

5.3 Litre Engine

GROUP A INTRODUCTION

VEHICLE IDENTIFICATION NUMBERING SYSTEM

The Vehicle Identification number comprisies of a series of letters and numbers presenting, in code such information as Marque, Model Range, Class, Engine and Emission type etc. etc. The code for the Jaguar 2.9 & 3.6 litre range is as follows;

Prefix 1 to 4
World Manufacture Identifier.
SAJJ Jaguar

Prefix 6
Class
A Base line
J Japanesse Spec
L Canadian Spec
V USA Spec

Prefix 9
Transmission and Steering
3 Auto RHS
4 Auto LHS
7 Manual RHS
8 Manual LHS

Prefix 11
Emission Control System
A UK/Eur. Emm B.
B ROW Emm.F.
C Middle East Emm.D.
D Germany Emm.D.
E 3.61 Canada Emm.E.
F Austratia Emm.D.
G 3.61 Aus/Germany/Switzerland Emm.A.
H 3.61 Can/J/Swed/USA Emm.A.

Prefix 5
Model Range.
N=XJS

Prefix 7
Body Variant.
C=Cabriolet
E=Coupe

Prefix 8
Engine Variant
C=3.6 Litre
W=5.3 Litre

Prefix 10
Model Change.
H Series 1 (1987 Models)

Prefix 12 to 17
Serial Number.
100001 on

GROUP A INTRODUCTION

MAJOR UNIT NUMBERING SYSTEM

ENGINE NUMBERING

3.6 LITRE

The 3.6 litre range utilises a 12 character code, similar to the VIN System and contains information such as engine size, compression ratio and emission levels etc. The Codes used are listed below:

Prifix	Description	Code Used
1	Marque	9=Jaguar
2	Engine Size	D=3.6 Litre
3	Fuel	P=Petrol
4	Compression Ratio	A=High
		B=Standard
5	Emission Specification	L=USA/J/Sweden/Australia/ Germany/Switzerland
		M=European
		P=Australia/Middle East
		R=Canada
		S=ROW
6	Speicification Change	E=1987 Model Year
7 to 12	Serial Number	100001 onwards

5.3 Litre

The 5.3 litre range engine numbers commence with the prefix 8S with a sequential number suffixed by one of the following:

HA=High Compression with grade A pistons & liners.

HB=High Compression with grade B pistons & liners

SA=Standard Compression with grade A pistons & liners.

SB=Standard Compression with grade B pistons & liners.

eg. 8S12345HA.

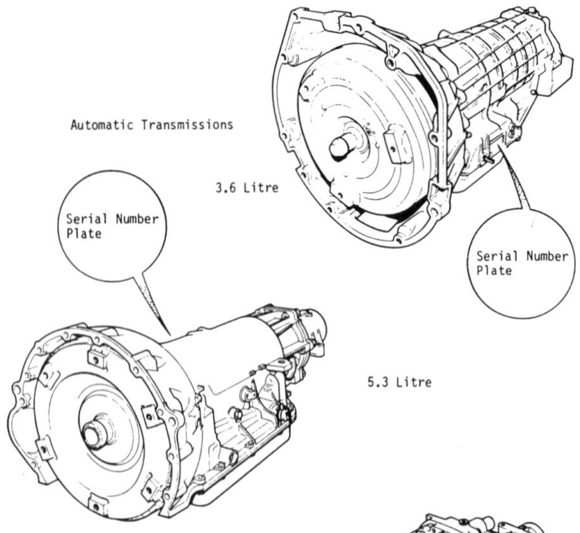

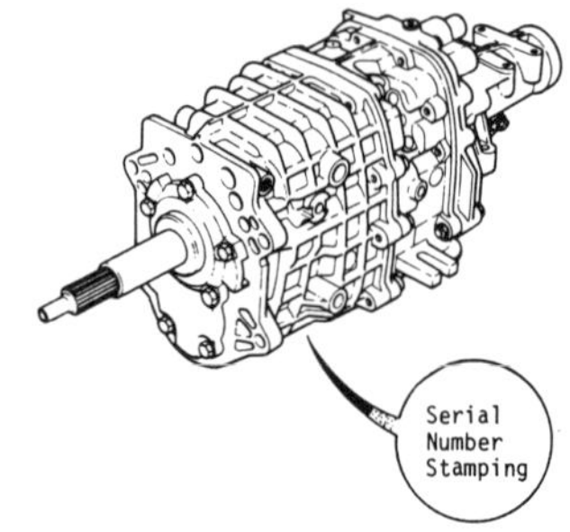

GROUP A INTRODUCTION

MAJOR UNIT NUMBERING SYSTEMS

Manual Gearbox

The serial number for the Getrag manual gearbox is located on the front lower edge and will be consecutively numbered. eg. 004221.

Automatic Gearbox

3.6 litre.

The serial number for the ZG Automatic gearbox is located on the left hand side of the transmission case and will be consecutively numbered. eg. 470216.

5.3 litre.

The serial number for the General Motors automatic gearbox is located on the right hand side of the transmission case and will be conecutively numbered prefixed with 86ZVA. eg. 86ZVA1234.

Final Drive Units

The final drive unit serial number will be located on the output shaft side of the main diff. case assembly and will be consecutively numbered.

Three other identification marks to look for are:

a. Part Number

b. Axle Ratio

c. Power Lok (Optional)

These are stamped on metal tags and secured to the back cover of the unit by means of the cover bolts. eg 86PO6563.

Final Drive Unit

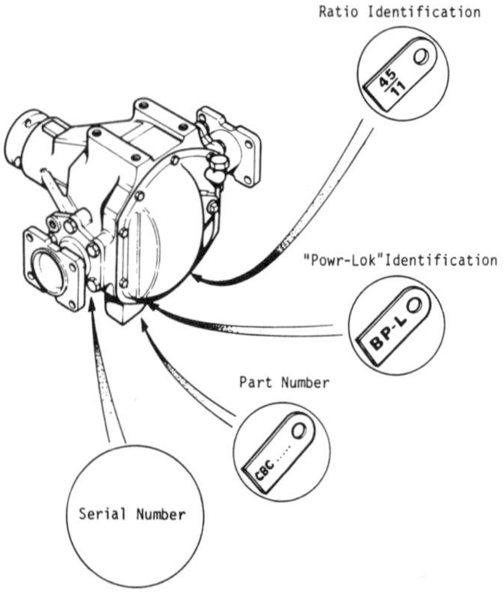

GROUP A INTRODUCTION

BODY COLOUR RANGE

BODY COLOUR SOLID	JBC No.	PAINT CODE	UCCS No.
NIMBUS WHITE	700	NDJ	JLM9055
GRENADIER RED	332	CEH	JLM9056
JAGUAR RACING GREEN	701	HEN	JLM9057
WESTMINSTER BLUE	712	JFG	JLM9058
BLACK	333	PDH	JLM9059
METALLIC			
BORDEAUX RED	340	CEK	JLM9060
CRIMSON	714	CEV	JLM9061
SOLENT BLUE	715	JEW	JLM9062
ARCTIC BLUE	337	JFE	JLM9063
TUNGSTEN	781	JEX	JLM9064
TALISMAN SILVER	336	MDF	JLM9065
DORCHESTER GREY	342	LDP	JLM9066
SOVEREIGN GOLD	341	GDF	JLM9067
SATIN BEIGE	711	AFV	JLM9068
SILVER BIRCH	716	MDJ	JLM9069
ALPINE GREEN	709	HES	JLM9070
MOORLAND GREEN	717	HET	JLM9071

GROUP A INTRODUCTION

TRIM COLOUR CROSS REFERENCE

COLOUR	MATERIAL	PROD CODE	UCCS CODE
BARLEY	LEATHER	AFW	AW
BISCUIT	LEATHER	ADE	AE
BUCKSKIN	LEATHER	AFR	AR
SAVILE GREY	LEATHER	LDY	LY
ISIS BLUE	LEATHER	JEF	JF
WARM CHARCOAL	LEATHER	LEG	LZ
MULBERRY	LEATHER	CEM	CM
DOESKIN	LEATHER	AEE	XE
MAGNOLIA	LEATHER	AEM	ZM
COTSWOLD TWEED	CLOTH	AFX	AX
CHEVIOT TWEED	CLOTH	LEF	YF
PENNINE TWEED	CLOTH	LEE	YE
CHILTERN TWEED	CLOTH	AFY	AY
RATTAN	CARPET	AEF	XF
MINK	CARPET	ADP	AP
SMOKE GREY	CARPET	LEA	XA
SLATE GREY	CARPET	LDE	LE
WINE RED	CARPET	CEN	CN
LIMESTONE	HEADLINING	NDD	ND

GROUP INTRODUCTION

TRIM COLOUR COMBINATION CHART

LEATHER TRIM

DIRECTIONS FOR USE

1 Establish the basic trim colour of the car. These are the colours which appear in the Jaguar sales and advertising literature..
2 Select the basic trim colour in the left hand colum of the chart.
3 Select the trim required from the description across the top of the chart.
4 By reading across and down the colour for the part can be obtained.

Basic trim Colour	Underscuttle casings, Windscreen finishers, Parcel shelf	Seats	Carpets	Door casings, Pockets, Armrests, Console	Rear quarter casings, pockets armrests	Flange finshers	Headlining, Cantrail Crash rolls, DE Posts	A Posts, Sunvisor, BC Posts, Coat Hook
Barley	LZ	AW	XF	AW	AW	PA	ND	XE
Biscuit	LZ	AE	AP	AE	AE	PA	ND	XE
Isis blue	LZ	JF	LE	JF	JF	PA	ND	XE
Buckskin	LZ	AR	XF	AR	AR	PA	ND	XE
Doeskin	LZ	XE	XF	XE	XE	PA	ND	XE
Magnolia	LZ	ZM	AP	ZM	ZM	PA	MD	XE
Mulberry	LZ	CM	CN	CM	CM	PA	ND	XE
Savile grey	LZ	LY	XA	LY	LY	PA	ND	XE
Warm CharcoaL	LZ	LZ	LE	LZ	LZ	PZ	ND	XE

GROUP A INTRODUCTION

TRIM COLOUR COMBINATION CHART

AMBLA/LEATHER/CLOTH TRIM

DIRECTIONS FOR USE

1 Establish the basic trim colour of the car. These are the colours which appear in the jaguar sales and advertising literature.
2 Select the basic trim colour in the left hand column of the chart.
3 Select the trim required from the description across the top of the chart.
4 By reading across and down the colour for the part can be obtained.

Basic Trim Colour	Underscuttle casings, Windscreen finishers	Seats	Carpets	Door casings, Pockets, Armrest Console	Rear quarter casings, Pockets, Armrests	Flange finishers	Headlining, Cantrail, Crash Roll, DE Posts	A Posts, Sunvisor, BC Posts, Coat hook
Cheviot Tweed	LZ	YF	XA	LY	LY	PA	ND	XE
Chiltern Tweed	LZ	AY	XF	XE	XE	PA	ND	XE
Coltswold Tweed	LZ	AX	XF	AW	AW	PA	ND	XE
Pennine Tweed	LZ	YE	LE	LZ	LZ	PA	ND	XE

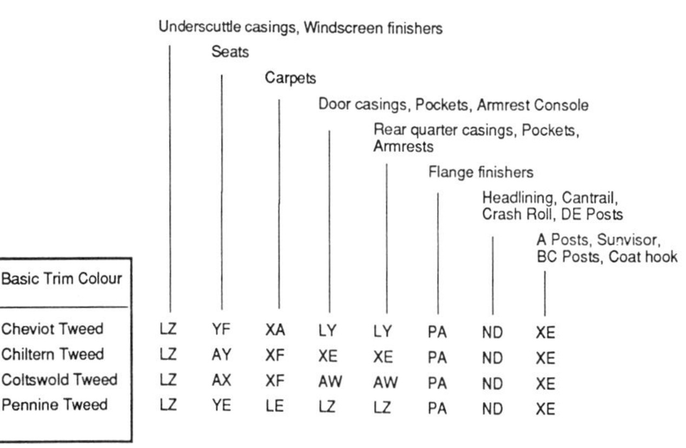

SEATS WITH CONTRASTING PIPING

Seat components trimmed with contrasting piping other than those shown on this fiche, will be special order only and the basic trim/piping combination will have to be quoted at the time of ordering.

JAGUAR SPORT MODELS

Parts which are unique to Jaguar Sport models have been allocated the following pre-fixes:-

SPB – Body Components
SPC – Chassis Components
SPD – Electrical Components
SPE – Engine Components

These parts will not appear on this fiche and any orders should be directed to Jaguar Sport Ltd or one of its nominated dealers.

XJ-S COACHLINE COMBINATION CHART

TRIM COLOUR \ BODY COLOUR	GLACIER WHITE	JET BLACK	WESTMINSTER BLUE	JAGUAR RACING GREEN	GRENADIER RED	SATIN BEIGE	SOLENT BLUE	TALISMAN SILVER	ARTIC BLUE	TUNGSTEN	DORCHESTER GREY	MOORLAND GREEN	BORDEAUX RED	ALPINE GREEN	CRIMSON	SILVER BIRCH	SIGNAL RED
ISIS BLUE	GM	GM	GM	O	GM	G	GM	GM	O	O	GM	O	O	O	O	G	GM
WARM CHARCOAL	GM	GM	GM	GM	GM	O	GM	GM	O	O	GM	GM	GM	GM	GM	GM	GM
MULBERRY	R	R	R	–	–	R	R	R	R	R	R	–	–	–	–	R	–
DOESKIN	O	O	O	O	O	G	O	GM	O	O	O	O	O	O	O	G	O
SAVILE GREY	GM	GM	GM	GM	GM	–	GM	GM	O	O	GM	GM	GM	GM	GM	–	GM
BUCKSKIN	O	O	O	O	O	G	O	GM	O	O	O	O	O	O	O	G	O
MAGNOLIA	O	O	O	O	O	G	O	GM	O	O	O	O	O	O	O	G	O
BARLEY	G	G	G	G	G	G	G	GM	O	O	G	O	G	O	G	G	G
COTSWOLD TWEED	G	G	G	G	G	G	G	GM	O	O	G	O	G	O	G	G	G
CHEVIOT TWEED	GM	GM	GM	GM	GM	–	GM	GM	O	O	GM	GM	GM	GM	GM	–	GM
PENNINE TWEED	GM	GM	GM	GM	GM	O	GM	GM	O	O	GM	GM	GM	GM	GM	GM	GM
CHILTERN TWEED	O	O	O	O	O	G	O	GM	O	O	O	O	O	O	O	G	O
3.6 SPORTSPACK (ALL TRIM COLOURS)	RQ	RB	RQ	RQ	AQ	RQ	RQ	RB	RB	RB	RQ	RQ	RQ	RQ	RQ	RQ	AQ

COACHLINE KEY

GM = GUNMETAL
O = OYSTER
G = GOLDLEAF
R = DARK RED
RB = FLAME RED – UPPER, PEWTER – LOWER
RQ = FLAME RED – UPPER, QUICKSILVER – LOWER
AQ = AMBER – UPPER, QUICKSILVER – LOWER

INTRODUCTION – Country Abbreviations

Index Mark	Office
A	Austria
ADN	Yemen People's Democratic Republic
AFG	Afghanistan
AL	Albania
AND	Andorra
AUS	Australia*
B	Belgium
BD	Bangladesh*
BDS	Barbados*
BG	Bulgaria
BH	Belize
BR	Brazil
BRN	Bahrain
BRU	Brunel*
BS	Bahamas*
BUR	Burma
C	Cuba
CAL	California
CDN	Canada
CH	Switzerland
CI	Ivory Coast
CL	Sri Lanka*
CO	Colombia
CR	Costa Rica
CS	Czechoslovakia
CY	Cyprus*
D	German Federal Republic
DDR	German Democratic Republic
DK	Denmark
DOM	Dominican Republic
DY	Benin
DZ	Algeria
E	Spain
EAK	Kenya*
EAT or EAZ	Tanzania*
EAU	Uganda*
EC	Ecuador
ES	El Salvador
ET	Egypt, Arab Republic of
ETH	Ethiopia
F	France
FJI	Fiji
FL	Liechtenstein
FR	Faroe Islands
GB	United Kingdom of Great Britain & Northern Ireland*
GBA	Alderney* Channel Islands
GBG	Guernsey* Channel Islands
GBJ	Jersey*
GBM	Isle of Man*
GBZ	Gibralter
GCA	Guatemala
GH	Ghana
GR	Greece
GUY	Guyana
H	Hungary
HK	Hong Kong*
HKJ	Jordan
I	Italy
IL	Israel
IND	India*
IR	Iran
IRL	Ireland*
IRQ	Iraq
IS	Iceland
J	Japan
JA	Jamaica
K	Kampuchea
KWT	Kuwait
L	Luxembourg
LAO	Lao People's Democratic Republic
LAR	Libya
LB	Liberia
LS	Lesotho*
M	Malta
MA	Morocco
MAL	Malaysia*
MC	Monaco
MEX	Mexico
MS	Mauritius*
MW	Malawi*
N	Norway
NA	Netherlands Antilles
NIC	Nicaragua
NL	Netherlands
NZ	New Zealand*
P	Portugal
PA	Panama
PAK	Pakistan*
PE	Peru
PL	Poland
PNG	Papua New Guinea*
PY	Paraguay
RA	Argentina
RB	Botswana*
RC	Taiwan
RCA	Central African Republic
RCB	Congo
RCH	Chile
RH	Haiti
RI	Indonesia*
RIM	Maruitania
RL	Lebanon
RM	Madagascar
RMM	Mali
RN	Niger
RO	Romania
ROK	Korea, Republic of
ROU	Uruguay
RP	Philippines
RSM	San Marino
RU	Burundi
RWA	Rwanda
S	Sweden
SD	Swaziland*
SF	Finland
SGP	Singapore*
SME	Suriname*
SN	Senegal
SU	Union of Soviet Socialist Republics
SWA or ZA	Namibia
SY	Seychelles
SYR	Syria
T	Thailand
TG	Togo
TN	Tunisia
TR	Turkey
TT	Trinidad & Tobago*
USA	United States of America
V	Vatican City
VN	Vietnam
WAG	Gambia
WAL	Sierra Leone
WAN	Nigeria
WD	Dominica*
WG	Grenada* Windward Islands
WL	St Lucia* Windward Islands
WS	Western Samoa
WV	St Vincent & the Grenadine
YU	Yugoslavia
YV	Venezuela
Z	Zambia
ZA	South Africa*
ZRE	Zaire
ZW	Zimbabwe*

NOTE: * In countries marked with an asterix the rule of the road is drive on the left; otherwise drive on the right.

JAGUAR XJS RANGE (JAN 1987 ON) — C01 fiche 1 — ENGINE

MASTER INDEX

Section	Ref
ENGINE / MOTEUR / MOTOR	1.C
FLYWHEEL AND CLUTCH / VOLANT ET EMBRAYAGE / SCHWUNGRAD UND KUPPLUNG	1.G
GEARBOX AND PROPSHAFT / BOITE DE VITESSES ET CARDAN / GETRIEBE UND KURDUMWELLE	1.H
AXLES, SUSPENSION, DRIVE SHAFTS, WHEELS / AXE, SUSPENSION, ARBRE DE COMMANDE, ROUES / ACHSE, AUFHANGUNG,ANTRIEBSWELLE, RADER	1.K
STEERING / DIRECTION / LENKUNG	1.M
BRAKES AND BRAKE CONTROLS / FREINS ET COMMANDES / BREMSEN UND BEDIENUNGSORGANE	1.N
FUEL, EXHAUST AND EMISSION SYSTEMS / ALIMENTATION ET ECHAPPEMENT / KRAFTSTOFF UND AUSPUFFANLAGE	2.C
COOLING, HEATING AND AIR CONDITIONING / REFROIDISSEMENT, CHAUFFAGE, AERATION / KUHLUNG, HEIZUNG, LUFTUNG	2.G
ELECTRICAL, WASHERS-WIPERS, INSTRUMENTS / ELECTRICITE, ESSUIE GLACE, INSTRUMENTS / ELELTRISCHE, SCHEIBENWISCHER, INSTRUMENTE	2.J
CHASSIS, SUBFRAMES, BODYSHELL, FITTINGS / CHASSIS, SOUBASSEMENT, CABINE / CHASSIS TEILE, HILFSRAHMEN, KABINE	3.C
FASCIA, TRIM, SEATS AND FIXINGS / TABLEAU DE BORD, GARNITURES INTERIEURE, SIEGES / ARMATURENBRETT, POLSTERUNG UND SITZ	3.H
ACCESSORIES AND PAINT / ACCESSOIRES ET PEINTURE / ZUBEHOREN UND FARBE	4.C
NUMERICAL INDEX / INDEX NUMERIQUE / NUMMERNVERZEICHNIS	4.D

GROUP INDEX

ENGINE 3.6 LITRE C02
CAM.-TAPPET BLOCK - 5.3 LITRE E04
CAMSHAFT & VALVES C12
CRANKSHAFT AND MAIN BEARINGS C05
CRANKSHAFT-MAIN BEAR.-5.3 LITRE D13
CYL. HEAD & VALVES - 5.3 LITRE E03
CYLINDER BLOCK D10
CYLINDER HEAD & GASKET C11
DRIVING BELTS-3.6 LITRE D05
EXHAUST MANIFOLD - 3.6 LITRE D03
EXHAUST MANIFOLDS - 5.3 LITRE F02
FRONT ENGINE MOUNTING 3.6 LITRE D08
FRONT ENGINE MOUNTING-5.3 LITRE F09
GASKET SETS - 3.6 LITRE D06
INDUCTION MANIFOLD-3.6 LITRE D02
INLET MANIFOLD-5.3 LITRE E17
LOWER ENG GASKET SET-5.3 LITRE F07
OIL COOLER & OIL PIPES 3.6 LITRE D07
OIL COOLER - 5.3 LITRE F04
OIL COOLER PIPES - 5.3 LITRE F05
OIL FILTER,FILLER CAP-3.6 LITRE C15
OIL FILTER-5.3 LITRE E08
OIL PUMP - 5.3 LITRE D15
OIL PUMP-3.6 LITRE C07
OIL SUMP & DIPSTICK - 5.3 LITRE D12
OIL SUMP & DIPSTICK 3.6 LITRE C04
PISTON & CONN RODS C06
PISTONS & CONN. RODS D14
UPPER ENG GASKET SET-5.3 LITRE F08
WATER PUMP C16
WATER PUMP - 5.3 LITRE E10

JAGUAR XJS RANGE (JAN 1987 ON)	C02 fiche 1	ENGINE

Illus	Part Number	Description	Quantity	Change Point	Remarks
		ENGINE 3.6 LITRE. ENGINE UNIT			
1	JLM 1290 N	STRIPPED ENGINE(HIGH COMP)	1		80 amp alternator
1	JLM 1338 N	STRIPPED ENGINE(HIGH COMP)	1	Up to Engine No. 128810	90 amp alternator
	JLM 1338 E	STRIPPED ENGINE(HIGH COMP)	1	Up to Engine No. 128810	90 amp alternator
1	JLM 1339 N	STRIPPED ENGINE (STD COMP)	1	Up to Engine No. 128810	90 amp alternator
	JLM 1339 E	STRIPPED ENGINE (STD COMP)	1	Up to Engine No. 128810	90 amp alternator
1	JLM 1564 N	STRIPPED ENGINE (HIGH COMP)	1	From Engine No. 128811	
1	JLM 1565 N	STRIPPED ENGINE (STD COMP)	1	From Engine No. 128811	

1C 02

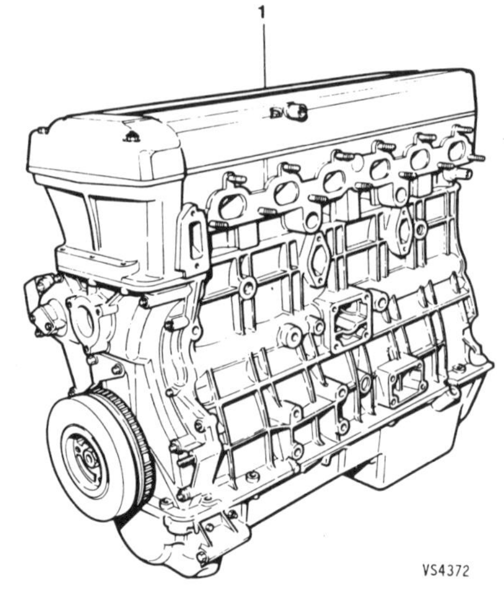

VS 4372

JAGUAR XJS RANGE (JAN 1987 ON)	C03 fiche 1	ENGINE

Illus	Part Number	Description	Quantity	Change Point	Remarks
		ENGINE 3.6 LITRE-continued. CYLINDER BLOCK-REAR HOUSING			
1	EAC 7002	CYLINDER BLOCK ASSEMBLY	1		
2	EAC 2522	Bush	1		
3	C 2373	Dowel-timing cover	1		
4	EAC 4119	Bolt-bearing caps	14		
5	EAC 3972	Drain plug	1		
6	EAC 3964	Washer	NLA		Use C2296/24
6	C 2296 24	Washer	1		
7	EAC 7769	Plug	1		
8	C 2296 23	Sealing washer	1		
9	EAC 2494	Plug-oil gallery	2		
10	EAC 2495	'O' ring	2		
11	SH 506051 J	Setscrew-oil gallery blanking	1		
12	C 2296 4	Sealing washer	1		
13	C 2373	Dowel-cylinder head	2		
14	EAC 4974	Rear oil seal	1		
15	EAC 1864	Housing-oil seal	1		
16	FS 108301 J	Setscrew-housing to block	6		
17	C 2373	Dowel	2		
18	EAC 4451	Windage tray	2		
19	FS 108201 J	Setscrew-tray to block	8		

B4AK12/10/88

1C 03

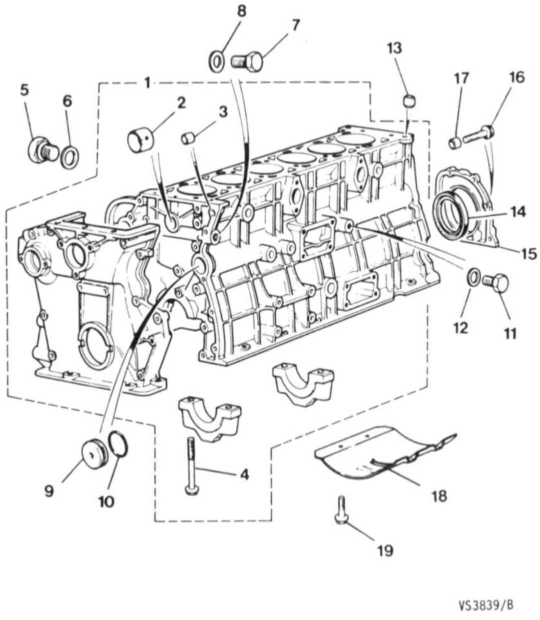

VS3839B

MASTER INDEX

ENGINE 1.C
FLYWHEEL AND CLUTCH 1.G
GEARBOX AND PROPSHAFT 1.H
AXLES, SUSPENSION, DRIVE SHAFTS, WHEELS 1.K
STEERING 1.M
BRAKES AND BRAKE CONTROLS 1.N
FUEL, EXHAUST AND EMISSION SYSTEMS 2.C
COOLING, HEATING AND AIR CONDITIONING 2.G
ELECTRICAL, WASHERS-WIPERS, INSTRUMENTS 2.J
CHASSIS, SUBFRAMES, BODYSHELL, FITTINGS 3.C
FASCIA, TRIM, SEATS AND FIXINGS 3.H
ACCESSORIES AND PAINT 4.C
NUMERICAL INDEX 4.D

GROUP INDEX

ENGINE 3.6 LITRE C02
CAM.-TAPPET BLOCK - 5.3 LITRE E04
CAMSHAFT & VALVES C12
CRANKSHAFT AND MAIN BEARINGS C05
CRANKSHAFT-MAIN BEAR.-5.3 LITRE D13
CYL. HEAD & VALVES - 5.3 LITRE E03
CYLINDER BLOCK D10
CYLINDER HEAD & GASKET C11
DRIVING BELTS-3.6 LITRE D05
EXHAUST MANIFOLD - 3.6 LITRE D03
EXHAUST MANIFOLDS - 5.3 LITRE F02
FRONT ENGINE MOUNTING 3.6 LITRE D08
FRONT ENGINE MOUNTING-5.3 LITRE F09
GASKET SETS - 3.6 LITRE D06
INDUCTION MANIFOLD-3.6 LITRE D02
INLET MANIFOLD-5.3 LITRE E17
LOWER ENG GASKET SET-5.3 LITRE F07
OIL COOLER & OIL PIPES 3.6 LITRE D07
OIL COOLER - 5.3 LITRE F04
OIL COOLER PIPES - 5.3 LITRE F05
OIL FILTER,FILLER CAP-3.6 LITRE C15
OIL FILTER-5.3 LITRE E08
OIL PUMP - 5.3 LITRE D15
OIL PUMP-3.6 LITRE C07
OIL SUMP & DIPSTICK - 5.3 LITRE D12
OIL SUMP & DIPSTICK 3.6 LITRE C04
PISTON & CONN RODS C06
PISTONS & CONN. RODS D14
UPPER ENG GASKET SET-5.3 LITRE F08
WATER PUMP C16
WATER PUMP - 5.3 LITRE E10

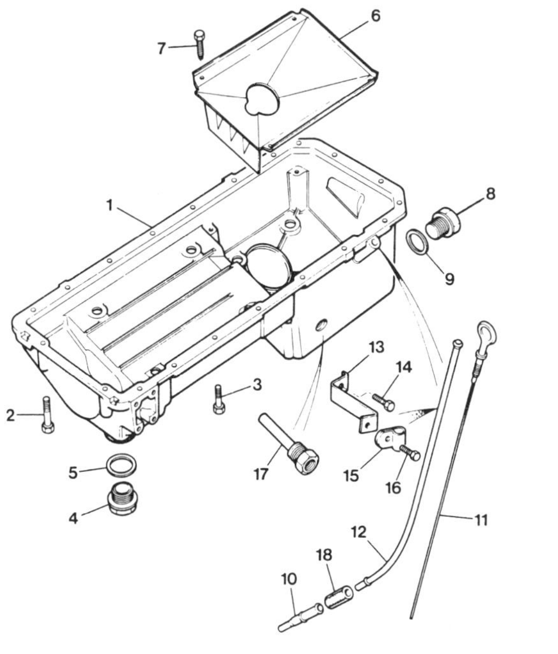

VS4477/B

VS4477/B

ENGINE 3.6 LITRE-continued.

OIL SUMP & DIPSTICK 3.6 LITRE

Illus	Part Number	123456 Description	Quantity	Change Point	Remarks
1	EAC 7436	Sump	1		
2	FB 108121 J	Bolt-sump to timing cover	2		
3	FS 108251 J	Bolt-sump to block	18		
4	EAC 5221	Access plug-oil relief valve	1		
5	C 2296 22	Washer	1		
6	EAC 5164	Baffle tray	1		Use EAC 7353
6	EAC 7353	Baffle tray	1		
7	FS 108161 J	Setscrew-tray to sump	4		
8	EAC 3972	Drain plug	1		
9	EAC 3964	Washer	NLA		Use C2296/24
9	C 2296 24	Washer	1		
10	EAC 2258	Location tube	1		
17	EAC 7428	Adaptor-oil drain pipe	1		
11	EAC 9276	Dipstick	1		
12	EAC 8958	Tube-dipstick	1		
13	EAC 4808	Bracket	1		
14	FS 106161 J	Setscrew	1		
15	C 44230	Clip	1		
16	C 35093 1	Screw	1		
18	EBC 2084	Sleeve - diptick tube	1		

B6AK12/10/88

MASTER INDEX

GROUP INDEX

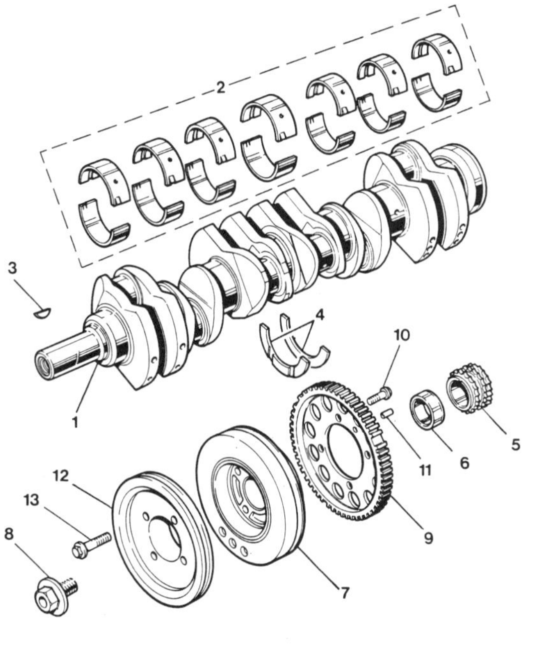

VS 4359A

VS4359/A

ENGINE 3.6 LITRE-continued.

CRANKSHAFT AND MAIN BEARINGS

Illus	Part Number	123456 Description	Quantity	Change Point	Remarks
1	EAC 7730	Crankshaft	1	Upto Eng No 128810	
1	EBC 1147	Crankshaft	1	From Eng No 128811	
2	JLM 764 S	Mainbearing kit-STD	1		
2	JLM 764 20	Main bearing kit-0.020"	1		
3	CAM 1978 J	Woodruff key	2		
4	C 34697	Thrust washer	2		
4	C 34697 1	Thrust washer	2		Oversize
5	EAC 2456	Sprocket	1		
6	EAC 2464	Distance piece	1		
7	EAC 3810	Damper	1		80 amp alternator
7	EAC 7005	Damper	1		90 amp alternator
8	C 32367 1	Bolt-damper to crankshaft	1		
9	EAC 6850	Timing ring	1		
10	EAC 7291	Bolt-timing ring to damper	4		
11	PA 108121 J	Roll pin	1		
12	EAC 4056	Crankshaft pulley	1		
13	FB 108091 J	Bolt-pulley to damper	4		
		CRANK BALANCE WEIGHTS			
	EAC 9921 A	4.0 gramme	A/R	Up to Engine No. 128810	
	EAC 9921 I	4.8 gramme	A/R		
	EAC 9921 B	5.3 gramme	A/R		
	EAC 9921 C	7.9 gramme	A/R		
	EAC 9921 D	10.5 gramme	A/R		
	EAC 9921 E	14.5 gramme	A/R		
	EAC 9921 F	17.1 gramme	A/R		
	EAC 9921 G	22.8 gramme	A/R		
	EAC 9921 H	28.1 gramme	A/R		

MASTER INDEX

GROUP INDEX

JAGUAR XJS RANGE (JAN 1987 ON) | C06 fiche 1 | ENGINE

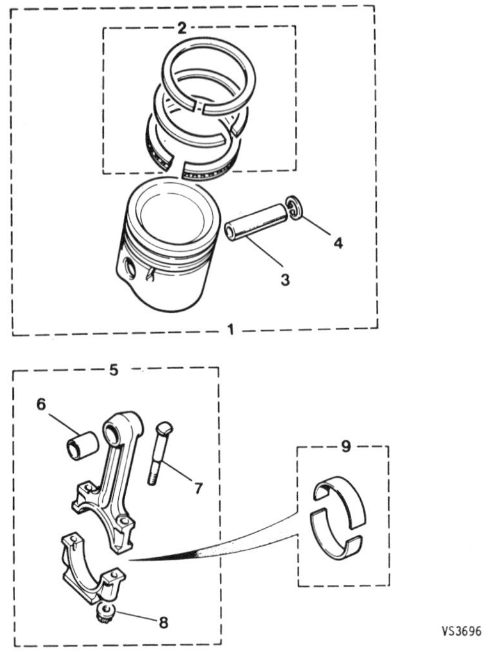

VS 3696

ENGINE 3.6 LITRE-continued.
PISTON & CONN RODS

Illus	Part Number	Description	Quantity	Change Point	Remarks
1	EAC 8744 S	PISTON-ASSY (HIGH COMPRESSION)	6	Upto VIN144343	
1	EAC 8744 20	PISTON-ASSY (HIGH COMPRESSION)	6		
2	JLM 581 S	Piston ring kit	1		
2	JLM 581 20	Piston ring kit	1		
3	EAC 4431	Gudgeon pin	1		
4	C 29699	Circlip	2		
1	EBC 1288 S	PISTON-ASSY (HIGH COMPRESSION)	6	From VIN144344	
1	EBC 1288 20	PISTON-ASSY (HIGH COMPRESSION)	6		
2	JLM 581 S	Piston ring kit	1		
2	JLM 581 20	Piston ring kit	1		
3	EAC 1620	Gudgeon pin	1		
4	C 29699	Circlip	2		
1	EAC 8745 S	PISTON ASSY (STD COMP)	6		
1	EAC 8746 20	PISTON ASSY (STD COMP)	6		
2	JLM 581 S	Piston ring kit	1		
2	JLM 581 20	Piston ring kit	1		
3	EAC 4431	Gudgeon pin	1		
4	C 29699	Circlip	2		
5	RTC 2996	CONNECTING ROD ASSY	1set		
6	EAC 2732	Bearing	6		
7	EAC 1027	Bolt	12		
8	EAC 5541	Nut	12		
9	RTC 2854 S	BIG END BEARING KIT	1SET		
	RTC 2854 20	BIG END BEARING KIT	1SET		

B10AK24/6/88

MASTER INDEX

GROUP INDEX

JAGUAR XJS RANGE (JAN 1987 ON) | C07 fiche 1 | ENGINE

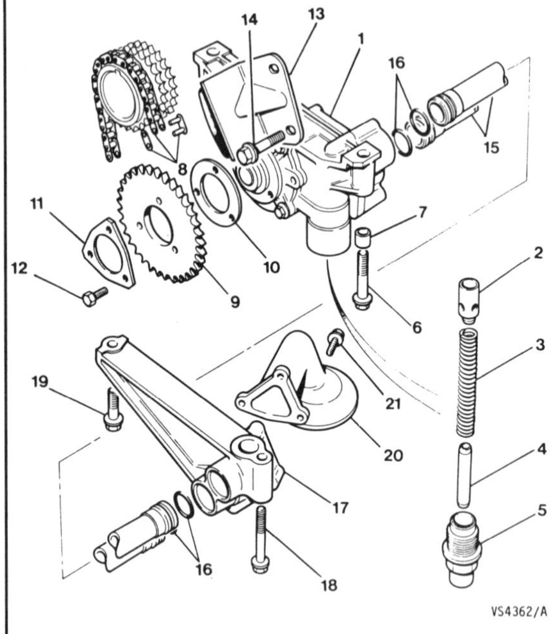

VS 4362A

ENGINE 3.6 LITRE-continued
OIL PUMP-3.6 LITRE

Illus	Part Number	Description	Quantity	Change Point	Remarks
1	EAC 7398	OIL PUMP ASSEMBLY	1		
2	EAC 4328	Plug-oil relief valve	1		
3	EAC 4327	Support mandrel	1		
4	EAC 4380	Spring-oil relief valve	1		
5	EAC 4326	Plunger-oil relief valve	1		
6	FB 108121 J	Bolt-pump to cyl block	1		
	FB 108101 J	Bolt-pump to cyl block	1		
7	C 2373	Dowel	2		
8	JLM 1279	Oil pump chain	1		
9	EAC 2577	Sprocket-oil pump	1		
10	EAC 4329 1	Shim(0.005")	A/R		
10	EAC 4329 2	Shim(0.010")	A/R		
10	EAC 4329 3	Shim(0.020")	A/R		
11	SH 106121 J	Setscrew-securing sprocket	3		
12	EAC 4505	Tabwasher	1		
13	EAC 6017	Tensioner blade	1		
14	FB 108201 J	Bolt	2		
15	EAC 3793	Oil pipe	2		
16	EAC 5536	'O' ring(oil pipes)	4		
17	EAC 5674	Oil pipe carrier	1		
18	FB 108061 J	Bolt-securing carrier	1		
19	FB 108121 J	Bolt-securing carrier	1		
20	EAC 4504	Oil suction pickup	1	Up to (E)164383	
20	EAC 7355	Oil suction pickup	1	From (E)164384	
21	FS 106201 J	Setscrew(pickup to carrier)	3		

B12AK23/9/88

MASTER INDEX

ENGINE 1.C
FLYWHEEL AND CLUTCH 1.G
GEARBOX AND PROPSHAFT 1.H
AXLES, SUSPENSION, DRIVE SHAFTS, WHEELS 1.K
STEERING 1.M
BRAKES AND BRAKE CONTROLS 1.N
FUEL, EXHAUST AND EMISSION SYSTEMS 2.C
COOLING, HEATING AND AIR CONDITIONING 2.G
ELECTRICAL, WASHERS-WIPERS, INSTRUMENTS 2.J
CHASSIS, SUBFRAMES, BODYSHELL, FITTINGS 3.C
FASCIA, TRIM, SEATS AND FIXINGS 3.H
ACCESSORIES AND PAINT 4.C
NUMERICAL INDEX 4.D

GROUP INDEX

ENGINE 3.6 LITRE C02
CAM.-TAPPET BLOCK - 5.3 LITRE E04
CAMSHAFT & VALVES C12
CRANKSHAFT AND MAIN BEARINGS C05
CRANKSHAFT-MAIN BEAR.-5.3 LITRE D13
CYL. HEAD & VALVES - 5.3 LITRE E03
CYLINDER BLOCK D10
CYLINDER HEAD & GASKET C11
DRIVING BELTS-3.6 LITRE D05
EXHAUST MANIFOLD - 3.6 LITRE D03
EXHAUST MANIFOLDS - 5.3 LITRE F02
FRONT ENGINE MOUNTING 3.6 LITRE D08
FRONT ENGINE MOUNTING-5.3 LITRE F09
GASKET SETS - 3.6 LITRE D06
INDUCTION MANIFOLD-3.6 LITRE D02
INLET MANIFOLD-5.3 LITRE E17
LOWER ENG GASKET SET-5.3 LITRE F07
OIL COOLER & OIL PIPES 3.6 LITRE D07
OIL COOLER - 5.3 LITRE F04
OIL COOLER PIPES - 5.3 LITRE F05
OIL FILTER,FILLER CAP-3.6 LITRE C15
OIL FILTER-5.3 LITRE E08
OIL PUMP - 5.3 LITRE D15
OIL PUMP-3.6 LITRE C07
OIL SUMP & DIPSTICK - 5.3 LITRE D12
OIL SUMP & DIPSTICK 3.6 LITRE C04
PISTON & CONN RODS C06
PISTONS & CONN. RODS D14
UPPER ENG GASKET SET-5.3 LITRE F08
WATER PUMP C16
WATER PUMP - 5.3 LITRE E10

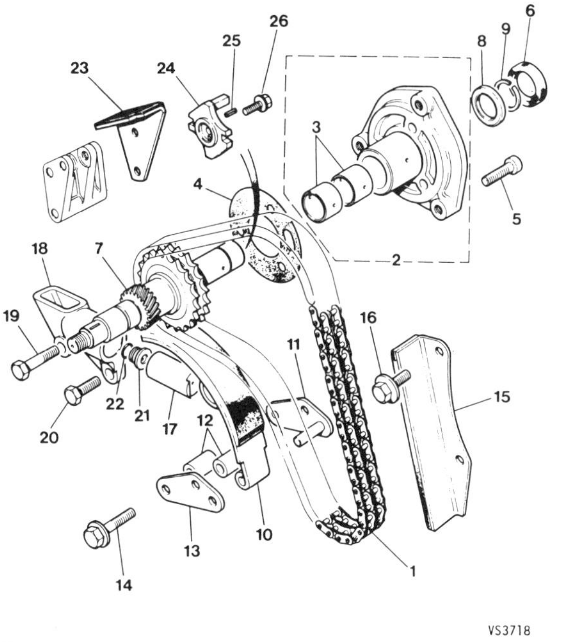

VS3718

ENGINE 3.6 LITRE-continued.
AUXILIARY SHAFT-PRIMARY CHAIN

Illus	Part Number	123456 Description	Quantity	Change Point	Remarks
1	EAC 4519	Primary timing chain	1		
2	EAC 3921	Housing-auxiliary shaft	1		
3	EAC 2522	Bush	2		
4	EAC 4038	Gasket	1		
5	EAC 8367	Capscrew	3		
6	EAC 7263	Oil seal	NLA	Upto 9DPAMA 123705	Use EAC 6156
6	EAC 6156	Oil seal	1	From 9DPAMA 123706	
7	EAC 6741	Auxiliary shaft assembly	1		
24	EAC 7080	Driving dog	1		
25	EAC 7643	Spirol pin	2		
26	FT 108255 J	Setscrew	1	....	Use EBC 1223
26	EBC 1223	Setscrew	1		
8	EAC 3455	Thrust washer	1		
9	CR 120225 J	Circlip	1		
10	EAC 4527	Chain tensioner assembly	1		
11	EAC 4528	Pivot plate	1		
12	EAC 4531	Distance piece	2		
13	EAC 4532	Support plate-pivot pin	1		
14	FB 108091 J	Bolt-securing plate	2		
15	EAC 4533	Chain damper	1		
16	AGU 1168	Setscrew-securing damper	2		
17	EAC 5317	Piston assembly tensioner	1		
18	EAC 6878	Housing piston	1		
19	FB 108901 J	Bolt-securing housing	1		
20	FB 108061 J	Setscrew-securing housing	1		
21	EAC 4647	Non return valve assembly	1		
22	EAC 6779	'O' ring non return valve	1		
23	EAC 6766	Upper damper-primary chain	1		

1C 08

MASTER INDEX

GROUP INDEX

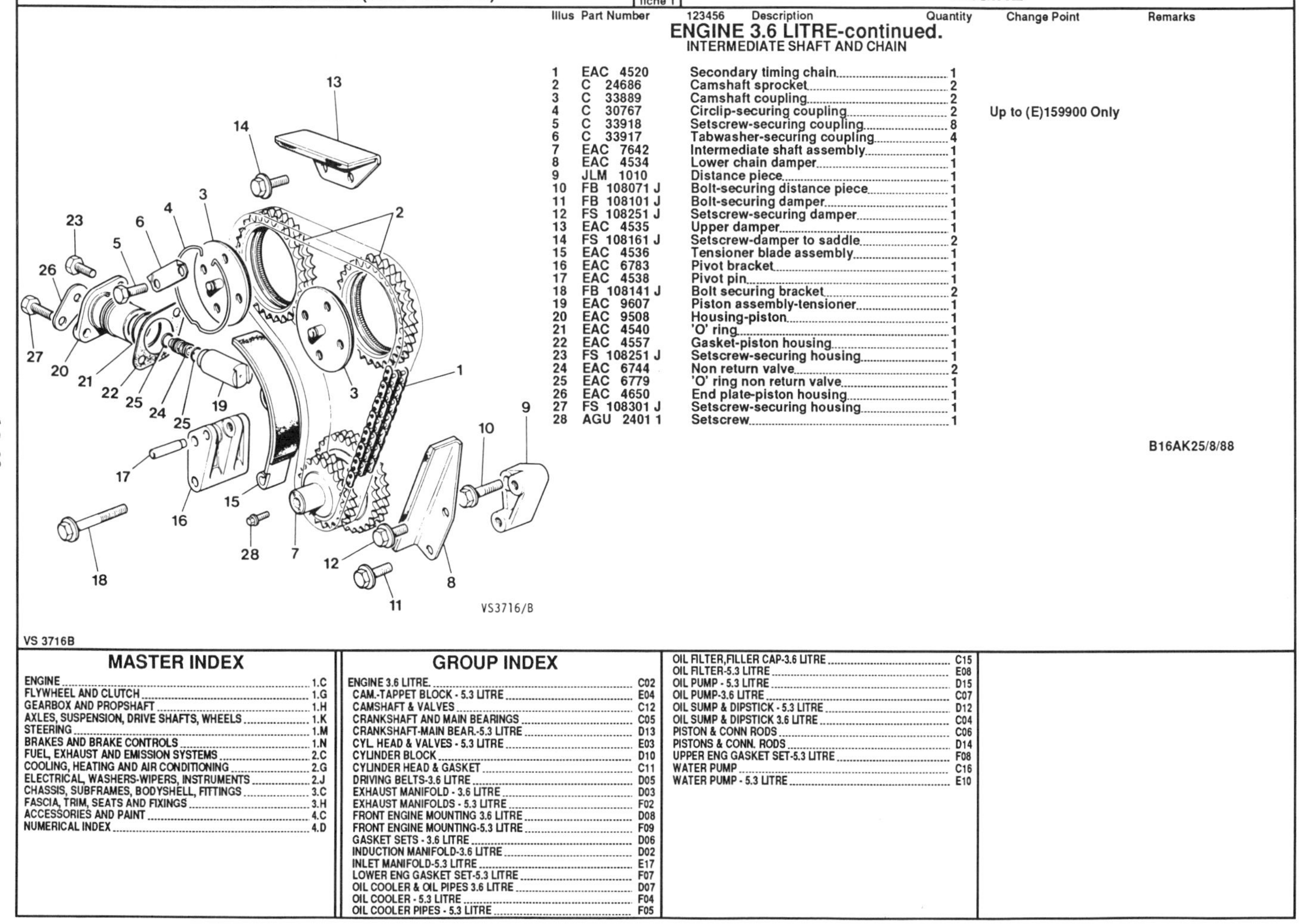

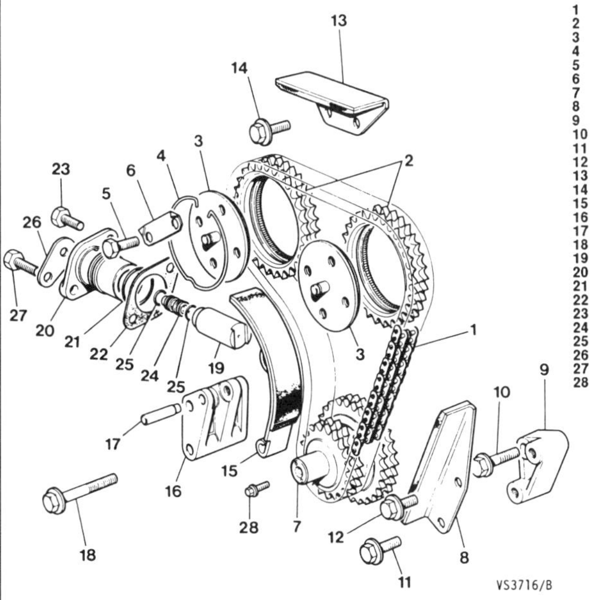

VS 3716B

ENGINE 3.6 LITRE-continued.
INTERMEDIATE SHAFT AND CHAIN

Illus	Part Number	123456 Description	Quantity	Change Point	Remarks
1	EAC 4520	Secondary timing chain	1		
2	C 24686	Camshaft sprocket	2		
3	C 33889	Camshaft coupling	2		
4	C 30767	Circlip-securing coupling	2	Up to (E)159900 Only	
5	C 33918	Setscrew-securing coupling	8		
6	C 33917	Tabwasher-securing coupling	4		
7	EAC 7642	Intermediate shaft assembly	1		
8	EAC 4534	Lower chain damper	1		
9	JLM 1010	Distance piece	1		
10	FB 108071 J	Bolt-securing distance piece	1		
11	FB 108101 J	Bolt-securing damper	1		
12	FS 108251 J	Setscrew-securing damper	1		
13	EAC 4535	Upper damper	1		
14	FS 108161 J	Setscrew-damper to saddle	2		
15	EAC 4536	Tensioner blade assembly	1		
16	EAC 6783	Pivot bracket	1		
17	EAC 4538	Pivot pin	1		
18	FB 108141 J	Bolt securing bracket	2		
19	EAC 9607	Piston assembly-tensioner	1		
20	EAC 9508	Housing-piston	1		
21	EAC 4540	'O' ring	1		
22	EAC 4557	Gasket-piston housing	1		
23	FS 108251 J	Setscrew-securing housing	1		
24	EAC 6744	Non return valve	2		
25	EAC 6779	'O' ring non return valve	1		
26	EAC 4650	End plate-piston housing	1		
27	FS 108301 J	Setscrew-securing housing	1		
28	AGU 2401 1	Setscrew	1		

B16AK25/8/88

1C 09

MASTER INDEX

ENGINE 1.C
FLYWHEEL AND CLUTCH 1.G
GEARBOX AND PROPSHAFT 1.H
AXLES, SUSPENSION, DRIVE SHAFTS, WHEELS 1.K
STEERING 1.M
BRAKES AND BRAKE CONTROLS 1.N
FUEL, EXHAUST AND EMISSION SYSTEMS 2.C
COOLING, HEATING AND AIR CONDITIONING 2.G
ELECTRICAL, WASHERS-WIPERS, INSTRUMENTS 2.J
CHASSIS, SUBFRAMES, BODYSHELL, FITTINGS 3.C
FASCIA, TRIM, SEATS AND FIXINGS 3.H
ACCESSORIES AND PAINT 4.C
NUMERICAL INDEX 4.D

GROUP INDEX

ENGINE 3.6 LITRE C02
CAM.-TAPPET BLOCK - 5.3 LITRE E04
CAMSHAFT & VALVES C12
CRANKSHAFT AND MAIN BEARINGS C05
CRANKSHAFT-MAIN BEAR.-5.3 LITRE D13
CYL. HEAD & VALVES - 5.3 LITRE E03
CYLINDER BLOCK D10
CYLINDER HEAD & GASKET C11
DRIVING BELTS-3.6 LITRE D05
EXHAUST MANIFOLD - 3.6 LITRE D03
EXHAUST MANIFOLDS - 5.3 LITRE F02
FRONT ENGINE MOUNTING 3.6 LITRE D08
FRONT ENGINE MOUNTING-5.3 LITRE F09
GASKET SETS - 3.6 LITRE D06
INDUCTION MANIFOLD-3.6 LITRE D02
INLET MANIFOLD-5.3 LITRE E17
LOWER ENG GASKET SET-5.3 LITRE F07
OIL COOLER & OIL PIPES 3.6 LITRE D07
OIL COOLER - 5.3 LITRE F04
OIL COOLER PIPES - 5.3 LITRE F05
OIL FILTER,FILLER CAP-3.6 LITRE C15
OIL FILTER-5.3 LITRE E08
OIL PUMP - 5.3 LITRE D15
OIL PUMP-3.6 LITRE C07
OIL SUMP & DIPSTICK - 5.3 LITRE D12
OIL SUMP & DIPSTICK 3.6 LITRE C04
PISTON & CONN RODS C06
PISTONS & CONN. RODS D14
UPPER ENG GASKET SET-5.3 LITRE F08
WATER PUMP C16
WATER PUMP - 5.3 LITRE E10

1C 10

ENGINE-3.6 LITRE-continued
TIMING COVER

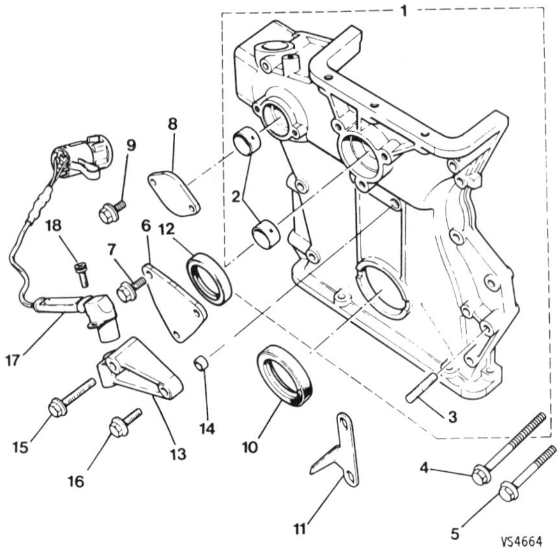

VS4664

Illus	Part Number	123456 Description	Quantity	Change Point	Remarks
1	EAC 5659	TIMING COVER ASSEMBLY	1	....	80 amp alternator
1	EAC 9657	TIMING COVER ASSEMBLY	1	....	90 amp alternator
2	EAC 2522	Bush	2		
3	TE 108051 J	Stud	1		
4	FB 108181 J	Bolt-securing cover	4		
5	FB 108131 J	Bolt-securing cover	4		
6	EAC 3977	Cover plate-interm.shaft	1		
7	FS 108201 J	Setscrew-securing plate	3		
8	EAC 3926	Cover plate-auxiliary shaft	1		
9	FS 106161 J	Setscrew-securing plate	2		
10	EAC 7954	Oil seal-crankshaft	1		
11	EAC 6811	Timing indicator	1		
12	EAC 7263	Oil seal-interm.shaft	NLA	Upto 9DPAMA 123705	Use EAC 5721
12	EAC 5721	Oil seal-intermediate shaft	1	From 9DPAMA 123706	
13	EAC 6001	Mounting bracket(sensor)	1	Up to Eng.No.108834	
13	EBC 1253	Mounting bracket(sensor)	1	From Eng.No.108835	
14	C 2373	Dowel	2		
15	FB 108111 J	Bolt	1		
16	FB 108071 J	Bolt	1	Up to Eng.No.108834	
16	FB 108061 J	Bolt	1	From Eng.No.108835	
17	DBC 2139	Crankshaft speed sensor	1		
18	SS 106121 J	Socket screw-sensor to bracket	1		

MASTER INDEX

GROUP INDEX

1C 11

ENGINE 3.6 LITRE-continued
CYLINDER HEAD & GASKET

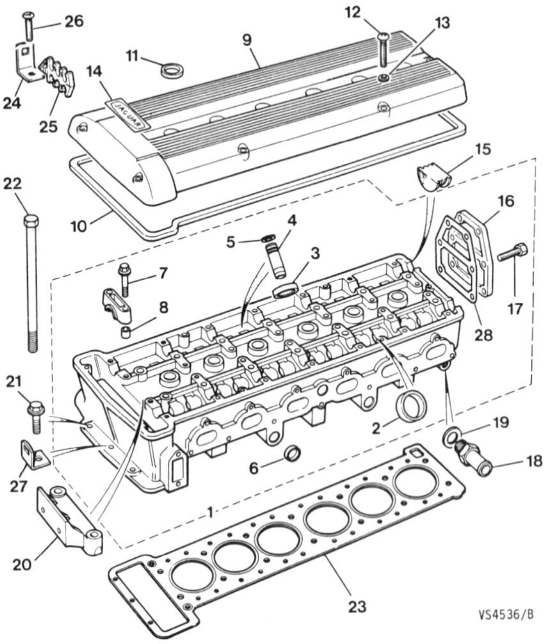

VS4536B

Illus	Part Number	123456 Description	Quantity	Change Point	Remarks
	JLM 741	CYLINDER HEAD ASSY(WITH VALVES/SPRINGS/CAMSHAFT	1		
1	EAC 7774	Cylinder head(less valves)	1	Up to (E)162265	
1	EBC 2925	Cylinder head (less valves)	1	From (E)162266	
2	EAC 6437	Insert-inlet valve	12		
3	EAC 6435	Insert-exhaust valve	12		
4	C 29390	Valve guide-(1st oversize)	24		
	C 29391	Valve guide-(2nd oversize)	24		
5	C 26199	Circlip-valve guide	24		
6	C 16862 3	Core plug	6		
7	FB 108101 J	Bolt-bearing caps	28		
8	C 2373	Dowel	28		
16	EAC 2911	Rear cover	1		
17	FS 108251 J	Setscrew	6		
28	EBC 1131	Gasket - rear cover	1	From (E)162266	
9	EAC 4989	Camshaft cover	1		
10	EAC 2912	Seal-camshaft cover	1		
11	EAC 4049	Seal-spark plug boss	6		
12	SE 106601 J	Screw-cover to head	7		
13	EBC 1284	Washer	7		
14	C 35732	Emblem	1		
15	C 30431	Rear sealing plug	2		
18	EAC 7189	Adaptor-heater feed	1		
19	C 2296 4	Washer	1		
20	EAC 4175	Saddle-upper chain damper	1		
21	FB 108071 J	Bolt-cyl head to timing cover	3		
22	EAC 4744	Bolt-cyl head to block	14	Up to (E)143668	
22	EBC 1771	Bolt-cyl head to block	14	From (E)143669	
23	EAC 7504	Cylinder head gasket	1		
24	EAC 5046	Bracket	3		
25	EAC 5045	Clip-distributor leads	4		
26	FS 106101 J	Setscrew	3		
27	DBC 1626	Bracket-harness attachment	1		

MASTER INDEX

ENGINE	1.C
FLYWHEEL AND CLUTCH	1.G
GEARBOX AND PROPSHAFT	1.H
AXLES, SUSPENSION, DRIVE SHAFTS, WHEELS	1.K
STEERING	1.M
BRAKES AND BRAKE CONTROLS	1.N
FUEL, EXHAUST AND EMISSION SYSTEMS	2.C
COOLING, HEATING AND AIR CONDITIONING	2.G
ELECTRICAL, WASHERS-WIPERS, INSTRUMENTS	2.J
CHASSIS, SUBFRAMES, BODYSHELL, FITTINGS	3.C
FASCIA, TRIM, SEATS AND FIXINGS	3.H
ACCESSORIES AND PAINT	4.C
NUMERICAL INDEX	4.D

GROUP INDEX

ENGINE 3.6 LITRE	C02
CAM.-TAPPET BLOCK - 5.3 LITRE	E04
CAMSHAFT & VALVES	C12
CRANKSHAFT AND MAIN BEARINGS	C05
CRANKSHAFT-MAIN BEAR.-5.3 LITRE	D13
CYL. HEAD & VALVES - 5.3 LITRE	E03
CYLINDER BLOCK	D10
CYLINDER HEAD & GASKET	C11
DRIVING BELTS-3.6 LITRE	D05
EXHAUST MANIFOLD - 3.6 LITRE	D03
EXHAUST MANIFOLDS - 5.3 LITRE	F02
FRONT ENGINE MOUNTING 3.6 LITRE	D08
FRONT ENGINE MOUNTING-5.3 LITRE	F09
GASKET SETS - 3.6 LITRE	D06
INDUCTION MANIFOLD-3.6 LITRE	D02
INLET MANIFOLD-5.3 LITRE	E17
LOWER ENG GASKET SET-5.3 LITRE	F07
OIL COOLER & OIL PIPES 3.6 LITRE	D07
OIL COOLER - 5.3 LITRE	F04
OIL COOLER PIPES - 5.3 LITRE	F05
OIL FILTER,FILLER CAP-3.6 LITRE	C15
OIL FILTER-5.3 LITRE	E08
OIL PUMP - 5.3 LITRE	D15
OIL PUMP-3.6 LITRE	C07
OIL SUMP & DIPSTICK - 5.3 LITRE	D12
OIL SUMP & DIPSTICK 3.6 LITRE	C04
PISTON & CONN RODS	C06
PISTONS & CONN. RODS	D14
UPPER ENG GASKET SET-5.3 LITRE	F08
WATER PUMP	C16
WATER PUMP - 5.3 LITRE	E10

1C 12

JAGUAR XJS RANGE (JAN 1987 ON) | C12 fiche 1 | ENGINE

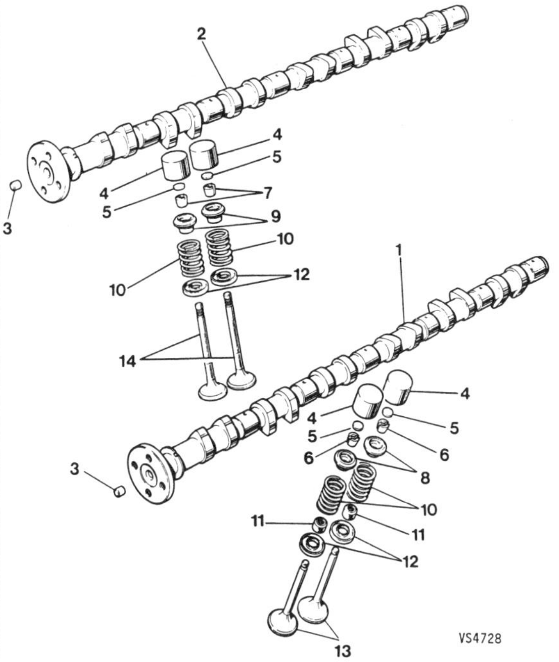

VS4728

ENGINE 3.6 LITRE-Continued
CAMSHAFT & VALVES

Illus	Part Number	123456	Description	Quantity	Change Point	Remarks
1	EAC 8523		Camshaft-inlet	1		Use EAC 9535
2	EAC 8525		Camshaft-exhaust	1		Use EAC 9534
1	EAC 9535		Camshaft-inlet	1	Up to Eng.No. 149789	
2	EAC 9534		Camshaft-exhaust	1		
1	EBC 2755		Camshaft-inlet	1	From Eng.No. 149790	
2	EBC 2756		Camshaft-exhaust	1		
3	C 33244		End plug	2		
4	EAC 8418		Tappet	24		
4	EBC 2197		Tappet	24		Oversize
			VALVE ADJUSTING SHIM			
	EAC 5013 2		0.0850	A/R		
	EAC 5013 3		0.0855	A/R		
	EAC 5013 4		0.0860	A/R		
	EAC 5013 5		0.0865	A/R		
	EAC 5013 6		0.0870	A/R		
	EAC 5013 7		0.0875	A/R		
	EAC 5013 8		0.0880	A/R		
	EAC 5013 9		0.0885	A/R		
	EAC 5013 10		0.0890	A/R		
	EAC 5013 11		0.0895	A/R		
	EAC 5013 12		0.0900	A/R		
	EAC 5013 13		0.0905	A/R		
	EAC 5013 14		0.0910	A/R		
	EAC 5013 15		0.0915	A/R		
	EAC 5013 16		0.0920	A/R		

MASTER INDEX

GROUP INDEX

1C 13

JAGUAR XJS RANGE (JAN 1987 ON) | C13 fiche 1 | ENGINE

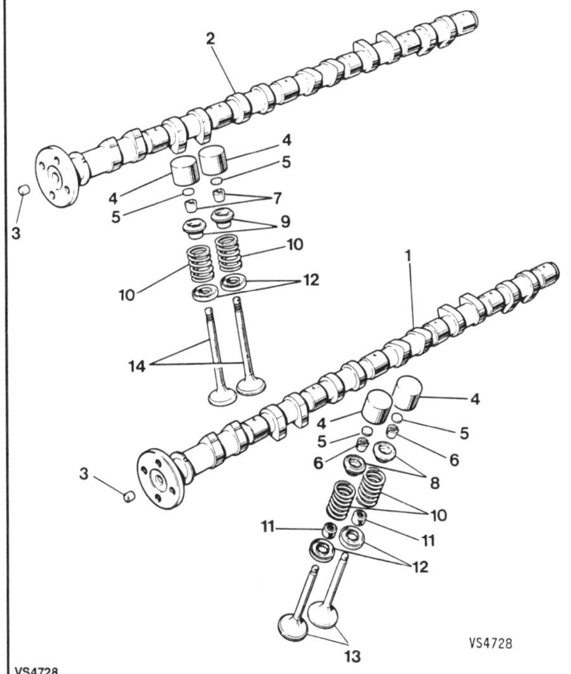

VS4728

ENGINE 3.6 LITRE-Continued.
CAMSHAFT & VALVES - 3.6 LITRE

Illus	Part Number	123456	Description	Quantity	Change Point	Remarks
			VALVE ADJUSTING SHIM			
	EAC 5013 17		0.0925	A/R		
	EAC 5013 18		0.0930	A/R		
	EAC 5013 19		0.0935	A/R		
	EAC 5013 20		0.0940	A/R		
	EAC 5013 21		0.0945	A/R		
	EAC 5013 22		0.0950	A/R		
	EAC 5013 23		0.0955	A/R		
	EAC 5013 24		0.0960	A/R		
	EAC 5013 25		0.0965	A/R		
	EAC 5013 26		0.0970	A/R		
	EAC 5013 27		0.0975	A/R		
	EAC 5013 28		0.0980	A/R		
	EAC 5013 29		0.0985	A/R		
	EAC 5013 30		0.0990	A/R		
	EAC 5013 31		0.0995	A/R		
	EAC 5013 32		0.1000	A/R		
	EAC 5013 33		0.1005	A/R		
	EAC 5013 34		0.1010	A/R		
	EAC 5013 35		0.1015	A/R		
	EAC 5013 36		0.1020	A/R		
	EAC 5013 37		0.1025	A/R		
	EAC 5013 38		0.1030	A/R		
	EAC 5013 39		0.1035	A/R		

MASTER INDEX
ENGINE 1.C
FLYWHEEL AND CLUTCH 1.G
GEARBOX AND PROPSHAFT 1.H
AXLES, SUSPENSION, DRIVE SHAFTS, WHEELS 1.K
STEERING 1.M
BRAKES AND BRAKE CONTROLS 1.N
FUEL, EXHAUST AND EMISSION SYSTEMS 2.C
COOLING, HEATING AND AIR CONDITIONING 2.G
ELECTRICAL, WASHERS-WIPERS, INSTRUMENTS 2.J
CHASSIS, SUBFRAMES, BODYSHELL, FITTINGS 3.C
FASCIA, TRIM, SEATS AND FIXINGS 3.H
ACCESSORIES AND PAINT 4.C
NUMERICAL INDEX 4.D
GROUP INDEX
ENGINE 3.6 LITRE. C02
CAM.-TAPPET BLOCK - 5.3 LITRE E04
CAMSHAFT & VALVES C12
CRANKSHAFT AND MAIN BEARINGS C05
CRANKSHAFT-MAIN BEAR.-5.3 LITRE D13
CYL. HEAD & VALVES - 5.3 LITRE E03
CYLINDER BLOCK D10
CYLINDER HEAD & GASKET C11
DRIVING BELTS-3.6 LITRE D05
EXHAUST MANIFOLD - 3.6 LITRE D03
EXHAUST MANIFOLDS - 5.3 LITRE F02
FRONT ENGINE MOUNTING 3.6 LITRE D08
FRONT ENGINE MOUNTING-5.3 LITRE F09
GASKET SETS - 3.6 LITRE D06
INDUCTION MANIFOLD-3.6 LITRE D02
INLET MANIFOLD-5.3 LITRE E17
LOWER ENG GASKET SET-5.3 LITRE F07
OIL COOLER & OIL PIPES 3.6 LITRE D07
OIL COOLER - 5.3 LITRE F04
OIL COOLER PIPES - 5.3 LITRE F05
OIL FILTER,FILLER CAP-3.6 LITRE C15
OIL FILTER-5.3 LITRE E08
OIL PUMP - 5.3 LITRE D15
OIL PUMP-3.6 LITRE C07
OIL SUMP & DIPSTICK - 5.3 LITRE D12
OIL SUMP & DIPSTICK 3.6 LITRE C04
PISTON & CONN RODS C06
PISTONS & CONN. RODS D14
UPPER ENG GASKET SET-5.3 LITRE F08
WATER PUMP C16
WATER PUMP - 5.3 LITRE E10

1C 14

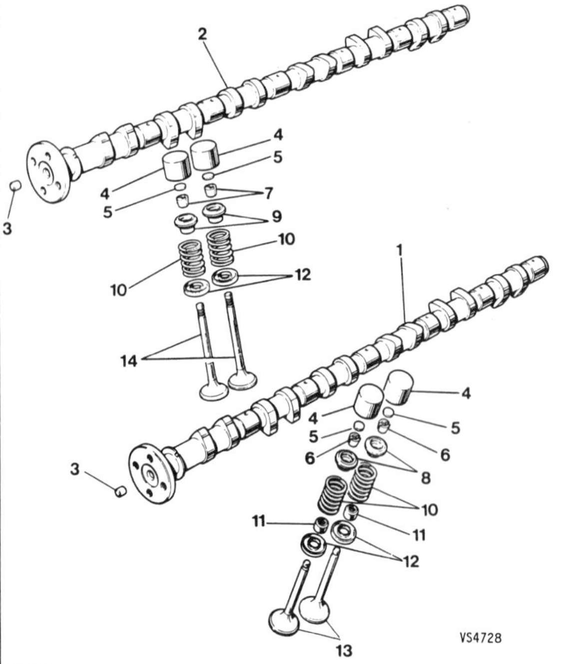

VS4728

ENGINE 3.6 LITRE-Continued
CAMSHAFT AND VALVES-Continued

Illus	Part Number	Description	Quantity	Change Point	Remarks
		VALVE ADJUSTING SHIM			
	EAC 5013 40	0.1040	24		
	EAC 5013 41	0.1045	24		
	EAC 5013 42	0.1050	24		
	EAC 5013 43	0.1055	24		
	EAC 5013 44	0.1060	24		
	EAC 5013 45	0.1065	24		
	EAC 5013 46	0.1070	24		
	EAC 5013 47	0.1075	24		
	EAC 5013 48	0.1080	24		
6	C 27482	Cotter-inlet valve	12PR		
7	EAC 7256	Cotter-exhaust valve	12PR		
8	EAC 4709	Collar-inlet valve	12		
9	EAC 7227	Collar-exhaust valve	12		
10	EAC 5875	Spring-valve	24		
11	EAC 5032	Seal-inlet valve	12		
12	C 27481	Seat-valve spring	24		
13	EAC 2909	Valve-inlet	12		
14	EAC 7255	Valve-exhaust	12		

MASTER INDEX

GROUP INDEX

1C 15

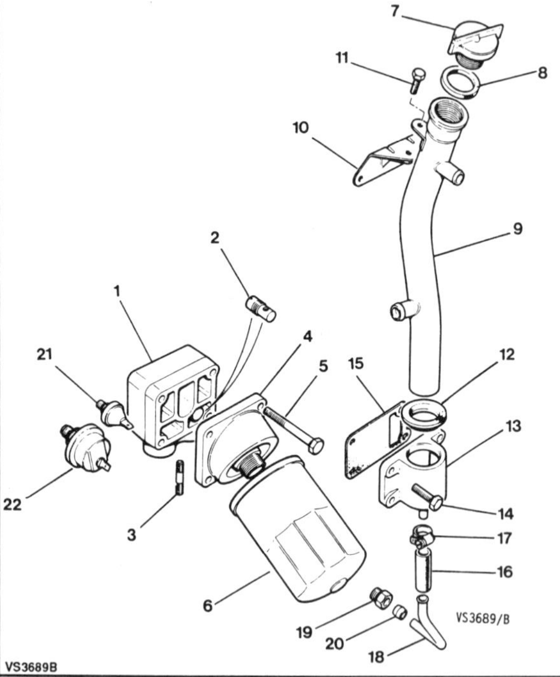

VS3689B

ENGINE 3.6 LITRE-Continued
OIL FILTER,FILLER CAP-3.6 LITRE

Illus	Part Number	Description	Quantity	Change Point	Remarks
1	EAC 3671	Adaptor-oil cooler	1		
2	EAC 2428	Bypass valve	1		
3	TE 108041 J	Stud-oil cooler pipes	1		
4	EAC 4512	Filter head	1		
5	FB 108141 J	Bolt-securing head and adaptor	4		
6	C 42797	OIL FILTER	1		
7	EAC 6523	Oil filler cap	1		
8	C 42511	'O' ring	1		
9	EAC 6434	Oil filler pipe	1		
10	EAC 5004	Support bracket	1		
11	FS 106121 J	Setscrew-pipe to bracket	1		
12	EAC 2632	Bush	1		
13	EAC 7443	Housing	1		
14	FS 108251 J	Setscrew-housing to block	4		
15	EAC 4886	Baffle plate	1		
16	EAC 7407	Hose-housing to drain pipe	1		
17	EAC 3215 1	Clip	2		
18	EAC 7441	Oil drain pipe	1		
19	EAC 4927	Nut	1		
20	EAC 4929	Olive(steel)	NLA		Use AGU2527/J
20	AGU 2527 J	Olive(brass)	1		
21	C 42200	OIL PRESSURE SWITCH	1		
22	C 46272	OIL PRESSURE TRANSMITTER	1		

B26AK13/7/88

MASTER INDEX

ENGINE 1.C
FLYWHEEL AND CLUTCH 1.G
GEARBOX AND PROPSHAFT 1.H
AXLES, SUSPENSION, DRIVE SHAFTS, WHEELS 1.K
STEERING 1.M
BRAKES AND BRAKE CONTROLS 1.N
FUEL, EXHAUST AND EMISSION SYSTEMS 2.C
COOLING, HEATING AND AIR CONDITIONING 2.G
ELECTRICAL, WASHERS-WIPERS, INSTRUMENTS 2.J
CHASSIS, SUBFRAMES, BODYSHELL, FITTINGS 3.C
FASCIA, TRIM, SEATS AND FIXINGS 3.H
ACCESSORIES AND PAINT 4.C
NUMERICAL INDEX 4.D

GROUP INDEX

ENGINE 3.6 LITRE C02
CAM.-TAPPET BLOCK - 5.3 LITRE E04
CAMSHAFT & VALVES C12
CRANKSHAFT AND MAIN BEARINGS C05
CRANKSHAFT-MAIN BEAR.-5.3 LITRE D13
CYL. HEAD & VALVES - 5.3 LITRE E03
CYLINDER BLOCK D10
CYLINDER HEAD & GASKET C11
DRIVING BELTS-3.6 LITRE D05
EXHAUST MANIFOLD - 3.6 LITRE D03
EXHAUST MANIFOLDS - 5.3 LITRE F02
FRONT ENGINE MOUNTING 3.6 LITRE D08
FRONT ENGINE MOUNTING-5.3 LITRE F09
GASKET SETS - 3.6 LITRE D06
INDUCTION MANIFOLD-3.6 LITRE D02
INLET MANIFOLD-5.3 LITRE E17
LOWER ENG GASKET SET-5.3 LITRE F07
OIL COOLER & OIL PIPES 3.6 LITRE D07
OIL COOLER - 5.3 LITRE F04
OIL COOLER PIPES - 5.3 LITRE F05
OIL FILTER,FILLER CAP-3.6 LITRE C15
OIL FILTER-5.3 LITRE E08
OIL PUMP - 5.3 LITRE D15
OIL PUMP-3.6 LITRE C07
OIL SUMP & DIPSTICK - 5.3 LITRE D12
OIL SUMP & DIPSTICK 3.6 LITRE C04
PISTON & CONN RODS C06
PISTONS & CONN. RODS D14
UPPER ENG GASKET SET-5.3 LITRE F08
WATER PUMP C16
WATER PUMP - 5.3 LITRE E10

1C 16

ENGINE 3.6 LITRE-continued
WATER PUMP

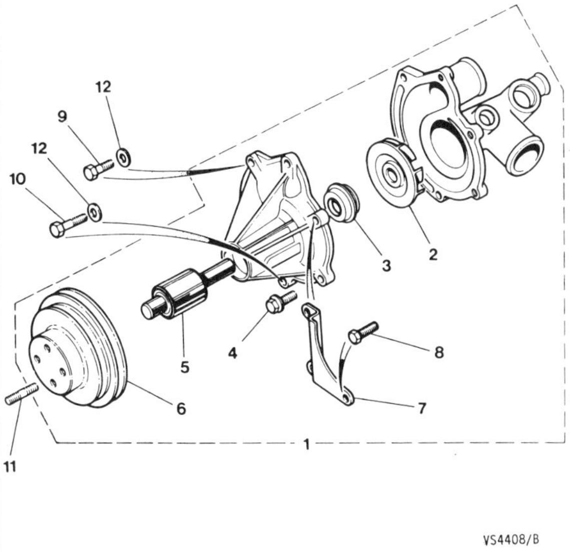

VS4408B

Illus	Part Number	Description	Quantity	Change Point	Remarks
1	EAC 7243	WATER PUMP ASSEMBLY	1		Non air injection 80 amp alternator
2	EAC 5442	Pulley	1		
5	EAC 6305	Bearing	1		
4	FS 108301 J	Setscrew	4		
3	EAC 6447	Seal	1		
6	EAC 7220	Impellor	1		
1	EAC 7246	WATER PUMP ASSEMBLY	1		Air injection only 80 amp alternator
2	EAC 5442	Pulley	1		
5	EAC 6305	Bearing	1		
4	FS 108301 J	Setscrew	4		
3	EAC 6447	Seal	1		
6	EAC 7220	Impellor	1		
7	EAC 7236	Adjusting bracket(air pump)	1		
8	FS 108301 J	Setscrew	4		
1	EAC 9211	WATER PUMP ASSEMBLY	1		Non air injection 90 amp alternator
2	EAC 6651	Pulley	1		
5	EAC 6305	Bearing	1		
4	FS 108301 J	Setscrew	4		
3	EAC 6447	Seal	1		
6	EAC 7220	Impellor	1		
1	EAC 9172	WATER PUMP ASSEMBLY	1		Air injection 90 amp alternator
2	EAC 6651	Pulley	1		
5	EAC 6305	Bearing	1		
4	FS 108301 J	Setscrew	4		
3	EAC 6447	Seal	1		
6	EAC 7220	Impellor	1		
7	EAC 7236	Adjusting bracket (air pump)	1		
8	FS 108301 J	Setscrew	4		
9	BH 108101 J	Bolt-pump to timing cover	1		Non air injection
10	BH 108141 J	Bolt-pump to cylinder block	2		
12	WB 108051 J	Washer	3		
11	TE 108031 J	Stud-pulley to drive unit	4		

AKB2811/7/88

1C 17

ENGINE 3.6 LITRE-continued.
WATER RAIL-HEATER RETURN PIPE

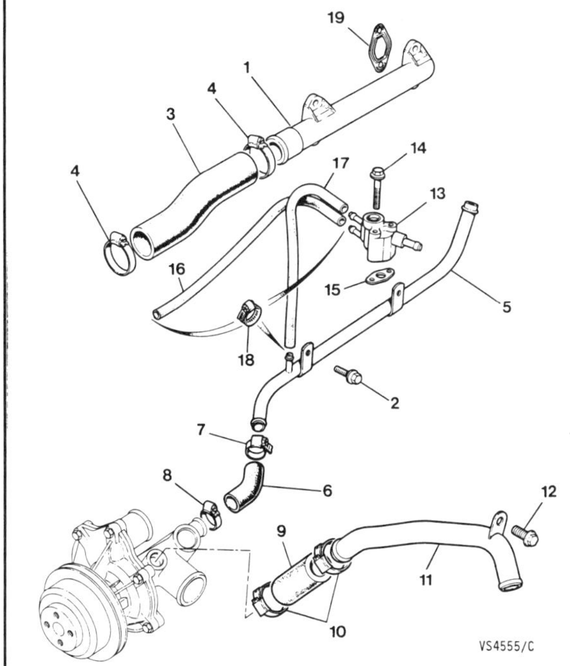

VS4555C

Illus	Part Number	Description	Quantity	Change Point	Remarks
1	EAC 4644	Water rail	1		
2	FS 108251 J	Setscrew-securing rail and pipe	4		
19	EAC 9745	Gasket - water rail	1		
3	EAC 3454	Hose-water rail to water pump	1		
4	EAC 3215 13	Hose clip	2		
5	EAC 7222	Heater return pipe	1		
6	EAC 7224	Hose-return pipe to water pump	1		
7	EAC 3215 8	Hose clip	1		
8	EAC 3215 7	Hose clip	1		
9	EAC 7343	Water hose	1		Air injection only
10	EAC 3215 8	Hose clip	2		
11	EAC 7339	Water pipe	1		
12	FS 108141 J	Setscrew	1		
13	EAC 6106	Restrictor	1	Up to Eng.No.135612	Use EAC 7019
13	EAC 7019	Restrictor	1	From Eng.No.135613	
14	FS 106111 J	Setscrew - restrictor to manifold	2		
15	EAC 5513	Gasket	1		
16	EAC 6124	Hose-restrictor to thermo hsg	1		
17	EAC 6123	Hose-restrictor to return pipe	!		
18	EAC 3215 1	Hose clip	3		

MASTER INDEX

ENGINE 1.C
FLYWHEEL AND CLUTCH 1.G
GEARBOX AND PROPSHAFT 1.H
AXLES, SUSPENSION, DRIVE SHAFTS, WHEELS 1.K
STEERING 1.M
BRAKES AND BRAKE CONTROLS 1.N
FUEL, EXHAUST AND EMISSION SYSTEMS 2.C
COOLING, HEATING AND AIR CONDITIONING 2.G
ELECTRICAL, WASHERS-WIPERS, INSTRUMENTS 2.J
CHASSIS, SUBFRAMES, BODYSHELL, FITTINGS 3.C
FASCIA, TRIM, SEATS AND FIXINGS 3.H
ACCESSORIES AND PAINT 4.C
NUMERICAL INDEX 4.D

GROUP INDEX

ENGINE 3.6 LITRE C02
CAM.-TAPPET BLOCK - 5.3 LITRE E04
CAMSHAFT & VALVES C12
CRANKSHAFT AND MAIN BEARINGS C05
CRANKSHAFT-MAIN BEAR.-5.3 LITRE D13
CYL. HEAD & VALVES - 5.3 LITRE E03
CYLINDER BLOCK D10
CYLINDER HEAD & GASKET C11
DRIVING BELTS-3.6 LITRE D05
EXHAUST MANIFOLD - 3.6 LITRE D03
EXHAUST MANIFOLDS - 5.3 LITRE F02
FRONT ENGINE MOUNTING 3.6 LITRE D08
FRONT ENGINE MOUNTING-5.3 LITRE F09
GASKET SETS - 3.6 LITRE D06
INDUCTION MANIFOLD-3.6 LITRE D02
INLET MANIFOLD-5.3 LITRE E17
LOWER ENG GASKET SET-5.3 LITRE F07
OIL COOLER & OIL PIPES 3.6 LITRE D07
OIL COOLER - 5.3 LITRE F04
OIL COOLER PIPES - 5.3 LITRE F05
OIL FILTER,FILLER CAP-3.6 LITRE C15
OIL FILTER-5.3 LITRE E08
OIL PUMP - 5.3 LITRE D15
OIL PUMP-3.6 LITRE C07
OIL SUMP & DIPSTICK - 5.3 LITRE D12
OIL SUMP & DIPSTICK 3.6 LITRE C04
PISTON & CONN RODS C06
PISTONS & CONN. RODS D14
UPPER ENG GASKET SET-5.3 LITRE F08
WATER PUMP C16
WATER PUMP - 5.3 LITRE E10

1C 18

JAGUAR XJS RANGE (JAN 1987 ON) — C18 fiche 1 — ENGINE

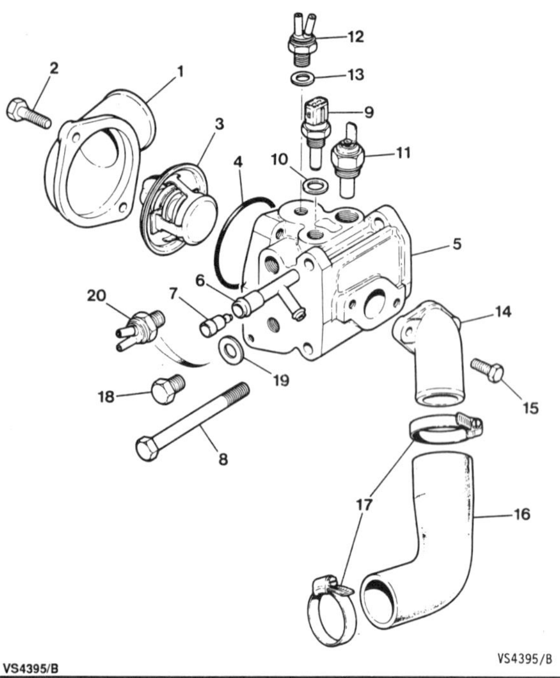

VS4395/B

ENGINE 3.6 LITRE-Continued
THERMOSTAT AND BYPASS HOSE

Illus	Part Number	123456 Description	Quantity	Change Point	Remarks
1	EAC 8209	Thermostat cover	1		
2	FS 108201 J	Setscrew-cover to housing	2		
3	JLM 322	THERMOSTAT	1		
4	EAC 8999	'O' ring	1		
5	EAC 6140	Thermostat housing	1		Emiss 'B/F'
5	EAC 6063	Thermostat housing	1		Emiss 'A'
6	EAC 6142	Tee piece	1		
7	EAC 4126	Water valve	1		
8	FB 108201 J	Bolt-housing to cylinder head	2		
9	EAC 3927	TEMPERATURE SENSOR	1		
10	C 2296 23	Washer	1		
11	DAC 2583	TEMPERATURE TRANSMITTER	1		
12	EAC 2630	Thermal valve	1		] Emiss 'A' only
13	C 2296 20	Washer	1		
20	EAC 2630	THERMAL VALVE	1	From Eng No 139500	
14	EAC 6813	Bypass elbow	1		
15	FS 108201 J	Setscrew-elbow to housing	2		
16	EAC 3803	Hose-elbow to water pump	1		
17	EAC 3215 10	Hose clip	2		
18	EAC 5063	Blanking plug	1		] Emiss 'A' w/out carbon canister purge
19	C 2296 20	Washer	2		

B30AK13/7/88

1D 02

JAGUAR XJS RANGE (JAN 1987 ON) — D02 fiche 1 — ENGINE

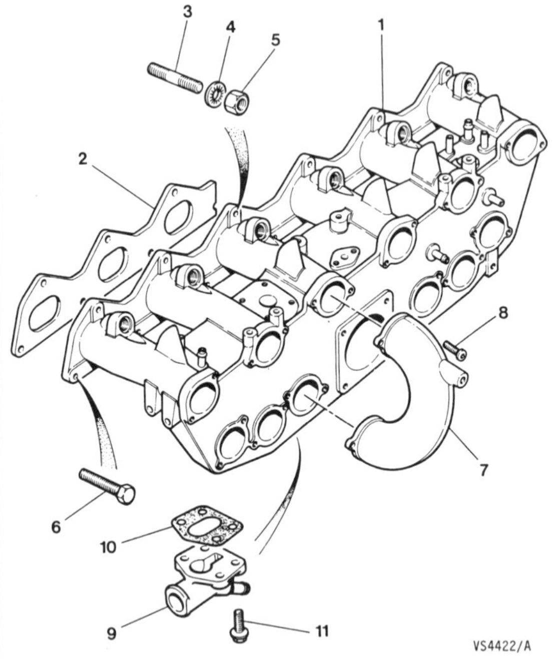

VS4422A

ENGINE 3.6 LITRE-continued.
INDUCTION MANIFOLD-3.6 LITRE

Illus	Part Number	123456 Description	Quantity	Change Point	Remarks
1	EAC 8989	Induction manifold	1	Up to Eng.No.139920	With Decel Valve
1	EBC 2057	Induction manifold	1	(E)139921to(E)156653	Less Decel Valve
1	EBC 2054	Induction manifold	1	From (E)156654	Teves ABS
2	EAC 4691	Gasket	2		
3	TE 108051 J	Stud-manifold to cyl head	2	] Up to (E)151396	
3	TE 108061 J	Stud	4		
3	TE 108061 J	Stud-manifold to cyl head	6	From (E)151397	
4	C 30075 2	Washer	6		
5	NH 108041 J	Nut	6		
6	FB 108061 J	Bolt-manifold to cyl head	7		
7	EAC 6382	Elbow	5		
7	EAC 4861	Elbow	1		
8	EAC 7544	Screw-securing elbow	24		
	CAC 6966	Adaptor (brake servo)	1	] Up to (E)156653	] Not required when ABS fitted
	C 2296 1	Washer (servo adaptor)	1		
9	EAC 5838	Adaptor-idle air	1	Up to (E)139499	
9	EBC 2381	Adaptor-idle air	NLA	From (E)139500	Use EBC 2735
9	EBC 2735	Adaptor-idle air	1		
9	EAC 7759	Adaptor-idle air	1	Up to (E)139499	] Purge engine Emiss A
9	EBC 2351	Adaptor-idle air	1	From (E)139500	
10	EAC 5869	Gasket-adaptor to manifold	1		
11	FS 106201 J	Setscrew-adaptor to manifold	4		

B32AK12/10/88

MASTER INDEX

ENGINE 1.C
FLYWHEEL AND CLUTCH 1.G
GEARBOX AND PROPSHAFT 1.H
AXLES, SUSPENSION, DRIVE SHAFTS, WHEELS 1.K
STEERING 1.M
BRAKES AND BRAKE CONTROLS 1.N
FUEL, EXHAUST AND EMISSION SYSTEMS 2.C
COOLING, HEATING AND AIR CONDITIONING 2.G
ELECTRICAL, WASHERS-WIPERS, INSTRUMENTS 2.J
CHASSIS, SUBFRAMES, BODYSHELL, FITTINGS 3.C
FASCIA, TRIM, SEATS AND FIXINGS 3.H
ACCESSORIES AND PAINT 4.C
NUMERICAL INDEX 4.D

GROUP INDEX

ENGINE 3.6 LITRE C02
CAM.-TAPPET BLOCK - 5.3 LITRE E04
CAMSHAFT & VALVES C12
CRANKSHAFT AND MAIN BEARINGS C05
CRANKSHAFT-MAIN BEAR.-5.3 LITRE D13
CYL. HEAD & VALVES - 5.3 LITRE E03
CYLINDER BLOCK D10
CYLINDER HEAD & GASKET C11
DRIVING BELTS-3.6 LITRE D05
EXHAUST MANIFOLD - 3.6 LITRE D03
EXHAUST MANIFOLDS - 5.3 LITRE F02
FRONT ENGINE MOUNTING 3.6 LITRE D08
FRONT ENGINE MOUNTING-5.3 LITRE F09
GASKET SETS - 3.6 LITRE D06
INDUCTION MANIFOLD-3.6 LITRE D02
INLET MANIFOLD-5.3 LITRE E17
LOWER ENG GASKET SET-5.3 LITRE F07
OIL COOLER & OIL PIPES 3.6 LITRE D07
OIL COOLER - 5.3 LITRE F04
OIL COOLER PIPES - 5.3 LITRE F05
OIL FILTER,FILLER CAP-3.6 LITRE C15
OIL FILTER-5.3 LITRE E08
OIL PUMP - 5.3 LITRE D15
OIL PUMP-3.6 LITRE C07
OIL SUMP & DIPSTICK - 5.3 LITRE D12
OIL SUMP & DIPSTICK 3.6 LITRE C04
PISTON & CONN RODS C06
PISTONS & CONN. RODS D14
UPPER ENG GASKET SET-5.3 LITRE F08
WATER PUMP C16
WATER PUMP - 5.3 LITRE E10

1D 03

ENGINE 3.6 LITRE-Continued

EXHAUST MANIFOLD - 3.6 LITRE

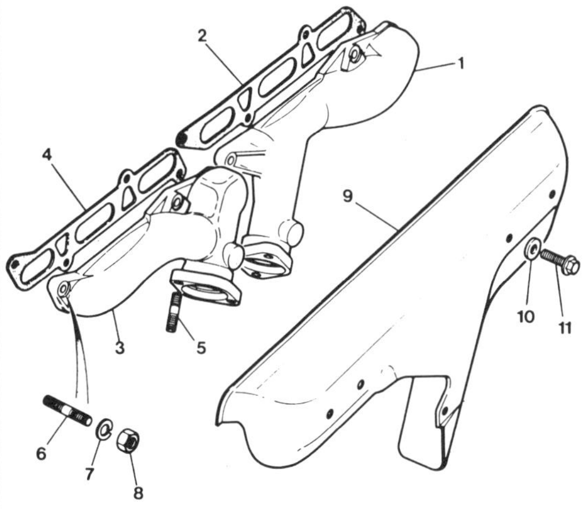

VS3691/B

VS 3691B

Illus	Part Number	123456	Description	Quantity	Change Point	Remarks
1	EAC 7264		Exhaust manifold-front	1		Non air injection
1	EAC 7275		Exhaust manifold-front	1		Air injection
2	EAC 8994		Gasket-front manifold	1		
3	EAC 7261		Exhaust manifold-rear	1		Non air injection
3	EAC 7276		Exhaust manifold-rear	1		Air injection
4	EAC 8995		Gasket-rear manifold	1		
5	EAC 3823		Stud-outlet flange	4		
6	TE 110071 J		Stud-manifold to cyl head	8		Use EBC1292
6	EBC 1292		Stud-manifold to cyl head	8		
7	WL 110001 J		Washer-manifold to cyl head	4		
8	EAC 5308		Nut-manifold to cyl head	4		
9	EAC 9675		Heatshield	1		
10	FS 108141 J		Bolt	5		
11	WA 108051 J		Washer	5		

1D 04

ENGINE 3.6 LITRE-Continued

TRANSMISSION ADAPTOR

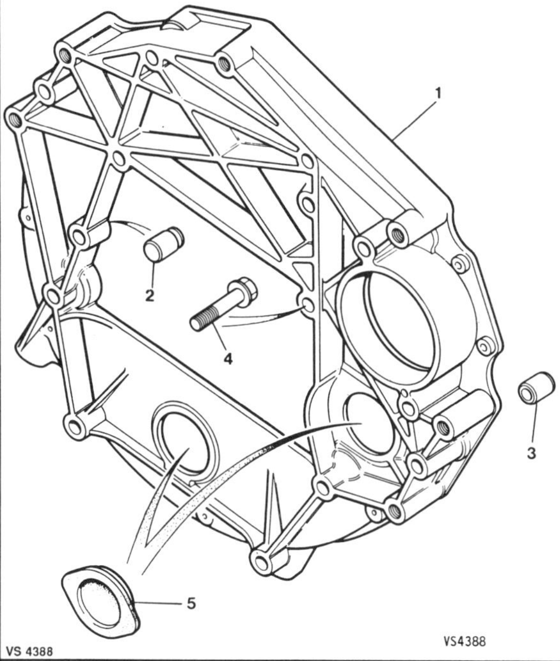

VS4388

VS 4388

Illus	Part Number	123456	Description	Quantity	Change Point	Remarks
1	EAC 7067		Adaptor-transmission	1	Up to 9DPAMA110668	
1	EBC 1647		Adaptor-transmission	1	From 9DPAMA110669	
2	C 13478		Dowel adaptor to cyl block	2		
3	C 46021		Dowel-adaptor to gearbox	2		
4	FS 110351 J		Setscrew-adaptor to cyl block	4		
5	EAC 4378		Seal-rubber	2		

MASTER INDEX

ENGINE ... 1.C
FLYWHEEL AND CLUTCH ... 1.G
GEARBOX AND PROPSHAFT ... 1.H
AXLES, SUSPENSION, DRIVE SHAFTS, WHEELS ... 1.K
STEERING ... 1.M
BRAKES AND BRAKE CONTROLS ... 1.N
FUEL, EXHAUST AND EMISSION SYSTEMS ... 2.C
COOLING, HEATING AND AIR CONDITIONING ... 2.G
ELECTRICAL, WASHERS-WIPERS, INSTRUMENTS ... 2.J
CHASSIS, SUBFRAMES, BODYSHELL, FITTINGS ... 3.C
FASCIA, TRIM, SEATS AND FIXINGS ... 3.H
ACCESSORIES AND PAINT ... 4.C
NUMERICAL INDEX ... 4.D

GROUP INDEX

ENGINE 3.6 LITRE ... C02
CAM.-TAPPET BLOCK - 5.3 LITRE ... E04
CAMSHAFT & VALVES ... C12
CRANKSHAFT AND MAIN BEARINGS ... C05
CRANKSHAFT-MAIN BEAR.-5.3 LITRE ... D13
CYL. HEAD & VALVES - 5.3 LITRE ... E03
CYLINDER BLOCK ... D10
CYLINDER HEAD & GASKET ... C11
DRIVING BELTS-3.6 LITRE ... D05
EXHAUST MANIFOLD - 3.6 LITRE ... D03
EXHAUST MANIFOLDS - 5.3 LITRE ... F02
FRONT ENGINE MOUNTING 3.6 LITRE ... D08
FRONT ENGINE MOUNTING-5.3 LITRE ... F09
GASKET SETS - 3.6 LITRE ... D06
INDUCTION MANIFOLD-3.6 LITRE ... D02
INLET MANIFOLD-5.3 LITRE ... E17
LOWER ENG GASKET SET-5.3 LITRE ... F07
OIL COOLER & OIL PIPES 3.6 LITRE ... D07
OIL COOLER - 5.3 LITRE ... F04
OIL COOLER PIPES - 5.3 LITRE ... F05
OIL FILTER,FILLER CAP-3.6 LITRE ... C15
OIL FILTER-5.3 LITRE ... E08
OIL PUMP - 5.3 LITRE ... D15
OIL PUMP-3.6 LITRE ... C07
OIL SUMP & DIPSTICK - 5.3 LITRE ... D12
OIL SUMP & DIPSTICK 3.6 LITRE ... C04
PISTON & CONN RODS ... C06
PISTONS & CONN. RODS ... D14
UPPER ENG GASKET SET-5.3 LITRE ... F08
WATER PUMP ... C16
WATER PUMP - 5.3 LITRE ... E10

Illus	Part Number	123456 Description	Quantity	Change Point	Remarks

ENGINE 3.6 LITRE-continued
DRIVING BELTS-3.6 LITRE

Illus	Part Number	Description	Quantity	Change Point	Remarks
1	EAC 3850	Fan belt	1		80 amp alternator
1	EAC 7033	Fan belt	1		90 amp alternator
2	EAC 3878	Compressor belt	1		
3	EAC 6335	Air pump belt	1		Air injection only

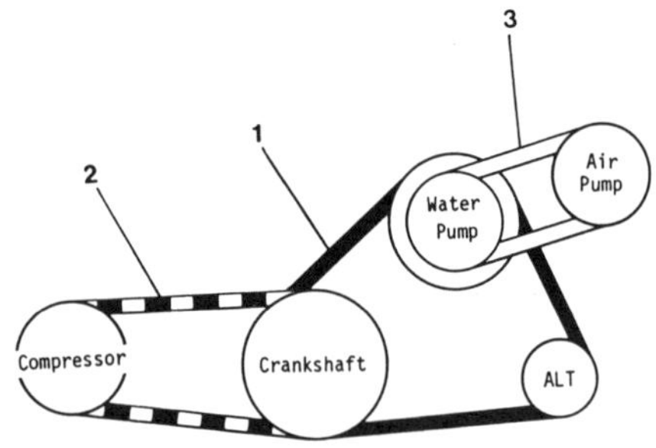

VS4721

1D 05

Illus	Part Number	123456 Description	Quantity	Change Point	Remarks

ENGINE 3.6 LITRE-continued.
GASKET SETS - 3.6 LITRE

Illus	Part Number	Description	Quantity
1	JLM 113	CYLINDER HEAD GASKET SET	1
		Gasket-cylinder head	1
		Gasket-camshaft cover	1
		Gasket-cyl head rear cover	1
		Gasket-front exhaust manifold	1
		Gasket-rear exhaust manifold	1
		Gasket-inlet manifold	2
		Gasket-secondary chain piston	1
		Seal-cam cover(rear)	2
		Seal-spark plug	6
		Seal-inlet valve	12
		'O' Ring-thermostat	1
		'O' Ring-secondary chain piston	1
		'O' Ring-secondary chain N.R.V.	1
		'O' Ring-air cleaner	1
		Ring-exhaust flange	2
2	JLM 835	SUPPLEMENTARY GASKET SET	1
		Gasket-throttle body	2
		Gasket-throttle trumpet	2
		Gasket-manifold adaptor	1
		Gasket-heated restrictor	1
		Gasket-idle air valve	1
		Gasket-auxiliary shaft	1
		Gasket-water rail	1
		Seal-crankshaft(front)	1
		Seal-crankshaft(rear)	1
		Seal-auxiliary shaft	1
		Seal-intermediate shaft	1
		Seal-throttle spindle	2
		Seal-transmission adaptor	2
		'O' Ring-oil gallery plug	2
		'O' Ring-oil pipes	4
		'O' Ring-oil cooler pipes	2
		'O' Ring-chain N/R valve	1
		'O' Ring-oil filler cap	1

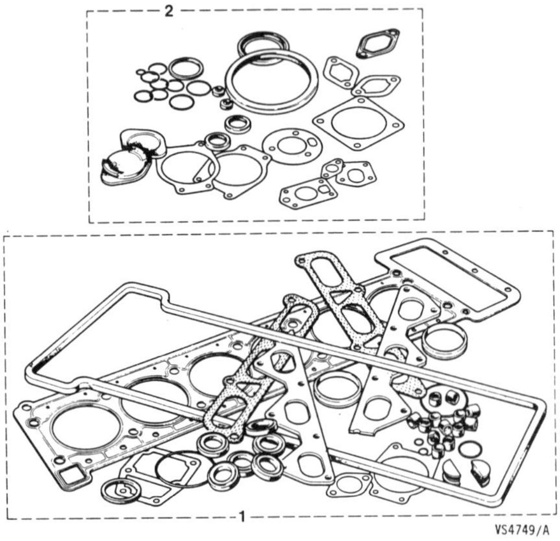

B40AK13/10/88

VS4749A

1D 06

MASTER INDEX
ENGINE 1.C
FLYWHEEL AND CLUTCH 1.G
GEARBOX AND PROPSHAFT 1.H
AXLES, SUSPENSION, DRIVE SHAFTS, WHEELS 1.K
STEERING 1.M
BRAKES AND BRAKE CONTROLS 1.N
FUEL, EXHAUST AND EMISSION SYSTEMS 2.C
COOLING, HEATING AND AIR CONDITIONING 2.G
ELECTRICAL, WASHERS-WIPERS, INSTRUMENTS 2.J
CHASSIS, SUBFRAMES, BODYSHELL, FITTINGS 3.C
FASCIA, TRIM, SEATS AND FIXINGS 3.H
ACCESSORIES AND PAINT 4.C
NUMERICAL INDEX 4.D

GROUP INDEX
ENGINE 3.6 LITRE C02
CAM.-TAPPET BLOCK - 5.3 LITRE E04
CAMSHAFT & VALVES C12
CRANKSHAFT AND MAIN BEARINGS C05
CRANKSHAFT-MAIN BEAR.-5.3 LITRE D13
CYL. HEAD & VALVES - 5.3 LITRE E03
CYLINDER BLOCK D10
CYLINDER HEAD & GASKET C11
DRIVING BELTS-3.6 LITRE D05
EXHAUST MANIFOLD - 3.6 LITRE D03
EXHAUST MANIFOLDS - 5.3 LITRE F02
FRONT ENGINE MOUNTING 3.6 LITRE D08
FRONT ENGINE MOUNTING-5.3 LITRE F09
GASKET SETS - 3.6 LITRE D06
INDUCTION MANIFOLD-3.6 LITRE D02
INLET MANIFOLD-5.3 LITRE E17
LOWER ENG GASKET SET-5.3 LITRE F07
OIL COOLER & OIL PIPES 3.6 LITRE D07
OIL COOLER - 5.3 LITRE F04
OIL COOLER PIPES - 5.3 LITRE F05
OIL FILTER,FILLER CAP-3.6 LITRE C15
OIL FILTER-5.3 LITRE E08
OIL PUMP - 5.3 LITRE D15
OIL PUMP-3.6 LITRE C07
OIL SUMP & DIPSTICK - 5.3 LITRE D12
OIL SUMP & DIPSTICK 3.6 LITRE C04
PISTON & CONN RODS C06
PISTONS & CONN. RODS D14
UPPER ENG GASKET SET-5.3 LITRE F08
WATER PUMP C16
WATER PUMP - 5.3 LITRE E10

JAGUAR XJS RANGE (JAN 1987 ON)

D07 fiche 1

ENGINE

1D 07

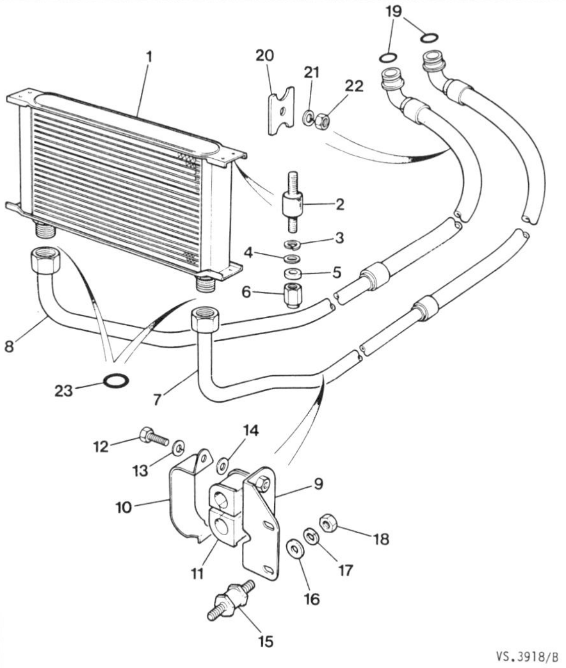

VS 3918B

ENGINE 3.6 LITRE-continued.
OIL COOLER & OIL PIPES 3.6 LITRE

Illus	Part Number	123456 Description	Quantity	Change Point	Remarks
1	CAC 5261	Oil cooler	1		
2	C 45801	Bolt	4		
3	WA 104001 J	Washer	8		
4	WL 104001 J	Lockwasher	8		
5	C 44649	Spacer	8		
6	DAC 2667	Nut	8		
7	CAC 5804	Oil feed pipe	1		
8	CAC 5805	Oil return pipe	1		
9	CBC 1416	Support bracket	1		
10	CAC 5477	Strap	1		
11	CAC 5537	Insulating bush	2		
12	SH 604041 J	Setscrew	3		
13	JLM 298	Washer	3		
14	JLM 302	Lockwasher	3		
15	CAC 8478	Bolt	2		
16	WA 106041 J	Washer	2		
17	WL 600041 J	Lockwasher	2		
18	NH 106041 J	Nut	2		
20	CAC 6647	Clamp plate	1		
19	CAC 5118	'O' ring	2		
21	JLM 9574	Lockwasher	1		
22	JLM 9688	Nut	1		
23	KRC 1160	"O" Ring	2		

MASTER INDEX

GROUP INDEX

JAGUAR XJS RANGE (JAN 1987 ON)

D08 fiche 1

ENGINE

1D 08

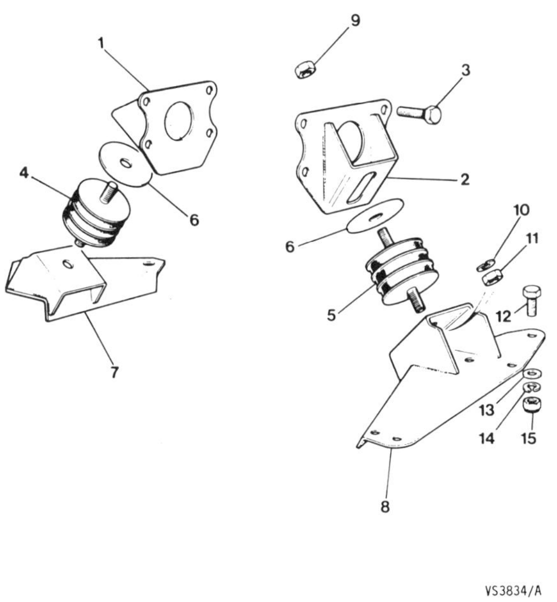

VS 3834A

ENGINE 3.6 LITRE-continued.
FRONT ENGINE MOUNTING 3.6 LITRE

Illus	Part Number	123456 Description	Quantity	Change Point	Remarks
1	CAC 6612	Engine mounting bracket RH	1		
2	CAC 6613	Engine mounting bracket LH	1		
3	FS 110251 J	Setscrew-brackets to engine	8		
4	JLM 1222	Mounting rubber RH	1		
5	JLM 1008	Mounting rubber LH	1		
6	C 30721	Insulator	2		Use CBC 5931
6	CBC 5931	Insulator	2		
7	CAC 6617	Mounting bracket RH	1		
8	CAC 6618	Mounting bracket LH	1		
9	CAC 9816	Nut	2		
10	JLM 304	Washer	2		
11	JLM 9685	Nut rubbers to mounting brackets	2		
12	SH 605071 J	Setscrew-mounting bracket to body	6		
13	JLM 298	Washer-plain	4		
14	JLM 304	Washer-lock	4		
15	NZ 605041 J	Nut-nyloc	2		

MASTER INDEX

ENGINE 1.C
FLYWHEEL AND CLUTCH 1.G
GEARBOX AND PROPSHAFT 1.H
AXLES, SUSPENSION, DRIVE SHAFTS, WHEELS 1.K
STEERING 1.M
BRAKES AND BRAKE CONTROLS 1.N
FUEL, EXHAUST AND EMISSION SYSTEMS 2.C
COOLING, HEATING AND AIR CONDITIONING 2.G
ELECTRICAL, WASHERS-WIPERS, INSTRUMENTS 2.J
CHASSIS, SUBFRAMES, BODYSHELL, FITTINGS 3.C
FASCIA, TRIM, SEATS AND FIXINGS 3.H
ACCESSORIES AND PAINT 4.C
NUMERICAL INDEX 4.D

GROUP INDEX

ENGINE 3.6 LITRE C02
CAM.-TAPPET BLOCK - 5.3 LITRE E04
CAMSHAFT & VALVES C12
CRANKSHAFT AND MAIN BEARINGS C05
CRANKSHAFT-MAIN BEAR.-5.3 LITRE D13
CYL. HEAD & VALVES - 5.3 LITRE E03
CYLINDER BLOCK D10
CYLINDER HEAD & GASKET C11
DRIVING BELTS-3.6 LITRE D05
EXHAUST MANIFOLD - 3.6 LITRE D03
EXHAUST MANIFOLDS - 5.3 LITRE F02
FRONT ENGINE MOUNTING 3.6 LITRE D08
FRONT ENGINE MOUNTING-5.3 LITRE F09
GASKET SETS - 3.6 LITRE D06
INDUCTION MANIFOLD-3.6 LITRE D02
INLET MANIFOLD-5.3 LITRE E17
LOWER ENG GASKET SET-5.3 LITRE F07
OIL COOLER & OIL PIPES 3.6 LITRE D07
OIL COOLER - 5.3 LITRE F04
OIL COOLER PIPES - 5.3 LITRE F05
OIL FILTER,FILLER CAP-3.6 LITRE C15
OIL FILTER-5.3 LITRE E08
OIL PUMP - 5.3 LITRE D15
OIL PUMP-3.6 LITRE C07
OIL SUMP & DIPSTICK - 5.3 LITRE D12
OIL SUMP & DIPSTICK 3.6 LITRE C04
PISTON & CONN RODS C06
PISTONS & CONN. RODS D14
UPPER ENG GASKET SET-5.3 LITRE F08
WATER PUMP C16
WATER PUMP - 5.3 LITRE E10

1D 09

JAGUAR XJS RANGE (JAN 1987 ON) | D09 fiche 1 | ENGINE

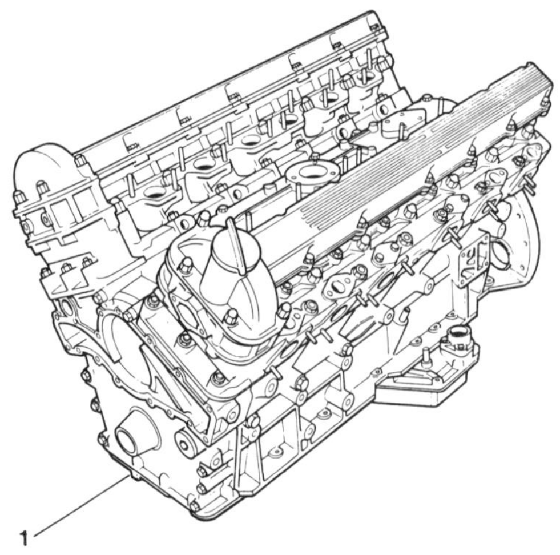

ENGINE 5.3 LITRE
ENGINE UNIT

Illus	Part Number	Description	Quantity	Change Point	Remarks
1	JLM 463 N	STRIPPED ENGINE (HIGH COMP)	1		
1	JLM 463 E	STRIPPED ENGINE (HIGH COMP)	1		
1	JLM 464 N	STRIPPED ENGINE (STD COMP)	1	Up to 1989 MY	Australia/ME
1	JLM 853 N	STRIPPED ENGINE (STD COMP)	1		Germany/Sweden/
1	JLM 853 E	STRIPPED ENGINE (STD COMP)	1		Switzerland
1	RTC 2983 N	STRIPPED ENGINE (STD COMP)	1		USA/CDN/ Japan (see 1)
	JLM 1634 N	STRIPPED ENGINE (STD COMP)	1		Europe (EmissB) Aus/Middle East
	JLM 1635 N	STRIPPED ENGINE (STD COMP)	1	From 1989 MY Digital Ignition	Germany/Sweden Switzerland
	JLM 1636 N	STRIPPED ENGINE (STD COMP)	1		USA/CDN Japan (see 1)

B100AK13/9/88

(1) USA/CDN/Japan use "Relief" Oil Cooling-all other countries use "Full Flow" Oil Cooling.

VS1615

1D 10

JAGUAR XJS RANGE (JAN 1987 ON) | D10 fiche 1 | ENGINE

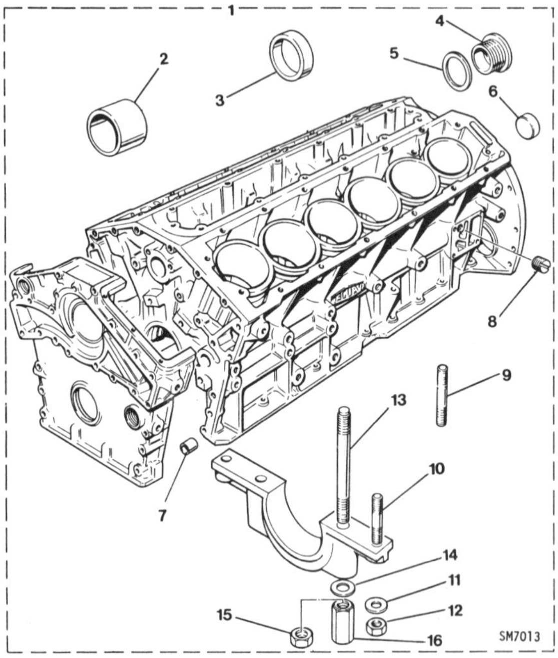

ENGINE 5.3 LITRE-Continued
CYLINDER BLOCK

Illus	Part Number	Description	Quantity	Change Point	Remarks
1	C 46064	CYLINDER BLOCK ASSEMBLY	1	Up to 1989MY	
1	EBC 2255	CYLINDER BLOCK ASSEMBLY	1	From 1989MY	
2	C 29273	Bearing(auxiliary shaft)	1		
3	C 29274	Bearing(auxiliary shaft)	2		
4	C 29268	Plug(oil gallery)	1		
5	C 2296 17	Sealing washer	1		
6	C 16862 2	Plug	2		
7	C 2373	Dowel(timing cover)	2		
8	C 29272	Plug-filter face	1		
9	C 32272	Stud	2		
10	C 32273	Stud	14		
11	WA 110061 J	Washer	16		
12	C 34734	Nut	16		
13	C 29283	Stud	14		
14	C 28082	Washer	14		
15	JLM 9568	Nut	8		
16	C 32726	Nut	6		

SM7013

MASTER INDEX

ENGINE	1.C
FLYWHEEL AND CLUTCH	1.G
GEARBOX AND PROPSHAFT	1.H
AXLES, SUSPENSION, DRIVE SHAFTS, WHEELS	1.K
STEERING	1.M
BRAKES AND BRAKE CONTROLS	1.N
FUEL, EXHAUST AND EMISSION SYSTEMS	2.C
COOLING, HEATING AND AIR CONDITIONING	2.G
ELECTRICAL, WASHERS-WIPERS, INSTRUMENTS	2.J
CHASSIS, SUBFRAMES, BODYSHELL, FITTINGS	3.C
FASCIA, TRIM, SEATS AND FIXINGS	3.H
ACCESSORIES AND PAINT	4.C
NUMERICAL INDEX	4.D

GROUP INDEX

ENGINE 3.6 LITRE	C02
CAM.-TAPPET BLOCK - 5.3 LITRE	E04
CAMSHAFT & VALVES	C12
CRANKSHAFT AND MAIN BEARINGS	C05
CRANKSHAFT-MAIN BEAR.-5.3 LITRE	D13
CYL. HEAD & VALVES - 5.3 LITRE	E03
CYLINDER BLOCK	D10
CYLINDER HEAD & GASKET	C11
DRIVING BELTS-3.6 LITRE	D05
EXHAUST MANIFOLD - 3.6 LITRE	D03
EXHAUST MANIFOLDS - 5.3 LITRE	F02
FRONT ENGINE MOUNTING 3.6 LITRE	D08
FRONT ENGINE MOUNTING-5.3 LITRE	F09
GASKET SETS - 3.6 LITRE	D06
INDUCTION MANIFOLD-3.6 LITRE	D02
INLET MANIFOLD-5.3 LITRE	E17
LOWER ENG GASKET SET-5.3 LITRE	F07
OIL COOLER & OIL PIPES 3.6 LITRE	D07
O'L COOLER - 5.3 LITRE	F04
OIL COOLER PIPES - 5.3 LITRE	F05
OIL FILTER,FILLER CAP-3.6 LITRE	C15
OIL FILTER-5.3 LITRE	E08
OIL PUMP - 5.3 LITRE	D15
OIL PUMP-3.6 LITRE	C07
OIL SUMP & DIPSTICK - 5.3 LITRE	D12
OIL SUMP & DIPSTICK 3.6 LITRE	C04
PISTON & CONN RODS	C06
PISTONS & CONN. RODS	D14
UPPER ENG GASKET SET-5.3 LITRE	F08
WATER PUMP	C16
WATER PUMP - 5.3 LITRE	E10

1D 11

ENGINE 5.3 LITRE-Continued
TOP COVER-CYLINDER LINER

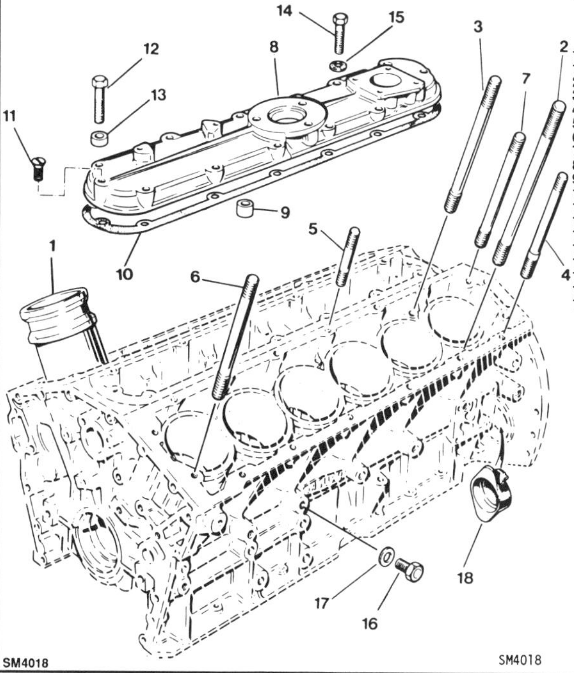

SM4018

Illus	Part Number	Description	Quantity	Change Point	Remarks
1	C 32544 A	Cylinder liner(grade A)	12		
1	C 32544 B	Cylinder liner(grade B)	12		
2	C 32534	Stud-cyl head (exh.side)	10		
3	C 32535	Stud-cyl head (inlet side)	10		
4	C 32679	Stud-cyl head (exh side)	12		
5	C 32680	Stud-cyl head (inlet side)	12		
6	C 32331	Stud-cyl head (exh side)	4		
7	C 32332	Stud-cyl head (inlet side)	4		
8	C 39655	Top cover	1		
9	C 2373	Dowel	2		
10	C 29485	Gasket	1		
11	SF 505101 J	Screw	1		
12	BH 505111 J	Bolt	3		
13	C 19706	Distance piece	2		
14	BH 505101 J	Bolt	10		
15	C 30075 2	Washer	13		
16	SH 508051 J	Plug-oil gallery	5		
17	C 2296 3	Sealing washer	5		
18	C 34016	Seal-rear flange	1		

MASTER INDEX

GROUP INDEX

1D 12

ENGINE 5.3 LITRE-Continued
OIL SUMP & DIPSTICK - 5.3 LITRE

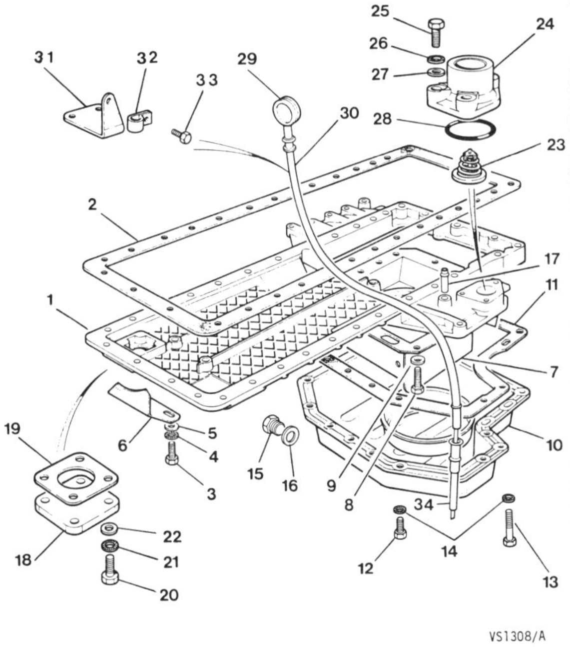

VS1308A

Illus	Part Number	Description	Quantity	Change Point	Remarks
1	C 35572	Oil sump sandwich	1		
2	EAC 7252	Gasket	1		
3	SH 505081 J	Setscrew	18		
4	C 30075 2	Washer	18		
5	WA 108051 J	Washer	2		
6	C 37355	Timing indicator	1	Up to 1989MY only	
7	EAC 8926	Sump baffle	1		
8	SH 505051 J	Setscrew	4		
9	C 30075 2	Washer	4		
10	EAC 3234	Oil sump	1		
11	EAC 7251	Gasket	1		
12	SH 505061 J	Setscrew	14		
13	BH 505141 J	Bolt	15		
14	C 30075 2	Washer	29		
15	C 23435	Oil drain plug	1		
16	C 2296 1	Sealing washer	1		
17	C 37999	Stub pipe	1		
18	EAC 6420	Blanking plate	1		Full flow oil cooling
19	EAC 6421	Gasket	1	....	Full flow oil cooling
20	SH 504051 J	Setscrew	4	....	Full flow oil cooling
21	C 30075 1	Washer	4		Full flow oil cooling
22	WA 106041 J	Washer	4		Full flow oil cooling
23	C 37888	Bypass valve	1	....	Relief oil cooling
24	C 37873	Adaptor	1		
25	SH 504081 J	Setscrew	3		
26	C 30075 1	Washer	3		
27	WA 106041 J	Washer	3		
28	EAC 8101	'O' ring	1		
29	C 44151	Dipstick assembly	1		
30	C 44152	Tube	1		
31	C 45467	Bracket	1		
32	C 44230	Clip	1		
33	C 35093 1	Screw-clip to bracket	1		
34	C 36520	Tube	1		

B106AK13/9/88

MASTER INDEX

ENGINE 1.C
FLYWHEEL AND CLUTCH 1.G
GEARBOX AND PROPSHAFT 1.H
AXLES, SUSPENSION, DRIVE SHAFTS, WHEELS 1.K
STEERING 1.M
BRAKES AND BRAKE CONTROLS 1.N
FUEL, EXHAUST AND EMISSION SYSTEMS 2.C
COOLING, HEATING AND AIR CONDITIONING 2.G
ELECTRICAL, WASHERS-WIPERS, INSTRUMENTS 2.J
CHASSIS, SUBFRAMES, BODYSHELL, FITTINGS 3.C
FASCIA, TRIM, SEATS AND FIXINGS 3.H
ACCESSORIES AND PAINT 4.C
NUMERICAL INDEX 4.D

GROUP INDEX

ENGINE 3.6 LITRE C02
CAM.-TAPPET BLOCK - 5.3 LITRE E04
CAMSHAFT & VALVES C12
CRANKSHAFT AND MAIN BEARINGS C05
CRANKSHAFT-MAIN BEAR.-5.3 LITRE D13
CYL. HEAD & VALVES - 5.3 LITRE E03
CYLINDER BLOCK D10
CYLINDER HEAD & GASKET C11
DRIVING BELTS-3.6 LITRE D05
EXHAUST MANIFOLD - 3.6 LITRE D03
EXHAUST MANIFOLDS - 5.3 LITRE F02
FRONT ENGINE MOUNTING 3.6 LITRE D08
FRONT ENGINE MOUNTING-5.3 LITRE F09
GASKET SETS - 3.6 LITRE D06
INDUCTION MANIFOLD-3.6 LITRE D02
INLET MANIFOLD-5.3 LITRE E17
LOWER ENG GASKET SET-5.3 LITRE F07
OIL COOLER & OIL PIPES 3.6 LITRE D07
OIL COOLER - 5.3 LITRE F04
OIL COOLER PIPES - 5.3 LITRE F05
OIL FILTER,FILLER CAP-3.6 LITRE C15
OIL FILTER-5.3 LITRE E08
OIL PUMP - 5.3 LITRE D15
OIL PUMP-3.6 LITRE C07
OIL SUMP & DIPSTICK - 5.3 LITRE D12
OIL SUMP & DIPSTICK 3.6 LITRE C04
PISTON & CONN RODS C06
PISTONS & CONN. RODS D14
UPPER ENG GASKET SET-5.3 LITRE F08
WATER PUMP C16
WATER PUMP - 5.3 LITRE E10

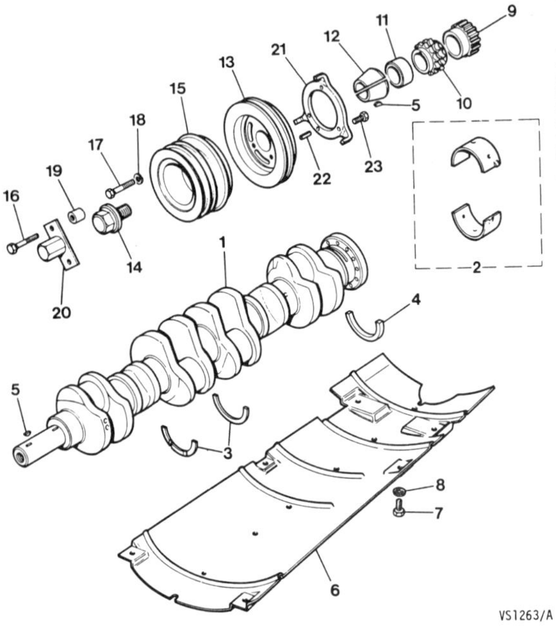

VS1263A

ENGINE 5.3 LITRE-Continued
CRANKSHAFT-MAIN BEAR.-5.3 LITRE

Illus	Part Number	Description	Quantity	Change Point	Remarks
1	EAC 2757	Crankshaft	1	Up to 8S.61038	
1	EBC 2871	Crankshaft	1	From 8S.61039	
2	RTC 2728	Main bearings set	1		
3	C 34697	Thrust washer	2		
3	C 34697 1	Thrust washer	2		Oversize
4	C 19654	Rear oil seal	2		
5	C 757	Key	4		
6	EAC 8927	Undershield	1		
7	SH 605041 J	Setscrew	2		
8	C 30075 2	Washer	2		
9	C 32185	Collar-oil pump driving	1		
10	C 29330	Sprocket	1		
11	C 32197	Distance piece	1		
12	C 13892	Damper cone	1		
13	C 36013	Damper	1	Up to 8S.57571	
13	EAC 9248	Damper	1	8S57572 to 1989 MY	With A.B.S Brakes
13	EAC 9693	Damper	1	From 1989 MY Digital Ignition	
21	EAC 9692	Timing disc	1		
22	PA 112121 J	Pin-disc to damper	1		
23	SH 106121 PJ	Screw-disc to damper	3		
14	C 32367 1	Bolt-securing damper	1		
15	C 37892	Crankshaft pulley	1		
16	BH 605171 J	Bolt	2	Up to 8S.87571	
17	SH 605091 J	Setscrew	2		
18	WM 600051 J	Washer	4		
16	BH 108111 J	Bolt	2	From 8S.57571	With A.B.S. Brakes
17	BH 108061 J	Bolt	2		
18	WL 108001 J	Washer	4		
19	C 19129	Spacer	2		
20	C 38549	Turning plate	1		

B108AK1/9/88

1D 13

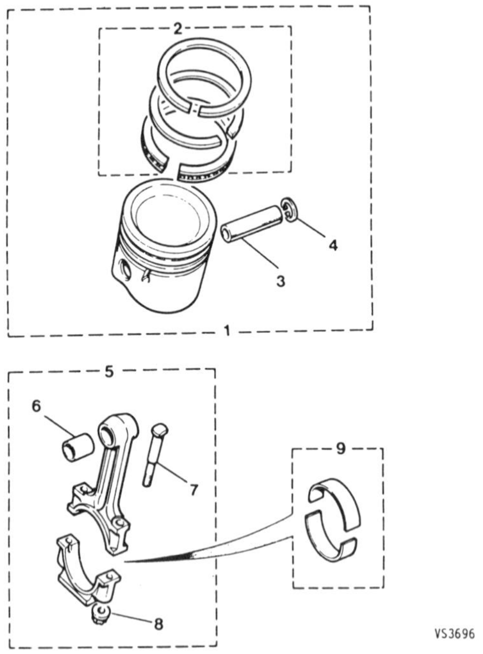

VS3696

ENGINE 5.3 LITRE-Continued
PISTONS & CONN. RODS

Illus	Part Number	Description	Quantity	Change Point	Remarks
1	EAC 6252 PA	PISTON ASSY(HIGH COMP)-GRADE A	12	Up to 1989MY only	
1	EAC 6252 PB	PISTON ASSY(HIGH COMP)-GRADE B	12		
2	RTC 2708	Piston ring kit	1		
3	EAC 3701	Gudgeon pin	1		
4	C 29699	Circlip	2		
1	EAC 6253 PA	PISTON ASSY(STD COMP)-GRADE A	12		
1	EAC 6253 PB	PISTON ASSY(STD COMP)-GRADE B	12		
2	RTC 2708	Piston ring kit	1		
3	EAC 3701	Gudgeon pin	1		
4	C 29699	Circlip	2		
5	RTC 2997	CONNECTING ROD ASSY	1SET		
6	C 29308	Small end bush	12		
7	C 43809	Bolt	24		
8	EAC 5541	Nut	24		
9	JS 571	BEARING KIT(BIG END)	1SET		

B110AK24/6/88

1D 14

1D 15

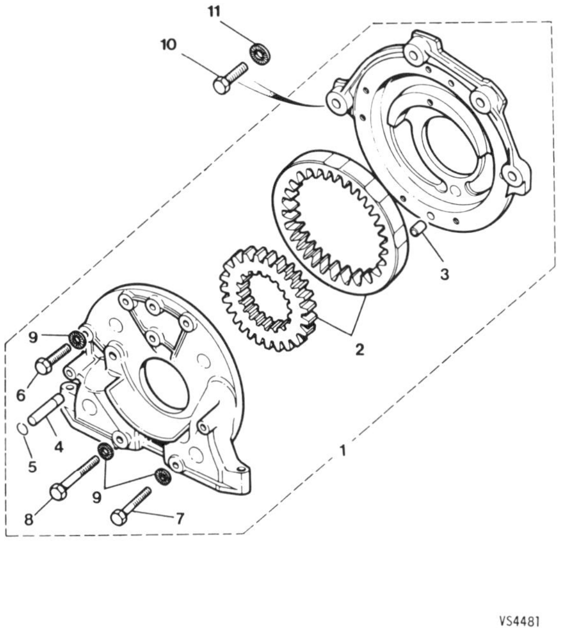

VS4481

ENGINE 5.3 LITRE-Continued
OIL PUMP - 5.3 LITRE

Illus	Part Number	Description	Quantity	Change Point	Remarks
1	C 40177	OIL PUMP ASSEMBLY	1		
2	RTC 2075	Pump gears kit	1		
3	C 31193	Dowel	2		
4	C 44250	Pivot pin	1		
5	C 44251	Circlip	1		
6	BH 604091 J	Bolt - cover to housing	4		
7	BH 604161 J	Bolt	2		
8	BH 604181 J	Bolt	2		
9	C 30075 1	Washer	8		
10	SH 505091 J	Setscrew-pump to block	4		
11	C 30075 2	Washer			

1D 16

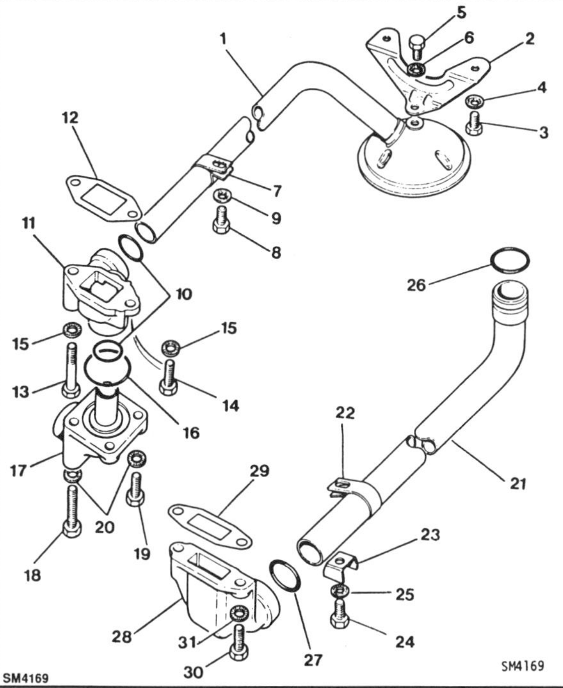

SM4169

ENGINE 5.3 LITRE-continued
OIL PUMP PIPES
USA CANADA JAPAN ONLY

Illus	Part Number	Description	Quantity	Change Point	Remarks
1	C 35512	Oil suction pipe	1		
2	C 35521	Bracket	1		
3	SH 605051 J	Setscrew-securing bracket	2		
4	C 30075 2	Washer	2		
5	SH 605041 J	Setscrew-pipe to bracket	2		
6	C 30075 2	Washer	2		
7	C 33851	Clip	1		
8	SH 605051 J	Setscrew-securing clip	1		
9	C 30075 2	Washer	1		
10	C 33855	'O' Ring	2		
11	C 33869	Oil suction elbow	1		
12	C 31063	Gasket	1		
13	BH 505151 J	Bolt-elbow to pump	1		
14	SH 505071 J	Setscrew-elbow to pump	1		
15	C 30075 2	Washer	2		
16	C 35657	'O' Ring	1		Relief oil cooling engine (USA,CDN, Japan only)
17	C 37882	Inlet elbow	1		
18	SH 504171 J	Setscrew-elbow to sump	2		
19	SH 504081 J	Setscrew-elbow to sump	2		
20	C 30075 1	Washer	4		
	WA 106041 J	Washer	4		
21	C 35568	Oil delivery pipe	1		
22	C 35570	Clip	1		
23	C 41028	Stiffener	1		
24	SH 605061 J	Setscrew	1		
25	C 30075 2	Washer	1		
26	EAC 8102	'O' Ring	1		
27	C 29828	'O' Ring	1		
28	C 35571	Oil delivery elbow	1		
29	C 31064	Gasket	1		
30	SH 505071 J	Setscrew-elbow to pump	2		
31	C 30075 2	Washer	2		

B114AK20/6/88

MASTER INDEX

ENGINE 1.C
FLYWHEEL AND CLUTCH 1.G
GEARBOX AND PROPSHAFT 1.H
AXLES, SUSPENSION, DRIVE SHAFTS, WHEELS 1.K
STEERING 1.M
BRAKES AND BRAKE CONTROLS 1.N
FUEL, EXHAUST AND EMISSION SYSTEMS 2.C
COOLING, HEATING AND AIR CONDITIONING 2.G
ELECTRICAL, WASHERS-WIPERS, INSTRUMENTS 2.J
CHASSIS, SUBFRAMES, BODYSHELL, FITTINGS 3.C
FASCIA, TRIM, SEATS AND FIXINGS 3.H
ACCESSORIES AND PAINT 4.C
NUMERICAL INDEX 4.D

GROUP INDEX

ENGINE 3.6 LITRE C02
CAM.-TAPPET BLOCK - 5.3 LITRE E04
CAMSHAFT & VALVES C12
CRANKSHAFT AND MAIN BEARINGS C05
CRANKSHAFT-MAIN BEAR.-5.3 LITRE D13
CYL. HEAD & VALVES - 5.3 LITRE E03
CYLINDER BLOCK D10
CYLINDER HEAD & GASKET C11
DRIVING BELTS-3.6 LITRE D05
EXHAUST MANIFOLD - 3.6 LITRE D03
EXHAUST MANIFOLDS - 5.3 LITRE F02
FRONT ENGINE MOUNTING 3.6 LITRE D08
FRONT ENGINE MOUNTING-5.3 LITRE F09
GASKET SETS - 3.6 LITRE D06
INDUCTION MANIFOLD-3.6 LITRE D02
INLET MANIFOLD-5.3 LITRE E17
LOWER ENG GASKET SET-5.3 LITRE F07
OIL COOLER & OIL PIPES 3.6 LITRE D07
OIL COOLER - 5.3 LITRE F04
OIL COOLER PIPES - 5.3 LITRE F05
OIL FILTER,FILLER CAP-3.6 LITRE C15
OIL FILTER-5.3 LITRE E08
OIL PUMP - 5.3 LITRE D15
OIL PUMP-3.6 LITRE C07
OIL SUMP & DIPSTICK - 5.3 LITRE D12
OIL SUMP & DIPSTICK 3.6 LITRE C04
PISTON & CONN RODS C06
PISTONS & CONN. RODS D14
UPPER ENG GASKET SET-5.3 LITRE F08
WATER PUMP C16
WATER PUMP - 5.3 LITRE E10

1D 17

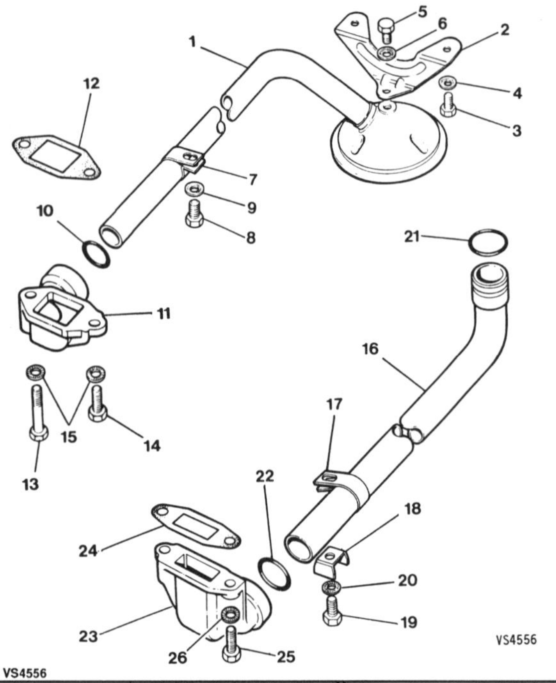

VS4556

ENGINE 5.3 LITRE-continued
OIL PUMP PIPES
EXCEPT USA CANADA JAPAN

Illus	Part Number	Description	Quantity	Change Point	Remarks
1	EAC 6424	Oil suction pipe	1		"Full flow" oil cooling engine (except USA, CDN & Japan)
2	C 35521	Bracket	1		
3	SH 605051 J	Setscrew-securing bracket	2		
4	C 30075 2	Washer	2		
5	SH 605041 J	Setscrew-pipe to bracket	2		
6	C 30075 2	Washer	2		
7	EAC 8841	Clip	1		
8	SH 605051 J	Setscrew-securing clip	1		
9	C 30075 2	Washer	1		
10	EAC 6423	'O' ring	1		
11	EAC 6422	Oil suction elbow	1		
12	C 31063	Gasket	1		
13	BH 505151 J	Bolt - elbow to pump	1		
14	SH 505071 J	Setscrew	1		
15	C 30075 2	Washer	2		
16	C 35568	Oil delivery pipe	1		
17	C 35570	Clip	1		
18	C 41028	Stiffener	1		
19	SH 605061 J	Setscrew	1		
20	C 30075 2	Washer	1		
21	EAC 8102	'O' ring	1		
22	C 29829	'O' ring	1		
23	C 35571	Oil delivery elbow	1		
24	C 31064	Gasket	1		
25	SH 505071 J	Setscrew - elbow to pump	2		
26	C 30075 2	Washer	2		

b116AK20/6/88

MASTER INDEX

GROUP INDEX

1D 18

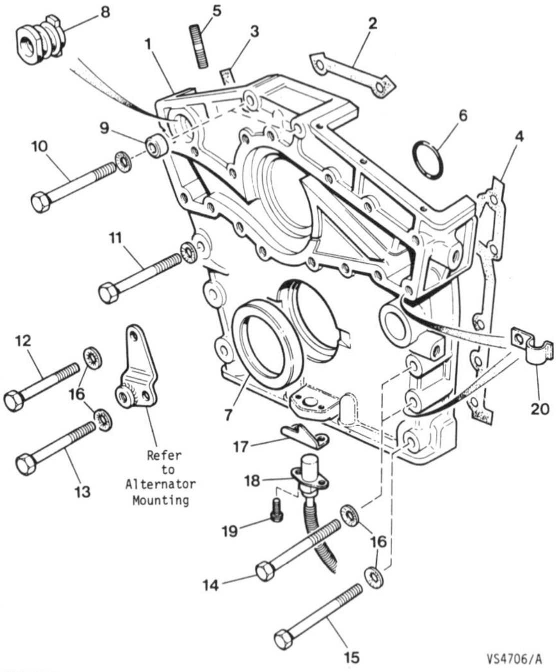

VS4706A

ENGINE 5.3 LITRE-continued
TIMING COVER

Illus	Part Number	Description	Quantity	Change Point	Remarks
1	EAC 1729	Timing cover	1	Up to 1989MY	
1	EBC 1416	Timing cover	1	From 1989MY	
2	C 29339	Gasket (top)	1		
3	C 29338	Gasket (RH)	1		
4	C 29337	Gasket (LH)	1		
5	C 29340	Stud	6		
6	EAC 8102	'O' ring	1		
7	C 30002	Oil seal	1		
8	C 37925	Plug (chain adjusting)	1		
9	BD 19203	Spacer	2		
10	BH 505291 J	Bolt - cover to block	2		
11	BH 505251 J	Bolt	1		
12	BH 505261 J	Bolt	1		
13	BH 505281 J	Bolt	1		
14	BH 505241 J	Bolt	2		
15	BH 505271 J	Bolt	1		
16	C 30075 2	Washer	8		
17	EAC 9848	Timing indicator	1	From 1989 MY Digital Ignition	
18	DAC 4606	Electromagnetic sensor	1		
19	SN 105101 J	Screw	2		
20	C 34920	Clip	2		
	ADU 9028	Cable tie	1		
	DAC 5998	Hose-sensor cable	1		

B118AK23/9/88

MASTER INDEX

ENGINE ... 1.C
FLYWHEEL AND CLUTCH ... 1.G
GEARBOX AND PROPSHAFT ... 1.H
AXLES, SUSPENSION, DRIVE SHAFTS, WHEELS ... 1.K
STEERING ... 1.M
BRAKES AND BRAKE CONTROLS ... 1.N
FUEL, EXHAUST AND EMISSION SYSTEMS ... 2.C
COOLING, HEATING AND AIR CONDITIONING ... 2.G
ELECTRICAL, WASHERS-WIPERS, INSTRUMENTS ... 2.J
CHASSIS, SUBFRAMES, BODYSHELL, FITTINGS ... 3.C
FASCIA, TRIM, SEATS AND FIXINGS ... 3.H
ACCESSORIES AND PAINT ... 4.C
NUMERICAL INDEX ... 4.D

GROUP INDEX

ENGINE 3.6 LITRE ... C02
CAM.-TAPPET BLOCK - 5.3 LITRE ... E04
CAMSHAFT & VALVES ... C12
CRANKSHAFT AND MAIN BEARINGS ... C05
CRANKSHAFT-MAIN BEAR.-5.3 LITRE ... D13
CYL. HEAD & VALVES - 5.3 LITRE ... E03
CYLINDER BLOCK ... D10
CYLINDER HEAD & GASKET ... C11
DRIVING BELTS-3.6 LITRE ... D05
EXHAUST MANIFOLD - 3.6 LITRE ... D03
EXHAUST MANIFOLDS - 5.3 LITRE ... F02
FRONT ENGINE MOUNTING 3.6 LITRE ... D08
FRONT ENGINE MOUNTING-5.3 LITRE ... F09
GASKET SETS - 3.6 LITRE ... D06
INDUCTION MANIFOLD-3.6 LITRE ... D02
INLET MANIFOLD-5.3 LITRE ... E17
LOWER ENG GASKET SET-5.3 LITRE ... F07
OIL COOLER & OIL PIPES 3.6 LITRE ... D07
OIL COOLER - 5.3 LITRE ... F04
OIL COOLER PIPES - 5.3 LITRE ... F05
OIL FILTER,FILLER CAP-3.6 LITRE ... C15
OIL FILTER-5.3 LITRE ... E08
OIL PUMP - 5.3 LITRE ... D15
OIL PUMP-3.6 LITRE ... C07
OIL SUMP & DIPSTICK - 5.3 LITRE ... D12
OIL SUMP & DIPSTICK 3.6 LITRE ... C04
PISTON & CONN RODS ... C06
PISTONS & CONN. RODS ... D14
UPPER ENG GASKET SET-5.3 LITRE ... F08
WATER PUMP ... C16
WATER PUMP - 5.3 LITRE ... E10

JAGUAR XJS RANGE (JAN 1987 ON) — E02 fiche 1 — ENGINE

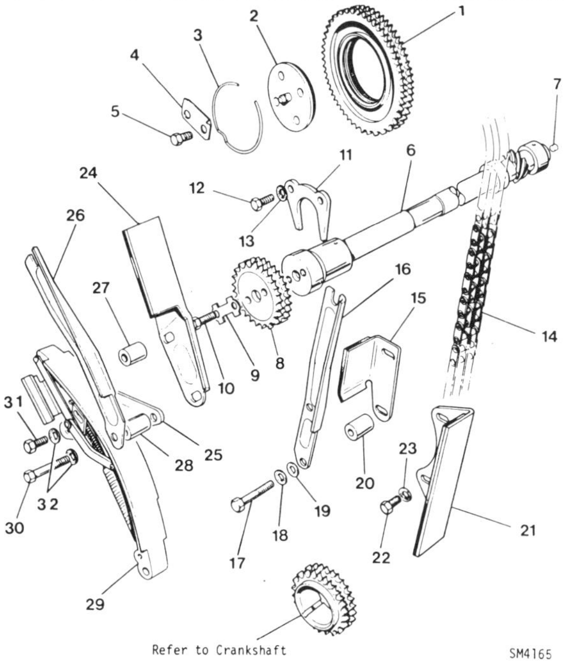

SM4165

ENGINE 5.3 LITRE-continued
TIMING CHAIN - AUXILIARY SHAFT

Illus	Part Number	Description	Quantity	Change Point	Remarks
1	C 29349	Camshaft sprocket	2		
2	C 33889	Coupling	2		
3	C 30767	Circlip	2		
4	C 33917	Tabwasher	4		
5	C 33918	Screw	8		
6	C 38421	AUXILIARY SHAFT	1		
7	C 33244	Plug	2		
8	C 32227	Sprocket	1		
9	C 31076	Lockplate	1		
10	SH 604061 J	Setscrew	2		
11	C 29335	Thrust plate	1		
12	SH 505061 J	Setscrew	2		
13	C 30075 2	Washer	2		
14	C 29590	Timing chain	1		
15	C 29635	Upper damper	1		
16	C 30427	Support bracket	1		
17	BH 505141 J	Bolt	2		
18	C 30075 2	Washer	2		
19	JLM 299	Washer	2		
20	C 30493	Distance piece	2		
21	C 29636	Lower damper	1		
22	SH 505061 J	Setscrew	2		
23	C 30075 2	Washer	2		
24	C 31068	Bracket-damper	1		
25	C 30484	Bracket-tensioner support	1		
26	C 31201	Bracket-sprocket support	1		
27	C 19129	Distance piece	1		
28	C 30493	Distance piece	1		
29	C 30485	Chain tensioner	1		
30	BH 505141 J	Bolt	2		
31	SH 505061 J	Setscrew	1		
32	C 30075 2	Washer	3		

JAGUAR XJS RANGE (JAN 1987 ON) — E03 fiche 1 — ENGINE

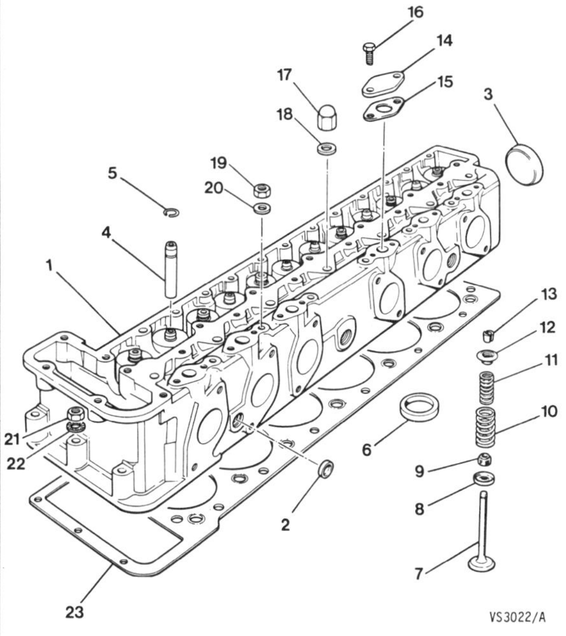

VS3022A

ENGINE 5.3 LITRE-Continued
CYL. HEAD & VALVES - 5.3 LITRE

Illus	Part Number	Description	Quantity	Change Point	Remarks
	RTC 2970	CYLINDER HEAD ASSEMBLY-RH	1		
	RTC 2971	CYLINDER HEAD ASSEMBLY-LH	1		
	JLM 564	CYLINDER HEAD ASSEMBLY-RH	1		See (1) below
	JLM 565	CYLINDER HEAD ASSEMBLY-LH	1		See (1) below
1	EAC 5112	CYLINDER HEAD RH(LESS VALVES)	1		
1	EAC 5113	CYLINDER HEAD LH(LESS VALVES)	1		
2	C 16862 2	Core plug	14		
3	C 16862 6	Core plug	2		
4	C 29390	Inlet valve guide	12		1st Oversize
4	C 29391	Inlet valve guide	12		2nd Oversize
4	EAC 3636	Exhaust valve guide	12		1st Oversize
4	EAC 3637	Exhaust valve guide	12		2nd Oversize
5	C 26199	Circlip	24		
6	EAC 3639	Insert-inlet valve	12		
6	EAC 3641	Insert-exhaust valve	12		
7	EAC 3191	Inlet valve	12		
	EAC 3192	Exhaust valve	12		
	EAC 7318	Exhaust valve	12		See (1) below
8	C 27481	Seat-valve spring	24		
9	EAC 5032	Seal-inlet valve	12		
10	C 14591	Valve spring(outer)	24		
11	C 14592	Valve spring(inner)	24		
12	C 27480	Valve collar	24		
13	C 27482	Valve cotter	24pr		
12	C 27480	Valve collar(inlet)	12		See (1) below
12	EAC 7309	Valve collar(exhaust)	12		See (1) below
13	C 27482	Valve cotter(inlet)	12pr		See (1) below
13	EAC 7256	Valve cotter(exhaust)	12pr		See (1) below
14	C 34283	Blanking plate	4		
15	C 30344	Gasket	4		
16	FB 108201 J	Bolt	8		
17	C 37523	Nut	28		
18	C 29419	Washer	28		
19	C 34734	Nut	24		
20	C 29420	Washer	24		
21	JLM 9684	Nut	6		
22	C 30075 2	Washer	6		
23	EAC 1954	Cylinder head gasket-RH	1		
	EAC 1955	Cylinder head gasket-LH	1		

B122AK13/9/88

NOTE:
(1).Germany (Emiss "F"/"G") - Sweden - Switzerland (Unleaded Fuel).

MASTER INDEX

ENGINE 1.C
FLYWHEEL AND CLUTCH 1.G
GEARBOX AND PROPSHAFT 1.H
AXLES, SUSPENSION, DRIVE SHAFTS, WHEELS 1.K
STEERING 1.M
BRAKES AND BRAKE CONTROLS 1.N
FUEL, EXHAUST AND EMISSION SYSTEMS 2.C
COOLING, HEATING AND AIR CONDITIONING 2.G
ELECTRICAL, WASHERS-WIPERS, INSTRUMENTS 2.J
CHASSIS, SUBFRAMES, BODYSHELL, FITTINGS 3.C
FASCIA, TRIM, SEATS AND FIXINGS 3.H
ACCESSORIES AND PAINT 4.C
NUMERICAL INDEX 4.D

GROUP INDEX

ENGINE 3.6 LITRE C02
CAM.-TAPPET BLOCK - 5.3 LITRE E04
CAMSHAFT & VALVES C12
CRANKSHAFT AND MAIN BEARINGS C05
CRANKSHAFT-MAIN BEAR.-5.3 LITRE D13
CYL HEAD & VALVES - 5.3 LITRE E03
CYLINDER BLOCK D10
CYLINDER HEAD & GASKET C11
DRIVING BELTS-3.6 LITRE D05
EXHAUST MANIFOLD - 3.6 LITRE D03
EXHAUST MANIFOLDS - 5.3 LITRE F02
FRONT ENGINE MOUNTING 3.6 LITRE D08
FRONT ENGINE MOUNTING-5.3 LITRE F09
GASKET SETS - 3.6 LITRE D06
INDUCTION MANIFOLD-3.6 LITRE D02
INLET MANIFOLD-5.3 LITRE E17
LOWER ENG GASKET SET-5.3 LITRE F07
OIL COOLER & OIL PIPES 3.6 LITRE D07
OIL COOLER - 5.3 LITRE F04
OIL COOLER PIPES - 5.3 LITRE F05
OIL FILTER,FILLER CAP-3.6 LITRE C15
OIL FILTER-5.3 LITRE E08
OIL PUMP - 5.3 LITRE D15
OIL PUMP-3.6 LITRE C07
OIL SUMP & DIPSTICK - 5.3 LITRE D12
OIL SUMP & DIPSTICK 3.6 LITRE C04
PISTON & CONN RODS C06
PISTONS & CONN. RODS D14
UPPER ENG GASKET SET-5.3 LITRE F08
WATER PUMP C16
WATER PUMP - 5.3 LITRE E10

1E 04

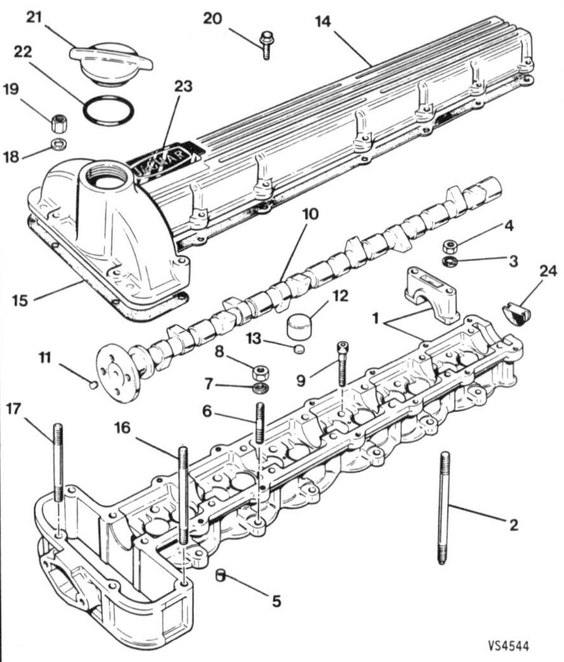

ENGINE 5.3 LITRE-continued
CAM.-TAPPET BLOCK - 5.3 LITRE

Illus	Part Number	123456 Description	Quantity	Change Point	Remarks
1	RTC 3005	Tappet block-RH	1		
1	RTC 3006	Tappet block-LH	1		
2	EAC 3653	Stud	28		
3	C 30075 2	Washer	28		
4	JLM 9688	Nut	28		
5	C 2373	Dowel	4		
6	TE 108051 J	Stud	12		
7	C 30075 2	Washer	12		
8	JLM 9688	Nut	12		
9	SS 108400 J	Socket screw	12		
10	C 42176	CAMSHAFT ASSEMBLY-RH	1		
10	C 42177	CAMSHAFT ASSEMBLY-LH	1		
11	C 33244	Plug	4		
12	C 44010	Tappet	24		
		TAPPET ADJUSTING SHIM			
	C 2243 A	0.085"	A/R		
	C 2243 B	0.086"	A/R		
	C 2243 C	0.087"	A/R		
	C 2243 D	0.088"	A/R		
	C 2243 E	0.089"	A/R		
	C 2243 F	0.090"	A/R		
	C 2243 G	0.091"	A/R		
	C 2243 H	0.092"	A/R		
	C 2243 I	0.093"	A/R		
	C 2243 J	0.094"	A/R		
	C 2243 K	0.095"	A/R		
	C 2243 L	0.096"	A/R		
	C 2243 M	0.097"	A/R		
	C 2243 N	0.098"	A/R		

MASTER INDEX

GROUP INDEX

1E 05

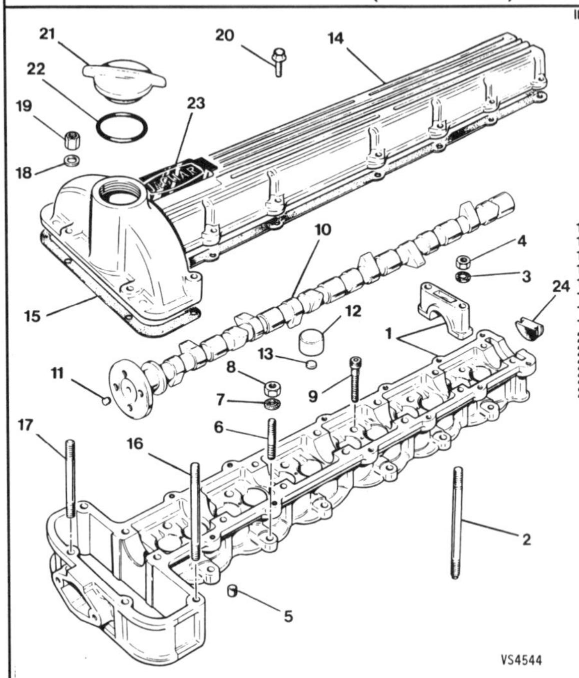

ENGINE 5.3 LITRE-continued
CAMSHAFT-TAPPET BLOCK

Illus	Part Number	123456 Description	Quantity	Change Point	Remarks
		TAPPET ADJUSTING SHIM			
	C 2243 O	0.0990"	A/R		
	C 2243 P	0.0100"	A/R		
	C 2243 Q	0.0101"	A/R		
	C 2243 R	0.0102"	A/R		
	C 2243 S	0.0103"	A/R		
	C 2243 T	0.0104"	A/R		
	C 2243 U	0.0105"	A/R		
	C 2243 V	0.0106"	A/R		
	C 2243 W	0.0107"	A/R		
	C 2243 X	0.0108"	A/R		
14	C 35800	Camshaft cover-RH	1		
14	C 35801	Camshaft cover-LH	1		
15	C 29428	Gasket-RH	1		
15	C 29429	Gasket-LH	1		
16	TE 108171 J	Stud	4		
17	TE 108151 J	Stud	4		
18	WA 108051 J	Washer	8		
19	JLM 9688	Nut	8		
20	SN 106251 J	Screw	24		
21	C 34757	Oil filler cap	1		
22	C 42511	'O' ring	1		
23	C 35732	Emblem	2		
24	C 30431	Rear sealing plug	2		

MASTER INDEX

ENGINE 1.C
FLYWHEEL AND CLUTCH 1.G
GEARBOX AND PROPSHAFT 1.H
AXLES, SUSPENSION, DRIVE SHAFTS, WHEELS 1.K
STEERING 1.M
BRAKES AND BRAKE CONTROLS 1.N
FUEL, EXHAUST AND EMISSION SYSTEMS 2.C
COOLING, HEATING AND AIR CONDITIONING 2.G
ELECTRICAL, WASHERS-WIPERS, INSTRUMENTS 2.J
CHASSIS, SUBFRAMES, BODYSHELL, FITTINGS 3.C
FASCIA, TRIM, SEATS AND FIXINGS 3.H
ACCESSORIES AND PAINT 4.C
NUMERICAL INDEX 4.D

GROUP INDEX

ENGINE 3.6 LITRE C02
CAM.-TAPPET BLOCK - 5.3 LITRE E04
CAMSHAFT & VALVES C12
CRANKSHAFT AND MAIN BEARINGS C05
CRANKSHAFT-MAIN BEAR.-5.3 LITRE D13
CYL. HEAD & VALVES - 5.3 LITRE E03
CYLINDER BLOCK D10
CYLINDER HEAD & GASKET C11
DRIVING BELTS-3.6 LITRE D05
EXHAUST MANIFOLD - 3.6 LITRE D03
EXHAUST MANIFOLDS - 5.3 LITRE F02
FRONT ENGINE MOUNTING 3.6 LITRE D08
FRONT ENGINE MOUNTING-5.3 LITRE F09
GASKET SETS - 3.6 LITRE D06
INDUCTION MANIFOLD-3.6 LITRE D02
INLET MANIFOLD-5.3 LITRE E17
LOWER ENG GASKET SET-5.3 LITRE F07
OIL COOLER & OIL PIPES 3.6 LITRE D07
OIL COOLER - 5.3 LITRE F04
OIL COOLER PIPES - 5.3 LITRE F05
OIL FILTER,FILLER CAP-3.6 LITRE C15
OIL FILTER-5.3 LITRE E08
OIL PUMP - 5.3 LITRE D15
OIL PUMP-3.6 LITRE C07
OIL SUMP & DIPSTICK - 5.3 LITRE D12
OIL SUMP & DIPSTICK 3.6 LITRE C04
PISTON & CONN RODS C06
PISTONS & CONN. RODS D14
UPPER ENG GASKET SET-5.3 LITRE F08
WATER PUMP C16
WATER PUMP - 5.3 LITRE E10

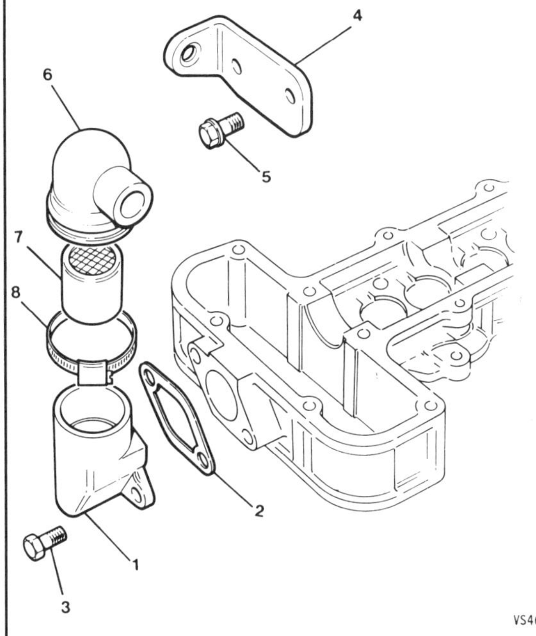

VS4630

ENGINE 5.3 LITRE-continued

BREATHER HOUSING

Illus	Part Number	Description	Quantity	Change Point	Remarks
1	C 44219	Breather housing	1		
2	C 32366	Gasket	1		
3	FS 108161 J	Screw-housing to tappet block	2		
4	C 41943	Adjuster bracket	1		
5	FS 108161 J	Setscrew-brkt to tappet block-RH	2		
6	C 44386	Breather cap	1		
7	C 44410	Capsule	1		
8	C 41116 29	Clip	1		

1E 06

MASTER INDEX

GROUP INDEX

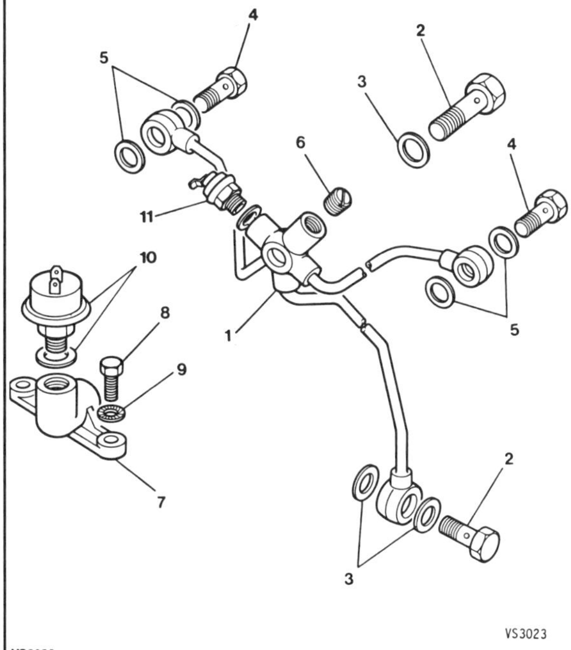

VS3023

ENGINE 5.3 LITRE-Continued

OIL FEED PIPE

Illus	Part Number	Description	Quantity	Change Point	Remarks
1	EAC 3489	Oil feed pipe	1		
2	C 29652	Banjo bolt	2		
3	C 2296 3	Washer	4		
4	C 5846	Banjo bolt	2		
5	C 4146	Washer	4		
6	EAC 3569	Plug	1		
7	C 42072	Adaptor	1		
8	SH 504071 J	Setscrew	2		
9	C 30075 1	Washer	2		
10	C 46272	Oil gauge transmitter	1		
11	C 42200	Oil pressure switch	1		

1E 07

MASTER INDEX

ENGINE 1.C
FLYWHEEL AND CLUTCH 1.G
GEARBOX AND PROPSHAFT 1.H
AXLES, SUSPENSION, DRIVE SHAFTS, WHEELS 1.K
STEERING 1.M
BRAKES AND BRAKE CONTROLS 1.N
FUEL, EXHAUST AND EMISSION SYSTEMS 2.C
COOLING, HEATING AND AIR CONDITIONING 2.G
ELECTRICAL, WASHERS-WIPERS, INSTRUMENTS 2.J
CHASSIS, SUBFRAMES, BODYSHELL, FITTINGS 3.C
FASCIA, TRIM, SEATS AND FIXINGS 3.H
ACCESSORIES AND PAINT 4.C
NUMERICAL INDEX 4.D

GROUP INDEX

ENGINE 3.6 LITRE C02
CAM.-TAPPET BLOCK - 5.3 LITRE E04
CAMSHAFT & VALVES C12
CRANKSHAFT AND MAIN BEARINGS C05
CRANKSHAFT-MAIN BEAR.-5.3 LITRE D13
CYL. HEAD & VALVES - 5.3 LITRE E03
CYLINDER BLOCK D10
CYLINDER HEAD & GASKET C11
DRIVING BELTS-3.6 LITRE D05
EXHAUST MANIFOLD - 3.6 LITRE D03
EXHAUST MANIFOLDS - 5.3 LITRE F02
FRONT ENGINE MOUNTING 3.6 LITRE D08
FRONT ENGINE MOUNTING-5.3 LITRE F09
GASKET SETS - 3.6 LITRE D06
INDUCTION MANIFOLD-3.6 LITRE D02
INLET MANIFOLD-5.3 LITRE E17
LOWER ENG GASKET SET-5.3 LITRE F07
OIL COOLER & OIL PIPES 3.6 LITRE D07
OIL COOLER - 5.3 LITRE F04
OIL COOLER PIPES - 5.3 LITRE F05
OIL FILTER,FILLER CAP-3.6 LITRE C15
OIL FILTER-5.3 LITRE E08
OIL PUMP - 5.3 LITRE D15
OIL PUMP-3.6 LITRE C07
OIL SUMP & DIPSTICK - 5.3 LITRE D12
OIL SUMP & DIPSTICK 3.6 LITRE C04
PISTON & CONN RODS C06
PISTONS & CONN. RODS D14
UPPER ENG GASKET SET-5.3 LITRE F08
WATER PUMP C16
WATER PUMP - 5.3 LITRE E10

1E 08

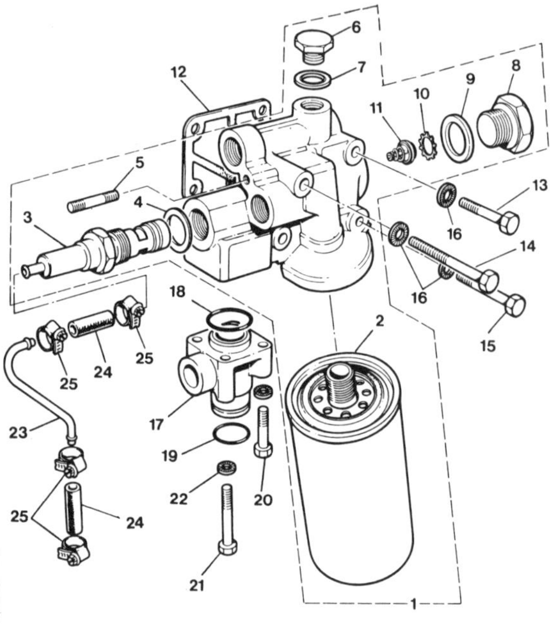

VS4619

VS4619

ENGINE 5.3 LITRE-Continued

OIL FILTER-5.3 LITRE
EXCEPT USA CANADA JAPAN

Illus	Part Number	123456 Description	Quantity	Change Point	Remarks
1	EAC 6398	OIL FILTER ASSEMBLY	1		
2	C 42797	Oil filter	1		
3	EAC 7149	Relief valve	1		
4	JS 224	Sealing washer	1		
5	TE 505111 J	Stud	1		
6	EAC 6402	Plug	1		
7	EAC 3964	Sealing washer	1		
8	EAC 6400	Plug	1		Full flow
9	EAC 6401	Sealing washer	1		oil cooling
10	EAC 6404	Lock ring	1		(except USA,
11	C 37888	Bypass valve	1		CDN,Japan)
12	EAC 6337	Gasket	1		
13	BH 505181 J	Bolt-filter to cyl block	2		
14	BH 505321 J	Bolt	1		
15	BH 505241 J	Bolt	1		
16	C 30075	Washer	4		
17	EAC 6789	Outlet elbow	1		
18	EAC 8101	'O' ring	1		
19	EAC 8102	'O' ring	1		
20	BH 504121 J	Bolt	2		
21	BH 504161 J	Bolt	2		
22	C 30075 1	Washer	4		
23	C 37874	Oil vent pipes	1		
24	CAC 5866	Hose	1		
25	EAC 3215 1	Hose clip	4		

B132AK20/6/88

MASTER INDEX

GROUP INDEX

1E 09

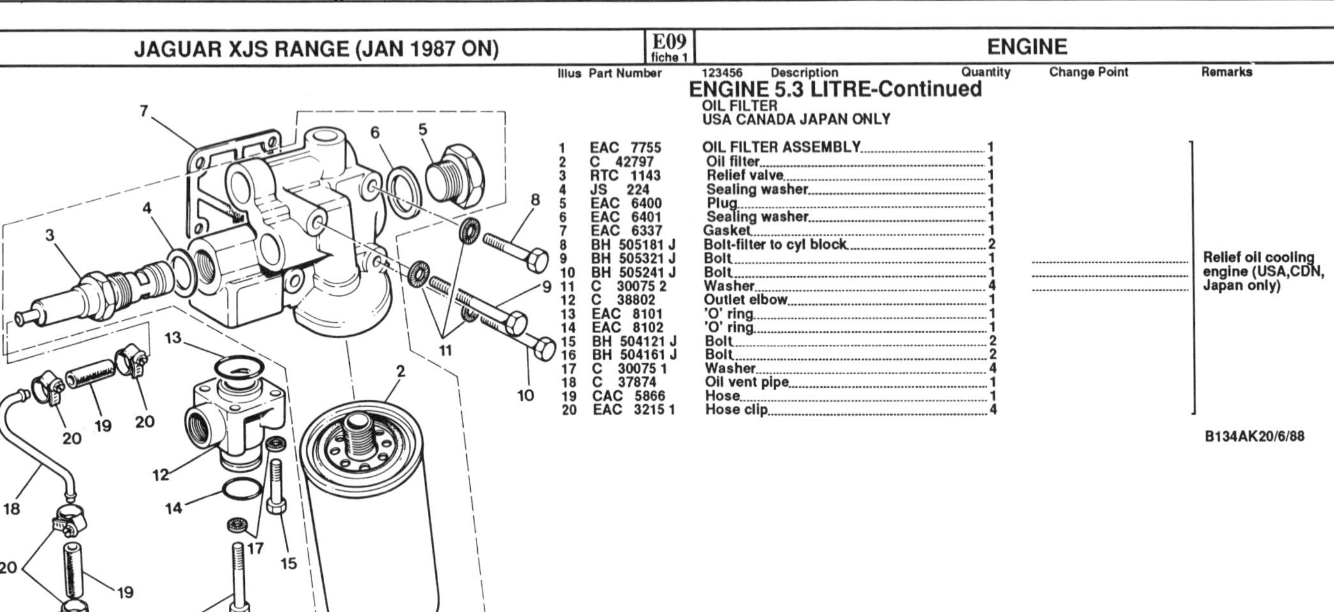

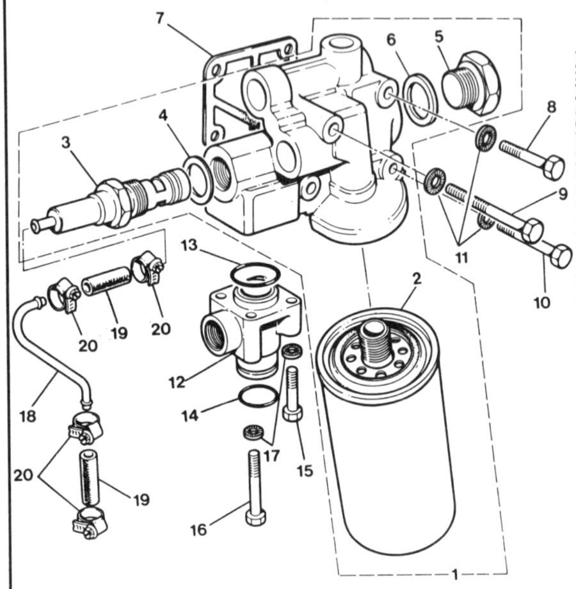

VS4557

VS4557

ENGINE 5.3 LITRE-Continued

OIL FILTER
USA CANADA JAPAN ONLY

Illus	Part Number	123456 Description	Quantity	Change Point	Remarks
1	EAC 7755	OIL FILTER ASSEMBLY	1		
2	C 42797	Oil filter	1		
3	RTC 1143	Relief valve	1		
4	JS 224	Sealing washer	1		
5	EAC 6400	Plug	1		
6	EAC 6401	Sealing washer	1		
7	EAC 6337	Gasket	1		
8	BH 505181 J	Bolt-filter to cyl block	2		
9	BH 505321 J	Bolt	1		Relief oil cooling
10	BH 505241 J	Bolt	1		engine (USA,CDN,
11	C 30075 2	Washer	4		Japan only)
12	C 38802	Outlet elbow	1		
13	EAC 8101	'O' ring	1		
14	EAC 8102	'O' ring	1		
15	BH 504121 J	Bolt	2		
16	BH 504161 J	Bolt	2		
17	C 30075 1	Washer	4		
18	C 37874	Oil vent pipe	1		
19	CAC 5866	Hose	1		
20	EAC 3215 1	Hose clip	4		

B134AK20/6/88

MASTER INDEX

ENGINE 1.C
FLYWHEEL AND CLUTCH 1.G
GEARBOX AND PROPSHAFT 1.H
AXLES, SUSPENSION, DRIVE SHAFTS, WHEELS 1.K
STEERING 1.M
BRAKES AND BRAKE CONTROLS 1.N
FUEL, EXHAUST AND EMISSION SYSTEMS 2.C
COOLING, HEATING AND AIR CONDITIONING 2.G
ELECTRICAL, WASHERS-WIPERS, INSTRUMENTS 2.J
CHASSIS, SUBFRAMES, BODYSHELL, FITTINGS 3.C
FASCIA, TRIM, SEATS AND FIXINGS 3.H
ACCESSORIES AND PAINT 4.C
NUMERICAL INDEX 4.D

GROUP INDEX

ENGINE 3.6 LITRE C02
CAM.-TAPPET BLOCK - 5.3 LITRE E04
CAMSHAFT & VALVES C12
CRANKSHAFT AND MAIN BEARINGS C05
CRANKSHAFT-MAIN BEAR.-5.3 LITRE D13
CYL HEAD & VALVES - 5.3 LITRE E03
CYLINDER BLOCK D10
CYLINDER HEAD & GASKET C11
DRIVING BELTS-3.6 LITRE D05
EXHAUST MANIFOLD - 3.6 LITRE D03
EXHAUST MANIFOLDS - 5.3 LITRE F02
FRONT ENGINE MOUNTING 3.6 LITRE D08
FRONT ENGINE MOUNTING-5.3 LITRE F09
GASKET SETS - 3.6 LITRE D06
INDUCTION MANIFOLD-3.6 LITRE D02
INLET MANIFOLD-5.3 LITRE E17
LOWER ENG GASKET SET-5.3 LITRE F07
OIL COOLER & OIL PIPES 3.6 LITRE D07
OIL COOLER - 5.3 LITRE F04
OIL COOLER PIPES - 5.3 LITRE F05
OIL FILTER,FILLER CAP-3.6 LITRE C15
OIL FILTER-5.3 LITRE E08
OIL PUMP - 5.3 LITRE D15
OIL PUMP-3.6 LITRE C07
OIL SUMP & DIPSTICK - 5.3 LITRE D12
OIL SUMP & DIPSTICK 3.6 LITRE C04
PISTON & CONN RODS C06
PISTONS & CONN. RODS D14
UPPER ENG GASKET SET-5.3 LITRE F08
WATER PUMP C16
WATER PUMP - 5.3 LITRE E10

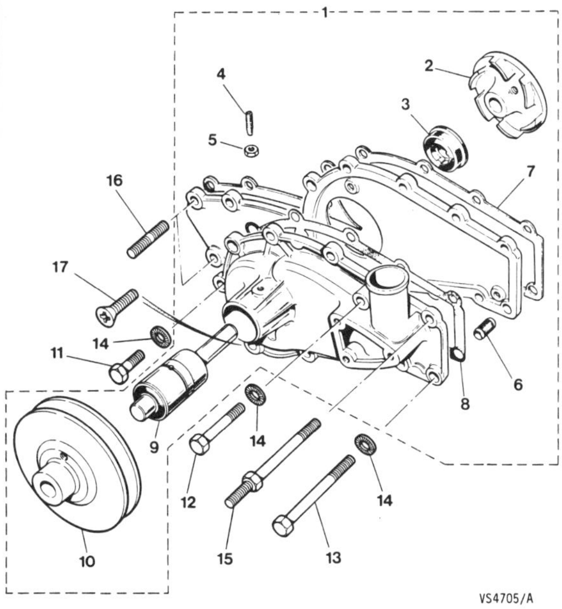

VS4705/A

ENGINE 5.3 LITRE-Continued
WATER PUMP - 5.3 LITRE

Illus	Part Number	Description	Quantity	Change Point	Remarks
1	JLM 314	WATER PUMP ASSEMBLY	1	Up to VIN144699	
1	JLM 1330	WATER PUMP ASSEMBLY	1	From VIN144700	
2	EAC 3212	Impellor	1		
3	EAC 6447	Seal	1		
4	C 23128	Retaining screw	1		
5	NT 604041 J	Locknut	1		
6	C 36298	Dowel	2		
7	C 29626	Gasket	1		
8	C 36542	Gasket	1		
9	EAC 6305	Bearing	1		
10	C 37817	Pulley	1		
11	BH 505111 J	Bolt	7	Up to VIN144699	
11	BH 505111 J	Bolt	6	From VIN144700	
17	SF 505101 J	Screw	1		
12	BH 505181 J	Bolt	1		
13	BH 505321 J	Bolt	2		
14	C 30075 2	Washer	10		
15	C 36196	Pivot Stud	1		
16	C 8137	Stud-Fan Housing	2		

MASTER INDEX

GROUP INDEX

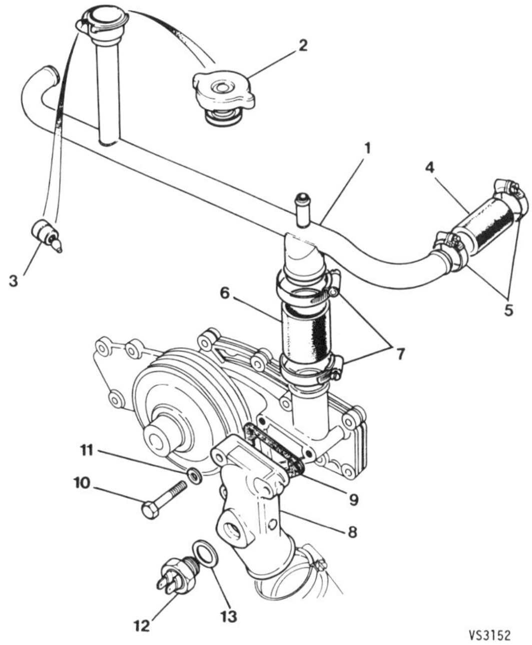

VS3152

ENGINE 5.3 LITRE-Continued
WATER INLET SPOUT-BYPASS PIPE

Illus	Part Number	Description	Quantity	Change Point	Remarks
1	EAC 4191	Water bypass pipe	1		
2	EAC 4192	Pressure cap	1		
3	EAC 4126	Water valve	1		
4	C 41373 1	Hose-pipe to water pipe	1		
5	EAC 3215 8	Hose clip	2		
6	C 41374	Hose-pipe to water pump	1		
7	EAC 3215 11	Hose clip	2		
8	EAC 3195	Water inlet spout	1		
9	C 36020	Gasket	1		
10	BH 505141 J	Bolt	3		
11	C 30075 2	Washer	3		
12	EAC 2510	Thermostat switch	1		
13	C 45759 3	Sealing washer	1		

MASTER INDEX
ENGINE 1.C
FLYWHEEL AND CLUTCH 1.G
GEARBOX AND PROPSHAFT 1.H
AXLES, SUSPENSION, DRIVE SHAFTS, WHEELS 1.K
STEERING 1.M
BRAKES AND BRAKE CONTROLS 1.N
FUEL, EXHAUST AND EMISSION SYSTEMS 2.C
COOLING, HEATING AND AIR CONDITIONING 2.G
ELECTRICAL, WASHERS-WIPERS, INSTRUMENTS 2.J
CHASSIS, SUBFRAMES, BODYSHELL, FITTINGS 3.C
FASCIA, TRIM, SEATS AND FIXINGS 3.H
ACCESSORIES AND PAINT 4.C
NUMERICAL INDEX 4.D
GROUP INDEX
ENGINE 3.6 LITRE C02
CAM.-TAPPET BLOCK - 5.3 LITRE E04
CAMSHAFT & VALVES C12
CRANKSHAFT AND MAIN BEARINGS C05
CRANKSHAFT-MAIN BEAR.-5.3 LITRE D13
CYL. HEAD & VALVES - 5.3 LITRE E03
CYLINDER BLOCK D10
CYLINDER HEAD & GASKET C11
DRIVING BELTS-3.6 LITRE D05
EXHAUST MANIFOLD - 3.6 LITRE D03
EXHAUST MANIFOLDS - 5.3 LITRE F02
FRONT ENGINE MOUNTING 3.6 LITRE D08
FRONT ENGINE MOUNTING-5.3 LITRE F09
GASKET SETS - 3.6 LITRE D06
INDUCTION MANIFOLD-3.6 LITRE D02
INLET MANIFOLD-5.3 LITRE E17
LOWER ENG GASKET SET-5.3 LITRE F07
OIL COOLER & OIL PIPES 3.6 LITRE D07
OIL COOLER - 5.3 LITRE F04
OIL COOLER PIPES - 5.3 LITRE F05
OIL FILTER,FILLER CAP-3.6 LITRE C15
OIL FILTER-5.3 LITRE E08
OIL PUMP - 5.3 LITRE D15
OIL PUMP-3.6 LITRE C07
OIL SUMP & DIPSTICK - 5.3 LITRE D12
OIL SUMP & DIPSTICK 3.6 LITRE C04
PISTON & CONN RODS C06
PISTONS & CONN. RODS D14
UPPER ENG GASKET SET-5.3 LITRE F08
WATER PUMP C16
WATER PUMP - 5.3 LITRE E10

1E 12

ENGINE 5.3 LITRE-continued
THERMOSTAT-WATER PIPES-RH

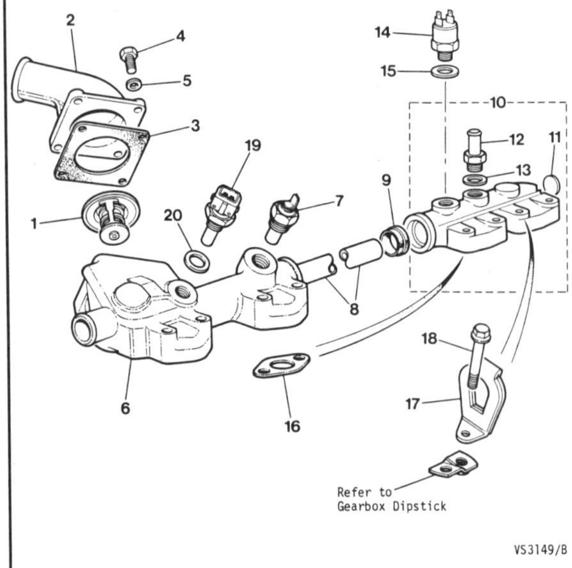

Illus	Part Number	123456 Description	Quantity	Change Point	Remarks
1	JLM 9634	Thermostat(82 degrees)	1		UK/Europe (Emiss B) Middle East
2	C 42606	Thermostat cover-RH	1	Up to VIN144699	
2	EAC 9903	Thermostat cover-RH	1	From VIN144700	
3	EAC 7048	Gasket-cover to water pipe	1		
4	SH 504071 J	Setscrew	3		
5	C 30075 2	Washer	3		
	C 42717	Stud-securing dump valve	1		
	NH 604041 J	Nut	1		
6	EAC 6821	Front water pipe-RH	1	Up to 1989MY	
6	EAC 9844	Front water pipe-LH	1	From 1989 MY Digital Ignition	
19	DAC 4737	TEMPERATURE SENSOR	1		
20	C 2296 3	Sealing washer	1		
7	DAC 2583	Temperature transmitter	1		
8	C 42595	Connecting pipe	1		
9	C 37990	Sealing bush	2		
10	EAC 4017	WATER PIPE ASSEMBLY-RH-REAR	1		
11	C 16862	Core plug	1		
12	C 17421	Connector	1		
13	C 2296 1	Sealing washer	1		
14	EAC 3762	Temperature switch	1		
15	C 2296 20	Sealing washer	1		
16	C 30344	Gasket-water pipes	4		
17	C 38720	Lifting eye	2		
18	FB 108091 J	Bolt	4		
	FB 108101 J	Bolt	4		

VS3149B

1E 13

ENGINE 5.3 LITRE-continued
THERMOSTAT-WATER PIPES RH

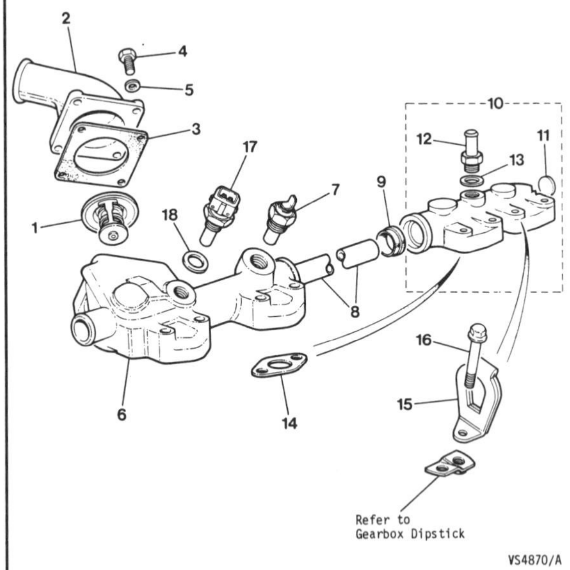

Illus	Part Number	123456 Description	Quantity	Change Point	Remarks
1	JLM 9635	Thermostat	1		
2	C 42606	Thermo Cover	1	Up to VIN144699	
2	EAC 9903	Thermo Cover	1	From VIN144700	
3	EAC 7048	Gasket-Cover to water pipe	1		
4	SH 504071 J	Setscrew	3		
5	C 30075 2	Washer	3		
	C 42717	Stud-securing dump valve	1		
	NH 604041 J	Nut	1		Australia
6	EAC 6821	Front Water Pipe RH	1	Up to 1989MY	Canada (Emiss C)
6	EAC 9844	Front water pipe RH	1	From 1989MY Digital Ignition	
17	DAC 4737	TEMPERATURE SENSOR	1		
18	C 2296 3	Sealing washer	1		
7	DAC 2583	TEMPERATURE TRANSMITTER	1		Japan
8	C 42595	Connecting Pipe	1		Germany (Emiss G)
9	C 37990	Sealing Bush	2		
10	C 42794	Rear Water Pipe RH	1	Up to 8S.57761	
11	C 16862	Core Plug	1	From 1989MY	
12	C 17421	Connector	1		
13	C 2296 1	Sealing Washer	1		
10	EBC 1202	Rear water pipe RH	1	From 8S57762 Up to 1989MY	
11	C 16862	Core plug	1		
12	C 17421	Connector	1		
13	C 2296 1	Sealing washer	1		
	EBC 1197	Vacuum control valve	1		
14	C 30344	Gasket-water pipes	4		
15	C 38720	Lifting Eye	2		
16	FB 108091 J	Bolt	4		
	FB 108101 J	Bolt	4		

B140.02AK1/9/88

VS4870A

MASTER INDEX

ENGINE 1.C
FLYWHEEL AND CLUTCH 1.G
GEARBOX AND PROPSHAFT 1.H
AXLES, SUSPENSION, DRIVE SHAFTS, WHEELS 1.K
STEERING 1.M
BRAKES AND BRAKE CONTROLS 1.N
FUEL, EXHAUST AND EMISSION SYSTEMS 2.C
COOLING, HEATING AND AIR CONDITIONING 2.G
ELECTRICAL, WASHERS-WIPERS, INSTRUMENTS 2.J
CHASSIS, SUBFRAMES, BODYSHELL, FITTINGS 3.C
FASCIA, TRIM, SEATS AND FIXINGS 3.H
ACCESSORIES AND PAINT 4.C
NUMERICAL INDEX 4.D

GROUP INDEX

ENGINE 3.6 LITRE C02
CAM.-TAPPET BLOCK - 5.3 LITRE E04
CAMSHAFT & VALVES C12
CRANKSHAFT AND MAIN BEARINGS C05
CRANKSHAFT-MAIN BEAR.-5.3 LITRE D13
CYL. HEAD & VALVES - 5.3 LITRE E03
CYLINDER BLOCK D10
CYLINDER HEAD & GASKET C11
DRIVING BELTS-3.6 LITRE D05
EXHAUST MANIFOLD - 3.6 LITRE D03
EXHAUST MANIFOLDS - 5.3 LITRE F02
FRONT ENGINE MOUNTING 3.6 LITRE D08
FRONT ENGINE MOUNTING-5.3 LITRE F09
GASKET SETS - 3.6 LITRE D06
INDUCTION MANIFOLD-3.6 LITRE D02
INLET MANIFOLD-5.3 LITRE E17
LOWER ENG GASKET SET-5.3 LITRE F07
OIL COOLER & OIL PIPES 3.6 LITRE D07
OIL COOLER - 5.3 LITRE F04
OIL COOLER PIPES - 5.3 LITRE F05
OIL FILTER,FILLER CAP-3.6 LITRE C15
OIL FILTER-5.3 LITRE E08
OIL PUMP - 5.3 LITRE D15
OIL PUMP-3.6 LITRE C07
OIL SUMP & DIPSTICK - 5.3 LITRE D12
OIL SUMP & DIPSTICK 3.6 LITRE C04
PISTON & CONN RODS C06
PISTONS & CONN. RODS D14
UPPER ENG GASKET SET-5.3 LITRE F08
WATER PUMP C16
WATER PUMP - 5.3 LITRE E10

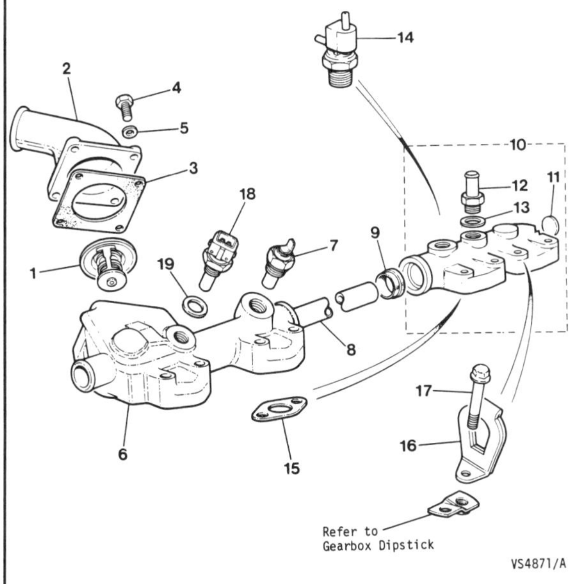

VS4871A

ENGINE 5.3 LITRE-continued

THERMOSTAT WATER PIPES RH

Illus	Part Number	Description	Quantity	Change Point	Remarks
1	JLM 9635	THERMOSTAT	1		
2	C 42606	Thermo Cover	1	Up to VIN144699	
2	EAC 9903	Thermo Cover	1	From VIN144700	
3	EAC 7048	Gasket-cover to water pipe	1		
4	SH 504071 J	Setscrew	3		
5	C 30075 2	Washer	3		
	C 42717	Stud-securing dump valve	1		
	NH 604041 J	Nut	1		
6	EAC 6821	Front Water Pipe-RH	1	Up to 1989MY	Air Injection
6	EAC 9844	Front water pipe LH	1	From 1989MY	
18	DAC 4737	Temperature sensor	1	Digital Ignition	
19	C 2296 3	Sealing washer	1		
7	DAC 2583	TEMPERATURE TRANSMITTER	1		USA
8	C 42595	Connecting Pipe	1		CDN (Emiss "A")
9	C 37990	Sealing Bush	2		Switzerland
10	EAC 4084	Rear Water Pipe-RH	1	Up to 8S57761	Germany (Emiss F)
11	C 16862	Core Plug	1	From 1989MY	
12	C 17421	Connector	1		
13	C 2296 L	Sealing Washer	1		
10	EBC 1200	Rear water pipe RH	1	From 8S57762	
11	C 16862	Core plug	1	Up to 1989MY	
12	C 17421	Connector	1		
13	C 2296 1	Sealing washer	1		
	EBC 1197	VACUUM CONTROL VALVE	1		
14	EAC 4083	THERMAL VACUUM VALVE	1		
15	C 30344	Gasket-water pipes	4		
16	C 38720	Lifting Eye	2		
17	FB 108091 J	Bolt	4		
	FB 108101 J	Bolt	4		

B140.04AK1/9/88

1E 14

MASTER INDEX

GROUP INDEX

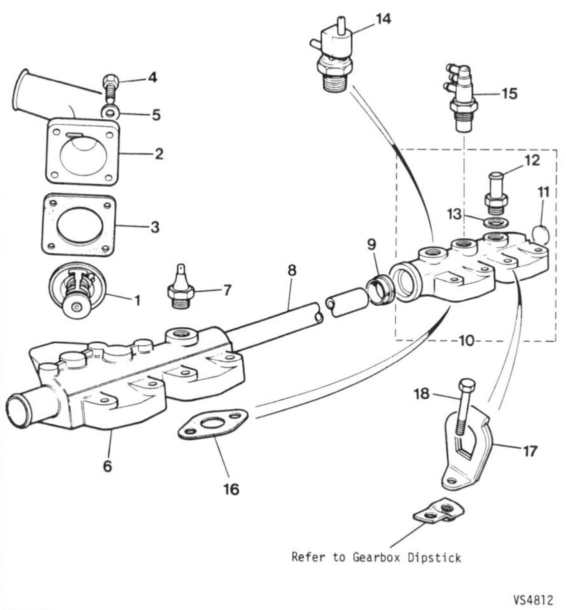

VS4812

ENGINE 5.3 LITRE-continued

THERMOSTAT-WATER PIPES RH

Illus	Part Number	Description	Quantity	Change Point	Remarks
1	JLM 9635	THERMOSTAT	1		
2	EAC 9903	Thermostat Cover-RH	1		
3	EAC 7048	Gasket	1		
4	SH 504071 J	Setscrew	3		
5	C 30075 2	Washer	3		
	C 42717	Stud-securing dump valve	1		
	NH 604041 J	Nut	1		USA/Canada/
6	EAC 6821	Front Water Pipe-RH	1		Sweden/
7	DAC 2583	TEMPERATURE TRANSMITTER	1		Switzerland/
8	C 42595	Connecting Pipe	1		Germany/Japan/
9	C 37990	Sealing Bush	2		Australia
16	C 30344	Gasket-water pipes	4		
17	C 38720	Lifting Eye	2		
18	FB 108091 J	Bolt	4		
	FB 108101 J	Bolt	4	From 8S 57762	
15	EBC 1197	VACUUM CONTROL VALVE	1	Up to 1989 MY only	
10	EBC 1200	Rear Water Pipe-RH	1		
11	C 16862	Core Plug	1		USA/CDN/Sweden/
12	C 17421	Connector	1		Switzerland
13	C 2296 1	Washer	1		with air injection
14	EAC 4083	THERMAL VACUUM VALVE	1		
10	EBC 1202	Rear Water Pipe-RH	1		Australia/
11	C 16862	Core Plug	1		Germany/
12	C 17421	Connector	1		Japan
13	C 2296 1	Washer	1		

B140.06AK1/9/88

1E 15

MASTER INDEX

ENGINE 1.C
FLYWHEEL AND CLUTCH 1.G
GEARBOX AND PROPSHAFT 1.H
AXLES, SUSPENSION, DRIVE SHAFTS, WHEELS 1.K
STEERING 1.M
BRAKES AND BRAKE CONTROLS 1.N
FUEL, EXHAUST AND EMISSION SYSTEMS 2.C
COOLING, HEATING AND AIR CONDITIONING 2.G
ELECTRICAL, WASHERS-WIPERS, INSTRUMENTS 2.J
CHASSIS, SUBFRAMES, BODYSHELL, FITTINGS 3.C
FASCIA, TRIM, SEATS AND FIXINGS 3.H
ACCESSORIES AND PAINT 4.C
NUMERICAL INDEX 4.D

GROUP INDEX

ENGINE 3.6 LITRE C02
CAM.-TAPPET BLOCK - 5.3 LITRE E04
CAMSHAFT & VALVES C12
CRANKSHAFT AND MAIN BEARINGS C05
CRANKSHAFT-MAIN BEAR.-5.3 LITRE D13
CYL. HEAD & VALVES - 5.3 LITRE E03
CYLINDER BLOCK D10
CYLINDER HEAD & GASKET C11
DRIVING BELTS-3.6 LITRE D05
EXHAUST MANIFOLD - 3.6 LITRE D03
EXHAUST MANIFOLDS - 5.3 LITRE F02
FRONT ENGINE MOUNTING 3.6 LITRE D08
FRONT ENGINE MOUNTING-5.3 LITRE F09
GASKET SETS - 3.6 LITRE D06
INDUCTION MANIFOLD-3.6 LITRE D02
INLET MANIFOLD-5.3 LITRE E17
LOWER ENG GASKET SET-5.3 LITRE F07
OIL COOLER & OIL PIPES 3.6 LITRE D07
OIL COOLER - 5.3 LITRE F04
OIL COOLER PIPES - 5.3 LITRE F05
OIL FILTER,FILLER CAP-3.6 LITRE C15
OIL FILTER-5.3 LITRE E08
OIL PUMP - 5.3 LITRE D15
OIL PUMP-3.6 LITRE C07
OIL SUMP & DIPSTICK - 5.3 LITRE D12
OIL SUMP & DIPSTICK 3.6 LITRE C04
PISTON & CONN RODS C06
PISTONS & CONN. RODS D14
UPPER ENG GASKET SET-5.3 LITRE F08
WATER PUMP C16
WATER PUMP - 5.3 LITRE E10

1E 16

JAGUAR XJS RANGE (JAN 1987 ON) | E16 fiche 1 | ENGINE

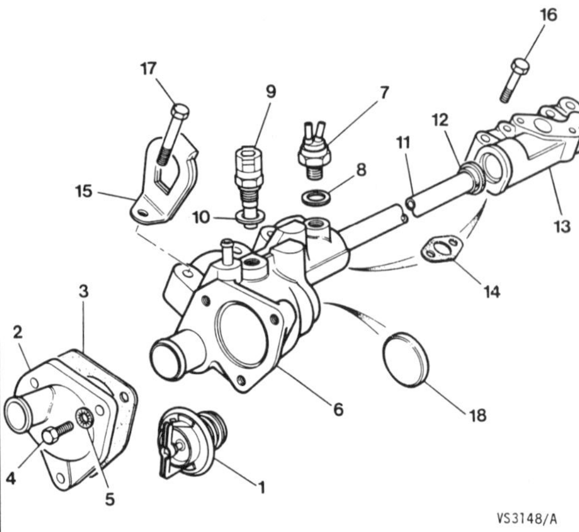

VS3148/A

ENGINE 5.3 LITRE-Continued
THERMOSTAT-WATER PIPES-LH

Illus	Part Number	123456 Description	Quantity	Change Point	Remarks
1	JLM 9634	Thermostat	1	....	82 Degree opening
1	JLM 9635	Thermostat	1	....	88 Degree opening
2	C 42599	Thermostat cover-LH	1		
3	EAC 7047	Gasket	1		
4	SH 505091 J	Setscrew	3		
5	C 30075 2	Washer	3		
6	EAC 4018	Front water pipe-LH	1		
18	C 16862 1	Core plug	1		
6	EAC 9382	Front water pipe-LH	1	....	USA,Canada, Japan,Australia Middle East
18	C 16862 1	Core plug	1	....	
7	EAC 2630	Thermal valve	1	....	
8	C 2246 8	Sealing washer	1		
9	EAC 3927	Water temperature sensor	1		
10	C 2296 3	Sealing washer	1		
11	C 42595	Connecting pipe	1		
12	C 37990	Sealing bush	2		
13	C 42795	Rear water pipe-LH	1		
14	C 30344	Gasket-water pipes	4		
15	C 38720	Lifting eye	2		
16	FB 108091 J	Bolt	4		
17	FB 108101 J	Bolt	4		

MASTER INDEX

GROUP INDEX

1E 17

JAGUAR XJS RANGE (JAN 1987 ON) | E17 fiche 1 | ENGINE

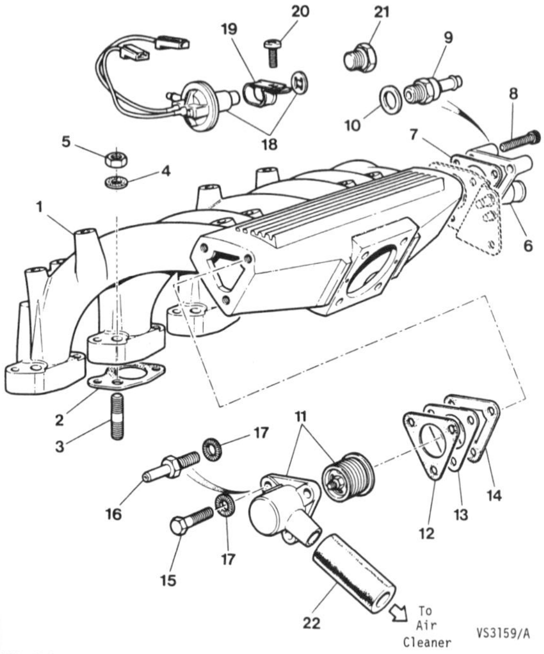

VS3159A

ENGINE 5.3 LITRE-Continued
INLET MANIFOLD-5.3 LITRE
UP TO 1989 MODEL YEAR ONLY

Illus	Part Number	123456 Description	Quantity	Change Point	Remarks
1	EAC 6816	Inlet manifold-LH	1	....	Less air injection
	EAC 6815	Inlet manifold-RH	1		
1	EAC 6819	Inlet manifold-LH	1	....	Air injection
	EAC 6818	Inlet manifold-RH	1		
2	C 43354	Gasket	12		
3	TE 108091 J	Stud-securing manifold	24		
4	C 30075 2	Washer	24		
5	JLM 9688	Nut	24		
6	EAC 3051	End connector-LH	1		
	EAC 3099	End connector-RH	1		
7	EAC 2650	Gasket	2		
8	C 42301 11	Screw (LH Connector)	1		
	C 42301 10	Screw (LH Connector)	1		
	C 42301 5	Screw (LH Connector)	1		
8	C 42301 11	Screw (RH Connector)	2		
	C 42301 6	Screw (RH Connector)	1		
9	CAC 6966	Adaptor (brake servo)	1	Up to 8S.57571	
21	C 23435	Blanking Plug	1	From 8S.57572	With A.B.S Brakes
10	C 2296 1	Washer	1		
11	EAC 8619	Deceleration valve	2		
22	C 41351 9	Hose-decel valve to air cleaner	2		
12	C 44190	Gasket	2		
13	C 44287	Sandwich plate	2		
14	C 41887	Gasket	2		
15	BH 505291 J	Bolt	2		
	C 42301 5	Screw	1		
16	EAC 2899	Screw	3		
17	C 30075 2	Washer	3		
18	JLM 473	Throttle control valve	1	....	USA/CDN/Japan Australia/Sweden Germany/Switz
19	CP 108161 J	Clip	1	....	
20	C 42301 1	Screw	1	....	

B144AK1/9/88

MASTER INDEX

ENGINE 1.C
FLYWHEEL AND CLUTCH 1.G
GEARBOX AND PROPSHAFT 1.H
AXLES, SUSPENSION, DRIVE SHAFTS, WHEELS 1.K
STEERING 1.M
BRAKES AND BRAKE CONTROLS 1.N
FUEL, EXHAUST AND EMISSION SYSTEMS 2.C
COOLING, HEATING AND AIR CONDITIONING 2.G
ELECTRICAL, WASHERS-WIPERS, INSTRUMENTS 2.J
CHASSIS, SUBFRAMES, BODYSHELL, FITTINGS 3.C
FASCIA, TRIM, SEATS AND FIXINGS 3.H
ACCESSORIES AND PAINT 4.C
NUMERICAL INDEX 4.D

GROUP INDEX

ENGINE 3.6 LITRE C02
CAM.-TAPPET BLOCK - 5.3 LITRE E04
CAMSHAFT & VALVES C12
CRANKSHAFT AND MAIN BEARINGS C05
CRANKSHAFT-MAIN BEAR.-5.3 LITRE D13
CYL. HEAD & VALVES - 5.3 LITRE E03
CYLINDER BLOCK D10
CYLINDER HEAD & GASKET C11
DRIVING BELTS-3.6 LITRE D05
EXHAUST MANIFOLD - 3.6 LITRE D03
EXHAUST MANIFOLDS - 5.3 LITRE F02
FRONT ENGINE MOUNTING 3.6 LITRE D08
FRONT ENGINE MOUNTING-5.3 LITRE F09
GASKET SETS - 3.6 LITRE D06
INDUCTION MANIFOLD-3.6 LITRE D02
INLET MANIFOLD-5.3 LITRE E17
LOWER ENG GASKET SET-5.3 LITRE F07
OIL COOLER & OIL PIPES 3.6 LITRE D07
OIL COOLER - 5.3 LITRE F04
OIL COOLER PIPES - 5.3 LITRE F05
OIL FILTER,FILLER CAP-3.6 LITRE C15
OIL FILTER-5.3 LITRE E08
OIL PUMP - 5.3 LITRE D15
OIL PUMP-3.6 LITRE C07
OIL SUMP & DIPSTICK - 5.3 LITRE D12
OIL SUMP & DIPSTICK 3.6 LITRE C04
PISTON & CONN RODS C06
PISTONS & CONN. RODS D14
UPPER ENG GASKET SET-5.3 LITRE F08
WATER PUMP C16
WATER PUMP - 5.3 LITRE E10

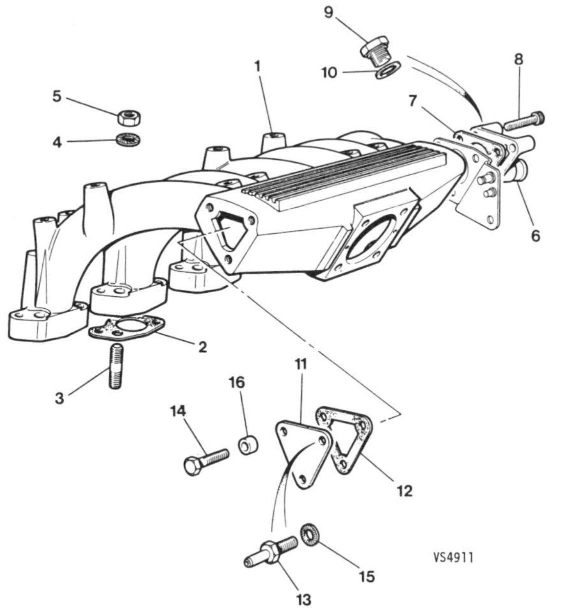

VS4911

ENGINE 5.3 LITRE - continued
INLET MANIFOLD - 5.3 LITRE
FROM 1989 MODEL YEAR

Illus	Part Number	123456 Description	Quantity	Change Point	Remarks
1	EAC 6816	Inlet manifold LH	1		Less Air Injection
	EAC 6815	Inlet manifold RH	1		
1	EAC 6819	Inlet manifold LH	1		With Air Injection
	EAC 6818	Inlet manifold RH	1		
2	C 43354	Gasket	12		
3	TE 108091 J	Stud	24		
4	C 30075 2	Washer	24		
5	JLM 9688	Nut	24		
6	EAC 3051	End connector LH	1		
	EAC 3099	End connector RH	!		
7	EAC 2650	Gasket	2		
8	C 42301 11	Screw - securing LH connector	1		
	C 42301 10	Screw - securing LH connector	1		
	C 42301 5	Screw - securing LH connector	1		
	C 42301 11	Screw - securing RH connector	2		
	C 42301 6	Screw - securing RH connector	1		
9	C 23435	Blanking plug	1		
10	C 2296 1	Washer	1		
11	C 41888	Blanking plate	2		
12	C 41887	Gasket	2		
13	EBC 2489	Screw - vacuum take-off	3		
14	BH 505291 J	Bolt	2		
	C 42301 2	Screw	1		
15	C 30075 2	Washer	3		
16	EBC 2479	Spacing boss	2		

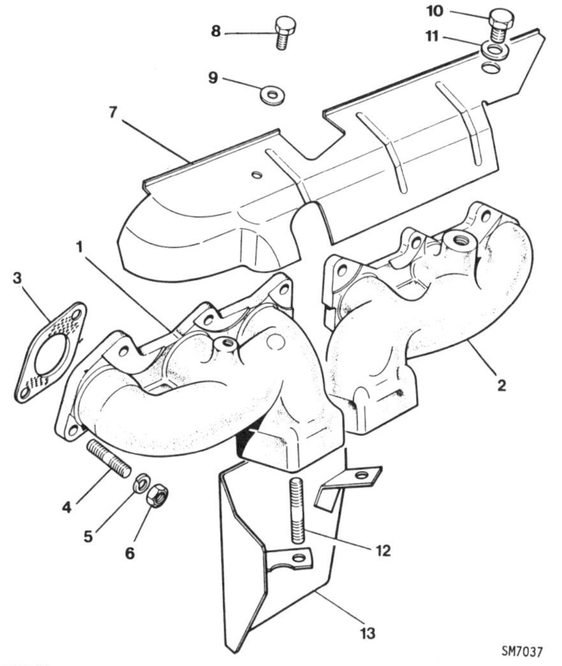

SM7037

ENGINE 5.3 litre-Continued
EXHAUST MANIFOLDS - 5.3 LITRE

Illus	Part Number	123456 Description	Quantity	Change Point	Remarks
1	C 44008	Front exhaust manifold RH	1		
1	C 44006	Front exhaust manifold LH	1		
2	C 44009	Rear exhaust manifold RH	1		
2	C 44007	Rear exhaust manifold LH	1		
3	C 33921	Gasket	12		
4	TE 108051 J	Stud	24		
5	JLM 9574	Washer	24		
6	EAC 5277	Nut	24		
7	C 39687	Heatshield RH	1		
7	C 39686	Heatshield LH	1		
8	SH 605041 J	Setscrew	2		
9	JLM 298	Washer	2		
10	SH 608051 J	Setscrew	2		
11	C 2296 3	Washer	2		
12	EAC 3823	Stud-exhaust flange	8		
13	C 44979	Heatshield (PAS hose)	1		
13	C 44979 1	Heatshield (PAS hose)	1		Japan/Australia

MASTER INDEX
ENGINE 1.C
FLYWHEEL AND CLUTCH 1.G
GEARBOX AND PROPSHAFT 1.H
AXLES, SUSPENSION, DRIVE SHAFTS, WHEELS 1.K
STEERING 1.M
BRAKES AND BRAKE CONTROLS 1.N
FUEL, EXHAUST AND EMISSION SYSTEMS 2.C
COOLING, HEATING AND AIR CONDITIONING 2.G
ELECTRICAL, WASHERS-WIPERS, INSTRUMENTS 2.J
CHASSIS, SUBFRAMES, BODYSHELL, FITTINGS 3.C
FASCIA, TRIM, SEATS AND FIXINGS 3.H
ACCESSORIES AND PAINT 4.C
NUMERICAL INDEX 4.D

GROUP INDEX
ENGINE 3.6 LITRE C02
CAM.-TAPPET BLOCK - 5.3 LITRE E04
CAMSHAFT & VALVES C12
CRANKSHAFT AND MAIN BEARINGS C05
CRANKSHAFT-MAIN BEAR.-5.3 LITRE D13
CYL. HEAD & VALVES - 5.3 LITRE E03
CYLINDER BLOCK D10
CYLINDER HEAD & GASKET C11
DRIVING BELTS-3.6 LITRE D05
EXHAUST MANIFOLD - 3.6 LITRE D03
EXHAUST MANIFOLDS - 5.3 LITRE F02
FRONT ENGINE MOUNTING 3.6 LITRE D08
FRONT ENGINE MOUNTING-5.3 LITRE F09
GASKET SETS - 3.6 LITRE D06
INDUCTION MANIFOLD-3.6 LITRE D02
INLET MANIFOLD-5.3 LITRE E17
LOWER ENG GASKET SET-5.3 LITRE F07
OIL COOLER & OIL PIPES 3.6 LITRE D07
OIL COOLER - 5.3 LITRE F04
OIL COOLER PIPES - 5.3 LITRE F05
OIL FILTER,FILLER CAP-3.6 LITRE C15
OIL FILTER-5.3 LITRE E08
OIL PUMP - 5.3 LITRE D15
OIL PUMP-3.6 LITRE C07
OIL SUMP & DIPSTICK - 5.3 LITRE D12
OIL SUMP & DIPSTICK 3.6 LITRE C04
PISTON & CONN RODS C06
PISTONS & CONN. RODS D14
UPPER ENG GASKET SET-5.3 LITRE F08
WATER PUMP C16
WATER PUMP - 5.3 LITRE E10

1F 03

ENGINE 5.3 LITRE-Continued
DRIVING BELTS

Illus	Part Number	123456 Description	Quantity	Change Point	Remarks
1	EAC 1605	Driving belt(compressor)	1		
2	EAC 1323	Driving belt(steering pump)	1		
3	C 43070	Driving belt(fan)	1		
4	EAC 1616	Driving belt(alternator)	1	Up to 8S.57571	
4	EAC 9369	Driving belt(alternator)	1	From 8S.57572	With A.B.S Brakes

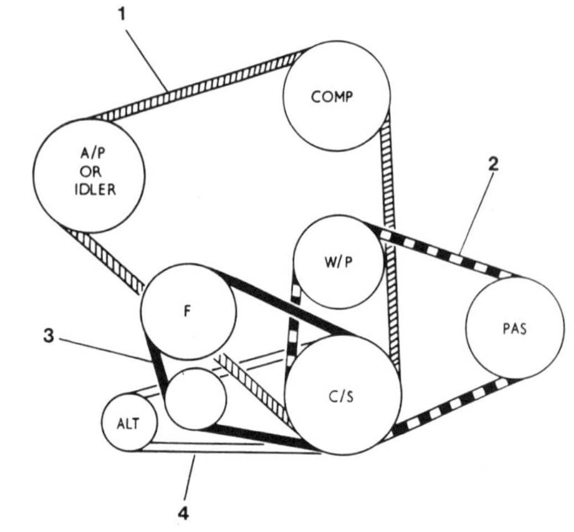

VS1259

MASTER INDEX

GROUP INDEX

1F 04

ENGINE 5.3 LITRE-Continued
OIL COOLER - 5.3 LITRE
EXCEPT USA CANADA JAPAN

Illus	Part Number	123456 Description	Quantity	Change Point	Remarks
1	CBC 2692	Oil cooler	1		Full flow oil cooling(except USA,CDN,Japan)
2	CBC 2691	Oil feed pipe	1		
3	CBC 2690	Oil return pipe	1		
4	CAC 3785	Mounting bracket	2		
5	SE 605051 J	Setscrew - cooler to bracket	4		
6	WM 600051 J	Lockwasher	4		
7	FW 105 T	Washer	4		
8	UFS 925 6R	Setscrew-securing bracket	4		

B150AK20/6/88

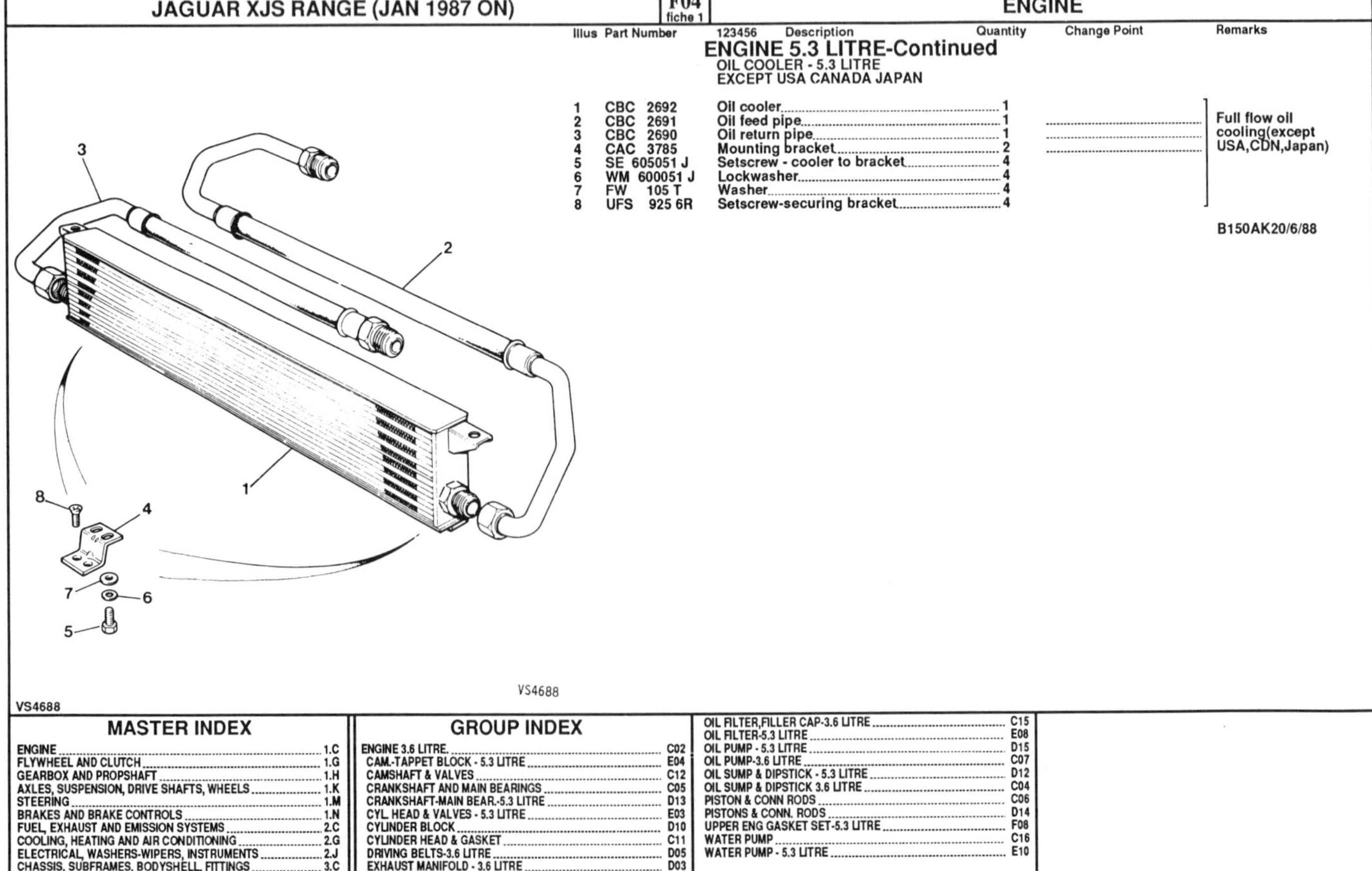

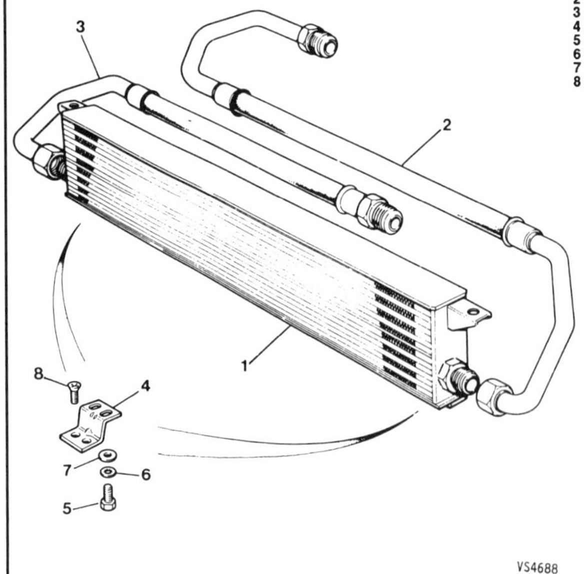

VS4688

MASTER INDEX

ENGINE 1.C
FLYWHEEL AND CLUTCH 1.G
GEARBOX AND PROPSHAFT 1.H
AXLES, SUSPENSION, DRIVE SHAFTS, WHEELS 1.K
STEERING 1.M
BRAKES AND BRAKE CONTROLS 1.N
FUEL, EXHAUST AND EMISSION SYSTEMS 2.C
COOLING, HEATING AND AIR CONDITIONING 2.G
ELECTRICAL, WASHERS-WIPERS, INSTRUMENTS 2.J
CHASSIS, SUBFRAMES, BODYSHELL, FITTINGS 3.C
FASCIA, TRIM, SEATS AND FIXINGS 3.H
ACCESSORIES AND PAINT 4.C
NUMERICAL INDEX 4.D

GROUP INDEX

ENGINE 3.6 LITRE C02
CAM.-TAPPET BLOCK - 5.3 LITRE E04
CAMSHAFT & VALVES C12
CRANKSHAFT AND MAIN BEARINGS C05
CRANKSHAFT-MAIN BEAR.-5.3 LITRE D13
CYL. HEAD & VALVES - 5.3 LITRE E03
CYLINDER BLOCK D10
CYLINDER HEAD & GASKET C11
DRIVING BELTS-3.6 LITRE D05
EXHAUST MANIFOLD - 3.6 LITRE D03
EXHAUST MANIFOLDS - 5.3 LITRE F02
FRONT ENGINE MOUNTING 3.6 LITRE D08
FRONT ENGINE MOUNTING-5.3 LITRE F09
GASKET SETS - 3.6 LITRE D06
INDUCTION MANIFOLD-3.6 LITRE D02
INLET MANIFOLD-5.3 LITRE E17
LOWER ENG GASKET SET-5.3 LITRE F07
OIL COOLER & OIL PIPES 3.6 LITRE D07
OIL COOLER - 5.3 LITRE F04
OIL COOLER PIPES - 5.3 LITRE F05
OIL FILTER,FILLER CAP-3.6 LITRE C15
OIL FILTER-5.3 LITRE E08
OIL PUMP - 5.3 LITRE D15
OIL PUMP-3.6 LITRE C07
OIL SUMP & DIPSTICK - 5.3 LITRE D12
OIL SUMP & DIPSTICK 3.6 LITRE C04
PISTON & CONN RODS C06
PISTONS & CONN. RODS D14
UPPER ENG GASKET SET-5.3 LITRE F08
WATER PUMP C16
WATER PUMP - 5.3 LITRE E10

JAGUAR XJS RANGE (JAN 1987 ON)	F05 fiche 1	ENGINE

1F 05

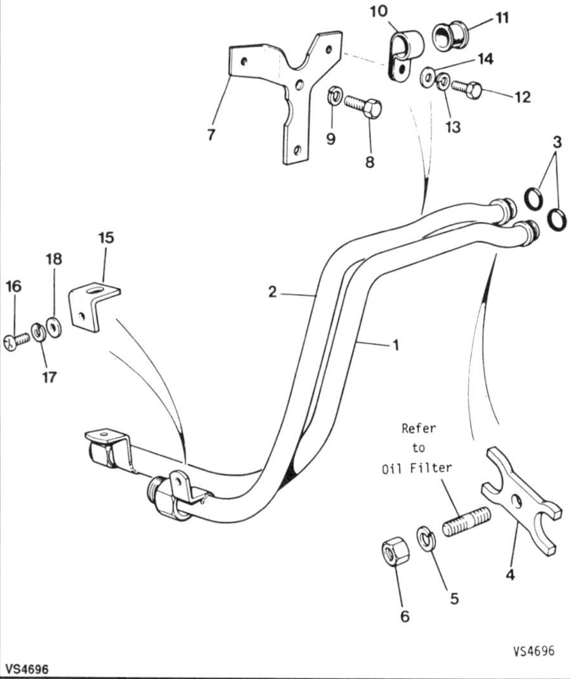

VS4696

ENGINE 5.3 LITRE-continued

OIL COOLER PIPES - 5.3 LITRE
EXCEPT USA CANADA JAPAN

Illus	Part Number	Description	Quantity	Change Point	Remarks
1	EAC 8956	Oil feed pipe	1		'Full flow' oil cooling (except USA, CDN,Japan)
2	EAC 8954	Oil return pipe	1		
3	EAC 6423	'O' ring	2		
4	EAC 6413	Clamp plate	1		
5	WL 600051 J	Lockwasher	1		
6	JLM 9684	Nut	1		
7	EAC 6414	Support bracket	1		
8	SH 507121 J	Setscrew-bracket to cyl block	1		
9	WL 600071 J	Lockwasher	1		
10	EAC 6800	Clip	2		
11	EAC 6790	Bush	2		
12	SH 605071 J	Setscrew - clip to bracket	2		
13	WL 600051 J	Lockwasher	2		
14	WA 108051 J	Washer	2		
15	EAC 6419	Bracket	1		
16	SL 106161 J	Screw - securing bracket	1		
17	WL 600041 J	Lockwasher	1		
18	WA 106041 J	Washer	1		

JAGUAR XJS RANGE (JAN 1987 ON)	F06 fiche 1	ENGINE

1F 06

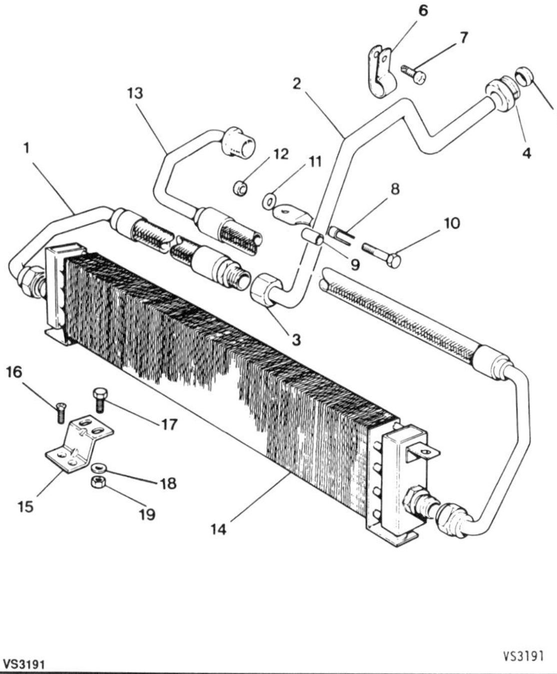

VS3191

ENGINE 5.3 LITRE-Continued

OIL COOLER AND PIPES
USA CANADA JAPAN ONLY

Illus	Part Number	Description	Quantity	Change Point	Remarks
1	C 38074	Oil feed pipe	1		'Relief' oil cooling (USA,CDN, Japan only)
2	EAC 1380	Oil relief pipe	1		
3	C 38189	Nut	1		
4	C 38801	Nut	1		
5	C 37133 09	Olive	1		
6	C 34608	Pipe clip	1		
7	BD 37824 1	Setscrew	1		
8	EAC 1381	Clip	1		
9	C 38522	Distance piece	1		
10	JLM 454	Setscrew	1		
11	WC 702101 J	Washer	1		
12	NH 910011 J	Nut	1		
13	C 38075	Oil return pipe	1		
14	C 43923	Oil cooler	1		
15	CAC 3785	Mounting bracket	2		
16	UFS 925 6R	Setscrew	4		
17	SE 605051 J	Setscrew	4		
18	JLM 303	Washer	4		
19	JLM 299	Nut	4		

B154AK20/6/88

MASTER INDEX

ENGINE	1.C
FLYWHEEL AND CLUTCH	1.G
GEARBOX AND PROPSHAFT	1.H
AXLES, SUSPENSION, DRIVE SHAFTS, WHEELS	1.K
STEERING	1.M
BRAKES AND BRAKE CONTROLS	1.N
FUEL, EXHAUST AND EMISSION SYSTEMS	2.C
COOLING, HEATING AND AIR CONDITIONING	2.G
ELECTRICAL, WASHERS-WIPERS, INSTRUMENTS	2.J
CHASSIS, SUBFRAMES, BODYSHELL, FITTINGS	3.C
FASCIA, TRIM, SEATS AND FIXINGS	3.H
ACCESSORIES AND PAINT	4.C
NUMERICAL INDEX	4.D

GROUP INDEX

ENGINE 3.6 LITRE	C02
CAM.-TAPPET BLOCK - 5.3 LITRE	E04
CAMSHAFT & VALVES	C12
CRANKSHAFT AND MAIN BEARINGS	C05
CRANKSHAFT-MAIN BEAR.-5.3 LITRE	D13
CYL. HEAD & VALVES - 5.3 LITRE	E03
CYLINDER BLOCK	D10
CYLINDER HEAD & GASKET	C11
DRIVING BELTS-3.6 LITRE	D05
EXHAUST MANIFOLD - 3.6 LITRE	D03
EXHAUST MANIFOLDS - 5.3 LITRE	F02
FRONT ENGINE MOUNTING 3.6 LITRE	D08
FRONT ENGINE MOUNTING-5.3 LITRE	F09
GASKET SETS - 3.6 LITRE	D06
INDUCTION MANIFOLD-3.6 LITRE	D02
INLET MANIFOLD-5.3 LITRE	E17
LOWER ENG GASKET SET-5.3 LITRE	F07
OIL COOLER & OIL PIPES 3.6 LITRE	D07
OIL COOLER - 5.3 LITRE	F04
OIL COOLER PIPES - 5.3 LITRE	F05
OIL FILTER,FILLER CAP-3.6 LITRE	C15
OIL FILTER-5.3 LITRE	E08
OIL PUMP - 5.3 LITRE	D15
OIL PUMP-3.6 LITRE	C07
OIL SUMP & DIPSTICK - 5.3 LITRE	D12
OIL SUMP & DIPSTICK 3.6 LITRE	C04
PISTON & CONN RODS	C06
PISTONS & CONN. RODS	D14
UPPER ENG GASKET SET-5.3 LITRE	F08
WATER PUMP	C16
WATER PUMP - 5.3 LITRE	E10

1F 07

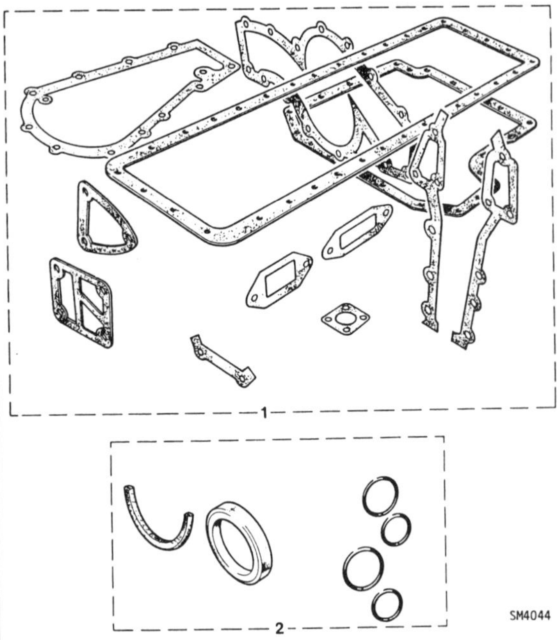

SM4044

ENGINE 5.3 LITRE-Continued
LOWER ENG GASKET SET-5.3 LITRE

Illus	Part Number	123456 Description	Quantity	Change Point	Remarks
1	JLM 447	LOWER ENGINE GASKET SET	1		
		Gasket-timing cover	3		
		Gasket-water pump	2		
		Gasket-inlet spout	1		
		Gasket-oil delivery elbow	1		
		Gasket-oil suction elbow	1		
		Gasket-oil filter	1		
		Gasket-sump	1		
		Gasket-sump sandwich	1		
		Gasket-sump plate	1		
2	JS 615	SEAL AND 'O' RING SET	1		'Relief' Oil Cooling USA/CDN/J only
		Seal-crankshaft(front)	1		
		Seal-crankshaft(rear)	2		
		'O' ring-oil suction pipe	2		
		'O' ring-oil delivery pipe	2		
		'O' ring-oil outlet elbow	2		
		'O' ring-oil inlet elbow	1		
		'O' ring-sump adaptor	1		
		'O' ring-timing cover	1		
2	JLM 399	SEAL AND 'O' RING SET	1		'Full Flow' Oil Cooling Not USA/CDN/J
		Seal-crankshaft(front)	1		
		Seal-crankshaft(rear)	2		
		'O' Ring-oil suction pipe	1		
		'O' Ring-oil delivery pipe	2		
		'O' ring-oil outlet elbow	2		
		'O' ring-oil cooler pipes	2		
		'O' ring-sump adaptor	1		
		'O' ring-timing cover	1		

1F 08

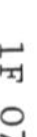

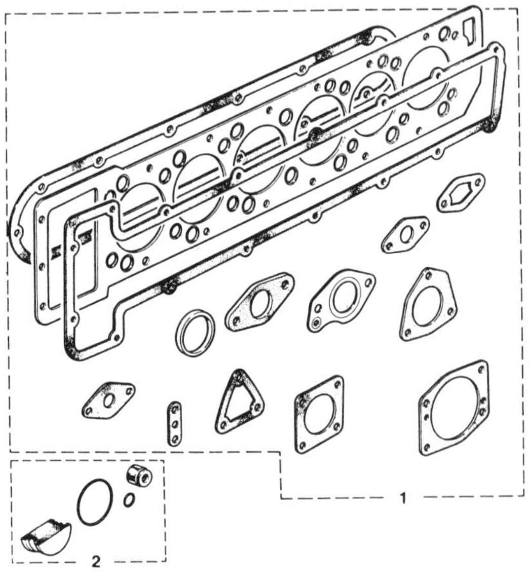

VS3151

ENGINE 5.3 LITRE-Continued
UPPER ENG GASKET SET-5.3 LITRE

Illus	Part Number	123456 Description	Quantity	Change Point	Remarks
1	JLM 9703	CYLINDER HEAD GASKET SET	1		
		Gasket-cylinder head	2		
		Gasket-camshaft cover	2		
		Gasket-exhaust manifold	12		
		Gasket-inlet manifold	12		
		Gasket-inlet manifold(front)	4		
		Gasket-inlet manifold(rear)	2		
		Gasket-thermostat	2		
		Gasket-water pipes	12		
		Gasket-throttle body	4		
		Gasket-top cover	1		
		Gasket-breather housing	1		
		Gasket-extra air valve	1		
		Gasket-cold start injector	2		
		'O' ring-exhaust flange	4		
2	12840	SEAL AND 'O' RING KIT	1		
		Seal-inlet valve	12		
		Seal-cam.cover(rear)	2		
		'O' ring-air pipe	24		
		'O' ring-distributor	1		

MASTER INDEX

ENGINE 1.C
FLYWHEEL AND CLUTCH 1.G
GEARBOX AND PROPSHAFT 1.H
AXLES, SUSPENSION, DRIVE SHAFTS, WHEELS 1.K
STEERING 1.M
BRAKES AND BRAKE CONTROLS 1.N
FUEL, EXHAUST AND EMISSION SYSTEMS 2.C
COOLING, HEATING AND AIR CONDITIONING 2.G
ELECTRICAL, WASHERS-WIPERS, INSTRUMENTS 2.J
CHASSIS, SUBFRAMES, BODYSHELL, FITTINGS 3.C
FASCIA, TRIM, SEATS AND FIXINGS 3.H
ACCESSORIES AND PAINT 4.C
NUMERICAL INDEX 4.D

GROUP INDEX

ENGINE 3.6 LITRE C02
CAM.-TAPPET BLOCK - 5.3 LITRE E04
CAMSHAFT & VALVES C12
CRANKSHAFT AND MAIN BEARINGS C05
CRANKSHAFT-MAIN BEAR.-5.3 LITRE D13
CYL HEAD & VALVES - 5.3 LITRE E03
CYLINDER BLOCK D10
CYLINDER HEAD & GASKET C11
DRIVING BELTS-3.6 LITRE D05
EXHAUST MANIFOLD - 3.6 LITRE D03
EXHAUST MANIFOLDS - 5.3 LITRE F02
FRONT ENGINE MOUNTING 3.6 LITRE D08
FRONT ENGINE MOUNTING-5.3 LITRE F09
GASKET SETS - 3.6 LITRE D06
INDUCTION MANIFOLD-3.6 LITRE D02
INLET MANIFOLD-5.3 LITRE E17
LOWER ENG GASKET SET-5.3 LITRE F07
OIL COOLER & OIL PIPES 3.6 LITRE D07
OIL COOLER - 5.3 LITRE F04
OIL COOLER PIPES - 5.3 LITRE F05
OIL FILTER,FILLER CAP-3.6 LITRE C15
OIL FILTER-5.3 LITRE E08
OIL PUMP - 5.3 LITRE D15
OIL PUMP-3.6 LITRE C07
OIL SUMP & DIPSTICK - 5.3 LITRE D12
OIL SUMP & DIPSTICK 3.6 LITRE C04
PISTON & CONN RODS C06
PISTONS & CONN. RODS D14
UPPER ENG GASKET SET-5.3 LITRE F08
WATER PUMP C16
WATER PUMP - 5.3 LITRE E10

1F 09

ENGINE-5.3 LITRE-continued.
FRONT ENGINE MOUNTING-5.3 LITRE

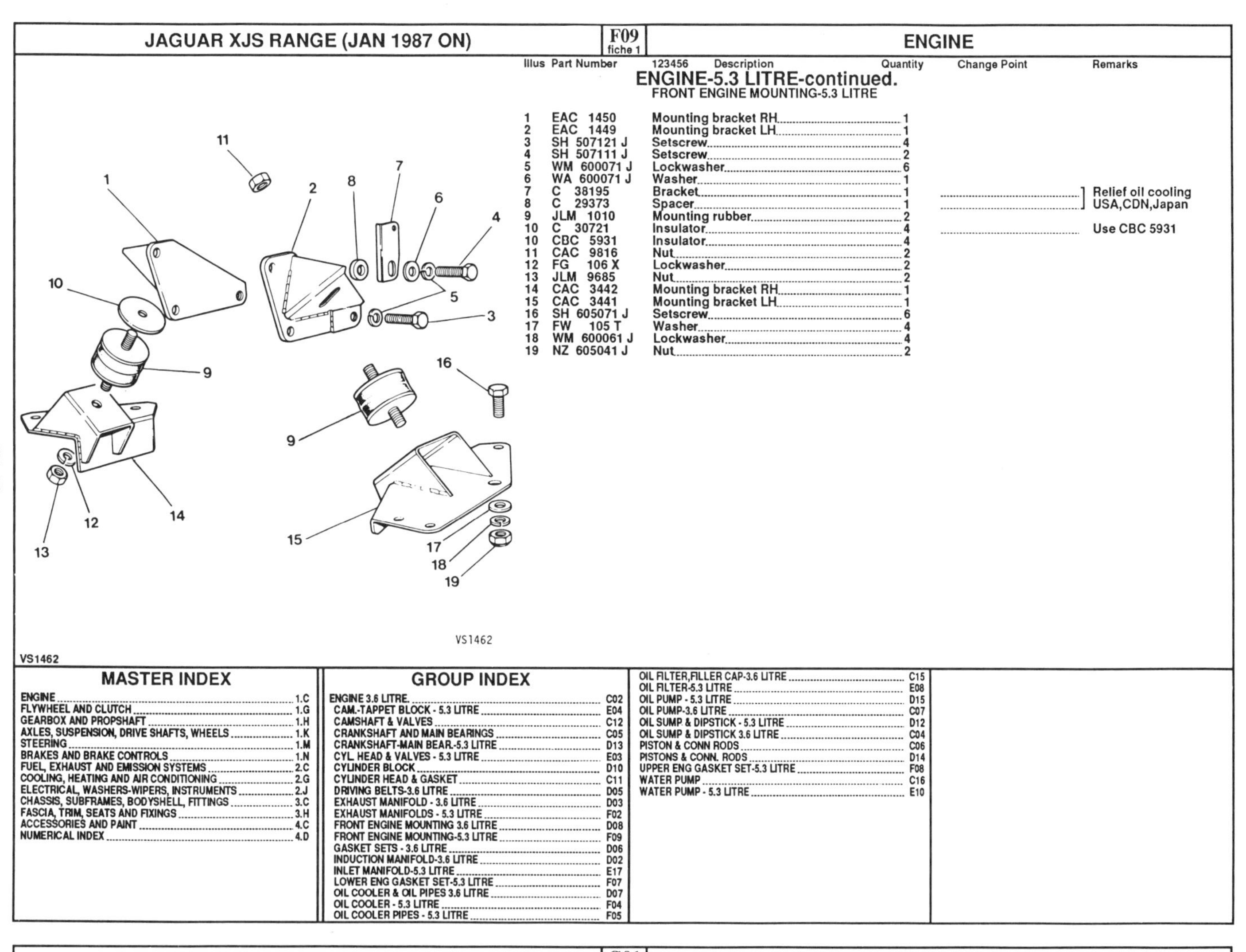

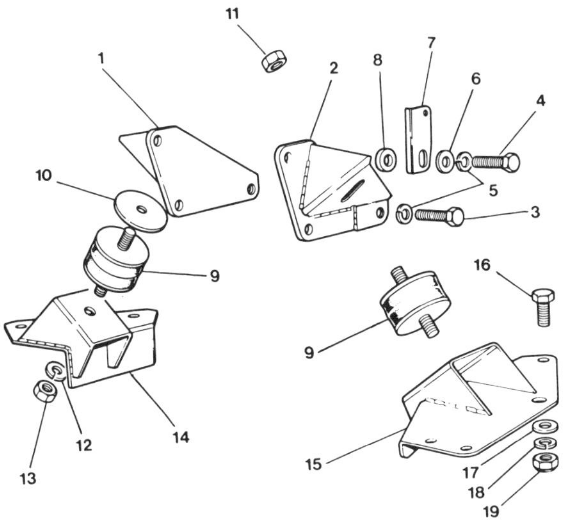

VS1462

Illus	Part Number	123456 Description	Quantity	Change Point	Remarks
1	EAC 1450	Mounting bracket RH	1		
2	EAC 1449	Mounting bracket LH	1		
3	SH 507121 J	Setscrew	4		
4	SH 507111 J	Setscrew	2		
5	WM 600071 J	Lockwasher	6		
6	WA 600071 J	Washer	1		
7	C 38195	Bracket	1		Relief oil cooling USA,CDN,Japan
8	C 29373	Spacer	1		Relief oil cooling USA,CDN,Japan
9	JLM 1010	Mounting rubber	2		
10	C 30721	Insulator	4		Use CBC 5931
10	CBC 5931	Insulator	4		
11	CAC 9816	Nut	2		
12	FG 106 X	Lockwasher	2		
13	JLM 9685	Nut	2		
14	CAC 3442	Mounting bracket RH	1		
15	CAC 3441	Mounting bracket LH	1		
16	SH 605071 J	Setscrew	6		
17	FW 105 T	Washer	4		
18	WM 600061 J	Lockwasher	4		
19	NZ 605041 J	Nut	2		

MASTER INDEX

GROUP INDEX

1G 01

MASTER INDEX

GROUP INDEX

DRIVEPLATE 3.6 LITRE.

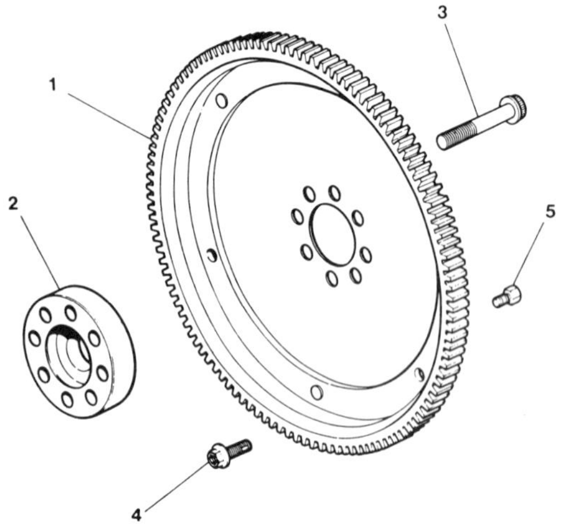

VS3684/A

Illus	Part Number	123456 Description	Quantity	Change Point	Remarks
1	EAC 9961	Driveplate	1		
5	EAC 9799	Bolt (balancing)	1	Upto Eng No 128810	
2	EAC 4619	Adaptor	1	Up to Eng No 139065	
3	EAC 4620	Screw-driveplate to crankshaft	8		
2	EBC 1908	Adaptor	1	From Eng No 139066	
3	EBC 1939	Screw-driveplate to crankshaft	8		
4	EAC 5580	Bolt-driveplate to converter	3		

MASTER INDEX

GROUP INDEX

1G 02

DRIVEPLATE 5.3 LITRE.

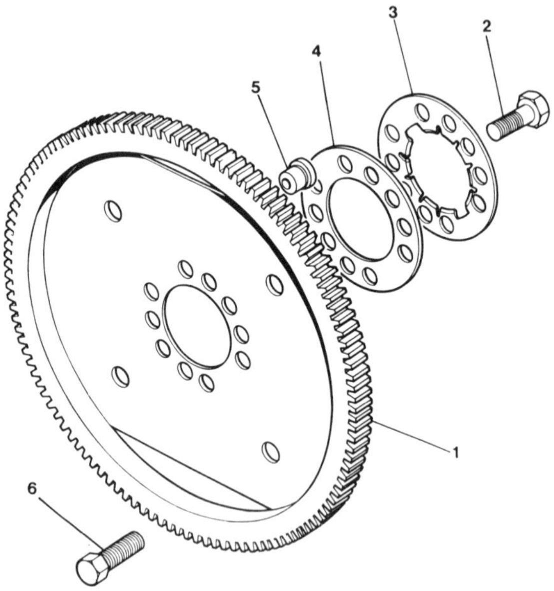

VS1519

Illus	Part Number	123456 Description	Quantity	Change Point	Remarks
1	EAC 4604	Driveplate	1		
2	C 6839	Screw - driveplate to crank	10		
3	C 4810	Lockplate	1		
4	C 6861	Stiffener plate	1		
5	C 6862	Dowel	2		
6	EAC 5580	Screw-driveplate to converter	6		

MASTER INDEX

GROUP INDEX

1G 03

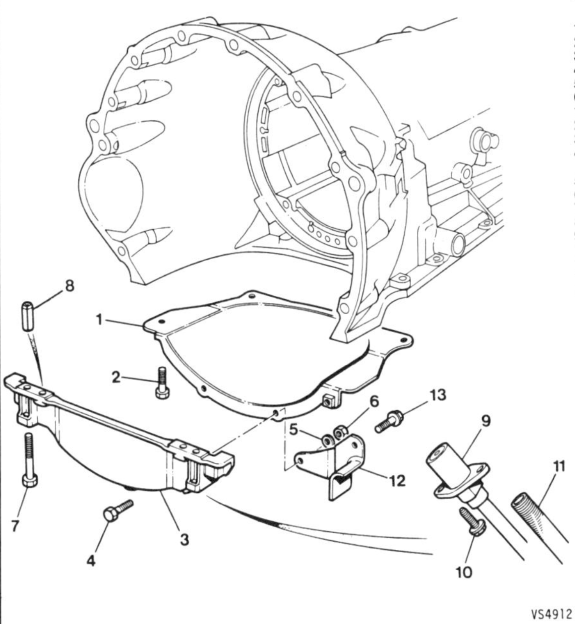

VS4912

FLYWHEEL COVERS 5.3 LITRE

Illus	Part Number	Description	Quantity	Change Point	Remarks
1	C 45232	BOTTOM COVER	1	Up to 1989 MY	
2	C 35093 1	Screw-cover to gearbox	4		
3	C 29462	FLYWHEEL COVER	1		
4	SH 605051 J	Setscrew	2		
5	WF 600040 J	Washer	2		
6	NT 604041 J	Nut	2		
1	EBC 1548	BOTTOM COVER	1	From 1989 Digital Ignition	
2	C 35093 1	Screw-cover to gearbox	3		
3	EBC 2995	FLYWHEEL COVER	1		
4	SH 604061 J	Setscrew	1		
	SH 604071 J	Setscrew	1		
5	WF 600040 J	Washer	2		
6	NT 604041 J	Nut	2		
7	BH 504181 J	Bolt-flywheel cover to engine	2		
8	PA 112121 J	Spring pin	2		
9	DAC 4606	ELECTROMAGNETIC SENSOR	1		
10	SN 105101 J	Screw	2		
11	DAC 5998	Hose-sensor cable	1		
12	EBC 1556	Heatshield	1		
13	FS 106121 J	Setscrew	1		

C4.02AK23/9/88

MASTER INDEX

GROUP INDEX

1G 04

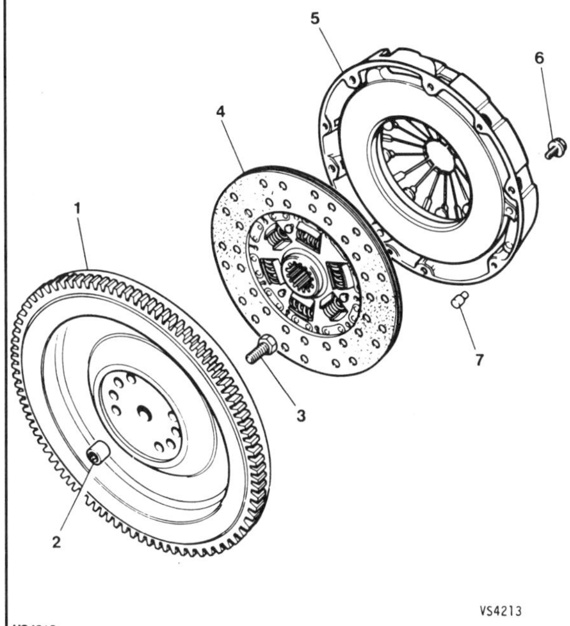

VS4213

FLYWHEEL AND CLUTCH 3.6 LITRE.

Illus	Part Number	Description	Quantity	Change Point	Remarks
1	EAC 8192	Flywheel	1	Up to Eng.No.128477	Use EBC 1832
1	EBC 1832	Flywheel	1	From Eng.No.128478	
2	UKC 8154 J	Bearing	1		
3	EAC 2547	Setscrew-flywheel to crank	8		
4	EAC 8595	Clutch plate	1		
5	EAC 9985	Clutch cover	1		
6	FS 108251 J	Setscrew-clutch to flywheel	6		
7	C 8631	Dowel-clutch to flywheel	3		

MASTER INDEX

GROUP INDEX

1G 05

1G 06

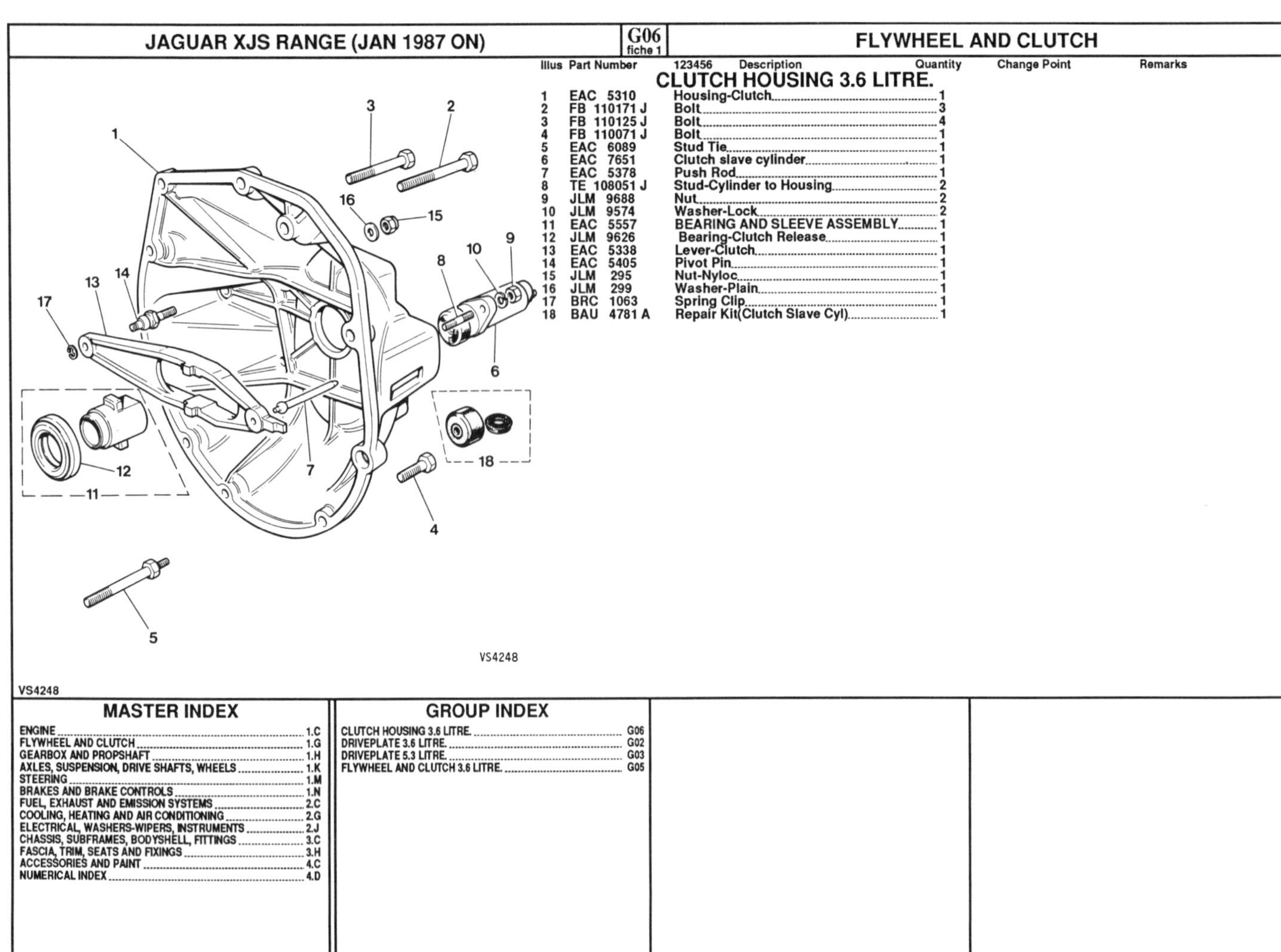

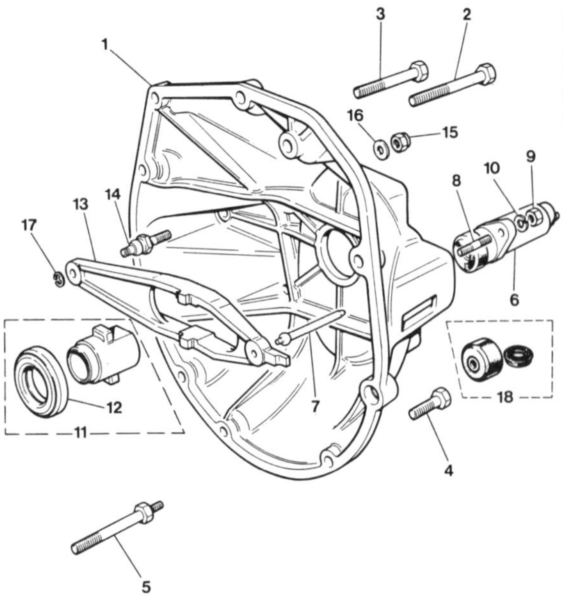

VS4248

CLUTCH HOUSING 3.6 LITRE.

Illus	Part Number	123456	Description	Quantity	Change Point	Remarks
1	EAC 5310		Housing-Clutch	1		
2	FB 110171 J		Bolt	3		
3	FB 110125 J		Bolt	4		
4	FB 110071 J		Bolt	1		
5	EAC 6089		Stud Tie	1		
6	EAC 7651		Clutch slave cylinder	1		
7	EAC 5378		Push Rod	1		
8	TE 108051 J		Stud-Cylinder to Housing	2		
9	JLM 9688		Nut	2		
10	JLM 9574		Washer-Lock	2		
11	EAC 5557		BEARING AND SLEEVE ASSEMBLY	1		
12	JLM 9626		Bearing-Clutch Release	1		
13	EAC 5338		Lever-Clutch	1		
14	EAC 5405		Pivot Pin	1		
15	JLM 295		Nut-Nyloc	1		
16	JLM 299		Washer-Plain	1		
17	BRC 1063		Spring Clip	1		
18	BAU 4781 A		Repair Kit(Clutch Slave Cyl)	1		

MASTER INDEX

GROUP INDEX

1H 01

MASTER INDEX

GROUP INDEX

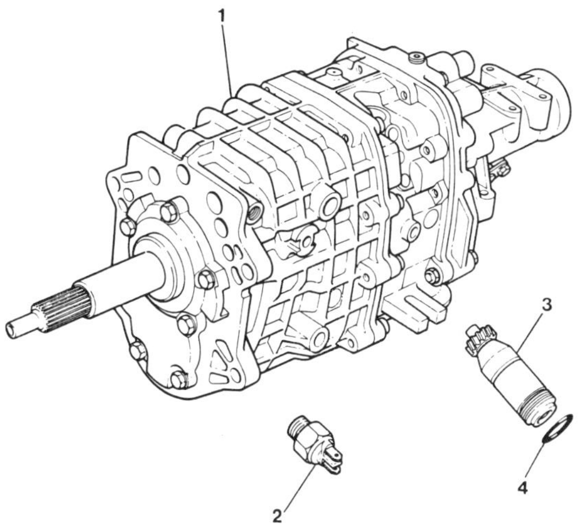

VS4272

VS 4272

MANUAL GEARBOX ASSEMBLY 3.6 LITRE

Illus	Part Number	Description	Quantity	Change Point	Remarks
1	EAC 8195	Gearbox assembly	1		Non speed sensor
1	EAC 8730	Gearbox assembly	1	From eng No124215	Speed sensor
2	EAC 6023	Switch reverse light	1		
	AEU 4005	Gearbox front seal	1		
	AEU 4006	Gearbox rear seal	1		
3	EAC 5681	Speedo driven gear	1		Non speed sensor
4	EAC 5683	Seal	1		Non speed sensor
	JLM 1180	Plug filler	1		
	JLM 1182	Plug magnetic	1		
	JLM 1181	Seal oil support tube	1		

1H 02

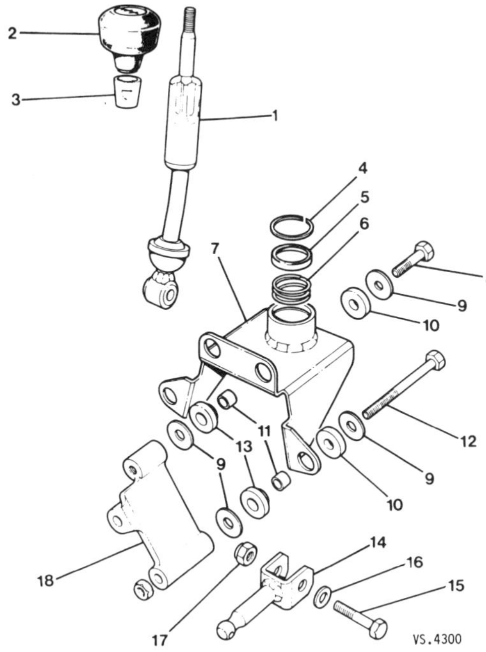

VS 4300

MANUAL G/BOX-LEVER & MOUNTINGS 3.6

Illus	Part Number	Description	Quantity	Change Point	Remarks
1	EAC 7657	Assembly c/speed lever	1		
2	EAC 6268	Knob assembly C/S lever	1		
3	EAC 1896	Locking cone	1		
4	AGU 1381	Spring ring	1		
5	EAC 5261	Spring retainer	1		
6	EAC 9855	Coil spring	1		
7	EAC 5753	Assembly housing C/S lever	1		
8	BH 108071 J	Bolt housing to pedestal	2		
9	WJ 108001 J	Washer	4		
10	UKC 1090	Washer	4		
11	EAC 4902	Tube spacer	4		
12	BH 108241 J	Bolt pedestal to g/box	2		
13	UKC 854	Bush	4		
14	EAC 5268	Assembly selector shaft	1		
15	BH 108081 J	Bolt selector shaft	1		
16	AGU 1533	Special washer	1		
17	NY 108041 J	Nut	1		
18	EAC 5751	Pedestal g/change mounting-LH	1		] Non speed sensor
18	EAC 5752	Pedestal g/change mounting-RH	1		]
	EAC 5259	Spherical washer C/S housing	1		
	EAC 5271	Sleeve C/S lever	1		
	EAC 6024	Joint-selector shaft	1		
	EAC 6025	Pin	1		
	EAC 6026	Foam disc	1		
	EAC 6027	Retainer clip	1		

MASTER INDEX

ENGINE 1.C
FLYWHEEL AND CLUTCH 1.G
GEARBOX AND PROPSHAFT 1.H
AXLES, SUSPENSION, DRIVE SHAFTS, WHEELS 1.K
STEERING 1.M
BRAKES AND BRAKE CONTROLS 1.N
FUEL, EXHAUST AND EMISSION SYSTEMS 2.C
COOLING, HEATING AND AIR CONDITIONING 2.G
ELECTRICAL, WASHERS-WIPERS, INSTRUMENTS 2.J
CHASSIS, SUBFRAMES, BODYSHELL, FITTINGS 3.C
FASCIA, TRIM, SEATS AND FIXINGS 3.H
ACCESSORIES AND PAINT 4.C
NUMERICAL INDEX 4.D

GROUP INDEX

'A' CLUTCH INPUT SHAFT 3.6 LITRE H11
AUTO GEARBOX 'B' CLUTCH 3.6 LITRE H12
AUTO GEARBOX 'C' CLUTCH 3.6 LITRE H13
AUTO GEARBOX 'D' CLUTCH 3.6 LITRE H14
AUTO GEARBOX CLUTCH 'E' 3.6 LITRE H16
AUTO GEARBOX MOUNTINGS-3.6 LITRE I04
AUTO TRANSMISSION SUMP PAN 3.6 LITRE I03
AUTOMATIC TRANSMISSION UNIT 3.6 L/T H07
CONVERTER ASSEMBLY H08
FRONT PUMP ASSEMBLY H10
AUTOMATIC TRANSMISSION UNIT 5.3 L/T I08
CLUTCH HOUSING I11
GEARBOX CASE ASSEMBLY I09
TORQUE CONVERTER I10
FRONT AND REAR SERVO 5.3 LITRE I14
GEARBOX MOUNTING PLATE 5.3 LITRE J08
GEARBOX MOUNTING-5.3 LITRE J07
MANUAL GEARBOX ASSEMBLY 3.6 LITRE H02
MANUAL GEARBOX MOUNTINGS 3.6 LITRE H04
PARKING BRAKE ASSEMBLY 5.3 LITRE I16
SELECTOR LEVER ASSEMBLY J04
SNAP RING KIT J02
PROPELLOR SHAFT 3.6 LITRE I06
PROPELLOR SHAFT 5.3 LITRE I07
REAR BAND SUNGEAR MAINSHAFT 5.3 LITRE I12
OUTPUT SHAFT I13
SELECTOR GATE & LEVER 3.6 AUTOMATIC H05
SPEEDO HOUSING GOVERNOR & MODULATOR I15
VALVE BODIES ASSEMBLY 3.6 LITRE H15

1H 03

MANUAL GEARBOX MOUNTINGS 3.6 LITRE

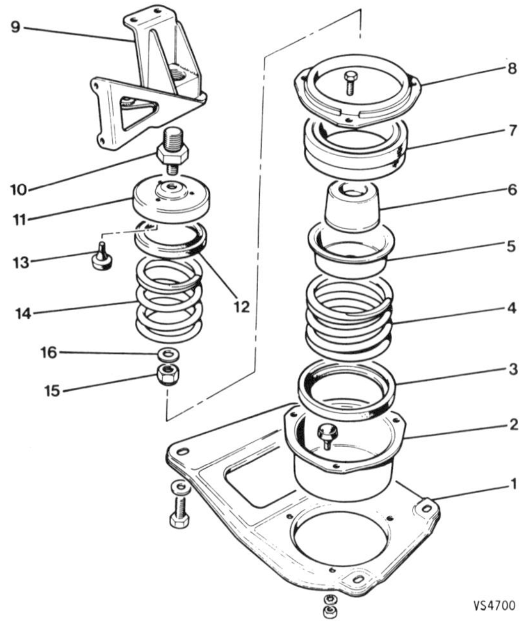

VS4700

Illus	Part Number	123456 Description	Quantity	Change Point	Remarks
1	CBC 4125	Gearbox mounting bracket	1		
2	CAC 9333	Spring cup	1		
3	CBC 1883	Spring seat lower	1		
4	CBC 4744	Outer coil spring	1		
5	CBC 1319	Spring cradle	1		
6	CBC 1318	Locating weight	1		
7	CBC 1324	Foam buffer	1		
8	CBC 1321	Buffer ring assembly	1		
9	CAC 7996	Mounting bracket	1		Not req'd on cars fitted with EAC 8730 G/box
			1		
10	CBC 1320	Adaptor pin	1		
11	CBC 1329	Upper spring seat	1		
12	CAC 9343	Spring seat rubber	1		
13	CAC 9344	Buffer	6		
14	CBC 4745	Inner coil spring	1		
15	NY 112041 J	Locknut	1		
16	WA 112041 J	Washer	1		
	SH 108201 J	Screw	2		
	SH 108351 J	Screw	2		
	BD 541 48	Special washer	4		
	WL 108001 J	Washer	4		

MASTER INDEX

GROUP INDEX

SELECTOR GATE & LEVER 3.6 AUTOMATIC

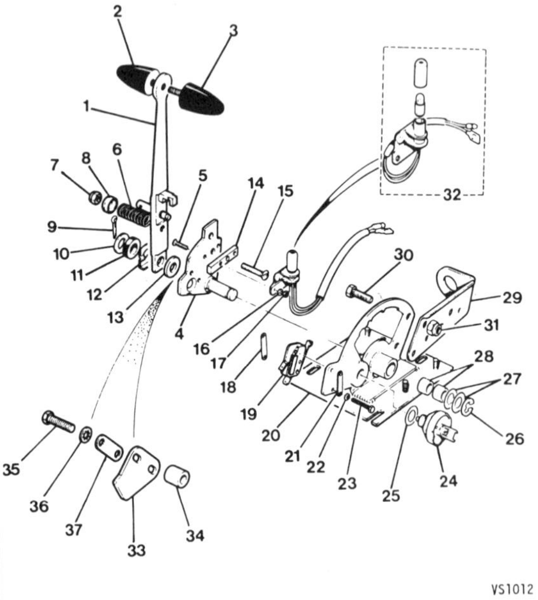

VS1012

Illus	Part Number	123456 Description	Quantity	Change Point	Remarks
1	CAC 6724	Selector lever assembly	1		
2	C 31228	Sel lever knob	1		
3	C 31232	Sel lever knob	1		
4	CAC 6012	Cam plate assembly	1		
5	UCS 511 5H	Setscrew	1		
6	CAC 5366	Return spring	1		
7	TN 3205	Nut	1		
8	CAC 5340	Spring retainer cup	1		
9	4900	Split pin	1		
10	C 24781	Washer	1		
11	C 25997	Damper rubber	1		
12	C 33050	Hinge washer	1		
13	FW 106 E	Washer	1		
14	CAC 6136	Tapped block	1		
15	UFS 519 10H	Setscrew	1		
16	WF 706041 J	Washer	2		
17	UCN 111 L	Locknut	2		
18	C 33784	Fixing plate	1		
19	C 33509	Kickdown switch	1		
20	CAC 5957	Sel gate & mounting plate assembly	1		
21	C 35840	Clamping plate	2		
22	WF 706041 J	Washer	2		
23	UCS 111 7R	Setscrew	2		
24	DAC 3844	Reverse lamp switch	1		
25	C 16626 3	Shim	A/R		
25	C 15896	Shim	A/R		
26	C 25981	Pivot pin circlip	1		
27	C 26644	Shim	A/R		
28	C 20953 3	Pivot pin bush	2		
29	CAC 4845	Abutment plate	1		
30	JLM 9561	Hose clip	1		
31	NY 604041 J	Nut	2		
32	DAC 4192	Gear indicator light assembly	1		
33	CAC 8341	Cam plate speed control	1		
34	C 45125	Spacer	2		
35	UCS 116 5R	Setscrew	2		
36	C 722	Washer	1		
37	CAC 2040	Clamp plate	1		

MASTER INDEX
ENGINE 1.C
FLYWHEEL AND CLUTCH 1.G
GEARBOX AND PROPSHAFT 1.H
AXLES, SUSPENSION, DRIVE SHAFTS, WHEELS 1.K
STEERING 1.M
BRAKES AND BRAKE CONTROLS 1.N
FUEL, EXHAUST AND EMISSION SYSTEMS 2.C
COOLING, HEATING AND AIR CONDITIONING 2.G
ELECTRICAL, WASHERS-WIPERS, INSTRUMENTS 2.J
CHASSIS, SUBFRAMES, BODYSHELL, FITTINGS 3.C
FASCIA, TRIM, SEATS AND FIXINGS 3.H
ACCESSORIES AND PAINT 4.C
NUMERICAL INDEX 4.D
GROUP INDEX
'A' CLUTCH INPUT SHAFT 3.6 LITRE H11
AUTO GEARBOX 'B' CLUTCH 3.6 LITRE H12
AUTO GEARBOX 'C' CLUTCH 3.6 LITRE H13
AUTO GEARBOX 'D' CLUTCH 3.6 LITRE H14
AUTO GEARBOX CLUTCH 'E' 3.6 LITRE H16
AUTO GEARBOX MOUNTINGS-3.6 LITRE I04
AUTO TRANSMISSION SUMP PAN 3.6 LITRE I03
AUTOMATIC TRANSMISSION UNIT 3.6 L/T H07
CONVERTER ASSEMBLY H08
FRONT PUMP ASSEMBLY H10
AUTOMATIC TRANSMISSION UNIT 5.3 L/T I08
CLUTCH HOUSING I11
GEARBOX CASE ASSEMBLY I09
TORQUE CONVERTER I10
FRONT AND REAR SERVO 5.3 LITRE I14
GEARBOX MOUNTING PLATE 5.3 LITRE J08
GEARBOX MOUNTING-5.3 LITRE J07
MANUAL GEARBOX ASSEMBLY 3.6 LITRE H02
MANUAL GEARBOX MOUNTINGS 3.6 LITRE H04
PARKING BRAKE ASSEMBLY 5.3 LITRE I16
SELECTOR LEVER ASSEMBLY J04
SNAP RING KIT J02
PROPELLOR SHAFT 3.6 LITRE I06
PROPELLOR SHAFT 5.3 LITRE I07
REAR BAND SUNGEAR MAINSHAFT 5.3 LITRE I12
OUTPUT SHAFT I13
SELECTOR GATE & LEVER 3.6 AUTOMATIC H05
SPEEDO HOUSING GOVERNOR & MODULATOR I15
VALVE BODIES ASSEMBLY 3.6 LITRE H15

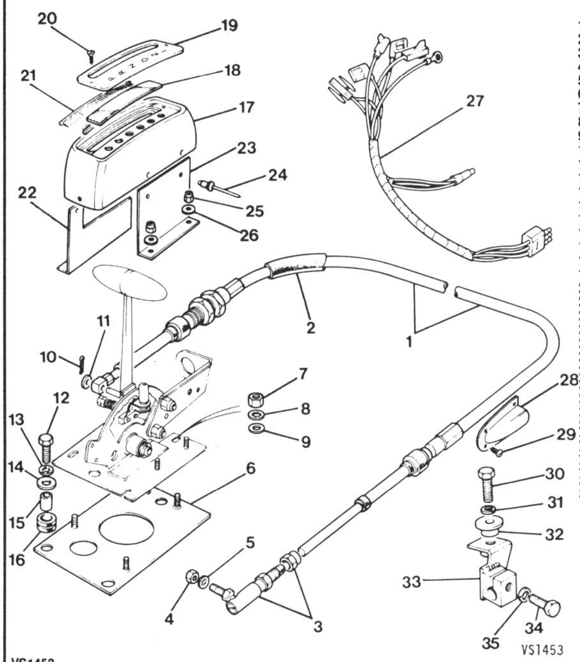

VS1453

GEAR CONTROL CABLE SELECTOR PANEL 3.6 LITRE

Illus	Part Number	123456	Description	Quantity	Change Point	Remarks
1	CBC 4721		Gear control cable	1		
2	C 38494 6		Insulation tube	2		
3	C 25152		Adjustable ball end	1		
4	JLM 9683		Nut cable to gearbox	1		
5	C 724		Washer	1		
6	C 42100		Base plate	1		
7	UFN 119 L		Nut	3		
8	WF 702101 J		Washer	3		
9	WC 702101 J		Washer	3		
10	4900		Split pin	1		
11	FW 104 T		Washer	1		
12	SH 604071 J		Setscrew	3		
13	C 724		Washer	3		
14	FW 104 T		Washer	3		
15	C 25949		Spacer	3		
16	C 25982		Mounting plate rubber	3		
17	CAC 8357		Selector cover	1		
18	C 26337		Filter	1		
19	CBC 2526		Selector cover panel	1		
20	AD 604031 J		Screw	2		
21	BD 31067		Seal	2		
22	CAC 5951		Bracket	1		
23	CAC 5952		Support bracket	1		
24	BD 38427 1		Rivet	4		
25	TN 3205		Nut	4		
26	WC 702101 J		Washer	4		
27	DAC 2163		Auto trans harness	1		
28	CAC 5893		Cover plate	1		
29	AB 606041 J		Screw	4		
30	SH 605101 J		Setscrew	1		
31	FG 105 X		Washer	1		
32	C 45248		Clamp washer	1		
33	CAC 6995		Cable abutment lower	1		
34	BH 108071 J		Bolt	1		
35	WM 600051 J		Washer	1		

MASTER INDEX

GROUP INDEX

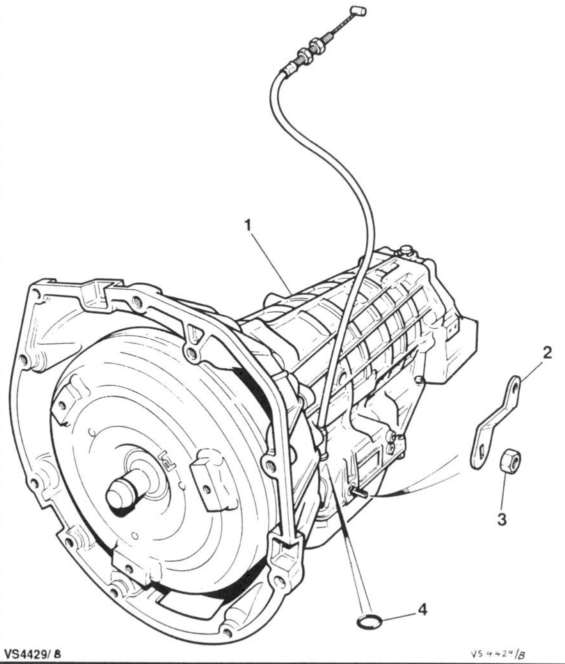

VS4429/ 8

AUTOMATIC TRANSMISSION UNIT 3.6 L/T

Illus	Part Number	123456	Description	Quantity	Change Point	Remarks
1	EBC 2195 N		Automatic trans assembly	1		
2	EAC 6439		Selector Lever	1		
3	NH 208041 J		Nut Securing Lever	1		
4	JLM 1672		'O' ring seal kickdown cable	1		

MASTER INDEX

ENGINE ... 1.C
FLYWHEEL AND CLUTCH ... 1.G
GEARBOX AND PROPSHAFT ... 1.H
AXLES, SUSPENSION, DRIVE SHAFTS, WHEELS ... 1.K
STEERING ... 1.M
BRAKES AND BRAKE CONTROLS ... 1.N
FUEL, EXHAUST AND EMISSION SYSTEMS ... 2.C
COOLING, HEATING AND AIR CONDITIONING ... 2.G
ELECTRICAL, WASHERS-WIPERS, INSTRUMENTS ... 2.J
CHASSIS, SUBFRAMES, BODYSHELL, FITTINGS ... 3.C
FASCIA, TRIM, SEATS AND FIXINGS ... 3.H
ACCESSORIES AND PAINT ... 4.C
NUMERICAL INDEX ... 4.D

GROUP INDEX

'A' CLUTCH INPUT SHAFT 3.6 LITRE ... H11
AUTO GEARBOX 'B' CLUTCH 3.6 LITRE ... H12
AUTO GEARBOX 'C' CLUTCH 3.6 LITRE ... H13
AUTO GEARBOX 'D' CLUTCH 3.6 LITRE ... H14
AUTO GEARBOX CLUTCH 'E' 3.6 LITRE ... H16
AUTO GEARBOX MOUNTINGS-3.6 LITRE ... I04
AUTO TRANSMISSION SUMP PAN 3.6 LITRE ... I03
AUTOMATIC TRANSMISSION UNIT 3.6 L/T ... H07
CONVERTER ASSEMBLY ... H08
FRONT PUMP ASSEMBLY ... H10
AUTOMATIC TRANSMISSION UNIT 5.3 L/T ... I08
CLUTCH HOUSING ... I11
GEARBOX CASE ASSEMBLY ... I09
TORQUE CONVERTER ... I10
FRONT AND REAR SERVO 5.3 LITRE ... I14
GEARBOX MOUNTING PLATE 5.3 LITRE ... J08
GEARBOX MOUNTING-5.3 LITRE ... J07
MANUAL GEARBOX ASSEMBLY 3.6 LITRE ... H02
MANUAL GEARBOX MOUNTINGS 3.6 LITRE ... H04
PARKING BRAKE ASSEMBLY 5.3 LITRE ... I16
SELECTOR LEVER ASSEMBLY ... J04
SNAP RING KIT ... J02
PROPELLOR SHAFT 3.6 LITRE ... I06
PROPELLOR SHAFT 5.3 LITRE ... I07
REAR BAND SUNGEAR MAINSHAFT 5.3 LITRE ... I12
OUTPUT SHAFT ... I13
SELECTOR GATE & LEVER 3.6 AUTOMATIC ... H05
SPEEDO HOUSING GOVERNOR & MODULATOR ... I15
VALVE BODIES ASSEMBLY 3.6 LITRE ... H15

1H 08

JAGUAR XJS RANGE (JAN 1987 ON)	H08 fiche 1	GEARBOX AND PROPSHAFT

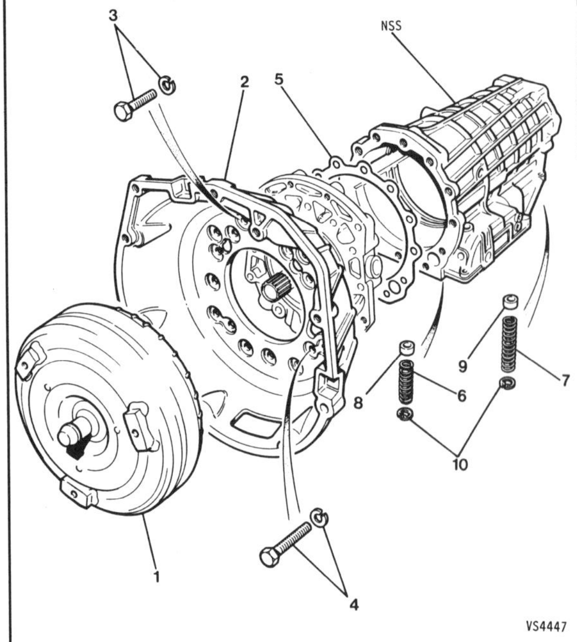

VS4447

AUTOMATIC TRANSMISSON CONVERTER & HOUSING 3.6 LITRE

Illus	Part Number	123456 Description	Quantity	Change Point	Remarks
1	EAC 6923	CONVERTER ASSEMBLY	1		
2	EAC 5371	Torque converter housing	1		
3	JLM 659	Bolt housing to gearbox	6		
4	JLM 666	Bolt housing to gearbox	12		
5	JLM 662	Gasket	1		
6	JLM 674	Compression spring	4		
7	JLM 675	Compression spring	5		
8	JLM 672	Sleeve	8		
9	JLM 863	Sleeve	1		
10	JLM 864	Ring			
	C 39246	Stoneguard Torque Conv Housing			

1H 09

JAGUAR XJS RANGE (JAN 1987 ON)	H09 fiche 1	GEARBOX AND PROPSHAFT

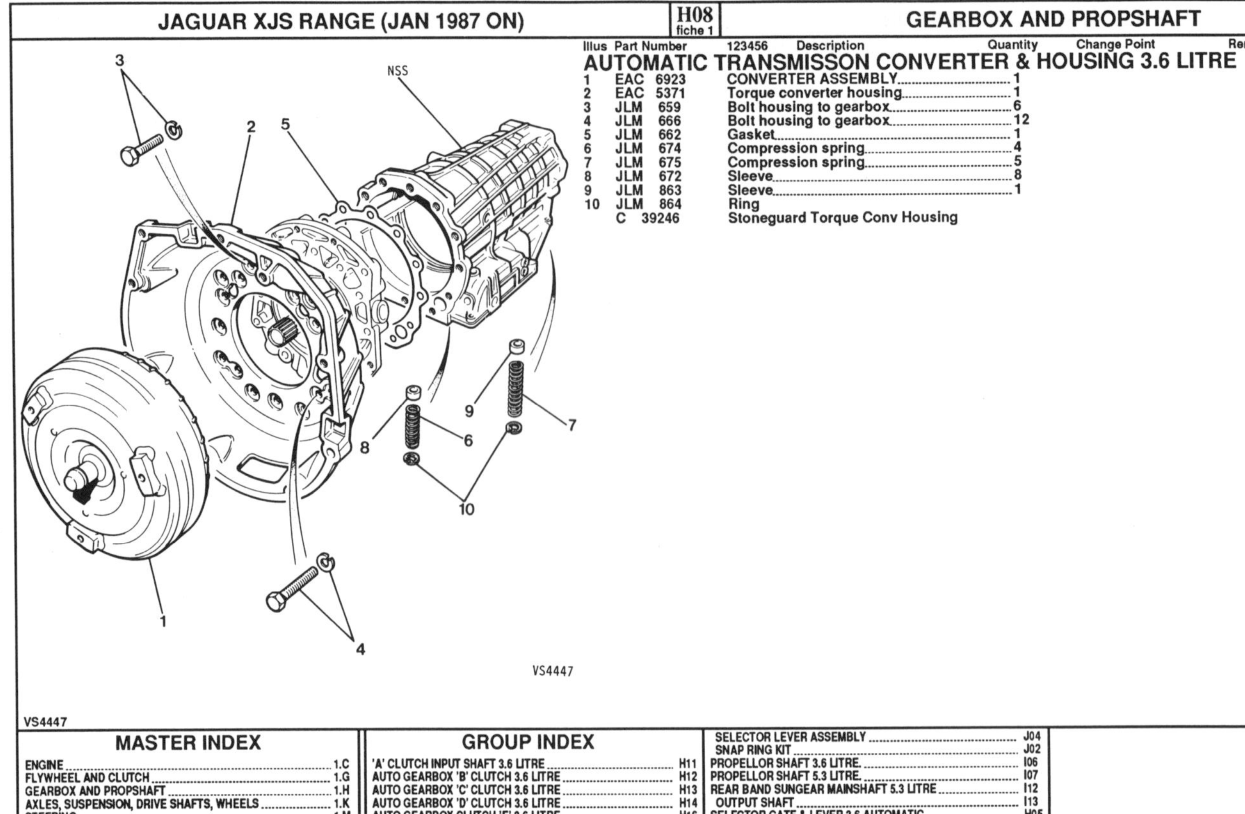

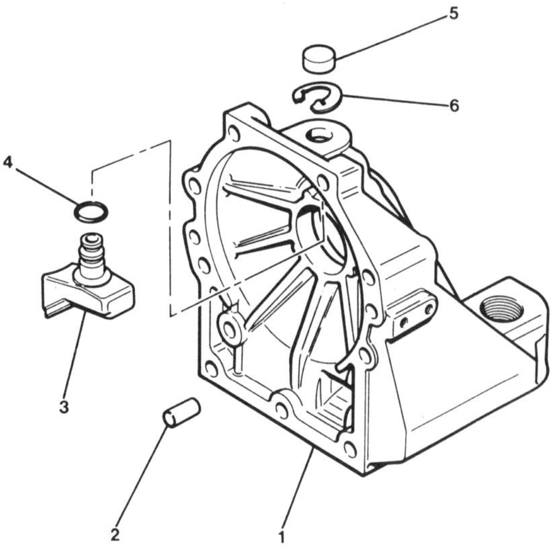

VS4703

REAR EXTENSION AND BREATHER 3.6 LITRE

Illus	Part Number	123456 Description	Quantity	Change Point	Remarks
1	JLM 693	Extension	1		Fitted up to Trans No. 891230
1	EBC 1178	Extension	1		Fitted from Trans No. 891231
2	JLM 697	Cylindrical pin	2		
3	JLM 705	Breather	1		
4	JLM 703	'O' ring	1		
5	JLM 706	Breather cover	1		
6	JLM 905	Locking washer	1		
	JLM 700	Setscrew	8		
	JLM 701	Setscrew	1		

MASTER INDEX

ENGINE 1.C
FLYWHEEL AND CLUTCH 1.G
GEARBOX AND PROPSHAFT 1.H
AXLES, SUSPENSION, DRIVE SHAFTS, WHEELS 1.K
STEERING 1.M
BRAKES AND BRAKE CONTROLS 1.N
FUEL, EXHAUST AND EMISSION SYSTEMS 2.C
COOLING, HEATING AND AIR CONDITIONING 2.G
ELECTRICAL, WASHERS-WIPERS, INSTRUMENTS 2.J
CHASSIS, SUBFRAMES, BODYSHELL, FITTINGS 3.C
FASCIA, TRIM, SEATS AND FIXINGS 3.H
ACCESSORIES AND PAINT 4.C
NUMERICAL INDEX 4.D

GROUP INDEX

'A' CLUTCH INPUT SHAFT 3.6 LITRE H11
AUTO GEARBOX 'B' CLUTCH 3.6 LITRE H12
AUTO GEARBOX 'C' CLUTCH 3.6 LITRE H13
AUTO GEARBOX 'D' CLUTCH 3.6 LITRE H14
AUTO GEARBOX CLUTCH 'E' 3.6 LITRE H16
AUTO GEARBOX MOUNTINGS-3.6 LITRE I04
AUTO TRANSMISSION SUMP PAN 3.6 LITRE I03
AUTOMATIC TRANSMISSION UNIT 3.6 L/T H07
CONVERTER ASSEMBLY H08
FRONT PUMP ASSEMBLY H10
AUTOMATIC TRANSMISSION UNIT 5.3 L/T I08
CLUTCH HOUSING I11
GEARBOX CASE ASSEMBLY I09
TORQUE CONVERTER I10
FRONT AND REAR SERVO 5.3 LITRE I14
GEARBOX MOUNTING PLATE 5.3 LITRE J08
GEARBOX MOUNTING-5.3 LITRE J07
MANUAL GEARBOX ASSEMBLY 3.6 LITRE H02
MANUAL GEARBOX MOUNTINGS 3.6 LITRE H04
PARKING BRAKE ASSEMBLY 5.3 LITRE I16
SELECTOR LEVER ASSEMBLY J04
SNAP RING KIT J02
PROPELLOR SHAFT 3.6 LITRE I06
PROPELLOR SHAFT 5.3 LITRE I07
REAR BAND SUNGEAR MAINSHAFT 5.3 LITRE I12
OUTPUT SHAFT I13
SELECTOR GATE & LEVER 3.6 AUTOMATIC H05
SPEEDO HOUSING GOVERNOR & MODULATOR I15
VALVE BODIES ASSEMBLY 3.6 LITRE H15

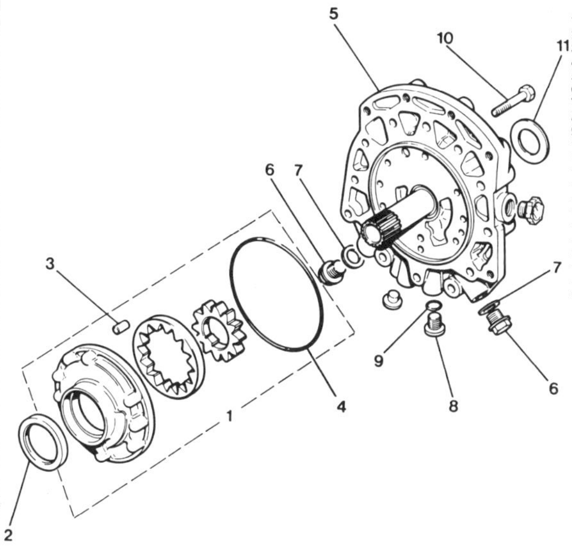

FRONT PUMP AND INTERMEDIATE PLATE 3.6 LITRE

Illus	Part Number	Description	Quantity	Change Point	Remarks
1	JLM 882	FRONT PUMP ASSEMBLY	1		
2	JLM 660	Seal front pump	1		
3	JLM 884	Cylindrical pin	1		
4	JLM 661	'O' ring	1		
5	JLM 885	Intermediate plate	1		
6	JLM 737	Screw plug	2		
7	JLM 734	Sealing ring	2		
8	JLM 738	Screw plug	2		
9	JLM 733	Sealing ring	2		
10	JLM 663	Hexagon screw	8		
11	JLM 702	Special washer	1		
11	JLM 692	Special washer	1		
11	JLM 689	Special washer	1		
11	JLM 685	Special washer	1		
11	JLM 684	Special washer	1		
11	JLM 679	Special washer	1		
11	JLM 678	Special washer	1		
11	JLM 673	Special washer	1		
11	JLM 670	Special washer	1		
11	JLM 668	Special washer	1		
11	JLM 739	Special washer	1		

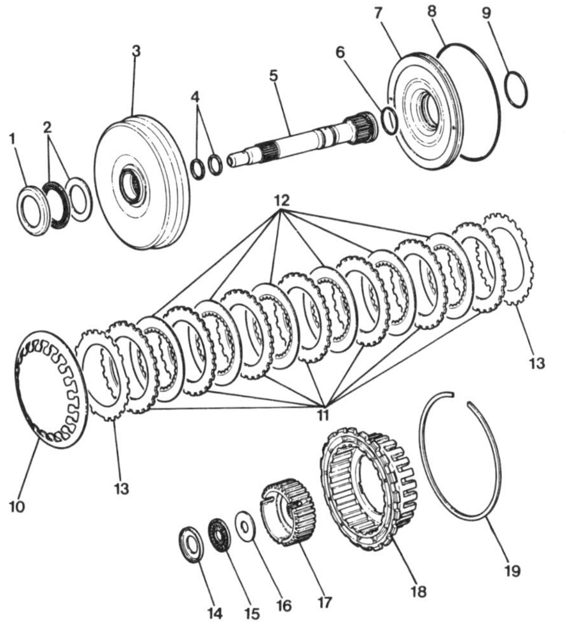

'A' CLUTCH INPUT SHAFT 3.6 LITRE

Illus	Part Number	Description	Quantity	Change Point	Remarks
1	JLM 1027	Angle disc	1		
2	JLM 1026	Needle cage	1		
3	JLM 1024	Cylinder	1		
4	JLM 1025	Piston ring	1		
5	JLM 1022	Input shaft	1		
6	JLM 1023	'O' ring	1		
7	JLM 1068	Piston	1		
8	JLM 1072	'O' ring	1		
9	JLM 1071	'O' ring	1		
10	JLM 1073	Cup spring	1		
11	JLM 1064	Clutch plate steel	A/R		
11	JLM 1065	Clutch plate steel	A/R		
12	JLM 1066	Clutch plate friction	6		Fitted up to serial no.524424
12	JLM 1332	Clutch plate friction	6		Fitted from serial no.524425
13	JLM 1067	Clutch plate waved	2		
14	JLM 944	Washer	1		
15	JLM 1029	Needle cage	1		
16	JLM 1030	Washer	1		
17	JLM 1031	Disc carrier	1		
18	JLM 1033	Disc carrier	1		
19	JLM 1036	Circlip	1		
	JLM 1004	'O' ring & sealing kit	1		
	JLM 1005	Needle bearing & piston ring kit	1		

MASTER INDEX

ENGINE 1.C
FLYWHEEL AND CLUTCH 1.G
GEARBOX AND PROPSHAFT 1.H
AXLES, SUSPENSION, DRIVE SHAFTS, WHEELS 1.K
STEERING 1.M
BRAKES AND BRAKE CONTROLS 1.N
FUEL, EXHAUST AND EMISSION SYSTEMS 2.C
COOLING, HEATING AND AIR CONDITIONING 2.G
ELECTRICAL, WASHERS-WIPERS, INSTRUMENTS 2.J
CHASSIS, SUBFRAMES, BODYSHELL, FITTINGS 3.C
FASCIA, TRIM, SEATS AND FIXINGS 3.H
ACCESSORIES AND PAINT 4.C
NUMERICAL INDEX 4.D

GROUP INDEX

'A' CLUTCH INPUT SHAFT 3.6 LITRE H11
AUTO GEARBOX 'B' CLUTCH 3.6 LITRE H12
AUTO GEARBOX 'C' CLUTCH 3.6 LITRE H13
AUTO GEARBOX 'D' CLUTCH 3.6 LITRE H14
AUTO GEARBOX CLUTCH 'E' 3.6 LITRE H16
AUTO GEARBOX MOUNTINGS-3.6 LITRE I04
AUTO TRANSMISSION SUMP PAN 3.6 LITRE I03
AUTOMATIC TRANSMISSION UNIT 3.6 L/T H07
CONVERTER ASSEMBLY H08
FRONT PUMP ASSEMBLY H10
AUTOMATIC TRANSMISSION UNIT 5.3 L/T I08
CLUTCH HOUSING I11
GEARBOX CASE ASSEMBLY I09
TORQUE CONVERTER I10
FRONT AND REAR SERVO 5.3 LITRE I14
GEARBOX MOUNTING PLATE 5.3 LITRE J08
GEARBOX MOUNTING-5.3 LITRE J07
MANUAL GEARBOX ASSEMBLY 3.6 LITRE H02
MANUAL GEARBOX MOUNTINGS 3.6 LITRE H04
PARKING BRAKE ASSEMBLY 5.3 LITRE I16
SELECTOR LEVER ASSEMBLY J04
SNAP RING KIT J02
PROPELLOR SHAFT 3.6 LITRE I06
PROPELLOR SHAFT 5.3 LITRE I07
REAR BAND SUNGEAR MAINSHAFT 5.3 LITRE I12
OUTPUT SHAFT I13
SELECTOR GATE & LEVER 3.6 AUTOMATIC H05
SPEEDO HOUSING GOVERNOR & MODULATOR I15
VALVE BODIES ASSEMBLY 3.6 LITRE H15

1H 12

JAGUAR XJS RANGE (JAN 1987 ON)	H12 fiche 1	GEARBOX AND PROPSHAFT

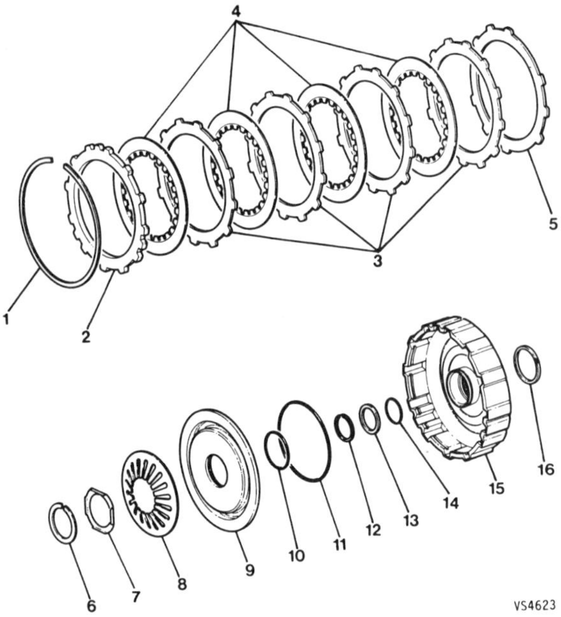

VS4623

AUTO GEARBOX 'B' CLUTCH 3.6 LITRE

Illus	Part Number	123456	Description	Quantity	Change Point	Remarks
1	JLM 1076		Snap ring	1		
2	JLM 1001		End disc	1		
3	JLM 999		Clutch disc	4		
4	JLM 1000		Friction disc	4		
5	JLM 1070		Clutch plate waved	1		
6	JLM 1075		Snap ring	1		
7	JLM 1074		Locking washer	1		
8	JLM 1087		Cup spring	1		
9	JLM 1088		Piston	1		
10	JLM 1071		'O' ring	1		
11	JLM 1089		'O' ring	1		
12	JLM 1036		Snap ring	1		
13	JLM 1035		Support ring	1		
14	JLM 1091		'O' ring	1		
15	JLM 1121		Cylinder	1		
16	JLM 1090		Piston ring	1		
	JLM 1004		'O' ring & seal kit	1		
	JLM 1005		Needle bearing & piston ring kit	1		

MASTER INDEX

GROUP INDEX

1H 13

JAGUAR XJS RANGE (JAN 1987 ON)	H13 fiche 1	GEARBOX AND PROPSHAFT

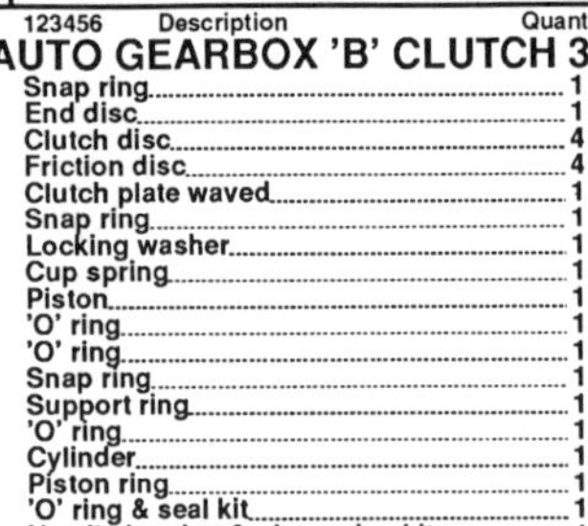

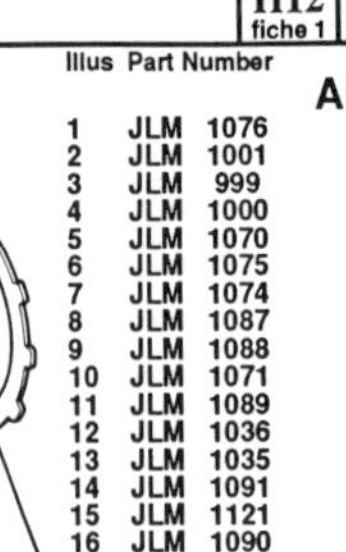

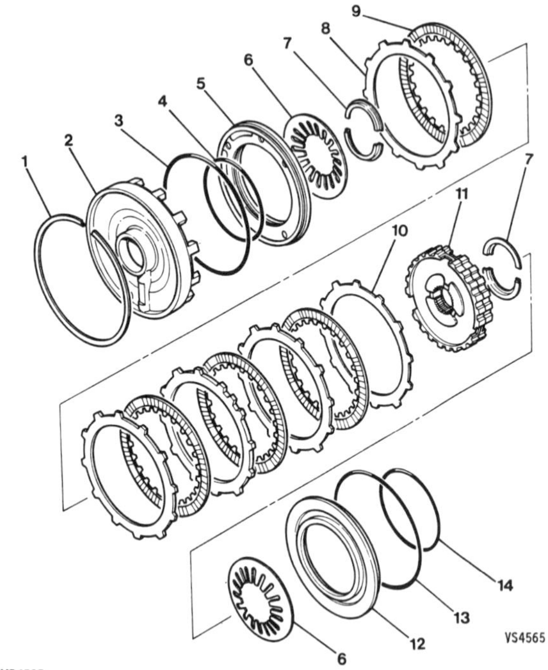

VS4565

AUTO GEARBOX 'C' CLUTCH 3.6 LITRE

Illus	Part Number	123456	Description	Quantity	Change Point	Remarks
1	JLM 916		Snap ring	1		
2	JLM 917		Centering plate	1		
3	JLM 878		'O' ring	1		
4	JLM 919		'O' ring	1		
5	JLM 918		Piston	1		
6	JLM 920		Cup spring	2		
7	JLM 921		Stop ring	2		
8	JLM 997		Clutch plate	2		
8	JLM 999		Clutch plate	2		
9	JLM 1000		Friction plate	4		
10	JLM 1001		End disc	1		
11	JLM 922		Free wheel	1		
12	JLM 923		Piston	1		
13	JLM 878		'O' ring	1		
14	JLM 919		'O' ring	1		

MASTER INDEX
ENGINE ... 1.C
FLYWHEEL AND CLUTCH ... 1.G
GEARBOX AND PROPSHAFT ... 1.H
AXLES, SUSPENSION, DRIVE SHAFTS, WHEELS ... 1.K
STEERING ... 1.M
BRAKES AND BRAKE CONTROLS ... 1.N
FUEL, EXHAUST AND EMISSION SYSTEMS ... 2.C
COOLING, HEATING AND AIR CONDITIONING ... 2.G
ELECTRICAL, WASHERS-WIPERS, INSTRUMENTS ... 2.J
CHASSIS, SUBFRAMES, BODYSHELL, FITTINGS ... 3.C
FASCIA, TRIM, SEATS AND FIXINGS ... 3.H
ACCESSORIES AND PAINT ... 4.C
NUMERICAL INDEX ... 4.D
GROUP INDEX
'A' CLUTCH INPUT SHAFT 3.6 LITRE ... H11
AUTO GEARBOX 'B' CLUTCH 3.6 LITRE ... H12
AUTO GEARBOX 'C' CLUTCH 3.6 LITRE ... H13
AUTO GEARBOX 'D' CLUTCH 3.6 LITRE ... H14
AUTO GEARBOX CLUTCH 'E' 3.6 LITRE ... H16
AUTO GEARBOX MOUNTINGS-3.6 LITRE ... I04
AUTO TRANSMISSION SUMP PAN 3.6 LITRE ... I03
AUTOMATIC TRANSMISSION UNIT 3.6 L/T ... H07
CONVERTER ASSEMBLY ... H08
FRONT PUMP ASSEMBLY ... H10
AUTOMATIC TRANSMISSION UNIT 5.3 L/T ... I08
CLUTCH HOUSING ... I11
GEARBOX CASE ASSEMBLY ... I09
TORQUE CONVERTER ... I10
FRONT AND REAR SERVO 5.3 LITRE ... I14
GEARBOX MOUNTING PLATE 5.3 LITRE ... J08
GEARBOX MOUNTING-5.3 LITRE ... J07
MANUAL GEARBOX ASSEMBLY 3.6 LITRE ... H02
MANUAL GEARBOX MOUNTINGS 3.6 LITRE ... H04
PARKING BRAKE ASSEMBLY 5.3 LITRE ... I16
SELECTOR LEVER ASSEMBLY ... J04
SNAP RING KIT ... J02
PROPELLOR SHAFT 3.6 LITRE ... I06
PROPELLOR SHAFT 5.3 LITRE ... I07
REAR BAND SUNGEAR MAINSHAFT 5.3 LITRE ... I12
OUTPUT SHAFT ... I13
SELECTOR GATE & LEVER 3.6 AUTOMATIC ... H05
SPEEDO HOUSING GOVERNOR & MODULATOR ... I15
VALVE BODIES ASSEMBLY 3.6 LITRE ... H15

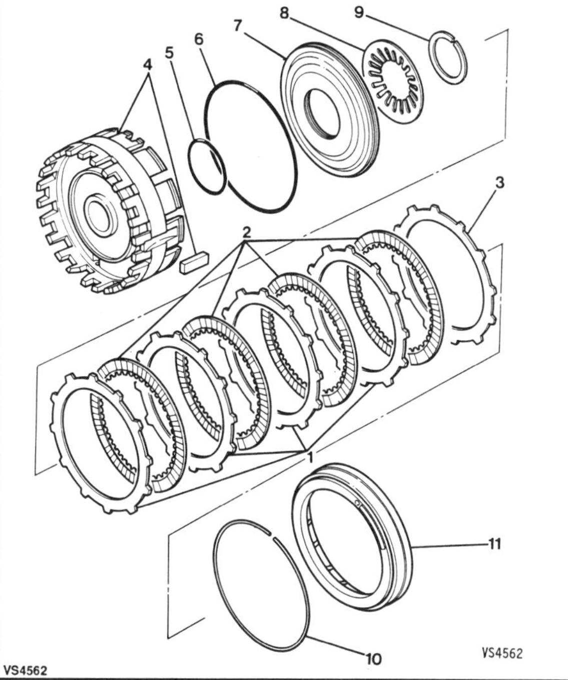

VS4562

AUTO GEARBOX 'D' CLUTCH 3.6 LITRE

Illus	Part Number	Description	Quantity	Change Point	Remarks
1	JLM 992	Clutch disc	4		
2	JLM 993	Clutch disc	4		
3	JLM 991	Clutch disc	1		
4	JLM 875	Cylinder	1		
5	JLM 877	'O' ring	1		
6	JLM 878	'O' ring	1		
7	JLM 876	Piston	1		
8	JLM 672	Cup spring	1		
9	JLM 879	Snap ring	1		
10	JLM 880	Snap ring	1		
11	JLM 881	Support ring	1		
	JLM 1004	'O' ring & seal ring kit	1		

MASTER INDEX

GROUP INDEX

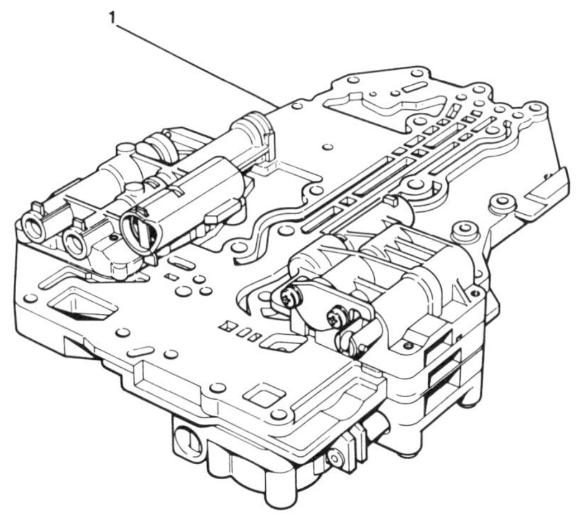

VS4595

VALVE BODIES ASSEMBLY 3.6 LITRE

Illus	Part Number	Description	Quantity	Change Point	Remarks
1	EBC 2196	Valve bodies assembly	1		
	JLM 1045	Cap screw	1		
	JLM 1078	Cap screw	1		
	JLM 1079	Cap screw	5		
	JLM 1080	Cap screw	4		
	JLM 1081	Cap screw	3		
	JLM 669	Cap screw	3		
	JLM 1083	Cap screw	7		
	JLM 1084	Cap screw	2		
	JLM 1085	Cap screw	1		

MASTER INDEX
ENGINE 1.C
FLYWHEEL AND CLUTCH 1.G
GEARBOX AND PROPSHAFT 1.H
AXLES, SUSPENSION, DRIVE SHAFTS, WHEELS 1.K
STEERING 1.M
BRAKES AND BRAKE CONTROLS 1.N
FUEL, EXHAUST AND EMISSION SYSTEMS 2.C
COOLING, HEATING AND AIR CONDITIONING 2.G
ELECTRICAL, WASHERS-WIPERS, INSTRUMENTS 2.J
CHASSIS, SUBFRAMES, BODYSHELL, FITTINGS 3.C
FASCIA, TRIM, SEATS AND FIXINGS 3.H
ACCESSORIES AND PAINT 4.C
NUMERICAL INDEX 4.D
GROUP INDEX
'A' CLUTCH INPUT SHAFT 3.6 LITRE H11
AUTO GEARBOX 'B' CLUTCH 3.6 LITRE H12
AUTO GEARBOX 'C' CLUTCH 3.6 LITRE H13
AUTO GEARBOX 'D' CLUTCH 3.6 LITRE H14
AUTO GEARBOX CLUTCH 'E' 3.6 LITRE H16
AUTO GEARBOX MOUNTINGS-3.6 LITRE I04
AUTO TRANSMISSION SUMP PAN 3.6 LITRE I03
AUTOMATIC TRANSMISSION UNIT 3.6 L/T H07
CONVERTER ASSEMBLY H08
FRONT PUMP ASSEMBLY H10
AUTOMATIC TRANSMISSION UNIT 5.3 L/T I08
CLUTCH HOUSING I11
GEARBOX CASE ASSEMBLY I09
TORQUE CONVERTER I10
FRONT AND REAR SERVO 5.3 LITRE I14
GEARBOX MOUNTING PLATE 5.3 LITRE J08
GEARBOX MOUNTING-5.3 LITRE J07
MANUAL GEARBOX ASSEMBLY 3.6 LITRE H02
MANUAL GEARBOX MOUNTINGS 3.6 LITRE H04
PARKING BRAKE ASSEMBLY 5.3 LITRE I16
SELECTOR LEVER ASSEMBLY J04
SNAP RING KIT J02
PROPELLOR SHAFT 3.6 LITRE I06
PROPELLOR SHAFT 5.3 LITRE I07
REAR BAND SUNGEAR MAINSHAFT 5.3 LITRE I12
OUTPUT SHAFT I13
SELECTOR GATE & LEVER 3.6 AUTOMATIC H05
SPEEDO HOUSING GOVERNOR & MODULATOR I15
VALVE BODIES ASSEMBLY 3.6 LITRE H15

JAGUAR XJS RANGE (JAN 1987 ON)	H16 fiche 1	GEARBOX AND PROPSHAFT

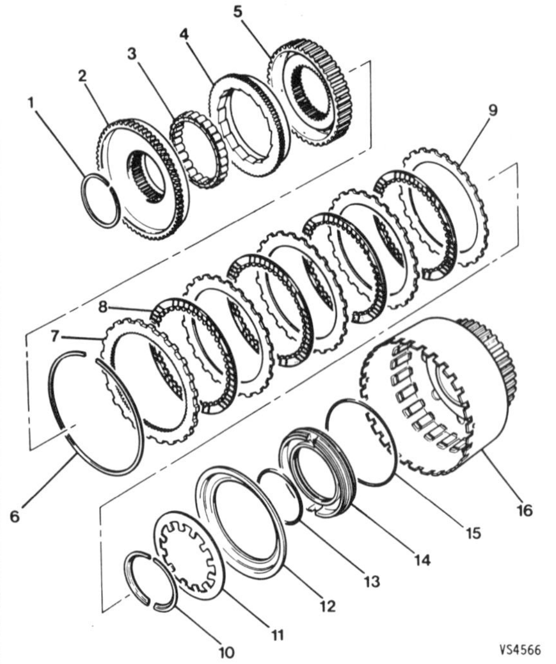

VS4566

1H 16

AUTO GEARBOX CLUTCH 'E' 3.6 LITRE

Illus	Part Number	Description	Quantity	Change Point	Remarks
1	JLM 914	Snap ring	1		
2	JLM 911	Free wheel ring	1		
3	JLM 912	Free wheel case	1		
4	JLM 913	Free wheel ring	1		
5	JLM 910	Disc carrier	1		
6	JLM 897	Snap ring	1		
7	JLM 1003	End disc	1		
8	JLM 995	Friction plate	4		
9	JLM 1002	Clutch disc	4		
10	JLM 909	Stop ring	1		
11	JLM 908	Cup spring	1		
12	JLM 907	Pressure	1		
13	JLM 904	'O' ring	1		
14	JLM 903	Piston	1		
15	JLM 1051	'O' ring	1		
16	JLM 898	Cylinder	1		

MASTER INDEX

GROUP INDEX

JAGUAR XJS RANGE (JAN 1987 ON)	H17 fiche 1	GEARBOX AND PROPSHAFT

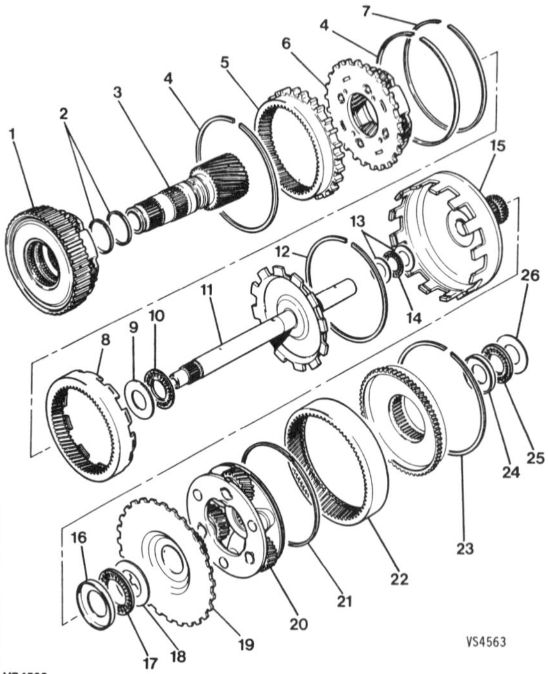

VS4563

1H 17

SUN GEARS & PLANETARY DRIVE 3.6 LITRE

Illus	Part Number	Description	Quantity	Change Point	Remarks
1	JLM 927	Planetary drive	1		
2	JLM 925	Piston ring	1		
3	JLM 1006	Sun gear shaft	1		
4	JLM 1069	Snap ring	2		
5	JLM 924	Ring gear	1		
6	JLM 934	Planetary drive	1		
7	JLM 937	Snap ring	1		
8	JLM 926	Ring gear	1		
9	JLM 929	Washer	1		
10	JLM 928	Needle cage	1		
11	JLM 930	Intermediate shaft	1		
12	JLM 931	Snap ring	1		
13	JLM 932	Washer	2		
14	JLM 933	Needle cage	1		
15	JLM 936	Spider shaft	1		
16	JLM 938	Angle disc	1		
17	JLM 902	Needle cage	1		
18	JLM 939	Washer	1		
19	JLM 940	Sun gear	1		
20	JLM 941	Planetary drive	1		
21		Snap ring	1		
22	JLM 942	Ring gear	1		
23	JLM 943	Snap ring	1		
24	JLM 944	Washer	1		
25	JLM 928	Needle cage	1		
26	JLM 929	Washer	1		

MASTER INDEX

ENGINE 1.C
FLYWHEEL AND CLUTCH 1.G
GEARBOX AND PROPSHAFT 1.H
AXLES, SUSPENSION, DRIVE SHAFTS, WHEELS 1.K
STEERING 1.M
BRAKES AND BRAKE CONTROLS 1.N
FUEL, EXHAUST AND EMISSION SYSTEMS 2.C
COOLING, HEATING AND AIR CONDITIONING 2.G
ELECTRICAL, WASHERS-WIPERS, INSTRUMENTS 2.J
CHASSIS, SUBFRAMES, BODYSHELL, FITTINGS 3.C
FASCIA, TRIM, SEATS AND FIXINGS 3.H
ACCESSORIES AND PAINT 4.C
NUMERICAL INDEX 4.D

GROUP INDEX

'A' CLUTCH INPUT SHAFT 3.6 LITRE H11
AUTO GEARBOX 'B' CLUTCH 3.6 LITRE H12
AUTO GEARBOX 'C' CLUTCH 3.6 LITRE H13
AUTO GEARBOX 'D' CLUTCH 3.6 LITRE H14
AUTO GEARBOX CLUTCH 'E' 3.6 LITRE H16
AUTO GEARBOX MOUNTINGS-3.6 LITRE I04
AUTO TRANSMISSION SUMP PAN 3.6 LITRE I03
AUTOMATIC TRANSMISSION UNIT 3.6 L/T H07
CONVERTER ASSEMBLY H08
FRONT PUMP ASSEMBLY H10
AUTOMATIC TRANSMISSION UNIT 5.3 L/T I08
CLUTCH HOUSING I11
GEARBOX CASE ASSEMBLY I09
TORQUE CONVERTER I10
FRONT AND REAR SERVO 5.3 LITRE I14
GEARBOX MOUNTING PLATE 5.3 LITRE J08
GEARBOX MOUNTING-5.3 LITRE J07
MANUAL GEARBOX ASSEMBLY 3.6 LITRE H02
MANUAL GEARBOX MOUNTINGS 3.6 LITRE H04
PARKING BRAKE ASSEMBLY 5.3 LITRE I16
SELECTOR LEVER ASSEMBLY J04
SNAP RING KIT J02
PROPELLOR SHAFT 3.6 LITRE I06
PROPELLOR SHAFT 5.3 LITRE I07
REAR BAND SUNGEAR MAINSHAFT 5.3 LITRE I12
OUTPUT SHAFT I13
SELECTOR GATE & LEVER 3.6 AUTOMATIC H05
SPEEDO HOUSING GOVERNOR & MODULATOR I15
VALVE BODIES ASSEMBLY 3.6 LITRE H15

OUTPUT SHAFT & 'F' CLUTCH 3.6 LITRE

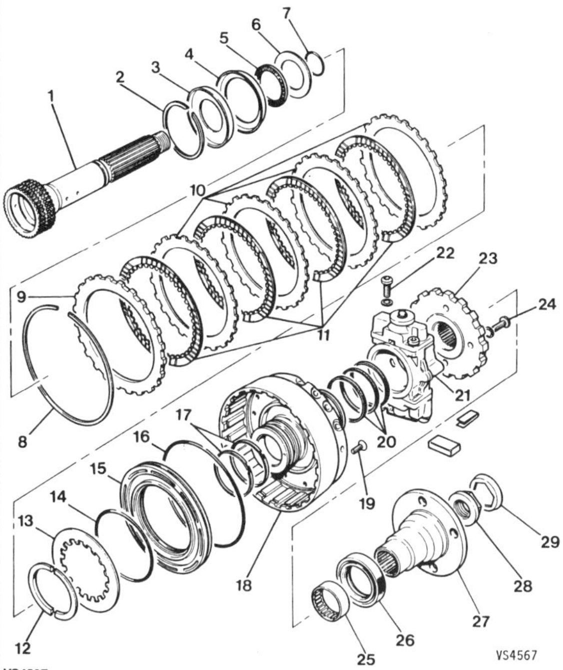

VS4567

Illus	Part Number	123456	Description	Quantity	Change Point	Remarks
1	JLM 873		Output shaft	1		
2	JLM 915		Snap ring	1		
3	JLM 900		Thrust washer	1		
4	JLM 901		Thrust washer	1		
5	JLM 902		Needle bearing	1		
6	JLM 899		Washer	1		
7	JLM 718		'O' ring	1		
8	JLM 897		Snap ring	1		
9	JLM 996		End disc	1		
10	JLM 994		Clutch disc	4		
11	JLM 995		Clutch disc lining	4		
12	JLM 909		Stop ring	1		
13	JLM 896		Cup spring	1		
14	JLM 894		'O' ring	1		
15	JLM 893		Piston	1		
16	JLM 895		'O' ring	1		
17	JLM 891		Piston ring	1		
18	JLM 889		Cylinder	1		
19	JLM 890		Screw	10		
20	JLM 892		Piston ring	3		
21	EAC 8106		Governor assembly	1		
22	JLM 945		Screw	2		
23	JLM 715		Park lock wheel	1		
24	JLM 671		Screw	2		
25	JLM 694		Needle sleeve	1		
26	JLM 686		Seal output flange	1		
27	JLM 874		Output flange	1		
28	JLM 711		Collar nut	1		
29	JLM 699		Tab washer	1		

MASTER INDEX

GROUP INDEX

GEARSHIFT & PARKING LOCK 3.6 LITRE

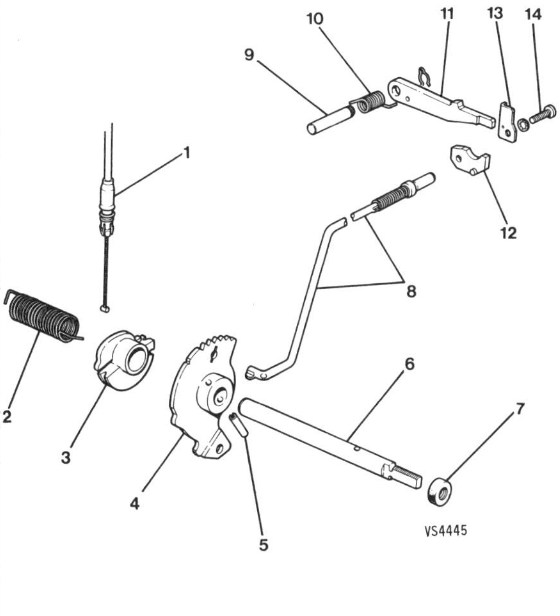

VS4445

Illus	Part Number	123456	Description	Quantity	Change Point	Remarks
1	JLM 872		Kickdown cable	1		
2	JLM 871		Leg spring	1		
3	JLM 870		Accelerator cam	1		
4	JLM 869		Detent disc	1		
5	JLM 1195		Stop pin	1		
6	JLM 867		Selector shaft	1		
7	JLM 866		Shaft seal	1		
8	JLM 887		Connecting bar	1		
9	JLM 712		Pin	1		
10	JLM 714		Spring	1		
11	JLM 713		Pawl	1		
12	JLM 717		Guide piece	1		
13	JLM 1054		Guide sheet	1		
14	JLM 888		Screw	1		

MASTER INDEX
ENGINE 1.C
FLYWHEEL AND CLUTCH 1.G
GEARBOX AND PROPSHAFT 1.H
AXLES, SUSPENSION, DRIVE SHAFTS, WHEELS 1.K
STEERING 1.M
BRAKES AND BRAKE CONTROLS 1.N
FUEL, EXHAUST AND EMISSION SYSTEMS 2.C
COOLING, HEATING AND AIR CONDITIONING 2.G
ELECTRICAL, WASHERS-WIPERS, INSTRUMENTS 2.J
CHASSIS, SUBFRAMES, BODYSHELL, FITTINGS 3.C
FASCIA, TRIM, SEATS AND FIXINGS 3.H
ACCESSORIES AND PAINT 4.C
NUMERICAL INDEX 4.D

GROUP INDEX
'A' CLUTCH INPUT SHAFT 3.6 LITRE H11
AUTO GEARBOX 'B' CLUTCH 3.6 LITRE H12
AUTO GEARBOX 'C' CLUTCH 3.6 LITRE H13
AUTO GEARBOX 'D' CLUTCH 3.6 LITRE H14
AUTO GEARBOX CLUTCH 'E' 3.6 LITRE H16
AUTO GEARBOX MOUNTINGS-3.6 LITRE I04
AUTO TRANSMISSION SUMP PAN 3.6 LITRE I03
AUTOMATIC TRANSMISSION UNIT 3.6 L/T H07
CONVERTER ASSEMBLY H08
FRONT PUMP ASSEMBLY H10
AUTOMATIC TRANSMISSION UNIT 5.3 L/T I08
CLUTCH HOUSING I11
GEARBOX CASE ASSEMBLY I09
TORQUE CONVERTER I10
FRONT AND REAR SERVO 5.3 LITRE I14
GEARBOX MOUNTING PLATE 5.3 LITRE J08
GEARBOX MOUNTING-5.3 LITRE J07
MANUAL GEARBOX ASSEMBLY 3.6 LITRE H02
MANUAL GEARBOX MOUNTINGS 3.6 LITRE H04
PARKING BRAKE ASSEMBLY 5.3 LITRE I16
SELECTOR LEVER ASSEMBLY J04
SNAP RING KIT J02
PROPELLOR SHAFT 3.6 LITRE I06
PROPELLOR SHAFT 5.3 LITRE I07
REAR BAND SUNGEAR MAINSHAFT 5.3 LITRE I12
OUTPUT SHAFT I13
SELECTOR GATE & LEVER 3.6 AUTOMATIC H05
SPEEDO HOUSING GOVERNOR & MODULATOR I15
VALVE BODIES ASSEMBLY 3.6 LITRE H15

1I 03

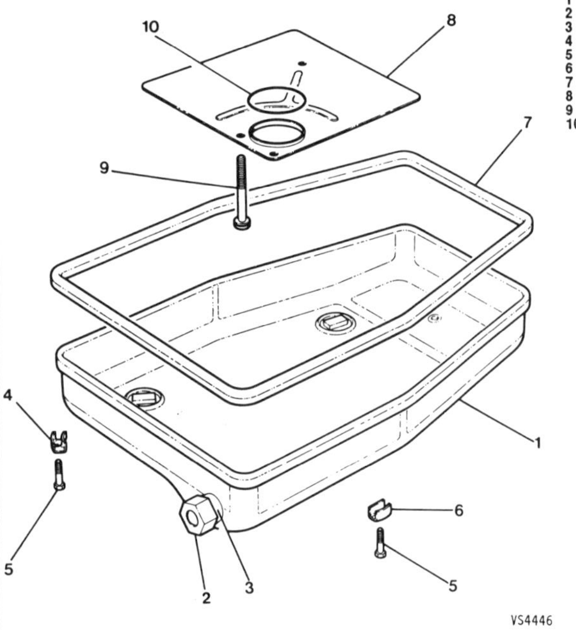

VS4446

AUTO TRANSMISSION SUMP PAN 3.6 LITRE

Illus	Part Number	123456 Description	Quantity	Change Point	Remarks
1	JLM 648	Sump pan	1		
2	JLM 655	Oil sump drain plug	1		
3	JLM 649	Drain plug washer	1		
4	JLM 652	Fixing plate	4		
5	JLM 651	Screw	6		
6	JLM 653	Fixing plate	2		
7	JLM 650	Gasket sump pan	1		
8	JLM 664	Oil screen	1		
9	JLM 667	Screw	2		
10	JLM 665	'O' ring	1		

1I 04

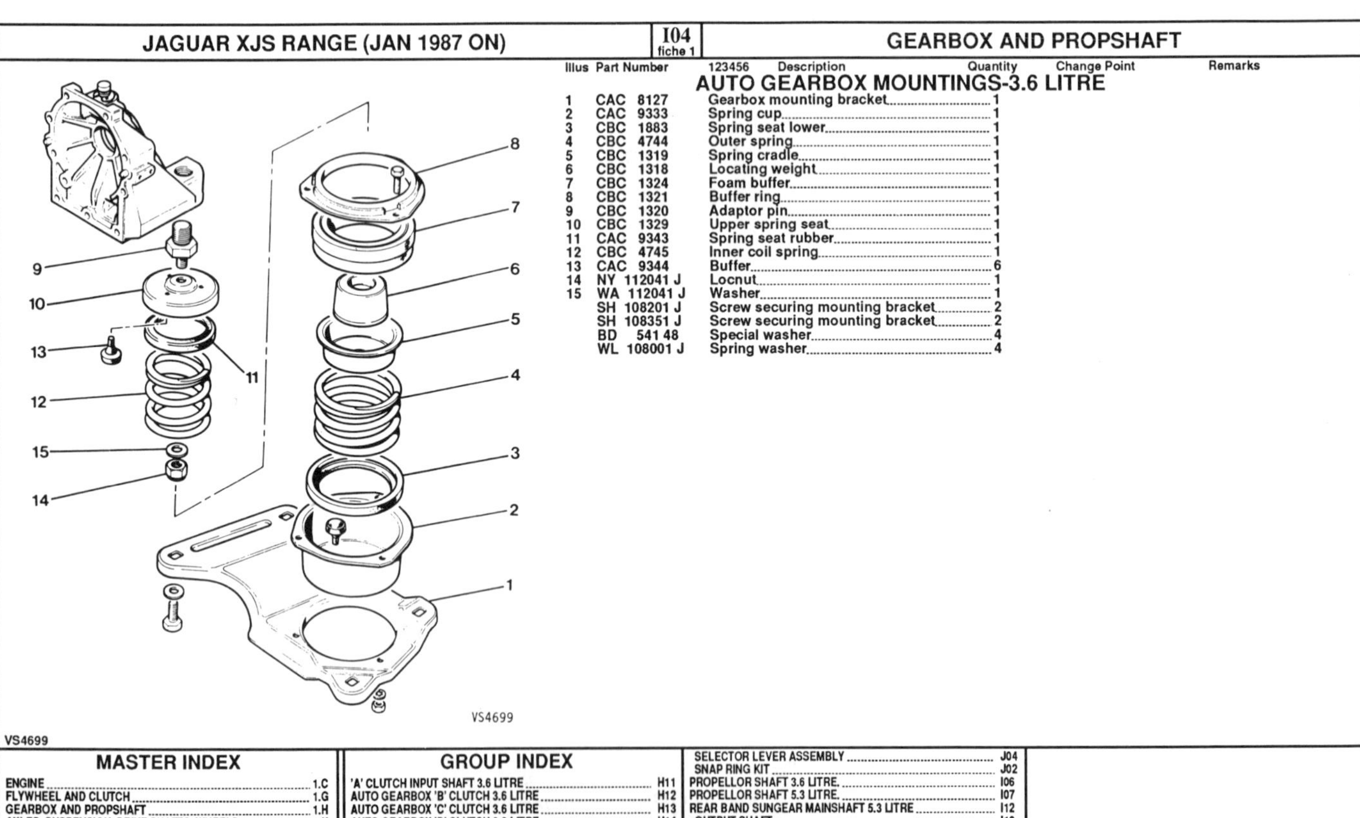

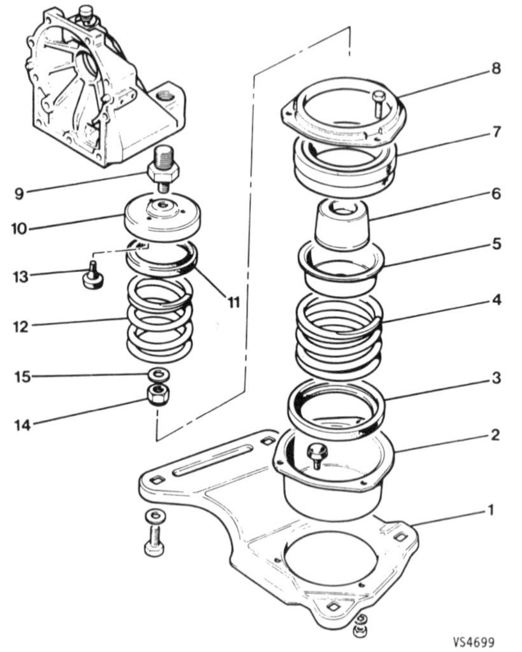

VS4699

AUTO GEARBOX MOUNTINGS-3.6 LITRE

Illus	Part Number	123456 Description	Quantity	Change Point	Remarks
1	CAC 8127	Gearbox mounting bracket	1		
2	CAC 9333	Spring cup	1		
3	CBC 1883	Spring seat lower	1		
4	CBC 4744	Outer spring	1		
5	CBC 1319	Spring cradle	1		
6	CBC 1318	Locating weight	1		
7	CBC 1324	Foam buffer	1		
8	CBC 1321	Buffer ring	1		
9	CBC 1320	Adaptor pin	1		
10	CBC 1329	Upper spring seat	1		
11	CAC 9343	Spring seat rubber	1		
12	CBC 4745	Inner coil spring	1		
13	CAC 9344	Buffer	6		
14	NY 112041 J	Locnut	1		
15	WA 112041 J	Washer	1		
	SH 108201 J	Screw securing mounting bracket	2		
	SH 108351 J	Screw securing mounting bracket	2		
	BD 541 48	Special washer	4		
	WL 108001 J	Spring washer	4		

MASTER INDEX

ENGINE 1.C
FLYWHEEL AND CLUTCH 1.G
GEARBOX AND PROPSHAFT 1.H
AXLES, SUSPENSION, DRIVE SHAFTS, WHEELS 1.K
STEERING 1.M
BRAKES AND BRAKE CONTROLS 1.N
FUEL, EXHAUST AND EMISSION SYSTEMS 2.C
COOLING, HEATING AND AIR CONDITIONING 2.G
ELECTRICAL, WASHERS-WIPERS, INSTRUMENTS 2.J
CHASSIS, SUBFRAMES, BODYSHELL, FITTINGS 3.C
FASCIA, TRIM, SEATS AND FIXINGS 3.H
ACCESSORIES AND PAINT 4.C
NUMERICAL INDEX 4.D

GROUP INDEX

'A' CLUTCH INPUT SHAFT 3.6 LITRE H11
AUTO GEARBOX 'B' CLUTCH 3.6 LITRE H12
AUTO GEARBOX 'C' CLUTCH 3.6 LITRE H13
AUTO GEARBOX 'D' CLUTCH 3.6 LITRE H14
AUTO GEARBOX CLUTCH 'E' 3.6 LITRE H16
AUTO GEARBOX MOUNTINGS-3.6 LITRE I04
AUTO TRANSMISSION SUMP PAN 3.6 LITRE I03
AUTOMATIC TRANSMISSION UNIT 3.6 L/T H07
CONVERTER ASSEMBLY H08
FRONT PUMP ASSEMBLY H10
AUTOMATIC TRANSMISSION UNIT 5.3 L/T I08
CLUTCH HOUSING I11
GEARBOX CASE ASSEMBLY I09
TORQUE CONVERTER I10
FRONT AND REAR SERVO 5.3 LITRE I14
GEARBOX MOUNTING PLATE 5.3 LITRE J08
GEARBOX MOUNTING-5.3 LITRE J07
MANUAL GEARBOX ASSEMBLY 3.6 LITRE H02
MANUAL GEARBOX MOUNTINGS 3.6 LITRE H04
PARKING BRAKE ASSEMBLY 5.3 LITRE I16
SELECTOR LEVER ASSEMBLY J04
SNAP RING KIT J02
PROPELLOR SHAFT 3.6 LITRE I06
PROPELLOR SHAFT 5.3 LITRE I07
REAR BAND SUNGEAR MAINSHAFT 5.3 LITRE I12
OUTPUT SHAFT I13
SELECTOR GATE & LEVER 3.6 AUTOMATIC H05
SPEEDO HOUSING GOVERNOR & MODULATOR I15
VALVE BODIES ASSEMBLY 3.6 LITRE H15

JAGUAR XJS RANGE (JAN 1987 ON) — I05 fiche 1 — GEARBOX AND PROPSHAFT

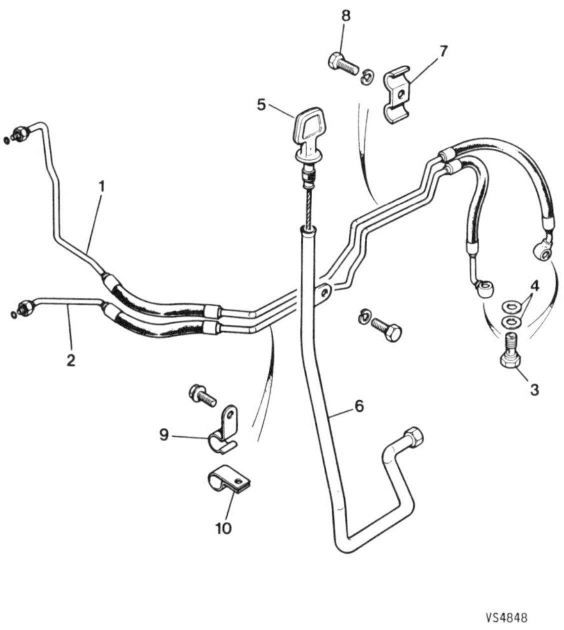

VS4848

DIPSTICK OIL COOLER PIPES-3.6 LITRE

Illus	Part Number	123456	Description	Quantity	Change Point	Remarks
1	EAC 9648		Oil cooler pipe assembly-output	1		
2	EAC 9649		Oil cooler pipe assembly-return	1		
	CAC 8968 2		'O' ring	1		
3	EAC 5625		Banjo bolt	2		
4	AGU 2093		Washer	4		
5	EAC 8065		Trans Dipstick Seal & Label Assembly	1		
6	EAC 8183		Dipstick Tube Assembly	1		
7	C 38894		Pipe Clip - Double	1		
8	UFS 819 4		Screw Securing Clip	1		
9	CP 108101		Clip	1		
10	C 43689		Clip	2		
	FS 11025 1		Setscrew	1		

D38.02DC16/8/88

MASTER INDEX

GROUP INDEX

JAGUAR XJS RANGE (JAN 1987 ON) — I06 fiche 1 — GEARBOX AND PROPSHAFT

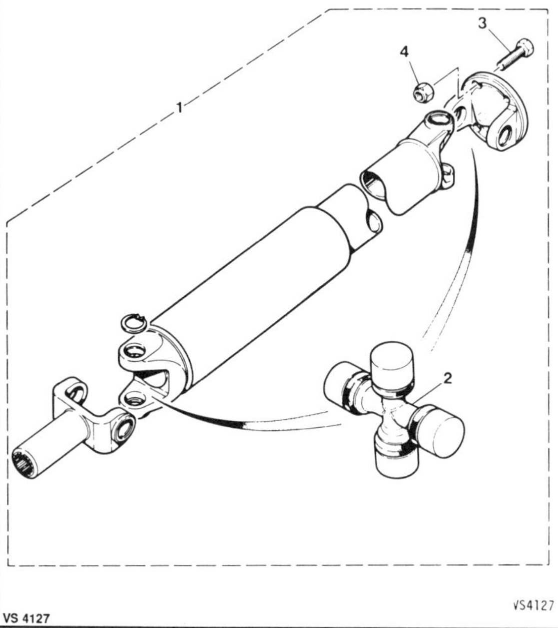

VS 4127

PROPELLOR SHAFT 3.6 LITRE.

Illus	Part Number	123456	Description	Quantity	Change Point	Remarks
1	CBC 1995		PROPELLOR SHAFT ASSEMBLY	1		Auto transmission
	AAU 6066		Centre bearing	1		
	CBC 5432		Mounting bracket	1		1988 Model Year Propshaft - centre
	SH 605111 MJ		Setscrew	2		
	FW 205 T		Plain washer	2		
	SH 605081 J		Screw	4		Centre bearing mtg brkt to body
	WA 108051 J		Plain washer	4		
	WM 108001 J		Spring washer	4		
	CAC 4560 2		PROPELLOR SHAFT ASSEMBLY	1		Manual Transmission
2	JLM 823		Kit universal joint	1		
3	C 19827		Bolt prop shaft coupling	8		
4	NZ 606041 J		Nut self locking	8		
	C 28654		Packer-propshaft alignment	A/R		
	C 28654 1		Packer-propshaft alignment	A/R		

MASTER INDEX

ENGINE 1.C
FLYWHEEL AND CLUTCH 1.G
GEARBOX AND PROPSHAFT 1.H
AXLES, SUSPENSION, DRIVE SHAFTS, WHEELS 1.K
STEERING 1.M
BRAKES AND BRAKE CONTROLS 1.N
FUEL, EXHAUST AND EMISSION SYSTEMS 2.C
COOLING, HEATING AND AIR CONDITIONING 2.G
ELECTRICAL, WASHERS-WIPERS, INSTRUMENTS 2.J
CHASSIS, SUBFRAMES, BODYSHELL, FITTINGS 3.C
FASCIA, TRIM, SEATS AND FIXINGS 3.H
ACCESSORIES AND PAINT 4.C
NUMERICAL INDEX 4.D

GROUP INDEX

'A' CLUTCH INPUT SHAFT 3.6 LITRE H11
AUTO GEARBOX 'B' CLUTCH 3.6 LITRE H12
AUTO GEARBOX 'C' CLUTCH 3.6 LITRE H13
AUTO GEARBOX 'D' CLUTCH 3.6 LITRE H14
AUTO GEARBOX CLUTCH 'E' 3.6 LITRE H16
AUTO GEARBOX MOUNTINGS-3.6 LITRE I04
AUTO TRANSMISSION SUMP PAN 3.6 LITRE I03
AUTOMATIC TRANSMISSION UNIT 3.6 L/T H07
CONVERTER ASSEMBLY H08
FRONT PUMP ASSEMBLY H10
AUTOMATIC TRANSMISSION UNIT 5.3 L/T I08
CLUTCH HOUSING I11
GEARBOX CASE ASSEMBLY I09
TORQUE CONVERTER I10
FRONT AND REAR SERVO 5.3 LITRE I14
GEARBOX MOUNTING PLATE 5.3 LITRE J08
GEARBOX MOUNTING-5.3 LITRE J07
MANUAL GEARBOX ASSEMBLY 3.6 LITRE H02
MANUAL GEARBOX MOUNTINGS 3.6 LITRE H04
PARKING BRAKE ASSEMBLY 5.3 LITRE I16
SELECTOR LEVER ASSEMBLY J04
SNAP RING KIT J02
PROPELLOR SHAFT 3.6 LITRE I06
PROPELLOR SHAFT 5.3 LITRE I07
REAR BAND SUNGEAR MAINSHAFT 5.3 LITRE I12
OUTPUT SHAFT I13
SELECTOR GATE & LEVER 3.6 AUTOMATIC H05
SPEEDO HOUSING GOVERNOR & MODULATOR I15
VALVE BODIES ASSEMBLY 3.6 LITRE H15

JAGUAR XJS RANGE (JAN 1987 ON) — I07 fiche 1 — GEARBOX AND PROPSHAFT

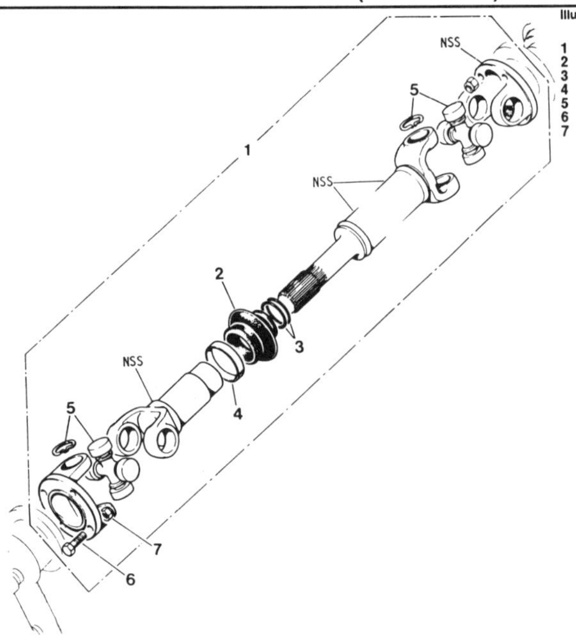

SM4083

SM 4083

PROPELLOR SHAFT 5.3 LITRE.

Illus	Part Number	123456	Description	Quantity	Change Point	Remarks
1	CAC 21911	1	PROPELLOR SHAFT ASSEMBLY	1		
2	9411		Gaiter-prop shaft	1		
3	9412		Ring-P.S. gaiter	2		
4	9413		Clip-P.S. gaiter	1		
5	JLM 823		Kit universal joint	1		
6	C 19827		Bolt prop shaft coupling	8		
7	C 8737	3	Nut self locking	8		

MASTER INDEX

GROUP INDEX

JAGUAR XJS RANGE (JAN 1987 ON) — I08 fiche 1 — GEARBOX AND PROPSHAFT

AUTOMATIC TRANSMISSION UNIT 5.3 L/T

Illus	Part Number	123456	Description	Quantity	Change Point	Remarks
1	EAC 8439	N	Automatic trans assembly	1		
2	BH 506201	J	Bolt	8		
3	WF 600060	J	Washer	8		
4	C 46021		Dowel	2		
	C 16862	15	Tapered plug for transducer aperture	1		

D50.00DC30/6/88

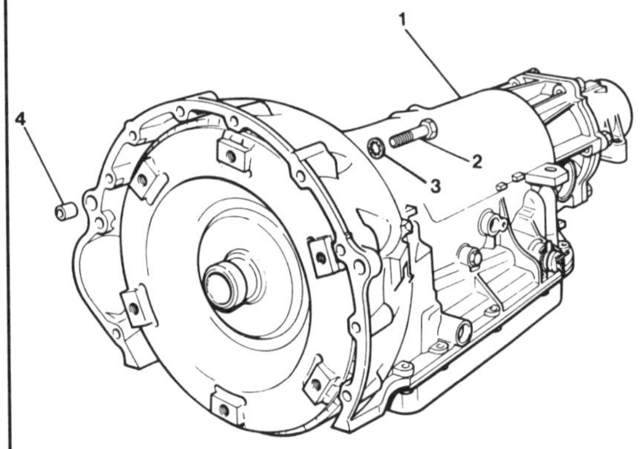

VS3218

VS3218

MASTER INDEX

ENGINE 1.C
FLYWHEEL AND CLUTCH 1.G
GEARBOX AND PROPSHAFT 1.H
AXLES, SUSPENSION, DRIVE SHAFTS, WHEELS 1.K
STEERING 1.M
BRAKES AND BRAKE CONTROLS 1.N
FUEL, EXHAUST AND EMISSION SYSTEMS 2.C
COOLING, HEATING AND AIR CONDITIONING 2.G
ELECTRICAL, WASHERS-WIPERS, INSTRUMENTS 2.J
CHASSIS, SUBFRAMES, BODYSHELL, FITTINGS 3.C
FASCIA, TRIM, SEATS AND FIXINGS 3.H
ACCESSORIES AND PAINT 4.C
NUMERICAL INDEX 4.D

GROUP INDEX

'A' CLUTCH INPUT SHAFT 3.6 LITRE H11
AUTO GEARBOX 'B' CLUTCH 3.6 LITRE H12
AUTO GEARBOX 'C' CLUTCH 3.6 LITRE H13
AUTO GEARBOX 'D' CLUTCH 3.6 LITRE H14
AUTO GEARBOX CLUTCH 'E' 3.6 LITRE H16
AUTO GEARBOX MOUNTINGS-3.6 LITRE I04
AUTO TRANSMISSION SUMP PAN 3.6 LITRE I03
AUTOMATIC TRANSMISSION UNIT 3.6 L/T H07
CONVERTER ASSEMBLY H08
FRONT PUMP ASSEMBLY H10
AUTOMATIC TRANSMISSION UNIT 5.3 L/T I08
CLUTCH HOUSING I11
GEARBOX CASE ASSEMBLY I09
TORQUE CONVERTER I10
FRONT AND REAR SERVO 5.3 LITRE I14
GEARBOX MOUNTING PLATE 5.3 LITRE J08
GEARBOX MOUNTING-5.3 LITRE J07
MANUAL GEARBOX ASSEMBLY 3.6 LITRE H02
MANUAL GEARBOX MOUNTINGS 3.6 LITRE H04
PARKING BRAKE ASSEMBLY 5.3 LITRE I16
SELECTOR LEVER ASSEMBLY J04
SNAP RING KIT J02
PROPELLOR SHAFT 3.6 LITRE I06
PROPELLOR SHAFT 5.3 LITRE I07
REAR BAND SUNGEAR MAINSHAFT 5.3 LITRE I12
OUTPUT SHAFT I13
SELECTOR GATE & LEVER 3.6 AUTOMATIC H05
SPEEDO HOUSING GOVERNOR & MODULATOR I15
VALVE BODIES ASSEMBLY 3.6 LITRE H15

JAGUAR XJS RANGE (JAN 1987 ON)	I09 fiche 1	GEARBOX AND PROPSHAFT

AUTOMATIC TRANSMISSION CASING 5.3 LITRE

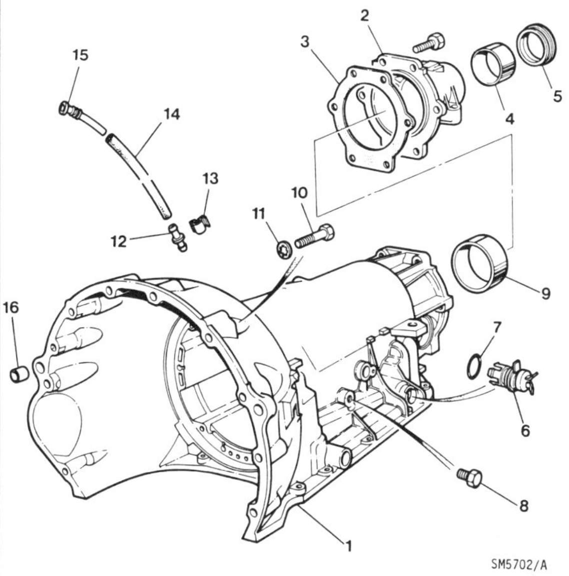

SM5702A

Illus	Part Number	123456 Description	Quantity	Change Point	Remarks
1	AAU 9595	GEARBOX CASE ASSEMBLY	1		
2	AAU 6603	Extension case	1		
3	AEU 1360	Gasket	1		See kit AAU 6639
4	AAU 6670	Ext case bushing	1		
5	AEU 1236	Seal	1		See kit AAU 6641
6	AAU 6677	Connector	1		
7	AAU 6639	O/haul kit-seals-gaskets	1		
8	AEU 1103	Plug	1		
9	AAU 6638	Case bushing	1		
10	BH 506201 J	Bolt	8		
11	C 726	Washer	8		
12	C 33835	Connector	1		
13	13H 9559	Edge clip	1		
14	C 38844 30	Hose	1		
15	AEU 1660	Vent pipe	1		
16	C 46021	Dowel	2		

JAGUAR XJS RANGE (JAN 1987 ON)	I10 fiche 1	GEARBOX AND PROPSHAFT

TORQUE CONVERTER-OIL PUMP 5.3 LITRE

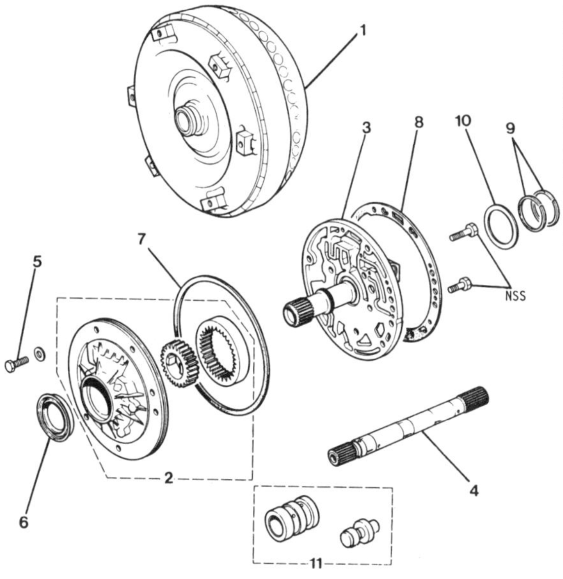

SM5673

Illus	Part Number	123456 Description	Quantity	Change Point	Remarks
1	EAC 7320	TORQUE CONVERTER	1		
2	AAU 6699	Oil pump body	1		
3	AAU 6698	Body cover assembly	1		
4	AAU 9528	Turbine shaft	1		
5	JLM 267	Bolt	1		
6	AEU 1065	Front seal	1		
7	AAU 6639	Gasket kit	1		
8	AAU 9525	Gasket	1		Part of AAU 6639
9	AAU 6641	Repair kit	1		
		THRUST WASHER			
10	AAU 6637	0.060/0.064	A/R		
10	AAU 6636	0.071/0.075	A/R		
10	AAU 6635	0.082/0.086	A/R		
10	AAU 6634	0.093/0.097	A/R		
10	AAU 6633	0.104/0.108	A/R		
10	AAU 6632	0.115/0.119	A/R		
10	AAU 6631	0.126/0.130	A/R		
11	AAU 6640	Kit-bushing and valve	1		

MASTER INDEX

ENGINE ... 1.C
FLYWHEEL AND CLUTCH ... 1.G
GEARBOX AND PROPSHAFT ... 1.H
AXLES, SUSPENSION, DRIVE SHAFTS, WHEELS ... 1.K
STEERING ... 1.M
BRAKES AND BRAKE CONTROLS ... 1.N
FUEL, EXHAUST AND EMISSION SYSTEMS ... 2.C
COOLING, HEATING AND AIR CONDITIONING ... 2.G
ELECTRICAL, WASHERS-WIPERS, INSTRUMENTS ... 2.J
CHASSIS, SUBFRAMES, BODYSHELL, FITTINGS ... 3.C
FASCIA, TRIM, SEATS AND FIXINGS ... 3.H
ACCESSORIES AND PAINT ... 4.C
NUMERICAL INDEX ... 4.D

GROUP INDEX

'A' CLUTCH INPUT SHAFT 3.6 LITRE ... H11
AUTO GEARBOX 'B' CLUTCH 3.6 LITRE ... H12
AUTO GEARBOX 'C' CLUTCH 3.6 LITRE ... H13
AUTO GEARBOX 'D' CLUTCH 3.6 LITRE ... H14
AUTO GEARBOX CLUTCH 'E' 3.6 LITRE ... H16
AUTO GEARBOX MOUNTINGS-3.6 LITRE ... I04
AUTO TRANSMISSION SUMP PAN 3.6 LITRE ... I03
AUTOMATIC TRANSMISSION UNIT 3.6 L/T ... H07
CONVERTER ASSEMBLY ... H08
FRONT PUMP ASSEMBLY ... H10
AUTOMATIC TRANSMISSION UNIT 5.3 L/T ... I08
CLUTCH HOUSING ... I11
GEARBOX CASE ASSEMBLY ... I09
TORQUE CONVERTER ... I10
FRONT AND REAR SERVO 5.3 LITRE ... I14
GEARBOX MOUNTING PLATE 5.3 LITRE ... J08
GEARBOX MOUNTING-5.3 LITRE ... J07
MANUAL GEARBOX ASSEMBLY 3.6 LITRE ... H02
MANUAL GEARBOX MOUNTINGS 3.6 LITRE ... H04
PARKING BRAKE ASSEMBLY 5.3 LITRE ... I16
SELECTOR LEVER ASSEMBLY ... J04
SNAP RING KIT ... J02
PROPELLOR SHAFT 3.6 LITRE ... I06
PROPELLOR SHAFT 5.3 LITRE ... I07
REAR BAND SUNGEAR MAINSHAFT 5.3 LITRE ... I12
OUTPUT SHAFT ... I13
SELECTOR GATE & LEVER 3.6 AUTOMATIC ... H05
SPEEDO HOUSING GOVERNOR & MODULATOR ... I15
VALVE BODIES ASSEMBLY 3.6 LITRE ... H15

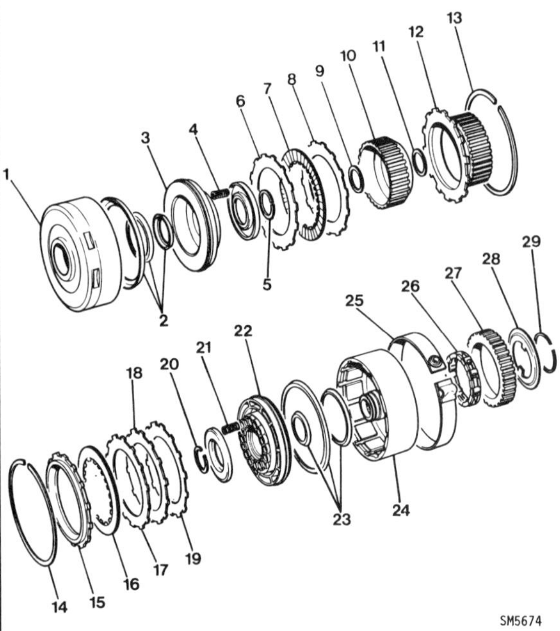

SM5674

I1 11

FORWARD AND DIRECT CLUTCH 5.3 LITRE

Illus	Part Number	Description	Quantity	Change Point	Remarks
1	AAU 6629	CLUTCH HOUSING	1		
2	AAU 9524	Seal kit	1		
3	AAU 6630	Piston-forward clutch	1		
4	AAU 9526	Spring-piston release	16		
5	AAU 6642	Kit-snap ring	1		
6	AAU 6667	Clutch plate	1		
7	AEU 1048	Clutch plate	5		
7	AAU 6668	Clutch plate	5		
8	AAU 6666	Clutch plate	5		
9	AEU 1039	Thrust washer	1		
10	AAU 6628	Hub forward clutch	1		
11	AAU 6626	Thrust washer	1		
12	AAU 6627	Hub-direct clutch	1		
13	AAU 6649	Snap ring	1		
14	AAU 6649	Snap ring	1		
15	AAU 6663	Direct clutch backing plate	1		
16	AAU 6668	Clutch plate	5		
17	AAU 6662	Clutch plate	4		
18	AAU 6666	Clutch plate	1		
19	AAU 6667	Clutch plate	1		
20	AAU 6642	Kit snap ring	1		
21	AAU 9526	Spring piston release	14		
22	AAU 6624	Piston direct clutch	1		
23	AAU 9524	Seal kit	1		
24	AEU 1040	Housing assembly-direct clutch	1		
25	AAU 6623	Front band assembly	1		
26	AAU 6625	Retainer	1		
27	AAU 6664	Race	1		
28	AAU 6625	Retainer	1		
29	AAU 6642	Kit snap ring	1		

MASTER INDEX

GROUP INDEX

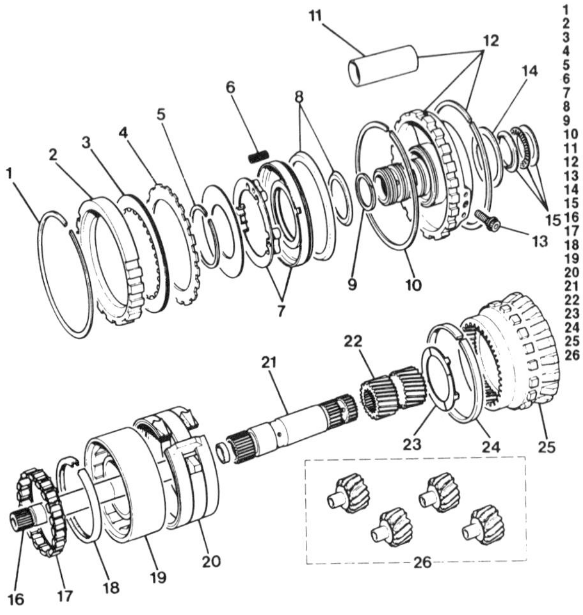

SM5675

I1 12

REAR BAND SUNGEAR MAINSHAFT 5.3 LITRE

Illus	Part Number	Description	Quantity	Change Point	Remarks
1	AAU 6650	Snap ring	1		
2	AAU 6659	Backing plate	1		
3	AAU 6658	Clutch plate	3		
4	AAU 6657	Clutch plate	3		
5	AAU 6642	Kit-snap ring	1		
6	AAU 9520	Spring-int clutch	3		
7	AAU 6619	Piston and return assembly	1		
8	AAU 6656	Seal kit	1		
9	AAU 6641	Repair kit	1		
10	AAU 9521	Snap ring	1		
11	AAU 6651	Bushing	1		
12	AAU 6643	Kit support and race assembly	1		
13	AEU 1038	Bolt-centre support	1		
14	AAU 6618	Thrust washer	1		
15	AAU 6644	Bearing	1		
16	AAU 6607	Mainshaft	1		
17	AAU 6617	Roller assembly	1		
18	AAU 6612	Spacer ring	1		
19	AAU 6613	Carrier assembly	1		
20	JLM 374	Brake band assembly	1		
21	AAU 6616	Sungear shaft	1		
22	AAU 6614	Sungear	1		
23	JLM 781	Washer	1		
24	AAU 6611	Ring	1		
25	AAU 6610	Carrier assembly	1		
26	AAU 6645	Kit & pinion replacement	1		

MASTER INDEX
ENGINE 1.C
FLYWHEEL AND CLUTCH 1.G
GEARBOX AND PROPSHAFT 1.H
AXLES, SUSPENSION, DRIVE SHAFTS, WHEELS 1.K
STEERING 1.M
BRAKES AND BRAKE CONTROLS 1.N
FUEL, EXHAUST AND EMISSION SYSTEMS 2.C
COOLING, HEATING AND AIR CONDITIONING 2.G
ELECTRICAL, WASHERS-WIPERS, INSTRUMENTS 2.J
CHASSIS, SUBFRAMES, BODYSHELL, FITTINGS 3.C
FASCIA, TRIM, SEATS AND FIXINGS 3.H
ACCESSORIES AND PAINT 4.C
NUMERICAL INDEX 4.D
GROUP INDEX
'A' CLUTCH INPUT SHAFT 3.6 LITRE H11
AUTO GEARBOX 'B' CLUTCH 3.6 LITRE H12
AUTO GEARBOX 'C' CLUTCH 3.6 LITRE H13
AUTO GEARBOX 'D' CLUTCH 3.6 LITRE H14
AUTO GEARBOX CLUTCH 'E' 3.6 LITRE H16
AUTO GEARBOX MOUNTINGS-3.6 LITRE I04
AUTO TRANSMISSION SUMP PAN 3.6 LITRE I03
AUTOMATIC TRANSMISSION UNIT 3.6 L/T H07
CONVERTER ASSEMBLY H08
FRONT PUMP ASSEMBLY H10
AUTOMATIC TRANSMISSION UNIT 5.3 L/T I08
CLUTCH HOUSING I11
GEARBOX CASE ASSEMBLY I09
TORQUE CONVERTER I10
FRONT AND REAR SERVO 5.3 LITRE I14
GEARBOX MOUNTING PLATE 5.3 LITRE J08
GEARBOX MOUNTING-5.3 LITRE J07
MANUAL GEARBOX ASSEMBLY 3.6 LITRE H02
MANUAL GEARBOX MOUNTINGS 3.6 LITRE H04
PARKING BRAKE ASSEMBLY 5.3 LITRE I16
SELECTOR LEVER ASSEMBLY J04
SNAP RING KIT J02
PROPELLOR SHAFT 3.6 LITRE. I06
PROPELLOR SHAFT 5.3 LITRE. I07
REAR BAND SUNGEAR MAINSHAFT 5.3 LITRE I12
OUTPUT SHAFT I13
SELECTOR GATE & LEVER 3.6 AUTOMATIC H05
SPEEDO HOUSING GOVERNOR & MODULATOR I15
VALVE BODIES ASSEMBLY 3.6 LITRE H15

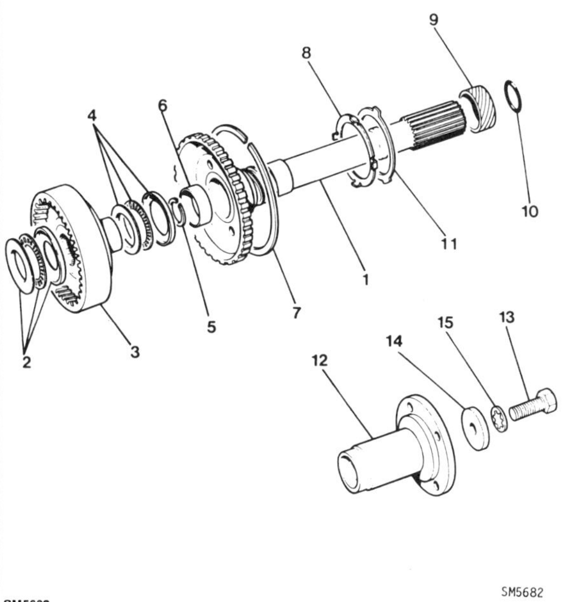

SM5682

OUTPUT SHAFT SPEEDO DRIVING GEAR 5.3 LITRE

Illus	Part Number	123456 Description	Quantity	Change Point	Remarks
1	AAU 6606	OUTPUT SHAFT	1		
2	AAU 6646	Thrust race	1		
3	AAU 6608	Rear internal gear	1		
4	AAU 6647	Kit-thrust brg and races	1		
5	AAU 6652	Snap ring	1		
6	AAU 6605	Bushing	1		
7	NSS	Snap ring	1		Part of AAU 6642 Kit
8	AAU 6694	Washer-output case	1		
9	EAC 4061	Speedo driving gear	1		18 teeth
10	AEU 1197	'O' ring			See kit AAU 6639
		THRUST WASHER			
11	AAU 9529	0.074 -0.078	A/R		Selective sizes
11	AAU 9530	0.082 -0.086	A/R		Selective sizes
11	AAU 9531	0.090 -0.094	A/R		Selective sizes
11	AAU 9532	0.098 -0.102	A/R		Selective sizes
11	AAU 9533	0.106 -0.110	A/R		Selective sizes
11	AAU 9534	0.114 -0.118	A/R		Selective sizes
12	C 45230	Output flange	1		
13	UFS 150 11R	Bolt	1		
14	C 45231	Washer	1		
15	C 728	Washer	1		

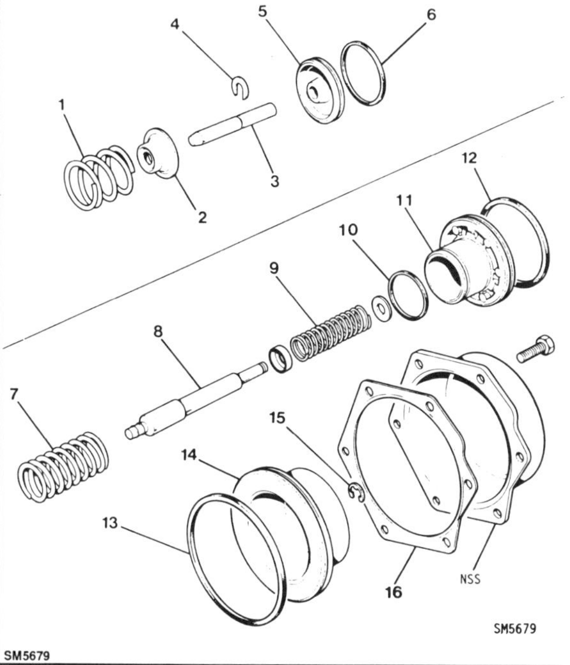

SM5679

FRONT AND REAR SERVO 5.3 LITRE

Illus	Part Number	123456 Description	Quantity	Change Point	Remarks
1	AAU 6622	Spring	1		
2	AAU 6621	Retainer	1		
3	AAU 6660	Servo piston pin	1		
4	AAU 6620	Front servo ring	1		
5	AAU 6661	Front servo piston	5		
6	AEU 1702	Oil seal	1		See kit AAU 6641
7	AAU 9527	Rear accumulator spring	1		
8	AAU 6696	Pin-rear band	1		
8	AAU 6697	Pin-rear band	1		
9	AAU 9523	Rear servo spring	1		
10	AAU 6653	Ring	1		
11	AAU 6600	Piston assembly	1		
12	AAU 6654	Oil seal acc-ring	1		
13	AAU 6639	Overhaul kit	1		
14	AAU 6601	Rear servo piston	1		
15	AAU 6642	Kit-snap ring	1		
16	AAU 6655	Cover gasket	1		

MASTER INDEX

ENGINE ... 1.C
FLYWHEEL AND CLUTCH ... 1.G
GEARBOX AND PROPSHAFT ... 1.H
AXLES, SUSPENSION, DRIVE SHAFTS, WHEELS ... 1.K
STEERING ... 1.M
BRAKES AND BRAKE CONTROLS ... 1.N
FUEL, EXHAUST AND EMISSION SYSTEMS ... 2.C
COOLING, HEATING AND AIR CONDITIONING ... 2.G
ELECTRICAL, WASHERS-WIPERS, INSTRUMENTS ... 2.J
CHASSIS, SUBFRAMES, BODYSHELL, FITTINGS ... 3.C
FASCIA, TRIM, SEATS AND FIXINGS ... 3.H
ACCESSORIES AND PAINT ... 4.C
NUMERICAL INDEX ... 4.D

GROUP INDEX

'A' CLUTCH INPUT SHAFT 3.6 LITRE ... H11
AUTO GEARBOX 'B' CLUTCH 3.6 LITRE ... H12
AUTO GEARBOX 'C' CLUTCH 3.6 LITRE ... H13
AUTO GEARBOX 'D' CLUTCH 3.6 LITRE ... H14
AUTO GEARBOX CLUTCH 'E' 3.6 LITRE ... H16
AUTO GEARBOX MOUNTINGS-3.6 LITRE ... I04
AUTO TRANSMISSION SUMP PAN 3.6 LITRE ... I03
AUTOMATIC TRANSMISSION UNIT 3.6 L/T ... H07
CONVERTER ASSEMBLY ... H08
FRONT PUMP ASSEMBLY ... H10
AUTOMATIC TRANSMISSION UNIT 5.3 L/T ... I08
CLUTCH HOUSING ... I11
GEARBOX CASE ASSEMBLY ... I09
TORQUE CONVERTER ... I10
FRONT AND REAR SERVO 5.3 LITRE ... I14
GEARBOX MOUNTING PLATE 5.3 LITRE ... J08
GEARBOX MOUNTING-5.3 LITRE ... J07
MANUAL GEARBOX ASSEMBLY 3.6 LITRE ... H02
MANUAL GEARBOX MOUNTINGS 3.6 LITRE ... H04
PARKING BRAKE ASSEMBLY 5.3 LITRE ... I16
SELECTOR LEVER ASSEMBLY ... J04
SNAP RING KIT ... J02
PROPELLOR SHAFT 3.6 LITRE ... I06
PROPELLOR SHAFT 5.3 LITRE ... I07
REAR BAND SUNGEAR MAINSHAFT 5.3 LITRE ... I12
OUTPUT SHAFT ... I13
SELECTOR GATE & LEVER 3.6 AUTOMATIC ... H05
SPEEDO HOUSING GOVERNOR & MODULATOR ... I15
VALVE BODIES ASSEMBLY 3.6 LITRE ... H15

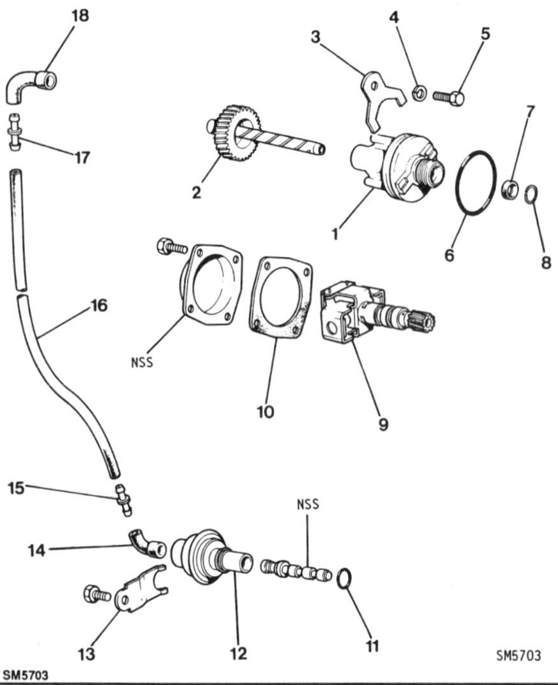

SM5703

SPEEDO HOUSING GOVERNOR & MODULATOR
5.3 LITRE

Illus	Part Number	123456 Description	Quantity	Change Point	Remarks
1	EAC 1424	Housing-driven gear	1		
2	C 46070	Speedo driven gear	1		39 teeth
3	EAC 1428	Retainer			
4	WL 106001 J	Washer	1		
5	SH 106161 J	Bolt	1		
6	EAC 1425	'O' ring	1		
7	EAC 1426	Oil seal	1		
8	EAC 1427	Retaining ring	1		
9	EAC 3185	Governor	1		
10	AEU 1049	Gasket	1		See kit AAU 6639
11	AAU 6639	Gasket kit	1		
12	EAC 1726	Vac modulator	1		
13	AAU 6676	Retainer	1		
14	C 34937	Elbow	1		
15	EAC 3094	Connector	1		
16	C 38844 30	Vac hose	1		
17	C 33835	Connector	1		
18	C 34937	Elbow	1		

II 15

MASTER INDEX

GROUP INDEX

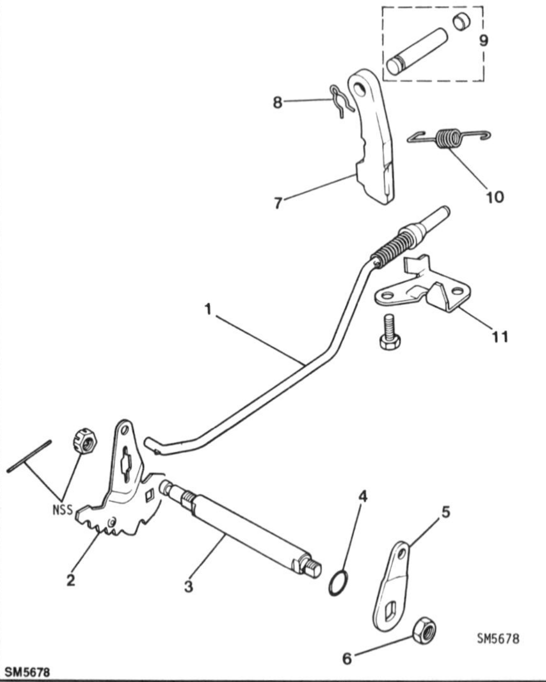

SM5678

PARKING BRAKE ASSEMBLY 5.3 LITRE

Illus	Part Number	123456 Description	Quantity	Change Point	Remarks
1	AAU 6673	Actuator assembly	Y		
2	AAU 6672	Lever & pin assembly	1		
3	AAU 6674	Manual shaft	1		
4	AAU 6639	Seal kit	1		
5	C 45247	Selector lever	1		
6	AAU 6841	Nut	1		
7	AEU 1659	Lock pawl	1		
8	AAU 8639	Retainer spring	1		
9	AAU 6648	Plug & shaft kit	1		
10	AAU 8641	Return spring	1		
11	AAU 6671	Pawl bracket	1		

II 16

MASTER INDEX

ENGINE 1.C
FLYWHEEL AND CLUTCH 1.G
GEARBOX AND PROPSHAFT 1.H
AXLES, SUSPENSION, DRIVE SHAFTS, WHEELS 1.K
STEERING 1.M
BRAKES AND BRAKE CONTROLS 1.N
FUEL, EXHAUST AND EMISSION SYSTEMS 2.C
COOLING, HEATING AND AIR CONDITIONING 2.G
ELECTRICAL, WASHERS-WIPERS, INSTRUMENTS 2.J
CHASSIS, SUBFRAMES, BODYSHELL, FITTINGS 3.C
FASCIA, TRIM, SEATS AND FIXINGS 3.H
ACCESSORIES AND PAINT 4.C
NUMERICAL INDEX 4.D

GROUP INDEX

'A' CLUTCH INPUT SHAFT 3.6 LITRE H11
AUTO GEARBOX 'B' CLUTCH 3.6 LITRE H12
AUTO GEARBOX 'C' CLUTCH 3.6 LITRE H13
AUTO GEARBOX 'D' CLUTCH 3.6 LITRE H14
AUTO GEARBOX CLUTCH 'E' 3.6 LITRE H16
AUTO GEARBOX MOUNTINGS-3.6 LITRE I04
AUTO TRANSMISSION SUMP PAN 3.6 LITRE I03
AUTOMATIC TRANSMISSION UNIT 3.6 L/T H07
CONVERTER ASSEMBLY H08
FRONT PUMP ASSEMBLY H10
AUTOMATIC TRANSMISSION UNIT 5.3 L/T I08
CLUTCH HOUSING I11
GEARBOX CASE ASSEMBLY I09
TORQUE CONVERTER I10
FRONT AND REAR SERVO 5.3 LITRE I14
GEARBOX MOUNTING PLATE 5.3 LITRE J08
GEARBOX MOUNTING-5.3 LITRE J07
MANUAL GEARBOX ASSEMBLY 3.6 LITRE H02
MANUAL GEARBOX MOUNTINGS 3.6 LITRE H04
PARKING BRAKE ASSEMBLY 5.3 LITRE I16
SELECTOR LEVER ASSEMBLY J04
SNAP RING KIT J02
PROPELLOR SHAFT 3.6 LITRE I06
PROPELLOR SHAFT 5.3 LITRE I07
REAR BAND SUNGEAR MAINSHAFT 5.3 LITRE I12
OUTPUT SHAFT I13
SELECTOR GATE & LEVER 3.6 AUTOMATIC H05
SPEEDO HOUSING GOVERNOR & MODULATOR I15
VALVE BODIES ASSEMBLY 3.6 LITRE H15

CONTROL VALVE SOLENOID GOVERNOR PIPES 5.3 LITRE

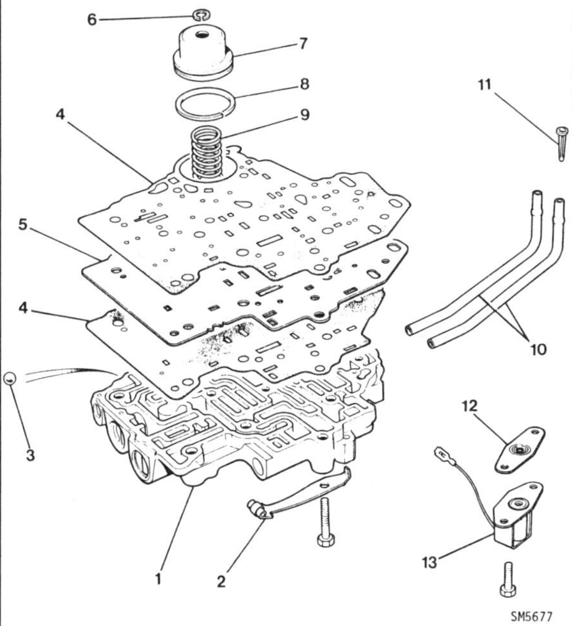

SM5677

Illus	Part Number	Description	Quantity	Change Point	Remarks
1	EAC 4048	Control valve assembly	1		
2	AAU 6685	Roller & spring-detent	1		
3	AAU 6680	Ball bearing	1		
4	AAU 6639	Gasket kit	1		
5	AAU 6679	Plate-valve body spacer	1		
6	AAU 6683	Retaining ring-acc piston	1		
7	AAU 6682	Piston-accumulator piston	1		
8	AAU 9522	Oil seal-acc piston	1		
9	AEU 1658	Spring-acc piston	1		
10	AAU 6687	Governor pipes	1		
11	AAU 6688	Screen governor	1		
12	AAU 6639	Gasket kit	1		
13	AAU 6686	Solenoid assembly	1		

MASTER INDEX

GROUP INDEX

OIL SUMP AND FILTER 5.3 LITRE

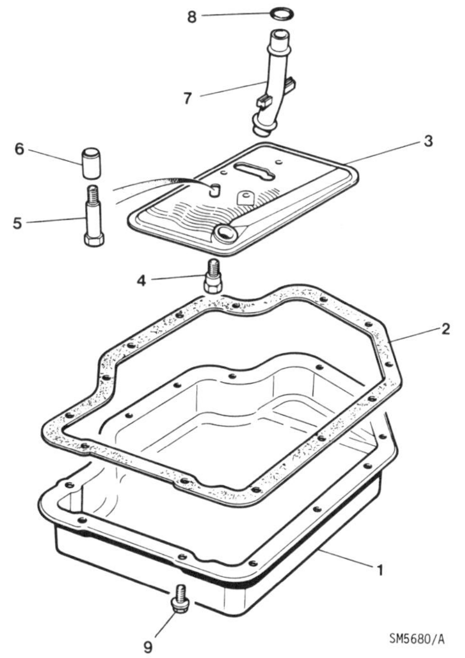

SM5680/A

Illus	Part Number	Description	Quantity	Change Point	Remarks
1	JLM 1189	Oil sump	1		
2	AEU 1050	Gasket oil sump	1		
3	AAU 6690	Filter assy	1		
4	JLM 1191	Bolt	1	Up to Eng No 7P56894	
5	JLM 1193	Bolt	1	From Eng No 7P56895	
6	JLM 1194	Spacer	1	From Eng No 7P56895	
7	AAU 6689	Oil pick up tube	1	Up to Eng No 7P56894	
7	JLM 1190	Oil pick up tube	1	From Eng No 7P56895	
8	AAU 6639	Seal kit	1		
9	AAU 6693	Screw	13		

D70DC27/7/88

MASTER INDEX
ENGINE 1.C
FLYWHEEL AND CLUTCH 1.G
GEARBOX AND PROPSHAFT 1.H
AXLES, SUSPENSION, DRIVE SHAFTS, WHEELS 1.K
STEERING 1.M
BRAKES AND BRAKE CONTROLS 1.N
FUEL, EXHAUST AND EMISSION SYSTEMS 2.C
COOLING, HEATING AND AIR CONDITIONING 2.G
ELECTRICAL, WASHERS-WIPERS, INSTRUMENTS 2.J
CHASSIS, SUBFRAMES, BODYSHELL, FITTINGS 3.C
FASCIA, TRIM, SEATS AND FIXINGS 3.H
ACCESSORIES AND PAINT 4.C
NUMERICAL INDEX 4.D
GROUP INDEX
'A' CLUTCH INPUT SHAFT 3.6 LITRE H11
AUTO GEARBOX 'B' CLUTCH 3.6 LITRE H12
AUTO GEARBOX 'C' CLUTCH 3.6 LITRE H13
AUTO GEARBOX 'D' CLUTCH 3.6 LITRE H14
AUTO GEARBOX CLUTCH 'E' 3.6 LITRE H16
AUTO GEARBOX MOUNTINGS-3.6 LITRE I04
AUTO TRANSMISSION SUMP PAN 3.6 LITRE I03
AUTOMATIC TRANSMISSION UNIT 3.6 L/T H07
CONVERTER ASSEMBLY H08
FRONT PUMP ASSEMBLY H10
AUTOMATIC TRANSMISSION UNIT 5.3 L/T I08
CLUTCH HOUSING I11
GEARBOX CASE ASSEMBLY I09
TORQUE CONVERTER I10
FRONT AND REAR SERVO 5.3 LITRE I14
GEARBOX MOUNTING PLATE 5.3 LITRE J08
GEARBOX MOUNTING-5.3 LITRE J07
MANUAL GEARBOX ASSEMBLY 3.6 LITRE H02
MANUAL GEARBOX MOUNTINGS 3.6 LITRE H04
PARKING BRAKE ASSEMBLY 5.3 LITRE I16
SELECTOR LEVER ASSEMBLY J04
SNAP RING KIT J02
PROPELLOR SHAFT 3.6 LITRE I06
PROPELLOR SHAFT 5.3 LITRE I07
REAR BAND SUNGEAR MAINSHAFT 5.3 LITRE I12
OUTPUT SHAFT I13
SELECTOR GATE & LEVER 3.6 AUTOMATIC H05
SPEEDO HOUSING GOVERNOR & MODULATOR I15
VALVE BODIES ASSEMBLY 3.6 LITRE H15

1J 02

JAGUAR XJS RANGE (JAN 1987 ON)	J02 fiche 1	GEARBOX AND PROPSHAFT

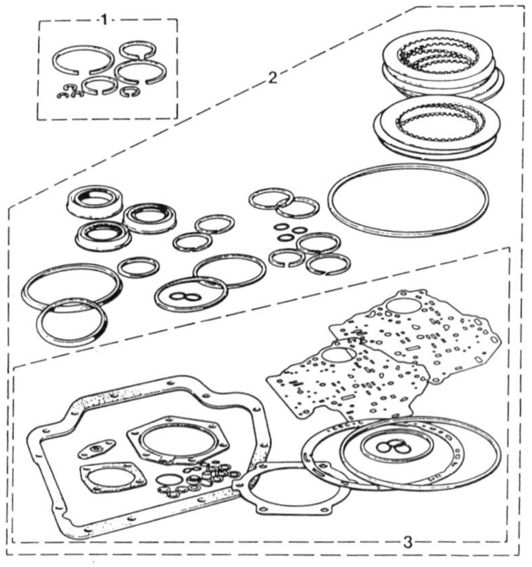

VS2094

AUTOMATIC TRANSMISSION SERVICE KITS-5.3 LITRE

Illus	Part Number	123456	Description	Quantity	Change Point	Remarks
1	AAU 6642		SNAP RING KIT	1		
2	AAU 6641		Overhaul repair kit	1		
3	AAU 6639		Gasket & seals kit	1		

1J 03

JAGUAR XJS RANGE (JAN 1987 ON)	J03 fiche 1	GEARBOX AND PROPSHAFT

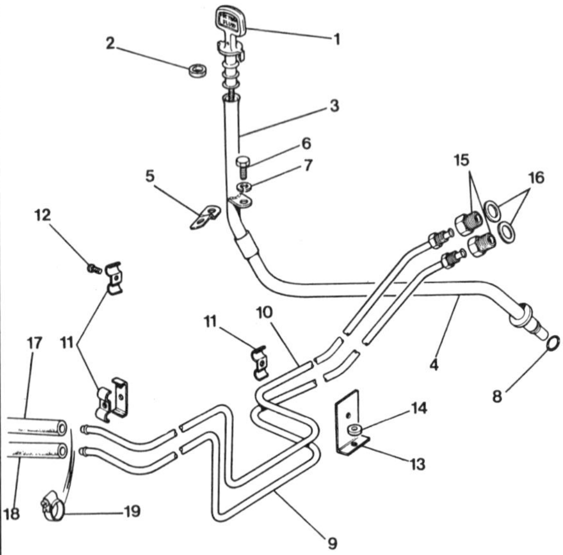

SM5681

TRANSMISSION DIPSTICK & OIL COOLER PIPES 5.3 LITRE.

Illus	Part Number	123456	Description	Quantity	Change Point	Remarks
1	EAC 8313		Dipstick assembly	1		
2	EAC 2978		Dipstick seal	1		
3	EAC 1884		Dipstick tube upper	1		
4	EAC 5706		Dipstick tube lower	1		
5	EAC 1675		Dipstick support bracket	1		
6	SH 604041 J		Setscrew	1		
7	FG 104 X		Washer	1		
8	EAC 5544		Seal	1		
9	EAC 7649		Oil cooler pipe outlet	1		
10	EAC 7650		Oil cooler pipe	1		
11	EAC 5910		Pipe clip double	3		
12	BD 34524 1		Screw	3		
13	C 43319		Bracket	1		
14	C 8136		Distance piece	1		
	AZ 610051		Screw	1		
	AH 610011		Flat speed nut	1		
	C 37175 7		Setscrew dist piece to block	1		
15	JLM 535		Cooler pipe connectors	2		
16	AAU 6669		Washer	2		
17	CAC 9734		Cooler pipe flexible	1	Up to VIN.139148	
18	CAC 9735		Cooler pipe flexible	1	Up to VIN.139148	
17	CBC 4491		Cooler pipe flexible	1	Fitted from VIN139149	
18	CBC 4492		Cooler pipe flexible	1	Fitted from VIN139149	
19	C 15886 3		Clip	4		

D72.02DC29/6/88

MASTER INDEX

ENGINE	1.C
FLYWHEEL AND CLUTCH	1.G
GEARBOX AND PROPSHAFT	1.H
AXLES, SUSPENSION, DRIVE SHAFTS, WHEELS	1.K
STEERING	1.M
BRAKES AND BRAKE CONTROLS	1.N
FUEL, EXHAUST AND EMISSION SYSTEMS	2.C
COOLING, HEATING AND AIR CONDITIONING	2.G
ELECTRICAL, WASHERS-WIPERS, INSTRUMENTS	2.J
CHASSIS, SUBFRAMES, BODYSHELL, FITTINGS	3.C
FASCIA, TRIM, SEATS AND FIXINGS	3.H
ACCESSORIES AND PAINT	4.C
NUMERICAL INDEX	4.D

GROUP INDEX

'A' CLUTCH INPUT SHAFT 3.6 LITRE	H11
AUTO GEARBOX 'B' CLUTCH 3.6 LITRE	H12
AUTO GEARBOX 'C' CLUTCH 3.6 LITRE	H13
AUTO GEARBOX 'D' CLUTCH 3.6 LITRE	H14
AUTO GEARBOX CLUTCH 'E' 3.6 LITRE	H16
AUTO GEARBOX MOUNTINGS-3.6 LITRE	I04
AUTO TRANSMISSION SUMP PAN 3.6 LITRE	I03
AUTOMATIC TRANSMISSION UNIT 3.6 L/T	H07
CONVERTER ASSEMBLY	H08
FRONT PUMP ASSEMBLY	H10
AUTOMATIC TRANSMISSION UNIT 5.3 L/T	I08
CLUTCH HOUSING	I11
GEARBOX CASE ASSEMBLY	I09
TORQUE CONVERTER	I10
FRONT AND REAR SERVO 5.3 LITRE	I14
GEARBOX MOUNTING PLATE 5.3 LITRE	J08
GEARBOX MOUNTING-5.3 LITRE	J07
MANUAL GEARBOX ASSEMBLY 3.6 LITRE	H02
MANUAL GEARBOX MOUNTINGS 3.6 LITRE	H04
PARKING BRAKE ASSEMBLY 5.3 LITRE	I16
SELECTOR LEVER ASSEMBLY	J04
SNAP RING KIT	J02
PROPELLOR SHAFT 3.6 LITRE	I06
PROPELLOR SHAFT 5.3 LITRE	I07
REAR BAND SUNGEAR MAINSHAFT 5.3 LITRE	I12
OUTPUT SHAFT	I13
SELECTOR GATE & LEVER 3.6 AUTOMATIC	H05
SPEEDO HOUSING GOVERNOR & MODULATOR	I15
VALVE BODIES ASSEMBLY 3.6 LITRE	H15

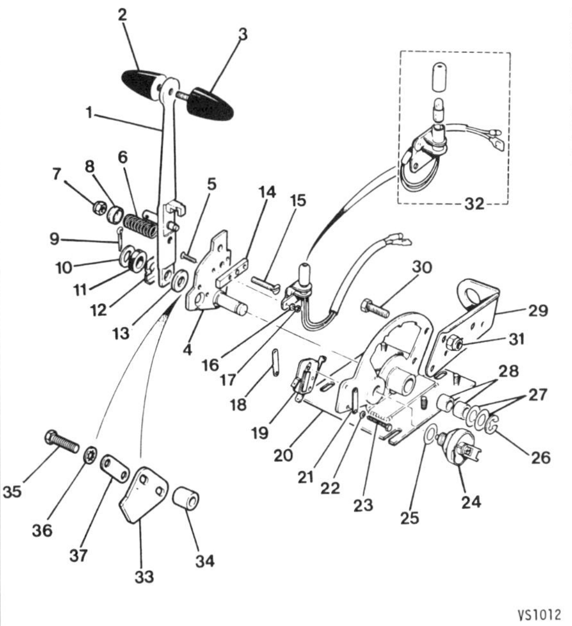

SELECTOR GATE & LEVER-5.3 LITRE

Illus	Part Number	Description	Quantity	Change Point	Remarks
1	CAC 4452	SELECTOR LEVER ASSEMBLY	1		
2	C 31228	Sel lever knob	1		
3	C 31232	Sel lever knob	1		
4	CAC 5304	Cam plate assembly	1		
5	UCS 511 5H	Setscrew	1		
6	CAC 5366	Return spring	1		
7	TN 3205	Nut	1		
8	CAC 5340	Spring retainer cup	1		
9	4900	Split pin	1		
10	C 24781	Washer	1		
11	C 25997	Damper rubber	1		
12	C 33050	Hinge washer	1		
13	FW 106 E	Washer	1		
14	C 35468	Tapped block	1		
15	UFS 519 10H	Setscrew	1		
16	WF 706041 J	Washer	2		
17	UCN 111 L	Locknut	2		
18	C 33784	Fixing plate	1		
19	C 33509	Kickdown switch	1		
20	CAC 4748	Sel gate & mounting plate assembly	1		
21	C 35840	Clamping plate	2		
22	WF 706041 J	Washer	2		
23	UCS 111 7R	Setscrew	2		
24	DAC 3844	Reverse lamp switch	1		
25	C 16626 3	Shim	A/R		
25	C 15896	Shim	A/R		
26	C 25981	Pivot pin circlip	1		
27	C 26644	Shim	A/R		
28	C 20953 3	Pivot pin bush	2		
29	C 28194	Abutment plate	1	Upto VIN 139051	
29	CAC 4845	Abutment plate	1	From VIN 139052	
30	JLM 9561	Hose clip	1		
31	NY 604041 J	Nut	2		
32	DAC 4192	Gear indicator light assembly	1		
33	CAC 5483	Cam plate speed control	1		
34	C 45125	Spacer	2		
35	UCS 116 5R	Setscrew	2		
36	C 722	Washer	1		
37	CAC 2040	Clamp plate	1		

VS1012

MASTER INDEX

GROUP INDEX

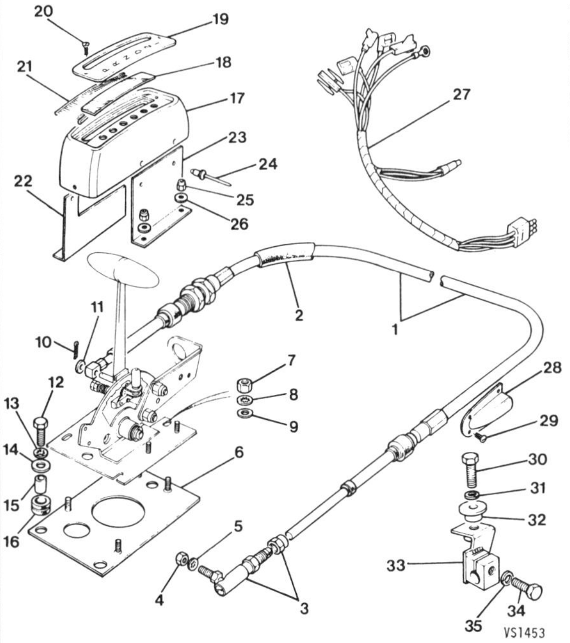

GEAR CONTROL CABLE SELECTOR PANEL 5.3 LITRE

Illus	Part Number	Description	Quantity	Change Point	Remarks
1	CBC 2482	Gear control cable	1	Upto VIN 135901	
1	CBC 5224	Gear control cable	1	From VIN 135902	
2	C 38494 6	Insulation tube	2		
3	C 25152	Adjustable ball end	1		
4	JLM 9683	Nut cable to gearbox	1		
5	C 724	Washer	1		
6	C 42100	Base plate	1		
7	C 8667 17	Nut	3		
8	WF 702101 J	Washer	3		
9	WC 702101 J	Washer	3		
10	4900	Split pin	1		
11	FW 104 T	Washer	1		
12	SH 604071 J	Setscrew	3		
13	C 724	Washer	3		
14	FW 104 T	Washer	3		
15	C 25949	Spacer	3		
16	C 25982	Mounting plate rubber	3		
17	CAC 5636	Selector cover	1		
18	C 26337	Filter	1		
19	CAC 5491	Selector cover panel	1		
20	AD 604031 J	Screw	2		
21	BD 31067	Seal	2		
22	BD 38393	Bracket	1		
23	CAC 5632	Support bracket	1		
24	BD 38427 1	Rivet	4		
25	TN 3205	Nut	4		
26	WC 702101 J	Washer	4		
27	DAC 2163	Auto trans harness	1		
28	CAC 5893	Cover plate	1		
29	AB 606041 J	Screw	4		
30	SH 605101 J	Setscrew	1		
31	FG 105 X	Washer	1		
32	C 45248	Clamp washer	1		
33	CBC 2605	Cable abutment lower	1		
34	BH 605101 J	Bolt	1		
35	WM 600051 J	Washer	1		

VS1453

MASTER INDEX
ENGINE ... 1.C
FLYWHEEL AND CLUTCH ... 1.G
GEARBOX AND PROPSHAFT ... 1.H
AXLES, SUSPENSION, DRIVE SHAFTS, WHEELS ... 1.K
STEERING ... 1.M
BRAKES AND BRAKE CONTROLS ... 1.N
FUEL, EXHAUST AND EMISSION SYSTEMS ... 2.C
COOLING, HEATING AND AIR CONDITIONING ... 2.G
ELECTRICAL, WASHERS-WIPERS, INSTRUMENTS ... 2.J
CHASSIS, SUBFRAMES, BODYSHELL, FITTINGS ... 3.C
FASCIA, TRIM, SEATS AND FIXINGS ... 3.H
ACCESSORIES AND PAINT ... 4.C
NUMERICAL INDEX ... 4.D
GROUP INDEX
'A' CLUTCH INPUT SHAFT 3.6 LITRE ... H11
AUTO GEARBOX 'B' CLUTCH 3.6 LITRE ... H12
AUTO GEARBOX 'C' CLUTCH 3.6 LITRE ... H13
AUTO GEARBOX 'D' CLUTCH 3.6 LITRE ... H14
AUTO GEARBOX CLUTCH 'E' 3.6 LITRE ... H16
AUTO GEARBOX MOUNTINGS-3.6 LITRE ... I04
AUTO TRANSMISSION SUMP PAN 3.6 LITRE ... I03
AUTOMATIC TRANSMISSION UNIT 3.6 L/T ... H07
CONVERTER ASSEMBLY ... H08
FRONT PUMP ASSEMBLY ... H10
AUTOMATIC TRANSMISSION UNIT 5.3 L/T ... I08
CLUTCH HOUSING ... I11
GEARBOX CASE ASSEMBLY ... I09
TORQUE CONVERTER ... I10
FRONT AND REAR SERVO 5.3 LITRE ... I14
GEARBOX MOUNTING PLATE 5.3 LITRE ... J08
GEARBOX MOUNTING-5.3 LITRE ... J07
MANUAL GEARBOX ASSEMBLY 3.6 LITRE ... H02
MANUAL GEARBOX MOUNTINGS 3.6 LITRE ... H04
PARKING BRAKE ASSEMBLY 5.3 LITRE ... I16
SELECTOR LEVER ASSEMBLY ... J04
SNAP RING KIT ... J02
PROPELLOR SHAFT 3.6 LITRE ... I06
PROPELLOR SHAFT 5.3 LITRE ... I07
REAR BAND SUNGEAR MAINSHAFT 5.3 LITRE ... I12
OUTPUT SHAFT ... I13
SELECTOR GATE & LEVER 3.6 AUTOMATIC ... H05
SPEEDO HOUSING GOVERNOR & MODULATOR ... I15
VALVE BODIES ASSEMBLY 3.6 LITRE ... H15

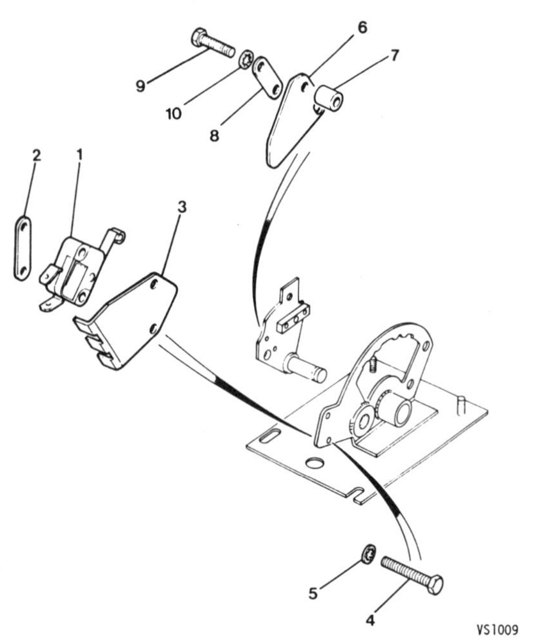

VS1009

1J 06

KICKDOWN INHIBITOR 5.3 LITRE

Illus	Part Number	123456 Description	Quantity	Change Point	Remarks
1	C 33509	Kickdown switch	1		
2	C 33784	Fixing plate	1		
3	C 37346	Bracket	1		
4	UCS 111 10R	Setscrew	2		
5	WF 706041 J	Washer	2		
6	CAC 5483	Cam plate	1		
7	C 45125	Distance piece	2		
8	CAC 2040	Clamp plate	1		
9	UCS 116 5R	Setscrew	2		
10	WF 703081	Washer	2		

MASTER INDEX

GROUP INDEX

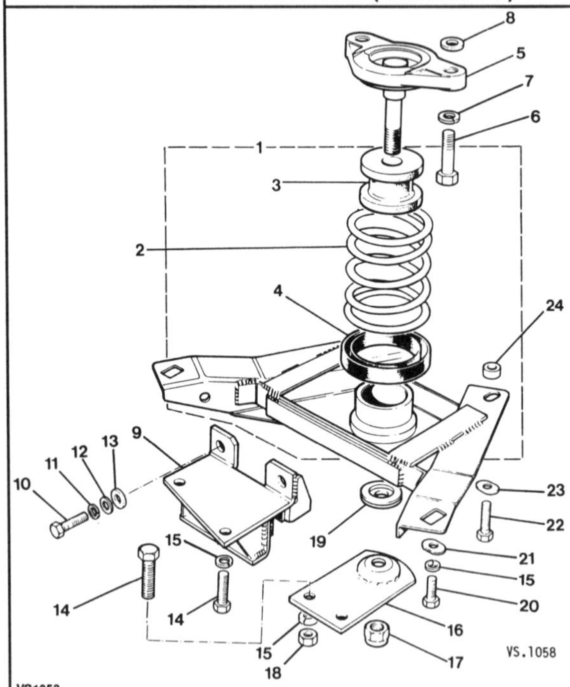

VS1058

1J 07

GEARBOX MOUNTING-5.3 LITRE

Illus	Part Number	123456 Description	Quantity	Change Point	Remarks
1	CAC 3242	Rear eng mtg assy	1		
2	CAC 2327	Coil spring	1		
3	CAC 3227	Centre bush	1		
4	C 12335	Rubber-seat lower	1		
5	CAC 6327	Spring retainer	1		
6	BH 110081 J	Bolt	2		
7	WL 110001 J	Washer	2		
8	C 1006	Washer	2		
9	CAC 2430	Buffer plate	1		
10	SH 505101 J	Setscrew	2		
11	JLM 303	Washer	2		
12	JLM 299	Washer	2		
13	CAC 2469	Spacer	2		
14	SH 606101 J	Setscrew	2		
15	JLM 304	Washer	6		
16	CAC 2462	Tie plate	1		
17	TN 3211	Nut	1		
18	JLM 9685	Nut	2		
19	CAC 2523	Spacer	1		
20	SH 606071 J	Setscrew	2		
21	BD 542 7	Washer	2		
22	SH 606121	Setscrew	2		
23	BD 542 6	Washer	2		
24	C 32458 7	Spacer	2		

MASTER INDEX

ENGINE 1.C
FLYWHEEL AND CLUTCH 1.G
GEARBOX AND PROPSHAFT 1.H
AXLES, SUSPENSION, DRIVE SHAFTS, WHEELS 1.K
STEERING 1.M
BRAKES AND BRAKE CONTROLS 1.N
FUEL, EXHAUST AND EMISSION SYSTEMS 2.C
COOLING, HEATING AND AIR CONDITIONING 2.G
ELECTRICAL, WASHERS-WIPERS, INSTRUMENTS 2.J
CHASSIS, SUBFRAMES, BODYSHELL, FITTINGS 3.C
FASCIA, TRIM, SEATS AND FIXINGS 3.H
ACCESSORIES AND PAINT 4.C
NUMERICAL INDEX 4.D

GROUP INDEX

'A' CLUTCH INPUT SHAFT 3.6 LITRE H11
AUTO GEARBOX 'B' CLUTCH 3.6 LITRE H12
AUTO GEARBOX 'C' CLUTCH 3.6 LITRE H13
AUTO GEARBOX 'D' CLUTCH 3.6 LITRE H14
AUTO GEARBOX CLUTCH 'E' 3.6 LITRE H16
AUTO GEARBOX MOUNTINGS-3.6 LITRE I04
AUTO TRANSMISSION SUMP PAN 3.6 LITRE I03
AUTOMATIC TRANSMISSION UNIT 3.6 L/T H07
CONVERTER ASSEMBLY H08
FRONT PUMP ASSEMBLY H10
AUTOMATIC TRANSMISSION UNIT 5.3 L/T I08
CLUTCH HOUSING I11
GEARBOX CASE ASSEMBLY I09
TORQUE CONVERTER I10
FRONT AND REAR SERVO 5.3 LITRE I14
GEARBOX MOUNTING PLATE 5.3 LITRE J08
GEARBOX MOUNTING-5.3 LITRE J07
MANUAL GEARBOX ASSEMBLY 3.6 LITRE H02
MANUAL GEARBOX MOUNTINGS 3.6 LITRE H04
PARKING BRAKE ASSEMBLY 5.3 LITRE I16
SELECTOR LEVER ASSEMBLY J04
SNAP RING KIT J02
PROPELLOR SHAFT 3.6 LITRE I06
PROPELLOR SHAFT 5.3 LITRE I07
REAR BAND SUNGEAR MAINSHAFT 5.3 LITRE I12
OUTPUT SHAFT I13
SELECTOR GATE & LEVER 3.6 AUTOMATIC H05
SPEEDO HOUSING GOVERNOR & MODULATOR I15
VALVE BODIES ASSEMBLY 3.6 LITRE H15

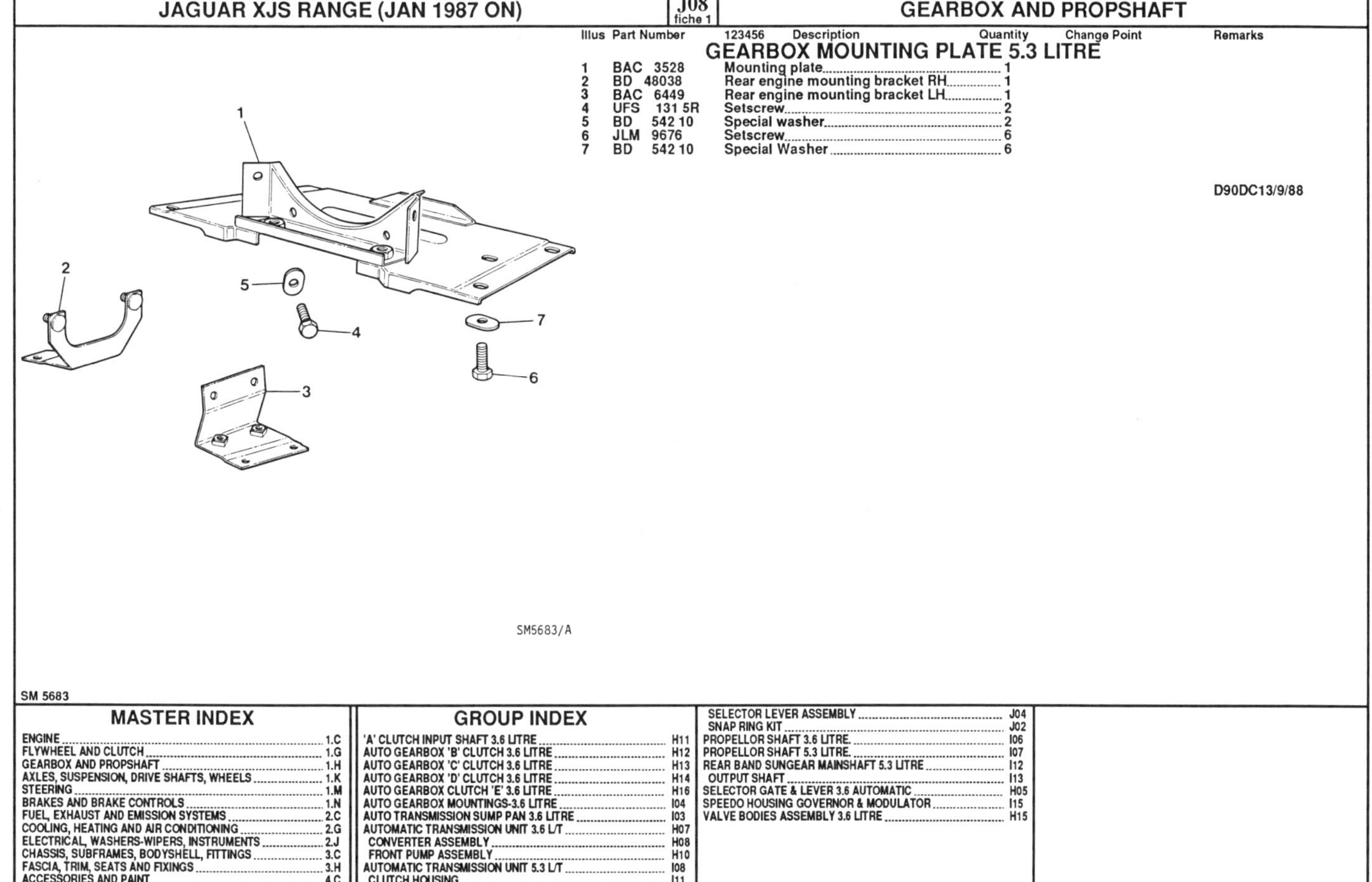

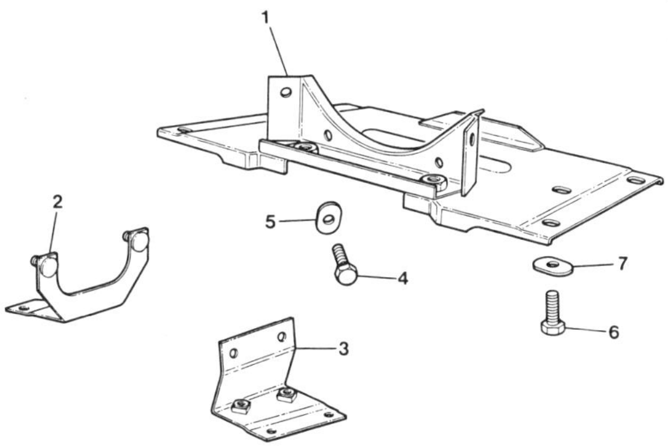

SM 5683

Illus	Part Number	123456 Description	Quantity	Change Point	Remarks
		GEARBOX MOUNTING PLATE 5.3 LITRE			
1	BAC 3528	Mounting plate	1		
2	BD 48038	Rear engine mounting bracket RH	1		
3	BAC 6449	Rear engine mounting bracket LH	1		
4	UFS 131 5R	Setscrew	2		
5	BD 542 10	Special washer	2		
6	JLM 9676	Setscrew	6		
7	BD 542 10	Special Washer	6		

D90DC13/9/88

MASTER INDEX

GROUP INDEX

MASTER INDEX

GROUP INDEX

1K 02

FRONT SUSPENSION - 3.6 LITRE & 5.3 LITRE
CROSSMEMBER & ANTI ROLL BAR

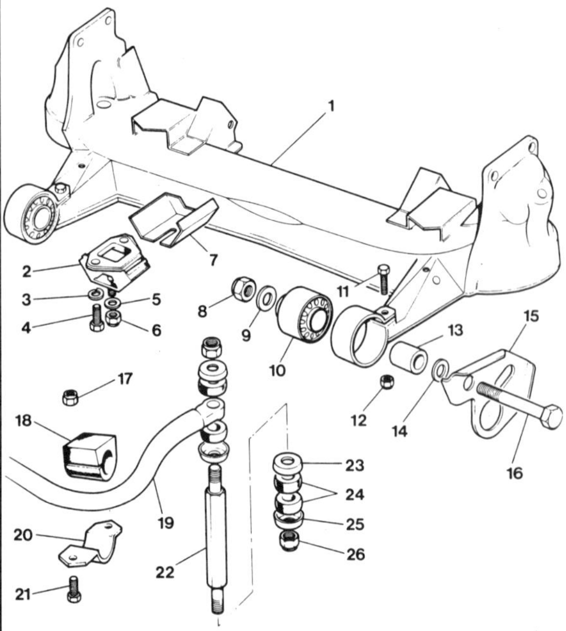

VS1116/A

Illus	Part Number	Description	Quantity	Change Point	Remarks
1	CAC 1162	Crossmember-front assembly	1		
2	C 45666	Vee mounting rear	2		
3	CBC 5735	Rear vee mounting	2		1988 1/2 MY
	FG 106 X	Washer spring	4		
4	JLM 9563	Bolt	4		
5	C 19466	Washer-plain	2		
6	C 8667 3	Nut nyloc	2		
7	C 45367	Heatshield RH	1		
	C 45368	Heatshield LH	1		
8	252167 J	Nut self locking	2		
9	WA 600121 J	Washer - pinion nut	2		
10	C 30314	Bush - mounting front	2		
	CBC 5736	Front mounting bush	2		1988 1/2 MY
11	SH 608101 J	Setscrew - 0.50" x 1.25"	2		
12	C 8667 5	Nut - self locking	2		
13	C 30371	Sleeve - mounting bush	2		
14	C 30370	Washer special	2		
15	CAC 3416	Tie bracket front RH	1		Use CBC2292
	CAC 3417	Tie bracket front LH	1		Use CBC2293
	CBC 2292	Tie bracket front RH	1		
	CBC 2293	Tie bracket front LH	1		
16	RTC 2584	Bolt - 0.75" x 6.50"	2		
17	NZ 606041 J	Self locknut	4		
18	CAC 4651 7	Bush anti roll bar	2		
19	C 36887 1	Anti roll bar	1		
	CBC 5579	Anti roll bar	1		XJS V12 Sportspack only
	CBC 5580	Bush anti roll bar	2		XJS V12 Sportspack only
20	CBC 5336	Support bracket - RH	1		
	CBC 5337	Support bracket - LH	1		
21	SH 606081 J	Setscrew 3/8 UNF	4		
22	C 46186	Link - A.R.B.	2		
23	C 33682	Pad - retaining washer	4		
24	C 10996	Rubber pad - ARB link	8		
25	C 11045	Pad retaining washer	4		
26	NZ 606041 J	Self locknut	4		

VS1116/A

MASTER INDEX

GROUP INDEX

1K 03

UPPER WISHBONES AND BALL JOINTS 3.6 LITRE & 5.3 LITRE
UPPER WISHBONES

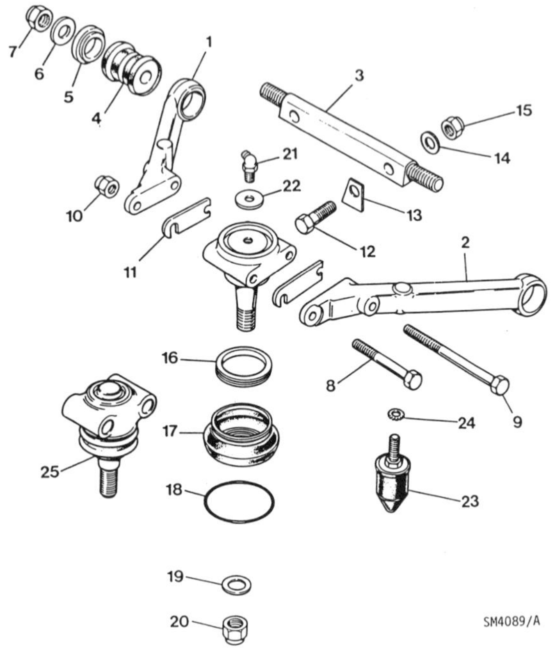

SM4089/A

Illus	Part Number	Description	Quantity	Change Point	Remarks
1	C 30611	Wishbone arm-upper front-RH	1		
	C 30612	Wishbone arm-upper front-LH	1		
2	C 30613	Wishbone arm-upper rear-RH	1		
	C 30614	Wishbone arm-upper rear-LH	1		
3	C 27859	Fulcrum shaft-upper	2		
4	C 33745 1	Bush-slipflex	4		
5	C 29975	Washer-slipflex bush	8		
6	C 30993	Washer-plain	4		
7	C 8737 5	Nut-self locking	4		
8	C 29890	Bolt special	2		Use CBC7361
	CBC 7361	Bolt special	2		
9	BH 606321 J	Bolt-0.75" x 4"	2		Use BX606321/J
	BX 606321 J	Bolt special	2		
10	C 8667 3	Nut-nyloc	4		
11	C 15308	Plate-spacing	8		Use CBC6416/1/2
	CBC 6416 1	Spacing plate	6		
	CBC 6416 2	Spacing plate	4		
12	BH 607181 J	Bolt	4		
13	C 30486	Washer 'D'	4		
14	C 1919 1	Shim-camber adjusting-0.031	A/R		
	C 1919 3	Shim-camber adjusting-0.128	A/R		
15	NY 608041 J	Nut-nyloc	4		
16	C 22970	Gaiter-moulded	2		Parts of upper ball joint kit C 23024
17	C 43216	Gaiter-lower ball pin	2		Parts of upper ball joint kit C 23024
18	C 22969	Clip-upper ball joint	2		Parts of upper ball joint kit C 23024
19	WC 112081 J	Washer-plain	2		Parts of upper ball joint kit C 23024
20	C 8737 5	Nut-nyloc	2		Parts of upper ball joint kit C 23024
21	LN 30041 J	Nipple-grease	2		Parts of upper ball joint kit C 23024
22	C 22971	Washer	2		Parts of upper ball joint kit C 23024
23	C 29979	Rubber-rebound	4		
24	WE 600050 J	Washer	4		
	C 23024	KIT-UPPER BALL JOINT	NLA		Use CAC 9938
25	CAC 9938	Ball joint assembly	2		

SM 4089A

MASTER INDEX

GROUP INDEX

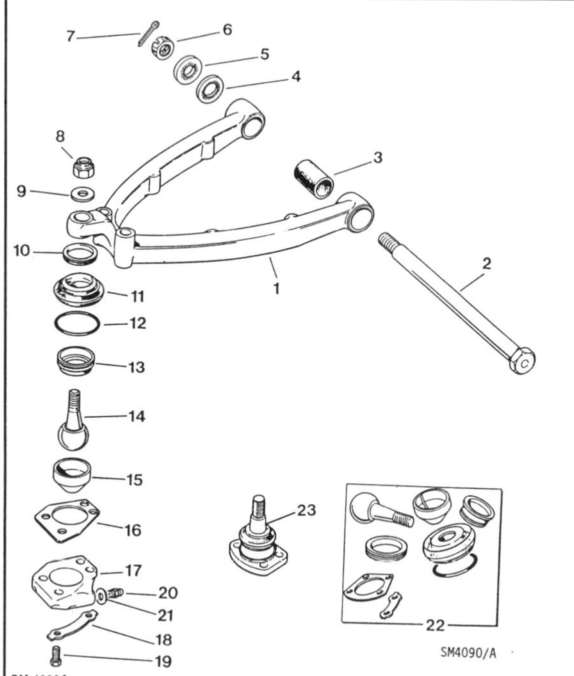

SM 4090A

LOWER WISHBONE AND BALL 3.6 & 5.3 LITRE

LOWER WIHSBONES

Illus	Part Number	123456 Description	Quantity	Change Point	Remarks
1	C 41314	Wishbone-lower-RH	1		
	C 41315	Wishbone-lower-LH	1		
2	C 30722	Shaft assembly-lower fulcrum	2		
3	C 8673	Bush-fulcrum	4		
4	C 30954	Washer-special	4		
5	C 30955	Washer-special	2		
6	NL 609041 J	Nut-slotted	2		
7	L 10410 U	Pin-split	2		
8	C 8737 6	Nut-self locking	2		
9	C 794	Washer-plain	2		
10	C 22970	Gaiter-moulded	2		
11	C 43216	Gaiter-lower ball pin	2		
12	C 22969	Clip-upper ball joint	2		
13	C 30952	Socket-lower ball pin	2		Upper
14	CBC 1380	Ball pin-lower	2		
15	C 30953	Socket-lower ball pin	2		Lower
16	C 15341	Shim-socket cap-0.002"	A/R		
	C 15342	Shim-socket cap-0.004"	A/R		
	C 17175	Shim-socket cap-0.010"	A/R		
17	C 24514	Socket cap-lower ball pin	2		
18	C 15343	Washer tab-lower ball joint	4		
19	C 15344	Bolt-socket cap	8		
20	234532 J	Nipple-grease	2		
21	C 22971	Washer-grease nipple	2		
22	12803	KIT-LOWER BALL JOINT	2		Use CAC9937 & CBC1805 4 off
23	CAC 9937	Lower ball joint assy	2		
	CBC 1805	Special bolt 5/16 UNF	8		

MASTER INDEX

GROUP INDEX

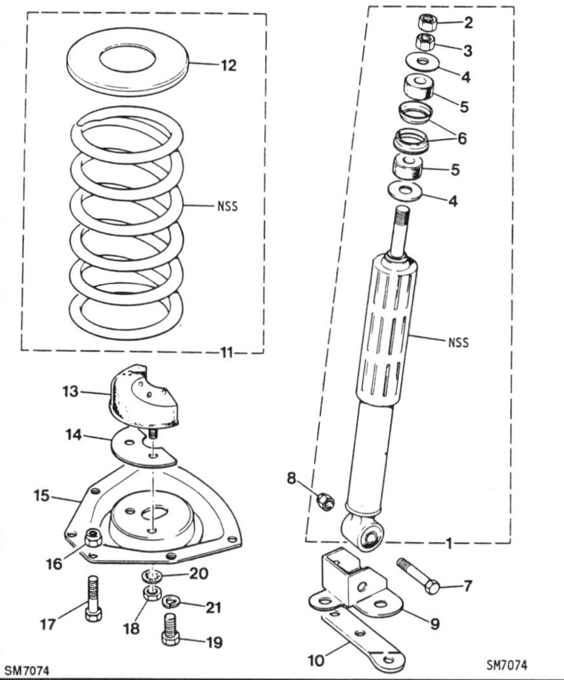

SM7074

FRONT SHOCK ABSORBER & SPRING-3.6 LT

Illus	Part Number	123456 Description	Quantity	Change Point	Remarks
1	C 27795	Shock absorber	2		Use CAC 9089
	CAC 9089	Shock absorber	2		
	CBC 5740	Shock absorber	2	Up to 19881/2 MY	
	CBC 6558	Shock absorber	2	From 19881/2 MY	
	CBC 5173	Upper washer	2		Use CBC 6608
	CBC 6608	Upper washer	2		
	CBC 5174	Lower washer	2		
2	UFN 237 L	Nut	1		
3	JLM 9685	Nut-0.375"	1		
5	C 30115	Bush-front damper	2		
6	C 30114	Washer-cup	2		
7	C 22827	Bolt	2		
8	TN 3210 J	Nyloc-nut	2		
	AGU 2484	Damper cap	2		
9	CBC 4102	Mounting bracket-RH	1		
	CBC 4103	Mounting bracket-LH	1		
10	C 27775	Anti roll bar-bracket	2		
11	RTC 660	Front road spring kit	1	Upto VIN 144699	
	JLM 1424	Front road spring kit	1	1988MY VIN144700	
	C 41271	Packing ring	A/R		See kit list for qty req'd 1988MY
12	C 41271	Packing ring-0.125"	A/R		
13	C 43790	Bump stop rubber	2		
14	C 43791	Spacer-bump rubber	4		
15	C 43058	Spring pan assembly	2		
16	C 8667 3	Nut-self locking	4		
17	BH 606261 J	Bolt-0.375" x 3.25"	4		
18	JLM 9684	Nut-0.3125"	4		
19	SH 606061 J	Setscrew-CSK-0.375 x 0.75	8		
20	C 725	Washer-shakeproof	4		
21	JLM 304	Washer-spring-0.375 ID	8		

MASTER INDEX

ENGINE ... 1.C
FLYWHEEL AND CLUTCH ... 1.G
GEARBOX AND PROPSHAFT ... 1.H
AXLES, SUSPENSION, DRIVE SHAFTS, WHEELS ... 1.K
STEERING ... 1.M
BRAKES AND BRAKE CONTROLS ... 1.N
FUEL, EXHAUST AND EMISSION SYSTEMS ... 2.C
COOLING, HEATING AND AIR CONDITIONING ... 2.G
ELECTRICAL, WASHERS-WIPERS, INSTRUMENTS ... 2.J
CHASSIS, SUBFRAMES, BODYSHELL, FITTINGS ... 3.C
FASCIA, TRIM, SEATS AND FIXINGS ... 3.H
ACCESSORIES AND PAINT ... 4.C
NUMERICAL INDEX ... 4.D

GROUP INDEX

FRONT SHOCK ABSORBER & SPRING-3.6 LT ... K05
FRONT SHOCK ABSORBER AND SPRING ... K06
FRONT SUSPENSION - 3.6 LITRE & 5.3 LITRE ... K02
KIT-LOWER BALL JOINT ... K04
KIT-UPPER BALL JOINT ... K03
LOWER WIHSBONES ... K04
UPPER WISHBONES ... K03
HUB AND STUB AXLE CARRIER ... K07
REAR SUSPENSION 3.6 & 5.3 LITRE ... K08
REAR SUSPENSION 3.6 LITRE ALL ... K13
HALF SHAFTS ... K14
REAR SUSPENSION-continued ... K09
DAMPER UNIT AND SPRING ... K10
RADIUS ARM AND MOUNTINGS ... K12
REAR HUB AND CARRIER ... K11
ROAD WHEEL-ALLOY-NOT USA/CDN. ... L05

1K 06

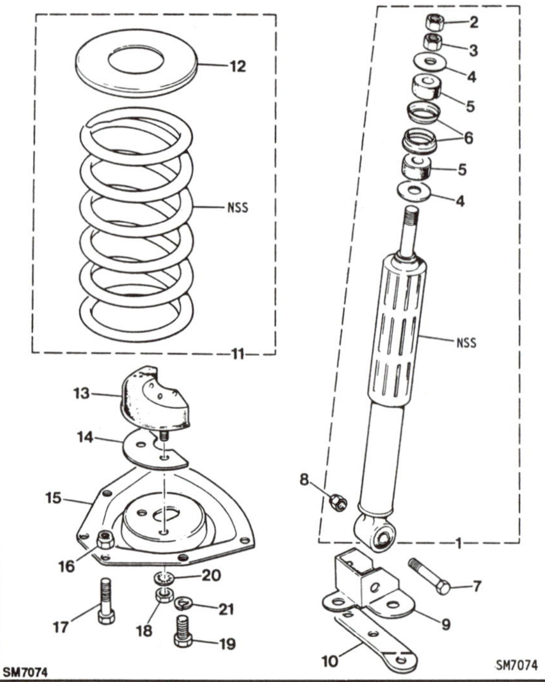

FRONT SHOCK ABSORBER AND SPRING

5.3 LITRE

Illus	Part Number	Description	Quantity	Change Point	Remarks
1	C 27795	Shock absorber	NLA		Use CAC 9089
	CAC 9089	Shock absorber	2		Also Convert.
2	UFN 237 L	Nut	1		
3	JLM 9685	Nut-0.375"	1		
4	C 30117	Washer	2		
5	C 30115	Bush-front damper	2		
6	C 30114	Washer-cup	2		
	CBC 6559	Shock Absorber	2		V12 Sportspack only
	CBC 5611	Cup Washer	4		
	CBC 5616	Bush	4		
	CBC 5173	Upper Washer	2		
	CBC 5174	Lower Washer	2		
	NY 606041 J	Nyloc Nut	2		
7	C 22827	Bolt	2		
8	TN 3210 J	Nyloc-nut	2		
9	C 38112	Mounting bracket-RH	1		
	C 38113	Mounting bracket-LH	1		
10	C 27775	Anti roll bar-bracket	2		
11	RTC 2522	Coil spring kit	2		19881/2 All V12 Not Sportspack
	JLM 1449	Coil spring kit	1		19881/2 XJ77
	JLM 1718	Coil spring kit	1		1989MY all V12 Sportspack
12	C 41271	Packing ring-0.125"	A/R		
13	C 43790	Bumper stop rubber	2		
14	C 43791	Spacer-bump rubber	4		
15	C 43058	Spring pan assembly	2		
16	C 8667 3	Nut-self locking	4		
17	BH 606261 J	Bolt-0.375" x 3.25	4		
18	JLM 9684	Nut-0.3125"	4		
19	SH 606061 J	Setscrew-CSK-0.375 x 0.75	8		
20	C 725	Washer-shakeproof	4		
21	JLM 304	Washer-spring-0.375 ID	8		
	CBC 4102	Damper mounting bracket-RH	1		
	CBC 4103	Damper mounting bracket-LH	1		

MASTER INDEX

GROUP INDEX

1K 07

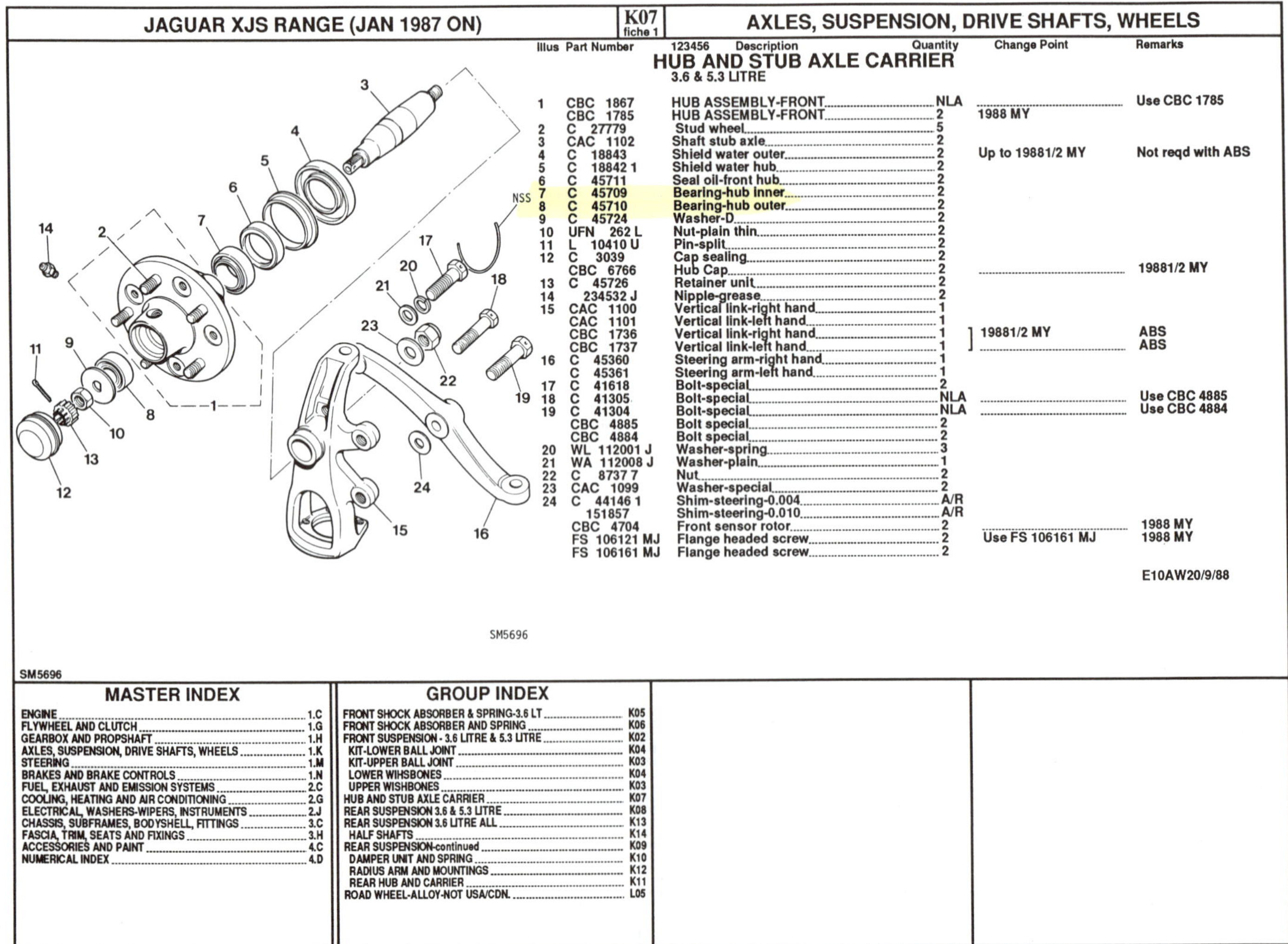

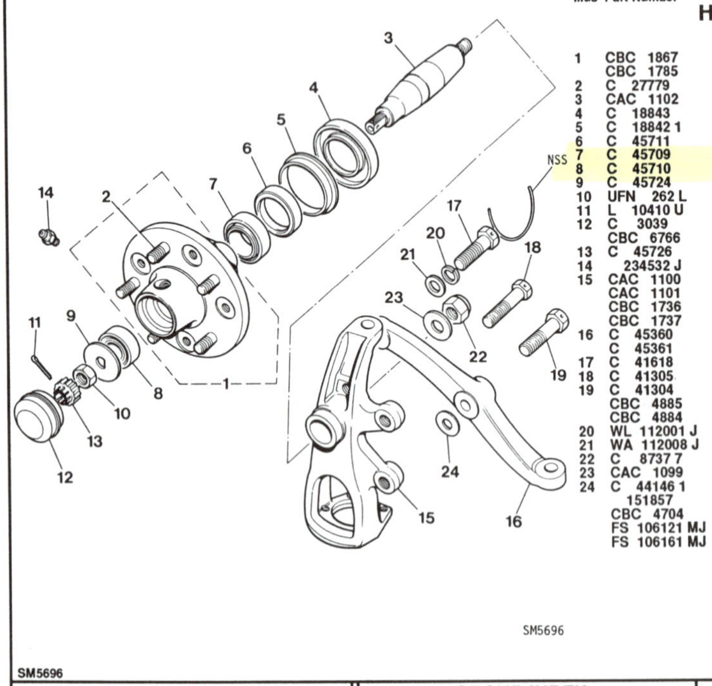

HUB AND STUB AXLE CARRIER

3.6 & 5.3 LITRE

Illus	Part Number	Description	Quantity	Change Point	Remarks
1	CBC 1867	HUB ASSEMBLY-FRONT	NLA		Use CBC 1785
	CBC 1785	HUB ASSEMBLY-FRONT	2	1988 MY	
2	C 27779	Stud wheel	5		
3	CAC 1102	Shaft stub axle	2		
4	C 18843	Shield water outer	2	Up to 19881/2 MY	Not reqd with ABS
5	C 18842 1	Shield water hub	2		
6	C 45711	Seal oil-front hub	2		
7	C 45709	Bearing-hub inner	2		
8	C 45710	Bearing-hub outer	2		
9	C 45724	Washer-D	2		
10	UFN 262 L	Nut-plain thin	2		
11	L 10410 U	Pin-split	2		
12	C 3039	Cap sealing	2		
	CBC 6766	Hub Cap	2		19881/2 MY
13	C 45726	Retainer unit	2		
14	234532 J	Nipple-grease	2		
15	CAC 1100	Vertical link-right hand	1		
	CAC 1101	Vertical link-left hand	1		
	CBC 1736	Vertical link-right hand	1	19881/2 MY	ABS
	CBC 1737	Vertical link-left hand	1		ABS
16	C 45360	Steering arm-right hand	1		
	C 45361	Steering arm-left hand	1		
17	C 41618	Bolt-special	2		
18	C 41305	Bolt-special	NLA		Use CBC 4885
19	C 41304	Bolt-special	NLA		Use CBC 4884
	CBC 4885	Bolt special	2		
	CBC 4884	Bolt special	2		
20	WL 112001 J	Washer-spring	3		
21	WA 112008 J	Washer-plain	1		
22	C 8737 7	Nut	2		
23	CAC 1099	Washer-special	2		
24	C 44146 1	Shim-steering-0.004	A/R		
	151857	Shim-steering-0.010	A/R		
	CBC 4704	Front sensor rotor	2		1988 MY
	FS 106121 MJ	Flange headed screw	2	Use FS 106161 MJ	1988 MY
	FS 106161 MJ	Flange headed screw	2		

E10AW20/9/88

MASTER INDEX

ENGINE	1.C
FLYWHEEL AND CLUTCH	1.G
GEARBOX AND PROPSHAFT	1.H
AXLES, SUSPENSION, DRIVE SHAFTS, WHEELS	1.K
STEERING	1.M
BRAKES AND BRAKE CONTROLS	1.N
FUEL, EXHAUST AND EMISSION SYSTEMS	2.C
COOLING, HEATING AND AIR CONDITIONING	2.G
ELECTRICAL, WASHERS-WIPERS, INSTRUMENTS	2.J
CHASSIS, SUBFRAMES, BODYSHELL, FITTINGS	3.C
FASCIA, TRIM, SEATS AND FIXINGS	3.H
ACCESSORIES AND PAINT	4.C
NUMERICAL INDEX	4.D

GROUP INDEX

FRONT SHOCK ABSORBER & SPRING-3.6 LT	K05
FRONT SHOCK ABSORBER AND SPRING	K06
FRONT SUSPENSION - 3.6 LITRE & 5.3 LITRE	K02
KIT-LOWER BALL JOINT	K04
KIT-UPPER BALL JOINT	K03
LOWER WIHSBONES	K04
UPPER WISHBONES	K03
HUB AND STUB AXLE CARRIER	K07
REAR SUSPENSION 3.6 & 5.3 LITRE	K08
REAR SUSPENSION 3.6 LITRE ALL	K13
HALF SHAFTS	K14
REAR SUSPENSION-continued	K09
DAMPER UNIT AND SPRING	K10
RADIUS ARM AND MOUNTINGS	K12
REAR HUB AND CARRIER	K11
ROAD WHEEL-ALLOY-NOT USA/CDN.	L05

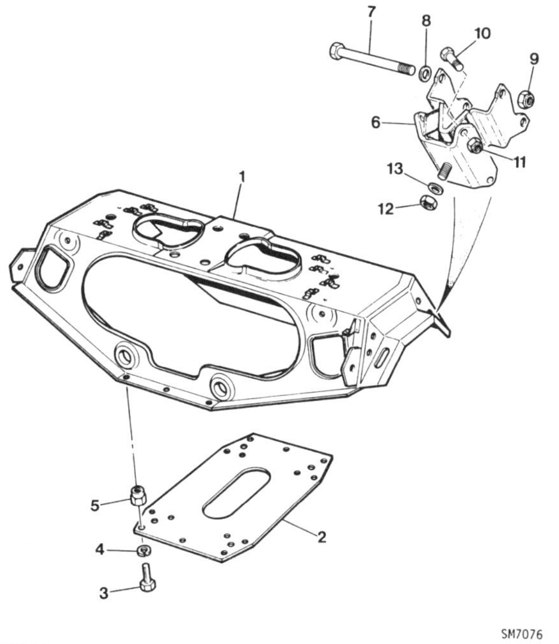

SM 7076

REAR SUSPENSION 3.6 & 5.3 LITRE
CROSSMEMBER AND MOUNTINGS

Illus	Part Number	123456 Description	Quantity	Change Point	Remarks
1	CAC 3301	Crossmember	1		
2	C 20651	Bottom plate	1		Drive unit MTG
3	UFS 131 5R	Setscrew 0.3125 x 0.625	14		
4	JLM 303	Washer spring 0.3125 10	8		
5	C 8737 2	Nut self locking 0.3125	6		
6	CAC 3067	Mounting vee	4		
	CBC 5737	Mounting vee	4		XJ77
7	BH 606301 J	Bolt 0.375 x 4"	8		
8	JLM 300	Washer plain	4		
9	NZ 606041 J	Nut self locking 0.375	8		
10	UFS 131 7R	Bolt 0.3125 x 0.875	8		
11	NY 605041 J	Nut nyloc 0.3125	8		
12	NY 605041 J	Nut nyloc 0.3125	4		
13	JLM 299	Washer plain 0.3125 10	4		

MASTER INDEX

GROUP INDEX

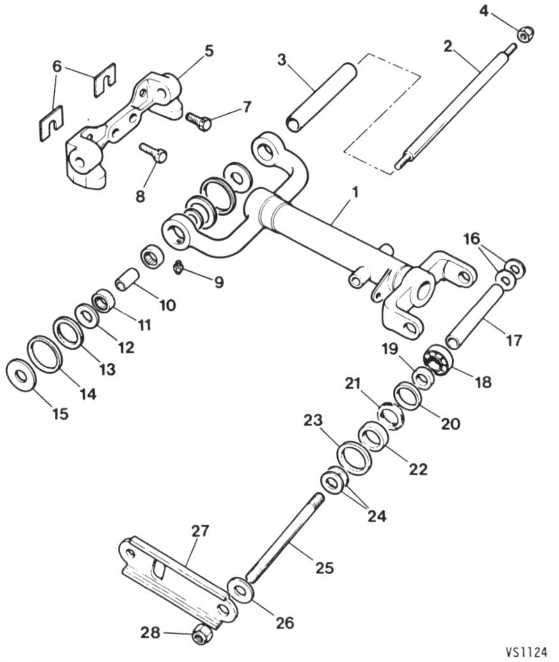

VS 1124

REAR SUSPENSION-continued
REAR WISHBONES AND MOUNTINGS- 3.6 & 5.3 LITRE

Illus	Part Number	123456 Description	Quantity	Change Point	Remarks
1	C 36137	Wishbone assembly-RH	1		
	C 36138	Wishbone assembly-LH	1		
2	C 17008	Fulcrum pin	2		
3	C 17663 1	Spacer tube	2		
4	C 8667 5	Nut self locking	4		
5	C 21977	Mounting fulcrum	2		
6	C 28427	Shim-mounting-0.005	A/R		Selective
	C 28428	Shim-mounting-0.007	A/R		Selective
7	C 17010	Setscrew-fulcrum mounting	2		
8	C 17009	Setscrew-fulcrum mounting	2		
9	234532 J	Nipple-grease	4		
10	C 17168 1	Tube-bearing	4		
11	C 17167	Bearing-needle	8		
12	C 17166	Washer-thrust	8		
13	C 17213	Ring-sealing	8		
14	C 17936	Retainer-seal	8		
15	C 17165	Washer-thrust	8		
16	C 16626	Shim-pivot pin 0.003	A/R		Selective
	C 16626 3	Shim-pivot pin 0.007	A/R		Selective
17	C 16623 1	Sleeve-pivot pin	2		
18	C 16029	Bearing-roller	4		
19	C 16628	Seal-oil-track	4		
20	C 20180	Spacer-seal	4		
21	C 20178	Seal-oil-felt	4		
22	C 20179	Shell-oil seal	4		
23	C 20182	Washer-retaining	4		
24	C 16626 1	Shim-pivot pin 0.003	A/R		Selective
	C 16626 2	Shim-pivot pin 0.007	A/R		Selective
25	CAC 2316	Pin-pivot	2		
26	JLM 9571	Washer	4		
27	CAC 2416	Bracket-rear tie down	2		
28	C 8667 7	Nut-self locking	4		

MASTER INDEX
ENGINE 1.C
FLYWHEEL AND CLUTCH 1.G
GEARBOX AND PROPSHAFT 1.H
AXLES, SUSPENSION, DRIVE SHAFTS, WHEELS 1.K
STEERING 1.M
BRAKES AND BRAKE CONTROLS 1.N
FUEL, EXHAUST AND EMISSION SYSTEMS 2.C
COOLING, HEATING AND AIR CONDITIONING 2.G
ELECTRICAL, WASHERS-WIPERS, INSTRUMENTS 2.J
CHASSIS, SUBFRAMES, BODYSHELL, FITTINGS 3.C
FASCIA, TRIM, SEATS AND FIXINGS 3.H
ACCESSORIES AND PAINT 4.C
NUMERICAL INDEX 4.D
GROUP INDEX
FRONT SHOCK ABSORBER & SPRING-3.6 LT K05
FRONT SHOCK ABSORBER AND SPRING K06
FRONT SUSPENSION - 3.6 LITRE & 5.3 LITRE K02
KIT-LOWER BALL JOINT K04
KIT-UPPER BALL JOINT K03
LOWER WIHSBONES K04
UPPER WISHBONES K03
HUB AND STUB AXLE CARRIER K07
REAR SUSPENSION 3.6 & 5.3 LITRE K08
REAR SUSPENSION 3.6 LITRE ALL K13
HALF SHAFTS K14
REAR SUSPENSION-continued K09
DAMPER UNIT AND SPRING K10
RADIUS ARM AND MOUNTINGS K12
REAR HUB AND CARRIER K11
ROAD WHEEL-ALLOY-NOT USA/CDN. L05

JAGUAR XJS RANGE (JAN 1987 ON)	K10 fiche 1	AXLES, SUSPENSION, DRIVE SHAFTS, WHEELS

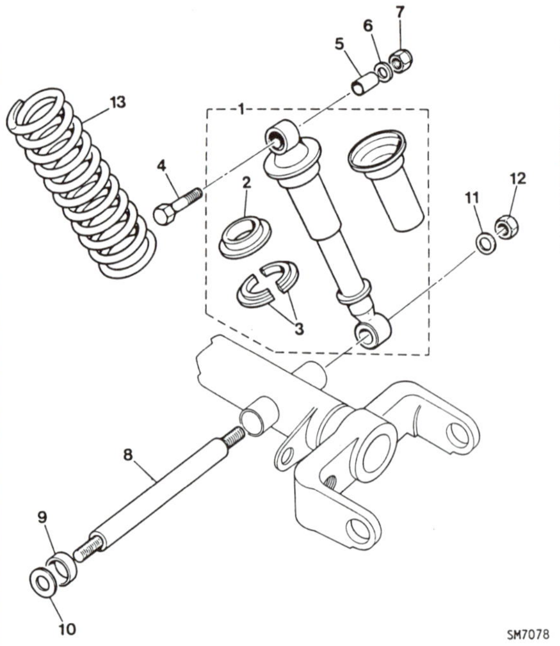

SM 7078

REAR SUSPENSION-continued 3.6 & 5.3 LITRE
DAMPER UNIT AND SPRING

Illus	Part Number	123456 Description	Quantity	Change Point	Remarks
1	CAC 9091 1	DAMPER UNIT ASSEMBLY	4		V12 Non Sportspack
	CBC 5742	Damper unit	4		3.6 & Sportspack V12
2	C 19821	Ring spring support	1		
3	C 37273	Retainer spring	2		
4	BH 607171 J	Damper pin upper	4		
5	C 17012	Sleeve damper	4		
6	WA 600071 J	Washer	4		
7	C 8737 4	Nut self locking 0.4375	4		
8	C 17013	Pin damper lower	2		
9	C 29374	Spacer damper	2		
10	C 3052	Washer special	4		
11	WA 600071 J	Washer	2		
12	C 8737 4	Nut self locking 0.4375	4		
13	C 39692	Spring road	4		
	CBC 2793	Spring road	4		3.6 & V12 Sportspack

MASTER INDEX

GROUP INDEX

1K 10

JAGUAR XJS RANGE (JAN 1987 ON)	K11 fiche 1	AXLES, SUSPENSION, DRIVE SHAFTS, WHEELS

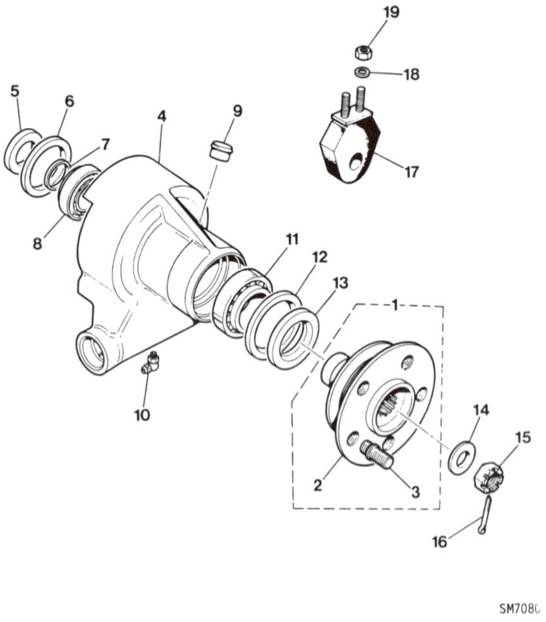

SM7080

REAR SUSPENSION 3.6 & 5.3 LITRE - continued
REAR HUB AND CARRIER

Illus	Part Number	123456 Description	Quantity	Change Point	Remarks
1	C 31657	HUB ASSEMBLY REAR	2		Use CBC 1784
	CBC 1784	Hub Assy Rear	2		1988 Model year
2	C 20813 1	Thrower water	1		
3	C 13365	Stud wheel	5		
	CBC 1740	Carrier hub	2		1988 Model year
4	C 26699	Carrier hub	2		
5	C 15232	Seal oil track	2		
6	C 15231	Seal oil hub	2		
	CBC 4706	Rear Rotor	2		
		SPACER REAR HUB			
	CAC 3818 10	0.110"	2		
	CAC 3818 12	0.112"	2		
	CAC 3818 14	0.114"	2		
	CAC 3818 16	0.116"	2		
	CAC 3818 18	0.118"	2		
	CAC 3818 20	0.120"	2		
	CAC 3818 22	0.122"	2		
	CAC 3818 24	0.124"	2		
	CAC 3818 26	0.126"	2	Selective	
	CAC 3818 28	0.128"	2		
	CAC 3818 30	0.130"	2		
	CAC 3818 32	0.132"	2		
	CAC 3818 34	0.134"	2		
	CAC 3818 36	0.136"	2		
	CAC 3818 38	0.138"	2		
	CAC 3818 40	0.140"	2		
	CAC 3818 42	0.142"	2		
	CAC 3818 44	0.144"	2		
	CAC 3818 46	0.146"	2		
	CAC 3818 48	0.148"	2		
	CAC 3818 50	0.150"	2		
8	C 15230	Bearing hub	2		
9	C 18124	Cap hub carrier	2		
10	LN 30041 J	Greaser	2		
11	C 19066	Bearing rear hub	2		
12	C 24789	Seal oil	2		
13	C 24791	Seal oil track	2		
14	C 24923	Washer plain	2		
15	ND 612041 J	Nut slotted	2		
16	LN 10513 UJ	Pin split	2		
17	C 40158	Stop bump	2		
18	WE 600051 J	Washer shakeproof	4		
19	JLM 9684	Nut 0.3125"	4		

MASTER INDEX

ENGINE 1.C
FLYWHEEL AND CLUTCH 1.G
GEARBOX AND PROPSHAFT 1.H
AXLES, SUSPENSION, DRIVE SHAFTS, WHEELS 1.K
STEERING 1.M
BRAKES AND BRAKE CONTROLS 1.N
FUEL, EXHAUST AND EMISSION SYSTEMS 2.C
COOLING, HEATING AND AIR CONDITIONING 2.G
ELECTRICAL, WASHERS-WIPERS, INSTRUMENTS 2.J
CHASSIS, SUBFRAMES, BODYSHELL, FITTINGS 3.C
FASCIA, TRIM, SEATS AND FIXINGS 3.H
ACCESSORIES AND PAINT 4.C
NUMERICAL INDEX 4.D

GROUP INDEX

FRONT SHOCK ABSORBER & SPRING-3.6 LT K05
FRONT SHOCK ABSORBER AND SPRING K06
FRONT SUSPENSION - 3.6 LITRE & 5.3 LITRE K02
KIT-LOWER BALL JOINT K04
KIT-UPPER BALL JOINT K03
LOWER WIHSBONES K04
UPPER WISHBONES K03
HUB AND STUB AXLE CARRIER K07
REAR SUSPENSION 3.6 & 5.3 LITRE K08
REAR SUSPENSION 3.6 LITRE ALL K13
HALF SHAFTS K14
REAR SUSPENSION-continued K09
DAMPER UNIT AND SPRING K10
RADIUS ARM AND MOUNTINGS K12
REAR HUB AND CARRIER K11
ROAD WHEEL-ALLOY-NOT USA/CDN. L05

1K 11

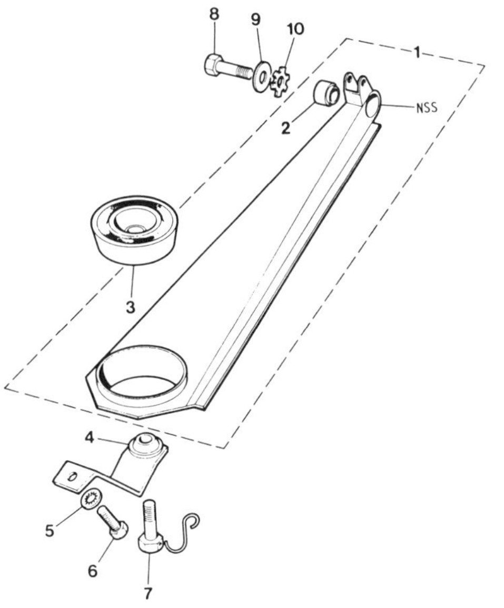

SM 7225

REAR SUSPENSION-3.6 LITRE & 5.3 LITRE - continued
RADIUS ARM AND MOUNTINGS

Illus	Part Number	Description	Quantity	Change Point	Remarks
1	C 29310	RADIUS ARM COMPLETE ASSEMBLY	2		V12 Non Sportspk
	C 41831	RADIUS ARM COMPLETE ASSEMBLY	2		3.6/V12 Sportspack
2	C 17146	Bush-rear-radius arm	1		
3	C 23782	Bush-front-radius arm	1		
4	C 42220	Strap assembly radius arm safety	2		
5	C 726	Washer shakeproof 0.375	2		
6	JLM 9563	Setscrew 0.375	2		
7	C 19045	Bolt	2		
8	C 25365	Bolt special	2		
9	C 36881	Washer plain	2		
10	C 25401	Washer-tab	2		

MASTER INDEX

GROUP INDEX

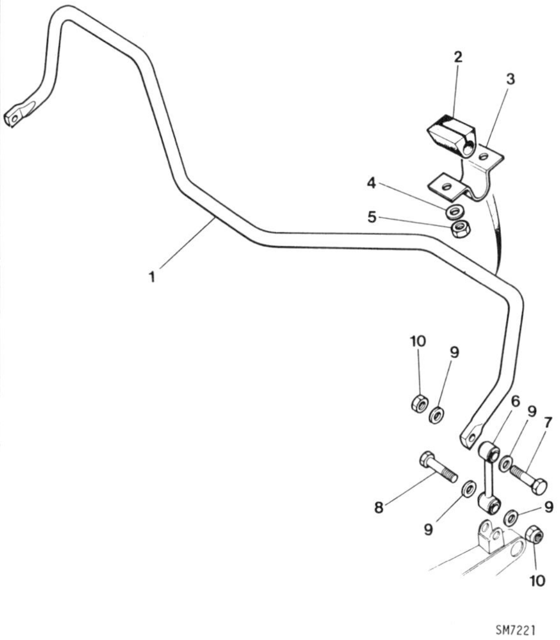

SM7221

REAR SUSPENSION 3.6 LITRE ALL
5.3 LITRE SPORTSPACK ONLY
ANTI ROLL BAR

Illus	Part Number	Description	Quantity	Change Point	Remarks
1	C 42178 3	Anti roll bar 16mm	1		
2	CBC 4901	Mounting bush-A.R.B.	2		
3	C 3054	Mounting bracket	2		
4	BD 541 9	Washer-radiator tie rod	4		
5	JLM 295	Self locknut	4		
6	C 42907	Link assembly-anti roll bar	2		
7	C 43807	Bolt-0.4375" UNF ARB to link	2		
8	BH 607161 J	Bolt-0.4375" UNF link to rad rod	2		
9	WA 600071 J	Plain washer	8		
10	TN 3210 J	Self locknut	4		

MASTER INDEX
ENGINE 1.C
FLYWHEEL AND CLUTCH 1.G
GEARBOX AND PROPSHAFT 1.H
AXLES, SUSPENSION, DRIVE SHAFTS, WHEELS 1.K
STEERING 1.M
BRAKES AND BRAKE CONTROLS 1.N
FUEL, EXHAUST AND EMISSION SYSTEMS 2.C
COOLING, HEATING AND AIR CONDITIONING 2.G
ELECTRICAL, WASHERS-WIPERS, INSTRUMENTS 2.J
CHASSIS, SUBFRAMES, BODYSHELL, FITTINGS 3.C
FASCIA, TRIM, SEATS AND FIXINGS 3.H
ACCESSORIES AND PAINT 4.C
NUMERICAL INDEX 4.D

GROUP INDEX
FRONT SHOCK ABSORBER & SPRING-3.6 LT K05
FRONT SHOCK ABSORBER AND SPRING K06
FRONT SUSPENSION - 3.6 LITRE & 5.3 LITRE K02
KIT-LOWER BALL JOINT K04
KIT-UPPER BALL JOINT K03
LOWER WIHSBONES K04
UPPER WISHBONES K03
HUB AND STUB AXLE CARRIER K07
REAR SUSPENSION 3.6 & 5.3 LITRE K08
REAR SUSPENSION 3.6 LITRE ALL K13
HALF SHAFTS K14
REAR SUSPENSION-continued K09
DAMPER UNIT AND SPRING K10
RADIUS ARM AND MOUNTINGS K12
REAR HUB AND CARRIER K11
ROAD WHEEL-ALLOY-NOT USA/CDN. L05

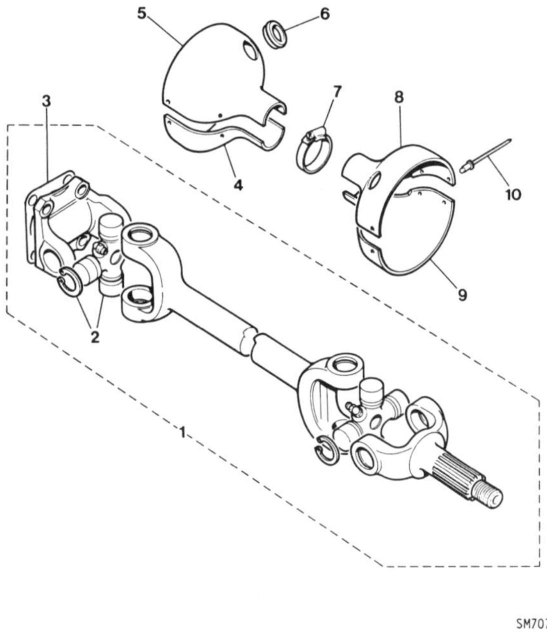

SM 7079

REAR SUSPENSION-continued 3.6 & 5.3 LITRE
HALF SHAFTS

Illus	Part Number	Description	Quantity	Change Point	Remarks
1	C 23655 1	HALF SHAFT ASSEMBLY	2		
2	JLM 9639	Universal joint assembly	2		
3	C 16621	Shim	A/R		
4	C 23615	Cover joint	2		
5	C 28843	Cover joint	2		
6	338023 J	Plug plastic	4		
7	C 43206 3	Clip hose	4		
8	C 28844	Cover joint	2		
9	C 23616	Cover joint	2		
10	BD 1814 4	Rivet pop	16		

1K 14

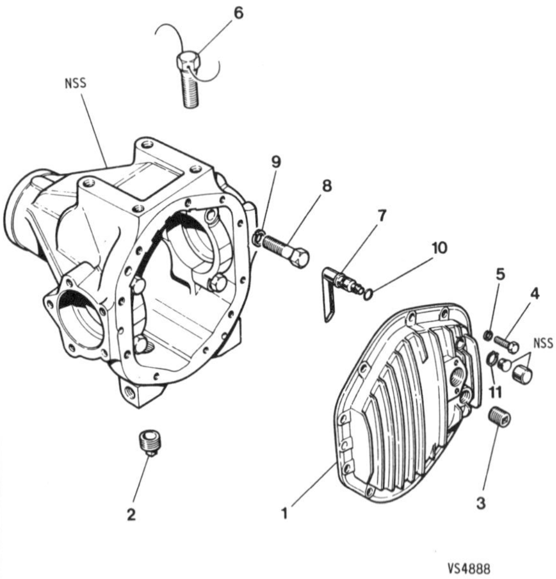

VS4888

FINAL DRIVE 3.54:1 3.6 Litre
FINAL DRIVE GKN POWR LOK

Illus	Part Number	Description	Quantity	Change Point	Remarks
	CBC 1792 35	Drive unit complete 3.54:1 ratio	1		
1	JLM 1351	Cover-alum	1		
2	JLM 1346	Magnetic drain plug	1		
3	JLM 632	Plug female 3/4 BSP	1		
4	JLM 1348	Bolt carrier cover	10		
5	JLM 613	Spring washer	10		
6	C 17024	Bolt, drive unit mounting	4		
7	JLM 631	Breather	1		
8	6114	Screw-diff bearing cap	4		
9	8489	Lock washer	4		
10	JLM 620	"O" ring - breather	1		
11	JLM 1710	Circlip, external	1		

MASTER INDEX

ENGINE ... 1.C
FLYWHEEL AND CLUTCH ... 1.G
GEARBOX AND PROPSHAFT ... 1.H
AXLES, SUSPENSION, DRIVE SHAFTS, WHEELS ... 1.K
STEERING ... 1.M
BRAKES AND BRAKE CONTROLS ... 1.N
FUEL, EXHAUST AND EMISSION SYSTEMS ... 2.C
COOLING, HEATING AND AIR CONDITIONING ... 2.G
ELECTRICAL, WASHERS-WIPERS, INSTRUMENTS ... 2.J
CHASSIS, SUBFRAMES, BODYSHELL, FITTINGS ... 3.C
FASCIA, TRIM, SEATS AND FIXINGS ... 3.H
ACCESSORIES AND PAINT ... 4.C
NUMERICAL INDEX ... 4.D

GROUP INDEX

FRONT SHOCK ABSORBER & SPRING-3.6 LT ... K05
FRONT SHOCK ABSORBER AND SPRING ... K06
FRONT SUSPENSION - 3.6 LITRE & 5.3 LITRE ... K02
KIT-LOWER BALL JOINT ... K04
KIT-UPPER BALL JOINT ... K03
LOWER WIHSBONES ... K04
UPPER WISHBONES ... K03
HUB AND STUB AXLE CARRIER ... K07
REAR SUSPENSION 3.6 & 5.3 LITRE ... K08
REAR SUSPENSION 3.6 LITRE ALL ... K13
HALF SHAFTS ... K14
REAR SUSPENSION-continued ... K09
DAMPER UNIT AND SPRING ... K10
RADIUS ARM AND MOUNTINGS ... K12
REAR HUB AND CARRIER ... K11
ROAD WHEEL-ALLOY-NOT USA/CDN. ... L05

1K 15

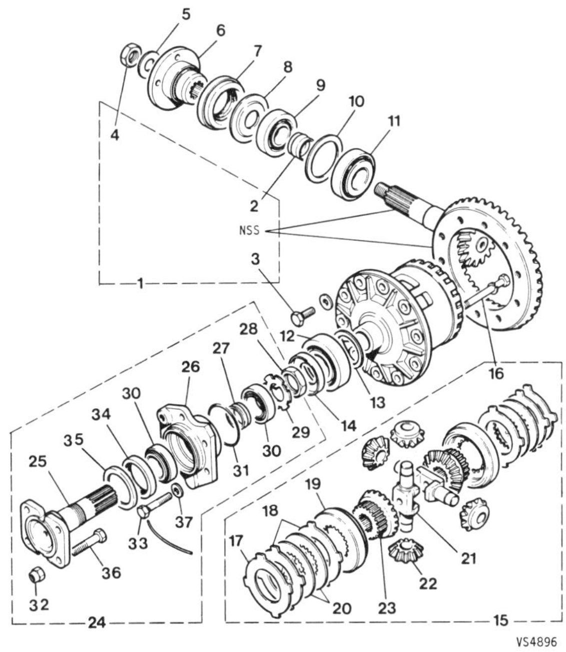

VS4896

FINAL DRIVE UNIT 3.6 LITRE BREAKDOWN

DIFF UNIT B/DOWN POWR/LOK
3.54

Illus	Part Number	123456 Description	Quantity	Change Point	Remarks
1	12458 1	Crown wheel/pinion 3.54:1	1		
2	12456	Spacer	1		
3	BAU 4604 J	Bolt drive gear	10		
4	11099	Nut pinion	1		
5	JLM 611	Washer pinion nut	1		
6	RTC 1009 A	Flange-companion	1		
7	JLM 527	Seal pinion-viton	1		
8	JLM 614	Pinion spacer	1		
9	ATA 7166	Bearing roller taper	1		
		SHIM PINION BEARING			
	JLM 1262 A	0.508mm	2		
	JLM 1262 B	0.533mm	2		
	JLM 1262 C	0.559mm	2		
	JLM 1262 D	0.584mm	2		
	JLM 1262 E	0.609mm	2		
	JLM 1262 F	0.635mm	2		
	JLM 1262 G	0.660mm	2		
	JLM 1262 H	0.686mm	2		
	JLM 1262 I	0.711mm	2		
	JLM 1262 J	0.737mm	2		
	JLM 1262 K	0.762mm	2		
	JLM 1262 L	0.787mm	2		
	JLM 1262 M	0.813mm	2		
	JLM 1262 N	0.838mm	2		
	JLM 1262 O	0.864mm	2		
	JLM 1262 P	0.889mm	2		
11	12252	Bearing-pinion-inner	1		
12	3845	Bearing roller	1		
		SHIM-DIFF BEARING			
	JLM 563 A	0.028mm	2		
	JLM 563 B	0.030mm	2		
	JLM 563 C	0.031mm	2		
	JLM 563 D	0.032mm	2		
	JLM 563 E	0.033mm	2		
	JLM 563 F	0.034mm	2		
	JLM 563 G	0.035mm	2		
	JLM 563 H	0.036mm	2		
	JLM 563 I	0.037mm	2		
	JLM 563 J	0.038mm	2		
	JLM 563 K	0.039mm	2		
	JLM 563 L	0.040mm	2		
	JLM 563 M	0.045mm	2		

MASTER INDEX

GROUP INDEX

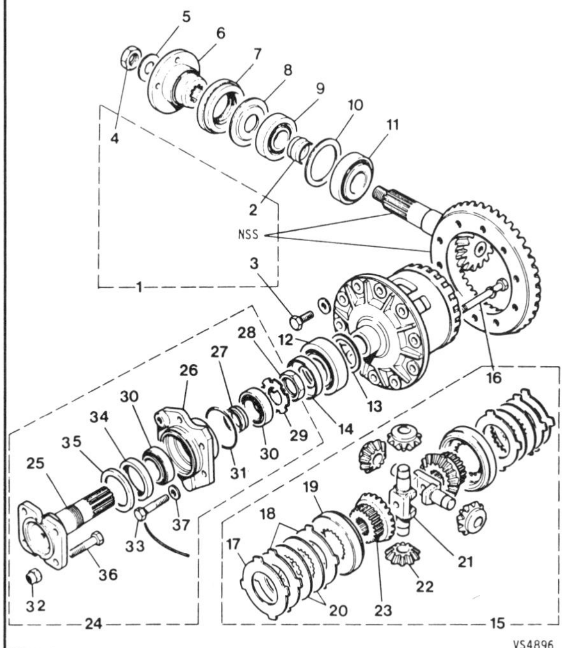

VS4896

DIFF UNIT BREAKDOWN 3.6-continued

Illus	Part Number	123456 Description	Quantity	Change Point	Remarks
14	JLM 1703	Oil-weir baffle	2		
15	JLM 1368	DIFFERENTIAL CASE ASSEMBLY	1		
16	7680	Bolt-diff case	8		
17	RTC 1350 A	Plates belleville clutch	2		
18	RTC 1349 A	Plates clutch tagged	4		
19	11290	Ring gear diff side	2		
20	RTC 1416 A	Disc friction	4		
21	7686	Shaft-cross-pinion mate	2		
22	7685	Bevel diff pinion mate	4		
23	11291	Bevel diff side gear	2		
24	JLM 558	AXLE SHAFT ASSEMBLY-RH	1		
	JLM 559	AXLE SHAFT ASSEMBLY-LH	1		
25	JLM 557	Axle shaft-stud & nut assembly	2		
26	AEU 1386	Housing-bearing-RH	1		
	AEU 1387	Housing-bearing-LH	1		
27	RTC 1348	Spacer collapsible	1		
28	RTC 1346	Nut shaft bearing	1		
29	JLM 556	Washer tab	1		
30	RTC 1347	BEARING ASSEMBLY	2		
31	RTC 1340	'O' ring bearing housing	1		
32	C 24119 4	Nut cleve lock	8		
33	JLM 1345	Bolt caliper mounting	10		
34	AAU 8271	Oil seal-output shaft	2		
35	AEU 1385	Dust shield	2		
36	JLM 442	Bolt-drive shaft	8		

E26.02AW31/8/88

MASTER INDEX

ENGINE 1.C
FLYWHEEL AND CLUTCH 1.G
GEARBOX AND PROPSHAFT 1.H
AXLES, SUSPENSION, DRIVE SHAFTS, WHEELS 1.K
STEERING 1.M
BRAKES AND BRAKE CONTROLS 1.N
FUEL, EXHAUST AND EMISSION SYSTEMS 2.C
COOLING, HEATING AND AIR CONDITIONING 2.G
ELECTRICAL, WASHERS-WIPERS, INSTRUMENTS 2.J
CHASSIS, SUBFRAMES, BODYSHELL, FITTINGS 3.C
FASCIA, TRIM, SEATS AND FIXINGS 3.H
ACCESSORIES AND PAINT 4.C
NUMERICAL INDEX 4.D

GROUP INDEX

FRONT SHOCK ABSORBER & SPRING-3.6 LT K05
FRONT SHOCK ABSORBER AND SPRING K06
FRONT SUSPENSION - 3.6 LITRE & 5.3 LITRE K02
KIT-LOWER BALL JOINT K04
KIT-UPPER BALL JOINT K03
LOWER WIHSBONES K04
UPPER WISHBONES K03
HUB AND STUB AXLE CARRIER K07
REAR SUSPENSION 3.6 & 5.3 LITRE K08
REAR SUSPENSION 3.6 LITRE ALL K13
HALF SHAFTS K14
REAR SUSPENSION-continued K09
DAMPER UNIT AND SPRING K10
RADIUS ARM AND MOUNTINGS K12
REAR HUB AND CARRIER K11
ROAD WHEEL-ALLOY-NOT USA/CDN. L05

1K 18

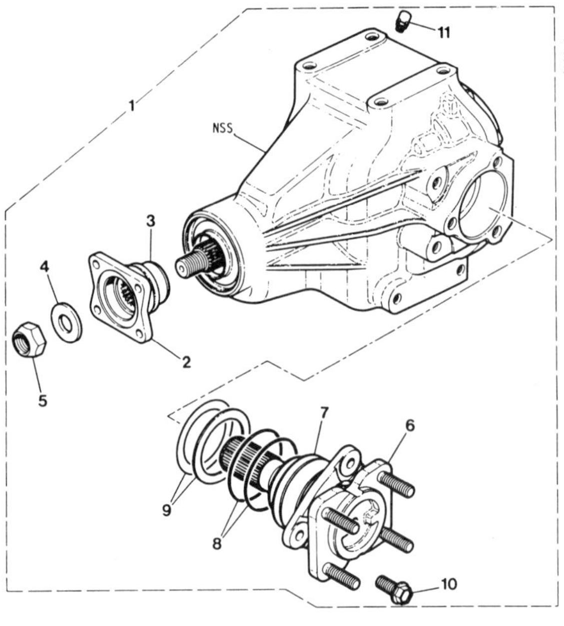

VS4473

DRIVE UNIT-5.3 LITRE

DANA AXLE L.S.D.

Illus	Part Number	123456	Description	Quantity	Change Point	Remarks
1	CAC 9281 H		DRIVE UNIT ASSEMBLY	1		5.3 litre only
2	JLM 340		Companion flange	1		
3	JLM 282		Tail bearing-input shaft	1		
	JLM 286		Shim adjusting-tail bearing	A/R		
	JLM 287		Shim adjusting-tail bearing	A/R		
	JLM 288		Shim adjusting-tail bearing	A/R		
	JLM 289		Shim adjusting-tail bearing	A/R		
	JLM 281		Head bearing	1		
	JLM 283		Shim adjusting head bearing	A/R		
	JLM 284		Shim adjusting head bearing	A/R		
	JLM 285		Shim adjusting head bearing	A/R		
	JLM 337		Oil seal-pinion	1		
4	JLM 339		Washer	1		
5	JLM 338		Nut	1		
6	JLM 313		Output shaft-complete	2		
7	JLM 290		Bearing-output shaft	2		
	JLM 291		Retainer bearing	2		
8	JLM 292		Sealing ring	4		
9	JLM 443		Shim adjusting 0.003	A/R		
	JLM 444		Shim adjusting 0.005	A/R		
	JLM 445		Shim adjusting 0.010	A/R		
	JLM 446		Shim adjusting 0.030	A/R		
10	JLM 408		Screw-flanged	6		
11	JLM 829		Breather	1		
	JLM 336		Slinger-pinion oil seal-outer	1		

E28AW12/10/88

1L 02

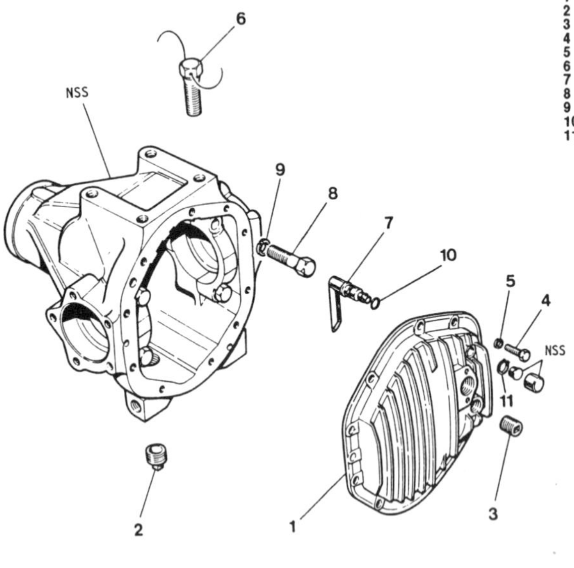

VS4888

FINAL DRIVE 2.88:1 5.3 Litre

FINAL DRIVE GKN POWR LOK

Illus	Part Number	123456	Description	Quantity	Change Point	Remarks
	CBC 1792 28		Drive unit complete 2.88:1 ratio	1		Alternative to Dana manufacture
1	JLM 1351		Cover-alum	1		
2	JLM 1346		Magnetic drain plug	1		
3	JLM 632		Plug female 3/4 BSP	1		
4	JLM 1348		Bolt carrier cover	10		
5	JLM 613		Spring washer	10		
6	C 17024		Bolt-drive unit mounting	4		
7	JLM 631		Breather	1		
8	6114		Screw-diff bearing cap	4		
9	8489		Lock washer	4		
10	JLM 620		'O' ring-breather	1		
11	JLM 1710		Circlip-external	1		

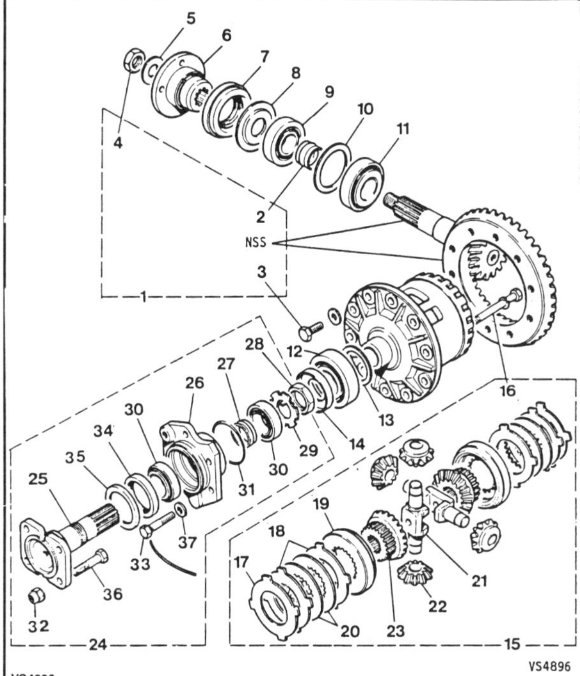

VS4896

FINAL DRIVE 2.88:1 5.3 Litre
FINAL DRIVE-COMPONENTS

Illus	Part Number	123456 Description	Quantity	Change Point	Remarks
1	JLM 635	CROWN WHEEL & PINION ASSEMBLY	1		2.88:1
2	12456	Spacer-collapsible	1		
3	BAU 4604 J	Bolt-drive gear	10		
4	11099	Nut-pinion	1		
5	JLM 611	Washer-pinion nut	1		
6	RTC 1009 A	Companion flange	1		
7	JLM 527	Seal final drive-viton	1		
8	JLM 614	Pinion spacer	1		
9	ATA 7166	Bearing-roller taper	1		
10	JLM 1262 A	Shim-pinion bearing	2		0.508mm
	JLM 1262 B	Shim-pinion bearing	2		0.533mm
	JLM 1262 C	Shim-pinion bearing	2		0.559mm
	JLM 1262 D	Shim-pinion bearing	2		0.584mm
	JLM 1262 E	Shim-pinion bearing	2		0.609mm
	JLM 1262 F	Shim-pinion bearing	2		0.635mm
	JLM 1262 G	Shim-pinion bearing	2		0.660mm
	JLM 1262 H	Shim-pinion bearing	2		0.686mm
	JLM 1262 I	Shim-pinion bearing	2		0.711mm
	JLM 1262 J	Shim-pinion bearing	2		0.737mm
	JLM 1262 K	Shim-pinion bearing	2		0.762mm
	JLM 1262 L	Shim-pinion bearing	2		0.787mm
	JLM 1262 M	Shim-pinion bearing	2		0.813mm
	JLM 1262 N	Shim-pinion bearing	2		0.838mm
	JLM 1262 O	Shim-pinion bearing	2		0.864mm
	JLM 1262 P	Shim-pinion bearing	2		0.889mm
11	12252	Bearing-pinion inner	1		
12	3845	Bearing roller	1		

MASTER INDEX

GROUP INDEX

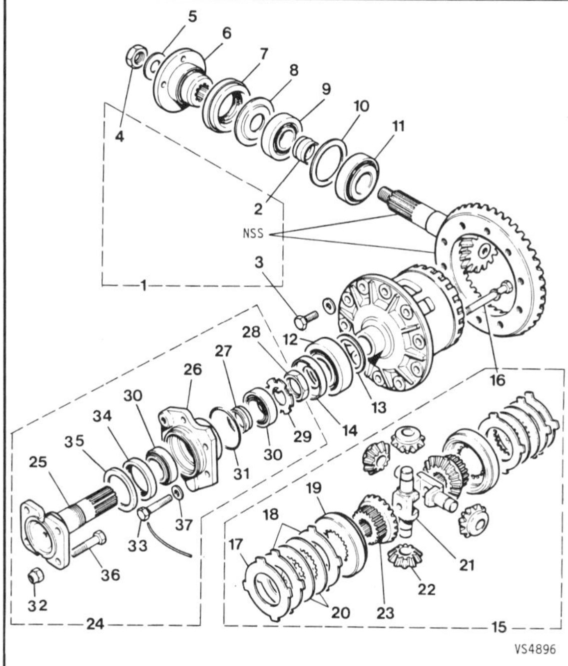

VS4896

FINAL DRIVE 2.88:1 5.3 Litre continued

Illus	Part Number	123456 Description	Quantity	Change Point	Remarks
13	JLM 563 A	Shim-diff bearing	2		0.028mm
	JLM 563 B	Shim-diff bearing	2		0.030mm
	JLM 563 C	Shim-diff bearing	2		0.031mm
	JLM 563 D	Shim-diff bearing	2		0.032mm
	JLM 563 E	Shim-diff bearing	2		0.033mm
	JLM 563 F	Shim-diff bearing	2		0.034mm
	JLM 563 G	Shim-diff bearing	2		0.035mm
	JLM 563 H	Shim-diff bearing	2		0.036mm
	JLM 563 I	Shim-diff bearing	2		0.037mm
	JLM 563 J	Shim-diff bearing	2		0.038mm
	JLM 563 K	Shim-diff bearing	2		0.039mm
	JLM 563 L	Shim-diff bearing	2		0.040mm
	JLM 563 M	Shim-diff bearing	2		0.045mm
14	JLM 1703	Oil weir baffle	2		
15	JLM 1352	DIFFERENTIAL ASSEMBLY	1		
16	7680	Bolt diff case	8		
17	RTC 1350 A	Plate belleville clutch	2		
18	RTC 1349 A	Plate clutch-tagged	4		
19	11290	Ring gear-diff side	2		
20	RTC 1416 A	Disc friction	4		
21	7686	Shaft cross pinion mate	2		
22	7685	Bevel diff pinion mate	4		
23	11291	Bevel diff side gear	2		
24	JLM 558	Shaft axle assembly-RH	1		
	JLM 559	Shaft axle assembly-LH	1		
25	JLM 557	Shaft and stud assembly	2		
26	AEU 1386	Housing bearing-RH	1		
	AEU 1387	Housing bearing-LH	1		
	AEU 1385	Oil thrower-brng housing	2		
27	RTC 1348	Spacer collapsible	1		
28	RTC 1346	Nut-shaft bearing	2		
29	JLM 556	Tag washer	2		
30	RTC 1347	Bearing assembly	4		
31	RTC 1340	O ring-bearing housing	2		
32	C 24119 4	Nut cleve lock	8		
33	JLM 1345	Bolt-caliper mounting	10		
34	AAU 8271	Oil Seal - Output Shaft	2		
35	AEU 1385	Oil thrower, bearing housing	2		
36	JLM 442	Bolt-drive shaft	8		

E32.02AW31/8/88

MASTER INDEX

ENGINE 1.C
FLYWHEEL AND CLUTCH 1.G
GEARBOX AND PROPSHAFT 1.H
AXLES, SUSPENSION, DRIVE SHAFTS, WHEELS 1.K
STEERING 1.M
BRAKES AND BRAKE CONTROLS 1.N
FUEL, EXHAUST AND EMISSION SYSTEMS 2.C
COOLING, HEATING AND AIR CONDITIONING 2.G
ELECTRICAL, WASHERS-WIPERS, INSTRUMENTS 2.J
CHASSIS, SUBFRAMES, BODYSHELL, FITTINGS 3.C
FASCIA, TRIM, SEATS AND FIXINGS 3.H
ACCESSORIES AND PAINT 4.C
NUMERICAL INDEX 4.D

GROUP INDEX

FRONT SHOCK ABSORBER & SPRING-3.6 LT K05
FRONT SHOCK ABSORBER AND SPRING K06
FRONT SUSPENSION - 3.6 LITRE & 5.3 LITRE K02
KIT-LOWER BALL JOINT K04
KIT-UPPER BALL JOINT K03
LOWER WIHSBONES K04
UPPER WISHBONES K03
HUB AND STUB AXLE CARRIER K07
REAR SUSPENSION 3.6 & 5.3 LITRE K08
REAR SUSPENSION 3.6 LITRE ALL K13
HALF SHAFTS K14
REAR SUSPENSION-continued K09
DAMPER UNIT AND SPRING K10
RADIUS ARM AND MOUNTINGS K12
REAR HUB AND CARRIER K11
ROAD WHEEL-ALLOY-NOT USA/CDN. L05

ROAD WHEEL-ALLOY-NOT USA/CDN.

3.6 LITRE AND 5.3 LITRE
ALLOY WHEEL OGLE PERFORATED

Illus	Part Number	Description	Quantity	Change Point	Remarks
1	CAC 5667	Wheel-road	5		
2	CBC 1262	Nut-wheel	20		
3	C 42191	Motif Jaguar	4		
4	CAC 6442	Retainer badge	4		
5	CBC 1491	Balance weight	a/r		
6	CAC 4123	Tyre valve	5		
	CAC 7218	Tyre valve	5		Alternative
	CAC 6377	Alloy wheel & tyre assy (Dunlop)	5		
	CAC 6380	Alloy wheel & tyre assy (Pirelli)	5		

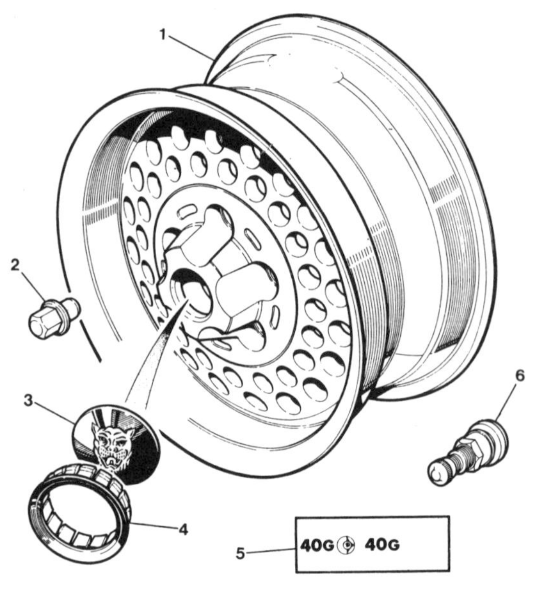

VS4347/A

VS 4347A

MASTER INDEX

GROUP INDEX

ROAD WHEELS-ALLOY-STARFISH WHEEL 3.6 & 5.3 LITRE

Illus	Part Number	Description	Quantity	Change Point	Remarks
1	CAC 4379	Road wheel	5		
2	CAC 4549	Nut wheel	20		
3	CAC 6502	Badge road wheel	4		
4	CAC 5073	Insert wheel	4		
5	CBC 1491	Balance weight	A/R		
6	CAC 4123	Valve	5		
	CAC 7218	Valve	5		Alternative
	CAC 4763	Alloy wheel & tyre assy(Dunlop)	5		Not CDN/USA
	CAC 4504	Alloy wheel & tyre assy(Pirelli)	5		

E38AW16/8/88

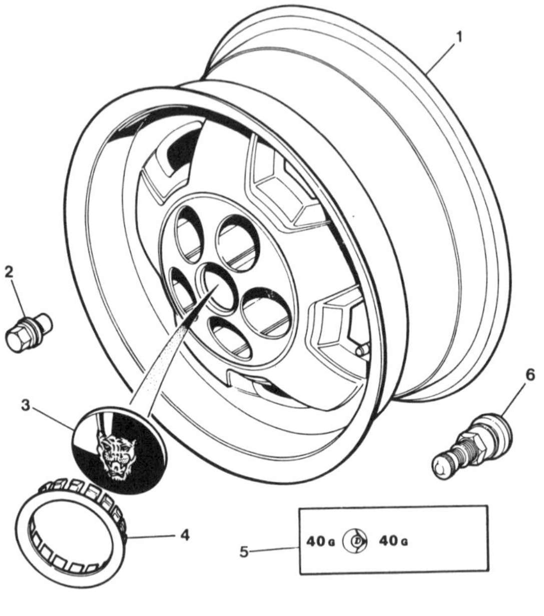

VS3526/A

VS3526A

MASTER INDEX

ENGINE 1.C
FLYWHEEL AND CLUTCH 1.G
GEARBOX AND PROPSHAFT 1.H
AXLES, SUSPENSION, DRIVE SHAFTS, WHEELS 1.K
STEERING 1.M
BRAKES AND BRAKE CONTROLS 1.N
FUEL, EXHAUST AND EMISSION SYSTEMS 2.C
COOLING, HEATING AND AIR CONDITIONING 2.G
ELECTRICAL, WASHERS-WIPERS, INSTRUMENTS 2.J
CHASSIS, SUBFRAMES, BODYSHELL, FITTINGS 3.C
FASCIA, TRIM, SEATS AND FIXINGS 3.H
ACCESSORIES AND PAINT 4.C
NUMERICAL INDEX 4.D

GROUP INDEX

FRONT SHOCK ABSORBER & SPRING-3.6 LT K05
FRONT SHOCK ABSORBER AND SPRING K06
FRONT SUSPENSION - 3.6 LITRE & 5.3 LITRE K02
KIT-LOWER BALL JOINT K04
KIT-UPPER BALL JOINT K03
LOWER WIHSBONES K04
UPPER WIHSBONES K03
HUB AND STUB AXLE CARRIER K07
REAR SUSPENSION 3.6 & 5.3 LITRE K08
REAR SUSPENSION 3.6 LITRE ALL K13
HALF SHAFTS K14
REAR SUSPENSION-continued K09
DAMPER UNIT AND SPRING K10
RADIUS ARM AND MOUNTINGS K12
REAR HUB AND CARRIER K11
ROAD WHEEL-ALLOY-NOT USA/CDN. L05

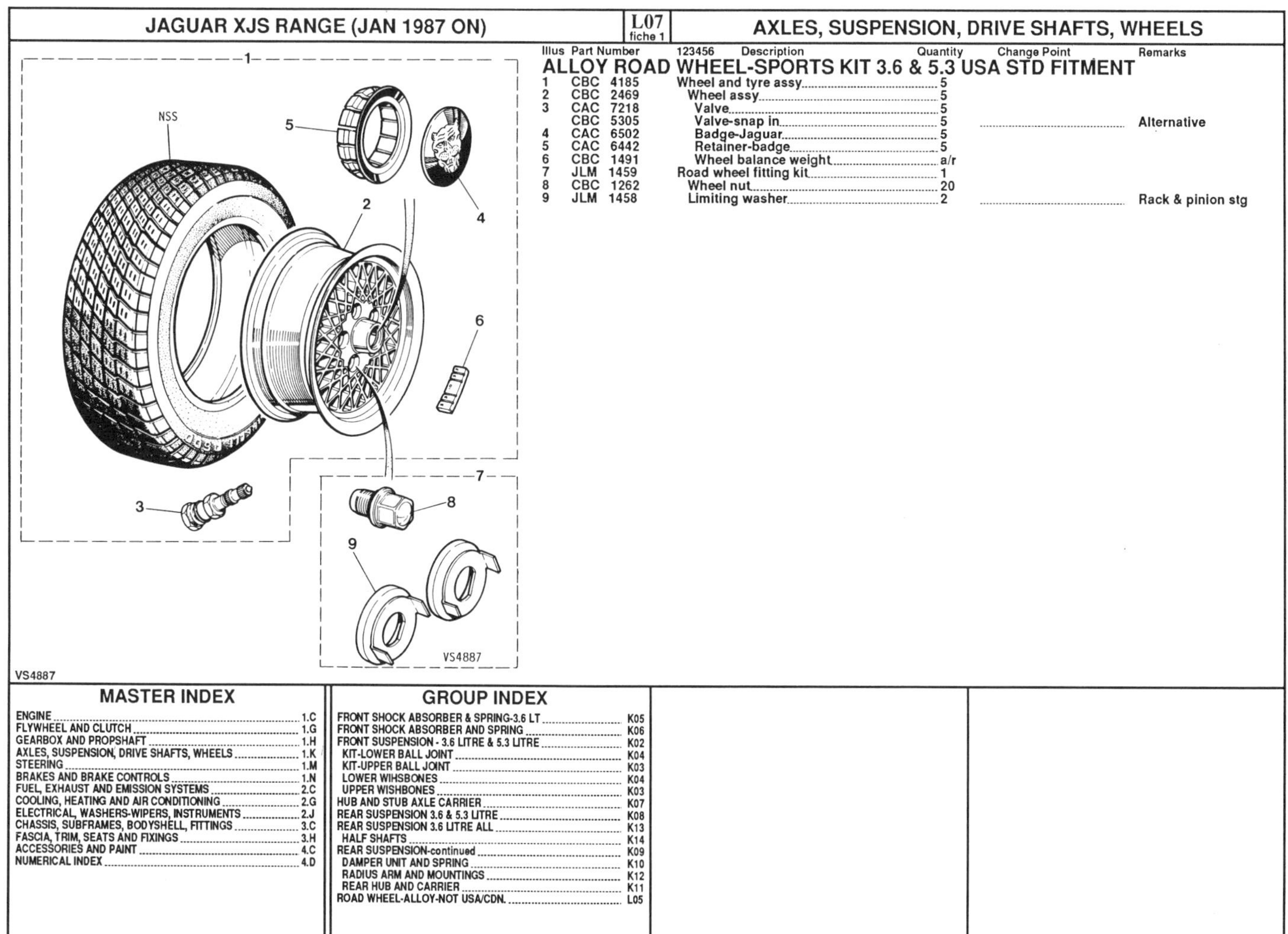

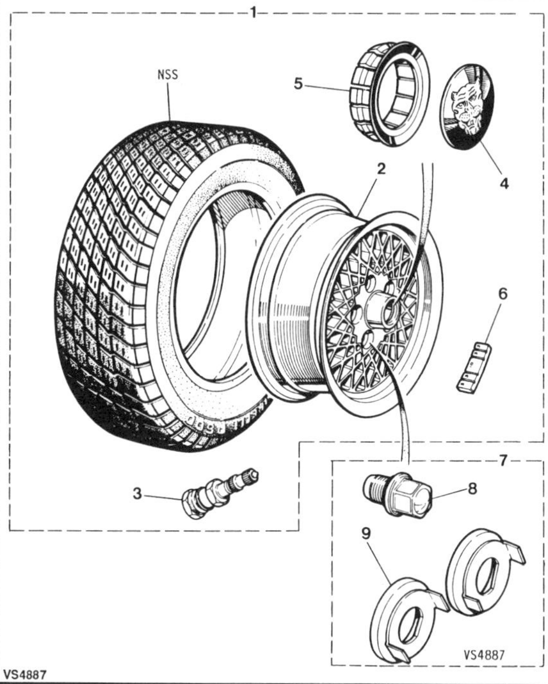

Illus	Part Number	Description	Quantity	Change Point	Remarks
		ALLOY ROAD WHEEL-SPORTS KIT 3.6 & 5.3 USA STD FITMENT			
1	CBC 4185	Wheel and tyre assy	5		
2	CBC 2469	Wheel assy	5		
3	CAC 7218	Valve	5		
	CBC 5305	Valve-snap in	5		Alternative
4	CAC 6502	Badge-Jaguar	5		
5	CAC 6442	Retainer-badge	5		
6	CBC 1491	Wheel balance weight	a/r		
7	JLM 1459	Road wheel fitting kit	1		
8	CBC 1262	Wheel nut	20		
9	JLM 1458	Limiting washer	2		Rack & pinion stg

MASTER INDEX

GROUP INDEX

MASTER INDEX

GROUP INDEX

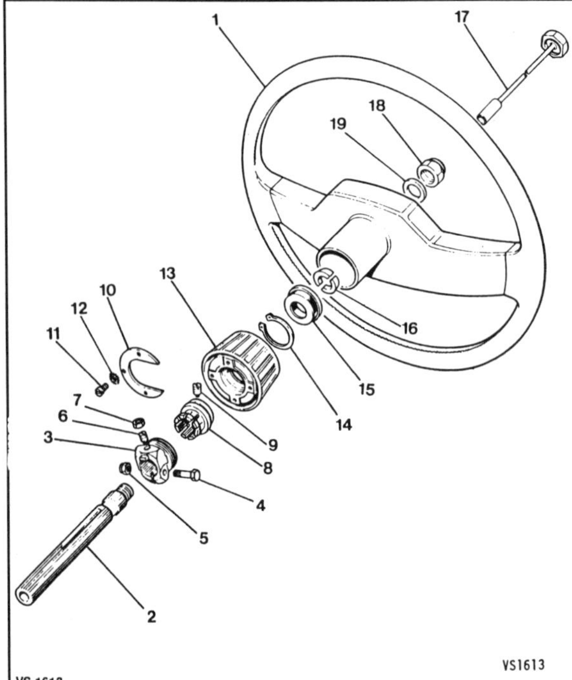

VS 1613

STEERING WHEEL AND SHAFT
3.6 & 5.3 LITRE

Illus	Part Number	123456 Description	Quantity	Change Point	Remarks
1	CAC 5885	Wheel steering	1	....	V12
	CBC 6333	Steering wheel	1	....	Use CBC2731
	CBC 2731	Steering wheel (less horn pad)	1		
	CBC 6345	Horn Pad	1	....	] 3.6 Standard V12 Sportspack
2	C 38526	Shaft steering wheel	1		
3	CAC 2970	Adaptor	1		
4	C 39559	Bolt collet	1		
5	C 8737	Nut self locking	1		
6	CAC 2967	Screw retaining-st collet	1		
7	UFN 231 L	Nut plain	1		
8	C 38524	Collet split	1		
9	C 35515	Button stop-st wheel shaft	1		
10	C 38527	Plate retaining hand locknut	1		
11	UCS 311 2H	Screw set	2		
12	WF 706041 J	Washer shakeproof	2		
13	C 38523	Locknut hand	1		
14	BD 38852 3	Circlip collet	1		
15	CAC 3321	Rubber impact assembly	1		
16	C 38528	Cone split	1		
17	DAC 2882	Guide horn & sleeve assembly	1		
18	C 8737 7	Nut nyloc	1		
19	C 38791	Washer plain	1		

MASTER INDEX

GROUP INDEX

1M 02

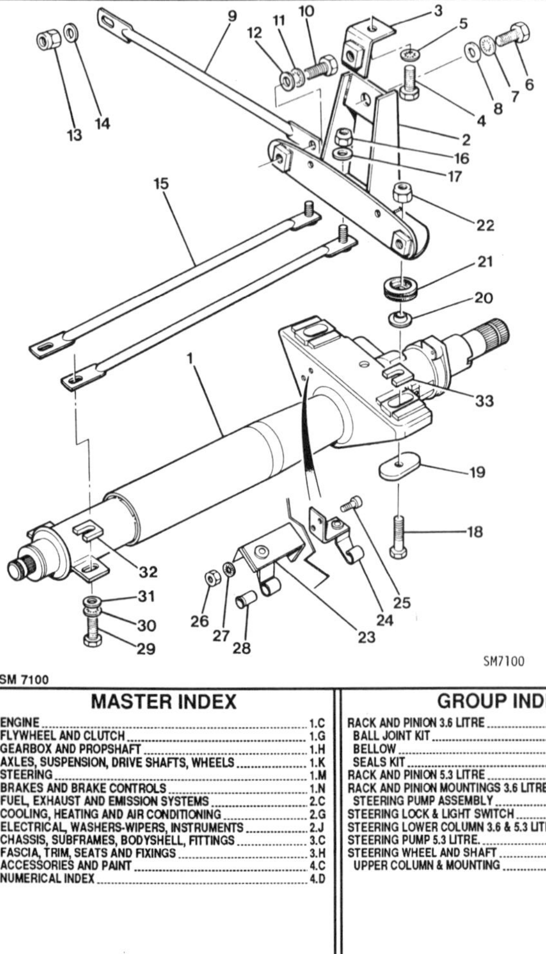

SM 7100

STEERING 3.6 & 5.3 LITRE
UPPER COLUMN & MOUNTING

Illus	Part Number	123456 Description	Quantity	Change Point	Remarks
1	CBC 4652	Steering upper column	1		
2	C 45804	Strut vertical	1		
3	C 44673	Bracket vertical strut	1		
4	UFS 131 5R	Setscrew 0.3125" x 0.625"	1		
5	C 725	Washer shakeproof	1		
6	UFS 131 6V	Setscrew 0.3125" x 0.75"	2		
7	C 725	Washer shakeproof	1		
8	JLM 299	Washer plain 0.3125" ID	1		
9	C 44674	Strut transverse	1		
10	UFS 131 5R	Setscrew 0.3125" x 0.625"	1		
11	C 725	Washer shakeproof	1		
12	JLM 299	Washer plain 0.3125" ID	1		
13	C 8737 2	Nut self locking 0.3125 UNF	1		
14	JLM 299	Washer shakeproof	1		
15	C 44676	Strut longitudinal	2		
16	JLM 9684	Nut 0.3125 UNF	1		
17	C 725	Washer shakeproof	2		
18	BH 605141 J	Bolt 0.3125" UNF x 1.75"	2		
19	BD 542 7	Washer oval	2		
20	C 30316	Cone st column mounting	4		
	CBC 5738	Cone - st column mounting	4	....	XJ77
21	C 30213	Grommet rubber	2		
22	C 8737 2	Nut self locking 0.3125" ID	2		
23	C 44551	Contact assembly feed horn	1		
24	C 44553	Contact assembly earth horn	1		
25	UCS 813 4H	Setscrew No.6 x 0.50"	2		
26	UCN 113 L	Nut plain No.6 UNC	2		
27	WF 704061 J	Washer shakeproof UBA No.6	2		
28	C 12194	Eyelet	1		
29	SH 605071 J	Setscrew 0.3125" UNC x 0.875"	2		
30	C 725	Washer shakeproof	2		
31	JLM 299	Washer plain 0.3125 ID	2		
32	C 16370	Packing	2		
33	C 16370	Packing	A/R		

MASTER INDEX

ENGINE 1.C
FLYWHEEL AND CLUTCH 1.G
GEARBOX AND PROPSHAFT 1.H
AXLES, SUSPENSION, DRIVE SHAFTS, WHEELS 1.K
STEERING 1.M
BRAKES AND BRAKE CONTROLS 1.N
FUEL, EXHAUST AND EMISSION SYSTEMS 2.C
COOLING, HEATING AND AIR CONDITIONING 2.G
ELECTRICAL, WASHERS-WIPERS, INSTRUMENTS 2.J
CHASSIS, SUBFRAMES, BODYSHELL, FITTINGS 3.C
FASCIA, TRIM, SEATS AND FIXINGS 3.H
ACCESSORIES AND PAINT 4.C
NUMERICAL INDEX 4.D

GROUP INDEX

RACK AND PINION 3.6 LITRE M06
BALL JOINT KIT M06
BELLOW M06
SEALS KIT M06
RACK AND PINION 5.3 LITRE M07
RACK AND PINION MOUNTINGS 3.6 LITRE M08
STEERING PUMP ASSEMBLY M11
STEERING LOCK & LIGHT SWITCH M04
STEERING LOWER COLUMN 3.6 & 5.3 LITRE M05
STEERING PUMP 5.3 LITRE M13
STEERING WHEEL AND SHAFT M02
UPPER COLUMN & MOUNTING M03

1M 03

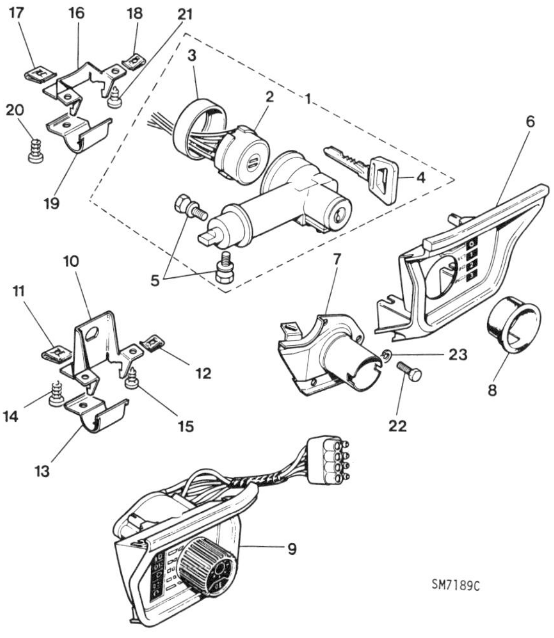

SM 7189C

STEERING LOCK & LIGHT SWITCH
3.6 & 5.3 LITRE

Illus	Part Number	123456	Description	Quantity	Change Point	Remarks
1	DAC 4151		LOCK AND IGNITION SWITCH ASSEMBLY	1		
2	DAC 2852		Inhibitor switch	1		
3	C 38037		Grommet	1		
4	37H 7193 J		Key blank	1		
5	DAC 4154		Bolt shear	2		
6	DAC 3170		Cover assembly-ignition switch	1		
7	C 44511		Support tube-switch cluster	1		
8	DAC 1054		Finisher-ignition switch	1		
9	DAC 5956		Switch lighting assembly	1		
	DAC 4775		Escutheon plate	1		
10	DAC 4781		Bracket light switch	1		
11	78723		Spire nut	1		
12	BD 8633 3		Spire nut	1		
13	C 44763		Clamp attachment bracket	1		
14	AB 608051 J		Screw self tapping No 8 x 0.625"	1		
15	DAZ 408 8C		Screw self tapping No 8 x 0.5"	1		
16	C 44760		Bracket ignition switch	1		
17	78723		Spire nut	1		
18	BD 8633 3		Spire nut	2		
19	C 44763		Clamp attachment bracket	1		
20	AB 608051 J		Screw self tapping No 8 x 0.625"	1		
21	DAZ 408 8C		Screw self tapping No 8 x 0.5"	2		
22	UCS 119/3R		Setscrew	3		
23	WF 702100 J		Shakeproof Washer	3		

MASTER INDEX

GROUP INDEX

STEERING LOWER COLUMN 3.6 & 5.3 LITRE

Illus	Part Number	123456	Description	Quantity	Change Point	Remarks
1	CAC 2765 3		Steering column lower	1		
2	CAC 8187 1		Bolt	1		
3	C 8737 2		Nut 0.3125" UNF	1		
4	CAC 3470		Universal joint	1		
5	CAC 8187 1		Bolt	1		
6	C 8737 2		Nut 0.3125 UNF	1		
7	CAC 8187 1		Bolt 0.3125 UNF	1		
8	JLM 299		Washer plain 0.3125 ID	1		
9	C 8737 2		Nut 0.3125 UNF	1		
10	C 42001		Gaiter	1		
11	C 42879		Sleeve gaiter	1		
12	C 41116 6		Clip	1		
13	C 43733		Retainer gaiter	1		
14	BD 36746 2		Screw self tapping	1		
15	C 30741		Sleeve sealing	1		
	CAC 7556		Cover - Univeral Joint	1		

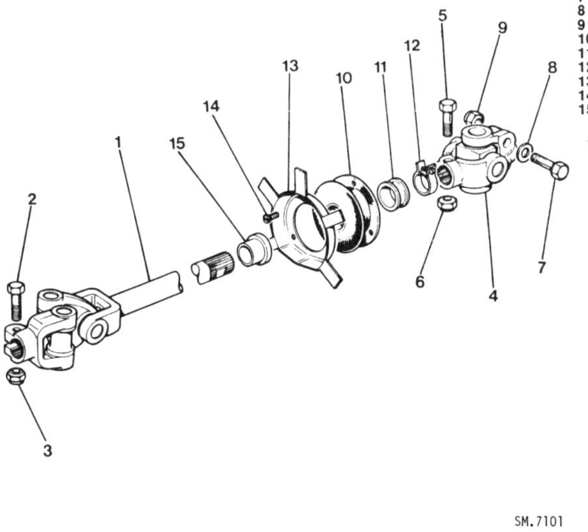

SM7101

MASTER INDEX
ENGINE 1.C
FLYWHEEL AND CLUTCH 1.G
GEARBOX AND PROPSHAFT 1.H
AXLES, SUSPENSION, DRIVE SHAFTS, WHEELS 1.K
STEERING 1.M
BRAKES AND BRAKE CONTROLS 1.N
FUEL, EXHAUST AND EMISSION SYSTEMS 2.C
COOLING, HEATING AND AIR CONDITIONING 2.G
ELECTRICAL, WASHERS-WIPERS, INSTRUMENTS 2.J
CHASSIS, SUBFRAMES, BODYSHELL, FITTINGS 3.C
FASCIA, TRIM, SEATS AND FIXINGS 3.H
ACCESSORIES AND PAINT 4.C
NUMERICAL INDEX 4.D
GROUP INDEX
RACK AND PINION 3.6 LITRE M06
BALL JOINT KIT M06
BELLOW M06
SEALS KIT M06
RACK AND PINION 5.3 LITRE M07
RACK AND PINION MOUNTINGS 3.6 LITRE M08
STEERING PUMP ASSEMBLY M11
STEERING LOCK & LIGHT SWITCH M04
STEERING LOWER COLUMN 3.6 & 5.3 LITRE M05
STEERING PUMP 5.3 LITRE M13
STEERING WHEEL AND SHAFT M02
UPPER COLUMN & MOUNTING M03

1M 06

JAGUAR XJS RANGE (JAN 1987 ON) — M06 fiche 1 — STEERING

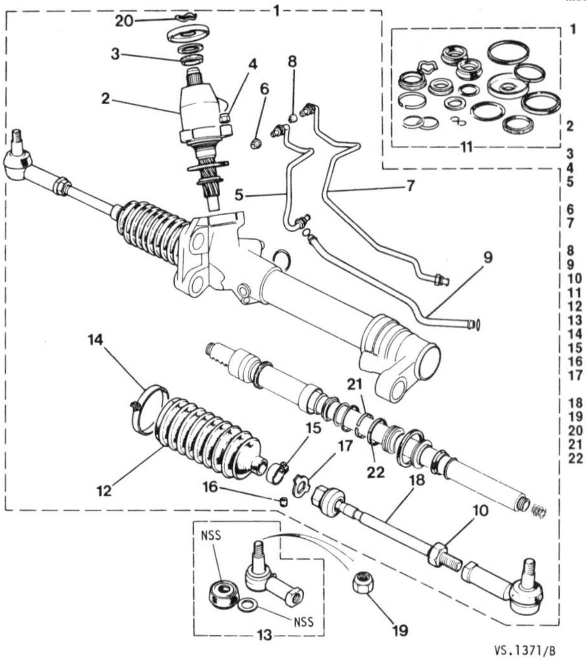

VS 1371B

RACK AND PINION 3.6 LITRE

Illus	Part Number	123456 Description	Quantity	Change Point	Remarks
1	CBC 1720 N	RACK AND PINION ASSEMBLY-RHD	1		
	CBC 1720 E	RACK AND PINION ASSY(ex.unit)RHD	1		
	CBC 1721	RACK AND PINION ASSEMBLY-LHD	1		
	CBC 4648 N	RACK AND PINION ASSEMBLY-RHD	1		1988 Model Year
	CBC 4648 E	RACK AND PINION ASSEMBLY-RHD EXCHANGE UNIT	1		1988 Model Year
	CBC 4649	RACK AND PINION ASSEMBLY-LHD	1		1988 Model Year
2	JLM 601	Valve body RHD	1		
	JLM 641 J	Valve body LHD	1		
3	AEU 1248 J	Seal	1		
4	AEU 1183	Nut self locking	3		
5	AEU 1445	Pipe short-LHD	1		
	JLM 405	Pipe short RHD	1		
6	11987	Seat inlet port	1		
7	AEU 2519	Pipe long-LHD	1		
	JLM 406	Pipe long-RHD	1		
8	C 45307	Seat inlet port	1		
9	AEU 1434	Air pipe	1		
10	AEU 2997	Nut lock	2		
11	JLM 533	SEALS KIT	1		
12	JLM 347	BELLOW	2		
13	JLM 9660	BALL JOINT KIT	2		
14	JLM 348	Clip large	2		
15	JLM 349	Clip small	2		
16	JLM 350	Cap screw thread	2		
17	C 39341	Tab washer	1		
	JLM 1458	Limiting washer	2		Sportspack
18	JLM 409	Inner ball joint assembly	1		
19	JLM 582	Self locking nut	2		
20	C 45068	Wave washer	1		
21	C 39456	Piston ring inner	1		
22	C 39457	Piston ring outer-(nylon)	1		

MASTER INDEX

GROUP INDEX

1M 07

JAGUAR XJS RANGE (JAN 1987 ON) — M07 fiche 1 — STEERING

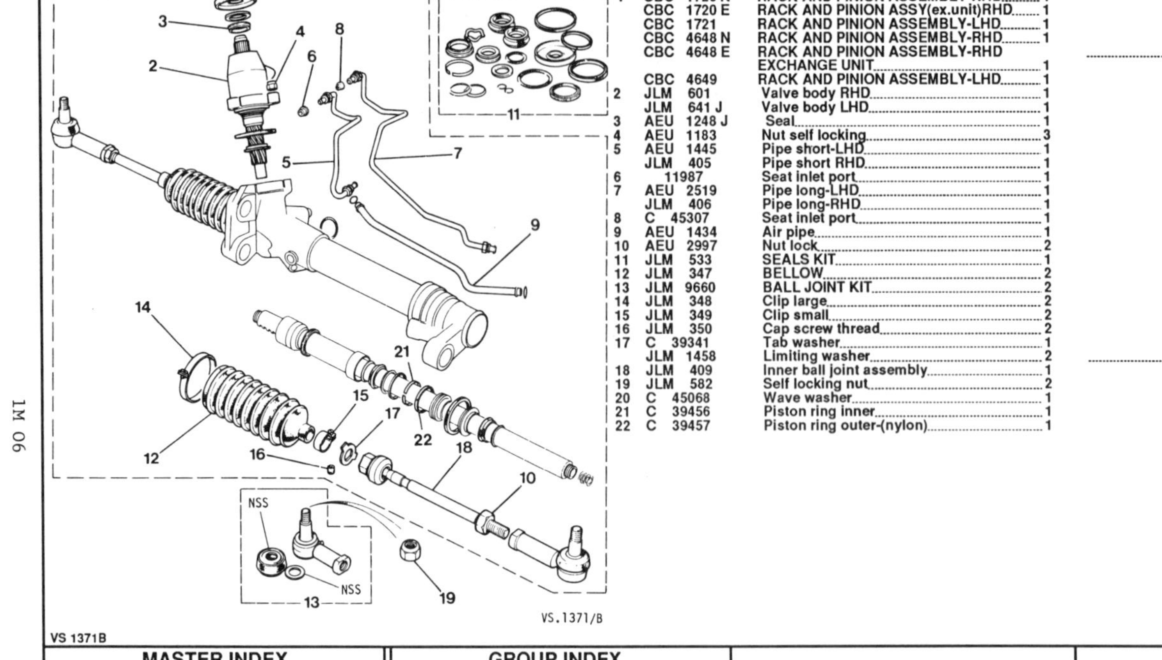

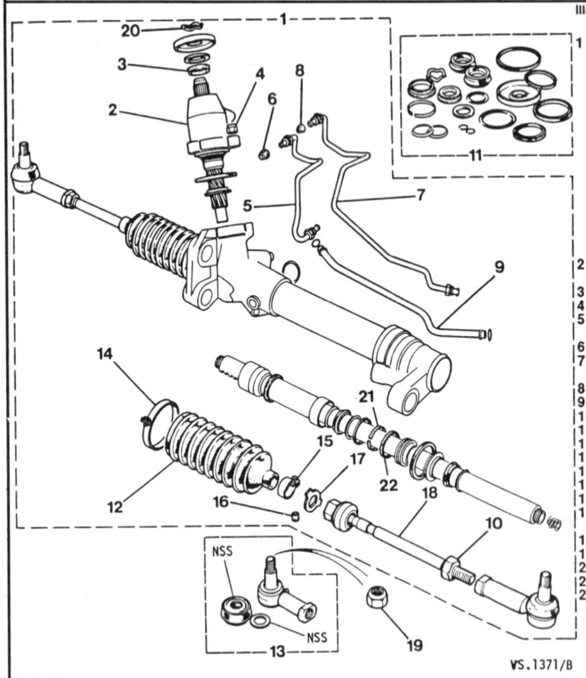

VS1371B

RACK AND PINION 5.3 LITRE

Illus	Part Number	123456 Description	Quantity	Change Point	Remarks
1	CBC 1717 N	RACK AND PINION ASSEMBLY-RHD	1		Not Japan
	CBC 1717 E	RACK AND PINION ASSY(ex.unit)RHD	1		
	CBC 1719 N	RACK AND PINION ASSY - RHD	1		Japan
	CBC 1718	RACK AND PINION ASSEMBLY-LHD	1		
		RACK & PINION FOR SPORTSPACK			
	CBC 5708 N	RACK & PINION ASSY RHD-NOT JAP	1		Use CBC 5682
	CBC 5682 N	RACK & PINION ASSY RHD-NOT JAP	1		
	CBC 5682 E	RACK & PINION ASSY RHD Exchange Unit-NOT JAPAN	1		
	CBC 5710	RACK & PINION ASSY RHD JAPAN ONLY	1		Use CBC 5684
	CBC 5684	RACK & PIN.ASSY RHD JAP ONLY	1		
	CBC 5709	RACK AND PINION ASSY LHD	1		Use CBC 5683
	CBC 5683	RACK & PINION ASSY LHD	1		
2	JLM 403	Valve body RHD	1		
	JLM 404	Valve body LHD	!		
3	AEU 1248 J	Seal	1		
4	AEU 1183	Nut self locking	3		
5	AEU 1445	Pipe short-LHD	1		
	JLM 405	Pipe long-RHD	1		
6	11987	Seat inlet port	1		
7	AEU 2519	Pipe long-LHD	1		
	JLM 406	Pipe short - RHD	1		
8	C 45307	Seat inlet port	1		
9	AEU 1434	Air pipe	1		
10	AEU 2997	Nut lock	2		
11	JLM 533	Seals kit	1		
12	JLM 347	Bellow	2		
13	JLM 9660	Ball joint kit	2		
14	JLM 348	Clip large	2		
15	JLM 349	Clip small	2		
16	JLM 350	Cap screw thread	2		
17	C 39341	Tab washer	1		
	JLM 1458	Limiting washer	2		Sportspack
18	JLM 409	Inner ball joint assembly	1		
19	JLM 582	Self locking nut	2		
20	C 45068	Wave washer	1		
21	C 39456	Piston ring inner	1		
22	C 39457	Piston ring outer (nylon)	1		
	JLM 462	Retaining cap,piston,seal housing	1		

MASTER INDEX

ENGINE 1.C
FLYWHEEL AND CLUTCH 1.G
GEARBOX AND PROPSHAFT 1.H
AXLES, SUSPENSION, DRIVE SHAFTS, WHEELS 1.K
STEERING 1.M
BRAKES AND BRAKE CONTROLS 1.N
FUEL, EXHAUST AND EMISSION SYSTEMS 2.C
COOLING, HEATING AND AIR CONDITIONING 2.G
ELECTRICAL, WASHERS-WIPERS, INSTRUMENTS 2.J
CHASSIS, SUBFRAMES, BODYSHELL, FITTINGS 3.C
FASCIA, TRIM, SEATS AND FIXINGS 3.H
ACCESSORIES AND PAINT 4.C
NUMERICAL INDEX 4.D

GROUP INDEX

RACK AND PINION 3.6 LITRE M06
BALL JOINT KIT M06
BELLOW M06
SEALS KIT M06
RACK AND PINION 5.3 LITRE M07
RACK AND PINION MOUNTINGS 3.6 LITRE M08
STEERING PUMP ASSEMBLY M11
STEERING LOCK & LIGHT SWITCH M04
STEERING LOWER COLUMN 3.6 & 5.3 LITRE M05
STEERING PUMP 5.3 LITRE M13
STEERING WHEEL AND SHAFT M02
UPPER COLUMN & MOUNTING M03

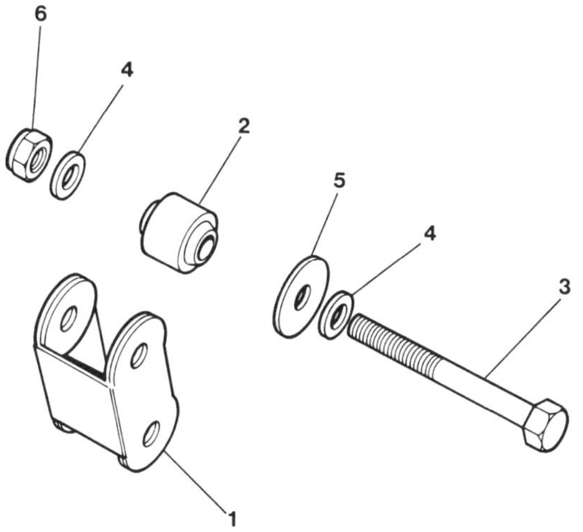

VS1372/A

VS 1372A

RACK AND PINION MOUNTINGS 3.6 LITRE

Illus	Part Number	123456 Description	Quantity	Change Point	Remarks
1	CAC 2036	Stiffener welded assembly RHD	1		1988 MY FOR PREVIOUS SEE 5.3 MODEL Non Sportspack
	CAC 2037	Stiffener welded assembly LHD	1		
2	CBC 5928	Mounting bush	2		
	CAC 1635	Mounting bush	1		
3	BH 605211 J	Bolt 0.3125" UNF x 2.75"	3		
4	JLM 299	Washer plain 0.3125" ID	6		
5	C 29407 1	Washer mounting	3		
6	C 8737 2	Nut self locking 0.3125" UNF	3		

MASTER INDEX

GROUP INDEX

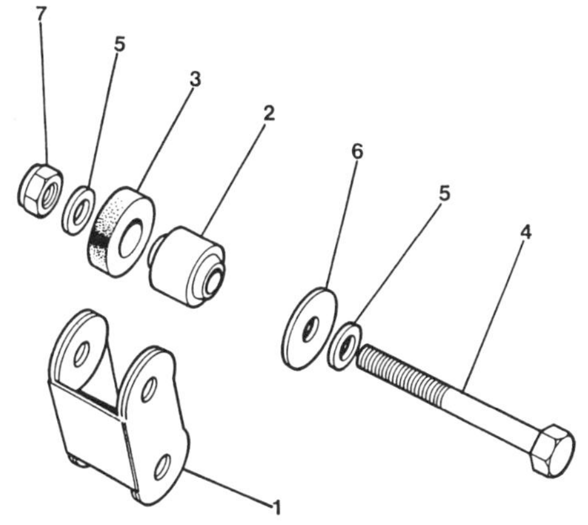

VS4904

VS 4904

RACK AND PINION MOUNTINGS 5.3 LITRE

Illus	Part Number	123456 Description	Quantity	Change Point	Remarks
1	CAC 2036	Stiffener welded assembly	1		RHD
	CAC 2037	Stiffener welded assembly	1		LHD
2	CAC 1635	Bush mountings	3		Not Sportspack
	CBC 5928	Mounting bush	2		Sportspack only
	CAC 1635	Mounting bush	1		
3	C 29406	Washer thrust	2		
4	BH 605211 J	Bolt 0.3125" UNF x 2.75"	3		
5	JLM 299	Washer plain 0.3125" ID	6		
6	C 29407 1	Washer mounting	3		
7	C 8737 2	Nut self locking 0.3125" UNF	3		

MASTER INDEX

ENGINE ... 1.C
FLYWHEEL AND CLUTCH ... 1.G
GEARBOX AND PROPSHAFT ... 1.H
AXLES, SUSPENSION, DRIVE SHAFTS, WHEELS ... 1.K
STEERING ... 1.M
BRAKES AND BRAKE CONTROLS ... 1.N
FUEL, EXHAUST AND EMISSION SYSTEMS ... 2.C
COOLING, HEATING AND AIR CONDITIONING ... 2.G
ELECTRICAL, WASHERS-WIPERS, INSTRUMENTS ... 2.J
CHASSIS, SUBFRAMES, BODYSHELL, FITTINGS ... 3.C
FASCIA, TRIM, SEATS AND FIXINGS ... 3.H
ACCESSORIES AND PAINT ... 4.C
NUMERICAL INDEX ... 4.D

GROUP INDEX

RACK AND PINION 3.6 LITRE ... M06
BALL JOINT KIT ... M06
BELLOW ... M06
SEALS KIT ... M06
RACK AND PINION 5.3 LITRE ... M07
RACK AND PINION MOUNTINGS 3.6 LITRE ... M08
STEERING PUMP ASSEMBLY ... M11
STEERING LOCK & LIGHT SWITCH ... M04
STEERING LOWER COLUMN 3.6 & 5.3 LITRE ... M05
STEERING PUMP 5.3 LITRE ... M13
STEERING WHEEL AND SHAFT ... M02
UPPER COLUMN & MOUNTING ... M03

1M 10

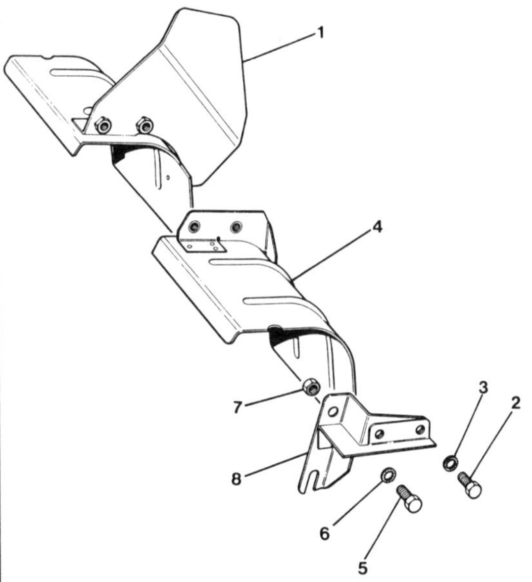

SM7117

STEERING HEATSHIELDS 5.3 LITRE

Illus	Part Number	Description	Quantity	Change Point	Remarks
1	C 45047	Heatshield assembly-RH-RHD	1		
	CBC 2550	Heatshield assembly-RH-RHD	1		Australia
	C 44703	Heatshield assembly-LH-RHD	1		
	CAC 2546	Heatshield assembly-LH-RHD	1		] Japan
	CAC 2549	Heatshield assembly-RH-RHD	1		
	C 44998	Bracket mounting heatshield-LH	1		] RHD
2	SH 504041 J	Setscrew 0.25 UNC x 0.50"	2		
	SH 604041 J	Setscrew 0.25 UNF x 0.50	2		LHD
3	JLM 302	Washer shakeproof	4		2 RHD 2 LHD
4	CAC 4330	Heatshield assembly-RH-LHD	1		
	CAC 3907	Heatshield assembly-LH-LHD	1		
5	UFS 131 5R	Setscrew 0.3125" UNC x 0.625	1		
6	C 725	Washer shakeproof	1		
7	JLM 9684	Nut 0.50 UNF	1		
8	C 44999	Bracket mounting heatshield-RH	1		RHD
	CAC 4328	Bracket mounting heatshield-RH	1		LHD

MASTER INDEX

GROUP INDEX

1M 11

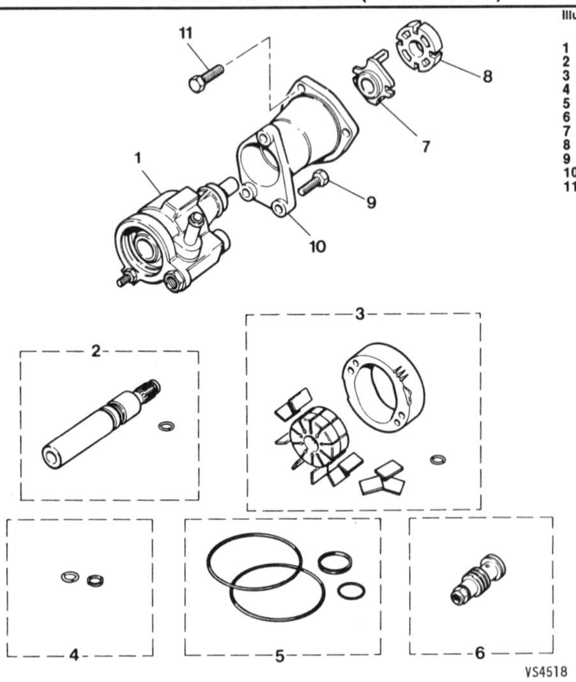

VS4518

STEERING PUMP 3.6 LITRE

Illus	Part Number	Description	Quantity	Change Point	Remarks
1	EAC 3970	STEERING PUMP ASSEMBLY	1		
2	BAU 2135 J	Shaft and ring	1		
3	BAU 2136 J	Rotor kit	1		
4	BAU 2137 J	Retaining rings kit	1		
5	AEU 2787	Seals kit	1		
6	AEU 2786	Control valve	1		
7	EAC 7081	Driving dog	1		
8	EAC 7107	Coupling	1		
9	FS 110251 J	Setscrew-pump to adaptor	3		
10	EAC 3968	Adaptor	3		
11	FS 108251 J	Setscrew-adaptor to engine	3		

MASTER INDEX

GROUP INDEX

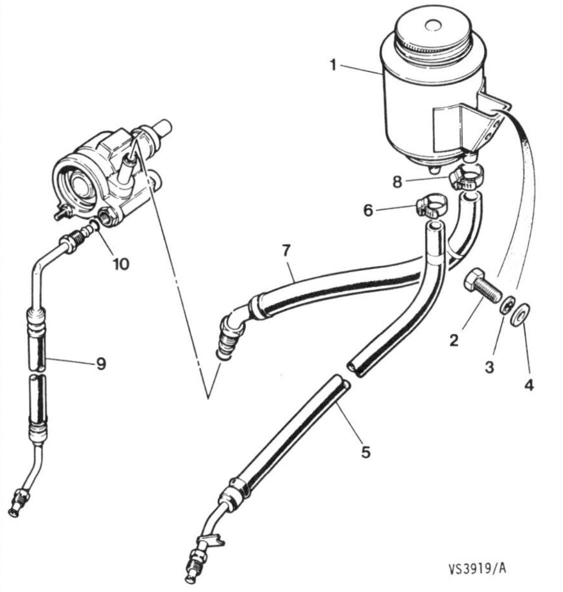

VS3919A

PAS OIL PIPES 3.6 LITRE

Illus	Part Number	123456 Description	Quantity	Change Point	Remarks
1	CAC 6853	Oil reservoir	1		
	CBC 5307	Support bracket	1		
2	SH 604041 J	Setscrew	2		Bracket to valance
3	WL 600041 J	Spring washer	2		
4	FW 104 T	Plain washer	1		Reservoir to valance
	NH 604041 J	Nut	1		
	SH 604051 J	Setscrew	1		Reservoir to bracket
	FW 104 T	Plain washer	1		
5	CBC 5357	Hose-LP reservoir to stg rack	1		RHD
	CBC 2616	Hose-Lp reservoir to stg rack	1		LHD
6	C 43206 1	Hose clip	1		
7	CBC 4918	Hose-reservoir to pump LP	1		
8	C 43206 2	Hose clip	1		
9	CAC 5509	Hose HP pump to stg rack	1		RHD
	CAC 6941	Hose HP pump to stg rack	1		LHD
10	CAC 3589 2	'O' ring	1		
	CAC 2522	Double pipe clip	2		Clipping to susp Crossmember
	C 31304 4	Taptite screw	2		

MASTER INDEX

GROUP INDEX

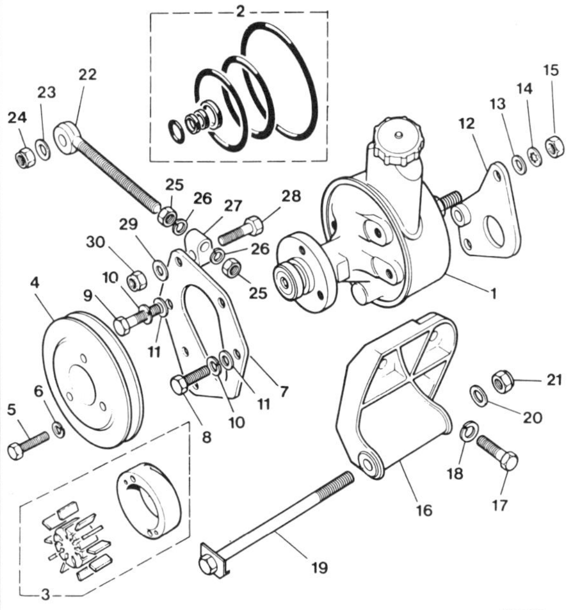

VS1490

STEERING PUMP 5.3 LITRE.

Illus	Part Number	123456 Description	Quantity	Change Point	Remarks
1	EAC 3167	Steering pump	1		
	37H 7627 J	Filler cap	1		
2	518564 J	Seals service kit	1		
3	JLM 1331	Rotor service kit	1		
4	C 45551	Pulley	1		
5	SH 605051 J	Setscrew-pulley to pump	3		
6	WM 600051 J	Lockwasher	3		
7	EAC 3170	Front support plate	1		
8	SH 110281 J	Setscrew-front plate	2		
9	BH 110071 J	Bolt-front plate	2		
10	WL 110001 J	Lockwasher-to pump	4		
11	WC 110061 J	Washer-to pump	4		
12	EAC 3171	Rear support plate	1		
13	WC 110061 J	Washer-rear plate to pump	2		
14	WF 110001 J	Washer-rear plate to pump	2		
15	NH 110041 J	Nut-rear plate to pump	2		
16	C 36189	Mounting bracket	1		
17	BH 507141 J	Bolt-bracket to engine	2		
18	WM 600071 J	Washer-bracket to engine	2		
19	C 36217	Pivot bolt-pump to bracket	1		
20	WA 110061 J	Washer-pump to bracket	1		
21	NY 606041 J	Nut-pump to bracket	1		
22	C 34241	Adjusting bolt	1		
23	WA 106041 J	Washer-bolt to engine	1		
24	NY 605041 J	Nut-bolt to engine	1		
25	NH 606041 J	Nut-bolt to sleeve	2		
26	WM 600061 J	Washer-bolt to sleeve	2		
27	C 34006	Sleeve	1		
28	BH 606121 J	Bolt-sleeve to bracket	1		
29	WA 110061 J	Washer-sleeve to bracket	1		
30	NY 606041 J	Nut-sleeve to bracket	1		

MASTER INDEX

ENGINE 1.C
FLYWHEEL AND CLUTCH 1.G
GEARBOX AND PROPSHAFT 1.H
AXLES, SUSPENSION, DRIVE SHAFTS, WHEELS 1.K
STEERING 1.M
BRAKES AND BRAKE CONTROLS 1.N
FUEL, EXHAUST AND EMISSION SYSTEMS 2.C
COOLING, HEATING AND AIR CONDITIONING 2.G
ELECTRICAL, WASHERS-WIPERS, INSTRUMENTS 2.J
CHASSIS, SUBFRAMES, BODYSHELL, FITTINGS 3.C
FASCIA, TRIM, SEATS AND FIXINGS 3.H
ACCESSORIES AND PAINT 4.C
NUMERICAL INDEX 4.D

GROUP INDEX

RACK AND PINION 3.6 LITRE M06
BALL JOINT KIT M06
BELLOW M06
SEALS KIT M06
RACK AND PINION 5.3 LITRE M07
RACK AND PINION MOUNTINGS 3.6 LITRE M08
STEERING PUMP ASSEMBLY M11
STEERING LOCK & LIGHT SWITCH M04
STEERING LOWER COLUMN 3.6 & 5.3 LITRE M05
STEERING PUMP 5.3 LITRE M13
STEERING WHEEL AND SHAFT M02
UPPER COLUMN & MOUNTING M03

1M 14

JAGUAR XJS RANGE (JAN 1987 ON)	M14 fiche 1	STEERING

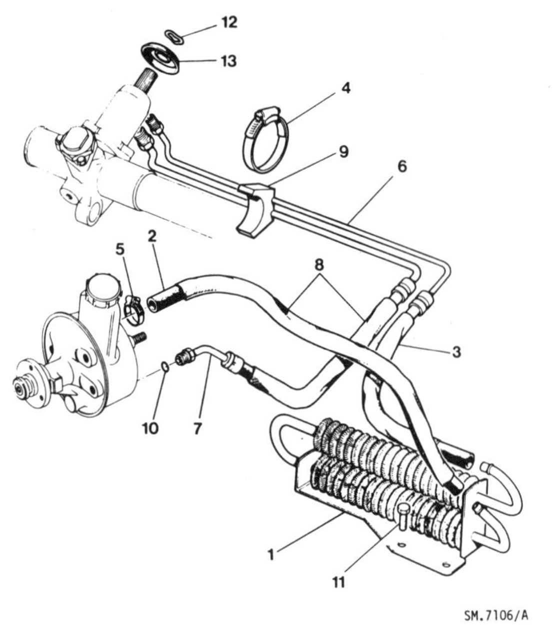

SM7106/A

STEERING 5.3 LITRE

PRESSURE HOSES & OIL COOLER

Illus	Part Number	123456 Description	Quantity	Change Point	Remarks
1	C 43719	PAS oil cooler	1		
2	C 43604 1	LP hose cooler to pump	1		
3	C 33139 1	Hose sleeve	1		
4	C 43206 1	Hose clip	2		Hose to cooler & pump LP hose to cooler
5	C 43206 1	Hose clip	1		
6	C 45906	LP hose assy RHD	1		Not Japan
7	CAC 3654	HP hose assy RHD	1		Not Japan
	CAC 3014	LP hose assy RHD	1		Japan only
	CAC 3656	HP Hose assy RHD	1		Japan only
8	C 33139 4	Sleeve LP and HP hoses	2	clipping PAS Hose to rack	Not Japan
9	C 45907	Packing block	2		Not Japan
10	CAC 3589 2	"O" Ring	1		
11	PFS 125 6E	Setscrew	2		Oil Cooler to Eng Mntg
12	C 45068	Wave washer	1		
13	C 45066	Heatshield, Pionion seal	1		
	C 38894	Double pipe clip	2		Japan only
	C 31304 4	Taptite screw	2		
	C 3169 5	Hose strap	1		
	C 43722	LP hose assy LHD	1		
	CAC 3655	HP hose assy LHD	1		
	C 33139 4	Sleeve for LP hose & HP hose	2		
	C 38894	Double pipe clip	1		
	C 1040 32	Single pipe clip	1		Clipping HP hose to IFS cross beam
	C 31304 4	Taptite screw	2		

MASTER INDEX

GROUP INDEX

1N 01

JAGUAR XJS RANGE (JAN 1987 ON)	N01 fiche 1	BRAKES AND BRAKE CONTROLS

MASTER INDEX

GROUP INDEX

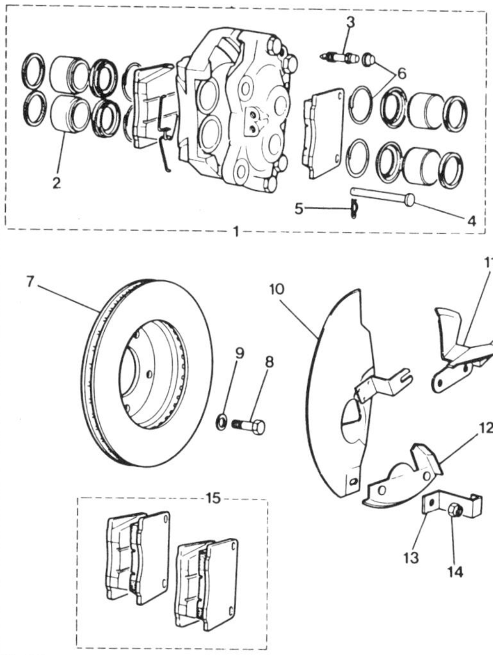

VS 4287

FRONT CALIPERS & DISCS

3.6 & 5.3 LITRE

Illus	Part Number	Description	Quantity	Change Point	Remarks
1	AAU 2102	Caliper front brake-RH	1		
	AAU 2103	Caliper front brake-LH	1		
2	RTC 1119	Piston caliper	4		
3	RTC 1115 J	Screw bleed	1		
4	12798	Pin retaining	2		
5	11369	Clip retaining pin	2		
6	RTC 1116	KIT SERVICE-FT CALIPER	1		
7	JLM 729	Disc brake kit - Axle set	1		Use JLM776
	JLM 727	Disc brake kit	1		Use JLM776
	JLM 776	Disc brake kit	1		
8	C 32533	Bolt 0.4375 UFN disc to hub	10		
9	WL 600071 J	Washer spring 0.4375 ID	10		
10	C 44198	Shield disc-RH	1		Use CBC 6466
	C 44199	Shield disc-LH	1		Use CBC 6467
	CBC 6466	Shield disc-RH	1		
	CBC 6467	Shield disc-LH	1		
11	C 44194	Bracket assembly shield and hose-RH	1		
	C 44195	Bracket assembly shield and hose-LH	1		
	CBC 5652	Bracket assy-shield & hose RH	1		1988 1/2 MY
	CBC 5653	Bracket assy-shield & hose LH	1		1988 1/2 MY
12	C 41606	Plate attachment-RH	1		
	C 41607	Plate attachment-LH	1		
	CBC 5190	Plate attachment - RH	1		1988 1/2 MY
	CBC 5191	Plate attachment - LH	1		1988 1/2 MY
13	C 30064	Clamp shield	4		
14	C 8737 1	Nut nyloc 0.25 UFN	4		
15	JLM 536	PADS FRONT BRAKE FERODO	1		Use JLM1157
	JLM 1157	Pads car set Jurid	1		
	JLM 1510	Pads front axle set Jurid	1		

MASTER INDEX

GROUP INDEX

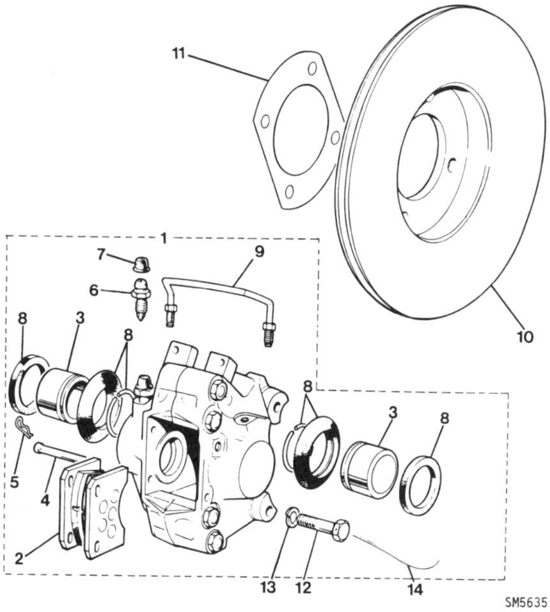

SM 5635

REAR CALIPERS & DISCS

CALIPERS AND DISCS
3.6 & 5.3 LITRE

Illus	Part Number	Description	Quantity	Change Point	Remarks
1	AAU 3378	Rear brake caliper RH	1		
	AAU 3379	Rear brake caliper LH	1		
2	JLM 9728	BRAKE PAD REAR	1		Use JLM 1157
	JLM 1157	Brake pad kit - car set Jurid	1		
	JLM 1513	Brake pad kit rear, axle set Jurid	1		
3	AAU 3377	Piston brake	2		
4	11368	Pin	2		
5	11369	Clip	2		
6	108756	Bleed screw	1		
7	1675	Dust cap	1		
8	AAU 3380	SEAL KIT	1		
9	AEU 1068	Pipe bridge	1		
10	JLM 731	Disc damped	2		
11	C 33440	Shim rear disc 0.020	A/R		
	C 33440 1	Shim rear disc 0.005	A/R		
	C 33440 2	Shim rear disc 0.010	A/R		
12	C 30251	Bolt 0.4375" UNF caliper to diff	4		
13	EAW 3502	Washer spring	4		
14	C 30251	Lockwire 12"	A/R		

G4.00AGW27/7/88

MASTER INDEX

ENGINE 1.C
FLYWHEEL AND CLUTCH 1.G
GEARBOX AND PROPSHAFT 1.H
AXLES, SUSPENSION, DRIVE SHAFTS, WHEELS 1.K
STEERING 1.M
BRAKES AND BRAKE CONTROLS 1.N
FUEL, EXHAUST AND EMISSION SYSTEMS 2.C
COOLING, HEATING AND AIR CONDITIONING 2.G
ELECTRICAL, WASHERS-WIPERS, INSTRUMENTS 2.J
CHASSIS, SUBFRAMES, BODYSHELL, FITTINGS 3.C
FASCIA, TRIM, SEATS AND FIXINGS 3.H
ACCESSORIES AND PAINT 4.C
NUMERICAL INDEX 4.D

GROUP INDEX

ACTUATION ASSY AND RESERVOIR TEVES N11
BRAKE AND CLUTCH PEDALS - TEVES BRAKES N07
BRAKE AND CLUTCH PEDALS 3.6 LITRE N04
BRAKE AND CLUTCH PEDALS 3.6 LITRE N05
SWITCH ASSEMBLY-STOPLIGHT N05
BRAKE PEDAL & HOUSING 5.3 LITRE N09
CLUTCH MASTER CYLINDER 3.6 LITRE N06
FRONT BRAKE PIPES-RHD 5.3 LITRE O02
FRONT BRAKE PIPES-TEVES 3.6 & 5.3 LITRE O05
FRONT CALIPERS & DISCS N02
KIT SERVICE-FT CALIPER N02
PADS FRONT BRAKE FERODO N02
MOTOR PUMP & ACCUMULATOR (TEVES BRAKES) N12
REAR CALIPERS & DISCS N03
BRAKE PAD REAR N03
SEAL KIT N03
SERVO AND MASTER CYLINDER N10
KIT MINOR SERVICE BRAKE SERVO N10
KIT REPAIR NON RETURN VALVE N10
KIT SERVICE BRAKE M/CYLINDER N10

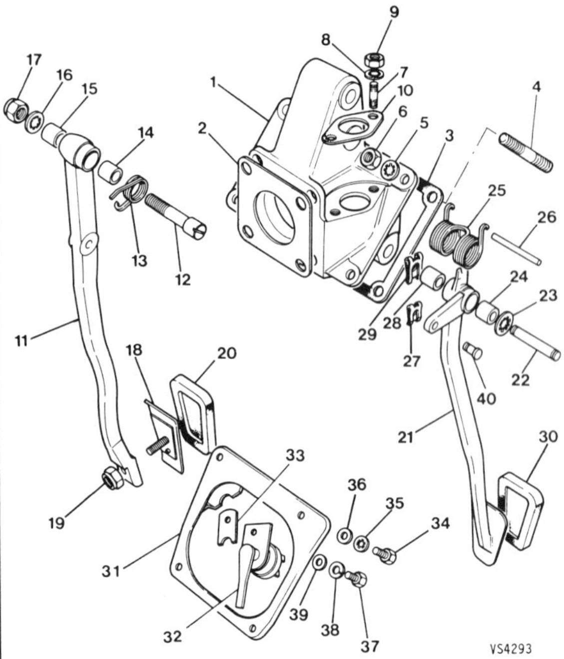

VS 4293

BRAKE AND CLUTCH PEDALS 3.6 LITRE

Illus	Part Number	123456 Description	Quantity	Change Point	Remarks
1	CAC 8471	Housing clutch and brake pedal	1		Speed control only
2	C 39597	Seal-brake booster & m/cyl	1		
3	C 44896	Gasket-base joint	1		
4	CBC 6083	Stud 0.3125"unf	4		
5	C 725	Washer-shakeproof	4		
6	JLM 9684	Nut 0.3125"unf	4		
7	C 8956	Stud-m/cyl to clutch housing	2		
8	C 725	Washer-shakeproof	2		
9	JLM 9684	Nut 0.3125"unf	2		
10	C 32202	Shim m/cyl to clutch housing	A/R		
11	CBC 5504	Pedal-brake assembly RHD	1		Manual
	CBC 5505	Pedal-brake assembly LHD	1		Manual
	CBC 5502	Pedal-brake assembly-RHD	1		Auto
	CBC 5503	Pedal-brake assembly-LHD	1		Auto
12	CAC 4347	Pin-pivot-brake pedal	1		
13	C 24689	Spring-return brake pedal	1		
14	C 39619	Bush-brake pedal	1		
15	C 39619	Bush-brake pedal	1		
15	C 39619	Bush-brake pedal	1		
16	JLM 300	Washer-shakeproof	1		
17	C 8737 3	Nut-self locking	1		
18	CAC 5355	Pad-brake pedal	1		
19	C 8737 2	Nut-self locking	1		
	FW 105 T	Washer	1		
20	CAC 5609	Rubber-pedal pad	1		

1N 04

MASTER INDEX

GROUP INDEX

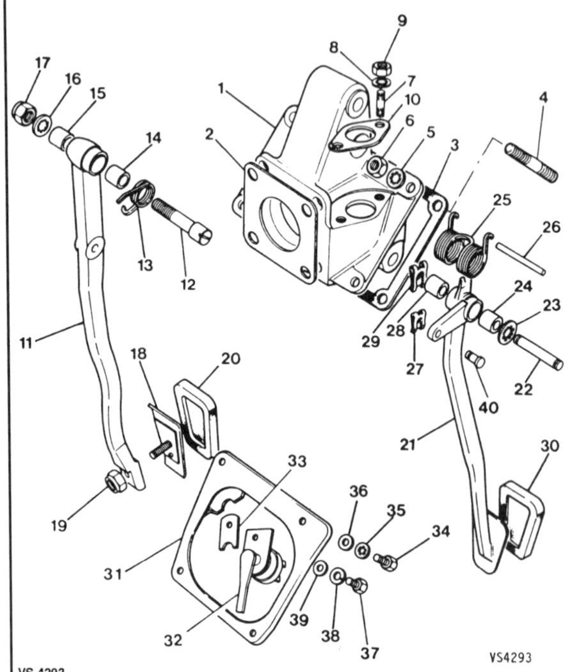

VS 4293

BRAKE AND CLUTCH PEDALS 3.6 LITRE

Illus	Part Number	123456 Description	Quantity	Change Point	Remarks
21	CAC 5613	Pedal-clutch-RHD	1		
	CAC 5612	Pedal-clutch-LHD	1		
22	C 39624	Pin-pivot	1		
23	WF 600081 J	Washer-shakeproof	1		
24	C 39619	Bush-clutch pedal	1		
25	C 39515	Spring-return-clutch pedal	1		
26	C 20725 40	Pin-pivot-clutch pedal ret spring	1		
27	C 27328 3Z	Clip-safety	1		
28	C 39619	Bush-clutch pedal	1		
29	C 27328 6Z	Clip-safety	1		
30	CAC 5608	Rubber-clutch pedal	1		
31	CBC 4176	Retainer assembly-pedal boot RHD	1		
	CBC 4175	Retainer assembly-pedal boot LHD	1		
	CBC 6044	Pedal boot retainer assy LHD	1		Manual G/Box
32	DBC 2134	SWITCH ASSEMBLY-STOPLIGHT	1		
33	C 43829	Plate-tapped-switch to MTG bkt	1		
34	UFS 119 5R	Setscrew No.10 x 0.625	4		
35	WF 702101 J	Washer-shakeproof	4		
36	WC 702101 J	Washer-0.500 x 0.202 x 0.048	4		
37	SH 604041 J	Setscrew 0.25 UNF x 0.50"	1		
38	JLM 302	Washer-spring 0.25 ID	1		
39	JLM 298	Washer-plain	1		
40	J 205 8S	Joint pin	1		

1N 05

MASTER INDEX

ENGINE 1.C
FLYWHEEL AND CLUTCH 1.G
GEARBOX AND PROPSHAFT 1.H
AXLES, SUSPENSION, DRIVE SHAFTS, WHEELS 1.K
STEERING 1.M
BRAKES AND BRAKE CONTROLS 1.N
FUEL, EXHAUST AND EMISSION SYSTEMS 2.C
COOLING, HEATING AND AIR CONDITIONING 2.G
ELECTRICAL, WASHERS-WIPERS, INSTRUMENTS 2.J
CHASSIS, SUBFRAMES, BODYSHELL, FITTINGS 3.C
FASCIA, TRIM, SEATS AND FIXINGS 3.H
ACCESSORIES AND PAINT 4.C
NUMERICAL INDEX 4.D

GROUP INDEX

ACTUATION ASSY AND RESERVOIR TEVES N11
BRAKE AND CLUTCH PEDALS - TEVES BRAKES N07
BRAKE AND CLUTCH PEDALS 3.6 LITRE N04
BRAKE AND CLUTCH PEDALS 3.6 LITRE N05
SWITCH ASSEMBLY-STOPLIGHT N05
BRAKE PEDAL & HOUSING 5.3 LITRE N09
CLUTCH MASTER CYLINDER 3.6 LITRE N06
FRONT BRAKE PIPES-RHD 5.3 LITRE O02
FRONT BRAKE PIPES-TEVES 3.6 & 5.3 LITRE O05
FRONT CALIPERS & DISCS N02
KIT SERVICE-FT CALIPER N02
PADS FRONT BRAKE FERODO N02
MOTOR PUMP & ACCUMULATOR (TEVES BRAKES) N12
REAR CALIPERS & DISCS N03
BRAKE PAD REAR N03
SEAL KIT N03
SERVO AND MASTER CYLINDER N10
KIT MINOR SERVICE BRAKE SERVO N10
KIT REPAIR NON RETURN VALVE N10
KIT SERVICE BRAKE M/CYLINDER N10

JAGUAR XJS RANGE (JAN 1987 ON)

N06 fiche 1

BRAKES AND BRAKE CONTROLS

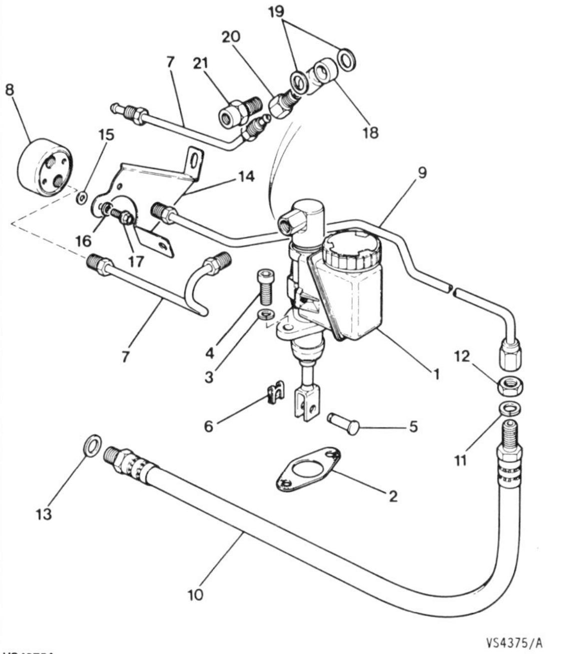

VS4375A

CLUTCH MASTER CYLINDER 3.6 LITRE

Illus	Part Number	Description	Quantity	Change Point	Remarks
1	CAC 8091	Master cylinder clutch	1		
2	C 32202	Shim	A/R		
3	WL 106001 J	Washer spring	2		
4	SS 505066 J	Screw-socket head	2		
5	J 205 8S	Pin-joint	1		
6	C 27328	Clip-safety	1		
7	CAC 8324	Pipe-clutch hydraulic LHD	1		Non speed control
7	CAC 9475	Pipe-clutch hydraulic RHD	1		Speed control
8	AAU 8008	Damper-clutch	1		
8	CAC 9508	Damper-clutch	1		Speed control
9	CAC 8106	Pipe-clutch-hydraulic-RHD	1		
	CAC 7461	Pipe-clutch-hydraulic-LHD	1		
	CAC 9476	Clutch pipe assembly-LHD	1		Banjo to B/head Flex hose
10	CAC 7876	Hose flexible assembly	1		
11	WE 112001 J	Washer spring	1		
12	NT 212041 J	Nut	1		
13	C 33622	Washer hose to bracket	1		
14	CAC 7455	Bracket-clutch damper mtg	1		
15	JLM 299	Washer	2		
16	WL 106001 J	Washer spring	2		
17	AGU 1198	Setscrew	2		
	BHM 7127 J	Seals kit-m/cyl	1		
18	CAC 9477	Banjo	1		Speed control (items 18–21)
19	CAC 9621 1	Washer copper	2		
20	CAC 9479	Adaptor	1		
21	CAC 9478	Banjo bolt switch adaptor	1		

MASTER INDEX

GROUP INDEX

JAGUAR XJS RANGE (JAN 1987 ON)

N07 fiche 1

BRAKES AND BRAKE CONTROLS

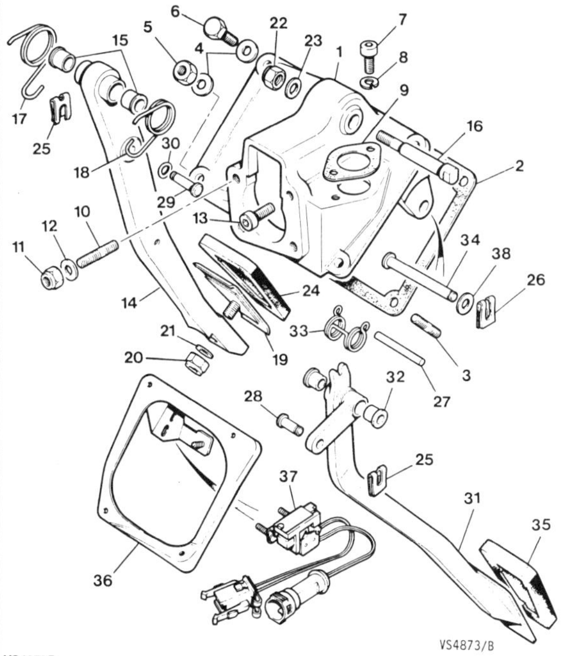

VS4873B

BRAKE AND CLUTCH PEDALS - TEVES BRAKES

3.6 &5.3 LITRE

Illus	Part Number	Description	Quantity	Change Point	Remarks
1	CBC 2141	Booster and Pedal Housing	1		Manual
	CBC 2142	Booster and Pedal Housing	1		Auto
2	C 44896	Base Joint	1		
3	CBC 6083	Stud Housing to Body	2		
4	WF 600050 J	Washer - Shakeproof	4		
5	JLM 9684	Nut	2		
6	JLM 9563	Setscrew	2		
7	SS 505066 J	Setscrew - Sockethead	2		
8	FG 105 X	Spring Washer	2		
9	C 32202	Shim, M/Cyl Mounting	1		
10	TE 108061 J	Stud	3		
11	NY 108041 J	Locknut	3		
12	FW 105 T	Washer	4		
13	SS 108250	Socket Head Screw	1		
14	CBC 7209	Brake Pedal Assy RHD	1		Manual
	CBC 5759	Brake Pedal Assy LHD	1		Manual
	CBC 7210	Brake Pedal Assy RHD	1		Auto
	CBC 5758	Brake Pedal Assy LHD	1		
15	CAC 1742	Bush, Pedal	2		
16	CBC 4681	Pivot Pin, Brake Pedal	1		
17	CBC 5253	Pedal Return Spring	1		
18	CBC 5255	Pedal Return Spring	1		
19	CAC 5355	Pedal Pad (Metal)	1		Manual
20	C 8737 2	Locknut, Pedal Pad	1		Manual
21	FW 105 T	Washer, Pedal Pad	1		Manual
22	NY 110041 J	Locknut, Pivot Pin	1		
23	FW 106 E	Washer Pivot Pin	1		

G6.06AW11/7/88

MASTER INDEX
ENGINE ... 1.C
FLYWHEEL AND CLUTCH ... 1.G
GEARBOX AND PROPSHAFT ... 1.H
AXLES, SUSPENSION, DRIVE SHAFTS, WHEELS ... 1.K
STEERING ... 1.M
BRAKES AND BRAKE CONTROLS ... 1.N
FUEL, EXHAUST AND EMISSION SYSTEMS ... 2.C
COOLING, HEATING AND AIR CONDITIONING ... 2.G
ELECTRICAL, WASHERS-WIPERS, INSTRUMENTS ... 2.J
CHASSIS, SUBFRAMES, BODYSHELL, FITTINGS ... 3.C
FASCIA, TRIM, SEATS AND FIXINGS ... 3.H
ACCESSORIES AND PAINT ... 4.C
NUMERICAL INDEX ... 4.D
GROUP INDEX
ACTUATION ASSY AND RESERVOIR TEVES ... N11
BRAKE AND CLUTCH PEDALS - TEVES BRAKES ... N07
BRAKE AND CLUTCH PEDALS 3.6 LITRE ... N04
BRAKE AND CLUTCH PEDALS 3.6 LITRE ... N05
SWITCH ASSEMBLY-STOPLIGHT ... N05
BRAKE PEDAL & HOUSING 5.3 LITRE ... N09
CLUTCH MASTER CYLINDER 3.6 LITRE ... N06
FRONT BRAKE PIPES-RHD 5.3 LITRE ... O02
FRONT BRAKE PIPES-TEVES 3.6 & 5.3 LITRE ... O05
FRONT CALIPERS & DISCS ... N02
KIT SERVICE-FT CALIPER ... N02
PADS FRONT BRAKE FERODO ... N02
MOTOR PUMP & ACCUMULATOR (TEVES BRAKES) ... N12
REAR CALIPERS & DISCS ... N03
BRAKE PAD REAR ... N03
SEAL KIT ... N03
SERVO AND MASTER CYLINDER ... N10
KIT MINOR SERVICE BRAKE SERVO ... N10
KIT REPAIR NON RETURN VALVE ... N10
KIT SERVICE BRAKE M/CYLINDER ... N10

1N 08

JAGUAR XJS RANGE (JAN 1987 ON) | N08 fiche 1 | BRAKES AND BRAKE CONTROLS

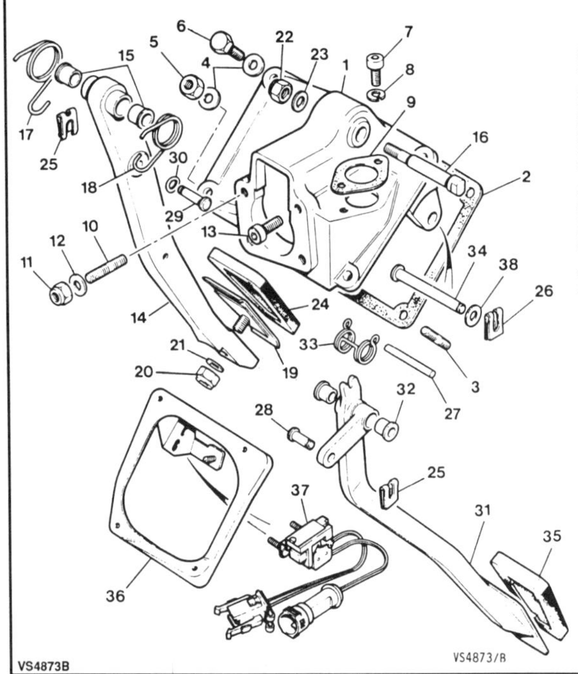

VS4873B

BRAKE AND CLUTCH PEDALS-TEVES BRAKES

3.6 & 5.3 LITRE

Illus	Part Number	123456 Description	Quantity	Change Point	Remarks
24	CAC 9760	Pedal Pad Rubber, Brake	1		
25	C 27328 3Z	Safety Clip	2		Manual
	C 27328 3Z	Safety Clip	1		Auto
26	C 27328 6	Safety Clip Clutch Pivot	1		Manual
27	C 20725 40	Roll Pin	1		Manual
28	J 205 8S	Joint Pin	1		
29	CBC 2600	Joint Pin	1		
30	WW 108001 J	Wave Spring Washer	1		
31	CAC 5613	Clutch Pedal Assy	1		Manual
	CAC 5612	Clutch Pedal Assy	1		Manual
32	C 39619	Bush	2		
33	C 39515	Return Spring - Clutch Pedal	1		
34	C 39624	Pivot Pin, Clutch Pedal	1		
35	CAC 9759	Pedal Pad Rubber Clutch	1		
36	CBC 7216	Pedal Boot Retainer Assy RHD	1		
	CBC 7217	Pedal Boot Retainer LHD	1		
37	DAC 6134	Brake/speed con.switch/mtg plate	1		
38	C 30993	Washer, Pivot Pin, Clutch	1		
	UFS 119 5R	Setscrew	4		Retainer to Body
	WC 702101 J	Plain Washer	4		Retainer to Body
	WF 702100 J	Shakeproof Washer	4		Retainer to Body
	AGU 2414	Cable Clamp	1		
	NH 104041 J	Nut	1		
	WC 703081 J	Washer	1		

G6.08AW11/7/88

MASTER INDEX

GROUP INDEX

1N 09

JAGUAR XJS RANGE (JAN 1987 ON) | N09 fiche 1 | BRAKES AND BRAKE CONTROLS

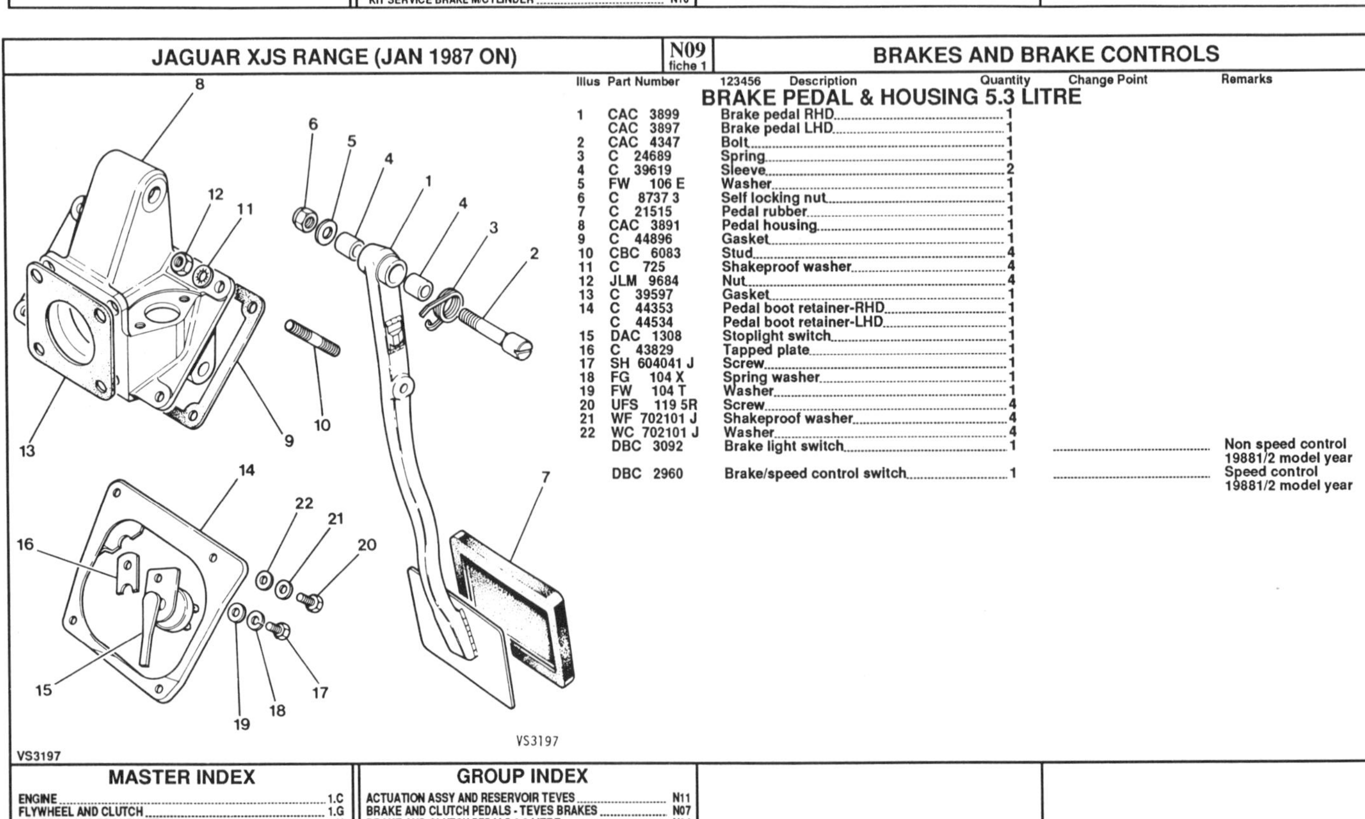

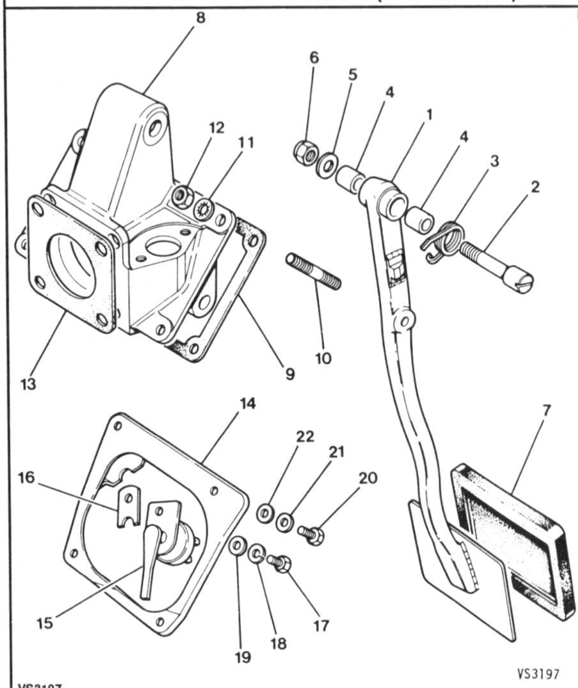

VS3197

BRAKE PEDAL & HOUSING 5.3 LITRE

Illus	Part Number	123456 Description	Quantity	Change Point	Remarks
1	CAC 3899	Brake pedal RHD	1		
	CAC 3897	Brake pedal LHD	1		
2	CAC 4347	Bolt	1		
3	C 24689	Spring	1		
4	C 39619	Sleeve	2		
5	FW 106 E	Washer	1		
6	C 8737 3	Self locking nut	1		
7	C 21515	Pedal rubber	1		
8	CAC 3891	Pedal housing	1		
9	C 44896	Gasket	1		
10	CBC 6083	Stud	4		
11	C 725	Shakeproof washer	4		
12	JLM 9684	Nut	4		
13	C 39597	Gasket	1		
14	C 44353	Pedal boot retainer-RHD	1		
	C 44534	Pedal boot retainer-LHD	1		
15	DAC 1308	Stoplight switch	1		
16	C 43829	Tapped plate	1		
17	SH 604041 J	Screw	1		
18	FG 104 X	Spring washer	1		
19	FW 104 T	Washer	1		
20	UFS 119 5R	Screw	4		
21	WF 702101 J	Shakeproof washer	4		
22	WC 702101 J	Washer	4		
	DBC 3092	Brake light switch	1		Non speed control 19881/2 model year
	DBC 2960	Brake/speed control switch	1		Speed control 19881/2 model year

MASTER INDEX

ENGINE 1.C
FLYWHEEL AND CLUTCH 1.G
GEARBOX AND PROPSHAFT 1.H
AXLES, SUSPENSION, DRIVE SHAFTS, WHEELS 1.K
STEERING 1.M
BRAKES AND BRAKE CONTROLS 1.N
FUEL, EXHAUST AND EMISSION SYSTEMS 2.C
COOLING, HEATING AND AIR CONDITIONING 2.G
ELECTRICAL, WASHERS-WIPERS, INSTRUMENTS 2.J
CHASSIS, SUBFRAMES, BODYSHELL, FITTINGS 3.C
FASCIA, TRIM, SEATS AND FIXINGS 3.H
ACCESSORIES AND PAINT 4.C
NUMERICAL INDEX 4.D

GROUP INDEX

ACTUATION ASSY AND RESERVOIR TEVES N11
BRAKE AND CLUTCH PEDALS - TEVES BRAKES N07
BRAKE AND CLUTCH PEDALS 3.6 LITRE N04
BRAKE AND CLUTCH PEDALS 3.6 LITRE N05
SWITCH ASSEMBLY-STOPLIGHT N05
BRAKE PEDAL & HOUSING 5.3 LITRE N09
CLUTCH MASTER CYLINDER 3.6 LITRE N06
FRONT BRAKE PIPES-RHD 5.3 LITRE O02
FRONT BRAKE PIPES-TEVES 3.6 & 5.3 LITRE O05
FRONT CALIPERS & DISCS N02
KIT SERVICE-FT CALIPER N02
PADS FRONT BRAKE FERODO N02
MOTOR PUMP & ACCUMULATOR (TEVES BRAKES) N12
REAR CALIPERS & DISCS N03
BRAKE PAD REAR N03
SEAL KIT N03
SERVO AND MASTER CYLINDER N10
KIT MINOR SERVICE BRAKE SERVO N10
KIT REPAIR NON RETURN VALVE N10
KIT SERVICE BRAKE M/CYLINDER N10

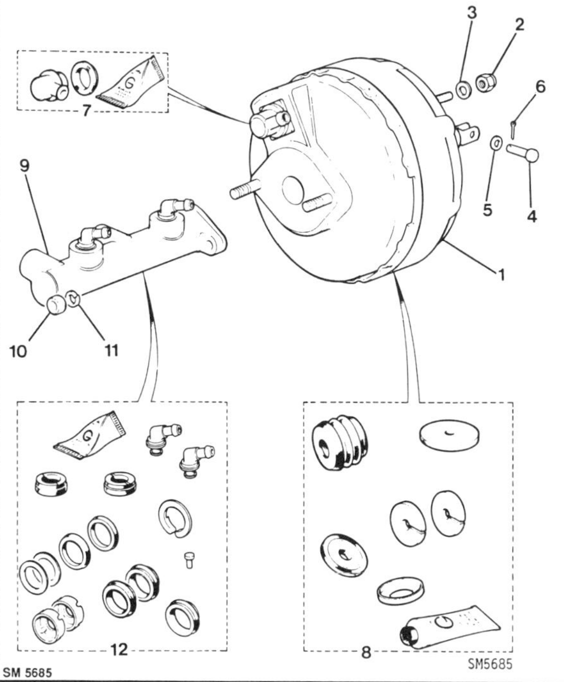

SM 5685

SERVO AND MASTER CYLINDER

Illus	Part Number	Description	Quantity	Change Point	Remarks
1	CAC 1227	Servo unit	1		
2	NY 108041 J	Nut nyloc 8mm	4		
3	JLM 299	Washer plain 0.312"	4		
4	J 10517 S	Pin joint	1		
5	JLM 299	Washer plain 0.312"	1		
6	2390	Pin split	1		
7	RTC 1128	KIT REPAIR NON RETURN VALVE	1		
8	12446	KIT MINOR SERVICE BRAKE SERVO	1		
9	CAC 1582	Master cylinder brake-RHD	1		
	CAC 1583	Master cylinder brake-LHD	1		
10	JLM 9689	Nut m/cylinder to servo	2		
11	AAU 6840	Washer m/cylinder to servo	2		
12	AAU 6839	KIT SERVICE BRAKE M/CYLINDER	1		
	CAC 7936	Seal-m/cylinder to servo unit	1		

MASTER INDEX

GROUP INDEX

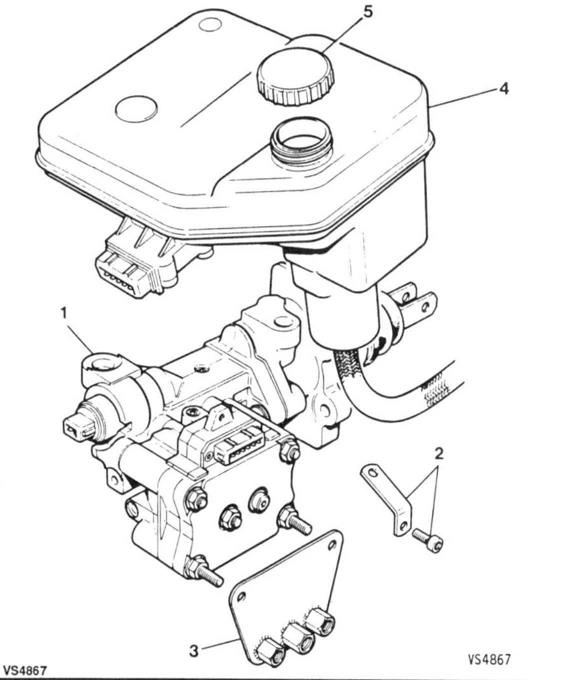

VS4867

ACTUATION ASSY AND RESERVOIR TEVES

3.6 & 5.3 LITRE

Illus	Part Number	Description	Quantity	Change Point	Remarks
1	JLM 1476	Actuation Assy Kit RHD	1		
2	JLM 1468	Bracket Kit RHD	1		
		Screw Reservoir Fixing	1		
	JLM 1475	Actuation Assy Kit LHD	1		
	JLM 1467	Bracket Kit LHD	1		
		Screw Reservoir Fixing	1		
3	CBC 5019	Connector Plate Assy	1		
4	JLM 1583	Reservoir Kit - RHD	1		
	JLM 1584	Reservoir Kit - LHD	1		
		Seal Reservoir to Booster	2		Part of Seal Kit JLM 1477
		Screw Reservoir Fixing	1		Part of Seal Kit JLM 1477
5	JLM 1469	Filler Cap (Reservoir)	1		

MASTER INDEX
ENGINE 1.C
FLYWHEEL AND CLUTCH 1.G
GEARBOX AND PROPSHAFT 1.H
AXLES, SUSPENSION, DRIVE SHAFTS, WHEELS 1.K
STEERING 1.M
BRAKES AND BRAKE CONTROLS 1.N
FUEL, EXHAUST AND EMISSION SYSTEMS 2.C
COOLING, HEATING AND AIR CONDITIONING 2.G
ELECTRICAL, WASHERS-WIPERS, INSTRUMENTS 2.J
CHASSIS, SUBFRAMES, BODYSHELL, FITTINGS 3.C
FASCIA, TRIM, SEATS AND FIXINGS 3.H
ACCESSORIES AND PAINT 4.C
NUMERICAL INDEX 4.D
GROUP INDEX
ACTUATION ASSY AND RESERVOIR TEVES N11
BRAKE AND CLUTCH PEDALS - TEVES BRAKES N07
BRAKE AND CLUTCH PEDALS 3.6 LITRE N04
BRAKE AND CLUTCH PEDALS 3.6 LITRE N05
SWITCH ASSEMBLY-STOPLIGHT N05
BRAKE PEDAL & HOUSING 5.3 LITRE N09
CLUTCH MASTER CYLINDER 3.6 LITRE N06
FRONT BRAKE PIPES-RHD 5.3 LITRE O02
FRONT BRAKE PIPES-TEVES 3.6 & 5.3 LITRE O05
FRONT CALIPERS & DISCS N02
KIT SERVICE-FT CALIPER N02
PADS FRONT BRAKE FERODO N02
MOTOR PUMP & ACCUMULATOR (TEVES BRAKES) N12
REAR CALIPERS & DISCS N03
BRAKE PAD REAR N03
SEAL KIT N03
SERVO AND MASTER CYLINDER N10
KIT MINOR SERVICE BRAKE SERVO N10
KIT REPAIR NON RETURN VALVE N10
KIT SERVICE BRAKE M/CYLINDER N10

MOTOR PUMP & ACCUMULATOR (TEVES BRAKES)

3.6 & 5.3 LITRE

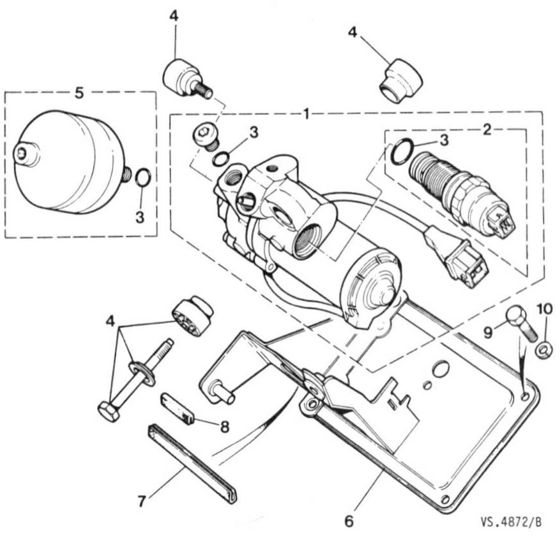

VS4872/B

Illus	Part Number	123456	Description	Quantity	Change Point	Remarks
1	JLM 1473		Motor and Pump Assy LHD	NLA		Use JLM 1663
	JLM 1474		Motor and Pump Assy RHD	NLA		Use JLM 1664
	JLM 1663		Motor,Pump & Pressure Warning Switch Kit LHD	1		
	JLM 1664		Motor,Pump & Pressure Warning Switch Kit RHD	1		
2	JLM 1465		Pressure & warning switch kit	1		
3	NSS		'O' ring	1		Part of seal kit JLM 1477
4	JLM 1464		Insulator kit	1		
	NSS		Insulator bushing	3		
	NSS		Sleeve pump body	2		
	NSS		Sleeve-front of pump	1		
5	JLM 1461		Accumulator kit	1		
3	NSS		"O" ring	1		Part of seal kit JLM 1477
6	CBC 6140		Cover plate-motor and pump	1		RHD 3.6 manual and Auto
	CBC 6141		Cover plate-motor and pump	1		LHD 3.6 manual and Auto
	CBC 6142		Cover plate-motor and pump	1		RHD V12 auto
	CBC 6143		Cover Plate-motor and pump	1		LHD V12 auto
7	BD 44643 12		Edge protection strip	1		
8	BD 44643 1		Flange finisher	1		
	JLM 1477		"O" ring and seals kit	1		
	NSS		"O" ring (accumulator)	1		
	NSS		"O" ring (pressure warning switch)	1		
	NSS		"O" ring (hollow bolt)	3		
	NSS		"O" ring (reservoir to booster)	1		
	NSS		Sleeve (reservoir to booster)	1		
	NSS		Seal (reservoir to booster/m/cyl)	2		
9	SH 605071 J		Bolt	4		
10	WM 600051 J		Spring washer	4		

G10.04AW23/9/88

1N 12

MASTER INDEX

GROUP INDEX

VACUUM TANK RHD 3.6 LITRE

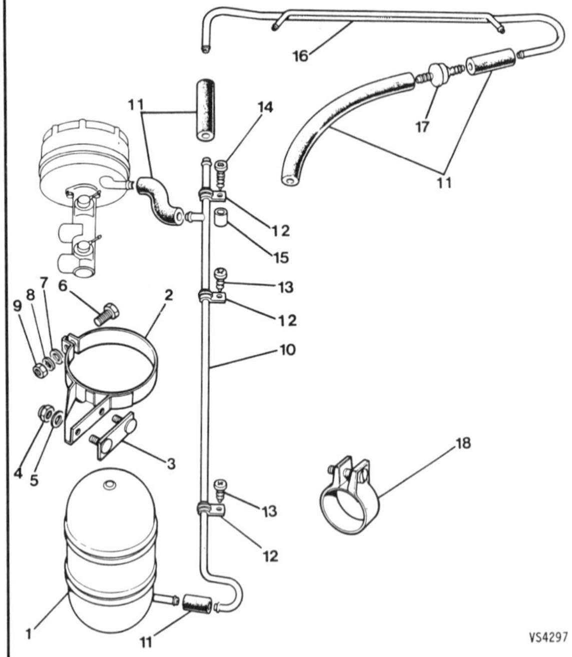

VS 4297

Illus	Part Number	123456	Description	Quantity	Change Point	Remarks
1	C 36471		Vacuum tank assembly	1		
2	C 43375		Mounting-vacuum tank	1		
3	C 43378		Plate-stud	1		
4	JLM 9684		Nut-0.3125 UNF	2		
5	FW 105 E		Washer-plain 0.3125 ID	2		
6	JLM 9675		Setscrew 0.25 x 1"	1		
7	FW 105 T		Washer-plain 0.25	1		
8	C 724		Washer-shakeproof 0.25	1		
9	JLM 9683		Nut-0.25 UNF	1		
10	C 43748		Pipe-vacuum	1		
11	JLM 1455		Hose-vacuum-0.343	5		
12	C 38894		Clip-pipe	3		
13	AB 610031 J		Screw-self tapping no.10 x 0.375	2		
14	JLM 9577		Screw-self tapping no.10 x 0.75	1		
15	C 33620		Spacer	1		
16	CAC 5409		Pipe vacuum assy	1		
17	CAC 3923		Valve non return	1		
18	C 43599 10		Clip hose	8		
	C 43599 12		Clip hose	2		
	C 43689		Clip vacuum pipe	2		

1N 13

MASTER INDEX
ENGINE 1.C
FLYWHEEL AND CLUTCH 1.G
GEARBOX AND PROPSHAFT 1.H
AXLES, SUSPENSION, DRIVE SHAFTS, WHEELS 1.K
STEERING 1.M
BRAKES AND BRAKE CONTROLS 1.N
FUEL, EXHAUST AND EMISSION SYSTEMS 2.C
COOLING, HEATING AND AIR CONDITIONING 2.G
ELECTRICAL, WASHERS-WIPERS, INSTRUMENTS 2.J
CHASSIS, SUBFRAMES, BODYSHELL, FITTINGS 3.C
FASCIA, TRIM, SEATS AND FIXINGS 3.H
ACCESSORIES AND PAINT 4.C
NUMERICAL INDEX 4.D
GROUP INDEX
ACTUATION ASSY AND RESERVOIR TEVES N11
BRAKE AND CLUTCH PEDALS - TEVES BRAKES N07
BRAKE AND CLUTCH PEDALS 3.6 LITRE N04
BRAKE AND CLUTCH PEDALS 3.6 LITRE N05
SWITCH ASSEMBLY-STOPLIGHT N05
BRAKE PEDAL & HOUSING 5.3 LITRE N09
CLUTCH MASTER CYLINDER 3.6 LITRE N06
FRONT BRAKE PIPES-RHD 5.3 LITRE O02
FRONT BRAKE PIPES-TEVES 3.6 & 5.3 LITRE O05
FRONT CALIPERS & DISCS N02
KIT SERVICE-FT CALIPER N02
PADS FRONT BRAKE FERODO N02
MOTOR PUMP & ACCUMULATOR (TEVES BRAKES) N12
REAR CALIPERS & DISCS N03
BRAKE PAD REAR N03
SEAL KIT N03
SERVO AND MASTER CYLINDER N10
KIT MINOR SERVICE BRAKE SERVO N10
KIT REPAIR NON RETURN VALVE N10
KIT SERVICE BRAKE M/CYLINDER N10

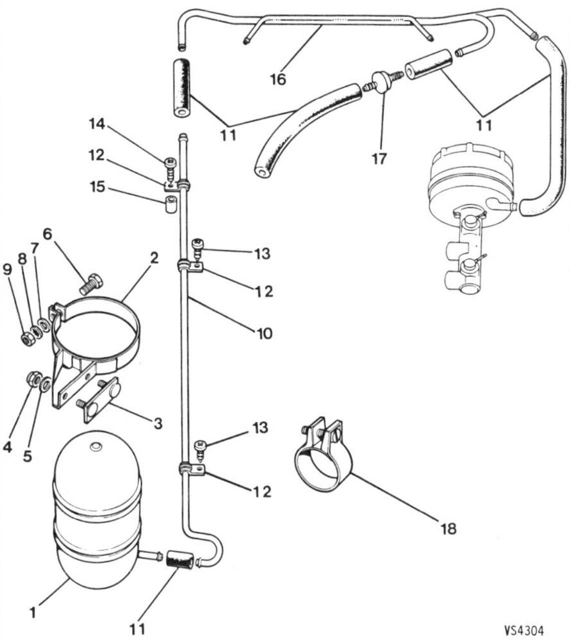

VS4304

VACUUM TANK LHD 3.6 LITRE

Illus	Part Number	123456 Description	Quantity	Change Point	Remarks
1	C 36471	Vacuum tank assembly	1		
2	C 43375	Mounting-vacuum tank	1		
3	C 43378	Plate stud	1		
4	JLM 9684	Nut-0.3125 UNF	2		
5	JLM 299	Washer-plain 0.3125 ID	2		
6	JLM 9675	Setscrew	1		
7	JLM 298	Washer-plain 0.25	1		
8	C 724	Washer-shakeproof 0.25	1		
9	JLM 9683	Nut-0.25 UNF	1		
10	C 43687	Pipe vacuum	1		
11	JLM 1455	Hose-vacuum-0.343	5		
12	C 38894	Clip-pipe	3		
13	AB 612061 J	Screw-self tapping no.10 x 0.375	2		
14	JLM 9577	Screw-self tapping no.10 x 0.75	1		
15	C 33620	Spacer	1		
16	CAC 5408	Pipe-vacuum assembly	1		
	C 43689	Clip-vacuum pipe	2		
17	CAC 4188	Valve-non return valve	1		
18	C 43599 10	Clip-hose	8		
	C 43599 12	Clip-hose	2		

MASTER INDEX

GROUP INDEX

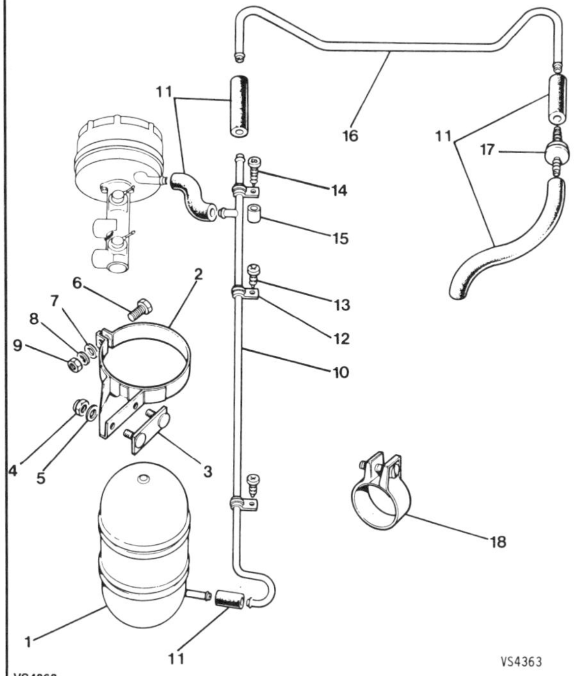

VS4363

VACUUM TANK RHD 5.3 LITRE

Illus	Part Number	123456 Description	Quantity	Change Point	Remarks
1	C 36471	Vacuum tank assy	1		
2	C 43375	Mounting vac tank	1		
3	C 43378	Plate stud	1		
4	JLM 9684	Nut-0.3125 UNF	2		
5	FW 105 E	Washer plain 0.3125 ID	2		
6	JLM 9675	Setscrew 0.25 x 1"	1		
7	FW 105 T	Washer plain	1		
8	C 724	Washer shakeproof	1		
9	JLM 9683	Nut 0.25 UNF			
10	C 43748	Pipe - vacuum	1		
11	JLM 1455	Hose - vacuum - 0.343	5		
12	C 38894	Clip - pipe	3		
13	AB 610031 J	Screw self tapping No 10 x 0.375	2		
14	JLM 9577	Screw self tapping No 10 x 0.75	1		
15	C 33620	Spacer	1		
16	CAC 3714	Pipe - vacuum assy	1		] AUS only
	CAC 1384	Pipe - vacuum assy	1		] AUS only
17	CAC 4188	Valve non return	1		
18	C 43599 10	Clip hose	8		
	C 43599 12	Clip hose	2		

MASTER INDEX

ENGINE	1.C
FLYWHEEL AND CLUTCH	1.G
GEARBOX AND PROPSHAFT	1.H
AXLES, SUSPENSION, DRIVE SHAFTS, WHEELS	1.K
STEERING	1.M
BRAKES AND BRAKE CONTROLS	1.N
FUEL, EXHAUST AND EMISSION SYSTEMS	2.C
COOLING, HEATING AND AIR CONDITIONING	2.G
ELECTRICAL, WASHERS-WIPERS, INSTRUMENTS	2.J
CHASSIS, SUBFRAMES, BODYSHELL, FITTINGS	3.C
FASCIA, TRIM, SEATS AND FIXINGS	3.H
ACCESSORIES AND PAINT	4.C
NUMERICAL INDEX	4.D

GROUP INDEX

ACTUATION ASSY AND RESERVOIR TEVES	N11
BRAKE AND CLUTCH PEDALS - TEVES BRAKES	N07
BRAKE AND CLUTCH PEDALS 3.6 LITRE	N04
BRAKE AND CLUTCH PEDALS 3.6 LITRE	N05
SWITCH ASSEMBLY-STOPLIGHT	N05
BRAKE PEDAL & HOUSING 5.3 LITRE	N09
CLUTCH MASTER CYLINDER 3.6 LITRE	N06
FRONT BRAKE PIPES-RHD 5.3 LITRE	O02
FRONT BRAKE PIPES-TEVES 3.6 & 5.3 LITRE	O05
FRONT CALIPERS & DISCS	N02
KIT SERVICE-FT CALIPER	N02
PADS FRONT BRAKE FERODO	N02
MOTOR PUMP & ACCUMULATOR (TEVES BRAKES)	N12
REAR CALIPERS & DISCS	N03
BRAKE PAD REAR	N03
SEAL KIT	N03
SERVO AND MASTER CYLINDER	N10
KIT MINOR SERVICE BRAKE SERVO	N10
KIT REPAIR NON RETURN VALVE	N10
KIT SERVICE BRAKE M/CYLINDER	N10

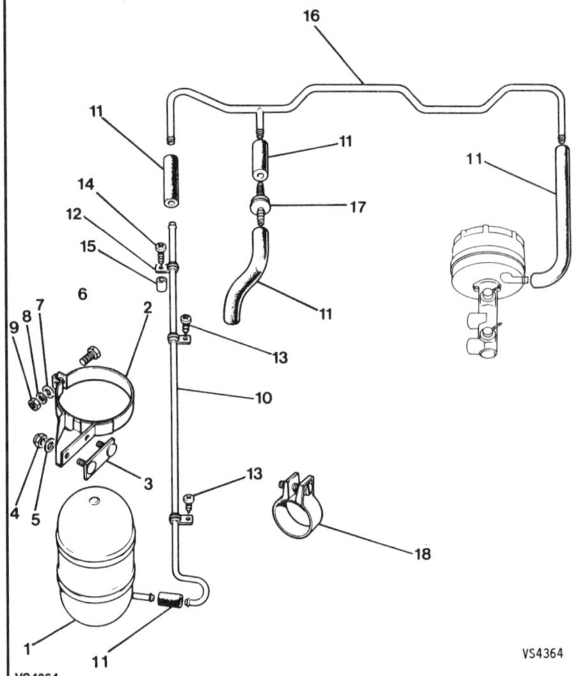

VS4364

VACUUM TANK LHD 5.3 LITRE

Illus	Part Number	Description (123456)	Quantity	Change Point	Remarks
1	C 36471	Vacuum tank assy	1		
2	C 43375	Mounting - vacuum tank	1		
3	C 43378	Plate stud	1		
4	JLM 9684	Nut-0.3125	2		
5	FW 105 E	Washer plain - 0.3125 ID	2		
6	JLM 9675	Setscrew	1		
7	FW 105 T	Washer plain 0.25	1		
8	C 724	Washer shakeproof 0.25	1		
9	JLM 9683	Nut 0.25 UNF	1		
10	C 43687	Pipe vacuum	1		
11	JLM 1455	Hose vacuum 0.343	5		
12	C 38894	Clip pipe	3		
13	AB 612061 J	Screw self tapping	2		
14	JLM 9577	Screw self tapping	1		
15	C 33620	Spacer	1		
16	CAC 3713	Pipe - vacuum assy	1		
	CAC 5574	Pipe - vacuum assy	1		
17	CAC 4188	Valve - non return	1		
18	C 43599 10	Clip - hose	8		
	C 43599 12	Clip - hose	2		

1N 16

MASTER INDEX

GROUP INDEX

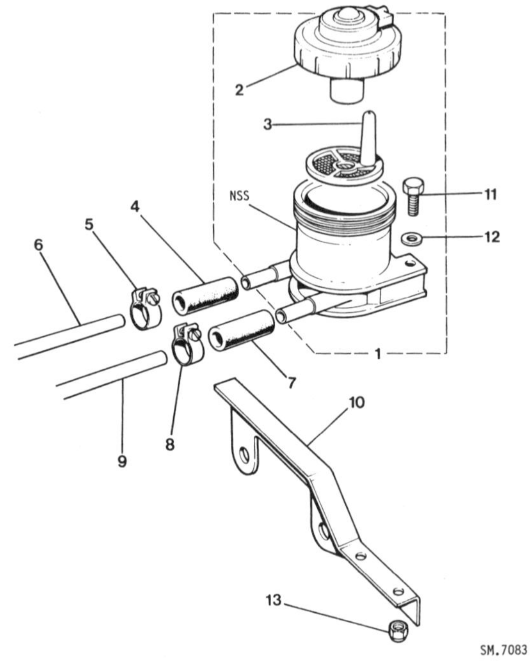

SM7083

BRAKE FLUID RESERVOIR 3.6 LITRE & 5.3 LITRE

Illus	Part Number	Description (123456)	Quantity	Change Point	Remarks
1	CAC 4578	Reservoir brake fluid	1		
2	C 45055	Cover and switch	1		
3	C 33081	Filter brake fluid	1		
4	C 45753 4	Hose low pressure pipe to reservoir	1		
5	JLM 9556	Clip hose	6		
6	CAC 4602	Pipe low pressure res to m/cyl	1		
	C 45088	Pipe low pressure res to m/cyl	NLA		Use CAC 4602
7	C 45090	Elbow LP pipe to reservoir	1		
8	C 15886 3	Clip hose	2		
9	CAC 4579	Pipe LP secondary res to m/cyl	1		
	C 45089	Pipe LP secondary res to m/cyl	NLA		Use CAC 4579
10	CAC 8073	Bracket mounting	1		RHD
	C 42421	Bracket mounting	1		LHD
11	JLM 9562	Bolt 0.25 x 1.50"	2		
12	JLM 298	Washer plain 0.25 ID	2		
13	C 8737 1	Nut nyloc 0.25 UNF	2		

1N 17

MASTER INDEX

GROUP INDEX

FRONT BRAKE PIPES RHD 3.6 LITRE

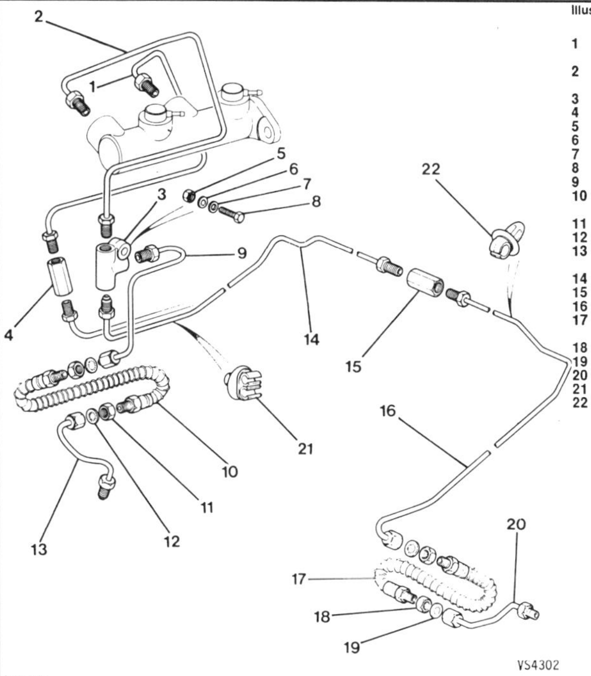

VS4302

Illus	Part Number	123456	Description	Quantity	Change Point	Remarks
1	CAC 5453		Hydraulic brake pipe m/cyl to connector	1		Primary
2	CBC 2671		Hydraulic brake pipe m/cyl to 3 way	1		Secondary
3	CAC 5466		Threeway M12 x 1	1		
4	C 42150		Connector	1		
5	JLM 9690		Nut-nyloc 0.25	1		
6	C 4999		Washer spacer	1		
7	JLM 296		Washer plain 0.25 ID	1		
8	JLM 9562		Bolt 0.25 x 1.50 UNF	1		
9	CAC 5437		Hydraulic brake pipe 3 way to flex	1		RH front flex
10	CAC 6140		Hose assembly-front-armoured	NLA		Use CBC 1390
	CBC 1390		Hose assembly-front armoured	1		
11	C 41712		Locknut 10mm	2		
12	C 726		Washer shakeproof 0.375 ID	2		
13	CAC 5434		Hyd pipe assembly-RH hose to ft caliper	1		
14	CAC 5438		Hyd pipe 3 way to connector	1		
15	C 42150		Connector	1		
16	CAC 5439		Connector to LH front flex	1		
17	CAC 6140		Hose assembly-front-armoured	NLA		Use CBC 1390
	CBC 1390		Hose assembly-front-armoured	1		
18	C 41712		Nut-locking 10mm	2		
19	C 726		Washer-shakeproof 0.375 ID	2		
20	CAC 5435		Hyd pipe LH hose to ft caliper	1		
21	CAC 9252		Clip double	A/R		
22	AAU 6367		Clip single	A/R		
	C 44778		Clip-CAC5438/9 to toe box	2		
	C 38887		Spacer	2		CAC5438/9 to bulkhead
	C 32746 1		Clip	2		
	AB 610061		Screw	2		

MASTER INDEX

GROUP INDEX

FRONT BRAKE PIPES-RHD 5.3 LITRE

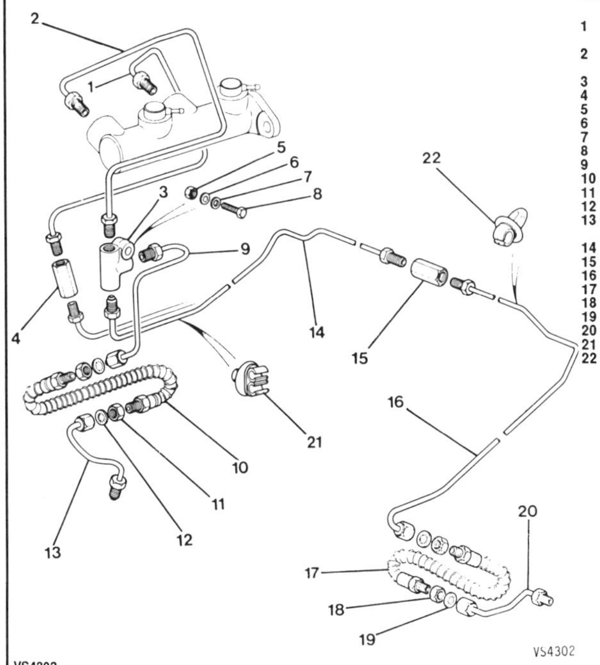

VS4302

Illus	Part Number	123456	Description	Quantity	Change Point	Remarks
1	CAC 5453		Hydraulic brake pipe m/cyl to connector	1		Primary
2	CAC 5436		Hydraulic brake pipe m/cyl to 3 way	1		Secondary
3	CAC 5466		Threeway M12 x 1	1		
4	C 42150		Connector	1		
5	JLM 9690		Nut-nyloc 0.25	1		
6	C 4999		Washer spacer	1		
7	JLM 296		Washer plain 0.25 ID	1		
8	JLM 9562		Bolt 0.25 x 1.50 UNF	1		
9	CAC 5437		Hydraulic brake pipe 3 way to flex	1		RH front flex
10	CAC 6140		Hose assembly-front-armoured	1		
11	C 41712		Locknut 10mm	2		
12	C 726		Washer shakeproof 0.375 ID	2		
13	CAC 5434		Hyd pipe assembly-RH hose to ft caliper	1		
14	CAC 5438		Hyd pipe 3 way to connector	1		
15	C 42150		Connector	1		
16	CAC 5439		Connector to LH front flex	1		
17	CAC 6140		Hose assembly-front-armoured	1		
18	C 41712		Nut-locking 10mm	2		
19	C 726		Washer-shakeproof 0.375 ID	2		
20	CAC 5435		Hyd pipe LH hose to ft caliper	1		
21	AAU 7383		Clip double	A/R		
22	AAU 6367		Clip single	A/R		

MASTER INDEX

ENGINE 1.C
FLYWHEEL AND CLUTCH 1.G
GEARBOX AND PROPSHAFT 1.H
AXLES, SUSPENSION, DRIVE SHAFTS, WHEELS 1.K
STEERING 1.M
BRAKES AND BRAKE CONTROLS 1.N
FUEL, EXHAUST AND EMISSION SYSTEMS 2.C
COOLING, HEATING AND AIR CONDITIONING 2.G
ELECTRICAL, WASHERS-WIPERS, INSTRUMENTS 2.J
CHASSIS, SUBFRAMES, BODYSHELL, FITTINGS 3.C
FASCIA, TRIM, SEATS AND FIXINGS 3.H
ACCESSORIES AND PAINT 4.C
NUMERICAL INDEX 4.D

GROUP INDEX

ACTUATION ASSY AND RESERVOIR TEVES N11
BRAKE AND CLUTCH PEDALS - TEVES BRAKES N07
BRAKE AND CLUTCH PEDALS 3.6 LITRE N04
BRAKE AND CLUTCH PEDALS 3.6 LITRE N05
SWITCH ASSEMBLY-STOPLIGHT N05
BRAKE PEDAL & HOUSING 5.3 LITRE N09
CLUTCH MASTER CYLINDER 3.6 LITRE N06
FRONT BRAKE PIPES-RHD 5.3 LITRE O02
FRONT BRAKE PIPES-TEVES 3.6 & 5.3 LITRE O05
FRONT CALIPERS & DISCS N02
KIT SERVICE-FT CALIPER N02
PADS FRONT BRAKE FERODO N02
MOTOR PUMP & ACCUMULATOR (TEVES BRAKES) N12
REAR CALIPERS & DISCS N03
BRAKE PAD REAR N03
SEAL KIT N03
SERVO AND MASTER CYLINDER N10
KIT MINOR SERVICE BRAKE SERVO N10
KIT REPAIR NON RETURN VALVE N10
KIT SERVICE BRAKE M/CYLINDER N10

10 03

JAGUAR XJS RANGE (JAN 1987 ON) | O03 fiche 1 | BRAKES AND BRAKE CONTROLS

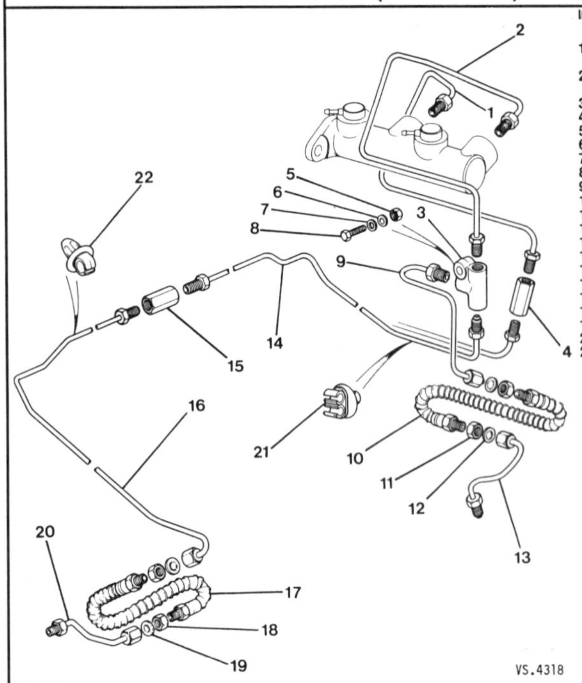

VS 4318

FRONT BRAKE PIPES LHD 3.6 LITRE

Illus	Part Number	123456 Description	Quantity	Change Point	Remarks
1	CAC 5455	Hydraulic brake pipe m/cyl to connector	1		Primary
2	CAC 5440	Hydraulic brake pipe m/cyl to 3 way	1		Secondary
3	CAC 5446	Threeway junction M12 x 1	1		
4	C 42150	Connector	1		
5	JLM 9690	Nut nyloc 0.25"	1		
6	C 4999	Washer-spacer	1		
7	JLM 296	Washer plain 0.25" ID	1		
8	JLM 9562	Bolt 0.25 x 1.50" UNF	1		
9	CAC 5441	Hyd brake pipe to LH ft flex	1		
10	CAC 6140	Hose assembly-front-armoured	1		
11	C 41712	Locknut 10mm	2		
12	C 726	Washer-shakeproof 0.375 ID	2		
13	CAC 5435	Hyd pipe LH hose to ft caliper	1		
14	CAC 5442	Hyd pipe 3 way to connector	1		
15	C 42150	Connector	1		
16	CAC 5443	Hyd pipe connector to RH ft flex	1		
17	CAC 6140	Hose assembly-front-armoured	1		
18	C 41712	Nut-lock 10mm	2		
19	C 726	Washer-shakeproof 0.375 ID	2		
20	CAC 5435	Hyd pipe LH hose to ft caliper	1		
21	CAC 9252	Clip double	A/R		
	AAU 6367	Clip single	A/R		

MASTER INDEX

GROUP INDEX

10 04

JAGUAR XJS RANGE (JAN 1987 ON) | O04 fiche 1 | BRAKES AND BRAKE CONTROLS

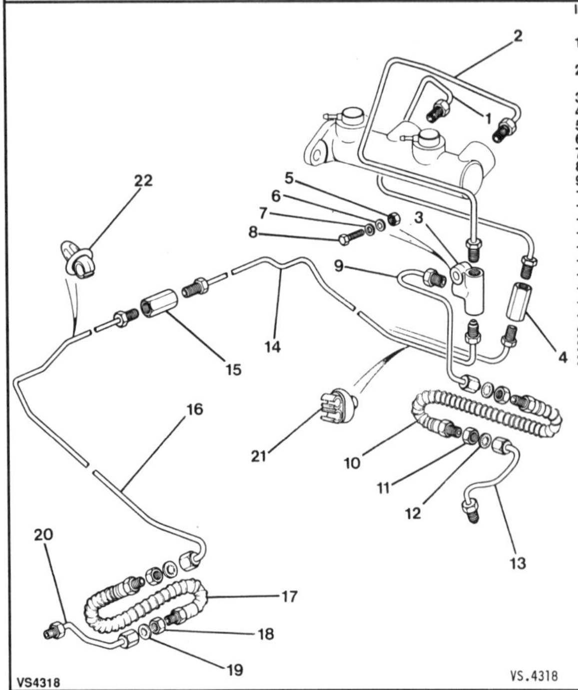

VS4318

FRONT BRAKE PIPES LHD 5.3 LITRE

Illus	Part Number	123456 Description	Quantity	Change Point	Remarks
1	CAC 5455	Hydraulic brake pipe m/cyl to connector	1		Primary
2	CAC 5440	Hydraulic brake pipe m/cyl to 3 way	1		Secondary
3	CAC 5446	Threeway junction M12 x 1	1		
4	C 42150	Connector	1		
5	JLM 9690	Nut nyloc 0.25"	1		
6	C 4999	Washer-spacer	1		
7	JLM 296	Washer plain 0.25" ID	1		
8	JLM 9562	Bolt 0.25 x 1.50" UNF	1		
9	CAC 5441	Hyd brake pipe to LH ft flex	1		
10	CAC 6140	Hose assembly-front-armoured	1		
11	C 41712	Locknut 10mm	2		
12	C 726	Washer-shakeproof 0.375 ID	2		
13	CAC 5435	Hyd pipe LH hose to ft caliper	1		
14	CAC 5442	Hyd pipe 3 way to connector	1		
15	C 42150	Connector	1		
16	CAC 5443	Hyd pipe connector to RH ft flex	1		
17	CAC 6140	Hose assembly-front-armoured	1		
18	C 41712	Nut-lock 10mm	2		
19	C 726	Washer-shakeproof 0.375 ID	2		
20	CAC 5435	Hyd pipe LH hose to ft caliper	1		
21	AAU 7383	Clip double	A/R		
22	AAU 6367	Clip single	A/R		

MASTER INDEX

ENGINE 1.C
FLYWHEEL AND CLUTCH 1.G
GEARBOX AND PROPSHAFT 1.H
AXLES, SUSPENSION, DRIVE SHAFTS, WHEELS 1.K
STEERING 1.M
BRAKES AND BRAKE CONTROLS 1.N
FUEL, EXHAUST AND EMISSION SYSTEMS 2.C
COOLING, HEATING AND AIR CONDITIONING 2.G
ELECTRICAL, WASHERS-WIPERS, INSTRUMENTS 2.J
CHASSIS, SUBFRAMES, BODYSHELL, FITTINGS 3.C
FASCIA, TRIM, SEATS AND FIXINGS 3.H
ACCESSORIES AND PAINT 4.C
NUMERICAL INDEX 4.D

GROUP INDEX

ACTUATION ASSY AND RESERVOIR TEVES N11
BRAKE AND CLUTCH PEDALS - TEVES BRAKES N07
BRAKE AND CLUTCH PEDALS 3.6 LITRE N04
BRAKE AND CLUTCH PEDALS 3.6 LITRE N05
SWITCH ASSEMBLY-STOPLIGHT N05
BRAKE PEDAL & HOUSING 5.3 LITRE N09
CLUTCH MASTER CYLINDER 3.6 LITRE N06
FRONT BRAKE PIPES-RHD 5.3 LITRE O02
FRONT BRAKE PIPES-TEVES 3.6 & 5.3 LITRE O05
FRONT CALIPERS & DISCS N02
KIT SERVICE-FT CALIPER N02
PADS FRONT BRAKE FERODO N02
MOTOR PUMP & ACCUMULATOR (TEVES BRAKES) N12
REAR CALIPERS & DISCS N03
BRAKE PAD REAR N03
SEAL KIT N03
SERVO AND MASTER CYLINDER N10
KIT MINOR SERVICE BRAKE SERVO N10
KIT REPAIR NON RETURN VALVE N10
KIT SERVICE BRAKE M/CYLINDER N10

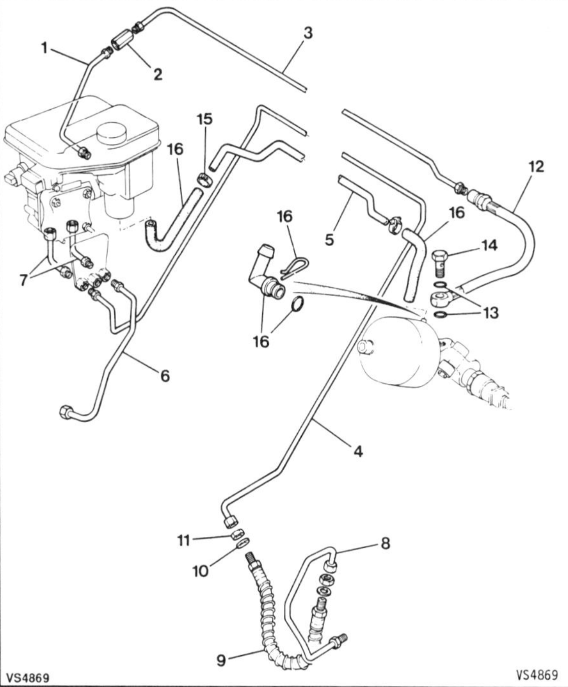

VS4869

VS4869

FRONT BRAKE PIPES-TEVES 3.6 & 5.3 LITRE

Illus	Part Number	123456 Description	Quantity	Change Point	Remarks
1	CBC 5428	Hyd pipe-connector to A/S Unit RHD	1		
	CBC 5429	Hyd pipe-connector to A/S Unit LHD	1		
2	C 42150	Connector	1		
3	CBC 2650	Hyd pipe pump to connector RHD	1		
	CBC 6174	Hyd pipe pump to connector LHD	1		
4	CBC 2357	Hyd pipe A/S unit to LH ft flex	1		RHD
	CBC 2358	Hyd pipe A/S unit to LH ft flex	1		LHD
5	CBC 6182	Hyd pipe reservoir to pump RHD	1		
	CBC 6183	Hyd pipe reservoir to pump LHD	1		
6	CBC 2352	Hyd pipe A/S to RH front flex RHD	1		
	CBC 2353	Hyd pipe A/S to RH front flex LHD	1		
7	CBC 4393	Brake pipe A/S unit to Items 4 & 6	2		
8	CAC 5434	Hyd pipe hose to ft cal RH	1		
	CAC 5435	Hyd pipe hose to ft cal LH	1		
9	CBC 1390	Front armoured hose	2		
10	WF 110001 J	Shakeproof washer	4		
11	C 41712	Locknut	4		
12	JLM 1462	High pressure hose kit	1		
		Hose	1		
13		"O" Ring	2		Part of Seals Kit JLM 1477
14	JLM 1463	Bolt, Banjo	1		
15	TBA	Hose clip	2		
16	JLM 1470	Low pressure hose kit LHD	1		
	JLM 1471	Low pressure hose kit RHD	1		
	JLM 1472	Pin-elbow to pump retainer	1		
	JLM 1477	"O" Ring and Seals Kit	1		Part of Seals Kit and Hose Kit

MASTER INDEX

GROUP INDEX

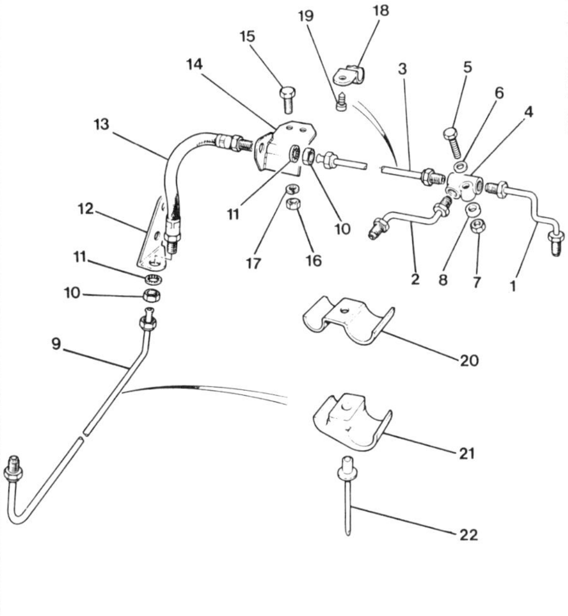

SM 7088

SM7088

REAR BRAKE PIPES 3.6 LITRE & 5.3 LITRE

Illus	Part Number	123456 Description	Quantity	Change Point	Remarks
1	CAC 5452	Hydraulic pipes 3 way to caliper	1		Left hand
2	CAC 5451	Hydraulic pipes 3 way to caliper	1		Right hand
3	CAC 5457	Hydraulic pipe rr hose to 3 way	1		RHD
	CAC 5458	Hydraulic pipe rr hose to 3 way	1		LHD
4	C 29888	3 way connector	1		
5	JLM 9562	Bolt-25 x 1.50	1		
6	JLM 296	Washer plain 0.25 ID	1		
7	JLM 9683	Nut 0.25 UNF	1		
8	C 38559	Spacer	1		
9	CAC 5454	Hydraulic pipe conn to rr flexhose	1		RHD
	CAC 5456	Hydraulic pipe conn to rr flexhose	1		5.3 only LHD
	CAC 5347	Hydraulic pipe conn to rr flex	1		3.6 only LHD
10	C 41712	Locknut 10mm	1		
11	WF 110001 J	Washer shakeproof	1		
12	C 43253	Bracket brake hose	1		
	C 43254	Bracket brake hose	1		LHD
13	CAC 6143	Hose rear flexible	1		Use CBC 1393
	CBC 1393	Hose rear flexible	1		
14	C 44014	Bracket brake hose	1		RHD
	C 44015	Bracket brake hose	1		LHD
15	UFS 131 6R	Setscrew 0.3125 x 0.75	1		
16	JLM 9684	Nut 0.3125 UNF	1		
17	FG 105 X	Washer spring 0.3125 ID	1		
18	C 32746	Clip single	2		
19	AB 610031 J	Screw self tap no.10 x 0.375	2		
20	C 44781	Clip double	1		
21	C 37627	Clip double	4		
22	BD 24978	Rivet pop	5		

G32AW

MASTER INDEX

ENGINE 1.C
FLYWHEEL AND CLUTCH 1.G
GEARBOX AND PROPSHAFT 1.H
AXLES, SUSPENSION, DRIVE SHAFTS, WHEELS 1.K
STEERING 1.M
BRAKES AND BRAKE CONTROLS 1.N
FUEL, EXHAUST AND EMISSION SYSTEMS 2.C
COOLING, HEATING AND AIR CONDITIONING 2.G
ELECTRICAL, WASHERS-WIPERS, INSTRUMENTS 2.J
CHASSIS, SUBFRAMES, BODYSHELL, FITTINGS 3.C
FASCIA, TRIM, SEATS AND FIXINGS 3.H
ACCESSORIES AND PAINT 4.C
NUMERICAL INDEX 4.D

GROUP INDEX

ACTUATION ASSY AND RESERVOIR TEVES N11
BRAKE AND CLUTCH PEDALS - TEVES BRAKES N07
BRAKE AND CLUTCH PEDALS 3.6 LITRE N04
BRAKE AND CLUTCH PEDALS 3.6 LITRE N05
SWITCH ASSEMBLY-STOPLIGHT N05
BRAKE PEDAL & HOUSING 5.3 LITRE N09
CLUTCH MASTER CYLINDER 3.6 LITRE N06
FRONT BRAKE PIPES-RHD 5.3 LITRE O02
FRONT BRAKE PIPES-TEVES 3.6 & 5.3 LITRE O05
FRONT CALIPERS & DISCS N02
KIT SERVICE-FT CALIPER N02
PADS FRONT BRAKE FERODO N02
MOTOR PUMP & ACCUMULATOR (TEVES BRAKES) N12
REAR CALIPERS & DISCS N03
BRAKE PAD REAR N03
SEAL KIT N03
SERVO AND MASTER CYLINDER N10
KIT MINOR SERVICE BRAKE SERVO N10
KIT REPAIR NON RETURN VALVE N10
KIT SERVICE BRAKE M/CYLINDER N10

JAGUAR XJS RANGE (JAN 1987 ON) — O07 fiche 1 — BRAKES AND BRAKE CONTROLS

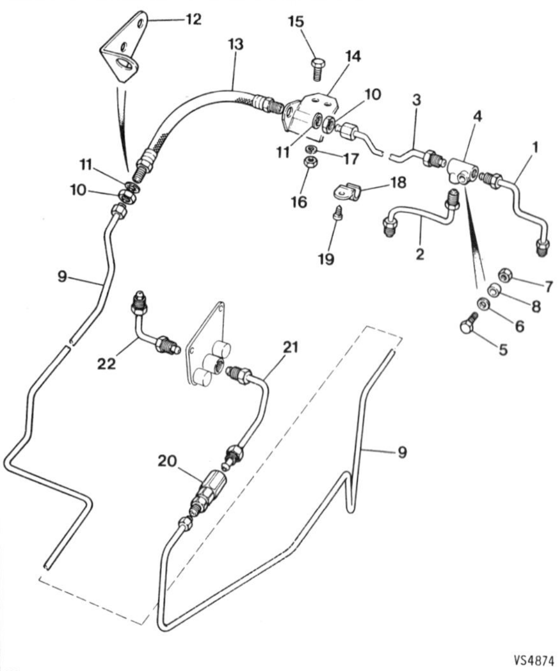

VS4874

REAR BRAKE PIPES - TEVES 3.6 LITRE & 5.3 LITRE

Illus	Part Number	123456 Description	Quantity	Change Point	Remarks
1	CAC 5452	Hydraulic pipe 3 way to LH Caliper	1		
2	CAC 5451	Hydraulic pipe 3 way to RH Caliper	1		
3	CAC 5457	Hydraulic pipe RR Hose to 3 way	1		RHD
	CAC 5458	Hydraulic pipe RR Hose to 3 way	1		LHD
4	C 29888	3 way connector	1		
5	JLM 9562	Bolt	1		
6	JLM 296	Washer plain	1		
7	JLM 9683	Nut 0.25 UNF	1		
8	C 38559	Spacer	1		
9	CBC 4234	Hydraulic pipe PD Valve to RR Flex	1		RHD All
	CBC 4235	Hydraulic pipe PD Valve to RR Flex	1		LHD V12
	CBC 4236	Hydraulic pipe PD Valve to RR Flex	1		LHD 3.6
10	C 41712	Locknut 10 mm	2		
11	C 726	Washer Shakeproof	2		
12	C 43253	Bracket Brake Hose	1		
13	CBC 1393	Rear flexible hose	2		
14	C 44014	Bracket Assy Flex Hose RHD	1		RHD
	C 44015	Bracket Assy Flex Hose LHD	1		LHD
15	UFS 131 6R	Setscrew 0.3125 x 0.75	1		
16	JLM 9684	Nut 0.3125	1		
17	FG 105 X	Spring washer	1		
18	C 32746 1	Clip, single	2		
19	AB 610031 J	Screw self tap No 10 x 0.375	2		
20	CBC 4233	Pressure delay valve	1		
21	CBC 2354	Hyd pipe m/cyl primary to PD Valve	1		RHD
	CBC 2355	Hyd pipe m/cyl primary to PD Valve	1		LHD
22	CBC 4392	Brake pipe	1		
	AGU 1317	Clip	1		
	FS 105161 J	Setscrew M6	1		
	NY 105041 J	Nut	1		
	WJ 105001 J	Washer	1		

MASTER INDEX

GROUP INDEX

JAGUAR XJS RANGE (JAN 1987 ON) — O08 fiche 1 — BRAKES AND BRAKE CONTROLS

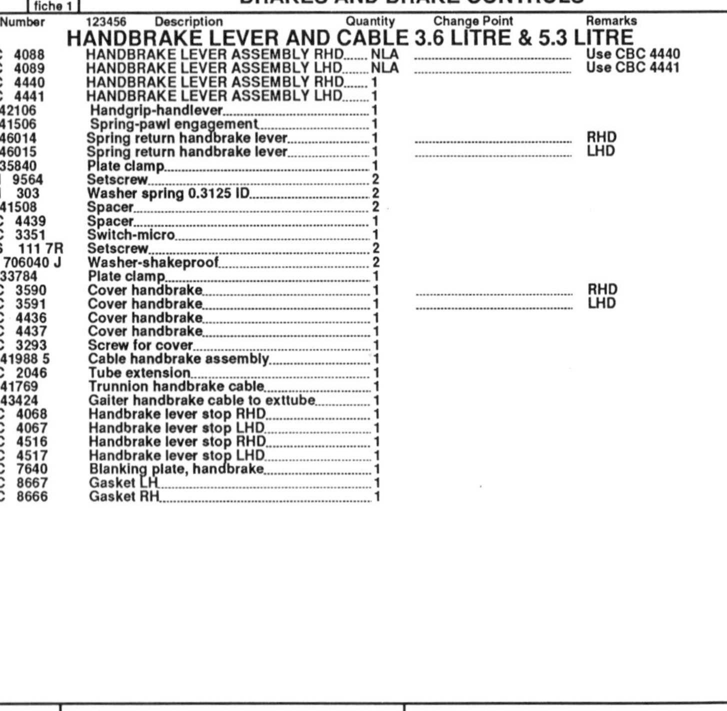

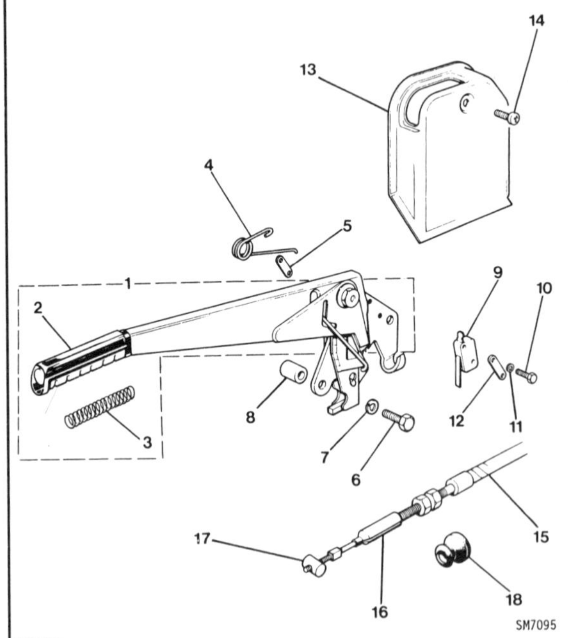

SM7095

HANDBRAKE LEVER AND CABLE 3.6 LITRE & 5.3 LITRE

Illus	Part Number	123456 Description	Quantity	Change Point	Remarks
1	CAC 4088	HANDBRAKE LEVER ASSEMBLY RHD	NLA		Use CBC 4440
	CAC 4089	HANDBRAKE LEVER ASSEMBLY LHD	NLA		Use CBC 4441
	CBC 4440	HANDBRAKE LEVER ASSEMBLY RHD	1		
	CBC 4441	HANDBRAKE LEVER ASSEMBLY LHD	1		
2	C 42106	Handgrip-handlever	1		
3	C 41506	Spring-pawl engagement	1		
4	C 46014	Spring return handbrake lever	1		RHD
	C 46015	Spring return handbrake lever	1		LHD
5	C 35840	Plate clamp	1		
6	JLM 9564	Setscrew	2		
7	JLM 303	Washer spring 0.3125 ID	2		
8	C 41508	Spacer	2		
	CBC 4439	Spacer	1		
9	DAC 3351	Switch-micro	1		
10	UCS 111 7R	Setscrew	2		
11	WF 706040 J	Washer-shakeproof	2		
12	C 33784	Plate clamp	1		
13	CAC 3590	Cover handbrake	1		RHD
	CAC 3591	Cover handbrake	1		LHD
	CBC 4436	Cover handbrake	1		
	CBC 4437	Cover handbrake	1		
14	CAC 3293	Screw for cover	1		
15	C 41988 5	Cable handbrake assembly	1		
16	CBC 2046	Tube extension	1		
17	C 41769	Trunnion handbrake cable	1		
18	C 43424	Gaiter handbrake cable to exttube	1		
	CAC 4068	Handbrake lever stop RHD	1		
	CAC 4067	Handbrake lever stop LHD	1		
	CBC 4516	Handbrake lever stop RHD	1		
	CBC 4517	Handbrake lever stop LHD	1		
	CAC 7640	Blanking plate, handbrake	1		
	CAC 8667	Gasket LH	1		
	CAC 8666	Gasket RH	1		

MASTER INDEX

GROUP INDEX

HANDBRAKE CONTROLS 3.6 LITRE & 5.3 LITRE

Illus	Part Number	123456 Description	Quantity	Change Point	Remarks
1	C 45083	Sleeve handbrake cable	1		
2	C 45700	Retainer h/brake cable	1		
3	C 22690	Spring-return-caliper lever	2		
4	C 27589	Bracket-return spring	2		
5	UFS 119 4R	Setscrew No10 UNF x 0.50"	2		
6	C 8667 17	Nut-self locking	2		

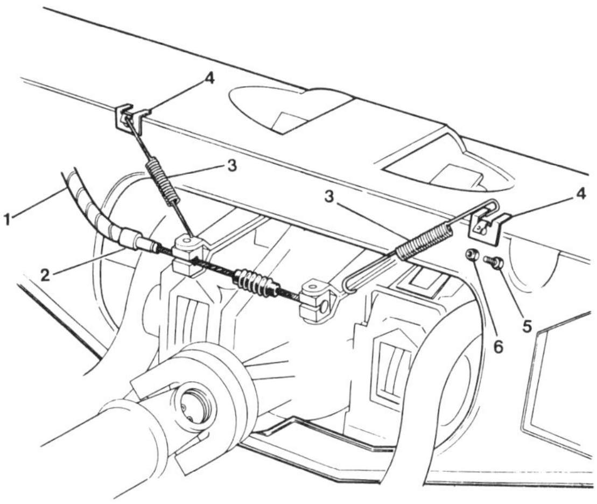

SM5167

SM 5167

MASTER INDEX

GROUP INDEX

HANDRBAKE MECHANISM 3.6 LITRE & 5.3 LITRE

Illus	Part Number	123456 Description	Quantity	Change Point	Remarks
1	CAC 2716	HANDBRAKE ASSEMBLY REAR	1		RH
	CAC 2717	HANDBRAKE ASSEMBLY REAR	1		LH
2	9336	Screw	2		
3	BD 2906 3	Washer shakeproof	2		
4	8836	Pawl assembly	2		
5	8980	Nut adjust	2		
6	8337	Spring	2		
7	8840	Spring friction	2		
8	8841	Pin hinge	2		
9	L 102 4U	Pin split	2		
10	8938	Pin anchor	2		
11	8940	Spring return	2		
12	9335	Cover protection	2		
13	8938	Pin anchor	2		
14	8845	Bolt	1		
15	4900	Pin split	2		
16	8016	Bolt	4		
17	9750	Plate	2		
18	9751	Lockplate	2		
19	JLM 9518	Brake pads	1SET		

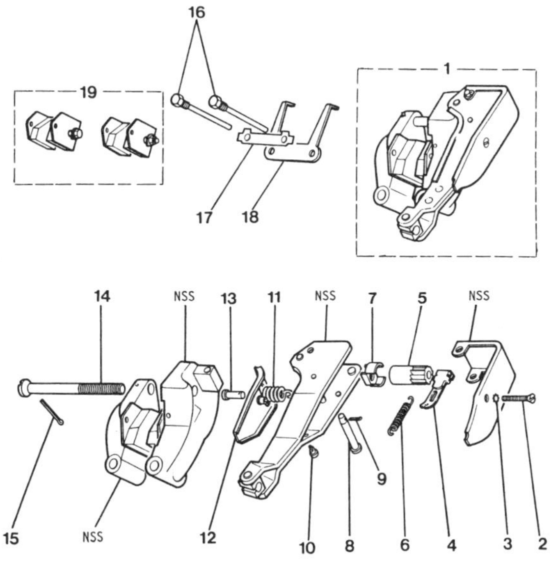

VS3108

VS 3108

MASTER INDEX

ENGINE	1.C
FLYWHEEL AND CLUTCH	1.G
GEARBOX AND PROPSHAFT	1.H
AXLES, SUSPENSION, DRIVE SHAFTS, WHEELS	1.K
STEERING	1.M
BRAKES AND BRAKE CONTROLS	1.N
FUEL, EXHAUST AND EMISSION SYSTEMS	2.C
COOLING, HEATING AND AIR CONDITIONING	2.G
ELECTRICAL, WASHERS-WIPERS, INSTRUMENTS	2.J
CHASSIS, SUBFRAMES, BODYSHELL, FITTINGS	3.C
FASCIA, TRIM, SEATS AND FIXINGS	3.H
ACCESSORIES AND PAINT	4.C
NUMERICAL INDEX	4.D

GROUP INDEX

ACTUATION ASSY AND RESERVOIR TEVES	N11
BRAKE AND CLUTCH PEDALS - TEVES BRAKES	N07
BRAKE AND CLUTCH PEDALS 3.6 LITRE	N04
BRAKE AND CLUTCH PEDALS 3.6 LITRE	N05
SWITCH ASSEMBLY-STOPLIGHT	N05
BRAKE PEDAL & HOUSING 5.3 LITRE	N09
CLUTCH MASTER CYLINDER 3.6 LITRE	N06
FRONT BRAKE PIPES-RHD 5.3 LITRE	O02
FRONT BRAKE PIPES-TEVES 3.6 & 5.3 LITRE	O05
FRONT CALIPERS & DISCS	N02
KIT SERVICE-FT CALIPER	N02
PADS FRONT BRAKE FERODO	N02
MOTOR PUMP & ACCUMULATOR (TEVES BRAKES)	N12
REAR CALIPERS & DISCS	N03
BRAKE PAD REAR	N03
SEAL KIT	N03
SERVO AND MASTER CYLINDER	N10
KIT MINOR SERVICE BRAKE SERVO	N10
KIT REPAIR NON RETURN VALVE	N10
KIT SERVICE BRAKE M/CYLINDER	N10

2C 01

MASTER INDEX

GROUP INDEX

2C 02

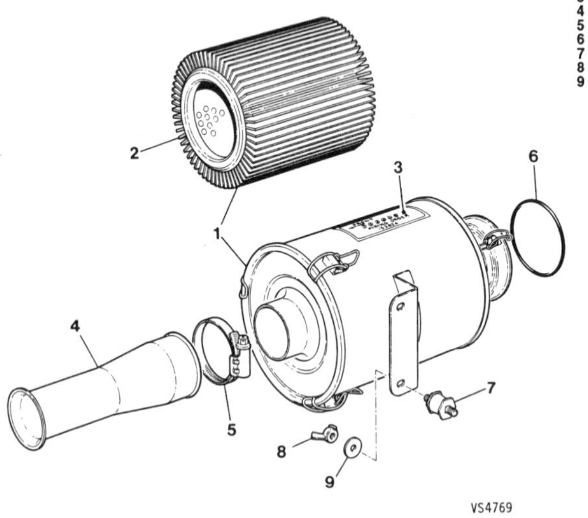

VS4769

AIR CLEANER-3.6 LITRE.

Illus	Part Number	123456 Description	Quantity	Change Point	Remarks
1	EAC 8448	AIR CLEANER ASSEMBLY	1		
1	EAC 8447	AIR CLEANER ASSEMBLY	1		Air inject.engine
2	EAC 4840	Air filter	1		
	EAC 5864	Clip - air cleaner cover	6		
3	EAC 5620	Firing order chart	1		
4	EAC 8963	Intake tube	1		
5	EAC 3215 13	Hose clip	2		
6	EAC 4714	'O' ring	1		
7	EAC 8130	Instrument	2		
8	NA 106011 J	Nut	2		
9	WJ 106001 J	Washer	2		

MASTER INDEX

ENGINE 1.C
FLYWHEEL AND CLUTCH 1.G
GEARBOX AND PROPSHAFT 1.H
AXLES, SUSPENSION, DRIVE SHAFTS, WHEELS 1.K
STEERING 1.M
BRAKES AND BRAKE CONTROLS 1.N
FUEL, EXHAUST AND EMISSION SYSTEMS 2.C
COOLING, HEATING AND AIR CONDITIONING 2.G
ELECTRICAL, WASHERS-WIPERS, INSTRUMENTS 2.J
CHASSIS, SUBFRAMES, BODYSHELL, FITTINGS 3.C
FASCIA, TRIM, SEATS AND FIXINGS 3.H
ACCESSORIES AND PAINT 4.C
NUMERICAL INDEX 4.D

GROUP INDEX

ACCELERATOR CABLE LHD 3.6 LITRE. C16
ACCELERATOR CABLE LHD-5.3 LITRE. C17
ACCELERATOR CABLE RHD 3.6 LITRE. C13
ACCELERATOR CABLE RHD 5.3 LITRE. C14
AIR CLEANER RH-5.3 LITRE. E02
AIR CLEANER-3.6 LITRE. C02
AIR INJECTION PUMP 3.6 LITRE. C09
AIR INJECTION PUMP-5.3 LITRE. F03
FRONT EXHAUST PIPE RH F10
FUEL FILLER CAP - CONVERTIBLE D11
FUEL FILLER CAP-CABRIOLET D10
FUEL PUMP AND FILTER. D04
FUEL RAIL-INJECTORS-3.6 LITRE. C05
FUEL TANK-CABRIOLET D09

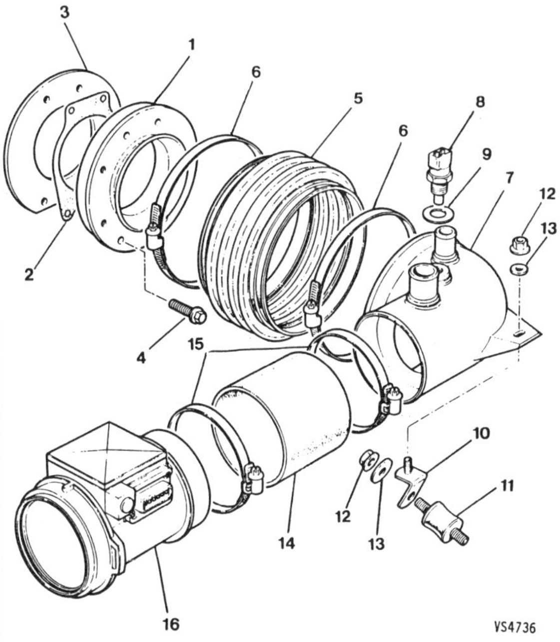

VS4736

AIR FLOWMETER-AIRBOX-3.6 LITRE.

Illus	Part Number	123456	Description	Quantity	Change Point	Remarks
1	EAC 5070		Trumpet-throttle body	1	Up to (E)163584	
1	EBC 3375		Trumpet-throttle body	1	From (E)163585	
2	EAC 6266		Gasket	1		
3	EAC 7561		Retaining plate	1		
4	FB 106071 M		Bolt-trumpet to throttle body	4		
5	EAC 6370		Hose-trumpet to airbox	1		
6	EAC 3215 20		Hose clip	2		
7	EAC 6962		Airbox	1		
7	EAC 9552		Airbox	1	]	Emiss 'A' only
8	EAC 2863		Air temperature sensor	1	]	
9	C 2296 4		Sealing washer	1		
10	EAC 8264		Mounting bracket	1		
11	CAC 8478		Instrument	1		
12	FN 106041 J		Nut	2		
13	WJ 106001		Washer	2		
14	EAC 1301		Hose-airbox to flowmeter	1		
15	EAC 3215 15		Hose clip	2		
16	EAC 6215		Air flowmeter	1		

H4AK23/9/88

MASTER INDEX

GROUP INDEX

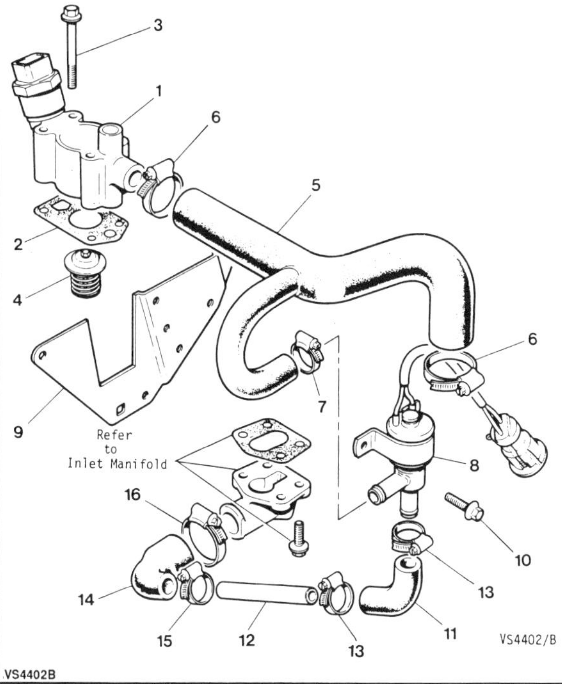

VS4402B

IDLE AIR VALVE-3.6 LITRE

Illus	Part Number	123456	Description	Quantity	Change Point	Remarks
1	EAC 6263		IDLE AIR VALVE	1	Upto Eng No 117231	Use EBC 1789
1	EBC 1789		IDLE AIR VALVE	1	(E)117232-(E)143565	
1	EBC 2675		IDLE AIR VALVE	1	From (E)143566	
	JLM 1741		Stepper motor	1		
2	EAC 5604		Gasket	1		
3	FB 106121 J		Bolt-securing air valve	4		
4	EAC 9440		Deceleration valve	1	Up to (E)139920	
5	EAC 8307		Hose-air valve to manifold	1		
6	EAC 3215 8		Hose clip	2		
7	EAC 3215 6		Hose clip	1		
8	EAC 7340		SUPPLEMENTARY AIR VALVE	1		
9	EAC 7922		Mounting bracket	1		
10	FS 106141 J		Setscrew-securing bracket	3		
11	EAC 6560		Hose-air valve to air pipe	1		
12	EAC 6255		Air pipe	1		
13	EAC 3215 6		Hose clip	2		
14	EAC 6254		Hose-air pipe to adaptor	1		
15	EAC 3215 6		Hose clip	1		
16	EAC 3215 9		Hose clip	1		

H6AK11/7/88

MASTER INDEX

ENGINE 1.C
FLYWHEEL AND CLUTCH 1.G
GEARBOX AND PROPSHAFT 1.H
AXLES, SUSPENSION, DRIVE SHAFTS, WHEELS 1.K
STEERING 1.M
BRAKES AND BRAKE CONTROLS 1.N
FUEL, EXHAUST AND EMISSION SYSTEMS 2.C
COOLING, HEATING AND AIR CONDITIONING 2.G
ELECTRICAL, WASHERS-WIPERS, INSTRUMENTS 2.J
CHASSIS, SUBFRAMES, BODYSHELL, FITTINGS 3.C
FASCIA, TRIM, SEATS AND FIXINGS 3.H
ACCESSORIES AND PAINT 4.C
NUMERICAL INDEX 4.D

GROUP INDEX

ACCELERATOR CABLE LHD 3.6 LITRE C16
ACCELERATOR CABLE LHD-5.3 LITRE C17
ACCELERATOR CABLE RHD 3.6 LITRE C13
ACCELERATOR CABLE RHD 5.3 LITRE C14
AIR CLEANER RH-5.3 LITRE E02
AIR CLEANER-3.6 LITRE C02
AIR INJECTION PUMP 3.6 LITRE C09
AIR INJECTION PUMP-5.3 LITRE F03
FRONT EXHAUST PIPE RH F10
FUEL FILLER CAP - CONVERTIBLE D11
FUEL FILLER CAP-CABRIOLET D10
FUEL PUMP AND FILTER D04
FUEL RAIL-INJECTORS-3.6 LITRE C05
FUEL TANK-CABRIOLET D09

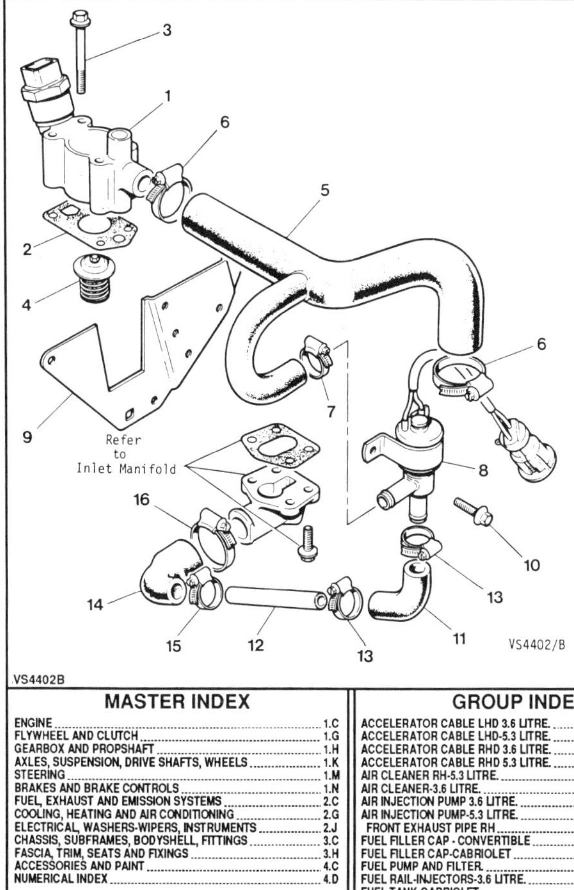

JAGUAR XJS RANGE (JAN 1987 ON) | C05 fiche 2 | FUEL, EXHAUST AND EMISSION SYSTEMS

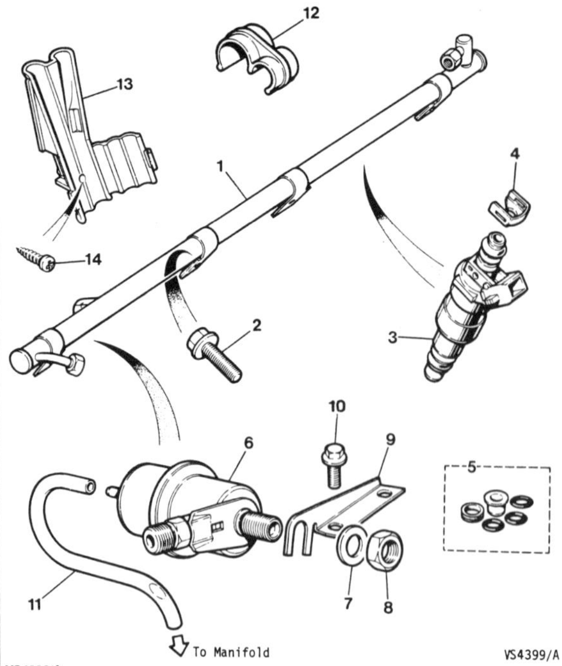

VS4399/A

FUEL RAIL-INJECTORS-3.6 LITRE.

Illus	Part Number	123456 Description	Quantity	Change Point	Remarks
1	EAC 9690	Fuel rail	1		
2	FS 106141 J	Setscrew-securing fuel rail	4		
3	EAC 4867	Fuel injector	6		
4	EAC 4579	Clip-injector to fuel rail	6		
5	EAC 5901	'O' ring kit(injectors)	6		
6	EAC 6314	Fuel pressure regulator	1		
7	WB 114001 J	Washer	1		
8	AGU 1329 J	Nut	1		
9	EAC 6278	Mounting bracket	1		
10	FS 106141 J	Setscrew-securing bracket	2		
11	EAC 8022 30	Hose-regulator to manifold	1		63" lengths
12	EAC 6995	Clip-fuel hose	2		
13	EAC 6996	Guide-fuel hose	1		
14	AB 612051 J	Screw-securing guide	1		

MASTER INDEX

GROUP INDEX

2C 05

JAGUAR XJS RANGE (JAN 1987 ON) | C06 fiche 2 | FUEL, EXHAUST AND EMISSION SYSTEMS

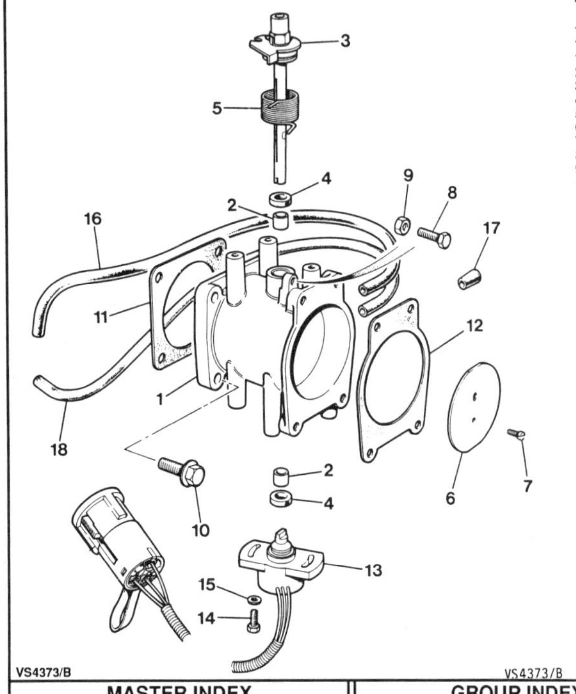

VS4373/B

THROTTLE BODY-3.6 LITRE

Illus	Part Number	123456 Description	Quantity	Change Point	Remarks
1	EBC 2108	Throttle body	1		
1	EBC 2107	Throttle body	1	Up to Eng No 139499	Emiss 'A' only
1	EBC 2344	Throttle body	1	From Eng No 139500	Emiss 'A' only
2	C 20952 3	Bush	2		
3	EAC 8011	Throttle spindle	1		
4	EAC 7329	Seal-throttle spindle	2		
5	EAC 4834	Torsion spring	1		
6	EAC 4820	Throttle disc	1		
7	AUC 1358 J	Screw-disc to spindle	2		
8	C 37939	Throttle stop screw	1		
9	UFN 119 L	Locknut	1		
10	FS 108251 J	Screw-throttle body to manifold	4		
11	EAC 5410	Gasket-throttle to manifold	1		
12	EAC 6266	Gasket-throttle to trumpet	1		
13	EAC 7341	Throttle potentiometer	1		
14	SN 104161 J	Screw	2		
15	WA 104001 J	Washer	2		
16	EAC 8022 30	Hose-throttle to thermal valve	1		
17	C 37021	Blanking cap	1		
18	EAC 8022 30	Hose-throttle to thermal valve	1	From Eng No 139500	

MASTER INDEX

ENGINE 1.C
FLYWHEEL AND CLUTCH 1.G
GEARBOX AND PROPSHAFT 1.H
AXLES, SUSPENSION, DRIVE SHAFTS, WHEELS 1.K
STEERING 1.M
BRAKES AND BRAKE CONTROLS 1.N
FUEL, EXHAUST AND EMISSION SYSTEMS 2.C
COOLING, HEATING AND AIR CONDITIONING 2.G
ELECTRICAL, WASHERS-WIPERS, INSTRUMENTS 2.J
CHASSIS, SUBFRAMES, BODYSHELL, FITTINGS 3.C
FASCIA, TRIM, SEATS AND FIXINGS 3.H
ACCESSORIES AND PAINT 4.C
NUMERICAL INDEX 4.D

GROUP INDEX

ACCELERATOR CABLE LHD 3.6 LITRE C16
ACCELERATOR CABLE LHD-5.3 LITRE C17
ACCELERATOR CABLE RHD 3.6 LITRE C13
ACCELERATOR CABLE RHD 5.3 LITRE C14
AIR CLEANER RH-5.3 LITRE E02
AIR CLEANER-3.6 LITRE C02
AIR INJECTION PUMP 3.6 LITRE C09
AIR INJECTION PUMP-5.3 LITRE F03
FRONT EXHAUST PIPE RH F10
FUEL FILLER CAP - CONVERTIBLE D11
FUEL FILLER CAP-CABRIOLET D10
FUEL PUMP AND FILTER D04
FUEL RAIL-INJECTORS-3.6 LITRE C05
FUEL TANK-CABRIOLET D09

2C 06

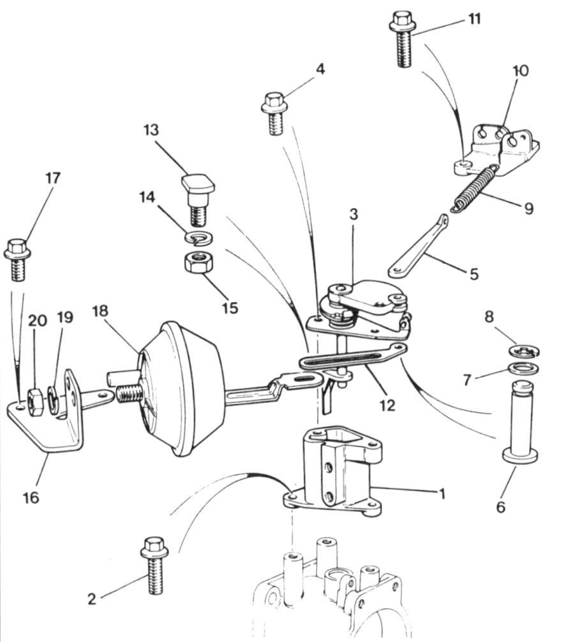

VS4625

THROTTLE LINKAGE-3.6 LITRE

Illus	Part Number	Description	Quantity	Change Point	Remarks
1	EAC 4888	Throttle pedestal	1		
2	FS 106201 J	Setscrew-securing pedestal	3		
3	EAC 9918	Throttle coupling assembly	1		
4	FS 106161 J	Setscrew-securing coupling	2		
5	EAC 8162	Retainer	1		
6	EAC 8164	Pin-retainer to coupling	1		
7	WA 105001 J	Washer	1		
8	C 14350	Circlip	1		
9	EAC 8616	Return spring	1		
10	EAC 9477	Abutment bracket	1		
11	FS 106201 J	Setscrew-securing bracket	2		
12	EAC 8778	Sliding link	1		
13	EAC 8167	Bolt-link to actuator	1		
14	WL 104001 J	Washer	1		
15	NH 104041 J	Nut	1		
16	EAC 8166	Bracket	1		
17	FS 106141 J	Setscrew-securing bracket	2		
18	DBC 2641	Actuator(speed control)	1		
19	FG 105 X	Washer	1		
20	JLM 9688	Nut-actuator to bracket	1		

MASTER INDEX

GROUP INDEX

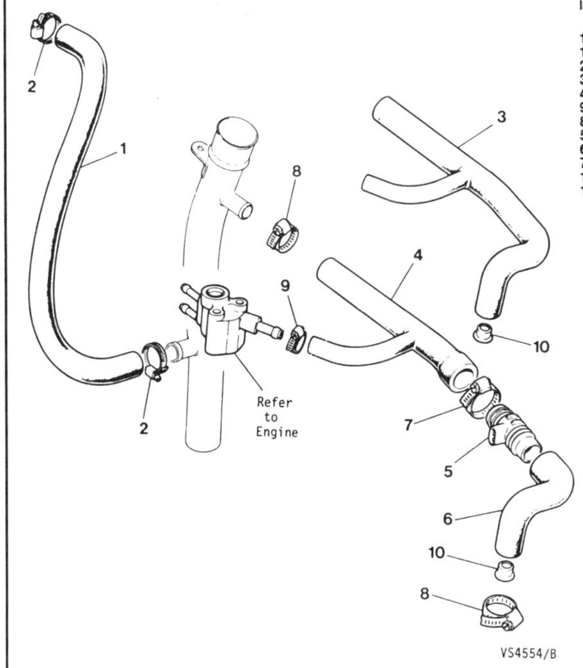

VS4554/B

ENGINE BREATHING 3.6 LITRE

Illus	Part Number	Description	Quantity	Change Point	Remarks
1	EAC 8997	Hose-filler pipe to cam cover	1	Up to VIN149810	Use EBC 1233
1	EBC 1233	Hose-filler pipe to cam cover	1	From VIN149811	
2	EAC 3215 6	Hose clip	2		
3	EAC 7478	Hose-filler pipe to restrictor	1		Non heated
4	EAC 7418	Hose-filler pipe to element	1		Heated breathing
9	EAC 3215 2	Hose clip	1		
8	EAC 3215 6	Hose clip	2		
5	EAC 7991	Heater element	1		Heated breathing Emiss "A" only
6	EAC 7447	Hose-element to restrictor	1		Heated breathing Emiss "A" only
7	EAC 3215 7	Hose clip	2		Heated breathing Emiss "A" only
10	EAC 7155	Breather restrictor	1		Use EAC 7557
10	EAC 7557	Breather restrictor	1		

MASTER INDEX

ENGINE 1.C
FLYWHEEL AND CLUTCH 1.G
GEARBOX AND PROPSHAFT 1.H
AXLES, SUSPENSION, DRIVE SHAFTS, WHEELS 1.K
STEERING 1.M
BRAKES AND BRAKE CONTROLS 1.N
FUEL, EXHAUST AND EMISSION SYSTEMS 2.C
COOLING, HEATING AND AIR CONDITIONING 2.G
ELECTRICAL, WASHERS-WIPERS, INSTRUMENTS 2.J
CHASSIS, SUBFRAMES, BODYSHELL, FITTINGS 3.C
FASCIA, TRIM, SEATS AND FIXINGS 3.H
ACCESSORIES AND PAINT 4.C
NUMERICAL INDEX 4.D

GROUP INDEX

ACCELERATOR CABLE LHD 3.6 LITRE. C16
ACCELERATOR CABLE LHD-5.3 LITRE. C17
ACCELERATOR CABLE RHD 3.6 LITRE. C13
ACCELERATOR CABLE RHD 5.3 LITRE. C14
AIR CLEANER RH-5.3 LITRE. E02
AIR CLEANER-3.6 LITRE. C02
AIR INJECTION PUMP 3.6 LITRE. C09
AIR INJECTION PUMP-5.3 LITRE. F03
FRONT EXHAUST PIPE RH F10
FUEL FILLER CAP - CONVERTIBLE D11
FUEL FILLER CAP-CABRIOLET D10
FUEL PUMP AND FILTER. D04
FUEL RAIL-INJECTORS-3.6 LITRE. C05
FUEL TANK-CABRIOLET D09

2C 09

AIR INJECTION PUMP 3.6 LITRE.

Illus	Part Number	123456 Description	Quantity	Change Point	Remarks
1	EAC 8809	AIR PUMP ASSEMBLY	1		
2	EAC 8810	Air pump	1		
3	EAC 8811	Clutch	1		
4	EAC 8951	Circlip(outer)	1		
5	EAC 8950	Circlip(inner)	1		Air injection engine only
6	AGU 1329 J	Nut	1		
7	EAC 5192	Vent cover	1		
8	EAC 8272	Hose-pump to air cleaner	1		
9	EAC 5911	Hose-pump to switch valve	1		
10	EAC 3215 8	Hose clip	4		

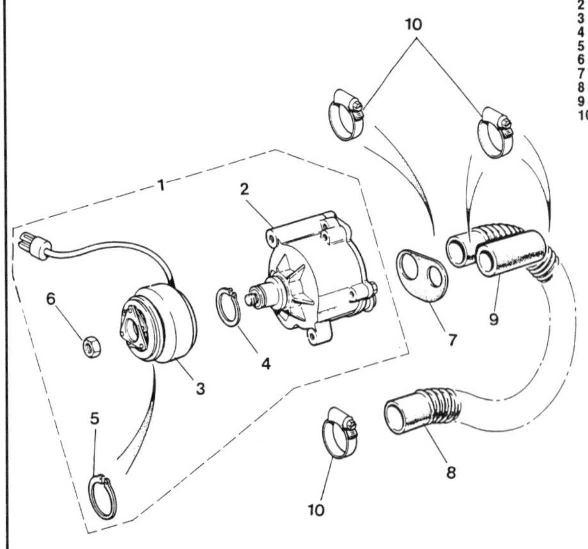

VS4588/A

VS4588/A

MASTER INDEX

GROUP INDEX

2C 10

AIR PUMP MOUNTING 3.6 LITRE

Illus	Part Number	123456 Description	Quantity	Change Point	Remarks
1	EAC 5958	Mounting bracket	1		
2	FB 108181 J	Bolt-bracket to water pump	1		
3	FB 108111 J	Bolt-bracket to water pump	1		
4	EAC 7237	Distance piece	2		
5	FS 108201 J	Screw-bracket to thermo hsg	1		
6	BH 108231 J	Pivot bolt	1		
7	EAC 5438	Distance piece	1		
8	EAC 4897	Distance piece	1		
9	EAC 4898	Clamping washer	1		Air injection engine only
10	NY 108041 J	Nut	1		
11	EAC 7508	Adjusting bolt	1		
12	EAC 5961	Adjusting sleeve	1		
13	WM 600061 J	Washer	2		
14	JLM 9685	Nut	2		
15	FB 108081 J	Bolt-sleeve to pump	1		
16	SH 606091 J	Setscrew	1		
17	WM 600061 J	Washer	1		
18	JLM 9685	Nut	1		

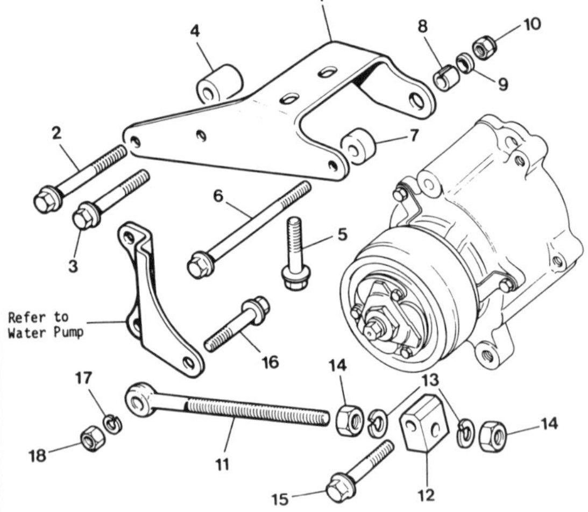

VS4428

VS 4428

MASTER INDEX

ENGINE ... 1.C
FLYWHEEL AND CLUTCH ... 1.G
GEARBOX AND PROPSHAFT ... 1.H
AXLES, SUSPENSION, DRIVE SHAFTS, WHEELS ... 1.K
STEERING ... 1.M
BRAKES AND BRAKE CONTROLS ... 1.N
FUEL, EXHAUST AND EMISSION SYSTEMS ... 2.C
COOLING, HEATING AND AIR CONDITIONING ... 2.G
ELECTRICAL, WASHERS-WIPERS, INSTRUMENTS ... 2.J
CHASSIS, SUBFRAMES, BODYSHELL, FITTINGS ... 3.C
FASCIA, TRIM, SEATS AND FIXINGS ... 3.H
ACCESSORIES AND PAINT ... 4.C
NUMERICAL INDEX ... 4.D

GROUP INDEX

ACCELERATOR CABLE LHD 3.6 LITRE. ... C16
ACCELERATOR CABLE LHD-5.3 LITRE. ... C17
ACCELERATOR CABLE RHD 3.6 LITRE. ... C13
ACCELERATOR CABLE RHD 5.3 LITRE. ... C14
AIR CLEANER RH-5.3 LITRE. ... E02
AIR CLEANER-3.6 LITRE. ... C02
AIR INJECTION PUMP 3.6 LITRE. ... C09
AIR INJECTION PUMP-5.3 LITRE. ... F03
FRONT EXHAUST PIPE RH ... F10
FUEL FILLER CAP - CONVERTIBLE ... D11
FUEL FILLER CAP-CABRIOLET ... D10
FUEL PUMP AND FILTER. ... D04
FUEL RAIL-INJECTORS-3.6 LITRE. ... C05
FUEL TANK-CABRIOLET ... D09

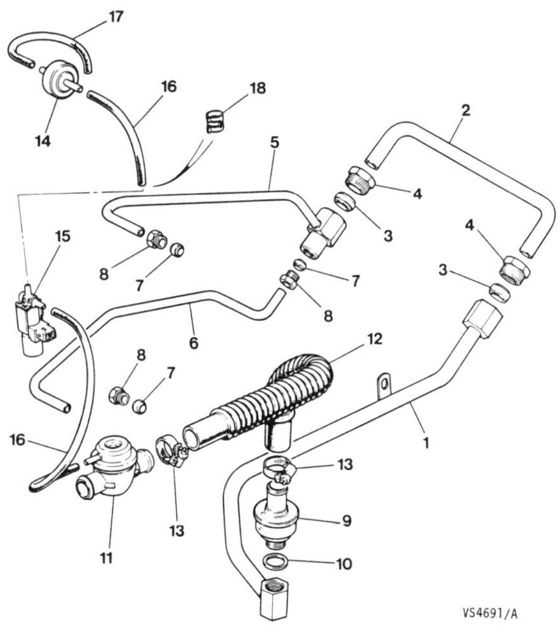

VS4691/A

AIR INJECTION PIPES 3.6 LITRE

Illus	Part Number	123456 Description	Quantity	Change Point	Remarks
1	EAC 6191	Air delivery pipe (primary)	1		Air injection engine only
2	EAC 5944	Air delivery pipe (secondary)	1		
3	EAC 5939	Sleeve	2		
4	EAC 5940	Nut	2		
5	EAC 5943	Air tube	1		
6	EAC 5945	Air tube	1		
7	EAC 4929	Sleeve	3		
8	EAC 4927	Nut	3		
9	EAC 5042	CHECK VALVE	1		
10	C 45759 3	Sealing washer	1		
11	EAC 5040	AIR SWITCHING VALVE	1		
12	EAC 5896	Air hose	1		
13	EAC 3215 8	Hose clip	2		
14	EAC 2206	Delay valve	1		
15	EAC 7905	SOLENOID VALVE	1		
16	EAC 8021 18	Hose (65" lengths)	A/R		
17	EAC 8022 30	Hose (63" lengths)	A/R		
18	AGU 2536 J	Hose Band	1		

MASTER INDEX

GROUP INDEX

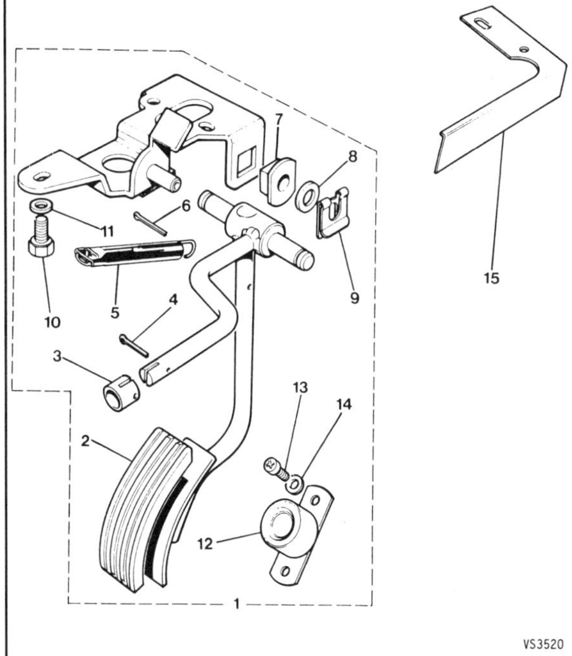

VS 3520

ACCELERATOR-PEDAL RHD 3.6 LITRE & 5.3 LITRE

Illus	Part Number	123456 Description	Quantity	Change Point	Remarks
1	CAC 5180	PEDAL ASSY ACCELERATOR	1		Automatic Non-speed control
	CAC 5181	PEDAL ASSY ACCELERATOR	1		Automatic Speed control
	CBC 4844	PEDAL ASSY ACCELERATOR	1		Manual Non Speed control
	CBC 2473	PEDAL ASSY ACCELERATOR	1		Manual Speed control
2	C 45287	Facing acc foot pad	1		Automatic
	CBC 2132	Facing acc foot pad	1		Manual
3	CAC 5328	Retainer acc cable	1		
4	2554	Pin split	1		
5	CAC 1429	Spring pedal return	1		
6	2554	Pin split	1		
7	C 28732	Bearing acc pedal rod	2		
8	BD 20013 5	Washer brass	1		
9	C 27328 4	Clip safety	1		
10	SH 604051 J	Setscrew	3		
11	JLM 296	Washer plain	3		
12	C 44616	Stop pedal unit assembly	1		
13	BD 27546 2	Screw paint clearing	2		
14	WE 600041 J	Washer	2		
15	CAC 5146	Guide sound deadening	1		

MASTER INDEX

GROUP INDEX

ACCELERATOR CABLE RHD 3.6 LITRE.

Illus	Part Number	Description 123456	Quantity	Change Point	Remarks
1	CBC 1904 1	Accelerator cable assy	1		
	CBC 4576 1	Accelerator cable assy	1		19871/2 MY
2	NT 605041 J	Nut	2		
	NT 108041 J	Locknut	1		
3	C 2850	Grommet	1		
4	FW 105 T	Washer	1		
5	SH 604051 J	Setscrew	2		
6	CAC 4316	Cable abutment bracket	1		
	AGU 2385 J	Spacer clip-cable to a/con pipe	2		

H22.02AW31/8/88

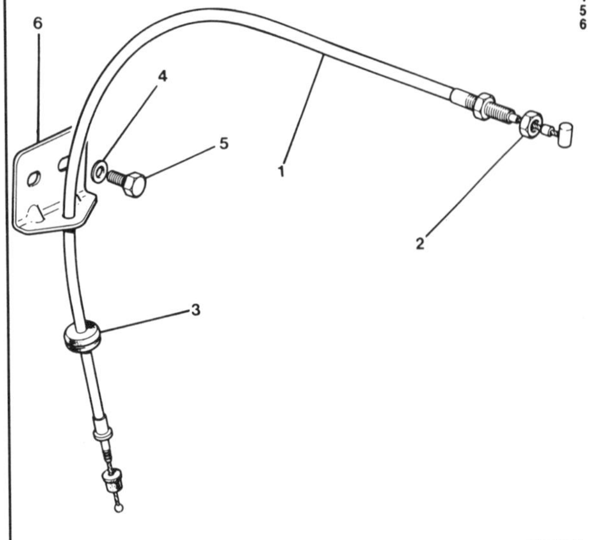

VS4344/A

VS 4344A

2C 13

MASTER INDEX

GROUP INDEX

ACCELERATOR CABLE RHD 5.3 LITRE.

Illus	Part Number	Description 123456	Quantity	Change Point	Remarks
1	CBC 1915 4	Accelerator cable	1		Use CBC5056
	CBC 5056	Accelerator cable	1		
2	NT 605041 J	Locknut	2		
	NT 108041 J	Locknut	1		
3	C 2850	Grommet	1		
4	FW 105 T	Washer	2		
5	SH 605061 J	Setscrew	2		
6	CAC 4316	Cable abutment bracket	1		
7	FW 105 E	Washer-copper	1		
8	C 725	Washer-shakeproof	1		
9	C 33509	Switch	1		
10	C 35840	Clamp plate	1		
11	C 33784	Fixing plate	1		
12	UCS 111 6RJ	Setscrew	2		

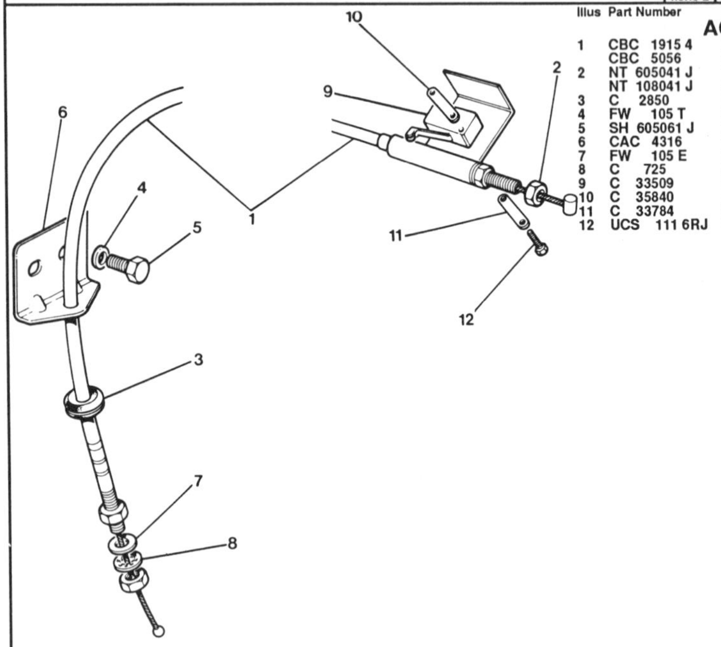

VS3521

VS3521

2C 14

MASTER INDEX

ENGINE ... 1.C
FLYWHEEL AND CLUTCH ... 1.G
GEARBOX AND PROPSHAFT ... 1.H
AXLES, SUSPENSION, DRIVE SHAFTS, WHEELS ... 1.K
STEERING ... 1.M
BRAKES AND BRAKE CONTROLS ... 1.N
FUEL, EXHAUST AND EMISSION SYSTEMS ... 2.C
COOLING, HEATING AND AIR CONDITIONING ... 2.G
ELECTRICAL, WASHERS-WIPERS, INSTRUMENTS ... 2.J
CHASSIS, SUBFRAMES, BODYSHELL, FITTINGS ... 3.C
FASCIA, TRIM, SEATS AND FIXINGS ... 3.H
ACCESSORIES AND PAINT ... 4.C
NUMERICAL INDEX ... 4.D

GROUP INDEX

ACCELERATOR CABLE LHD 3.6 LITRE. ... C16
ACCELERATOR CABLE LHD-5.3 LITRE. ... C17
ACCELERATOR CABLE RHD 3.6 LITRE. ... C13
ACCELERATOR CABLE RHD 5.3 LITRE. ... C14
AIR CLEANER RH-5.3 LITRE. ... E02
AIR CLEANER-3.6 LITRE. ... C02
AIR INJECTION PUMP 3.6 LITRE. ... C09
AIR INJECTION PUMP-5.3 LITRE. ... F03
FRONT EXHAUST PIPE RH ... F10
FUEL FILLER CAP - CONVERTIBLE ... D11
FUEL FILLER CAP-CABRIOLET ... D10
FUEL PUMP AND FILTER. ... D04
FUEL RAIL-INJECTORS-3.6 LITRE. ... C05
FUEL TANK-CABRIOLET ... D09

2C 15

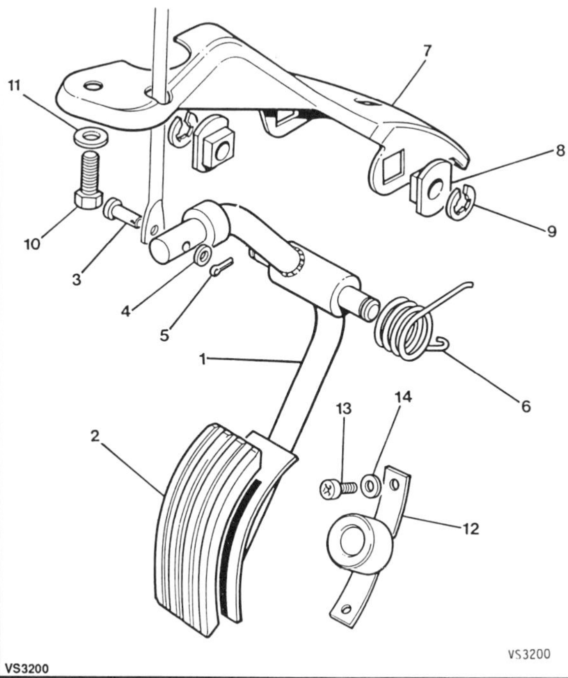

VS3200

ACCELERATPR PEDAL LHD 3.6 LITRE & 5.3 LITRE

Illus	Part Number	123456	Description	Quantity	Change Point	Remarks
1	CAC 3858		Pedal accelerator assy	1		
2	C 45287		Facing-acc foot pad	1		
3	J 103 9S		Pin-joint	1		
4	BD 541 28		Washer-plain	1		
5	L 102 4U		Pin split	1		
6	C 46063		Spring torsion	1		
7	CAC 3842		Bracket acc pivot bearing	1		
8	C 28732		Bearing pivot	2		
9	C 33600 1		Circlip for bearing 0.375 ID	2		
10	JLM 9566		Setscrew-recessed 0.25UNF x 0.75	2		
11	JLM 296		Washer 0.25" ID	2		
12	CAC 1610		Stop-pedal-assembly	1		
13	BD 27546		Screw-paint clearing	2		
14	WE 600041 J		Washer-shakeproof 0.25" ID	2		

MASTER INDEX

GROUP INDEX

2C 16

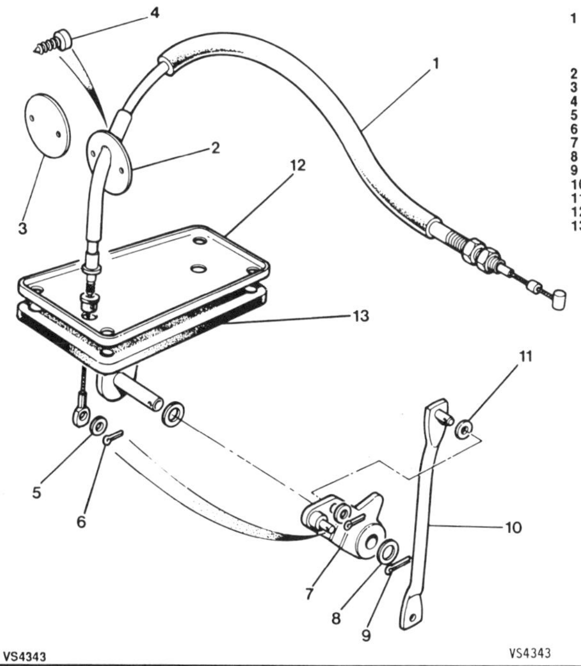

VS4343

ACCELERATOR CABLE LHD 3.6 LITRE.

Illus	Part Number	123456	Description	Quantity	Change Point	Remarks
1	CAC 6254 1		Cable accelerator	NLA		WSE Use CBC 1925 1
	CBC 1925 1		Cable accelerator	1		
	CBC 4577 1		Accelerator cable	1		1987 1/2 MY
2	C 44630		Plate blanking	1		
3	C 44602		Gasket	1		
4	DAZ 408 8C		Screw self tapping	2		
5	BD 541 28		Washer plain	1		
6	2390		Pin split	1		
7	CAC 3827		Lever accelerator/bush assembly	1		
8	BD 20013 7		Washer plain	2		
9	2390		Pin split	1		
10	CAC 3835		Rod operating assembly	1		
11	BD 20013 9		Washer plain	2		
12	CAC 3823		Plate attachment assembly	1		
13	CAC 3826		Seal attachment plate	1		
	CAC 6200		Steady bracket, acc cable	1		
	DRC 1449 J		Grommet	1		

MASTER INDEX

ENGINE 1.C
FLYWHEEL AND CLUTCH 1.G
GEARBOX AND PROPSHAFT 1.H
AXLES, SUSPENSION, DRIVE SHAFTS, WHEELS 1.K
STEERING 1.M
BRAKES AND BRAKE CONTROLS 1.N
FUEL, EXHAUST AND EMISSION SYSTEMS 2.C
COOLING, HEATING AND AIR CONDITIONING 2.G
ELECTRICAL, WASHERS-WIPERS, INSTRUMENTS 2.J
CHASSIS, SUBFRAMES, BODYSHELL, FITTINGS 3.C
FASCIA, TRIM, SEATS AND FIXINGS 3.H
ACCESSORIES AND PAINT 4.C
NUMERICAL INDEX 4.D

GROUP INDEX

ACCELERATOR CABLE LHD 3.6 LITRE C16
ACCELERATOR CABLE LHD-5.3 LITRE C17
ACCELERATOR CABLE RHD 3.6 LITRE C13
ACCELERATOR CABLE RHD 5.3 LITRE C14
AIR CLEANER RH-5.3 LITRE E02
AIR CLEANER-3.6 LITRE C02
AIR INJECTION PUMP 3.6 LITRE C09
AIR INJECTION PUMP-5.3 LITRE F03
FRONT EXHAUST PIPE RH F10
FUEL FILLER CAP - CONVERTIBLE D11
FUEL FILLER CAP-CABRIOLET D10
FUEL PUMP AND FILTER D04
FUEL RAIL-INJECTORS-3.6 LITRE C05
FUEL TANK-CABRIOLET D09

2C 17

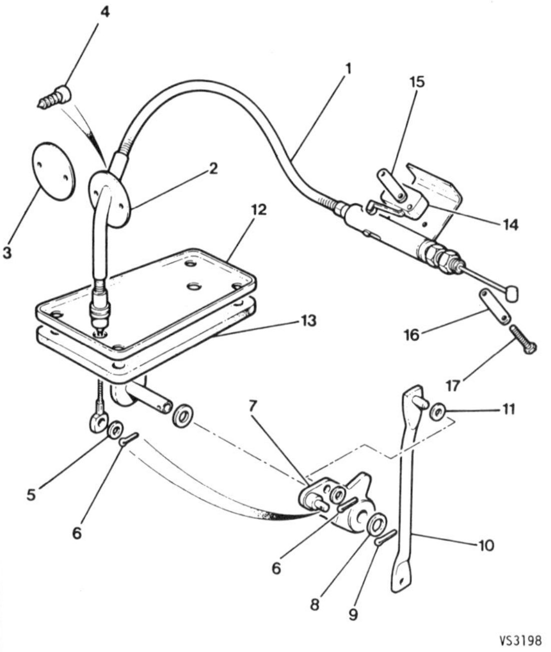

VS3198

ACCELERATOR CABLE LHD-5.3 LITRE.

Illus	Part Number	123456 Description	Quantity	Change Point	Remarks
1	CBC 1910 1	Cable accelerator	1		
2	C 44630	Plate blanking	1		
3	C 44602	Gasket	1		
4	DAZ 408 8C	Screw self tapping	2		
5	BD 541 28	Washer plain	1		
6	L 102 4U	Pin split	1		
7	CAC 3827	Lever accelerator/bush assembly	1		
8	BD 20013 7	Washer plain	1		
9	L 102 4U	Pin split	1		
10	CAC 3835	Rod operating assembly	1		
11	BD 20013 9	Washer plain	2		
12	CAC 3823	Plate attachment assembly	1		
13	CAC 3826	Seal attachment plate	1		
14	C 33509	Switch	1		
15	C 35840	Clamp plate	1		
16	C 33784	Fixing plate	1		
17	UCS 111 6R	Setscrew	2		

MASTER INDEX

GROUP INDEX

2C 18

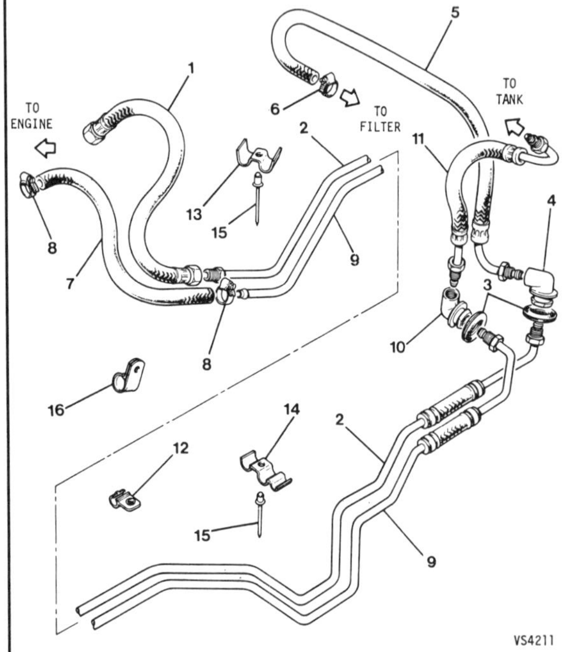

VS 4211

FUEL FEED AND RETURN PIPES 3.6 LITRE

Illus	Part Number	123456 Description	Quantity	Change Point	Remarks
1	CAC 6241	Fuel feed hose-fuel rail to pipes	NLA		Use CBC 5355
	CBC 5355	Fuel hose assembly (feed)	1		
2	CAC 5343	Fuel feed pipe assembly	1		
3	BD 37522	Grommet	2		
4	CAC 5406	Connector-feed hose	1		
5	CBC 4897	Hose-connector to filter	1		
6	C 43599 3	Clip-hose to filter	1		
7	CAC 5867 22	Hose-fuel rail to return pipe	NLA		Use CBC 1976
	CBC 1976	Fuel hose assembly (return)	1		
8	C 43206 1	Clip	2		
9	CBC 4743	Fuel return pipe assembly	1		
10	C 41770	Connector-return hose	1		
11	CAC 1646	Fuel return hose	NLA		Use CBC 8189
	CBC 8189	Fuel return hose	1		
12	C 32746 1	Pipe clip (single)	7		
13	C 24599	Pipe clip (double)	8		
14	CAC 5652	Pipe clip (treble)	5		
15	BD 24978 2	Rivet	20		
	CAC 8662	Bracket-fuel hose retention	1		
	CAC 8635	Double clip	1		
	C 17432	Grommet - forward bulkhead	2		
	AGU 2098 J	Clip - CBC 5355 to rail	2		
	FB 106061 J	Bolt-flange head (for clip)	2		

MASTER INDEX

ENGINE 1.C
FLYWHEEL AND CLUTCH 1.G
GEARBOX AND PROPSHAFT 1.H
AXLES, SUSPENSION, DRIVE SHAFTS, WHEELS 1.K
STEERING 1.M
BRAKES AND BRAKE CONTROLS 1.N
FUEL, EXHAUST AND EMISSION SYSTEMS 2.C
COOLING, HEATING AND AIR CONDITIONING 2.G
ELECTRICAL, WASHERS-WIPERS, INSTRUMENTS 2.J
CHASSIS, SUBFRAMES, BODYSHELL, FITTINGS 3.C
FASCIA, TRIM, SEATS AND FIXINGS 3.H
ACCESSORIES AND PAINT 4.C
NUMERICAL INDEX 4.D

GROUP INDEX

ACCELERATOR CABLE LHD 3.6 LITRE C16
ACCELERATOR CABLE LHD-5.3 LITRE C17
ACCELERATOR CABLE RHD 3.6 LITRE C13
ACCELERATOR CABLE RHD 5.3 LITRE C14
AIR CLEANER RH-5.3 LITRE E02
AIR CLEANER-3.6 LITRE C02
AIR INJECTION PUMP 3.6 LITRE C09
AIR INJECTION PUMP-5.3 LITRE F03
FRONT EXHAUST PIPE RH F10
FUEL FILLER CAP - CONVERTIBLE D11
FUEL FILLER CAP-CABRIOLET D10
FUEL PUMP AND FILTER D04
FUEL RAIL-INJECTORS-3.6 LITRE C05
FUEL TANK-CABRIOLET D09

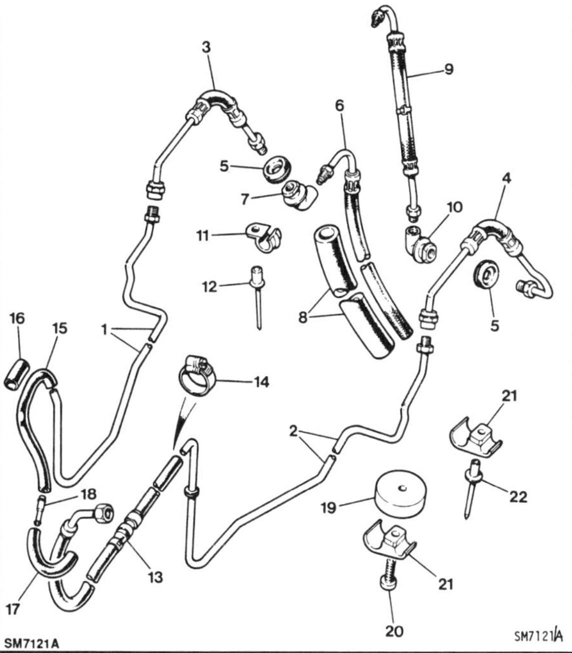

SM7121A

FUEL FEED AND RETURN PIPES-5.3 LITRE

Illus	Part Number	123456 Description	Quantity	Change Point	Remarks
		FUEL PIPE UNDERFLOOR			
1	CAC 7877	RH	1		Use CBC 4940
2	CAC 7878	LH	1		Use CBC 4941
	CBC 4940	Fuel pipe underfloor feed	1		
	CBC 4941	Fuel pipe underflorr return	1		
3	CAC 2878	Fuel pipe overaxle-RH	1		Use CBC 4938
4	CAC 2879	Fuel pipe overaxle-LH	1		Use CBC 4939
	CBC 4938	Fuel pipe overaxle-RH	1		
	CBC 4939	Fuel pipe overaxle-LH	1		
5	BD 37522	Grommet	2		
	BD 21899	Rubber plug	1		
6	CBC 4896	Fuel pipe	1		
7	C 43200	Elbow-fuel pipe	1		
8	CAC 4307 2	Sleeve	1		
9	CAC 1646	Fuel return pipe assembly	NLA		Use CBC 8189
	CBC 8189	Return hose assembly	1		
10	C 41770	Elbow fuel pipe	1		
11	C 32746 3	Pipe clip	2		
12	BD 24978 1	Rivet	2		
13	CBC 2179	Feed pipe, underfloor to engine	1		
	CBC 6710	Feed pipe, underfloor to engine	NLA		Use CBC 7979
	CBC 7979	Feed pipe, underfloor to engine	1		
14	C 43599 10	Hose clip	2		
	C 43599 9	Hose clip	4		
15	CAC 5868	Fuel hose	1		
16	C 38494 3	Sleeve	1		
17	CAC 5867	Fuel hose	1		
18	C 44479	Reducer-fuel feed	1		
19	C 38887	Spacer	2		
20	DBZ 80812 C	Screw	2		
21	C 44781	Clip	16		
22	BD 24978 1	Rivet	14		
	C 17432	Grommet-forward bulkhead	2		

H34JP13/7/88

MASTER INDEX

GROUP INDEX

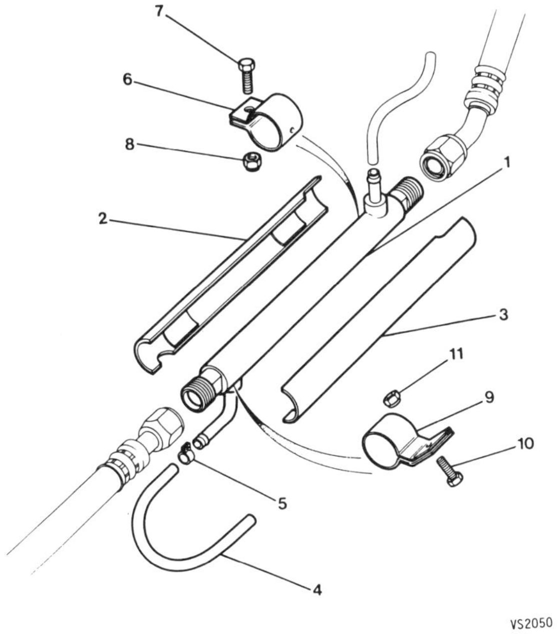

VS2050

FUEL COOLER 5.3 LITRE

Illus	Part Number	123456 Description	Quantity	Change Point	Remarks
1	CAC 7912	Fuel cooler	1		
2	CAC 3916	Insulation sleeve RH	1		
3	CAC 3917	Insulation sleeve LH	1		
4	CAC 5868	Fuel hose	1		
5	C 43599 9	Clip	1		
6	CAC 3723	Bracket	1		
7	UFS 119 4R	Bolt	1		
8	TN 3205 J	Nut	1		
9	CAC 3722	Bracket	1		
10	UFS 119 4R	Bolt	1		
11	TN 3205 J	Nut			

MASTER INDEX

ENGINE 1.C
FLYWHEEL AND CLUTCH 1.G
GEARBOX AND PROPSHAFT 1.H
AXLES, SUSPENSION, DRIVE SHAFTS, WHEELS 1.K
STEERING 1.M
BRAKES AND BRAKE CONTROLS 1.N
FUEL, EXHAUST AND EMISSION SYSTEMS 2.C
COOLING, HEATING AND AIR CONDITIONING 2.G
ELECTRICAL, WASHERS-WIPERS, INSTRUMENTS 2.J
CHASSIS, SUBFRAMES, BODYSHELL, FITTINGS 3.C
FASCIA, TRIM, SEATS AND FIXINGS 3.H
ACCESSORIES AND PAINT 4.C
NUMERICAL INDEX 4.D

GROUP INDEX

ACCELERATOR CABLE LHD 3.6 LITRE C16
ACCELERATOR CABLE LHD-5.3 LITRE C17
ACCELERATOR CABLE RHD 3.6 LITRE C13
ACCELERATOR CABLE RHD 5.3 LITRE C14
AIR CLEANER RH-5.3 LITRE E02
AIR CLEANER-3.6 LITRE C02
AIR INJECTION PUMP 3.6 LITRE C09
AIR INJECTION PUMP-5.3 LITRE F03
FRONT EXHAUST PIPE RH F10
FUEL FILLER CAP - CONVERTIBLE D11
FUEL FILLER CAP-CABRIOLET D10
FUEL PUMP AND FILTER D04
FUEL RAIL-INJECTORS-3.6 LITRE C05
FUEL TANK-CABRIOLET D09

2D 04

FUEL PUMP AND FILTER.
5.3 LITRE AND 3.6 LITRE

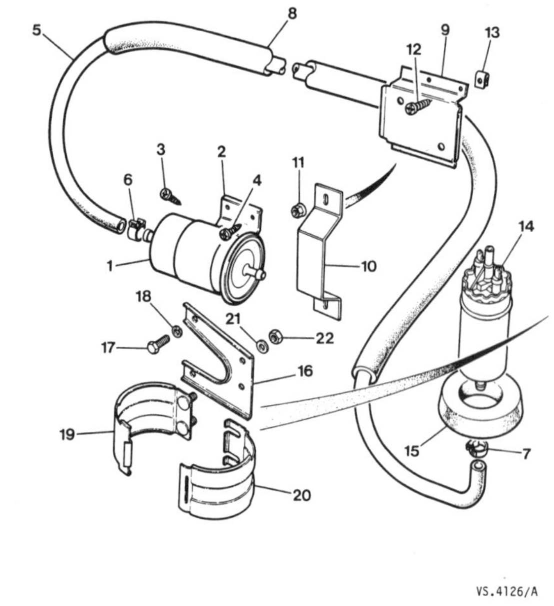

VS4126/A

Illus	Part Number	123456 Description	Quantity	Change Point	Remarks
1	EAC 3112	Fuel filter	1		
2	CAC 3684	Mounting bracket(fuel filter)	1		
3	AB 610051 J	Screw-clamping bracket	2		
4	AB 610031 J	Screw-bracket to body	2		
5	CBC 4895	Fuel hose filter to pump	1		
6	C 43599 9	Clip-hose to filter	1		
7	C 43206 10	Clip-hose to pump	1		
8	C 38494 3	Insulation sleeve	1		
9	CAC 3685	Hose guard	1		
10	BCC 7689	Hose guard	1	....	Convertible only
11	AGU 2436	Nut plastic push on	2	....	Convertible only
12	AB 610051 J	Screw-guard to body	2		
13	FU 2545	Spirenut-guard to body	2		
14	CAC 4269	Fuel pump	1	....	Use CBC5657
	CBC 5657	Fuel pump	1		
15	CAC 2481	Mounting rubber	2		
16	CAC 3132	Mounting bracket (fuel pump)	1		
17	SH 604041 J	Setscrew	2		
18	WM 600041 J	Lockwasher	2		
19	CAC 3477	Strap (fuel pump)	1		
20	CAC 1706	Strap (fuel pump)	1		
21	JLM 298 T	Washer-strap to bracket	2		
22	C 8737 1	Locknut-strap to bracket	2		

H40AW5/7/88

MASTER INDEX

GROUP INDEX

2D 05

FUEL SUMP TANK 3.6 LITRE

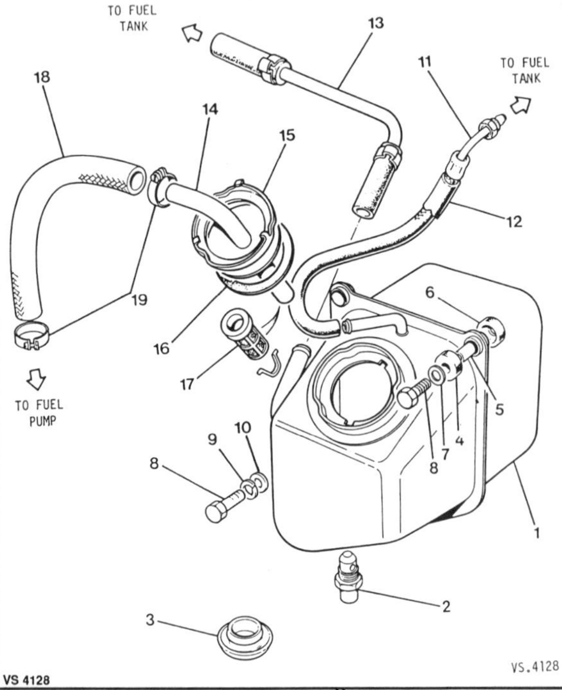

VS 4128

Illus	Part Number	123456 Description	Quantity	Change Point	Remarks
1	C 41794	Fuel sump tank	1		
2	C 41803	Drain plug-sump tank	NLA	....	Use CBC4518
	CBC 4518	Magnetic drain plug	1		
3	BD 15588 8	Sealing plug-drain plug	1		
4	C 41802	Spacer	2		
5	C 41801	Bush	2		
6	C 41856	Spacer	2		
7	FW 104 T	Washer	2		
8	BH 604091 J	Bolt	3		
9	WM 600041 J	Lockwasher	3		
10	C 21930	Washer	1		
11	CAC 1648	Breather hose	NLA	....	Use CBC6346
	CBC 6346	Breather hose	1		
12	C 38494 3	Insulating tube	1		
13	CBC 4722	Fuel feed hose	1		
	EAC 3215 5	Clip-feed hose to sump tank	2		
14	C 43196	Fuel suction pipe	1		
15	ARA 1501 J	Locking ring	1		
16	ARA 1502 J	Sealing ring	1		
17	CAC 3762	Fuel filter	NLA	....	Use CBC5649
	CBC 5649	Fuel filter	1		
18	C 46109 1	Hose-sump tank to fuel pump	1		
19	C 43599 14	Hose clip	2		

MASTER INDEX

GROUP INDEX

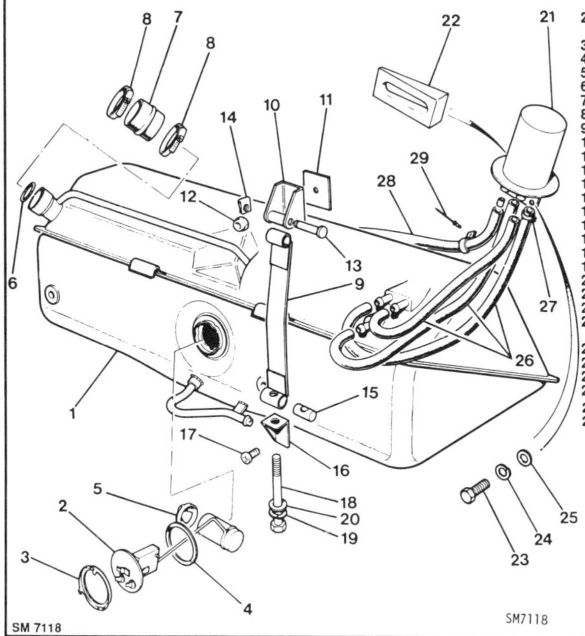

SM 7118

FUEL TANK COUPE 3.6 LITRE & 5.3 LITRE

Illus	Part Number	123456 Description	Quantity	Change Point	Remarks
1	CBC 4760 1	Fuel tank assembly	1		
	CBC 5525 1	Fuel tank assembly	1		1988 1/2 MY
2	C 42123	Tank gauge unit	NLA		Use DAC 5500
	DAC 5500	Tank gauge unit	1		
3	ARA 1501 J	Locking ring	1		
4	ARA 1502 J	Sealing ring	1		
5	C 31131	End cap	1		
6	C 46157	'O' ring-fuel tank neck	1		
7	C 43958	Fuel filler hose	1		
8	C 43206 7	Hose clip	2		
9	C 42806	Fuel tank strap	2		
10	C 42804	Upper mounting bracket	2		
11	C 42803	Reinforcement plate	2		
12	NY 606041 J	Locknut	2		
13	J 20829 S	Joint pin	2		
14	C 273286 6Z	Clip	2		
15	C 42817	Trunnion	2		
16	C 42805	Lower mounting bracket	2		
17	UFS 937 8R	Setscrew	2		
18	BH 606481 J	Bolt	2		
19	WM 600061 J	Lockwasher	2		
20	FW 106 T	Washer	2		
21	C 44144	Vapour seperator	1		
22	C 27737	Mounting block	1		
23	JLM 9675	Setscrew	NLA		Use CAC 2015
	CAC 2015	Paint clearing screw 0.25 UNF	2		
24	WM 600041 J	Lockwasher	2		
25	FW 104 T	Washer	2		
26	CAC 5866	Hose	3		1 metre length
27	C 43599 6	Clip	3		
28	CAC 5867	Hose	1		
29	BD 24978/1	Rivet	1		

MASTER INDEX

GROUP INDEX

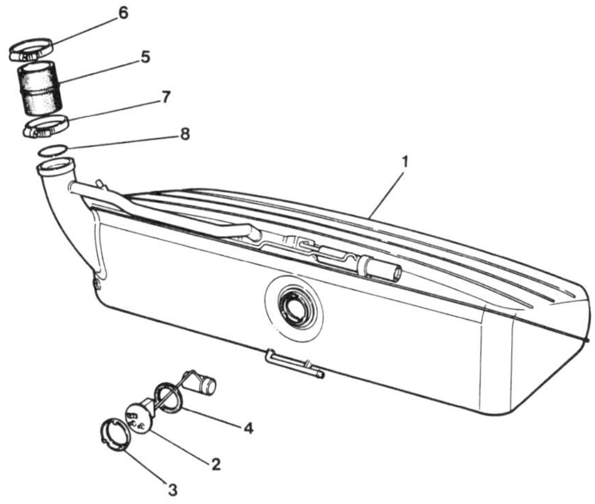

VS.4793

VS4793

FUEL TANK - CONVERTIBLE 5.3

Illus	Part Number	123456 Description	Quantity	Change Point	Remarks
1	CBC 6072	Fuel tank	NLA		Use CBC6579/1
	CBC 6579 1	Fuel tank	NLA		Use CBC6644/1
2	DAC 4405	Gauge unit-fuel tank	1		
	CBC 6644 1	Fuel tank	1		
	DAC 6103	Gauge unit-fuel tank	1		
3	ARA 1501 J	Locking ring	1		
4	ARA 1502 J	Sealing ring	1		
5	CBC 2887	Gaiter-fuel filler pipe	1		
6	EAC 3215 13	Clip-gaiter to filler cap	1		
7	EAC 3215 14	Clip-gaiter to tank nect	1		
8	CAC 7509	'O' ring-fuel tank neck	1		
	CBC 4238	Filler breather hose	1		
	CAC 8772	Hose clip	2		
	C 15150 3	Drain tube	1		To fuel filler box

MASTER INDEX

ENGINE 1.C
FLYWHEEL AND CLUTCH 1.G
GEARBOX AND PROPSHAFT 1.H
AXLES, SUSPENSION, DRIVE SHAFTS, WHEELS 1.K
STEERING 1.M
BRAKES AND BRAKE CONTROLS 1.N
FUEL, EXHAUST AND EMISSION SYSTEMS 2.C
COOLING, HEATING AND AIR CONDITIONING 2.G
ELECTRICAL, WASHERS-WIPERS, INSTRUMENTS 2.J
CHASSIS, SUBFRAMES, BODYSHELL, FITTINGS 3.C
FASCIA, TRIM, SEATS AND FIXINGS 3.H
ACCESSORIES AND PAINT 4.C
NUMERICAL INDEX 4.D

GROUP INDEX

ACCELERATOR CABLE LHD 3.6 LITRE C16
ACCELERATOR CABLE LHD-5.3 LITRE C17
ACCELERATOR CABLE RHD 3.6 LITRE C13
ACCELERATOR CABLE RHD 5.3 LITRE C14
AIR CLEANER RH-5.3 LITRE E02
AIR CLEANER-3.6 LITRE C02
AIR INJECTION PUMP 3.6 LITRE C09
AIR INJECTION PUMP-5.3 LITRE F03
FRONT EXHAUST PIPE RH F10
FUEL FILLER CAP - CONVERTIBLE D11
FUEL FILLER CAP-CABRIOLET D10
FUEL PUMP AND FILTER D04
FUEL RAIL-INJECTORS-3.6 LITRE C05
FUEL TANK-CABRIOLET D09

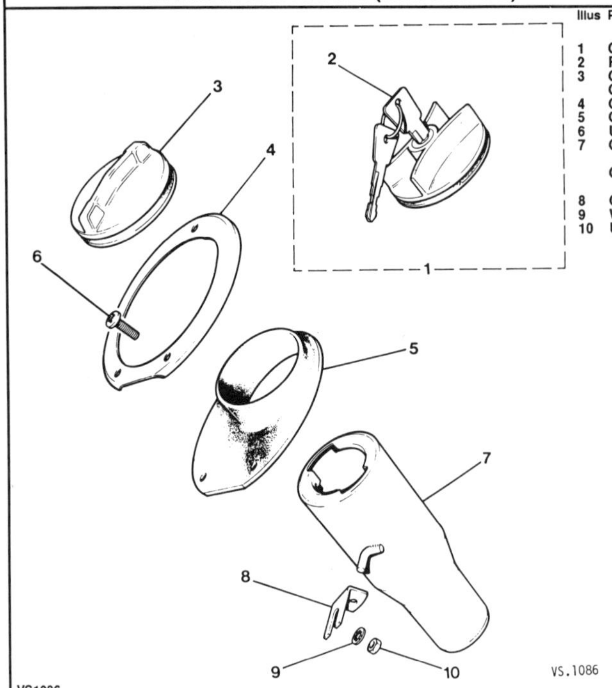

VS1086

FUEL FILLER CAP COUPE 3.6 LITRE & 5.3 LITRE

Illus	Part Number	123456 Description	Quantity	Change Point	Remarks
1	CAC 3565	Fuel filler cap (lockable)	1		Less USA
2	RTC 2182 B	Key (blank)	1		
3	CAC 2961	Fuel filler cap (non lockable)	1		USA
	CBC 6027	Fuel filler cap (non lockable)	1		All markets
4	CAC 2404	Clamp ring	1		
5	CAC 2403	Grommet	1		
6	UFS 819 5H	Setscrew-securing ring/grommet	3		
7	CAC 2395	Fuel filler pipe	1		Not USA/CDN/ Switzerland
	CAC 2398	Fuel filler pipe	1		5.3 USA/CDN/ Switz Unleaded
8	CAC 2405	Support bracket	1		
9	WF 702101 J	Washer	1		
10	UFN 119 L	Nut	1		

2D 08

MASTER INDEX

GROUP INDEX

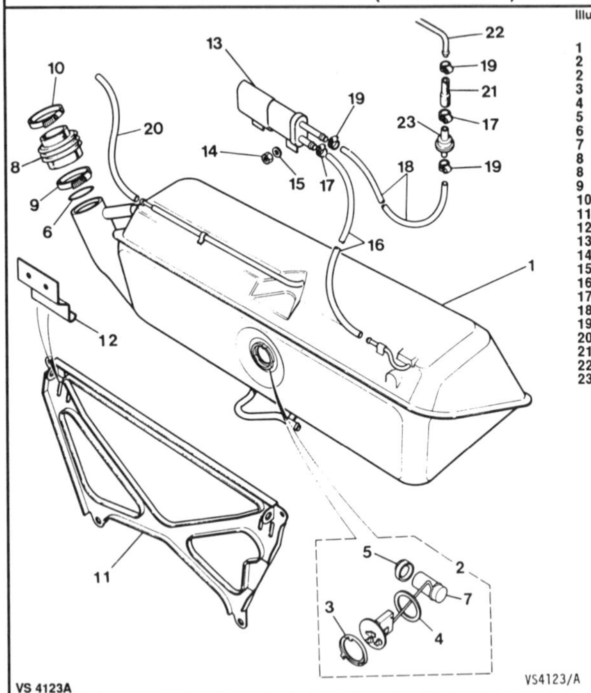

VS 4123A

FUEL TANK-CABRIOLET

Illus	Part Number	123456 Description	Quantity	Change Point	Remarks
1	CBC 4762 1	Tank-fuel	1		
2	C 42123	Element-tank	1		Use DAC 5525
2	DAC 5525	Element-tank	1		
3	ARA 1501 J	Ring-locking	1		
4	ARA 1502 J	Washer-sealing	1		
5	C 31131	Cap-anti-rattle	1		
6	CAC 7509	'O' ring-filler	1		
7	JLM 772	Float-Element	1		
8	CAC 7508	Gaiter-filler	1		Use CBC 2887
8	CBC 2887	Gaiter-filler	1		
9	EAC 3215 14	Clip-tank	1		
10	EAC 3215 13	Clip-filler cap	1		
11	CAC 7203	Stiffener-boot	1		
12	CAC 7204	Clamp	2		
13	CAC 7599	Separator-vapour	1		
14	UFN 119 L	Nut-plain	2		
15	JLM 9572	Washer-plain	2		
16	CAC 5867	Hose	1		1 Metre length
17	EAC 3215 2	Clip	2		
18	CAC 5866	Hose	1		1 Metre length
19	EAC 3215 1	Clip	1		
20	CAC 8900 11	Hose	1		
21	CAC 7586	Connector	1		
22	C 41993	Vapour pipe	1		
23	CAC 3101	Vacuum valve	1		

2D 09

MASTER INDEX

ENGINE ... 1.C
FLYWHEEL AND CLUTCH ... 1.G
GEARBOX AND PROPSHAFT ... 1.H
AXLES, SUSPENSION, DRIVE SHAFTS, WHEELS ... 1.K
STEERING ... 1.M
BRAKES AND BRAKE CONTROLS ... 1.N
FUEL, EXHAUST AND EMISSION SYSTEMS ... 2.C
COOLING, HEATING AND AIR CONDITIONING ... 2.G
ELECTRICAL, WASHERS-WIPERS, INSTRUMENTS ... 2.J
CHASSIS, SUBFRAMES, BODYSHELL, FITTINGS ... 3.C
FASCIA, TRIM, SEATS AND FIXINGS ... 3.H
ACCESSORIES AND PAINT ... 4.C
NUMERICAL INDEX ... 4.D

GROUP INDEX

ACCELERATOR CABLE LHD 3.6 LITRE ... C16
ACCELERATOR CABLE LHD-5.3 LITRE ... C17
ACCELERATOR CABLE RHD 3.6 LITRE ... C13
ACCELERATOR CABLE RHD 5.3 LITRE ... C14
AIR CLEANER RH-5.3 LITRE ... E02
AIR CLEANER-3.6 LITRE ... C02
AIR INJECTION PUMP 3.6 LITRE ... C09
AIR INJECTION PUMP-5.3 LITRE ... F03
FRONT EXHAUST PIPE RH ... F10
FUEL FILLER CAP - CONVERTIBLE ... D11
FUEL FILLER CAP-CABRIOLET ... D10
FUEL PUMP AND FILTER ... D04
FUEL RAIL-INJECTORS-3.6 LITRE ... C05
FUEL TANK-CABRIOLET ... D09

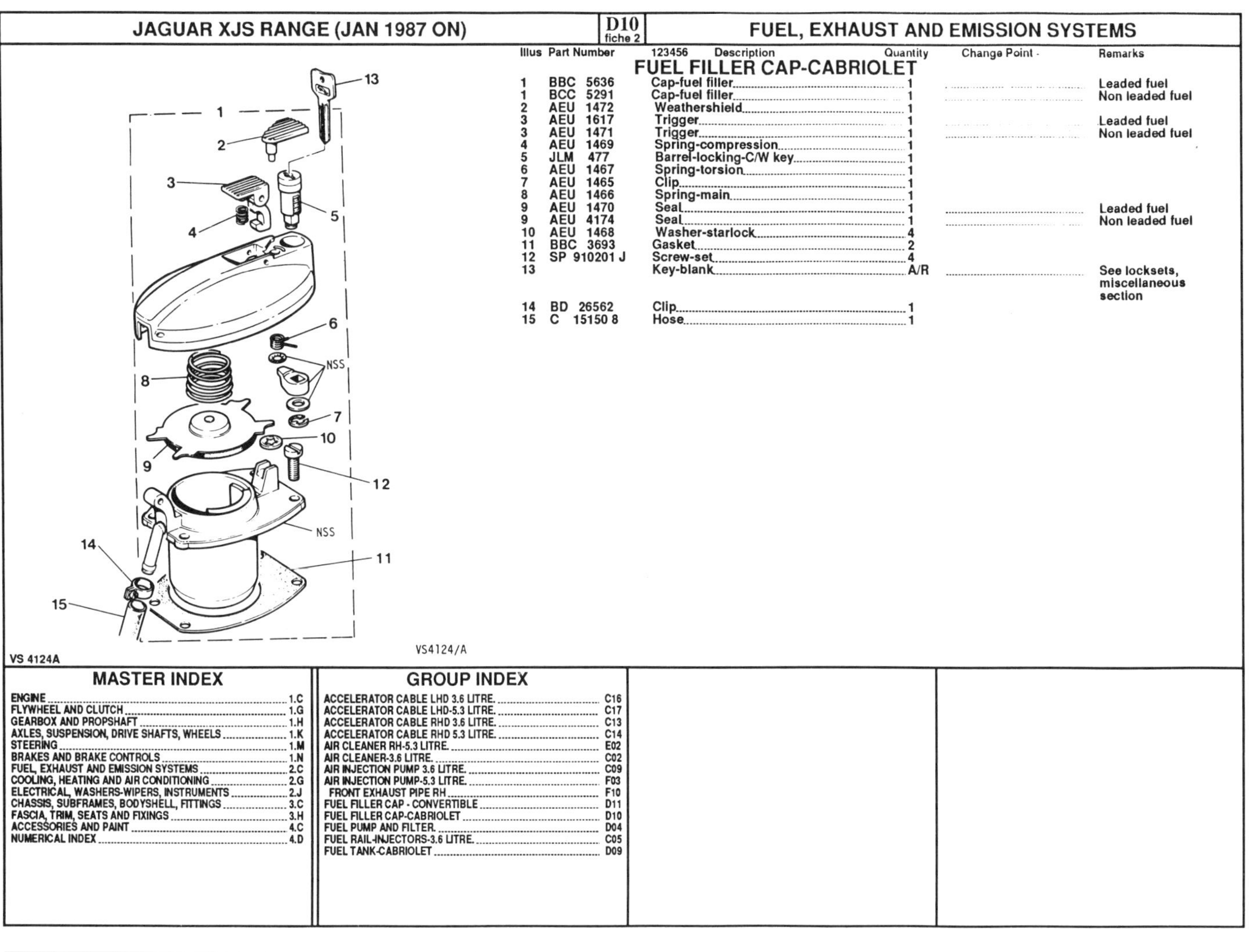

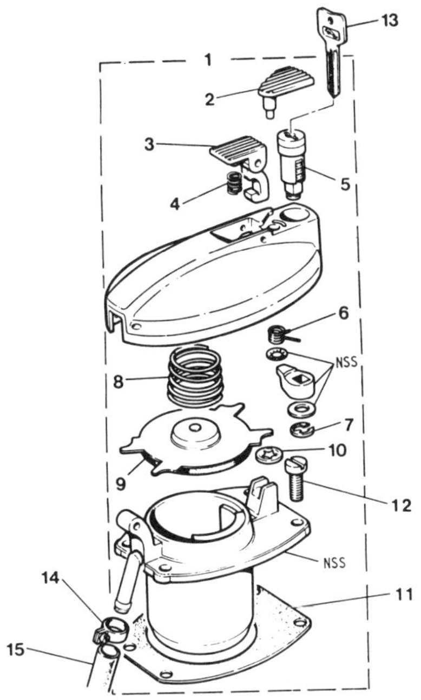

VS 4124A

FUEL FILLER CAP-CABRIOLET

Illus	Part Number	Description	Quantity	Change Point	Remarks
1	BBC 5636	Cap-fuel filler	1		Leaded fuel
1	BCC 5291	Cap-fuel filler	1		Non leaded fuel
2	AEU 1472	Weathershield	1		
3	AEU 1617	Trigger	1		Leaded fuel
3	AEU 1471	Trigger	1		Non leaded fuel
4	AEU 1469	Spring-compression	1		
5	JLM 477	Barrel-locking-C/W key	1		
6	AEU 1467	Spring-torsion	1		
7	AEU 1465	Clip	1		
8	AEU 1466	Spring-main	1		
9	AEU 1470	Seal	1		Leaded fuel
9	AEU 4174	Seal	1		Non leaded fuel
10	AEU 1468	Washer-starlock	4		
11	BBC 3693	Gasket	2		
12	SP 910201 J	Screw-set	4		
13		Key-blank	A/R		See locksets, miscellaneous section
14	BD 26562	Clip	1		
15	C 15150 8	Hose	1		

MASTER INDEX

GROUP INDEX

FUEL FILLER CAP - CONVERTIBLE

Illus	Part Number	Description	Quantity	Change Point	Remarks
1	BDC 7628	Cap - fuel filler	1		Leaded fuel
1	BDC 7627	Cap - fuel filler	1		Non leaded fuel
2	BBC 3693	Gasket - fuel cap	1		

ILLUSTRATION TO FOLLOW

ILLUSTRATION SUIT

ABBILDUNG WIRD NACHGEREIGHT

SEGUE ILLUSTRAZIONE

TF

MASTER INDEX

ENGINE 1.C
FLYWHEEL AND CLUTCH 1.G
GEARBOX AND PROPSHAFT 1.H
AXLES, SUSPENSION, DRIVE SHAFTS, WHEELS 1.K
STEERING 1.M
BRAKES AND BRAKE CONTROLS 1.N
FUEL, EXHAUST AND EMISSION SYSTEMS 2.C
COOLING, HEATING AND AIR CONDITIONING 2.G
ELECTRICAL, WASHERS-WIPERS, INSTRUMENTS 2.J
CHASSIS, SUBFRAMES, BODYSHELL, FITTINGS 3.C
FASCIA, TRIM, SEATS AND FIXINGS 3.H
ACCESSORIES AND PAINT 4.C
NUMERICAL INDEX 4.D

GROUP INDEX

ACCELERATOR CABLE LHD 3.6 LITRE C16
ACCELERATOR CABLE LHD-5.3 LITRE C17
ACCELERATOR CABLE RHD 3.6 LITRE C13
ACCELERATOR CABLE RHD 5.3 LITRE C14
AIR CLEANER RH-5.3 LITRE E02
AIR CLEANER-3.6 LITRE C02
AIR INJECTION PUMP 3.6 LITRE C09
AIR INJECTION PUMP-5.3 LITRE F03
FRONT EXHAUST PIPE RH F10
FUEL FILLER CAP - CONVERTIBLE D11
FUEL FILLER CAP-CABRIOLET D10
FUEL PUMP AND FILTER D04
FUEL RAIL-INJECTORS-3.6 LITRE C05
FUEL TANK-CABRIOLET D09

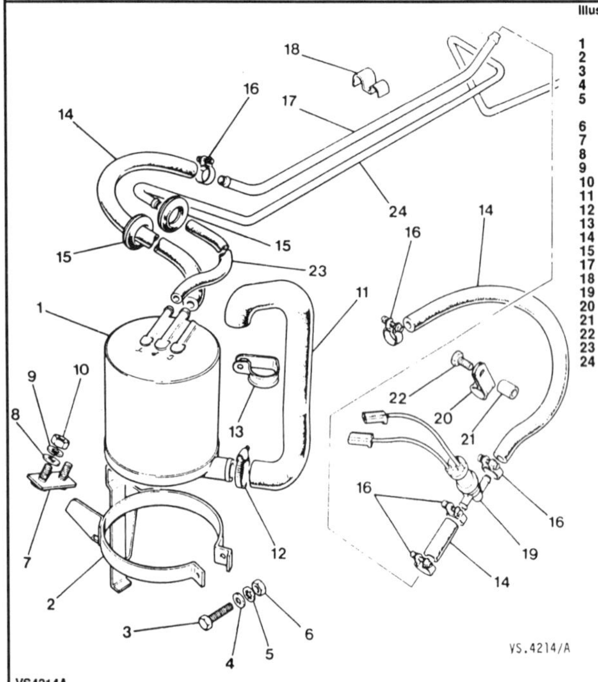

PURGE VALVE AND CANISTER USA 3.6 LITRE

Illus	Part Number	123456 Description	Quantity	Change Point	Remarks
1	CAC 3814	Charcoal canister	1		
2	CAC 3603	Mounting bracket	1		
3	JLM 9675	Setscrew canister to bracket	1		
4	JLM 296	Plain washer canister to bracket	1		
5	WF 600040 J	Shakeproof washer canister to bracket	1		
6	JLM 9683	Nut	1		
7	C 43378	Stud plate bracket to body	1		
8	JLM 298	Plain washer bracket to body	2		
9	WM 600051 J	Lockwasher bracket to body	2		
10	JLM 9684	Nut bracket to body	2		
11	CAC 2571	Breather hose	1		
12	EAC 3215 8	Hose clip	1		
13	C 1040 21	'P' clip	1		
14	CAC 8900 11	Hose	A/R		1 metre length
15	C 976	Grommet	2		
17	CBC 2705	Purge pipe	1		
18	CAC 6659	Pipe clip	2		
19	CAC 6874	Purge valve	1		
20	CAC 1040 3	'P' clip	1		
21	C 33620	Spacer	1		
22	AR 610061 J	Setscrew	1		
23	CAC 1853 1	Hose-canister to vapour pipe	1		
24	CAC 5345	Vapour pipe assembly	1		

2D 12

MASTER INDEX

GROUP INDEX

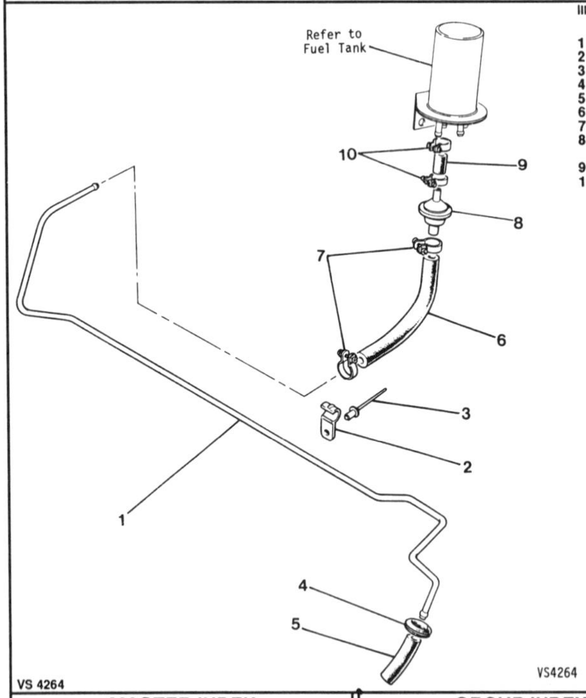

REAR VAPOUR USA 3.6 LITRE

Illus	Part Number	123456 Description	Quantity	Change Point	Remarks
1	CAC 8285	Rear vapour pipe	1		
2	C 32746 1	Pipe clip	3		
3	BD 24978 1	Rivet-securing clip	3		
4	C 42013	Grommet	1		
5	CAC 5866	Hose-vapour pipe to vap pipe	1		1 metre lengths
6	CBC 4118 1	Hose-vapour pipe to valve	1		
7	C 43599 8	Hose clip	2		
8	CAC 3101	Vacuum valve	NLA		Use CBC6280
	CBC 6280	Vacuum valve	1		
9	CAC 5867	Hose-valve to seperator	1		1 metre lengths
10	C 43599 6	Hose clip	2		

2D 13

MASTER INDEX

ENGINE 1.C
FLYWHEEL AND CLUTCH 1.G
GEARBOX AND PROPSHAFT 1.H
AXLES, SUSPENSION, DRIVE SHAFTS, WHEELS 1.K
STEERING 1.M
BRAKES AND BRAKE CONTROLS 1.N
FUEL, EXHAUST AND EMISSION SYSTEMS 2.C
COOLING, HEATING AND AIR CONDITIONING 2.G
ELECTRICAL, WASHERS-WIPERS, INSTRUMENTS 2.J
CHASSIS, SUBFRAMES, BODYSHELL, FITTINGS 3.C
FASCIA, TRIM, SEATS AND FIXINGS 3.H
ACCESSORIES AND PAINT 4.C
NUMERICAL INDEX 4.D

GROUP INDEX

ACCELERATOR CABLE LHD 3.6 LITRE C16
ACCELERATOR CABLE LHD-5.3 LITRE C17
ACCELERATOR CABLE RHD 3.6 LITRE C13
ACCELERATOR CABLE RHD 5.3 LITRE C14
AIR CLEANER RH-5.3 LITRE E02
AIR CLEANER-3.6 LITRE C02
AIR INJECTION PUMP 3.6 LITRE C09
AIR INJECTION PUMP-5.3 LITRE F03
FRONT EXHAUST PIPE RH F10
FUEL FILLER CAP - CONVERTIBLE D11
FUEL FILLER CAP-CABRIOLET D10
FUEL PUMP AND FILTER D04
FUEL RAIL-INJECTORS-3.6 LITRE C05
FUEL TANK-CABRIOLET D09

BLANK

MASTER INDEX

GROUP INDEX

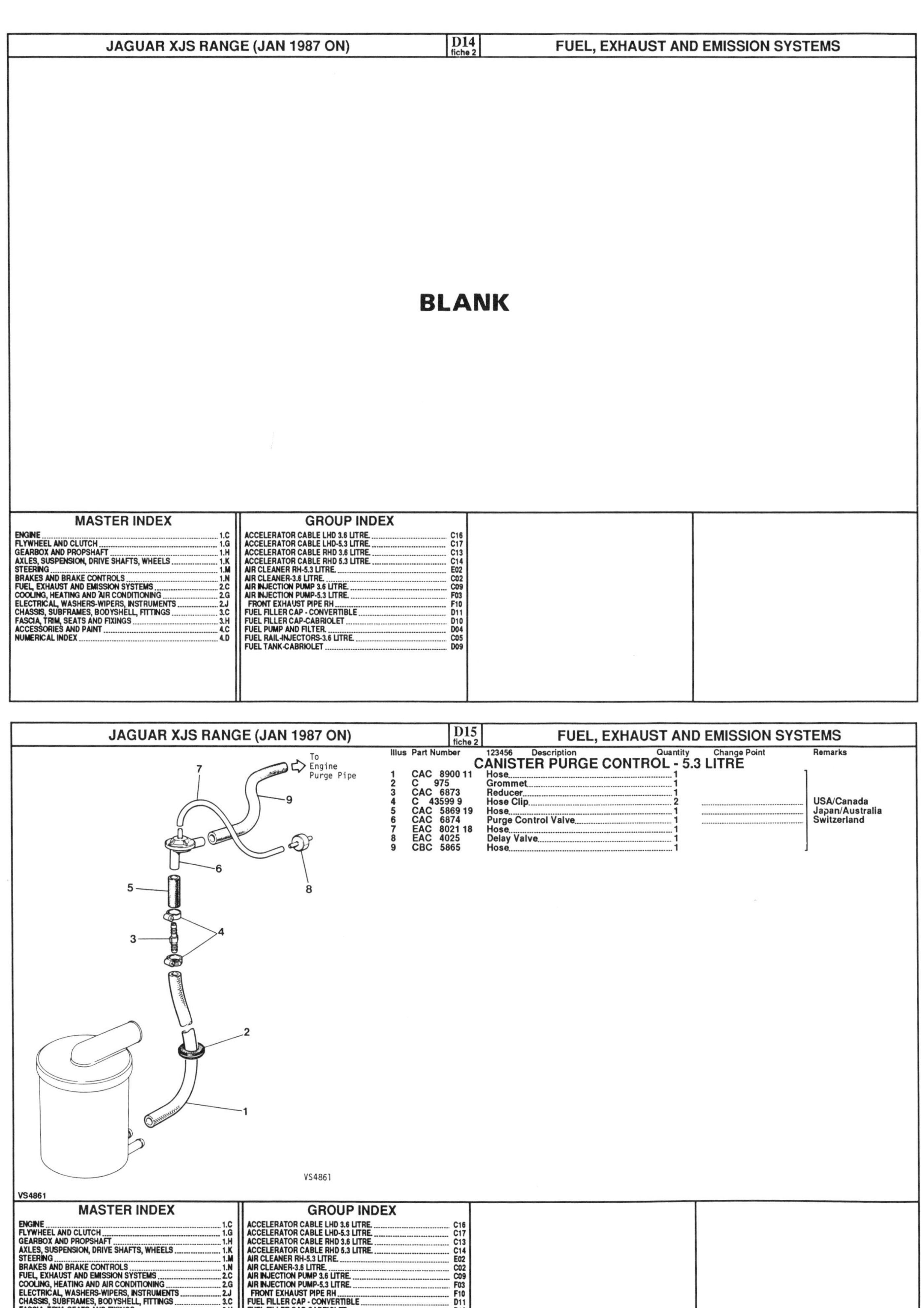

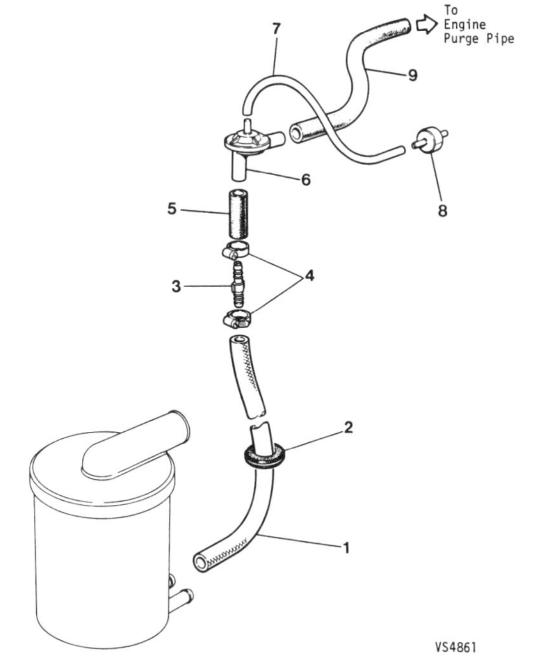

VS4861

CANISTER PURGE CONTROL - 5.3 LITRE

Illus	Part Number	123456	Description	Quantity	Change Point	Remarks
1	CAC 8900 11		Hose	1		
2	C 975		Grommet	1		
3	CAC 6873		Reducer	1		
4	C 43599 9		Hose Clip	2		USA/Canada
5	CAC 5869 19		Hose	1		Japan/Australia
6	CAC 6874		Purge Control Valve	1		Switzerland
7	EAC 8021 18		Hose	1		
8	EAC 4025		Delay Valve	1		
9	CBC 5865		Hose	1		

MASTER INDEX
ENGINE 1.C
FLYWHEEL AND CLUTCH 1.G
GEARBOX AND PROPSHAFT 1.H
AXLES, SUSPENSION, DRIVE SHAFTS, WHEELS 1.K
STEERING 1.M
BRAKES AND BRAKE CONTROLS 1.N
FUEL, EXHAUST AND EMISSION SYSTEMS 2.C
COOLING, HEATING AND AIR CONDITIONING 2.G
ELECTRICAL, WASHERS-WIPERS, INSTRUMENTS 2.J
CHASSIS, SUBFRAMES, BODYSHELL, FITTINGS 3.C
FASCIA, TRIM, SEATS AND FIXINGS 3.H
ACCESSORIES AND PAINT 4.C
NUMERICAL INDEX 4.D
GROUP INDEX
ACCELERATOR CABLE LHD 3.6 LITRE C16
ACCELERATOR CABLE LHD-5.3 LITRE C17
ACCELERATOR CABLE RHD 3.6 LITRE C13
ACCELERATOR CABLE RHD 5.3 LITRE C14
AIR CLEANER RH-5.3 LITRE E02
AIR CLEANER-3.6 LITRE C02
AIR INJECTION PUMP 3.6 LITRE C09
AIR INJECTION PUMP-5.3 LITRE F03
FRONT EXHAUST PIPE RH F10
FUEL FILLER CAP - CONVERTIBLE D11
FUEL FILLER CAP-CABRIOLET D10
FUEL PUMP AND FILTER D04
FUEL RAIL-INJECTORS-3.6 LITRE C05
FUEL TANK-CABRIOLET D09

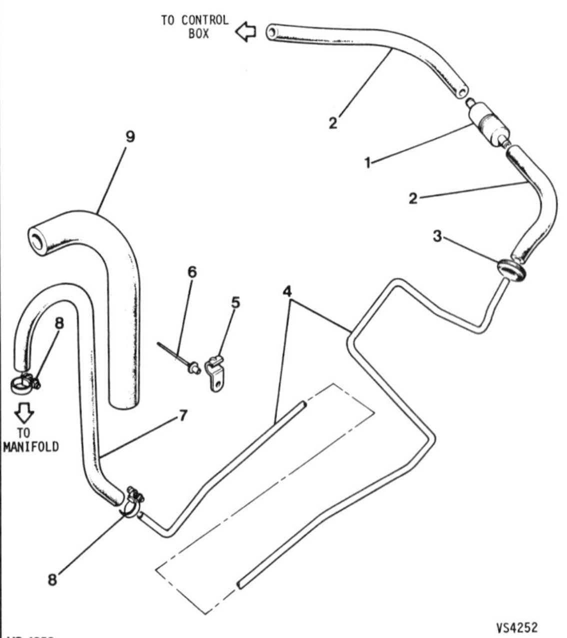

VS 4252

PRESSURE SENSOR PIPE-5.3 LITRE

Illus	Part Number	123456	Description	Quantity	Change Point	Remarks
1	CAC 4968		Reservoir	1		
2	CAC 3231 4		Hose	2		24 Inch lengths
3	C 5574		Grommet	1		
4	CAC 7849		Pressure sensor pipe	1		Manual gearbox
4	CAC 3677		Pressure sensor pipe	1		Automatic gearbox
5	C 32746 3		Pipe clip	5		
6	BD 24978 1		Rivet-securing clip	5		
7	CAC 4033 22		Hose	1		51 Inch lengths
8	C 43599 7		Hose clip	2		
9	C 38494 5		Sleeve	1		
	CBC 2591 4		Hose-pipe to reservoir	A/R		
	CBC 2591 4		Hose-reservoir to control box	A/R		

2D 16

MASTER INDEX

GROUP INDEX

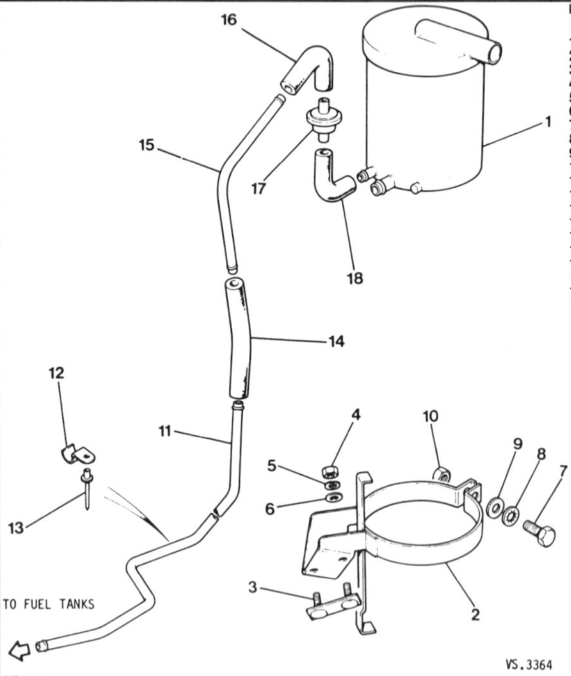

VS3364

CARBON CANISTER USA/CDN 5.3 LITRE

Illus	Part Number	123456	Description	Quantity	Change Point	Remarks
1	CAC 3814		Carbon canister	1		
2	CAC 3603		Mounting bracket assembly	1		
3	C 43378		Stud plate	1		
4	JLM 9684		Nut	2		
5	FG 105 X		Spring washer	2		
6	FW 105 E		Plain washer	2		
7	SH 604091 J		Setscrew	1		
8	C 724		Shakeproof washer	1		
9	FW 105 T		Plain washer	1		
10	JLM 9683		Nut	1		
11	C 41994		Vapour pipe intermediate	1		
12	C 32746 2		Pipe clip	9		
13	RU 610183		Imex rivet	9		
14	CAC 5866 4		Petrol hose	1		
15	C 41995		Vapour pipe-front	1		
16	CAC 5866 4		Petrol hose	1		
17	CAC 3101		Pressure reducing valve	1		Use CBC 6280
	CBC 6280		Pressure/Vacuum Valve	1		
18	CAC 4040		Fuel hose	1		

2D 17

MASTER INDEX

ENGINE 1.C
FLYWHEEL AND CLUTCH 1.G
GEARBOX AND PROPSHAFT 1.H
AXLES, SUSPENSION, DRIVE SHAFTS, WHEELS 1.K
STEERING 1.M
BRAKES AND BRAKE CONTROLS 1.N
FUEL, EXHAUST AND EMISSION SYSTEMS 2.C
COOLING, HEATING AND AIR CONDITIONING 2.G
ELECTRICAL, WASHERS-WIPERS, INSTRUMENTS 2.J
CHASSIS, SUBFRAMES, BODYSHELL, FITTINGS 3.C
FASCIA, TRIM, SEATS AND FIXINGS 3.H
ACCESSORIES AND PAINT 4.C
NUMERICAL INDEX 4.D

GROUP INDEX

ACCELERATOR CABLE LHD 3.6 LITRE C16
ACCELERATOR CABLE LHD-5.3 LITRE C17
ACCELERATOR CABLE RHD 3.6 LITRE C13
ACCELERATOR CABLE RHD 5.3 LITRE C14
AIR CLEANER RH-5.3 LITRE E02
AIR CLEANER-3.6 LITRE C02
AIR INJECTION PUMP 3.6 LITRE C09
AIR INJECTION PUMP-5.3 LITRE F03
FRONT EXHAUST PIPE RH F10
FUEL FILLER CAP - CONVERTIBLE D11
FUEL FILLER CAP-CABRIOLET D10
FUEL PUMP AND FILTER D04
FUEL RAIL-INJECTORS-3.6 LITRE C05
FUEL TANK-CABRIOLET D09

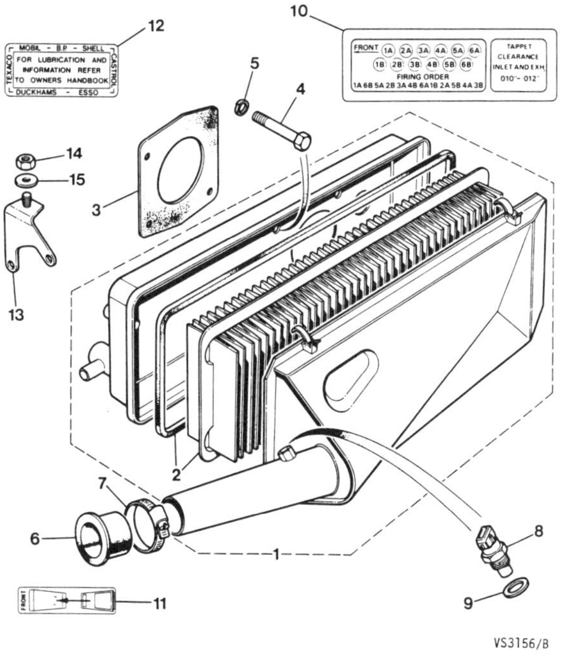

VS3156B

AIR CLEANER LH-5.3 LITRE

Illus	Part Number	123456 Description	Quantity	Change Point	Remarks
1	EAC 2946	AIR CLEANER ASSEMBLY LH	1	Up to 8S.57571	
	EAC 1828	Air filter	1		
	EBC 1077	AIR CLEANER ASSEMBLY LH	1	From 8S.57572	
2	EBC 1084	Air filter	1	To 1989 MY	
1	EBC 2525	AIR CLEANER ASSEMBLY LH	1	From 1989 MY	
2	EBC 1084	Air filter	1	Digital Ignition	
3	C 33280	Gasket	1		
4	BH 505181 J	Bolt	4		
5	WM 600051 J	Washer	4		
6	C 31422	Flare end	1		
7	C 34646 23	Clip	1		
8	EAC 2863	AIR TEMPERATURE SENSOR	1		
9	C 2296 4	Sealing washer	1		
10	EAC 4262	Firing order chart	1		
11	EAC 1133	Element label	1	Up to 8S.57571	
11	EBC 2241	Element label	1	From 8S.57572	
12	C 33884	Lubrication chart	1		
13	EBC 1180	Bracket	1	From 8S.57572	
14	NH 605041 J	Nut	1		
15	WL 600051 J	Washer	1		

H64AK1/9/88

MASTER INDEX

GROUP INDEX

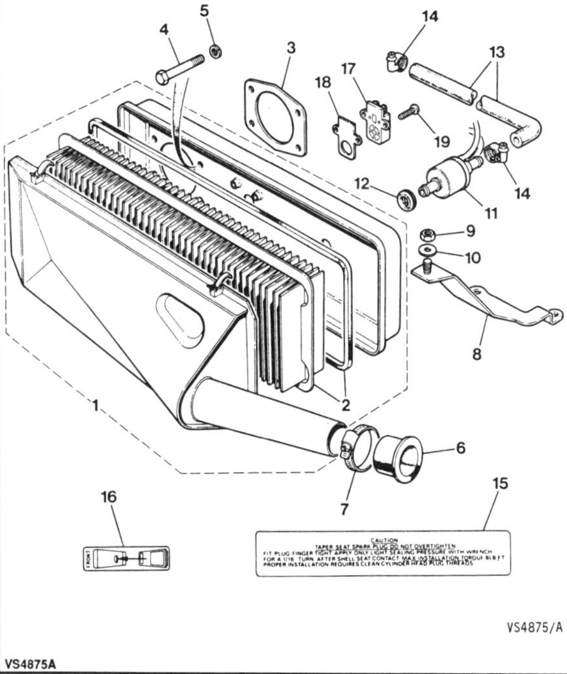

VS4875A

AIR CLEANER RH-5.3 LITRE.

Illus	Part Number	123456 Description	Quantity	Change Point	Remarks
1	EAC 3164	AIR CLEANER ASSEMBLY RH	1		Non air injection
1	EAC 4057	AIR CLEANER ASSEMBLY RH	1	Up to 8S.57571	Air Injection only
2	EAC 1828	Air filter	1		
1	EBC 9846	AIR CLEANER ASSEMBLY RH	1	From 8S.57572	
2	EBC 1084	Air filter	1	Up to 1989 MY	
1	EBC 2524	AIR CLEANER ASSEMBLY RH	1	From 1989 MY	
2	EBC 1084	Air filter	1		
3	C 33280	Gasket	1		
4	BH 505181 J	Bolt	4		
5	WM 600051 J	Washer	4		
6	C 31422	Flare end	1		
7	C 34646 23	Clip	1		
8	EBC 1181	Bracket	1		
9	NH 605041 J	Nut	1	From 8S.57572	
10	WL 600051 J	Washer	1		
11	EAC 4161	SUPPLEMENTARY AIR VALVE	1		
12	C 16393	Grommet	1		
13	EAC 4096 1	Hose-air valve to manifold	1		
14	EAC 3215 2	Clip	1		
15	EAC 4271	Label	1		
16	EAC 1133	Label	1	Up to 8S.57571	
16	EBC 2241	Label	1	From 8S.57572	
17	DAC 5061	AIR TEMPERATURE SWITCH	1	From 1989MY	
18	EBC 1974	Gasket	1		
19	SL 104061 J	Screw	1		

H64.02AK13/9/88

MASTER INDEX

GROUP INDEX

2E 03

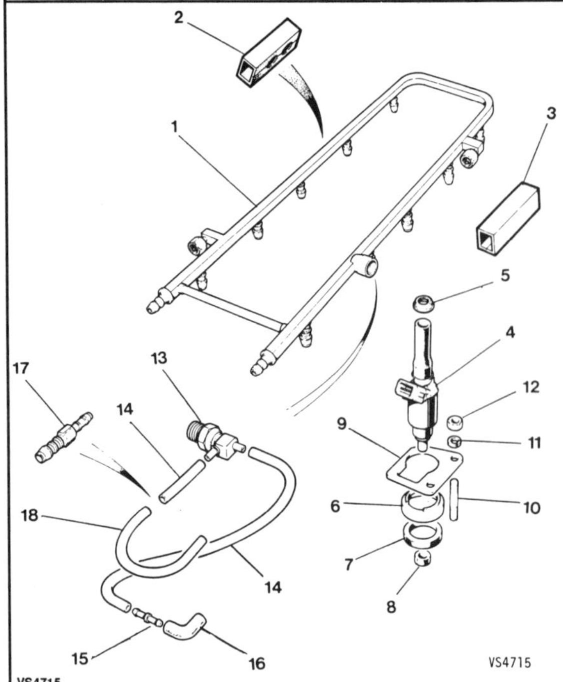

VS4715

FUEL RAIL-INJECTORS-5.3 LITRE.

Illus	Part Number	Description	Quantity	Change Point	Remarks
1	EAC 8930	Fuel rail	1		
2	EAC 8275	Separator	2		
3	EAC 7973	Separator	3		
4	EAC 7874	Fuel injector (short hose)	10		
5	EAC 7875	Fuel injector (long hose)	2		
	EAC 7876	Sleeve	12		
6	C 42077	Collar	12		
7	EAC 2414	Upper mounting ring	12		
8	EAC 2415	Lower mounting ring	12		
9	EAC 2644	Clamp plate	12		
10	C 16757	Stud	24		
11	FG 104 X	Lockwasher	24		
12	JLM 9683	Nut	24		
13	EAC 5086	Thermal valve	1		
14	EAC 8022 30	Vacuum hose	1		63" lengths
15	C 33835	Connector	1		
16	CAC 8050	Elbow	1		
17	EAC 9286	Reducer	1		
18	CBC 2591 4	Hose-reducer to manifold	1		

MASTER INDEX

GROUP INDEX

2E 04

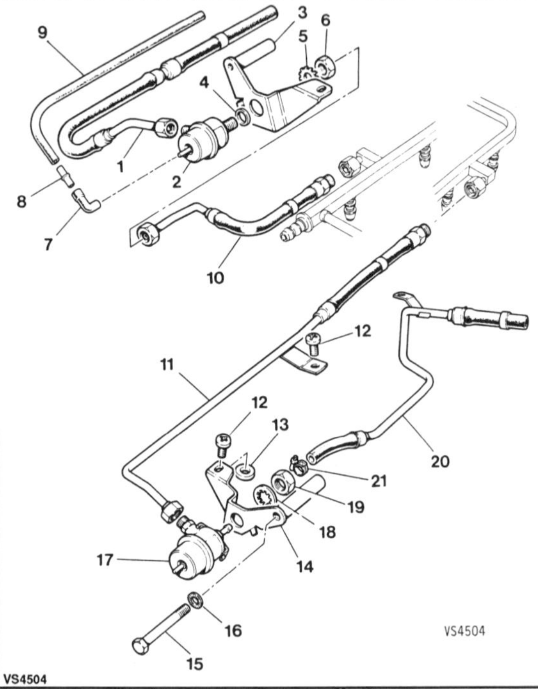

VS4504

FUEL PIPES-REGULATORS-5.3 LITRE.

Illus	Part Number	Description	Quantity	Change Point	Remarks
1	CBC 2179	Fuel feed hose	1		
2	EAC 6314	Pressure regulator	1		
3	EAC 8043	Mounting bracket	1		
4	EAC 8055	Spacing ring	1		
5	WE 114001 J	Washer	1		
6	AGU 1329 J	Nut	1		
7	EAC 7965	Elbow	1		
8	EAC 7982	Adaptor	1		
9	CBC 2591 4	Vacuum hose	1		24" lengths
10	EAC 8122	Fuel feed pipe	NLA		Use EAC 9987
10	EAC 9987	Fuel feed pipe	1		
11	EAC 8088	Fuel return pipe	NLA		Use EAC 9986
11	EAC 9986	Fuel return pipe	1		
12	SE 505051 J	Screw	2		
13	EAC 7972	Spacer	2		
14	EAC 7877	Mounting bracket	1		
15	BH 505291 J	Bolt	1		
16	C 30075 2	Washer	1		
17	EAC 4864	Pressure regulator	1		
18	WF 116001 J	Washer	1		
19	AGU 1287 J	Nut	1		
20	EAC 7983	Fuel pipe	1		
21	EAC 3215 2	Hose clip	1		

MASTER INDEX

GROUP INDEX

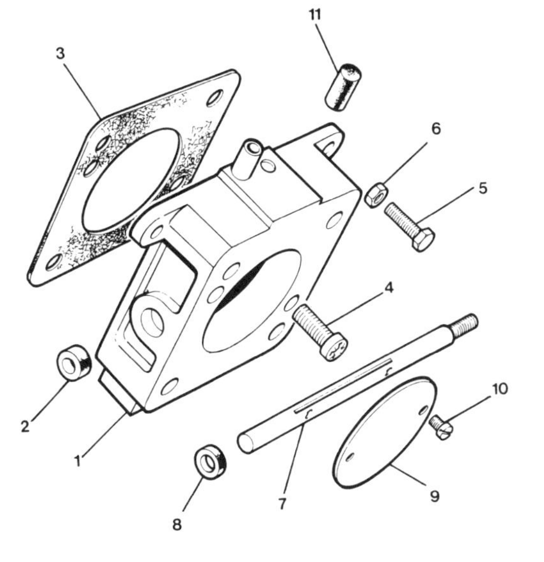

VS1059

VS1059

THROTTLE HOUSING 5.3 LITRE.

Illus	Part Number	123456 Description	Quantity	Change Point	Remarks
1	EAC 4029	Throttle housing RH	1		UK/Europe
1	C 44097	Throttle housing LH	1	Up to 1989MY	(Emiss B)
1	EAC 5226	Throttle housing RH	1		Less Europe
	EAC 5599	Throttle housing LH	1		Emiss B
1	EBC 2554	Throttle housing RH	1		UK/Europe
1	EAC 2556	Throttle housing LH	1		Emiss B
1	EBC 2556	Throttle housing	2	From 1989MY	Germany Emiss G
1	EAC 5226	Throttle housing RH	1	Digital Ignition	Middle East
1	EAC 5599	Throttle housing LH	1		
1	EAC 5599	Throttle housing	2		USA/CDN Japan/Aust Swe/Switz
2	C 20952 4	Bush	2		
3	C 33280	Gasket	2		
4	C 42325 12	Screw-housing attachment	4		
5	UFS 319 5H	Throttle screw	2		
6	NH 910011	Locknut	2		
7	C 44069	Throttle spindle	2		
8	C 43265 1	Spindle seal	4		
9	C 43280	Throttle disc	2		
10	AUC 1358 J	Screw-disc to spindle	4		
11	C 34946	Blanking Plug	1		

MASTER INDEX

GROUP INDEX

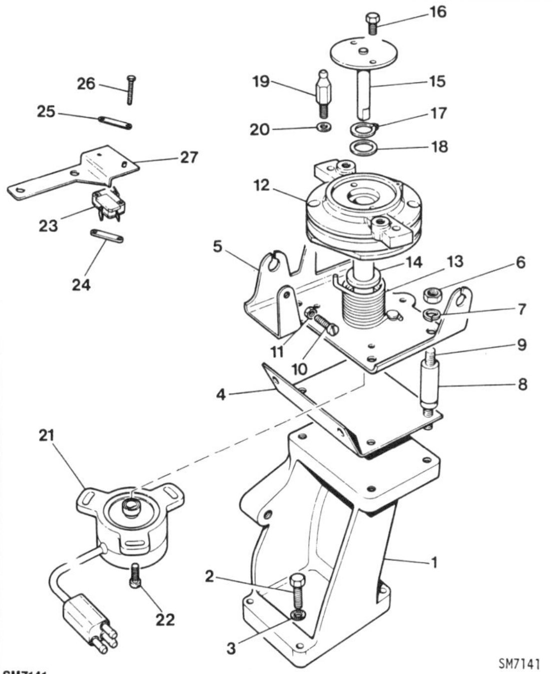

SM7141

SM7141

THROTTLE POTENTIOMETER 5.3 LITRE

UP TO ENGINE NO 8S61794 ONLY

Illus	Part Number	123456 Description	Quantity	Change Point	Remarks
1	EAC 2929	Throttle pedestal	1		
2	SH 504071 J	Setscrew	4		
3	C 30075 1	Washer	4		
4	EAC 5044	Bracket-coil mounting	1		
5	EAC 8415	Bracket-cable abutment	1		
6	JLM 9685	Nut	4		
7	FG 104 X	Washer	4		
8	EAC 8114	Spacer	4		
9	TE 504251 J	Stud	4		
10	UFS 319 6H	Stop screw	1		
11	NH 910011	Locknut	1		
12	EAC 3047	Pulley-throttle cable	1		
13	C 43208	Throttle spring	1		
14	EAC 3610	Washer	A/R		
14	EAC 3610 1	Washer	A/R		
14	EAC 3610 2	Washer	A/R		
15	EAC 2815	Drive shaft	1		
16	C 44477 1	Screw	2		
17	C 2261	Circlip	1		
18	EAC 3178	Washer	1		
19	EAC 7920	Ball end	2		
20	C 30075 1	Washer	2		
21	EAC 2670	Throttle potentiometer	1		
22	SN 104101 J	Screw	3		
23	C 35509	Throttle switch	1		USA/Canada/
24	C 33784	Fixing plate	1		Japan/Australia/
25	C 35840	Clamp plate	1		Sweden/Switz./
26	UCS 111 6R	Screw	2		Germany
27	EAC 3032	Mounting bracket	1		

H72AK13/9/88

MASTER INDEX

ENGINE 1.C
FLYWHEEL AND CLUTCH 1.G
GEARBOX AND PROPSHAFT 1.H
AXLES, SUSPENSION, DRIVE SHAFTS, WHEELS 1.K
STEERING 1.M
BRAKES AND BRAKE CONTROLS 1.N
FUEL, EXHAUST AND EMISSION SYSTEMS 2.C
COOLING, HEATING AND AIR CONDITIONING 2.G
ELECTRICAL, WASHERS-WIPERS, INSTRUMENTS 2.J
CHASSIS, SUBFRAMES, BODYSHELL, FITTINGS 3.C
FASCIA, TRIM, SEATS AND FIXINGS 3.H
ACCESSORIES AND PAINT 4.C
NUMERICAL INDEX 4.D

GROUP INDEX

ACCELERATOR CABLE LHD 3.6 LITRE C16
ACCELERATOR CABLE LHD-5.3 LITRE C17
ACCELERATOR CABLE RHD 3.6 LITRE C13
ACCELERATOR CABLE RHD 5.3 LITRE C14
AIR CLEANER RH-5.3 LITRE E02
AIR CLEANER-3.6 LITRE C02
AIR INJECTION PUMP 3.6 LITRE C09
AIR INJECTION PUMP-5.3 LITRE F03
FRONT EXHAUST PIPE RH F10
FUEL FILLER CAP - CONVERTIBLE D11
FUEL FILLER CAP-CABRIOLET D10
FUEL PUMP AND FILTER D04
FUEL RAIL-INJECTORS-3.6 LITRE C05
FUEL TANK-CABRIOLET D09

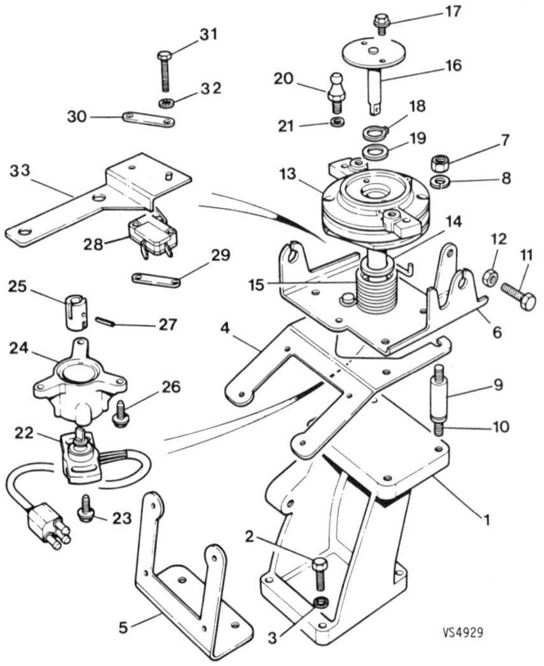

VS4929

THROTTLE POTENTIOMETER 5.3 LITRE

FROM ENGINE NO 8S61795

Illus	Part Number	Description	Quantity	Change Point	Remarks
1	EAC 2929	Throttle pedestal	1		
2	SH 504071 J	Setscrew-pedestal to top cover	4		
3	C 30075 1	Washer	4		
4	EAC 5044	Bracket-coil mounting	1	Up to 1989MY	
4	EAC 9842	Bracket-coil mounting	1	From 1989MY	
5	EBC 1979	Bracket-coil mounting	1		
6	EAC 9797	Bracket-cable abutment	1		
7	NH 606041 J	Nut	4		
8	WM 600041 J	Washer	4		
9	EAC 8114	Spacer	4		
10	TE 504251 J	Stud	4		
11	UFS 319 6H	Stop screw	1		
12	NH 910011	Locknut	1		
13	EAC 3047	Pulley - throttle cable	1		
14	C 43208	Throttle spring	1		
15	EAC 3610	Washer	A/R		
15	EAC 3610 1	Washer	A/R		
15	EAC 3610 2	Washer	A/R		
16	EAC 9566	Drive shaft - throttle switch	1		
17	C 44477 1	Screw-drive shaft to pulley	2		
18	C 2261	Circlip	1		
19	EAC 3178	Washer	1		
20	EAC 7920	Ball end	2		
21	C 30075 1	Washer	2		
22	EAC 9634	THROTTLE POTENTIOMETER	!		
23	SN 104101 J	Screw	2		
24	EAC 9567	Adaptor	1		
25	EAC 9568	Coupling	1		
26	SN 104101 J	Screw	3		
27	PA 103161 J	Spring pin	2		
28	C 35509	THROTTLE SWITCH	1		Not UK & Europe
29	C 33784	Fixing plate	1		(Emiss B)
30	C 35840	Clamp plate	1		Middle East
31	UCS 111 6R	Screw	2		
32	C 720	Washer	2		
33	EAC 3032	Bracket - switch mounting	1		

H72.02AK13/9/88

2E 07

MASTER INDEX

GROUP INDEX

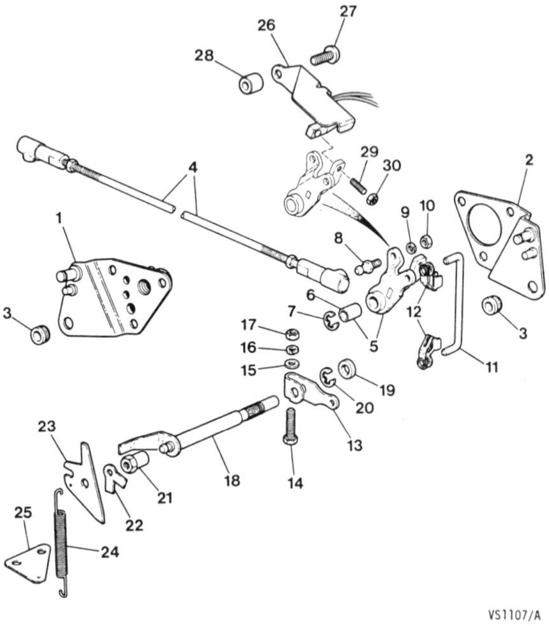

VS1107A

THROTTLE LINKAGE 5.3 LITRE

Illus	Part Number	Description	Quantity	Change Point	Remarks
1	EAC 2658	Mounting bracket RH	1		
2	EAC 2648	Mounting bracket LH	1		
3	C 34388	Bush	2		
4	EAC 1942	Connecting rod	2		
5	EAC 7921	Idler lever RH	1		
5	C 41884	Idler lever LH	1	Up to 1989MY	
5	EBC 2203	Idler lever LH	1	From 1989MY	
6	C 20952 4	Bush	2		
7	C 17380	Circlip	2		
8	C 37435	Ball end	2		
9	WM 702001 J	Lockwasher	2		
10	NH 910011	Locknut	2		
11	C 42618	Throttle link	2		
12	C 13824	Clip	4		
13	C 41902	Throttle lever	2		
14	UFS 119 7R	Setscrew	2		
15	C 28992	Washer	2		
16	WM 702001 J	Lockwasher	2		
17	NH 910011	Locknut	2		
18	C 44067	Throttle coupling shaft	2		
19	C 19706	Distance piece	2		
20	C 17380	Circlip	2		
21	C 43232	Nut-throttle spindle	2		
22	C 43231	Tabwasher	2		
23	C 43708	Throttle stop lever	2		
24	C 45981	Throttle return spring	2	Up to 1989MY	
24	C 45981	Throttle return spring RH	1	From 1989MY	
24	EBC 3003	Throttle return spring LH	1		
25	C 44478	Anchor plate	2		
26	DAC 6049	Idler switch	1	From 1989 MY	
27	C 42301 11	Screw	1	Digital Ignition	
28	EBC 2209	Boss	1		
29	GS 104161 J	Grub screw	1		
30	NH 104041 J	Locknut	1		

H74AK1/9/88

2E 08

MASTER INDEX

ENGINE ... 1.C
FLYWHEEL AND CLUTCH ... 1.G
GEARBOX AND PROPSHAFT ... 1.H
AXLES, SUSPENSION, DRIVE SHAFTS, WHEELS ... 1.K
STEERING ... 1.M
BRAKES AND BRAKE CONTROLS ... 1.N
FUEL, EXHAUST AND EMISSION SYSTEMS ... 2.C
COOLING, HEATING AND AIR CONDITIONING ... 2.G
ELECTRICAL, WASHERS-WIPERS, INSTRUMENTS ... 2.J
CHASSIS, SUBFRAMES, BODYSHELL, FITTINGS ... 3.C
FASCIA, TRIM, SEATS AND FIXINGS ... 3.H
ACCESSORIES AND PAINT ... 4.C
NUMERICAL INDEX ... 4.D

GROUP INDEX

ACCELERATOR CABLE LHD 3.6 LITRE ... C16
ACCELERATOR CABLE LHD-5.3 LITRE ... C17
ACCELERATOR CABLE RHD 3.6 LITRE ... C13
ACCELERATOR CABLE RHD 5.3 LITRE ... C14
AIR CLEANER RH-5.3 LITRE ... E02
AIR CLEANER-3.6 LITRE ... C02
AIR INJECTION PUMP 3.6 LITRE ... C09
AIR INJECTION PUMP-5.3 LITRE ... F03
FRONT EXHAUST PIPE RH ... F10
FUEL FILLER CAP - CONVERTIBLE ... D11
FUEL FILLER CAP-CABRIOLET ... D10
FUEL PUMP AND FILTER ... D04
FUEL RAIL-INJECTORS-3.6 LITRE ... C05
FUEL TANK-CABRIOLET ... D09

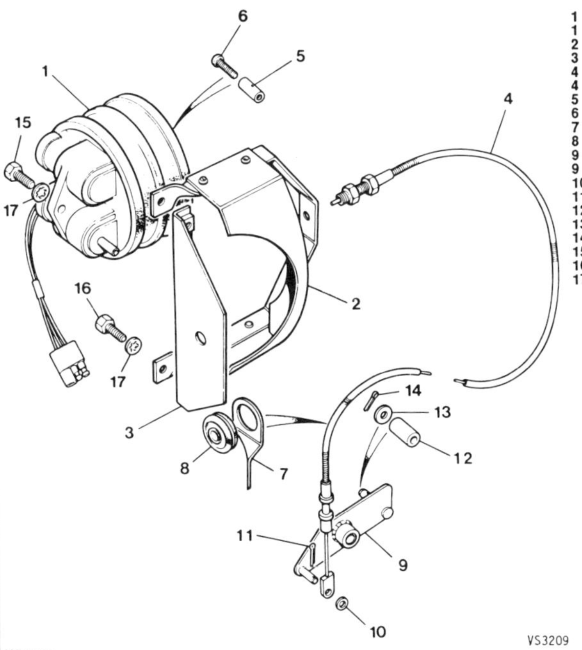

VS 3209

SPEED CONTROL ACTUATOR ASSEMBLY-5.3 LITRE

Illus	Part Number	123456	Description	Quantity	Change Point	Remarks
1	DAC 1994		Actuator assembly	NLA		Use DAC5964
1	DAC 5964		Actuator assembly	1		
2	CBC 2054		Bracket assembly	1		
3	CAC 4268		Bracket	1		
4	CBC 1382 2		Actuator cable-RHD	1		
4	CBC 1382 1		Actuator cable-LHD	1		
5	CAC 2552		Nipple-solderless	1		
6	UCS 811 3H		Setscrew	1		
7	CAC 4407		Cable bracket-LHD	1		
8	C 29302		Grommet-LHD	1		
9	CAC 4029		Lever and bush-RHD	1		
9	CAC 3832		Lever and bush-LHD	1		
10	BD 541 28		Washer-plain	1		
11	L 102 4H		Pin-split	1		
12	CAC 3039		Spacer-RHD	1		
13	BD 541 32		Washer-plain	1		
14	2390		Pin-split	1		
15	JLM 9566		Setscrew	1		
16	SH 604051 J		Setscrew	2		
17	C 724		Washer-locking	3		

MASTER INDEX

GROUP INDEX

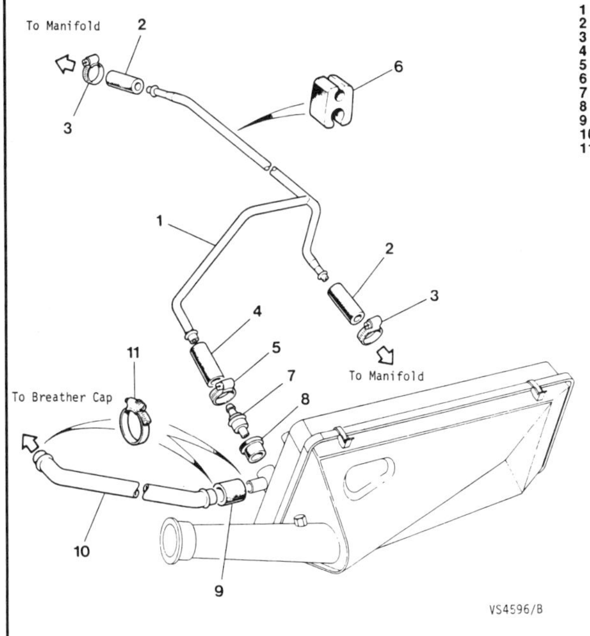

VS4596B

ENGINE BREATHER PIPE 5.3 LITRE

Illus	Part Number	123456	Description	Quantity	Change Point	Remarks
1	EAC 3389		Breather pipe	1		
2	CAC 5867 22		Hose-pipe to manifold	1		
3	EAC 3215 2		Hose clip	4		
4	CAC 8900 11		Hose-pipe to vent valve	1		
5	C 43599 10		Hose clip	2		
6	EAC 3179		Retainer	1		
7	C 44231		Vent valve	1		
8	C 37990		Bush	1		
9	C 44412		Hose	1		
10	C 44392		Breather pipe	1		
11	EAC 3215 7		Clip	3		

MASTER INDEX

GROUP INDEX

2E 11

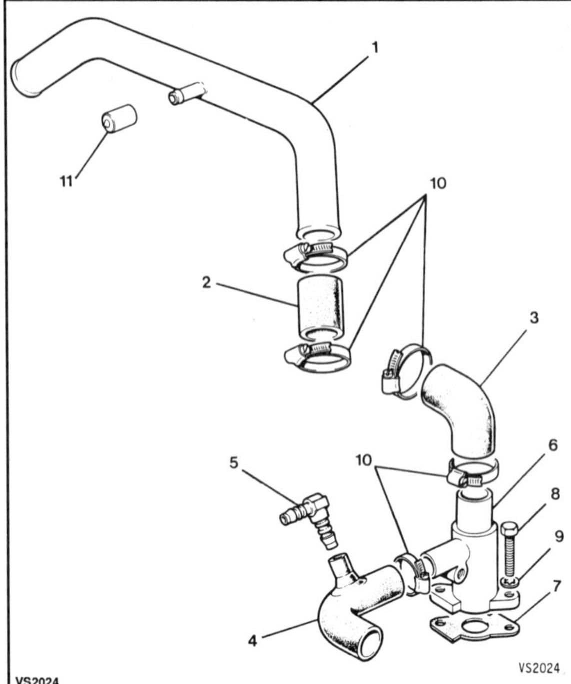

VS2024

EXTRA AIR VALVE 5.3 LITRE

Illus	Part Number 123456	Description	Quantity	Change Point	Remarks
1	EAC 3090	Balance pipe	1		
2	C 41373	Hose-pipe to manifold	2		
1	EAC 3634	Balance pipe	1		Air injection only
2	EAC 2654	Hose-pipe to manifold	2		Air injection only
3	EAC 2655	Hose-air valve to manifold	1		
	EAC 7968	Hose-air valve to air cleaner	1	Up to VIN144699	
4	EAC 9866	Hose-air valve to air cleaner	1	From VIN144700	
5	EAC 8036 5	Connector	1		
6	EAC 2273	Extra air valve	1	Up to 8S.57761	Emission B/C/D/E/G
6	EAC 4438	Extra air valve	1		Emission A/F
6	EAC 2273	Extra air valve	1	From 8S.57762	Emission B/E
6	EBC 1198	Extra air valve	1		Emission A/C/F/G
7	C 42165	Gasket	1		
8	C 42325 4	Screw	2		
9	C 30075 1	Washer	2		
10	EAC 3215 8	Hose clip	7		
11	C 34946	Blanking cap	1		

2E 12

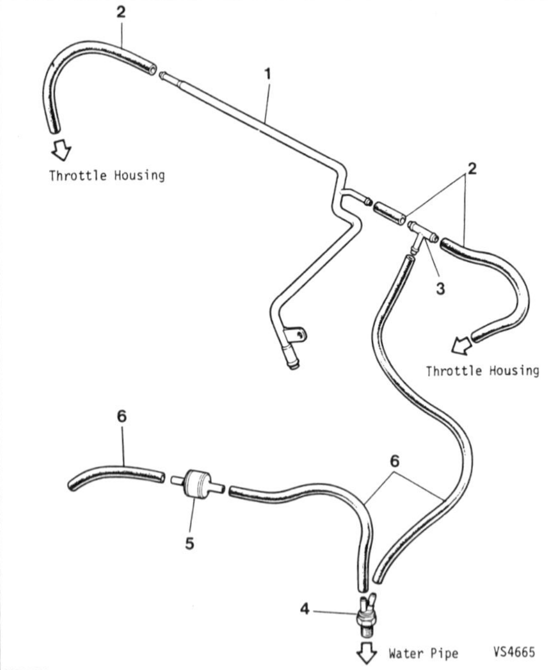

VS4665

ENGINE PURGE 5.3 LITRE

Illus	Part Number 123456	Description	Quantity	Change Point	Remarks
1	EAC 3419	Purge pipe	1		Vehicles with carbon canister
2	CBC 2591 4	Hose (24" lengths)	A/R		Vehicles with carbon canister
3	EAC 9299	Tee piece	1		Vehicles with carbon canister
4	EAC 2630	Thermal valve	1		Vehicles with carbon canister
5	EAC 9270	Delay valve	1		Vehicles with carbon canister
6	EAC 8022 30	Hose (63" lengths)	A/R		Vehicles with carbon canister

MASTER INDEX

ENGINE 1.C
FLYWHEEL AND CLUTCH 1.G
GEARBOX AND PROPSHAFT 1.H
AXLES, SUSPENSION, DRIVE SHAFTS, WHEELS 1.K
STEERING 1.M
BRAKES AND BRAKE CONTROLS 1.N
FUEL, EXHAUST AND EMISSION SYSTEMS 2.C
COOLING, HEATING AND AIR CONDITIONING 2.G
ELECTRICAL, WASHERS-WIPERS, INSTRUMENTS 2.J
CHASSIS, SUBFRAMES, BODYSHELL, FITTINGS 3.C
FASCIA, TRIM, SEATS AND FIXINGS 3.H
ACCESSORIES AND PAINT 4.C
NUMERICAL INDEX 4.D

GROUP INDEX

ACCELERATOR CABLE LHD 3.6 LITRE C16
ACCELERATOR CABLE LHD-5.3 LITRE C17
ACCELERATOR CABLE RHD 3.6 LITRE C13
ACCELERATOR CABLE RHD 5.3 LITRE C14
AIR CLEANER RH-5.3 LITRE E02
AIR CLEANER-3.6 LITRE C02
AIR INJECTION PUMP 3.6 LITRE C09
AIR INJECTION PUMP-5.3 LITRE F03
FRONT EXHAUST PIPE RH F10
FUEL FILLER CAP - CONVERTIBLE D11
FUEL FILLER CAP-CABRIOLET D10
FUEL PUMP AND FILTER D04
FUEL RAIL-INJECTORS-3.6 LITRE C05
FUEL TANK-CABRIOLET D09

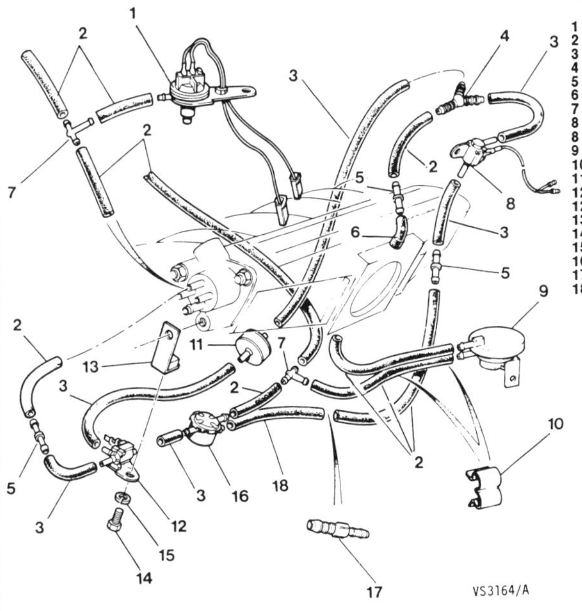

VACUUM ADVANCE CONTROL 5.3 LITRE

Illus	Part Number	123456	Description	Quantity	Change Point	Remarks
1	EAC 4371		Throttle control switch	1		
2	EAC 8022 30		Vacuum hose	1		
3	EAC 8021 18		Vacuum hose	1		
4	EAC 2249 2		Connector	1		
5	C 33835		Connector	2	Up to 1989MY	
6	EAC 7965		Elbow	1		
7	C 15644		'T' piece	2		UK/Europe(Emiss B)
8	EAC 4013		Solenoid valve (bullet connector)	1		Middle East
8	EAC 7934		Solenoid valve (moulded connector)	1		
9	EAC 4069		Dump valve	1		
10	C 37330 1		Clip	1		
11	EAC 4025		Delay valve	1		
12	EAC 3488		Solenoid valve (bullet connector)	1		
12	EAC 7904		Solenoid valve (moulded connector)	1		
13	EAC 4123		Bracket	1		
14	SH 605041 J		Setscrew-valve to bracket	1		
15	JLM 303		Lockwasher	1		
16	EAC 4012		Vacuum regulator	1		
17	EAC 9286		Reducer	1		
18	CBC 2591 4		Vacuum hose	1		

VS3164/A

MASTER INDEX

GROUP INDEX

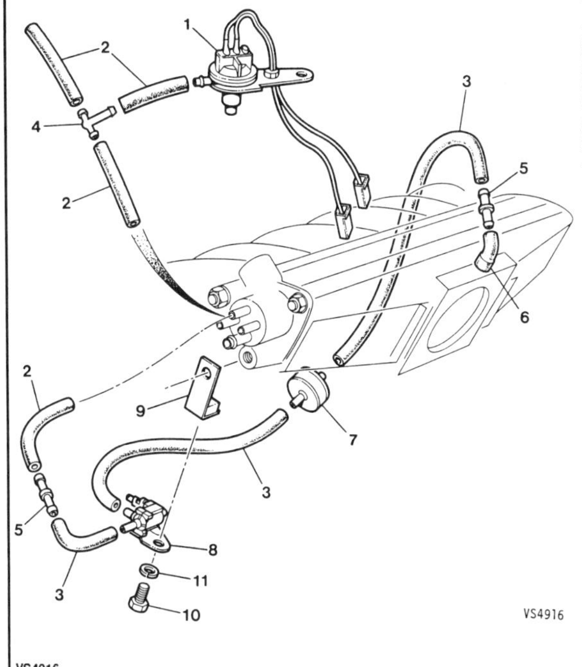

Illus	Part Number	123456	Description	Quantity	Change Point	Remarks
1	EAC 4371		THROTTLE CONTROL SWITCH	1		
2	EAC 8022 30		Vacuum hose	A/R		
3	EAC 8021 18		Vacuum hose	A/R		
4	C 15644		"T" Piece	1		
5	C 33835		Connector	2		
6	EAC 7965		Elbow	1	From 1989MY	UK/Europe
7	EAC 4025		Delay valve	1		Emiss B
8	EAC 7904		SOLENOID VALVE	1		Middle East
9	EAC 4123		Bracket	1		
10	SH 605041 J		Setscrew-valve to brkt	1		
11	JLM 303		Lockwasher	1		

VS4916

MASTER INDEX

ENGINE 1.C
FLYWHEEL AND CLUTCH 1.G
GEARBOX AND PROPSHAFT 1.H
AXLES, SUSPENSION, DRIVE SHAFTS, WHEELS 1.K
STEERING 1.M
BRAKES AND BRAKE CONTROLS 1.N
FUEL, EXHAUST AND EMISSION SYSTEMS 2.C
COOLING, HEATING AND AIR CONDITIONING 2.G
ELECTRICAL, WASHERS-WIPERS, INSTRUMENTS 2.J
CHASSIS, SUBFRAMES, BODYSHELL, FITTINGS 3.C
FASCIA, TRIM, SEATS AND FIXINGS 3.H
ACCESSORIES AND PAINT 4.C
NUMERICAL INDEX 4.D

GROUP INDEX

ACCELERATOR CABLE LHD 3.6 LITRE C16
ACCELERATOR CABLE LHD-5.3 LITRE C17
ACCELERATOR CABLE RHD 3.6 LITRE C13
ACCELERATOR CABLE RHD 5.3 LITRE C14
AIR CLEANER RH-5.3 LITRE E02
AIR CLEANER-3.6 LITRE C02
AIR INJECTION PUMP 3.6 LITRE C09
AIR INJECTION PUMP-5.3 LITRE F03
FRONT EXHAUST PIPE RH F10
FUEL FILLER CAP - CONVERTIBLE D11
FUEL FILLER CAP-CABRIOLET D10
FUEL PUMP AND FILTER D04
FUEL RAIL-INJECTORS-3.6 LITRE C05
FUEL TANK-CABRIOLET D09

2E 15

VACUUM ADVANCE CONTROL 5.3 LITRE

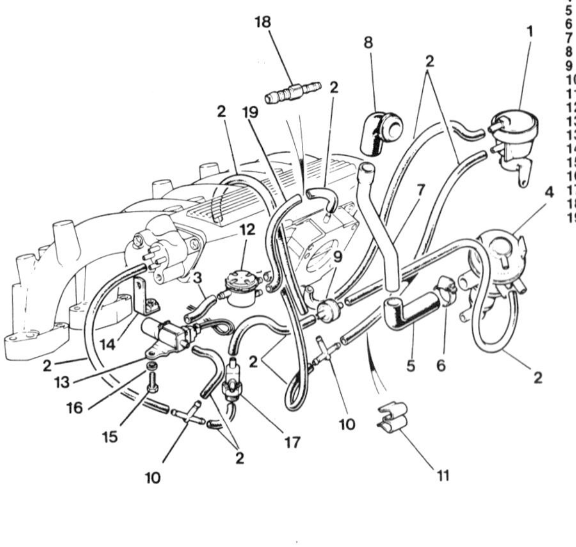

VS4002

VS4002

Illus	Part Number	123456 Description	Quantity	Change Point	Remarks
1	EAC 4069	Dump valve	1		
2	EAC 8022 30	Vacuum hose	1		
3	EAC 8021 30	Vacuum hose	1		
4	EAC 3827	Air switching valve	1		
5	EAC 3954 3	Hose-valve to air pipe	1		
6	EAC 3215 6	Clip	2		
7	EAC 4082	Air pipe	1		Air injection USA/Sweden/ Switz/Germany (optional) and from VIN144700 for Canada
8	C 35125	Hose	1		
9	EAC 4025	Delay valve	1	Up to Eng No 8S.57761	
10	C 15644	'T' piece	2		
11	C 37330 1	Clip	2		
12	EAC 5157	Vacuum regulator	1		
13	EAC 4100	Solenoid valve (bullet connector)	1		
13	EAC 7935	Solenoid valve (moulded connector)	1		
14	EAC 4123	Bracket	1		
15	SH 605041 J	Setscrew-valve to bracket	1		
16	JLM 303	Lockwasher	1		
17	EAC 4083	Thermal valve	1		
18	EAC 9286	Reducer	1		
19	CBC 2591 4	Vacuum hose	1		

H84AK13/9/88

MASTER INDEX

GROUP INDEX

2E 16

VACUUM ADVANCE CONTROL-5.3 LITRE

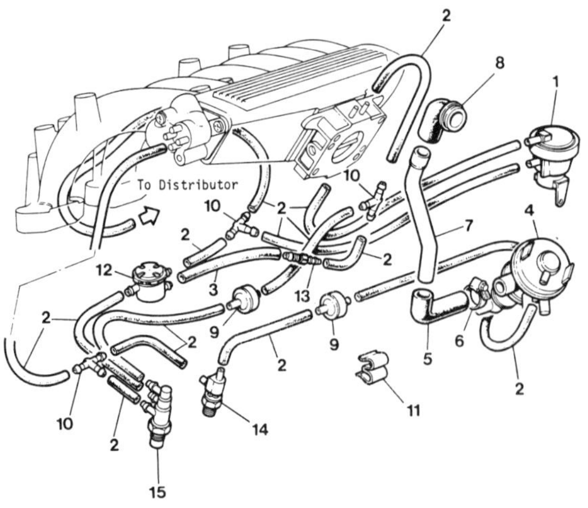

VS4880

VS4880

Illus	Part Number	123456 Description	Quantity	Change Point	Remarks
1	EAC 4069	Dump valve	1		
2	EAC 8022 30	Vacuum hose	1		
3	CBC 2591 4	Vacuum hose	1		
4	EAC 3827	Air switching valve	1		
5	EAC 3954 3	Hose-valve to air pipe	1		
6	EAC 3215 6	Clip	2		
7	EAC 4082	Air Pipe	1	From Eng No 8S.57762 Up to 1989 MY	Air injection USA/CDN/Sweden/ Switzerland
8	C 35125	Hose	1		
9	EAC 4025	Delay valve	2		
10	C 15644	"T" piece	3		
11	C 37330 1	Clip	2		
12	EAC 5157	Vacuum regulator	1		
13	EAC 9286	Reducer	1		
14	EAC 4083	Thermal valve	1		
15	EBC 1197	Vacuum control valve	1		

H84.02AK13/9/88

MASTER INDEX

ENGINE ... 1.C
FLYWHEEL AND CLUTCH ... 1.G
GEARBOX AND PROPSHAFT ... 1.H
AXLES, SUSPENSION, DRIVE SHAFTS, WHEELS ... 1.K
STEERING ... 1.M
BRAKES AND BRAKE CONTROLS ... 1.N
FUEL, EXHAUST AND EMISSION SYSTEMS ... 2.C
COOLING, HEATING AND AIR CONDITIONING ... 2.G
ELECTRICAL, WASHERS-WIPERS, INSTRUMENTS ... 2.J
CHASSIS, SUBFRAMES, BODYSHELL, FITTINGS ... 3.C
FASCIA, TRIM, SEATS AND FIXINGS ... 3.H
ACCESSORIES AND PAINT ... 4.C
NUMERICAL INDEX ... 4.D

GROUP INDEX

ACCELERATOR CABLE LHD 3.6 LITRE ... C16
ACCELERATOR CABLE LHD-5.3 LITRE ... C17
ACCELERATOR CABLE RHD 3.6 LITRE ... C13
ACCELERATOR CABLE RHD 5.3 LITRE ... C14
AIR CLEANER RH-5.3 LITRE ... E02
AIR CLEANER-3.6 LITRE ... C02
AIR INJECTION PUMP 3.6 LITRE ... C09
AIR INJECTION PUMP-5.3 LITRE ... F03
FRONT EXHAUST PIPE RH ... F10
FUEL FILLER CAP - CONVERTIBLE ... D11
FUEL FILLER CAP-CABRIOLET ... D10
FUEL PUMP AND FILTER ... D04
FUEL RAIL-INJECTORS-3.6 LITRE ... C05
FUEL TANK-CABRIOLET ... D09

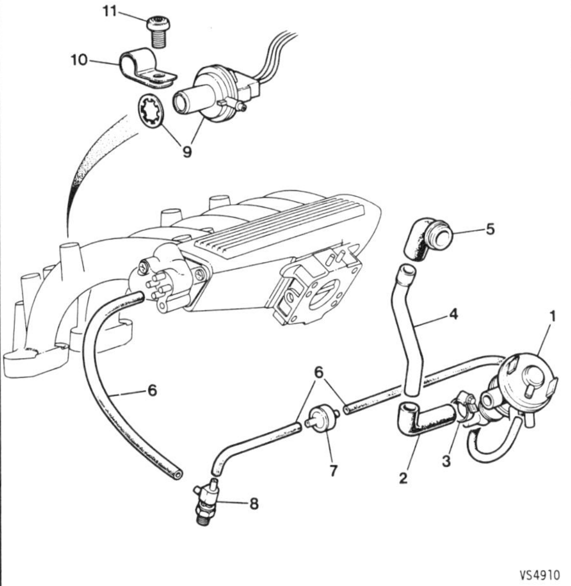

VS4910

VACUUM ADVANCE CONTROL-5.3 LITRE

Illus	Part Number	123456 Description	Quantity	Change Point	Remarks
1	EAC 3827	AIR SWITCHING VALVE	1		
2	EAC 3954 3	Hose-valve to air pipe	1		
3	EAC 3215 6	Clip	2		
4	EAC 4082	Air pipe	1	From 1989 MY	Air Injection
5	C 35125	Hose	1	Digital Ignition	USA/CDN/Sweden/
6	EAC 8022 30	Vacuum hose	A/R	....	Switzerland
7	EAC 4025	Delay valve	1		
8	EAC 4083	THERMAL VALVE	1		
9	JLM 473	THROTTLE CONTROL VALVE	1	....	USA/CDN/
10	CP 108161 J	Clip	1	....	Australia/Japan/
11	C 42301 1	Screw	1	....	Sweden/Germany/ Switzerland

H84.04AK13/9/88

MASTER INDEX

GROUP INDEX

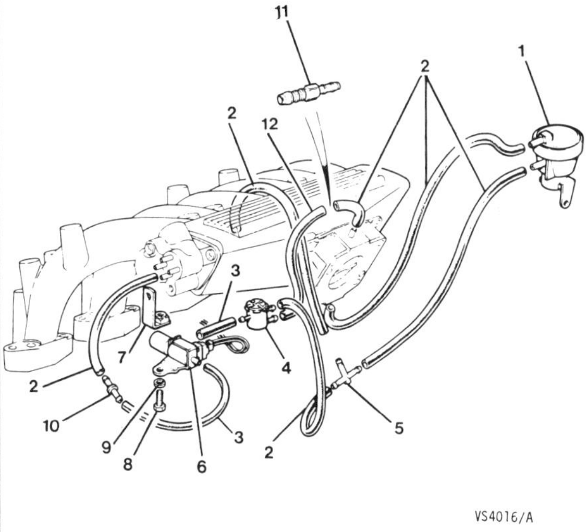

VS4016/A

VACUUM ADVANCE CONTROL 5.3 LITRE

Illus	Part Number	123456 Description	Quantity	Change Point	Remarks
1	EAC 4069	Dump valve	1		
2	EAC 8022 30	Vacuum hose	1		
3	EAC 8021 18	Vacuum hose	1		
4	EAC 5157	Vacuum regulator	1		
5	C 15644	'T' piece	1	Up to Eng No	Emission 'C'
6	EAC 4100	Solenoid valve (bullet connector)	1	8S.57761	Canada (up to
6	EAC 7935	Solenoid valve (moulded connector)	1	....	VIN144699 only)
7	EAC 4123	Bracket	1	....	Japan/Australia
8	SH 605041 J	Setscrew-valve to bracket	1		
9	JLM 303	Lockwasher	1		
10	C 33835	Connector	1		
11	EAC 9286	Reducer	1		
12	CBC 2591 4	Vacuum hose	1		

MASTER INDEX

ENGINE 1.C
FLYWHEEL AND CLUTCH 1.G
GEARBOX AND PROPSHAFT 1.H
AXLES, SUSPENSION, DRIVE SHAFTS, WHEELS 1.K
STEERING 1.M
BRAKES AND BRAKE CONTROLS 1.N
FUEL, EXHAUST AND EMISSION SYSTEMS 2.C
COOLING, HEATING AND AIR CONDITIONING 2.G
ELECTRICAL, WASHERS-WIPERS, INSTRUMENTS 2.J
CHASSIS, SUBFRAMES, BODYSHELL, FITTINGS 3.C
FASCIA, TRIM, SEATS AND FIXINGS 3.H
ACCESSORIES AND PAINT 4.C
NUMERICAL INDEX 4.D

GROUP INDEX

ACCELERATOR CABLE LHD 3.6 LITRE C16
ACCELERATOR CABLE LHD-5.3 LITRE C17
ACCELERATOR CABLE RHD 3.6 LITRE C13
ACCELERATOR CABLE RHD 5.3 LITRE C14
AIR CLEANER RH-5.3 LITRE E02
AIR CLEANER-3.6 LITRE C02
AIR INJECTION PUMP 3.6 LITRE C09
AIR INJECTION PUMP-5.3 LITRE F03
FRONT EXHAUST PIPE RH F10
FUEL FILLER CAP - CONVERTIBLE D11
FUEL FILLER CAP-CABRIOLET D10
FUEL PUMP AND FILTER D04
FUEL RAIL-INJECTORS-3.6 LITRE C05
FUEL TANK-CABRIOLET D09

2F 02

VACUUM ADVANCE CONTROL 5.3 LITRE

Illus	Part Number	Description (123456)	Quantity	Change Point	Remarks
1	EAC 4069	Dump valve	1	From Eng No 8S.57762	Australia/ Japan/ Germany
2	EAC 8022 30	Vacuum hose	1		
3	CBC 2591 4	Vacuum hose	1		
4	EAC 5157	Vacuum regulator	1		
5	C 15644	"T" piece	2		
6	EAC 4025	Delay valve	1		
7	EAC 9286	Reducer	1		
8	C 37330 1	Clip	2		
9	EBC 1197	Vacuum control valve	1		

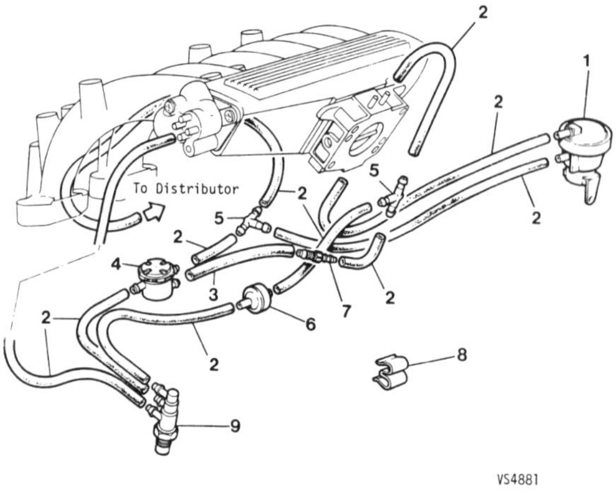

VS4881

MASTER INDEX

GROUP INDEX

2F 03

AIR INJECTION PUMP-5.3 LITRE.

Illus	Part Number	Description (123456)	Quantity	Change Point	Remarks
1	EAC 8839	Air injection pump	1		Air injection engine only USA/Sweden/ Switzerland/ Germany(optional) and from VIN144700 for Canada
2	EAC 8858	Pulley	1		
3	SH 108201 J	Setscrew-pulley to pump	3		
4	C 30075 2	Washer-pulley to pump	3		
5	C 36143	Mounting bracket	1		
6	SH 505071 J	Setscrew-securing bracket	4		
7	WM 600051 J	Lockwasher	4		
8	BH 606421 J	Pivot bolt-pump to bracket	1		
9	C 35557	Clamp washer	4		
10	WA 110061 J	Washer	2		
11	NZ 606041 J	Nut	1		
12	C 36119	Adjusting bolt	1		
13	SH 110301 J	Bolt-adjuster bolt to pump	1		
14	WF 110061 J	Washer-adjuster bolt to pump	1		
15	C 37278	Washer	2		
16	WM 600061 J	Washer	2		
17	JLM 9685	Nut	2		
18	EAC 3833	Outlet adaptor	1		
19	EAC 3909	Gasket	1		
20	TE 108041 J	Stud	2		
21	C 30075 2	Washer	2		
22	JLM 9688	Nut	2		
23	C 37990	Sealing bush	1		

900H88AK17/6/88

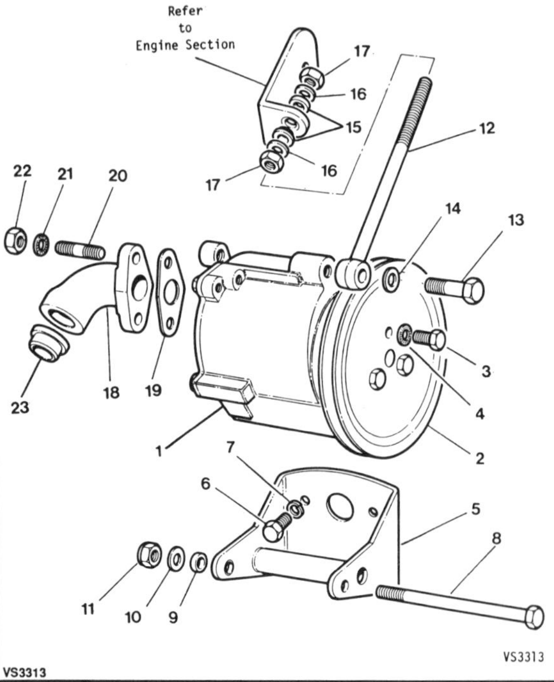

VS3313

MASTER INDEX

ENGINE ... 1.C
FLYWHEEL AND CLUTCH ... 1.G
GEARBOX AND PROPSHAFT ... 1.H
AXLES, SUSPENSION, DRIVE SHAFTS, WHEELS ... 1.K
STEERING ... 1.M
BRAKES AND BRAKE CONTROLS ... 1.N
FUEL, EXHAUST AND EMISSION SYSTEMS ... 2.C
COOLING, HEATING AND AIR CONDITIONING ... 2.G
ELECTRICAL, WASHERS-WIPERS, INSTRUMENTS ... 2.J
CHASSIS, SUBFRAMES, BODYSHELL, FITTINGS ... 3.C
FASCIA, TRIM, SEATS AND FIXINGS ... 3.H
ACCESSORIES AND PAINT ... 4.C
NUMERICAL INDEX ... 4.D

GROUP INDEX

ACCELERATOR CABLE LHD 3.6 LITRE ... C16
ACCELERATOR CABLE LHD-5.3 LITRE ... C17
ACCELERATOR CABLE RHD 3.6 LITRE ... C13
ACCELERATOR CABLE RHD 5.3 LITRE ... C14
AIR CLEANER RH-5.3 LITRE ... E02
AIR CLEANER-3.6 LITRE ... C02
AIR INJECTION PUMP 3.6 LITRE ... C09
AIR INJECTION PUMP-5.3 LITRE ... F03
FRONT EXHAUST PIPE RH ... F10
FUEL FILLER CAP - CONVERTIBLE ... D11
FUEL FILLER CAP-CABRIOLET ... D10
FUEL PUMP AND FILTER ... D04
FUEL RAIL-INJECTORS-3.6 LITRE ... C05
FUEL TANK-CABRIOLET ... D09

AIR INJECTION PIPES-5.3 LITRE

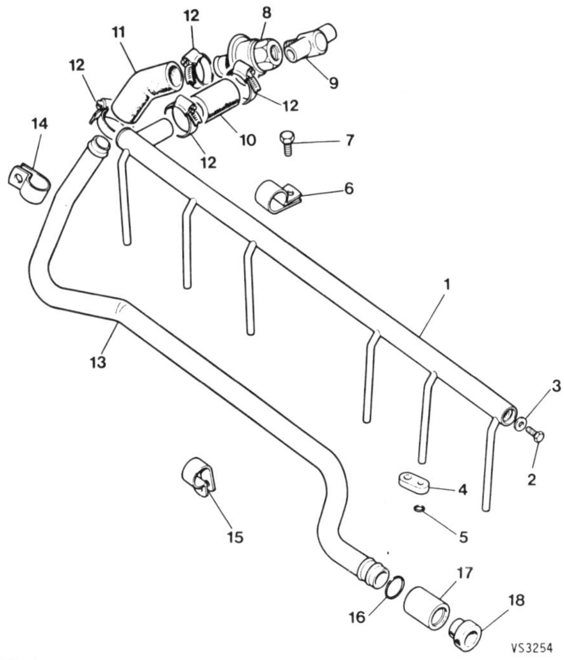

VS3254

Illus	Part Number	123456 Description	Quantity	Change Point	Remarks
1	C 39930	Air pipe RH	1		
1	C 39929	Air pipe LH	1		
2	SH 605041 J	Setscrew	2		
3	C 2296 7	Washer	2		
4	C 35770	Clamp plate	12		
5	C 35779	'O' ring	24		
6	C 42566	Clip	4		
7	SE 505051 J	Setscrew	4		
8	C 34017	Check valve	1		
9	EAC 3714	Tee piece	1		Air injection
10	C 34632	Hose-tee piece to air pipe	2	Up to 8S58990	Engine only
10	EBC 1488	Hose-tee piece to air pipe	2	From 8S58991	USA/Sweden/
11	C 35099	Hose-check valve to air pipe	1		Switzerland/
12	EAC 3215 6	Hose clip	4		Germany(optional)
13	C 46337	Air pipe	1		and from
14	EAC 1621	Clip	1		VIN144700
15	C 43608	Clip	1		for Canada
16	C 21058	'O' ring	1		
17	C 44543	Sealing sleeve	1		
18	C 37990	Sealing bush	1		

MASTER INDEX

GROUP INDEX

JAGUAR XJS RANGE (JAN 1987 ON) | F05 fiche 2 | FUEL, EXHAUST AND EMISSION SYSTEMS

FRT EXHAUST NON CATALYST - 3.6 LITRE

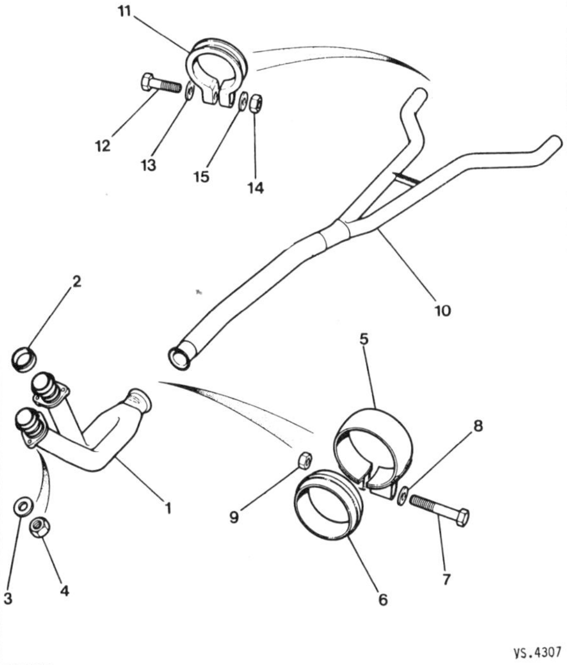

VS 4307

Illus	Part Number	123456 Description	Quantity	Change Point	Remarks
1	CBC 2521	Downpipe	1		
2	CBC 4843	Sealing ring	2		
3	FW 106 T	Washer	4		
4	C 17916 1	Locknut	4		
5	CAC 4073	Clamp	1		
6	CAC 4336	Olive	1		
7	BH 605201 J	Bolt	1		
8	FW 105 T	Washer	1		
9	C 17914	Locknut-helicoil	1		
10	CAC 5081	Intermediate pipe	1		
	CBC 4179	Inter exhaust pipe-European Mkts	NLA		Use CBC4888
	CBC 4888	Inter exhaust pipe-European Mkts	1		
11	C 13063	Clip	2		
12	BH 605221 J	Bolt	2		
13	JLM 298	Washer	2		
14	JLM 9684	Nut	2		
15	WF 600050 J	Washer	2		
	CAC 1765	Clip	2		Japan only

MASTER INDEX

ENGINE ... 1.C
FLYWHEEL AND CLUTCH ... 1.G
GEARBOX AND PROPSHAFT ... 1.H
AXLES, SUSPENSION, DRIVE SHAFTS, WHEELS ... 1.K
STEERING ... 1.M
BRAKES AND BRAKE CONTROLS ... 1.N
FUEL, EXHAUST AND EMISSION SYSTEMS ... 2.C
COOLING, HEATING AND AIR CONDITIONING ... 2.G
ELECTRICAL, WASHERS-WIPERS, INSTRUMENTS ... 2.J
CHASSIS, SUBFRAMES, BODYSHELL, FITTINGS ... 3.C
FASCIA, TRIM, SEATS AND FIXINGS ... 3.H
ACCESSORIES AND PAINT ... 4.C
NUMERICAL INDEX ... 4.D

GROUP INDEX

ACCELERATOR CABLE LHD 3.6 LITRE ... C16
ACCELERATOR CABLE LHD-5.3 LITRE ... C17
ACCELERATOR CABLE RHD 3.6 LITRE ... C13
ACCELERATOR CABLE RHD 5.3 LITRE ... C14
AIR CLEANER RH-5.3 LITRE ... E02
AIR CLEANER-3.6 LITRE ... C02
AIR INJECTION PUMP 3.6 LITRE ... C09
AIR INJECTION PUMP-5.3 LITRE ... F03
FRONT EXHAUST PIPE RH ... F10
FUEL FILLER CAP - CONVERTIBLE ... D11
FUEL FILLER CAP-CABRIOLET ... D10
FUEL PUMP AND FILTER ... D04
FUEL RAIL-INJECTORS-3.6 LITRE ... C05
FUEL TANK-CABRIOLET ... D09

2F 06

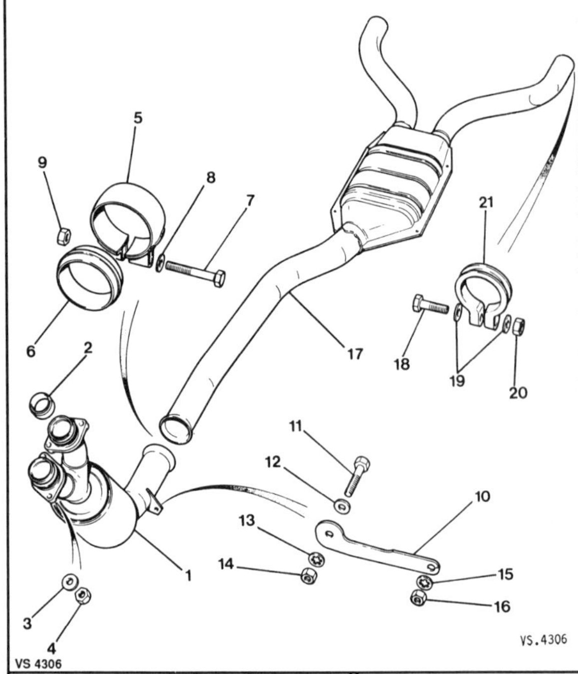

FRT EXHAUST CATALYST-3.6 LITRE

Illus	Part Number	123456 Description	Quantity	Change Point	Remarks
1	CBC 2630	Downpipe and catalyst	NLA		Use CBC 8104
	CBC 5728	Front downpipe and catalyst	1		Germany only
	CBC 8104	Front downpipe and catalyst	1		Not Germany
2	CBC 4834	Joint-manifold to downpipe	2		Downpipe to manifold
3	FW 106 E	Washer	4		
4	C 17916 1	Locknut	4		
5	CAC 4073	Clamp	1		Downpipe to inter pipe
6	CAC 4336	Olive	1		
7	FB 108121 J	Bolt	1		
8	FW 105 T	Washer	1		
9	ACU 6289	Locknut-helicoil	1		
10	CAC 7670	Stay	1		
11	SH 604041 J	Setscrew	1		
12	WF 600040 J	Washer	2		
13	FW 104 T	Washer-plain	2		Manual g/box
	WA 106041 J	Washer-plain	1		Auto trans
14	JLM 9683	Nut	1		
15	WF 106001 J	Washer	2		Manual only
16	NH 106041 J	Nut	2		
	SH 604041 J	Setscrew	1		Auto trans
	WA 106041 J	Washer plain	1		
17	CBC 1401	Inter pipe and catalyst assembly	NLA		Use CBC 8106
	CBC 6014	Inter pipe and catalyst assembly	1		Germany only
	CBC 8106	Inter pipe and catalyst assembly	1		Not Germany
18	CAC 3802	Bolt	2		
19	FW 104 T	Washer	4		
20	JLM 9684	Nut	2		
21	CAC 3801 4	Clamp	2		
	CAC 8535	Stay-exhaust downpipe	1		
	SH 106121 J	Setscrew	1		
	FN 106041 J	Flanged nut	2		

H162AW13/9/88

MASTER INDEX

GROUP INDEX

2F 07

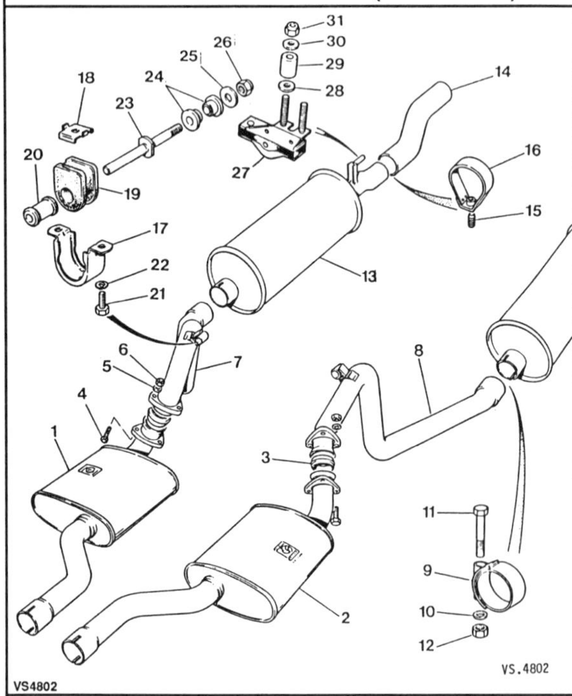

FRT AND REAR SILENCERS 3.6 LITRE

Illus	Part Number	123456 Description	Quantity	Change Point	Remarks
1	CAC 2236	Silencer-inter RH	1		
2	CAC 2237	Silencer-inter LH	1		
3	C 43831	Olive	2		
4	BH 605161 J	Bolt	6		
5	JLM 298	Washer	6		
6	C 17914	Locknut	6		
7	CAC 5620	Exhaust pipe-over axle RH	1		
8	CAC 5621	Exhaust pipe-over axle LH	1		
9	CAC 3801 3	Clamp	2		
10	JLM 296	Washer	4		
11	CAC 3802	Bolt	2		
12	CAC 3803	Locknut	2		
13	CAC 2494	Silencer-rear	2		
14	CAC 6068	Tail pipe-RH	1		
	CAC 6069	Tail pipe-LH	1		
15	AGU 1004	Setscrew	2		
16	CAC 4333	Clamp	2		
17	CAC 2753	Cradle	2		
18	CAC 2823	Keeper plate	2		
19	CAC 2751	Mounting rubber	2		
20	CAC 2750	Bush	2		
21	UFS 131 5R	Setscrew	4		
22	JLM 303	Washer	4		
23	CAC 3103	Mounting bar	2		
24	C 44083	Collar	4		
25	FW 206 T	Washer	2		
26	C 31075	Locknut	2		
27	C 45084	Mounting bracket	2		
28	BD 541 41	Washer	4		
29	C 19706	Distance piece	4		
30	C 724	Washer	4		
31	JLM 9684	Nut	4		

MASTER INDEX

ENGINE ... 1.C
FLYWHEEL AND CLUTCH ... 1.G
GEARBOX AND PROPSHAFT ... 1.H
AXLES, SUSPENSION, DRIVE SHAFTS, WHEELS ... 1.K
STEERING ... 1.M
BRAKES AND BRAKE CONTROLS ... 1.N
FUEL, EXHAUST AND EMISSION SYSTEMS ... 2.C
COOLING, HEATING AND AIR CONDITIONING ... 2.G
ELECTRICAL, WASHERS-WIPERS, INSTRUMENTS ... 2.J
CHASSIS, SUBFRAMES, BODYSHELL, FITTINGS ... 3.C
FASCIA, TRIM, SEATS AND FIXINGS ... 3.H
ACCESSORIES AND PAINT ... 4.C
NUMERICAL INDEX ... 4.D

GROUP INDEX

ACCELERATOR CABLE LHD 3.6 LITRE ... C16
ACCELERATOR CABLE LHD-5.3 LITRE ... C17
ACCELERATOR CABLE RHD 3.6 LITRE ... C13
ACCELERATOR CABLE RHD 5.3 LITRE ... C14
AIR CLEANER RH-5.3 LITRE ... E02
AIR CLEANER-3.6 LITRE ... C02
AIR INJECTION PUMP 3.6 LITRE ... C09
AIR INJECTION PUMP-5.3 LITRE ... F03
FRONT EXHAUST PIPE RH ... F10
FUEL FILLER CAP - CONVERTIBLE ... D11
FUEL FILLER CAP-CABRIOLET ... D10
FUEL PUMP AND FILTER ... D04
FUEL RAIL-INJECTORS-3.6 LITRE ... C05
FUEL TANK-CABRIOLET ... D09

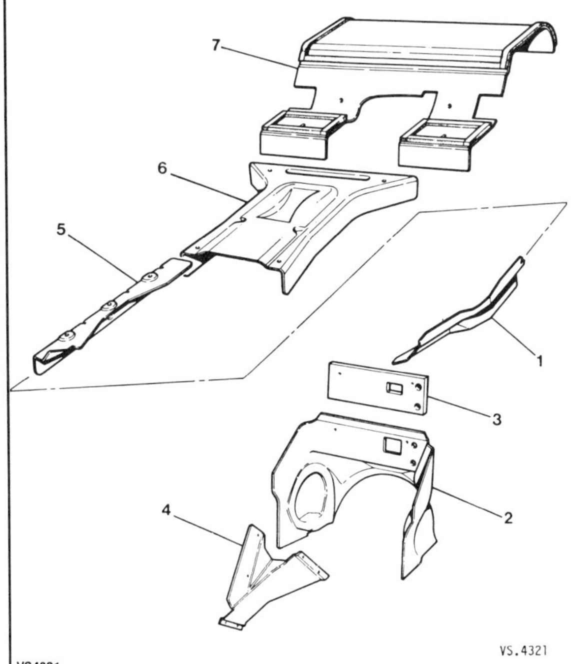

VS4321

HEATSHIELDS 3.6 LITRE

Illus	Part Number	123456 Description	Quantity	Change Point	Remarks
1	CAC 5970	Forward upperfloor h/shield LH	1		
	DAZ 810 8C	Screw-shield to body	2		
	BD 541 34	Washer-shield to body	2		
	DAZ 810 8C	Screw-shield to inter shield	1		
	BD 541 34	Washer-shield to inter shield	1		
	BD 8633 10	Nut-shield to inter shield	1		
2	CAC 5423	Heatshield-bulkhead	1		
	BD 35306 1	Rivet	2		
3	C 43686 1	Insulation pad	1		
4	CAC 5468	Heatshield-side RH	1		
	CAC 5469	Heatshield-side LH	1		
	BD 33506 1	Rivet	2		
	JLM 9577	Screw	1		
	BD 8633 15	Nut	2		
	AR 608051 J	Screw	1		
	BD 541 34	Washer	2		
5	CAC 4408 1	Heatshield-underfloor	1		
	CBC 2181	Heatshield-underfloor (Auto)	1	XJ57 Up to 1988 MY	
	CBC 5688	Heatshield-underfloor (Auto)	1	XJ57 1988 MY	
	DAZ 810 8C	Screw	3		
	BD 541 34	Washer	3		
6	CAC 7657	Heatshield-underfloor (Manual)	1	XJ57 Up to 1988MY	
	CBC 5691	Heatshield-underfloor (Manual)	1	XJ57 1988 MY	
	AB 610031 J	Screw	NLA		Use AR610071
	AR 610071	Self tapping screw	4		
	C 38559	Spacer	4		
	BD 542 9	Washer	4		
7	CAC 3915	Heatshield-overaxle	NLA		Use CBC5361
	CBC 5361	Heatshield-overaxle	1		
	BD 542 9	Washer	10		
	DAZ 810 8C	Screw	10		

MASTER INDEX

GROUP INDEX

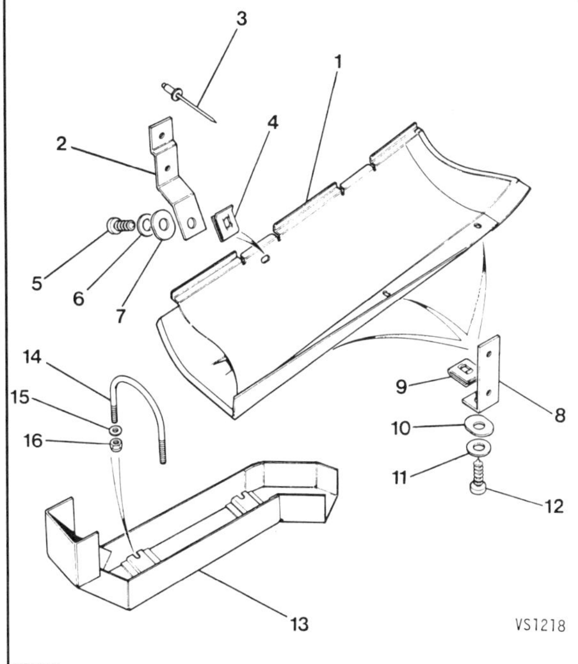

VS1218

EXHAUST-GRASS SHIELDS-JAPAN

Illus	Part Number	123456 Description	Quantity	Change Point	Remarks
1	CAC 2582	Grass shield-rear silencer RH	1		
	CAC 2583	Grass shield-rear silencer LH	1		
2	CAC 2589	Attachment bkt-grass shield	2		
3	BD 33506 1	Imex rivet	16		
4	BD 8633 3	Spire nut	2		Shield to attachment brackets
5	DBB 808 8C	Selt tapping screw	2		
6	BD 541 51	Plain washer	2		
7	BD 542 12	Oval washer	2		
8	CAC 2590	Attachment bkt-grass shield	6		
9	BD 8633 3	Spire nut	6		
10	BD 542 12	Oval washer	6		
11	BD 541 51	Plain washer	6		
12	DBB 808 8C	Self tapping screw	6		
13	CAC 2588 4	Grass shield-o/axle pipe RH	1		
	CAC 2588 5	Grass shield-o/axle pipe LH	1		
14	CAC 2631	'U' bolt-grass shield	4		
15	WC 702101 J	Plain washer	8		
16	C 24119 27	Cleveloc nut	8		

MASTER INDEX
ENGINE 1.C
FLYWHEEL AND CLUTCH 1.G
GEARBOX AND PROPSHAFT 1.H
AXLES, SUSPENSION, DRIVE SHAFTS, WHEELS 1.K
STEERING 1.M
BRAKES AND BRAKE CONTROLS 1.N
FUEL, EXHAUST AND EMISSION SYSTEMS 2.C
COOLING, HEATING AND AIR CONDITIONING 2.G
ELECTRICAL, WASHERS-WIPERS, INSTRUMENTS 2.J
CHASSIS, SUBFRAMES, BODYSHELL, FITTINGS 3.C
FASCIA, TRIM, SEATS AND FIXINGS 3.H
ACCESSORIES AND PAINT 4.C
NUMERICAL INDEX 4.D
GROUP INDEX
ACCELERATOR CABLE LHD 3.6 LITRE C16
ACCELERATOR CABLE LHD-5.3 LITRE C17
ACCELERATOR CABLE RHD 3.6 LITRE C13
ACCELERATOR CABLE RHD 5.3 LITRE C14
AIR CLEANER RH-5.3 LITRE E02
AIR CLEANER-3.6 LITRE C02
AIR INJECTION PUMP 3.6 LITRE C09
AIR INJECTION PUMP-5.3 LITRE F03
FRONT EXHAUST PIPE RH F10
FUEL FILLER CAP - CONVERTIBLE D11
FUEL FILLER CAP-CABRIOLET D10
FUEL PUMP AND FILTER D04
FUEL RAIL-INJECTORS-3.6 LITRE C05
FUEL TANK-CABRIOLET D09

FRONT EXHAUST PIPES-NOT USA/CDN/J 5.3 LITRE.

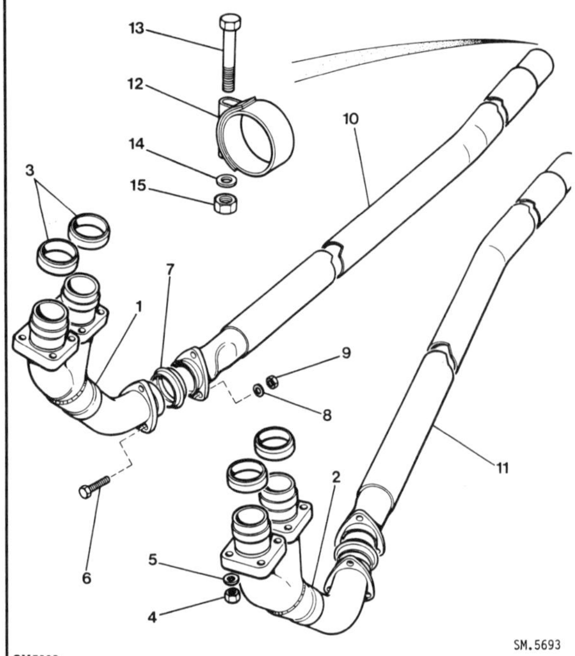

SM5693

Illus	Part Number	123456 Description	Quantity	Change Point	Remarks
1	CAC 1974	FRONT EXHAUST PIPE RH	1		
2	CAC 1975	Front exhaust pipe LH	1		
3	CBC 4834	Manifold to downpipe joint	4		
4	C 17916	Nut	8		
5	FW 106 T	Washer	8		
6	BH 605141 J	Bolt	6		
7	C 35401	Olive	2		
8	FW 105 E	Washer	6		
9	C 17914	Nut	6		
10	CAC 1402	Intermediate exhaust pipe RH	1		
11	CAC 1403	Intermediate exhaust pipe LH	1		
12	CAC 3801 1	Exhaust strap	4		
13	CAC 3802	Bolt	4		
14	FW 105 T	Washer	4		
15	CAC 3803	Nut	4		

2F 10

MASTER INDEX

GROUP INDEX

FRONT EXHAUST PIPES 5.3 LITRE.

USA,CDN, AUSTRALIA, GERMANY &
OPT.EUROPEAN LEVEL 1

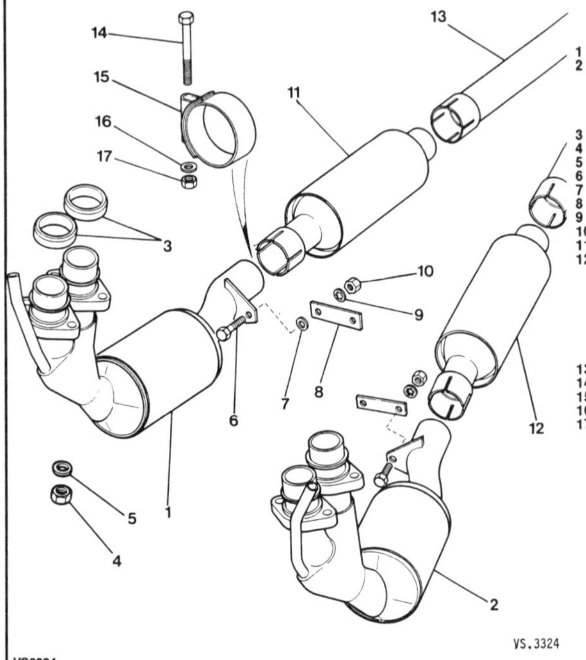

VS3324

Illus	Part Number	123456 Description	Quantity	Change Point	Remarks
1	CBC 2072	Exhaust catalyst & downpipe-RH	1	] USA/CDN Up to	
2	CBC 2073	Exhaust catalyst & downpipe-LH	1	] 1989 MY	
	CBC 5320	Exhaust catalyst & downpipe-RH	1	] Germany/USA/CDN	19881/2 MY
	CBC 5321	Exhaust catalyst & downpipe-LH	1	] Up to 1989 MY	19881/2 MY
	CBC 7958	Exhaust downpipe & catalyst-RH	1	] USA/CDN/Not	
	CBC 7959	Exhaust downpipe & catalyst-LH	1	] Germany 1989 MY	
3	CBC 4834	Joint-manifold to downpipe	4		
4	C 17916 1	Nut	8		
5	FW 106 T	Washer	8		
6	SH 604041 J	Setscrew	2		
7	FW 105 T	Washer	2		
8	CAC 4267	Exhaust stay	2		
9	C 724	Shakeproof washer	4		
10	JLM 9683	Nut	4		
11	CBC 2074	Secondary exhaust catalyst-RH	1		
12	CBC 2075	Secondary exhaust catalyst-LH	1		
	CBC 5324	Secondary exhaust catalyst-RH	1	....	19881/2 MY
	CBC 5325	Secondary exhaust catalyst-LH	1	....	19881/2 MY
	CBC 4850	Secondary exhaust catalyst-RH	1	....	] 1989MY
	CBC 4851	Secondary exhaust catalyst-LH	1		]
	CBC 8212	Secondary exhaust catalyst-RH	1	....	] Aust/USA/CDN/
	CBC 8213	Secondary exhaust catalyst-LH	1	....	] Not Germany
13	CAC 4319	Intermediate exhaust pipe	2		
14	CAC 3802	Bolt	4		
15	CAC 3801 1	Exhaust strap	4		
16	FW 105 T	Washer	4		
17	CAC 3803	Nut	4		

H174AW25/8/88

2F 11

MASTER INDEX

GROUP INDEX

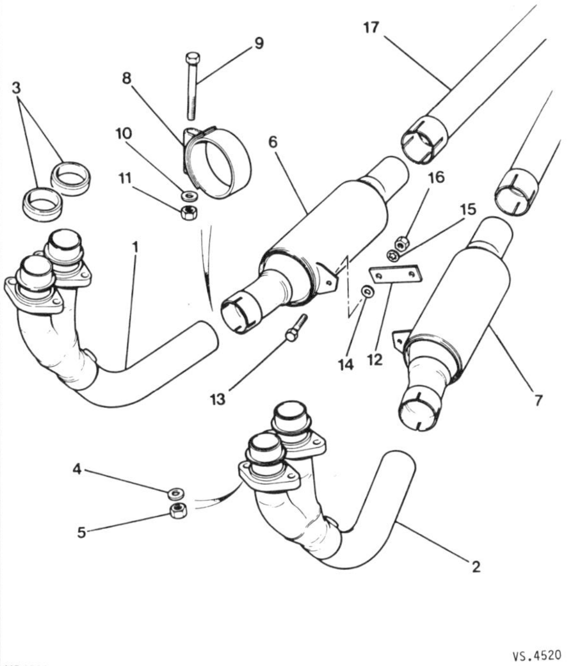

VS4520

FRONT EXHAUST PIPES-AUSTRALIA/CDN/USA/Germ/Eur Level 1

FRONT EXHAUST PIPES

Illus	Part Number	123456 Description	Quantity	Change Point	Remarks
1	CBC 1530	Front downpipe RH	1		
2	CBC 1531	Front downpipe LH	1		
3	CBC 4843	Joint-manifold to downpipe	4		
4	FW 106 T	Plain washer	8		
5	C 17916 1	Helicoil nut	8		
6	CBC 2074	Secondary catalyst-RH	NLA		Use CBC5324
7	CBC 2075	Secondary catalyst-LH	NLA		Use CBC5325
	CBC 5324	Secondary catalyst-RH	1	Up to 1989 MY	
	CBC 5325	Secondary catalyst-LH	1	Up to 1989 MY	
	CBC 4850	Secondary catalyst-RH	1	Germany only from 1989 MY	
	CBC 4851	Secondary catalyst-LH	1	Germany only from 1989 MY	
	CBC 8212	Secondary catalyst-RH	1	1989 MY Not Germany	
	CBC 8213	Secondary catalyst-LH	1	1989 MY Not Germany	
8	CAC 3801 1	Clamp-pipe band	2		
9	CAC 3802	Bolt 0.025 UNF	2		
10	FW 104 T	Plain washer	2		
11	CAC 3803	Locknut 1.4 UNF	2		
12	CAC 4267	Stay	2		
13	SH 604041 J	Setscrew	2		Stay to exhaust tie
14	FW 104 T	Plain washer	2		Stay to exhaust tie
15	WF 600040 J	Shakeproof washer	2		Stay to exhaust tie
16	NH 604041 J	Nut	2		Stay to exhaust tie
17	CAC 4319	Intermediate exhaust pipe	2		

H176AW31/8/88

MASTER INDEX

GROUP INDEX

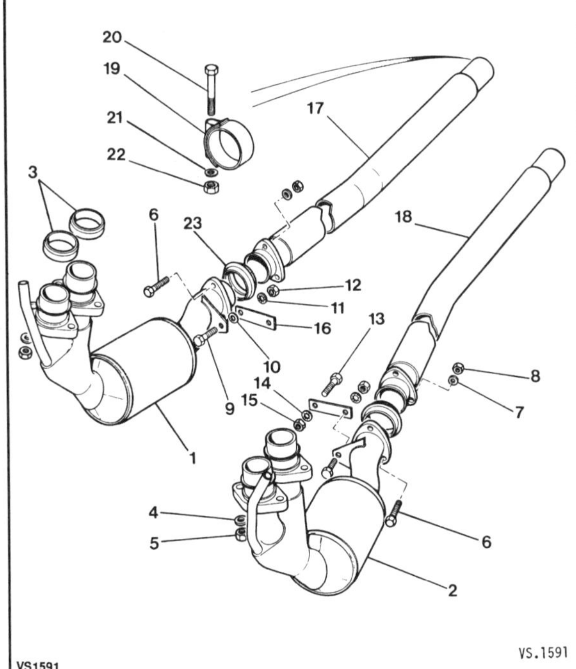

VS1591

FRONT EXHAUST PIPES-5.3 LITRE JAPAN

Illus	Part Number	123456 Description	Quantity	Change Point	Remarks
1	CBC 2388	Exhaust catalyst & downpipe-RH	NLA		Use CBC5542
2	CBC 2389	Exhaust catalyst & downpipe-LH	NLA		Use CBC5543
	CBC 5542	Exhaust catalyst & downpipe-RH	1		
	CBC 5543	Exhaust catalyst & downpipe-LH	1		
3	CBC 4834	Sealing ring	4		
4	FW 106 T	Washer	8		
5	C 17916 1	Nut-helicoil	8		
6	BH 605141 J	Bolt	6		
7	FW 105 T	Washer	6		
8	C 17914	Nut	6		
9	SH 604041 J	Setscrew	2		
10	FW 104 T	Washer	2		
11	WF 600040 J	Shakeproof washer	4		
12	JLM 9683	Nut	4		
13	SH 504081 J	Setscrew	1		
14	C 724	Shakeproof washer	1		
15	JLM 9683	Nut	1		
16	C 45173	Stay	2		
17	CAC 1928	Intermediate exhaust pipe-RH	1		
18	CAC 1929	Intermediate exhaust pipe-LH	1		
19	CAC 3801 1	Exhaust pipe clamp	4		
20	CAC 3802	Bolt	4		
21	FW 104 T	Washer	4		
22	CAC 3803	Nut	4		
23	C 35401	Olive	2		

H178AW21/7/88

MASTER INDEX

ENGINE 1.C
FLYWHEEL AND CLUTCH 1.G
GEARBOX AND PROPSHAFT 1.H
AXLES, SUSPENSION, DRIVE SHAFTS, WHEELS 1.K
STEERING 1.M
BRAKES AND BRAKE CONTROLS 1.N
FUEL, EXHAUST AND EMISSION SYSTEMS 2.C
COOLING, HEATING AND AIR CONDITIONING 2.G
ELECTRICAL, WASHERS-WIPERS, INSTRUMENTS 2.J
CHASSIS, SUBFRAMES, BODYSHELL, FITTINGS 3.C
FASCIA, TRIM, SEATS AND FIXINGS 3.H
ACCESSORIES AND PAINT 4.C
NUMERICAL INDEX 4.D

GROUP INDEX

ACCELERATOR CABLE LHD 3.6 LITRE C16
ACCELERATOR CABLE LHD-5.3 LITRE C17
ACCELERATOR CABLE RHD 3.6 LITRE C13
ACCELERATOR CABLE RHD 5.3 LITRE C14
AIR CLEANER RH-5.3 LITRE E02
AIR CLEANER-3.6 LITRE C02
AIR INJECTION PUMP 3.6 LITRE C09
AIR INJECTION PUMP-5.3 LITRE F03
FRONT EXHAUST PIPE RH F10
FUEL FILLER CAP - CONVERTIBLE D11
FUEL FILLER CAP-CABRIOLET D10
FUEL PUMP AND FILTER D04
FUEL RAIL-INJECTORS-3.6 LITRE C05
FUEL TANK-CABRIOLET D09

2F 14

EXHAUST DOWNPIPES EUROPEAN OPTIONAL LEVEL 2-5.3 LITRE

FRONT DOWNPIPE & CATALYST

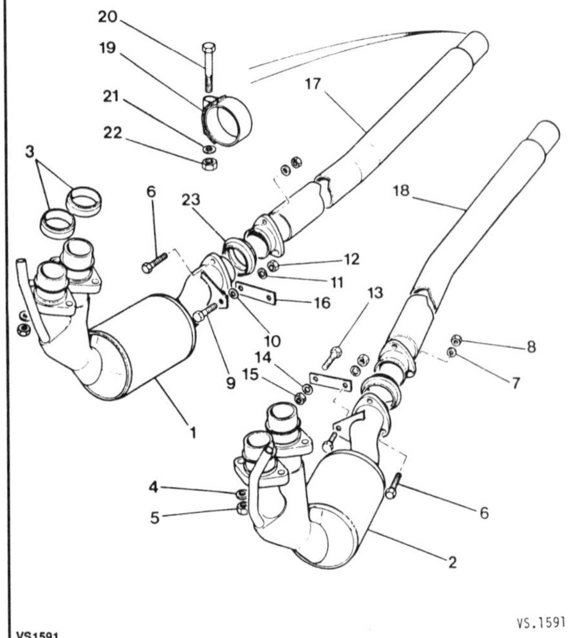

VS1591

Illus	Part Number	Description	Quantity	Change Point	Remarks
1	CBC 4060	Frt downpipe & catalyst assy RH	NLA		Use CBC5328
	CBC 5328	Frt downpipe & catalyst assy RH	1		
2	CBC 4061	Frt downpipe & catalyst assy LH	NLA		Use CBC5329
	CBC 5329	Frt downpipe & catalyst assy LH	1		
3	CBC 4834	Joint-manifold to downpipe	4		
4	FW 106 T	Washer	9		
5	C 17916 1	Helicoil locknut	8		
6	BH 605141 J	Bolt	6		
7	FW 105 T	Washer	6		
8	C 17914	Helicoil locknut	6		
9	SH 604041 J	Setscrew	2		
10	FW 104 T	Washer	2		
11	WF 600040 J	Shakeproof washer	4		
12	JLM 9683	Nut	4		
13	SH 604041 J	Setscrew	2		
14	C 724	Shakeproof washer	2		
15	JLM 9683	Nut	2		
16	C 45173	Stay	2		
17	CAC 1410	Exhaust pipe	1		
18	CAC 1410	Exhaust pipe	1		
19	CAC 3801 1	Pipe band clamp	4		
20	CAC 3802	Bolt	4		
21	FW 104 T	Plain washer	4		
22	CAC 3803	Nut	4		
23	C 35401	Olive	2		

MASTER INDEX

GROUP INDEX

2F 15

INTERMEDIATE & REAR SILENCERS 5.3 LITRE

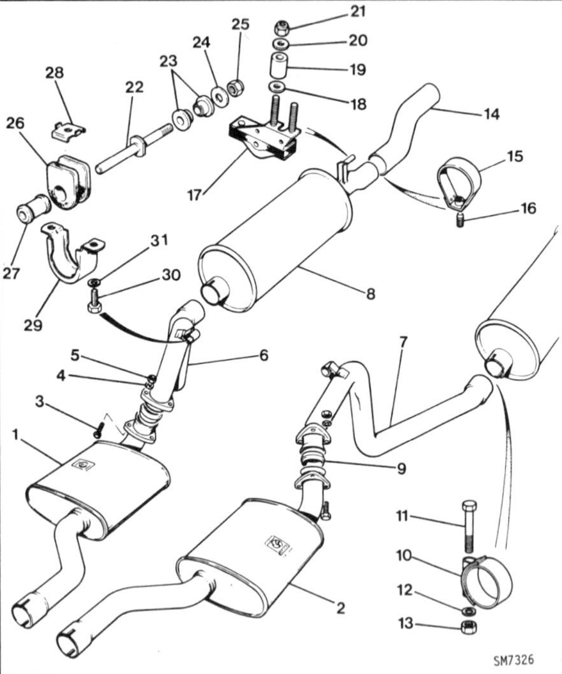

SM7326

Illus	Part Number	Description	Quantity	Change Point	Remarks
1	CAC 1660	Intermediate exhaust silencer-RH	1		
	CAC 1930	Intermediate exhaust silencer-RH	1		Japan only
2	CAC 1661	Intermediate exhaust silencer-LH	1		
	CAC 1931	Intermediate exhaust silencer-LH	1		Japan only
3	BH 605141 J	Bolt	6		
4	FW 105 T	Washer	6		
5	C 17914	Nut	6		
6	CAC 5620	Over axle pipe-RH	1		
7	CAC 5621	Over axle pipe-LH	1		
8	CAC 2494	Rear exhaust silencer	2		
9	C 43831	Olive	2		
10	CAC 3801 3	Exhaust pipe clamp	2		
11	CAC 3802	Bolt	2		
12	FW 105 T	Washer	2		
13	CAC 3803	Nut	2		
14	CAC 6068	Tail pipe-RH	1		
	CAC 6069	Tail pipe-LH	1		
15	CAC 4333	Exhaust clamp	2		Use CBC8388
	CBC 8388	Exhaust clamp	2		
16	AGU 1004	Hexagon socket cup point screw	2		
17	C 45084	Tail pipe mounting assembly	2		
18	BD 541 41	Washer	4		
19	C 19706	Distance piece	4		
20	WF 600050 J	Shakeproof washer	4		
21	NH 605041 J	Nut	4		
22	CAC 3103	Mounting bar	2		
23	C 44083	Collar	4		
24	FW 206 T	Washer	2		
25	C 31075	Nut	2		
26	CAC 2751	Mounting rubber	2		
27	CAC 2750	Bush	2		
28	CAC 2823	Keeper plate	2		
29	CAC 2753	Cradle	2		
30	UFS 131 5R	Setscrew	4		
31	JLM 303	Washer-spring	4		

MASTER INDEX

ENGINE 1.C
FLYWHEEL AND CLUTCH 1.G
GEARBOX AND PROPSHAFT 1.H
AXLES, SUSPENSION, DRIVE SHAFTS, WHEELS 1.K
STEERING 1.M
BRAKES AND BRAKE CONTROLS 1.N
FUEL, EXHAUST AND EMISSION SYSTEMS 2.C
COOLING, HEATING AND AIR CONDITIONING 2.G
ELECTRICAL, WASHERS-WIPERS, INSTRUMENTS 2.J
CHASSIS, SUBFRAMES, BODYSHELL, FITTINGS 3.C
FASCIA, TRIM, SEATS AND FIXINGS 3.H
ACCESSORIES AND PAINT 4.C
NUMERICAL INDEX 4.D

GROUP INDEX

ACCELERATOR CABLE LHD 3.6 LITRE C16
ACCELERATOR CABLE LHD-5.3 LITRE C17
ACCELERATOR CABLE RHD 3.6 LITRE C13
ACCELERATOR CABLE RHD 5.3 LITRE C14
AIR CLEANER RH-5.3 LITRE E02
AIR CLEANER-3.6 LITRE C02
AIR INJECTION PUMP 3.6 LITRE C09
AIR INJECTION PUMP-5.3 LITRE F03
FRONT EXHAUST PIPE RH F10
FUEL FILLER CAP - CONVERTIBLE D11
FUEL FILLER CAP-CABRIOLET D10
FUEL PUMP AND FILTER D04
FUEL RAIL-INJECTORS-3.6 LITRE C05
FUEL TANK-CABRIOLET D09

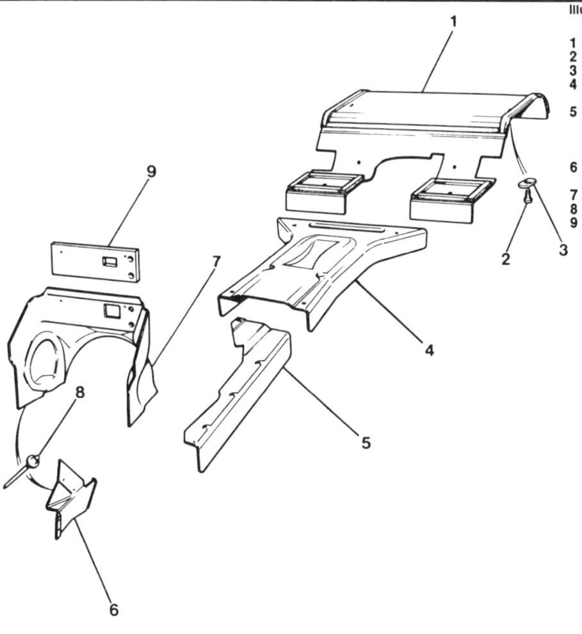

SM.7116

SM7116

EXHAUST HEATSHIELDS-5.3 LITRE

Illus	Part Number	Description	Quantity	Change Point	Remarks
1	CAC 3915	Heatshield-overaxle	1		
2	DAZ 810 8C	Self tapping screw	10		
3	BD 542 9	Washer	10		
4	CAC 7657	Heatshield-underfloor	1		
	CAC 7207	Heatshield-underfloor	1		Cabriolet
5	CAC 4410	H/shield underfloor interpipes RH	1		Use CBC 6888
	CAC 4411	H/shield underfloor interpipes LH	1		Use CBC 6889
	CBC 6888	Heatshield u/floor interpipes RH	1		
	CBC 6889	Heatshield u/floor interpipes LH	1		
6	CAC 5468	H/shield bulkhead side RH	1		
	CAC 5469	H/shield bulkhead side LH	1		
7	CAC 5109 1	Heatshield bulkhead	1		
8	BD 33506 1	Pop rivet	16		
9	C 43686 1	Insulation pad	1		
	CAC 4375	Heatshield LH-ex downpipe	1		N America only
	CAC 4374	Heatshield LH-ex downpipe	1		N America only
	CAC 4373 1	Heatshield LH-ex downpipe	1		Japan only

MASTER INDEX

GROUP INDEX

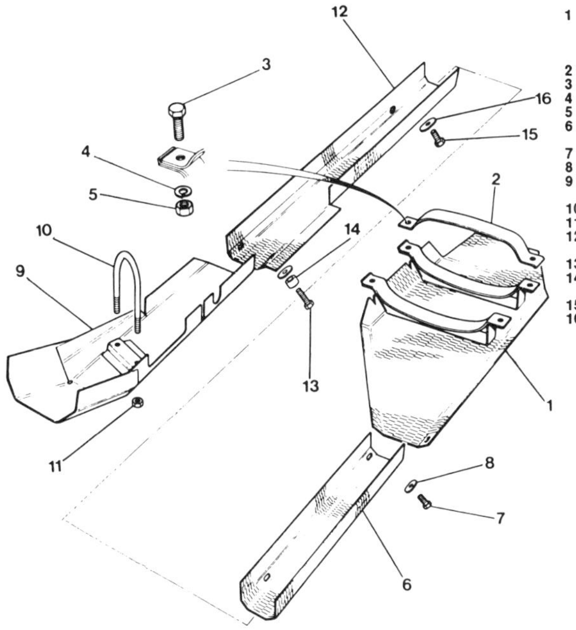

SM.5655

SM5655

GRASS HEATSHIELDS(JAPAN)-5.3 LITRE

Illus	Part Number	Description	Quantity	Change Point	Remarks
1	CAC 3144	Heatshield inter silencer RH	1		XJS
	CAC 3145	Heatshield inter silencer LH	1		XJS
	CBC 2368	Heatshield inter silencer RH	1		Cabriolet
	CBC 2369	Heatshield inter silencer LH	1		Cabriolet
2	C 31343	Shield support clip	4		
3	SH 604071 J	Screw	8		
4	C 725	Spring washer	8		
5	JLM 9684	Nut	8		
6	CAC 1922	H/shield inter pipe RH	1		
	CAC 1923	H/shield inter pipe LH	1		
7	SH 604041 J	Screw	8		
8	BD 542 8	Washer	8		
9	CAC 2924	Heatshield-catalyst RH	1		
	CAC 2995	Heatshield-catalyst LH	1		
10	CAC 1018	Bolt	2		
11	C 17916 2	Nut	4		
12	CAC 1792	H/shield inter pipe front RH	1		
	CAC 1793	H/shield inter pipe front LH	1		
13	SH 604071 J	Setscrew	4		Catalyst &
14	C 16145	Spacer	4		intershield to
	C 16941	D Washer	4		inter ex pipes
15	SH 604041 J	Setscrew	6		Intershield to
16	BD 542 8	D Washer	6		inter ex pipes

MASTER INDEX

GROUP INDEX

MASTER INDEX

GROUP INDEX

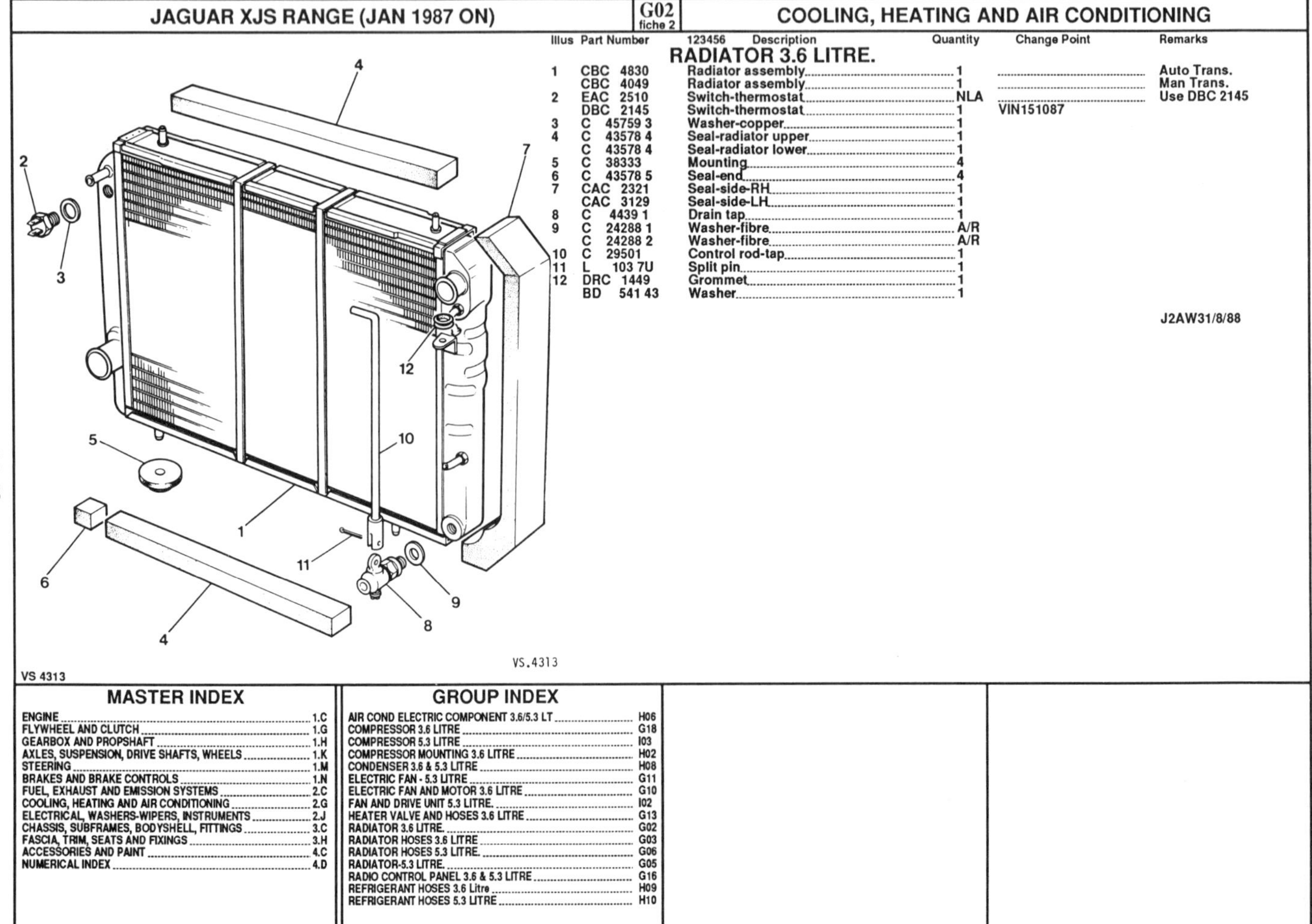

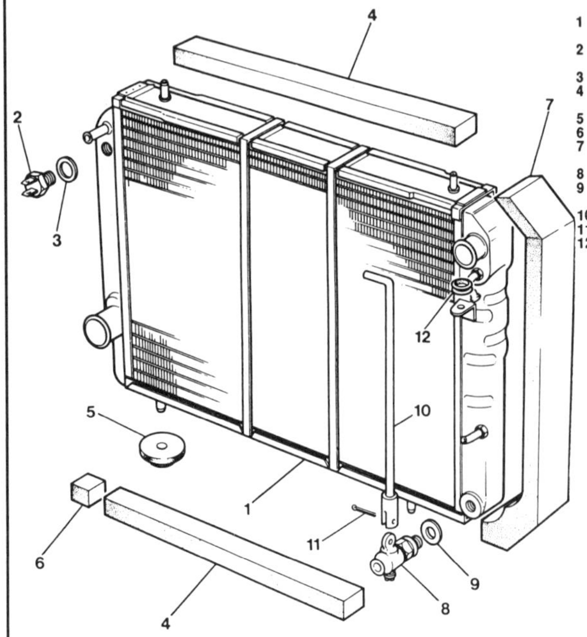

VS.4313

VS 4313

RADIATOR 3.6 LITRE.

Illus	Part Number	123456 Description	Quantity	Change Point	Remarks
1	CBC 4830	Radiator assembly	1		Auto Trans.
	CBC 4049	Radiator assembly	1		Man Trans.
2	EAC 2510	Switch-thermostat	NLA		Use DBC 2145
	DBC 2145	Switch-thermostat	1	VIN151087	
3	C 45759 3	Washer-copper	1		
4	C 43578 4	Seal-radiator upper	1		
	C 43578 4	Seal-radiator lower	1		
5	C 38333	Mounting	4		
6	C 43578 5	Seal-end	4		
7	CAC 2321	Seal-side-RH	1		
	CAC 3129	Seal-side-LH	1		
8	C 4439 1	Drain tap	1		
9	C 24288 1	Washer-fibre	A/R		
	C 24288 2	Washer-fibre	A/R		
10	C 29501	Control rod-tap	1		
11	L 103 7U	Split pin	1		
12	DRC 1449	Grommet	1		
	BD 541 43	Washer	1		

J2AW31/8/88

MASTER INDEX

GROUP INDEX

AIR COND ELECTRIC COMPONENT 3.6/5.3 LT H06
COMPRESSOR 3.6 LITRE G18
COMPRESSOR 5.3 LITRE I03
COMPRESSOR MOUNTING 3.6 LITRE H02
CONDENSER 3.6 & 5.3 LITRE H08
ELECTRIC FAN - 5.3 LITRE G11
ELECTRIC FAN AND MOTOR 3.6 LITRE G10
FAN AND DRIVE UNIT 5.3 LITRE. I02
HEATER VALVE AND HOSES 3.6 LITRE G13
RADIATOR 3.6 LITRE. G02
RADIATOR HOSES 3.6 LITRE G03
RADIATOR HOSES 5.3 LITRE. G06
RADIATOR-5.3 LITRE. G05
RADIO CONTROL PANEL 3.6 & 5.3 LITRE G16
REFRIGERANT HOSES 3.6 Litre H09
REFRIGERANT HOSES 5.3 LITRE H10

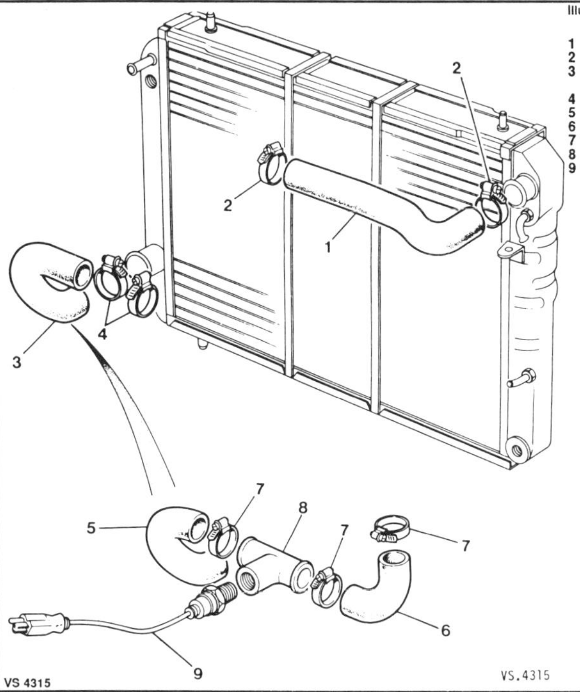

VS 4315

RADIATOR HOSES 3.6 LITRE

Illus	Part Number	123456 Description	Quantity	Change Point	Remarks
1	CAC 5100	Water hose-top	1		
2	C 43206 4	Clip-hose	2		
3	CAC 5225	Water hose bottom	1		Cars without water heater
	CBC 1864	Water hose-bottom	1		
4	C 43206 5	Clip-hose	2		
5	CAC 6593	Water hose-bottom	1		Cars with water heater
6	CAC 6594	Water hose-bottom	1		
7	C 43206 5	Clip-hose	4		
8	EAC 5038	Water pipe	1		
9	C 30381	Bray heater(240 volt)	1		
	C 34403	Heater(110 volt)	1		Canada only

MASTER INDEX

GROUP INDEX

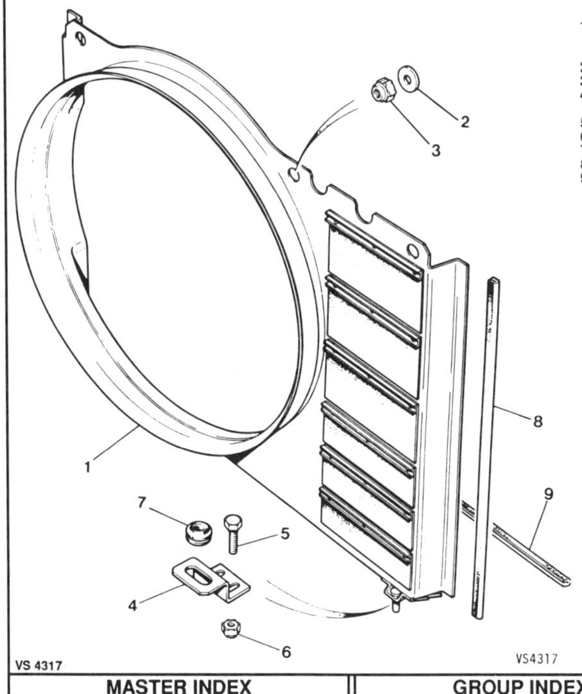

VS 4317

FAN COWL MECHANICAL-3.6 LITRE

Illus	Part Number	123456 Description	Quantity	Change Point	Remarks
1	CAC 5201	Fan cowl assembly	NLA		Use CBC7297 & CBC6209
	CBC 7297	Fan cowl assembly	1		
2	C 16941	Washer-plain	3		
3	NZ 604041 J	Locknut	3		
4	CAC 4797	Bracket-cowl	2		
	CBC 6209	Bracket-cowl	2		
5	SH 604041 J	Setscrew	2		
6	NZ 604041 J	Locknut	2		
7	C 16393	Grommet	2		
8	C 37941 1	Sealing strip-upper	2		
9	C 37941 7	Sealing strip-lower	2		

MASTER INDEX

ENGINE 1.C
FLYWHEEL AND CLUTCH 1.G
GEARBOX AND PROPSHAFT 1.H
AXLES, SUSPENSION, DRIVE SHAFTS, WHEELS 1.K
STEERING 1.M
BRAKES AND BRAKE CONTROLS 1.N
FUEL, EXHAUST AND EMISSION SYSTEMS 2.C
COOLING, HEATING AND AIR CONDITIONING 2.G
ELECTRICAL, WASHERS-WIPERS, INSTRUMENTS 2.J
CHASSIS, SUBFRAMES, BODYSHELL, FITTINGS 3.C
FASCIA, TRIM, SEATS AND FIXINGS 3.H
ACCESSORIES AND PAINT 4.C
NUMERICAL INDEX 4.D

GROUP INDEX

AIR COND ELECTRIC COMPONENT 3.6/5.3 LT H06
COMPRESSOR 3.6 LITRE G18
COMPRESSOR 5.3 LITRE I03
COMPRESSOR MOUNTING 3.6 LITRE H02
CONDENSER 3.6 & 5.3 LITRE H08
ELECTRIC FAN - 5.3 LITRE G11
ELECTRIC FAN AND MOTOR 3.6 LITRE G10
FAN AND DRIVE UNIT 5.3 LITRE I02
HEATER VALVE AND HOSES 3.6 LITRE G13
RADIATOR 3.6 LITRE G02
RADIATOR HOSES 3.6 LITRE G03
RADIATOR HOSES 5.3 LITRE G06
RADIATOR-5.3 LITRE G05
RADIO CONTROL PANEL 3.6 & 5.3 LITRE G16
REFRIGERANT HOSES 3.6 Litre H09
REFRIGERANT HOSES 5.3 LITRE H10

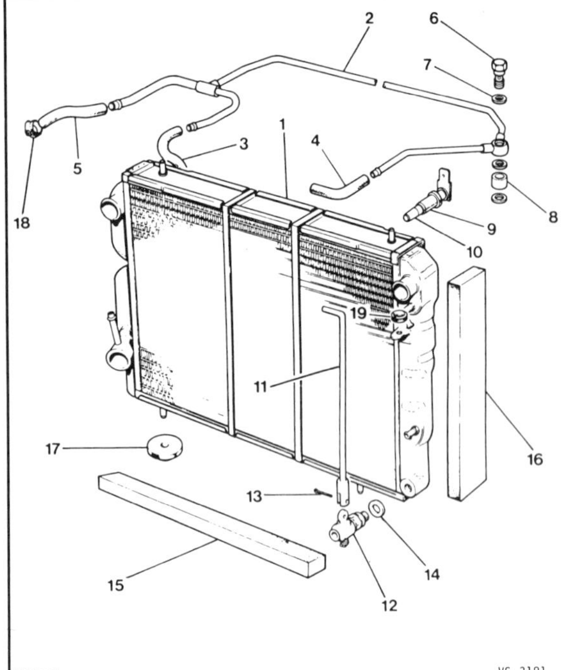

VS3181

RADIATOR-5.3 LITRE.

Illus	Part Number	123456 Description	Quantity	Change Point	Remarks
1	CBC 4047	Radiator	1		CDN-Not Germ/UK/ Euro temp zones
	CBC 4048	Radiator	1		UK/Germ/Euro temp zones
2	CAC 4474	Bleed pipe-radiator	1		
3	CAC 5092 1	Bleed hose	1		
4	CAC 5093 1	Bleed hose-thermo to header tank	1		
5	CAC 4572 1	Hose-thermo vent to header	1		
6	C 44085 2	Banjo bolt	1		
7	C 4146	Washer	3		
8	CAC 4559	Spacer	1		
9	C 43221	Probe bush	1		
10	C 43222	Low coolant probe	1		
11	C 29501	Control rod	1		
12	C 4439 1	Drain tube	1		
13	L 103 7U	Split pin	1		
14	C 24288 1	Fibre washer 0.03125"	A/R		
	C 24288 2	Fibre washer 0.01515	A/R		
15	CAC 3821	Seal	2		
16	CAC 2321	Seal side	2		
17	C 38333	Radiator mounting rubber	2		
18	EAC 3215 2	Hose clip	10		
19	C 1042	Grommet-control rod	2		
	CAC 4723	Plug 0.125 BSP	1		
	C 930	Fibre washer	1		

J6.02AW31/8/88

2G 05

MASTER INDEX

GROUP INDEX

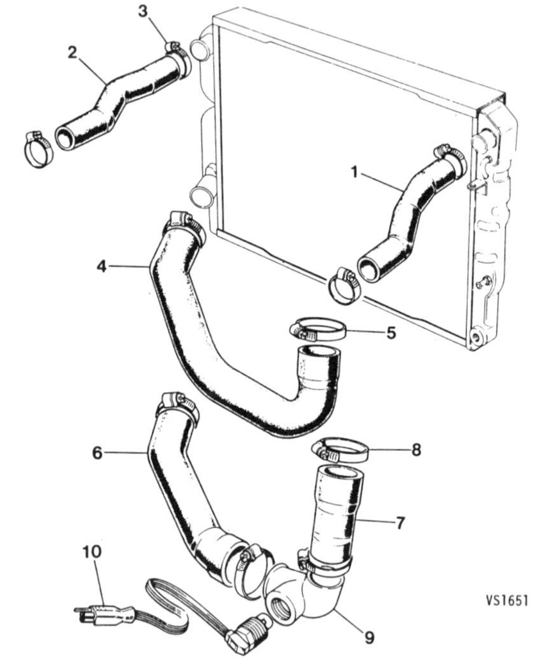

VS 1651

RADIATOR HOSES 5.3 LITRE.

Illus	Part Number	123456 Description	Quantity	Change Point	Remarks
1	C 43266	Top hose radiator RH	NLA		Use CBC4838
	CBC 4838	Water hose top RH	1		
2	C 43267	Top hose radiator LH	1		
3	C 43206 3	Hose clip	4		
4	C 41102	Bottom hose	1		
5	C 43206 5	Hose clip	2		
6	C 41103	Bottom hose	1		For cars fitted with heater element
7	C 41099	Bottom hose	1		For cars fitted with heater element
8	C 43206 5	Hose clip	4		For cars fitted with heater element
9	C 36996	Elbow-bottom water hose	1		For cars fitted with heater element
10	EAC 9587	Heater plug(less cable)	1		For cars fitted with heater element
	EAC 9588	Cable-heater plug	1		For cars fitted with heater element

2G 06

MASTER INDEX

ENGINE 1.C
FLYWHEEL AND CLUTCH 1.G
GEARBOX AND PROPSHAFT 1.H
AXLES, SUSPENSION, DRIVE SHAFTS, WHEELS 1.K
STEERING 1.M
BRAKES AND BRAKE CONTROLS 1.N
FUEL, EXHAUST AND EMISSION SYSTEMS 2.C
COOLING, HEATING AND AIR CONDITIONING 2.G
ELECTRICAL, WASHERS-WIPERS, INSTRUMENTS 2.J
CHASSIS, SUBFRAMES, BODYSHELL, FITTINGS 3.C
FASCIA, TRIM, SEATS AND FIXINGS 3.H
ACCESSORIES AND PAINT 4.C
NUMERICAL INDEX 4.D

GROUP INDEX

AIR COND ELECTRIC COMPONENT 3.6/5.3 LT H06
COMPRESSOR 3.6 LITRE G18
COMPRESSOR 5.3 LITRE I03
COMPRESSOR MOUNTING 3.6 LITRE H02
CONDENSER 3.6 & 5.3 LITRE H08
ELECTRIC FAN - 5.3 LITRE G11
ELECTRIC FAN AND MOTOR 3.6 LITRE G10
FAN AND DRIVE UNIT 5.3 LITRE I02
HEATER VALVE AND HOSES 3.6 LITRE G13
RADIATOR 3.6 LITRE G02
RADIATOR HOSES 3.6 LITRE G03
RADIATOR HOSES 5.3 LITRE G06
RADIATOR-5.3 LITRE G05
RADIO CONTROL PANEL 3.6 & 5.3 LITRE G16
REFRIGERANT HOSES 3.6 Litre H09
REFRIGERANT HOSES 5.3 LITRE H10

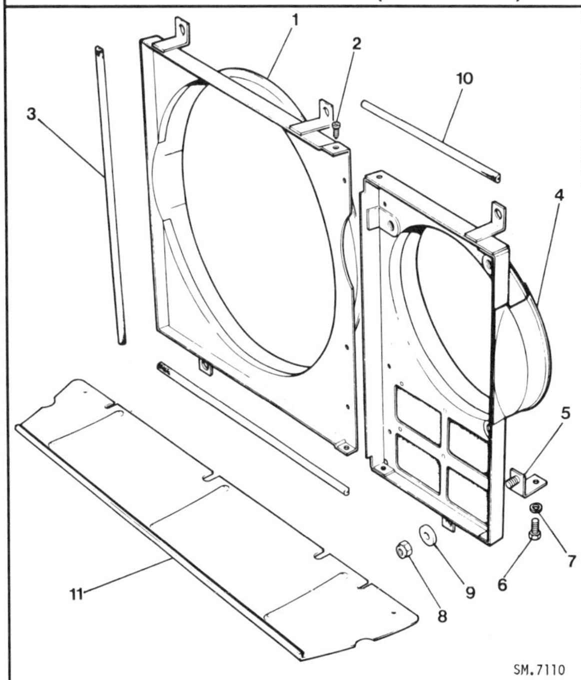

SM.7110

SM7110

COOLING-FAN COWLS-5.3 LITRE

Illus	Part Number	123456 Description	Quantity	Change Point	Remarks
1	C 40187	Fan cowl	1		
2	DAZ 810 8C	Self tapping screw	5		
3	C 37941 1	Sealing strip	5		
4	CAC 3794	Fan cowl	1		
5	C 36389	Fixing bracket	2		
6	SH 604041 J	Setscrew	2		
7	FG 104 X	Spring washer	2		
8	C 8737 1	Nut	5		
9	C 16941	Plain washer	5		
10	C 37941 1	Sealing strip	2		
11	C 45019	Undershield	1		Japan only

MASTER INDEX

GROUP INDEX

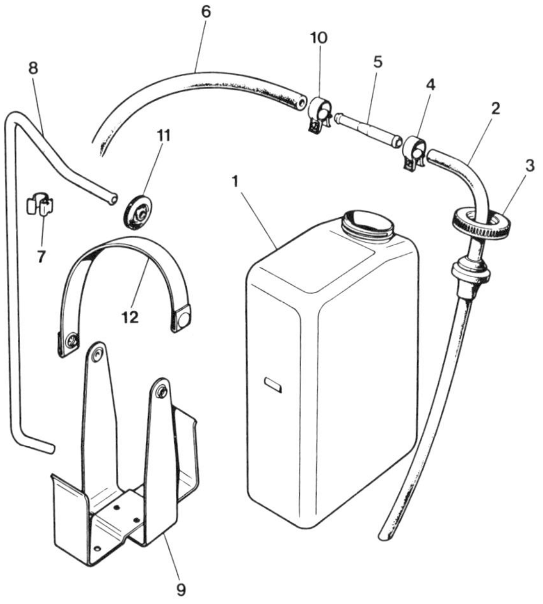

VS.2002/B

VS2002B

ATMOSPHERIC RECOVERY 3.6 LITRE

Illus	Part Number	123456 Description	Quantity	Change Point	Remarks
1	CBC 4889	Atmospheric recovery bottle	1		
2	CAC 3730	Assembly closure and tube	1		
3	CAC 3674	Cap bottle	1		
4	C 15886 3	Clip	1		
5	C 44479	Pipe adaptor	1		
6	EAC 2928 7	Hose pipe to tank	1		
7	CRC 1977	Clip-edge	2		
8	CBC 1931	Pipe overflow	1		
9	CAC 3678	Bracket mounting recovery bottle	1		
10	C 15886 4	Clip	2		
11	C 17432	Grommet	1		
12	CAC 3672	Strap flexible	1		

MASTER INDEX

ENGINE 1.C
FLYWHEEL AND CLUTCH 1.G
GEARBOX AND PROPSHAFT 1.H
AXLES, SUSPENSION, DRIVE SHAFTS, WHEELS 1.K
STEERING 1.M
BRAKES AND BRAKE CONTROLS 1.N
FUEL, EXHAUST AND EMISSION SYSTEMS 2.C
COOLING, HEATING AND AIR CONDITIONING 2.G
ELECTRICAL, WASHERS-WIPERS, INSTRUMENTS 2.J
CHASSIS, SUBFRAMES, BODYSHELL, FITTINGS 3.C
FASCIA, TRIM, SEATS AND FIXINGS 3.H
ACCESSORIES AND PAINT 4.C
NUMERICAL INDEX 4.D

GROUP INDEX

AIR COND ELECTRIC COMPONENT 3.6/5.3 LT H06
COMPRESSOR 3.6 LITRE G18
COMPRESSOR 5.3 LITRE I03
COMPRESSOR MOUNTING 3.6 LITRE H02
CONDENSER 3.6 & 5.3 LITRE H08
ELECTRIC FAN - 5.3 LITRE G11
ELECTRIC FAN AND MOTOR 3.6 LITRE G10
FAN AND DRIVE UNIT 5.3 LITRE I02
HEATER VALVE AND HOSES 3.6 LITRE G13
RADIATOR 3.6 LITRE G02
RADIATOR HOSES 3.6 LITRE G03
RADIATOR HOSES 5.3 LITRE G06
RADIATOR-5.3 LITRE G05
RADIO CONTROL PANEL 3.6 & 5.3 LITRE G16
REFRIGERANT HOSES 3.6 Litre H09
REFRIGERANT HOSES 5.3 LITRE H10

ATMOSPHERIC RECOVERY BOTTLE 5.3 LITRE

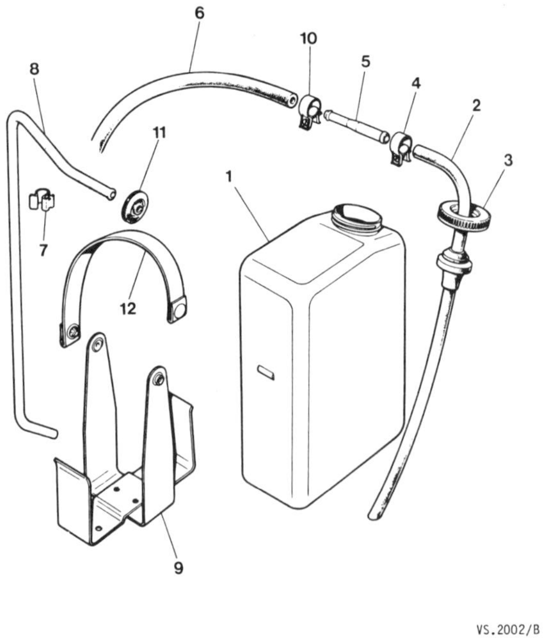

VS2002B

Illus	Part Number	123456	Description	Quantity	Change Point	Remarks
1	CAC 3673 9		Atmospheric recovery bottle	1		
	CBC 5942		Mounting bracket	1		
	CBC 5984		Atmospheric recovery bottle	1		Convertible
	CBC 6336		Overflow pipe	1		Convertible
2	CAC 3730		Closure and tube assembly	1		
3	CAC 3674		Cap-bottle	1		
4	C 15886 3		Clip-hose	1		
5	C 44479		Pipe adaptor	1		
6	EAC 2928 7		Hose pipe to tank	1		
7	CRC 1977		Edge clip	2		
8	CAC 3682		Pipe-overflow	1		
9	CAC 3678		Bracket-mounting	1		
10	C 15886 4		Clip-hose to pipe & hdr tank	2		
11	C 17432		Grommet	1		
12	CAC 3672		Strap-flexible	1		
	CBC 5945		Strap-flexible	1		Convertible

J8.02AW31/8/88

2G 09

MASTER INDEX

GROUP INDEX

ELECTRIC FAN AND MOTOR 3.6 LITRE

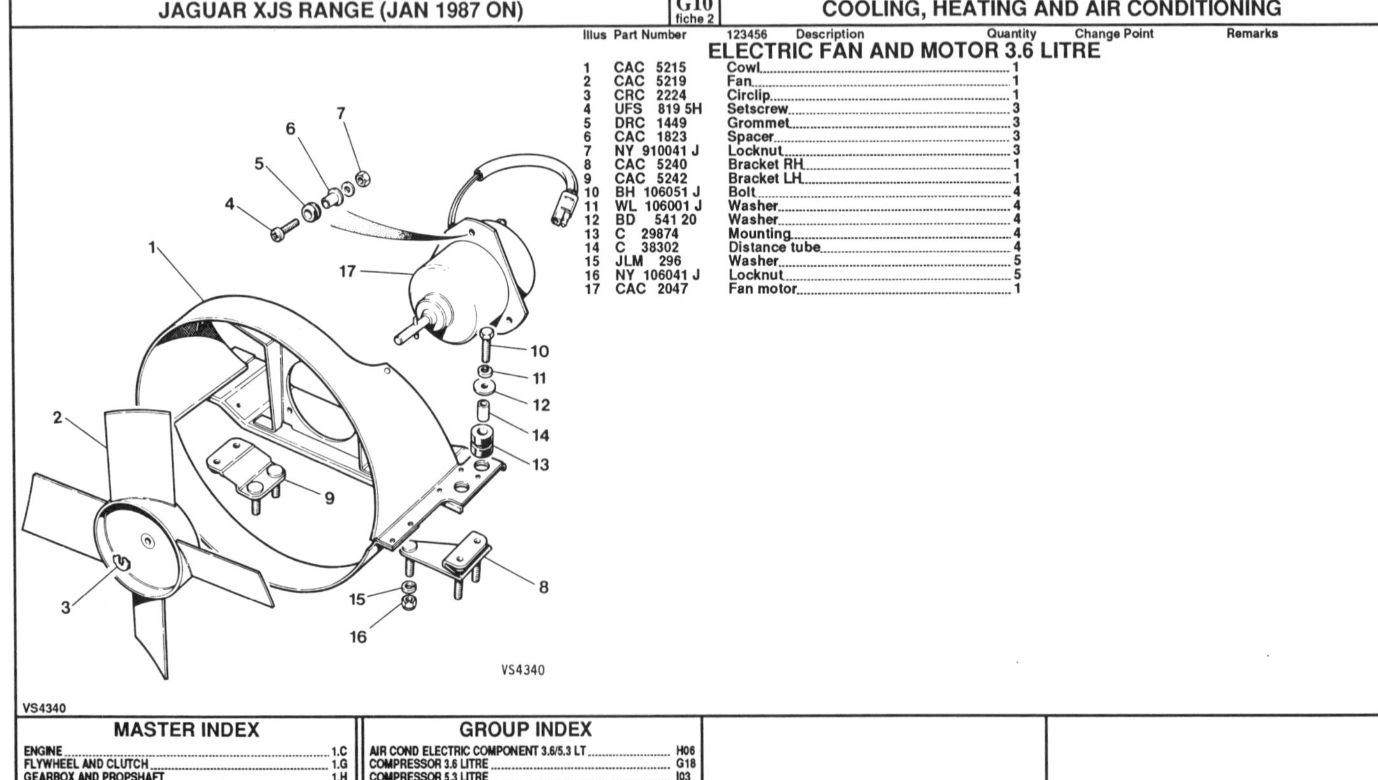

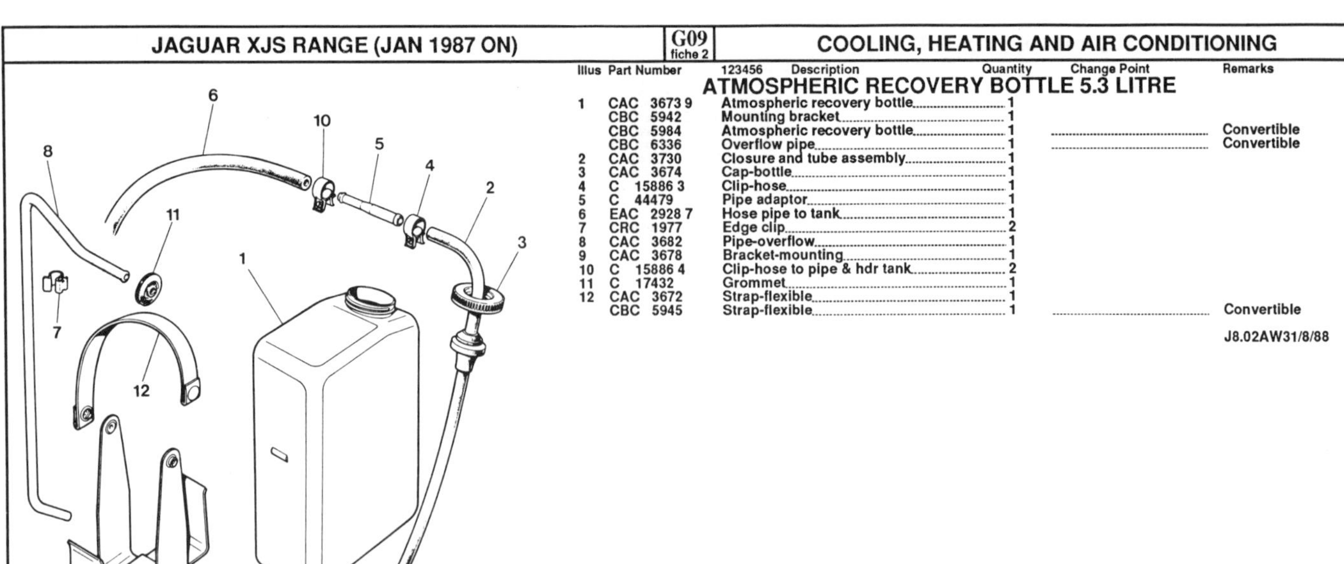

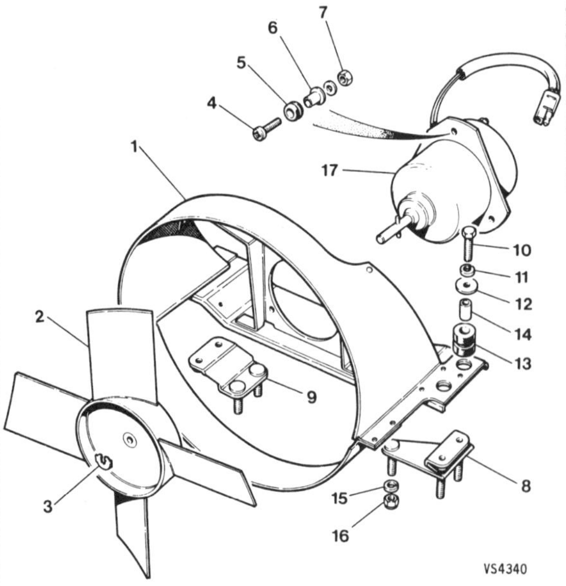

VS4340

Illus	Part Number	123456	Description	Quantity	Change Point	Remarks
1	CAC 5215		Cowl	1		
2	CAC 5219		Fan	1		
3	CRC 2224		Circlip	1		
4	UFS 819 5H		Setscrew	3		
5	DRC 1449		Grommet	3		
6	CAC 1823		Spacer	3		
7	NY 910041 J		Locknut	3		
8	CAC 5240		Bracket RH	1		
9	CAC 5242		Bracket LH	1		
10	BH 106051 J		Bolt	4		
11	WL 106001 J		Washer	4		
12	BD 541 20		Washer	4		
13	C 29874		Mounting	4		
14	C 38302		Distance tube	4		
15	JLM 296		Washer	5		
16	NY 106041 J		Locknut	5		
17	CAC 2047		Fan motor	1		

2G 10

MASTER INDEX

ENGINE 1.C
FLYWHEEL AND CLUTCH 1.G
GEARBOX AND PROPSHAFT 1.H
AXLES, SUSPENSION, DRIVE SHAFTS, WHEELS 1.K
STEERING 1.M
BRAKES AND BRAKE CONTROLS 1.N
FUEL, EXHAUST AND EMISSION SYSTEMS 2.C
COOLING, HEATING AND AIR CONDITIONING 2.G
ELECTRICAL, WASHERS-WIPERS, INSTRUMENTS 2.J
CHASSIS, SUBFRAMES, BODYSHELL, FITTINGS 3.C
FASCIA, TRIM, SEATS AND FIXINGS 3.H
ACCESSORIES AND PAINT 4.C
NUMERICAL INDEX 4.D

GROUP INDEX

AIR COND ELECTRIC COMPONENT 3.6/5.3 LT H06
COMPRESSOR 3.6 LITRE G18
COMPRESSOR 5.3 LITRE I03
COMPRESSOR MOUNTING 3.6 LITRE H02
CONDENSER 3.6 & 5.3 LITRE H08
ELECTRIC FAN - 5.3 LITRE G11
ELECTRIC FAN AND MOTOR 3.6 LITRE G10
FAN AND DRIVE UNIT 5.3 LITRE I02
HEATER VALVE AND HOSES 3.6 LITRE G13
RADIATOR 3.6 LITRE G02
RADIATOR HOSES 3.6 LITRE G03
RADIATOR HOSES 5.3 LITRE G06
RADIATOR-5.3 LITRE G05
RADIO CONTROL PANEL 3.6 & 5.3 LITRE G16
REFRIGERANT HOSES 3.6 Litre H09
REFRIGERANT HOSES 5.3 LITRE H10

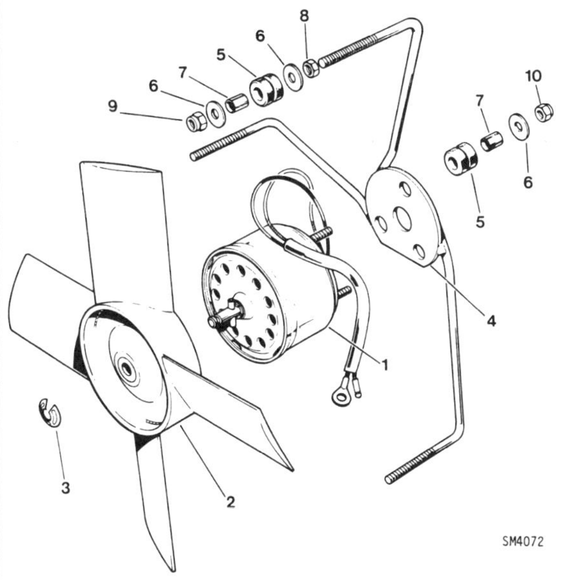

SM4072

ELECTRIC FAN - 5.3 LITRE

Illus	Part Number	123456 Description	Quantity	Change Point	Remarks
1	C 38009	Motor - fan	1		
2	C 37880	Blade - fan	1		
3	AAU 2019	Circlip - fan blade	1		
4	C 37921	Bracket - fan mtg	1		
5	CBC 1448	Grommet	6		
6	C 21930	Washer	9		
7	C 38302	Sleeve	6		
8	JLM 9683	Nut	3		
9	C 8337 1	Nut - locking	3		
10	C 41115 1	Nut - locking	3		

MASTER INDEX

GROUP INDEX

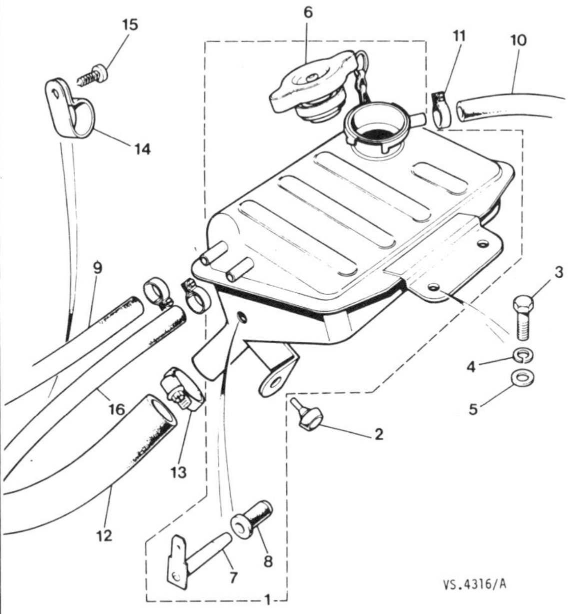

VS4316A

HEADER TANK 3.6 LITRE

Illus	Part Number	123456 Description	Quantity	Change Point	Remarks
1	CBC 2096	Header tank	1		
2	BD 36436	Buffer	1		
3	UFS 131 5R	Setscrew	2		
4	JLM 9574	Lockwasher	2		
5	JLM 298	Washer-plain	2		
6	CAC 5095	Filler cap	1		
7	C 43222	Probe-low coolant	1		
8	C 43221	Grommet	1		
9	C 36175 20	Hose radiator to"T" Piece	1		
10	EAC 2928 4	Hose-tank to recovery tank	1		
11	EAC 3215 2	Clip-hose	6		
12	CBC 4266	Hose-tank to water pump	1		
13	EAC 3215 7	Clip-hose	4		
14	C 1040 3	Clip	1		
15	AB 610031 J	Screw	1		
16	CAC 4572 1	Bleed hose Thermo to "T" Piece	1		
	C 36175 11	Bleed Hose "T" Piece to H/Tank	1		
	CBC 1948	Filler pipe-pump to Filler Hose	1		
	CBC 1958	"T" Piece	1		
	CBC 5201	Hose-Filler Pipe to Water Pump	1		
	EAC 3215 7	Hose Clip	4		

MASTER INDEX
ENGINE 1.C
FLYWHEEL AND CLUTCH 1.G
GEARBOX AND PROPSHAFT 1.H
AXLES, SUSPENSION, DRIVE SHAFTS, WHEELS 1.K
STEERING 1.M
BRAKES AND BRAKE CONTROLS 1.N
FUEL, EXHAUST AND EMISSION SYSTEMS 2.C
COOLING, HEATING AND AIR CONDITIONING 2.G
ELECTRICAL, WASHERS-WIPERS, INSTRUMENTS 2.J
CHASSIS, SUBFRAMES, BODYSHELL, FITTINGS 3.C
FASCIA, TRIM, SEATS AND FIXINGS 3.H
ACCESSORIES AND PAINT 4.C
NUMERICAL INDEX 4.D
GROUP INDEX
AIR COND ELECTRIC COMPONENT 3.6/5.3 LT H06
COMPRESSOR 3.6 LITRE G18
COMPRESSOR 5.3 LITRE I03
COMPRESSOR MOUNTING 3.6 LITRE H02
CONDENSER 3.6 & 5.3 LITRE H08
ELECTRIC FAN - 5.3 LITRE G11
ELECTRIC FAN AND MOTOR 3.6 LITRE G10
FAN AND DRIVE UNIT 5.3 LITRE I02
HEATER VALVE AND HOSES 3.6 LITRE G13
RADIATOR 3.6 LITRE G02
RADIATOR HOSES 3.6 LITRE G03
RADIATOR HOSES 5.3 LITRE G06
RADIATOR-5.3 LITRE G05
RADIO CONTROL PANEL 3.6 & 5.3 LITRE G16
REFRIGERANT HOSES 3.6 Litre H09
REFRIGERANT HOSES 5.3 LITRE H10

HEATER VALVE AND HOSES 3.6 LITRE

Illus	Part Number	123456	Description	Quantity	Change Point	Remarks
1	C 41051		Valve heater	1		
2	C 45016		Hose - valve to heater	1		
3	CAC 9800		Hose - Heater to Engine	1		
4	CAC 9801		Hose - Engine to W/Valve	1		
5	C 43206 8		Clip	6		
6	CBC 2590 8		Vacuum Pipe	1		

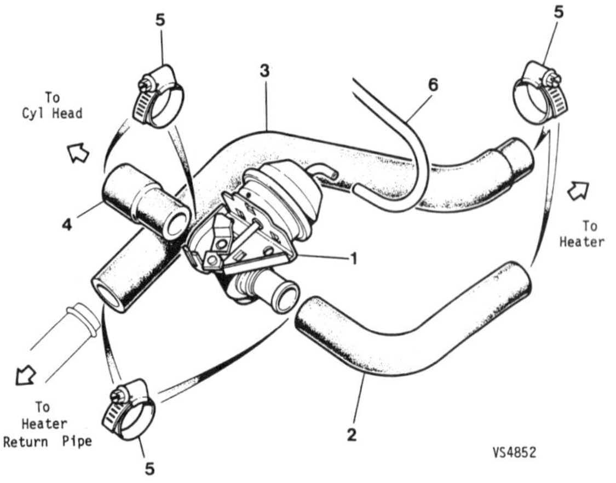

VS4852

MASTER INDEX

GROUP INDEX

EXPANSION TANK &WATER PIPES 5.3 LITRE

Illus	Part Number	123456	Description	Quantity	Change Point	Remarks
1	CAC 8614		Expansion Tank & Probe Assy	1		
2	CAC 5095		Expansion Tank Cap Assy	1		
3	C 19798 3		Drain Tube	1		
4	CAC 4572 1		Bleed Pipe	1		
5	C 1040 2		Clip	1		
6	CAC 1786 8		Vacuum Pipe	1		
7	C 18066		Grommet	1		
8	JLM 9577		Screw	2		
9	C 33620		Spacer	2		
10	SH 605061 J		Setscrew	1		
11	C 43206 1		Clip	1		
12	C 41051		Heater Water Valve	1		
13	C 42452		Bracket	1		
14	UFS 819 4H		Screw	2		
15	C 724		Washer	2		
16	JLM 297		Nut	2		
17	JLM 9566		Setscrew	2		
18	C 724		Washer	2		
19	JLM 297		Washer	2		
20	C 43999		Hose valve to water pipe	2		
21	C 43206 8		Clip	4		
22	CAC 2917		Hose valve to heater	1		
23	C 42967		Hose heater to engine	1		
24	C 43994 16		Clip	1		
25	CAC 4508		Water Pipe	1		
26	CAC 5125		Hose	1		
27	CAC 4605/2		Hose	1		
28	CAC 4564		Pipe	1		
29	C 43206/1		Clip	4		
30	CAC 4605/1		Hose	1		

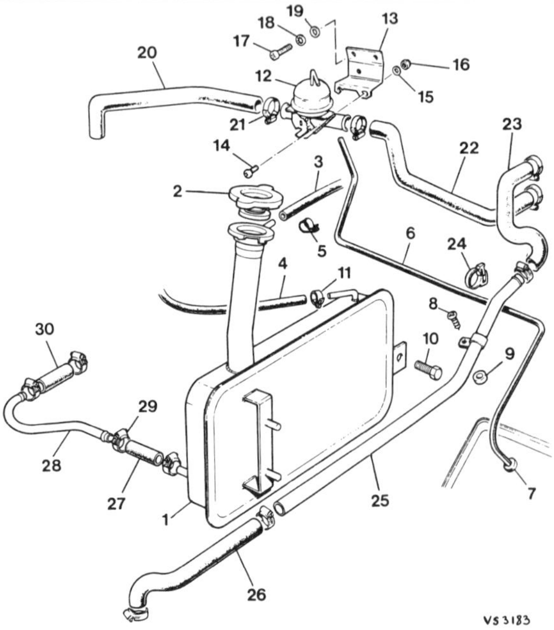

VS3183

MASTER INDEX

GROUP INDEX

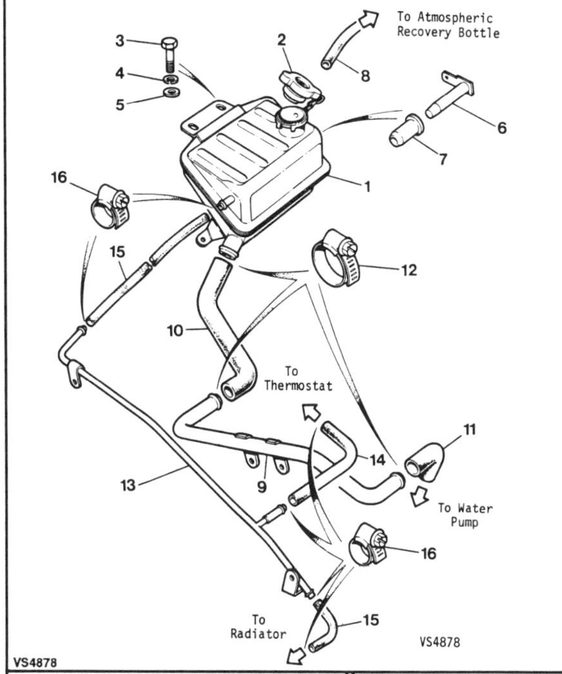

VS4878

EXPANSION TANK 3.6 LITRE

Illus	Part Number	123456	Description	Quantity	Change Point	Remarks
1	CBC 2096		Expansion tank & cap assy	1		
2	CAC 4568		Expansion Tank Cap	1		
3	SH 605051 J		Setscrew securing tank	2		
3	SH 604091 J		Setscrew securing tank	1		
4	WL 108001 J		Washer	2		
4	WA 108051 J		Washer	1		
4	WA 106041 J		Washer	1		
6	C 43222		Low coolant probe	1		
7	C 43221		Grommet	1		
8	EAC 2928 4		Hose pipe to recovery bottle	1		
9	CBC 1948		Filler pipe	1		
10	CBC 4266		Hose filler pipe to tank	1		
11	CBC 5201		Hose pipe to water pump	1		
12	EAC 3215 7		Clip	4		
13	CBC 1958		T Piece	1		
14	CBC 4572 1		Hose T Piece to thermostat	1		
15	C 36175 11		Bleed Hose	A/R		
16	EAC 3215 2		Clip	6		

J16.04DC25/8/88

MASTER INDEX

GROUP INDEX

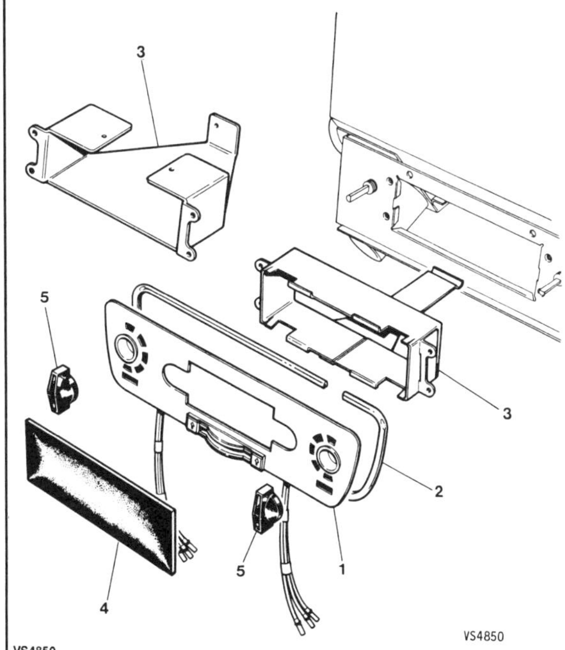

VS4850

RADIO CONTROL PANEL 3.6 & 5.3 LITRE

Illus	Part Number	123456	Description	Quantity	Change Point	Remarks
1	DAC 4310		Panel a/con control & radio	1		Clarion E920 radio 3.6 & 5.3 litre
1	DAC 4311		Panel a/con control & radio	1		Clarion E950 radio 87 1/2 MY CDN/USA 3.6/5.3 Litre
1	DAC 5511		Panel a/con control & radio			CDN/USA 88 1/2 MY 3.6/5.3 litre Clarion E950 & 9021
2	DAC 2804		Panel edge finisher	1		
3	DAC 4259		Radio support bracket	1		3.6/5.3 E950 Clarion not Aus/CDN/USA
3	DAC 3475		Radio support bracket	1		3.6 & 5.3 litre countries including AUS/ CDN/USA
4	DAC 4434		Radio blanking plate (trimmed)	1		
5	C 39413		Control knob - first condition	2		White rectangle
	CBC 6443		Control knob	2		Pull/Manual instruction

MASTER INDEX

ENGINE ... 1.C
FLYWHEEL AND CLUTCH ... 1.G
GEARBOX AND PROPSHAFT ... 1.H
AXLES, SUSPENSION, DRIVE SHAFTS, WHEELS ... 1.K
STEERING ... 1.M
BRAKES AND BRAKE CONTROLS ... 1.N
FUEL, EXHAUST AND EMISSION SYSTEMS ... 2.C
COOLING, HEATING AND AIR CONDITIONING ... 2.G
ELECTRICAL, WASHERS-WIPERS, INSTRUMENTS ... 2.J
CHASSIS, SUBFRAMES, BODYSHELL, FITTINGS ... 3.C
FASCIA, TRIM, SEATS AND FIXINGS ... 3.H
ACCESSORIES AND PAINT ... 4.C
NUMERICAL INDEX ... 4.D

GROUP INDEX

AIR COND ELECTRIC COMPONENT 3.6/5.3 LT ... H06
COMPRESSOR 3.6 LITRE ... G18
COMPRESSOR 5.3 LITRE ... I03
COMPRESSOR MOUNTING 3.6 LITRE ... H02
CONDENSER 3.6 & 5.3 LITRE ... H08
ELECTRIC FAN - 5.3 LITRE ... G11
ELECTRIC FAN AND MOTOR 3.6 LITRE ... G10
FAN AND DRIVE UNIT 5.3 LITRE ... I02
HEATER VALVE AND HOSES 3.6 LITRE ... G13
RADIATOR 3.6 LITRE ... G02
RADIATOR HOSES 3.6 LITRE ... G03
RADIATOR HOSES 5.3 LITRE ... G06
RADIATOR-5.3 LITRE ... G05
RADIO CONTROL PANEL 3.6 & 5.3 LITRE ... G16
REFRIGERANT HOSES 3.6 Litre ... H09
REFRIGERANT HOSES 5.3 LITRE ... H10

2G 17

AIR CON CONTROLS 3.6 & 5.3 LITRE

Illus	Part Number	123456	Description	Quantity	Change Point	Remarks
1	JLM 1166		Temp setting potentiometer	1		
2	JLM 1161		Control switch off/high/low/ auto/fan	1		
3	JLM 1162		Differential control	1		

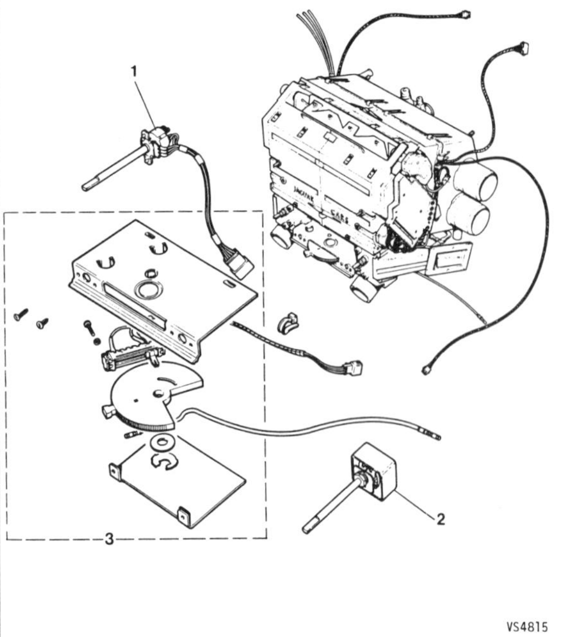

VS4815

MASTER INDEX

GROUP INDEX

2G 18

COMPRESSOR 3.6 LITRE

Illus	Part Number	123456	Description	Quantity	Change Point	Remarks
1	EAC 9616		COMPRESSOR ASSEMBLY	1		
2	JLM 1271		Compressor	1		
3	AEU 1690 J		'O' ring	1		
4	JLM 1165		Pressure switch	1		
5	AEU 1689 J		Relief valve	1		
6	AEU 1695		Clutch housing	1		
7	AEU 1696		Pulley	1		
8	AEU 1687		Bearing	1		
9	AEU 1858		Retaining ring	1		
10	AEU 1692		Drive clutch	1		
11	AEU 1694 J		Key	1		
12	AEU 1693		Nut	1		
13	AEU 1699		Seal kit	1		
14	AEU 1697		Retaining ring kit	1		

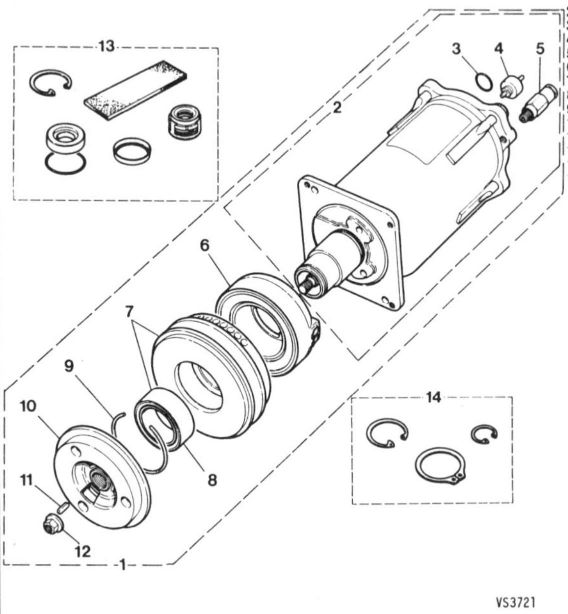

VS3721

MASTER INDEX

ENGINE 1.C
FLYWHEEL AND CLUTCH 1.G
GEARBOX AND PROPSHAFT 1.H
AXLES, SUSPENSION, DRIVE SHAFTS, WHEELS 1.K
STEERING 1.M
BRAKES AND BRAKE CONTROLS 1.N
FUEL, EXHAUST AND EMISSION SYSTEMS 2.C
COOLING, HEATING AND AIR CONDITIONING 2.G
ELECTRICAL, WASHERS-WIPERS, INSTRUMENTS 2.J
CHASSIS, SUBFRAMES, BODYSHELL, FITTINGS 3.C
FASCIA, TRIM, SEATS AND FIXINGS 3.H
ACCESSORIES AND PAINT 4.C
NUMERICAL INDEX 4.D

GROUP INDEX

AIR COND ELECTRIC COMPONENT 3.6/5.3 LT H06
COMPRESSOR 3.6 LITRE G18
COMPRESSOR 5.3 LITRE I03
COMPRESSOR MOUNTING 3.6 LITRE H02
CONDENSER 3.6 & 5.3 LITRE H08
ELECTRIC FAN - 5.3 LITRE G11
ELECTRIC FAN AND MOTOR 3.6 LITRE G10
FAN AND DRIVE UNIT 5.3 LITRE I02
HEATER VALVE AND HOSES 3.6 LITRE G13
RADIATOR 3.6 LITRE G02
RADIATOR HOSES 3.6 LITRE G03
RADIATOR HOSES 5.3 LITRE G06
RADIATOR-5.3 LITRE G05
RADIO CONTROL PANEL 3.6 & 5.3 LITRE G16
REFRIGERANT HOSES 3.6 Litre H09
REFRIGERANT HOSES 5.3 LITRE H10

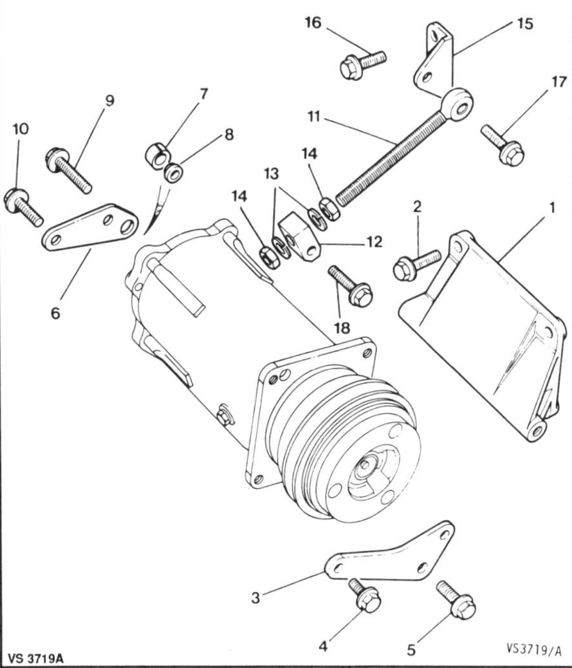

VS 3719A

COMPRESSOR MOUNTING 3.6 LITRE

Illus	Part Number	123456	Description	Quantity	Change Point	Remarks
1	EAC 6601		Mounting bracket	1		
2	FB 108061 J		Setscrew-securing bracket	4		
3	EAC 3864		Front pivot bracket	1		
4	FS 110301 J		Setscrew	1		
5	FS 108121 J		Setscrew	2		
6	EAC 6642		Rear pivot bracket	1		
7	EAC 3866		Distance tube	1		
8	EAC 3867		Clamping washer	1		
9	FS 110351 J		Setscrew	2		
10	FS 110251 J		Setscrew	2		
11	C 34241		Adjusting bolt	1		
12	EAC 7301		Adjusting sleeve	1		
13	JLM 304		Lockwasher	2		
14	NH 606041 J		Nut	2		
15	EAC 7786		Bracket (adjusting bolt)	1		
16	FS 108201 J		Setscrew	2		
17	FB 108071 J		Bolt	1		
18	FS 108251 J		Bolt	1		

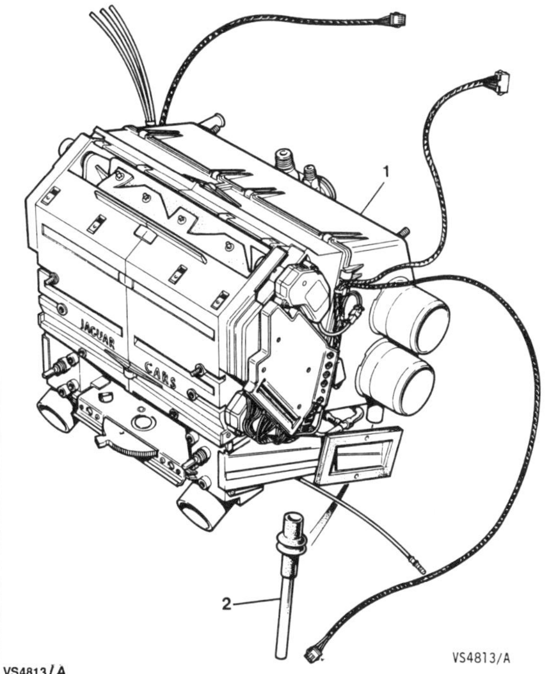

VS4813/A

AIR CONDITIONING UNIT 3.6 & 5.3 LITRE

Illus	Part Number	123456	Description	Quantity	Change Point	Remarks
1	CBC 4444		AIR CONDITIONING ASSY	1		Up to 1989 MY
1	CBC 7463		AIR CONDITIONING ASSY	1		From 1989 MY
1	CBC 6607		AIR CONDITIONING ASSY	1		Convertible only
	CBC 6606		LINK LEAD ASSY	1		Convertible only
2	CAC 2299		Condensate drain	2		Up to 1989 MY
2	CBC 7462		Condensate drain	2		From 1989 MY

J26DC4/8/88

MASTER INDEX

ENGINE 1.C
FLYWHEEL AND CLUTCH 1.G
GEARBOX AND PROPSHAFT 1.H
AXLES, SUSPENSION, DRIVE SHAFTS, WHEELS 1.K
STEERING 1.M
BRAKES AND BRAKE CONTROLS 1.N
FUEL, EXHAUST AND EMISSION SYSTEMS 2.C
COOLING, HEATING AND AIR CONDITIONING 2.G
ELECTRICAL, WASHERS-WIPERS, INSTRUMENTS 2.J
CHASSIS, SUBFRAMES, BODYSHELL, FITTINGS 3.C
FASCIA, TRIM, SEATS AND FIXINGS 3.H
ACCESSORIES AND PAINT 4.C
NUMERICAL INDEX 4.D

GROUP INDEX

AIR COND ELECTRIC COMPONENT 3.6/5.3 LT H06
COMPRESSOR 3.6 LITRE G18
COMPRESSOR 5.3 LITRE I03
COMPRESSOR MOUNTING 3.6 LITRE H02
CONDENSER 3.6 & 5.3 LITRE H08
ELECTRIC FAN - 5.3 LITRE G11
ELECTRIC FAN AND MOTOR 3.6 LITRE G10
FAN AND DRIVE UNIT 5.3 LITRE I02
HEATER VALVE AND HOSES 3.6 LITRE G13
RADIATOR 3.6 LITRE G02
RADIATOR HOSES 3.6 LITRE G03
RADIATOR HOSES 5.3 LITRE G06
RADIATOR-5.3 LITRE G05
RADIO CONTROL PANEL 3.6 & 5.3 LITRE G16
REFRIGERANT HOSES 3.6 Litre H09
REFRIGERANT HOSES 5.3 LITRE H10

2H 04

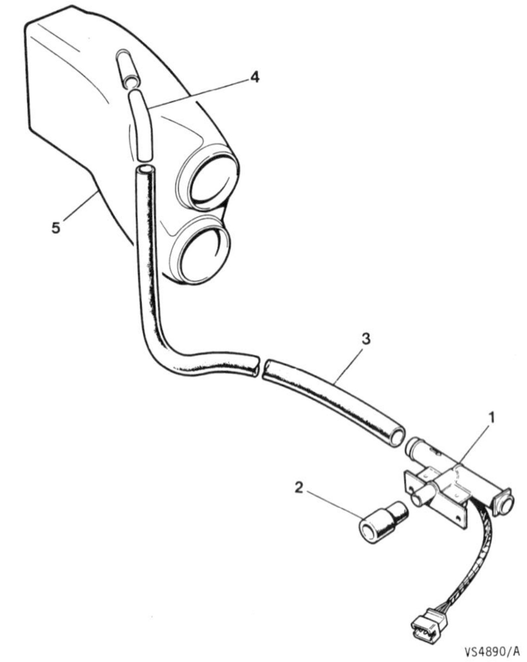

VS4890A

IN CAR SENSOR 3.6 & 5.3 LITRE

Illus	Part Number	123456	Description	Quantity	Change Point	Remarks
1	JLM 1248		In car sensor	1		
2	CBC 2855		Stepped hose	1		
3	CAC 80881 1		Connecting hose to LH air duct	1		
4	CAC 3398		Tube	1		
5	C 41920		Air duct-RH	1		
5	C 41921		Air duct-LH	1		
			Ambient sensor(see blower motors)	1		

MASTER INDEX

GROUP INDEX

2H 05

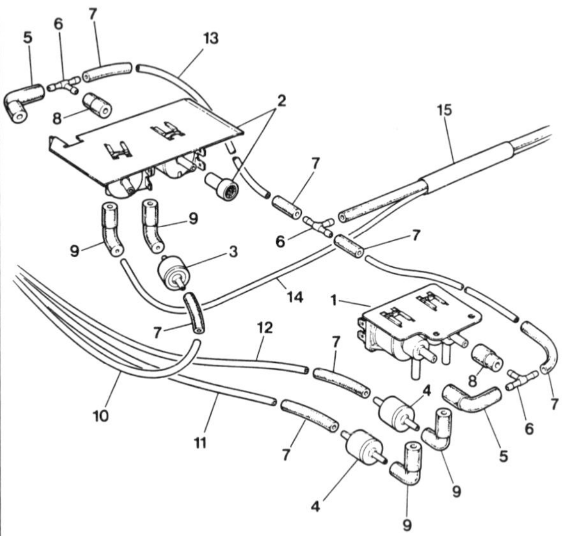

VS4824

VS4824

VACUUM SOLENOIDS 3.6 & 5.3 LITRE

Illus	Part Number	123456	Description	Quantity	Change Point	Remarks
1	JLM 1125		Solenoid assy RH	1		
2	JLM 1126		Solenoid assy LH	1		
3	JLM 1164		Vac restrictor - brown	1		
4	JLM 750		Vac restrictor - blue	2		
5	JLM 1459		Vac elbow	2		
6	JLM 1236		Vac "T" connector	3		
7	JLM 1133		Vac hose	7		
8	JLM 1132		Vac hose stepped	2		
9	JLM 1238		Vac elbow	4		
10	JLM 1234		Vac pipe - blue	1		
11	JLM 1233		Vac pipe - black	1		
12	JLM 1232		Vac pipe - green	1		
13	JLM 1231		Vac pipe - white	1		
14	JLM 1235		Vac pipe - red	1		
15	JLM 1127		Vac pipe	1		

MASTER INDEX

ENGINE 1.C
FLYWHEEL AND CLUTCH 1.G
GEARBOX AND PROPSHAFT 1.H
AXLES, SUSPENSION, DRIVE SHAFTS, WHEELS 1.K
STEERING 1.M
BRAKES AND BRAKE CONTROLS 1.N
FUEL, EXHAUST AND EMISSION SYSTEMS 2.C
COOLING, HEATING AND AIR CONDITIONING 2.G
ELECTRICAL, WASHERS-WIPERS, INSTRUMENTS 2.J
CHASSIS, SUBFRAMES, BODYSHELL, FITTINGS 3.C
FASCIA, TRIM, SEATS AND FIXINGS 3.H
ACCESSORIES AND PAINT 4.C
NUMERICAL INDEX 4.D

GROUP INDEX

AIR COND ELECTRIC COMPONENT 3.6/5.3 LT H06
COMPRESSOR 3.6 LITRE G18
COMPRESSOR 5.3 LITRE I03
COMPRESSOR MOUNTING 3.6 LITRE H02
CONDENSER 3.6 & 5.3 LITRE H08
ELECTRIC FAN - 5.3 LITRE G11
ELECTRIC FAN AND MOTOR 3.6 LITRE G10
FAN AND DRIVE UNIT 5.3 LITRE I02
HEATER VALVE AND HOSES 3.6 LITRE G13
RADIATOR 3.6 LITRE G02
RADIATOR HOSES 3.6 LITRE G03
RADIATOR HOSES 5.3 LITRE G06
RADIATOR-5.3 LITRE G05
RADIO CONTROL PANEL 3.6 & 5.3 LITRE G16
REFRIGERANT HOSES 3.6 Litre H09
REFRIGERANT HOSES 5.3 LITRE H10

AIR COND ELECTRIC COMPONENT 3.6/5.3 LT

Illus	Part Number	Description	Quantity	Change Point	Remarks
1	CAC 8032	Electronic control unit	1		
2	JLM 755	Upper & lower servo motor	2		
3	JLM 763	Water temperature switch	1		
4	JLM 1229	Feedback pot upper	1		
4	JLM 1230	Feedback pot lower	1		
5	JLM 748	Evaporator sensor	1		

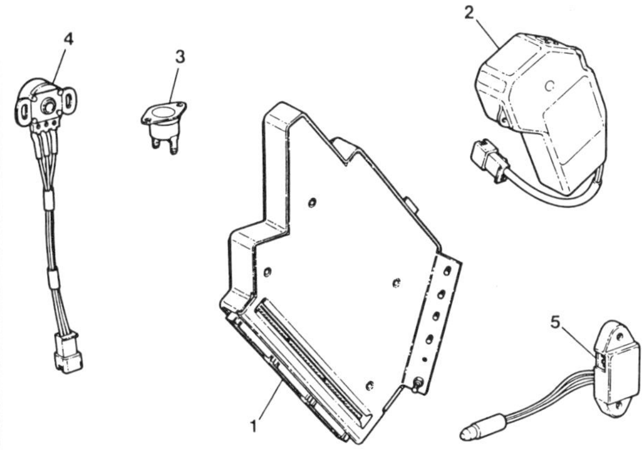

VS4825

AIR CON BLOWER UNITS 3.6 & 5.3 LITRE

Illus	Part Number	Description	Quantity	Change Point	Remarks
	CAC 6746	Blower assembly RH	1		
1	CAC 6747	Blower assembly LH	1		
2	JLM 1179	Vacuum actuator	1		
3	JLM 771	Air con relay	1		
4	JLM 1167	Ambient sensor	1		

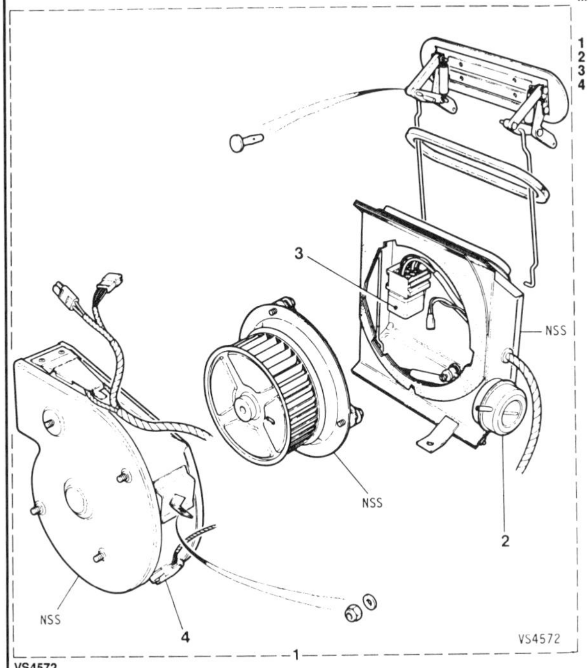

VS4572

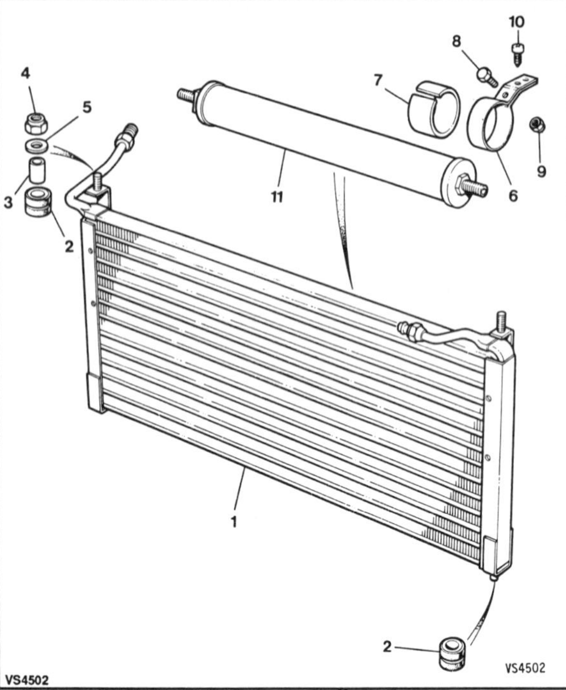

CONDENSER 3.6 & 5.3 LITRE

Illus	Part Number	123456	Description	Quantity	Change Point	Remarks
1	CBC 2185		Condenser	1		3.6 litre
1	CBC 2846		Condenser	1		5.3 litre
1	CBC 6037		Condenser	1		19881/2 MY 5.3l
2	CBC 1448		Mounting	4		
3	CAC 1825		Distance piece	2		
3	C 38302		Distance piece	2		
4	NZ 604041		Nyloc nut	2		
5	BD 541 20		Washer	2		
6	C 41013		Bracket	2		
7	C 41014		Mounting Pad	2		
8	UFS 119 4R		Setscrew	2		
9	TN 3205		Nut	2		
10	AB 100031		Setscrew	4		
11	CAC 1881		Drier Bottle	1		

2H 08

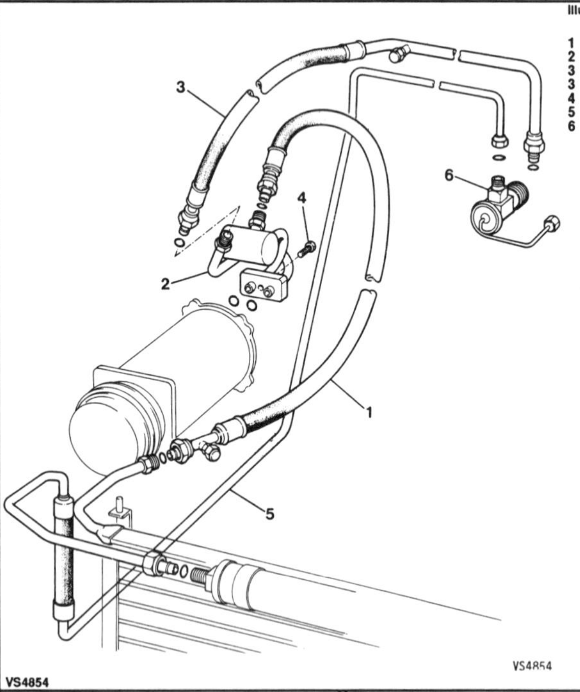

REFRIGERANT HOSES 3.6 Litre

Illus	Part Number	123456	Description	Quantity	Change Point	Remarks
1	CBC 1956		Hose assy comp to condenser	1		
2	CBC 2366		Air con muffler	1		
3	CBC 1955		Hose assy comp to evaporator	1	Up to 19881/2 MY	
3	CBC 5110		Hose assy comp to evaporator	1	19881/2 MY	
4	AGU 1135		Screw	1		
5	CAC 5596		Hose assy drier to evaporator	1		
6	JLM 1228		Expansion valve	1		
	KRC 1154		"O" Ring	A/R		
	KRC 1156		"O" Ring	A/R		
	KRC 1158		"O" Ring	A/R		

2H 09

MASTER INDEX

ENGINE 1.C
FLYWHEEL AND CLUTCH 1.G
GEARBOX AND PROPSHAFT 1.H
AXLES, SUSPENSION, DRIVE SHAFTS, WHEELS 1.K
STEERING 1.M
BRAKES AND BRAKE CONTROLS 1.N
FUEL, EXHAUST AND EMISSION SYSTEMS 2.C
COOLING, HEATING AND AIR CONDITIONING 2.G
ELECTRICAL, WASHERS-WIPERS, INSTRUMENTS 2.J
CHASSIS, SUBFRAMES, BODYSHELL, FITTINGS 3.C
FASCIA, TRIM, SEATS AND FIXINGS 3.H
ACCESSORIES AND PAINT 4.C
NUMERICAL INDEX 4.D

GROUP INDEX

AIR COND ELECTRIC COMPONENT 3.6/5.3 LT H06
COMPRESSOR 3.6 LITRE G18
COMPRESSOR 5.3 LITRE I03
COMPRESSOR MOUNTING 3.6 LITRE H02
CONDENSER 3.6 & 5.3 LITRE H08
ELECTRIC FAN - 5.3 LITRE G11
ELECTRIC FAN AND MOTOR 3.6 LITRE G10
FAN AND DRIVE UNIT 5.3 LITRE I02
HEATER VALVE AND HOSES 3.6 LITRE G13
RADIATOR 3.6 LITRE G02
RADIATOR HOSES 3.6 LITRE G03
RADIATOR HOSES 5.3 LITRE G06
RADIATOR-5.3 LITRE G05
RADIO CONTROL PANEL 3.6 & 5.3 LITRE G16
REFRIGERANT HOSES 3.6 Litre H09
REFRIGERANT HOSES 5.3 LITRE H10

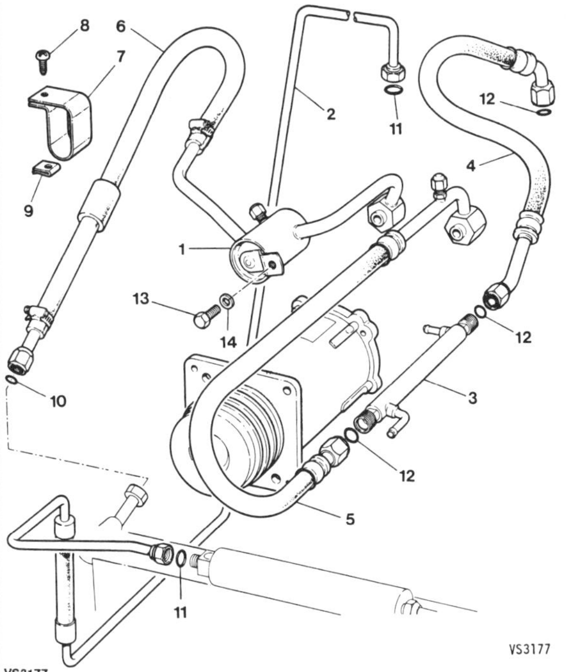

VS3177

REFRIGERANT HOSES 5.3 LITRE

Illus	Part Number	Description	Quantity	Change Point	Remarks
1	CAC 2341	Hose-comp to condenser	1	Up to 19881/2 MY	
1	CBC 6036	Hose-comp to condenser	1	From 19881/2 MY	
2	CAC 5596	Hose-drier to evap	1		
3	CAC 7912	Fuel cooler	1		
4	CAC 3696	Hose-cooler to evap	1		
5	CAC 3697	Hose-cooler to comp	1		
6	C 38494 3	Insulation	1		
7	C 38518	Bracket	1		
8	JLM 308	Screw	1		
9	FU 2545 4	Distance Piece	1		
10	KRC 1156	"O" Ring	1		
11	KRC 1154	"O" Ring	2		
12	KRC 1158	"O" Ring	2		
13	SH 604051	Setscrew	1		
14	WF 600041	Washer	1		

MASTER INDEX

GROUP INDEX

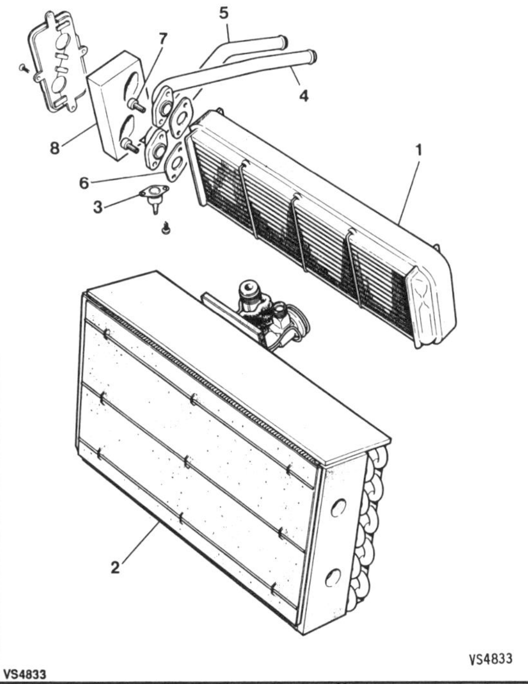

VS4833

AIR COND EVAPORATOR-3.6 & 5.3 litre

Illus	Part Number	Description	Quantity	Change Point	Remarks
1	JLM 1169	Heater matrix	1		
2	JLM 1197	Evaporator & seals	1		
3	JLM 763	Water temperature switch	1		
4	JLM 1515	Inlet pipe	1		
5	JLM 1516	Outlet pipe	1		
6	JLM 759	Gasket fitting	2		
6	CBC 4967	"O" ring fitting	2		
7	JLM 760	Screw	4		
8	JLM 761	Seal	1		

J38.00DC30/6/88

MASTER INDEX

ENGINE 1.C
FLYWHEEL AND CLUTCH 1.G
GEARBOX AND PROPSHAFT 1.H
AXLES, SUSPENSION, DRIVE SHAFTS, WHEELS 1.K
STEERING 1.M
BRAKES AND BRAKE CONTROLS 1.N
FUEL, EXHAUST AND EMISSION SYSTEMS 2.C
COOLING, HEATING AND AIR CONDITIONING 2.G
ELECTRICAL, WASHERS-WIPERS, INSTRUMENTS 2.J
CHASSIS, SUBFRAMES, BODYSHELL, FITTINGS 3.C
FASCIA, TRIM, SEATS AND FIXINGS 3.H
ACCESSORIES AND PAINT 4.C
NUMERICAL INDEX 4.D

GROUP INDEX

AIR COND ELECTRIC COMPONENT 3.6/5.3 LT H06
COMPRESSOR 3.6 LITRE G18
COMPRESSOR 5.3 LITRE I03
COMPRESSOR MOUNTING 3.6 LITRE H02
CONDENSER 3.6 & 5.3 LITRE H08
ELECTRIC FAN - 5.3 LITRE G11
ELECTRIC FAN AND MOTOR 3.6 LITRE G10
FAN AND DRIVE UNIT 5.3 LITRE I02
HEATER VALVE AND HOSES 3.6 LITRE G13
RADIATOR 3.6 LITRE G02
RADIATOR HOSES 3.6 LITRE G03
RADIATOR HOSES 5.3 LITRE G06
RADIATOR-5.3 LITRE G05
RADIO CONTROL PANEL 3.6 & 5.3 LITRE G16
REFRIGERANT HOSES 3.6 Litre H09
REFRIGERANT HOSES 5.3 LITRE H10

2H 12

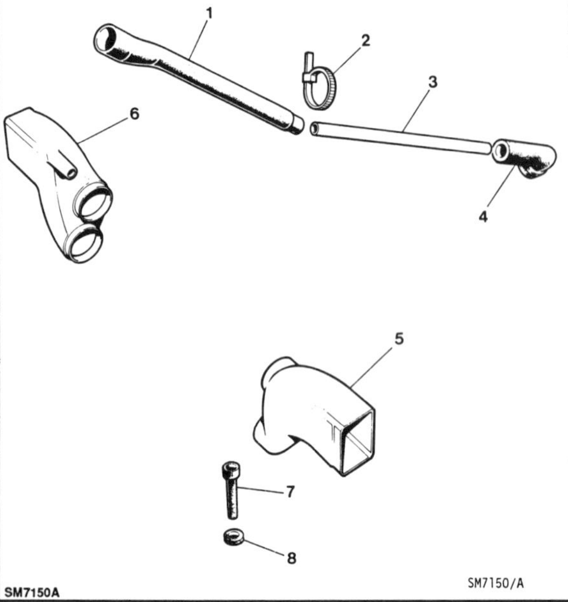

SM7150A

PLENUM DRAIN-3.6 and 5.3 litre

Illus	Part Number	123456	Description	Quantity	Change Point	Remarks
1	BAC 3411		Tube drain	2		
2	C 44468 2		Clip	2		
3	BD 46664 1		Drain plenum	2		
4	C 35125 1		Hose elbow	2		
5	C 41920		Duct air-RH	1		
6	C 44221		Duct air-LH	1		
7	CAC 2299		Condensate Drain Assy	1		
8	C 29302		Grommet	1		

MASTER INDEX

GROUP INDEX

2H 13

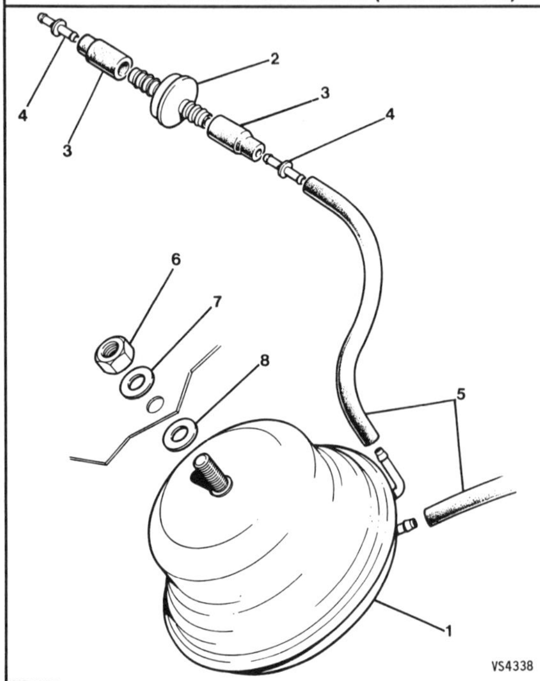

VS4338

VACUUM TANK AND HOSES 3.6 & 5.3 LITRE

Illus	Part Number	123456	Description	Quantity	Change Point	Remarks
1	CAC 5009		Tank vacuum	1		
2	CAC 4188		Valve non return	1		
3	C 39520		Hose stepped	2		
4	C 33835		Connector	2		
5	EAC 8022 30		Hose vac to resv	1		
5	EAC 8022 30		Hose to ac unit	1		
5	EAC 8022 30		Hose conn to manifold	1		
6	NH 604041 J		Nut fixing vac tank	1		
7	JLM 304		Washer spring	1		
8	C 42308		Washer sealing	1		

MASTER INDEX

GROUP INDEX

AIR DISTRIBUTION DUCTS 3.6 & 5.3 LITRE

Illus	Part Number	123456	Description	Quantity	Change Point	Remarks
1	C 42736		Duct assembly	1		RHD
1	C 42737		Duct assembly	1		LHD
2	CBC 5212		End cover-RH	1		
2	CBC 5213		End cover-LH	1		
3	BD 44130		Fascia support bracket	1		
4	SH 605051		Setscrew	2		
5	JLM 303		Washer	2		
6	C 44117		Seal centre outlet	1		
7	C 44969		Vacuum motor	1		
8	AB 606021		Screw	1		
9	CAC 3797		Centre outlet duct	1		
10	C 44142		Bracket-RH-duct	1		
11	C 44143		Bracket-LH-duct	1		
12	C 41920		Air duct-RH	1		
12	C 44221		Air duct-LH	1		
13	BD 46143		Clip	1		
14	CZG 4707 J		Screw	1		
15	CAC 1786 8		Vacuum hose	1		
16	C 42799 1		Vacuum pipe	1		
17	BAC 4255		Centre vent	1	Up to 871/2 MY	
18	BAC 2953		Outer vent	2	Up to 871/2 MY	
17	BAC 3748		Centre vent	1	From 871/2 MY	
18	BAC 2953		Outer vent	1	From 871/2 MY	

J44DC5/7/88

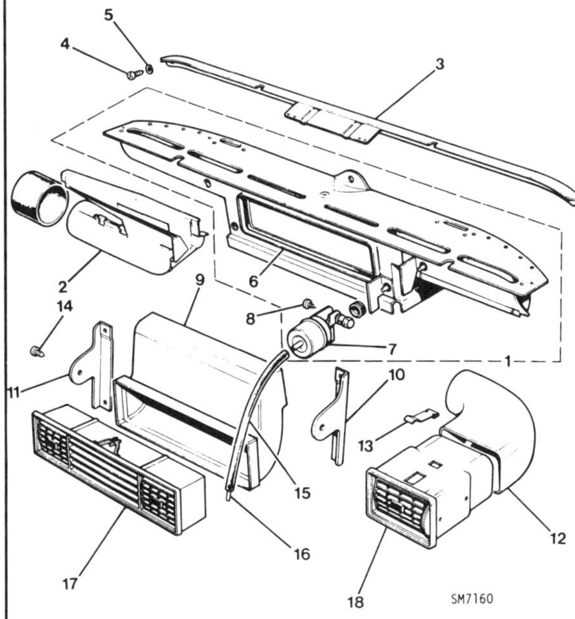

SM7160

MASTER INDEX

GROUP INDEX

AIRFLOW REAR COMPARTMENT 3.6 & 5.3 Litre

Illus	Part Number	123456	Description	Quantity	Change Point	Remarks
1	BD 48281		Duct rear	1		
2	DAZ 810 8C		Screw	4		
3	C 31973 11		Nut plastic	4		
4	BCC 5430		Hose RH	1		
4	BCC 5431		Hose LH	1		

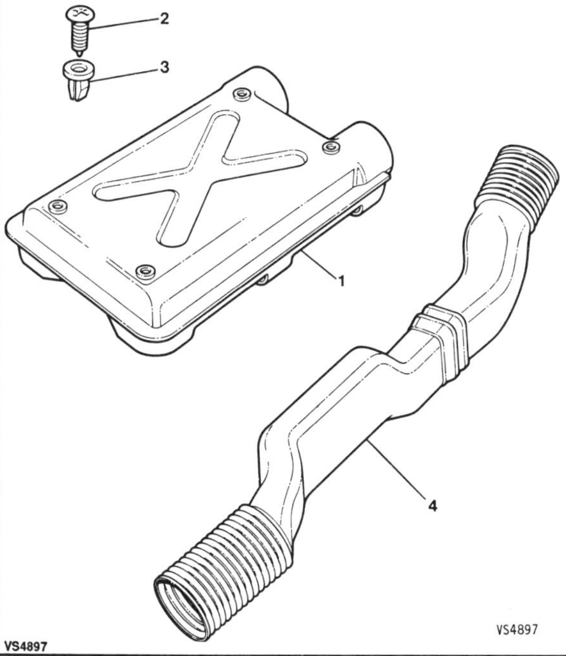

VS4897

MASTER INDEX

ENGINE 1.C
FLYWHEEL AND CLUTCH 1.G
GEARBOX AND PROPSHAFT 1.H
AXLES, SUSPENSION, DRIVE SHAFTS, WHEELS 1.K
STEERING 1.M
BRAKES AND BRAKE CONTROLS 1.N
FUEL, EXHAUST AND EMISSION SYSTEMS 2.C
COOLING, HEATING AND AIR CONDITIONING 2.G
ELECTRICAL, WASHERS-WIPERS, INSTRUMENTS 2.J
CHASSIS, SUBFRAMES, BODYSHELL, FITTINGS 3.C
FASCIA, TRIM, SEATS AND FIXINGS 3.H
ACCESSORIES AND PAINT 4.C
NUMERICAL INDEX 4.D

GROUP INDEX

AIR COND ELECTRIC COMPONENT 3.6/5.3 LT H06
COMPRESSOR 3.6 LITRE G18
COMPRESSOR 5.3 LITRE I03
COMPRESSOR MOUNTING 3.6 LITRE H02
CONDENSER 3.6 & 5.3 LITRE H08
ELECTRIC FAN - 5.3 LITRE G11
ELECTRIC FAN AND MOTOR 3.6 LITRE G10
FAN AND DRIVE UNIT 5.3 LITRE I02
HEATER VALVE AND HOSES 3.6 LITRE G13
RADIATOR 3.6 LITRE G02
RADIATOR HOSES 3.6 LITRE G03
RADIATOR HOSES 5.3 LITRE G06
RADIATOR-5.3 LITRE G05
RADIO CONTROL PANEL 3.6 & 5.3 LITRE G16
REFRIGERANT HOSES 3.6 Litre H09
REFRIGERANT HOSES 5.3 LITRE H10

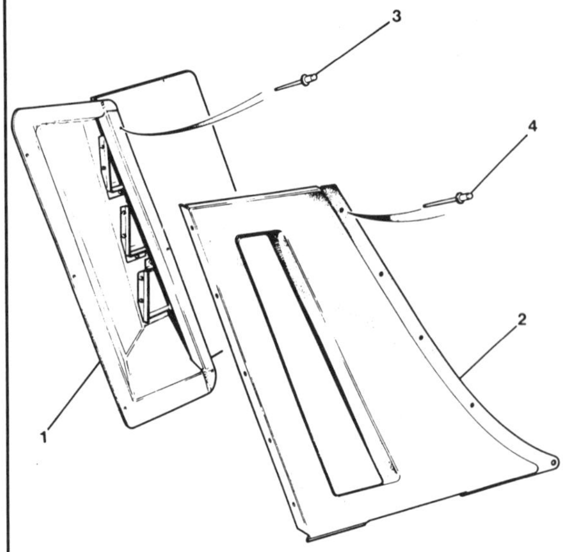

SM7149

AIR EXTRACTOR. 3.6 & 5.3 LITRE

Illus	Part Number	123456	Description	Quantity	Change Point	Remarks
1	BDC 7186		Extractor rear RH	1		
1	BDC 7187		Extractor rear LH	1		
2	BDC 3562		Panel extractor RH	1		
2	BDC 3563		Panel extractor LH	1		
3	BD 11712 4		Rivet	22		
4	BD 26706 1		Rivet	8		

J50DC31/8/88

MASTER INDEX

GROUP INDEX

2H 16

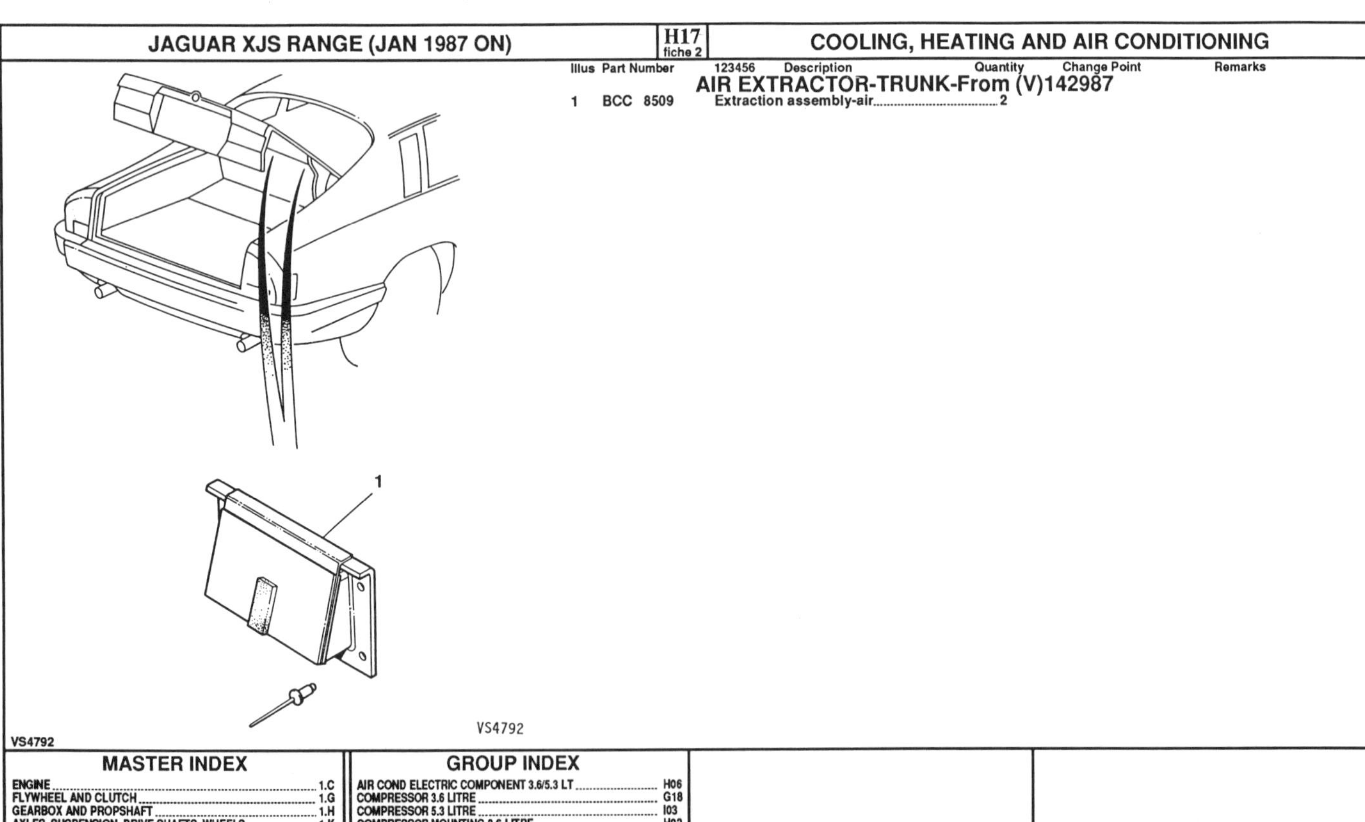

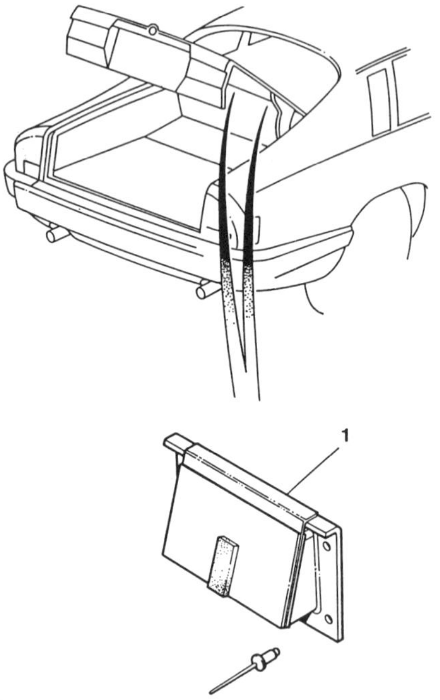

VS4792

AIR EXTRACTOR-TRUNK-From (V)142987

Illus	Part Number	123456	Description	Quantity	Change Point	Remarks
1	BCC 8509		Extraction assembly-air	2		

MASTER INDEX

ENGINE ... 1.C
FLYWHEEL AND CLUTCH ... 1.G
GEARBOX AND PROPSHAFT ... 1.H
AXLES, SUSPENSION, DRIVE SHAFTS, WHEELS ... 1.K
STEERING ... 1.M
BRAKES AND BRAKE CONTROLS ... 1.N
FUEL, EXHAUST AND EMISSION SYSTEMS ... 2.C
COOLING, HEATING AND AIR CONDITIONING ... 2.G
ELECTRICAL, WASHERS-WIPERS, INSTRUMENTS ... 2.J
CHASSIS, SUBFRAMES, BODYSHELL, FITTINGS ... 3.C
FASCIA, TRIM, SEATS AND FIXINGS ... 3.H
ACCESSORIES AND PAINT ... 4.C
NUMERICAL INDEX ... 4.D

GROUP INDEX

AIR COND ELECTRIC COMPONENT 3.6/5.3 LT ... H06
COMPRESSOR 3.6 LITRE ... G18
COMPRESSOR 5.3 LITRE ... I03
COMPRESSOR MOUNTING 3.6 LITRE ... H02
CONDENSER 3.6 & 5.3 LITRE ... H08
ELECTRIC FAN - 5.3 LITRE ... G11
ELECTRIC FAN AND MOTOR 3.6 LITRE ... G10
FAN AND DRIVE UNIT 5.3 LITRE ... I02
HEATER VALVE AND HOSES 3.6 LITRE ... G13
RADIATOR 3.6 LITRE ... G02
RADIATOR HOSES 3.6 LITRE ... G03
RADIATOR HOSES 5.3 LITRE ... G06
RADIATOR-5.3 LITRE ... G05
RADIO CONTROL PANEL 3.6 & 5.3 LITRE ... G16
REFRIGERANT HOSES 3.6 Litre ... H09
REFRIGERANT HOSES 5.3 LITRE ... H10

2H 17

AIR EXTRACTOR - "B" POST-CONVERTIBLE 3.6 & 5.3 Litre

Illus	Part Number	123456	Description	Quantity	Change Point	Remarks
1	BBC 7743		Extractor - air	2		

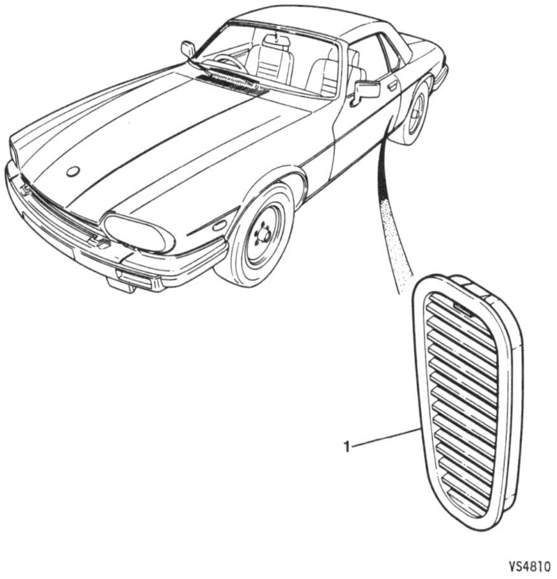

VS4810

MASTER INDEX

GROUP INDEX

FAN AND DRIVE UNIT 5.3 LITRE.

Illus	Part Number	123456	Description	Quantity	Change Point	Remarks
1	EAC 3265		Fan assembly	1		
2	EAC 4751		Drive unit	1		
3	SH 505051 J		Setscrew	4		
4	C 30075 2		Washer	4		
5	C 36546		Stud	4		
6	WF 600050 J		Washer	4		
7	JLM 9684		Nut	4		
8	EAC 3438		Fan pulley	1		
9	EAC 3437		Bearing	1		
10	EAC 3038		Bearing housing	1		
11	EAC 8097		Jockey pulley	1		
12	BH 508162 J		Bolt	1		
13	C 30075 5		Washer	1		
14	C 34230		Adjuster bolt	1		
15	BH 606131 J		Bolt	1		
16	JLM 300		Washer	1		
17	NZ 606041 J		Nut	1		
18	JLM 304		Washer	2		
19	JLM 9685		Nut	2		
20	C 34006		Sleeve	1		
21	C 9655		Stud	1		
22	NZ 606041 J		Nut	1		

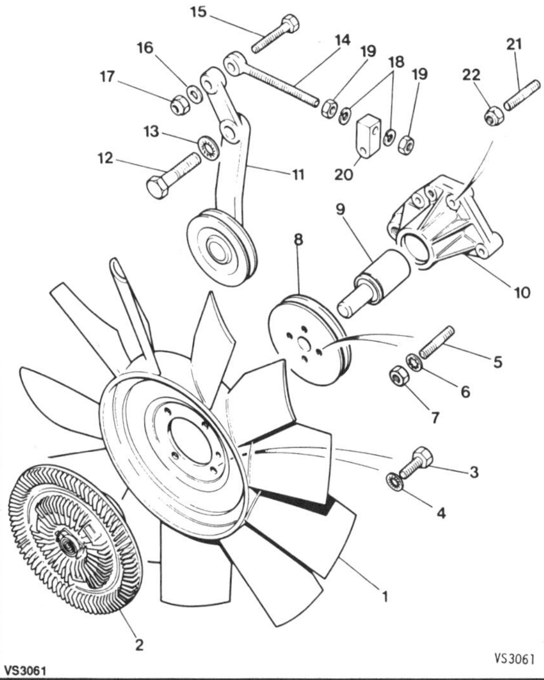

VS3061

MASTER INDEX
ENGINE 1.C
FLYWHEEL AND CLUTCH 1.G
GEARBOX AND PROPSHAFT 1.H
AXLES, SUSPENSION, DRIVE SHAFTS, WHEELS 1.K
STEERING 1.M
BRAKES AND BRAKE CONTROLS 1.N
FUEL, EXHAUST AND EMISSION SYSTEMS 2.C
COOLING, HEATING AND AIR CONDITIONING 2.G
ELECTRICAL, WASHERS-WIPERS, INSTRUMENTS 2.J
CHASSIS, SUBFRAMES, BODYSHELL, FITTINGS 3.C
FASCIA, TRIM, SEATS AND FIXINGS 3.H
ACCESSORIES AND PAINT 4.C
NUMERICAL INDEX 4.D
GROUP INDEX
AIR COND ELECTRIC COMPONENT 3.6/5.3 LT H06
COMPRESSOR 3.6 LITRE G18
COMPRESSOR 5.3 LITRE I03
COMPRESSOR MOUNTING 3.6 LITRE H02
CONDENSER 3.6 & 5.3 LITRE H08
ELECTRIC FAN - 5.3 LITRE G11
ELECTRIC FAN AND MOTOR 3.6 LITRE G10
FAN AND DRIVE UNIT 5.3 LITRE. I02
HEATER VALVE AND HOSES 3.6 LITRE G13
RADIATOR 3.6 LITRE. G02
RADIATOR HOSES 3.6 LITRE G03
RADIATOR HOSES 5.3 LITRE. G06
RADIATOR-5.3 LITRE. G05
RADIO CONTROL PANEL 3.6 & 5.3 LITRE G16
REFRIGERANT HOSES 3.6 Litre H09
REFRIGERANT HOSES 5.3 LITRE H10

2I 03

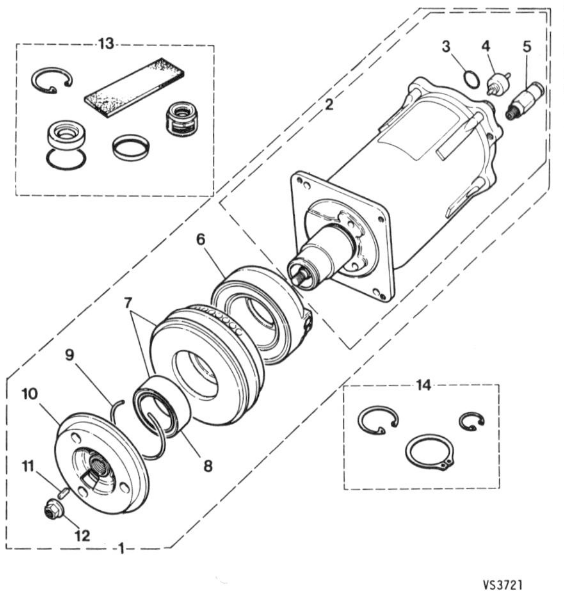

VS3721

COMPRESSOR 5.3 LITRE

Illus	Part Number	123456 Description	Quantity	Change Point	Remarks
1	EAC 9616	COMPRESSOR ASSEMBLY	1		
2	JLM 1271	Compressor	1		
3	AEU 1690 J	'O' ring	2		
4	JLM 1165	Pressure switch	1		
5	AEU 1689 J	Pressure relief valve	1		
6	AEU 1695	Clutch housing	1		
7	AEU 1696	PULLEY ASSEMBLY	1		
8	AEU 1687	Bearing	1		
9	AEU 1858	Retaining ring	1		
10	AEU 1692	Drive clutch	1		
11	AEU 1694 J	Key	1		
12	AEU 1693	Nut	1		
13	AEU 1699	Seal kit	1		
14	AEU 1697	Retaining ring kit	1		

MASTER INDEX

GROUP INDEX

2I 04

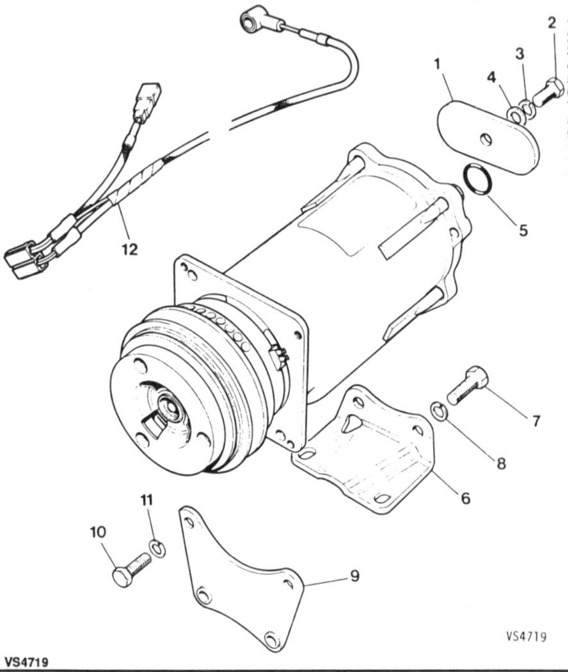

VS4719

COMPRESSOR MOUNTING 5.3 LITRE

Illus	Part Number	123456 Description	Quantity	Change Point	Remarks
1	C 36983	Clamp plate	1		
2	BH 110091 J	Bolt	1		
3	WL 110001 J	Lockwasher	1		
4	WA 110061 J	Washer	1		
5	C 35864	"O" ring	2		
6	C 37566	Rear mounting bracket	1		
7	SH 110251 J	Setscrew	2		
8	WL 110001 J	Lockwasher	2		
9	C 37576	Front mounting bracket	1		
10	SH 108161 J	Setscrew	2		
11	WL 108001 J	Lockwasher	2		
12	DAC 4651	Compressor cable	1		

MASTER INDEX
ENGINE 1.C
FLYWHEEL AND CLUTCH 1.G
GEARBOX AND PROPSHAFT 1.H
AXLES, SUSPENSION, DRIVE SHAFTS, WHEELS 1.K
STEERING 1.M
BRAKES AND BRAKE CONTROLS 1.N
FUEL, EXHAUST AND EMISSION SYSTEMS 2.C
COOLING, HEATING AND AIR CONDITIONING 2.G
ELECTRICAL, WASHERS-WIPERS, INSTRUMENTS 2.J
CHASSIS, SUBFRAMES, BODYSHELL, FITTINGS 3.C
FASCIA, TRIM, SEATS AND FIXINGS 3.H
ACCESSORIES AND PAINT 4.C
NUMERICAL INDEX 4.D
GROUP INDEX
AIR COND ELECTRIC COMPONENT 3.6/5.3 LT H06
COMPRESSOR 3.6 LITRE G18
COMPRESSOR 5.3 LITRE I03
COMPRESSOR MOUNTING 3.6 LITRE H02
CONDENSER 3.6 & 5.3 LITRE H08
ELECTRIC FAN - 5.3 LITRE G11
ELECTRIC FAN AND MOTOR 3.6 LITRE G10
FAN AND DRIVE UNIT 5.3 LITRE I02
HEATER VALVE AND HOSES 3.6 LITRE G13
RADIATOR 3.6 LITRE G02
RADIATOR HOSES 3.6 LITRE G03
RADIATOR HOSES 5.3 LITRE G06
RADIATOR-5.3 LITRE G05
RADIO CONTROL PANEL 3.6 & 5.3 LITRE G16
REFRIGERANT HOSES 3.6 Litre H09
REFRIGERANT HOSES 5.3 LITRE H10

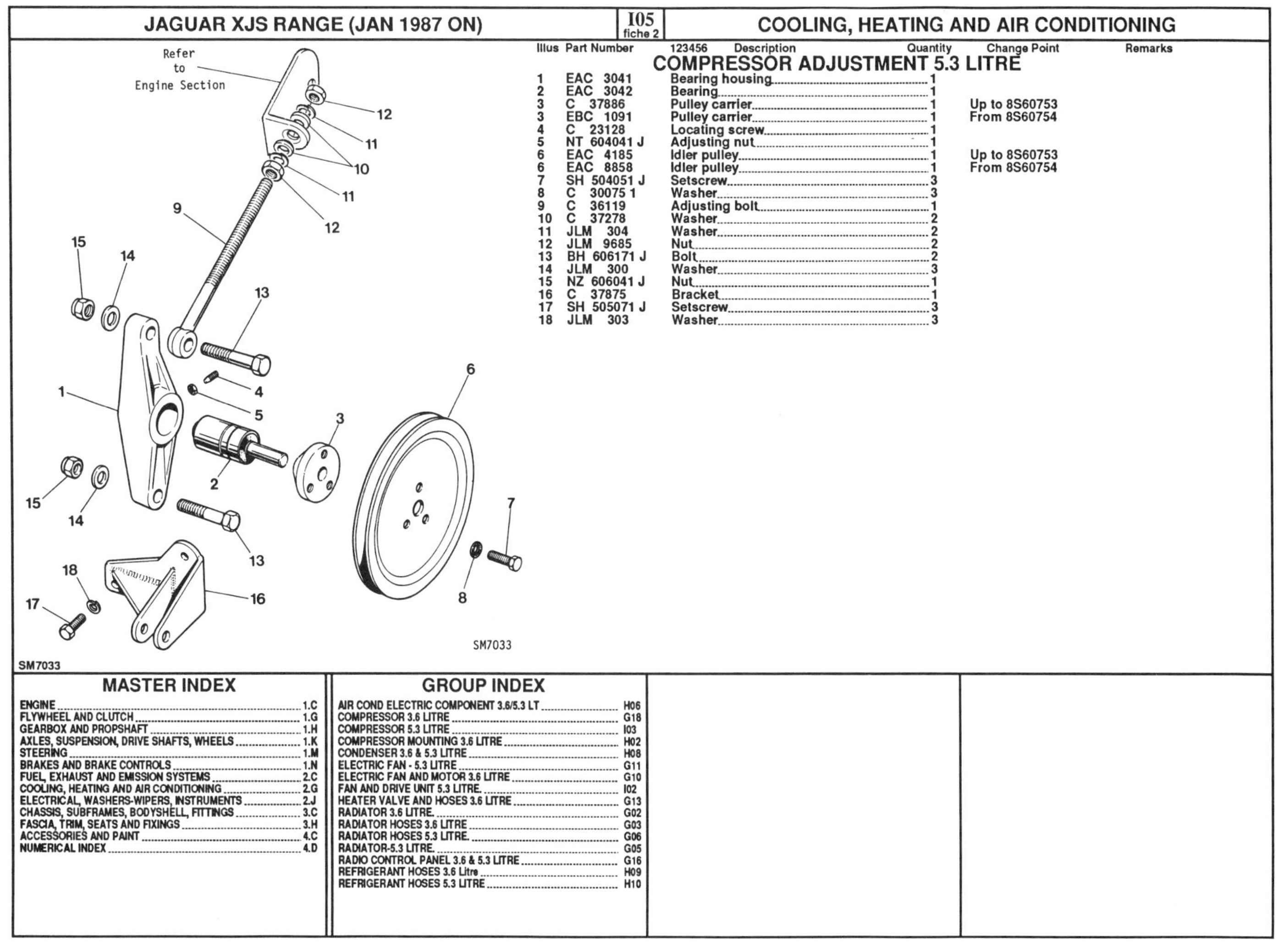

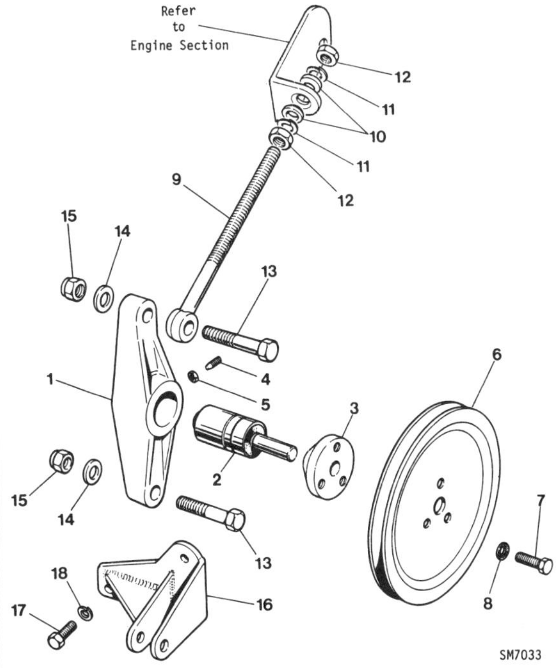

SM7033

COMPRESSOR ADJUSTMENT 5.3 LITRE

Illus	Part Number	Description	Quantity	Change Point	Remarks
1	EAC 3041	Bearing housing	1		
2	EAC 3042	Bearing	1		
3	C 37886	Pulley carrier	1	Up to 8S60753	
3	EBC 1091	Pulley carrier	1	From 8S60754	
4	C 23128	Locating screw	1		
5	NT 604041 J	Adjusting nut	1		
6	EAC 4185	Idler pulley	1	Up to 8S60753	
6	EAC 8858	Idler pulley	1	From 8S60754	
7	SH 504051 J	Setscrew	3		
8	C 30075 1	Washer	3		
9	C 36119	Adjusting bolt	1		
10	C 37278	Washer	2		
11	JLM 304	Washer	2		
12	JLM 9685	Nut	2		
13	BH 606171 J	Bolt	2		
14	JLM 300	Washer	3		
15	NZ 606041 J	Nut	1		
16	C 37875	Bracket	1		
17	SH 505071 J	Setscrew	3		
18	JLM 303	Washer	3		

MASTER INDEX

GROUP INDEX

MASTER INDEX

GROUP INDEX

2J 02

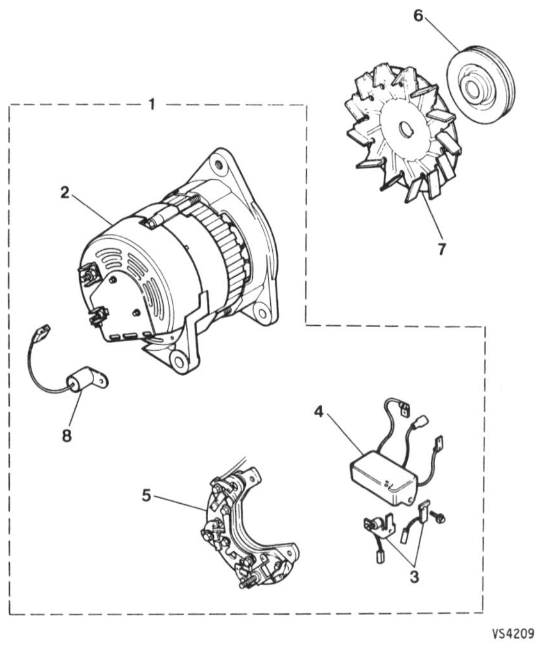

VS4209

ALTERNATOR 75 AMP (UP TO (E)109969).

Illus	Part Number	123456	Description	Quantity	Change Point	Remarks
1	AEU 1930 N		ALTERNATOR	1		
2	AEU 1672		Cover	1		
3	JLM 9555		Brush set	1		
4	JLM 277		Regulator	1		
5	JLM 278		Rectifier	1		
6	EAC 3765		Pulley	1		
7	605410 J		Cooling Fan	1		
8	AEU 1616 J		Capacitor	1		

MASTER INDEX

GROUP INDEX

2J 03

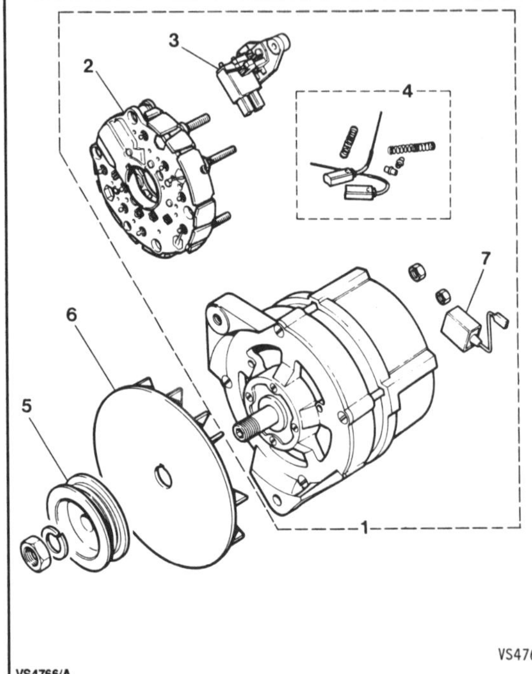

VS4766/A

ALTERNATOR-BOSCH-90 AMP-3.6 LITRE
FROM (E)109970 UPTO (E)141909

Illus	Part Number	123456	Description	Quantity	Change Point	Remarks
1	DBC 3102		ALTERNATOR	1		
2	JLM 1300		Rectifier-alternator	1		
3	JLM 1301		Regulator-alternator	1		
4	JLM 1302		Brush set-alternator	1		
5	DBC 3231		Pulley	1		
6	DBC 3230		Fan-alternator	1		
7	DBC 3232		Supressor-alternator	1		

MASTER INDEX
ENGINE 1.C
FLYWHEEL AND CLUTCH 1.G
GEARBOX AND PROPSHAFT 1.H
AXLES, SUSPENSION, DRIVE SHAFTS, WHEELS 1.K
STEERING 1.M
BRAKES AND BRAKE CONTROLS 1.N
FUEL, EXHAUST AND EMISSION SYSTEMS 2.C
COOLING, HEATING AND AIR CONDITIONING 2.G
ELECTRICAL, WASHERS-WIPERS, INSTRUMENTS 2.J
CHASSIS, SUBFRAMES, BODYSHELL, FITTINGS 3.C
FASCIA, TRIM, SEATS AND FIXINGS 3.H
ACCESSORIES AND PAINT 4.C
NUMERICAL INDEX 4.D
GROUP INDEX
A.B.S EQUIPMENT N09
AERIAL AND FIXINGS N12
ALTERNATOR 75 AMP (UP TO (E)109969) J02
BATTERY AND LEADS N17
CONSOLE SWITCHES AND CIGAR LIGHTER M08
CONTROL UNIT-PJ-V12 COUPE L11
FROM 89 MY L13
CONTROL UNITS-PI/CRUISE CONTROL 3.6LITRE L10
DISTRIBUTOR-3.6 LITRE J11
DISTRIBUTOR-5.3 LITRE J12
FROM 89 MY J13
DOOR WARNING LAMP K15
FOG REAR GUARD LAMP K14
FRONT FLASHER LAMPS K06
FRONT FOG LAMPS K13
FUEL INTERFACE UNIT AND RELAYS L15
HARNESSES-FRONT 3.6 LITRE-FROM (V)148945 .... O04
HARNESSES-FRONT 5.3 LITRE-FROM (V)148782 .... O06
HARNESSES-FRONT-3.6 LITRE-UPTO (V)148944 .... O03
HEADLAMP WASH WIPE RESERVOIR L07
HEADLAMP WASHWIPE EQUIPMENT L06
HEADLAMPS-NOT USA K05
HEADLAMPS-USA K03
HIGH MOUNTED STOP LAMP-USA/CDN ONLY K12
HOOD CONTROL MODULE - CONVERTIBLE N10
HORNS AND HORN PAD M05
IGNITION COIL & MODULE-3.6 LITRE J14
IGNITION COILS AND AMPLIFIER-5.3 LITRE J15
INERTIA SWITCH AND COVER M13
INSTRUMENT PACK-3.6 LITRE L17
INSTRUMENT PACK-5.3 LITRE L18
INTERIOR LAMPS-CABRIOLET K09
INTERIOR LAMPS-COUPE K08
LOUDSPEAKERS AND MOUNTINGS N15
MAIN FUSE BLOCK AND MOUNTING PLATE M15
NUMBER PLATE AND REVERSE LAMP K16
PASSIVE RESTRAINT MODULE & FIXINGS N08
PLUGS AND RELAYS N07
POWER RESISTOR AND LAMBDA SENSOR L16
RADIO CASSETTE PLAYER-FROM (V)148782 .... N14
SHOULDER BAG-REMOVEABLE RADIO N14
RADIO'S-UP TO (V)148781 .... N13
REAR LAMPS K11
SEAT CUSHION SENSOR N11
SENSORS,TRANSMITTERS AND BRACKET M12
SIDE LAMPS K07
SPEED CONTROL PUMP & DUMP VALVE O11
STARTER MOTOR - 3.6 LITRE J08
STARTER MOTOR 5.3 LITRE J09
STEERING COLUMN SWITCHES M04
TRIP COMPUTER AND PANEL SWITCHES M06
WASH/WIPE RESERVOIR AND FIXINGS L08
WINDSCREEN WASHER RESERVOIR L04
WINDSCREEN WIPER MOTOR-LUCAS TYPE K17
WINDSCREEN WIPERS L03

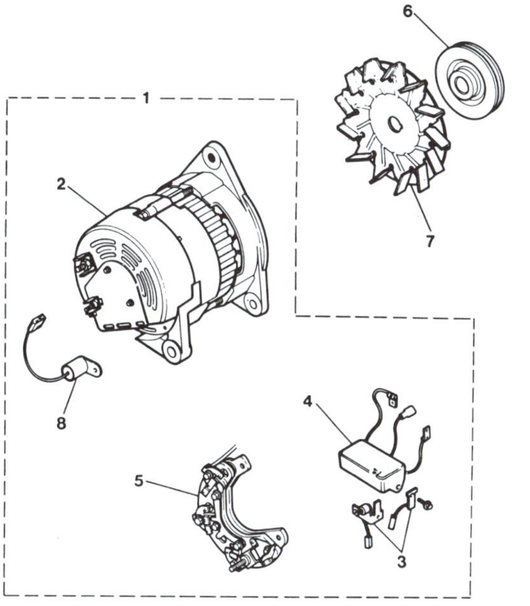

VS4209

VS4209

ALTERNATOR-LUCAS-75 AMP-5.3 LITRE - UP TO (E) 8S.57571

Illus	Part Number	123456 Description	Quantity	Change Point	Remarks
1	AEU 1930 N	ALTERNATOR	1		
2	AEU 1672	Cover	1		
3	JLM 9555	Brush set	1		
4	JLM 277	Regulator	1		
5	JLM 278	Rectifier	1		
6	EAC 3765	Pulley	1		
7	605410 J	Cooling fan	1		
8	AEU 1616 J	Capacitor	1		

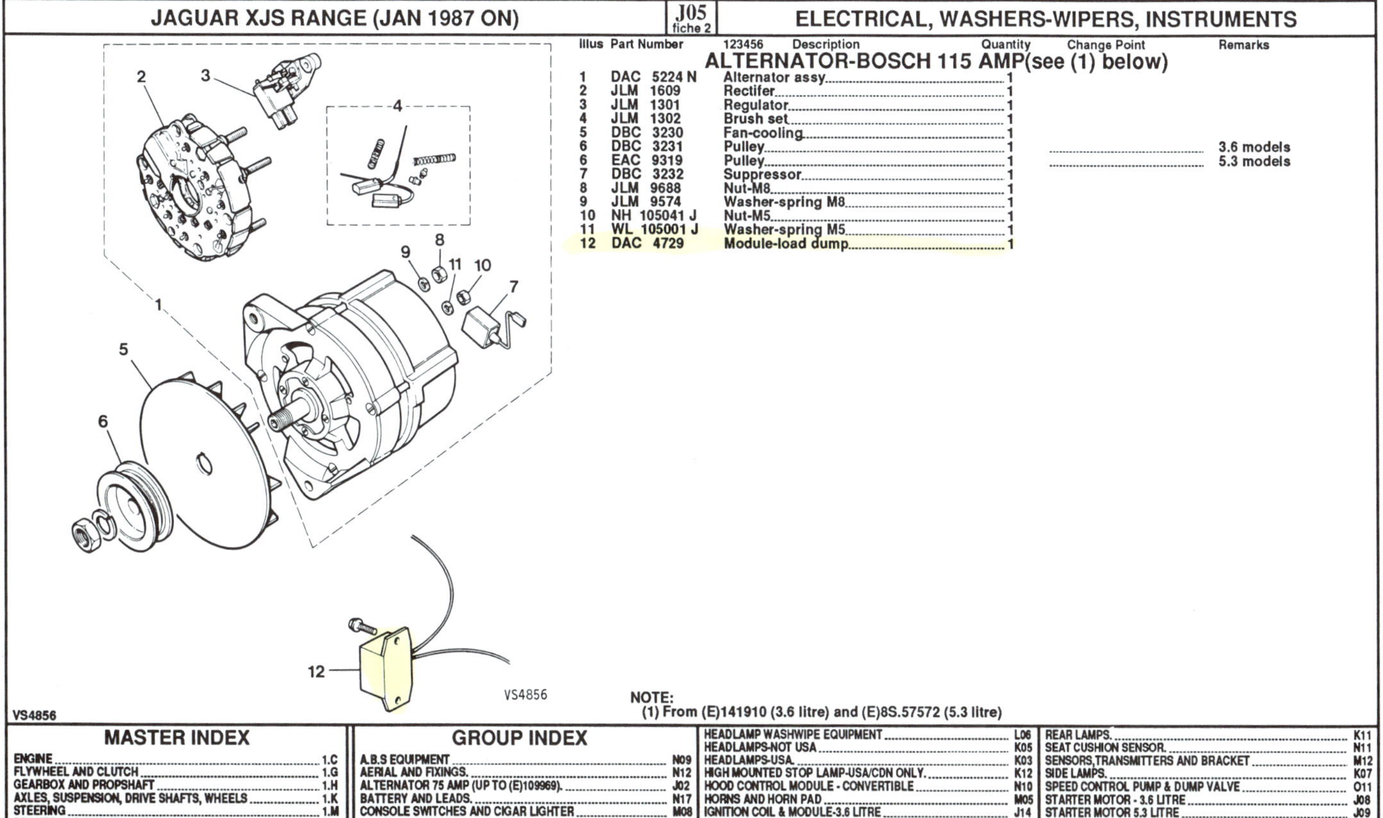

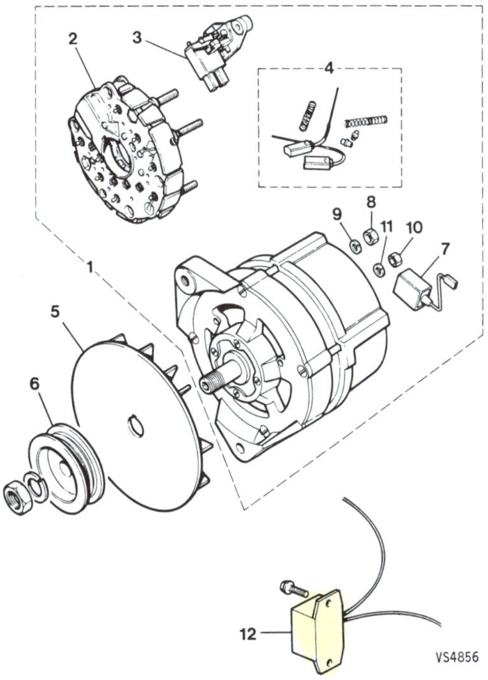

VS4856

VS4856

ALTERNATOR-BOSCH 115 AMP(see (1) below)

Illus	Part Number	123456 Description	Quantity	Change Point	Remarks
1	DAC 5224 N	Alternator assy	1		
2	JLM 1609	Rectifer	1		
3	JLM 1301	Regulator	1		
4	JLM 1302	Brush set	1		
5	DBC 3230	Fan-cooling	1		
6	DBC 3231	Pulley	1	....	3.6 models
6	EAC 9319	Pulley	1	....	5.3 models
7	DBC 3232	Suppressor	1		
8	JLM 9688	Nut-M8	1		
9	JLM 9574	Washer-spring M8	1		
10	NH 105041 J	Nut-M5	1		
11	WL 105001 J	Washer-spring M5	1		
12	DAC 4729	Module-load dump	1		

NOTE:
(1) From (E)141910 (3.6 litre) and (E)8S.57572 (5.3 litre)

MASTER INDEX

ENGINE 1.C
FLYWHEEL AND CLUTCH 1.G
GEARBOX AND PROPSHAFT 1.H
AXLES, SUSPENSION, DRIVE SHAFTS, WHEELS 1.K
STEERING 1.M
BRAKES AND BRAKE CONTROLS 1.N
FUEL, EXHAUST AND EMISSION SYSTEMS 2.C
COOLING, HEATING AND AIR CONDITIONING 2.G
ELECTRICAL, WASHERS-WIPERS, INSTRUMENTS 2.J
CHASSIS, SUBFRAMES, BODYSHELL, FITTINGS 3.C
FASCIA, TRIM, SEATS AND FIXINGS 3.H
ACCESSORIES AND PAINT 4.C
NUMERICAL INDEX 4.D

GROUP INDEX

A.B.S EQUIPMENT N09
AERIAL AND FIXINGS. N12
ALTERNATOR 75 AMP (UP TO (E)109969). J02
BATTERY AND LEADS. N17
CONSOLE SWITCHES AND CIGAR LIGHTER M08
CONTROL UNIT-P.J-V12 COUPE L11
FROM 89 MY L13
CONTROL UNITS-PI/CRUISE CONTROL 3.6LITRE L10
DISTRIBUTOR-3.6 LITRE. J11
DISTRIBUTOR-5.3 LITRE. J12
FROM 89 MY J13
DOOR WARNING LAMP. K15
FOG REAR GUARD LAMP. K14
FRONT FLASHER LAMPS. K06
FRONT FOG LAMPS. K13
FUEL INTERFACE UNIT AND RELAYS L15
HARNESSES-FRONT 3.6 LITRE-FROM (V)148945 .... O04
HARNESSES-FRONT 5.3 LITRE-FROM (V)148782 .... O06
HARNESSES-FRONT-3.6 LITRE-UPTO (V)148944 .... O03
HEADLAMP WASH WIPE RESERVOIR L07
HEADLAMP WASHWIPE EQUIPMENT L06
HEADLAMPS-NOT USA K05
HEADLAMPS-USA. K03
HIGH MOUNTED STOP LAMP-USA/CDN ONLY. K12
HOOD CONTROL MODULE - CONVERTIBLE N10
HORNS AND HORN PAD M05
IGNITION COIL & MODULE-3.6 LITRE J14
IGNITION COILS AND AMPLIFIER-5.3 LITRE J15
INERTIA SWITCH AND COVER M13
INSTRUMENT PACK-3.6 LITRE L17
INSTRUMENT PACK-5.3 LITRE L18
INTERIOR LAMPS-CABRIOLET. K09
INTERIOR LAMPS-COUPE. K08
LOUDSPEAKERS AND MOUNTINGS. N15
MAIN FUSE BLOCK AND MOUNTING PLATE M15
NUMBER PLATE AND REVERSE LAMP. K16
PASSIVE RESTRAINT MODULE & FIXINGS N08
PLUGS AND RELAYS. N07
POWER RESISTOR AND LAMBDA SENSOR. L16
RADIO CASSETTE PLAYER-FROM (V)148782 .... N14
SHOULDER BAG-REMOVEABLE RADIO N14
RADIO'S-UP TO (V)148781 .... N13
REAR LAMPS. K11
SEAT CUSHION SENSOR. N11
SENSORS,TRANSMITTERS AND BRACKET M12
SIDE LAMPS. K07
SPEED CONTROL PUMP & DUMP VALVE O11
STARTER MOTOR - 3.6 LITRE J08
STARTER MOTOR 5.3 LITRE J09
STEERING COLUMN SWITCHES. M04
TRIP COMPUTER AND PANEL SWITCHES. M06
WASH/WIPE RESERVOIR AND FIXINGS L08
WINDSCREEN WASHER RESERVOIR L04
WINDSCREEN WIPER MOTOR-LUCAS TYPE K17
WINDSCREEN WIPERS L03

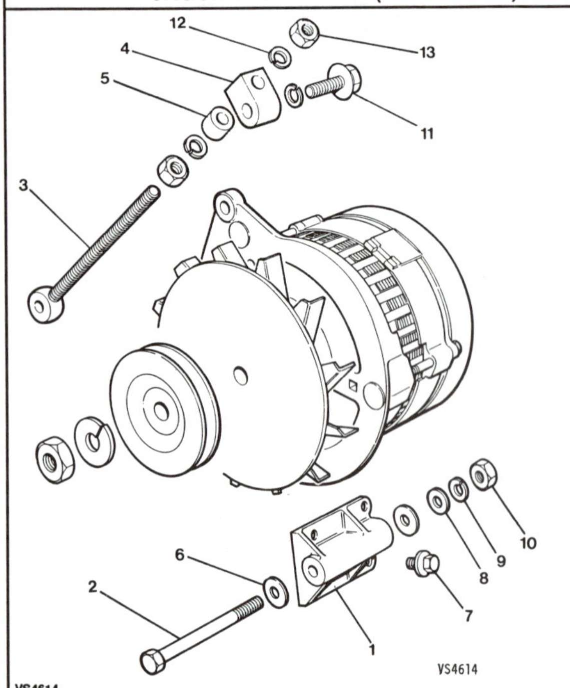

VS4614

ALTERNATOR MOUNTINGS-3.6 LITRE.

Illus	Part Number	123456 Description	Quantity	Change Point	Remarks
1	EAC 3385	Bracket-alternator pivot	1	Up to (E) 109969	
1	EAC 9510	Bracket-alternator pivot	1	From (E) 109970 to (E) 141909	
1	EBC 1052	Bracket-alternator pivot	1	From (E) 141910	
2	BH 108261 J	Bolt-pivot	1	Up to (E) 141909	
2	BH 110281	Bolt-pivot	1	From (E) 141910	
3	C 34241	Bolt-adjustment	1		
4	C 36120	Sleeve-adjustment	1		
5	EAC 4542	Spacer-adjustment bolt	1		
6	C 15183	Washer-plain	2	Up to (E) 141909	
6	WC 110061 J	Washer-plain	1	From (E) 141910	
7	AGU 1169	Setscrew	4	Up to (E) 109969	
7	FS 108251 J	Setscrew	2	From (E) 109970	
7	FB 108091 J	Setscrew	2		
8	WA 108051 J	Washer-plain	1	Up to (E) 141909	
8	WK 110061 J	Washer-plain	1	From (E) 141910	
9	WL 108001 J	Washer-spring	1		
10	NH 108041 J	Nut	1	Up to (E) 141909	
10	JLM 9691	Nut	1	From (E) 141910	
11	FB 108061 J	Bolt	1		
12	WM 600061 J	Washer-spring	2		
13	NH 606041 J	Nut	2		

2J 06

JAGUAR XJS RANGE (JAN 1987 ON)

J07 fiche 2

ELECTRICAL, WASHERS-WIPERS, INSTRUMENTS

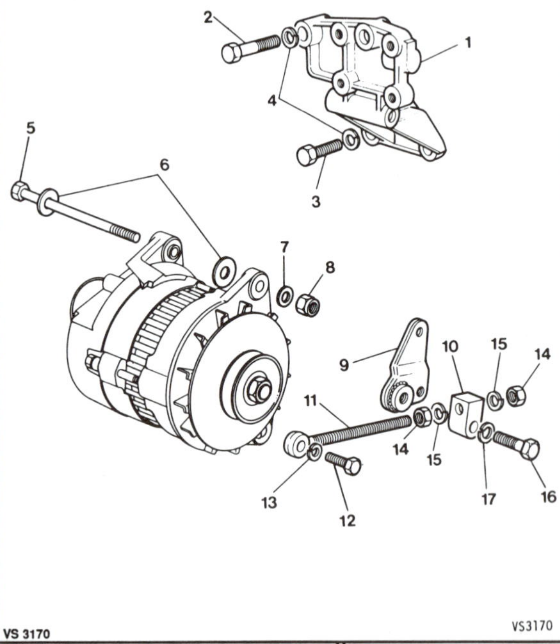

VS 3170

ALTERNATOR MOUNTINGS-5.3 LITRE

Illus	Part Number	123456 Description	Quantity	Change Point	Remarks
1	EAC 4181	Bracket-Alternator Pivot	1	See (1) below	
1	EAC 9320	Bracket-Alternator Pivot	1	See (2) below	
2	UCB 143 16R	Bolt-Bracket to Block	2		
3	UCB 143 12R	Bolt-Bracket to Block	2		
4	WL 600071 J	Washer-Spring	4		
5	BH 605401	Bolt-Pivot	1	See (1) below	
5	BH 110281	Bolt-Pivot	1	See (2) below	
6	C 15183	Washer-Plain	2	] See (1) below	
7	JLM 299	Washer-Plain	1	]	
7	WA 110061 J	Washer-Plain	3	See (2) below	
8	C 8667 2	Nut-Locking	1	See (1) below	
8	JLM 9691	Nut-Locking	1	See (2) below	
9	C 36151	Bracket-Adjustment Sleeve	1	See (1) below	
9	EBC 2149	Bracket-Adjustment Sleeve	1	See (2) below	
10	C 36120	Sleeve-Adjustment	1		
11	EAC 2127	Bolt-Adjustment	1		
12	JLM 9679	Setscrew-Bolt to Alternator	1	See (1) below	
12	SH 108301	Setscrew-Bolt to Alternator	1	See (2) below	
13	JLM 303	Washer-Spring	1	See (1) below	
13	WM 600051 J	Washer-Spring	1	See (2) below	
14	JLM 9685	Nut-Adjustment	2		
15	JLM 304	Washer-Spring	2		
16	BH 605101 J	Bolt-Sleeve to Bracket	1		
17	WM 600051 J	Washer-Spring	1		

(1) Up to 8S.57571
(2) From 8S.57572

2J 07

MASTER INDEX

ENGINE 1.C
FLYWHEEL AND CLUTCH 1.G
GEARBOX AND PROPSHAFT 1.H
AXLES, SUSPENSION, DRIVE SHAFTS, WHEELS 1.K
STEERING 1.M
BRAKES AND BRAKE CONTROLS 1.N
FUEL, EXHAUST AND EMISSION SYSTEMS 2.C
COOLING, HEATING AND AIR CONDITIONING 2.G
ELECTRICAL, WASHERS-WIPERS, INSTRUMENTS 2.J
CHASSIS, SUBFRAMES, BODYSHELL, FITTINGS 3.C
FASCIA, TRIM, SEATS AND FIXINGS 3.H
ACCESSORIES AND PAINT 4.C
NUMERICAL INDEX 4.D

GROUP INDEX

A.B.S EQUIPMENT N09
AERIAL AND FIXINGS. N12
ALTERNATOR 75 AMP (UP TO (E)109969). J02
BATTERY AND LEADS. N17
CONSOLE SWITCHES AND CIGAR LIGHTER M08
CONTROL UNIT-PJ-V12 COUPE L11
FROM 89 MY L13
CONTROL UNITS-PI/CRUISE CONTROL 3.6LITRE L10
DISTRIBUTOR-3.6 LITRE. J11
DISTRIBUTOR-5.3 LITRE. J12
FROM 89 MY J13
DOOR WARNING LAMP. K15
FOG REAR GUARD LAMP. K14
FRONT FLASHER LAMPS. K06
FRONT FOG LAMPS. K13
FUEL INTERFACE UNIT AND RELAYS L15
HARNESSES-FRONT 3.6 LITRE-FROM (V)148945 O04
HARNESSES-FRONT 5.3 LITRE-FROM (V)148782 O06
HARNESSES-FRONT-3.6 LITRE-UPTO (V)148944 O03
HEADLAMP WASH WIPE RESERVOIR L07
HEADLAMP WASHWIPE EQUIPMENT L06
HEADLAMPS-NOT USA K05
HEADLAMPS-USA. K03
HIGH MOUNTED STOP LAMP-USA/CDN ONLY. K12
HOOD CONTROL MODULE - CONVERTIBLE N10
HORNS AND HORN PAD M05
IGNITION COIL & MODULE-3.6 LITRE J14
IGNITION COILS AND AMPLIFIER-5.3 LITRE J15
INERTIA SWITCH AND COVER M13
INSTRUMENT PACK-3.6 LITRE L17
INSTRUMENT PACK-5.3 LITRE L18
INTERIOR LAMPS-CABRIOLET. K09
INTERIOR LAMPS-COUPE. K08
LOUDSPEAKERS AND MOUNTINGS. N15
MAIN FUSE BLOCK AND MOUNTING PLATE M15
NUMBER PLATE AND REVERSE LAMP. K16
PASSIVE RESTRAINT MODULE & FIXINGS N08
PLUGS AND RELAYS. N07
POWER RESISTOR AND LAMBDA SENSOR. L16
RADIO CASSETTE PLAYER-FROM (V)148782 N14
SHOULDER BAG-REMOVEABLE RADIO N14
RADIO'S-UP TO (V)148781 N13
REAR LAMPS. K11
SEAT CUSHION SENSOR. N11
SENSORS,TRANSMITTERS AND BRACKET M12
SIDE LAMPS. K07
SPEED CONTROL PUMP & DUMP VALVE O11
STARTER MOTOR - 3.6 LITRE J08
STARTER MOTOR 5.3 LITRE J09
STEERING COLUMN SWITCHES. M04
TRIP COMPUTER AND PANEL SWITCHES. M06
WASH/WIPE RESERVOIR AND FIXINGS L08
WINDSCREEN WASHER RESERVOIR L04
WINDSCREEN WIPER MOTOR-LUCAS TYPE K17
WINDSCREEN WIPERS L03

STARTER MOTOR - 3.6 LITRE

Illus	Part Number	Description	Quantity	Change Point	Remarks
1	DBC 2937 N	STARTER MOTOR ASSEMBLY	1		
	JLM 1312	Solenoid	1		
2	EBC 1632	Shim	1	Up to (E)110668	
3	FB 110071 J	Bolt-lower	1		
	FB 110171 J	Bolt-upper	1		

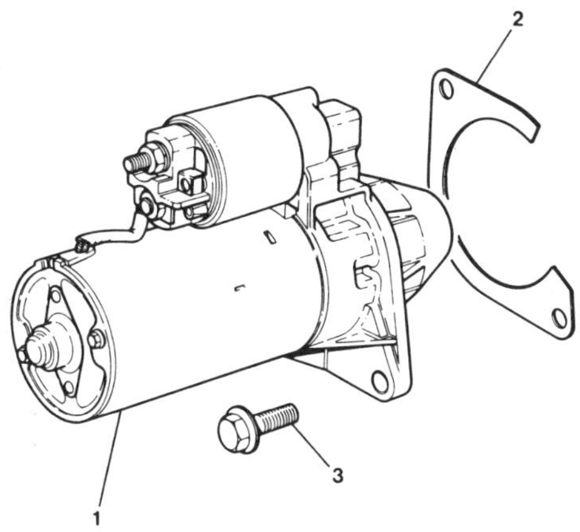

VS 4831

STARTER MOTOR 5.3 LITRE

Illus	Part Number	Description	Quantity	Change Point	Remarks
1	DAC 3269 N	STARTER MOTOR ASSEMBLY	1		Use DAC 4334 N
1	DAC 4334 N	STARTER MOTOR ASSEMBLY	!		
2	JLM 279	Solenoid	1		Use with
3	AEU 1768	Brush set	1		Starter motor
4	JLM 280	Drive assembly	1		DAC 3269 N
2	JLM 1521	Solenoid	1		Use with
3	JLM 1520	Brush set	1		Starter motor
4	JLM 1522	Drive assembly	1		DAC 4334 N
5	BH 506271	Bolt-upper fixing	1		
5	C 46269	Bolt-lower fixing	1		

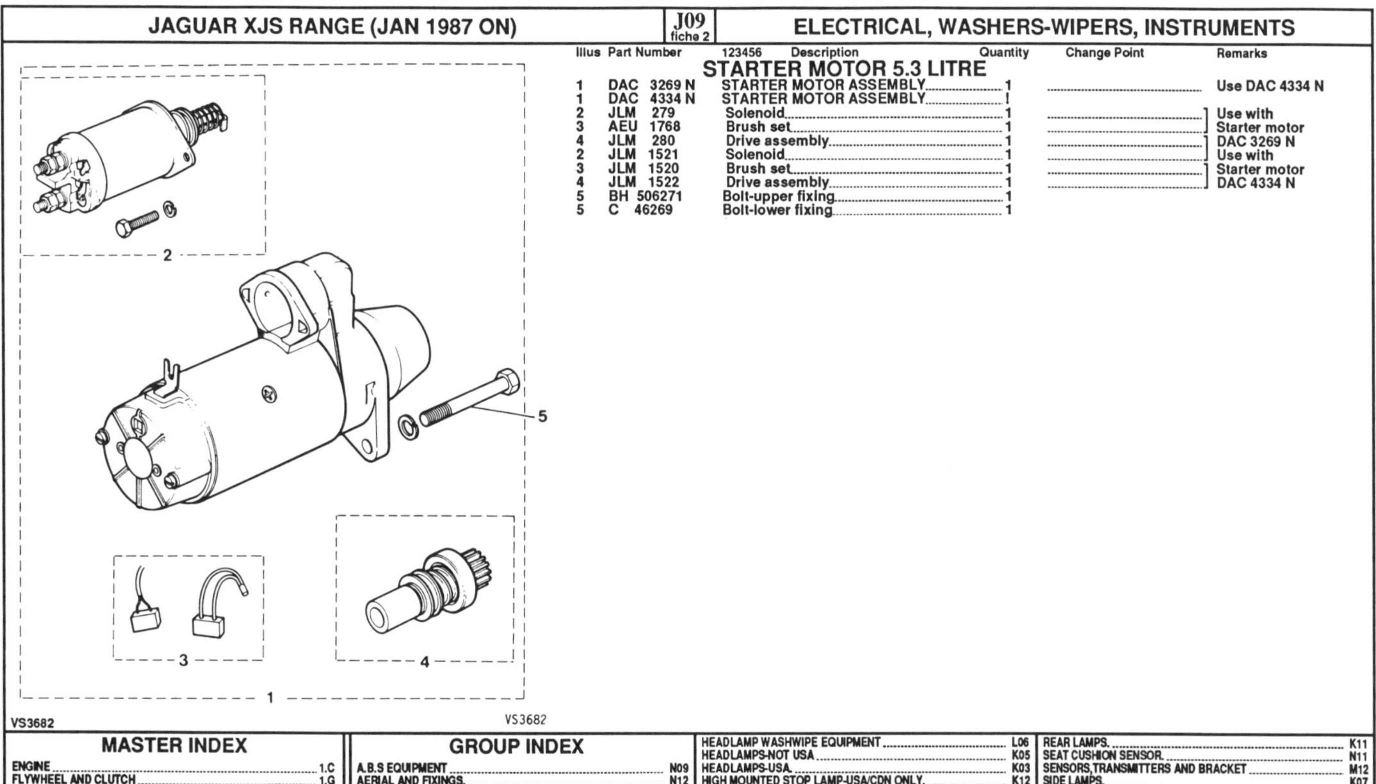

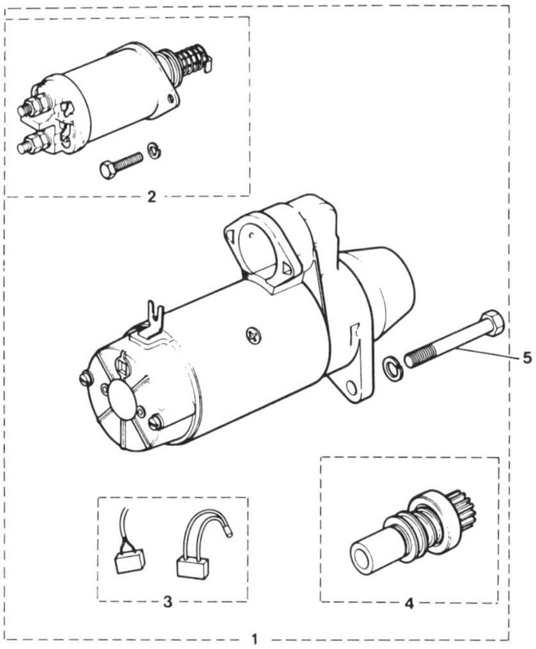

VS3682

MASTER INDEX

ENGINE ... 1.C
FLYWHEEL AND CLUTCH ... 1.G
GEARBOX AND PROPSHAFT ... 1.H
AXLES, SUSPENSION, DRIVE SHAFTS, WHEELS ... 1.K
STEERING ... 1.M
BRAKES AND BRAKE CONTROLS ... 1.N
FUEL, EXHAUST AND EMISSION SYSTEMS ... 2.C
COOLING, HEATING AND AIR CONDITIONING ... 2.G
ELECTRICAL, WASHERS-WIPERS, INSTRUMENTS ... 2.J
CHASSIS, SUBFRAMES, BODYSHELL, FITTINGS ... 3.C
FASCIA, TRIM, SEATS AND FIXINGS ... 3.H
ACCESSORIES AND PAINT ... 4.C
NUMERICAL INDEX ... 4.D

GROUP INDEX

A.B.S EQUIPMENT ... N09
AERIAL AND FIXINGS. ... N12
ALTERNATOR 75 AMP (UP TO (E)109969). ... J02
BATTERY AND LEADS. ... N17
CONSOLE SWITCHES AND CIGAR LIGHTER ... M08
CONTROL UNIT-P.J-V12 COUPE ... L11
FROM 89 MY ... L13
CONTROL UNITS-PI/CRUISE CONTROL 3.6LITRE ... L10
DISTRIBUTOR-3.6 LITRE. ... J11
DISTRIBUTOR-5.3 LITRE. ... J12
FROM 89 MY ... J13
DOOR WARNING LAMP. ... K15
FOG REAR GUARD LAMP. ... K14
FRONT FLASHER LAMPS. ... K06
FRONT FOG LAMPS. ... K13
FUEL INTERFACE UNIT AND RELAYS ... L15
HARNESSES-FRONT 3.6 LITRE-FROM (V)148945 ... O04
HARNESSES-FRONT 5.3 LITRE-FROM (V)148782 ... O06
HARNESSES-FRONT-3.6 LITRE-UPTO (V)148944 ... O03
HEADLAMP WASH WIPE RESERVOIR ... L07
HEADLAMP WASHWIPE EQUIPMENT ... L06
HEADLAMPS-NOT USA ... K05
HEADLAMPS-USA. ... K03
HIGH MOUNTED STOP LAMP-USA/CDN ONLY. ... K12
HOOD CONTROL MODULE - CONVERTIBLE ... N10
HORNS AND HORN PAD ... M05
IGNITION COIL & MODULE-3.6 LITRE ... J14
IGNITION COILS AND AMPLIFIER-5.3 LITRE ... J15
INERTIA SWITCH AND COVER ... M13
INSTRUMENT PACK-3.6 LITRE ... L17
INSTRUMENT PACK-5.3 LITRE ... L18
INTERIOR LAMPS-CABRIOLET. ... K09
INTERIOR LAMPS-COUPE. ... K08
LOUDSPEAKERS AND MOUNTINGS. ... N15
MAIN FUSE BLOCK AND MOUNTING PLATE ... M15
NUMBER PLATE AND REVERSE LAMP. ... K16
PASSIVE RESTRAINT MODULE & FIXINGS ... N08
PLUGS AND RELAYS. ... N07
POWER RESISTOR AND LAMBDA SENSOR. ... L16
RADIO CASSETTE PLAYER-FROM (V)148782 ... N14
SHOULDER BAG-REMOVEABLE RADIO ... N14
RADIO'S-UP TO (V)148781 ... N13
REAR LAMPS. ... K11
SEAT CUSHION SENSOR. ... N11
SENSORS,TRANSMITTERS AND BRACKET ... M12
SIDE LAMPS. ... K07
SPEED CONTROL PUMP & DUMP VALVE ... O11
STARTER MOTOR - 3.6 LITRE ... J08
STARTER MOTOR 5.3 LITRE ... J09
STEERING COLUMN SWITCHES. ... M04
TRIP COMPUTER AND PANEL SWITCHES. ... M06
WASH/WIPE RESERVOIR AND FIXINGS ... L08
WINDSCREEN WASHER RESERVOIR ... L04
WINDSCREEN WIPER MOTOR-LUCAS TYPE ... K17
WINDSCREEN WIPERS ... L03

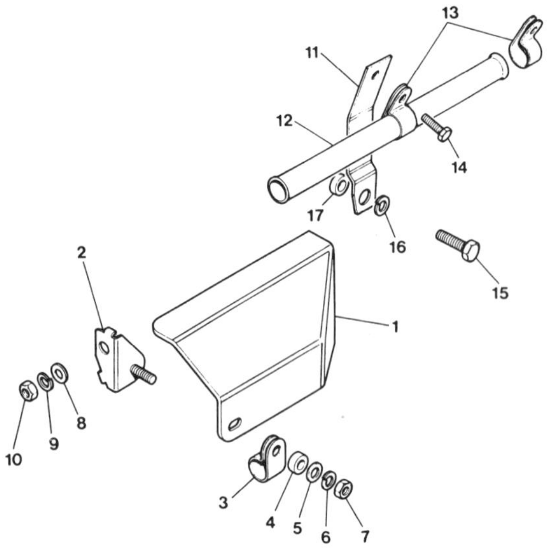

SM7213

SM 7213

2J 10

STARTER MOTOR HEATSHIELD-5.3 LITRE

Illus	Part Number	123456 Description	Quantity	Change Point	Remarks
1	C 43317	Heatshield	1		Not required with starter motor DAC 4334 N
2	DAC 4179	Bracket-mounting	1		
3	AGU 1117	P-clip	1		
4	C 16541	Spacer	1		
5	WC 702101 J	Washer-plain	1		
6	WF 702101 J	Washer-shakeproof	1		
7	UFN 119 L	Nut	1		
8	JLM 296	Washer-plain	1		
9	JLM 302	Washer-spring	1		
10	JLM 9683	Nut	1		
11	C 43318	Bracket	1		
12	C 43340	Conduit-cable	1		
13	C 34608	P-clip	2		
14	C 35093 2	Setscrew	2		
15	SH 507101 J	Setscrew	1		
16	WM 600071 J	Washer-spring	1		
17	BD 23149	Spacer	1		

MASTER INDEX

GROUP INDEX

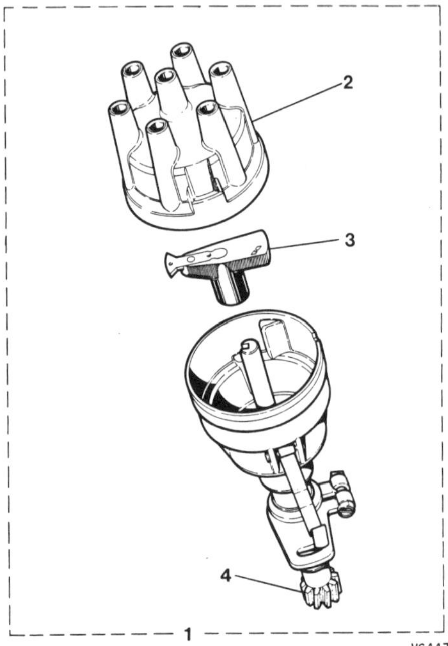

VS4475

VS4475

2J 11

DISTRIBUTOR-3.6 LITRE.

Illus	Part Number	123456 Description	Quantity	Change Point	Remarks
1	DBC 2690	DISTRIBUTOR ASSEMBLY	1		
2	JLM 150	Cap-distributor	1		
	608094 J	Brush and spring	1		
3	JLM 381	Rotor arm-distributor	1		
4	EAC 4674	Gear-driving	1		
	513682 J	'O'Ring oil seal	1		

MASTER INDEX
ENGINE 1.C
FLYWHEEL AND CLUTCH 1.G
GEARBOX AND PROPSHAFT 1.H
AXLES, SUSPENSION, DRIVE SHAFTS, WHEELS 1.K
STEERING 1.M
BRAKES AND BRAKE CONTROLS 1.N
FUEL, EXHAUST AND EMISSION SYSTEMS 2.C
COOLING, HEATING AND AIR CONDITIONING 2.G
ELECTRICAL, WASHERS-WIPERS, INSTRUMENTS 2.J
CHASSIS, SUBFRAMES, BODYSHELL, FITTINGS 3.C
FASCIA, TRIM, SEATS AND FIXINGS 3.H
ACCESSORIES AND PAINT 4.C
NUMERICAL INDEX 4.D
GROUP INDEX
A.B.S EQUIPMENT N09
AERIAL AND FIXINGS. N12
ALTERNATOR 75 AMP (UP TO (E)109969). J02
BATTERY AND LEADS. N17
CONSOLE SWITCHES AND CIGAR LIGHTER M08
CONTROL UNIT-PJ-V12 COUPE L11
FROM 89 MY L13
CONTROL UNITS-PI/CRUISE CONTROL 3.6LITRE L10
DISTRIBUTOR-3.6 LITRE. J11
DISTRIBUTOR-5.3 LITRE. J12
FROM 89 MY J13
DOOR WARNING LAMP. K15
FOG REAR GUARD LAMP. K14
FRONT FLASHER LAMPS. K06
FRONT FOG LAMPS. K13
FUEL INTERFACE UNIT AND RELAYS L15
HARNESSES-FRONT 3.6 LITRE-FROM (V)148945 .... O04
HARNESSES-FRONT 5.3 LITRE-FROM (V)148782 .... O06
HARNESSES-FRONT-3.6 LITRE-UPTO (V)148944 .... O03
HEADLAMP WASH WIPE RESERVOIR L07
HEADLAMP WASHWIPE EQUIPMENT L06
HEADLAMPS-NOT USA K05
HEADLAMPS-USA. K03
HIGH MOUNTED STOP LAMP-USA/CDN ONLY. K12
HOOD CONTROL MODULE - CONVERTIBLE N10
HORNS AND HORN PAD M05
IGNITION COIL & MODULE-3.6 LITRE J14
IGNITION COILS AND AMPLIFIER-5.3 LITRE J15
INERTIA SWITCH AND COVER M13
INSTRUMENT PACK-3.6 LITRE L17
INSTRUMENT PACK-5.3 LITRE L18
INTERIOR LAMPS-CABRIOLET. K09
INTERIOR LAMPS-COUPE. K08
LOUDSPEAKERS AND MOUNTINGS. N15
MAIN FUSE BLOCK AND MOUNTING PLATE M15
NUMBER PLATE AND REVERSE LAMP. K16
PASSIVE RESTRAINT MODULE & FIXINGS N08
PLUGS AND RELAYS. N07
POWER RESISTOR AND LAMBDA SENSOR. L16
RADIO CASSETTE PLAYER-FROM (V)148782 .... N14
SHOULDER BAG-REMOVEABLE RADIO N14
RADIO'S-UP TO (V)148781 .... N13
REAR LAMPS. K11
SEAT CUSHION SENSOR. N11
SENSORS,TRANSMITTERS AND BRACKET M12
SIDE LAMPS. K07
SPEED CONTROL PUMP & DUMP VALVE O11
STARTER MOTOR - 3.6 LITRE J08
STARTER MOTOR 5.3 LITRE J09
STEERING COLUMN SWITCHES. M04
TRIP COMPUTER AND PANEL SWITCHES. M06
WASH/WIPE RESERVOIR AND FIXINGS L08
WINDSCREEN WASHER RESERVOIR L04
WINDSCREEN WIPER MOTOR-LUCAS TYPE K17
WINDSCREEN WIPERS L03

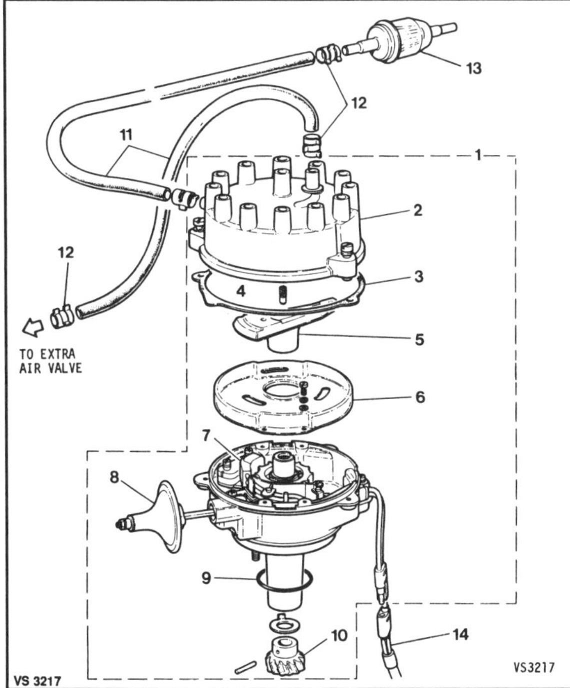

VS 3217

DISTRIBUTOR-5.3 LITRE.

UP TO 89 MY

Illus	Part Number	123456 Description	Quantity	Change Point	Remarks
1	DAC 4329	DISTRIBUTOR ASSEMBLY	NLA		Use DAC 4379
1	DAC 4379	DISTRIBUTOR ASSEMBLY	1		
2	DAC 4168	Distributor cap	1		
3	DAC 4063	Gasket	1		
4	608 094 J	Centre brush	1		
5	JLM 9624	Rotor arm	1		
6	AEU 1722	Dust cover	1		
7	AEU 1721	Pick-up and base plate	1		
8	JLM 519	Vacuum unit	1		
9	JS 407	'O' Ring	1		
10	AEU 1723	Drive gear	1		
11	DAC 4087 2	Hose	2		
12	CAC 8770	Pipe clip	4		
13	DAC 4062	Air filter	1		
14	DAC 3247	Screen link lead	1		

K8.02JP31/8/88

MASTER INDEX

GROUP INDEX

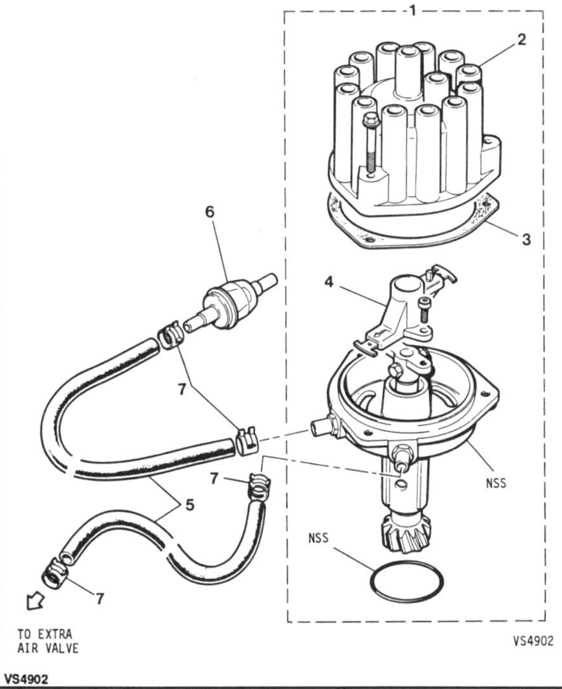

VS4902

DISTRIBUTOR - 5.3 LITRE

FROM 89 MY

Illus	Part Number	123456 Description	Quantity	Change Point	Remarks
1	DAC 4758	DISTRIBUTOR ASSEMBLY	1		
2	TBA	Cap - distributor	1		
3	TBA	Gasket	1		
4	TBA	Arm - rotor	1		
5	DAC 4087 2	Hose	2		
6	DAC 4062	Filter - air	1		
7	CAC 8770	Clip - pipe	4		

K8.04JP1/9/88

MASTER INDEX
ENGINE 1.C
FLYWHEEL AND CLUTCH 1.G
GEARBOX AND PROPSHAFT 1.H
AXLES, SUSPENSION, DRIVE SHAFTS, WHEELS 1.K
STEERING 1.M
BRAKES AND BRAKE CONTROLS 1.N
FUEL, EXHAUST AND EMISSION SYSTEMS 2.C
COOLING, HEATING AND AIR CONDITIONING 2.G
ELECTRICAL, WASHERS-WIPERS, INSTRUMENTS 2.J
CHASSIS, SUBFRAMES, BODYSHELL, FITTINGS 3.C
FASCIA, TRIM, SEATS AND FIXINGS 3.H
ACCESSORIES AND PAINT 4.C
NUMERICAL INDEX 4.D
GROUP INDEX
A.B.S EQUIPMENT N09
AERIAL AND FIXINGS N12
ALTERNATOR 75 AMP (UP TO (E)109969) J02
BATTERY AND LEADS N17
CONSOLE SWITCHES AND CIGAR LIGHTER M08
CONTROL UNIT-PJ-V12 COUPE L11
FROM 89 MY L13
CONTROL UNITS-PI/CRUISE CONTROL 3.6LITRE L10
DISTRIBUTOR-3.6 LITRE J11
DISTRIBUTOR-5.3 LITRE J12
FROM 89 MY J13
DOOR WARNING LAMP K15
FOG REAR GUARD LAMP K14
FRONT FLASHER LAMPS K06
FRONT FOG LAMPS K13
FUEL INTERFACE UNIT AND RELAYS L15
HARNESSES-FRONT 3.6 LITRE-FROM (V)148945 O04
HARNESSES-FRONT 5.3 LITRE-FROM (V)148782 O06
HARNESSES-FRONT-3.6 LITRE-UPTO (V)148944 O03
HEADLAMP WASH WIPE RESERVOIR L07
HEADLAMP WASHWIPE EQUIPMENT L06
HEADLAMPS-NOT USA K05
HEADLAMPS-USA K03
HIGH MOUNTED STOP LAMP-USA/CDN ONLY K12
HOOD CONTROL MODULE - CONVERTIBLE N10
HORNS AND HORN PAD M05
IGNITION COIL & MODULE-3.6 LITRE J14
IGNITION COILS AND AMPLIFIER-5.3 LITRE J15
INERTIA SWITCH AND COVER M13
INSTRUMENT PACK-3.6 LITRE L17
INSTRUMENT PACK-5.3 LITRE L18
INTERIOR LAMPS-CABRIOLET K09
INTERIOR LAMPS-COUPE K08
LOUDSPEAKERS AND MOUNTINGS N15
MAIN FUSE BLOCK AND MOUNTING PLATE M15
NUMBER PLATE AND REVERSE LAMP K16
PASSIVE RESTRAINT MODULE & FIXINGS N08
PLUGS AND RELAYS N07
POWER RESISTOR AND LAMBDA SENSOR L16
RADIO CASSETTE PLAYER-FROM (V)148782 N14
SHOULDER BAG-REMOVEABLE RADIO N14
RADIO'S-UP TO (V)148781 N13
REAR LAMPS K11
SEAT CUSHION SENSOR N11
SENSORS,TRANSMITTERS AND BRACKET M12
SIDE LAMPS K07
SPEED CONTROL PUMP & DUMP VALVE O11
STARTER MOTOR - 3.6 LITRE J08
STARTER MOTOR 5.3 LITRE J09
STEERING COLUMN SWITCHES M04
TRIP COMPUTER AND PANEL SWITCHES M06
WASH/WIPE RESERVOIR AND FIXINGS L08
WINDSCREEN WASHER RESERVOIR L04
WINDSCREEN WIPER MOTOR-LUCAS TYPE K17
WINDSCREEN WIPERS L03

JAGUAR XJS RANGE (JAN 1987 ON) | J14 fiche 2 | ELECTRICAL, WASHERS-WIPERS, INSTRUMENTS

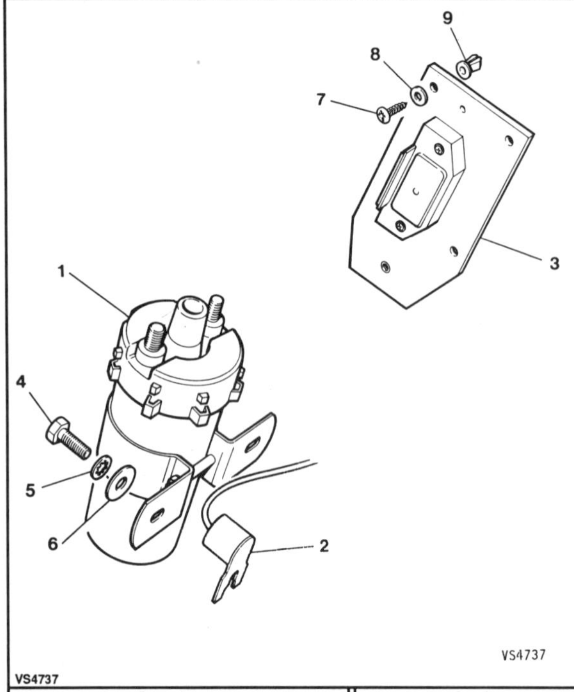

VS4737

IGNITION COIL & MODULE-3.6 LITRE

Illus	Part Number	123456	Description	Quantity	Change Point	Remarks
1	DBC 1140		Coil-ignition	1		
2	DBC 2000		Suppressor-ign coil	1		
3	DAC 4731		Module-ignition control	1		
4	SH 604041 J		Screw-set	2		
5	C 724		Washer-shakeproof	2		
6	BD 541 9		Washer-plain	2		
7	AB 610081 J		Screw-self tap	3		
8	WA 105001 J		Washer-plain	3		
9	C 31973 11		Nut-plastic	3		

2J 14

JAGUAR XJS RANGE (JAN 1987 ON) | J15 fiche 2 | ELECTRICAL, WASHERS-WIPERS, INSTRUMENTS

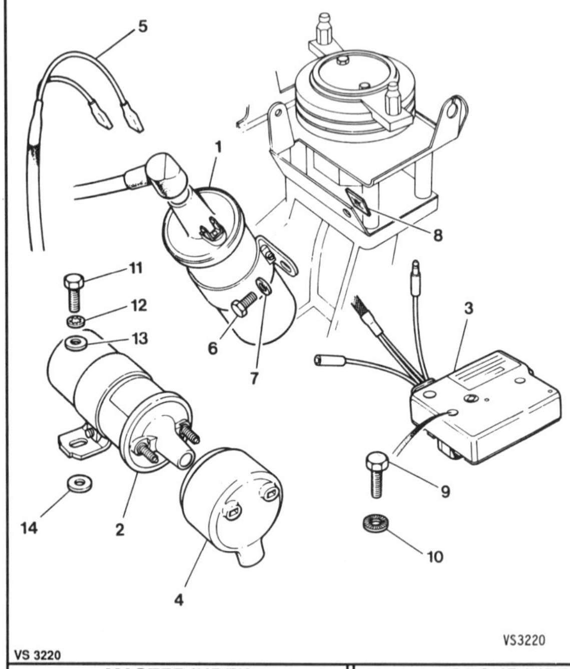

VS3220

IGNITION COILS AND AMPLIFIER-5.3 LITRE

UP TO 89 MY

Illus	Part Number	123456	Description	Quantity	Change Point	Remarks
1	DAC 2945		Main coil	1		
2	DAC 2693		Auxiliary coil	1		
3	DAC 4104		Amplifier	1		
4	DAC 2173		Boot-auxiliary coil	1		
5	DAC 2691		Link lead-coils	1		
6	SH 504041 J		Setscrew	2		
7	JLM 296		Washer-plain	2		
8	BD 8633 12		Spire nut	2		
9	SH 505141 J		Setscrew	2		
10	C 30075 2		Washer-locking	2		
11	GHF 117		Setscrew	2		
12	WF 600040 J		Washer-shakeproof	2		
13	BD 541 9		Washer-plain	2		
14	BD 541 47		Washer-plain	2		

K10.02JP31/8/88

MASTER INDEX

ENGINE ... 1.C
FLYWHEEL AND CLUTCH ... 1.G
GEARBOX AND PROPSHAFT ... 1.H
AXLES, SUSPENSION, DRIVE SHAFTS, WHEELS ... 1.K
STEERING ... 1.M
BRAKES AND BRAKE CONTROLS ... 1.N
FUEL, EXHAUST AND EMISSION SYSTEMS ... 2.C
COOLING, HEATING AND AIR CONDITIONING ... 2.G
ELECTRICAL, WASHERS-WIPERS, INSTRUMENTS ... 2.J
CHASSIS, SUBFRAMES, BODYSHELL, FITTINGS ... 3.C
FASCIA, TRIM, SEATS AND FIXINGS ... 3.H
ACCESSORIES AND PAINT ... 4.C
NUMERICAL INDEX ... 4.D

GROUP INDEX

A.B.S EQUIPMENT ... N09
AERIAL AND FIXINGS ... N12
ALTERNATOR 75 AMP (UP TO (E)109969) ... J02
BATTERY AND LEADS ... N17
CONSOLE SWITCHES AND CIGAR LIGHTER ... M08
CONTROL UNIT-PJ-V12 COUPE ... L11
FROM 89 MY ... L13
CONTROL UNITS-PI/CRUISE CONTROL 3.6LITRE ... L10
DISTRIBUTOR-3.6 LITRE ... J11
DISTRIBUTOR-5.3 LITRE ... J12
FROM 89 MY ... J13
DOOR WARNING LAMP ... K15
FOG REAR GUARD LAMP ... K14
FRONT FLASHER LAMPS ... K06
FRONT FOG LAMPS ... K13
FUEL INTERFACE UNIT AND RELAYS ... L15
HARNESSES-FRONT 3.6 LITRE-FROM (V)148945 ... O04
HARNESSES-FRONT 5.3 LITRE-FROM (V)148782 ... O06
HARNESSES-FRONT-3.6 LITRE-UPTO (V)148944 ... O03
HEADLAMP WASH WIPE RESERVOIR ... L07
HEADLAMP WASHWIPE EQUIPMENT ... L06
HEADLAMPS-NOT USA ... K05
HEADLAMPS-USA ... K03
HIGH MOUNTED STOP LAMP-USA/CDN ONLY ... K12
HOOD CONTROL MODULE - CONVERTIBLE ... N10
HORNS AND HORN PAD ... M05
IGNITION COIL & MODULE-3.6 LITRE ... J14
IGNITION COILS AND AMPLIFIER-5.3 LITRE ... J15
INERTIA SWITCH AND COVER ... M13
INSTRUMENT PACK-3.6 LITRE ... L17
INSTRUMENT PACK-5.3 LITRE ... L18
INTERIOR LAMPS-CABRIOLET ... K09
INTERIOR LAMPS-COUPE ... K08
LOUDSPEAKERS AND MOUNTINGS ... N15
MAIN FUSE BLOCK AND MOUNTING PLATE ... M15
NUMBER PLATE AND REVERSE LAMP ... K16
PASSIVE RESTRAINT MODULE & FIXINGS ... N08
PLUGS AND RELAYS ... N07
POWER RESISTOR AND LAMBDA SENSOR ... L16
RADIO CASSETTE PLAYER-FROM (V)148782 ... N14
SHOULDER BAG-REMOVEABLE RADIO ... N14
RADIO'S-UP TO (V)148781 ... N13
REAR LAMPS ... K11
SEAT CUSHION SENSOR ... N11
SENSORS,TRANSMITTERS AND BRACKET ... M12
SIDE LAMPS ... K07
SPEED CONTROL PUMP & DUMP VALVE ... O11
STARTER MOTOR - 3.6 LITRE ... J08
STARTER MOTOR 5.3 LITRE ... J09
STEERING COLUMN SWITCHES ... M04
TRIP COMPUTER AND PANEL SWITCHES ... M06
WASH/WIPE RESERVOIR AND FIXINGS ... L08
WINDSCREEN WASHER RESERVOIR ... L04
WINDSCREEN WIPER MOTOR-LUCAS TYPE ... K17
WINDSCREEN WIPERS ... L03

2J 15

Illus	Part Number	123456 Description	Quantity	Change Point	Remarks
		IGNITION COILS AND MODULES - 5.3 LITRE			
		FROM 89 MY			
1	DAC 4608	Coil - Ignition	2		
2	DAC 4607	Module - Ignition	2		Mtd on rad crossmember
					K10.04JP2/9/88

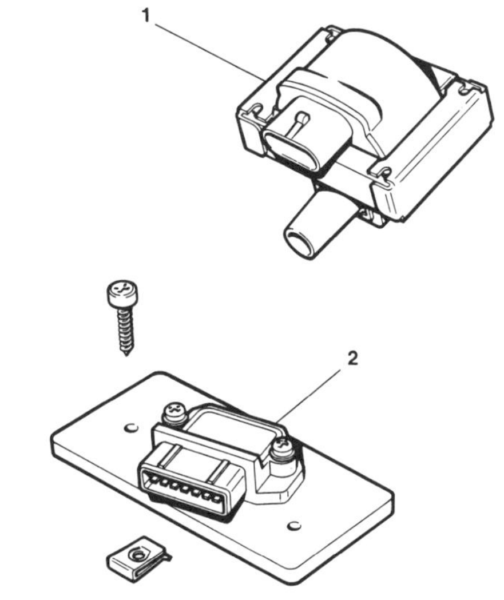

VS4901

VS4901

Illus	Part Number	123456 Description	Quantity	Change Point	Remarks
		SPARK PLUG,LEADS AND FIXINGS-3.6 LITRE			
1	EAC 8295	Plug-sparking	NLA		Use EBC2230
	EBC 2230	Plug-sparking	6		
2	DBC 3229	Kit-plug leads	1		Use DAC 6040
2	DAC 6040	Kit-plug leads	1		
3	DAC 5583	Cover - coil	1		
4	EAC 5045	Clip-H.T. leads	4		
5	EAC 5046	Bracket-H.T. lead clip	3		

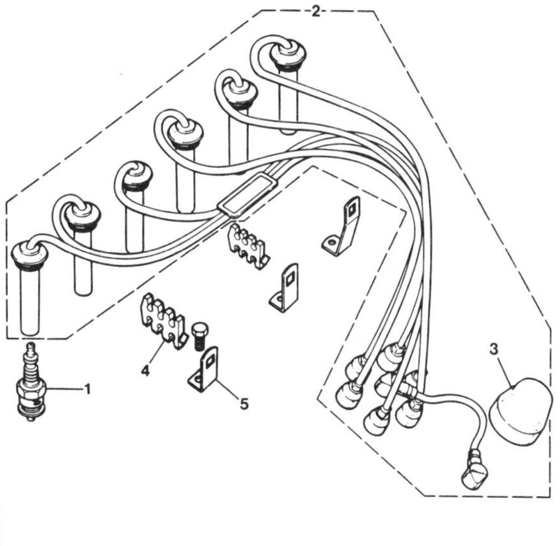

VS4859

VS4859

MASTER INDEX

ENGINE 1.C
FLYWHEEL AND CLUTCH 1.G
GEARBOX AND PROPSHAFT 1.H
AXLES, SUSPENSION, DRIVE SHAFTS, WHEELS 1.K
STEERING 1.M
BRAKES AND BRAKE CONTROLS 1.N
FUEL, EXHAUST AND EMISSION SYSTEMS 2.C
COOLING, HEATING AND AIR CONDITIONING 2.G
ELECTRICAL, WASHERS-WIPERS, INSTRUMENTS 2.J
CHASSIS, SUBFRAMES, BODYSHELL, FITTINGS 3.C
FASCIA, TRIM, SEATS AND FIXINGS 3.H
ACCESSORIES AND PAINT 4.C
NUMERICAL INDEX 4.D

GROUP INDEX

A.B.S EQUIPMENT N09
AERIAL AND FIXINGS N12
ALTERNATOR 75 AMP (UP TO (E)109969) J02
BATTERY AND LEADS N17
CONSOLE SWITCHES AND CIGAR LIGHTER M08
CONTROL UNIT-P.J-V12 COUPE L11
FROM 89 MY L13
CONTROL UNITS-PI/CRUISE CONTROL 3.6LITRE L10
DISTRIBUTOR-3.6 LITRE J11
DISTRIBUTOR-5.3 LITRE J12
FROM 89 MY J13
DOOR WARNING LAMP K15
FOG REAR GUARD LAMP K14
FRONT FLASHER LAMPS K06
FRONT FOG LAMPS K13
FUEL INTERFACE UNIT AND RELAYS L15
HARNESSES-FRONT 3.6 LITRE-FROM (V)148945 O04
HARNESSES-FRONT 5.3 LITRE-FROM (V)148782 O06
HARNESSES-FRONT-3.6 LITRE-UPTO (V)148944 O03
HEADLAMP WASH WIPE RESERVOIR L07
HEADLAMP WASHWIPE EQUIPMENT L06
HEADLAMPS-NOT USA K05
HEADLAMPS-USA K03
HIGH MOUNTED STOP LAMP-USA/CDN ONLY K12
HOOD CONTROL MODULE - CONVERTIBLE N10
HORNS AND HORN PAD M05
IGNITION COIL & MODULE-3.6 LITRE J14
IGNITION COILS AND AMPLIFIER-5.3 LITRE J15
INERTIA SWITCH AND COVER M13
INSTRUMENT PACK-3.6 LITRE L17
INSTRUMENT PACK-5.3 LITRE L18
INTERIOR LAMPS-CABRIOLET K09
INTERIOR LAMPS-COUPE K08
LOUDSPEAKERS AND MOUNTINGS N15
MAIN FUSE BLOCK AND MOUNTING PLATE M15
NUMBER PLATE AND REVERSE LAMP K16
PASSIVE RESTRAINT MODULE & FIXINGS N08
PLUGS AND RELAYS N07
POWER RESISTOR AND LAMBDA SENSOR L16
RADIO CASSETTE PLAYER-FROM (V)148782 N14
SHOULDER BAG-REMOVEABLE RADIO N14
RADIO'S-UP TO (V)148781 N13
REAR LAMPS K11
SEAT CUSHION SENSOR N11
SENSORS,TRANSMITTERS AND BRACKET M12
SIDE LAMPS K07
SPEED CONTROL PUMP & DUMP VALVE O11
STARTER MOTOR - 3.6 LITRE J08
STARTER MOTOR 5.3 LITRE J09
STEERING COLUMN SWITCHES M04
TRIP COMPUTER AND PANEL SWITCHES M06
WASH/WIPE RESERVOIR AND FIXINGS L08
WINDSCREEN WASHER RESERVOIR L04
WINDSCREEN WIPER MOTOR-LUCAS TYPE K17
WINDSCREEN WIPERS L03

2J 18

JAGUAR XJS RANGE (JAN 1987 ON) | J18 fiche 2 | ELECTRICAL, WASHERS-WIPERS, INSTRUMENTS

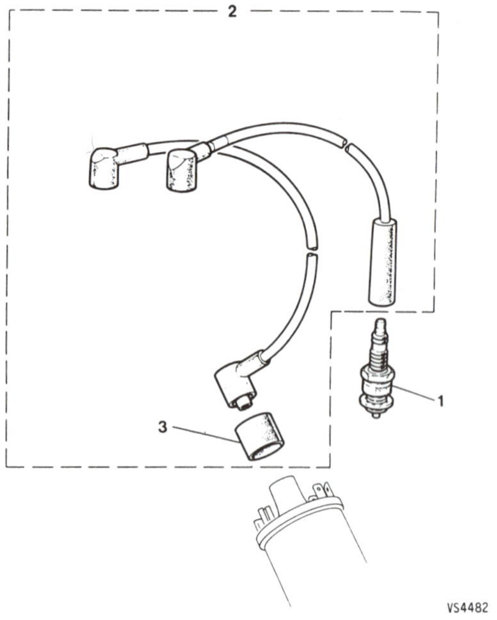

VS4482

SPARK PLUG, LEADS AND FIXINGS-5.3 LITRE

UP TO 89 MY

Illus	Part Number	Description	Quantity	Change Point	Remarks
1	EAC 9186	Plug - sparking	12		USA/CDN/J/AUS
1	EAC 8554	Plug - sparking	12		EEC/S/ME
2	JLM 726	Kit - HT leads	1		
3	DAC 4197	Sleeve - shrinklok	1		

K10.08JP1/9/88

2K 02

JAGUAR XJS RANGE (JAN 1987 ON) | K02 fiche 2 | ELECTRICAL, WASHERS-WIPERS, INSTRUMENTS

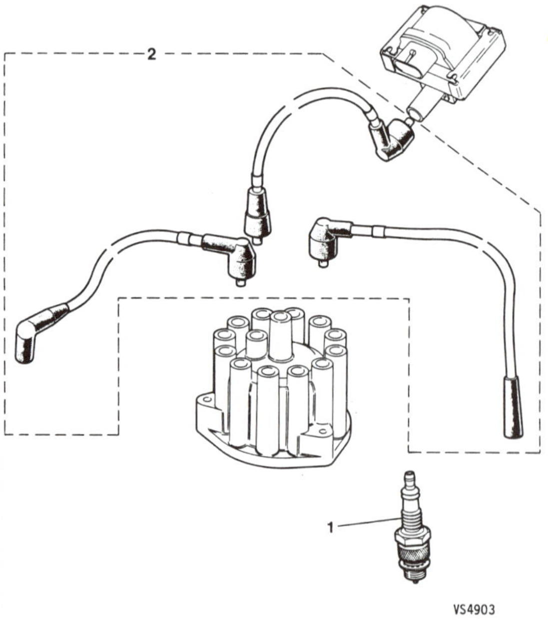

VS4903

SPARK PLUG AND LEADS - 5.3 LITRE

FROM 89 MY

Illus	Part Number	Description	Quantity	Change Point	Remarks
1	EAC 9186	Plug - sparking	12		USA/CDN,J,AUS
1	EAC 8554	Plug - sparking	12		EEC/S/ME
2	JLM 1714	Kit - HT leads	1		

K10.10JP31/8/88

MASTER INDEX

ENGINE 1.C
FLYWHEEL AND CLUTCH 1.G
GEARBOX AND PROPSHAFT 1.H
AXLES, SUSPENSION, DRIVE SHAFTS, WHEELS 1.K
STEERING 1.M
BRAKES AND BRAKE CONTROLS 1.N
FUEL, EXHAUST AND EMISSION SYSTEMS 2.C
COOLING, HEATING AND AIR CONDITIONING 2.G
ELECTRICAL, WASHERS-WIPERS, INSTRUMENTS 2.J
CHASSIS, SUBFRAMES, BODYSHELL, FITTINGS 3.C
FASCIA, TRIM, SEATS AND FIXINGS 3.H
ACCESSORIES AND PAINT 4.C
NUMERICAL INDEX 4.D

GROUP INDEX

A.B.S EQUIPMENT N09
AERIAL AND FIXINGS N12
ALTERNATOR 75 AMP (UP TO (E)109969) J02
BATTERY AND LEADS N17
CONSOLE SWITCHES AND CIGAR LIGHTER M08
CONTROL UNIT-PJ-V12 COUPE L11
FROM 89 MY L13
CONTROL UNITS-PI/CRUISE CONTROL 3.6LITRE L10
DISTRIBUTOR-3.6 LITRE J11
DISTRIBUTOR-5.3 LITRE J12
FROM 89 MY J13
DOOR WARNING LAMP K15
FOG REAR GUARD LAMP K14
FRONT FLASHER LAMPS K06
FRONT FOG LAMPS K13
FUEL INTERFACE UNIT AND RELAYS L15
HARNESSES-FRONT 3.6 LITRE-FROM (V)148945 O04
HARNESSES-FRONT 5.3 LITRE-FROM (V)148782 O06
HARNESSES-FRONT-3.6 LITRE-UPTO (V)148944 O03
HEADLAMP WASH WIPE RESERVOIR L07
HEADLAMP WASHWIPE EQUIPMENT L06
HEADLAMPS-NOT USA K05
HEADLAMPS-USA K03
HIGH MOUNTED STOP LAMP-USA/CDN ONLY K12
HOOD CONTROL MODULE - CONVERTIBLE N10
HORNS AND HORN PAD M05
IGNITION COIL & MODULE-3.6 LITRE J14
IGNITION COILS AND AMPLIFIER-5.3 LITRE J15
INERTIA SWITCH AND COVER M13
INSTRUMENT PACK-3.6 LITRE L17
INSTRUMENT PACK-5.3 LITRE L18
INTERIOR LAMPS-CABRIOLET K09
INTERIOR LAMPS-COUPE K08
LOUDSPEAKERS AND MOUNTINGS N15
MAIN FUSE BLOCK AND MOUNTING PLATE M15
NUMBER PLATE AND REVERSE LAMP K16
PASSIVE RESTRAINT MODULE & FIXINGS N08
PLUGS AND RELAYS N07
POWER RESISTOR AND LAMBDA SENSOR L16
RADIO CASSETTE PLAYER-FROM (V)148782 N14
SHOULDER BAG-REMOVEABLE RADIO N14
RADIO'S-UP TO (V)148781 N13
REAR LAMPS K11
SEAT CUSHION SENSOR N11
SENSORS,TRANSMITTERS AND BRACKET M12
SIDE LAMPS K07
SPEED CONTROL PUMP & DUMP VALVE O11
STARTER MOTOR - 3.6 LITRE J08
STARTER MOTOR 5.3 LITRE J09
STEERING COLUMN SWITCHES M04
TRIP COMPUTER AND PANEL SWITCHES M06
WASH/WIPE RESERVOIR AND FIXINGS L08
WINDSCREEN WASHER RESERVOIR L04
WINDSCREEN WIPER MOTOR-LUCAS TYPE K17
WINDSCREEN WIPERS L03

HEADLAMPS-USA.

Illus	Part Number	Description	Quantity	Change Point	Remarks
1	JLM 215	Light unit-inner	1		
2	JLM 216	Light unit-outer	1		
3	511598	Rim-headlamp	4		
	AEU 1364	Rim-headlamp seating-inner	2		
4	AEU 1365	Rim-headlamp seating-outer	2		
5	AEU 1366	Screw kit-headlamp trimmer	2		
6	AB 606021 J	Screw	12		
7	C 43863	Cable assembly-headlamp	2		

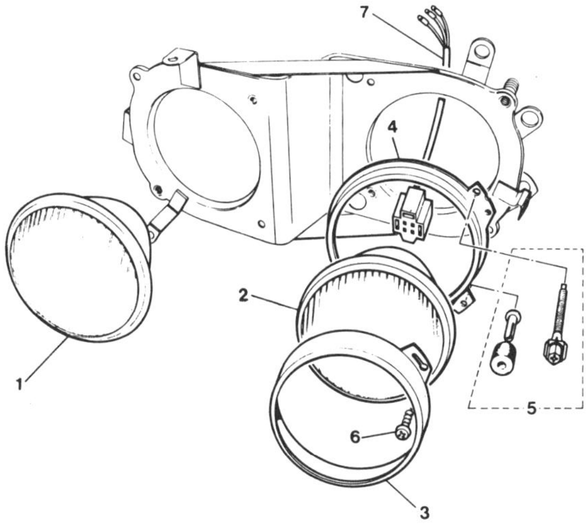

SM7185

SM7185

HEADLAMP MOUNTING-USA.

Illus	Part Number	Description	Quantity	Change Point	Remarks
1	AEU 2594	Bracket-headlamp mounting-RH	1		
1	AEU 2593	Bracket-headlamp mounting-LH	1		
2	DAC 1633	Fixing plate-headlamp bracket	2		
3	BD 48375	Base plate-headlamp bracket	2		
4	UFS 119 4R	Setscrew	4		
5	WF 702101 J	Washer-shakeproof	4		
6	WC 702101 J	Washer-plain	4		
7	SE 604041 J	Setscrew	6		
8	C 724	Washer-plain	4		
9	BD 541 9	Washer-shakeproof	4		
10	UFS 825 4H	Setscrew	2		
11	JLM 296	Washer-plain	2		
12	78319 J	Nut-spire	6		

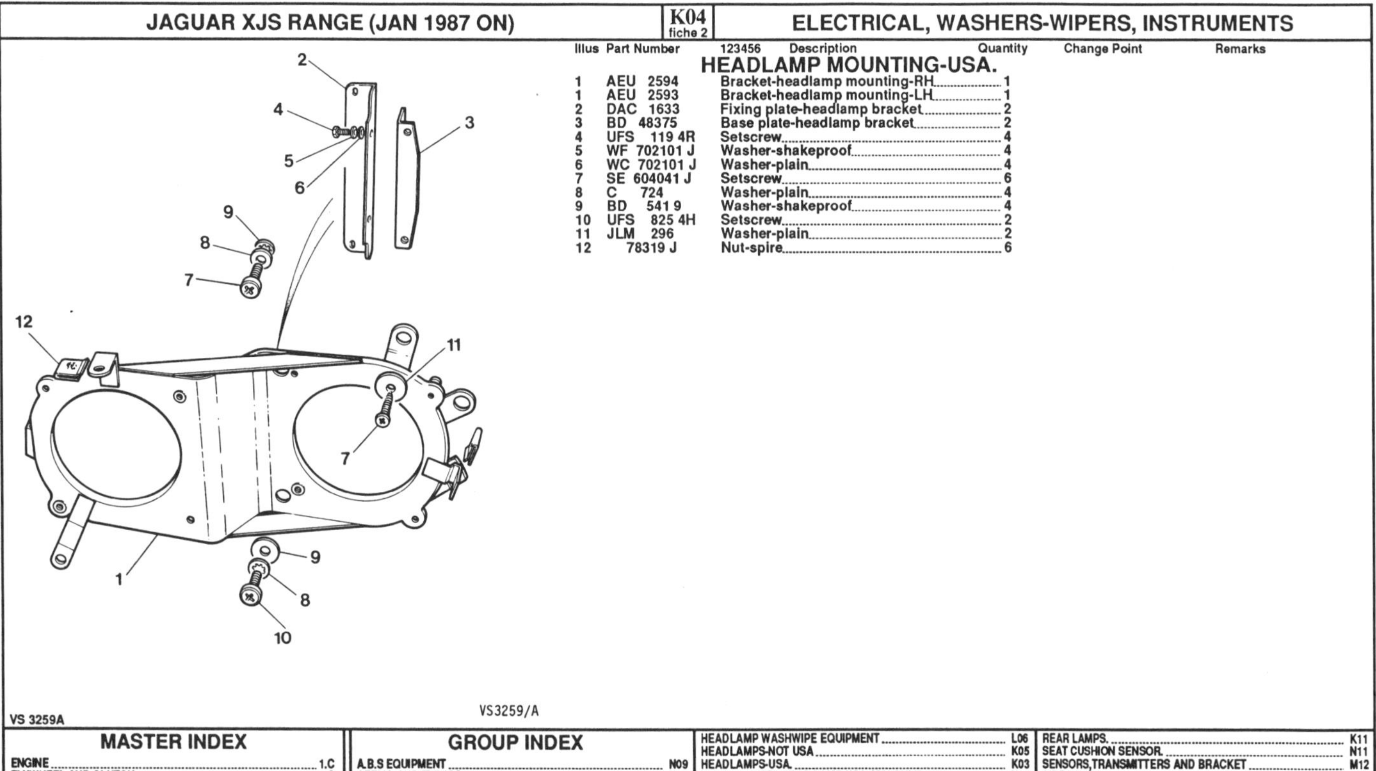

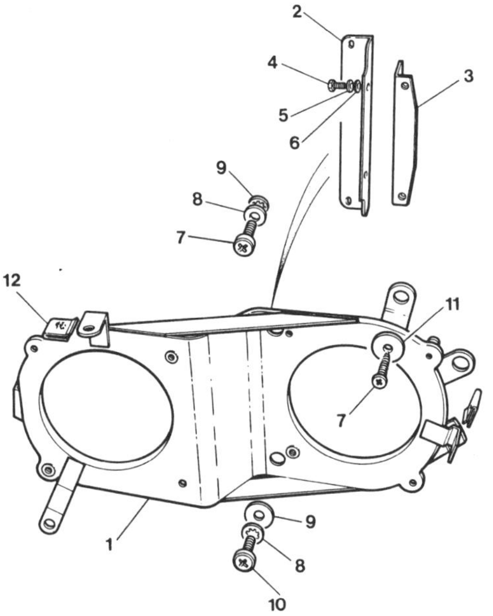

VS3259/A

VS 3259A

MASTER INDEX

ENGINE 1.C
FLYWHEEL AND CLUTCH 1.G
GEARBOX AND PROPSHAFT 1.H
AXLES, SUSPENSION, DRIVE SHAFTS, WHEELS 1.K
STEERING 1.M
BRAKES AND BRAKE CONTROLS 1.N
FUEL, EXHAUST AND EMISSION SYSTEMS 2.C
COOLING, HEATING AND AIR CONDITIONING 2.G
ELECTRICAL, WASHERS-WIPERS, INSTRUMENTS 2.J
CHASSIS, SUBFRAMES, BODYSHELL, FITTINGS 3.C
FASCIA, TRIM, SEATS AND FIXINGS 3.H
ACCESSORIES AND PAINT 4.C
NUMERICAL INDEX 4.D

GROUP INDEX

A.B.S EQUIPMENT N09
AERIAL AND FIXINGS. N12
ALTERNATOR 75 AMP (UP TO (E)109969). J02
BATTERY AND LEADS. N17
CONSOLE SWITCHES AND CIGAR LIGHTER M08
CONTROL UNIT-PJ-V12 COUPE L11
FROM 89 MY L13
CONTROL UNITS-PI/CRUISE CONTROL 3.6LITRE L10
DISTRIBUTOR-3.6 LITRE. J11
DISTRIBUTOR-5.3 LITRE. J12
FROM 89 MY J13
DOOR WARNING LAMP. K15
FOG REAR GUARD LAMP. K14
FRONT FLASHER LAMPS. K06
FRONT FOG LAMPS. K13
FUEL INTERFACE UNIT AND RELAYS L15
HARNESSES-FRONT 3.6 LITRE-FROM (V)148945 .... O04
HARNESSES-FRONT 5.3 LITRE-FROM (V)148782 .... O06
HARNESSES-FRONT-3.6 LITRE-UPTO (V)148944 .... O03
HEADLAMP WASH WIPE RESERVOIR L07
HEADLAMP WASHWIPE EQUIPMENT L06
HEADLAMPS-NOT USA K05
HEADLAMPS-USA. K03
HIGH MOUNTED STOP LAMP-USA/CDN ONLY. K12
HOOD CONTROL MODULE - CONVERTIBLE N10
HORNS AND HORN PAD M05
IGNITION COIL & MODULE-3.6 LITRE J14
IGNITION COILS AND AMPLIFIER-5.3 LITRE J15
INERTIA SWITCH AND COVER M13
INSTRUMENT PACK-3.6 LITRE L17
INSTRUMENT PACK-5.3 LITRE L18
INTERIOR LAMPS-CABRIOLET. K09
INTERIOR LAMPS-COUPE. K08
LOUDSPEAKERS AND MOUNTINGS. N15
MAIN FUSE BLOCK AND MOUNTING PLATE M15
NUMBER PLATE AND REVERSE LAMP. K16
PASSIVE RESTRAINT MODULE & FIXINGS N08
PLUGS AND RELAYS. N07
POWER RESISTOR AND LAMBDA SENSOR. L16
RADIO CASSETTE PLAYER-FROM (V)148782 .... N14
SHOULDER BAG-REMOVEABLE RADIO N14
RADIO'S-UP TO (V)148781 .... N13
REAR LAMPS. K11
SEAT CUSHION SENSOR. N11
SENSORS,TRANSMITTERS AND BRACKET M12
SIDE LAMPS. K07
SPEED CONTROL PUMP & DUMP VALVE O11
STARTER MOTOR - 3.6 LITRE J08
STARTER MOTOR 5.3 LITRE J09
STEERING COLUMN SWITCHES. M04
TRIP COMPUTER AND PANEL SWITCHES. M06
WASH/WIPE RESERVOIR AND FIXINGS L08
WINDSCREEN WASHER RESERVOIR L04
WINDSCREEN WIPER MOTOR-LUCAS TYPE K17
WINDSCREEN WIPERS L03

HEADLAMPS-NOT USA

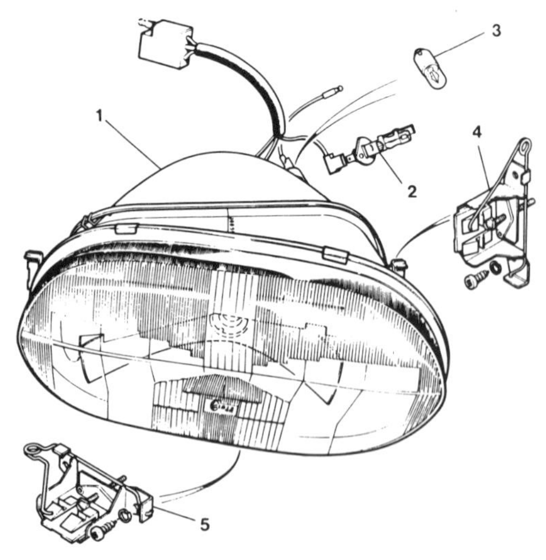

SM 7211A

Illus	Part Number	123456 Description	Quantity	Change Point	Remarks
1	DAC 2500	Headlamp assembly-LH	1		RHD
1	DAC 2501	Headlamp assembly-RH	1		RHD
1	DAC 2507	Headlamp assembly-LH	1		LHD
1	DAC 2508	Headlamp assembly-RH	1		LHD
2	JLM 9599	Bulb-headlamp	2		France only
2	JLM 9597	Bulb-headlamp	2		Japan only
2	JLM 9598	Bulb-headlamp	2		Other markets
3	JLM 9589	Bulb-pilot	2		
4	AAU 1312	Mounting bracket-headlamp	2		
5	AAU 1313	Mounting bracket-headlamp	2		

MASTER INDEX

GROUP INDEX

2K 05

FRONT FLASHER LAMPS.

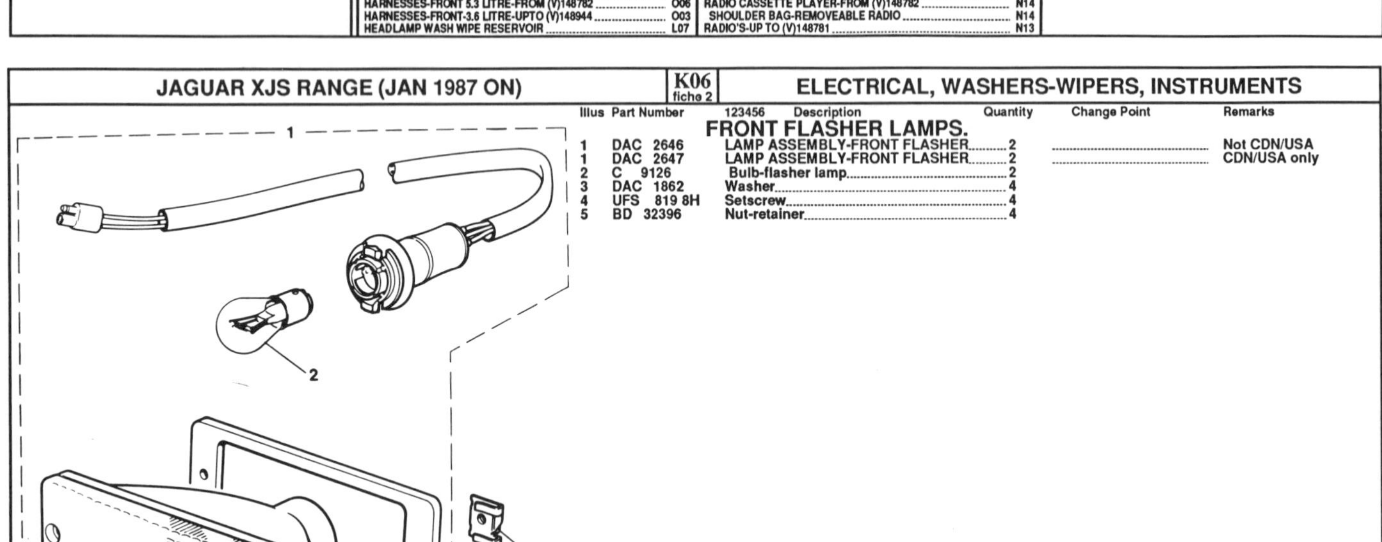

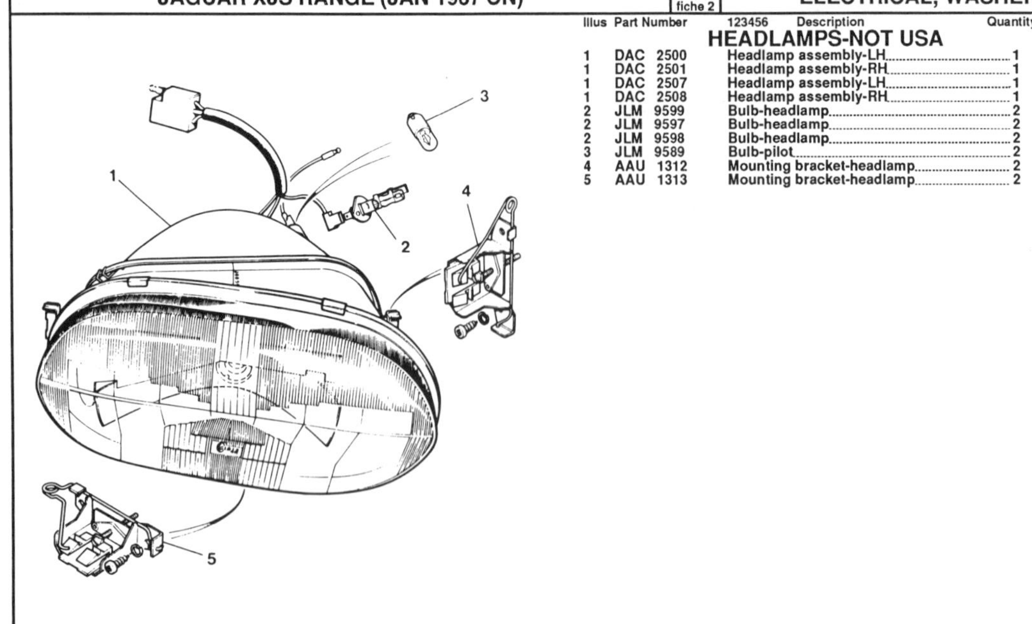

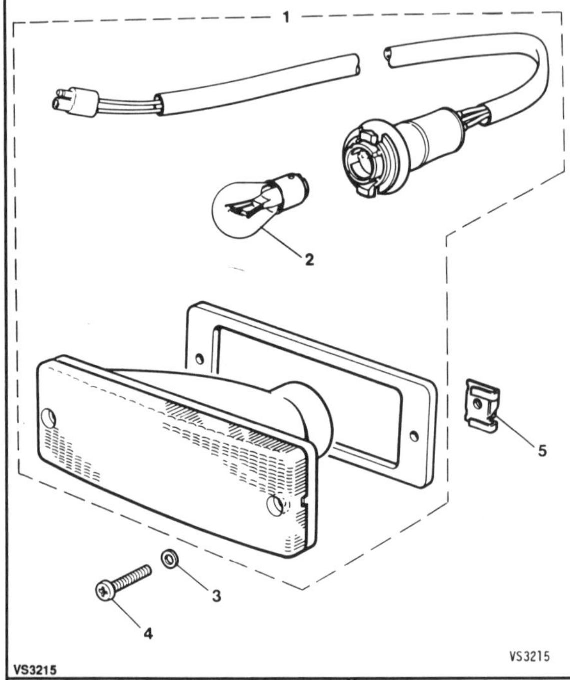

VS3215

Illus	Part Number	123456 Description	Quantity	Change Point	Remarks
1	DAC 2646	LAMP ASSEMBLY-FRONT FLASHER	2		Not CDN/USA
1	DAC 2647	LAMP ASSEMBLY-FRONT FLASHER	2		CDN/USA only
2	C 9126	Bulb-flasher lamp	2		
3	DAC 1862	Washer	4		
4	UFS 819 8H	Setscrew	4		
5	BD 32396	Nut-retainer	4		

MASTER INDEX
ENGINE 1.C
FLYWHEEL AND CLUTCH 1.G
GEARBOX AND PROPSHAFT 1.H
AXLES, SUSPENSION, DRIVE SHAFTS, WHEELS 1.K
STEERING 1.M
BRAKES AND BRAKE CONTROLS 1.N
FUEL, EXHAUST AND EMISSION SYSTEMS 2.C
COOLING, HEATING AND AIR CONDITIONING 2.G
ELECTRICAL, WASHERS-WIPERS, INSTRUMENTS 2.J
CHASSIS, SUBFRAMES, BODYSHELL, FITTINGS 3.C
FASCIA, TRIM, SEATS AND FIXINGS 3.H
ACCESSORIES AND PAINT 4.C
NUMERICAL INDEX 4.D
GROUP INDEX
A.B.S EQUIPMENT N09
AERIAL AND FIXINGS N12
ALTERNATOR 75 AMP (UP TO (E)109969) J02
BATTERY AND LEADS N17
CONSOLE SWITCHES AND CIGAR LIGHTER M08
CONTROL UNIT-PJ-V12 COUPE L11
FROM 89 MY L13
CONTROL UNITS-PI/CRUISE CONTROL 3.6LITRE L10
DISTRIBUTOR-3.6 LITRE J11
DISTRIBUTOR-5.3 LITRE J12
FROM 89 MY J13
DOOR WARNING LAMP K15
FOG REAR GUARD LAMP K14
FRONT FLASHER LAMPS K06
FRONT FOG LAMPS K13
FUEL INTERFACE UNIT AND RELAYS L15
HARNESSES-FRONT 3.6 LITRE-FROM (V)148945 O04
HARNESSES-FRONT 5.3 LITRE-FROM (V)148782 O06
HARNESSES-FRONT-3.6 LITRE-UPTO (V)148944 O03
HEADLAMP WASH WIPE RESERVOIR L07
HEADLAMP WASHWIPE EQUIPMENT L06
HEADLAMPS-NOT USA K05
HEADLAMPS-USA K03
HIGH MOUNTED STOP LAMP-USA/CDN ONLY K12
HOOD CONTROL MODULE - CONVERTIBLE N10
HORNS AND HORN PAD M05
IGNITION COIL & MODULE-3.6 LITRE J14
IGNITION COILS AND AMPLIFIER-5.3 LITRE J15
INERTIA SWITCH AND COVER M13
INSTRUMENT PACK-3.6 LITRE L17
INSTRUMENT PACK-5.3 LITRE L18
INTERIOR LAMPS-CABRIOLET K09
INTERIOR LAMPS-COUPE K08
LOUDSPEAKERS AND MOUNTINGS N15
MAIN FUSE BLOCK AND MOUNTING PLATE M15
NUMBER PLATE AND REVERSE LAMP K16
PASSIVE RESTRAINT MODULE & FIXINGS N08
PLUGS AND RELAYS N07
POWER RESISTOR AND LAMBDA SENSOR L16
RADIO CASSETTE PLAYER-FROM (V)148782 N14
SHOULDER BAG-REMOVEABLE RADIO N14
RADIO'S-UP TO (V)148781 N13
REAR LAMPS K11
SEAT CUSHION SENSOR N11
SENSORS,TRANSMITTERS AND BRACKET M12
SIDE LAMPS K07
SPEED CONTROL PUMP & DUMP VALVE O11
STARTER MOTOR - 3.6 LITRE J08
STARTER MOTOR 5.3 LITRE J09
STEERING COLUMN SWITCHES M04
TRIP COMPUTER AND PANEL SWITCHES M06
WASH/WIPE RESERVOIR AND FIXINGS L08
WINDSCREEN WASHER RESERVOIR L04
WINDSCREEN WIPER MOTOR-LUCAS TYPE K17
WINDSCREEN WIPERS L03

2K 06

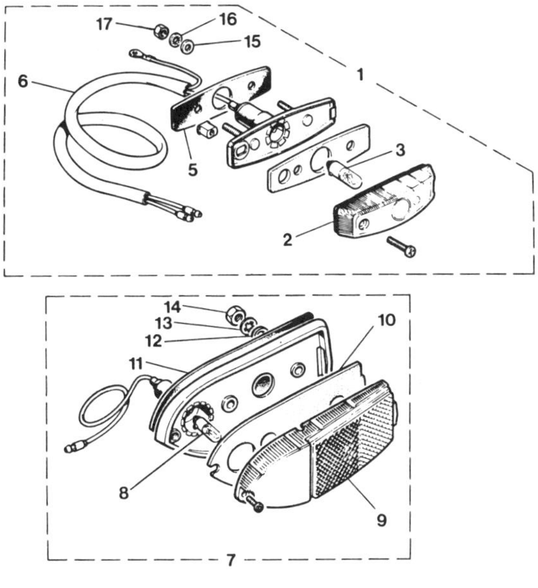

VS1892A

VS 1892A

SIDE LAMPS.

Illus	Part Number	123456	Description	Quantity	Change Point	Remarks
1	DAC 1373		LAMP ASSEMBLY-FLASHER SIDE REPEATER	NLA		Use DAC 4820
1	DAC 4820		LAMP ASSEMBLY-FLASHER SIDE REPEATER	2		Not CDN/USA
2	27H 2403 J		Lens	2		
3	JLM 9589		Bulb	2		
4	608311 J		Gasket	2		
5	608312 J		Gasket	2		
6	11356		Bulb holder	2		
7	C 40091		Lamp-sidemarker-RH	1		CDN/USA only
7	C 40092		Lamp-sidemarker-LH	1		CDN/USA only
8	JLM 9589		Bulb	2		
9	11727		Lens-RH	1		
9	11730		Lens-LH	1		
10	11728		Gasket	2		
11	BD 46717		Gasket-lamp to body-RH	1		
11	BD 46718		Gasket-lamp to body-LH	1		
12	WC 702101 J		Washer-plain	4		
13	WF 702100 J		Washer-shakeproof	4		
14	UFN 119 L		Nut	4		
15	WC 704061 J		Washer-plain	4		
16	WF 704060 J		Washer-shakeproof	4		
17	UCN 113 L		Nut	4		

MASTER INDEX

GROUP INDEX

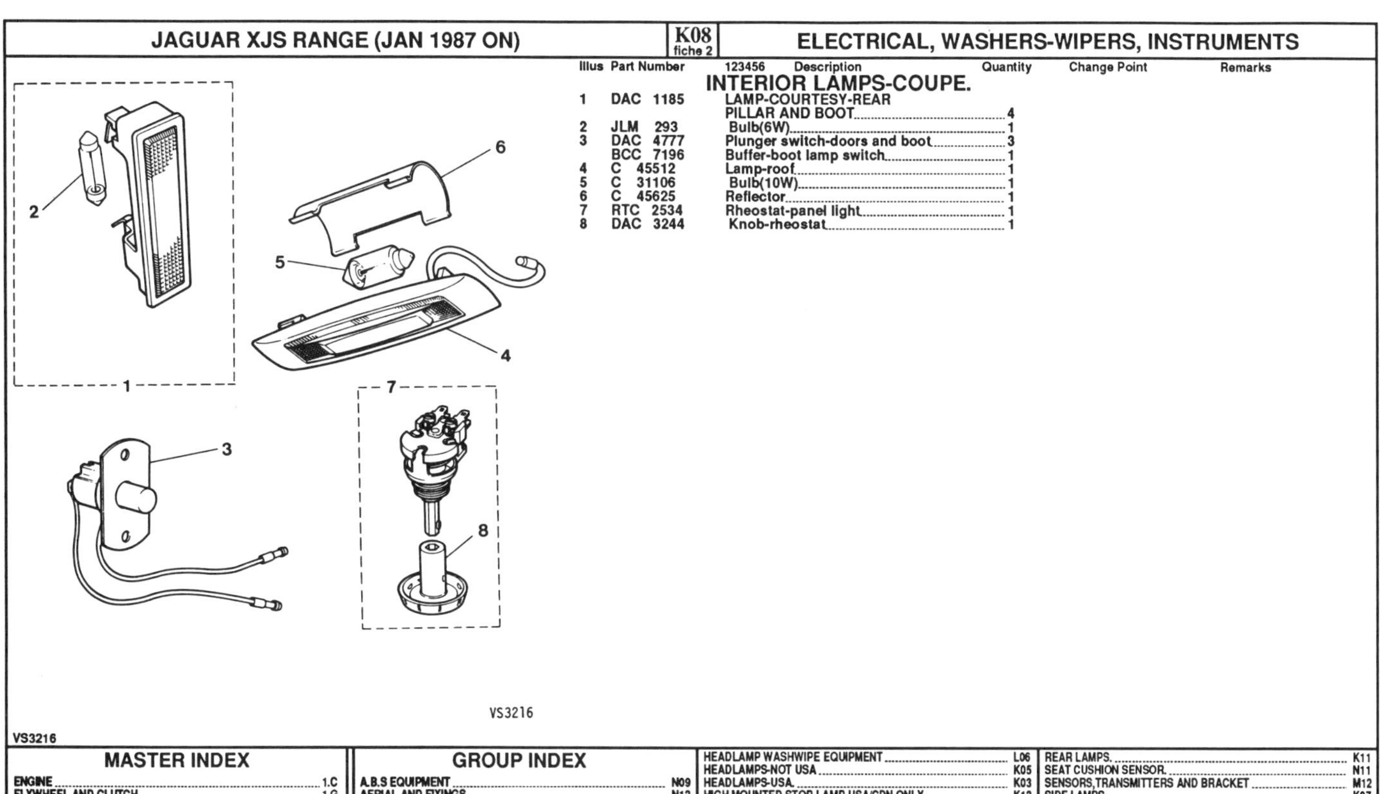

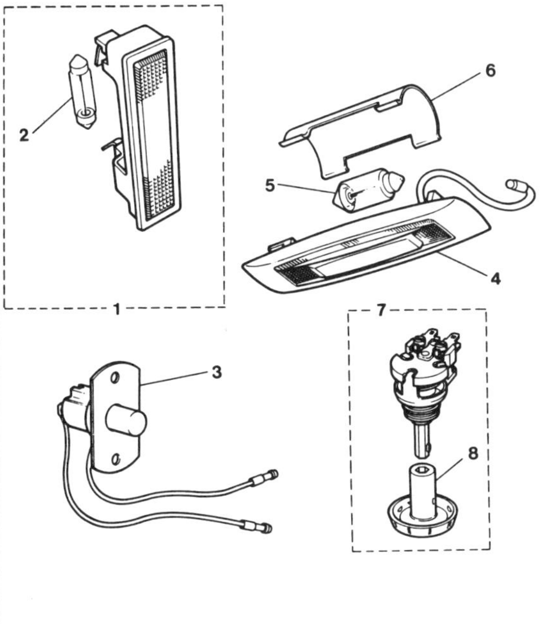

VS3216

VS3216

INTERIOR LAMPS-COUPE.

Illus	Part Number	123456	Description	Quantity	Change Point	Remarks
1	DAC 1185		LAMP-COURTESY-REAR PILLAR AND BOOT	4		
2	JLM 293		Bulb(6W)	1		
3	DAC 4777		Plunger switch-doors and boot	3		
	BCC 7196		Buffer-boot lamp switch	1		
4	C 45512		Lamp-roof	1		
5	C 31106		Bulb(10W)	1		
6	C 45625		Reflector	1		
7	RTC 2534		Rheostat-panel light	1		
8	DAC 3244		Knob-rheostat	1		

MASTER INDEX
ENGINE 1.C
FLYWHEEL AND CLUTCH 1.G
GEARBOX AND PROPSHAFT 1.H
AXLES, SUSPENSION, DRIVE SHAFTS, WHEELS 1.K
STEERING 1.M
BRAKES AND BRAKE CONTROLS 1.N
FUEL, EXHAUST AND EMISSION SYSTEMS 2.C
COOLING, HEATING AND AIR CONDITIONING 2.G
ELECTRICAL, WASHERS-WIPERS, INSTRUMENTS 2.J
CHASSIS, SUBFRAMES, BODYSHELL, FITTINGS 3.C
FASCIA, TRIM, SEATS AND FIXINGS 3.H
ACCESSORIES AND PAINT 4.C
NUMERICAL INDEX 4.D
GROUP INDEX
A.B.S EQUIPMENT N09
AERIAL AND FIXINGS N12
ALTERNATOR 75 AMP (UP TO (E)109969) J02
BATTERY AND LEADS N17
CONSOLE SWITCHES AND CIGAR LIGHTER M08
CONTROL UNIT-PJ-V12 COUPE L11
FROM 89 MY L13
CONTROL UNITS-PI/CRUISE CONTROL 3.6LITRE L10
DISTRIBUTOR-3.6 LITRE J11
DISTRIBUTOR-5.3 LITRE J12
FROM 89 MY J13
DOOR WARNING LAMP K15
FOG REAR GUARD LAMP K14
FRONT FLASHER LAMPS K06
FRONT FOG LAMPS K13
FUEL INTERFACE UNIT AND RELAYS L15
HARNESSES-FRONT 3.6 LITRE-FROM (V)148945 O04
HARNESSES-FRONT 5.3 LITRE-FROM (V)148782 O06
HARNESSES-FRONT-3.6 LITRE-UPTO (V)148944 O03
HEADLAMP WASH WIPE RESERVOIR L07
HEADLAMP WASHWIPE EQUIPMENT L06
HEADLAMPS-NOT USA K05
HEADLAMPS-USA K03
HIGH MOUNTED STOP LAMP-USA/CDN ONLY K12
HOOD CONTROL MODULE - CONVERTIBLE N10
HORNS AND HORN PAD M05
IGNITION COIL & MODULE-3.6 LITRE J14
IGNITION COILS AND AMPLIFIER-5.3 LITRE J15
INERTIA SWITCH AND COVER M13
INSTRUMENT PACK-3.6 LITRE L17
INSTRUMENT PACK-5.3 LITRE L18
INTERIOR LAMPS-CABRIOLET K09
INTERIOR LAMPS-COUPE K08
LOUDSPEAKERS AND MOUNTINGS N15
MAIN FUSE BLOCK AND MOUNTING PLATE M15
NUMBER PLATE AND REVERSE LAMP K16
PASSIVE RESTRAINT MODULE & FIXINGS N08
PLUGS AND RELAYS N07
POWER RESISTOR AND LAMBDA SENSOR L16
RADIO CASSETTE PLAYER-FROM (V)148782 N14
SHOULDER BAG-REMOVEABLE RADIO N14
RADIO'S-UP TO (V)148781 N13
REAR LAMPS K11
SEAT CUSHION SENSOR N11
SENSORS,TRANSMITTERS AND BRACKET M12
SIDE LAMPS K07
SPEED CONTROL PUMP & DUMP VALVE O11
STARTER MOTOR - 3.6 LITRE J08
STARTER MOTOR 5.3 LITRE J09
STEERING COLUMN SWITCHES M04
TRIP COMPUTER AND PANEL SWITCHES M06
WASH/WIPE RESERVOIR AND FIXINGS L08
WINDSCREEN WASHER RESERVOIR L04
WINDSCREEN WIPER MOTOR-LUCAS TYPE K17
WINDSCREEN WIPERS L03

2K 09

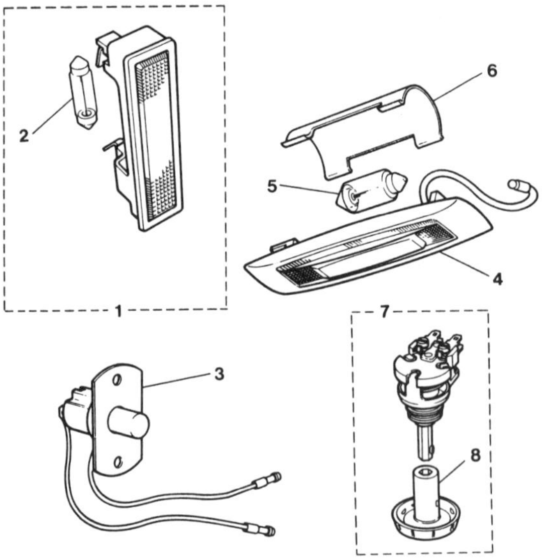

VS 3216

INTERIOR LAMPS-CABRIOLET.

Illus	Part Number	Description	Quantity	Change Point	Remarks
1	DAC 1185	Lamp-courtesy and boot	2		
2	JLM 293	Bulb (6W)	2		
3	DAC 4777	Plunger switch-doors and boot	3		
	BCC 7196	Buffer-boot lamp switch	1		
4	C 45512	Lamp-roof and rear quarter	3		
5	C 31106	Bulb-(10W)-roof	1		
	JLM 9590	Bulb-(5W)-rear quarter	2		
6	C 45625	Reflector	3		
7	RTC 2534	Rheostat-panel light	1		
8	DAC 3244	Knob-rheostat			

2K 10

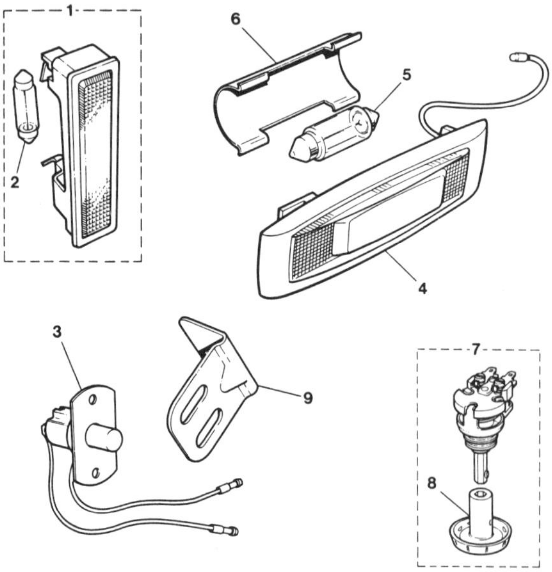

VS 4799

INTERIOR LAMPS-CONVERTIBLE.

Illus	Part Number	Description	Quantity	Change Point	Remarks
1	DAC 1185	Lamp-boot	2		
2	JLM 293	Bulb (6w)	2		
3	DAC 4777	Plunger switch-doors, boot and heated backlight disable	4		
	BCC 7196	Buffer-boot lamp switch	1		
4	DAC 2660	Lamp-rear quarter	2		
5	JLM 9591	Bulb (10w)	2		
6	C 45625	Reflector	2		
7	RTC 2534	Rheostat-panel light	1		
8	DAC 3244	Knob-rheostat	1		
9	BCC 8003	Striker-backlight disable switch	1		

MASTER INDEX

ENGINE 1.C
FLYWHEEL AND CLUTCH 1.G
GEARBOX AND PROPSHAFT 1.H
AXLES, SUSPENSION, DRIVE SHAFTS, WHEELS 1.K
STEERING 1.M
BRAKES AND BRAKE CONTROLS 1.N
FUEL, EXHAUST AND EMISSION SYSTEMS 2.C
COOLING, HEATING AND AIR CONDITIONING 2.G
ELECTRICAL, WASHERS-WIPERS, INSTRUMENTS 2.J
CHASSIS, SUBFRAMES, BODYSHELL, FITTINGS 3.C
FASCIA, TRIM, SEATS AND FIXINGS 3.H
ACCESSORIES AND PAINT 4.C
NUMERICAL INDEX 4.D

GROUP INDEX

A.B.S EQUIPMENT N09
AERIAL AND FIXINGS. N12
ALTERNATOR 75 AMP (UP TO (E)109969). J02
BATTERY AND LEADS. N17
CONSOLE SWITCHES AND CIGAR LIGHTER M08
CONTROL UNIT-PJ-V12 COUPE L11
FROM 89 MY L13
CONTROL UNITS-PI/CRUISE CONTROL 3.6LITRE L10
DISTRIBUTOR-3.6 LITRE. J11
DISTRIBUTOR-5.3 LITRE. J12
FROM 89 MY J13
DOOR WARNING LAMP. K15
FOG REAR GUARD LAMP. K14
FRONT FLASHER LAMPS. K06
FRONT FOG LAMPS. K13
FUEL INTERFACE UNIT AND RELAYS L15
HARNESSES-FRONT 3.6 LITRE-FROM (V)148945 .... O04
HARNESSES-FRONT 5.3 LITRE-FROM (V)148782 .... O06
HARNESSES-FRONT-3.6 LITRE-UPTO (V)148944 .... O03
HEADLAMP WASH WIPE RESERVOIR L07
HEADLAMP WASHWIPE EQUIPMENT L06
HEADLAMPS-NOT USA K05
HEADLAMPS-USA. K03
HIGH MOUNTED STOP LAMP-USA/CDN ONLY. K12
HOOD CONTROL MODULE - CONVERTIBLE N10
HORNS AND HORN PAD M05
IGNITION COIL & MODULE-3.6 LITRE J14
IGNITION COILS AND AMPLIFIER-5.3 LITRE J15
INERTIA SWITCH AND COVER M13
INSTRUMENT PACK-3.6 LITRE L17
INSTRUMENT PACK-5.3 LITRE L18
INTERIOR LAMPS-CABRIOLET. K09
INTERIOR LAMPS-COUPE. K08
LOUDSPEAKERS AND MOUNTINGS. N15
MAIN FUSE BLOCK AND MOUNTING PLATE M15
NUMBER PLATE AND REVERSE LAMP. K16
PASSIVE RESTRAINT MODULE & FIXINGS N08
PLUGS AND RELAYS. N07
POWER RESISTOR AND LAMBDA SENSOR. L16
RADIO CASSETTE PLAYER-FROM (V)148782 .... N14
SHOULDER BAG-REMOVEABLE RADIO N14
RADIO'S-UP TO (V)148781 .... N13
REAR LAMPS. K11
SEAT CUSHION SENSOR. N11
SENSORS,TRANSMITTERS AND BRACKET M12
SIDE LAMPS. K07
SPEED CONTROL PUMP & DUMP VALVE O11
STARTER MOTOR - 3.6 LITRE J08
STARTER MOTOR 5.3 LITRE J09
STEERING COLUMN SWITCHES. M04
TRIP COMPUTER AND PANEL SWITCHES. M06
WASH/WIPE RESERVOIR AND FIXINGS L08
WINDSCREEN WASHER RESERVOIR L04
WINDSCREEN WIPER MOTOR-LUCAS TYPE K17
WINDSCREEN WIPERS L03

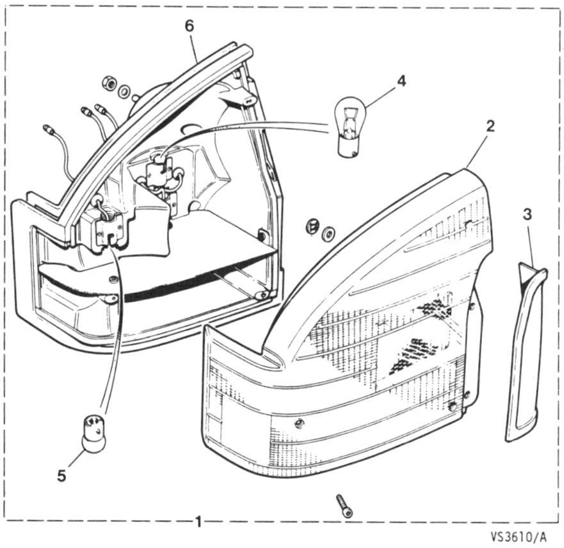

VS3610A

REAR LAMPS.

Illus	Part Number	123456 Description	Quantity	Change Point	Remarks
1	DAC 3800	Lamp-stop/tail/flasher-RH	1		Not CDN/USA
1	DAC 3801	Lamp-stop/tail/flasher-LH	1		
	DAC 3802	Lamp-stop/tail/flasher-RH	1		CDN/USA
1	DAC 3803	Lamp-stop/tail/flasher-LH	1		
	AEU 2556	Lens-RH	NLA		Use JLM 1551
2	AEU 2555	Lens-LH	NLA		Use JLM 1552
	AEU 2554	Lens-RH	NLA		Use JLM 1553
2	AEU 2553	Lens-LH	NLA		Use JLM 1554
	JLM 1551	Lens-RH	1		Not CDN/USA
2	JLM 1552	Lens-LH	1		
	JLM 1553	Lens-RH	1		CDN/USA
2	JLM 1554	Lens-LH	1		
	JLM 1549	Infill-chrome RH	1		
3	JLM 1550	Infill-chrome LH	1		
4	C 9126	Bulb	2		
5	JLM 9587	Bulb	2		
6	RTC 1208	Gasket-lamp to body-RH	1		
6	RTC 1209	Gasket-lamp to body-LH	1		

MASTER INDEX

GROUP INDEX

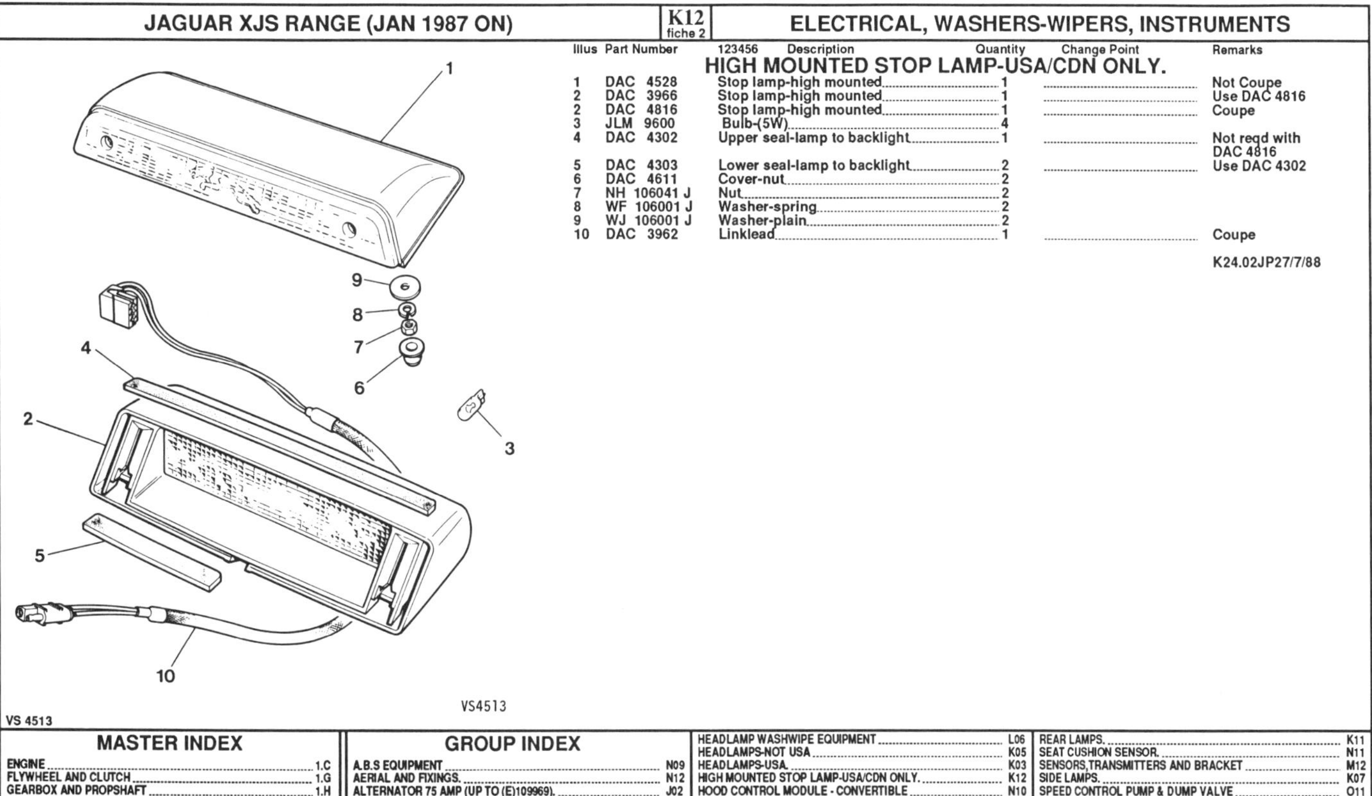

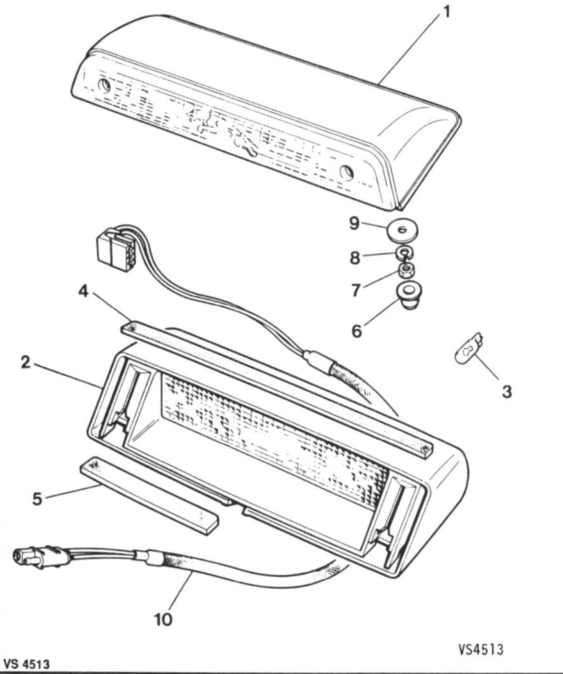

VS 4513

HIGH MOUNTED STOP LAMP-USA/CDN ONLY.

Illus	Part Number	123456 Description	Quantity	Change Point	Remarks
1	DAC 4528	Stop lamp-high mounted	1		Not Coupe
2	DAC 3966	Stop lamp-high mounted	1		Use DAC 4816
2	DAC 4816	Stop lamp-high mounted	1		Coupe
3	JLM 9600	Bulb-(5W)	4		
4	DAC 4302	Upper seal-lamp to backlight	1		Not reqd with DAC 4816
5	DAC 4303	Lower seal-lamp to backlight	2		Use DAC 4302
6	DAC 4611	Cover-nut	2		
7	NH 106041 J	Nut	2		
8	WF 106001 J	Washer-spring	2		
9	WJ 106001 J	Washer-plain	2		
10	DAC 3962	Linklead	1		Coupe

K24.02JP27/7/88

MASTER INDEX
ENGINE ... 1.C
FLYWHEEL AND CLUTCH ... 1.G
GEARBOX AND PROPSHAFT ... 1.H
AXLES, SUSPENSION, DRIVE SHAFTS, WHEELS ... 1.K
STEERING ... 1.M
BRAKES AND BRAKE CONTROLS ... 1.N
FUEL, EXHAUST AND EMISSION SYSTEMS ... 2.C
COOLING, HEATING AND AIR CONDITIONING ... 2.G
ELECTRICAL, WASHERS-WIPERS, INSTRUMENTS ... 2.J
CHASSIS, SUBFRAMES, BODYSHELL, FITTINGS ... 3.C
FASCIA, TRIM, SEATS AND FIXINGS ... 3.H
ACCESSORIES AND PAINT ... 4.C
NUMERICAL INDEX ... 4.D
GROUP INDEX
A.B.S EQUIPMENT ... N09
AERIAL AND FIXINGS. ... N12
ALTERNATOR 75 AMP (UP TO (E)109969). ... J02
BATTERY AND LEADS. ... N17
CONSOLE SWITCHES AND CIGAR LIGHTER ... M08
CONTROL UNIT-PJ-V12 COUPE ... L11
FROM 89 MY ... L13
CONTROL UNITS-PI/CRUISE CONTROL 3.6LITRE ... L10
DISTRIBUTOR-3.6 LITRE. ... J11
DISTRIBUTOR-5.3 LITRE. ... J12
FROM 89 MY ... J13
DOOR WARNING LAMP. ... K15
FOG REAR GUARD LAMP. ... K14
FRONT FLASHER LAMPS. ... K06
FRONT FOG LAMPS. ... K13
FUEL INTERFACE UNIT AND RELAYS ... L15
HARNESSES-FRONT 3.6 LITRE-FROM (V)148945 ... O04
HARNESSES-FRONT 5.3 LITRE-FROM (V)148782 ... O06
HARNESSES-FRONT-3.6 LITRE-UPTO (V)148944 ... O03
HEADLAMP WASH WIPE RESERVOIR ... L07
HEADLAMP WASHWIPE EQUIPMENT ... L06
HEADLAMPS-NOT USA ... K05
HEADLAMPS-USA. ... K03
HIGH MOUNTED STOP LAMP-USA/CDN ONLY. ... K12
HOOD CONTROL MODULE - CONVERTIBLE ... N10
HORNS AND HORN PAD ... M05
IGNITION COIL & MODULE-3.6 LITRE ... J14
IGNITION COILS AND AMPLIFIER-5.3 LITRE ... J15
INERTIA SWITCH AND COVER ... M13
INSTRUMENT PACK-3.6 LITRE ... L17
INSTRUMENT PACK-5.3 LITRE ... L18
INTERIOR LAMPS-CABRIOLET. ... K09
INTERIOR LAMPS-COUPE. ... K08
LOUDSPEAKERS AND MOUNTINGS. ... N15
MAIN FUSE BLOCK AND MOUNTING PLATE ... M15
NUMBER PLATE AND REVERSE LAMP. ... K16
PASSIVE RESTRAINT MODULE & FIXINGS ... N08
PLUGS AND RELAYS. ... N07
POWER RESISTOR AND LAMBDA SENSOR. ... L16
RADIO CASSETTE PLAYER-FROM (V)148782 ... N14
SHOULDER BAG-REMOVEABLE RADIO ... N14
RADIO'S-UP TO (V)148781 ... N13
REAR LAMPS. ... K11
SEAT CUSHION SENSOR. ... N11
SENSORS,TRANSMITTERS AND BRACKET ... M12
SIDE LAMPS. ... K07
SPEED CONTROL PUMP & DUMP VALVE ... O11
STARTER MOTOR - 3.6 LITRE ... J08
STARTER MOTOR 5.3 LITRE ... J09
STEERING COLUMN SWITCHES. ... M04
TRIP COMPUTER AND PANEL SWITCHES. ... M06
WASH/WIPE RESERVOIR AND FIXINGS ... L08
WINDSCREEN WASHER RESERVOIR ... L04
WINDSCREEN WIPER MOTOR-LUCAS TYPE ... K17
WINDSCREEN WIPERS ... L03

2K 13

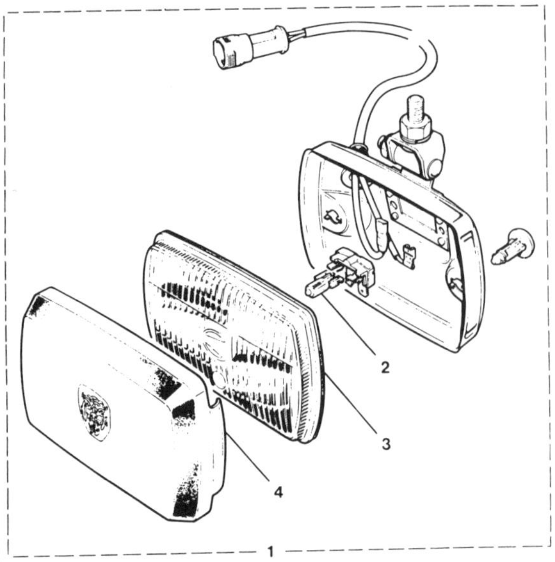

VS4747A

FRONT FOG LAMPS.

Illus	Part Number	Description	Quantity	Change Point	Remarks
1	DAC 4241	Foglamp-front	2		White
	DAC 4242	Foglamp-front	2		Yellow
2	JLM 9588	Bulb-front foglamp	2		
3	JLM 1378	Lens/reflector assembly	2		White
3	JLM 1379	Lens/reflector assembly	2		Yellow
4	JLM 1410	Cover-lens	2		

MASTER INDEX

GROUP INDEX

2K 14

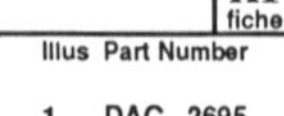

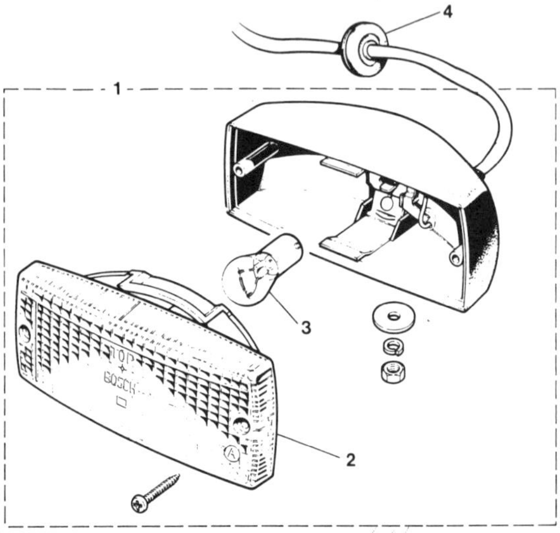

VS 1927

FOG REAR GUARD LAMP.

Illus	Part Number	Description	Quantity	Change Point	Remarks
1	DAC 2695	LAMP ASSY-FOG REAR GUARD	2		Lucas type
1	DAC 1855	LAMP ASSY-FOG REAR GUARD	2		Bosch type
2	AEU 4004	Lens	1		Lucas type
2	AEU 1590	Lens	1		Bosch type
3	C 9126	Bulb	1		
4	C 18066	Grommet	2		

MASTER INDEX
ENGINE ... 1.C
FLYWHEEL AND CLUTCH ... 1.G
GEARBOX AND PROPSHAFT ... 1.H
AXLES, SUSPENSION, DRIVE SHAFTS, WHEELS ... 1.K
STEERING ... 1.M
BRAKES AND BRAKE CONTROLS ... 1.N
FUEL, EXHAUST AND EMISSION SYSTEMS ... 2.C
COOLING, HEATING AND AIR CONDITIONING ... 2.G
ELECTRICAL, WASHERS-WIPERS, INSTRUMENTS ... 2.J
CHASSIS, SUBFRAMES, BODYSHELL, FITTINGS ... 3.C
FASCIA, TRIM, SEATS AND FIXINGS ... 3.H
ACCESSORIES AND PAINT ... 4.C
NUMERICAL INDEX ... 4.D
GROUP INDEX
A.B.S EQUIPMENT ... N09
AERIAL AND FIXINGS. ... N12
ALTERNATOR 75 AMP (UP TO (E)109969). ... J02
BATTERY AND LEADS. ... N17
CONSOLE SWITCHES AND CIGAR LIGHTER ... M08
CONTROL UNIT-PJ-V12 COUPE ... L11
FROM 89 MY ... L13
CONTROL UNITS-PI/CRUISE CONTROL 3.6LITRE ... L10
DISTRIBUTOR-3.6 LITRE. ... J11
DISTRIBUTOR-5.3 LITRE. ... J12
FROM 89 MY ... J13
DOOR WARNING LAMP. ... K15
FOG REAR GUARD LAMP. ... K14
FRONT FLASHER LAMPS. ... K06
FRONT FOG LAMPS. ... K13
FUEL INTERFACE UNIT AND RELAYS ... L15
HARNESSES-FRONT 3.6 LITRE-FROM (V)148945 ... O04
HARNESSES-FRONT 5.3 LITRE-FROM (V)148782 ... O06
HARNESSES-FRONT-3.6 LITRE-UPTO (V)148944 ... O03
HEADLAMP WASH WIPE RESERVOIR ... L07
HEADLAMP WASHWIPE EQUIPMENT ... L06
HEADLAMPS-NOT USA ... K05
HEADLAMPS-USA. ... K03
HIGH MOUNTED STOP LAMP-USA/CDN ONLY. ... K12
HOOD CONTROL MODULE - CONVERTIBLE ... N10
HORNS AND HORN PAD ... M05
IGNITION COIL & MODULE-3.6 LITRE ... J14
IGNITION COILS AND AMPLIFIER-5.3 LITRE ... J15
INERTIA SWITCH AND COVER ... M13
INSTRUMENT PACK-3.6 LITRE ... L17
INSTRUMENT PACK-5.3 LITRE ... L18
INTERIOR LAMPS-CABRIOLET. ... K09
INTERIOR LAMPS-COUPE. ... K08
LOUDSPEAKERS AND MOUNTINGS. ... N15
MAIN FUSE BLOCK AND MOUNTING PLATE ... M15
NUMBER PLATE AND REVERSE LAMP. ... K16
PASSIVE RESTRAINT MODULE & FIXINGS ... N08
PLUGS AND RELAYS. ... N07
POWER RESISTOR AND LAMBDA SENSOR. ... L16
RADIO CASSETTE PLAYER-FROM (V)148782 ... N14
SHOULDER BAG-REMOVEABLE RADIO ... N14
RADIO'S-UP TO (V)148781 ... N13
REAR LAMPS. ... K11
SEAT CUSHION SENSOR. ... N11
SENSORS,TRANSMITTERS AND BRACKET ... M12
SIDE LAMPS. ... K07
SPEED CONTROL PUMP & DUMP VALVE ... O11
STARTER MOTOR - 3.6 LITRE ... J08
STARTER MOTOR 5.3 LITRE ... J09
STEERING COLUMN SWITCHES. ... M04
TRIP COMPUTER AND PANEL SWITCHES. ... M06
WASH/WIPE RESERVOIR AND FIXINGS ... L08
WINDSCREEN WASHER RESERVOIR ... L04
WINDSCREEN WIPER MOTOR-LUCAS TYPE ... K17
WINDSCREEN WIPERS ... L03

DOOR WARNING LAMP.

Illus	Part Number	123456 Description	Quantity	Change Point	Remarks
1	150653	LAMP ASSY-DOOR WARNING-RH	1		
1	150654	LAMP ASSY-DOOR WARNING-LH	1		
2	519542	Lens assembly-RH	1		
2	519544	Lens assembly-LH	1		
3	JLM 9600	Bulb	1		

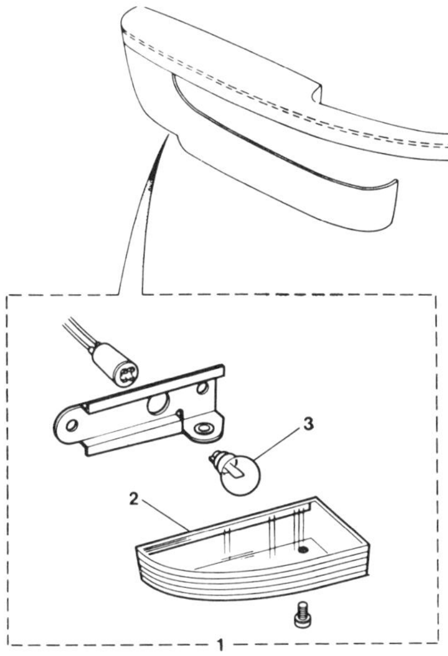

VS 3246

VS3246

MASTER INDEX

GROUP INDEX

NUMBER PLATE AND REVERSE LAMP.

Illus	Part Number	123456 Description	Quantity	Change Point	Remarks
1	DAC 2090	Lamp-number plate/reverse	1		
2	AAU 2043	Hinge assy	2		
3	AAU 2044	Lens-reverse lamp-RH	1		
4	AAU 2045	Lens-reverse lamp-LH	1		
5	JLM 293	Bulb	2		
6	JLM 9592	Bulb	2		
7	C 17432	Grommet	2		
8	GT 604071	Setscrew	6		
9	WF 702101 J	Washer-plain	6		
10	UFN 119 L	Nut	6		

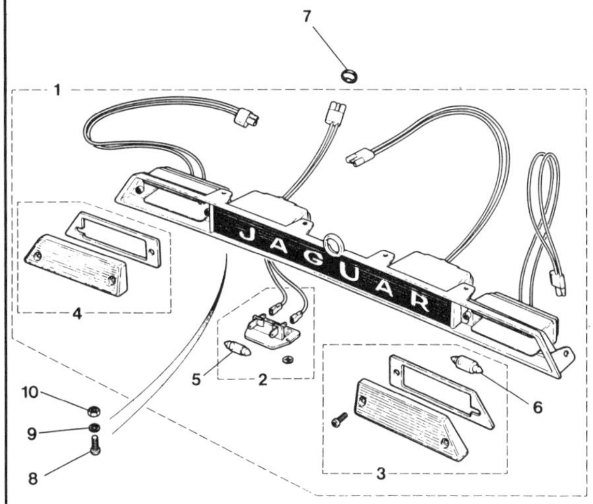

VS3228/A

VS 3228A

MASTER INDEX
ENGINE 1.C
FLYWHEEL AND CLUTCH 1.G
GEARBOX AND PROPSHAFT 1.H
AXLES, SUSPENSION, DRIVE SHAFTS, WHEELS 1.K
STEERING 1.M
BRAKES AND BRAKE CONTROLS 1.N
FUEL, EXHAUST AND EMISSION SYSTEMS 2.C
COOLING, HEATING AND AIR CONDITIONING 2.G
ELECTRICAL, WASHERS-WIPERS, INSTRUMENTS 2.J
CHASSIS, SUBFRAMES, BODYSHELL, FITTINGS 3.C
FASCIA, TRIM, SEATS AND FIXINGS 3.H
ACCESSORIES AND PAINT 4.C
NUMERICAL INDEX 4.D
GROUP INDEX
A.B.S EQUIPMENT N09
AERIAL AND FIXINGS. N12
ALTERNATOR 75 AMP (UP TO (E)109969). J02
BATTERY AND LEADS. N17
CONSOLE SWITCHES AND CIGAR LIGHTER M08
CONTROL UNIT-PJ-V12 COUPE L11
FROM 89 MY L13
CONTROL UNITS-PI/CRUISE CONTROL 3.6LITRE L10
DISTRIBUTOR-3.6 LITRE. J11
DISTRIBUTOR-5.3 LITRE. J12
FROM 89 MY J13
DOOR WARNING LAMP. K15
FOG REAR GUARD LAMP. K14
FRONT FLASHER LAMPS. K06
FRONT FOG LAMPS. K13
FUEL INTERFACE UNIT AND RELAYS L15
HARNESSES-FRONT 3.6 LITRE-FROM (V)148945 .... O04
HARNESSES-FRONT 5.3 LITRE-FROM (V)148782 .... O06
HARNESSES-FRONT-3.6 LITRE-UPTO (V)148944 .... O03
HEADLAMP WASH WIPE RESERVOIR L07
HEADLAMP WASHWIPE EQUIPMENT L06
HEADLAMPS-NOT USA K05
HEADLAMPS-USA. K03
HIGH MOUNTED STOP LAMP-USA/CDN ONLY. K12
HOOD CONTROL MODULE - CONVERTIBLE N10
HORNS AND HORN PAD M05
IGNITION COIL & MODULE-3.6 LITRE J14
IGNITION COILS AND AMPLIFIER-5.3 LITRE J15
INERTIA SWITCH AND COVER M13
INSTRUMENT PACK-3.6 LITRE L17
INSTRUMENT PACK-5.3 LITRE L18
INTERIOR LAMPS-CABRIOLET. K09
INTERIOR LAMPS-COUPE. K08
LOUDSPEAKERS AND MOUNTINGS. N15
MAIN FUSE BLOCK AND MOUNTING PLATE M15
NUMBER PLATE AND REVERSE LAMP. K16
PASSIVE RESTRAINT MODULE & FIXINGS N08
PLUGS AND RELAYS. N07
POWER RESISTOR AND LAMBDA SENSOR. L16
RADIO CASSETTE PLAYER-FROM (V)148782 .... N14
SHOULDER BAG-REMOVEABLE RADIO N14
RADIO'S-UP TO (V)148781 .... N13
REAR LAMPS. K11
SEAT CUSHION SENSOR. N11
SENSORS,TRANSMITTERS AND BRACKET M12
SIDE LAMPS. K07
SPEED CONTROL PUMP & DUMP VALVE O11
STARTER MOTOR - 3.6 LITRE J08
STARTER MOTOR 5.3 LITRE J09
STEERING COLUMN SWITCHES. M04
TRIP COMPUTER AND PANEL SWITCHES. M06
WASH/WIPE RESERVOIR AND FIXINGS L08
WINDSCREEN WASHER RESERVOIR L04
WINDSCREEN WIPER MOTOR-LUCAS TYPE K17
WINDSCREEN WIPERS L03

2K 17

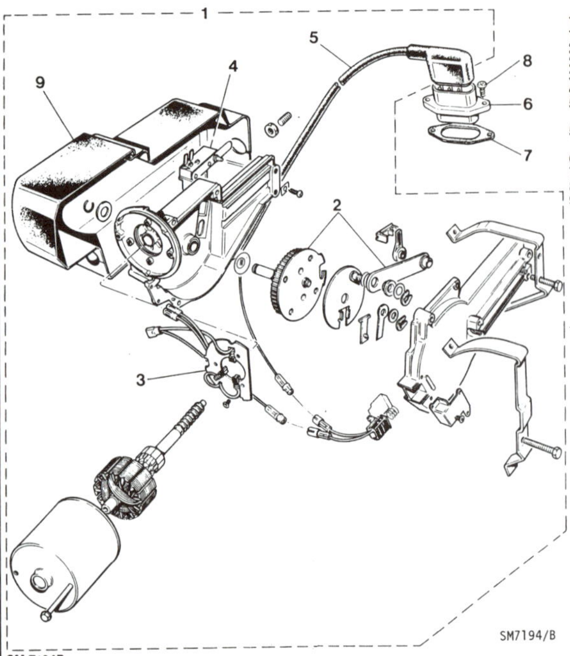

SM 7194B

WINDSCREEN WIPER MOTOR-LUCAS TYPE

Illus	Part Number	123456 Description	Quantity	Change Point	Remarks
1	AEU 1738	WIPER MOTOR ASSEMBLY	1		RHD
1	AEU 1774	WIPER MOTOR ASSEMBLY	1		LHD
2	AEU 1737	Gear assembly	1		RHD
2	AEU 1773	Gear assembly	1		LHD
3	AEU 1736	Brush gear	1		
4	AEU 1427	Park switch	1		
5	DAC 2644	Link lead	1		Not heated washer jets
5	JLM 1698	Link Lead	1		With heated washer jets
6	C 42373	Connecting plug			
7	C 42372	Gasket	1		
8	AB 608031 J	Screw	2		
9	DAC 2641	Cover-wiper motor assembly	1		

K32JP20/7/88

2K 18

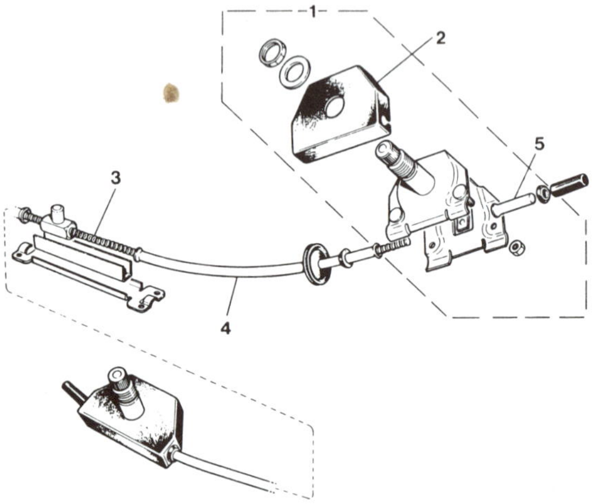

SM 7200

WIPER WHEELBOX-LUCAS TYPE MOTOR

Illus	Part Number	123456 Description	Quantity	Change Point	Remarks
1	AEU 1763	Wiper wheelbox assembly	2		
2	AAU 1497	Cover-wheelbox	1		
3	AAU 1498	Rack assembly	1		
4	AAU 1909 J	Bundy tube-to LH wheelbox	1		
5	575047 J	Bundy tube to RH wheelbox	1		

K34JP20/7/88

MASTER INDEX

ENGINE	1.C
FLYWHEEL AND CLUTCH	1.G
GEARBOX AND PROPSHAFT	1.H
AXLES, SUSPENSION, DRIVE SHAFTS, WHEELS	1.K
STEERING	1.M
BRAKES AND BRAKE CONTROLS	1.N
FUEL, EXHAUST AND EMISSION SYSTEMS	2.C
COOLING, HEATING AND AIR CONDITIONING	2.G
ELECTRICAL, WASHERS-WIPERS, INSTRUMENTS	2.J
CHASSIS, SUBFRAMES, BODYSHELL, FITTINGS	3.C
FASCIA, TRIM, SEATS AND FIXINGS	3.H
ACCESSORIES AND PAINT	4.C
NUMERICAL INDEX	4.D

GROUP INDEX

A.B.S EQUIPMENT	N09
AERIAL AND FIXINGS.	N12
ALTERNATOR 75 AMP (UP TO (E)109969).	J02
BATTERY AND LEADS.	N17
CONSOLE SWITCHES AND CIGAR LIGHTER	M08
CONTROL UNIT-PJ-V12 COUPE	L11
FROM 89 MY	L13
CONTROL UNITS-PI/CRUISE CONTROL 3.6LITRE	L10
DISTRIBUTOR-3.6 LITRE.	J11
DISTRIBUTOR-5.3 LITRE.	J12
FROM 89 MY	J13
DOOR WARNING LAMP.	K15
FOG REAR GUARD LAMP.	K14
FRONT FLASHER LAMPS.	K06
FRONT FOG LAMPS.	K13
FUEL INTERFACE UNIT AND RELAYS	L15
HARNESSES-FRONT 3.6 LITRE-FROM (V)148945	O04
HARNESSES-FRONT 5.3 LITRE-FROM (V)148782	O06
HARNESSES-FRONT-3.6 LITRE-UPTO (V)148944	O03
HEADLAMP WASH WIPE RESERVOIR	L07
HEADLAMP WASHWIPE EQUIPMENT	L06
HEADLAMPS-NOT USA	K05
HEADLAMPS-USA.	K03
HIGH MOUNTED STOP LAMP-USA/CDN ONLY.	K12
HOOD CONTROL MODULE - CONVERTIBLE	N10
HORNS AND HORN PAD	M05
IGNITION COIL & MODULE-3.6 LITRE	J14
IGNITION COILS AND AMPLIFIER-5.3 LITRE	J15
INERTIA SWITCH AND COVER	M13
INSTRUMENT PACK-3.6 LITRE	L17
INSTRUMENT PACK-5.3 LITRE	L18
INTERIOR LAMPS-CABRIOLET.	K09
INTERIOR LAMPS-COUPE.	K08
LOUDSPEAKERS AND MOUNTINGS.	N15
MAIN FUSE BLOCK AND MOUNTING PLATE	M15
NUMBER PLATE AND REVERSE LAMP.	K16
PASSIVE RESTRAINT MODULE & FIXINGS	N08
PLUGS AND RELAYS.	N07
POWER RESISTOR AND LAMBDA SENSOR.	L16
RADIO CASSETTE PLAYER-FROM (V)148782	N14
SHOULDER BAG-REMOVEABLE RADIO	N14
RADIO'S-UP TO (V)148781	N13
REAR LAMPS.	K11
SEAT CUSHION SENSOR.	N11
SENSORS,TRANSMITTERS AND BRACKET	M12
SIDE LAMPS.	K07
SPEED CONTROL PUMP & DUMP VALVE	O11
STARTER MOTOR - 3.6 LITRE	J08
STARTER MOTOR 5.3 LITRE	J09
STEERING COLUMN SWITCHES.	M04
TRIP COMPUTER AND PANEL SWITCHES.	M06
WASH/WIPE RESERVOIR AND FIXINGS	L08
WINDSCREEN WASHER RESERVOIR	L04
WINDSCREEN WIPER MOTOR-LUCAS TYPE	K17
WINDSCREEN WIPERS	L03

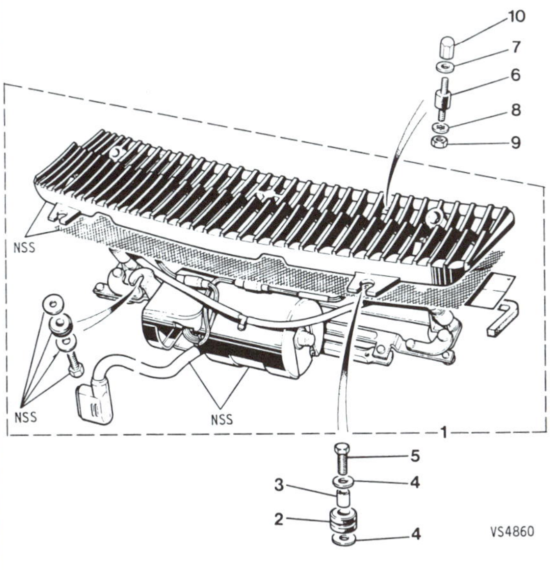

VS4860

WINDSCREEN WIPER MOTOR-ELECTROLUX TYPE

Illus	Part Number	123456	Description	Quantity	Change Point	Remarks
1	DAC 5504		Wiper motor assembly	1		RHD
	DAC 5505		Wiper motor assembly	1		LHD
2	DAC 6347		Grommet	2		
3	C 29875		Spacer	2		
4	BD 541 20		Washer	4		
5	SH 604101 J		Screw-set	2		
6	DAC 6348		Mounting	2		
7	C 44649		Spacer	2		
8	WF 702101 J		Washer	4		
9	NH 104041 J		Nut	2		
10	DAC 2667		Nut-dome	2		

K34.02JP16/8/88

MASTER INDEX

GROUP INDEX

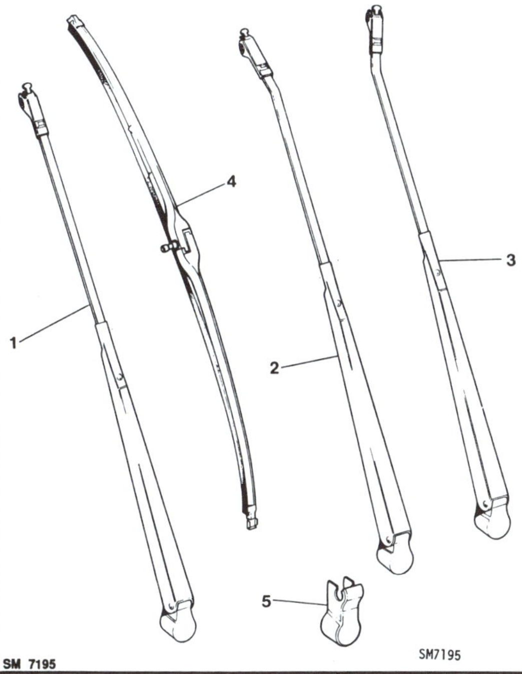

SM 7195

WINDSCREEN WIPERS

Illus	Part Number	123456	Description	Quantity	Change Point	Remarks
1	DAC 3438		Wiper arm-driver	1		With Lucas type wiper motor
2	DAC 3436		Wiper arm-passenger-RHD	1		
3	DAC 3437		Wiper arm-passenger-LHD	1		
4	DAC 3439		Wiper blade	2		
5	BAU 1124		Cover-wiper arm boss	2		
	DAC 6148		Wiper arm-driver-RHD	1		With Electrolux type wiper motor
	DAC 6149		Wiper arm-driver-LHD	1		
	DAC 6150		Wiper arm-passenger	1		
	NH 110041 J		Nut-arm fixing	2		
4	DAC 3439		Wiper blade	2		

K36JP20/7/88

MASTER INDEX
ENGINE 1.C
FLYWHEEL AND CLUTCH 1.G
GEARBOX AND PROPSHAFT 1.H
AXLES, SUSPENSION, DRIVE SHAFTS, WHEELS 1.K
STEERING 1.M
BRAKES AND BRAKE CONTROLS 1.N
FUEL, EXHAUST AND EMISSION SYSTEMS 2.C
COOLING, HEATING AND AIR CONDITIONING 2.G
ELECTRICAL, WASHERS-WIPERS, INSTRUMENTS 2.J
CHASSIS, SUBFRAMES, BODYSHELL, FITTINGS 3.C
FASCIA, TRIM, SEATS AND FIXINGS 3.H
ACCESSORIES AND PAINT 4.C
NUMERICAL INDEX 4.D
GROUP INDEX
A.B.S EQUIPMENT N09
AERIAL AND FIXINGS. N12
ALTERNATOR 75 AMP (UP TO (E)109969). J02
BATTERY AND LEADS. N17
CONSOLE SWITCHES AND CIGAR LIGHTER M08
CONTROL UNIT-PJ-V12 COUPE L11
FROM 89 MY L13
CONTROL UNITS-PI/CRUISE CONTROL 3.6LITRE L10
DISTRIBUTOR-3.6 LITRE. J11
DISTRIBUTOR-5.3 LITRE. J12
FROM 89 MY J13
DOOR WARNING LAMP. K15
FOG REAR GUARD LAMP. K14
FRONT FLASHER LAMPS. K06
FRONT FOG LAMPS. K13
FUEL INTERFACE UNIT AND RELAYS L15
HARNESSES-FRONT 3.6 LITRE-FROM (V)148945 .... O04
HARNESSES-FRONT 5.3 LITRE-FROM (V)148782 .... O06
HARNESSES-FRONT-3.6 LITRE-UPTO (V)148944 .... O03
HEADLAMP WASH WIPE RESERVOIR L07
HEADLAMP WASHWIPE EQUIPMENT L06
HEADLAMPS-NOT USA K05
HEADLAMPS-USA. K03
HIGH MOUNTED STOP LAMP-USA/CDN ONLY. K12
HOOD CONTROL MODULE - CONVERTIBLE N10
HORNS AND HORN PAD M05
IGNITION COIL & MODULE-3.6 LITRE J14
IGNITION COILS AND AMPLIFIER-5.3 LITRE J15
INERTIA SWITCH AND COVER M13
INSTRUMENT PACK-3.6 LITRE L17
INSTRUMENT PACK-5.3 LITRE L18
INTERIOR LAMPS-CABRIOLET. K09
INTERIOR LAMPS-COUPE. K08
LOUDSPEAKERS AND MOUNTINGS. N15
MAIN FUSE BLOCK AND MOUNTING PLATE M15
NUMBER PLATE AND REVERSE LAMP. K16
PASSIVE RESTRAINT MODULE & FIXINGS N08
PLUGS AND RELAYS. N07
POWER RESISTOR AND LAMBDA SENSOR. L16
RADIO CASSETTE PLAYER-FROM (V)148782 .... N14
SHOULDER BAG-REMOVEABLE RADIO N14
RADIO'S-UP TO (V)148781 .... N13
REAR LAMPS. K11
SEAT CUSHION SENSOR. N11
SENSORS,TRANSMITTERS AND BRACKET M12
SIDE LAMPS. K07
SPEED CONTROL PUMP & DUMP VALVE O11
STARTER MOTOR - 3.6 LITRE J08
STARTER MOTOR 5.3 LITRE J09
STEERING COLUMN SWITCHES. M04
TRIP COMPUTER AND PANEL SWITCHES. M06
WASH/WIPE RESERVOIR AND FIXINGS L08
WINDSCREEN WASHER RESERVOIR L04
WINDSCREEN WIPER MOTOR-LUCAS TYPE K17
WINDSCREEN WIPERS L03

WINDSCREEN WASHER RESERVOIR

UP TO (V)148781

Illus	Part Number	Description	Quantity	Change Point	Remarks
1	RTC 3009	Reservoir-windscreen washer	1		
2	JLM 9645	Cap assembly-filler	1		
3	DAC 2876	Pump assembly-washer	1		
4	BAU 2043	Grommet	1		
5	DAC 4244	Jet assembly (Heated)	2		
6	DAC 1234 3	Tubing	A/R		
7	DAC 2963	Valve-non return	1		
8	DAC 4808	Mounting bracket assembly-reservoir	1		3.6 litre
8	DAC 3386	Mounting bracket assembly-reservoir	1		5.3 litre
9	C 44896	Gasket-mounting bracket	1		
10	C 16393	Grommet	1		
11	SH 605071 J	Setscrew	4		
12	WM 600051 J	Washer-spring	4		
13	JLM 298	Washer-plain	4		
14	JLM 9566	Setscrew	2		
15	C 10193	Washer	2		
16	JLM 296	Washer	2		
17	NZ 604041 J	Nut	2		

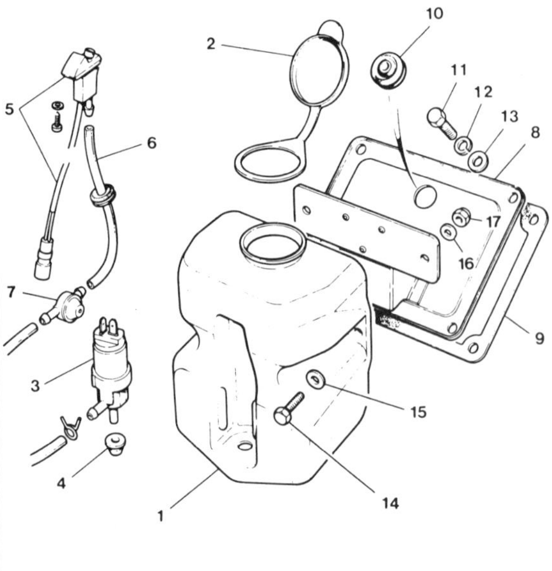

VS 4122

2L 04

MASTER INDEX

GROUP INDEX

THERMAL JET SENSOR

Illus	Part Number	Description	Quantity	Change Point	Remarks
1	DBC 2640	Sensor-thermal-heated jets	1		
2	C 15749	Grommet	1		

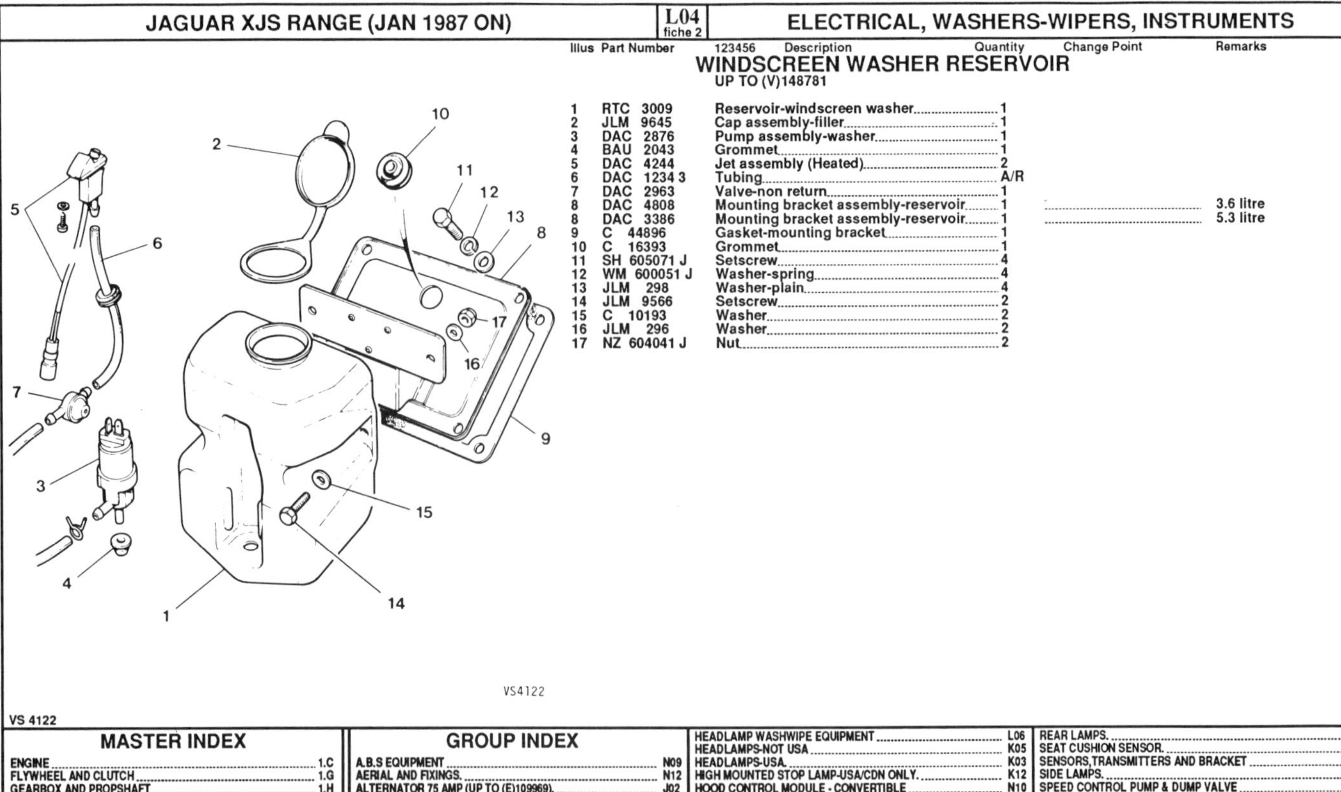

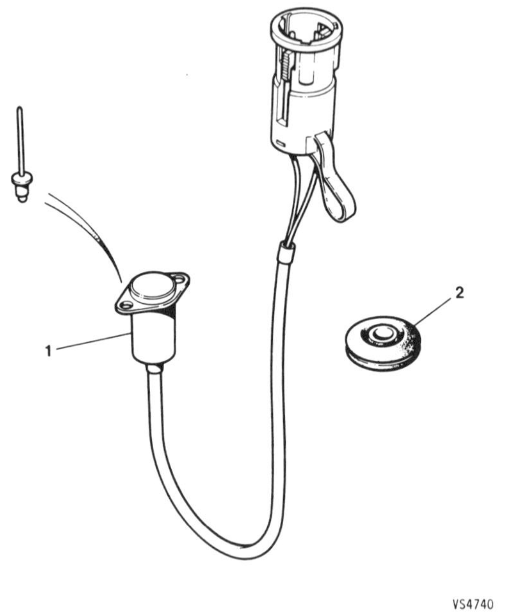

VS4740

2L 05

MASTER INDEX
ENGINE ... 1.C
FLYWHEEL AND CLUTCH ... 1.G
GEARBOX AND PROPSHAFT ... 1.H
AXLES, SUSPENSION, DRIVE SHAFTS, WHEELS ... 1.K
STEERING ... 1.M
BRAKES AND BRAKE CONTROLS ... 1.N
FUEL, EXHAUST AND EMISSION SYSTEMS ... 2.C
COOLING, HEATING AND AIR CONDITIONING ... 2.G
ELECTRICAL, WASHERS-WIPERS, INSTRUMENTS ... 2.J
CHASSIS, SUBFRAMES, BODYSHELL, FITTINGS ... 3.C
FASCIA, TRIM, SEATS AND FIXINGS ... 3.H
ACCESSORIES AND PAINT ... 4.C
NUMERICAL INDEX ... 4.D
GROUP INDEX
A.B.S EQUIPMENT ... N09
AERIAL AND FIXINGS. ... N12
ALTERNATOR 75 AMP (UP TO (E)109969). ... J02
BATTERY AND LEADS. ... N17
CONSOLE SWITCHES AND CIGAR LIGHTER ... M08
CONTROL UNIT-PJ-V12 COUPE ... L11
FROM 89 MY ... L13
CONTROL UNITS-PI/CRUISE CONTROL 3.6LITRE ... L10
DISTRIBUTOR-3.6 LITRE. ... J11
DISTRIBUTOR-5.3 LITRE. ... J12
FROM 89 MY ... J13
DOOR WARNING LAMP. ... K15
FOG REAR GUARD LAMP. ... K14
FRONT FLASHER LAMPS. ... K06
FRONT FOG LAMPS. ... K13
FUEL INTERFACE UNIT AND RELAYS ... L15
HARNESSES-FRONT 3.6 LITRE-FROM (V)148945 ... O04
HARNESSES-FRONT 5.3 LITRE-FROM (V)148782 ... O06
HARNESSES-FRONT-3.6 LITRE-UPTO (V)148944 ... O03
HEADLAMP WASH WIPE RESERVOIR ... L07
HEADLAMP WASHWIPE EQUIPMENT ... L06
HEADLAMPS-NOT USA ... K05
HEADLAMPS-USA. ... K03
HIGH MOUNTED STOP LAMP-USA/CDN ONLY. ... K12
HOOD CONTROL MODULE - CONVERTIBLE ... N10
HORNS AND HORN PAD ... M05
IGNITION COIL & MODULE-3.6 LITRE ... J14
IGNITION COILS AND AMPLIFIER-5.3 LITRE ... J15
INERTIA SWITCH AND COVER ... M13
INSTRUMENT PACK-3.6 LITRE ... L17
INSTRUMENT PACK-5.3 LITRE ... L18
INTERIOR LAMPS-CABRIOLET. ... K09
INTERIOR LAMPS-COUPE. ... K08
LOUDSPEAKERS AND MOUNTINGS. ... N15
MAIN FUSE BLOCK AND MOUNTING PLATE ... M15
NUMBER PLATE AND REVERSE LAMP. ... K16
PASSIVE RESTRAINT MODULE & FIXINGS ... N08
PLUGS AND RELAYS. ... N07
POWER RESISTOR AND LAMBDA SENSOR. ... L16
RADIO CASSETTE PLAYER-FROM (V)148782 ... N14
SHOULDER BAG-REMOVEABLE RADIO ... N14
RADIO'S-UP TO (V)148781 ... N13
REAR LAMPS. ... K11
SEAT CUSHION SENSOR. ... N11
SENSORS,TRANSMITTERS AND BRACKET ... M12
SIDE LAMPS. ... K07
SPEED CONTROL PUMP & DUMP VALVE ... O11
STARTER MOTOR - 3.6 LITRE ... J08
STARTER MOTOR 5.3 LITRE ... J09
STEERING COLUMN SWITCHES. ... M04
TRIP COMPUTER AND PANEL SWITCHES. ... M06
WASH/WIPE RESERVOIR AND FIXINGS ... L08
WINDSCREEN WASHER RESERVOIR ... L04
WINDSCREEN WIPER MOTOR-LUCAS TYPE ... K17
WINDSCREEN WIPERS ... L03

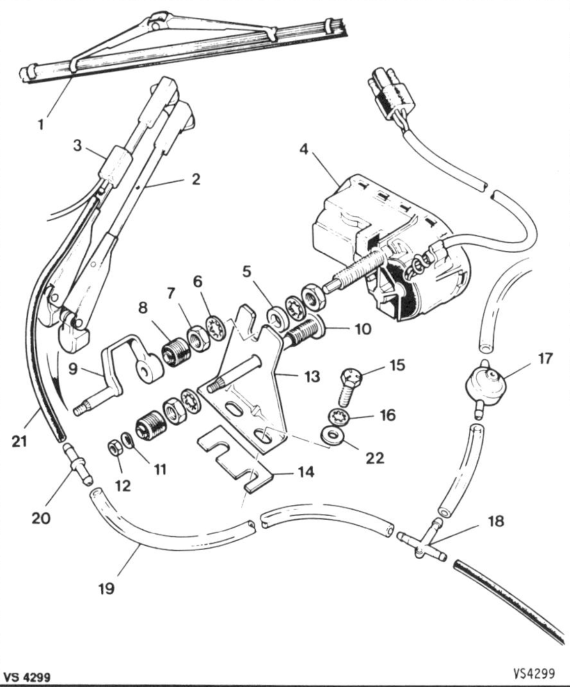

VS 4299

HEADLAMP WASHWIPE EQUIPMENT

Illus	Part Number	Description	Quantity	Change Point	Remarks
1	DAC 2488	Wiper blade	2		
2	DAC 4798	Wiper arm and jet-RH	1		
	DAC 4799	Wiper arm and jet-LH	1		
3	DAC 4532	Jet-heated-RH	1		
	DAC 4533	Jet-heated-LH	1		
4	DAC 2940	Wiper motor-RH	1		
	DAC 2941	Wiper motor-LH	1		
5	DAC 2993	Spacer	2		
6	C 30075 3	Washer shakeproof	6		
7	DAC 2899	Nut	6		
8	DAC 2007	Seal	2		
9	DAC 2898	Spindle assembly	2		
10	DAC 2432	Shaft secondary	2		
11	DAC 2005	Washer spring	2		
12	NH 105041 J	Nut	2		
13	BAC 4252	Bracket-RH	1		
	BAC 4253	Bracket-LH	1		
14	BAC 4333	Shim	A/R		
15	SH 504051 J	Setscrew	4		
16	WF 600040 J	Washer	4		
17	DAC 2963	Non return valve	1		
18	C 15644	'T' piece	1		
19	DAC 1234 3	Tubing	A/R		180" long
20	C 33835	Connector	1		
21	DAC 1847 5	Tubing	2		15" long
	C 39213	Setscrew spindle	2		
22	JLM 296	Washer-plain	4		

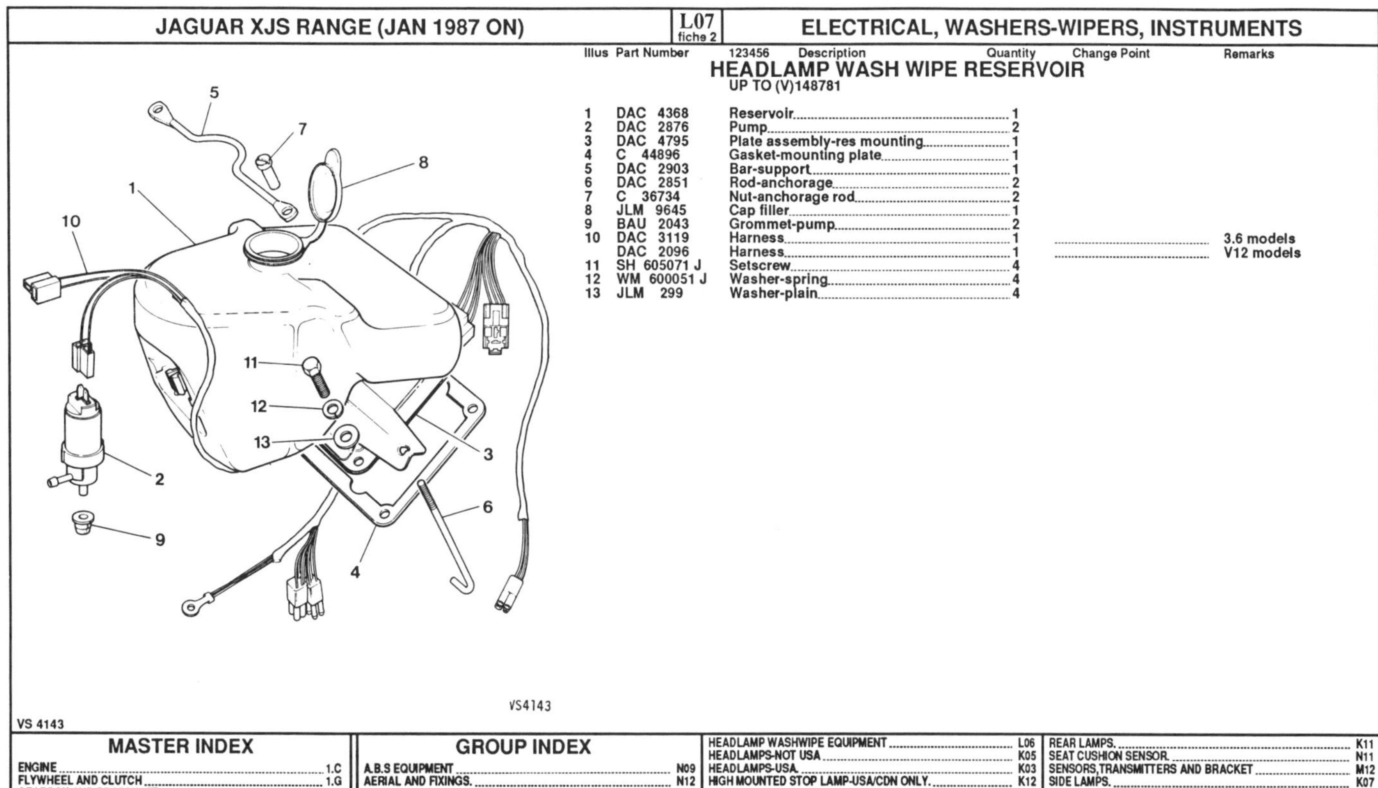

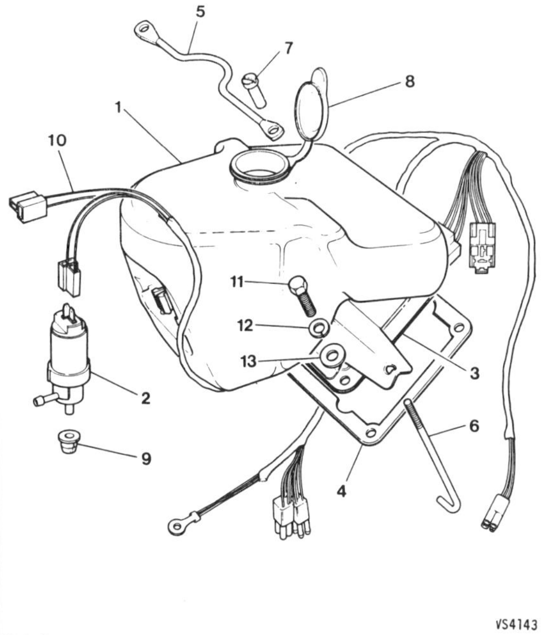

VS 4143

HEADLAMP WASH WIPE RESERVOIR

UP TO (V)148781

Illus	Part Number	Description	Quantity	Change Point	Remarks
1	DAC 4368	Reservoir	1		
2	DAC 2876	Pump	2		
3	DAC 4795	Plate assembly-res mounting	1		
4	C 44896	Gasket-mounting plate	1		
5	DAC 2903	Bar-support	1		
6	DAC 2851	Rod-anchorage	2		
7	C 36734	Nut-anchorage rod	2		
8	JLM 9645	Cap filler	1		
9	BAU 2043	Grommet-pump	2		
10	DAC 3119	Harness	1		3.6 models
	DAC 2096	Harness	1		V12 models
11	SH 605071 J	Setscrew	4		
12	WM 600051 J	Washer-spring	4		
13	JLM 299	Washer-plain	4		

MASTER INDEX

ENGINE 1.C
FLYWHEEL AND CLUTCH 1.G
GEARBOX AND PROPSHAFT 1.H
AXLES, SUSPENSION, DRIVE SHAFTS, WHEELS 1.K
STEERING 1.M
BRAKES AND BRAKE CONTROLS 1.N
FUEL, EXHAUST AND EMISSION SYSTEMS 2.C
COOLING, HEATING AND AIR CONDITIONING 2.G
ELECTRICAL, WASHERS-WIPERS, INSTRUMENTS 2.J
CHASSIS, SUBFRAMES, BODYSHELL, FITTINGS 3.C
FASCIA, TRIM, SEATS AND FIXINGS 3.H
ACCESSORIES AND PAINT 4.C
NUMERICAL INDEX 4.D

GROUP INDEX

A.B.S EQUIPMENT N09
AERIAL AND FIXINGS N12
ALTERNATOR 75 AMP (UP TO (E)109969) J02
BATTERY AND LEADS N17
CONSOLE SWITCHES AND CIGAR LIGHTER M08
CONTROL UNIT-PJ-V12 COUPE L11
FROM 89 MY L13
CONTROL UNITS-PI/CRUISE CONTROL 3.6LITRE L10
DISTRIBUTOR-3.6 LITRE J11
DISTRIBUTOR-5.3 LITRE J12
FROM 89 MY J13
DOOR WARNING LAMP K15
FOG REAR GUARD LAMP K14
FRONT FLASHER LAMPS K06
FRONT FOG LAMPS K13
FUEL INTERFACE UNIT AND RELAYS L15
HARNESSES-FRONT 3.6 LITRE-FROM (V)148945 O04
HARNESSES-FRONT 5.3 LITRE-FROM (V)148782 O06
HARNESSES-FRONT-3.6 LITRE-UPTO (V)148944 O03
HEADLAMP WASH WIPE RESERVOIR L07
HEADLAMP WASHWIPE EQUIPMENT L06
HEADLAMPS-NOT USA K05
HEADLAMPS-USA K03
HIGH MOUNTED STOP LAMP-USA/CDN ONLY K12
HOOD CONTROL MODULE - CONVERTIBLE N10
HORNS AND HORN PAD M05
IGNITION COIL & MODULE-3.6 LITRE J14
IGNITION COILS AND AMPLIFIER-5.3 LITRE J15
INERTIA SWITCH AND COVER M13
INSTRUMENT PACK-3.6 LITRE L17
INSTRUMENT PACK-5.3 LITRE L18
INTERIOR LAMPS-CABRIOLET K09
INTERIOR LAMPS-COUPE K08
LOUDSPEAKERS AND MOUNTINGS N15
MAIN FUSE BLOCK AND MOUNTING PLATE M15
NUMBER PLATE AND REVERSE LAMP K16
PASSIVE RESTRAINT MODULE & FIXINGS N08
PLUGS AND RELAYS N07
POWER RESISTOR AND LAMBDA SENSOR L16
RADIO CASSETTE PLAYER-FROM (V)148782 N14
SHOULDER BAG-REMOVEABLE RADIO N14
RADIO'S-UP TO (V)148781 N13
REAR LAMPS K11
SEAT CUSHION SENSOR N11
SENSORS,TRANSMITTERS AND BRACKET M12
SIDE LAMPS K07
SPEED CONTROL PUMP & DUMP VALVE O11
STARTER MOTOR - 3.6 LITRE J08
STARTER MOTOR 5.3 LITRE J09
STEERING COLUMN SWITCHES M04
TRIP COMPUTER AND PANEL SWITCHES M06
WASH/WIPE RESERVOIR AND FIXINGS L08
WINDSCREEN WASHER RESERVOIR L04
WINDSCREEN WIPER MOTOR-LUCAS TYPE K17
WINDSCREEN WIPERS L03

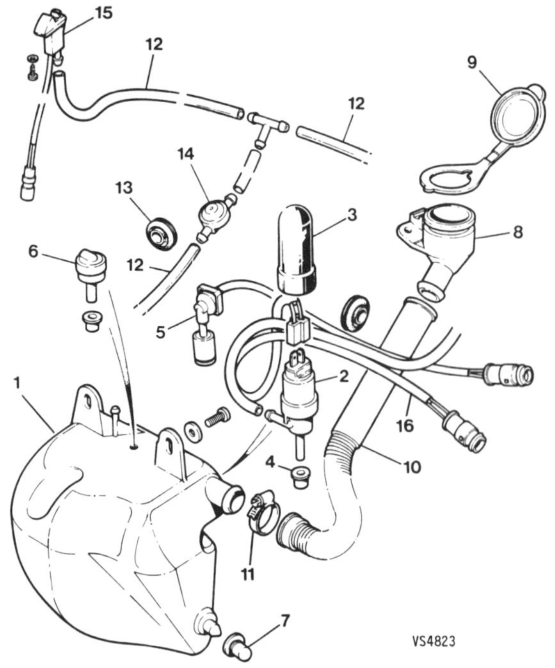

VS4823

WASH/WIPE RESERVOIR AND FIXINGS

FROM (V)148782

Illus	Part Number	123456 Description	Quantity	Change Point	Remarks
		RESERVOIR			
1	DAC 4732 2	Not H/lamp wash wipe	1		
	DAC 4732 1	With H/lamp wash wipe	1		Not Japan
	DAC 4732 3	With H/lamp wash wipe	1		Japan only
2	DAC 2876	Pump	A/R		
3	DAC 2877	Boot-protective	A/R		
4	BAU 2043	Grommet	A/R		
5	DAC 4772	Sensor-low level	1		
6	DAC 5942	Valve-air bleed	1		
7	DAC 4773	Grommet	1		
8	DAC 4734	Neck-filler	1		
9	DAC 4888	Cap-filler	1		
10	DAC 4859	Tube-connector	1		3.6 models
	DAC 5556	Tube-connector	1		5.3 models
11	EAC 3215 9	Clip	2		
12	DAC 1234 3	Tube	A/R		Cut to length
13	C 16393	Grommet	1		
14	DAC 2963	Valve-non return	1		
15	DAC 4244	Jet assy (Heated)	2		
16	DAC 4792	Harness-pump	1		Not H/lamp wash
	DAC 4793	Harness-pump	1		With H/lamp wash
	DAC 4484	Harness-Wash/wipe	1		5.3 models
	DAC 4485	Harness-Wash/wipe	1		3.6 models

2L 08

HEADLAMP WASH WIPE AND ANTI-STALL RELAYS.

Illus	Part Number	123456 Description	Quantity	Change Point	Remarks
1	DAC 1996	Module-wash/wipe	1		
2	DAC 1027	Relay-wash/wipe	1		
3	AGU 1068	Relay-anti-stall norm open	1		5.3 litre only
4	AGU 1109	Relay-anti-stall changeover	1		
5	DAC 4838	Bracket-relay mounting	1		
5	DAC 4543	Bracket-relay mounting	1		5.3 litre
6	UFS 119 3R	Screw	2		
7	WF 702100 J	Washer	2		
8	UFN 119 L	Nut	1		
7	C 31973 11	Nut-lokut	2		
8	DAC 4524	Module-dim dip	1		Use DAC 5322
8	DAC 5322	Module-dim dip	1		

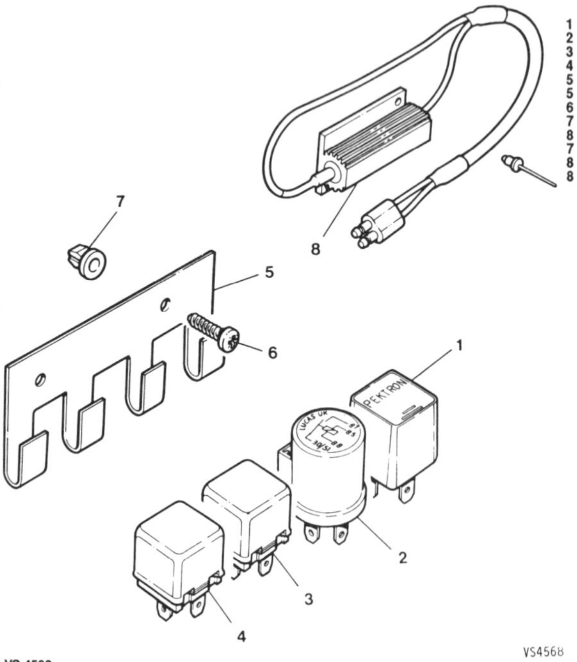

VS 4568

2L 09

MASTER INDEX

ENGINE 1.C
FLYWHEEL AND CLUTCH 1.G
GEARBOX AND PROPSHAFT 1.H
AXLES, SUSPENSION, DRIVE SHAFTS, WHEELS 1.K
STEERING 1.M
BRAKES AND BRAKE CONTROLS 1.N
FUEL, EXHAUST AND EMISSION SYSTEMS 2.C
COOLING, HEATING AND AIR CONDITIONING 2.G
ELECTRICAL, WASHERS-WIPERS, INSTRUMENTS 2.J
CHASSIS, SUBFRAMES, BODYSHELL, FITTINGS 3.C
FASCIA, TRIM, SEATS AND FIXINGS 3.H
ACCESSORIES AND PAINT 4.C
NUMERICAL INDEX 4.D

GROUP INDEX

A.B.S EQUIPMENT N09
AERIAL AND FIXINGS. N12
ALTERNATOR 75 AMP (UP TO (E)109969). J02
BATTERY AND LEADS. N17
CONSOLE SWITCHES AND CIGAR LIGHTER M08
CONTROL UNIT-P.J-V12 COUPE L11
FROM 89 MY L13
CONTROL UNITS-PI/CRUISE CONTROL 3.6LITRE L10
DISTRIBUTOR-3.6 LITRE. J11
DISTRIBUTOR-5.3 LITRE. J12
FROM 89 MY J13
DOOR WARNING LAMP. K15
FOG REAR GUARD LAMP. K14
FRONT FLASHER LAMPS. K06
FRONT FOG LAMPS. K13
FUEL INTERFACE UNIT AND RELAYS L15
HARNESSES-FRONT 3.6 LITRE-FROM (V)148945 O04
HARNESSES-FRONT 5.3 LITRE-FROM (V)148782 O06
HARNESSES-FRONT-3.6 LITRE-UPTO (V)148944 O03
HEADLAMP WASH WIPE RESERVOIR L07
HEADLAMP WASHWIPE EQUIPMENT L06
HEADLAMPS-NOT USA K05
HEADLAMPS-USA. K03
HIGH MOUNTED STOP LAMP-USA/CDN ONLY. K12
HOOD CONTROL MODULE - CONVERTIBLE N10
HORNS AND HORN PAD M05
IGNITION COIL & MODULE-3.6 LITRE J14
IGNITION COILS AND AMPLIFIER-5.3 LITRE J15
INERTIA SWITCH AND COVER M13
INSTRUMENT PACK-3.6 LITRE L17
INSTRUMENT PACK-5.3 LITRE L18
INTERIOR LAMPS-CABRIOLET. K09
INTERIOR LAMPS-COUPE. K08
LOUDSPEAKERS AND MOUNTINGS. N15
MAIN FUSE BLOCK AND MOUNTING PLATE M15
NUMBER PLATE AND REVERSE LAMP. K16
PASSIVE RESTRAINT MODULE & FIXINGS N08
PLUGS AND RELAYS. N07
POWER RESISTOR AND LAMBDA SENSOR. L16
RADIO CASSETTE PLAYER-FROM (V)148782 N14
SHOULDER BAG-REMOVEABLE RADIO N14
RADIO'S-UP TO (V)148781 N13
REAR LAMPS. K11
SEAT CUSHION SENSOR. N11
SENSORS,TRANSMITTERS AND BRACKET M12
SIDE LAMPS. K07
SPEED CONTROL PUMP & DUMP VALVE O11
STARTER MOTOR - 3.6 LITRE J08
STARTER MOTOR 5.3 LITRE J09
STEERING COLUMN SWITCHES. M04
TRIP COMPUTER AND PANEL SWITCHES. M06
WASH/WIPE RESERVOIR AND FIXINGS L08
WINDSCREEN WASHER RESERVOIR L04
WINDSCREEN WIPER MOTOR-LUCAS TYPE K17
WINDSCREEN WIPERS L03

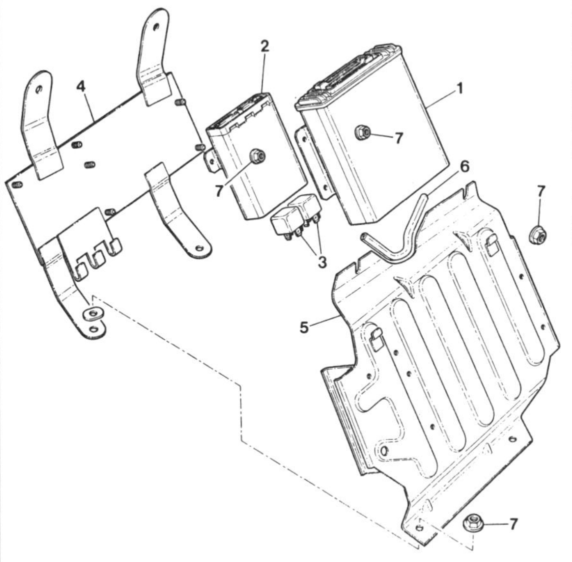

VS4743

CONTROL UNITS-PI/CRUISE CONTROL 3.6LITRE

Illus	Part Number	123456	Description	Quantity	Change Point	Remarks
			PI CONTROL UNIT			
1	DBC 2911		Auto with catalyst	1	Up to (V)144596	
1	DBC 3711		Auto with catalyst	1	From (V)144597	Use DBC 4411
1	DBC 4411		Auto with catalyst	1		
1	DBC 2735		Manual with catalyst	1		
1	DBC 2910		Non catalyst	1	Up to (V)144596	
1	DBC 3710		Non catalyst	1	From (V)144597	Use DBC 4410
1	DBC 4410		Non catalyst	1		
2	DBC 1169		Control unit-speed control	1		
3	AGU 1070		Relay-main	2		
3	AGU 1068		Relay-fuel pump	1		
4	BCC 5642		Plate-mounting-RH-LHD	1		Use BCC 7900
4	BCC 7900		Plate mounting-RH-LHD	1		
	BCC 5643		Plate-mounting-LH-RHD	1		Use BCC 7901
	BCC 7901		Plate-mounting-LH-RHD	1		
5	BCC 8098		Panel-toeboard-RH-LHD	1		
	BCC 8099		Panel-toeboard-LH-RHD	1		
6	BCC 2712	3	Strip-protection	3		
7	FN 105041		Nut-flanged	10		

VS4743

MASTER INDEX

GROUP INDEX

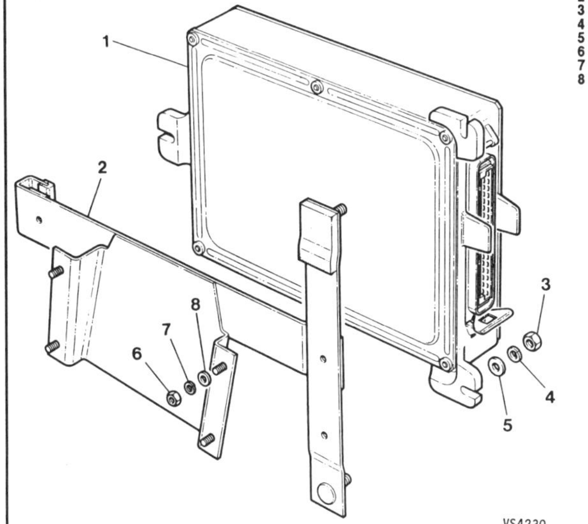

VS4230

CONTROL UNIT-P.I-V12 COUPE

Illus	Part Number	123456	Description	Quantity	Change Point	Remarks
1	DAC 4118		Control unit	1		Use DAC 4585
1	DAC 4585		Control unit	1	Up to 89 MY	With catalyst
1	DAC 6337		Control unit	1	From 89 MY	With catalyst
1	DAC 4478		Control unit	1		Use DAC 4586
1	DAC 4586		Control unit	1	Up to 89 MY	Not catalyst
1	DAC 6338		Control unit	1	From 89 My	Not catalyst
2	DAC 4750		Mounting bracket	1		
3	JLM 9683		Nut-unit to bracket	2		
4	JLM 302		Washer-spring	2		
5	JLM 296		Washer-plain	2		
6	UFN 119 L		Nut-bracket to drain channel	4		
7	WM 702001 J		Washer-spring	4		
8	WC 702101 J		Washer-plain	4		

K46JP1/9/88

VS 4230

MASTER INDEX
ENGINE 1.C
FLYWHEEL AND CLUTCH 1.G
GEARBOX AND PROPSHAFT 1.H
AXLES, SUSPENSION, DRIVE SHAFTS, WHEELS 1.K
STEERING 1.M
BRAKES AND BRAKE CONTROLS 1.N
FUEL, EXHAUST AND EMISSION SYSTEMS 2.C
COOLING, HEATING AND AIR CONDITIONING 2.G
ELECTRICAL, WASHERS-WIPERS, INSTRUMENTS 2.J
CHASSIS, SUBFRAMES, BODYSHELL, FITTINGS 3.C
FASCIA, TRIM, SEATS AND FIXINGS 3.H
ACCESSORIES AND PAINT 4.C
NUMERICAL INDEX 4.D
GROUP INDEX
A.B.S EQUIPMENT N09
AERIAL AND FIXINGS N12
ALTERNATOR 75 AMP (UP TO (E)109969) J02
BATTERY AND LEADS N17
CONSOLE SWITCHES AND CIGAR LIGHTER M08
CONTROL UNIT-P.I-V12 COUPE L11
FROM 89 MY L13
CONTROL UNITS-PI/CRUISE CONTROL 3.6LITRE L10
DISTRIBUTOR-3.6 LITRE J11
DISTRIBUTOR-5.3 LITRE J12
FROM 89 MY J13
DOOR WARNING LAMP K15
FOG REAR GUARD LAMP K14
FRONT FLASHER LAMPS K06
FRONT FOG LAMPS K13
FUEL INTERFACE UNIT AND RELAYS L15
HARNESSES-FRONT 3.6 LITRE-FROM (V)148945 O04
HARNESSES-FRONT 5.3 LITRE-FROM (V)148782 O06
HARNESSES-FRONT-3.6 LITRE-UPTO (V)148944 O03
HEADLAMP WASH WIPE RESERVOIR L07
HEADLAMP WASHWIPE EQUIPMENT L06
HEADLAMPS-NOT USA K05
HEADLAMPS-USA K03
HIGH MOUNTED STOP LAMP-USA/CDN ONLY K12
HOOD CONTROL MODULE - CONVERTIBLE N10
HORNS AND HORN PAD M05
IGNITION COIL & MODULE-3.6 LITRE J14
IGNITION COILS AND AMPLIFIER-5.3 LITRE J15
INERTIA SWITCH AND COVER M13
INSTRUMENT PACK-3.6 LITRE L17
INSTRUMENT PACK-5.3 LITRE L18
INTERIOR LAMPS-CABRIOLET K09
INTERIOR LAMPS-COUPE K08
LOUDSPEAKERS AND MOUNTINGS N15
MAIN FUSE BLOCK AND MOUNTING PLATE M15
NUMBER PLATE AND REVERSE LAMP K16
PASSIVE RESTRAINT MODULE & FIXINGS N08
PLUGS AND RELAYS N07
POWER RESISTOR AND LAMBDA SENSOR L16
RADIO CASSETTE PLAYER-FROM (V)148782 N14
SHOULDER BAG-REMOVEABLE RADIO N14
RADIO'S-UP TO (V)148781 N13
REAR LAMPS K11
SEAT CUSHION SENSOR N11
SENSORS,TRANSMITTERS AND BRACKET M12
SIDE LAMPS K07
SPEED CONTROL PUMP & DUMP VALVE O11
STARTER MOTOR - 3.6 LITRE J08
STARTER MOTOR 5.3 LITRE J09
STEERING COLUMN SWITCHES M04
TRIP COMPUTER AND PANEL SWITCHES M06
WASH/WIPE RESERVOIR AND FIXINGS L08
WINDSCREEN WASHER RESERVOIR L04
WINDSCREEN WIPER MOTOR-LUCAS TYPE K17
WINDSCREEN WIPERS L03

2L 12

JAGUAR XJS RANGE (JAN 1987 ON) | L12 fiche 2 | ELECTRICAL, WASHERS-WIPERS, INSTRUMENTS

CONTROL UNIT-P.I.-5.3 LITRE CABRIOLET & CONVERTIBLE

Illus	Part Number	Description	Quantity	Change Point	Remarks
1	DAC 4118	Control unit	1		Use DAC 4585
1	DAC 4585	Control unit	1	Up to 89 MY	With catalyst
1	DAC 6337	Control unit	1	From 89 MY	With catalyst
1	DAC 4478	Control unit	1		Use DAC 4586
1	DAC 4586	Control unit	1	Up to 89 MY	Not catalyst
1	DAC 6338	Control unit	1	From 89 MY	Not catalyst
2	BAC 7735	Mounting bracket	1		
3	JLM 296	Washer-plain	2		
4	JLM 302	Washer-spring	2		
5	JLM 9685	Nut-plain	2		

K46.02JP31/8/88

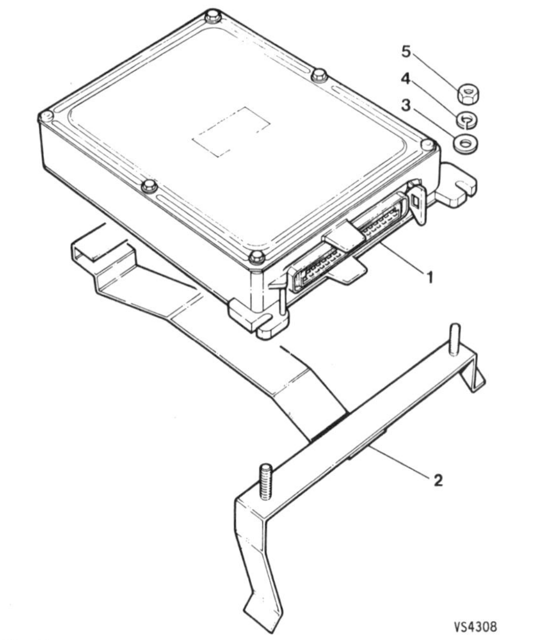

VS 4308

2L 13

JAGUAR XJS RANGE (JAN 1987 ON) | L13 fiche 2 | ELECTRICAL, WASHERS-WIPERS, INSTRUMENTS

CONTROL UNIT - IGNITION - 5.3 LITRE

FROM 89 MY

Illus	Part Number	Description	Quantity	Change Point	Remarks
1	DAC 5869	Control unit - Ign	1		Non catalyst
1	DAC 5870	Control unit - Ign	1		Single catalyst
1	DAC 5871	Control unit - Ign	1		Full catalyst
2	BDC 6801	Elbow	2		
3	C 42799 17	Hose - vacuum	1		ECU to manifold
4	FS 105121 J	Nut - flanged	3		ECU to bracket
5	BDC 6092	Mounting bracket	1		LHD only
6	FN 105041	Nut - flanged	2		Brkt to cover

K46.04JP1/9/88

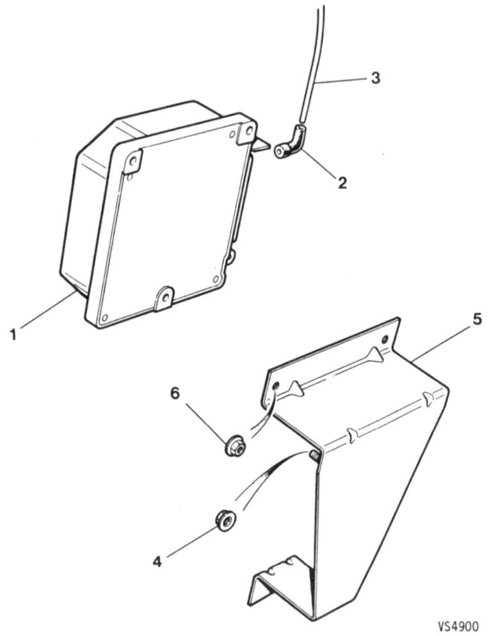

VS4900

MASTER INDEX

ENGINE 1.C
FLYWHEEL AND CLUTCH 1.G
GEARBOX AND PROPSHAFT 1.H
AXLES, SUSPENSION, DRIVE SHAFTS, WHEELS 1.K
STEERING 1.M
BRAKES AND BRAKE CONTROLS 1.N
FUEL, EXHAUST AND EMISSION SYSTEMS 2.C
COOLING, HEATING AND AIR CONDITIONING 2.G
ELECTRICAL, WASHERS-WIPERS, INSTRUMENTS 2.J
CHASSIS, SUBFRAMES, BODYSHELL, FITTINGS 3.C
FASCIA, TRIM, SEATS AND FIXINGS 3.H
ACCESSORIES AND PAINT 4.C
NUMERICAL INDEX 4.D

GROUP INDEX

A.B.S EQUIPMENT N09
AERIAL AND FIXINGS. N12
ALTERNATOR 75 AMP (UP TO (E)109969). J02
BATTERY AND LEADS. N17
CONSOLE SWITCHES AND CIGAR LIGHTER M08
CONTROL UNIT-P.I.-V12 COUPE L11
FROM 89 MY L13
CONTROL UNITS-PI/CRUISE CONTROL 3.6LITRE L10
DISTRIBUTOR-3.6 LITRE. J11
DISTRIBUTOR-5.3 LITRE. J12
FROM 89 MY J13
DOOR WARNING LAMP. K15
FOG REAR GUARD LAMP. K14
FRONT FLASHER LAMPS. K06
FRONT FOG LAMPS. K13
FUEL INTERFACE UNIT AND RELAYS L15
HARNESSES-FRONT 3.6 LITRE-FROM (V)148945 .... O04
HARNESSES-FRONT 5.3 LITRE-FROM (V)148782 .... O06
HARNESSES-FRONT-3.6 LITRE-UPTO (V)148944 .... O03
HEADLAMP WASH WIPE RESERVOIR L07
HEADLAMP WASHWIPE EQUIPMENT L06
HEADLAMPS-NOT USA K05
HEADLAMPS-USA. K03
HIGH MOUNTED STOP LAMP-USA/CDN ONLY. K12
HOOD CONTROL MODULE - CONVERTIBLE N10
HORNS AND HORN PAD M05
IGNITION COIL & MODULE-3.6 LITRE J14
IGNITION COILS AND AMPLIFIER-5.3 LITRE J15
INERTIA SWITCH AND COVER M13
INSTRUMENT PACK-3.6 LITRE L17
INSTRUMENT PACK-5.3 LITRE L18
INTERIOR LAMPS-CABRIOLET. K09
INTERIOR LAMPS-COUPE. K08
LOUDSPEAKERS AND MOUNTINGS. N15
MAIN FUSE BLOCK AND MOUNTING PLATE M15
NUMBER PLATE AND REVERSE LAMP. K16
PASSIVE RESTRAINT MODULE & FIXINGS N08
PLUGS AND RELAYS. N07
POWER RESISTOR AND LAMBDA SENSOR. L16
RADIO CASSETTE PLAYER-FROM (V)148782 .... N14
SHOULDER BAG-REMOVEABLE RADIO N14
RADIO'S-UP TO (V)148781 .... N13
REAR LAMPS. K11
SEAT CUSHION SENSOR. N11
SENSORS,TRANSMITTERS AND BRACKET M12
SIDE LAMPS. K07
SPEED CONTROL PUMP & DUMP VALVE O11
STARTER MOTOR - 3.6 LITRE J08
STARTER MOTOR 5.3 LITRE J09
STEERING COLUMN SWITCHES. M04
TRIP COMPUTER AND PANEL SWITCHES. M06
WASH/WIPE RESERVOIR AND FIXINGS L08
WINDSCREEN WASHER RESERVOIR L04
WINDSCREEN WIPER MOTOR-LUCAS TYPE K17
WINDSCREEN WIPERS L03

SPEED INTERFACE UNIT-WITH DIFF MTD SPEED SENSOR

Illus	Part Number	123456	Description	Quantity	Change Point	Remarks
1	DAC 4864		Interface unit	1		
1	DAC 4591		Interface unit	1		Sportspack only

K46.06JP31/8/88

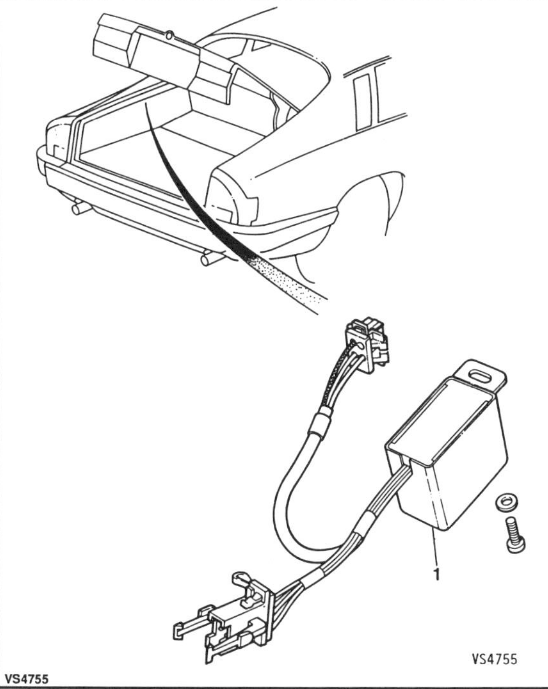

VS4755

FUEL INTERFACE UNIT AND RELAYS

5.3 LITRE

Illus	Part Number	123456	Description	Quantity	Change Point	Remarks
1	DAC 2589		Interface unit	1		
2	AGU 1068		Relay-fuel pump	1		
3	AGU 1070		Relay-main	1		
4	DAC 3113		Bracket-mounting	1		
5	UFS 819 3H		Setscrew	2		
6	WF 702101 J		Washer-shakeproof	2		

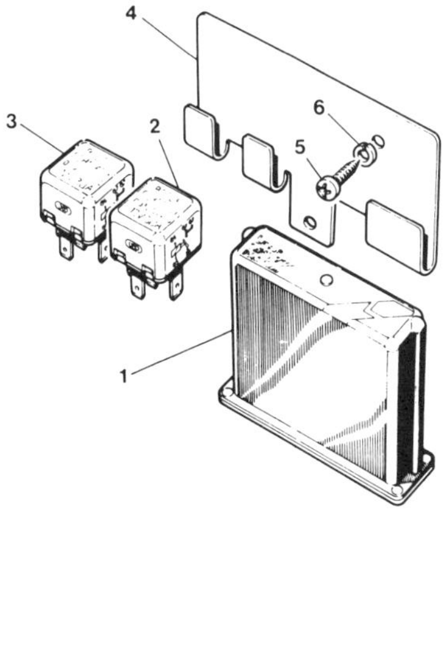

VS 4263

MASTER INDEX

ENGINE ... 1.C
FLYWHEEL AND CLUTCH ... 1.G
GEARBOX AND PROPSHAFT ... 1.H
AXLES, SUSPENSION, DRIVE SHAFTS, WHEELS ... 1.K
STEERING ... 1.M
BRAKES AND BRAKE CONTROLS ... 1.N
FUEL, EXHAUST AND EMISSION SYSTEMS ... 2.C
COOLING, HEATING AND AIR CONDITIONING ... 2.G
ELECTRICAL, WASHERS-WIPERS, INSTRUMENTS ... 2.J
CHASSIS, SUBFRAMES, BODYSHELL, FITTINGS ... 3.C
FASCIA, TRIM, SEATS AND FIXINGS ... 3.H
ACCESSORIES AND PAINT ... 4.C
NUMERICAL INDEX ... 4.D

GROUP INDEX

A.B.S EQUIPMENT ... N09
AERIAL AND FIXINGS ... N12
ALTERNATOR 75 AMP (UP TO (E)109969) ... J02
BATTERY AND LEADS ... N17
CONSOLE SWITCHES AND CIGAR LIGHTER ... M08
CONTROL UNIT-PJ-V12 COUPE ... L11
FROM 89 MY ... L13
CONTROL UNITS-PI/CRUISE CONTROL 3.6LITRE ... L10
DISTRIBUTOR-3.6 LITRE ... J11
DISTRIBUTOR-5.3 LITRE ... J12
FROM 89 MY ... J13
DOOR WARNING LAMP ... K15
FOG REAR GUARD LAMP ... K14
FRONT FLASHER LAMPS ... K06
FRONT FOG LAMPS ... K13
FUEL INTERFACE UNIT AND RELAYS ... L15
HARNESSES-FRONT 3.6 LITRE-FROM (V)148945 ... O04
HARNESSES-FRONT 5.3 LITRE-FROM (V)148782 ... O06
HARNESSES-FRONT-3.6 LITRE-UPTO (V)148944 ... O03
HEADLAMP WASH WIPE RESERVOIR ... L07
HEADLAMP WASHWIPE EQUIPMENT ... L06
HEADLAMPS-NOT USA ... K05
HEADLAMPS-USA ... K03
HIGH MOUNTED STOP LAMP-USA/CDN ONLY ... K12
HOOD CONTROL MODULE - CONVERTIBLE ... N10
HORNS AND HORN PAD ... M05
IGNITION COIL & MODULE-3.6 LITRE ... J14
IGNITION COILS AND AMPLIFIER-5.3 LITRE ... J15
INERTIA SWITCH AND COVER ... M13
INSTRUMENT PACK-3.6 LITRE ... L17
INSTRUMENT PACK-5.3 LITRE ... L18
INTERIOR LAMPS-CABRIOLET ... K09
INTERIOR LAMPS-COUPE ... K08
LOUDSPEAKERS AND MOUNTINGS ... N15
MAIN FUSE BLOCK AND MOUNTING PLATE ... M15
NUMBER PLATE AND REVERSE LAMP ... K16
PASSIVE RESTRAINT MODULE & FIXINGS ... N08
PLUGS AND RELAYS ... N07
POWER RESISTOR AND LAMBDA SENSOR ... L16
RADIO CASSETTE PLAYER-FROM (V)148782 ... N14
SHOULDER BAG-REMOVEABLE RADIO ... N14
RADIO'S-UP TO (V)148781 ... N13
REAR LAMPS ... K11
SEAT CUSHION SENSOR ... N11
SENSORS,TRANSMITTERS AND BRACKET ... M12
SIDE LAMPS ... K07
SPEED CONTROL PUMP & DUMP VALVE ... O11
STARTER MOTOR - 3.6 LITRE ... J08
STARTER MOTOR 5.3 LITRE ... J09
STEERING COLUMN SWITCHES ... M04
TRIP COMPUTER AND PANEL SWITCHES ... M06
WASH/WIPE RESERVOIR AND FIXINGS ... L08
WINDSCREEN WASHER RESERVOIR ... L04
WINDSCREEN WIPER MOTOR-LUCAS TYPE ... K17
WINDSCREEN WIPERS ... L03

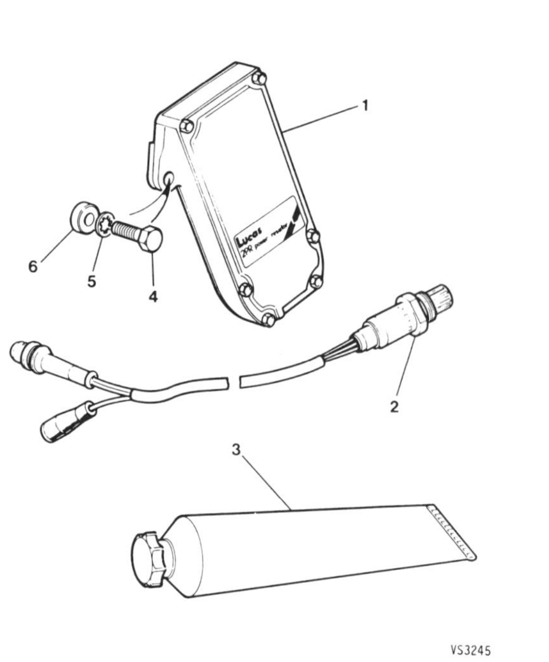

VS 3245

POWER RESISTOR AND LAMBDA SENSOR.

Illus	Part Number	123456 Description	Quantity	Change Point	Remarks
1	DAC 2044	Power resistor	1		
2	AGU 1870	Lambda sensor(Heated)	1		
2	AGU 1108	Lambda Sensor	1		Germany(optional)
3	AEU 1412	Compound-anti seizure	1		
4	UFS 119 7R	Screw	2		
5	WF 702101 J	Washer-shakeproof	2		
6	BD 541 34	Spacer	2		

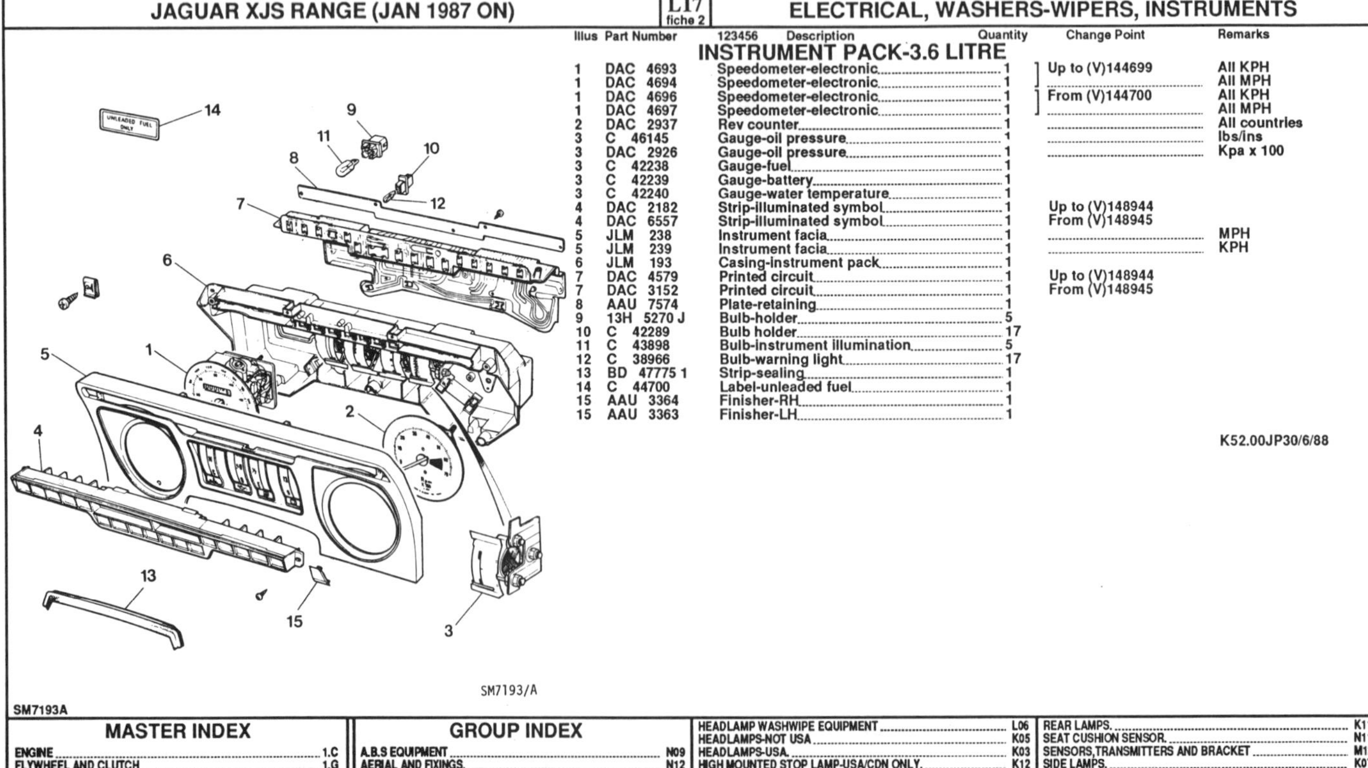

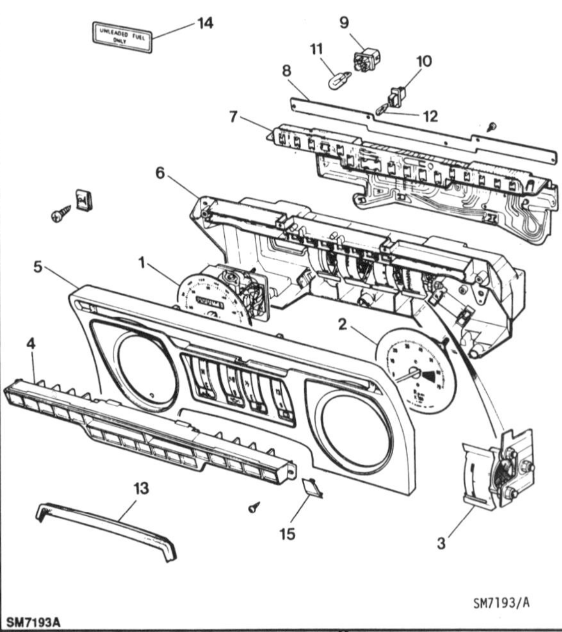

SM7193A

INSTRUMENT PACK-3.6 LITRE

Illus	Part Number	123456 Description	Quantity	Change Point	Remarks
1	DAC 4693	Speedometer-electronic	1	Up to (V)144699	All KPH
1	DAC 4694	Speedometer-electronic	1		All MPH
1	DAC 4696	Speedometer-electronic	1	From (V)144700	All KPH
1	DAC 4697	Speedometer-electronic	1		All MPH
2	DAC 2937	Rev counter	1		All countries
3	C 46145	Gauge-oil pressure	1		lbs/ins
3	DAC 2926	Gauge-oil pressure	1		Kpa x 100
3	C 42238	Gauge-fuel	1		
3	C 42239	Gauge-battery	1		
3	C 42240	Gauge-water temperature	1		
4	DAC 2182	Strip-illuminated symbol	1	Up to (V)148944	
4	DAC 6557	Strip-illuminated symbol	1	From (V)148945	
5	JLM 238	Instrument facia	1		MPH
5	JLM 239	Instrument facia	1		KPH
6	JLM 193	Casing-instrument pack	1		
7	DAC 4579	Printed circuit	1	Up to (V)148944	
7	DAC 3152	Printed circuit	1	From (V)148945	
8	AAU 7574	Plate-retaining	1		
9	13H 5270 J	Bulb-holder	5		
10	C 42289	Bulb holder	17		
11	C 43898	Bulb-instrument illumination	5		
12	C 38966	Bulb-warning light	17		
13	BD 47775 1	Strip-sealing	1		
14	C 44700	Label-unleaded fuel	1		
15	AAU 3364	Finisher-RH	1		
15	AAU 3363	Finisher-LH	1		

K52.00JP30/6/88

MASTER INDEX

ENGINE ... 1.C
FLYWHEEL AND CLUTCH ... 1.G
GEARBOX AND PROPSHAFT ... 1.H
AXLES, SUSPENSION, DRIVE SHAFTS, WHEELS ... 1.K
STEERING ... 1.M
BRAKES AND BRAKE CONTROLS ... 1.N
FUEL, EXHAUST AND EMISSION SYSTEMS ... 2.C
COOLING, HEATING AND AIR CONDITIONING ... 2.G
ELECTRICAL, WASHERS-WIPERS, INSTRUMENTS ... 2.J
CHASSIS, SUBFRAMES, BODYSHELL, FITTINGS ... 3.C
FASCIA, TRIM, SEATS AND FIXINGS ... 3.H
ACCESSORIES AND PAINT ... 4.C
NUMERICAL INDEX ... 4.D

GROUP INDEX

A.B.S EQUIPMENT ... N09
AERIAL AND FIXINGS. ... N12
ALTERNATOR 75 AMP (UP TO (E)109969). ... J02
BATTERY AND LEADS. ... N17
CONSOLE SWITCHES AND CIGAR LIGHTER ... M08
CONTROL UNIT-P.J-V12 COUPE ... L11
FROM 89 MY ... L13
CONTROL UNITS-PI/CRUISE CONTROL 3.6LITRE ... L10
DISTRIBUTOR-3.6 LITRE. ... J11
DISTRIBUTOR-5.3 LITRE. ... J12
FROM 89 MY ... J13
DOOR WARNING LAMP. ... K15
FOG REAR GUARD LAMP. ... K14
FRONT FLASHER LAMPS. ... K06
FRONT FOG LAMPS. ... K13
FUEL INTERFACE UNIT AND RELAYS ... L15
HARNESSES-FRONT 3.6 LITRE-FROM (V)148945 ... O04
HARNESSES-FRONT 5.3 LITRE-FROM (V)148782 ... O06
HARNESSES-FRONT-3.6 LITRE-UPTO (V)148944 ... O03
HEADLAMP WASH WIPE RESERVOIR ... L07
HEADLAMP WASHWIPE EQUIPMENT ... L06
HEADLAMPS-NOT USA ... K05
HEADLAMPS-USA. ... K03
HIGH MOUNTED STOP LAMP-USA/CDN ONLY. ... K12
HOOD CONTROL MODULE - CONVERTIBLE ... N10
HORNS AND HORN PAD ... M05
IGNITION COIL & MODULE-3.6 LITRE ... J14
IGNITION COILS AND AMPLIFIER-5.3 LITRE ... J15
INERTIA SWITCH AND COVER ... M13
INSTRUMENT PACK-3.6 LITRE ... L17
INSTRUMENT PACK-5.3 LITRE ... L18
INTERIOR LAMPS-CABRIOLET. ... K09
INTERIOR LAMPS-COUPE. ... K08
LOUDSPEAKERS AND MOUNTINGS. ... N15
MAIN FUSE BLOCK AND MOUNTING PLATE ... M15
NUMBER PLATE AND REVERSE LAMP. ... K16
PASSIVE RESTRAINT MODULE & FIXINGS ... N08
PLUGS AND RELAYS. ... N07
POWER RESISTOR AND LAMBDA SENSOR. ... L16
RADIO CASSETTE PLAYER-FROM (V)148782 ... N14
SHOULDER BAG-REMOVEABLE RADIO ... N14
RADIO'S-UP TO (V)148781 ... N13
REAR LAMPS. ... K11
SEAT CUSHION SENSOR. ... N11
SENSORS,TRANSMITTERS AND BRACKET ... M12
SIDE LAMPS. ... K07
SPEED CONTROL PUMP & DUMP VALVE ... O11
STARTER MOTOR - 3.6 LITRE ... J08
STARTER MOTOR 5.3 LITRE ... J09
STEERING COLUMN SWITCHES. ... M04
TRIP COMPUTER AND PANEL SWITCHES. ... M06
WASH/WIPE RESERVOIR AND FIXINGS ... L08
WINDSCREEN WASHER RESERVOIR ... L04
WINDSCREEN WIPER MOTOR-LUCAS TYPE ... K17
WINDSCREEN WIPERS ... L03

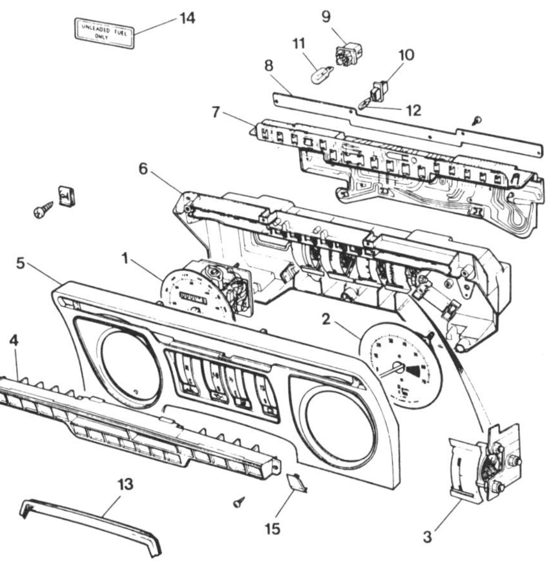

SM7193A

SM7193/A

INSTRUMENT PACK-5.3 LITRE

Illus	Part Number	123456 Description	Quantity	Change Point	Remarks
1	DAC 3408	Speedometer-electronic	1	] Up to (V) 144262	All KPH
1	DAC 3407	Speedometer-electronic	1	]	All MPH
1	DAC 4693	Speedometer-electronic	1	] From (V) 144263	All KPH
1	DAC 4694	Speedometer-electronic	1	] See (1) below	All MPH
1	DAC 4695	Speedometer-electronic	1	From (V)148782	MPH-not USA
1	DAC 4699	Speedometer-electronic	1	From (V)148782	MPH-USA only
1	DAC 4696	Speedometer-electronic(sportspack)	1		All KPH
1	DAC 4697	Speedometer-electronic(sportspack)	1		All MPH - not USA
2	DAC 1750	Rev counter	1	Up to 89 MY	
2	DAC 5531	Rev counter	1	From 89 MY	
3	C 46145	Gauge-oil pressure	1		lbs/ins
3	DAC 2926	Gauge-oil pressure	1		Kpa x 100
3	C 42238	Gauge-fuel	1		
3	C 42239	Gauge-battery	1		
3	C 42240	Gauge-water temperature	1		
4	DAC 2181	Strip-illuminated symbol	1	] Up to (V)148781	AUS/CDN/USA
4	DAC 2182	Strip-illuminated symbol	1	]	All other Countries
4	DAC 6558	Strip-illuminated symbol	1	]	USA only
4	DAC 6556	Strip-illuminated symbol	1	] From (V)148782	Japan only
4	DAC 6557	Strip-illuminated symbol	1	]	UK/R.O.W
5	JLM 238	Instrument facia	1		MPH
5	JLM 239	Instrument facia	1		KPH
6	JLM 193	Casing-instrument pack	1		
7	DAC 4579	Printed circuit	1	Up to (V)148781	
7	DAC 3152	Printed circuit	1	From (V)148782	
8	AAU 7574	Plate-retaining	1		
9	13H 5270 J	Bulb-holder	5		
10	C 42289	Bulb holder	17		
11	C 43898	Bulb-instrument illumination	5		
12	C 38966	Bulb-warning light	17		
13	BD 47775 1	Strip-sealing	1		
14	C 44700	Label-"unleaded fuel only"	1		
	DAC 6204	Label-"Premium unleaded"	1		
15	AAU 3364	Finisher-RH	1		
15	AAU 3363	Finisher-LH	1		

K52.02JP31/8/88

NOTE:
(1). From (V) 144263 to (V)148781

MASTER INDEX

GROUP INDEX

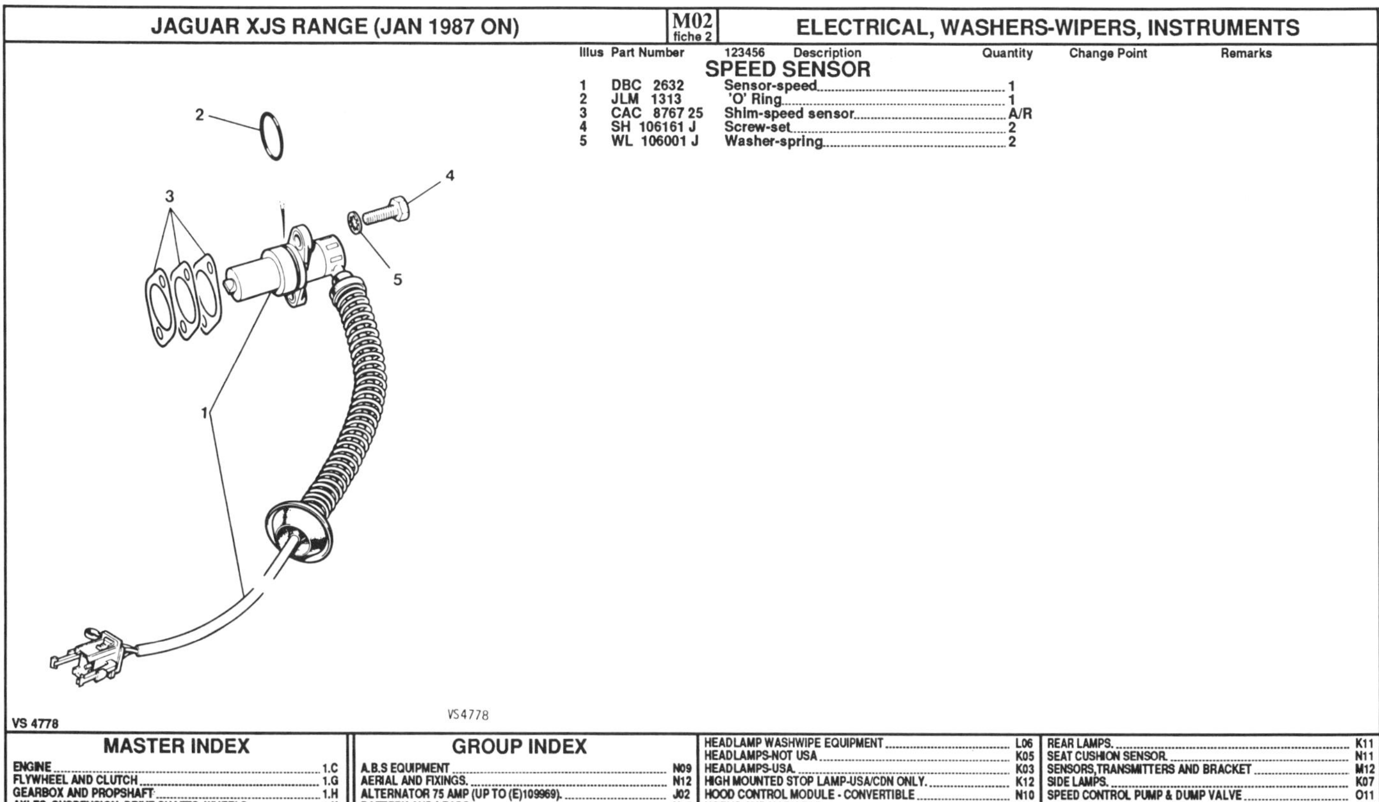

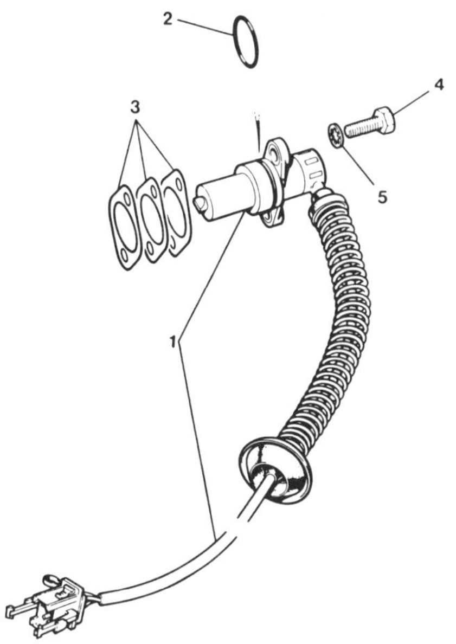

VS 4778

VS4778

SPEED SENSOR

Illus	Part Number	123456 Description	Quantity	Change Point	Remarks
1	DBC 2632	Sensor-speed	1		
2	JLM 1313	'O' Ring	1		
3	CAC 8767 25	Shim-speed sensor	A/R		
4	SH 106161 J	Screw-set	2		
5	WL 106001 J	Washer-spring	2		

MASTER INDEX

ENGINE 1.C
FLYWHEEL AND CLUTCH 1.G
GEARBOX AND PROPSHAFT 1.H
AXLES, SUSPENSION, DRIVE SHAFTS, WHEELS 1.K
STEERING 1.M
BRAKES AND BRAKE CONTROLS 1.N
FUEL, EXHAUST AND EMISSION SYSTEMS 2.C
COOLING, HEATING AND AIR CONDITIONING 2.G
ELECTRICAL, WASHERS-WIPERS, INSTRUMENTS 2.J
CHASSIS, SUBFRAMES, BODYSHELL, FITTINGS 3.C
FASCIA, TRIM, SEATS AND FIXINGS 3.H
ACCESSORIES AND PAINT 4.C
NUMERICAL INDEX 4.D

GROUP INDEX

A.B.S EQUIPMENT N09
AERIAL AND FIXINGS. N12
ALTERNATOR 75 AMP (UP TO (E)109969). J02
BATTERY AND LEADS. N17
CONSOLE SWITCHES AND CIGAR LIGHTER M08
CONTROL UNIT-PJ-V12 COUPE L11
FROM 89 MY L13
CONTROL UNITS-PI/CRUISE CONTROL 3.6LITRE L10
DISTRIBUTOR-3.6 LITRE. J11
DISTRIBUTOR-5.3 LITRE. J12
FROM 89 MY J13
DOOR WARNING LAMP. K15
FOG REAR GUARD LAMP. K14
FRONT FLASHER LAMPS. K06
FRONT FOG LAMPS. K13
FUEL INTERFACE UNIT AND RELAYS L15
HARNESSES-FRONT 3.6 LITRE-FROM (V)148945 .... O04
HARNESSES-FRONT 5.3 LITRE-FROM (V)148782 .... O06
HARNESSES-FRONT-3.6 LITRE-UPTO (V)148944 .... O03
HEADLAMP WASH WIPE RESERVOIR L07
HEADLAMP WASHWIPE EQUIPMENT L06
HEADLAMPS-NOT USA K05
HEADLAMPS-USA. K03
HIGH MOUNTED STOP LAMP-USA/CDN ONLY. K12
HOOD CONTROL MODULE - CONVERTIBLE N10
HORNS AND HORN PAD M05
IGNITION COIL & MODULE-3.6 LITRE J14
IGNITION COILS AND AMPLIFIER-5.3 LITRE J15
INERTIA SWITCH AND COVER M13
INSTRUMENT PACK-3.6 LITRE L17
INSTRUMENT PACK-5.3 LITRE L18
INTERIOR LAMPS-CABRIOLET. K09
INTERIOR LAMPS-COUPE. K08
LOUDSPEAKERS AND MOUNTINGS. N15
MAIN FUSE BLOCK AND MOUNTING PLATE M15
NUMBER PLATE AND REVERSE LAMP. K16
PASSIVE RESTRAINT MODULE & FIXINGS N08
PLUGS AND RELAYS. N07
POWER RESISTOR AND LAMBDA SENSOR. L16
RADIO CASSETTE PLAYER-FROM (V)148782 .... N14
SHOULDER BAG-REMOVEABLE RADIO N14
RADIO'S-UP TO (V)148781 .... N13
REAR LAMPS. K11
SEAT CUSHION SENSOR. N11
SENSORS,TRANSMITTERS AND BRACKET M12
SIDE LAMPS. K07
SPEED CONTROL PUMP & DUMP VALVE O11
STARTER MOTOR - 3.6 LITRE J08
STARTER MOTOR 5.3 LITRE J09
STEERING COLUMN SWITCHES. M04
TRIP COMPUTER AND PANEL SWITCHES. M06
WASH/WIPE RESERVOIR AND FIXINGS L08
WINDSCREEN WASHER RESERVOIR L04
WINDSCREEN WIPER MOTOR-LUCAS TYPE K17
WINDSCREEN WIPERS L03

2M 03

TRANSDUCER - UP TO (V) 144262

Illus	Part Number	123456 Description	Quantity	Change Point	Remarks
1	DAC 4569	Transducer	1		

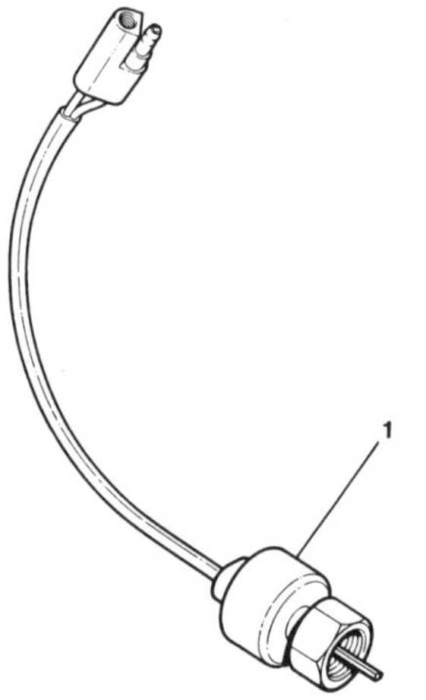

VS 3638A

2M 04

STEERING COLUMN SWITCHES.

Illus	Part Number	123456 Description	Quantity	Change Point	Remarks
		COLUMN SWITCHES			
1	AEU 2525	Direction indicator	1		Speed control
1	AEU 2526	Direction indicator	1		Not speed control
2	AEU 2527	Windscreen wiper	1		With Lucas wiper motor
2	JLM 1670	Windscreen wiper	1		With Electrolux wiper motor
3	CAC 2884	Switch cover upper	1		
4	CAC 2885	Switch cover lower	1		
5	CAC 2886	Blanking plug	1		
6	UCS 416 2H	Screw set	3		
7	RTC 2678	Screw	2		
8	AB 608031 J	Screw-set	1		

K58JP20/7/88

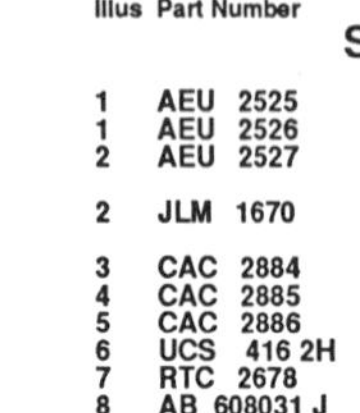

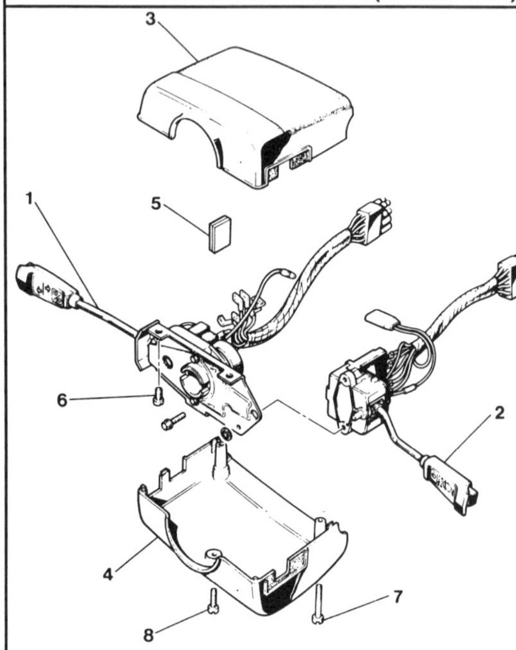

SM 3611

SM3611

MASTER INDEX

ENGINE ... 1.C
FLYWHEEL AND CLUTCH ... 1.G
GEARBOX AND PROPSHAFT ... 1.H
AXLES, SUSPENSION, DRIVE SHAFTS, WHEELS ... 1.K
STEERING ... 1.M
BRAKES AND BRAKE CONTROLS ... 1.N
FUEL, EXHAUST AND EMISSION SYSTEMS ... 2.C
COOLING, HEATING AND AIR CONDITIONING ... 2.G
ELECTRICAL, WASHERS-WIPERS, INSTRUMENTS ... 2.J
CHASSIS, SUBFRAMES, BODYSHELL, FITTINGS ... 3.C
FASCIA, TRIM, SEATS AND FIXINGS ... 3.H
ACCESSORIES AND PAINT ... 4.C
NUMERICAL INDEX ... 4.D

GROUP INDEX

A.B.S EQUIPMENT ... N09
AERIAL AND FIXINGS ... N12
ALTERNATOR 75 AMP (UP TO (E)109969) ... J02
BATTERY AND LEADS ... N17
CONSOLE SWITCHES AND CIGAR LIGHTER ... M08
CONTROL UNIT-PJ-V12 COUPE ... L11
FROM 89 MY ... L13
CONTROL UNITS-PI/CRUISE CONTROL 3.6LITRE ... L10
DISTRIBUTOR-3.6 LITRE ... J11
DISTRIBUTOR-5.3 LITRE ... J12
FROM 89 MY ... J13
DOOR WARNING LAMP ... K15
FOG REAR GUARD LAMP ... K14
FRONT FLASHER LAMPS ... K06
FRONT FOG LAMPS ... K13
FUEL INTERFACE UNIT AND RELAYS ... L15
HARNESSES-FRONT 3.6 LITRE-FROM (V)148945 ... O04
HARNESSES-FRONT 5.3 LITRE-FROM (V)148782 ... O06
HARNESSES-FRONT-3.6 LITRE-UPTO (V)148944 ... O03
HEADLAMP WASH WIPE RESERVOIR ... L07
HEADLAMP WASHWIPE EQUIPMENT ... L06
HEADLAMPS-NOT USA ... K05
HEADLAMPS-USA ... K03
HIGH MOUNTED STOP LAMP-USA/CDN ONLY ... K12
HOOD CONTROL MODULE - CONVERTIBLE ... N10
HORNS AND HORN PAD ... M05
IGNITION COIL & MODULE-3.6 LITRE ... J14
IGNITION COILS AND AMPLIFIER-5.3 LITRE ... J15
INERTIA SWITCH AND COVER ... M13
INSTRUMENT PACK-3.6 LITRE ... L17
INSTRUMENT PACK-5.3 LITRE ... L18
INTERIOR LAMPS-CABRIOLET ... K09
INTERIOR LAMPS-COUPE ... K08
LOUDSPEAKERS AND MOUNTINGS ... N15
MAIN FUSE BLOCK AND MOUNTING PLATE ... M15
NUMBER PLATE AND REVERSE LAMP ... K16
PASSIVE RESTRAINT MODULE & FIXINGS ... N08
PLUGS AND RELAYS ... N07
POWER RESISTOR AND LAMBDA SENSOR ... L16
RADIO CASSETTE PLAYER-FROM (V)148782 ... N14
SHOULDER BAG-REMOVEABLE RADIO ... N14
RADIO'S-UP TO (V)148781 ... N13
REAR LAMPS ... K11
SEAT CUSHION SENSOR ... N11
SENSORS,TRANSMITTERS AND BRACKET ... M12
SIDE LAMPS ... K07
SPEED CONTROL PUMP & DUMP VALVE ... O11
STARTER MOTOR - 3.6 LITRE ... J08
STARTER MOTOR 5.3 LITRE ... J09
STEERING COLUMN SWITCHES ... M04
TRIP COMPUTER AND PANEL SWITCHES ... M06
WASH/WIPE RESERVOIR AND FIXINGS ... L08
WINDSCREEN WASHER RESERVOIR ... L04
WINDSCREEN WIPER MOTOR-LUCAS TYPE ... K17
WINDSCREEN WIPERS ... L03

HORNS AND HORN PAD

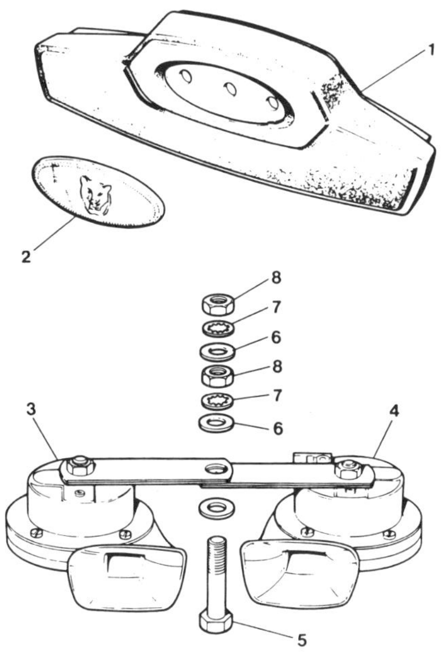

VS 4339

Illus	Part Number	123456 Description	Quantity	Change Point	Remarks
1	CAC 2418	Horn pad	1		Not Sportspack
	CBC 6345	Horn pad	1		Sportspack only
2	C 40184	Centre-horn pad	1		
3	DAC 4537	Horn-high note	1		
4	DAC 4536	Horn-low note	1		
5	BH 605101 J	Bolt-horn fixing	1		
7	C 725	Washer-shakeproof	3		
8	JLM 9684	Nut	2		
6	JLM 301	Washer-plain	3		

TRIP COMPUTER AND PANEL SWITCHES.

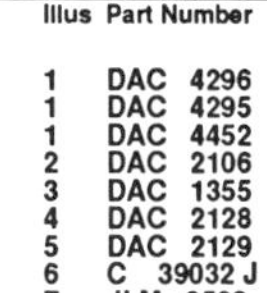

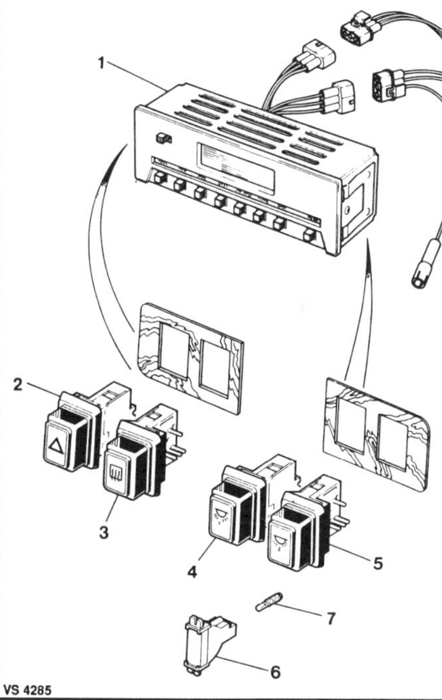

VS 4285

Illus	Part Number	123456 Description	Quantity	Change Point	Remarks
1	DAC 4296	Trip computer	1		5.3 litre USA
1	DAC 4295	Trip computer	1		5.3 litre not USA
1	DAC 4452	Trip computer	1		3.6 litre
2	DAC 2106	Switch-hazard	1		
3	DAC 1355	Switch-heated backlight	1		
4	DAC 2128	Switch-map light	1		
5	DAC 2129	Switch-interior light	1		
6	C 39032 J	Illumination unit-switch	2		
7	JLM 9593	Bulb	2		

MASTER INDEX

ENGINE 1.C
FLYWHEEL AND CLUTCH 1.G
GEARBOX AND PROPSHAFT 1.H
AXLES, SUSPENSION, DRIVE SHAFTS, WHEELS 1.K
STEERING 1.M
BRAKES AND BRAKE CONTROLS 1.N
FUEL, EXHAUST AND EMISSION SYSTEMS 2.C
COOLING, HEATING AND AIR CONDITIONING 2.G
ELECTRICAL, WASHERS-WIPERS, INSTRUMENTS 2.J
CHASSIS, SUBFRAMES, BODYSHELL, FITTINGS 3.C
FASCIA, TRIM, SEATS AND FIXINGS 3.H
ACCESSORIES AND PAINT 4.C
NUMERICAL INDEX 4.D

GROUP INDEX

A.B.S EQUIPMENT N09
AERIAL AND FIXINGS N12
ALTERNATOR 75 AMP (UP TO (E)109969) J02
BATTERY AND LEADS N17
CONSOLE SWITCHES AND CIGAR LIGHTER M08
CONTROL UNIT-PJ-V12 COUPE L11
FROM 89 MY L13
CONTROL UNITS-PI/CRUISE CONTROL 3.6LITRE L10
DISTRIBUTOR-3.6 LITRE J11
DISTRIBUTOR-5.3 LITRE J12
FROM 89 MY J13
DOOR WARNING LAMP K15
FOG REAR GUARD LAMP K14
FRONT FLASHER LAMPS K06
FRONT FOG LAMPS K13
FUEL INTERFACE UNIT AND RELAYS L15
HARNESSES-FRONT 3.6 LITRE-FROM (V)148945 O04
HARNESSES-FRONT 5.3 LITRE-FROM (V)148782 O06
HARNESSES-FRONT-3.6 LITRE-UPTO (V)148944 O03
HEADLAMP WASH WIPE RESERVOIR L07
HEADLAMP WASHWIPE EQUIPMENT L06
HEADLAMPS-NOT USA K05
HEADLAMPS-USA K03
HIGH MOUNTED STOP LAMP-USA/CDN ONLY K12
HOOD CONTROL MODULE - CONVERTIBLE N10
HORNS AND HORN PAD M05
IGNITION COIL & MODULE-3.6 LITRE J14
IGNITION COILS AND AMPLIFIER-5.3 LITRE J15
INERTIA SWITCH AND COVER M13
INSTRUMENT PACK-3.6 LITRE L17
INSTRUMENT PACK-5.3 LITRE L18
INTERIOR LAMPS-CABRIOLET K09
INTERIOR LAMPS-COUPE K08
LOUDSPEAKERS AND MOUNTINGS N15
MAIN FUSE BLOCK AND MOUNTING PLATE M15
NUMBER PLATE AND REVERSE LAMP K16
PASSIVE RESTRAINT MODULE & FIXINGS N08
PLUGS AND RELAYS N07
POWER RESISTOR AND LAMBDA SENSOR L16
RADIO CASSETTE PLAYER-FROM (V)148782 N14
SHOULDER BAG-REMOVEABLE RADIO N14
RADIO'S-UP TO (V)148781 N13
REAR LAMPS K11
SEAT CUSHION SENSOR N11
SENSORS,TRANSMITTERS AND BRACKET M12
SIDE LAMPS K07
SPEED CONTROL PUMP & DUMP VALVE O11
STARTER MOTOR - 3.6 LITRE J08
STARTER MOTOR 5.3 LITRE J09
STEERING COLUMN SWITCHES M04
TRIP COMPUTER AND PANEL SWITCHES M06
WASH/WIPE RESERVOIR AND FIXINGS L08
WINDSCREEN WASHER RESERVOIR L04
WINDSCREEN WIPER MOTOR-LUCAS TYPE K17
WINDSCREEN WIPERS L03

2M 07

CLOCK AND PANEL SWITCHES - 3.6 Litre

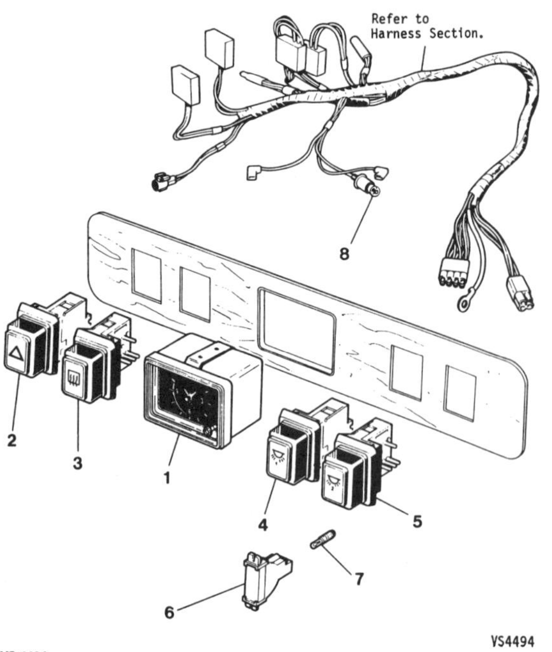

VS 4494

Illus	Part Number	Description	Quantity	Change Point	Remarks
1	DAC 1571	Clock	1		
2	DAC 2106	Switch-hazard	1		
3	DAC 1355	Switch-heated backlight	1		
4	DAC 2128	Switch-map light	1		
5	DAC 2129	Switch-interior light	1		
6	C 39032	Illumination unit-switch	2		
7	JLM 9593	Bulb	2		
8	C 15788	Bulb-clock	1		

2M 08

CONSOLE SWITCHES AND CIGAR LIGHTER

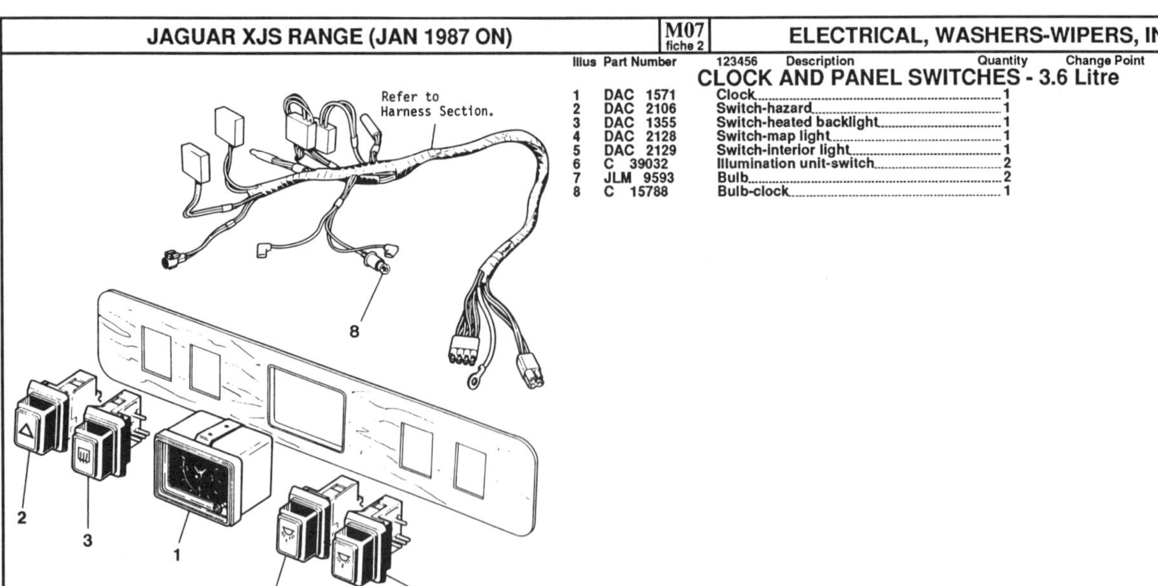

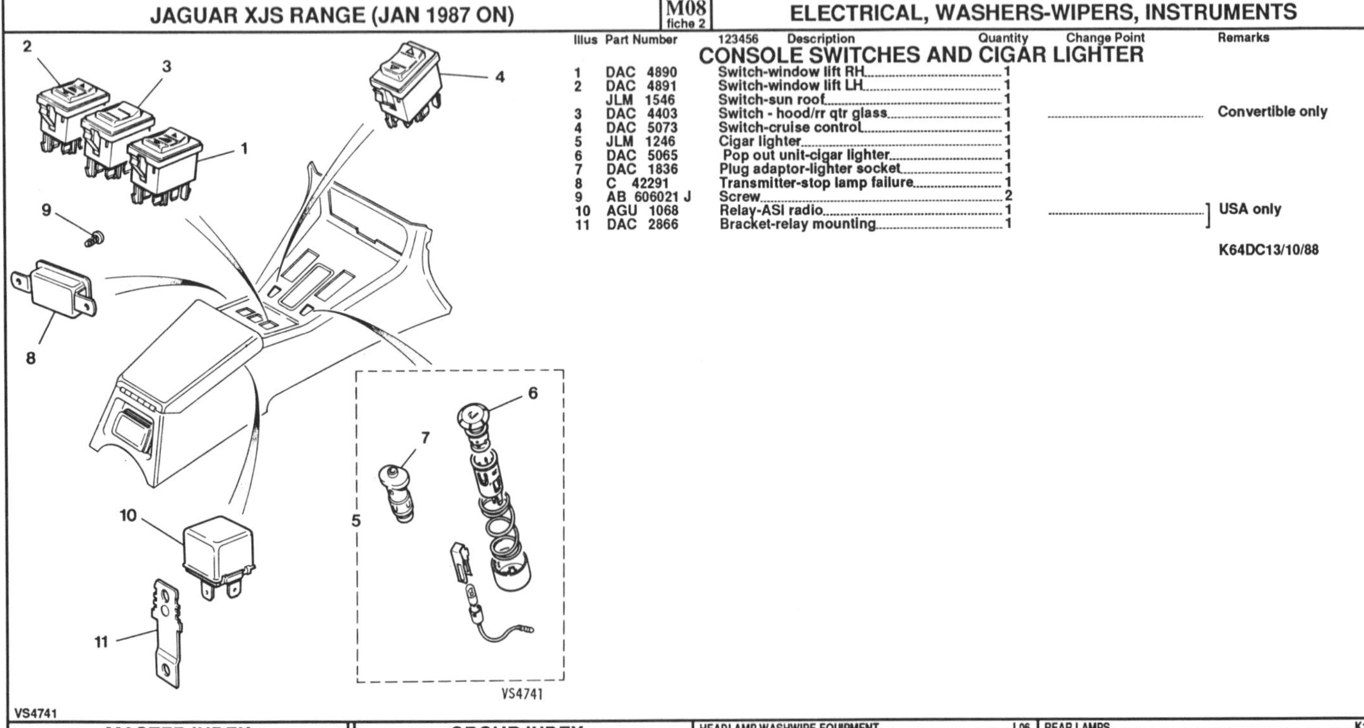

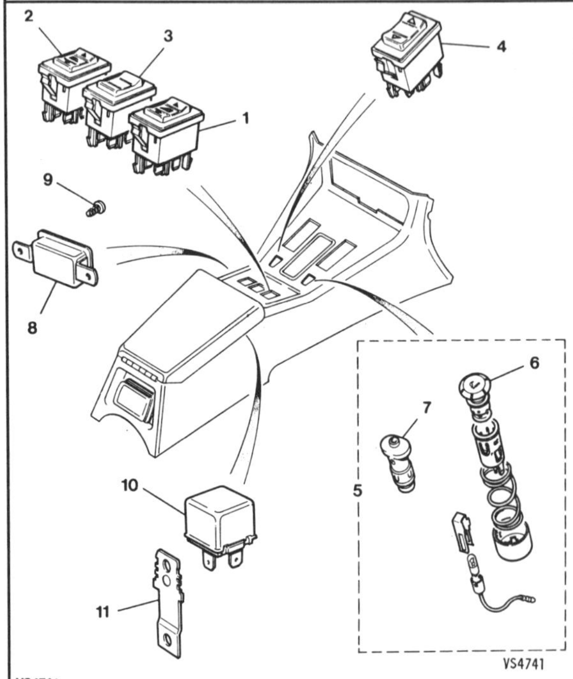

VS4741

Illus	Part Number	Description	Quantity	Change Point	Remarks
1	DAC 4890	Switch-window lift RH	1		
2	DAC 4891	Switch-window lift LH	1		
	JLM 1546	Switch-sun roof	1		
3	DAC 4403	Switch - hood/rr qtr glass	1		Convertible only
4	DAC 5073	Switch-cruise control	1		
5	JLM 1246	Cigar lighter	1		
6	DAC 5065	Pop out unit-cigar lighter	1		
7	DAC 1836	Plug adaptor-lighter socket	1		
8	C 42291	Transmitter-stop lamp failure	1		
9	AB 606021 J	Screw	2		
10	AGU 1068	Relay-ASI radio	1		] USA only
11	DAC 2866	Bracket-relay mounting	1		] USA only

K64DC13/10/88

MASTER INDEX

ENGINE 1.C
FLYWHEEL AND CLUTCH 1.G
GEARBOX AND PROPSHAFT 1.H
AXLES, SUSPENSION, DRIVE SHAFTS, WHEELS 1.K
STEERING 1.M
BRAKES AND BRAKE CONTROLS 1.N
FUEL, EXHAUST AND EMISSION SYSTEMS 2.C
COOLING, HEATING AND AIR CONDITIONING 2.G
ELECTRICAL, WASHERS-WIPERS, INSTRUMENTS 2.J
CHASSIS, SUBFRAMES, BODYSHELL, FITTINGS 3.C
FASCIA, TRIM, SEATS AND FIXINGS 3.H
ACCESSORIES AND PAINT 4.C
NUMERICAL INDEX 4.D

GROUP INDEX

A.B.S EQUIPMENT N09
AERIAL AND FIXINGS. N12
ALTERNATOR 75 AMP (UP TO (E)109969). J02
BATTERY AND LEADS. N17
CONSOLE SWITCHES AND CIGAR LIGHTER M08
CONTROL UNIT-P.J-V12 COUPE L11
FROM 89 MY L13
CONTROL UNITS-PI/CRUISE CONTROL 3.6LITRE L10
DISTRIBUTOR-3.6 LITRE. J11
DISTRIBUTOR-5.3 LITRE. J12
FROM 89 MY J13
DOOR WARNING LAMP. K15
FOG REAR GUARD LAMP. K14
FRONT FLASHER LAMPS. K06
FRONT FOG LAMPS. K13
FUEL INTERFACE UNIT AND RELAYS L15
HARNESSES-FRONT 3.6 LITRE-FROM (V)148945 .... O04
HARNESSES-FRONT 5.3 LITRE-FROM (V)148782 .... O06
HARNESSES-FRONT-3.6 LITRE-UPTO (V)148944 .... O03
HEADLAMP WASH WIPE RESERVOIR L07
HEADLAMP WASHWIPE EQUIPMENT L06
HEADLAMPS-NOT USA K05
HEADLAMPS-USA. K03
HIGH MOUNTED STOP LAMP-USA/CDN ONLY. K12
HOOD CONTROL MODULE - CONVERTIBLE N10
HORNS AND HORN PAD M05
IGNITION COIL & MODULE-3.6 LITRE J14
IGNITION COILS AND AMPLIFIER-5.3 LITRE J15
INERTIA SWITCH AND COVER M13
INSTRUMENT PACK-3.6 LITRE L17
INSTRUMENT PACK-5.3 LITRE L18
INTERIOR LAMPS-CABRIOLET. K09
INTERIOR LAMPS-COUPE. K08
LOUDSPEAKERS AND MOUNTINGS. N15
MAIN FUSE BLOCK AND MOUNTING PLATE M15
NUMBER PLATE AND REVERSE LAMP. K16
PASSIVE RESTRAINT MODULE & FIXINGS N08
PLUGS AND RELAYS. N07
POWER RESISTOR AND LAMBDA SENSOR. L16
RADIO CASSETTE PLAYER-FROM (V)148782 .... N14
SHOULDER BAG-REMOVEABLE RADIO N14
RADIO'S-UP TO (V)148781 .... N13
REAR LAMPS. K11
SEAT CUSHION SENSOR. N11
SENSORS,TRANSMITTERS AND BRACKET M12
SIDE LAMPS. K07
SPEED CONTROL PUMP & DUMP VALVE O11
STARTER MOTOR - 3.6 LITRE J08
STARTER MOTOR 5.3 LITRE J09
STEERING COLUMN SWITCHES. M04
TRIP COMPUTER AND PANEL SWITCHES. M06
WASH/WIPE RESERVOIR AND FIXINGS L08
WINDSCREEN WASHER RESERVOIR L04
WINDSCREEN WIPER MOTOR-LUCAS TYPE K17
WINDSCREEN WIPERS L03

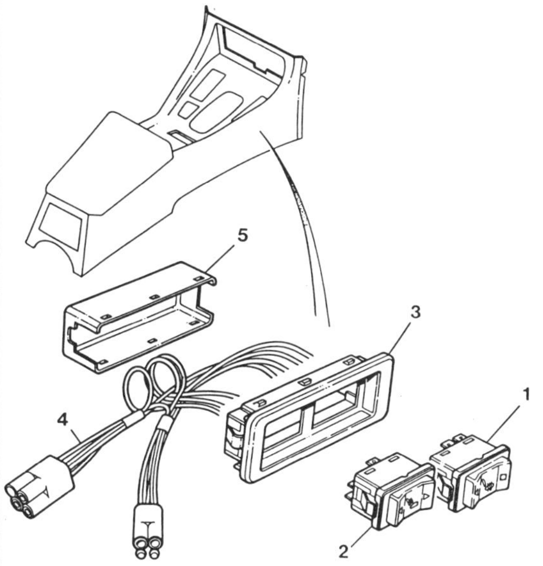

VS 4790

CONSOLE SEAT SWITCHES

Illus	Part Number	123456	Description	Quantity	Change Point	Remarks
1	DAC 5076		Switch-seat heater RH	1		
	DAC 5077		Switch-seat heater LH	1		
2	DAC 4436		Switch-lumbar adjust RH	1		
	DAC 4435		Switch-lumbar adjust LH	1		
3	DAC 4377		Moulding-plinth-double switch	2		
	DAC 4438		Moulding-plinth-single switch	2		
4	DAC 4522		Harness-seat heater & lumbar switch	2		
	DAC 4523		Harness-seat heater only	2		
	DAC 4559		Harness-lumbar adjust only	2		
5	DAC 4519		Cover moulding-console switch	2		

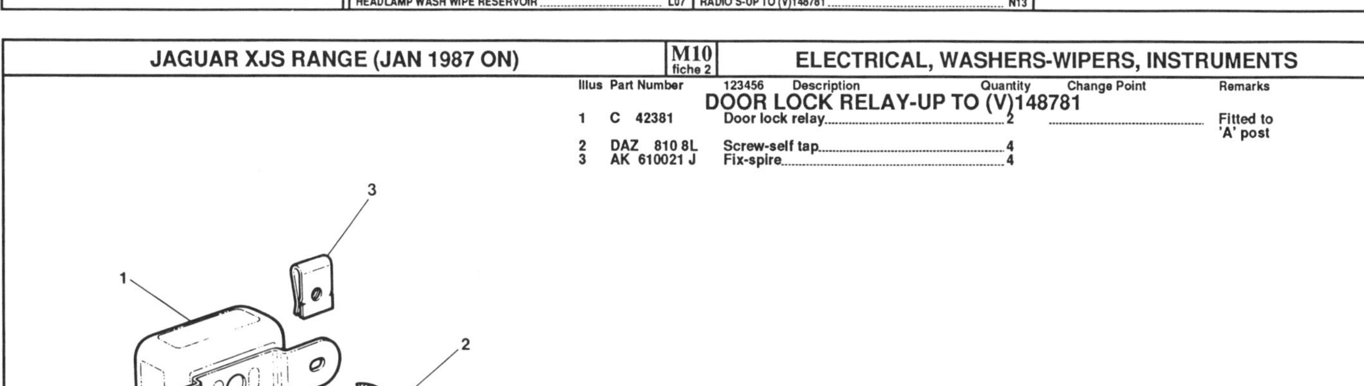

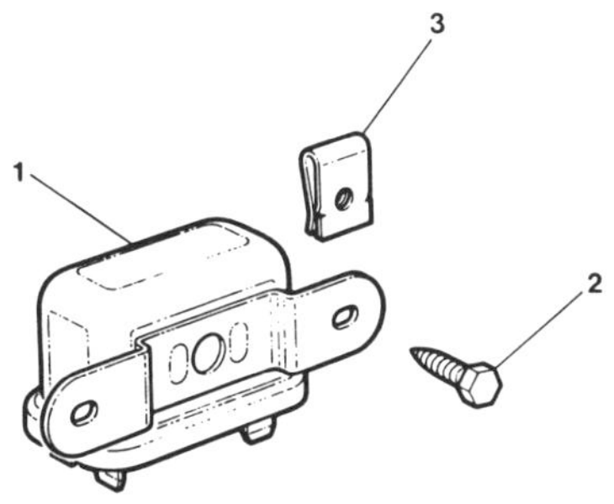

VS 4529

DOOR LOCK RELAY-UP TO (V)148781

Illus	Part Number	123456	Description	Quantity	Change Point	Remarks
1	C 42381		Door lock relay	2	....	Fitted to 'A' post
2	DAZ 810 8L		Screw-self tap	4		
3	AK 610021 J		Fix-spire	4		

MASTER INDEX

ENGINE 1.C
FLYWHEEL AND CLUTCH 1.G
GEARBOX AND PROPSHAFT 1.H
AXLES, SUSPENSION, DRIVE SHAFTS, WHEELS 1.K
STEERING 1.M
BRAKES AND BRAKE CONTROLS 1.N
FUEL, EXHAUST AND EMISSION SYSTEMS 2.C
COOLING, HEATING AND AIR CONDITIONING 2.G
ELECTRICAL, WASHERS-WIPERS, INSTRUMENTS 2.J
CHASSIS, SUBFRAMES, BODYSHELL, FITTINGS 3.C
FASCIA, TRIM, SEATS AND FIXINGS 3.H
ACCESSORIES AND PAINT 4.C
NUMERICAL INDEX 4.D

GROUP INDEX

A.B.S EQUIPMENT N09
AERIAL AND FIXINGS N12
ALTERNATOR 75 AMP (UP TO (E)109969) J02
BATTERY AND LEADS N17
CONSOLE SWITCHES AND CIGAR LIGHTER M08
CONTROL UNIT-PJ-V12 COUPE L11
FROM 89 MY L13
CONTROL UNITS-PI/CRUISE CONTROL 3.6LITRE L10
DISTRIBUTOR-3.6 LITRE J11
DISTRIBUTOR-5.3 LITRE J12
FROM 89 MY J13
DOOR WARNING LAMP K15
FOG REAR GUARD LAMP K14
FRONT FLASHER LAMPS K06
FRONT FOG LAMPS K13
FUEL INTERFACE UNIT AND RELAYS L15
HARNESSES-FRONT 3.6 LITRE-FROM (V)148945 O04
HARNESSES-FRONT 5.3 LITRE-FROM (V)148782 O06
HARNESSES-FRONT-3.6 LITRE-UPTO (V)148944 O03
HEADLAMP WASH WIPE RESERVOIR L07
HEADLAMP WASHWIPE EQUIPMENT L06
HEADLAMPS-NOT USA K05
HEADLAMPS-USA K03
HIGH MOUNTED STOP LAMP-USA/CDN ONLY K12
HOOD CONTROL MODULE - CONVERTIBLE N10
HORNS AND HORN PAD M05
IGNITION COIL & MODULE-3.6 LITRE J14
IGNITION COILS AND AMPLIFIER-5.3 LITRE J15
INERTIA SWITCH AND COVER M13
INSTRUMENT PACK-3.6 LITRE L17
INSTRUMENT PACK-5.3 LITRE L18
INTERIOR LAMPS-CABRIOLET K09
INTERIOR LAMPS-COUPE K08
LOUDSPEAKERS AND MOUNTINGS N15
MAIN FUSE BLOCK AND MOUNTING PLATE M15
NUMBER PLATE AND REVERSE LAMP K16
PASSIVE RESTRAINT MODULE & FIXINGS N08
PLUGS AND RELAYS N07
POWER RESISTOR AND LAMBDA SENSOR L16
RADIO CASSETTE PLAYER-FROM (V)148782 N14
SHOULDER BAG-REMOVEABLE RADIO N14
RADIO'S-UP TO (V)148781 N13
REAR LAMPS K11
SEAT CUSHION SENSOR N11
SENSORS,TRANSMITTERS AND BRACKET M12
SIDE LAMPS K07
SPEED CONTROL PUMP & DUMP VALVE O11
STARTER MOTOR - 3.6 LITRE J08
STARTER MOTOR 5.3 LITRE J09
STEERING COLUMN SWITCHES M04
TRIP COMPUTER AND PANEL SWITCHES M06
WASH/WIPE RESERVOIR AND FIXINGS L08
WINDSCREEN WASHER RESERVOIR L04
WINDSCREEN WIPER MOTOR-LUCAS TYPE K17
WINDSCREEN WIPERS L03

2M 11

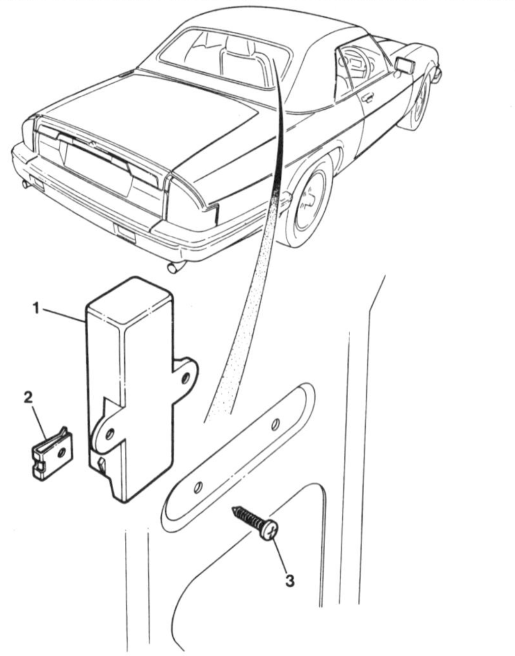

VS4817

VS 4817

CENTRAL LOCKING CONTROL UNIT-FROM (V)148782

Illus	Part Number	123456 Description	Quantity	Change Point	Remarks
1	DAC 4862	Control unit-central locking	1		
2	DAC 4870	Nut-spire	2		
3	AB 604044 J	Screw	2		

2M 12

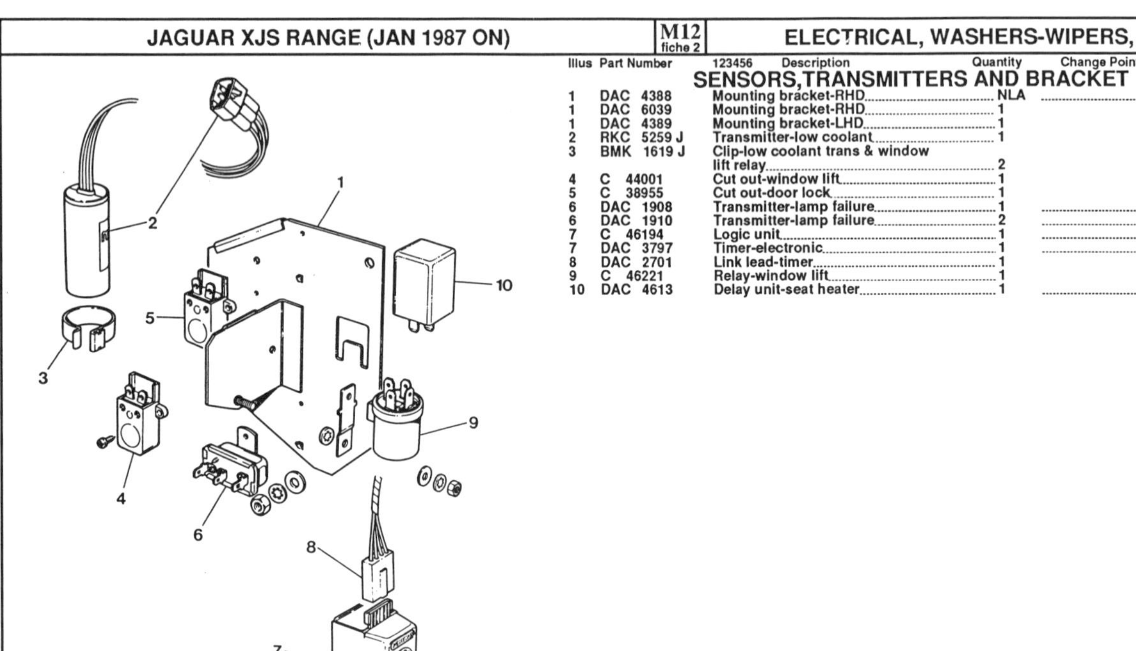

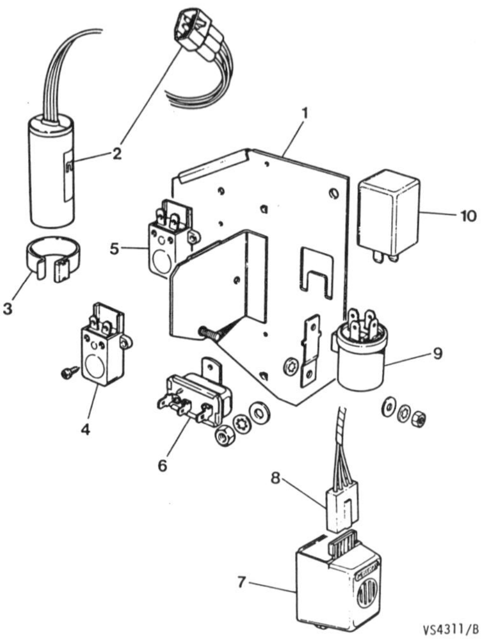

VS4311/B

VS 4311B

SENSORS,TRANSMITTERS AND BRACKET

Illus	Part Number	123456 Description	Quantity	Change Point	Remarks
1	DAC 4388	Mounting bracket-RHD	NLA		Use DAC6039
1	DAC 6039	Mounting bracket-RHD	1		
1	DAC 4389	Mounting bracket-LHD	1		
2	RKC 5259 J	Transmitter-low coolant	1		
3	BMK 1619 J	Clip-low coolant trans & window lift relay	2		
4	C 44001	Cut out-window lift	1		
5	C 38955	Cut out-door lock	1		
6	DAC 1908	Transmitter-lamp failure	1		Green Disc
6	DAC 1910	Transmitter-lamp failure	2		Red disc
7	C 46194	Logic unit	1		CDN/J/USA/ME
7	DAC 3797	Timer-electronic	1		Not catalyst
8	DAC 2701	Link lead-timer	1		
9	C 46221	Relay-window lift	1		
10	DAC 4613	Delay unit-seat heater	1		Passenger seat

MASTER INDEX

ENGINE 1.C
FLYWHEEL AND CLUTCH 1.G
GEARBOX AND PROPSHAFT 1.H
AXLES, SUSPENSION, DRIVE SHAFTS, WHEELS 1.K
STEERING 1.M
BRAKES AND BRAKE CONTROLS 1.N
FUEL, EXHAUST AND EMISSION SYSTEMS 2.C
COOLING, HEATING AND AIR CONDITIONING 2.G
ELECTRICAL, WASHERS-WIPERS, INSTRUMENTS 2.J
CHASSIS, SUBFRAMES, BODYSHELL, FITTINGS 3.C
FASCIA, TRIM, SEATS AND FIXINGS 3.H
ACCESSORIES AND PAINT 4.C
NUMERICAL INDEX 4.D

GROUP INDEX

A.B.S EQUIPMENT N09
AERIAL AND FIXINGS. N12
ALTERNATOR 75 AMP (UP TO (E)109969). J02
BATTERY AND LEADS. N17
CONSOLE SWITCHES AND CIGAR LIGHTER M08
CONTROL UNIT-PJ-V12 COUPE L11
FROM 89 MY L13
CONTROL UNITS-PI/CRUISE CONTROL 3.6LITRE L10
DISTRIBUTOR-3.6 LITRE. J11
DISTRIBUTOR-5.3 LITRE. J12
FROM 89 MY J13
DOOR WARNING LAMP. K15
FOG REAR GUARD LAMP. K14
FRONT FLASHER LAMPS. K06
FRONT FOG LAMPS. K13
FUEL INTERFACE UNIT AND RELAYS L15
HARNESSES-FRONT 3.6 LITRE-FROM (V)148945 ..... O04
HARNESSES-FRONT 5.3 LITRE-FROM (V)148782 ..... O06
HARNESSES-FRONT-3.6 LITRE-UPTO (V)148944 ..... O03
HEADLAMP WASH WIPE RESERVOIR L07
HEADLAMP WASHWIPE EQUIPMENT L06
HEADLAMPS-NOT USA K05
HEADLAMPS-USA. K03
HIGH MOUNTED STOP LAMP-USA/CDN ONLY. K12
HOOD CONTROL MODULE - CONVERTIBLE N10
HORNS AND HORN PAD M05
IGNITION COIL & MODULE-3.6 LITRE J14
IGNITION COILS AND AMPLIFIER-5.3 LITRE J15
INERTIA SWITCH AND COVER M13
INSTRUMENT PACK-3.6 LITRE L17
INSTRUMENT PACK-5.3 LITRE L18
INTERIOR LAMPS-CABRIOLET. K09
INTERIOR LAMPS-COUPE. K08
LOUDSPEAKERS AND MOUNTINGS. N15
MAIN FUSE BLOCK AND MOUNTING PLATE M15
NUMBER PLATE AND REVERSE LAMP. K16
PASSIVE RESTRAINT MODULE & FIXINGS N08
PLUGS AND RELAYS. N07
POWER RESISTOR AND LAMBDA SENSOR. L16
RADIO CASSETTE PLAYER-FROM (V)148782 ..... N14
SHOULDER BAG-REMOVEABLE RADIO N14
RADIO'S-UP TO (V)148781 ..... N13
REAR LAMPS. K11
SEAT CUSHION SENSOR. N11
SENSORS,TRANSMITTERS AND BRACKET M12
SIDE LAMPS. K07
SPEED CONTROL PUMP & DUMP VALVE O11
STARTER MOTOR - 3.6 LITRE J08
STARTER MOTOR 5.3 LITRE J09
STEERING COLUMN SWITCHES. M04
TRIP COMPUTER AND PANEL SWITCHES. M06
WASH/WIPE RESERVOIR AND FIXINGS L08
WINDSCREEN WASHER RESERVOIR L04
WINDSCREEN WIPER MOTOR-LUCAS TYPE K17
WINDSCREEN WIPERS L03

INERTIA SWITCH AND COVER

Illus	Part Number	Description	Quantity	Change Point	Remarks
1	DBC 2022	Switch-inertia	1		
2	DAC 1762	Bracket-mounting	1		
3	DAC 1821	Cover-inertia switch	1		
4	RU 610183	Rivet-pop	2		
5	AB 604021 J	Screw	2		

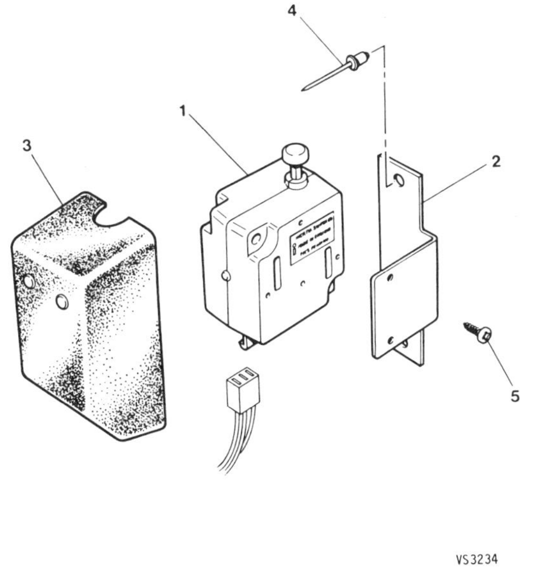

VS 3234

FIBRE OPTIC UNIT.

Illus	Part Number	Description	Quantity	Change Point	Remarks
1	DAC 4307	Opticell unit	1		
2	DAC 2662 1	Optic strand-fibre-Red	1		830mm
2	DAC 2662 2	Optic strand-fibre-Red	1		1120mm
2	DAC 2662 8	Optic strand-fibre-Red	1		490mm
2	DAC 2662 9	Optic strand-fibre-Blue	1		535mm
2	DAC 2663 2	Optic strand-fibre-long ferule	1		1120mm
3	DAC 2095 1	Tubing	1		
4	JLM 9601	Bulb	1		
5	DAC 2621	Cover-opticell unit	1		
6	DAC 5064	Bracket-opticell mounting	1		Auto
6	DAC 4889	Bracket-opticell mounting	1		Manual

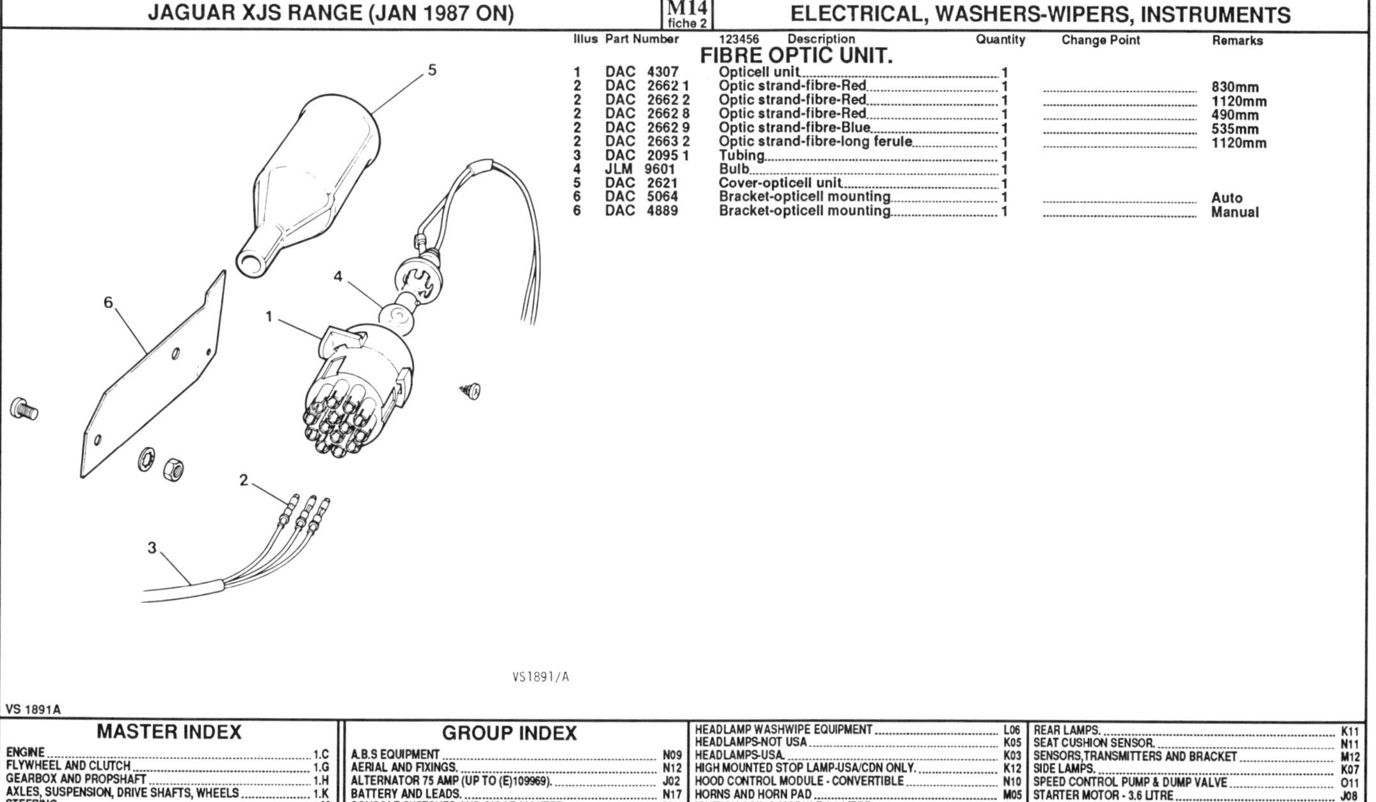

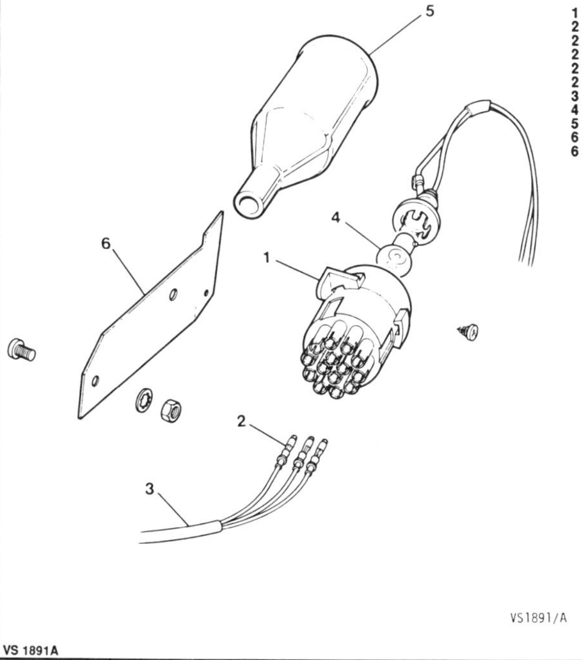

VS 1891A

MASTER INDEX

ENGINE 1.C
FLYWHEEL AND CLUTCH 1.G
GEARBOX AND PROPSHAFT 1.H
AXLES, SUSPENSION, DRIVE SHAFTS, WHEELS 1.K
STEERING 1.M
BRAKES AND BRAKE CONTROLS 1.N
FUEL, EXHAUST AND EMISSION SYSTEMS 2.C
COOLING, HEATING AND AIR CONDITIONING 2.G
ELECTRICAL, WASHERS-WIPERS, INSTRUMENTS 2.J
CHASSIS, SUBFRAMES, BODYSHELL, FITTINGS 3.C
FASCIA, TRIM, SEATS AND FIXINGS 3.H
ACCESSORIES AND PAINT 4.C
NUMERICAL INDEX 4.D

GROUP INDEX

A.B.S EQUIPMENT N09
AERIAL AND FIXINGS N12
ALTERNATOR 75 AMP (UP TO (E)109969) J02
BATTERY AND LEADS N17
CONSOLE SWITCHES AND CIGAR LIGHTER M08
CONTROL UNIT-PJ-V12 COUPE L11
FROM 89 MY L13
CONTROL UNITS-PI/CRUISE CONTROL 3.6LITRE L10
DISTRIBUTOR-3.6 LITRE J11
DISTRIBUTOR-5.3 LITRE J12
FROM 89 MY J13
DOOR WARNING LAMP K15
FOG REAR GUARD LAMP K14
FRONT FLASHER LAMPS K06
FRONT FOG LAMPS K13
FUEL INTERFACE UNIT AND RELAYS L15
HARNESSES-FRONT 3.6 LITRE-FROM (V)148945 O04
HARNESSES-FRONT 5.3 LITRE-FROM (V)148782 O06
HARNESSES-FRONT-3.6 LITRE-UPTO (V)148944 O03
HEADLAMP WASH WIPE RESERVOIR L07
HEADLAMP WASHWIPE EQUIPMENT L06
HEADLAMPS-NOT USA K05
HEADLAMPS-USA K03
HIGH MOUNTED STOP LAMP-USA/CDN ONLY K12
HOOD CONTROL MODULE - CONVERTIBLE N10
HORNS AND HORN PAD M05
IGNITION COIL & MODULE-3.6 LITRE J14
IGNITION COILS AND AMPLIFIER-5.3 LITRE J15
INERTIA SWITCH AND COVER M13
INSTRUMENT PACK-3.6 LITRE L17
INSTRUMENT PACK-5.3 LITRE L18
INTERIOR LAMPS-CABRIOLET K09
INTERIOR LAMPS-COUPE K08
LOUDSPEAKERS AND MOUNTINGS N15
MAIN FUSE BLOCK AND MOUNTING PLATE M15
NUMBER PLATE AND REVERSE LAMP K16
PASSIVE RESTRAINT MODULE & FIXINGS N08
PLUGS AND RELAYS N07
POWER RESISTOR AND LAMBDA SENSOR L16
RADIO CASSETTE PLAYER-FROM (V)148782 N14
SHOULDER BAG-REMOVEABLE RADIO N14
RADIO'S-UP TO (V)148781 N13
REAR LAMPS K11
SEAT CUSHION SENSOR N11
SENSORS,TRANSMITTERS AND BRACKET M12
SIDE LAMPS K07
SPEED CONTROL PUMP & DUMP VALVE O11
STARTER MOTOR - 3.6 LITRE J08
STARTER MOTOR 5.3 LITRE J09
STEERING COLUMN SWITCHES M04
TRIP COMPUTER AND PANEL SWITCHES M06
WASH/WIPE RESERVOIR AND FIXINGS L08
WINDSCREEN WASHER RESERVOIR L04
WINDSCREEN WIPER MOTOR-LUCAS TYPE K17
WINDSCREEN WIPERS L03

2M 15

Illus	Part Number	123456 Description	Quantity	Change Point	Remarks
		MAIN FUSE BLOCK AND MOUNTING PLATE			
1	DAC 4664	Mounting plate-main fuse block RHD	1		
1	DAC 4665	Mounting plate-main fuse block LHD	1		
2	DBC 2967	Fuse block-main	1		
3	DAC 4423	Mounting bracket-fuse box	1		
4	DGC 2490	Fuse-2 amp	A/R		
4	DGC 2491	Fuse-3 amp	A/R		
4	DGC 2492	Fuse-4 amp	A/R		
4	DGC 2493	Fuse-5 amp	A/R		
4	DGC 2494	Fuse-7.5 amp	A/R		
4	DGC 2495	Fuse-10 amp	A/R		
4	DGC 2496	Fuse-15 amp	A/R		
4	DGC 2497	Fuse-20 amp	A/R		
4	DGC 2498	Fuse-25 amp	A/R		
4	DGC 2499	Fuse-30 amp	A/R		
5	UFS 81910 H	Setscrew	2		
6	WF 702100 J	Washer-shakeproof	2		
7	DBC 3085	Extractor-fuse	1		

K72JP31/8/88

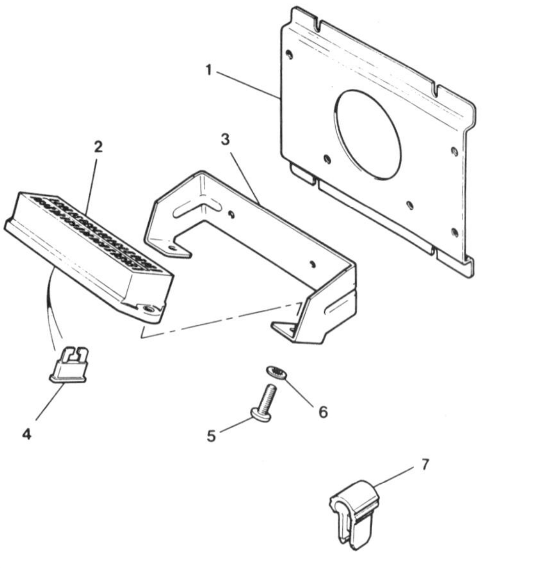

VS4742

2M 16

Illus	Part Number	123456 Description	Quantity	Change Point	Remarks
		MAIN FUSE BOX MTG PLATE RELAYS.			
1	DAC 4660	Plate-mounting	1		RHD
1	DAC 4661	Plate-mounting	1		LHD
2	DAC 1328	Delay unit-heated backlight	1		
3	DAC 1027	Relay-fog rear guard	1		
4	AGU 1068	Relay-ignition switch	1		
5	DAC 1562	Bulb check unit	1		
6	DAC 1731	Flasher unit-standard	1		
6	C 38850	Flasher unit-for towing	1		
7	AGU 1068	Relay-seat load	1		
8	DBC 2483	Relay-front fog lamp	1		
9	DAC 4613	Delay unit-seat heater	1		Driver seat
10	C 43773	Indicator unit-over voltage	1		
11	UFS 819 3H	Setscrew	2		
12	WF 702100 J	Washer	1		

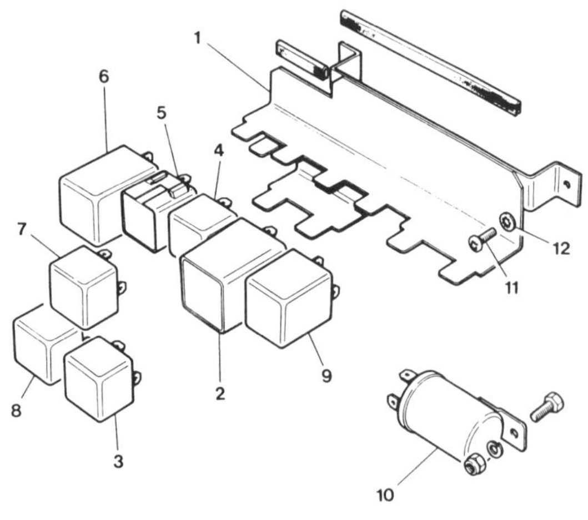

VS 4832

MASTER INDEX

ENGINE 1.C
FLYWHEEL AND CLUTCH 1.G
GEARBOX AND PROPSHAFT 1.H
AXLES, SUSPENSION, DRIVE SHAFTS, WHEELS 1.K
STEERING 1.M
BRAKES AND BRAKE CONTROLS 1.N
FUEL, EXHAUST AND EMISSION SYSTEMS 2.C
COOLING, HEATING AND AIR CONDITIONING 2.G
ELECTRICAL, WASHERS-WIPERS, INSTRUMENTS 2.J
CHASSIS, SUBFRAMES, BODYSHELL, FITTINGS 3.C
FASCIA, TRIM, SEATS AND FIXINGS 3.H
ACCESSORIES AND PAINT 4.C
NUMERICAL INDEX 4.D

GROUP INDEX

A.B.S EQUIPMENT N09
AERIAL AND FIXINGS. N12
ALTERNATOR 75 AMP (UP TO (E)109969). J02
BATTERY AND LEADS. N17
CONSOLE SWITCHES AND CIGAR LIGHTER M08
CONTROL UNIT-P.I-V12 COUPE L11
FROM 89 MY L13
CONTROL UNITS-PI/CRUISE CONTROL 3.6LITRE L10
DISTRIBUTOR-3.6 LITRE. J11
DISTRIBUTOR-5.3 LITRE. J12
FROM 89 MY J13
DOOR WARNING LAMP. K15
FOG REAR GUARD LAMP. K14
FRONT FLASHER LAMPS. K06
FRONT FOG LAMPS. K13
FUEL INTERFACE UNIT AND RELAYS L15
HARNESSES-FRONT 3.6 LITRE-FROM (V)148945 O04
HARNESSES-FRONT 5.3 LITRE-FROM (V)148782 O06
HARNESSES-FRONT-3.6 LITRE-UPTO (V)148944 O03
HEADLAMP WASH WIPE RESERVOIR L07
HEADLAMP WASHWIPE EQUIPMENT L06
HEADLAMPS-NOT USA K05
HEADLAMPS-USA. K03
HIGH MOUNTED STOP LAMP-USA/CDN ONLY. K12
HOOD CONTROL MODULE - CONVERTIBLE N10
HORNS AND HORN PAD M05
IGNITION COIL & MODULE-3.6 LITRE J14
IGNITION COILS AND AMPLIFIER-5.3 LITRE J15
INERTIA SWITCH AND COVER M13
INSTRUMENT PACK-3.6 LITRE L17
INSTRUMENT PACK-5.3 LITRE L18
INTERIOR LAMPS-CABRIOLET. K09
INTERIOR LAMPS-COUPE. K08
LOUDSPEAKERS AND MOUNTINGS. N15
MAIN FUSE BLOCK AND MOUNTING PLATE M15
NUMBER PLATE AND REVERSE LAMP. K16
PASSIVE RESTRAINT MODULE & FIXINGS N08
PLUGS AND RELAYS. N07
POWER RESISTOR AND LAMBDA SENSOR. L16
RADIO CASSETTE PLAYER-FROM (V)148782 N14
SHOULDER BAG-REMOVEABLE RADIO N14
RADIO'S-UP TO (V)148781 N13
REAR LAMPS. K11
SEAT CUSHION SENSOR. N11
SENSORS,TRANSMITTERS AND BRACKET M12
SIDE LAMPS. K07
SPEED CONTROL PUMP & DUMP VALVE O11
STARTER MOTOR - 3.6 LITRE J08
STARTER MOTOR 5.3 LITRE J09
STEERING COLUMN SWITCHES. M04
TRIP COMPUTER AND PANEL SWITCHES. M06
WASH/WIPE RESERVOIR AND FIXINGS L08
WINDSCREEN WASHER RESERVOIR L04
WINDSCREEN WIPER MOTOR-LUCAS TYPE K17
WINDSCREEN WIPERS L03

AUX FUSE BOX AND MOUNTING BRACKET

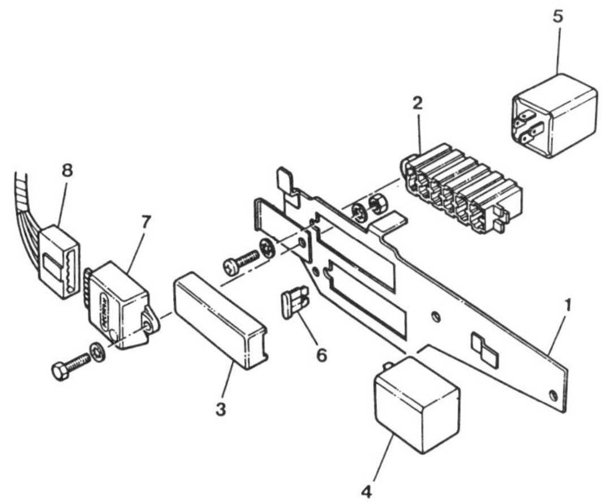

VS4788

Illus	Part Number	123456 Description	Quantity	Change Point	Remarks
1	DAC 4380	Bracket-mounting	1	Up to (V)148782	R.H.D.
1	DAC 4381	Bracket-mounting	1		L.H.D.
1	DAC 5320	Bracket - mounting	1		RHD Japan only
1	DAC 5316	Bracket - mounting	1	See (1) below	RHD not Japan
1	DAC 5317	Bracket - mounting	1		LHD
1	DAC 6222	Bracket - mounting	1	From 89 MY	RHD
1	DAC 6221	Bracket - mounting	1		LHD
2	DBC 2972	Fuse block-(Black)	1		
3	DBC 3060	Cover-fuse block-(Black)	1		
4	DBC 2760	Delay unit-interior light	1		
5	DAC 1329	Delay unit-wiper	1		With Lucas type wiper motor
5	DAC 6053	Delay unit-wiper	1		With Electrolux type wiper motor
6	DGC 2490	Fuse-2 amp	A/R		
6	DGC 2491	Fuse-3 amp	A/R		
6	DGC 2492	Fuse-4 amp	A/R		
6	DGC 2493	Fuse-5 amp	A/R		
6	DGC 2494	Fuse-7.5 amp	A/R		
6	DGC 2495	Fuse-10 amp	A/R		
6	DGC 2496	Fuse-15 amp	A/R		
6	DGC 2497	Fuse-20 amp	A/R		
6	DGC 2498	Fuse-25 amp	A/R		
6	DGC 2499	Fuse-35 amp	A/R		
7	DAC 3796	Timer-Electronic	1		With catalyst
8	DAC 2711	Link Lead-Timer	1		

K76JP31/8/88

(1) From (V) 148783 up to 89 MY.

VS 4788

HEADLAMP FUSE BLOCK & RELAY MTG BRKT-3.6 LITRE

UP TO (V)148838

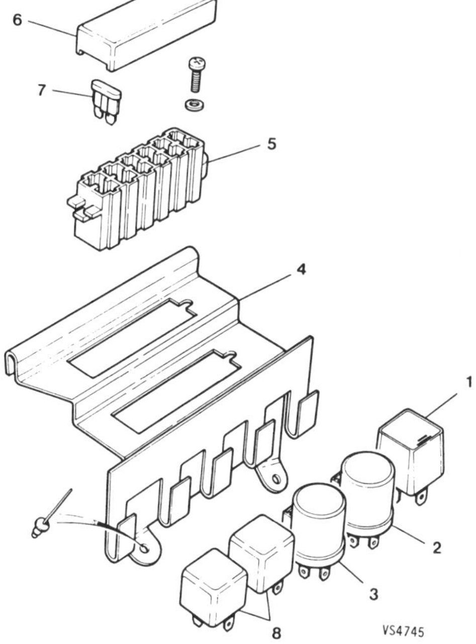

VS4745

Illus	Part Number	123456 Description	Quantity	Change Point	Remarks
1	DAC 1996	Pektron unit-fan run on	1		
2	DAC 1027	Relay-horn	1		
3	DAC 1028	Relay-fan	1		
4	DAC 5051	Bracket-headlamp fuse box	1		Not air inject
4	TBA	Bracket-headlamp fuse box	1		With air inject
5	DBC 2972	Fuse block-(Black)	1		
5	DBC 2973	Fuse block-(natural)	1		
6	DBC 3060	Cover-fuse block-(Black)	1		
6	DBC 3778	Cover-fuse block-(natural)	1		
7	DGC 2490	Fuse-2 amp	A/R		
7	DGC 2491	Fuse-3 amp	A/R		
7	DGC 2492	Fuse-4 amp	A/R		
7	DGC 2493	Fuse-5 amp	A/R		
7	DGC 2494	Fuse-7.5 amp	A/R		
7	DGC 2495	Fuse-10 amp	A/R		
7	DGC 2496	Fuse-15 amp	A/R		
7	DGC 2497	Fuse-20 amp	A/R		
	DGC 2498	Fuse-25 amp	A/R		
	DGC 2499	Fuse-30 amp	A/R		
8	DBC 2484	Relay-heated breather	1		With air inject
8	DBC 2484	Relay-air pump clutch	1		

VS4745

MASTER INDEX

ENGINE 1.C
FLYWHEEL AND CLUTCH 1.G
GEARBOX AND PROPSHAFT 1.H
AXLES, SUSPENSION, DRIVE SHAFTS, WHEELS 1.K
STEERING 1.M
BRAKES AND BRAKE CONTROLS 1.N
FUEL, EXHAUST AND EMISSION SYSTEMS 2.C
COOLING, HEATING AND AIR CONDITIONING 2.G
ELECTRICAL, WASHERS-WIPERS, INSTRUMENTS 2.J
CHASSIS, SUBFRAMES, BODYSHELL, FITTINGS 3.C
FASCIA, TRIM, SEATS AND FIXINGS 3.H
ACCESSORIES AND PAINT 4.C
NUMERICAL INDEX 4.D

GROUP INDEX

A.B.S EQUIPMENT N09
AERIAL AND FIXINGS. N12
ALTERNATOR 75 AMP (UP TO (E)109969). J02
BATTERY AND LEADS. N17
CONSOLE SWITCHES AND CIGAR LIGHTER M08
CONTROL UNIT-P.J-V12 COUPE L11
FROM 89 MY L13
CONTROL UNITS-PI/CRUISE CONTROL 3.6LITRE L10
DISTRIBUTOR-3.6 LITRE. J11
DISTRIBUTOR-5.3 LITRE. J12
FROM 89 MY J13
DOOR WARNING LAMP. K15
FOG REAR GUARD LAMP. K14
FRONT FLASHER LAMPS. K06
FRONT FOG LAMPS. K13
FUEL INTERFACE UNIT AND RELAYS L15
HARNESSES-FRONT 3.6 LITRE-FROM (V)148945 .... O04
HARNESSES-FRONT 5.3 LITRE-FROM (V)148782 .... O06
HARNESSES-FRONT-3.6 LITRE-UPTO (V)148944 .... O03
HEADLAMP WASH WIPE RESERVOIR L07
HEADLAMP WASHWIPE EQUIPMENT L06
HEADLAMPS-NOT USA K05
HEADLAMPS-USA. K03
HIGH MOUNTED STOP LAMP-USA/CDN ONLY. K12
HOOD CONTROL MODULE - CONVERTIBLE N10
HORNS AND HORN PAD M05
IGNITION COIL & MODULE-3.6 LITRE J14
IGNITION COILS AND AMPLIFIER-5.3 LITRE J15
INERTIA SWITCH AND COVER M13
INSTRUMENT PACK-3.6 LITRE L17
INSTRUMENT PACK-5.3 LITRE L18
INTERIOR LAMPS-CABRIOLET. K09
INTERIOR LAMPS-COUPE. K08
LOUDSPEAKERS AND MOUNTINGS. N15
MAIN FUSE BLOCK AND MOUNTING PLATE M15
NUMBER PLATE AND REVERSE LAMP. K16
PASSIVE RESTRAINT MODULE & FIXINGS N08
PLUGS AND RELAYS. N07
POWER RESISTOR AND LAMBDA SENSOR. L16
RADIO CASSETTE PLAYER-FROM (V)148782 .... N14
SHOULDER BAG-REMOVEABLE RADIO N14
RADIO'S-UP TO (V)148781 .... N13
REAR LAMPS. K11
SEAT CUSHION SENSOR. N11
SENSORS,TRANSMITTERS AND BRACKET M12
SIDE LAMPS. K07
SPEED CONTROL PUMP & DUMP VALVE O11
STARTER MOTOR - 3.6 LITRE J08
STARTER MOTOR 5.3 LITRE J09
STEERING COLUMN SWITCHES. M04
TRIP COMPUTER AND PANEL SWITCHES. M06
WASH/WIPE RESERVOIR AND FIXINGS L08
WINDSCREEN WASHER RESERVOIR L04
WINDSCREEN WIPER MOTOR-LUCAS TYPE K17
WINDSCREEN WIPERS L03

2N 02

HEADLAMP FUSE BLOCK AND RELAY BRKT - 3.6 LITRE

FROM (V)148839

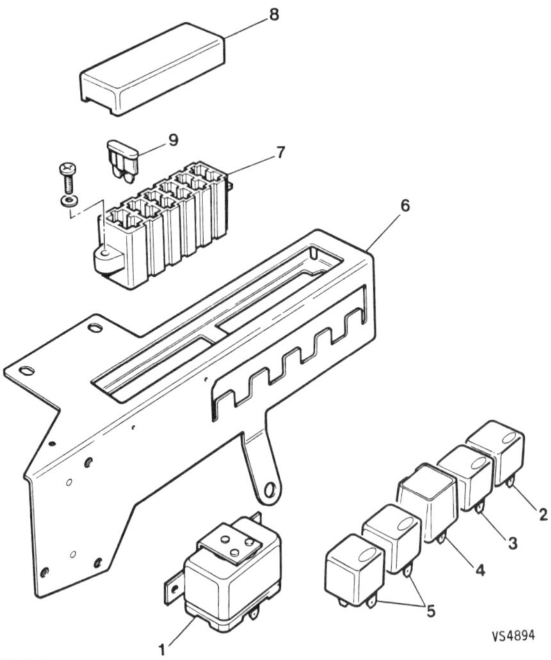

VS4894

Illus	Part Number	Description	Quantity	Change Point	Remarks
1	C 38616	Relay - headlamp	1		
2	BBC 2483	Relay - fan	1		
3	DBC 2484	Relay - horn	1		
4	DAC 1996	Pektron unit - fan run on	1		
5	DBC 2484	Relay - dim/dip	2		Where fitted
5	DBC 2484	Relay - heated breather	1		Where fitted
5	DBC 2484	Relay - air pump clutch	1		Where fitted
6	DAC 5517	Bracket - fuse box & speed control	1		
	DBC 2972	Fuse block - (black)	1		
7	DBC 2973	Fuse block - (natural)	1		
	DBC 3060	Cover - fuse block (black)	1		
8	DBC 3778	Cover - fuse block (natural)	1		
9	DGC 2490	Fuse - 2 amp	A/R		
9	DGC 2491	Fuse - 3 amp	A/R		
9	DGC 2492	Fuse - 4 amp	A/R		
9	DGC 2493	Fuse - 5 amp	A/R		
9	DGC 2494	Fuse - 7.5 amp	A/R		
9	DGC 2495	Fuse - 10 amp	A/R		
9	DGC 2496	Fuse - 15 amp	A/R		
9	DGC 2497	Fuse - 20 amp	A/R		
9	DGC 2498	Fuse - 25 amp	A/R		
9	DGC 2499	Fuse - 30 amp	A/R		

MASTER INDEX

GROUP INDEX

2N 03

HEADLAMP FUSE BLOCK & RELAY MTG BRKT-5.3 LITRE

UP TO (V)148838

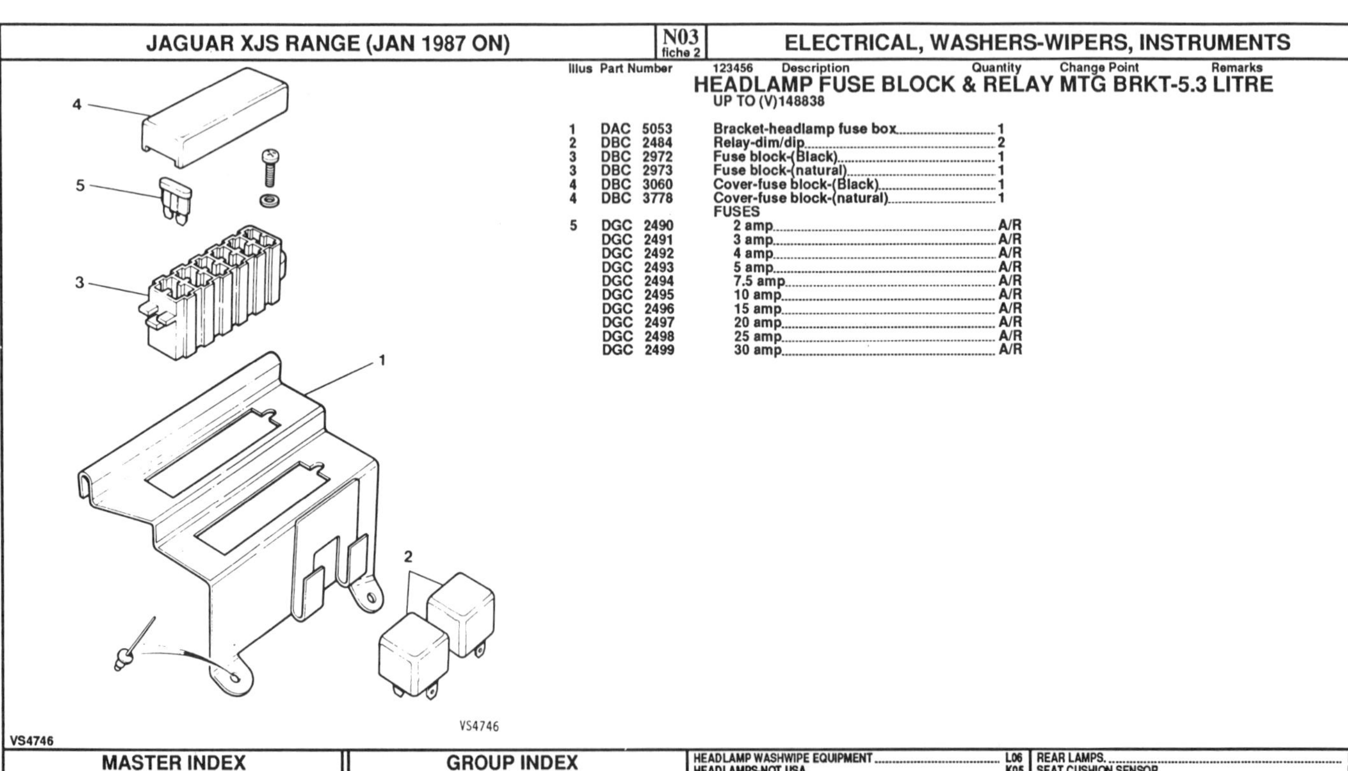

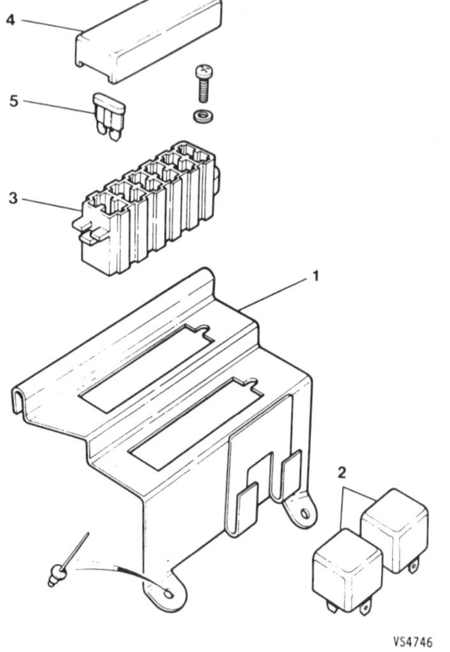

VS4746

Illus	Part Number	Description	Quantity	Change Point	Remarks
1	DAC 5053	Bracket-headlamp fuse box	1		
2	DBC 2484	Relay-dim/dip	2		
3	DBC 2972	Fuse block-(Black)	1		
3	DBC 2973	Fuse block-(natural)	1		
4	DBC 3060	Cover-fuse block-(Black)	1		
4	DBC 3778	Cover-fuse block-(natural)	1		
		FUSES			
5	DGC 2490	2 amp	A/R		
	DGC 2491	3 amp	A/R		
	DGC 2492	4 amp	A/R		
	DGC 2493	5 amp	A/R		
	DGC 2494	7.5 amp	A/R		
	DGC 2495	10 amp	A/R		
	DGC 2496	15 amp	A/R		
	DGC 2497	20 amp	A/R		
	DGC 2498	25 amp	A/R		
	DGC 2499	30 amp	A/R		

MASTER INDEX
ENGINE 1.C
FLYWHEEL AND CLUTCH 1.G
GEARBOX AND PROPSHAFT 1.H
AXLES, SUSPENSION, DRIVE SHAFTS, WHEELS 1.K
STEERING 1.M
BRAKES AND BRAKE CONTROLS 1.N
FUEL, EXHAUST AND EMISSION SYSTEMS 2.C
COOLING, HEATING AND AIR CONDITIONING 2.G
ELECTRICAL, WASHERS-WIPERS, INSTRUMENTS 2.J
CHASSIS, SUBFRAMES, BODYSHELL, FITTINGS 3.C
FASCIA, TRIM, SEATS AND FIXINGS 3.H
ACCESSORIES AND PAINT 4.C
NUMERICAL INDEX 4.D
GROUP INDEX
A.B.S EQUIPMENT N09
AERIAL AND FIXINGS. N12
ALTERNATOR 75 AMP (UP TO (E)109969). J02
BATTERY AND LEADS. N17
CONSOLE SWITCHES AND CIGAR LIGHTER M08
CONTROL UNIT-P.J-V12 COUPE L11
FROM 89 MY L13
CONTROL UNITS-PI/CRUISE CONTROL 3.6LITRE L10
DISTRIBUTOR-3.6 LITRE. J11
DISTRIBUTOR-5.3 LITRE. J12
FROM 89 MY J13
DOOR WARNING LAMP. K15
FOG REAR GUARD LAMP. K14
FRONT FLASHER LAMPS. K06
FRONT FOG LAMPS. K13
FUEL INTERFACE UNIT AND RELAYS L15
HARNESSES-FRONT 3.6 LITRE-FROM (V)148945 .... O04
HARNESSES-FRONT 5.3 LITRE-FROM (V)148782 .... O06
HARNESSES-FRONT-3.6 LITRE-UPTO (V)148944 .... O03
HEADLAMP WASH WIPE RESERVOIR L07
HEADLAMP WASHWIPE EQUIPMENT L06
HEADLAMPS-NOT USA K05
HEADLAMPS-USA. K03
HIGH MOUNTED STOP LAMP-USA/CDN ONLY. K12
HOOD CONTROL MODULE - CONVERTIBLE N10
HORNS AND HORN PAD M05
IGNITION COIL & MODULE-3.6 LITRE J14
IGNITION COILS AND AMPLIFIER-5.3 LITRE J15
INERTIA SWITCH AND COVER M13
INSTRUMENT PACK-3.6 LITRE L17
INSTRUMENT PACK-5.3 LITRE L18
INTERIOR LAMPS-CABRIOLET. K09
INTERIOR LAMPS-COUPE. K08
LOUDSPEAKERS AND MOUNTINGS. N15
MAIN FUSE BLOCK AND MOUNTING PLATE M15
NUMBER PLATE AND REVERSE LAMP. K16
PASSIVE RESTRAINT MODULE & FIXINGS N08
PLUGS AND RELAYS. N07
POWER RESISTOR AND LAMBDA SENSOR. L16
RADIO CASSETTE PLAYER-FROM (V)148782 .... N14
SHOULDER BAG-REMOVEABLE RADIO N14
RADIO'S-UP TO (V)148781 .... N13
REAR LAMPS. K11
SEAT CUSHION SENSOR. N11
SENSORS,TRANSMITTERS AND BRACKET M12
SIDE LAMPS. K07
SPEED CONTROL PUMP & DUMP VALVE O11
STARTER MOTOR - 3.6 LITRE J08
STARTER MOTOR 5.3 LITRE J09
STEERING COLUMN SWITCHES. M04
TRIP COMPUTER AND PANEL SWITCHES. M06
WASH/WIPE RESERVOIR AND FIXINGS L08
WINDSCREEN WASHER RESERVOIR L04
WINDSCREEN WIPER MOTOR-LUCAS TYPE K17
WINDSCREEN WIPERS L03

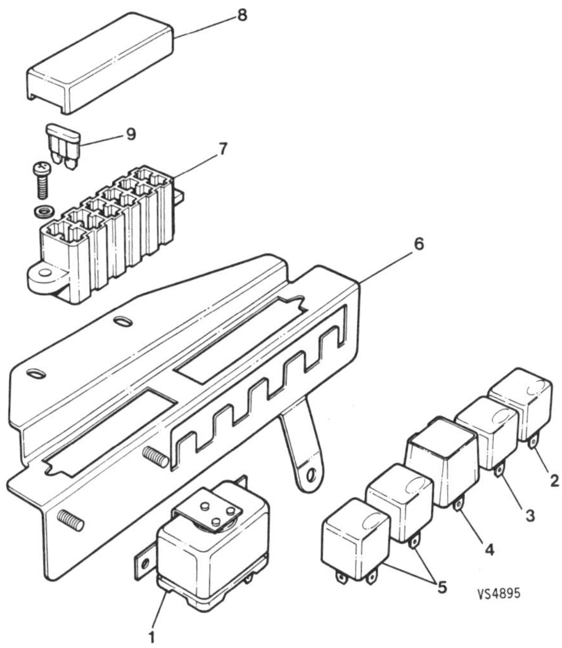

VS 4895

HEADLAMP FUSE BLOCK AND RELAY BRKT 5.3 LITRE

FROM (V)148839

Illus	Part Number	123456	Description	Quantity	Change Point	Remarks
1	C 38616		Relay - headlamp	1		
2	DBC 2483		Relay - fan	1		
3	DBC 2484		Relay - horn	1		
4	DAC 1996		Pektron unit - fan run on	1		
5	DBC 2484		Relay - dim/dip	2		
6	DAC 5092		Bracket - headlamp fuse box	1		
	DBC 2972		Fuse block - (black)	1		
7	DBC 2973		Fuse block - (natural)	1		
	DBC 3060		Cover - fuse block - (black)	1		
8	DBC 3778		Cover - fuse block - (natural)	1		
9	DGC 2490		Fuse - 2 amp	A/R		
9	DGC 2491		Fuse - 3 amp	A/R		
9	DGC 2492		Fuse - 4 amp	A/R		
9	DGC 2493		Fuse - 5 amp	A/R		
9	DGC 2494		Fuse - 7.5 amp	A/R		
9	DGC 2495		Fuse - 10 amp	A/R		
9	DGC 2496		Fuse - 15 amp	A/R		
9	DGC 2497		Fuse - 20 amp	A/R		
9	DGC 2498		Fuse - 30 amp	A/R		

MASTER INDEX

GROUP INDEX

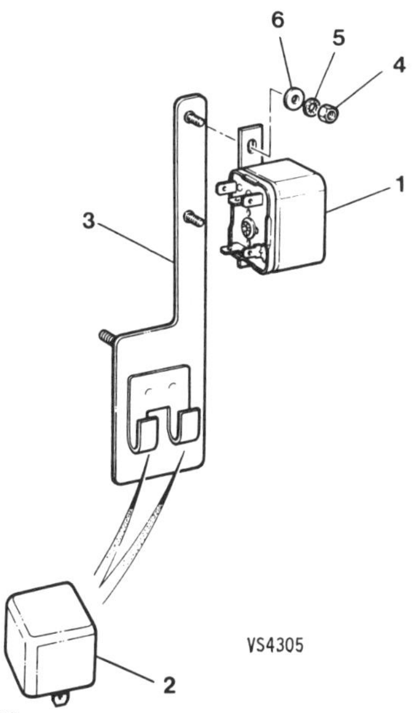

VS 4305

HEADLAMP RELAY & FIXINGS 3.6 LITRE-UP TO (V)148838

Illus	Part Number	123456	Description	Quantity	Change Point	Remarks
1	C 38616		Relay-headlamp	1		
2	DBC 2484		Relay-dim/dip	2		
3	DAC 3978		Bracket-mounting	1		
4	UFN 119 L		Nut	5		
5	WF 702100 J		Washer-shakeproof	5		
6	WC 702101 J		Washer-plain	5		

MASTER INDEX
ENGINE 1.C
FLYWHEEL AND CLUTCH 1.G
GEARBOX AND PROPSHAFT 1.H
AXLES, SUSPENSION, DRIVE SHAFTS, WHEELS 1.K
STEERING 1.M
BRAKES AND BRAKE CONTROLS 1.N
FUEL, EXHAUST AND EMISSION SYSTEMS 2.C
COOLING, HEATING AND AIR CONDITIONING 2.G
ELECTRICAL, WASHERS-WIPERS, INSTRUMENTS 2.J
CHASSIS, SUBFRAMES, BODYSHELL, FITTINGS 3.C
FASCIA, TRIM, SEATS AND FIXINGS 3.H
ACCESSORIES AND PAINT 4.C
NUMERICAL INDEX 4.D
GROUP INDEX
A.B.S EQUIPMENT N09
AERIAL AND FIXINGS N12
ALTERNATOR 75 AMP (UP TO (E)109969) J02
BATTERY AND LEADS N17
CONSOLE SWITCHES AND CIGAR LIGHTER M08
CONTROL UNIT-PJ-V12 COUPE L11
FROM 89 MY L13
CONTROL UNITS-PI/CRUISE CONTROL 3.6LITRE L10
DISTRIBUTOR-3.6 LITRE J11
DISTRIBUTOR-5.3 LITRE J12
FROM 89 MY J13
DOOR WARNING LAMP K15
FOG REAR GUARD LAMP K14
FRONT FLASHER LAMPS K06
FRONT FOG LAMPS K13
FUEL INTERFACE UNIT AND RELAYS L15
HARNESSES-FRONT 3.6 LITRE-FROM (V)148945 O04
HARNESSES-FRONT 5.3 LITRE-FROM (V)148782 O06
HARNESSES-FRONT-3.6 LITRE-UPTO (V)148944 O03
HEADLAMP WASH WIPE RESERVOIR L07
HEADLAMP WASHWIPE EQUIPMENT L06
HEADLAMPS-NOT USA K05
HEADLAMPS-USA K03
HIGH MOUNTED STOP LAMP-USA/CDN ONLY K12
HOOD CONTROL MODULE - CONVERTIBLE N10
HORNS AND HORN PAD M05
IGNITION COIL & MODULE-3.6 LITRE J14
IGNITION COILS AND AMPLIFIER-5.3 LITRE J15
INERTIA SWITCH AND COVER M13
INSTRUMENT PACK-3.6 LITRE L17
INSTRUMENT PACK-5.3 LITRE L18
INTERIOR LAMPS-CABRIOLET K09
INTERIOR LAMPS-COUPE K08
LOUDSPEAKERS AND MOUNTINGS N15
MAIN FUSE BLOCK AND MOUNTING PLATE M15
NUMBER PLATE AND REVERSE LAMP K16
PASSIVE RESTRAINT MODULE & FIXINGS N08
PLUGS AND RELAYS N07
POWER RESISTOR AND LAMBDA SENSOR L16
RADIO CASSETTE PLAYER-FROM (V)148782 N14
SHOULDER BAG-REMOVEABLE RADIO N14
RADIO'S-UP TO (V)148781 N13
REAR LAMPS K11
SEAT CUSHION SENSOR N11
SENSORS,TRANSMITTERS AND BRACKET M12
SIDE LAMPS K07
SPEED CONTROL PUMP & DUMP VALVE O11
STARTER MOTOR - 3.6 LITRE J08
STARTER MOTOR 5.3 LITRE J09
STEERING COLUMN SWITCHES M04
TRIP COMPUTER AND PANEL SWITCHES M06
WASH/WIPE RESERVOIR AND FIXINGS L08
WINDSCREEN WASHER RESERVOIR L04
WINDSCREEN WIPER MOTOR-LUCAS TYPE K17
WINDSCREEN WIPERS L03

HEADLAMP RELAY & FIXINGS 5.3 LITRE-UP TO (V)148838

Illus	Part Number	123456 Description	Quantity	Change Point	Remarks
1	C 38616	Relay-headlamp	1		
2	DAC 1028	Relay-fan	1		
3	DAC 1027	Relay-horn	1		
4	DAC 2592	Bracket-mounting	1		
5	DAC 2593	Cover-relay	1		
6	AR 606031	Screw-selftap	1		
7	UFN 119 L	Nut	5		
8	WF 702100 J	Washer-locking	5		
9	WC 702101 J	Washer-plain	5		

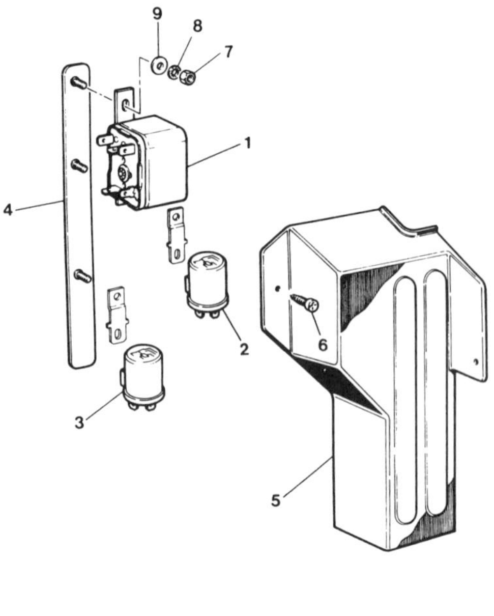

VS 3224

2N 06

PLUGS AND RELAYS.

Illus	Part Number	123456 Description	Quantity	Change Point	Remarks
1	DAC 4878	Bracket assembly-relay mounting	1		On RH valance
2	AGU 1097	Relay-feedback inhibit	1	Up to 89 MY	
3	DBC 2484	Relay-stop lamp	1		
4	C 36611	Relay-starter	1		
5	DBC 2484	Relay-air conditioning	1		
6	DAC 3636	Cover-relay	1		Use DAC 5884
6	DAC 5884	Cover-relay	1		
7	C 39314	Plug-bulkhead connector	2		
8	C 39315	Gasket-plug	2		

K80JP31/8/88

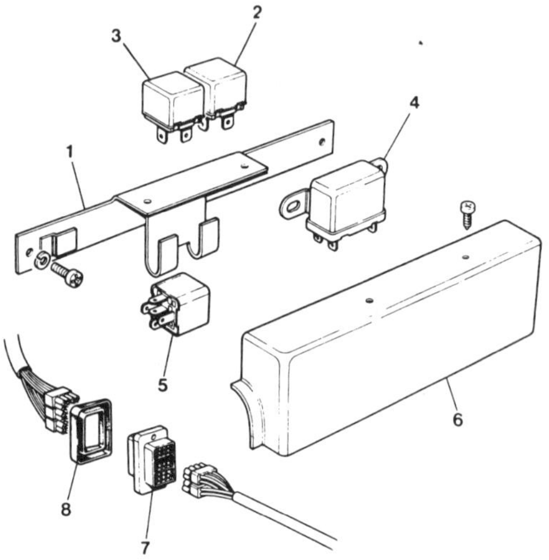

VS 2023B

2N 07

MASTER INDEX

ENGINE 1.C
FLYWHEEL AND CLUTCH 1.G
GEARBOX AND PROPSHAFT 1.H
AXLES, SUSPENSION, DRIVE SHAFTS, WHEELS 1.K
STEERING 1.M
BRAKES AND BRAKE CONTROLS 1.N
FUEL, EXHAUST AND EMISSION SYSTEMS 2.C
COOLING, HEATING AND AIR CONDITIONING 2.G
ELECTRICAL, WASHERS-WIPERS, INSTRUMENTS 2.J
CHASSIS, SUBFRAMES, BODYSHELL, FITTINGS 3.C
FASCIA, TRIM, SEATS AND FIXINGS 3.H
ACCESSORIES AND PAINT 4.C
NUMERICAL INDEX 4.D

GROUP INDEX

A.B.S EQUIPMENT N09
AERIAL AND FIXINGS N12
ALTERNATOR 75 AMP (UP TO (E)109969) J02
BATTERY AND LEADS N17
CONSOLE SWITCHES AND CIGAR LIGHTER M08
CONTROL UNIT-PJ-V12 COUPE L11
FROM 89 MY L13
CONTROL UNITS-PI/CRUISE CONTROL 3.6LITRE L10
DISTRIBUTOR-3.6 LITRE J11
DISTRIBUTOR-5.3 LITRE J12
FROM 89 MY J13
DOOR WARNING LAMP K15
FOG REAR GUARD LAMP K14
FRONT FLASHER LAMPS K06
FRONT FOG LAMPS K13
FUEL INTERFACE UNIT AND RELAYS L15
HARNESSES-FRONT 3.6 LITRE-FROM (V)148945 O04
HARNESSES-FRONT 5.3 LITRE-FROM (V)148782 O06
HARNESSES-FRONT-3.6 LITRE-UPTO (V)148944 O03
HEADLAMP WASH WIPE RESERVOIR L07
HEADLAMP WASHWIPE EQUIPMENT L06
HEADLAMPS-NOT USA K05
HEADLAMPS-USA K03
HIGH MOUNTED STOP LAMP-USA/CDN ONLY K12
HOOD CONTROL MODULE - CONVERTIBLE N10
HORNS AND HORN PAD M05
IGNITION COIL & MODULE-3.6 LITRE J14
IGNITION COILS AND AMPLIFIER-5.3 LITRE J15
INERTIA SWITCH AND COVER M13
INSTRUMENT PACK-3.6 LITRE L17
INSTRUMENT PACK-5.3 LITRE L18
INTERIOR LAMPS-CABRIOLET K09
INTERIOR LAMPS-COUPE K08
LOUDSPEAKERS AND MOUNTINGS N15
MAIN FUSE BLOCK AND MOUNTING PLATE M15
NUMBER PLATE AND REVERSE LAMP K16
PASSIVE RESTRAINT MODULE & FIXINGS N08
PLUGS AND RELAYS N07
POWER RESISTOR AND LAMBDA SENSOR L16
RADIO CASSETTE PLAYER-FROM (V)148782 N14
SHOULDER BAG-REMOVEABLE RADIO N14
RADIO'S-UP TO (V)148781 N13
REAR LAMPS K11
SEAT CUSHION SENSOR N11
SENSORS,TRANSMITTERS AND BRACKET M12
SIDE LAMPS K07
SPEED CONTROL PUMP & DUMP VALVE O11
STARTER MOTOR - 3.6 LITRE J08
STARTER MOTOR 5.3 LITRE J09
STEERING COLUMN SWITCHES M04
TRIP COMPUTER AND PANEL SWITCHES M06
WASH/WIPE RESERVOIR AND FIXINGS L08
WINDSCREEN WASHER RESERVOIR L04
WINDSCREEN WIPER MOTOR-LUCAS TYPE K17
WINDSCREEN WIPERS L03

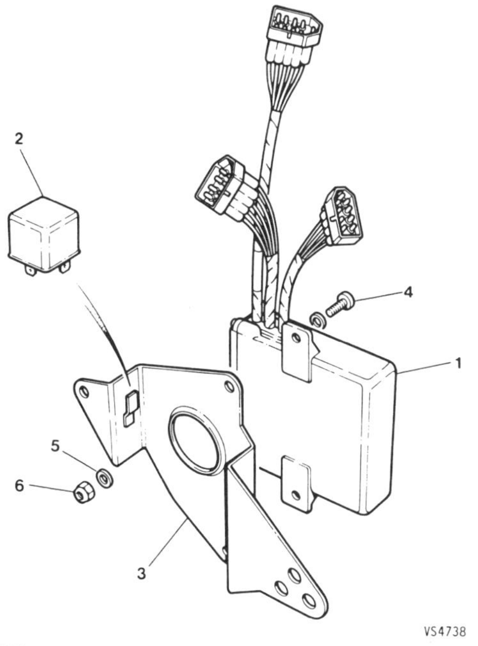

VS4738

PASSIVE RESTRAINT MODULE & FIXINGS

Illus	Part Number	123456	Description	Quantity	Change Point	Remarks
1	DAC 4292		Module-passive restraint	1		
2	DBC 2483		Relay-reverse inhibit	1		
3	DAC 5205		Bracket-module mounting	1		
4	SE 105161		Screw-set	2		
5	WC 105001 J		Washer-plain	2		
6	NY 105041 J		Nut-nyloc	2		

MASTER INDEX

GROUP INDEX

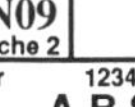

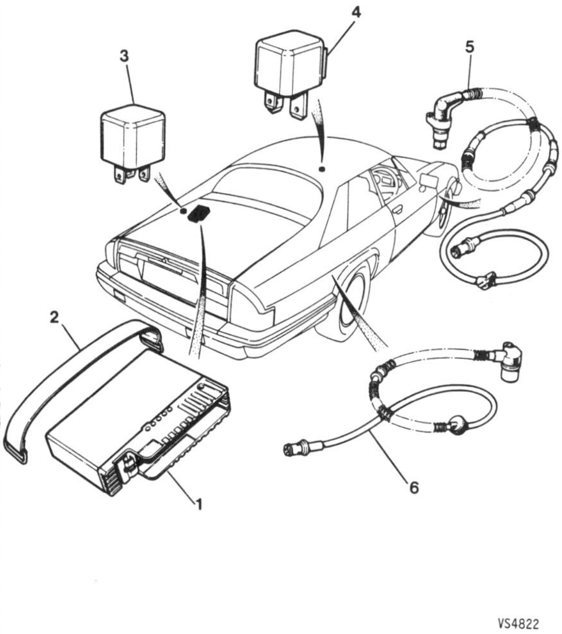

VS4822

A.B.S EQUIPMENT

Illus	Part Number	123456	Description	Quantity	Change Point	Remarks
1	DAC 5863		Controller-anti skid	1		
2	DAC 4653		Strap-retaining	1		
3	DAC 5096		Relay-main	1	.....	Situated in boot
4	DAC 4506		Relay-motor	1	.....	Situated on bulkhead
			WHEEL SPEED SENSORS			
5	DAC 4842		Front-RH	1		
	DAC 4841		Front-LH	1		
6	DAC 4840		Rear-RH	1		
	DAC 4839		Rear-LH	1		

MASTER INDEX
ENGINE 1.C
FLYWHEEL AND CLUTCH 1.G
GEARBOX AND PROPSHAFT 1.H
AXLES, SUSPENSION, DRIVE SHAFTS, WHEELS 1.K
STEERING 1.M
BRAKES AND BRAKE CONTROLS 1.N
FUEL, EXHAUST AND EMISSION SYSTEMS 2.C
COOLING, HEATING AND AIR CONDITIONING 2.G
ELECTRICAL, WASHERS-WIPERS, INSTRUMENTS 2.J
CHASSIS, SUBFRAMES, BODYSHELL, FITTINGS 3.C
FASCIA, TRIM, SEATS AND FIXINGS 3.H
ACCESSORIES AND PAINT 4.C
NUMERICAL INDEX 4.D
GROUP INDEX
A.B.S EQUIPMENT N09
AERIAL AND FIXINGS. N12
ALTERNATOR 75 AMP (UP TO (E)109969). J02
BATTERY AND LEADS. N17
CONSOLE SWITCHES AND CIGAR LIGHTER M08
CONTROL UNIT-P.I-V12 COUPE L11
FROM 89 MY L13
CONTROL UNITS-PI/CRUISE CONTROL 3.6LITRE L10
DISTRIBUTOR-3.6 LITRE. J11
DISTRIBUTOR-5.3 LITRE. J12
FROM 89 MY J13
DOOR WARNING LAMP. K15
FOG REAR GUARD LAMP. K14
FRONT FLASHER LAMPS. K06
FRONT FOG LAMPS. K13
FUEL INTERFACE UNIT AND RELAYS L15
HARNESSES-FRONT 3.6 LITRE-FROM (V)148945 ..... O04
HARNESSES-FRONT 5.3 LITRE-FROM (V)148782 ..... O06
HARNESSES-FRONT-3.6 LITRE-UPTO (V)148944 ..... O03
HEADLAMP WASH WIPE RESERVOIR L07
HEADLAMP WASHWIPE EQUIPMENT L06
HEADLAMPS-NOT USA K05
HEADLAMPS-USA. K03
HIGH MOUNTED STOP LAMP-USA/CDN ONLY. K12
HOOD CONTROL MODULE - CONVERTIBLE N10
HORNS AND HORN PAD M05
IGNITION COIL & MODULE-3.6 LITRE J14
IGNITION COILS AND AMPLIFIER-5.3 LITRE J15
INERTIA SWITCH AND COVER M13
INSTRUMENT PACK-3.6 LITRE L17
INSTRUMENT PACK-5.3 LITRE L18
INTERIOR LAMPS-CABRIOLET. K09
INTERIOR LAMPS-COUPE. K08
LOUDSPEAKERS AND MOUNTINGS. N15
MAIN FUSE BLOCK AND MOUNTING PLATE M15
NUMBER PLATE AND REVERSE LAMP. K16
PASSIVE RESTRAINT MODULE & FIXINGS N08
PLUGS AND RELAYS. N07
POWER RESISTOR AND LAMBDA SENSOR. L16
RADIO CASSETTE PLAYER-FROM (V)148782 ..... N14
SHOULDER BAG-REMOVEABLE RADIO N14
RADIO'S-UP TO (V)148781 ..... N13
REAR LAMPS. K11
SEAT CUSHION SENSOR. N11
SENSORS,TRANSMITTERS AND BRACKET M12
SIDE LAMPS. K07
SPEED CONTROL PUMP & DUMP VALVE O11
STARTER MOTOR - 3.6 LITRE J08
STARTER MOTOR 5.3 LITRE J09
STEERING COLUMN SWITCHES. M04
TRIP COMPUTER AND PANEL SWITCHES. M06
WASH/WIPE RESERVOIR AND FIXINGS L08
WINDSCREEN WASHER RESERVOIR L04
WINDSCREEN WIPER MOTOR-LUCAS TYPE K17
WINDSCREEN WIPERS L03

2N 10

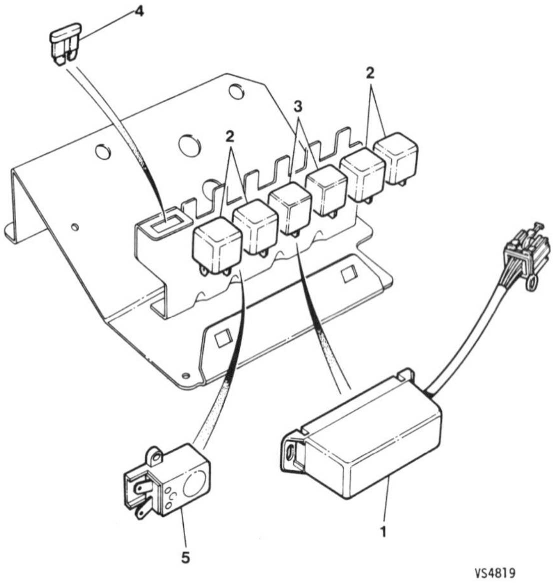

VS4819

VS 4819

HOOD CONTROL MODULE - CONVERTIBLE

Illus	Part Number	123456	Description	Quantity	Change Point	Remarks
1	DAC 4404		Module-hood control	1		
2	DAC 5881		Relay-rear qtr motor	4		
3	DAC 4408		Relay-changeover	2		
4	DGC 2499		Fuse-30 AMP	1		
5	C 44001		Circuit breaker	2		

2N 11

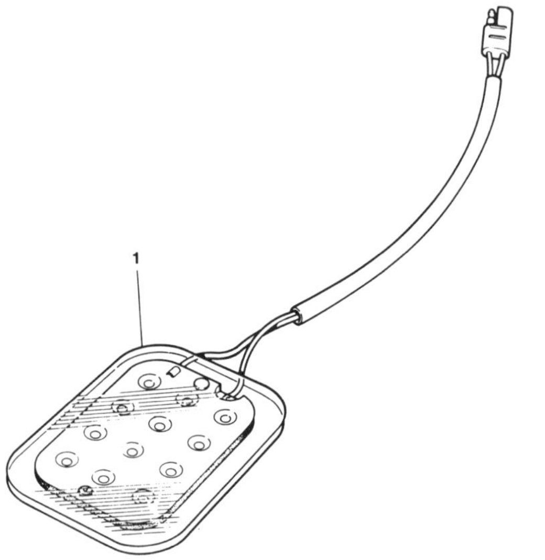

VS3232

VS 3232

SEAT CUSHION SENSOR.

Illus	Part Number	123456	Description	Quantity	Change Point	Remarks
1	DAC 1911		Sensor-seat cushion	1		

MASTER INDEX

ENGINE 1.C
FLYWHEEL AND CLUTCH 1.G
GEARBOX AND PROPSHAFT 1.H
AXLES, SUSPENSION, DRIVE SHAFTS, WHEELS 1.K
STEERING 1.M
BRAKES AND BRAKE CONTROLS 1.N
FUEL, EXHAUST AND EMISSION SYSTEMS 2.C
COOLING, HEATING AND AIR CONDITIONING 2.G
ELECTRICAL, WASHERS-WIPERS, INSTRUMENTS 2.J
CHASSIS, SUBFRAMES, BODYSHELL, FITTINGS 3.C
FASCIA, TRIM, SEATS AND FIXINGS 3.H
ACCESSORIES AND PAINT 4.C
NUMERICAL INDEX 4.D

GROUP INDEX

A.B.S EQUIPMENT N09
AERIAL AND FIXINGS. N12
ALTERNATOR 75 AMP (UP TO (E)109969). J02
BATTERY AND LEADS. N17
CONSOLE SWITCHES AND CIGAR LIGHTER M08
CONTROL UNIT-PJ-V12 COUPE L11
FROM 89 MY L13
CONTROL UNITS-PI/CRUISE CONTROL 3.6LITRE L10
DISTRIBUTOR-3.6 LITRE. J11
DISTRIBUTOR-5.3 LITRE. J12
FROM 89 MY J13
DOOR WARNING LAMP. K15
FOG REAR GUARD LAMP. K14
FRONT FLASHER LAMPS. K06
FRONT FOG LAMPS. K13
FUEL INTERFACE UNIT AND RELAYS L15
HARNESSES-FRONT 3.6 LITRE-FROM (V)148945 .... O04
HARNESSES-FRONT 5.3 LITRE-FROM (V)148782 .... O06
HARNESSES-FRONT-3.6 LITRE-UPTO (V)148944 .... O03
HEADLAMP WASH WIPE RESERVOIR L07
HEADLAMP WASHWIPE EQUIPMENT L06
HEADLAMPS-NOT USA K05
HEADLAMPS-USA. K03
HIGH MOUNTED STOP LAMP-USA/CDN ONLY. K12
HOOD CONTROL MODULE - CONVERTIBLE N10
HORNS AND HORN PAD M05
IGNITION COIL & MODULE-3.6 LITRE J14
IGNITION COILS AND AMPLIFIER-5.3 LITRE J15
INERTIA SWITCH AND COVER M13
INSTRUMENT PACK-3.6 LITRE L17
INSTRUMENT PACK-5.3 LITRE L18
INTERIOR LAMPS-CABRIOLET. K09
INTERIOR LAMPS-COUPE. K08
LOUDSPEAKERS AND MOUNTINGS. N15
MAIN FUSE BLOCK AND MOUNTING PLATE M15
NUMBER PLATE AND REVERSE LAMP. K16
PASSIVE RESTRAINT MODULE & FIXINGS N08
PLUGS AND RELAYS. N07
POWER RESISTOR AND LAMBDA SENSOR. L16
RADIO CASSETTE PLAYER-FROM (V)148782 .... N14
SHOULDER BAG-REMOVEABLE RADIO N14
RADIO'S-UP TO (V)148781 .... N13
REAR LAMPS. K11
SEAT CUSHION SENSOR. N11
SENSORS,TRANSMITTERS AND BRACKET M12
SIDE LAMPS. K07
SPEED CONTROL PUMP & DUMP VALVE O11
STARTER MOTOR - 3.6 LITRE J08
STARTER MOTOR 5.3 LITRE J09
STEERING COLUMN SWITCHES. M04
TRIP COMPUTER AND PANEL SWITCHES. M06
WASH/WIPE RESERVOIR AND FIXINGS L08
WINDSCREEN WASHER RESERVOIR L04
WINDSCREEN WIPER MOTOR-LUCAS TYPE K17
WINDSCREEN WIPERS L03

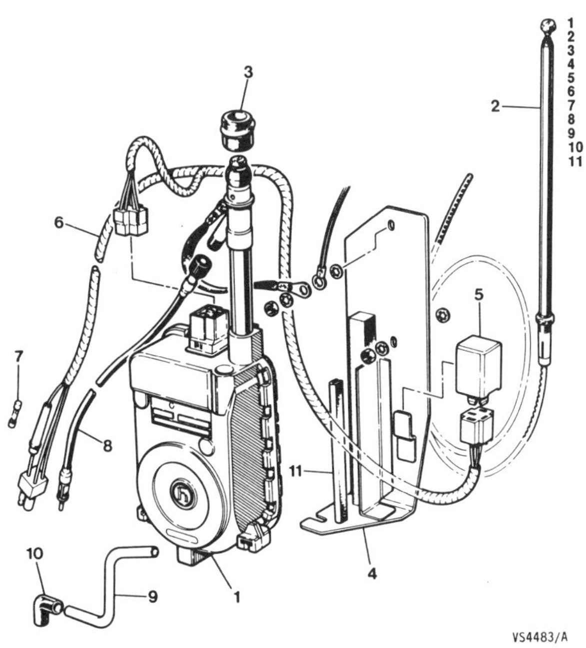

AERIAL AND FIXINGS.

Illus	Part Number	123456	Description	Quantity	Change Point	Remarks
1	DAC 3579		AERIAL ASSEMBLY-ELECTRIC	1		
2	DBC 2200		Assembly-mast and perlon	1		
3	DAC 3588		Grommet	1		
4	DAC 3516		Mounting bracket	1		
5	DAC 1820		Relay-aerial	1		
6	DAC 3518		Link harness	1		
7	C 39462		Fuse-1 AMP	1		
8	DAC 3726		Aerial lead	1		
9	DAC 3255 3		Hose drain	1		
10	AGU 1299		Outlet drain	1		
11	BD 44643 1		Strip-edging	1		

VS4483/A

MASTER INDEX

GROUP INDEX

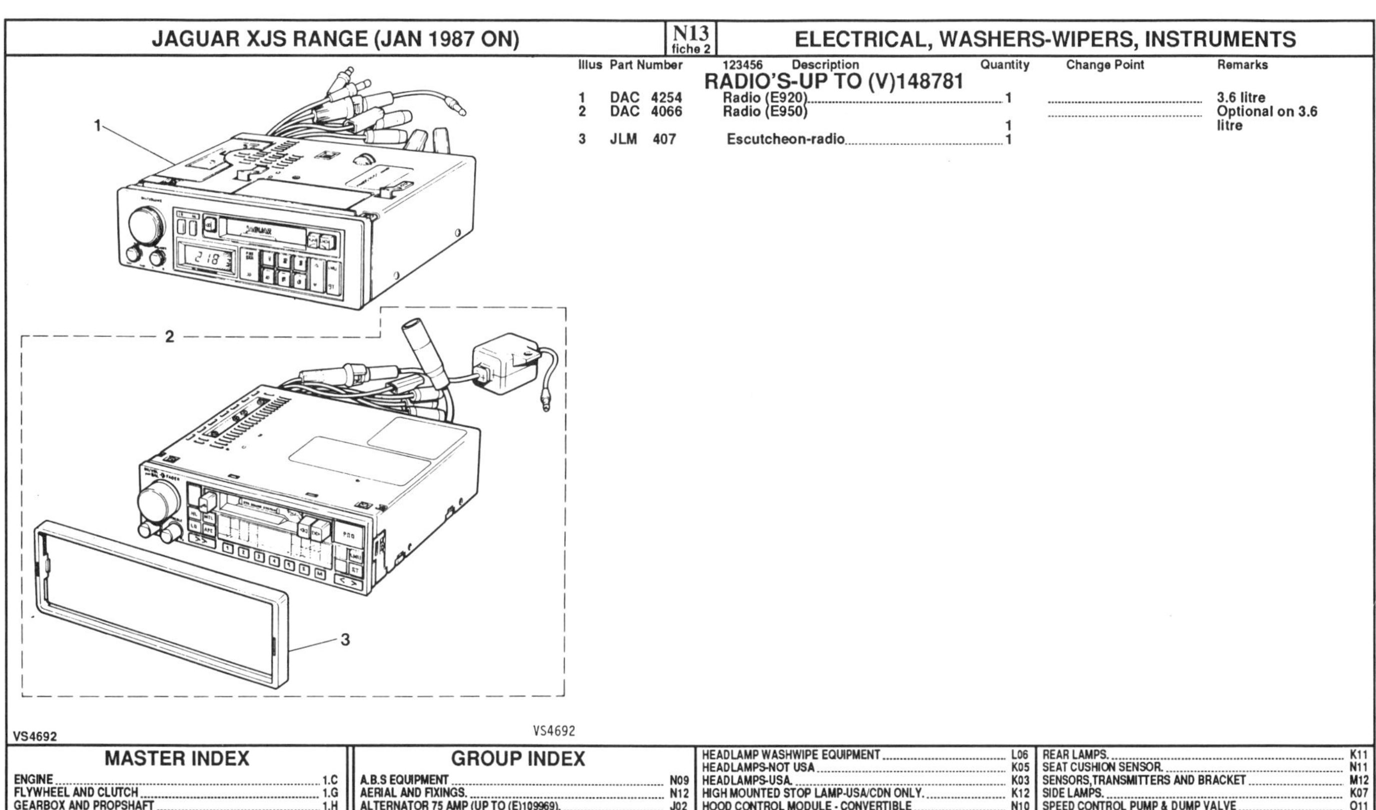

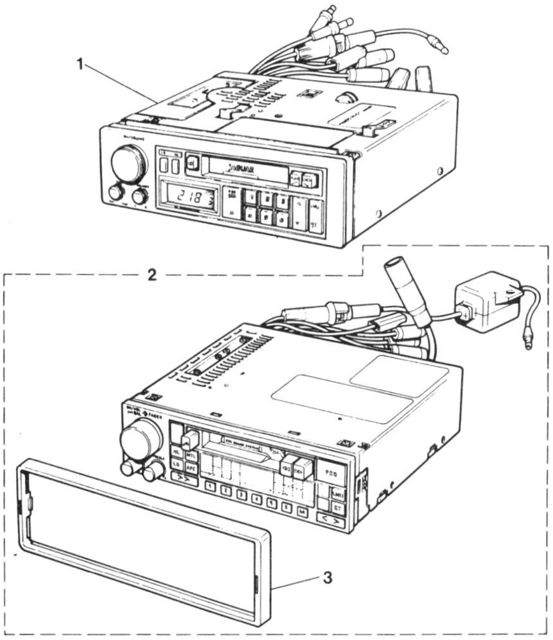

JAGUAR XJS RANGE (JAN 1987 ON) | N13 fiche 2 | ELECTRICAL, WASHERS-WIPERS, INSTRUMENTS

RADIO'S-UP TO (V)148781

Illus	Part Number	123456	Description	Quantity	Change Point	Remarks
1	DAC 4254		Radio (E920)	1		3.6 litre
2	DAC 4066		Radio (E950)	1		Optional on 3.6 litre
3	JLM 407		Escutcheon-radio	1		

VS4692

MASTER INDEX
ENGINE ... 1.C
FLYWHEEL AND CLUTCH ... 1.G
GEARBOX AND PROPSHAFT ... 1.H
AXLES, SUSPENSION, DRIVE SHAFTS, WHEELS ... 1.K
STEERING ... 1.M
BRAKES AND BRAKE CONTROLS ... 1.N
FUEL, EXHAUST AND EMISSION SYSTEMS ... 2.C
COOLING, HEATING AND AIR CONDITIONING ... 2.G
ELECTRICAL, WASHERS-WIPERS, INSTRUMENTS ... 2.J
CHASSIS, SUBFRAMES, BODYSHELL, FITTINGS ... 3.C
FASCIA, TRIM, SEATS AND FIXINGS ... 3.H
ACCESSORIES AND PAINT ... 4.C
NUMERICAL INDEX ... 4.D
GROUP INDEX
A.B.S EQUIPMENT ... N09
AERIAL AND FIXINGS. ... N12
ALTERNATOR 75 AMP (UP TO (E)109969). ... J02
BATTERY AND LEADS. ... N17
CONSOLE SWITCHES AND CIGAR LIGHTER ... M08
CONTROL UNIT-P.I-V12 COUPE ... L11
FROM 89 MY ... L13
CONTROL UNITS-PI/CRUISE CONTROL 3.6LITRE ... L10
DISTRIBUTOR-3.6 LITRE. ... J11
DISTRIBUTOR-5.3 LITRE. ... J12
FROM 89 MY ... J13
DOOR WARNING LAMP. ... K15
FOG REAR GUARD LAMP. ... K14
FRONT FLASHER LAMPS. ... K06
FRONT FOG LAMPS. ... K13
FUEL INTERFACE UNIT AND RELAYS ... L15
HARNESSES-FRONT 3.6 LITRE-FROM (V)148945 ... O04
HARNESSES-FRONT 5.3 LITRE-FROM (V)148782 ... O06
HARNESSES-FRONT-3.6 LITRE-UPTO (V)148944 ... O03
HEADLAMP WASH WIPE RESERVOIR ... L07
HEADLAMP WASHWIPE EQUIPMENT ... L06
HEADLAMPS-NOT USA ... K05
HEADLAMPS-USA. ... K03
HIGH MOUNTED STOP LAMP-USA/CDN ONLY. ... K12
HOOD CONTROL MODULE - CONVERTIBLE ... N10
HORNS AND HORN PAD ... M05
IGNITION COIL & MODULE-3.6 LITRE ... J14
IGNITION COILS AND AMPLIFIER-5.3 LITRE ... J15
INERTIA SWITCH AND COVER ... M13
INSTRUMENT PACK-3.6 LITRE ... L17
INSTRUMENT PACK-5.3 LITRE ... L18
INTERIOR LAMPS-CABRIOLET. ... K09
INTERIOR LAMPS-COUPE. ... K08
LOUDSPEAKERS AND MOUNTINGS. ... N15
MAIN FUSE BLOCK AND MOUNTING PLATE ... M15
NUMBER PLATE AND REVERSE LAMP. ... K16
PASSIVE RESTRAINT MODULE & FIXINGS ... N08
PLUGS AND RELAYS. ... N07
POWER RESISTOR AND LAMBDA SENSOR. ... L16
RADIO CASSETTE PLAYER-FROM (V)148782 ... N14
SHOULDER BAG-REMOVEABLE RADIO ... N14
RADIO'S-UP TO (V)148781 ... N13
REAR LAMPS. ... K11
SEAT CUSHION SENSOR. ... N11
SENSORS,TRANSMITTERS AND BRACKET ... M12
SIDE LAMPS. ... K07
SPEED CONTROL PUMP & DUMP VALVE ... O11
STARTER MOTOR - 3.6 LITRE ... J08
STARTER MOTOR 5.3 LITRE ... J09
STEERING COLUMN SWITCHES. ... M04
TRIP COMPUTER AND PANEL SWITCHES. ... M06
WASH/WIPE RESERVOIR AND FIXINGS ... L08
WINDSCREEN WASHER RESERVOIR ... L04
WINDSCREEN WIPER MOTOR-LUCAS TYPE ... K17
WINDSCREEN WIPERS ... L03

RADIO CASSETTE PLAYER-FROM (V)148782

Illus	Part Number	123456	Description	Quantity	Change Point	Remarks
			RADIO CASSETTE PLAYER			
1	DAC 5574		Clarion 950HX	1		Std UK coupe
	DAC 5575		Clarion 959HX	1		Optional-not UK
2	JLM 407		Escutcheon	1		
3	DAC 5937		Alpine 7283L	1		Std UK convertible
	DAC 5931		Alpine 7283M	1		Optional-not UK
4	DAC 5936		Bracket-removable radio	1		
	DAC 5933		SHOULDER BAG-REMOVEABLE RADIO	1		

K84.06JP23/9/88

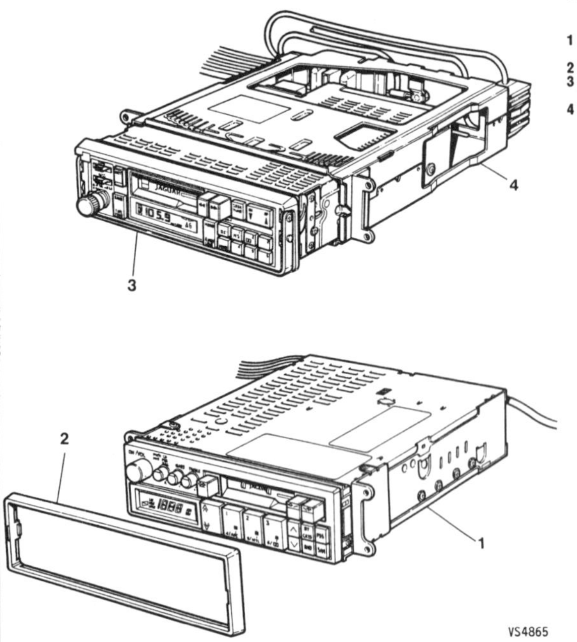

VS4865

2N 14

MASTER INDEX

GROUP INDEX

LOUDSPEAKERS AND MOUNTINGS.

Illus	Part Number	123456	Description	Quantity	Change Point	Remarks
			NOT CONVERTIBLE REAR			
1	DAC 3416		Loudspeaker (4 OHM)	4		
	DAC 3013		Loudspeaker (8 OHM)	4	Up to (V)148945	3.6 Litre with Clarion E920 radio
2	DAC 2237		Water shield	2		
3	C 45917		Grille	4		
4	BBC 7429		Bezel assembly	4		
5	C 38005		Ring-card	4		

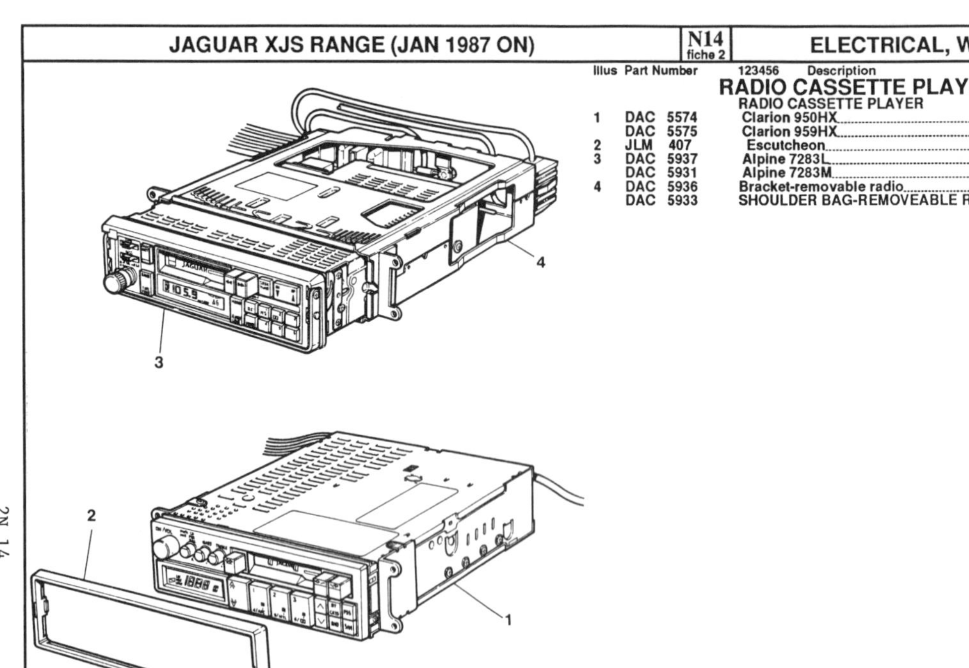

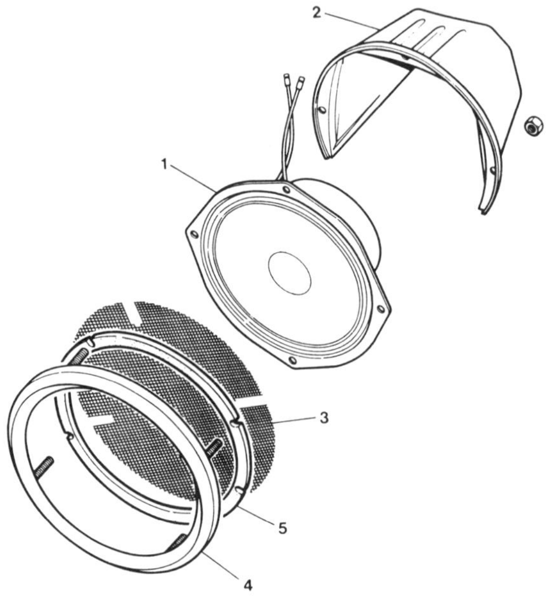

VS 1980A

2N 15

MASTER INDEX
ENGINE ... 1.C
FLYWHEEL AND CLUTCH ... 1.G
GEARBOX AND PROPSHAFT ... 1.H
AXLES, SUSPENSION, DRIVE SHAFTS, WHEELS ... 1.K
STEERING ... 1.M
BRAKES AND BRAKE CONTROLS ... 1.N
FUEL, EXHAUST AND EMISSION SYSTEMS ... 2.C
COOLING, HEATING AND AIR CONDITIONING ... 2.G
ELECTRICAL, WASHERS-WIPERS, INSTRUMENTS ... 2.J
CHASSIS, SUBFRAMES, BODYSHELL, FITTINGS ... 3.C
FASCIA, TRIM, SEATS AND FIXINGS ... 3.H
ACCESSORIES AND PAINT ... 4.C
NUMERICAL INDEX ... 4.D
GROUP INDEX
A.B.S EQUIPMENT ... N09
AERIAL AND FIXINGS. ... N12
ALTERNATOR 75 AMP (UP TO (E)109969). ... J02
BATTERY AND LEADS. ... N17
CONSOLE SWITCHES AND CIGAR LIGHTER ... M08
CONTROL UNIT-P.J-V12 COUPE ... L11
FROM 89 MY ... L13
CONTROL UNITS-PI/CRUISE CONTROL 3.6LITRE ... L10
DISTRIBUTOR-3.6 LITRE. ... J11
DISTRIBUTOR-5.3 LITRE. ... J12
FROM 89 MY ... J13
DOOR WARNING LAMP. ... K15
FOG REAR GUARD LAMP. ... K14
FRONT FLASHER LAMPS. ... K06
FRONT FOG LAMPS. ... K13
FUEL INTERFACE UNIT AND RELAYS ... L15
HARNESSES-FRONT 3.6 LITRE-FROM (V)148945 ... O04
HARNESSES-FRONT 5.3 LITRE-FROM (V)148782 ... O06
HARNESSES-FRONT-3.6 LITRE-UPTO (V)148944 ... O03
HEADLAMP WASH WIPE RESERVOIR ... L07
HEADLAMP WASHWIPE EQUIPMENT ... L06
HEADLAMPS-NOT USA ... K05
HEADLAMPS-USA. ... K03
HIGH MOUNTED STOP LAMP-USA/CDN ONLY. ... K12
HOOD CONTROL MODULE - CONVERTIBLE ... N10
HORNS AND HORN PAD ... M05
IGNITION COIL & MODULE-3.6 LITRE ... J14
IGNITION COILS AND AMPLIFIER-5.3 LITRE ... J15
INERTIA SWITCH AND COVER ... M13
INSTRUMENT PACK-3.6 LITRE ... L17
INSTRUMENT PACK-5.3 LITRE ... L18
INTERIOR LAMPS-CABRIOLET. ... K09
INTERIOR LAMPS-COUPE. ... K08
LOUDSPEAKERS AND MOUNTINGS. ... N15
MAIN FUSE BLOCK AND MOUNTING PLATE ... M15
NUMBER PLATE AND REVERSE LAMP. ... K16
PASSIVE RESTRAINT MODULE & FIXINGS ... N08
PLUGS AND RELAYS. ... N07
POWER RESISTOR AND LAMBDA SENSOR. ... L16
RADIO CASSETTE PLAYER-FROM (V)148782 ... N14
SHOULDER BAG-REMOVEABLE RADIO ... N14
RADIO'S-UP TO (V)148781 ... N13
REAR LAMPS. ... K11
SEAT CUSHION SENSOR. ... N11
SENSORS,TRANSMITTERS AND BRACKET ... M12
SIDE LAMPS. ... K07
SPEED CONTROL PUMP & DUMP VALVE ... O11
STARTER MOTOR - 3.6 LITRE ... J08
STARTER MOTOR 5.3 LITRE ... J09
STEERING COLUMN SWITCHES. ... M04
TRIP COMPUTER AND PANEL SWITCHES. ... M06
WASH/WIPE RESERVOIR AND FIXINGS ... L08
WINDSCREEN WASHER RESERVOIR ... L04
WINDSCREEN WIPER MOTOR-LUCAS TYPE ... K17
WINDSCREEN WIPERS ... L03

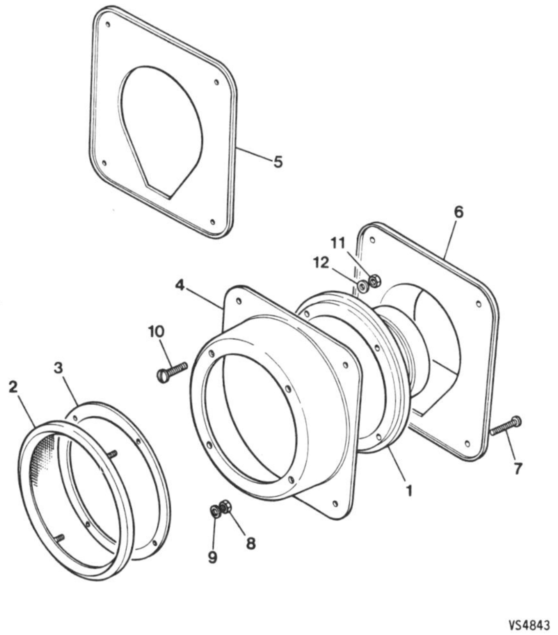

VS4843

REAR LOUDSPEAKERS - CONVERTIBLE

Illus	Part Number	123456 Description	Quantity	Change Point	Remarks
1	DAC 3416	Loudspeaker	2		
2	DAC 4418	Bezel	2		
3	C 38005	Spacer	2		
4	DAC 5966	Plinth	2		
5	DAC 5962	Cover - speaker RH	1		
6	DAC 5963	Cover - speaker LH	1		
7	DA 606084	Screw	8		Plinth & cover to panel
8	VCN 113 L	Nut	8		Bezel to panel
9	WF 704061 J	Washer	8		
10	SE 104201	Screw	8		Speaker to plinth
11	JLM 9686	Nut	8		
12	WF 104001 J	Washer	8		

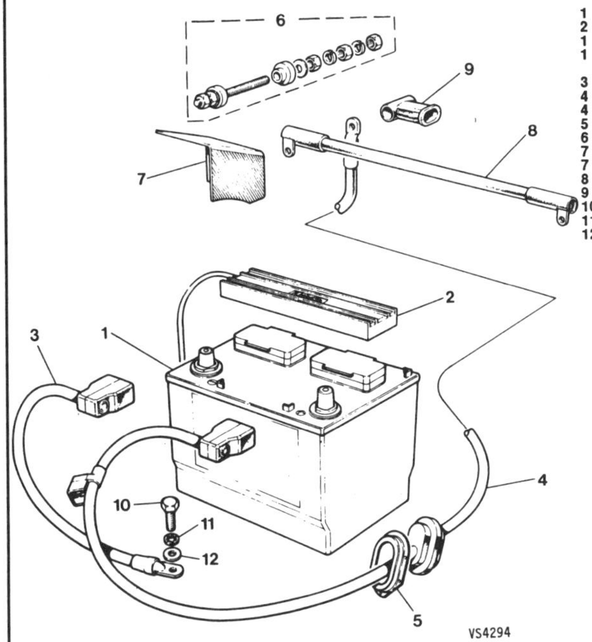

VS 4294

BATTERY AND LEADS.

Illus	Part Number	123456 Description	Quantity	Change Point	Remarks
1	JLM 9896	Battery	1	Upto (V)145729	
2	DAC 2846	Cover-breather	1		
1	DAC 4545	Battery	1	From (V)145730	Wet charged
1	JLM 9992	Battery	1		Dry charged
	C 43502	Pipe-vent	1		
3	C 43861	Lead-battery-negative	1		
4	C 43865	Lead-battery-positive	1	Up to (V)145729	
4	DAC 5608	Lead-battery-positive	1	From (V)145730	
5	C 44163	Clamping plate-cable	1		
6	C 45021	Post-terminal	2		
7	C 44718	Cover-terminal post-RH	1		
7	C 44719	Cover-terminal post-LH	1		
8	C 44183	Cable-terminal post link	1		
9	AGU 2074	Boot-rubber	2		
10	SH 606071 J	Setscrew-earth lead	1		
11	WF 600060 J	Washer shakeproof	1		
12	JLM 301	Washer-plain	1		

MASTER INDEX

ENGINE 1.C
FLYWHEEL AND CLUTCH 1.G
GEARBOX AND PROPSHAFT 1.H
AXLES, SUSPENSION, DRIVE SHAFTS, WHEELS 1.K
STEERING 1.M
BRAKES AND BRAKE CONTROLS 1.N
FUEL, EXHAUST AND EMISSION SYSTEMS 2.C
COOLING, HEATING AND AIR CONDITIONING 2.G
ELECTRICAL, WASHERS-WIPERS, INSTRUMENTS 2.J
CHASSIS, SUBFRAMES, BODYSHELL, FITTINGS 3.C
FASCIA, TRIM, SEATS AND FIXINGS 3.H
ACCESSORIES AND PAINT 4.C
NUMERICAL INDEX 4.D

GROUP INDEX

A.B.S EQUIPMENT N09
AERIAL AND FIXINGS N12
ALTERNATOR 75 AMP (UP TO (E)109969) J02
BATTERY AND LEADS N17
CONSOLE SWITCHES AND CIGAR LIGHTER M08
CONTROL UNIT-PJ-V12 COUPE L11
FROM 89 MY L13
CONTROL UNITS-PI/CRUISE CONTROL 3.6LITRE L10
DISTRIBUTOR-3.6 LITRE J11
DISTRIBUTOR-5.3 LITRE J12
FROM 89 MY J13
DOOR WARNING LAMP K15
FOG REAR GUARD LAMP K14
FRONT FLASHER LAMPS K06
FRONT FOG LAMPS K13
FUEL INTERFACE UNIT AND RELAYS L15
HARNESSES-FRONT 3.6 LITRE-FROM (V)148945 O04
HARNESSES-FRONT 5.3 LITRE-FROM (V)148782 O06
HARNESSES-FRONT-3.6 LITRE-UPTO (V)148944 O03
HEADLAMP WASH WIPE RESERVOIR L07
HEADLAMP WASHWIPE EQUIPMENT L06
HEADLAMPS-NOT USA K05
HEADLAMPS-USA K03
HIGH MOUNTED STOP LAMP-USA/CDN ONLY K12
HOOD CONTROL MODULE - CONVERTIBLE N10
HORNS AND HORN PAD M05
IGNITION COIL & MODULE-3.6 LITRE J14
IGNITION COILS AND AMPLIFIER-5.3 LITRE J15
INERTIA SWITCH AND COVER M13
INSTRUMENT PACK-3.6 LITRE L17
INSTRUMENT PACK-5.3 LITRE L18
INTERIOR LAMPS-CABRIOLET K09
INTERIOR LAMPS-COUPE K08
LOUDSPEAKERS AND MOUNTINGS N15
MAIN FUSE BLOCK AND MOUNTING PLATE M15
NUMBER PLATE AND REVERSE LAMP K16
PASSIVE RESTRAINT MODULE & FIXINGS N08
PLUGS AND RELAYS N07
POWER RESISTOR AND LAMBDA SENSOR L16
RADIO CASSETTE PLAYER-FROM (V)148782 N14
SHOULDER BAG-REMOVEABLE RADIO N14
RADIO'S-UP TO (V)148781 N13
REAR LAMPS K11
SEAT CUSHION SENSOR N11
SENSORS,TRANSMITTERS AND BRACKET M12
SIDE LAMPS K07
SPEED CONTROL PUMP & DUMP VALVE O11
STARTER MOTOR - 3.6 LITRE J08
STARTER MOTOR 5.3 LITRE J09
STEERING COLUMN SWITCHES M04
TRIP COMPUTER AND PANEL SWITCHES M06
WASH/WIPE RESERVOIR AND FIXINGS L08
WINDSCREEN WASHER RESERVOIR L04
WINDSCREEN WIPER MOTOR-LUCAS TYPE K17
WINDSCREEN WIPERS L03

2N 18

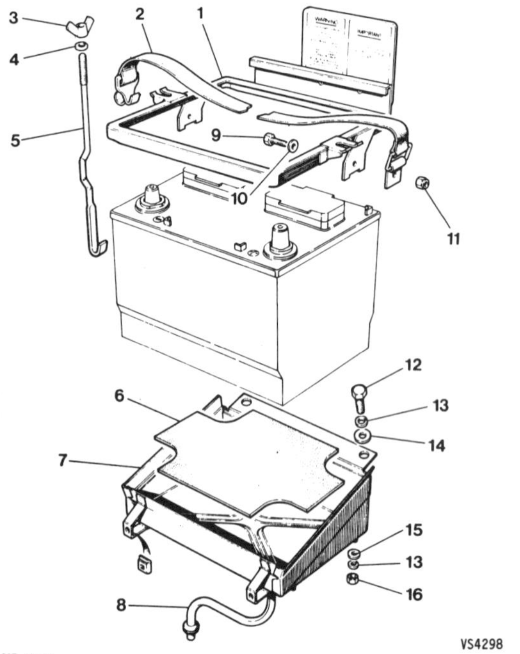

VS 4298

BATTERY FIXINGS - Upto (V)145729

Illus	Part Number	123456	Description	Quantity	Change Point	Remarks
1	C 45950		Strap and label assembly	1		
2	C 43943		Strap-cover	1		
3	C 31352		Wing nut	2		
4	WS 600041 J		Washer-spring	2		
5	C 43760		Rod-battery-long	1		
5	C 43761		Rod-battery-short	1		
6	C 42853		Spacer-battery	1		
7	BD 46151		Tray-battery	1		
8	C 43762		Drain tube-battery	1		
9	UFS 119 4R		Screw	1		
10	WC 702101 J		Washer-plain	1		
11	C 8667 17		Locknut	1		
12	SH 604051 J		Screw	2		
13	C 724		Washer-shakeproof	2		
14	BD 541 20		Washer-plain	2		
15	JLM 296		Washer-plain	4		
16	JLM 9683		Nut	4		

2O 02

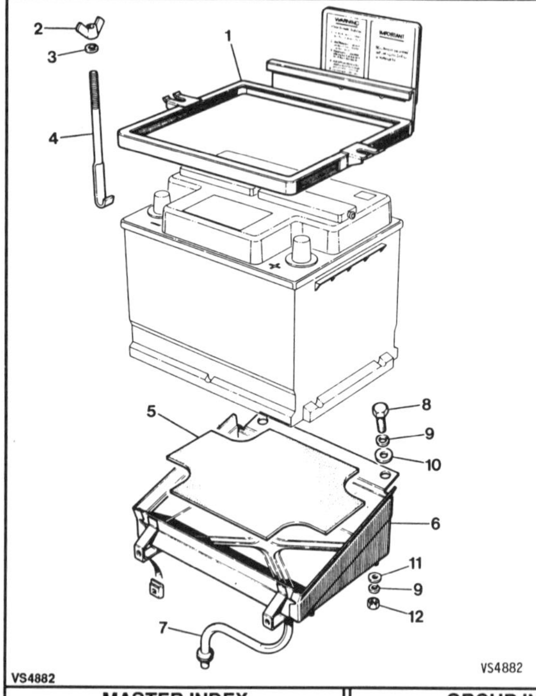

VS4882

BATTERY FIXINGS - FROM (V)145730

Illus	Part Number	123456	Description	Quantity	Change Point	Remarks
1	DAC 4727		CLAMP AND LABEL ASSEMBLY	1		
2	C 31352		Wing nut	2		
3	WS 600041 J		Washer-spring	2		
4	DAC 4754		Rod-battery	2		
5	C 42853		Spacer-battery	1		
6	BCC 9561		Tray-battery	1		
6	BDC 3536		Battery tray	1		Convertible only
7	C 43762		Drain tube-tray	1		
8	SH 604051 J		Screw	2		
9	C 724		Washer-shakeproof	2		
10	BD 541 20		Washer-plain	2		
11	JLM 296		Washer-plain	4		
12	JLM 9683		Nut	4		

MASTER INDEX

ENGINE 1.C
FLYWHEEL AND CLUTCH 1.G
GEARBOX AND PROPSHAFT 1.H
AXLES, SUSPENSION, DRIVE SHAFTS, WHEELS 1.K
STEERING 1.M
BRAKES AND BRAKE CONTROLS 1.N
FUEL, EXHAUST AND EMISSION SYSTEMS 2.C
COOLING, HEATING AND AIR CONDITIONING 2.G
ELECTRICAL, WASHERS-WIPERS, INSTRUMENTS 2.J
CHASSIS, SUBFRAMES, BODYSHELL, FITTINGS 3.C
FASCIA, TRIM, SEATS AND FIXINGS 3.H
ACCESSORIES AND PAINT 4.C
NUMERICAL INDEX 4.D

GROUP INDEX

A.B.S EQUIPMENT N09
AERIAL AND FIXINGS. N12
ALTERNATOR 75 AMP (UP TO (E)109969). J02
BATTERY AND LEADS. N17
CONSOLE SWITCHES AND CIGAR LIGHTER M08
CONTROL UNIT-P.J-V12 COUPE L11
FROM 89 MY L13
CONTROL UNITS-PI/CRUISE CONTROL 3.6LITRE L10
DISTRIBUTOR-3.6 LITRE. J11
DISTRIBUTOR-5.3 LITRE. J12
FROM 89 MY J13
DOOR WARNING LAMP. K15
FOG REAR GUARD LAMP. K14
FRONT FLASHER LAMPS. K06
FRONT FOG LAMPS. K13
FUEL INTERFACE UNIT AND RELAYS L15
HARNESSES-FRONT 3.6 LITRE-FROM (V)148945 .... O04
HARNESSES-FRONT 5.3 LITRE-FROM (V)148782 .... O06
HARNESSES-FRONT-3.6 LITRE-UPTO (V)148944 .... O03
HEADLAMP WASH WIPE RESERVOIR L07
HEADLAMP WASHWIPE EQUIPMENT L06
HEADLAMPS-NOT USA K05
HEADLAMPS-USA. K03
HIGH MOUNTED STOP LAMP-USA/CDN ONLY. K12
HOOD CONTROL MODULE - CONVERTIBLE N10
HORNS AND HORN PAD M05
IGNITION COIL & MODULE-3.6 LITRE J14
IGNITION COILS AND AMPLIFIER-5.3 LITRE J15
INERTIA SWITCH AND COVER M13
INSTRUMENT PACK-3.6 LITRE L17
INSTRUMENT PACK-5.3 LITRE L18
INTERIOR LAMPS-CABRIOLET. K09
INTERIOR LAMPS-COUPE. K08
LOUDSPEAKERS AND MOUNTINGS. N15
MAIN FUSE BLOCK AND MOUNTING PLATE M15
NUMBER PLATE AND REVERSE LAMP. K16
PASSIVE RESTRAINT MODULE & FIXINGS N08
PLUGS AND RELAYS. N07
POWER RESISTOR AND LAMBDA SENSOR. L16
RADIO CASSETTE PLAYER-FROM (V)148782 .... N14
SHOULDER BAG-REMOVEABLE RADIO N14
RADIO'S-UP TO (V)148781 .... N13
REAR LAMPS. K11
SEAT CUSHION SENSOR. N11
SENSORS,TRANSMITTERS AND BRACKET M12
SIDE LAMPS. K07
SPEED CONTROL PUMP & DUMP VALVE O11
STARTER MOTOR - 3.6 LITRE J08
STARTER MOTOR 5.3 LITRE J09
STEERING COLUMN SWITCHES. M04
TRIP COMPUTER AND PANEL SWITCHES. M06
WASH/WIPE RESERVOIR AND FIXINGS L08
WINDSCREEN WASHER RESERVOIR L04
WINDSCREEN WIPER MOTOR-LUCAS TYPE K17
WINDSCREEN WIPERS L03

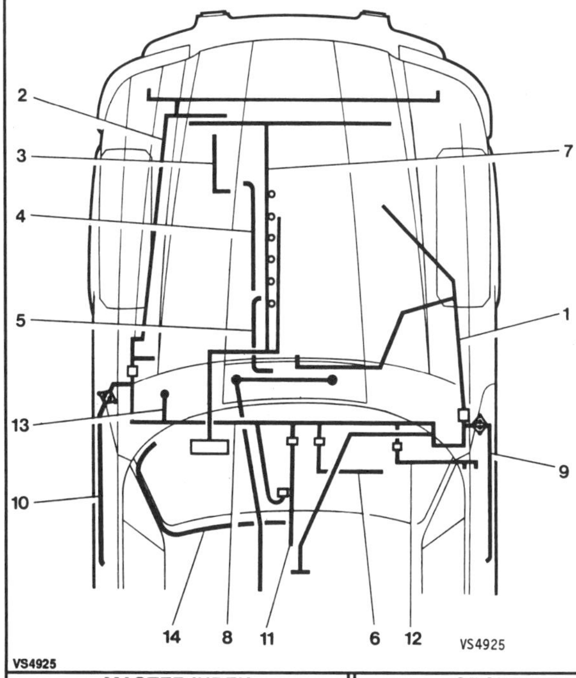

HARNESSES-FRONT-3.6 LITRE-UPTO (V)148944

Illus	Part Number	123456 Description	Quantity	Change Point	Remarks
1	DAC 4282	Harness-right forward	1		
2	DAC 4280	Harness-left forward	1		Not dim/dip
	DAC 4278	Harness-left forward	1		With dim/dip
	DAC 3541	Suppressor-headlamp relay	1		
3	C 44184	Engine/body earth	2		
4	DBC 2175	Harness-alternator	1		
5	DAC 4886	Cable-starter	1		
6	DAC 4619	Harness-switch panel	1		
	DAC 4849	Harness-clock link	1		
7	DAC 4286	Harness-engine/main cable	1		Non catalyst
	DAC 4287	Harness-engine/main cable	1		With catalyst
8	DAC 4270	Harness-bulkhead-RHD	1		
	DAC 4273	Harness-bulkhead-LHD	1		
	DAC 1915	Shorting plug	1		
	DAC 5482	Harness-lighting switch link	1		3.6 Litre without front foglamps
9	DAC 4289	Harness-drivers door	1		
10	DAC 4288	Harness-passenger door	1		
	DAC 2050	Wiring sheath-door upper	1		
	DAC 2854	Wiring sheath-door lower	1		
11	DAC 4385	Harness-console	1		
	DAC 4595	Harness-'hands free' CRT	1		
12	DAC 2650	Harness-window lift/sensor	1		
13	DAC 2081	Harness-seat belt warning	1		
14	DAC 2097	Harness-roof lamp	1		
	DAC 4214	Harness-auto trans	1		

K92JP27/6/88

MASTER INDEX

GROUP INDEX

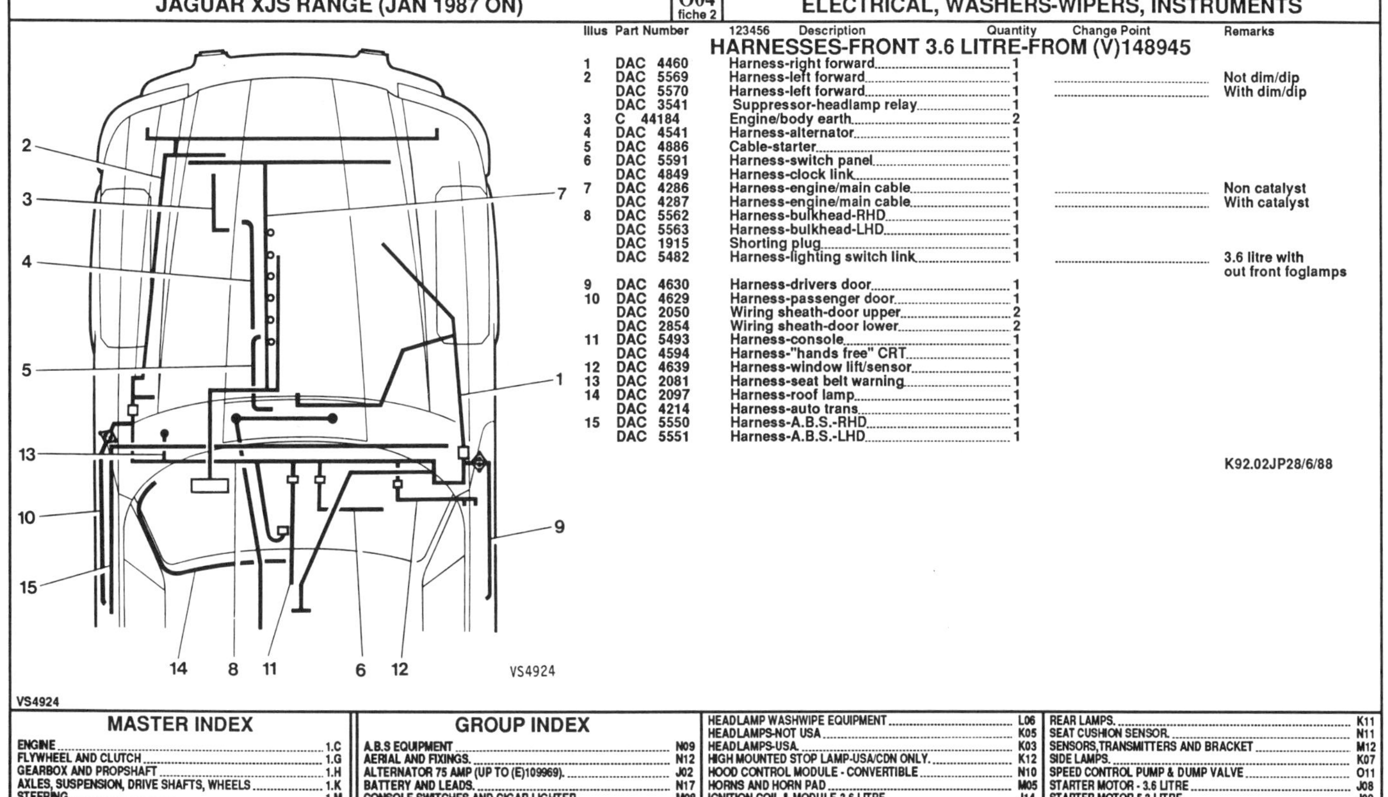

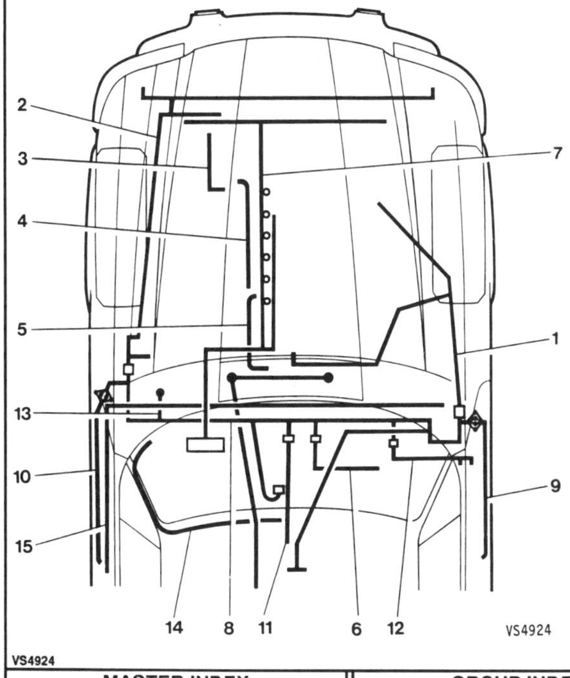

HARNESSES-FRONT 3.6 LITRE-FROM (V)148945

Illus	Part Number	123456 Description	Quantity	Change Point	Remarks
1	DAC 4460	Harness-right forward	1		
2	DAC 5569	Harness-left forward	1		Not dim/dip
	DAC 5570	Harness-left forward	1		With dim/dip
	DAC 3541	Suppressor-headlamp relay	1		
3	C 44184	Engine/body earth	2		
4	DAC 4541	Harness-alternator	1		
5	DAC 4886	Cable-starter	1		
6	DAC 5591	Harness-switch panel	1		
	DAC 4849	Harness-clock link	1		
7	DAC 4286	Harness-engine/main cable	1		Non catalyst
	DAC 4287	Harness-engine/main cable	1		With catalyst
8	DAC 5562	Harness-bulkhead-RHD	1		
	DAC 5563	Harness-bulkhead-LHD	1		
	DAC 1915	Shorting plug	1		
	DAC 5482	Harness-lighting switch link	1		3.6 litre with out front foglamps
9	DAC 4630	Harness-drivers door	1		
10	DAC 4629	Harness-passenger door	1		
	DAC 2050	Wiring sheath-door upper	2		
	DAC 2854	Wiring sheath-door lower	2		
11	DAC 5493	Harness-console	1		
	DAC 4594	Harness-"hands free" CRT	1		
12	DAC 4639	Harness-window lift/sensor	1		
13	DAC 2081	Harness-seat belt warning	1		
14	DAC 2097	Harness-roof lamp	1		
	DAC 4214	Harness-auto trans	1		
15	DAC 5550	Harness-A.B.S.-RHD	1		
	DAC 5551	Harness-A.B.S.-LHD	1		

K92.02JP28/6/88

MASTER INDEX
ENGINE 1.C
FLYWHEEL AND CLUTCH 1.G
GEARBOX AND PROPSHAFT 1.H
AXLES, SUSPENSION, DRIVE SHAFTS, WHEELS 1.K
STEERING 1.M
BRAKES AND BRAKE CONTROLS 1.N
FUEL, EXHAUST AND EMISSION SYSTEMS 2.C
COOLING, HEATING AND AIR CONDITIONING 2.G
ELECTRICAL, WASHERS-WIPERS, INSTRUMENTS 2.J
CHASSIS, SUBFRAMES, BODYSHELL, FITTINGS 3.C
FASCIA, TRIM, SEATS AND FIXINGS 3.H
ACCESSORIES AND PAINT 4.C
NUMERICAL INDEX 4.D
GROUP INDEX
A.B.S EQUIPMENT N09
AERIAL AND FIXINGS. N12
ALTERNATOR 75 AMP (UP TO (E)109969). J02
BATTERY AND LEADS. N17
CONSOLE SWITCHES AND CIGAR LIGHTER M08
CONTROL UNIT-P.J-V12 COUPE L11
FROM 89 MY L13
CONTROL UNITS-PI/CRUISE CONTROL 3.6LITRE L10
DISTRIBUTOR-3.6 LITRE. J11
DISTRIBUTOR-5.3 LITRE. J12
FROM 89 MY J13
DOOR WARNING LAMP. K15
FOG REAR GUARD LAMP. K14
FRONT FLASHER LAMPS. K06
FRONT FOG LAMPS. K13
FUEL INTERFACE UNIT AND RELAYS L15
HARNESSES-FRONT 3.6 LITRE-FROM (V)148945 O04
HARNESSES-FRONT 5.3 LITRE-FROM (V)148782 O06
HARNESSES-FRONT-3.6 LITRE-UPTO (V)148944 O03
HEADLAMP WASH WIPE RESERVOIR L07
HEADLAMP WASHWIPE EQUIPMENT L06
HEADLAMPS-NOT USA K05
HEADLAMPS-USA. K03
HIGH MOUNTED STOP LAMP-USA/CDN ONLY. K12
HOOD CONTROL MODULE - CONVERTIBLE N10
HORNS AND HORN PAD M05
IGNITION COIL & MODULE-3.6 LITRE J14
IGNITION COILS AND AMPLIFIER-5.3 LITRE J15
INERTIA SWITCH AND COVER M13
INSTRUMENT PACK-3.6 LITRE L17
INSTRUMENT PACK-5.3 LITRE L18
INTERIOR LAMPS-CABRIOLET. K09
INTERIOR LAMPS-COUPE. K08
LOUDSPEAKERS AND MOUNTINGS. N15
MAIN FUSE BLOCK AND MOUNTING PLATE M15
NUMBER PLATE AND REVERSE LAMP. K16
PASSIVE RESTRAINT MODULE & FIXINGS N08
PLUGS AND RELAYS. N07
POWER RESISTOR AND LAMBDA SENSOR. L16
RADIO CASSETTE PLAYER-FROM (V)148782 N14
SHOULDER BAG-REMOVEABLE RADIO N14
RADIO'S-UP TO (V)148781 N13
REAR LAMPS. K11
SEAT CUSHION SENSOR. N11
SENSORS,TRANSMITTERS AND BRACKET M12
SIDE LAMPS. K07
SPEED CONTROL PUMP & DUMP VALVE O11
STARTER MOTOR - 3.6 LITRE J08
STARTER MOTOR 5.3 LITRE J09
STEERING COLUMN SWITCHES. M04
TRIP COMPUTER AND PANEL SWITCHES. M06
WASH/WIPE RESERVOIR AND FIXINGS L08
WINDSCREEN WASHER RESERVOIR L04
WINDSCREEN WIPER MOTOR-LUCAS TYPE K17
WINDSCREEN WIPERS L03

JAGUAR XJS RANGE (JAN 1987 ON)

O05 fiche 2

ELECTRICAL, WASHERS-WIPERS, INSTRUMENTS

20 05

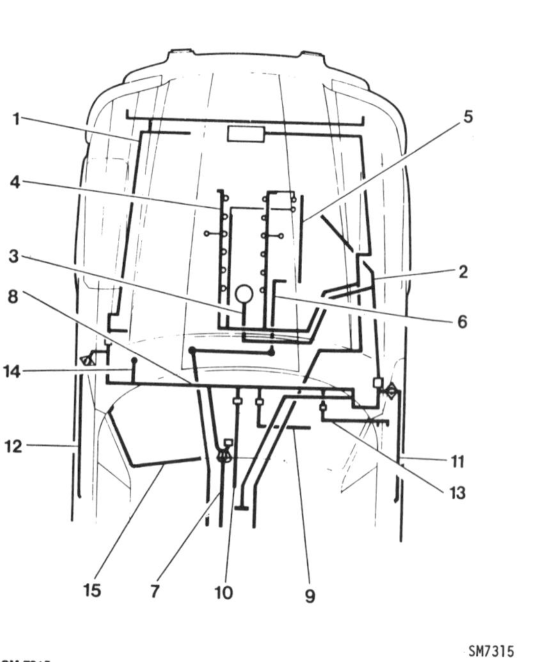

SM 7315

HARNESSES-FRONT-5.3 LITRE-UPTO (V)148781

Illus	Part Number	123456 Description	Quantity	Change Point	Remarks
2	DAC 4550	Harness-right forward	1		
	DAC 4552	Anti stall relays harness	1		
1	DAC 4281	Harness-left forward	1		Not dim/dip
1	DAC 4279	Harness-left forward	1		With dim/dip
	DAC 3541	Suppressor-headlamp relay	1		
	C 44184	Engine/body earth	1		
3	DAC 2689	Harness engine	1		
4	DAC 3718	Harness-engine injector	1		
5	DAC 3489	Harness-alternator	1		
6	DAC 4372	Harness-starter motor	1		
	C 44183	Terminal post/cable	1		
7	DAC 1673	Harness-gearbox	1		
8	DAC 4272	Harness-bulkhead RHD	1		
8	DAC 4271	Harness-bulkhead LHD	1		Not Pass rest
8	DAC 4274	Harness-bulkhead LHD	1		With Passive
	DAC 1915	Shorting plug	1		Not USA/CDN/J/ Middle East
9	DAC 4619	Harness-switch panel RHD	1		
9	DAC 3120	Harness-switch panel LHD	1		
10	DAC 4384	Harness-console	1		
	DAC 4595	Harness-'Hands Free' CRT	1		
	DAC 4246	Harness-ASI radio link	1		] USA only
	DAC 3674	Harness-speaker link	1		]
11	DAC 4289	Harness-drivers door	1		
12	DAC 4288	Harness-passenger door	1		
	DAC 2050	Wiring sheath-door upper	2		
	DAC 2854	Wiring sheath-door lower	2		
13	DAC 2650	Harness-windowlift/sensor	1		
14	DAC 2081	Harness-seat belt warning	1		Not USA/CDN
15	DAC 2097	Harness-roof lamp	1		

MASTER INDEX

GROUP INDEX

JAGUAR XJS RANGE (JAN 1987 ON)

O06 fiche 2

ELECTRICAL, WASHERS-WIPERS, INSTRUMENTS

20 06

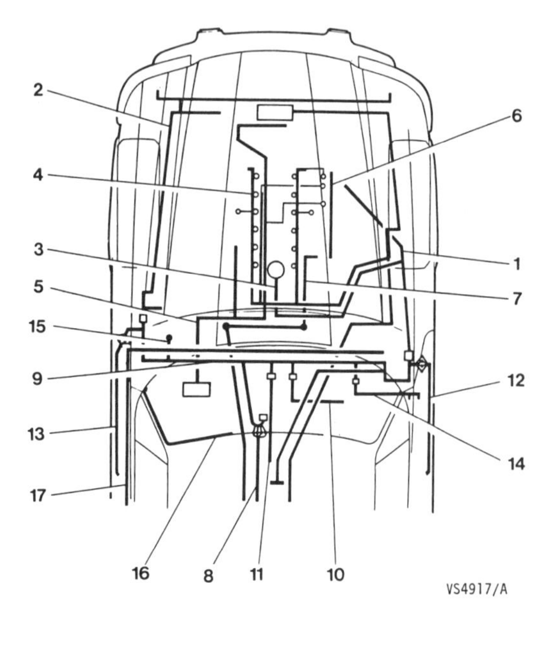

VS4917A

HARNESSES-FRONT 5.3 LITRE-FROM (V)148782

Illus	Part Number	123456 Description	Quantity	Change Point	Remarks
1	DAC 4550	Harness-Right forward	1		
	DAC 4552	Anti Stall relays harness	1		
2	DAC 5567	Harness-left forward	1		Not dim/dip
2	DAC 5568	Harness-left forward	1		With dim/dip
	DAC 3541	Suppressor-headlamp relay	1		
	C 44184	Engine/Body earth	1		
3	DAC 2689	Harness-engine	1	Up to 89 MY	
3	DAC 5895	Harness-engine	1	From 89 MY	
4	DAC 3718	Harness-engine injector	1		
5	DAC 4876	Harness-ignition	1	From 89 MY	
6	DAC 6074	Harness-alternator	1		
7	DAC 6072	Harness-starter motor	1		
	C 44183	Terminal Post/Cable	1		
8	DAC 1673	Harness-Gearbox	1		
9	DAC 5560	Harness-bulkhead-RHD	1		
9	DAC 5561	Harness-bulkhead-LHD	1		Not pass rest
9	DAC 5559	Harness-bulkhead-LHD	1		With Pass rest
	DAC 1915	Shorting plug	1		Not USA/CDN/J Middle East
10	DAC 5591	Harness-switch panel RHD	1		
10	DAC 3120	Harness-switch panel LHD	1		
11	DAC 5492	Harness-console	1		
	DAC 4595	Harness-"Hands free" CRT	1		
	DAC 4246	Harness-ASI radio link	1		
	DAC 3674	Harness-speaker link	1		
12	DAC 4630	Harness-drivers door	1		
13	DAC 4629	Harness-passenger door	1		
	DAC 2050	Wiring sheath-door upper	2		
	DAC 2854	Wiring sheath-door lower	2		
13	DAC 4639	Harness-Window lift/sensor	1		
15	DAC 2081	Harness-Seat belt warning	1		
16	DAC 2097	Harness-roof lamp	1		
17	DAC 5550	Harness-A.B.S-RHD	1		] Not Convertible
	DAC 5551	Harness-A.B.S-LHD	1		]
	DAC 5552	Harness-A.B.S-RHD	1		] Convertible
	DAC 5553	Harness-A.B.S-LHD	1		]

K92.06JP31/8/88

MASTER INDEX
ENGINE 1.C
FLYWHEEL AND CLUTCH 1.G
GEARBOX AND PROPSHAFT 1.H
AXLES, SUSPENSION, DRIVE SHAFTS, WHEELS 1.K
STEERING 1.M
BRAKES AND BRAKE CONTROLS 1.N
FUEL, EXHAUST AND EMISSION SYSTEMS 2.C
COOLING, HEATING AND AIR CONDITIONING 2.G
ELECTRICAL, WASHERS-WIPERS, INSTRUMENTS 2.J
CHASSIS, SUBFRAMES, BODYSHELL, FITTINGS 3.C
FASCIA, TRIM, SEATS AND FIXINGS 3.H
ACCESSORIES AND PAINT 4.C
NUMERICAL INDEX 4.D
GROUP INDEX
A.B.S EQUIPMENT N09
AERIAL AND FIXINGS. N12
ALTERNATOR 75 AMP (UP TO (E)109969). J02
BATTERY AND LEADS. N17
CONSOLE SWITCHES AND CIGAR LIGHTER M08
CONTROL UNIT-PJ-V12 COUPE L11
FROM 89 MY L13
CONTROL UNITS-PI/CRUISE CONTROL 3.6LITRE L10
DISTRIBUTOR-3.6 LITRE. J11
DISTRIBUTOR-5.3 LITRE. J12
FROM 89 MY J13
DOOR WARNING LAMP. K15
FOG REAR GUARD LAMP. K14
FRONT FLASHER LAMPS. K06
FRONT FOG LAMPS. K13
FUEL INTERFACE UNIT AND RELAYS L15
HARNESSES-FRONT 3.6 LITRE-FROM (V)148945 .... O04
HARNESSES-FRONT 5.3 LITRE-FROM (V)148782 .... O06
HARNESSES-FRONT-3.6 LITRE-UPTO (V)148944 .... O03
HEADLAMP WASH WIPE RESERVOIR L07
HEADLAMP WASHWIPE EQUIPMENT L06
HEADLAMPS-NOT USA K05
HEADLAMPS-USA. K03
HIGH MOUNTED STOP LAMP-USA/CDN ONLY. K12
HOOD CONTROL MODULE - CONVERTIBLE N10
HORNS AND HORN PAD M05
IGNITION COIL & MODULE-3.6 LITRE J14
IGNITION COILS AND AMPLIFIER-5.3 LITRE J15
INERTIA SWITCH AND COVER M13
INSTRUMENT PACK-3.6 LITRE L17
INSTRUMENT PACK-5.3 LITRE L18
INTERIOR LAMPS-CABRIOLET. K09
INTERIOR LAMPS-COUPE. K08
LOUDSPEAKERS AND MOUNTINGS. N15
MAIN FUSE BLOCK AND MOUNTING PLATE M15
NUMBER PLATE AND REVERSE LAMP. K16
PASSIVE RESTRAINT MODULE & FIXINGS N08
PLUGS AND RELAYS. N07
POWER RESISTOR AND LAMBDA SENSOR. L16
RADIO CASSETTE PLAYER-FROM (V)148782 .... N14
SHOULDER BAG-REMOVEABLE RADIO N14
RADIO'S-UP TO (V)148781 .... N13
REAR LAMPS. K11
SEAT CUSHION SENSOR. N11
SENSORS,TRANSMITTERS AND BRACKET M12
SIDE LAMPS. K07
SPEED CONTROL PUMP & DUMP VALVE O11
STARTER MOTOR - 3.6 LITRE J08
STARTER MOTOR 5.3 LITRE J09
STEERING COLUMN SWITCHES. M04
TRIP COMPUTER AND PANEL SWITCHES. M06
WASH/WIPE RESERVOIR AND FIXINGS L08
WINDSCREEN WASHER RESERVOIR L04
WINDSCREEN WIPER MOTOR-LUCAS TYPE K17
WINDSCREEN WIPERS L03

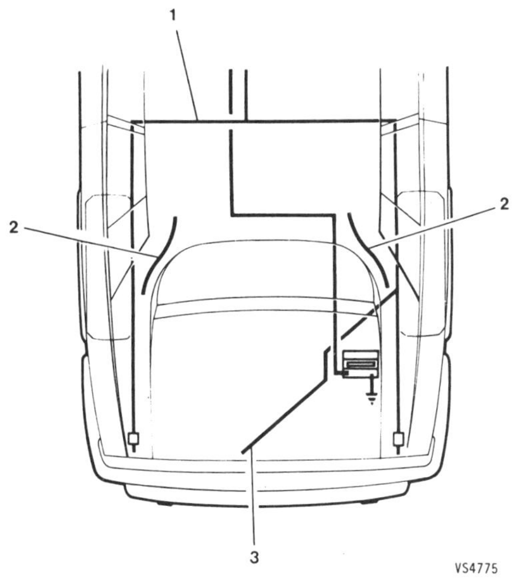

VS 4775

HARNESS-REAR-3.6 LITRE

Illus	Part Number	Description	Quantity	Change Point	Remarks
1	DAC 4285	Harness-rearward	1	Up to (V)148944	
	DAC 5565	Harness-rearward	1	From (V)148945	
2	C 42701	Lead-interior light	2		Rear Pillar
3	DAC 2084	Harness-boot lid	1	Up to (V)148944	
	DAC 4628	Harness-boot lid	1	From (V)148945	
	DAC 3551	Link lead-heated rear screen	1		Cabriolet with hardtop only

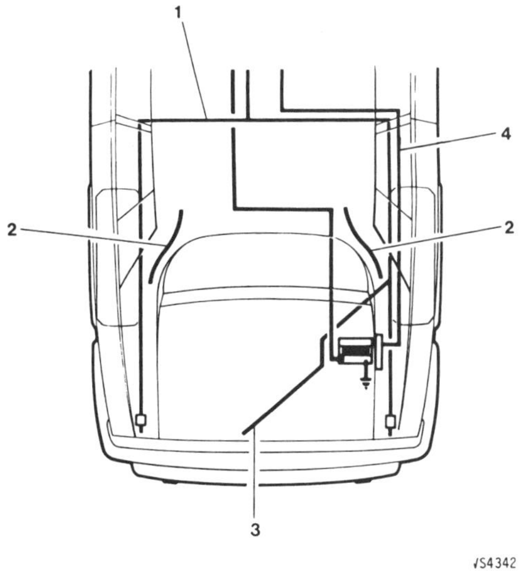

VS 4342

HARNESSES-REAR-5.3 LITRE-UP TO (V) 148781

Illus	Part Number	Description	Quantity	Change Point	Remarks
1	DAC 4285	Harness-rearward	1		Not pas rest.
1	DAC 4277	Harness-rearward	1		With pas.rest.
	DAC 3567	Harness-weight switch	1		With pas.rest.
2	C 42701	Lead-interior light	1		Rear pillar
3	DAC 2084	Harness-boot lid	1		
4	DAC 4131	Harness-main cable	1		USA/CDN/AUS
4	DAC 4094	Harness-main cable	1		EUR./M.East
4	DAC 4093	Harness-main cable	1		J/Germany
	DAC 2654	Harness-extra air valve	1	Upto (V)141532	USA/CDN/J/AUS
	DAC 4480	Harness-extra air valve	1	From (V)141533	Germany
	DAC 2712	Harness-extra air valve	1	Upto (V)141532	Europe/M.East
	DAC 4481	Harness-extra air valve	1	From (V)141533	Europe/M.East
	DAC 3551	Link lead-heated rear screen	1		Cabriolet with hardtop only

MASTER INDEX

ENGINE 1.C
FLYWHEEL AND CLUTCH 1.G
GEARBOX AND PROPSHAFT 1.H
AXLES, SUSPENSION, DRIVE SHAFTS, WHEELS 1.K
STEERING 1.M
BRAKES AND BRAKE CONTROLS 1.N
FUEL, EXHAUST AND EMISSION SYSTEMS 2.C
COOLING, HEATING AND AIR CONDITIONING 2.G
ELECTRICAL, WASHERS-WIPERS, INSTRUMENTS 2.J
CHASSIS, SUBFRAMES, BODYSHELL, FITTINGS 3.C
FASCIA, TRIM, SEATS AND FIXINGS 3.H
ACCESSORIES AND PAINT 4.C
NUMERICAL INDEX 4.D

GROUP INDEX

A.B.S EQUIPMENT N09
AERIAL AND FIXINGS N12
ALTERNATOR 75 AMP (UP TO (E)109969) J02
BATTERY AND LEADS N17
CONSOLE SWITCHES AND CIGAR LIGHTER M08
CONTROL UNIT-PJ-V12 COUPE L11
FROM 89 MY L13
CONTROL UNITS-PI/CRUISE CONTROL 3.6LITRE L10
DISTRIBUTOR-3.6 LITRE J11
DISTRIBUTOR-5.3 LITRE J12
FROM 89 MY J13
DOOR WARNING LAMP K15
FOG REAR GUARD LAMP K14
FRONT FLASHER LAMPS K06
FRONT FOG LAMPS K13
FUEL INTERFACE UNIT AND RELAYS L15
HARNESSES-FRONT 3.6 LITRE-FROM (V)148945 O04
HARNESSES-FRONT 5.3 LITRE-FROM (V)148782 O06
HARNESSES-FRONT-3.6 LITRE-UPTO (V)148944 O03
HEADLAMP WASH WIPE RESERVOIR L07
HEADLAMP WASHWIPE EQUIPMENT L06
HEADLAMPS-NOT USA K05
HEADLAMPS-USA K03
HIGH MOUNTED STOP LAMP-USA/CDN ONLY K12
HOOD CONTROL MODULE - CONVERTIBLE N10
HORNS AND HORN PAD M05
IGNITION COIL & MODULE-3.6 LITRE J14
IGNITION COILS AND AMPLIFIER-5.3 LITRE J15
INERTIA SWITCH AND COVER M13
INSTRUMENT PACK-3.6 LITRE L17
INSTRUMENT PACK-5.3 LITRE L18
INTERIOR LAMPS-CABRIOLET K09
INTERIOR LAMPS-COUPE K08
LOUDSPEAKERS AND MOUNTINGS N15
MAIN FUSE BLOCK AND MOUNTING PLATE M15
NUMBER PLATE AND REVERSE LAMP K16
PASSIVE RESTRAINT MODULE & FIXINGS N08
PLUGS AND RELAYS N07
POWER RESISTOR AND LAMBDA SENSOR L16
RADIO CASSETTE PLAYER-FROM (V)148782 N14
SHOULDER BAG-REMOVEABLE RADIO N14
RADIO'S-UP TO (V)148781 N13
REAR LAMPS K11
SEAT CUSHION SENSOR N11
SENSORS,TRANSMITTERS AND BRACKET M12
SIDE LAMPS K07
SPEED CONTROL PUMP & DUMP VALVE O11
STARTER MOTOR - 3.6 LITRE J08
STARTER MOTOR 5.3 LITRE J09
STEERING COLUMN SWITCHES M04
TRIP COMPUTER AND PANEL SWITCHES M06
WASH/WIPE RESERVOIR AND FIXINGS L08
WINDSCREEN WASHER RESERVOIR L04
WINDSCREEN WIPER MOTOR-LUCAS TYPE K17
WINDSCREEN WIPERS L03

HARNESSES-REAR 5.3 LITRE-FROM (V) 148782

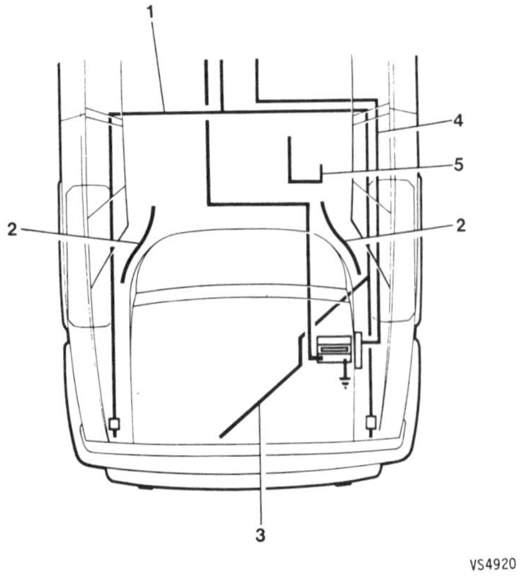

VS4920

Illus	Part Number	123456 Description	Quantity	Change Point	Remarks
1	DAC 5565	Harness-rearward	1		Not pas rest
1	DAC 5564	Harness-rearward	1		With pas rest
	DAC 5566	Harness-rearward	1		Convertible
	DAC 3567	Harness-weight switch	1		With pas rest
2	C 42701	Lead-interior light	1		Rear pillar
3	DAC 4628	Harness-boot lid	1		
4	DAC 5592	Harness-main cable	1		USA/CDN/AUS/J Germany
4	DAC 5594	Harness-main cable	1		EUR/M East
	DAC 4480	Harness-extra air valve	1		USA/CDN/AUS/J Germany
	DAC 4481	Harness-extra air valve	1		EUR/M East
5	DAC 5580	Link harness-hood module	1		Convertible

20 09

SPEED CONTROL CAM PLATE & INHIBITOR 3.6 MODELS

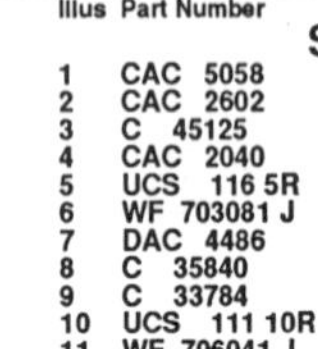

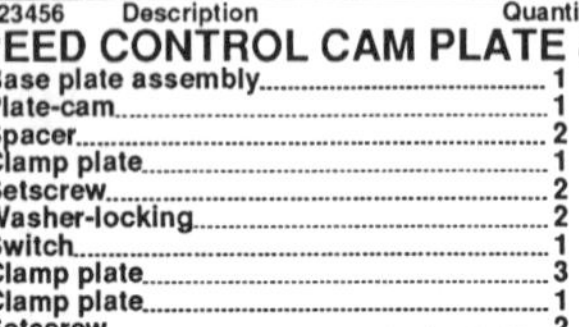

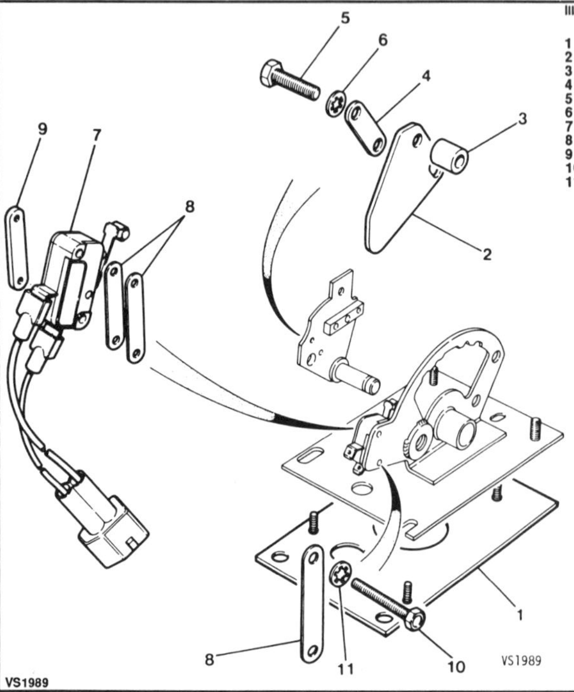

VS1989

Illus	Part Number	123456 Description	Quantity	Change Point	Remarks
1	CAC 5058	Base plate assembly	1		
2	CAC 2602	Plate-cam	1		
3	C 45125	Spacer	2		
4	CAC 2040	Clamp plate	1		
5	UCS 116 5R	Setscrew	2		
6	WF 703081 J	Washer-locking	2		
7	DAC 4486	Switch	1		
8	C 35840	Clamp plate	3		
9	C 33784	Clamp plate	1		
10	UCS 111 10R	Setscrew	2		
11	WF 706041 J	Washer-locking	2		

20 10

MASTER INDEX
ENGINE 1.C; FLYWHEEL AND CLUTCH 1.G; GEARBOX AND PROPSHAFT 1.H; AXLES, SUSPENSION, DRIVE SHAFTS, WHEELS 1.K; STEERING 1.M; BRAKES AND BRAKE CONTROLS 1.N; FUEL, EXHAUST AND EMISSION SYSTEMS 2.C; COOLING, HEATING AND AIR CONDITIONING 2.G; ELECTRICAL, WASHERS-WIPERS, INSTRUMENTS 2.J; CHASSIS, SUBFRAMES, BODYSHELL, FITTINGS 3.C; FASCIA, TRIM, SEATS AND FIXINGS 3.H; ACCESSORIES AND PAINT 4.C; NUMERICAL INDEX 4.D
GROUP INDEX
A.B.S EQUIPMENT N09; AERIAL AND FIXINGS N12; ALTERNATOR 75 AMP (UP TO (E)109969) J02; BATTERY AND LEADS N17; CONSOLE SWITCHES AND CIGAR LIGHTER M08; CONTROL UNIT-PJ-V12 COUPE L11; FROM 89 MY L13; CONTROL UNITS-PI/CRUISE CONTROL 3.6LITRE L10; DISTRIBUTOR-3.6 LITRE J11; DISTRIBUTOR-5.3 LITRE J12; FROM 89 MY J13; DOOR WARNING LAMP K15; FOG REAR GUARD LAMP K14; FRONT FLASHER LAMPS K06; FRONT FOG LAMPS K13; FUEL INTERFACE UNIT AND RELAYS L15; HARNESSES-FRONT 3.6 LITRE-FROM (V)148945 O04; HARNESSES-FRONT 5.3 LITRE-FROM (V)148782 O06; HARNESSES-FRONT-3.6 LITRE-UPTO (V)148944 O03; HEADLAMP WASH WIPE RESERVOIR L07; HEADLAMP WASHWIPE EQUIPMENT L06; HEADLAMPS-NOT USA K05; HEADLAMPS-USA K03; HIGH MOUNTED STOP LAMP-USA/CDN ONLY K12; HOOD CONTROL MODULE - CONVERTIBLE N10; HORNS AND HORN PAD M05; IGNITION COIL & MODULE-3.6 LITRE J14; IGNITION COILS AND AMPLIFIER-5.3 LITRE J15; INERTIA SWITCH AND COVER M13; INSTRUMENT PACK-3.6 LITRE L17; INSTRUMENT PACK-5.3 LITRE L18; INTERIOR LAMPS-CABRIOLET K09; INTERIOR LAMPS-COUPE K08; LOUDSPEAKERS AND MOUNTINGS N15; MAIN FUSE BLOCK AND MOUNTING PLATE M15; NUMBER PLATE AND REVERSE LAMP K16; PASSIVE RESTRAINT MODULE & FIXINGS N08; PLUGS AND RELAYS N07; POWER RESISTOR AND LAMBDA SENSOR L16; RADIO CASSETTE PLAYER-FROM (V)148782 N14; SHOULDER BAG-REMOVEABLE RADIO N14; RADIO'S-UP TO (V)148781 N13; REAR LAMPS K11; SEAT CUSHION SENSOR N11; SENSORS,TRANSMITTERS AND BRACKET M12; SIDE LAMPS K07; SPEED CONTROL PUMP & DUMP VALVE O11; STARTER MOTOR - 3.6 LITRE J08; STARTER MOTOR 5.3 LITRE J09; STEERING COLUMN SWITCHES M04; TRIP COMPUTER AND PANEL SWITCHES M06; WASH/WIPE RESERVOIR AND FIXINGS L08; WINDSCREEN WASHER RESERVOIR L04; WINDSCREEN WIPER MOTOR-LUCAS TYPE K17; WINDSCREEN WIPERS L03

SPEED CONTROL PUMP & DUMP VALVE

3.6 LITRE

Illus	Part Number	Description	Quantity	Change Point	Remarks
1	CRC 4787 J	Pump-speed control actuator	NLA		Use DBC 3722
1	DBC 3722	Pump-speed control actuator	1		
2	DBC 2182	Valve-dump	1		
3	DBC 3056	Bracket-mounting	1	See (1) below	
4	SH 605051 J	Screw-set	2	See (1) below	
5	WL 108001 J	Washer-lock	2	See (1) below	
6	FW 105 T	Washer-plain	2	See (1) below	
7	NY 105041 J	Nut-nyloc	2		
8	SG 105161 J	Screw	4		
9	DBC 3047	Hose-vacuum	1		Pump to tee piece
10	DBC 1167	Tee piece	1		
11	DBC 3048	Hose-vacuum	1		To dump valve
12	DBC 1168 1	Hose-vacuum	1		To actuator(cut to length)

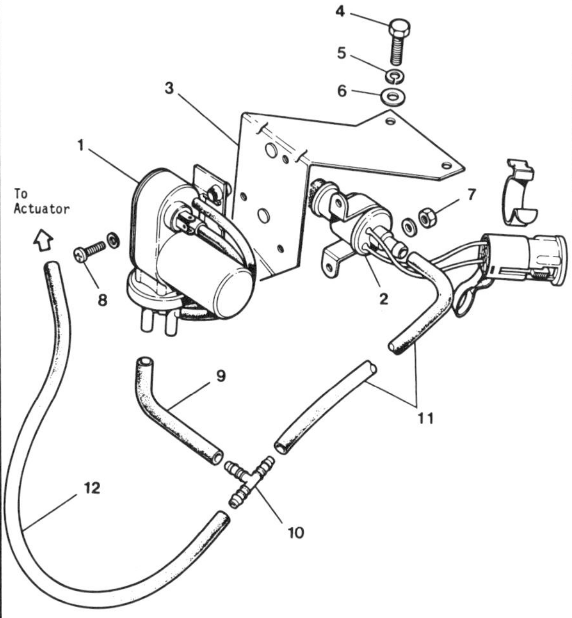

VS4744

(1) Fitted up to (V)148838-For later condtion see Aux fuse box mtg brkt

SPEED CONTROL SWITCHES 3.6 MODELS

Illus	Part Number	Description	Quantity	Change Point	Remarks
1	DAC 3719	Switch-clutch	1		Use DAC 5295
1	DAC 5295	Switch-clutch	1		

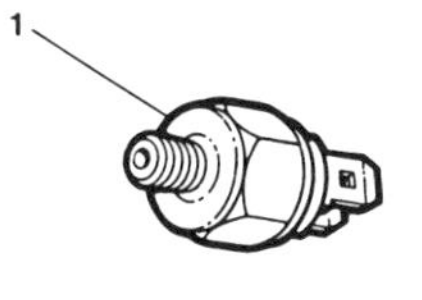

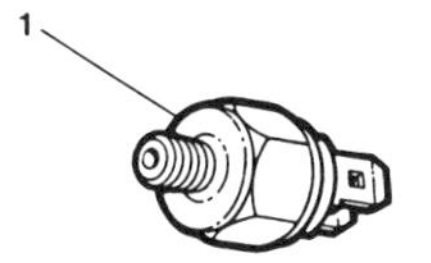

VS4543

VS4543

MASTER INDEX

ENGINE 1.C
FLYWHEEL AND CLUTCH 1.G
GEARBOX AND PROPSHAFT 1.H
AXLES, SUSPENSION, DRIVE SHAFTS, WHEELS 1.K
STEERING 1.M
BRAKES AND BRAKE CONTROLS 1.N
FUEL, EXHAUST AND EMISSION SYSTEMS 2.C
COOLING, HEATING AND AIR CONDITIONING 2.G
ELECTRICAL, WASHERS-WIPERS, INSTRUMENTS 2.J
CHASSIS, SUBFRAMES, BODYSHELL, FITTINGS 3.C
FASCIA, TRIM, SEATS AND FIXINGS 3.H
ACCESSORIES AND PAINT 4.C
NUMERICAL INDEX 4.D

GROUP INDEX

A.B.S EQUIPMENT N09
AERIAL AND FIXINGS. N12
ALTERNATOR 75 AMP (UP TO (E)109969). J02
BATTERY AND LEADS. N17
CONSOLE SWITCHES AND CIGAR LIGHTER M08
CONTROL UNIT-PJ-V12 COUPE L11
FROM 89 MY L13
CONTROL UNITS-PI/CRUISE CONTROL 3.6LITRE L10
DISTRIBUTOR-3.6 LITRE. J11
DISTRIBUTOR-5.3 LITRE. J12
FROM 89 MY J13
DOOR WARNING LAMP. K15
FOG REAR GUARD LAMP. K14
FRONT FLASHER LAMPS. K06
FRONT FOG LAMPS. K13
FUEL INTERFACE UNIT AND RELAYS L15
HARNESSES-FRONT 3.6 LITRE-FROM (V)148945 O04
HARNESSES-FRONT 5.3 LITRE-FROM (V)148782 O06
HARNESSES-FRONT-3.6 LITRE-UPTO (V)148944 O03
HEADLAMP WASH WIPE RESERVOIR L07
HEADLAMP WASHWIPE EQUIPMENT L06
HEADLAMPS-NOT USA K05
HEADLAMPS-USA. K03
HIGH MOUNTED STOP LAMP-USA/CDN ONLY. K12
HOOD CONTROL MODULE - CONVERTIBLE N10
HORNS AND HORN PAD M05
IGNITION COIL & MODULE-3.6 LITRE J14
IGNITION COILS AND AMPLIFIER-5.3 LITRE J15
INERTIA SWITCH AND COVER M13
INSTRUMENT PACK-3.6 LITRE L17
INSTRUMENT PACK-5.3 LITRE L18
INTERIOR LAMPS-CABRIOLET. K09
INTERIOR LAMPS-COUPE. K08
LOUDSPEAKERS AND MOUNTINGS. N15
MAIN FUSE BLOCK AND MOUNTING PLATE M15
NUMBER PLATE AND REVERSE LAMP. K16
PASSIVE RESTRAINT MODULE & FIXINGS N08
PLUGS AND RELAYS. N07
POWER RESISTOR AND LAMBDA SENSOR. L16
RADIO CASSETTE PLAYER-FROM (V)148782 N14
SHOULDER BAG-REMOVEABLE RADIO N14
RADIO'S-UP TO (V)148781 N13
REAR LAMPS. K11
SEAT CUSHION SENSOR. N11
SENSORS,TRANSMITTERS AND BRACKET M12
SIDE LAMPS. K07
SPEED CONTROL PUMP & DUMP VALVE O11
STARTER MOTOR - 3.6 LITRE J08
STARTER MOTOR 5.3 LITRE J09
STEERING COLUMN SWITCHES. M04
TRIP COMPUTER AND PANEL SWITCHES. M06
WASH/WIPE RESERVOIR AND FIXINGS L08
WINDSCREEN WASHER RESERVOIR L04
WINDSCREEN WIPER MOTOR-LUCAS TYPE K17
WINDSCREEN WIPERS L03

20 13

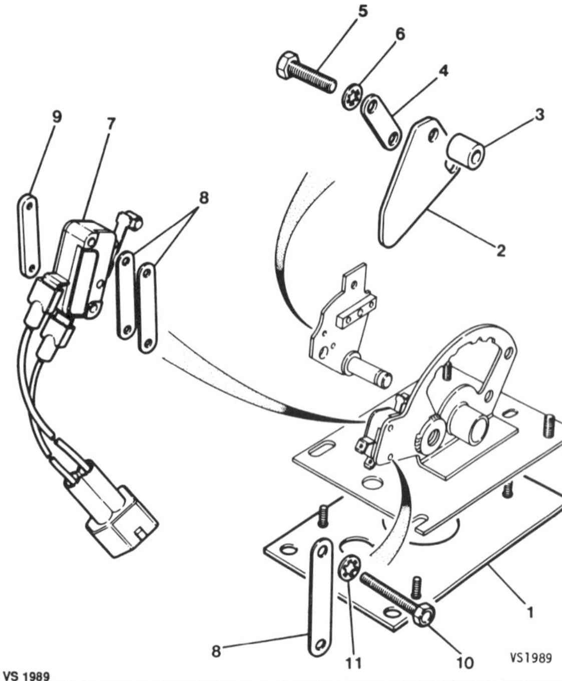

VS 1989

SPEED CONTROL CAM PLATE AND INHIBITOR-V-12 MODELS.

Illus	Part Number	Description	Quantity	Change Point	Remarks
1	CAC 5058	Assembly-base plate	1		
2	CAC 2602	Plate-cam	1		
3	C 45125	Spacer	2		
4	CAC 2040	Clamp plate	1		
5	UCS 116 5R	Setscrew	2		
6	WF 703081 J	Washer-locking	2		
7	DAC 1345	Switch	1		
8	C 35840	Clamp plate	3		
9	C 33784	Clamp plate	1		
10	UCS 111 10R	Setscrew	2		
11	WF 706041 J	Washer-locking	2		

MASTER INDEX

GROUP INDEX

20 14

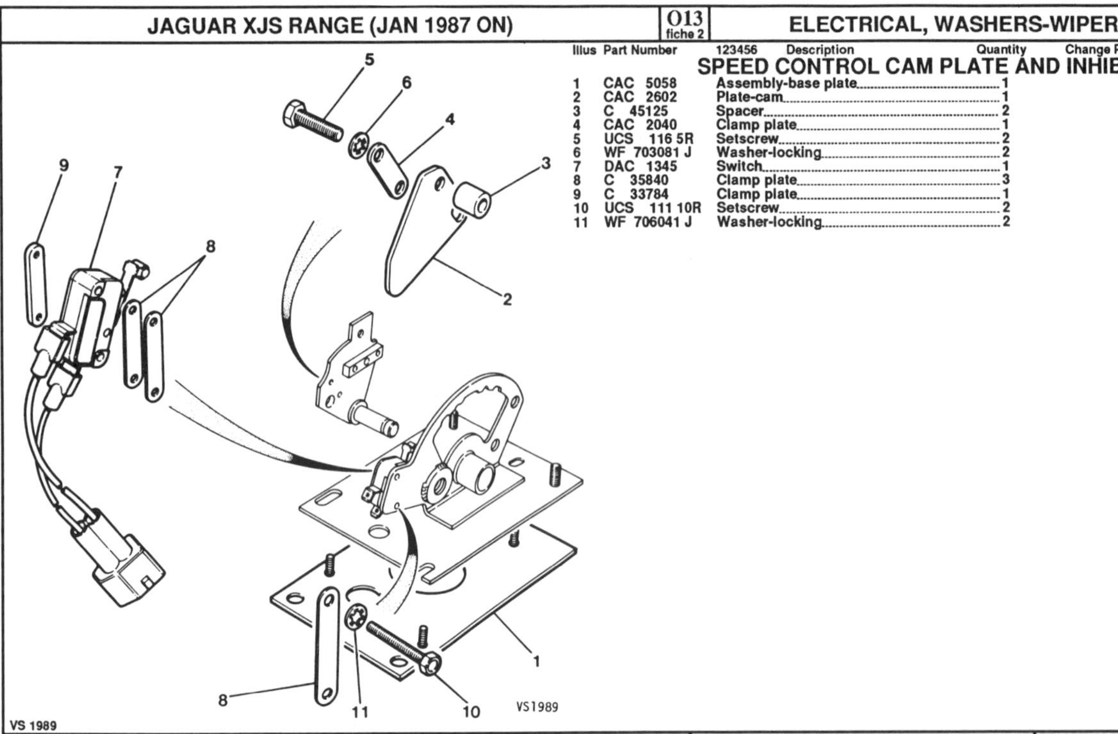

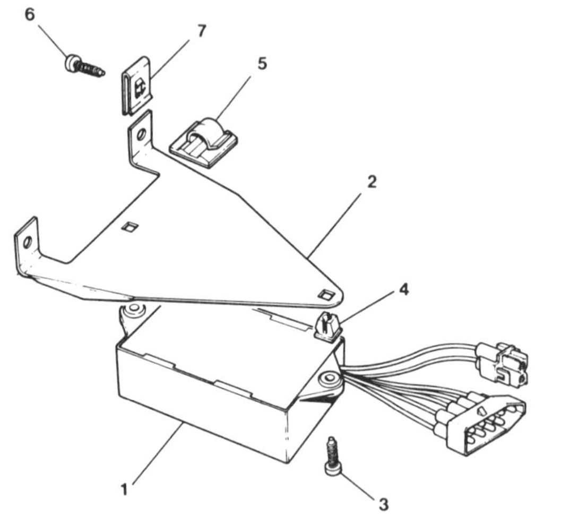

VS 1955

SPEED CONTROL-CONTROL UNIT-V-12 MODELS.

Illus	Part Number	Description	Quantity	Change Point	Remarks
1	DAC 4293	Control unit	1		
2	DAC 1594	Mounting bracket	1		
3	DAZ 408 8C	Screw-self tap	2		
4	C 31973 6	Nut-lokut	2		
5	JLM 9580	Clip-self adhesive	1		
6	DJZ 810 10C	Screw-self tap	2		
7	BD 8633 2	Spire-fix	2		

MASTER INDEX
ENGINE 1.C
FLYWHEEL AND CLUTCH 1.G
GEARBOX AND PROPSHAFT 1.H
AXLES, SUSPENSION, DRIVE SHAFTS, WHEELS 1.K
STEERING 1.M
BRAKES AND BRAKE CONTROLS 1.N
FUEL, EXHAUST AND EMISSION SYSTEMS 2.C
COOLING, HEATING AND AIR CONDITIONING 2.G
ELECTRICAL, WASHERS-WIPERS, INSTRUMENTS 2.J
CHASSIS, SUBFRAMES, BODYSHELL, FITTINGS 3.C
FASCIA, TRIM, SEATS AND FIXINGS 3.H
ACCESSORIES AND PAINT 4.C
NUMERICAL INDEX 4.D
GROUP INDEX
A.B.S EQUIPMENT N09
AERIAL AND FIXINGS N12
ALTERNATOR 75 AMP (UP TO (E)109969) J02
BATTERY AND LEADS N17
CONSOLE SWITCHES AND CIGAR LIGHTER M08
CONTROL UNIT-P.J-V12 COUPE L11
FROM 89 MY L13
CONTROL UNITS-P/CRUISE CONTROL 3.6LITRE L10
DISTRIBUTOR-3.6 LITRE J11
DISTRIBUTOR-5.3 LITRE J12
FROM 89 MY J13
DOOR WARNING LAMP K15
FOG REAR GUARD LAMP K14
FRONT FLASHER LAMPS K06
FRONT FOG LAMPS K13
FUEL INTERFACE UNIT AND RELAYS L15
HARNESSES-FRONT 3.6 LITRE-FROM (V)148945 O04
HARNESSES-FRONT 5.3 LITRE-FROM (V)148782 O06
HARNESSES-FRONT-3.6 LITRE-UPTO (V)148944 O03
HEADLAMP WASH WIPE RESERVOIR L07
HEADLAMP WASHWIPE EQUIPMENT L06
HEADLAMPS-NOT USA K05
HEADLAMPS-USA K03
HIGH MOUNTED STOP LAMP-USA/CDN ONLY K12
HOOD CONTROL MODULE - CONVERTIBLE N10
HORNS AND HORN PAD M05
IGNITION COIL & MODULE-3.6 LITRE J14
IGNITION COILS AND AMPLIFIER-5.3 LITRE J15
INERTIA SWITCH AND COVER M13
INSTRUMENT PACK-3.6 LITRE L17
INSTRUMENT PACK-5.3 LITRE L18
INTERIOR LAMPS-CABRIOLET K09
INTERIOR LAMPS-COUPE K08
LOUDSPEAKERS AND MOUNTINGS N15
MAIN FUSE BLOCK AND MOUNTING PLATE M15
NUMBER PLATE AND REVERSE LAMP K16
PASSIVE RESTRAINT MODULE & FIXINGS N08
PLUGS AND RELAYS N07
POWER RESISTOR AND LAMBDA SENSOR L16
RADIO CASSETTE PLAYER-FROM (V)148782 N14
SHOULDER BAG-REMOVEABLE RADIO N14
RADIO'S-UP TO (V)148781 N13
REAR LAMPS K11
SEAT CUSHION SENSOR N11
SENSORS,TRANSMITTERS AND BRACKET M12
SIDE LAMPS K07
SPEED CONTROL PUMP & DUMP VALVE O11
STARTER MOTOR - 3.6 LITRE J08
STARTER MOTOR 5.3 LITRE J09
STEERING COLUMN SWITCHES M04
TRIP COMPUTER AND PANEL SWITCHES M06
WASH/WIPE RESERVOIR AND FIXINGS L08
WINDSCREEN WASHER RESERVOIR L04
WINDSCREEN WIPER MOTOR-LUCAS TYPE K17
WINDSCREEN WIPERS L03

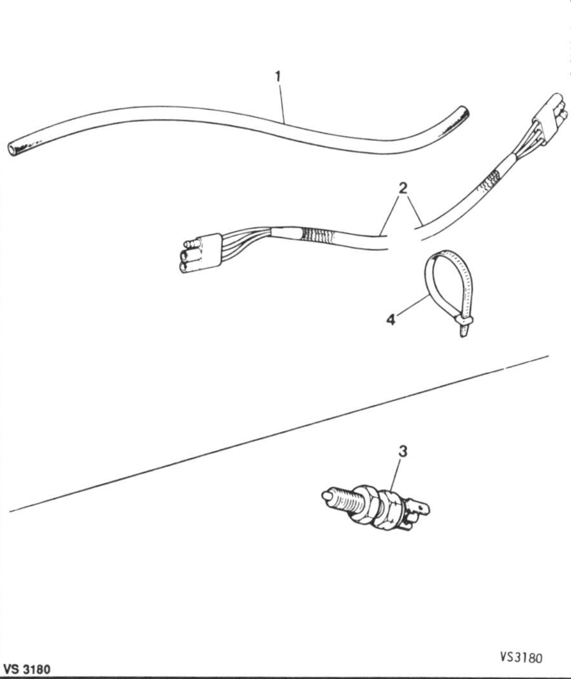

VS 3180

SPEED CONTROL HARNESS AND BRAKE-SWITCH-V12 MODELS

Illus	Part Number	123456	Description	Quantity	Change Point	Remarks
1	CAC 3231 4		Hose-actuator	1		
2	DAC 1995		Harness-speed control	1		
3	C 42525		Switch-brake pedal	1		
4	C 25969		Cable tie	3		

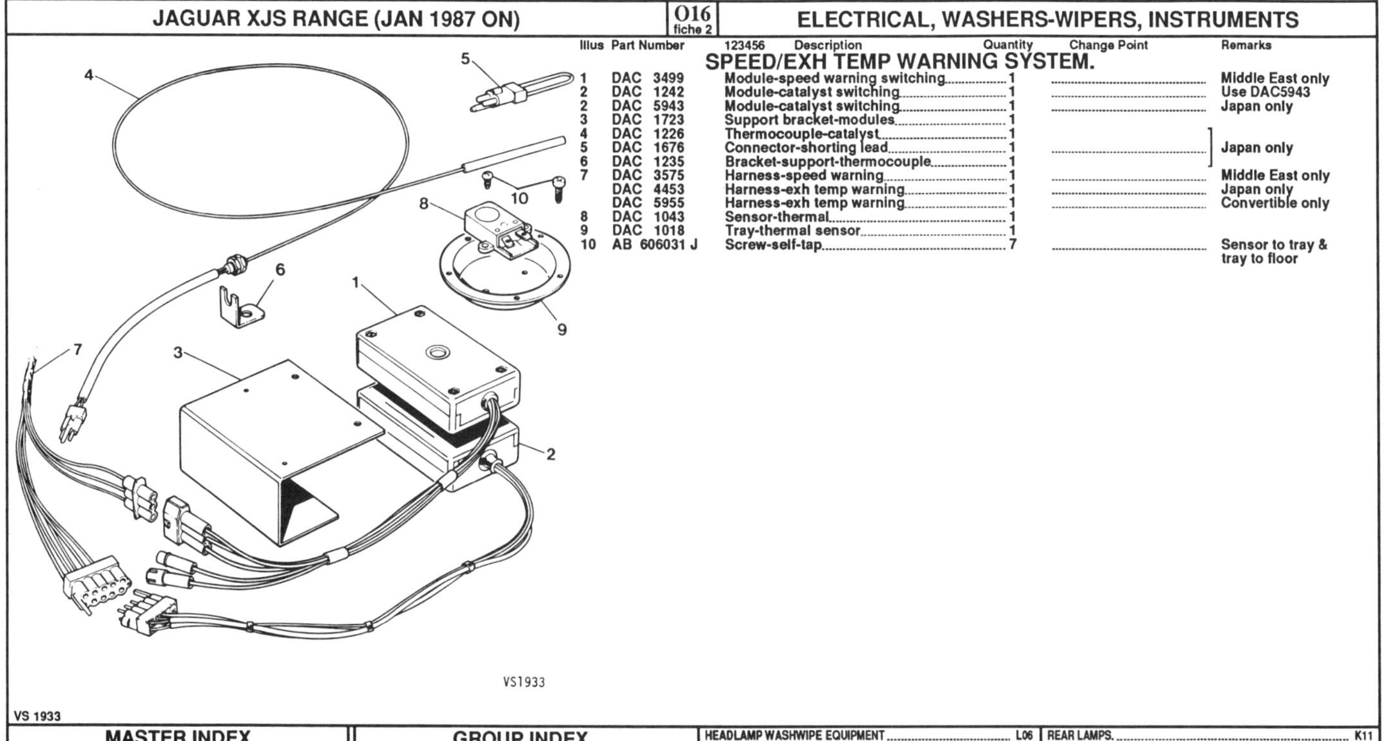

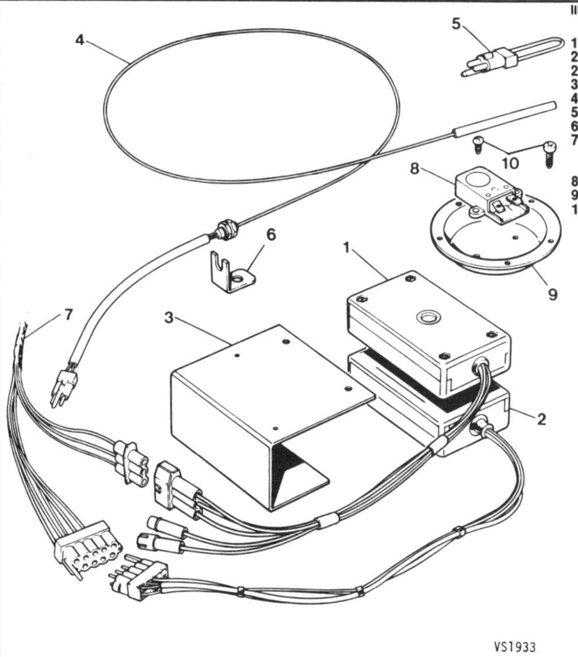

VS 1933

SPEED/EXH TEMP WARNING SYSTEM.

Illus	Part Number	123456	Description	Quantity	Change Point	Remarks
1	DAC 3499		Module-speed warning switching	1		Middle East only
2	DAC 1242		Module-catalyst switching	1		Use DAC5943
2	DAC 5943		Module-catalyst switching	1		Japan only
3	DAC 1723		Support bracket-modules	1		
4	DAC 1226		Thermocouple-catalyst	1		Japan only
5	DAC 1676		Connector-shorting lead	1		Japan only
6	DAC 1235		Bracket-support-thermocouple	1		Japan only
7	DAC 3575		Harness-speed warning	1		Middle East only
	DAC 4453		Harness-exh temp warning	1		Japan only
	DAC 5955		Harness-exh temp warning	1		Convertible only
8	DAC 1043		Sensor-thermal	1		
9	DAC 1018		Tray-thermal sensor	1		
10	AB 606031 J		Screw-self-tap	7		Sensor to tray & tray to floor

MASTER INDEX

ENGINE ... 1.C
FLYWHEEL AND CLUTCH ... 1.G
GEARBOX AND PROPSHAFT ... 1.H
AXLES, SUSPENSION, DRIVE SHAFTS, WHEELS ... 1.K
STEERING ... 1.M
BRAKES AND BRAKE CONTROLS ... 1.N
FUEL, EXHAUST AND EMISSION SYSTEMS ... 2.C
COOLING, HEATING AND AIR CONDITIONING ... 2.G
ELECTRICAL, WASHERS-WIPERS, INSTRUMENTS ... 2.J
CHASSIS, SUBFRAMES, BODYSHELL, FITTINGS ... 3.C
FASCIA, TRIM, SEATS AND FIXINGS ... 3.H
ACCESSORIES AND PAINT ... 4.C
NUMERICAL INDEX ... 4.D

GROUP INDEX

A.B.S EQUIPMENT ... N09
AERIAL AND FIXINGS ... N12
ALTERNATOR 75 AMP (UP TO (E)109969) ... J02
BATTERY AND LEADS ... N17
CONSOLE SWITCHES AND CIGAR LIGHTER ... M08
CONTROL UNIT-PJ-V12 COUPE ... L11
FROM 89 MY ... L13
CONTROL UNITS-PI/CRUISE CONTROL 3.6LITRE ... L10
DISTRIBUTOR-3.6 LITRE ... J11
DISTRIBUTOR-5.3 LITRE ... J12
FROM 89 MY ... J13
DOOR WARNING LAMP ... K15
FOG REAR GUARD LAMP ... K14
FRONT FLASHER LAMPS ... K06
FRONT FOG LAMPS ... K13
FUEL INTERFACE UNIT AND RELAYS ... L15
HARNESSES-FRONT 3.6 LITRE-FROM (V)148945 ... O04
HARNESSES-FRONT 5.3 LITRE-FROM (V)148782 ... O06
HARNESSES-FRONT-3.6 LITRE-UPTO (V)148944 ... O03
HEADLAMP WASH WIPE RESERVOIR ... L07
HEADLAMP WASHWIPE EQUIPMENT ... L06
HEADLAMPS-NOT USA ... K05
HEADLAMPS-USA ... K03
HIGH MOUNTED STOP LAMP-USA/CDN ONLY ... K12
HOOD CONTROL MODULE - CONVERTIBLE ... N10
HORNS AND HORN PAD ... M05
IGNITION COIL & MODULE-3.6 LITRE ... J14
IGNITION COILS AND AMPLIFIER-5.3 LITRE ... J15
INERTIA SWITCH AND COVER ... M13
INSTRUMENT PACK-3.6 LITRE ... L17
INSTRUMENT PACK-5.3 LITRE ... L18
INTERIOR LAMPS-CABRIOLET ... K09
INTERIOR LAMPS-COUPE ... K08
LOUDSPEAKERS AND MOUNTINGS ... N15
MAIN FUSE BLOCK AND MOUNTING PLATE ... M15
NUMBER PLATE AND REVERSE LAMP ... K16
PASSIVE RESTRAINT MODULE & FIXINGS ... N08
PLUGS AND RELAYS ... N07
POWER RESISTOR AND LAMBDA SENSOR ... L16
RADIO CASSETTE PLAYER-FROM (V)148782 ... N14
SHOULDER BAG-REMOVEABLE RADIO ... N14
RADIO'S-UP TO (V)148781 ... N13
REAR LAMPS ... K11
SEAT CUSHION SENSOR ... N11
SENSORS,TRANSMITTERS AND BRACKET ... M12
SIDE LAMPS ... K07
SPEED CONTROL PUMP & DUMP VALVE ... O11
STARTER MOTOR - 3.6 LITRE ... J08
STARTER MOTOR 5.3 LITRE ... J09
STEERING COLUMN SWITCHES ... M04
TRIP COMPUTER AND PANEL SWITCHES ... M06
WASH/WIPE RESERVOIR AND FIXINGS ... L08
WINDSCREEN WASHER RESERVOIR ... L04
WINDSCREEN WIPER MOTOR-LUCAS TYPE ... K17
WINDSCREEN WIPERS ... L03

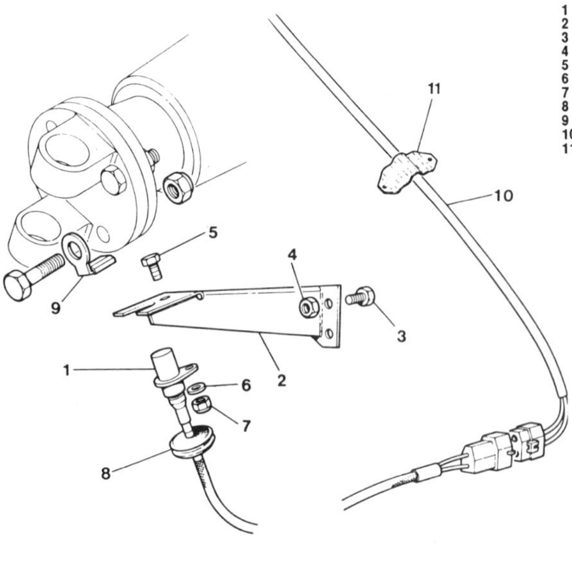

VS 1354

SPEED WARNING SYSTEM-TRANSDUCER-Middle East only

Illus	Part Number	123456	Description	Quantity	Change Point	Remarks
1	DAC 1391		Transducer assembly	1		
2	CAC 2591		Bracket-transducer mounting	1		
3	SH 604051 J		Screw-set	3		
4	C 87371		Nut	3		
5	UFS 119 4R		Screw-set	1		
6	WC 702101 J		Washer-plain	1		
7	TN 3205 J		Nut	1		
8	C 32341		Grommet	1		
9	DAC 2035		Tab-magnetic	2		
10	DAC 3650		Cable-transducer	1		
11	DAC 1394		Protector-transducer cable	1		Under rear seat

20 17

3C 01

Illus	Part Number	123456 Description	Quantity	Change Point	Remarks
		BODY SHELL-COUPE.			
1	JLM 1444	Body shell assembly (Less doors,front wings,bonnet and boot lid)	1		

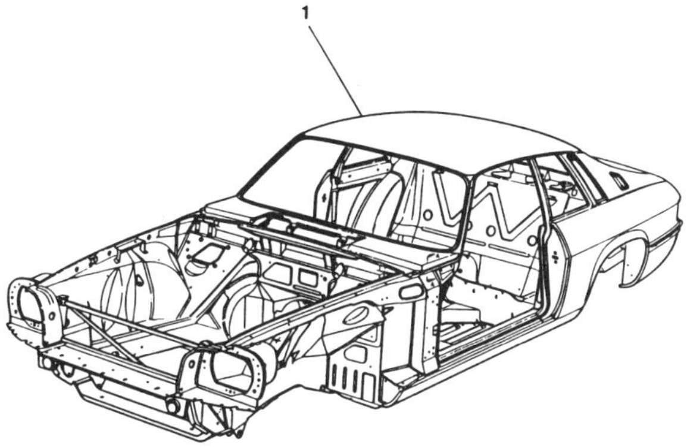

SM7011

SM 7011

MASTER INDEX

GROUP INDEX

BODY SHELL-COUPE-continued.
FRONT END PANELS

Illus	Part Number	123456 Description	Quantity	Change Point	Remarks
1	CBP 314	Crossmember-bumper	1		
2	RTC 1581	Bracket-support RH	1		
	RTC 1582	Bracket-support LH	1		
	RTC 1552	Sill-inner RH	1		
3	RTC 1553	Sill-inner LH	1		
	RTC 1555	Reinforcement RH	1		
4	RTC 1556	Reinforcement LH	1		
5	CBP 313	Crossmember-lower	1		
	RTC 1565	VALANCE ASSEMBLY-RH	1		
6	RTC 1548	Mounting RH	1		
	RTC 1563	Mounting-damper RH	1		
	RTC 1561	Panel-filler RH	NLA		Use CBP224
	CBP 224	Panel-filler RH	1		
	RTC 1557	Panel-closing RH	1		
	RTC 1559	Panel-mounting RH	1		
7	RTC 1566	VALANCE ASSEMBLY-LH	1		
	RTC 1549	Mounting LH	1		
8	RTC 1564	Mounting-damper LH	1		
9	RTC 1562	Panel-filler LH	1		
10	RTC 1558	Panel-closing LH	1		
11	RTC 1560	Panel-mounting LH	1		

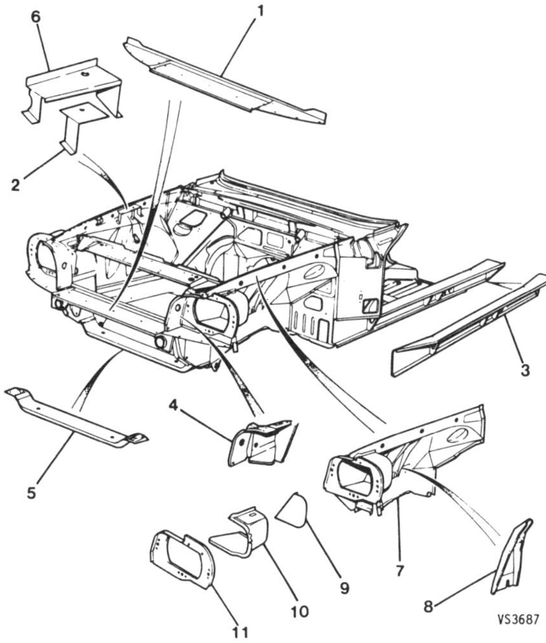

VS3687

VS 3687

MASTER INDEX
ENGINE 1.C
FLYWHEEL AND CLUTCH 1.G
GEARBOX AND PROPSHAFT 1.H
AXLES, SUSPENSION, DRIVE SHAFTS, WHEELS 1.K
STEERING 1.M
BRAKES AND BRAKE CONTROLS 1.N
FUEL, EXHAUST AND EMISSION SYSTEMS 2.C
COOLING, HEATING AND AIR CONDITIONING 2.G
ELECTRICAL, WASHERS-WIPERS, INSTRUMENTS 2.J
CHASSIS, SUBFRAMES, BODYSHELL, FITTINGS 3.C
FASCIA, TRIM, SEATS AND FIXINGS 3.H
ACCESSORIES AND PAINT 4.C
NUMERICAL INDEX 4.D
GROUP INDEX
BADGES - EMBLEMS - CONVERTIBLE G05
BADGES-EMBLEMS-CABRIOLET G04
BADGES-EMBLEMS-COUPE. G03
BODY SHELL - CONVERTIBLE D03
BODY SHELL-CABRIOLET C10
BODY SHELL-COUPE. C02
REAR WINGS (TONNEAU) C08
BONNET LOCK MECHANISM D15
BONNET SAFETY CATCH AND HINGES D14
BONNET-WINGS-DOORS-BOOT LID. D12
BOOT HINGES-SEAL-LAMP HOUSING. E14
BRIGHT FINISHERS-NOT CONVERTIBLE F14
COACHLINES(TAPE) V12 MODELS. F17
COACHLINES(TAPE)-3.6 MODELS. F16
CRUCIFORM AND HEATSHIELD-CABRIOLET F02
DOOR GLASS-SEALS - NOT CONVERTIBLE F06
DOOR HANDLES-LOCKS-SOLENOIDS E07
DOOR HINGES D18
DOOR SEALS AND TREAD PLATES. E02
FRONT BUMPER. G07
FRONT DOOR GLASS-SEALS-CONVERTIBLE F07
FRONT WINDOW REGULATORS - CONVERTIBLE E05
FUEL FILLER LID AND LOCK-COUPE C09
GEARBOX COVER PLATES. E13
GEARBOX TUNNEL COVER-AUTO G/BOX MODELS E11
GEARBOX TUNNEL COVER. E10
HEADLAMP SURROUNDS. F12
HEATED BACKLIGHT-SEAL-COUPE. F11
MIRRORS. G02
REAR BUMPER. G09
REAR QTR GLASS AND SEALS-CONVERTIBLE F10
REAR QTR WINDOW REGULATORS - CONVERTIBLE E06
REAR QUARTER GLASS-SEALS-CABRIOLET F09
BADGE RH-B POST F09
REAR QUARTER GLASS-SEALS-COUPE. F08
SPOILER-UNDERTRAY. G06
STONE GUARDS. D17
TRUNK REINFORCEMENT - CONVERTIBLE F03
WEATHERSTRIPS AND TREADPLATES E03
WINDOW REGULATORS-NOT CONVERTIBLE E04
WINDSCREEN - SEALS - CONVERTIBLE F05
WINDSCREEN-SEALS - NOT CONVERTIBLE F04

3C 04

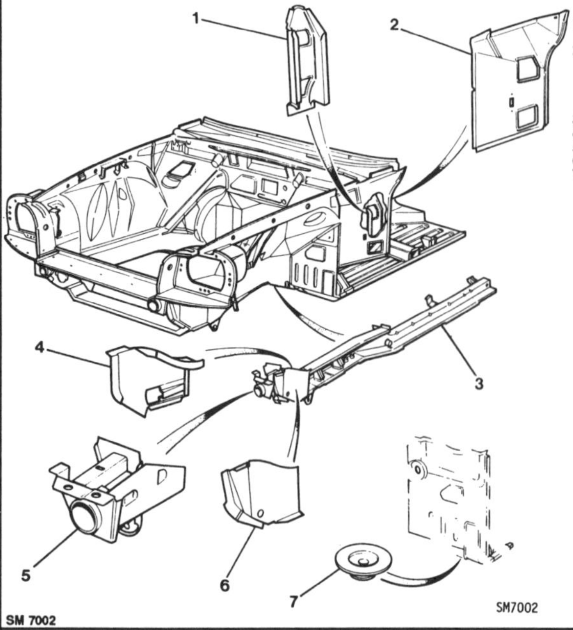

SM 7002

BODY SHELL-COUPE-continued.
PANELS AND SIDE MEMBERS

Illus	Part Number	123456	Description	Quantity	Change Point	Remarks
	RTC 1567		A post-inner RH	1		
1	RTC 1568		A post-inner LH	1		
	RTC 1569		Panel-dash side RH	1		
2	RTC 1570		Panel-dash side LH	1		
	RTC 1577		Sidemember-front RH	1		
	RTC 1573		Extension-inner RH	1		
	RTC 1575		Mounting-bumper RH	1		
	RTC 1579		Extension-outer RH	1		
3	RTC 1578		Sidemember-front LH	1		
4	RTC 1574		Extension-inner LH	1		
5	RTC 1576		Mounting-bumper LH	1		
6	RTC 1580		Extension-outer LH	1		
7	BD 28658		Bracket-front jacking	2		

MASTER INDEX

GROUP INDEX

3C 05

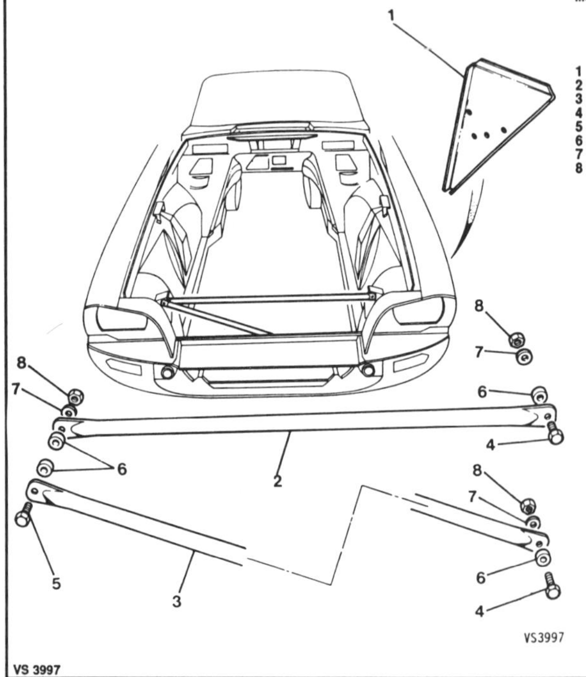

VS 3997

BODY SHELL-COUPE-continued.
BRACING STRUTS

Illus	Part Number	123456	Description	Quantity	Change Point	Remarks
	RTC 1542		Support-bumper RH	1		
1	RTC 1543		Support-bumper LH	1		
2	BD 44991		Brace-horizontal	1		
3	BD 44992		Brace-diagonal	1		
4	JLM 9676		Screw-set	2		
5	SH 606021 J		Screw-set	1		
6	C 31301		Distance piece	4		
7	JLM 301		Washer-plain	3		
8	JLM 9685		Nut-plain	3		

MASTER INDEX

ENGINE 1.C
FLYWHEEL AND CLUTCH 1.G
GEARBOX AND PROPSHAFT 1.H
AXLES, SUSPENSION, DRIVE SHAFTS, WHEELS 1.K
STEERING 1.M
BRAKES AND BRAKE CONTROLS 1.N
FUEL, EXHAUST AND EMISSION SYSTEMS 2.C
COOLING, HEATING AND AIR CONDITIONING 2.G
ELECTRICAL, WASHERS-WIPERS, INSTRUMENTS 2.J
CHASSIS, SUBFRAMES, BODYSHELL, FITTINGS 3.C
FASCIA, TRIM, SEATS AND FIXINGS 3.H
ACCESSORIES AND PAINT 4.C
NUMERICAL INDEX 4.D

GROUP INDEX

BADGES - EMBLEMS - CONVERTIBLE G05
BADGES-EMBLEMS-CABRIOLET G04
BADGES-EMBLEMS-COUPE G03
BODY SHELL - CONVERTIBLE D03
BODY SHELL-CABRIOLET C10
BODY SHELL-COUPE C02
REAR WINGS (TONNEAU) C08
BONNET LOCK MECHANISM D15
BONNET SAFETY CATCH AND HINGES D14
BONNET-WINGS-DOORS-BOOT LID D12
BOOT HINGES-SEAL-LAMP HOUSING E14
BRIGHT FINISHERS-NOT CONVERTIBLE F14
COACHLINES(TAPE) V12 MODELS F17
COACHLINES(TAPE)-3.6 MODELS F16
CRUCIFORM AND HEATSHIELD-CABRIOLET F02
DOOR GLASS-SEALS - NOT CONVERTIBLE F06
DOOR HANDLES-LOCKS-SOLENOIDS E07
DOOR HINGES D18
DOOR SEALS AND TREAD PLATES E02
FRONT BUMPER G07
FRONT DOOR GLASS-SEALS-CONVERTIBLE F07
FRONT WINDOW REGULATORS - CONVERTIBLE E05
FUEL FILLER LID AND LOCK-COUPE C09
GEARBOX COVER PLATES E13
GEARBOX TUNNEL COVER-AUTO G/BOX MODELS E11
GEARBOX TUNNEL COVER E10
HEADLAMP SURROUNDS F12
HEATED BACKLIGHT-SEAL-COUPE F11
MIRRORS G02
REAR BUMPER G09
REAR QTR GLASS AND SEALS-CONVERTIBLE F10
REAR QTR WINDOW REGULATORS - CONVERTIBLE E06
REAR QUARTER GLASS-SEALS-CABRIOLET F09
BADGE RH-B POST F09
REAR QUARTER GLASS-SEALS-COUPE F08
SPOILER-UNDERTRAY G06
STONE GUARDS D17
TRUNK REINFORCEMENT - CONVERTIBLE F03
WEATHERSTRIPS AND TREADPLATES E03
WINDOW REGULATORS-NOT CONVERTIBLE E04
WINDSCREEN - SEALS - CONVERTIBLE F05
WINDSCREEN-SEALS - NOT CONVERTIBLE F04

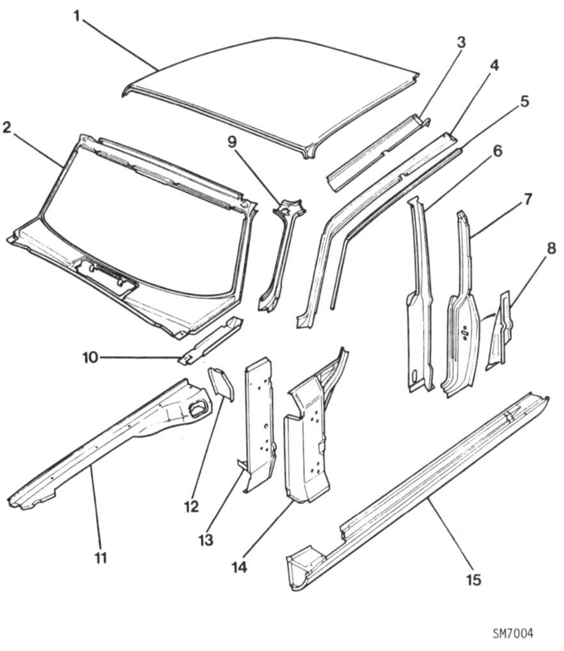

SM 7004

BODY SHELL-COUPE-continued.
ROOF WINDSCREEN PANELS

Illus	Part Number	123456 Description	Quantity	Change Point	Remarks
1	RTC 1601	Panel-roof	1		
2	RTC 1602	Panel-windscreen	1		
	RTC 1605	Panel-cantrail closing RH	1		
3	RTC 1606	Panel-cantrail closing LH	1		
	RTC 1607	Panel-cantrail RH	1		
4	RTC 1608	Panel-cantrail LH	1		
	RTC 1609	Moulding-drip RH	1		
5	RTC 1610	Moulding-drip LH	1		
	RTC 1611	B post-inner RH	1		
6	RTC 1612	B post-inner LH	1		
	RTC 1613	B post RH	1		
7	RTC 1614	B post LH	1		
	RTC 1615	B post-lower RH	1		
8	RTC 1616	B post-lower LH	1		
	RTC 1603	A post-upper RH	1		
9	RTC 1604	A post-upper LH	1		
	RTC 1619	A post-closing RH	1		
10	RTC 1620	A post-closing LH	1		
	RTC 1623	Reinforcement-upper RH	1		
11	RTC 1624	Reinforcement-upper LH	1		
	RTC 1617	Baffle-lower RH	1		
12	RTC 1618	Baffle-lower LH	1		
	RTC 1625	Reinforecement-lower RH	1		
13	RTC 1626	Reinforecement-lower LH	1		
	RTC 1621	A post-lower RH	1		
14	RTC 1622	A post-lower LH	1		
	RTC 1627	Sill-outer RH	1		
15	RTC 1628	Sill-outer LH	1		

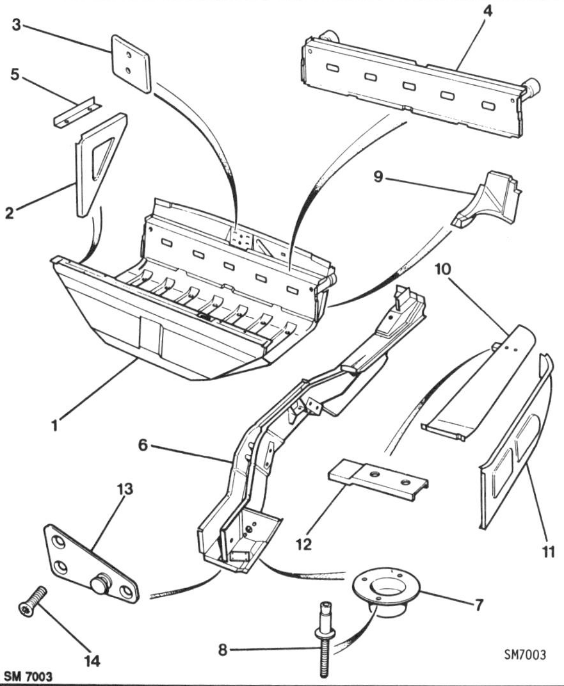

SM 7003

BODY SHELL-COUPE-continued.
REAR END PANELS

Illus	Part Number	123456 Description	Quantity	Change Point	Remarks
1	RTC 1594	Floor-boot	1		
2	RTC 1588	Support-battery tray	1		
3	RTC 1589	Mounting-striker	1		
4	RTC 1590	Crossmember-bumper	1		
	RTC 1599	Sidemember-rear RH	1		
5	RTC 1587	Angle support-battery	1		
7	JS 254	Mounting-radius arm	1		
8	BD 47948	Bolt-blind	1		
6	RTC 1600	Sidemember-rear LH	1		
7	JS 254	Mounting-radius arm	1		
8	BD 47948	Bolt-blind	1		
	RTC 1591	Panel-closing RH	1		
9	RTC 1592	Panel-closing LH	1		
	RTC 1595	Panel-boot side RH	1		
10	RTC 1596	Panel-boot side LH	1		
	RTC 1597	Panel-closing RH	1		
11	RTC 1598	Panel-closing LH	1		
12	RTC 1593	Reinforcement-exhaust	2		
	BD 44956	Plate-jack RH	1		
13	BD 44957	Plate-jack LH	1		
14	SF 606061 J	Screw-set	6		

MASTER INDEX

ENGINE 1.C
FLYWHEEL AND CLUTCH 1.G
GEARBOX AND PROPSHAFT 1.H
AXLES, SUSPENSION, DRIVE SHAFTS, WHEELS 1.K
STEERING 1.M
BRAKES AND BRAKE CONTROLS 1.N
FUEL, EXHAUST AND EMISSION SYSTEMS 2.C
COOLING, HEATING AND AIR CONDITIONING 2.G
ELECTRICAL, WASHERS-WIPERS, INSTRUMENTS 2.J
CHASSIS, SUBFRAMES, BODYSHELL, FITTINGS 3.C
FASCIA, TRIM, SEATS AND FIXINGS 3.H
ACCESSORIES AND PAINT 4.C
NUMERICAL INDEX 4.D

GROUP INDEX

BADGES - EMBLEMS - CONVERTIBLE G05
BADGES-EMBLEMS-CABRIOLET G04
BADGES-EMBLEMS-COUPE G03
BODY SHELL - CONVERTIBLE D03
BODY SHELL-CABRIOLET C10
BODY SHELL-COUPE C02
REAR WINGS (TONNEAU) C08
BONNET LOCK MECHANISM D15
BONNET SAFETY CATCH AND HINGES D14
BONNET-WINGS-DOORS-BOOT LID D12
BOOT HINGES-SEAL-LAMP HOUSING E14
BRIGHT FINISHERS-NOT CONVERTIBLE F14
COACHLINES(TAPE) V12 MODELS F17
COACHLINES(TAPE)-3.6 MODELS F16
CRUCIFORM AND HEATSHIELD-CABRIOLET F02
DOOR GLASS-SEALS - NOT CONVERTIBLE F06
DOOR HANDLES-LOCKS-SOLENOIDS E07
DOOR HINGES D18
DOOR SEALS AND TREAD PLATES E02
FRONT BUMPER G07
FRONT DOOR GLASS-SEALS-CONVERTIBLE F07
FRONT WINDOW REGULATORS - CONVERTIBLE E05
FUEL FILLER LID AND LOCK-COUPE C09
GEARBOX COVER PLATES E13
GEARBOX TUNNEL COVER-AUTO G/BOX MODELS E11
GEARBOX TUNNEL COVER E10
HEADLAMP SURROUNDS F12
HEATED BACKLIGHT-SEAL-COUPE F11
MIRRORS G02
REAR BUMPER G09
REAR QTR GLASS AND SEALS-CONVERTIBLE F10
REAR QTR WINDOW REGULATORS - CONVERTIBLE E06
REAR QUARTER GLASS-SEALS-CABRIOLET F09
BADGE RH-B POST F09
REAR QUARTER GLASS-SEALS-COUPE F08
SPOILER-UNDERTRAY G06
STONE GUARDS D17
TRUNK REINFORCEMENT - CONVERTIBLE F03
WEATHERSTRIPS AND TREADPLATES E03
WINDOW REGULATORS-NOT CONVERTIBLE E04
WINDSCREEN - SEALS - CONVERTIBLE F05
WINDSCREEN-SEALS - NOT CONVERTIBLE F04

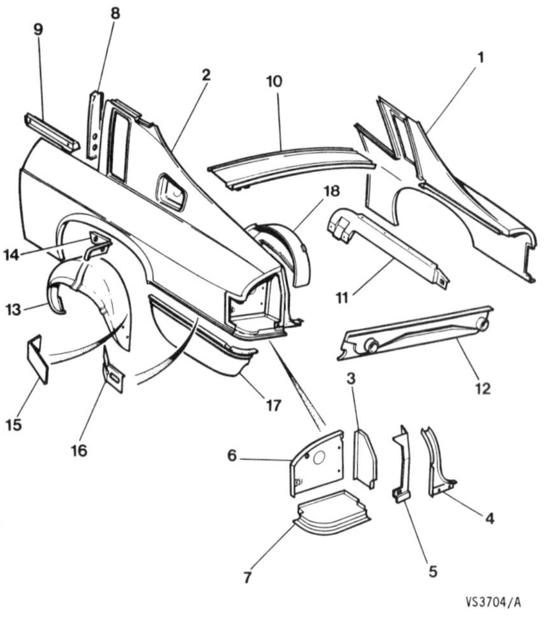

BODY SHELL-COUPE-continued.
REAR WINGS (TONNEAU)

Illus	Part Number	123456 Description	Quantity	Change Point	Remarks
1	RTC 1635	Wing-rear RH (tonneau)	1		
	RTC 1651	Reinforcement RH	1		
	RTC 1655	Gusset RH	1		
	RTC 1653	Panel-lamp RH	1		
	RTC 1647	Panel-closing RH	1		
	RTC 1649	Panel-lower RH	1		
	RTC 1631	Extension-cantrail RH	1		
2	RTC 1636	Wing-rear LH (tonneau)	1		
3	RTC 1652	Reinforcement LH	1		
4	RTC 1656	Gusset LH	1		
5	RTC 1654	Panel-lamp LH	1		
6	RTC 1648	Panel-closing LH	1		
7	RTC 1650	Panel-lower LH	1		
8	RTC 1632	Extension-cantrail LH	1		
	RTC 1629	Waistrail RH	1		
9	RTC 1630	Waistrail LH	1		
10	RTC 1634	Decking-rear	1		
11	BCC 3416	Reinforcement RH	1		
	BCC 3417	Reinforcemnt LH	1		
12	RTC 1637	Panel-lower rear	1		
	RTC 1638	Wheelarch-outer RH	1		
13	RTC 1639	Wheelarch-outer LH	1		
	RTC 1640	Baffle-rear RH	1		
14	RTC 1641	Baffle-rear LH	1		
15	RTC 1540	Support-bumper	2		
16	RTC 1644	Bracket-bumper	2		
	RTC 1645	Panel-lower rear RH	1		
17	RTC 1646	Panel-lower rear LH	1		
	JLM 742	Wheelarch-inner RH	1		
18	JLM 743	Wheelarch-inner LH	1		

VS 3704A

3C 08

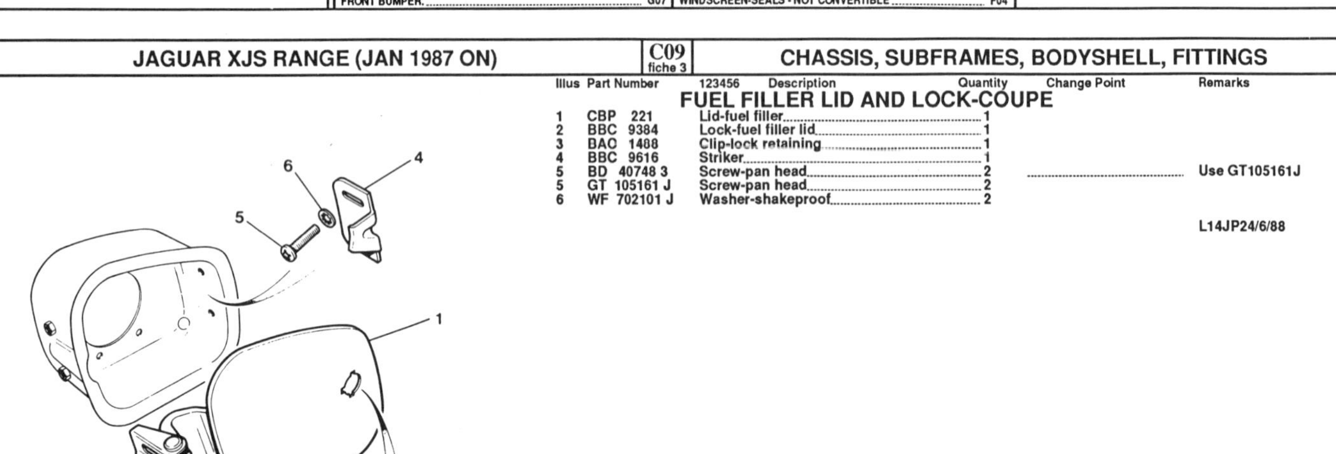

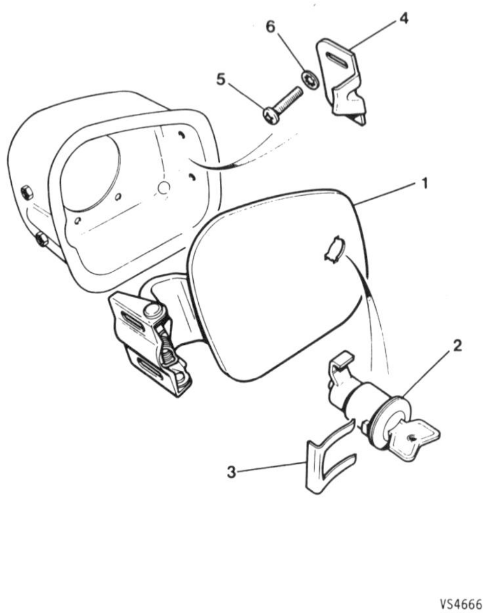

FUEL FILLER LID AND LOCK-COUPE

Illus	Part Number	123456 Description	Quantity	Change Point	Remarks
1	CBP 221	Lid-fuel filler	1		
2	BBC 9384	Lock-fuel filler lid	1		
3	BAC 1488	Clip-lock retaining	1		
4	BBC 9616	Striker	1		
5	BD 40748 3	Screw-pan head	2		Use GT105161J
5	GT 105161 J	Screw-pan head	2		
6	WF 702101 J	Washer-shakeproof	2		

L14JP24/6/88

VS4666

3C 09

MASTER INDEX

ENGINE	1.C
FLYWHEEL AND CLUTCH	1.G
GEARBOX AND PROPSHAFT	1.H
AXLES, SUSPENSION, DRIVE SHAFTS, WHEELS	1.K
STEERING	1.M
BRAKES AND BRAKE CONTROLS	1.N
FUEL, EXHAUST AND EMISSION SYSTEMS	2.C
COOLING, HEATING AND AIR CONDITIONING	2.G
ELECTRICAL, WASHERS-WIPERS, INSTRUMENTS	2.J
CHASSIS, SUBFRAMES, BODYSHELL, FITTINGS	3.C
FASCIA, TRIM, SEATS AND FIXINGS	3.H
ACCESSORIES AND PAINT	4.C
NUMERICAL INDEX	4.D

GROUP INDEX

BADGES - EMBLEMS - CONVERTIBLE	G05
BADGES-EMBLEMS-CABRIOLET	G04
BADGES-EMBLEMS-COUPE	G03
BODY SHELL - CONVERTIBLE	D03
BODY SHELL-CABRIOLET	C10
BODY SHELL-COUPE	C02
REAR WINGS (TONNEAU)	C08
BONNET LOCK MECHANISM	D15
BONNET SAFETY CATCH AND HINGES	D14
BONNET-WINGS-DOORS-BOOT LID	D12
BOOT HINGES-SEAL-LAMP HOUSING	E14
BRIGHT FINISHERS-NOT CONVERTIBLE	F14
COACHLINES(TAPE) V12 MODELS	F17
COACHLINES(TAPE)-3.6 MODELS	F16
CRUCIFORM AND HEATSHIELD-CABRIOLET	F02
DOOR GLASS-SEALS - NOT CONVERTIBLE	F06
DOOR HANDLES-LOCKS-SOLENOIDS	E07
DOOR HINGES	D18
DOOR SEALS AND TREAD PLATES	E02
FRONT BUMPER	G07
FRONT DOOR GLASS-SEALS-CONVERTIBLE	F07
FRONT WINDOW REGULATORS - CONVERTIBLE	E05
FUEL FILLER LID AND LOCK-COUPE	C09
GEARBOX COVER PLATES	E13
GEARBOX TUNNEL COVER-AUTO G/BOX MODELS	E11
GEARBOX TUNNEL COVER	E10
HEADLAMP SURROUNDS	F12
HEATED BACKLIGHT-SEAL-COUPE	F11
MIRRORS	G02
REAR BUMPER	G09
REAR QTR GLASS AND SEALS-CONVERTIBLE	F10
REAR QTR WINDOW REGULATORS - CONVERTIBLE	E06
REAR QUARTER GLASS-SEALS-CABRIOLET	F09
BADGE RH-B POST	F09
REAR QUARTER GLASS-SEALS-COUPE	F08
SPOILER-UNDERTRAY	G06
STONE GUARDS	D17
TRUNK REINFORCEMENT - CONVERTIBLE	F03
WEATHERSTRIPS AND TREADPLATES	E03
WINDOW REGULATORS-NOT CONVERTIBLE	E04
WINDSCREEN - SEALS - CONVERTIBLE	F05
WINDSCREEN-SEALS - NOT CONVERTIBLE	F04

Illus	Part Number	123456	Description	Quantity	Change Point	Remarks
			BODY SHELL-CABRIOLET			
1	JLM 1445		Body shell assembly(less doors, front wings,bonnet and boot lid)	1		

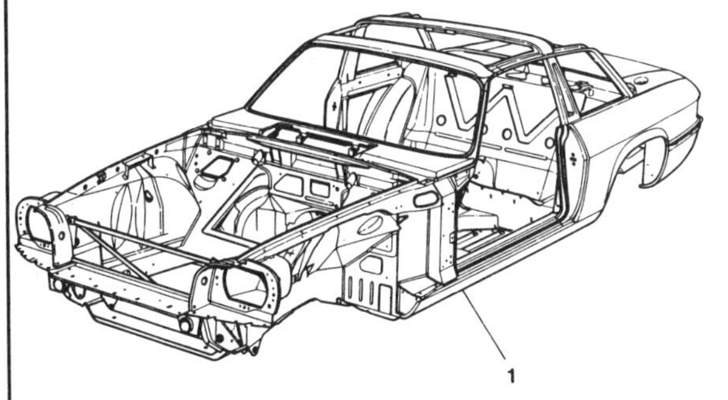

VS 4115

MASTER INDEX

GROUP INDEX

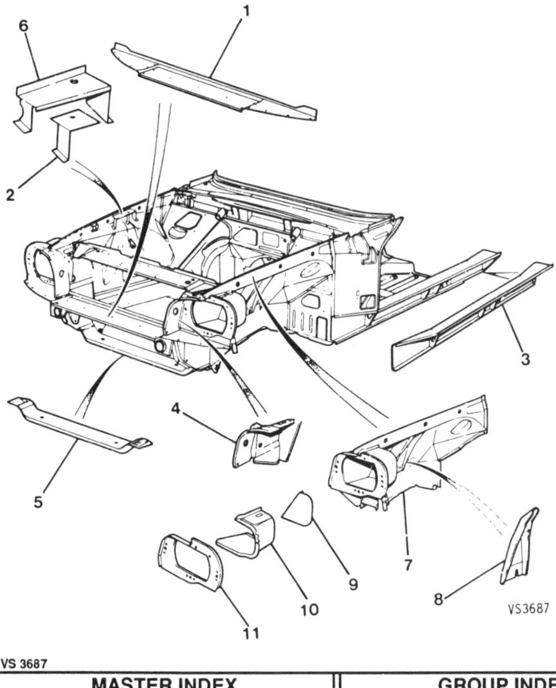

Illus	Part Number	123456	Description	Quantity	Change Point	Remarks
			BODY SHELL-CABRIOLET-Continued			
1	CBP 314		Crossmember-bumper	1		
2	RTC 1581		Bracket-support RH	1		
	RTC 1582		Bracket-support LH	1		
	RTC 1552		Sill-inner RH	1		
3	RTC 1553		Sill-inner LH	1		
	RTC 1555		Reinforcement RH	1		
4	RTC 1556		Reinforcement LH	1		
5	CBP 313		Crossmember-lower	1		
	RTC 1565		VALANCE ASSEMBLY RH	1		
6	RTC 1548		Mounting RH	1		
	RTC 1563		Mounting-damper RH	1		
	RTC 1561		Panel-filler RH	1		Use CBP224
	CBP 224		Panel-filler RH	1		
	RTC 1557		Panel-closing RH	1		
	RTC 1559		Panel-mounting RH	1		
7	RTC 1566		VALANCE ASSEMBLY LH	1		
	RTC 1549		Mounting LH	1		
8	RTC 1564		Mounting-damper LH	1		
9	RTC 1562		Panel-filler LH	1		
10	RTC 1558		Panel-closing LH	1		
11	RTC 1560		Panel-mounting LH	1		

VS 3687

MASTER INDEX

ENGINE	1.C
FLYWHEEL AND CLUTCH	1.G
GEARBOX AND PROPSHAFT	1.H
AXLES, SUSPENSION, DRIVE SHAFTS, WHEELS	1.K
STEERING	1.M
BRAKES AND BRAKE CONTROLS	1.N
FUEL, EXHAUST AND EMISSION SYSTEMS	2.C
COOLING, HEATING AND AIR CONDITIONING	2.G
ELECTRICAL, WASHERS-WIPERS, INSTRUMENTS	2.J
CHASSIS, SUBFRAMES, BODYSHELL, FITTINGS	3.C
FASCIA, TRIM, SEATS AND FIXINGS	3.H
ACCESSORIES AND PAINT	4.C
NUMERICAL INDEX	4.D

GROUP INDEX

BADGES - EMBLEMS - CONVERTIBLE	G05
BADGES-EMBLEMS-CABRIOLET	G04
BADGES-EMBLEMS-COUPE	G03
BODY SHELL - CONVERTIBLE	D03
BODY SHELL-CABRIOLET	C10
BODY SHELL-COUPE	C02
REAR WINGS (TONNEAU)	C08
BONNET LOCK MECHANISM	D15
BONNET SAFETY CATCH AND HINGES	D14
BONNET-WINGS-DOORS-BOOT LID	D12
BOOT HINGES-SEAL-LAMP HOUSING	E14
BRIGHT FINISHERS-NOT CONVERTIBLE	F14
COACHLINES(TAPE) V12 MODELS	F17
COACHLINES(TAPE)-3.6 MODELS	F16
CRUCIFORM AND HEATSHIELD-CABRIOLET	F02
DOOR GLASS-SEALS - NOT CONVERTIBLE	F06
DOOR HANDLES-LOCKS-SOLENOIDS	E07
DOOR HINGES	D18
DOOR SEALS AND TREAD PLATES	E02
FRONT BUMPER	G07
FRONT DOOR GLASS-SEALS-CONVERTIBLE	F07
FRONT WINDOW REGULATORS - CONVERTIBLE	E05
FUEL FILLER LID AND LOCK-COUPE	C09
GEARBOX COVER PLATES	E13
GEARBOX TUNNEL COVER-AUTO G/BOX MODELS	E11
GEARBOX TUNNEL COVER	E10
HEADLAMP SURROUNDS	F12
HEATED BACKLIGHT-SEAL-COUPE	F11
MIRRORS	G02
REAR BUMPER	G09
REAR QTR GLASS AND SEALS-CONVERTIBLE	F10
REAR QTR WINDOW REGULATORS - CONVERTIBLE	E06
REAR QUARTER GLASS-SEALS-CABRIOLET	F09
BADGE RH-B POST	F09
REAR QUARTER GLASS-SEALS-COUPE	F08
SPOILER-UNDERTRAY	G06
STONE GUARDS	D17
TRUNK REINFORCEMENT - CONVERTIBLE	F03
WEATHERSTRIPS AND TREADPLATES	E03
WINDOW REGULATORS-NOT CONVERTIBLE	E04
WINDSCREEN - SEALS - CONVERTIBLE	F05
WINDSCREEN-SEALS - NOT CONVERTIBLE	F04

3C 12

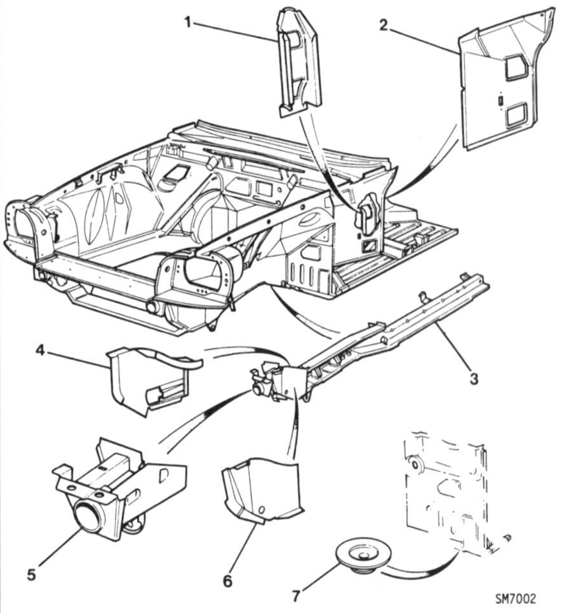

SM 7002

BODY SHELL-CABRIOLET-Continued

Illus	Part Number	123456	Description	Quantity	Change Point	Remarks
	RTC 1567		A post-inner RH	1		
1	RTC 1568		A post-inner LH	1		
	RTC 1569		Panel-dash side RH	1		
2	RTC 1570		Panel-dash side LH	1		
	RTC 1577		Sidemember-front RH	1		
	RTC 1573		Extension-inner RH	1		
	RTC 1575		Mounting-bumper RH	1		
	RTC 1579		Extension-outer RH	1		
3	RTC 1578		Sidemember-front LH	1		
4	RTC 1574		Extension-inner LH	1		
5	RTC 1576		Mounting-bumper LH	1		
6	RTC 1580		Extension-outer LH	1		
7	BD 28658		Bracket-front jacking	2		

MASTER INDEX

GROUP INDEX

3C 13

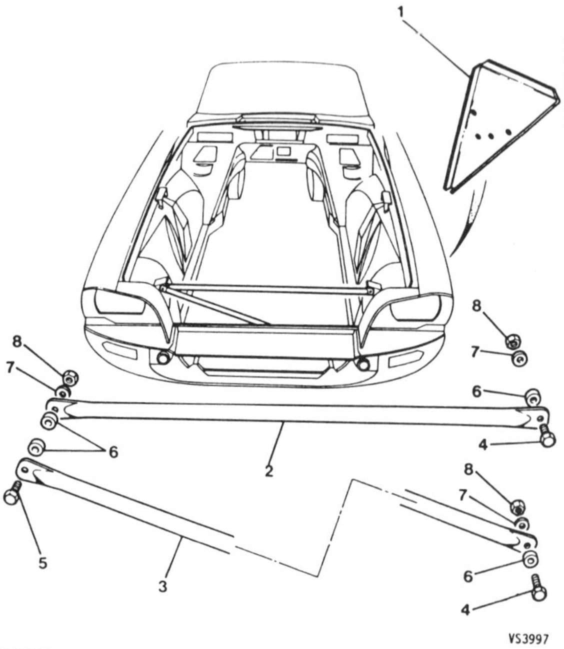

VS 3997

BODY SHELL-CABRIOLET-Continued

Illus	Part Number	123456	Description	Quantity	Change Point	Remarks
	RTC 1542		Support-bumper RH	1		
1	RTC 1543		Support-bumper LH	1		
2	BD 44991		Brace-horizontal	1		
3	BD 44992		Brace-diagonal	1		
4	JLM 9676		Screw-set	2		
5	SH 606021 J		Screw-set	1		
6	C 31302		Distance piece	4		
7	JLM 301		Washer-plain	3		
8	JLM 9685		Nut-plain	3		

MASTER INDEX
ENGINE ... 1.C
FLYWHEEL AND CLUTCH ... 1.G
GEARBOX AND PROPSHAFT ... 1.H
AXLES, SUSPENSION, DRIVE SHAFTS, WHEELS ... 1.K
STEERING ... 1.M
BRAKES AND BRAKE CONTROLS ... 1.N
FUEL, EXHAUST AND EMISSION SYSTEMS ... 2.C
COOLING, HEATING AND AIR CONDITIONING ... 2.G
ELECTRICAL, WASHERS-WIPERS, INSTRUMENTS ... 2.J
CHASSIS, SUBFRAMES, BODYSHELL, FITTINGS ... 3.C
FASCIA, TRIM, SEATS AND FIXINGS ... 3.H
ACCESSORIES AND PAINT ... 4.C
NUMERICAL INDEX ... 4.D
GROUP INDEX
BADGES - EMBLEMS - CONVERTIBLE ... G05
BADGES-EMBLEMS-CABRIOLET ... G04
BADGES-EMBLEMS-COUPE ... G03
BODY SHELL - CONVERTIBLE ... D03
BODY SHELL-CABRIOLET ... C10
BODY SHELL-COUPE ... C02
REAR WINGS (TONNEAU) ... C08
BONNET LOCK MECHANISM ... D15
BONNET SAFETY CATCH AND HINGES ... D14
BONNET-WINGS-DOORS-BOOT LID ... D12
BOOT HINGES-SEAL-LAMP HOUSING ... E14
BRIGHT FINISHERS-NOT CONVERTIBLE ... F14
COACHLINES(TAPE) V12 MODELS ... F17
COACHLINES(TAPE)-3.6 MODELS ... F16
CRUCIFORM AND HEATSHIELD-CABRIOLET ... F02
DOOR GLASS-SEALS - NOT CONVERTIBLE ... F06
DOOR HANDLES-LOCKS-SOLENOIDS ... E07
DOOR HINGES ... D18
DOOR SEALS AND TREAD PLATES ... E02
FRONT BUMPER ... G07
FRONT DOOR GLASS-SEALS-CONVERTIBLE ... F07
FRONT WINDOW REGULATORS - CONVERTIBLE ... E05
FUEL FILLER LID AND LOCK-COUPE ... C09
GEARBOX COVER PLATES ... E13
GEARBOX TUNNEL COVER-AUTO G/BOX MODELS ... E11
GEARBOX TUNNEL COVER ... E10
HEADLAMP SURROUNDS ... F12
HEATED BACKLIGHT-SEAL-COUPE ... F11
MIRRORS ... G02
REAR BUMPER ... G09
REAR QTR GLASS AND SEALS-CONVERTIBLE ... F10
REAR QTR WINDOW REGULATORS - CONVERTIBLE ... E06
REAR QUARTER GLASS-SEALS-CABRIOLET ... F09
BADGE RH-B POST ... F09
REAR QUARTER GLASS-SEALS-COUPE ... F08
SPOILER-UNDERTRAY ... G06
STONE GUARDS ... D17
TRUNK REINFORCEMENT - CONVERTIBLE ... F03
WEATHERSTRIPS AND TREADPLATES ... E03
WINDOW REGULATORS-NOT CONVERTIBLE ... E04
WINDSCREEN - SEALS - CONVERTIBLE ... F05
WINDSCREEN-SEALS - NOT CONVERTIBLE ... F04

BODY SHELL-CABRIOLET-Continued

Illus	Part Number	123456 Description	Quantity	Change Point	Remarks
1	RTC 1602	Panel-windscreen	1		
	RTC 1603	A post-upper RH	1		Requires modifying
2	RTC 1604	A post-upper LH	1		
	RTC 1619	A post-closing RH	1		
3	RTC 1620	A post-closing LH	1		
	RTC 1623	Reinforcement-upper RH	1		
4	RTC 1624	Reinforcement-upper LH	1		
	RTC 1617	Baffle-lower RH	1		
5	RTC 1618	Baffle-lower LH	1		
	RTC 1625	Reinforcement-lower RH	1		
6	RTC 1626	Reinforcement-lower LH	1		
	RTC 1621	A post-lower RH	1		
7	RTC 1622	A post-lower LH	1		
	RTC 1627	Sill-outer RH	1		
8	RTC 1628	Sill-outer LH	1		
	RTC 1611	B post-inner RH	1		Requires modifying
9	RTC 1612	B post-inner LH	1		
	RTC 1613	B post RH	1		
10	RTC 1614	B post LH	1		
	RTC 1615	B post-lower RH	1		
11	RTC 1616	B post-lower LH	1		

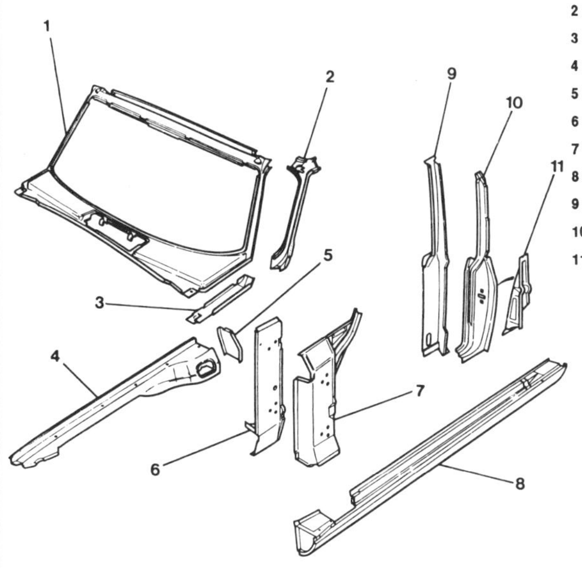

VS4149

VS 4149

BODY SHELL-CABRIOLET-Continued

Illus	Part Number	123456 Description	Quantity	Change Point	Remarks
1	BAC 9496	Header-windscreen	1		
2	BAC 7853	Catchplate-targa-front	1		
3	BAC 7736	Panel-roll bar-outer	1		
4	BAC 7854	Catchplate-targa-rear	1		
5	BAC 7897	Catchplate-hood	2		
6	BAC 8287	Tube-roll bar-front	1		
7	BAC 8288	Tube-roll bar-rear	1		
8	BAC 7820	Tube-cantrail RH	1		
9	BAC 7821	Tube-cantrail LH	1		
10	BAC 8285	Panel-roll bar-inner	1		

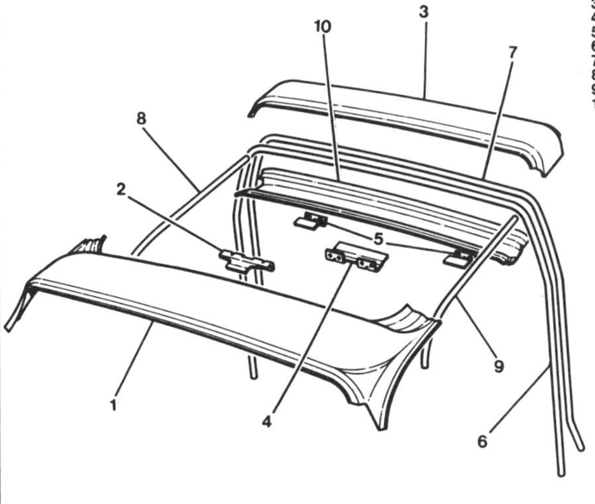

VS4117

VS 4117

MASTER INDEX

ENGINE 1.C
FLYWHEEL AND CLUTCH 1.G
GEARBOX AND PROPSHAFT 1.H
AXLES, SUSPENSION, DRIVE SHAFTS, WHEELS 1.K
STEERING 1.M
BRAKES AND BRAKE CONTROLS 1.N
FUEL, EXHAUST AND EMISSION SYSTEMS 2.C
COOLING, HEATING AND AIR CONDITIONING 2.G
ELECTRICAL, WASHERS-WIPERS, INSTRUMENTS 2.J
CHASSIS, SUBFRAMES, BODYSHELL, FITTINGS 3.C
FASCIA, TRIM, SEATS AND FIXINGS 3.H
ACCESSORIES AND PAINT 4.C
NUMERICAL INDEX 4.D

GROUP INDEX

BADGES - EMBLEMS - CONVERTIBLE G05
BADGES-EMBLEMS-CABRIOLET G04
BADGES-EMBLEMS-COUPE G03
BODY SHELL - CONVERTIBLE D03
BODY SHELL-CABRIOLET C10
BODY SHELL-COUPE C02
REAR WINGS (TONNEAU) C08
BONNET LOCK MECHANISM D15
BONNET SAFETY CATCH AND HINGES D14
BONNET-WINGS-DOORS-BOOT LID D12
BOOT HINGES-SEAL-LAMP HOUSING E14
BRIGHT FINISHERS-NOT CONVERTIBLE F14
COACHLINES(TAPE) V12 MODELS F17
COACHLINES(TAPE)-3.6 MODELS F16
CRUCIFORM AND HEATSHIELD-CABRIOLET F02
DOOR GLASS-SEALS - NOT CONVERTIBLE F06
DOOR HANDLES-LOCKS-SOLENOIDS E07
DOOR HINGES D18
DOOR SEALS AND TREAD PLATES E02
FRONT BUMPER G07
FRONT DOOR GLASS-SEALS-CONVERTIBLE F07
FRONT WINDOW REGULATORS - CONVERTIBLE E05
FUEL FILLER LID AND LOCK-COUPE C09
GEARBOX COVER PLATES E13
GEARBOX TUNNEL COVER-AUTO G/BOX MODELS E11
GEARBOX TUNNEL COVER E10
HEADLAMP SURROUNDS F12
HEATED BACKLIGHT-SEAL-COUPE F11
MIRRORS G02
REAR BUMPER G09
REAR QTR GLASS AND SEALS-CONVERTIBLE F10
REAR QTR WINDOW REGULATORS - CONVERTIBLE E06
REAR QUARTER GLASS-SEALS-CABRIOLET F09
BADGE RH-B POST F09
REAR QUARTER GLASS-SEALS-COUPE F08
SPOILER-UNDERTRAY G06
STONE GUARDS D17
TRUNK REINFORCEMENT - CONVERTIBLE F03
WEATHERSTRIPS AND TREADPLATES E03
WINDOW REGULATORS-NOT CONVERTIBLE E04
WINDSCREEN - SEALS - CONVERTIBLE F05
WINDSCREEN-SEALS - NOT CONVERTIBLE F04

3C 16

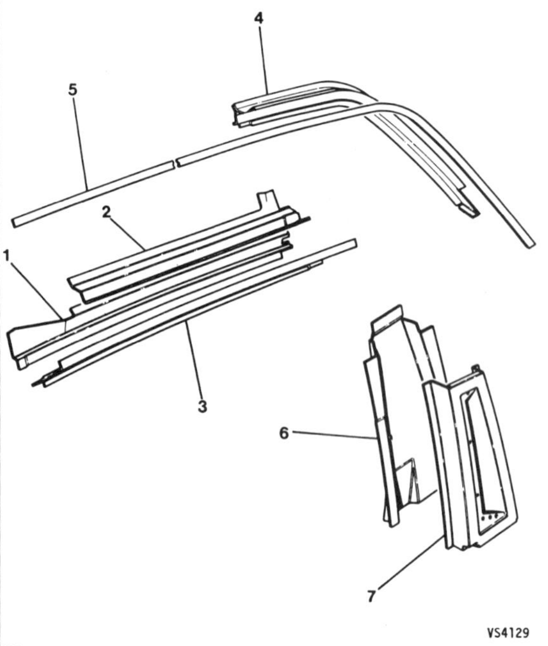

VS 4129

BODY SHELL-CABRIOLET-Continued

Illus	Part Number	123456	Description	Quantity	Change Point	Remarks
	BCC 4808		Cantrail-inner RH	1		
1	BCC 4809		Cantrail-inner LH	1		
	BAC 8294		Cantrail-closer RH	1		
2	BAC 8295		Cantrail-closer LH	1		
	BAC 7700		Cantrail-outer RH	1		
3	BAC 7701		Cantrail-outer LH	1		
	BCC 6084		Cantrail-rear RH	1		
4	BCC 6085		Cantrail-rear LH	1		
5	BAC 7728		Cantrail-strip-outer	2		
	BCC 7626		B post-inner RH	1		
6	BCC 7627		B post-inner LH	1		
	BCC 6082		B Post-outer RH	NLA		Use BBC 3916
	BBC 3916		B Post-outer RH	1		
7	BCC 6083		B Post-outer LH	NLA		Use BBC 3917
7	BBC 3917		B Post-outer LH	1		

L14.14JP25/8/88

3C 17

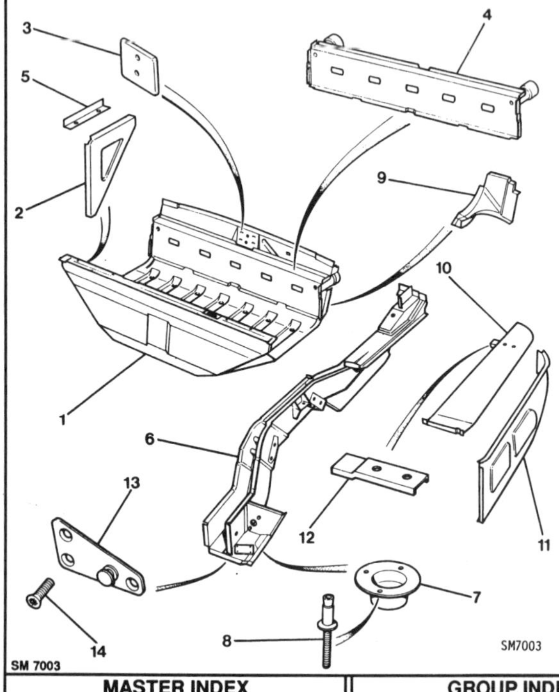

SM 7003

BODY SHELL-CABRIOLET-Continued

Illus	Part Number	123456	Description	Quantity	Change Point	Remarks
1	RTC 1594		Floor-boot	1		
2	RTC 1588		Support-battery tray	1		
3	RTC 1589		Mounting-striker	1		
4	RTC 1590		Crossmember-bumper	1		
	RTC 1599		Sidemember-rear RH	1		
5	RTC 1587		Angle support-battery	1		
7	JS 254		Mounting-radius arm	1		
8	BD 47948		Bolt-blind	1		
6	RTC 1600		Sidemember-rear LH	1		
7	JS 254		Mounting-radius arm	1		
8	BD 47948		Bolt-blind	1		
	RTC 1591		Panel-closing RH	1		
9	RTC 1592		Panel-closing LH	1		
	RTC 1595		Panel-boot side RH	1		
10	RTC 1596		Panel-boot side LH	1		
	RTC 1597		Panel-closing RH	1		
11	RTC 1598		Panel-closing LH	1		
12	RTC 1593		Reinforcement-exhaust	2		
	BD 44956		Plate-jack RH	1		
13	BD 44957		Plate-jack LH	1		
14	SF 606061 J		Screw-set	6		

MASTER INDEX

ENGINE 1.C
FLYWHEEL AND CLUTCH 1.G
GEARBOX AND PROPSHAFT 1.H
AXLES, SUSPENSION, DRIVE SHAFTS, WHEELS 1.K
STEERING 1.M
BRAKES AND BRAKE CONTROLS 1.N
FUEL, EXHAUST AND EMISSION SYSTEMS 2.C
COOLING, HEATING AND AIR CONDITIONING 2.G
ELECTRICAL, WASHERS-WIPERS, INSTRUMENTS 2.J
CHASSIS, SUBFRAMES, BODYSHELL, FITTINGS 3.C
FASCIA, TRIM, SEATS AND FIXINGS 3.H
ACCESSORIES AND PAINT 4.C
NUMERICAL INDEX 4.D

GROUP INDEX

BADGES - EMBLEMS - CONVERTIBLE G05
BADGES-EMBLEMS-CABRIOLET G04
BADGES-EMBLEMS-COUPE G03
BODY SHELL - CONVERTIBLE D03
BODY SHELL-CABRIOLET C10
BODY SHELL-COUPE C02
REAR WINGS (TONNEAU) C08
BONNET LOCK MECHANISM D15
BONNET SAFETY CATCH AND HINGES D14
BONNET-WINGS-DOORS-BOOT LID D12
BOOT HINGES-SEAL-LAMP HOUSING E14
BRIGHT FINISHERS-NOT CONVERTIBLE F14
COACHLINES(TAPE) V12 MODELS F17
COACHLINES(TAPE)-3.6 MODELS F16
CRUCIFORM AND HEATSHIELD-CABRIOLET F02
DOOR GLASS-SEALS - NOT CONVERTIBLE F06
DOOR HANDLES-LOCKS-SOLENOIDS E07
DOOR HINGES D18
DOOR SEALS AND TREAD PLATES E02
FRONT BUMPER G07
FRONT DOOR GLASS-SEALS-CONVERTIBLE F07
FRONT WINDOW REGULATORS - CONVERTIBLE E05
FUEL FILLER LID AND LOCK-COUPE C09
GEARBOX COVER PLATES E13
GEARBOX TUNNEL COVER-AUTO G/BOX MODELS E11
GEARBOX TUNNEL COVER E10
HEADLAMP SURROUNDS F12
HEATED BACKLIGHT-SEAL-COUPE F11
MIRRORS G02
REAR BUMPER G09
REAR QTR GLASS AND SEALS-CONVERTIBLE F10
REAR QTR WINDOW REGULATORS - CONVERTIBLE E06
REAR QUARTER GLASS-SEALS-CABRIOLET F09
BADGE RH-B POST F09
REAR QUARTER GLASS-SEALS-COUPE F08
SPOILER-UNDERTRAY G06
STONE GUARDS D17
TRUNK REINFORCEMENT - CONVERTIBLE F03
WEATHERSTRIPS AND TREADPLATES E03
WINDOW REGULATORS-NOT CONVERTIBLE E04
WINDSCREEN - SEALS - CONVERTIBLE F05
WINDSCREEN-SEALS - NOT CONVERTIBLE F04

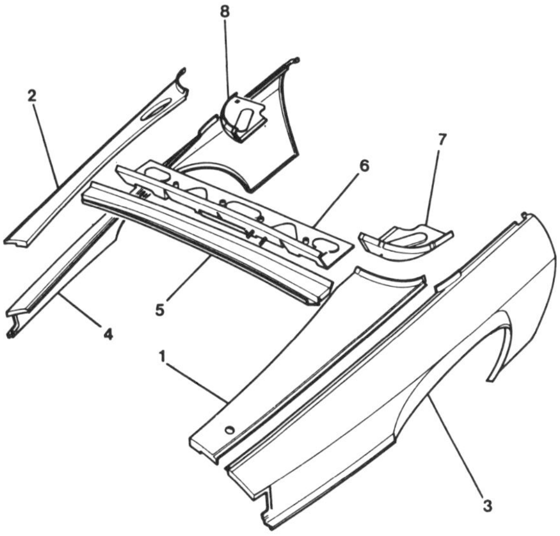

VS 4125

BODY SHELL-CABRIOLET-Continued

Illus	Part Number	123456 Description	Quantity	Change Point	Remarks
1	BAC 8274	Tonneau-upper RH	1		
2	BAC 7879	Tonneau-upper LH	1		
3	BAC 8276	Tonneau-lower RH	1		
4	BAC 8277	Tonneau-lower LH	1		
5	BAC 8279	Saddle	1		
6	BAC 8976	Shelf-parcel-centre	1		Use BCC 6659
6	BCC 6659	Shelf-parcel-centre	1		
7	BBC 3922	Shelf-parcel-side RH	1		
8	BBC 3923	Shelf-parcel-side LH	1		

MASTER INDEX

GROUP INDEX

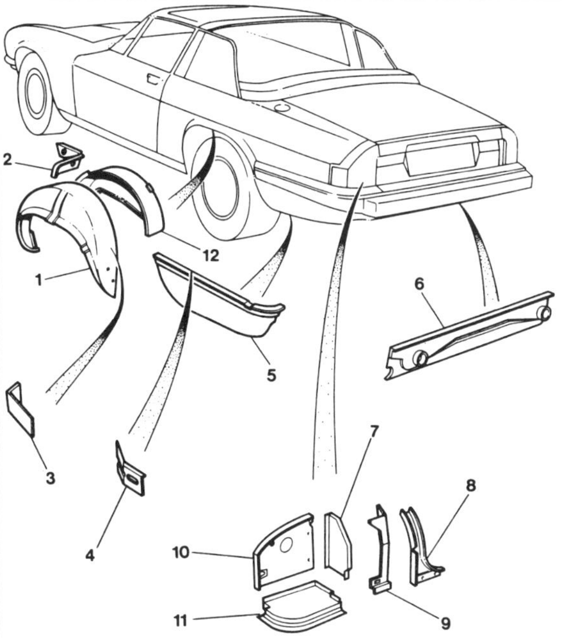

VS 4212A

BODY SHELL-CABRIOLET-Continued

Illus	Part Number	123456 Description	Quantity	Change Point	Remarks
	RTC 1638	Wheelarch-outer RH	1		
1	RTC 1639	Wheelarch-outer LH	1		
	RTC 1640	Baffle-rear RH	1		
2	RTC 1641	Baffle-rear LH	1		
3	RTC 1540	Support-bumper	2		
4	RTC 1644	Bracket-bumper	2		
	RTC 1645	Panel-lower rear RH	1		
5	RTC 1646	Panel-lower rear LH	1		
6	RTC 1637	Panel-lower rear	1		
	RTC 1651	Reinforcement RH	1		
7	RTC 1652	Reinforcement LH	1		
	RTC 1655	Gusset RH	1		
8	RTC 1656	Gusset LH	1		
	RTC 1653	Panel-lamp RH	1		
9	RTC 1654	Panel-lamp LH	1		
	RTC 1647	Panel-closing RH	1		
10	RTC 1648	Panel-closing LH	1		
	RTC 1649	Panel-lower RH	1		
11	RTC 1650	Panel-lower LH	1		
	JLM 742	Wheelarch-inner RH	1		
12	JLM 743	Wheelarch-inner LH	1		

MASTER INDEX
ENGINE 1.C
FLYWHEEL AND CLUTCH 1.G
GEARBOX AND PROPSHAFT 1.H
AXLES, SUSPENSION, DRIVE SHAFTS, WHEELS 1.K
STEERING 1.M
BRAKES AND BRAKE CONTROLS 1.N
FUEL, EXHAUST AND EMISSION SYSTEMS 2.C
COOLING, HEATING AND AIR CONDITIONING 2.G
ELECTRICAL, WASHERS-WIPERS, INSTRUMENTS 2.J
CHASSIS, SUBFRAMES, BODYSHELL, FITTINGS 3.C
FASCIA, TRIM, SEATS AND FIXINGS 3.H
ACCESSORIES AND PAINT 4.C
NUMERICAL INDEX 4.D
GROUP INDEX
BADGES - EMBLEMS - CONVERTIBLE G05
BADGES-EMBLEMS-CABRIOLET G04
BADGES-EMBLEMS-COUPE G03
BODY SHELL - CONVERTIBLE D03
BODY SHELL-CABRIOLET C10
BODY SHELL-COUPE C02
REAR WINGS (TONNEAU) C08
BONNET LOCK MECHANISM D15
BONNET SAFETY CATCH AND HINGES D14
BONNET-WINGS-DOORS-BOOT LID D12
BOOT HINGES-SEAL-LAMP HOUSING E14
BRIGHT FINISHERS-NOT CONVERTIBLE F14
COACHLINES(TAPE) V12 MODELS F17
COACHLINES(TAPE)-3.6 MODELS F16
CRUCIFORM AND HEATSHIELD-CABRIOLET F02
DOOR GLASS-SEALS - NOT CONVERTIBLE F06
DOOR HANDLES-LOCKS-SOLENOIDS E07
DOOR HINGES D18
DOOR SEALS AND TREAD PLATES E02
FRONT BUMPER G07
FRONT DOOR GLASS-SEALS-CONVERTIBLE F07
FRONT WINDOW REGULATORS - CONVERTIBLE E05
FUEL FILLER LID AND LOCK-COUPE C09
GEARBOX COVER PLATES E13
GEARBOX TUNNEL COVER-AUTO G/BOX MODELS E11
GEARBOX TUNNEL COVER E10
HEADLAMP SURROUNDS F12
HEATED BACKLIGHT-SEAL-COUPE F11
MIRRORS G02
REAR BUMPER G09
REAR QTR GLASS AND SEALS-CONVERTIBLE F10
REAR QTR WINDOW REGULATORS - CONVERTIBLE E06
REAR QUARTER GLASS-SEALS-CABRIOLET F09
BADGE RH-B POST F09
REAR QUARTER GLASS-SEALS-COUPE F08
SPOILER-UNDERTRAY G06
STONE GUARDS D17
TRUNK REINFORCEMENT - CONVERTIBLE F03
WEATHERSTRIPS AND TREADPLATES E03
WINDOW REGULATORS-NOT CONVERTIBLE E04
WINDSCREEN - SEALS - CONVERTIBLE F05
WINDSCREEN-SEALS - NOT CONVERTIBLE F04

3D 03

BODY SHELL - CONVERTIBLE

Illus	Part Number	123456	Description	Quantity	Change Point	Remarks
1	BJC 1502		Body shell assy (less doors,front wings bonnet and boot lid)	1		

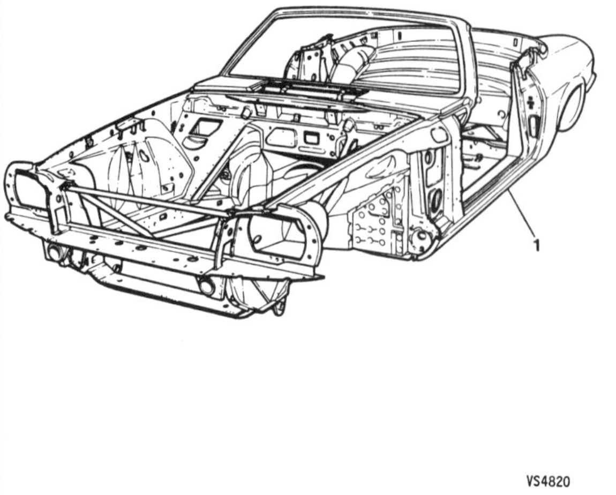

VS4820

3D 04

BODY SHELL CONVERTIBLE - continued

Illus	Part Number	123456	Description	Quantity	Change Point	Remarks
1	CBP 314		Crossmember-bumper	1		
2	RTC 1581		Bracket-support RH	1		
	RTC 1582		Bracket-support LH	1		
	JLM 1431		Sill-inner RH	1		
3	JLM 1432		Sill-inner LH	1		
	RTC 1555		Reinforcement RH	1		
4	RTC 1556		Reinforcement LH	1		
5	CBP 313		Crossmember-lower	1		
	RTC 1565		VALANCE ASSEMBLY RH	1		
6	RTC 1548		Mounting RH	1		
	RTC 1563		Mounting-damper RH	1		
	CBP 224		Panel-filler RH	1		
	RTC 1557		Panel-closing RH	1		
	RTC 1559		Panel-closing RH	1		
7	RTC 1566		VALANCE ASSEMBLY LH	1		
	RTC 1549		Mounting LH	1		
8	RTC 1564		Mounting-damper LH	1		
9	RTC 1562		Panel-filler LH	1		
10	RTC 1558		Panel-closing LH	1		
11	RTC 1560		Panel-closing LH	1		

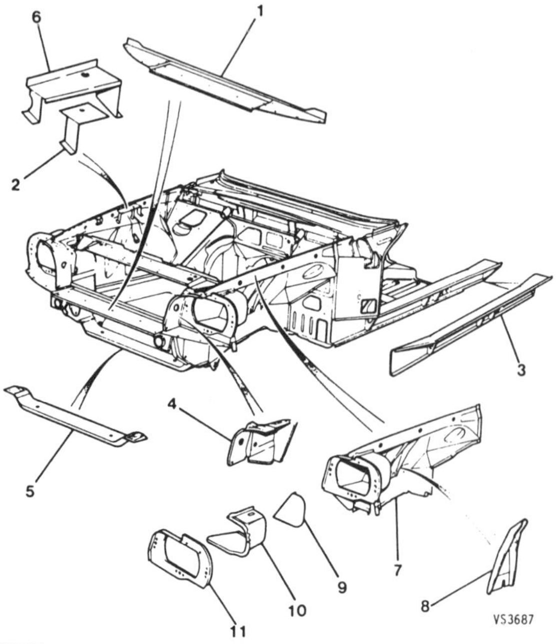

VS3687

MASTER INDEX

ENGINE ... 1.C
FLYWHEEL AND CLUTCH ... 1.G
GEARBOX AND PROPSHAFT ... 1.H
AXLES, SUSPENSION, DRIVE SHAFTS, WHEELS ... 1.K
STEERING ... 1.M
BRAKES AND BRAKE CONTROLS ... 1.N
FUEL, EXHAUST AND EMISSION SYSTEMS ... 2.C
COOLING, HEATING AND AIR CONDITIONING ... 2.G
ELECTRICAL, WASHERS-WIPERS, INSTRUMENTS ... 2.J
CHASSIS, SUBFRAMES, BODYSHELL, FITTINGS ... 3.C
FASCIA, TRIM, SEATS AND FIXINGS ... 3.H
ACCESSORIES AND PAINT ... 4.C
NUMERICAL INDEX ... 4.D

GROUP INDEX

BADGES - EMBLEMS - CONVERTIBLE ... G05
BADGES-EMBLEMS-CABRIOLET ... G04
BADGES-EMBLEMS-COUPE ... G03
BODY SHELL - CONVERTIBLE ... D03
BODY SHELL-CABRIOLET ... C10
BODY SHELL-COUPE ... C02
REAR WINGS (TONNEAU) ... C08
BONNET LOCK MECHANISM ... D15
BONNET SAFETY CATCH AND HINGES ... D14
BONNET-WINGS-DOORS-BOOT LID ... D12
BOOT HINGES-SEAL-LAMP HOUSING ... E14
BRIGHT FINISHERS-NOT CONVERTIBLE ... F14
COACHLINES(TAPE) V12 MODELS ... F17
COACHLINES(TAPE)-3.6 MODELS ... F16
CRUCIFORM AND HEATSHIELD-CABRIOLET ... F02
DOOR GLASS-SEALS - NOT CONVERTIBLE ... F06
DOOR HANDLES-LOCKS-SOLENOIDS ... E07
DOOR HINGES ... D18
DOOR SEALS AND TREAD PLATES ... E02
FRONT BUMPER ... G07
FRONT DOOR GLASS-SEALS-CONVERTIBLE ... F07
FRONT WINDOW REGULATORS - CONVERTIBLE ... E05
FUEL FILLER LID AND LOCK-COUPE ... C09
GEARBOX COVER PLATES ... E13
GEARBOX TUNNEL COVER-AUTO G/BOX MODELS ... E11
GEARBOX TUNNEL COVER ... E10
HEADLAMP SURROUNDS ... F12
HEATED BACKLIGHT-SEAL-COUPE ... F11
MIRRORS ... G02
REAR BUMPER ... G09
REAR QTR GLASS AND SEALS-CONVERTIBLE ... F10
REAR QTR WINDOW REGULATORS - CONVERTIBLE ... E06
REAR QUARTER GLASS-SEALS-CABRIOLET ... F09
BADGE RH-B POST ... F09
REAR QUARTER GLASS-SEALS-COUPE ... F08
SPOILER-UNDERTRAY ... G06
STONE GUARDS ... D17
TRUNK REINFORCEMENT - CONVERTIBLE ... F03
WEATHERSTRIPS AND TREADPLATES ... E03
WINDOW REGULATORS-NOT CONVERTIBLE ... E04
WINDSCREEN - SEALS - CONVERTIBLE ... F05
WINDSCREEN-SEALS - NOT CONVERTIBLE ... F04

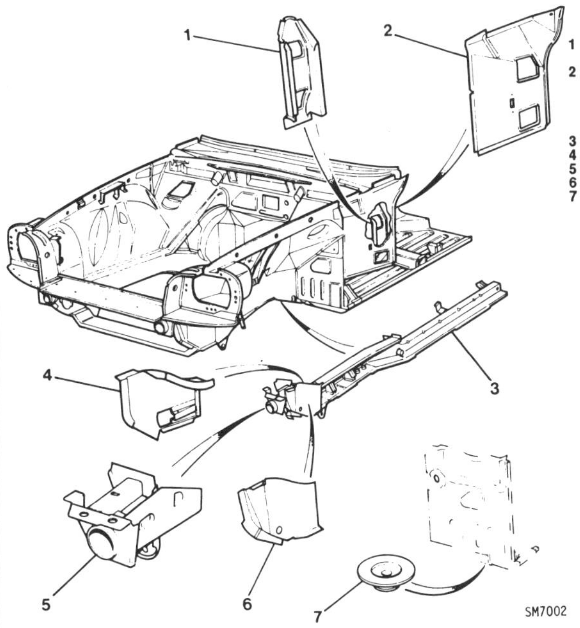

SM 7002

BODY SHELL-CONVERTIBLE - continued

PANELS AND SIDE MEMBERS

Illus	Part Number	123456	Description	Quantity	Change Point	Remarks
	RTC 1567		A post-inner RH	1		
1	RTC 1568		A post-inner LH	1		
	RTC 1569		Panel-dash side RH	1		
2	RTC 1570		Panel-dash side LH	1		
	RTC 1577		Sidemember-front RH	1		
	RTC 1573		Extension-inner RH	1		
	RTC 1575		Mounting-bumper RH	1		
	RTC 1579		Extension-outer RH	1		
3	RTC 1578		Sidemember-front LH	1		
4	RTC 1574		Extension-inner LH	1		
5	RTC 1576		Mounting-bumper LH	1		
6	RTC 1580		Extension-outer LH	1		
7	BD 28658		Bracket-front jacking	2		

MASTER INDEX

GROUP INDEX

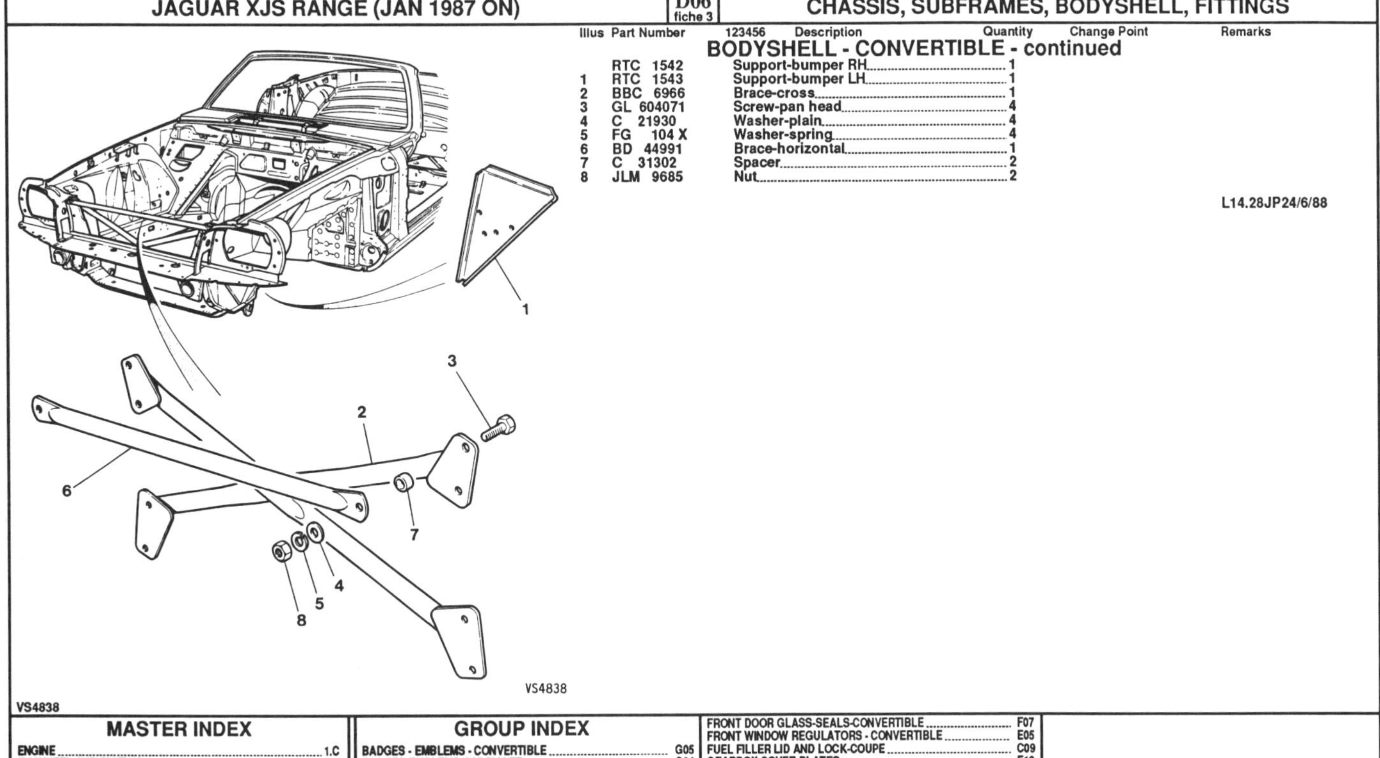

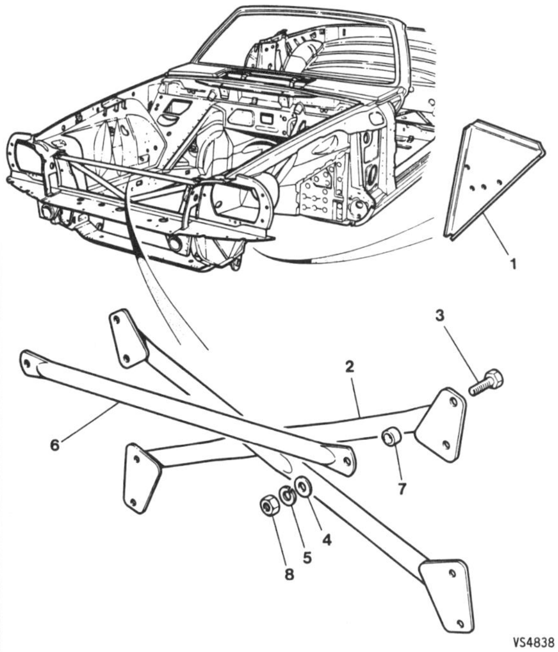

VS4838

BODYSHELL - CONVERTIBLE - continued

Illus	Part Number	123456	Description	Quantity	Change Point	Remarks
	RTC 1542		Support-bumper RH	1		
1	RTC 1543		Support-bumper LH	1		
2	BBC 6966		Brace-cross	1		
3	GL 604071		Screw-pan head	4		
4	C 21930		Washer-plain	4		
5	FG 104 X		Washer-spring	4		
6	BD 44991		Brace-horizontal	1		
7	C 31302		Spacer	2		
8	JLM 9685		Nut	2		

L14.28JP24/6/88

MASTER INDEX

ENGINE 1.C
FLYWHEEL AND CLUTCH 1.G
GEARBOX AND PROPSHAFT 1.H
AXLES, SUSPENSION, DRIVE SHAFTS, WHEELS 1.K
STEERING 1.M
BRAKES AND BRAKE CONTROLS 1.N
FUEL, EXHAUST AND EMISSION SYSTEMS 2.C
COOLING, HEATING AND AIR CONDITIONING 2.G
ELECTRICAL, WASHERS-WIPERS, INSTRUMENTS 2.J
CHASSIS, SUBFRAMES, BODYSHELL, FITTINGS 3.C
FASCIA, TRIM, SEATS AND FIXINGS 3.H
ACCESSORIES AND PAINT 4.C
NUMERICAL INDEX 4.D

GROUP INDEX

BADGES - EMBLEMS - CONVERTIBLE G05
BADGES-EMBLEMS-CABRIOLET G04
BADGES-EMBLEMS-COUPE G03
BODY SHELL - CONVERTIBLE D03
BODY SHELL-CABRIOLET C10
BODY SHELL-COUPE C02
REAR WINGS (TONNEAU) C08
BONNET LOCK MECHANISM D15
BONNET SAFETY CATCH AND HINGES D14
BONNET-WINGS-DOORS-BOOT LID D12
BOOT HINGES-SEAL-LAMP HOUSING E14
BRIGHT FINISHERS-NOT CONVERTIBLE F14
COACHLINES(TAPE) V12 MODELS F17
COACHLINES(TAPE)-3.6 MODELS F16
CRUCIFORM AND HEATSHIELD-CABRIOLET F02
DOOR GLASS-SEALS - NOT CONVERTIBLE F06
DOOR HANDLES-LOCKS-SOLENOIDS E07
DOOR HINGES D18
DOOR SEALS AND TREAD PLATES E02
FRONT BUMPER G07
FRONT DOOR GLASS-SEALS-CONVERTIBLE F07
FRONT WINDOW REGULATORS - CONVERTIBLE E05
FUEL FILLER LID AND LOCK-COUPE C09
GEARBOX COVER PLATES E13
GEARBOX TUNNEL COVER-AUTO G/BOX MODELS E11
GEARBOX TUNNEL COVER E10
HEADLAMP SURROUNDS F12
HEATED BACKLIGHT-SEAL-COUPE F11
MIRRORS G02
REAR BUMPER G09
REAR QTR GLASS AND SEALS-CONVERTIBLE F10
REAR QTR WINDOW REGULATORS - CONVERTIBLE E06
REAR QUARTER GLASS-SEALS-CABRIOLET F09
BADGE RH-B POST F09
REAR QUARTER GLASS-SEALS-COUPE F08
SPOILER-UNDERTRAY G06
STONE GUARDS D17
TRUNK REINFORCEMENT - CONVERTIBLE F03
WEATHERSTRIPS AND TREADPLATES E03
WINDOW REGULATORS-NOT CONVERTIBLE E04
WINDSCREEN - SEALS - CONVERTIBLE F05
WINDSCREEN-SEALS - NOT CONVERTIBLE F04

3D 07

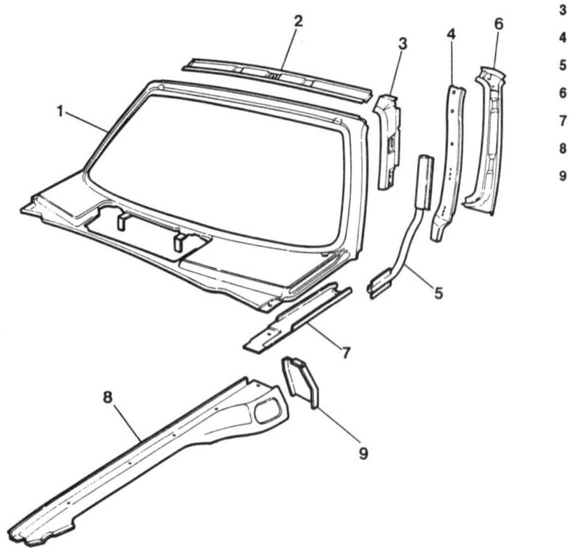

VS4862

BODY SHELL - CONVERTIBLE - continued

Illus	Part Number	123456 Description	Quantity	Change Point	Remarks
1	BJC 1518	Panel-windscreen	1		
2	BJC 1519	Windscreen panel-inner	1		
	BJC 1078	Panel-upper outer "A" post RH	1		
3	BJC 1079	Panel-upper outer "A" post LH	1		
	BJC 1052	Reinforcement-"A" Post RH	1		
4	BJC 1053	Reinforcement-"A" Post RH	1		
	BJC 1884	Tube assy-"A" Post RH	1		
5	BJC 1885	Tube assy-"A" Post LH	1		
	BJC 1830	Panel-closing upper "A" Post RH	1		
6	BJC 1831	Panel-closing upper "A" Post LH	1		
	BJC 1844	"A" Post-closing RH	1		
7	BJC 1845	"A" Post-closing LH	1		
	RTC 1623	Reinforcement-upper RH	1		
8	RTC 1624	Reinforcement-upper LH	1		
	RTC 1617	Baffle-RH	1		
9	RTC 1618	Baffle-LH	1		

MASTER INDEX

GROUP INDEX

3D 08

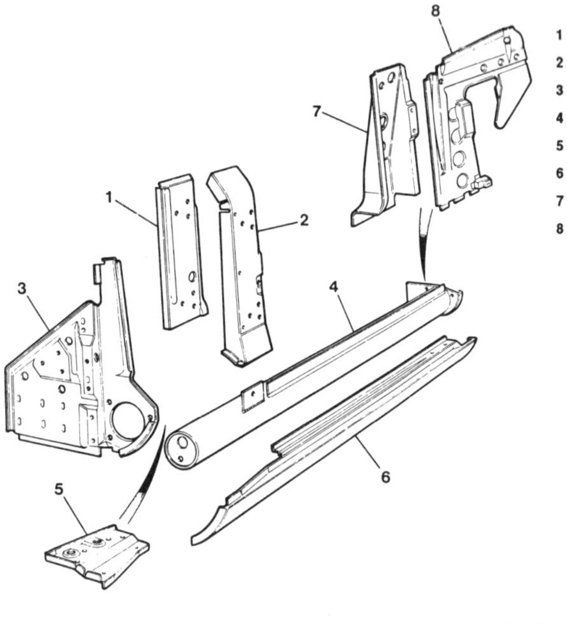

VS4840

BODYSHELL - CONVERTIBLE - continued

Illus	Part Number	123456 Description	Quantity	Change Point	Remarks
	BJC 1404	Reinforcement-lower Rh	1		
1	BJC 1405	Reinforcement-lower LH	1		
	BJC 1828	Panel-"A" Post RH	1		
2	BJC 1829	Panel-"A" Post LH	1		
	BJC 1824	Reinforcement-Wheelarch RH	1		
3	BJC 1825	Reinforcement-Wheelarch LH	1		
	BJC 1508	Tube-Sill RH	1		
4	BJC 1509	Tube-Sill LH	1		
	BJC 1858	Extension-Sill RH	1		
5	BJC 1859	Extension-Sill Lh	1		
	BJC 1412	Sill-outer RH	1		
6	BJC 1413	Sill-outer LH	1		
	JLM 1501	"B" Post inner RH	1		
7	JLM 1502	"B" Post inner LH	1		
	BJC 1538	Panel-inner rear qtr RH	1		
8	BJC 1539	Panel-inner rear qtr LH	1		

MASTER INDEX

ENGINE 1.C
FLYWHEEL AND CLUTCH 1.G
GEARBOX AND PROPSHAFT 1.H
AXLES, SUSPENSION, DRIVE SHAFTS, WHEELS 1.K
STEERING 1.M
BRAKES AND BRAKE CONTROLS 1.N
FUEL, EXHAUST AND EMISSION SYSTEMS 2.C
COOLING, HEATING AND AIR CONDITIONING 2.G
ELECTRICAL, WASHERS-WIPERS, INSTRUMENTS 2.J
CHASSIS, SUBFRAMES, BODYSHELL, FITTINGS 3.C
FASCIA, TRIM, SEATS AND FIXINGS 3.H
ACCESSORIES AND PAINT 4.C
NUMERICAL INDEX 4.D

GROUP INDEX

BADGES - EMBLEMS - CONVERTIBLE G05
BADGES-EMBLEMS-CABRIOLET G04
BADGES-EMBLEMS-COUPE G03
BODY SHELL - CONVERTIBLE D03
BODY SHELL-CABRIOLET C10
BODY SHELL-COUPE C02
REAR WINGS (TONNEAU) C08
BONNET LOCK MECHANISM D15
BONNET SAFETY CATCH AND HINGES D14
BONNET-WINGS-DOORS-BOOT LID D12
BOOT HINGES-SEAL-LAMP HOUSING E14
BRIGHT FINISHERS-NOT CONVERTIBLE F14
COACHLINES(TAPE) V12 MODELS F17
COACHLINES(TAPE)-3.6 MODELS F16
CRUCIFORM AND HEATSHIELD-CABRIOLET F02
DOOR GLASS-SEALS - NOT CONVERTIBLE F06
DOOR HANDLES-LOCKS-SOLENOIDS E07
DOOR HINGES D18
DOOR SEALS AND TREAD PLATES E02
FRONT BUMPER G07
FRONT DOOR GLASS-SEALS-CONVERTIBLE F07
FRONT WINDOW REGULATORS - CONVERTIBLE E05
FUEL FILLER LID AND LOCK-COUPE C09
GEARBOX COVER PLATES E13
GEARBOX TUNNEL COVER-AUTO G/BOX MODELS E11
GEARBOX TUNNEL COVER E10
HEADLAMP SURROUNDS F12
HEATED BACKLIGHT-SEAL-COUPE F11
MIRRORS G02
REAR BUMPER G09
REAR QTR GLASS AND SEALS-CONVERTIBLE F10
REAR QTR WINDOW REGULATORS - CONVERTIBLE E06
REAR QUARTER GLASS-SEALS-CABRIOLET F09
BADGE RH-B POST F09
REAR QUARTER GLASS-SEALS-COUPE F08
SPOILER-UNDERTRAY G06
STONE GUARDS D17
TRUNK REINFORCEMENT - CONVERTIBLE F03
WEATHERSTRIPS AND TREADPLATES E03
WINDOW REGULATORS-NOT CONVERTIBLE E04
WINDSCREEN - SEALS - CONVERTIBLE F05
WINDSCREEN-SEALS - NOT CONVERTIBLE F04

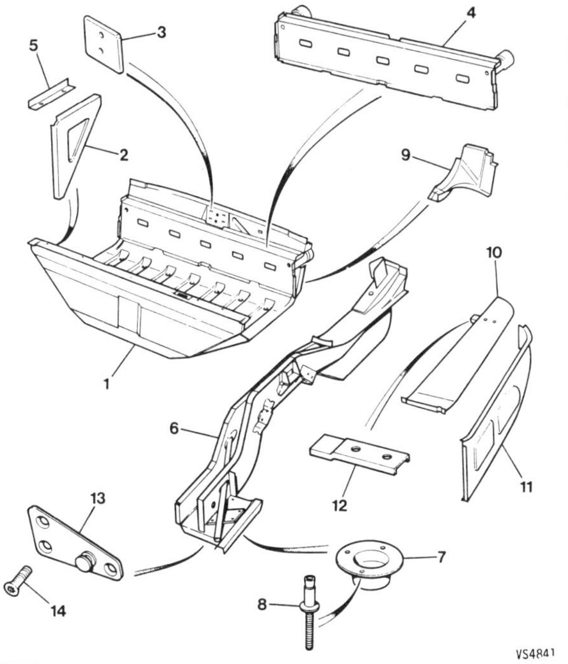

VS4841

BODY SHELL-CONVERTIBLE - continued

Illus	Part Number	123456	Description	Quantity	Change Point	Remarks
1	RTC 1594		Floor-boot	1		
2	RTC 1588		Support-battery tray	1		
3	RTC 1589		Mounting-striker	1		
4	RTC 1590		Crossmember-bumper	1		
	BJC 1506		Sidemember-rear RH	1		
5	RTC 1587		Angle support-battery	1		
7	JS 254		Mounting-radius arm	1		
8	BD 47948		Bolt-blind	1		
6	BJC 1507		Sidemember-rear LH	1		
7	JS 254		Mounting-radius arm	1		
8	BD 47948		Bolt-blind	1		
	RTC 1591		Panel-closing RH	1		
9	RTC 1592		Panel-closing LH	1		
	RTC 1595		Panel-boot side RH	1		
10	RTC 1596		Panel-boot side LH	1		
	RTC 1597		Panel-closing RH	1		
11	RTC 1598		Panel-closing LH	1		
12	RTC 1593		Reinforcement-exhaust	2		
	BD 44956		Plate-jack RH	1		
13	BD 44957		Plate-jack LH	1		
14	SF 606061 J		Screw-set	6		

MASTER INDEX

GROUP INDEX

BODY SHELL - CONVERTIBLE - continued

Illus	Part Number	123456	Description	Quantity	Change Point	Remarks
1	BJC 1516		Tonneau RH	1		
	BJC 1517		Tonneau Lh	1		
2	BJC 1073		Saddle	1		
3	BJC 1827		Panel-inner closing	1		

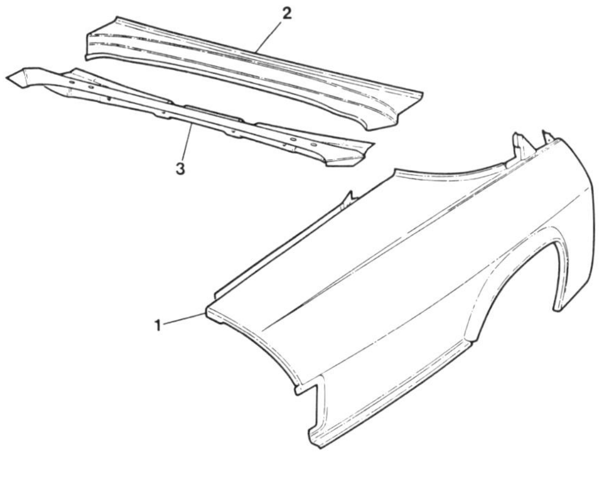

VS4837

MASTER INDEX

ENGINE 1.C
FLYWHEEL AND CLUTCH 1.G
GEARBOX AND PROPSHAFT 1.H
AXLES, SUSPENSION, DRIVE SHAFTS, WHEELS 1.K
STEERING 1.M
BRAKES AND BRAKE CONTROLS 1.N
FUEL, EXHAUST AND EMISSION SYSTEMS 2.C
COOLING, HEATING AND AIR CONDITIONING 2.G
ELECTRICAL, WASHERS-WIPERS, INSTRUMENTS 2.J
CHASSIS, SUBFRAMES, BODYSHELL, FITTINGS 3.C
FASCIA, TRIM, SEATS AND FIXINGS 3.H
ACCESSORIES AND PAINT 4.C
NUMERICAL INDEX 4.D

GROUP INDEX

BADGES - EMBLEMS - CONVERTIBLE G05
BADGES-EMBLEMS-CABRIOLET G04
BADGES-EMBLEMS-COUPE G03
BODY SHELL - CONVERTIBLE D03
BODY SHELL-CABRIOLET C10
BODY SHELL-COUPE C02
REAR WINGS (TONNEAU) C08
BONNET LOCK MECHANISM D15
BONNET SAFETY CATCH AND HINGES D14
BONNET-WINGS-DOORS-BOOT LID D12
BOOT HINGES-SEAL-LAMP HOUSING E14
BRIGHT FINISHERS-NOT CONVERTIBLE F14
COACHLINES(TAPE) V12 MODELS F17
COACHLINES(TAPE)-3.6 MODELS F16
CRUCIFORM AND HEATSHIELD-CABRIOLET F02
DOOR GLASS-SEALS - NOT CONVERTIBLE F06
DOOR HANDLES-LOCKS-SOLENOIDS E07
DOOR HINGES D18
DOOR SEALS AND TREAD PLATES E02
FRONT BUMPER G07
FRONT DOOR GLASS-SEALS-CONVERTIBLE F07
FRONT WINDOW REGULATORS - CONVERTIBLE E05
FUEL FILLER LID AND LOCK-COUPE C09
GEARBOX COVER PLATES E13
GEARBOX TUNNEL COVER-AUTO G/BOX MODELS E11
GEARBOX TUNNEL COVER E10
HEADLAMP SURROUNDS F12
HEATED BACKLIGHT-SEAL-COUPE F11
MIRRORS G02
REAR BUMPER G09
REAR QTR GLASS AND SEALS-CONVERTIBLE F10
REAR QTR WINDOW REGULATORS - CONVERTIBLE E06
REAR QUARTER GLASS-SEALS-CABRIOLET F09
BADGE RH-B POST F09
REAR QUARTER GLASS-SEALS-COUPE F08
SPOILER-UNDERTRAY G06
STONE GUARDS D17
TRUNK REINFORCEMENT - CONVERTIBLE F03
WEATHERSTRIPS AND TREADPLATES E03
WINDOW REGULATORS-NOT CONVERTIBLE E04
WINDSCREEN - SEALS - CONVERTIBLE F05
WINDSCREEN-SEALS - NOT CONVERTIBLE F04

JAGUAR XJS RANGE (JAN 1987 ON)	D11 fiche 3	CHASSIS, SUBFRAMES, BODYSHELL, FITTINGS

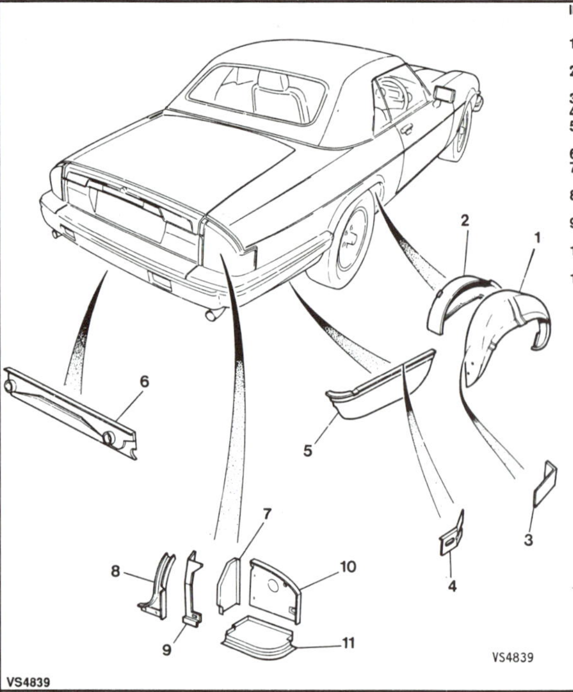

VS4839

BODYSHELL - CONVERTIBLE - continued

Illus	Part Number	123456	Description	Quantity	Change Point	Remarks
1	BJC 1414		Wheelarch-outer RH	1		
	BJC 1415		Wheelarch-outer LH	1		
2	BJC 1436		Wheelarch-inner Rh	1		
	BJC 1437		Wheelarch-inner LH	1		
3	RTC 1540		Support-bumper	2		
4	RTC 1644		Bracket-bumper	2		
5	RTC 1645		Panel-lower rear RH	1		
	RTC 1646		Panel-lower rear Lh	1		
6	RTC 1637		Panel-lower rear	1		
7	RTC 1651		Reinforcement RH	1		
	RTC 1652		Reinforcement LH	1		
8	RTC 1655		Russet RH	1		
	RTC 1656		Gusset LH	1		
9	RTC 1653		Panel-lamp RH	1		
	RTC 1654		Panel-lamp LH	1		
10	BCC 6804		Panel-closing RH	1		
	BCC 6805		Panel-closing LH	1		
11	RTC 1649		Panel-lower RH	1		
	RTC 1650		Panel-lower LH	1		

MASTER INDEX

GROUP INDEX

3D 11

JAGUAR XJS RANGE (JAN 1987 ON)	D12 fiche 3	CHASSIS, SUBFRAMES, BODYSHELL, FITTINGS

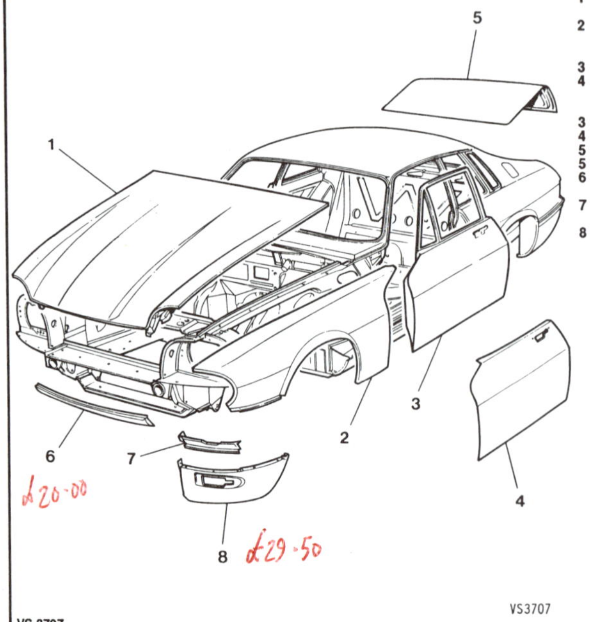

VS 3707

BONNET-WINGS-DOORS-BOOT LID.

Illus	Part Number	123456	Description	Quantity	Change Point	Remarks
1	CBP 303		Bonnet assembly	1	.....	3.6 litre
1	RTC 1530		Bonnet Assembly	1	.....	5.3 litre
	RTC 1536		Wing-front RH	1		
2	RTC 1537		Wing-front LH	1		
	RTC 1532		Door-RH	1		
	RTC 1534		Skin-door RH	1		
3	RTC 1533		Door-LH	1	.....	Not Convertible
4	RTC 1535		Skin-door LH	1		
	BJC 1536		Door RH	1	.....	Convertible only
	BJC 1834		Skin RH	1		
3	BJC 1537		Door LH	1		
4	BJC 1835		Skin LH	1		
5	CBP 318		Lid-boot	1	.....	Not Convertible
5	BJC 1524		Lid-boot	1	.....	Convertible
6	RTC 1531		Panel-lower front	1		
	RTC 1546		Extension-lower RH	1		
7	RTC 1547		Extension-lower LH	1		
	RTC 1544		Wing-lower RH	1		
8	RTC 1545		Wing-lower LH	1		

MASTER INDEX

ENGINE 1.C
FLYWHEEL AND CLUTCH 1.G
GEARBOX AND PROPSHAFT 1.H
AXLES, SUSPENSION, DRIVE SHAFTS, WHEELS 1.K
STEERING 1.M
BRAKES AND BRAKE CONTROLS 1.N
FUEL, EXHAUST AND EMISSION SYSTEMS 2.C
COOLING, HEATING AND AIR CONDITIONING 2.G
ELECTRICAL, WASHERS-WIPERS, INSTRUMENTS 2.J
CHASSIS, SUBFRAMES, BODYSHELL, FITTINGS 3.C
FASCIA, TRIM, SEATS AND FIXINGS 3.H
ACCESSORIES AND PAINT 4.C
NUMERICAL INDEX 4.D

GROUP INDEX

BADGES - EMBLEMS - CONVERTIBLE G05
BADGES-EMBLEMS-CABRIOLET G04
BADGES-EMBLEMS-COUPE. G03
BODY SHELL - CONVERTIBLE D03
BODY SHELL-CABRIOLET C10
BODY SHELL-COUPE. C02
REAR WINGS (TONNEAU) C08
BONNET LOCK MECHANISM D15
BONNET SAFETY CATCH AND HINGES D14
BONNET-WINGS-DOORS-BOOT LID. D12
BOOT HINGES-SEAL-LAMP HOUSING. E14
BRIGHT FINISHERS-NOT CONVERTIBLE F14
COACHLINES(TAPE) V12 MODELS. F17
COACHLINES(TAPE)-3.6 MODELS. F16
CRUCIFORM AND HEATSHIELD-CABRIOLET F02
DOOR GLASS-SEALS - NOT CONVERTIBLE F06
DOOR HANDLES-LOCKS-SOLENOIDS E07
DOOR HINGES D18
DOOR SEALS AND TREAD PLATES. E02
FRONT BUMPER. G07
FRONT DOOR GLASS-SEALS-CONVERTIBLE F07
FRONT WINDOW REGULATORS - CONVERTIBLE E05
FUEL FILLER LID AND LOCK-COUPE C09
GEARBOX COVER PLATES. E13
GEARBOX TUNNEL COVER-AUTO G/BOX MODELS E11
GEARBOX TUNNEL COVER. E10
HEADLAMP SURROUNDS. F12
HEATED BACKLIGHT-SEAL-COUPE. F11
MIRRORS. G02
REAR BUMPER. G09
REAR QTR GLASS AND SEALS-CONVERTIBLE F10
REAR QTR WINDOW REGULATORS - CONVERTIBLE E06
REAR QUARTER GLASS-SEALS-CABRIOLET F09
BADGE RH-B POST F09
REAR QUARTER GLASS-SEALS-COUPE. F08
SPOILER-UNDERTRAY. G06
STONE GUARDS. D17
TRUNK REINFORCEMENT - CONVERTIBLE F03
WEATHERSTRIPS AND TREADPLATES E03
WINDOW REGULATORS-NOT CONVERTIBLE E04
WINDSCREEN - SEALS - CONVERTIBLE F05
WINDSCREEN-SEALS - NOT CONVERTIBLE F04

3D 12

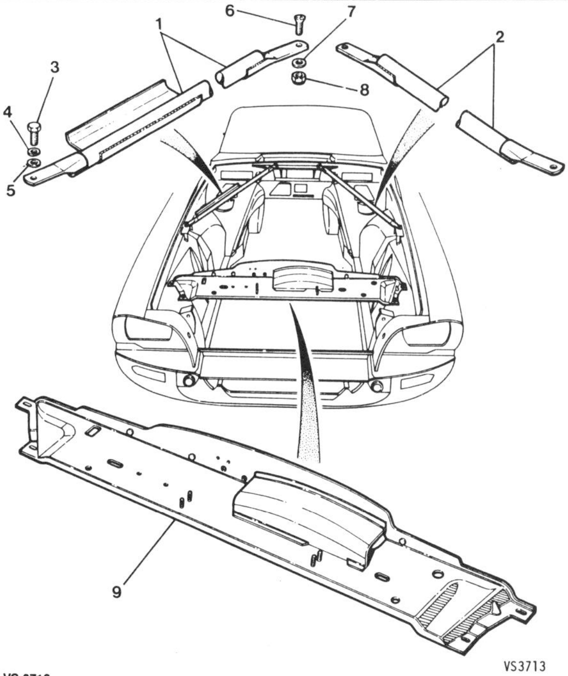

VS 3713

BRACING STRUTS-UPPER CROSSMEMBER.

Illus	Part Number	123456	Description	Quantity	Change Point	Remarks
1	BD 46085		Brace-RH	1		Use BDC 5512
1	BDC 5512		Brace-RH	1		
2	BAC 3477		Brace-LH	1		
3	SH 606062 J		Screw-set	2		
4	JLM 304		Washer-spring	2		
5	JLM 301		Washer-plain	2		
6	SE 606081 J		Screw-set	2		
7	JLM 301		Washer-plain	2		
8	C 8667 3		Nut-self-lock	2		
9	BAC 6306		Crossmember-front upper	1		3.6 litre
	BAC 4343		Crossmember-front upper	1	Up to 89 MY	5.3 litre
	BDC 8708		Crossmember-front upper	1	From 89 MY	5.3 litre
	SH 604071 J		Screw-set	4		
	C 21930		Washer-plain	4		Use WC600046J
	WC 600046 J		Washer-plain	4		
	JLM 302		Washer-spring	4		

L18JP31/8/88

MASTER INDEX

GROUP INDEX

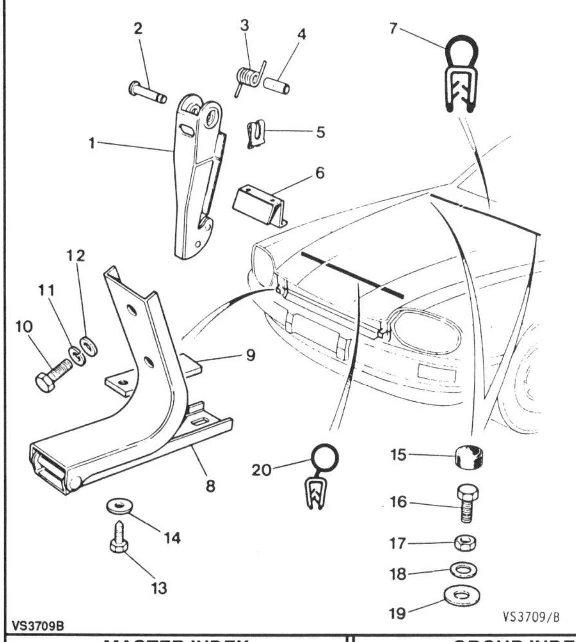

VS3709B

BONNET SAFETY CATCH AND HINGES

Illus	Part Number	123456	Description	Quantity	Change Point	Remarks
1	BD 31316		Catch-safety	1		
2	BAC 4300		Pin-pivot	1		
3	BD 48396		Spring-return	1		
4	BD 48393		Sleeve	1		
5	C 27328 3		Clip-safety	1		
6	BD 48203		Bracket-striker	1		
7	BBC 4411		Seal-rubber	1		
8	RTC 1675		Hinge-bonnet	2		
9	AEU 1057		Plate-tapped	2		
10	JLM 9676		Screw-set	4		
11	JLM 303		Washer-spring	4		
12	BD 541 37		Washer-plain	4		
13	BD 13565 3		Screw-set	4		
14	BD 541 5		Washer-plain	4		
15	BD 10224		Buffer-bonnet stop	2		
16	JLM 9563		Screw-set	2		
17	UFN 231 L		Nut-plain	2		
18	JLM 298		Washer-plain	2		
19	AGU 1079		Washer-rubber	2		
20	BBC 4454		Seal-rad crossmember	1		

MASTER INDEX
ENGINE ... 1.C
FLYWHEEL AND CLUTCH ... 1.G
GEARBOX AND PROPSHAFT ... 1.H
AXLES, SUSPENSION, DRIVE SHAFTS, WHEELS ... 1.K
STEERING ... 1.M
BRAKES AND BRAKE CONTROLS ... 1.N
FUEL, EXHAUST AND EMISSION SYSTEMS ... 2.C
COOLING, HEATING AND AIR CONDITIONING ... 2.G
ELECTRICAL, WASHERS-WIPERS, INSTRUMENTS ... 2.J
CHASSIS, SUBFRAMES, BODYSHELL, FITTINGS ... 3.C
FASCIA, TRIM, SEATS AND FIXINGS ... 3.H
ACCESSORIES AND PAINT ... 4.C
NUMERICAL INDEX ... 4.D
GROUP INDEX
BADGES - EMBLEMS - CONVERTIBLE ... G05
BADGES-EMBLEMS-CABRIOLET ... G04
BADGES-EMBLEMS-COUPE ... G03
BODY SHELL - CONVERTIBLE ... D03
BODY SHELL-CABRIOLET ... C10
BODY SHELL-COUPE ... C02
REAR WINGS (TONNEAU) ... C08
BONNET LOCK MECHANISM ... D15
BONNET SAFETY CATCH AND HINGES ... D14
BONNET-WINGS-DOORS-BOOT LID ... D12
BOOT HINGES-SEAL-LAMP HOUSING ... E14
BRIGHT FINISHERS-NOT CONVERTIBLE ... F14
COACHLINES(TAPE) V12 MODELS ... F17
COACHLINES(TAPE)-3.6 MODELS ... F16
CRUCIFORM AND HEATSHIELD-CABRIOLET ... F02
DOOR GLASS-SEALS - NOT CONVERTIBLE ... F06
DOOR HANDLES-LOCKS-SOLENOIDS ... E07
DOOR HINGES ... D18
DOOR SEALS AND TREAD PLATES ... E02
FRONT BUMPER ... G07
FRONT DOOR GLASS-SEALS-CONVERTIBLE ... F07
FRONT WINDOW REGULATORS - CONVERTIBLE ... E05
FUEL FILLER LID AND LOCK-COUPE ... C09
GEARBOX COVER PLATES ... E13
GEARBOX TUNNEL COVER-AUTO G/BOX MODELS ... E11
GEARBOX TUNNEL COVER ... E10
HEADLAMP SURROUNDS ... F12
HEATED BACKLIGHT-SEAL-COUPE ... F11
MIRRORS ... G02
REAR BUMPER ... G09
REAR QTR GLASS AND SEALS-CONVERTIBLE ... F10
REAR QTR WINDOW REGULATORS - CONVERTIBLE ... E06
REAR QUARTER GLASS-SEALS-CABRIOLET ... F09
BADGE RH-B POST ... F09
REAR QUARTER GLASS-SEALS-COUPE ... F08
SPOILER-UNDERTRAY ... G06
STONE GUARDS ... D17
TRUNK REINFORCEMENT - CONVERTIBLE ... F03
WEATHERSTRIPS AND TREADPLATES ... E03
WINDOW REGULATORS-NOT CONVERTIBLE ... E04
WINDSCREEN - SEALS - CONVERTIBLE ... F05
WINDSCREEN-SEALS - NOT CONVERTIBLE ... F04

3D 15

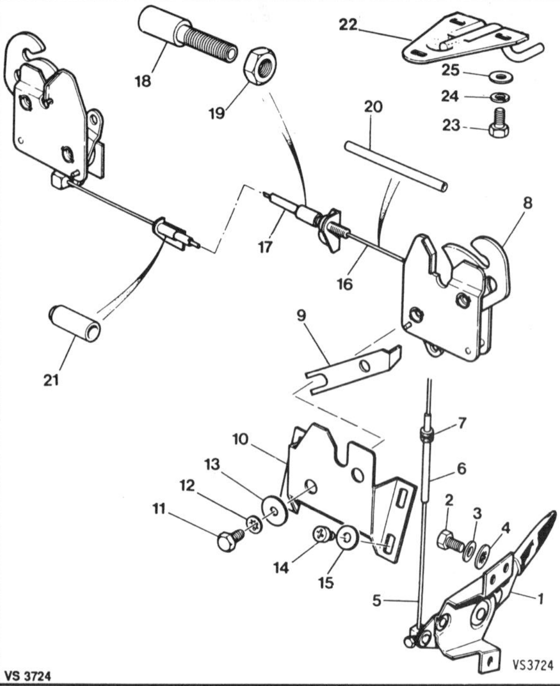

VS 3724

BONNET LOCK MECHANISM

Illus	Part Number	123456 Description	Quantity	Change Point	Remarks
1	BCC 2541	Lever-bonnet catch	1		LHD only From 89 MY
1	BDC 9064	Lever-bonnet catch	1	From 89 MY	RHD only
2	SH 604041 J	Screw-set	3		
3	JLM 296	Washer-plain	3		
4	C 724	Washer-shakeproof	3		
5	BD 47980	Cable-10.5"	1		
6	BD 47923	Sleeve	1		
7	BD 39893	Grommet	1		
	BD 43246	Catch-bonnet RH	1		
8	BD 43247	Catch-bonnet LH	1		
	BD 48322	Stop-catch RH	1		
9	BD 48323	Stop-catch LH	1		
	RTC 1538	Bracket-catch RH	1		
10	RTC 1539	Bracket-catch LH	1		
11	SH 604041 J	Screw-set	4		
12	C 724	Washer-shakeproof	4		
13	C 21930	Washer-plain	4		
14	UFS 825 3R	Screw-set	8		
15	JLM 296	Washer-plain	8		
16	BD 40755	Cable-inner-49.5"	1		
17	BD 46139	Cable-outer-26.125"	1		
	BD 46767	Sleeve-25.00"	1		
18	BD 48029	Adjuster-cable	1		
19	UCN 225 L	Nut-plain	1		
20	BD 47923	Sleeve-nylon	1		
21	BD 34982	Guide-outer cable	1		
22	RTC 1541	Striker	2		Three bolt fixing
	BJC 1899	Striker	2		Four bolt fixing
23	SH 604041 J	Screw-set	6		
24	JLM 302	Washer-spring	6		
25	JLM 296	Washer-plain	6		

L22JP1/9/88

3D 16

BONNET STAY GAS STRUTS

Illus	Part Number	123456 Description	Quantity	Change Point	Remarks
1	BD 48027	Stay-gas strut	2		
2	BH 605101 J	Bolt	4		
3	BD 48028	Spacer	4		
4	UFN 231 L	Nut-plain	4		
5	JLM 298	Washer-plain	8		

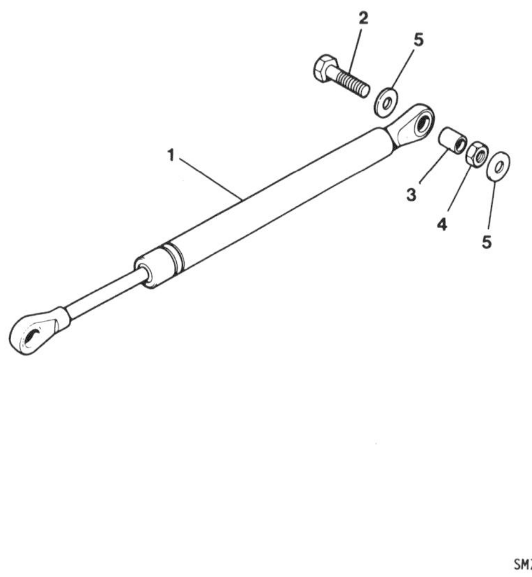

SM 7259

MASTER INDEX

ENGINE	1.C
FLYWHEEL AND CLUTCH	1.G
GEARBOX AND PROPSHAFT	1.H
AXLES, SUSPENSION, DRIVE SHAFTS, WHEELS	1.K
STEERING	1.M
BRAKES AND BRAKE CONTROLS	1.N
FUEL, EXHAUST AND EMISSION SYSTEMS	2.C
COOLING, HEATING AND AIR CONDITIONING	2.G
ELECTRICAL, WASHERS-WIPERS, INSTRUMENTS	2.J
CHASSIS, SUBFRAMES, BODYSHELL, FITTINGS	3.C
FASCIA, TRIM, SEATS AND FIXINGS	3.H
ACCESSORIES AND PAINT	4.C
NUMERICAL INDEX	4.D

GROUP INDEX

BADGES - EMBLEMS - CONVERTIBLE	G05
BADGES-EMBLEMS-CABRIOLET	G04
BADGES-EMBLEMS-COUPE	G03
BODY SHELL - CONVERTIBLE	D03
BODY SHELL-CABRIOLET	C10
BODY SHELL-COUPE	C02
REAR WINGS (TONNEAU)	C08
BONNET LOCK MECHANISM	D15
BONNET SAFETY CATCH AND HINGES	D14
BONNET-WINGS-DOORS-BOOT LID.	D12
BOOT HINGES-SEAL-LAMP HOUSING.	E14
BRIGHT FINISHERS-NOT CONVERTIBLE	F14
COACHLINES(TAPE) V12 MODELS.	F17
COACHLINES(TAPE)-3.6 MODELS.	F16
CRUCIFORM AND HEATSHIELD-CABRIOLET	F02
DOOR GLASS-SEALS - NOT CONVERTIBLE	F06
DOOR HANDLES-LOCKS-SOLENOIDS	E07
DOOR HINGES	D18
DOOR SEALS AND TREAD PLATES.	E02
FRONT BUMPER.	G07
FRONT DOOR GLASS-SEALS-CONVERTIBLE	F07
FRONT WINDOW REGULATORS - CONVERTIBLE	E05
FUEL FILLER LID AND LOCK-COUPE	C09
GEARBOX COVER PLATES.	E13
GEARBOX TUNNEL COVER-AUTO G/BOX MODELS	E11
GEARBOX TUNNEL COVER.	E10
HEADLAMP SURROUNDS.	F12
HEATED BACKLIGHT-SEAL-COUPE.	F11
MIRRORS.	G02
REAR BUMPER.	G09
REAR QTR GLASS AND SEALS-CONVERTIBLE	F10
REAR QTR WINDOW REGULATORS - CONVERTIBLE	E06
REAR QUARTER GLASS-SEALS-CABRIOLET	F09
BADGE RH-B POST	F09
REAR QUARTER GLASS-SEALS-COUPE.	F08
SPOILER-UNDERTRAY.	G06
STONE GUARDS.	D17
TRUNK REINFORCEMENT - CONVERTIBLE	F03
WEATHERSTRIPS AND TREADPLATES	E03
WINDOW REGULATORS-NOT CONVERTIBLE	E04
WINDSCREEN - SEALS - CONVERTIBLE	F05
WINDSCREEN-SEALS - NOT CONVERTIBLE	F04

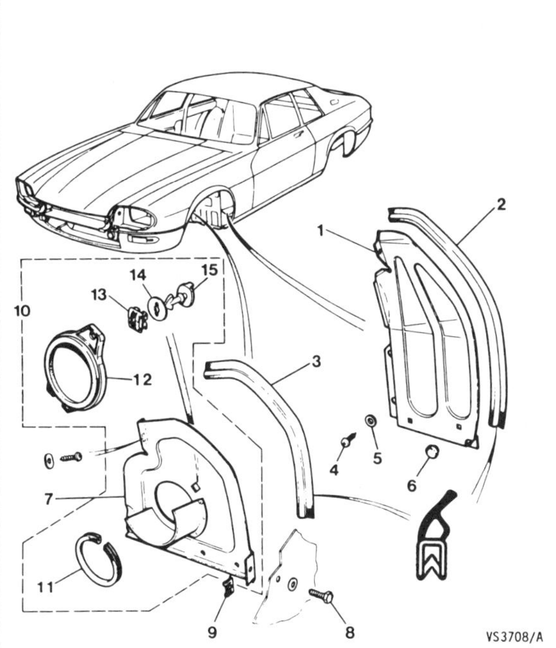

VS 3708A

STONE GUARDS.

Illus	Part Number	123456 Description	Quantity	Change Point	Remarks
	BCC 7073	Panel-stoneguard RH	1		
1	BD 46404	Panel-stoneguard LH	1		
	BBC 7684	Panel-stoneguard RH	1		Convertible
1	BBC 7685	Panel-stoneguard LH	1		
2	BCC 2614	Seal-rubber-stoneguard	2		
3	BBC 4444	Seal-rubber-baffle	2		
4	DAZ 810 8C	Screw-self-tap	18		
5	WC 702101 J	Washer-plain	16		
6	BD 30541 1	Plug-rubber	4		
	BD 46407	Baffle-RH	1		Not Bulb Access
7	BD 46408	Baffle-LH	1		
8	DBZ 11010 C	Screw-self-tap	2		
9	BD 8633 6	Nut-spire	2		
	BCC 4048	ASSY-HEADLAMP BULB ACCESS RH	1		Germany only
10	BCC 4049	ASSY-HEADLAMP BULB ACCESS LH	1		
11	BCC 2712 1	Strip-protection	2		
12	BBC 8710	Cover assembly-bulb access	2		
13	BBC 8679	Cam-fastener	2		
14	BD 46140 3	Washer-Fastener	2		
15	BCC 3594	Stud-Fastener	2		

MASTER INDEX

GROUP INDEX

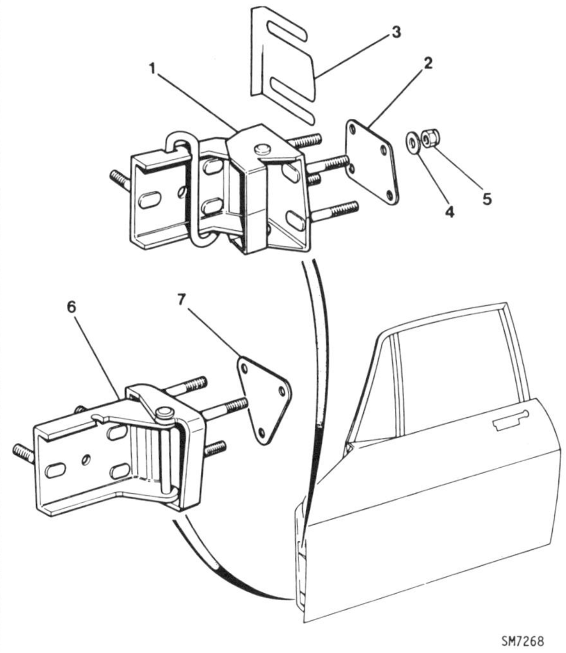

SM 7268

DOOR HINGES

Illus	Part Number	123456 Description	Quantity	Change Point	Remarks
	RTC 1667	Hinge-upper RH	1		
1	RTC 1668	Hinge-upper LH	1		
2	RTC 1671	Plate-upper	A/R		
	RTC 1676	Shim RH	2		
3	RTC 1682	Shim LH	2		
4	JLM 298	Washer-plain	14		
5	C 8737 2	Nut-self-lock	14		
	RTC 1669	Hinge-lower RH	1		
6	RTC 1670	Hinge-lower LH	1		
7	RTC 1672	Plate-lower	A/R		

MASTER INDEX
ENGINE 1.C
FLYWHEEL AND CLUTCH 1.G
GEARBOX AND PROPSHAFT 1.H
AXLES, SUSPENSION, DRIVE SHAFTS, WHEELS 1.K
STEERING 1.M
BRAKES AND BRAKE CONTROLS 1.N
FUEL, EXHAUST AND EMISSION SYSTEMS 2.C
COOLING, HEATING AND AIR CONDITIONING 2.G
ELECTRICAL, WASHERS-WIPERS, INSTRUMENTS 2.J
CHASSIS, SUBFRAMES, BODYSHELL, FITTINGS 3.C
FASCIA, TRIM, SEATS AND FIXINGS 3.H
ACCESSORIES AND PAINT 4.C
NUMERICAL INDEX 4.D
GROUP INDEX
BADGES - EMBLEMS - CONVERTIBLE G05
BADGES-EMBLEMS-CABRIOLET G04
BADGES-EMBLEMS-COUPE G03
BODY SHELL - CONVERTIBLE D03
BODY SHELL-CABRIOLET C10
BODY SHELL-COUPE C02
REAR WINGS (TONNEAU) C08
BONNET LOCK MECHANISM D15
BONNET SAFETY CATCH AND HINGES D14
BONNET-WINGS-DOORS-BOOT LID D12
BOOT HINGES-SEAL-LAMP HOUSING E14
BRIGHT FINISHERS-NOT CONVERTIBLE F14
COACHLINES(TAPE) V12 MODELS F17
COACHLINES(TAPE)-3.6 MODELS F16
CRUCIFORM AND HEATSHIELD-CABRIOLET F02
DOOR GLASS-SEALS - NOT CONVERTIBLE F06
DOOR HANDLES-LOCKS-SOLENOIDS E07
DOOR HINGES D18
DOOR SEALS AND TREAD PLATES E02
FRONT BUMPER G07
FRONT DOOR GLASS-SEALS-CONVERTIBLE F07
FRONT WINDOW REGULATORS - CONVERTIBLE E05
FUEL FILLER LID AND LOCK-COUPE C09
GEARBOX COVER PLATES E13
GEARBOX TUNNEL COVER-AUTO G/BOX MODELS E11
GEARBOX TUNNEL COVER E10
HEADLAMP SURROUNDS F12
HEATED BACKLIGHT-SEAL-COUPE F11
MIRRORS G02
REAR BUMPER G09
REAR QTR GLASS AND SEALS-CONVERTIBLE F10
REAR QTR WINDOW REGULATORS - CONVERTIBLE E06
REAR QUARTER GLASS-SEALS-CABRIOLET F09
BADGE RH-B POST F09
REAR QUARTER GLASS-SEALS-COUPE F08
SPOILER-UNDERTRAY G06
STONE GUARDS D17
TRUNK REINFORCEMENT - CONVERTIBLE F03
WEATHERSTRIPS AND TREADPLATES E03
WINDOW REGULATORS-NOT CONVERTIBLE E04
WINDSCREEN - SEALS - CONVERTIBLE F05
WINDSCREEN-SEALS - NOT CONVERTIBLE F04

JAGUAR XJS RANGE (JAN 1987 ON) | E02 fiche 3 | CHASSIS, SUBFRAMES, BODYSHELL, FITTINGS

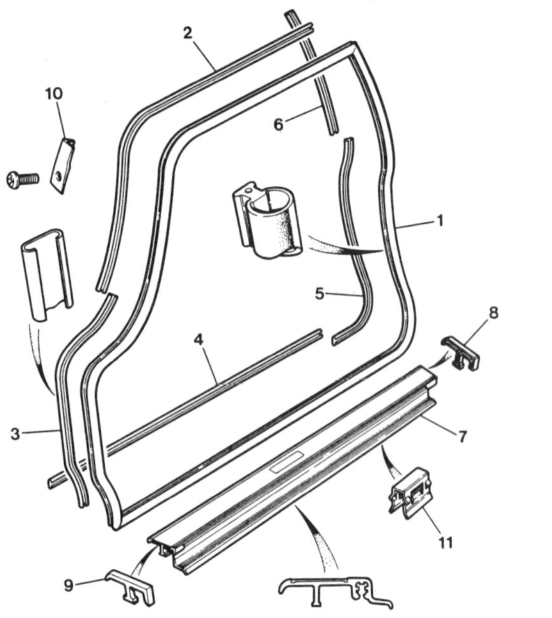

VS3838/A

DOOR SEALS AND TREAD PLATES.
NOT CONVERTIBLE

Illus	Part Number	123456	Description	Quantity	Change Point	Remarks
1	BAC 6019		Seal-door	2	Up to (V)153036	
	BCC 6690		Seal-door RH	1	From (V)153037	
1	BCC 6691		Seal-door LH	1		
	RTC 1659		Retainer-top-RH	1		
2	RTC 1660		Retainer-top-LH	1		
	RTC 1662		Retainer-front-RH	1		
3	RTC 1663		Retainer-front-LH	1		
4	RTC 1666		Retainer-lower	2		
	RTC 1664		Retainer-rear lower-RH	1		
5	RTC 1665		Retainer-rear lower-LH	1		
6	RTC 1661		Retainer-rear upper	2		
	BCC 4478		Tread plate-RH	1		Use BDC 4642
	BDC 4642		Tread plate-RH	1		
7	BCC 4479		Tread plate-LH	1		Use BDC 4643
7	BDC 4643		Tread plate-LH	1		
8	BCC 9742		Finisher-end	2		RH front LH rear
9	BCC 9743		Finisher-end	2		LH front RH rear
	BBC 5534		Finisher-RH	2		
10	BBC 5535		Finisher-LH	1		
11	BBC 9853		Clip-tread plate	10		

L30JP16/8/88

JAGUAR XJS RANGE (JAN 1987 ON) | E03 fiche 3 | CHASSIS, SUBFRAMES, BODYSHELL, FITTINGS

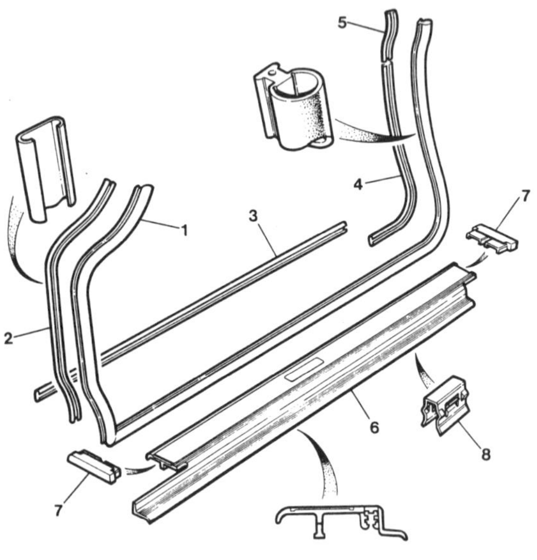

VS4836

WEATHERSTRIPS AND TREADPLATES
CONVERTIBLE

Illus	Part Number	123456	Description	Quantity	Change Point	Remarks
	BDC 8502		Weatherstrip RH	1		
1	BDC 8503		Weatherstrip LH	1		
	BJC 1426		Retainer-front RH	1		
2	BJC 1427		Retainer-front LH	1		
3	RTC 1666		Retainer-lower	2		
	BJC 1424		Retainer-rear lower RH	1		
4	BJC 1425		Retainer-rear lower LH	1		
	BJC 1082		Retainer-rear upper RH	1		
5	BJC 1083		Retainer-rear upper LH	1		
	BCC 6070		Treadplate RH	1		
6	BCC 6071		Treadplate LH	1		
	BCC 6886		Finisher RH	2		
7	BCC 6887		Finisher LH	2		
8	BCC 7408		Clip-tread plate	8		

MASTER INDEX

ENGINE 1.C
FLYWHEEL AND CLUTCH 1.G
GEARBOX AND PROPSHAFT 1.H
AXLES, SUSPENSION, DRIVE SHAFTS, WHEELS 1.K
STEERING 1.M
BRAKES AND BRAKE CONTROLS 1.N
FUEL, EXHAUST AND EMISSION SYSTEMS 2.C
COOLING, HEATING AND AIR CONDITIONING 2.G
ELECTRICAL, WASHERS-WIPERS, INSTRUMENTS 2.J
CHASSIS, SUBFRAMES, BODYSHELL, FITTINGS 3.C
FASCIA, TRIM, SEATS AND FIXINGS 3.H
ACCESSORIES AND PAINT 4.C
NUMERICAL INDEX 4.D

GROUP INDEX

BADGES - EMBLEMS - CONVERTIBLE G05
BADGES-EMBLEMS-CABRIOLET G04
BADGES-EMBLEMS-COUPE G03
BODY SHELL - CONVERTIBLE D03
BODY SHELL-CABRIOLET C10
BODY SHELL-COUPE C02
REAR WINGS (TONNEAU) C08
BONNET LOCK MECHANISM D15
BONNET SAFETY CATCH AND HINGES D14
BONNET-WINGS-DOORS-BOOT LID D12
BOOT HINGES-SEAL-LAMP HOUSING E14
BRIGHT FINISHERS-NOT CONVERTIBLE F14
COACHLINES(TAPE) V12 MODELS F17
COACHLINES(TAPE)-3.6 MODELS F16
CRUCIFORM AND HEATSHIELD-CABRIOLET F02
DOOR GLASS-SEALS - NOT CONVERTIBLE F06
DOOR HANDLES-LOCKS-SOLENOIDS E07
DOOR HINGES D18
DOOR SEALS AND TREAD PLATES E02
FRONT BUMPER G07
FRONT DOOR GLASS-SEALS-CONVERTIBLE F07
FRONT WINDOW REGULATORS - CONVERTIBLE E05
FUEL FILLER LID AND LOCK-COUPE C09
GEARBOX COVER PLATES E13
GEARBOX TUNNEL COVER-AUTO G/BOX MODELS E11
GEARBOX TUNNEL COVER E10
HEADLAMP SURROUNDS F12
HEATED BACKLIGHT-SEAL-COUPE F11
MIRRORS G02
REAR BUMPER G09
REAR QTR GLASS AND SEALS-CONVERTIBLE F10
REAR QTR WINDOW REGULATORS - CONVERTIBLE E06
REAR QUARTER GLASS-SEALS-CABRIOLET F09
BADGE RH-B POST F09
REAR QUARTER GLASS-SEALS-COUPE F08
SPOILER-UNDERTRAY G06
STONE GUARDS D17
TRUNK REINFORCEMENT - CONVERTIBLE F03
WEATHERSTRIPS AND TREADPLATES E03
WINDOW REGULATORS-NOT CONVERTIBLE E04
WINDSCREEN - SEALS - CONVERTIBLE F05
WINDSCREEN-SEALS - NOT CONVERTIBLE F04

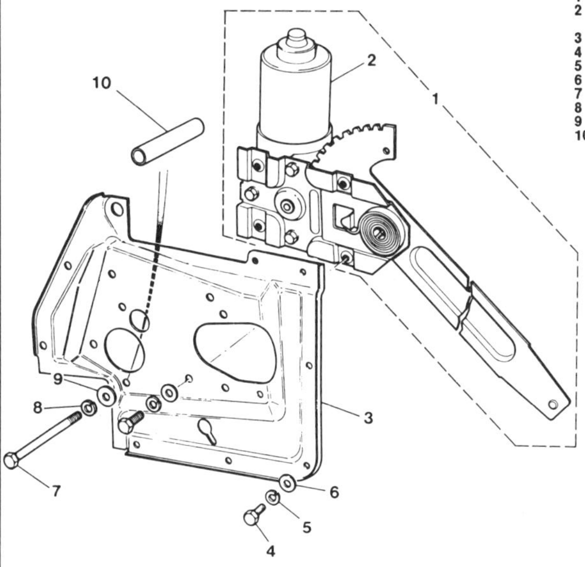

SM7270

WINDOW REGULATORS-NOT CONVERTIBLE

Illus	Part Number	123456	Description	Quantity	Change Point	Remarks
	BCC 5774		Regulator RH	1		
	JLM 974		Motor RH	1		
1	BCC 5775		Regulator LH	1		
2	JLM 975		Motor LH	1		
	BD 46405		Plate-mounting RH	1		
3	BD 46406		Plate-mounting LH	1		
4	SH 604041 J		Screw-set	22		
5	JLM 302		Washer-spring	22		
6	JLM 296		Washer-plain	22		
7	BH 604261 J		Bolt	2		
8	JLM 302		Washer-spring	2		
9	BD 541 9		Washer-plain	2		
10	BD 48286		Tube-spacing	2		

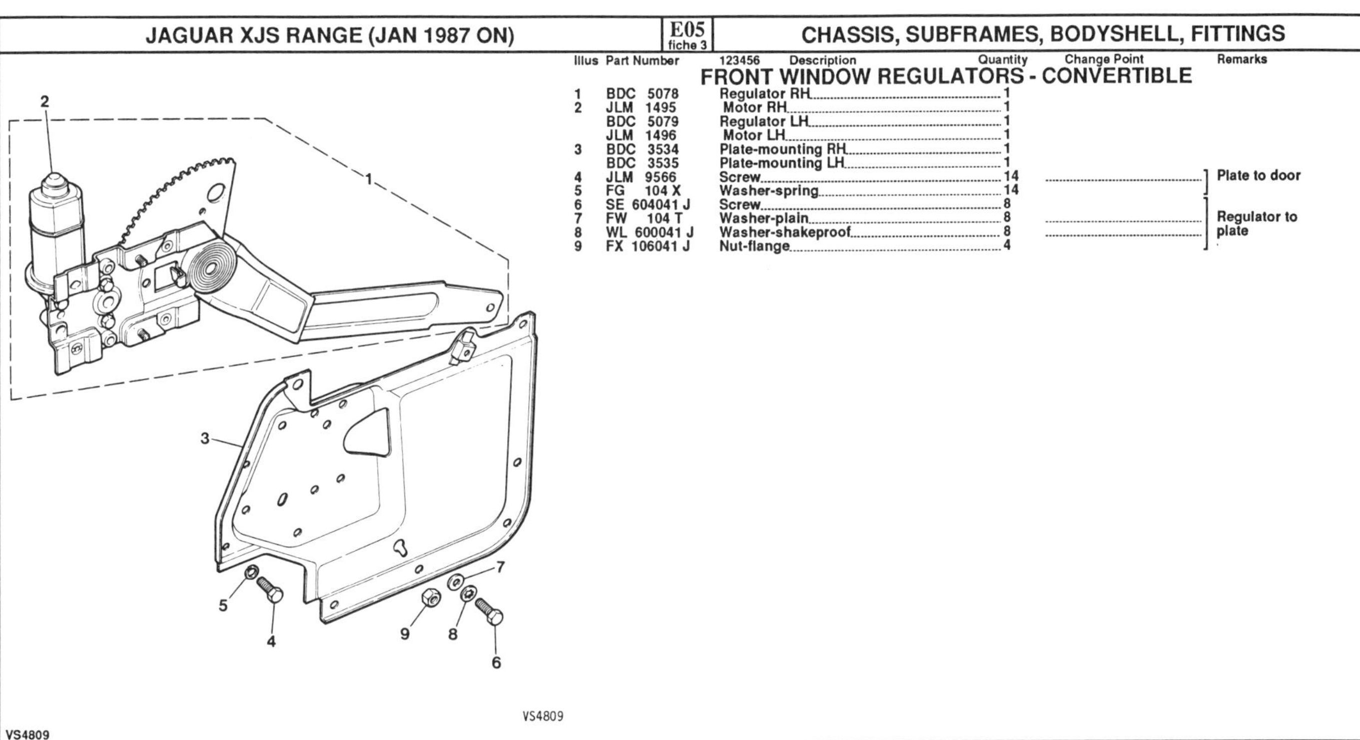

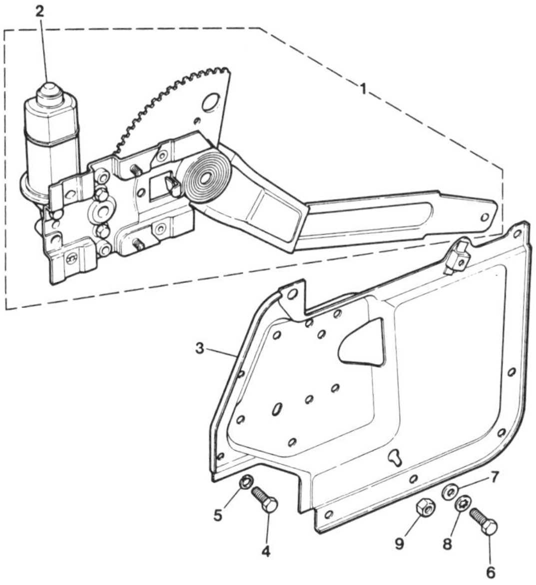

VS4809

FRONT WINDOW REGULATORS - CONVERTIBLE

Illus	Part Number	123456	Description	Quantity	Change Point	Remarks
1	BDC 5078		Regulator RH	1		
2	JLM 1495		Motor RH	1		
	BDC 5079		Regulator LH	1		
	JLM 1496		Motor LH	1		
3	BDC 3534		Plate-mounting RH	1		
	BDC 3535		Plate-mounting LH	1		
4	JLM 9566		Screw	14		Plate to door
5	FG 104 X		Washer-spring	14		
6	SE 604041 J		Screw	8		Regulator to plate
7	FW 104 T		Washer-plain	8		
8	WL 600041 J		Washer-shakeproof	8		
9	FX 106041 J		Nut-flange	4		

MASTER INDEX

ENGINE 1.C
FLYWHEEL AND CLUTCH 1.G
GEARBOX AND PROPSHAFT 1.H
AXLES, SUSPENSION, DRIVE SHAFTS, WHEELS 1.K
STEERING 1.M
BRAKES AND BRAKE CONTROLS 1.N
FUEL, EXHAUST AND EMISSION SYSTEMS 2.C
COOLING, HEATING AND AIR CONDITIONING 2.G
ELECTRICAL, WASHERS-WIPERS, INSTRUMENTS 2.J
CHASSIS, SUBFRAMES, BODYSHELL, FITTINGS 3.C
FASCIA, TRIM, SEATS AND FIXINGS 3.H
ACCESSORIES AND PAINT 4.C
NUMERICAL INDEX 4.D

GROUP INDEX

BADGES - EMBLEMS - CONVERTIBLE G05
BADGES-EMBLEMS-CABRIOLET G04
BADGES-EMBLEMS-COUPE G03
BODY SHELL - CONVERTIBLE D03
BODY SHELL-CABRIOLET C10
BODY SHELL-COUPE C02
REAR WINGS (TONNEAU) C08
BONNET LOCK MECHANISM D15
BONNET SAFETY CATCH AND HINGES D14
BONNET-WINGS-DOORS-BOOT LID D12
BOOT HINGES-SEAL-LAMP HOUSING E14
BRIGHT FINISHERS-NOT CONVERTIBLE F14
COACHLINES(TAPE) V12 MODELS F17
COACHLINES(TAPE)-3.6 MODELS F16
CRUCIFORM AND HEATSHIELD-CABRIOLET F02
DOOR GLASS-SEALS - NOT CONVERTIBLE F06
DOOR HANDLES-LOCKS-SOLENOIDS E07
DOOR HINGES D18
DOOR SEALS AND TREAD PLATES E02
FRONT BUMPER G07
FRONT DOOR GLASS-SEALS-CONVERTIBLE F07
FRONT WINDOW REGULATORS - CONVERTIBLE E05
FUEL FILLER LID AND LOCK-COUPE C09
GEARBOX COVER PLATES E13
GEARBOX TUNNEL COVER-AUTO G/BOX MODELS E11
GEARBOX TUNNEL COVER E10
HEADLAMP SURROUNDS F12
HEATED BACKLIGHT-SEAL-COUPE F11
MIRRORS G02
REAR BUMPER G09
REAR QTR GLASS AND SEALS-CONVERTIBLE F10
REAR QTR WINDOW REGULATORS - CONVERTIBLE E06
REAR QUARTER GLASS-SEALS-CABRIOLET F09
BADGE RH-B POST F09
REAR QUARTER GLASS-SEALS-COUPE F08
SPOILER-UNDERTRAY G06
STONE GUARDS D17
TRUNK REINFORCEMENT - CONVERTIBLE F03
WEATHERSTRIPS AND TREADPLATES E03
WINDOW REGULATORS-NOT CONVERTIBLE E04
WINDSCREEN - SEALS - CONVERTIBLE F05
WINDSCREEN-SEALS - NOT CONVERTIBLE F04

3E 06

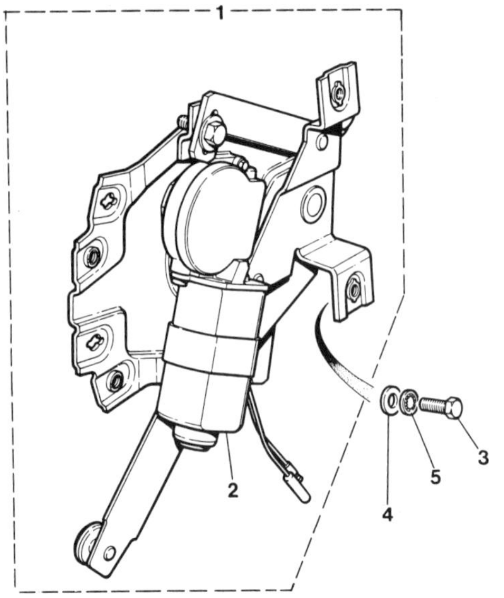

VS4794

VS4794

REAR QTR WINDOW REGULATORS - CONVERTIBLE

Illus	Part Number	123456	Description	Quantity	Change Point	Remarks
1	BDC 4096		Regulator RH	1		
2	JLM 1497		Motor RH	1		
	BDC 4097		Regulator LH	1		
	JLM 1498		Motor LH	1		
3	JLM 9566		Bolt	8		
4	WA 106041 J		Washer-plain	8		
5	C 724		Washer-shakeproof	8		

3E 07

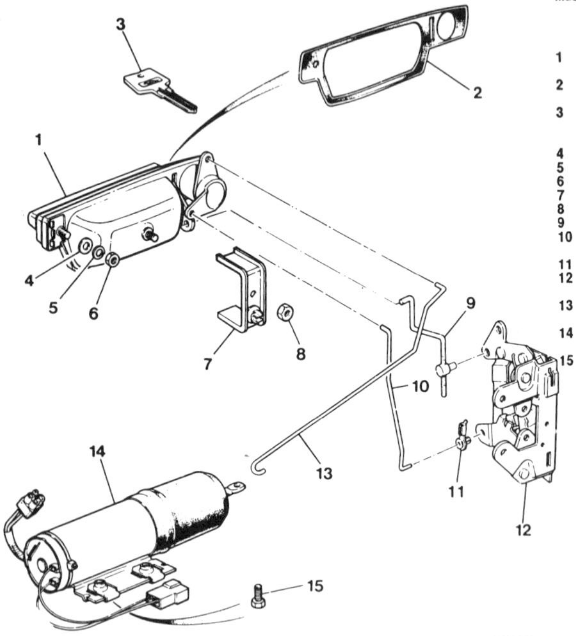

VS3221

VS 3221

DOOR HANDLES-LOCKS-SOLENOIDS

LUCAS SOLENOID

Illus	Part Number	123456	Description	Quantity	Change Point	Remarks
1	BBC 5932		Handle-door-RH	1		
	BBC 5933		Handle-door-LH	1		
2	BD 45850		Gasket-RH	1		
	BD 45851		Gasket-LH	1		
3			Key-blank	A/R		See locksets-Miscellaneous section
4	WC 702101 J		Washer-plain	2		
5	WM 702101 J		Washer-spring	2		
6	UNF 119 L		Nut-plain	2		
7	BD 47899		Clamp	2		
8	UNF 119 L		Nut-plain	2		
9	BD 45663		Link-release	2		
10	BD 45664		Link-locking-RH	1		
	BD 45665		Link-locking-LH	1		
11	BD 44650		Clip-linkage	6		
12	BD 43266		Lock-inner-RH	1		
	BD 43267		Lock-inner-LH	1		
13	BAC 3596		Link-operating-RH	1		
	BAC 3597		Link-operating-LH	1		
14	DAC 3046		Solenoid-RH	1		
	DAC 3047		Solenoid-LH	1		
15	SH 504061 J		Screw-set	6		

MASTER INDEX

ENGINE 1.C
FLYWHEEL AND CLUTCH 1.G
GEARBOX AND PROPSHAFT 1.H
AXLES, SUSPENSION, DRIVE SHAFTS, WHEELS 1.K
STEERING 1.M
BRAKES AND BRAKE CONTROLS 1.N
FUEL, EXHAUST AND EMISSION SYSTEMS 2.C
COOLING, HEATING AND AIR CONDITIONING 2.G
ELECTRICAL, WASHERS-WIPERS, INSTRUMENTS 2.J
CHASSIS, SUBFRAMES, BODYSHELL, FITTINGS 3.C
FASCIA, TRIM, SEATS AND FIXINGS 3.H
ACCESSORIES AND PAINT 4.C
NUMERICAL INDEX 4.D

GROUP INDEX

BADGES - EMBLEMS - CONVERTIBLE G05
BADGES-EMBLEMS-CABRIOLET G04
BADGES-EMBLEMS-COUPE G03
BODY SHELL - CONVERTIBLE D03
BODY SHELL-CABRIOLET C10
BODY SHELL-COUPE C02
REAR WINGS (TONNEAU) C08
BONNET LOCK MECHANISM D15
BONNET SAFETY CATCH AND HINGES D14
BONNET-WINGS-DOORS-BOOT LID D12
BOOT HINGES-SEAL-LAMP HOUSING E14
BRIGHT FINISHERS-NOT CONVERTIBLE F14
COACHLINES(TAPE) V12 MODELS F17
COACHLINES(TAPE)-3.6 MODELS F16
CRUCIFORM AND HEATSHIELD-CABRIOLET F02
DOOR GLASS-SEALS - NOT CONVERTIBLE F06
DOOR HANDLES-LOCKS-SOLENOIDS E07
DOOR HINGES D18
DOOR SEALS AND TREAD PLATES E02
FRONT BUMPER G07
FRONT DOOR GLASS-SEALS-CONVERTIBLE F07
FRONT WINDOW REGULATORS - CONVERTIBLE E05
FUEL FILLER LID AND LOCK-COUPE C09
GEARBOX COVER PLATES E13
GEARBOX TUNNEL COVER-AUTO G/BOX MODELS E11
GEARBOX TUNNEL COVER E10
HEADLAMP SURROUNDS F12
HEATED BACKLIGHT-SEAL-COUPE F11
MIRRORS G02
REAR BUMPER G09
REAR QTR GLASS AND SEALS-CONVERTIBLE F10
REAR QTR WINDOW REGULATORS - CONVERTIBLE E06
REAR QUARTER GLASS-SEALS-CABRIOLET F09
BADGE RH-B POST F09
REAR QUARTER GLASS-SEALS-COUPE F08
SPOILER-UNDERTRAY G06
STONE GUARDS D17
TRUNK REINFORCEMENT - CONVERTIBLE F03
WEATHERSTRIPS AND TREADPLATES E03
WINDOW REGULATORS-NOT CONVERTIBLE E04
WINDSCREEN - SEALS - CONVERTIBLE F05
WINDSCREEN-SEALS - NOT CONVERTIBLE F04

DOOR HANDLES-LOCKS-ACTUATORS-(KIEKERT ACTUATOR)

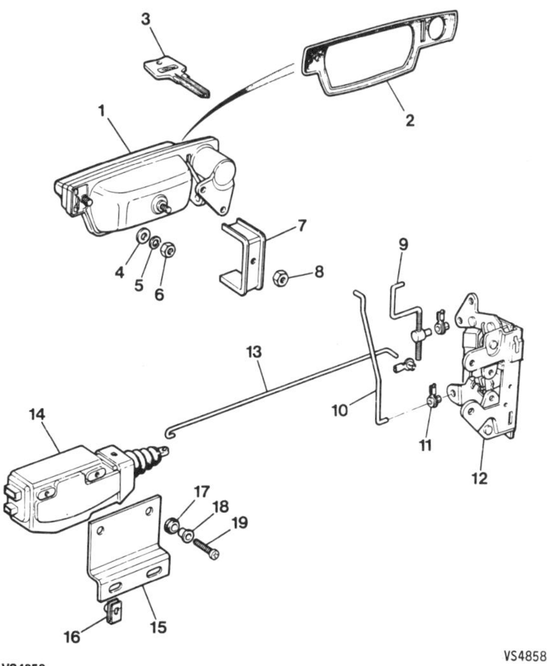

VS4858

Illus	Part Number	Description	Quantity	Change Point	Remarks
1	BDC 7600	Handle-door RH	1		
	BDC 7601	Handle-door LH	1		
2	BD 45850	Gasket-RH	1		
	BD 45851	Gasket-LH	1		
3		Key-blank	1		See miscellaneous section
4	WC 702101 J	Washer-plain	2		
5	WM 702101 J	Washer-spring	2		
6	UNF 119 L	Nut-plain	2		
7	BD 47899	Clamp	2		
8	UNF 119 L	Nut-plain	2		
9	BDC 4981	Link-release	2		
10	BDC 4982	Link-locking RH	1		
	BDC 4983	Link-locking LH	1		
11	BD 44650	Clip-linkage	6		
12	BD 43266	Lock-inner RH	1		
	BD 43267	Lock-inner LH	1		
13	BDC 4992	Link-actuator RH	1		
	BDC 4993	Link-actuator LH	1		
14	BCC 7685	Actuator	2		
15	BDC 4021	Mounting bracket-actuator	2		
16	BBC 8554	Lungnut	4		
17	DRC 1449 J	Grommet	4		
18	CAC 1823	Spacer	4		
19	BD 48986	Screw-plastite	4		

VS4858

DOOR LOCKS-REMOTE CONTROL

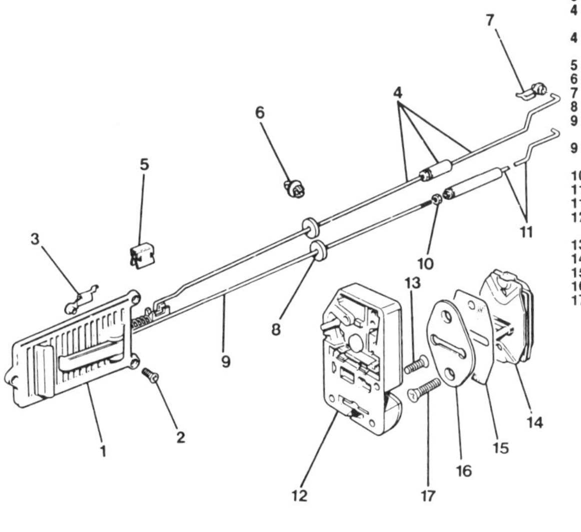

VS3310

Illus	Part Number	Description	Quantity	Change Point	Remarks
1	BCC 2392	Handle-remote control RH	1		
	BCC 2393	Handle remote control LH	1		
2	SF 604054 J	Screw-set	6		
3	BD 32396	Clip	2		
4	BD 47755	Link-release RH	1		Not Convertible
	BD 47756	Link-release LH	1		
4	BCC 8032	Link-release RH	1		Convertible only
	BCC 8033	Link-release LH	1		
5	BD 48049	Clip	2		
6	BD 34540	Clip	8		
7	BD 44650	Clip	8		
8	C 34059	Washer-rubber	4		
9	BD 47679	Link-locking RH	1		Not Convertible
	BD 47680	Link-locking LH	1		
9	BDC 4526	Link-locking RH	1		Convertible only
	BDC 4527	Link-locking LH	1		
10	JLM 9686	Nut-plain	2		
11	BD 47684	Device-overload	2		Not Convertible
11	BBC 6942	Device-overload	2		Convertible
12	BBC 7224	Lock-outer RH	1		
	BBC 7225	Lock-outer LH	1		
13	BBC 7221	Screw-set	8		
14	BD 48128	Striker	2		
15	BD 45974	Pad-friction	2		
16	BCC 3756	Cover	2		
17	BBC 7395	Screw-set	4		

VS 3310

MASTER INDEX

ENGINE 1.C
FLYWHEEL AND CLUTCH 1.G
GEARBOX AND PROPSHAFT 1.H
AXLES, SUSPENSION, DRIVE SHAFTS, WHEELS 1.K
STEERING 1.M
BRAKES AND BRAKE CONTROLS 1.N
FUEL, EXHAUST AND EMISSION SYSTEMS 2.C
COOLING, HEATING AND AIR CONDITIONING 2.G
ELECTRICAL, WASHERS-WIPERS, INSTRUMENTS 2.J
CHASSIS, SUBFRAMES, BODYSHELL, FITTINGS 3.C
FASCIA, TRIM, SEATS AND FIXINGS 3.H
ACCESSORIES AND PAINT 4.C
NUMERICAL INDEX 4.D

GROUP INDEX

BADGES - EMBLEMS - CONVERTIBLE G05
BADGES-EMBLEMS-CABRIOLET G04
BADGES-EMBLEMS-COUPE G03
BODY SHELL - CONVERTIBLE D03
BODY SHELL-CABRIOLET C10
BODY SHELL-COUPE C02
REAR WINGS (TONNEAU) C08
BONNET LOCK MECHANISM D15
BONNET SAFETY CATCH AND HINGES D14
BONNET-WINGS-DOORS-BOOT LID D12
BOOT HINGES-SEAL-LAMP HOUSING E14
BRIGHT FINISHERS-NOT CONVERTIBLE F14
COACHLINES(TAPE) V12 MODELS F17
COACHLINES(TAPE)-3.6 MODELS F16
CRUCIFORM AND HEATSHIELD-CABRIOLET F02
DOOR GLASS-SEALS - NOT CONVERTIBLE F06
DOOR HANDLES-LOCKS-SOLENOIDS E07
DOOR HINGES D18
DOOR SEALS AND TREAD PLATES E02
FRONT BUMPER G07
FRONT DOOR GLASS-SEALS-CONVERTIBLE F07
FRONT WINDOW REGULATORS - CONVERTIBLE E05
FUEL FILLER LID AND LOCK-COUPE C09
GEARBOX COVER PLATES E13
GEARBOX TUNNEL COVER-AUTO G/BOX MODELS E11
GEARBOX TUNNEL COVER E10
HEADLAMP SURROUNDS F12
HEATED BACKLIGHT-SEAL-COUPE F11
MIRRORS G02
REAR BUMPER G09
REAR QTR GLASS AND SEALS-CONVERTIBLE F10
REAR QTR WINDOW REGULATORS - CONVERTIBLE E06
REAR QUARTER GLASS-SEALS-CABRIOLET F09
BADGE RH-B POST F09
REAR QUARTER GLASS-SEALS-COUPE F08
SPOILER-UNDERTRAY G06
STONE GUARDS D17
TRUNK REINFORCEMENT - CONVERTIBLE F03
WEATHERSTRIPS AND TREADPLATES E03
WINDOW REGULATORS-NOT CONVERTIBLE E04
WINDSCREEN - SEALS - CONVERTIBLE F05
WINDSCREEN-SEALS - NOT CONVERTIBLE F04

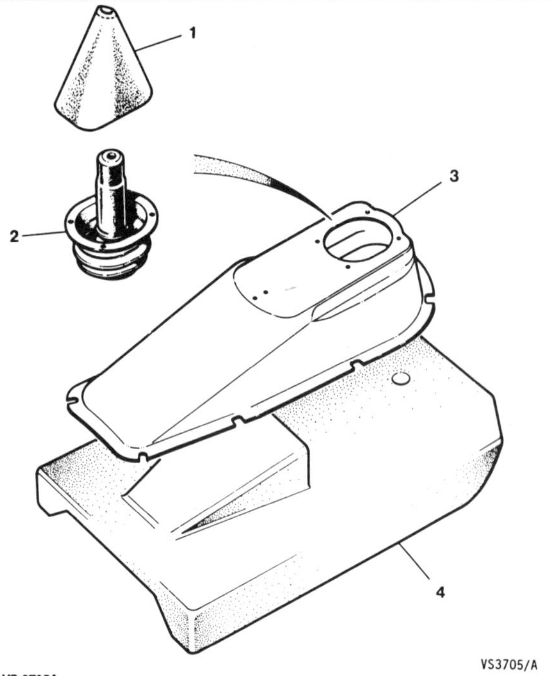

VS 3705A

GEARBOX TUNNEL COVER.

3.6 MANUAL G/BOX MODELS

Illus	Part Number	123456 Description	Quantity	Change Point	Remarks
1	BCC 2287	Gaiter-gear lever	1		
2	EAC 6620	Grommet-gear lever	1		
3	BBC 1305	Cover-gearbox tunnel	1		
4	BD 46958	Pad-insulation	1		

L38JP24/6/88

3E 10

MASTER INDEX

GROUP INDEX

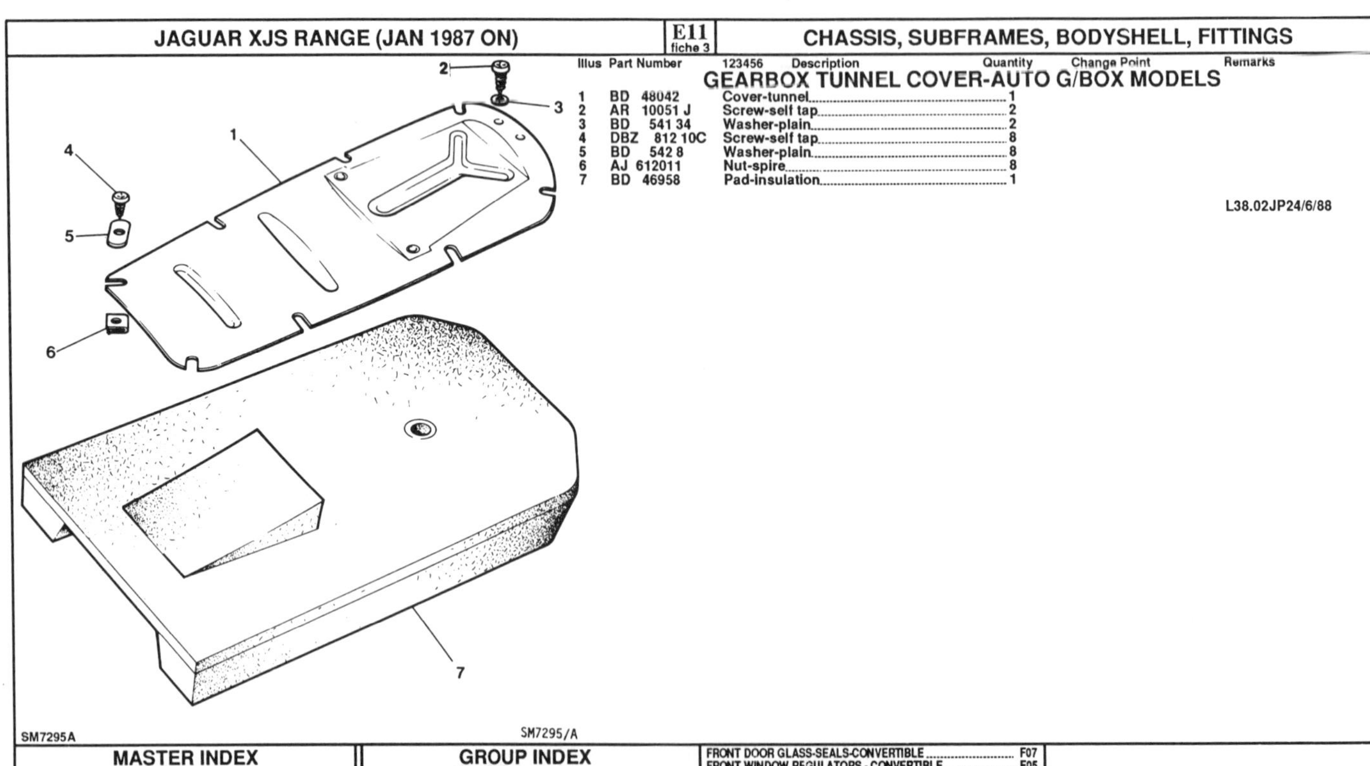

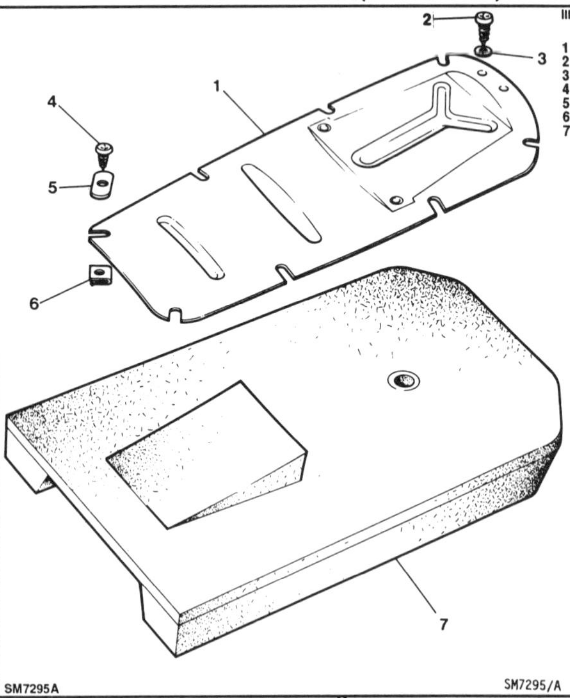

SM7295A

GEARBOX TUNNEL COVER-AUTO G/BOX MODELS

Illus	Part Number	123456 Description	Quantity	Change Point	Remarks
1	BD 48042	Cover-tunnel	1		
2	AR 10051 J	Screw-self tap	2		
3	BD 541 34	Washer-plain	2		
4	DBZ 812 10C	Screw-self tap	8		
5	BD 542 8	Washer-plain	8		
6	AJ 612011	Nut-spire	8		
7	BD 46958	Pad-insulation	1		

L38.02JP24/6/88

3E 11

MASTER INDEX
ENGINE 1.C
FLYWHEEL AND CLUTCH 1.G
GEARBOX AND PROPSHAFT 1.H
AXLES, SUSPENSION, DRIVE SHAFTS, WHEELS 1.K
STEERING 1.M
BRAKES AND BRAKE CONTROLS 1.N
FUEL, EXHAUST AND EMISSION SYSTEMS 2.C
COOLING, HEATING AND AIR CONDITIONING 2.G
ELECTRICAL, WASHERS-WIPERS, INSTRUMENTS 2.J
CHASSIS, SUBFRAMES, BODYSHELL, FITTINGS 3.C
FASCIA, TRIM, SEATS AND FIXINGS 3.H
ACCESSORIES AND PAINT 4.C
NUMERICAL INDEX 4.D
GROUP INDEX
BADGES - EMBLEMS - CONVERTIBLE G05
BADGES-EMBLEMS-CABRIOLET G04
BADGES-EMBLEMS-COUPE. G03
BODY SHELL - CONVERTIBLE D03
BODY SHELL-CABRIOLET C10
BODY SHELL-COUPE. C02
REAR WINGS (TONNEAU) C08
BONNET LOCK MECHANISM D15
BONNET SAFETY CATCH AND HINGES D14
BONNET-WINGS-DOORS-BOOT LID. D12
BOOT HINGES-SEAL-LAMP HOUSING. E14
BRIGHT FINISHERS-NOT CONVERTIBLE F14
COACHLINES(TAPE) V12 MODELS. F17
COACHLINES(TAPE)-3.6 MODELS. F16
CRUCIFORM AND HEATSHIELD-CABRIOLET F02
DOOR GLASS-SEALS - NOT CONVERTIBLE F06
DOOR HANDLES-LOCKS-SOLENOIDS E07
DOOR HINGES D18
DOOR SEALS AND TREAD PLATES. E02
FRONT BUMPER. G07
FRONT DOOR GLASS-SEALS-CONVERTIBLE F07
FRONT WINDOW REGULATORS - CONVERTIBLE E05
FUEL FILLER LID AND LOCK-COUPE C09
GEARBOX COVER PLATES. E13
GEARBOX TUNNEL COVER-AUTO G/BOX MODELS E11
GEARBOX TUNNEL COVER. E10
HEADLAMP SURROUNDS. F12
HEATED BACKLIGHT-SEAL-COUPE. F11
MIRRORS. G02
REAR BUMPER. G09
REAR QTR GLASS AND SEALS-CONVERTIBLE F10
REAR QTR WINDOW REGULATORS - CONVERTIBLE E06
REAR QUARTER GLASS-SEALS-CABRIOLET F09
BADGE RH-B POST F09
REAR QUARTER GLASS-SEALS-COUPE. F08
SPOILER-UNDERTRAY. G06
STONE GUARDS. D17
TRUNK REINFORCEMENT - CONVERTIBLE F03
WEATHERSTRIPS AND TREADPLATES E03
WINDOW REGULATORS-NOT CONVERTIBLE E04
WINDSCREEN - SEALS - CONVERTIBLE F05
WINDSCREEN-SEALS - NOT CONVERTIBLE F04

TUNNEL CLOSING PLATE - CONVERTIBLE

Illus	Part Number	123456	Description	Quantity	Change Point	Remarks
1	BDC 3174		Closing plate - tunnel	1		

L38.04JP1/9/88

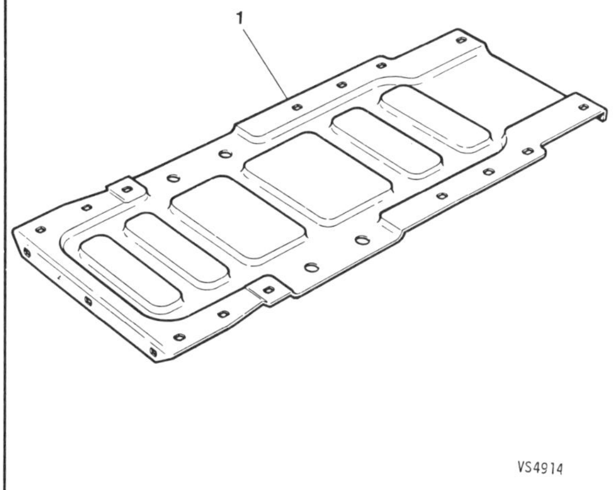

VS4914

MASTER INDEX

GROUP INDEX

GEARBOX COVER PLATES.

Illus	Part Number	123456	Description	Quantity	Change Point	Remarks
1	BD 34709 1		Plate	1		
2	BD 42123		Gasket	1		
3	SH 604041 J		Screw-set	2		
4	C 724		Washer-shakeproof	2		
5	BD 45619		Plate	1		
6	BD 45620		Gasket	1		
7	SH 604041 J		Screw-set	2		
	SH 604071 J		Screw-set	1		
8	C 724		Washer-shakeproof	3		
9	CAC 6272		Plate	1		3.6 litre
10	RU 608183 J		Pop-rivet	4		

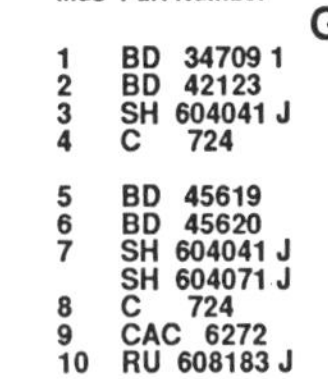

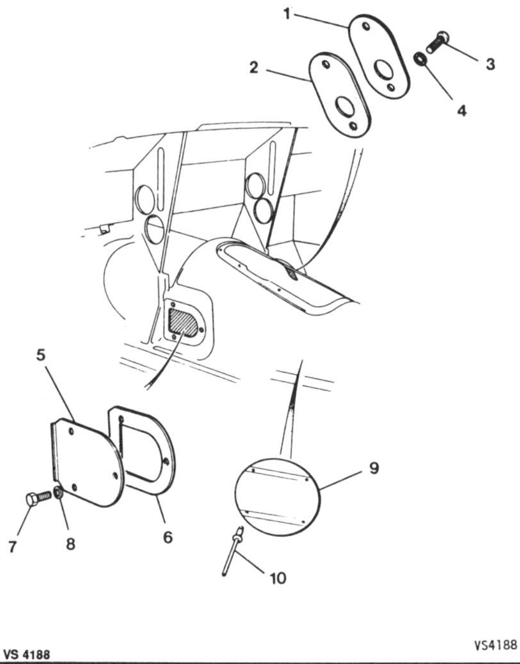

VS 4188

MASTER INDEX
ENGINE ... 1.C
FLYWHEEL AND CLUTCH ... 1.G
GEARBOX AND PROPSHAFT ... 1.H
AXLES, SUSPENSION, DRIVE SHAFTS, WHEELS ... 1.K
STEERING ... 1.M
BRAKES AND BRAKE CONTROLS ... 1.N
FUEL, EXHAUST AND EMISSION SYSTEMS ... 2.C
COOLING, HEATING AND AIR CONDITIONING ... 2.G
ELECTRICAL, WASHERS-WIPERS, INSTRUMENTS ... 2.J
CHASSIS, SUBFRAMES, BODYSHELL, FITTINGS ... 3.C
FASCIA, TRIM, SEATS AND FIXINGS ... 3.H
ACCESSORIES AND PAINT ... 4.C
NUMERICAL INDEX ... 4.D
GROUP INDEX
BADGES - EMBLEMS - CONVERTIBLE ... G05
BADGES-EMBLEMS-CABRIOLET ... G04
BADGES-EMBLEMS-COUPE ... G03
BODY SHELL - CONVERTIBLE ... D03
BODY SHELL-CABRIOLET ... C10
BODY SHELL-COUPE ... C02
REAR WINGS (TONNEAU) ... C08
BONNET LOCK MECHANISM ... D15
BONNET SAFETY CATCH AND HINGES ... D14
BONNET-WINGS-DOORS-BOOT LID ... D12
BOOT HINGES-SEAL-LAMP HOUSING ... E14
BRIGHT FINISHERS-NOT CONVERTIBLE ... F14
COACHLINES(TAPE) V12 MODELS ... F17
COACHLINES(TAPE)-3.6 MODELS ... F16
CRUCIFORM AND HEATSHIELD-CABRIOLET ... F02
DOOR GLASS-SEALS - NOT CONVERTIBLE ... F06
DOOR HANDLES-LOCKS-SOLENOIDS ... E07
DOOR HINGES ... D18
DOOR SEALS AND TREAD PLATES ... E02
FRONT BUMPER ... G07
FRONT DOOR GLASS-SEALS-CONVERTIBLE ... F07
FRONT WINDOW REGULATORS - CONVERTIBLE ... E05
FUEL FILLER LID AND LOCK-COUPE ... C09
GEARBOX COVER PLATES ... E13
GEARBOX TUNNEL COVER-AUTO G/BOX MODELS ... E11
GEARBOX TUNNEL COVER ... E10
HEADLAMP SURROUNDS ... F12
HEATED BACKLIGHT-SEAL-COUPE ... F11
MIRRORS ... G02
REAR BUMPER ... G09
REAR QTR GLASS AND SEALS-CONVERTIBLE ... F10
REAR QTR WINDOW REGULATORS - CONVERTIBLE ... E06
REAR QUARTER GLASS-SEALS-CABRIOLET ... F09
BADGE RH-B POST ... F09
REAR QUARTER GLASS-SEALS-COUPE ... F08
SPOILER-UNDERTRAY ... G06
STONE GUARDS ... D17
TRUNK REINFORCEMENT - CONVERTIBLE ... F03
WEATHERSTRIPS AND TREADPLATES ... E03
WINDOW REGULATORS-NOT CONVERTIBLE ... E04
WINDSCREEN - SEALS - CONVERTIBLE ... F05
WINDSCREEN-SEALS - NOT CONVERTIBLE ... F04

3E 14

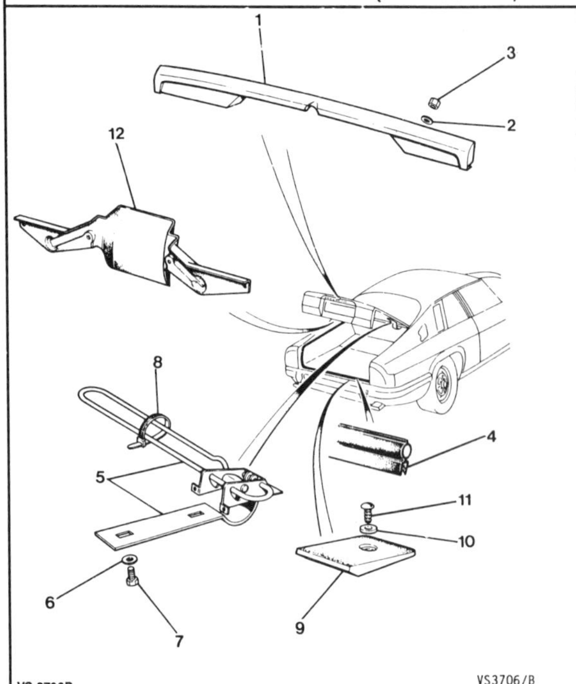

VS 3706B

VS3706/B

BOOT HINGES-SEAL-LAMP HOUSING.

NOT CONVERTIBLE

Illus	Part Number	123456 Description	Quantity	Change Point	Remarks
1	BD 44315	Housing-number plate lamp	1		
2	WE 702101 J	Washer-plain	10		
3	NV 105041 J	Nut-self lock	10		
4	BBC 4441	Seal boot	1		
5	RTC 1673	Hinge-RH	1		
	RTC 1674	Hinge-LH	1		
6	JLM 298	Washer-plain	4		
7	UFS 131 6V	Screw-set	4		
8	C 25969	Tie-cable	2		
9	BAC 5427	Pad-rubber	2		
10	BD 37943	Washer-plain	2		
11	AB 606031 J	Screw-self-tap	2		
12	DAC 3673	Shield-harness	1		Upto (V)148781
12	DAC 4834	Shield-harness	1		From (V)148782

3E 15

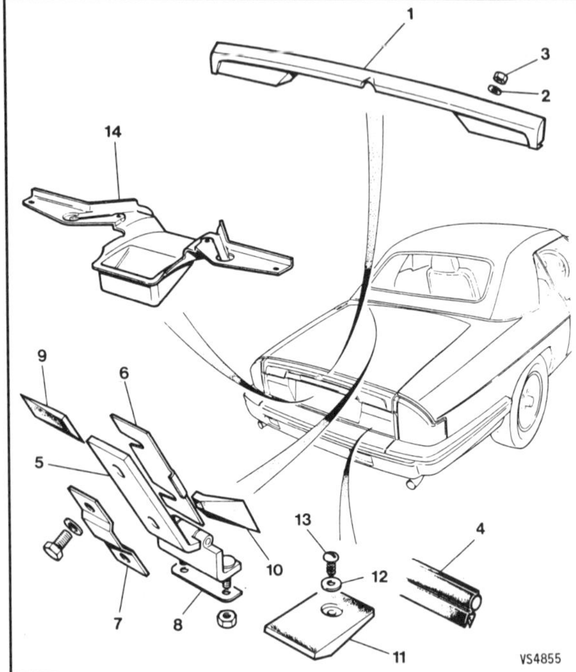

VS4855

VS4855

BOOT HINGES - SEAL - LAMP HOUSING - CONVERTIBLE

Illus	Part Number	123456 Description	Quantity	Change Point	Remarks
1	BD 44315	Housing-number plate lamp	1		
2	WE 702101 J	Washer-plain	10		
3	NV 105041 J	Nut-self lock	10		
4	BBC 4441	Seal-boot	1		
5	BBC 7654	Hinge-boot	2		
6	BCC 2841	Shim-boot hinge	2		
7	BJC 1182	Plate-hinge	2		
8	BBC 6991	Gasket-hinge	2		
	BDC 3573	Sealing ramp-outer RH	1		
9	BDC 3572	Sealing ramp-outer LH	1		
	BDC 3575	Sealing ramp-inner RH	1		
10	BDC 3574	Sealing ramp-inner LH	1		
11	BAC 5427	Pad-rubber	2		
12	BD 37943	Washer-plain	2		
13	AB 606031 J	Screw-self tap	2		
14	DAC 4834	Sheild-harness	1		

MASTER INDEX

ENGINE 1.C
FLYWHEEL AND CLUTCH 1.G
GEARBOX AND PROPSHAFT 1.H
AXLES, SUSPENSION, DRIVE SHAFTS, WHEELS 1.K
STEERING 1.M
BRAKES AND BRAKE CONTROLS 1.N
FUEL, EXHAUST AND EMISSION SYSTEMS 2.C
COOLING, HEATING AND AIR CONDITIONING 2.G
ELECTRICAL, WASHERS-WIPERS, INSTRUMENTS 2.J
CHASSIS, SUBFRAMES, BODYSHELL, FITTINGS 3.C
FASCIA, TRIM, SEATS AND FIXINGS 3.H
ACCESSORIES AND PAINT 4.C
NUMERICAL INDEX 4.D

GROUP INDEX

BADGES - EMBLEMS - CONVERTIBLE G05
BADGES-EMBLEMS-CABRIOLET G04
BADGES-EMBLEMS-COUPE G03
BODY SHELL - CONVERTIBLE D03
BODY SHELL-CABRIOLET C10
BODY SHELL-COUPE C02
REAR WINGS (TONNEAU) C08
BONNET LOCK MECHANISM D15
BONNET SAFETY CATCH AND HINGES D14
BONNET-WINGS-DOORS-BOOT LID D12
BOOT HINGES-SEAL-LAMP HOUSING E14
BRIGHT FINISHERS-NOT CONVERTIBLE F14
COACHLINES(TAPE) V12 MODELS F17
COACHLINES(TAPE)-3.6 MODELS F16
CRUCIFORM AND HEATSHIELD-CABRIOLET F02
DOOR GLASS-SEALS - NOT CONVERTIBLE F06
DOOR HANDLES-LOCKS-SOLENOIDS E07
DOOR HINGES D18
DOOR SEALS AND TREAD PLATES E02
FRONT BUMPER G07
FRONT DOOR GLASS-SEALS-CONVERTIBLE F07
FRONT WINDOW REGULATORS - CONVERTIBLE E05
FUEL FILLER LID AND LOCK-COUPE C09
GEARBOX COVER PLATES E13
GEARBOX TUNNEL COVER-AUTO G/BOX MODELS E11
GEARBOX TUNNEL COVER E10
HEADLAMP SURROUNDS F12
HEATED BACKLIGHT-SEAL-COUPE F11
MIRRORS G02
REAR BUMPER G09
REAR QTR GLASS AND SEALS-CONVERTIBLE F10
REAR QTR WINDOW REGULATORS - CONVERTIBLE E06
REAR QUARTER GLASS-SEALS-CABRIOLET F09
BADGE RH-B POST F09
REAR QUARTER GLASS-SEALS-COUPE F08
SPOILER-UNDERTRAY G06
STONE GUARDS D17
TRUNK REINFORCEMENT - CONVERTIBLE F03
WEATHERSTRIPS AND TREADPLATES E03
WINDOW REGULATORS-NOT CONVERTIBLE E04
WINDSCREEN - SEALS - CONVERTIBLE F05
WINDSCREEN-SEALS - NOT CONVERTIBLE F04

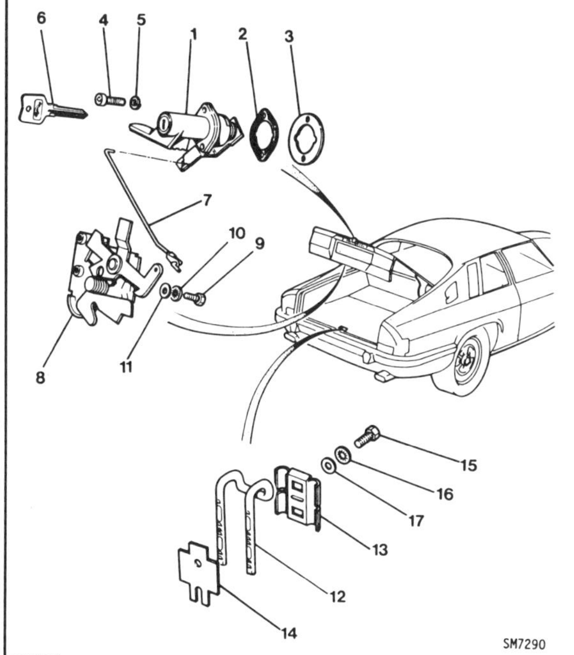

SM 7290

BOOT LOCK-STRIKER-UP TO (V)148781

Illus	Part Number	Description	Quantity	Change Point	Remarks
1	BD 44162	Lock-boot	1		
2	BD 44164	Seal-rubber	1		
3	BD 44163	Plate-tapped	1		
4	UFS 819 4H	Screw-set	2		
5	WM 702001 J	Washer-spring	2		
6		Key-blank	A/R		See locksets, miscellaneous group
7	BD 46985	Link	1		
8	BD 31101	Latch-boot	1		
9	SH 604041 J	Screw-set	3		
10	C 724	Washer-shakeproof	3		
11	BD 541 9	Washer-plain	3		
12	BD 47322	Striker-boot	1		
13	BD 31104	Retainer	1		
14	BD 36002	Plate-friction	1		
15	SH 604051 J	Screw-set	2		
16	C 724	Washer-shakeproof	2		
17	JLM 296	Washer-plain	2		

(Column header also shows "123456".)

MASTER INDEX

GROUP INDEX

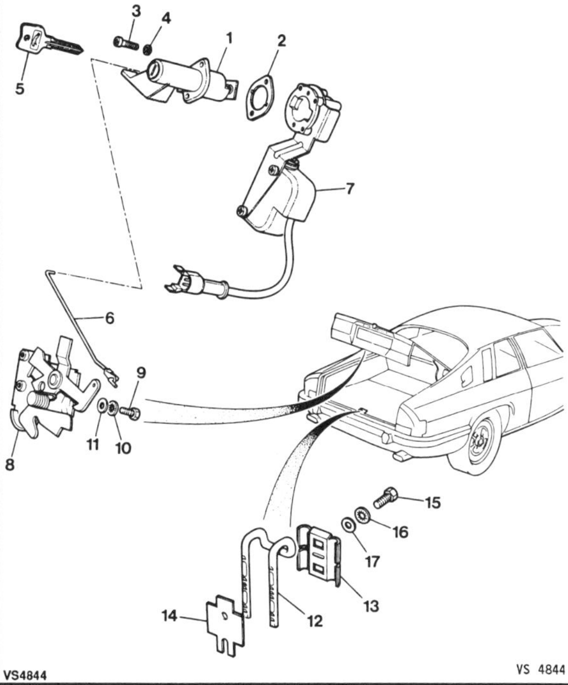

VS4844

BOOT LOCK-STRIKER-FROM (V)148782

Illus	Part Number	Description	Quantity	Change Point	Remarks
1	BDC 9038	Lock-boot	1		
2	BD 44164	Seal-rubber	1		
3	UFS 819 4H	Screw-set	2		
4	WM 702001 J	Washer-spring	2		
5		Key blank	A/R		See locksets miscellaneous group
6	BD 46985	Link	1		
7	BDC 8817	Base assy	1		
8	BD 31101	Latch-boot	1		Use BDC7153
8	BDC 7153	Latch-boot	1		
9	SH 604041 J	Screw-set	3		
10	C 724	Washer-shakeproof	3		
11	BD 541 9	Washer- plain	3		
12	BD 47322	Sinker- boot	1		
13	BD 31104	Retainer	1		
14	BD 36002	Plate-friction	1		
15	SH 604051 J	Screw-set	2		
16	C 724	Washer-shakeproof	2		
17	JLM 296	Washer-plain	2		

(Column header also shows "123456".)

MASTER INDEX

ENGINE 1.C
FLYWHEEL AND CLUTCH 1.G
GEARBOX AND PROPSHAFT 1.H
AXLES, SUSPENSION, DRIVE SHAFTS, WHEELS 1.K
STEERING 1.M
BRAKES AND BRAKE CONTROLS 1.N
FUEL, EXHAUST AND EMISSION SYSTEMS 2.C
COOLING, HEATING AND AIR CONDITIONING 2.G
ELECTRICAL, WASHERS-WIPERS, INSTRUMENTS 2.J
CHASSIS, SUBFRAMES, BODYSHELL, FITTINGS 3.C
FASCIA, TRIM, SEATS AND FIXINGS 3.H
ACCESSORIES AND PAINT 4.C
NUMERICAL INDEX 4.D

GROUP INDEX

BADGES - EMBLEMS - CONVERTIBLE G05
BADGES-EMBLEMS-CABRIOLET G04
BADGES-EMBLEMS-COUPE G03
BODY SHELL - CONVERTIBLE D03
BODY SHELL-CABRIOLET C10
BODY SHELL-COUPE C02
REAR WINGS (TONNEAU) C08
BONNET LOCK MECHANISM D15
BONNET SAFETY CATCH AND HINGES D14
BONNET-WINGS-DOORS-BOOT LID D12
BOOT HINGES-SEAL-LAMP HOUSING E14
BRIGHT FINISHERS-NOT CONVERTIBLE F14
COACHLINES(TAPE) V12 MODELS F17
COACHLINES(TAPE)-3.6 MODELS F16
CRUCIFORM AND HEATSHIELD-CABRIOLET F02
DOOR GLASS-SEALS - NOT CONVERTIBLE F06
DOOR HANDLES-LOCKS-SOLENOIDS E07
DOOR HINGES D18
DOOR SEALS AND TREAD PLATES E02
FRONT BUMPER G07
FRONT DOOR GLASS-SEALS-CONVERTIBLE F07
FRONT WINDOW REGULATORS - CONVERTIBLE E05
FUEL FILLER LID AND LOCK-COUPE C09
GEARBOX COVER PLATES E13
GEARBOX TUNNEL COVER-AUTO G/BOX MODELS E11
GEARBOX TUNNEL COVER E10
HEADLAMP SURROUNDS F12
HEATED BACKLIGHT-SEAL-COUPE F11
MIRRORS G02
REAR BUMPER G09
REAR QTR GLASS AND SEALS-CONVERTIBLE F10
REAR QTR WINDOW REGULATORS - CONVERTIBLE E06
REAR QUARTER GLASS-SEALS-CABRIOLET F09
BADGE RH-B POST F09
REAR QUARTER GLASS-SEALS-COUPE F08
SPOILER-UNDERTRAY G06
STONE GUARDS D17
TRUNK REINFORCEMENT - CONVERTIBLE F03
WEATHERSTRIPS AND TREADPLATES E03
WINDOW REGULATORS-NOT CONVERTIBLE E04
WINDSCREEN - SEALS - CONVERTIBLE F05
WINDSCREEN-SEALS - NOT CONVERTIBLE F04

3E 18

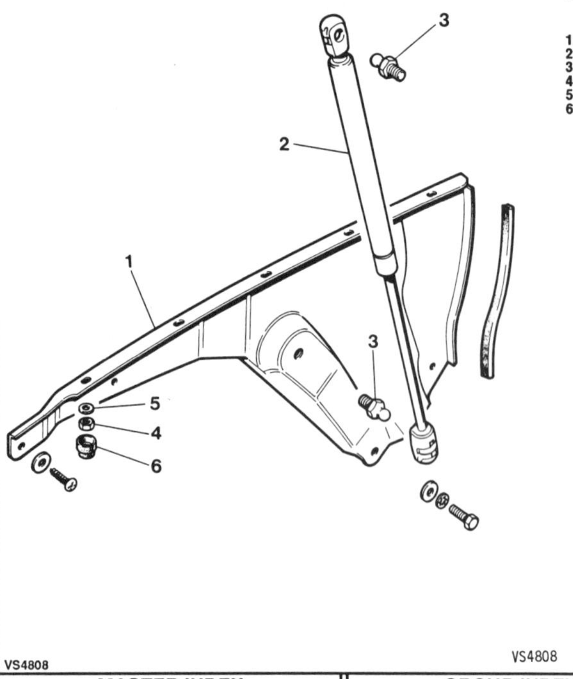

DRAIN CHANNEL REINFORCEMENT - CONVERTIBLE

Illus	Part Number	123456	Description	Quantity	Change Point	Remarks
	BCC 9550		Reinforcement-drain channel-RH	1		
1	BCC 9551		Reinforcement-drain channel-LH	1		
2	BBC 7000		Spring-gas	2		
3	BCC 3664		Pin-ball	4		
4	NH 105041 J		Nut-nyloc	10		
5	WC 105001 J		Washer-plain	10		
6	BCC 6778		Screw-cap	10		

MASTER INDEX

GROUP INDEX

3F 02

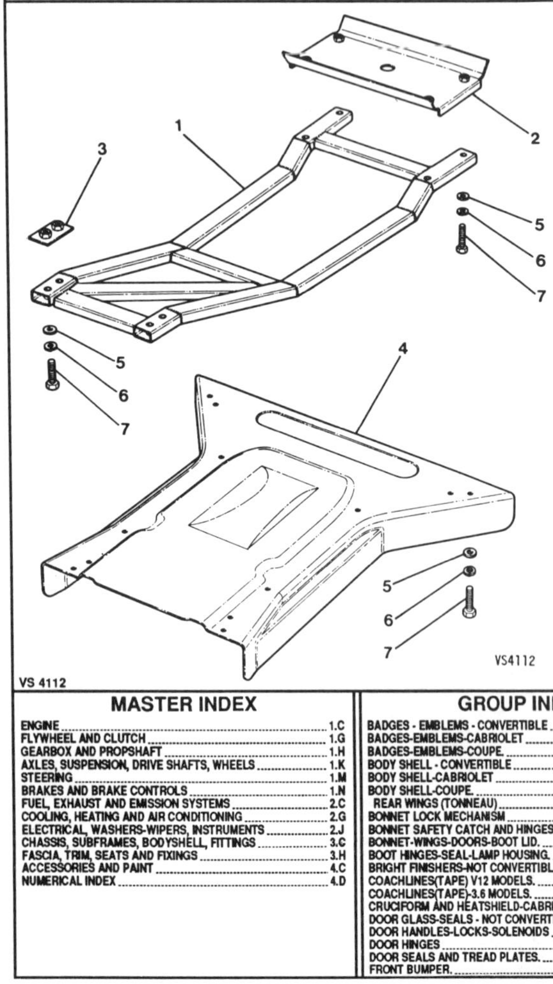

CRUCIFORM AND HEATSHIELD-CABRIOLET

Illus	Part Number	123456	Description	Quantity	Change Point	Remarks
1	CAC 7196		Cruciform	1		
2	CAC 7193		Bracket	1		
3	CAC 7191		Plate	2		
4	CAC 7207		Heatshield	1		
5	JLM 298		Washer-plain	16		
6	WM 600051 J		Washer-spring	16		
7	SH 605121 J		Screw-set	16		

MASTER INDEX
ENGINE 1.C
FLYWHEEL AND CLUTCH 1.G
GEARBOX AND PROPSHAFT 1.H
AXLES, SUSPENSION, DRIVE SHAFTS, WHEELS 1.K
STEERING 1.M
BRAKES AND BRAKE CONTROLS 1.N
FUEL, EXHAUST AND EMISSION SYSTEMS 2.C
COOLING, HEATING AND AIR CONDITIONING 2.G
ELECTRICAL, WASHERS-WIPERS, INSTRUMENTS 2.J
CHASSIS, SUBFRAMES, BODYSHELL, FITTINGS 3.C
FASCIA, TRIM, SEATS AND FIXINGS 3.H
ACCESSORIES AND PAINT 4.C
NUMERICAL INDEX 4.D
GROUP INDEX
BADGES - EMBLEMS - CONVERTIBLE G05
BADGES-EMBLEMS-CABRIOLET G04
BADGES-EMBLEMS-COUPE G03
BODY SHELL - CONVERTIBLE D03
BODY SHELL-CABRIOLET C10
BODY SHELL-COUPE C02
REAR WINGS (TONNEAU) C08
BONNET LOCK MECHANISM D15
BONNET SAFETY CATCH AND HINGES D14
BONNET-WINGS-DOORS-BOOT LID D12
BOOT HINGES-SEAL-LAMP HOUSING E14
BRIGHT FINISHERS-NOT CONVERTIBLE F14
COACHLINES(TAPE) V12 MODELS F17
COACHLINES(TAPE)-3.6 MODELS F16
CRUCIFORM AND HEATSHIELD-CABRIOLET F02
DOOR GLASS-SEALS - NOT CONVERTIBLE F06
DOOR HANDLES-LOCKS-SOLENOIDS E07
DOOR HINGES D18
DOOR SEALS AND TREAD PLATES E02
FRONT BUMPER G07
FRONT DOOR GLASS-SEALS-CONVERTIBLE F07
FRONT WINDOW REGULATORS - CONVERTIBLE E05
FUEL FILLER LID AND LOCK-COUPE C09
GEARBOX COVER PLATES E13
GEARBOX TUNNEL COVER-AUTO G/BOX MODELS E11
GEARBOX TUNNEL COVER E10
HEADLAMP SURROUNDS F12
HEATED BACKLIGHT-SEAL-COUPE F11
MIRRORS G02
REAR BUMPER G09
REAR QTR GLASS AND SEALS-CONVERTIBLE F10
REAR QTR WINDOW REGULATORS - CONVERTIBLE E06
REAR QUARTER GLASS-SEALS-CABRIOLET F09
BADGE RH-B POST F09
REAR QUARTER GLASS-SEALS-COUPE F08
SPOILER-UNDERTRAY G06
STONE GUARDS D17
TRUNK REINFORCEMENT - CONVERTIBLE F03
WEATHERSTRIPS AND TREADPLATES E03
WINDOW REGULATORS-NOT CONVERTIBLE E04
WINDSCREEN - SEALS - CONVERTIBLE F05
WINDSCREEN-SEALS - NOT CONVERTIBLE F04

JAGUAR XJS RANGE (JAN 1987 ON) — F03 fiche 3 — CHASSIS, SUBFRAMES, BODYSHELL, FITTINGS

TRUNK REINFORCEMENT - CONVERTIBLE

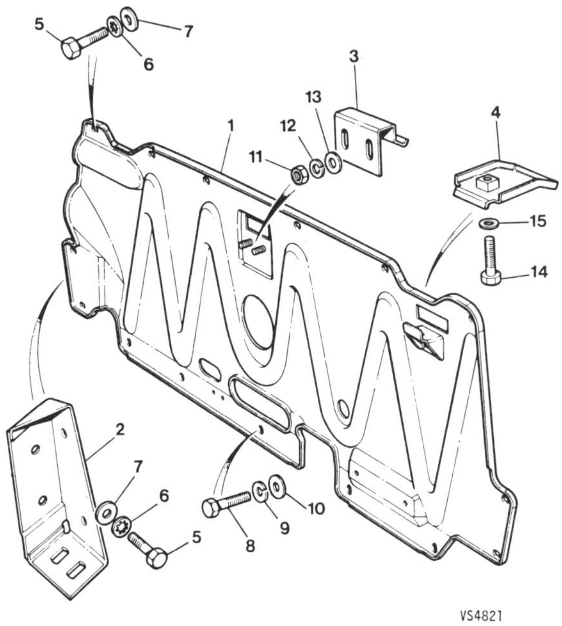

VS4821

Illus	Part Number	Description	Quantity	Change Point	Remarks
1	BDC 3725	Reinforcement-trunk	1		
	BDC 5708	Bracket-trunk reinf RH	1		
2	BDC 5709	Bracket-trunk reinf LH	1		
3	BCC 6831	Clamp-fuel tank	1		
4	BDC 4060	Clamp-fuel tank	1		
5	SH 605061 J	Bolt	18		Upper reinf & brkts to body
6	C 725	Washer-spring	18		
7	C 15183	Washer-plain	18		
8	SH 60601 J	Bolt	5		Lower reinf to body
9	WF 600060 J	Washer-spring	5		
10	FW 206 T	Washer-plain	5		
11	NH 604041 J	Nut	2		Clamp to reinf
12	WF 60041 J	Washer-spring	2		
13	WA 106041 J	Washer-plain	2		
14	BH 106051 J	Bolt	1		Clamp to reinf
15	WA 106041 J	Washer	1		

JAGUAR XJS RANGE (JAN 1987 ON) — F04 fiche 3 — CHASSIS, SUBFRAMES, BODYSHELL, FITTINGS

WINDSCREEN-SEALS - NOT CONVERTIBLE

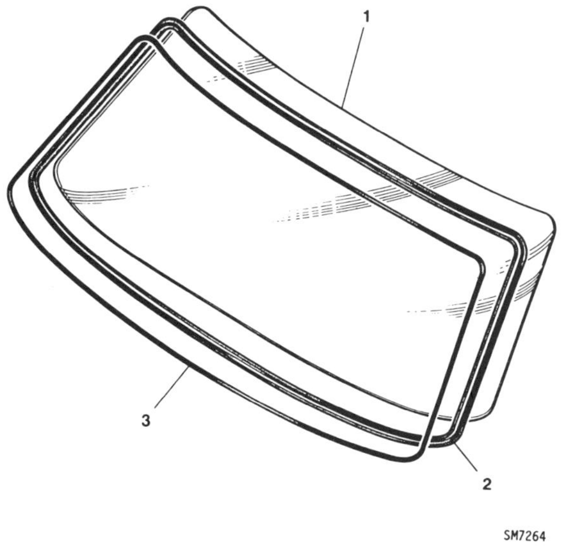

SM 7264

Illus	Part Number	Description	Quantity	Change Point	Remarks
1	BD 44231	Windscreen-clear	1		
	BD 44232	Windscreen-tinted	1		
2	BD 44236	Seal-windscreen	1		Use BDC 7178
2	BDC 7178	Seal-windscreen	1		
3	BD 15466 2	Insert-156 ins	1		
	BD 21181	Holder-licence	1		

L46JP28/7/88

MASTER INDEX

ENGINE 1.C
FLYWHEEL AND CLUTCH 1.G
GEARBOX AND PROPSHAFT 1.H
AXLES, SUSPENSION, DRIVE SHAFTS, WHEELS 1.K
STEERING 1.M
BRAKES AND BRAKE CONTROLS 1.N
FUEL, EXHAUST AND EMISSION SYSTEMS 2.C
COOLING, HEATING AND AIR CONDITIONING 2.G
ELECTRICAL, WASHERS-WIPERS, INSTRUMENTS 2.J
CHASSIS, SUBFRAMES, BODYSHELL, FITTINGS 3.C
FASCIA, TRIM, SEATS AND FIXINGS 3.H
ACCESSORIES AND PAINT 4.C
NUMERICAL INDEX 4.D

GROUP INDEX

BADGES - EMBLEMS - CONVERTIBLE G05
BADGES-EMBLEMS-CABRIOLET G04
BADGES-EMBLEMS-COUPE G03
BODY SHELL - CONVERTIBLE D03
BODY SHELL-CABRIOLET C10
BODY SHELL-COUPE C02
REAR WINGS (TONNEAU) C08
BONNET LOCK MECHANISM D15
BONNET SAFETY CATCH AND HINGES D14
BONNET-WINGS-DOORS-BOOT LID D12
BOOT HINGES-SEAL-LAMP HOUSING E14
BRIGHT FINISHERS-NOT CONVERTIBLE F14
COACHLINES(TAPE) V12 MODELS F17
COACHLINES(TAPE)-3.6 MODELS F16
CRUCIFORM AND HEATSHIELD-CABRIOLET F02
DOOR GLASS-SEALS - NOT CONVERTIBLE F06
DOOR HANDLES-LOCKS-SOLENOIDS E07
DOOR HINGES D18
DOOR SEALS AND TREAD PLATES E02
FRONT BUMPER G07
FRONT DOOR GLASS-SEALS-CONVERTIBLE F07
FRONT WINDOW REGULATORS - CONVERTIBLE E05
FUEL FILLER LID AND LOCK-COUPE C09
GEARBOX COVER PLATES E13
GEARBOX TUNNEL COVER-AUTO G/BOX MODELS E11
GEARBOX TUNNEL COVER E10
HEADLAMP SURROUNDS F12
HEATED BACKLIGHT-SEAL-COUPE F11
MIRRORS G02
REAR BUMPER G09
REAR QTR GLASS AND SEALS-CONVERTIBLE F10
REAR QTR WINDOW REGULATORS - CONVERTIBLE E06
REAR QUARTER GLASS-SEALS-CABRIOLET F09
BADGE RH-B POST F09
REAR QUARTER GLASS-SEALS-COUPE F08
SPOILER-UNDERTRAY G06
STONE GUARDS D17
TRUNK REINFORCEMENT - CONVERTIBLE F03
WEATHERSTRIPS AND TREADPLATES E03
WINDOW REGULATORS-NOT CONVERTIBLE E04
WINDSCREEN - SEALS - CONVERTIBLE F05
WINDSCREEN-SEALS - NOT CONVERTIBLE F04

3F 05

JAGUAR XJS RANGE (JAN 1987 ON) | F05 fiche 3 | CHASSIS, SUBFRAMES, BODYSHELL, FITTINGS

WINDSCREEN - SEALS - CONVERTIBLE

Illus	Part Number	123456	Description	Quantity	Change Point	Remarks
1	BBC 6907		Windscreen-clear	1		
1	BBC 6908		Windscreen-tinted	1		
2	BDC 3035		Rubber-retaining	1		
3	BCC 7335		Dam-mastic retention	1		
4	BDC 3036		Moulding-edge RH	1		
5	BDC 3037		Moulding-edge LH	1		
6	ACU 3010		Nut-Lokut	12		
7	DAC 706 8C		Screw-self tap	12		
8	BBC 6996		Finisher-upper	1		
9	BBC 6997		Finisher-lower	1		
	BBC 6998		Finisher-"A" Post RH	1		
10	BBC 6999		Finisher-"A" Post LH	1		

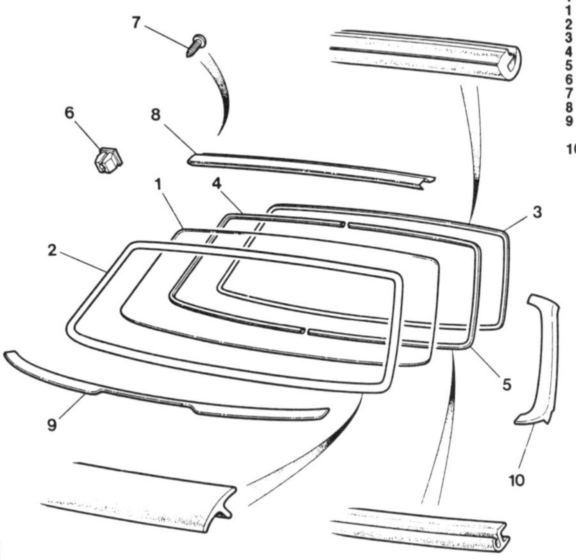

VS4835

VS4835

3F 06

JAGUAR XJS RANGE (JAN 1987 ON) | F06 fiche 3 | CHASSIS, SUBFRAMES, BODYSHELL, FITTINGS

DOOR GLASS-SEALS - NOT CONVERTIBLE

Illus	Part Number	123456	Description	Quantity	Change Point	Remarks
	BBC 2996		Door glass RH-tinted	1		
1	BBC 2997		Door glass LH-tinted	1		
	BD 45934		Channel-guide-RH	1		
2	BD 45935		Channel-guide-LH	1		
3	BBC 6292		Channel run-glass	2		
4	BD 46440 3		Seal-door waist	2		
5	BD 47779		Bracket-stop	2		
6	SH 604041 J		Screw-set	4		
7	C 724		Washer-shakeproof	4		
8	JLM 296		Washer-plain	4		
9	BD 21468		Block-stop	2		
10	JLM 307		Screw-self-tap	2		
11	C 18111		Washer-plain	2		
	BD 44093 1		Quarter glass RH-tinted	1		
12	BD 44094 1		Quarter glass LH-tinted	1		
13	BD 44773		Seal-quarter glass	2		
	BD 45088		Support-RH-glass	1		
14	BD 45089		Support-LH-glass	1		
15	AC 604042 J		Screw-self-tap	6		

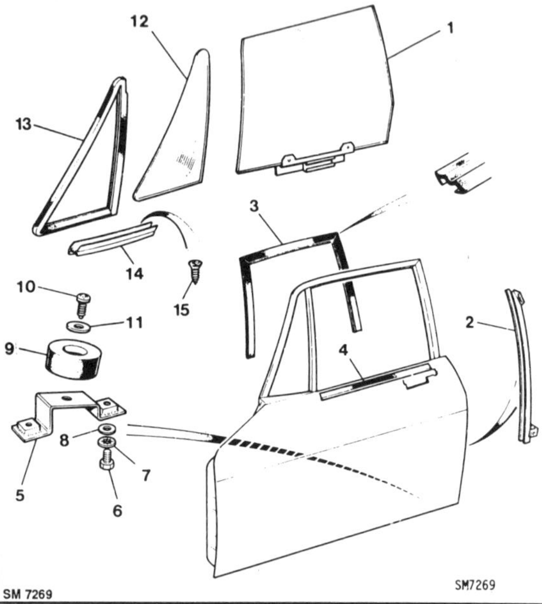

SM7269

SM 7269

MASTER INDEX

ENGINE ... 1.C
FLYWHEEL AND CLUTCH ... 1.G
GEARBOX AND PROPSHAFT ... 1.H
AXLES, SUSPENSION, DRIVE SHAFTS, WHEELS ... 1.K
STEERING ... 1.M
BRAKES AND BRAKE CONTROLS ... 1.N
FUEL, EXHAUST AND EMISSION SYSTEMS ... 2.C
COOLING, HEATING AND AIR CONDITIONING ... 2.G
ELECTRICAL, WASHERS-WIPERS, INSTRUMENTS ... 2.J
CHASSIS, SUBFRAMES, BODYSHELL, FITTINGS ... 3.C
FASCIA, TRIM, SEATS AND FIXINGS ... 3.H
ACCESSORIES AND PAINT ... 4.C
NUMERICAL INDEX ... 4.D

GROUP INDEX

BADGES - EMBLEMS - CONVERTIBLE ... G05
BADGES-EMBLEMS-CABRIOLET ... G04
BADGES-EMBLEMS-COUPE. ... G03
BODY SHELL - CONVERTIBLE ... D03
BODY SHELL-CABRIOLET ... C10
BODY SHELL-COUPE. ... C02
REAR WINGS (TONNEAU) ... C08
BONNET LOCK MECHANISM ... D15
BONNET SAFETY CATCH AND HINGES ... D14
BONNET-WINGS-DOORS-BOOT LID. ... D12
BOOT HINGES-SEAL-LAMP HOUSING. ... E14
BRIGHT FINISHERS-NOT CONVERTIBLE ... F14
COACHLINES(TAPE) V12 MODELS. ... F17
COACHLINES(TAPE)-3.6 MODELS. ... F16
CRUCIFORM AND HEATSHIELD-CABRIOLET ... F02
DOOR GLASS-SEALS - NOT CONVERTIBLE ... F06
DOOR HANDLES-LOCKS-SOLENOIDS ... E07
DOOR HINGES ... D18
DOOR SEALS AND TREAD PLATES. ... E02
FRONT BUMPER. ... G07
FRONT DOOR GLASS-SEALS-CONVERTIBLE ... F07
FRONT WINDOW REGULATORS - CONVERTIBLE ... E05
FUEL FILLER LID AND LOCK-COUPE ... C09
GEARBOX COVER PLATES. ... E13
GEARBOX TUNNEL COVER-AUTO G/BOX MODELS ... E11
GEARBOX TUNNEL COVER. ... E10
HEADLAMP SURROUNDS. ... F12
HEATED BACKLIGHT-SEAL-COUPE. ... F11
MIRRORS. ... G02
REAR BUMPER. ... G09
REAR QTR GLASS AND SEALS-CONVERTIBLE ... F10
REAR QTR WINDOW REGULATORS - CONVERTIBLE ... E06
REAR QUARTER GLASS-SEALS-CABRIOLET ... F09
BADGE RH-B POST ... F09
REAR QUARTER GLASS-SEALS-COUPE. ... F08
SPOILER-UNDERTRAY. ... G06
STONE GUARDS. ... D17
TRUNK REINFORCEMENT - CONVERTIBLE ... F03
WEATHERSTRIPS AND TREADPLATES ... E03
WINDOW REGULATORS-NOT CONVERTIBLE ... E04
WINDSCREEN - SEALS - CONVERTIBLE ... F05
WINDSCREEN-SEALS - NOT CONVERTIBLE ... F04

FRONT DOOR GLASS-SEALS-CONVERTIBLE

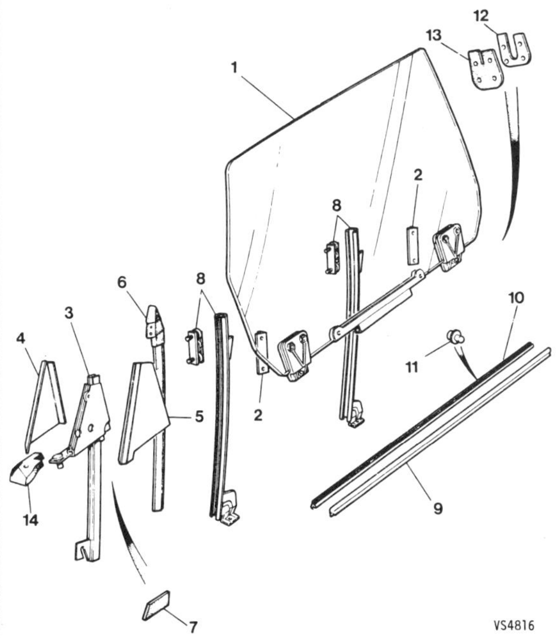

VS4816

Illus	Part Number	123456	Description	Quantity	Change Point	Remarks
	BDC 7910		Glass and carrier-front door RH	1		
1	BDC 7911		Glass and carrier-front door LH	1		
	BDC 8364		Angle - attachment RH	2		
2	BDC 8365		Angle-attachment LH	2		
	BDC 3556		Cheater-front RH	1		
3	BDC 3557		Cheater-front LH	1		
	BDC 8928		Bezel-inner RH	1		
4	BDC 8929		Bezel-inner LH	1		
	BCC 2816		Bezel-outer RH	1		
5	BCC 2817		Bezel-outer LH	1		
	BCC 2844		Run channel-glass RH	1		
6	BCC 2845		Run channel-glass LH	1		
	BCC 2838		Seal-outer lower RH	1		
7	BCC 2839		Seal-outer lower LH	1		
	BBC 7680		Guide rail RH	2		
8	BBC 7681		Guide rail LH	2		
	BBC 6934		Moulding-seal retainer RH	1		
9	BBC 6935		Moulding-seal retainer LH	1		
	BDC 4878		Seal-outer RH	1		
10	BDC 4879		Seal-outer LH	1		
11	AGU 2321		Rivet-plastic	16		
	BBC 6918		Finisher-RH	1		
12	BBC 6919		Finisher-LH	1		
	BBC 7730		Rubber plate RH	1		Door to "B" post
13	BBC 7731		Rubber plate LH	1		Door to "B" post
	BDC 5566		Moulding-door front RH	1		
14	BDC 5567		Moulding-door front LH	1		

REAR QUARTER GLASS-SEALS-COUPE.

Illus	Part Number	123456	Description	Quantity	Change Point	Remarks
	BD 44487 1		Quarter glass RH-tinted	1		
1	BD 44488 1		Quarter glass LH-tinted	1		
	BD 46763		Seal RH-quarter glass	1		
2	BD 46764		Seal LH-quarter glass	1		

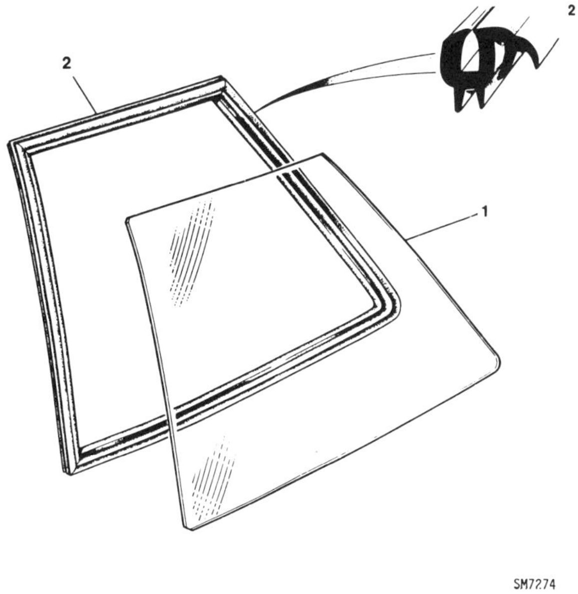

SM 7274

MASTER INDEX

ENGINE 1.C
FLYWHEEL AND CLUTCH 1.G
GEARBOX AND PROPSHAFT 1.H
AXLES, SUSPENSION, DRIVE SHAFTS, WHEELS 1.K
STEERING 1.M
BRAKES AND BRAKE CONTROLS 1.N
FUEL, EXHAUST AND EMISSION SYSTEMS 2.C
COOLING, HEATING AND AIR CONDITIONING 2.G
ELECTRICAL, WASHERS-WIPERS, INSTRUMENTS 2.J
CHASSIS, SUBFRAMES, BODYSHELL, FITTINGS 3.C
FASCIA, TRIM, SEATS AND FIXINGS 3.H
ACCESSORIES AND PAINT 4.C
NUMERICAL INDEX 4.D

GROUP INDEX

BADGES - EMBLEMS - CONVERTIBLE G05
BADGES-EMBLEMS-CABRIOLET G04
BADGES-EMBLEMS-COUPE G03
BODY SHELL - CONVERTIBLE D03
BODY SHELL-CABRIOLET C10
BODY SHELL-COUPE C02
REAR WINGS (TONNEAU) C08
BONNET LOCK MECHANISM D15
BONNET SAFETY CATCH AND HINGES D14
BONNET-WINGS-DOORS-BOOT LID D12
BOOT HINGES-SEAL-LAMP HOUSING E14
BRIGHT FINISHERS-NOT CONVERTIBLE F14
COACHLINES(TAPE) V12 MODELS F17
COACHLINES(TAPE)-3.6 MODELS F16
CRUCIFORM AND HEATSHIELD-CABRIOLET F02
DOOR GLASS-SEALS - NOT CONVERTIBLE F06
DOOR HANDLES-LOCKS-SOLENOIDS E07
DOOR HINGES D18
DOOR SEALS AND TREAD PLATES E02
FRONT BUMPER G07
FRONT DOOR GLASS-SEALS-CONVERTIBLE F07
FRONT WINDOW REGULATORS - CONVERTIBLE E05
FUEL FILLER LID AND LOCK-COUPE C09
GEARBOX COVER PLATES E13
GEARBOX TUNNEL COVER-AUTO G/BOX MODELS E11
GEARBOX TUNNEL COVER E10
HEADLAMP SURROUNDS F12
HEATED BACKLIGHT-SEAL-COUPE F11
MIRRORS G02
REAR BUMPER G09
REAR QTR GLASS AND SEALS-CONVERTIBLE F10
REAR QTR WINDOW REGULATORS - CONVERTIBLE E06
REAR QUARTER GLASS-SEALS-CABRIOLET F09
BADGE RH-B POST F09
REAR QUARTER GLASS-SEALS-COUPE F08
SPOILER-UNDERTRAY G06
STONE GUARDS D17
TRUNK REINFORCEMENT - CONVERTIBLE F03
WEATHERSTRIPS AND TREADPLATES E03
WINDOW REGULATORS-NOT CONVERTIBLE E04
WINDSCREEN - SEALS - CONVERTIBLE F05
WINDSCREEN-SEALS - NOT CONVERTIBLE F04

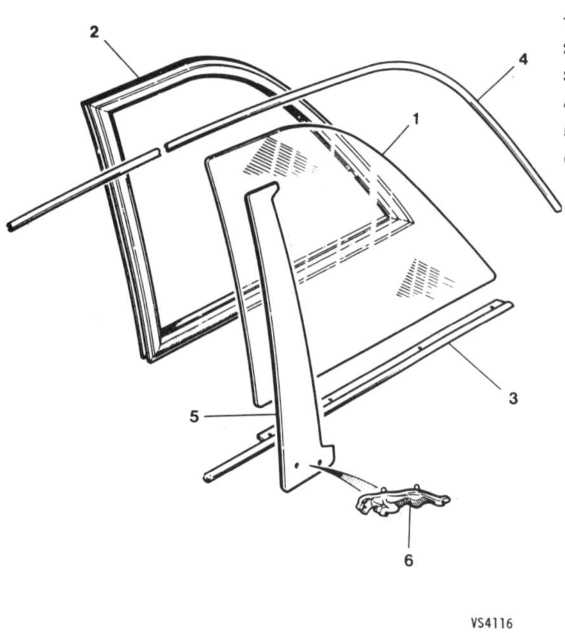

VS 4116

REAR QUARTER GLASS-SEALS-CABRIOLET

Illus	Part Number	Description	Quantity	Change Point	Remarks
	BAC 8272	Quarter glass RH-tinted	1		
1	BAC 8273	Quarter glass LH-tinted	1		
	BAC 7742	Seal RH-quarter glass	1		
2	BAC 7743	Seal LH-quarter glass	1		
	BAC 7738	Moulding RH-waist	1		
3	BAC 7739	Moulding LH-waist	1		
	BAC 7740	Moulding RH-drip	1		
4	BAC 7741	Moulding LH-drip	1		
	BAC 7788	Finisher-RH-B post	1		
5	BAC 7789	Finisher-LH-B post	1		
	BAC 7802	BADGE RH-B POST	1		
6	BAC 7803	Badge LH-B post	1		

3F 09

MASTER INDEX

GROUP INDEX

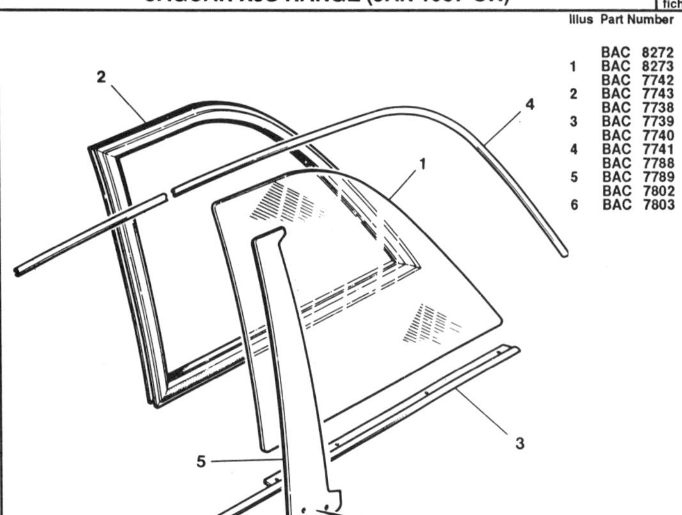

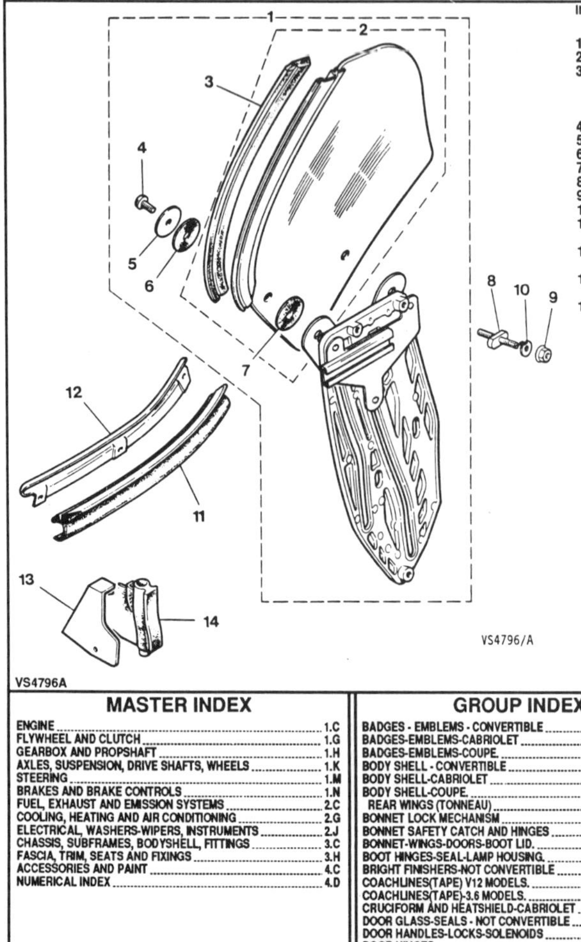

VS4796A

REAR QTR GLASS AND SEALS-CONVERTIBLE

Illus	Part Number	Description	Quantity	Change Point	Remarks
1	BDC 9444	GLASS AND GUIDE PLATE ASSY RH	1		
2	BDC 9442	Glass-rear qtr RH	1		
3	BDC 8676	Seal RH	1		
	BDC 9445	GLASS AND GUIDE PLATE ASSY LH	1		
	BDC 9443	Glass-rear qtr LH	1		
	BDC 8677	Seal LH	1		
4	BCC 8437	Screw	6		
5	BCC 8438	Washer-steel	6		
6	BCC 8439	Washer-rubber	6		
7	BCC 8440	Spacer-rubber	6		
8	BCC 5518	Screw-adjustment	6		
9	FN 108041 J	Nut-flanged	6		
10	BDC 4912	Washer	6		
11	BDC 4556	Seal-outer RH	1		
	BDC 4557	Seal-outer LH	1		
12	BDC 4554	Retainer-seal RH	1		
	BDC 4555	Retainer-seal LH	1		
13	BDC 6274	Finisher-RH	1		"B" Post to door
	BDC 6275	Finisher-LH	1		"B" Post to door
14	BDC 5360	Seal-RH	1		"B" Post to door
	BDC 5361	Seal-LH	1		"B" Post to door

3F 10

MASTER INDEX

ENGINE 1.C
FLYWHEEL AND CLUTCH 1.G
GEARBOX AND PROPSHAFT 1.H
AXLES, SUSPENSION, DRIVE SHAFTS, WHEELS 1.K
STEERING 1.M
BRAKES AND BRAKE CONTROLS 1.N
FUEL, EXHAUST AND EMISSION SYSTEMS 2.C
COOLING, HEATING AND AIR CONDITIONING 2.G
ELECTRICAL, WASHERS-WIPERS, INSTRUMENTS 2.J
CHASSIS, SUBFRAMES, BODYSHELL, FITTINGS 3.C
FASCIA, TRIM, SEATS AND FIXINGS 3.H
ACCESSORIES AND PAINT 4.C
NUMERICAL INDEX 4.D

GROUP INDEX

BADGES - EMBLEMS - CONVERTIBLE G05
BADGES-EMBLEMS-CABRIOLET G04
BADGES-EMBLEMS-COUPE G03
BODY SHELL - CONVERTIBLE D03
BODY SHELL-CABRIOLET C10
BODY SHELL-COUPE C02
REAR WINGS (TONNEAU) C08
BONNET LOCK MECHANISM D15
BONNET SAFETY CATCH AND HINGES D14
BONNET-WINGS-DOORS-BOOT LID D12
BOOT HINGES-SEAL-LAMP HOUSING E14
BRIGHT FINISHERS-NOT CONVERTIBLE F14
COACHLINES(TAPE) V12 MODELS F17
COACHLINES(TAPE)-3.6 MODELS F16
CRUCIFORM AND HEATSHIELD-CABRIOLET F02
DOOR GLASS-SEALS - NOT CONVERTIBLE F06
DOOR HANDLES-LOCKS-SOLENOIDS E07
DOOR HINGES D18
DOOR SEALS AND TREAD PLATES E02
FRONT BUMPER G07
FRONT DOOR GLASS-SEALS-CONVERTIBLE F07
FRONT WINDOW REGULATORS - CONVERTIBLE E05
FUEL FILLER LID AND LOCK-COUPE C09
GEARBOX COVER PLATES E13
GEARBOX TUNNEL COVER-AUTO G/BOX MODELS E11
GEARBOX TUNNEL COVER E10
HEADLAMP SURROUNDS F12
HEATED BACKLIGHT-SEAL-COUPE F11
MIRRORS G02
REAR BUMPER G09
REAR QTR GLASS AND SEALS-CONVERTIBLE F10
REAR QTR WINDOW REGULATORS - CONVERTIBLE E06
REAR QUARTER GLASS-SEALS-CABRIOLET F09
BADGE RH-B POST F09
REAR QUARTER GLASS-SEALS-COUPE F08
SPOILER-UNDERTRAY G06
STONE GUARDS D17
TRUNK REINFORCEMENT - CONVERTIBLE F03
WEATHERSTRIPS AND TREADPLATES E03
WINDOW REGULATORS-NOT CONVERTIBLE E04
WINDSCREEN - SEALS - CONVERTIBLE F05
WINDSCREEN-SEALS - NOT CONVERTIBLE F04

HEATED BACKLIGHT-SEAL-COUPE.

Illus	Part Number	123456 Description	Quantity	Change Point	Remarks
1	BD 44391	Glass-backlight-tinted	1		Not USA/CDN
1	BBC 6619	Glass-backlight-tinted	1		Use BDC 8074
1	BDC 8074	Glass-backlight-tinted	1		USA/CDN
2	BD 44393	Seal-backlight glass	1		

L52JP27/7/88

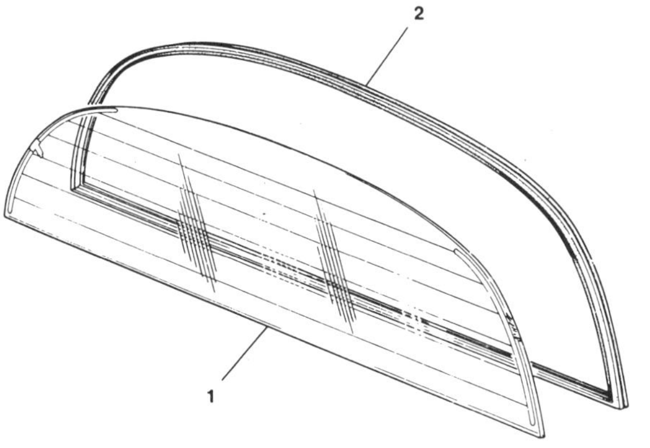

SM 7265

HEADLAMP SURROUNDS.

Illus	Part Number	123456 Description	Quantity	Change Point	Remarks
	DAC 2136	Surround RH-headlamp	1		CDN/USA
1	DAC 2137	Surround LH-headlamp	1		
2	AB 610052 J	Screw-self-tap	4		
3	AD 610052 J	Screw-self-tap	2		
4	BD 541 34	Washer-plain	6		
5	78319 J	Nut-spire	6		
	BAC 5416	Surround RH-headlamp	1		
6	BAC 5417	Surround LH-headlamp	1		
	BAC 4486	Surround RH-headlamp	1		Headlamp Wash/wipe
	BAC 4487	Surround LH-headlamp	1		
7	AB 610051 J	Screw-self-tap	2		
8	BD 541 34	Washer-plain	2		
9	C 31973 6	Nut-plastic	2		

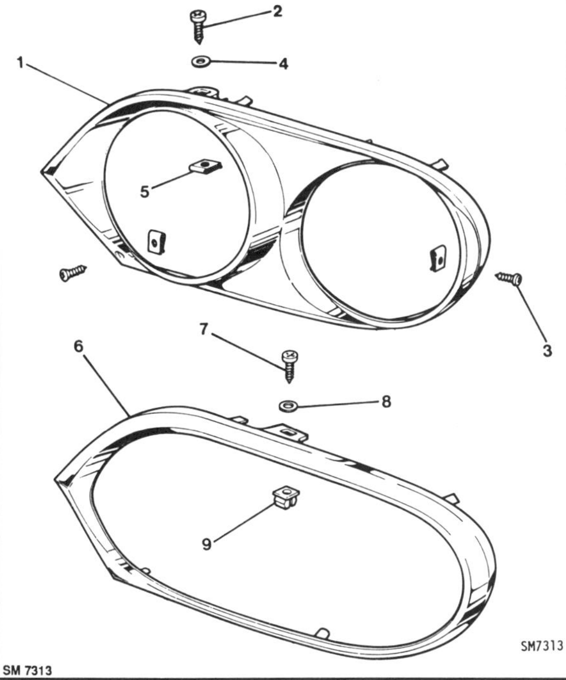

SM 7313

MASTER INDEX

ENGINE 1.C
FLYWHEEL AND CLUTCH 1.G
GEARBOX AND PROPSHAFT 1.H
AXLES, SUSPENSION, DRIVE SHAFTS, WHEELS 1.K
STEERING 1.M
BRAKES AND BRAKE CONTROLS 1.N
FUEL, EXHAUST AND EMISSION SYSTEMS 2.C
COOLING, HEATING AND AIR CONDITIONING 2.G
ELECTRICAL, WASHERS-WIPERS, INSTRUMENTS 2.J
CHASSIS, SUBFRAMES, BODYSHELL, FITTINGS 3.C
FASCIA, TRIM, SEATS AND FIXINGS 3.H
ACCESSORIES AND PAINT 4.C
NUMERICAL INDEX 4.D

GROUP INDEX

BADGES - EMBLEMS - CONVERTIBLE G05
BADGES-EMBLEMS-CABRIOLET G04
BADGES-EMBLEMS-COUPE G03
BODY SHELL - CONVERTIBLE D03
BODY SHELL-CABRIOLET C10
BODY SHELL-COUPE C02
REAR WINGS (TONNEAU) C08
BONNET LOCK MECHANISM D15
BONNET SAFETY CATCH AND HINGES D14
BONNET-WINGS-DOORS-BOOT LID D12
BOOT HINGES-SEAL-LAMP HOUSING E14
BRIGHT FINISHERS-NOT CONVERTIBLE F14
COACHLINES(TAPE) V12 MODELS F17
COACHLINES(TAPE)-3.6 MODELS F16
CRUCIFORM AND HEATSHIELD-CABRIOLET F02
DOOR GLASS-SEALS - NOT CONVERTIBLE F06
DOOR HANDLES-LOCKS-SOLENOIDS E07
DOOR HINGES D18
DOOR SEALS AND TREAD PLATES E02
FRONT BUMPER G07
FRONT DOOR GLASS-SEALS-CONVERTIBLE F07
FRONT WINDOW REGULATORS - CONVERTIBLE E05
FUEL FILLER LID AND LOCK-COUPE C09
GEARBOX COVER PLATES E13
GEARBOX TUNNEL COVER-AUTO G/BOX MODELS E11
GEARBOX TUNNEL COVER E10
HEADLAMP SURROUNDS F12
HEATED BACKLIGHT-SEAL-COUPE F11
MIRRORS G02
REAR BUMPER G09
REAR QTR GLASS AND SEALS-CONVERTIBLE F10
REAR QTR WINDOW REGULATORS - CONVERTIBLE E06
REAR QUARTER GLASS-SEALS-CABRIOLET F09
BADGE RH-B POST F09
REAR QUARTER GLASS-SEALS-COUPE F08
SPOILER-UNDERTRAY G06
STONE GUARDS D17
TRUNK REINFORCEMENT - CONVERTIBLE F03
WEATHERSTRIPS AND TREADPLATES E03
WINDOW REGULATORS-NOT CONVERTIBLE E04
WINDSCREEN - SEALS - CONVERTIBLE F05
WINDSCREEN-SEALS - NOT CONVERTIBLE F04

3F 13

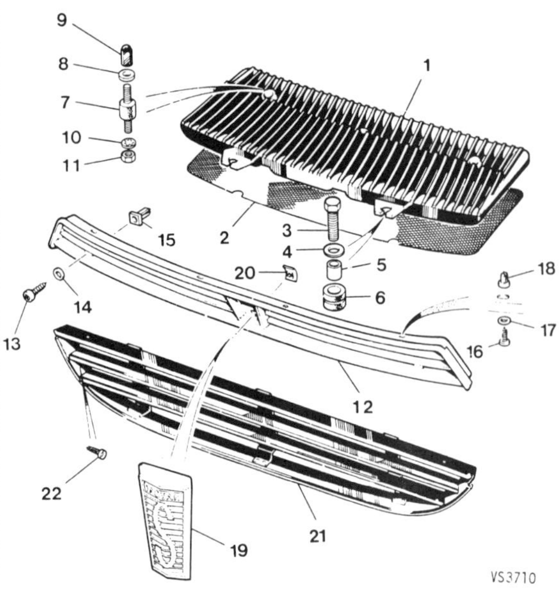

SCUTTLE VENT-RADIATOR GRILLES

Illus	Part Number	123456	Description	Quantity	Change Point	Remarks
1	BCC 2695		Grille-scuttle vent	1		See (1) below
2	BCC 2696		Mesh-grille	1		
3	SH 604101 J		Screw-set	2		With Lucas type wiper motor
4	BD 541 20		Washer-plain	2		
5	C 29875		Tube-distance	2		
6	CBC 1448		Grommet	2		
7	C 45801		Mounting	2		
8	C 44649		Spacer	2		
9	DAC 2667		Nut-dome	2		
10	WF 702101 J		Washer-shakeproof	2		
11	JLM 9686		Nut-plain	2		
12	BD 48339		Grille-radiator-upper	1		
13	JLM 308		Screw-self-tap	2		
14	BD 541 27		Washer-plain	2		
15	C 31973 6		Nut-plastic	2		
16	DA 81012 C		Screw-self-tap	4		
17	BD 541 27		Washer-plain	4		
18	C 31973 5		Nut-plastic	4		
19	BAC 5764		Badge-grille-'S'	1		
	BD 44533		Badge-grille-'V12'	1		
20	DBC 2013		Clip-push on	1		
21	BCC 2934		Grille-radiator-lower	1		
22	DAB 80810 C		Screw-self-tap	2		

L56JP20/7/88

VS3710

(1) Lucas wiper motor only. For cars with Electrolux wiper motor.

Mesh and grille are part of motor assembly.

VS3710

MASTER INDEX

GROUP INDEX

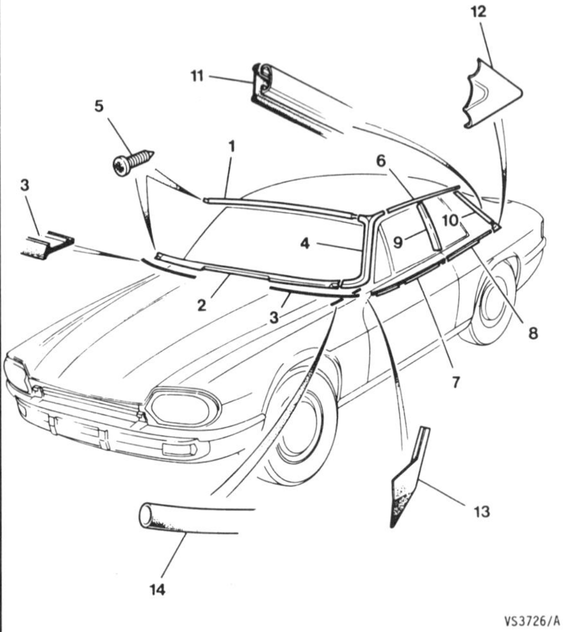

BRIGHT FINISHERS-NOT CONVERTIBLE

Illus	Part Number	123456	Description	Quantity	Change Point	Remarks
1	BD 44237		Finisher-upper	1		
2	BD 44243		Finisher-lower	1		
3	BD 46908 3		Seal-finisher-148ins	2		Cut to size
	BAC 4256		Finisher RH-"A" post	1		
4	BAC 4257		Finisher LH-"A" post	1		
5	DA 706 6S		Screw-self-tap	11		
	BD 44240		Finisher RH-gutter	1		Not Cabriolet
6	BD 44241		Finisher LH-gutter	1		
	BD 45628		Finisher RH-door waist	1		
7	BD 45629		Finisher LH-door waist	1		
	BD 45672		Finisher RH-rear quarter	1		Not Cabriolet
8	BD 45673		Finisher LH-rear quarter	1		
	BD 48752		Finisher RH-"B" post	1		
9	BD 48753		Finisher LH-"B" post	1		
	BD 45170		Finisher RH-rear air vent	1		
10	BD 45171		Finisher LH-rear air vent	1		
	BD 26706 1		Rivet-vent finisher fixing	8		
	BD 45175		Seal RH-finisher	1		
11	BD 45176		Seal LH-finisher	1		
	BD 45172		Finisher RH-corner	1		
12	BD 45173		Finisher LH-corner	1		
	BD 48015		Finisher RH-'A' post closing	NLA		Use BDC5780
	BDC 5780		Finisher RH-'A' post closing	1		
13	BD 48016		Finisher LH-'A' post closing	NLA		Use BDC5781
13	BDC 5781		Finisher LH-'A' post closing	1		
14	BAC 6236		Filler strip-wing to w/screen panel	2		

L58JP31/8/88

VS3726/A

VS 3726A

MASTER INDEX

ENGINE 1.C
FLYWHEEL AND CLUTCH 1.G
GEARBOX AND PROPSHAFT 1.H
AXLES, SUSPENSION, DRIVE SHAFTS, WHEELS 1.K
STEERING 1.M
BRAKES AND BRAKE CONTROLS 1.N
FUEL, EXHAUST AND EMISSION SYSTEMS 2.C
COOLING, HEATING AND AIR CONDITIONING 2.G
ELECTRICAL, WASHERS-WIPERS, INSTRUMENTS 2.J
CHASSIS, SUBFRAMES, BODYSHELL, FITTINGS 3.C
FASCIA, TRIM, SEATS AND FIXINGS 3.H
ACCESSORIES AND PAINT 4.C
NUMERICAL INDEX 4.D

GROUP INDEX

BADGES - EMBLEMS - CONVERTIBLE G05
BADGES-EMBLEMS-CABRIOLET G04
BADGES-EMBLEMS-COUPE G03
BODY SHELL - CONVERTIBLE D03
BODY SHELL-CABRIOLET C10
BODY SHELL-COUPE C02
REAR WINGS (TONNEAU) C08
BONNET LOCK MECHANISM D15
BONNET SAFETY CATCH AND HINGES D14
BONNET-WINGS-DOORS-BOOT LID D12
BOOT HINGES-SEAL-LAMP HOUSING E14
BRIGHT FINISHERS-NOT CONVERTIBLE F14
COACHLINES(TAPE) V12 MODELS F17
COACHLINES(TAPE)-3.6 MODELS F16
CRUCIFORM AND HEATSHIELD-CABRIOLET F02
DOOR GLASS-SEALS - NOT CONVERTIBLE F06
DOOR HANDLES-LOCKS-SOLENOIDS E07
DOOR HINGES D18
DOOR SEALS AND TREAD PLATES E02
FRONT BUMPER G07
FRONT DOOR GLASS-SEALS-CONVERTIBLE F07
FRONT WINDOW REGULATORS - CONVERTIBLE E05
FUEL FILLER LID AND LOCK-COUPE C09
GEARBOX COVER PLATES E13
GEARBOX TUNNEL COVER-AUTO G/BOX MODELS E11
GEARBOX TUNNEL COVER E10
HEADLAMP SURROUNDS F12
HEATED BACKLIGHT-SEAL-COUPE F11
MIRRORS G02
REAR BUMPER G09
REAR QTR GLASS AND SEALS-CONVERTIBLE F10
REAR QTR WINDOW REGULATORS - CONVERTIBLE E06
REAR QUARTER GLASS-SEALS-CABRIOLET F09
BADGE RH-B POST F09
REAR QUARTER GLASS-SEALS-COUPE F08
SPOILER-UNDERTRAY G06
STONE GUARDS D17
TRUNK REINFORCEMENT - CONVERTIBLE F03
WEATHERSTRIPS AND TREADPLATES E03
WINDOW REGULATORS-NOT CONVERTIBLE E04
WINDSCREEN - SEALS - CONVERTIBLE F05
WINDSCREEN-SEALS - NOT CONVERTIBLE F04

JAGUAR XJS RANGE (JAN 1987 ON) — F15 fiche 3 — CHASSIS, SUBFRAMES, BODYSHELL, FITTINGS

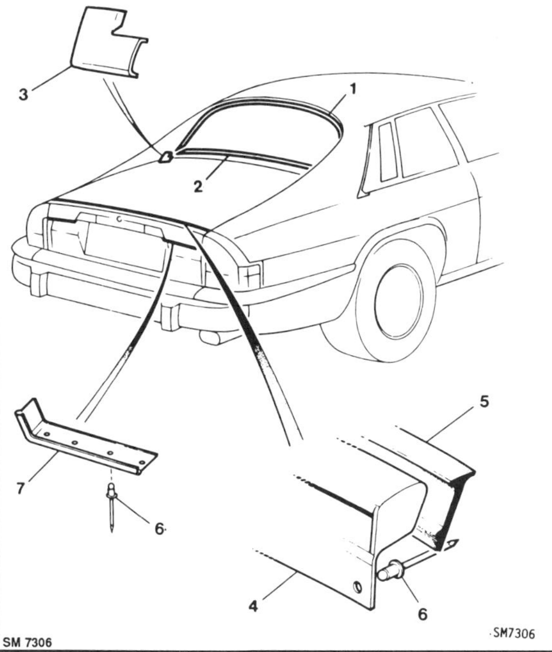

SM 7306

REAR END BRIGHT FINISHERS.

Illus	Part Number	123456	Description	Quantity	Change Point	Remarks
1	BD 44394		Finisher-upper	1		Coupe only
2	BD 44385		Finisher-lower	1		Coupe only
3	BD 47931		Finisher-corner	2		Coupe only
4	BD 45640		Finisher-boot upper	1		
5	BD 46908	1	Seal-finisher-41.75"	1		
6	BD 1818	1	Rivet	12		
7	BD 45919		Finisher RH-boot lower	1		
	BD 45920		Finisher LH-boot lower	1		

L60JP27/7/88

JAGUAR XJS RANGE (JAN 1987 ON) — F16 fiche 3 — CHASSIS, SUBFRAMES, BODYSHELL, FITTINGS

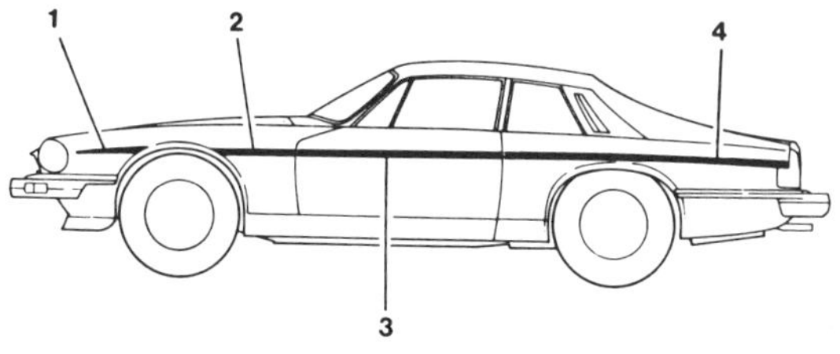

VS 3182

COACHLINES(TAPE)-3.6 MODELS.

UPTO (V)144699

Illus	Part Number	123456	Description	Quantity	Change Point	Remarks
			COACHLINE-FRONT WING-FRONT			
	BCC 5656		Goldleaf-RH	1		
1	BCC 5657		Goldleaf-LH	1		
	BCC 5658		Gunmetal-RH	1		
1	BCC 5659		Gunmetal-LH	1		
	BCC 5660		Oyster-RH	1		
1	BCC 5661		Oyster-LH	1		
	BCC 5662		Dark Red-RH	1		
1	BCC 5663		Dark Red-LH	1		
			COACHLINE-FRONT WING-REAR			
	BBC 9716		Goldleaf-RH	1		
2	BBC 9717		Goldleaf-LH	1		
	BBC 9718		Gunmetal-RH	1		
2	BBC 9719		Gunmetal-LH	1		
	BBC 9720		Oyster-RH	1		
2	BBC 9721		Oyster-LH	1		
	BCC 4578		Dark Red-RH	1		
2	BCC 4579		Dark Red-LH	1		
			COACHLINE-DOORS			
3	BBC 9722		Goldleaf	2		
3	BBC 9723		Gunmetal	2		
3	BBC 9724		Oyster	2		
3	BCC 4580		Dark Red	2		
			COACHLINE-REAR WING(TONNEAU)			
	BBC 9726		Goldleaf-RH	1		
4	BBC 9727		Goldleaf-LH	1		
	BBC 9728		Gunmetal-RH	1		
4	BBC 9729		Gunmetal-LH	1		
	BBC 9730		Oyster-RH	1		
4	BBC 9731		Oyster-LH	1		
	BCC 4582		Dark Red-RH	1		
4	BCC 4583		Dark Red-LH	1		

MASTER INDEX

ENGINE 1.C
FLYWHEEL AND CLUTCH 1.G
GEARBOX AND PROPSHAFT 1.H
AXLES, SUSPENSION, DRIVE SHAFTS, WHEELS 1.K
STEERING 1.M
BRAKES AND BRAKE CONTROLS 1.N
FUEL, EXHAUST AND EMISSION SYSTEMS 2.C
COOLING, HEATING AND AIR CONDITIONING 2.G
ELECTRICAL, WASHERS-WIPERS, INSTRUMENTS 2.J
CHASSIS, SUBFRAMES, BODYSHELL, FITTINGS 3.C
FASCIA, TRIM, SEATS AND FIXINGS 3.H
ACCESSORIES AND PAINT 4.C
NUMERICAL INDEX 4.D

GROUP INDEX

BADGES - EMBLEMS - CONVERTIBLE G05
BADGES-EMBLEMS-CABRIOLET G04
BADGES-EMBLEMS-COUPE G03
BODY SHELL - CONVERTIBLE D03
BODY SHELL-CABRIOLET C10
BODY SHELL-COUPE C02
REAR WINGS (TONNEAU) C08
BONNET LOCK MECHANISM D15
BONNET SAFETY CATCH AND HINGES D14
BONNET-WINGS-DOORS-BOOT LID D12
BOOT HINGES-SEAL-LAMP HOUSING E14
BRIGHT FINISHERS-NOT CONVERTIBLE F14
COACHLINES(TAPE) V12 MODELS F17
COACHLINES(TAPE)-3.6 MODELS F16
CRUCIFORM AND HEATSHIELD-CABRIOLET F02
DOOR GLASS-SEALS - NOT CONVERTIBLE F06
DOOR HANDLES-LOCKS-SOLENOIDS E07
DOOR HINGES D18
DOOR SEALS AND TREAD PLATES E02
FRONT BUMPER G07
FRONT DOOR GLASS-SEALS-CONVERTIBLE F07
FRONT WINDOW REGULATORS - CONVERTIBLE E05
FUEL FILLER LID AND LOCK-COUPE C09
GEARBOX COVER PLATES E13
GEARBOX TUNNEL COVER-AUTO G/BOX MODELS E11
GEARBOX TUNNEL COVER E10
HEADLAMP SURROUNDS F12
HEATED BACKLIGHT-SEAL-COUPE F11
MIRRORS G02
REAR BUMPER G09
REAR QTR GLASS AND SEALS-CONVERTIBLE F10
REAR QTR WINDOW REGULATORS - CONVERTIBLE E06
REAR QUARTER GLASS-SEALS-CABRIOLET F09
BADGE RH-B POST F09
REAR QUARTER GLASS-SEALS-COUPE F08
SPOILER-UNDERTRAY G06
STONE GUARDS D17
TRUNK REINFORCEMENT - CONVERTIBLE F03
WEATHERSTRIPS AND TREADPLATES E03
WINDOW REGULATORS-NOT CONVERTIBLE E04
WINDSCREEN - SEALS - CONVERTIBLE F05
WINDSCREEN-SEALS - NOT CONVERTIBLE F04

COACHLINES(TAPE) V12 MODELS.

UP TO (V)148781

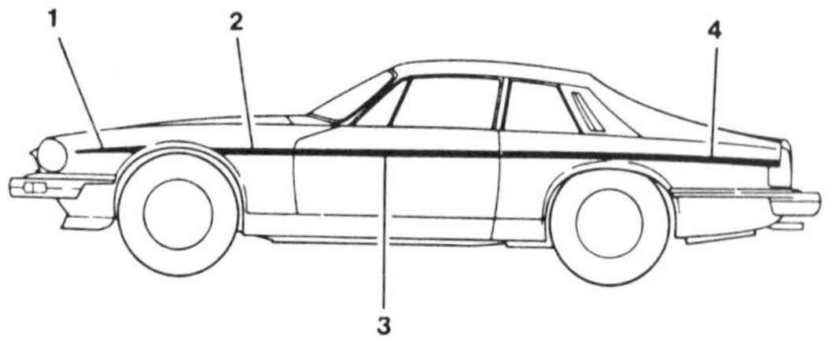

Illus	Part Number	123456 Description	Quantity	Change Point	Remarks
		COACHLINE-FRONT WING-FRONT			
	BCC 4604	Goldleaf RH	1		
1	BCC 4605	Goldleaf LH	1		
	BCC 4606	Gunmetal RH	1		
1	BCC 4607	Gunmetal LH	1		
	BCC 4608	Oyster RH	1		
1	BCC 4609	Oyster LH	1		
	BCC 4610	Dark red RH	1		
1	BCC 4611	Dark red LH	1		
		COACHLINE-FRONT WING-REAR			
	BCC 4612	Goldleaf RH	1		
2	BCC 4613	Goldleaf LH	1		
	BCC 4614	Gunmetal RH	1		
2	BCC 4615	Gunmetal LH	1		
	BCC 4616	Oyster RH	1		
2	BCC 4617	Oyster LH	1		
	BCC 4618	Dark red RH	1		
2	BCC 4619	Dark red LH	1		
		COACHLINE-DOORS			
3	BCC 4620	Goldleaf	2		
3	BCC 4621	Gunmetal	2		
3	BCC 4622	Oyster	2		
3	BCC 4623	Dark red	2		
		COACHLINE-REAR WING(TONNEAU)			
	BCC 4624	Goldleaf RH	1		
4	BCC 4625	Goldleaf LH	1		
	BCC 4626	Gunmetal RH	1		
4	BCC 4627	Gunmetal LH	1		
	BCC 4628	Oyster RH	1		
4	BCC 4629	Oyster LH	1		
	BCC 4630	Dark red RH	1		
4	BCC 4631	Dark red LH	1		

L62.02JP22/9/88

VS3182

VS3182

3F 17

MASTER INDEX

GROUP INDEX

COACHLINES (TAPE) - SPORTSPACK

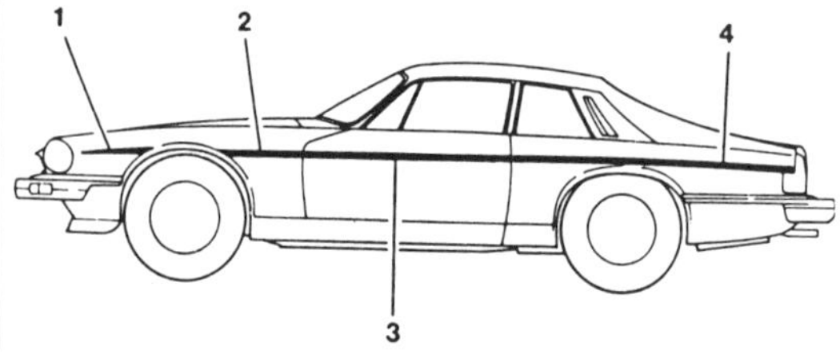

Illus	Part Number	123456 Description	Quantity	Change Point	Remarks
		COACHLINE-FRONT WING-FRONT			
	BCC 4632	Flame Red/Pewter RH	1		
1	BCC 4633	Flame Red/Pewter LH	1		
	BCC 4634	Flame Red/Quicksilver RH	1		
1	BCC 4635	Flame Red/Quicksilver LH	1		
	BDC 7252	Amber/Quicksilver RH	1		
1	BDC 7253	Amber/Quicksilver LH	1		
		COACHLINE-FRONT WING-REAR			
	BCC 4636	Flame Red/Pewter RH	1		
2	BCC 4637	Flame Red/Pewter LH	1		
	BCC 4638	Flame Red/Quicksilver RH	1		
2	BCC 4639	Flame Red/Quicksilver LH	1		
	BDC 7254	Amber/Quicksilver RH	1		
2	BDC 7255	Amber/Quicksilver LH	1		
		COACHLINES-DOORS			
3	BCC 4640	Flame Red/Pewter	2		
3	BCC 4641	Flame Red/Quicksilver	2		
3	BDC 7251	Amber/Quicksilver	2		
		COACHLINE-REAR WING (TONNEAU)			
	BCC 4642	Flame Red/Pewter RH	1		
4	BCC 4643	Flame Red/Pewter LH	1		
	BCC 4644	Flame Red/Quicksilver RH	1		
4	BCC 4645	Flame Red/Quicksilver LH	1		
	BDC 7256	Amber/Quicksilver RH	1		
4	BDC 7257	Amber/Quicksilver LH	1		

L62.04JP22/9/88

Std fit on 3.6 litre models from (V)144700, std.fit on 5.3 litre models from (V)148782

VS3182

VS3182

3F 18

MASTER INDEX

ENGINE 1.C
FLYWHEEL AND CLUTCH 1.G
GEARBOX AND PROPSHAFT 1.H
AXLES, SUSPENSION, DRIVE SHAFTS, WHEELS 1.K
STEERING 1.M
BRAKES AND BRAKE CONTROLS 1.N
FUEL, EXHAUST AND EMISSION SYSTEMS 2.C
COOLING, HEATING AND AIR CONDITIONING 2.G
ELECTRICAL, WASHERS-WIPERS, INSTRUMENTS 2.J
CHASSIS, SUBFRAMES, BODYSHELL, FITTINGS 3.C
FASCIA, TRIM, SEATS AND FIXINGS 3.H
ACCESSORIES AND PAINT 4.C
NUMERICAL INDEX 4.D

GROUP INDEX

BADGES - EMBLEMS - CONVERTIBLE G05
BADGES-EMBLEMS-CABRIOLET G04
BADGES-EMBLEMS-COUPE G03
BODY SHELL - CONVERTIBLE D03
BODY SHELL-CABRIOLET C10
BODY SHELL-COUPE C02
REAR WINGS (TONNEAU) C08
BONNET LOCK MECHANISM D15
BONNET SAFETY CATCH AND HINGES D14
BONNET-WINGS-DOORS-BOOT LID D12
BOOT HINGES-SEAL-LAMP HOUSING E14
BRIGHT FINISHERS-NOT CONVERTIBLE F14
COACHLINES(TAPE) V12 MODELS F17
COACHLINES(TAPE)-3.6 MODELS F16
CRUCIFORM AND HEATSHIELD-CABRIOLET F02
DOOR GLASS-SEALS - NOT CONVERTIBLE F06
DOOR HANDLES-LOCKS-SOLENOIDS E07
DOOR HINGES D18
DOOR SEALS AND TREAD PLATES E02
FRONT BUMPER G07
FRONT DOOR GLASS-SEALS-CONVERTIBLE F07
FRONT WINDOW REGULATORS - CONVERTIBLE E05
FUEL FILLER LID AND LOCK-COUPE C09
GEARBOX COVER PLATES E13
GEARBOX TUNNEL COVER-AUTO G/BOX MODELS E11
GEARBOX TUNNEL COVER E10
HEADLAMP SURROUNDS F12
HEATED BACKLIGHT-SEAL-COUPE F11
MIRRORS G02
REAR BUMPER G09
REAR QTR GLASS AND SEALS-CONVERTIBLE F10
REAR QTR WINDOW REGULATORS - CONVERTIBLE E06
REAR QUARTER GLASS-SEALS-CABRIOLET F09
BADGE RH-B POST F09
REAR QUARTER GLASS-SEALS-COUPE F08
SPOILER-UNDERTRAY G06
STONE GUARDS D17
TRUNK REINFORCEMENT - CONVERTIBLE F03
WEATHERSTRIPS AND TREADPLATES E03
WINDOW REGULATORS-NOT CONVERTIBLE E04
WINDSCREEN - SEALS - CONVERTIBLE F05
WINDSCREEN-SEALS - NOT CONVERTIBLE F04

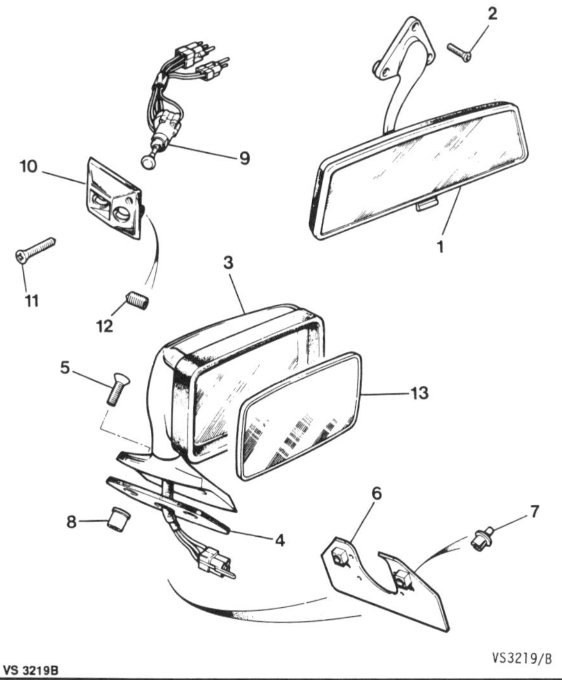

MIRRORS.

Illus	Part Number	Description	Quantity	Change Point	Remarks
1	BCC 4091	Mirror-interior	1		
2	BD 43278 3	Screw-set	1		
	BD 43278 4	Screw-set	2		
		DOOR MIRRORS-HEATED			
3	BBC 8606	RH-RHD	1		
	BBC 8607	LH-RHD	1		
	BCC 7784	RH-RHD	1		J
	BCC 7785	LH-LHD	1		J
	BBC 8608	RH-LHD	1		Not USA/CDN/J/N/S
	BBC 8609	LH-LHD	1		Not USA/CDN/J/N/S
	BBC 8610	RH-LHD	1		J/N/S
	BBC 8611	LH-LHD	1		J/N/S
	BBC 8610	RH-LHD	1		USA/CDN
	BBC 8609	LH-LHD	1		USA/CDN
4	BD 48757	Pad-mounting	2		Not Convertible
5	BD 48684 1	Screw-set	4		Not Convertible
	BDC 3014	Plinth-door mirror	2		Convertible
	BDC 3603	Screw-c/sunk	4		Convertible
6	BDC 8155	Reinforcement - mirror	2		Convertible
7	C 32312 3	Rivet - plastic	4		
8	ACU 2415	Nut-bear	4		
9	DAC 2648	Switch-door mirror	1		RHD
	DAC 2649	Switch-door mirror	1		LHD
10	BAC 3075	Escutcheon-door mirror	1		
11	AD 606042 J	Screw-self tap	2		
12	SK 104061 J	Screw-grub	2		
13	JLM 1114	Glass-mirror(flat)	A/R		
13	JLM 1115	Glass-mirror(convex)-RH	1		
	JLM 1116	Glass-mirror(convex)-LH	1		

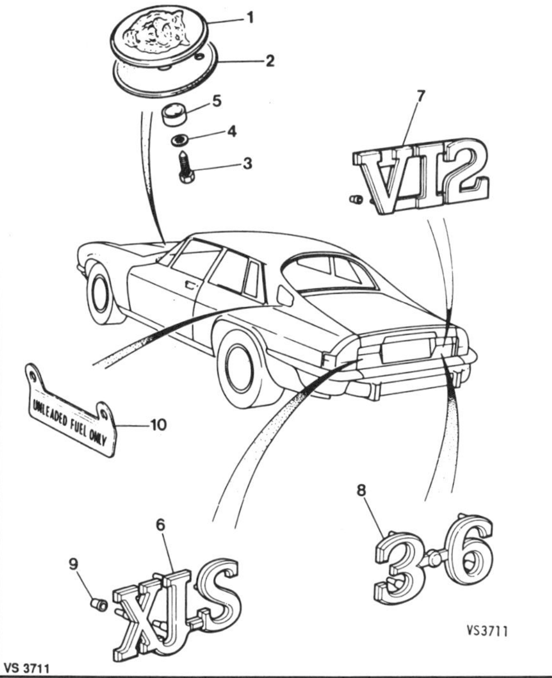

BADGES-EMBLEMS-COUPE.

Illus	Part Number	Description	Quantity	Change Point	Remarks
1	BAC 3838	Badge-bonnet	1		
2	BAC 3839	Mat-badge	1		
3	UCS 113 3R	Screw-set	1		
4	WF 702101 J	Washer-shakeproof	1		
5	BAC 3840	Washer-cup	1		
6	BCC 3665	Emblem-XJS	1		
7	BCC 3671	Emblem-V-12	1		
8	BCC 3668	Emblem-3.6	1		
9	BAC 1681	Pushnut-emblems	8		
10	BAC 6481	Badge-unleaded-fuel	1		

MASTER INDEX

ENGINE 1.C
FLYWHEEL AND CLUTCH 1.G
GEARBOX AND PROPSHAFT 1.H
AXLES, SUSPENSION, DRIVE SHAFTS, WHEELS 1.K
STEERING 1.M
BRAKES AND BRAKE CONTROLS 1.N
FUEL, EXHAUST AND EMISSION SYSTEMS 2.C
COOLING, HEATING AND AIR CONDITIONING 2.G
ELECTRICAL, WASHERS-WIPERS, INSTRUMENTS 2.J
CHASSIS, SUBFRAMES, BODYSHELL, FITTINGS 3.C
FASCIA, TRIM, SEATS AND FIXINGS 3.H
ACCESSORIES AND PAINT 4.C
NUMERICAL INDEX 4.D

GROUP INDEX

BADGES - EMBLEMS - CONVERTIBLE G05
BADGES-EMBLEMS-CABRIOLET G04
BADGES-EMBLEMS-COUPE G03
BODY SHELL - CONVERTIBLE D03
BODY SHELL-CABRIOLET C10
BODY SHELL-COUPE C02
REAR WINGS (TONNEAU) C08
BONNET LOCK MECHANISM D15
BONNET SAFETY CATCH AND HINGES D14
BONNET-WINGS-DOORS-BOOT LID D12
BOOT HINGES-SEAL-LAMP HOUSING E14
BRIGHT FINISHERS-NOT CONVERTIBLE F14
COACHLINES(TAPE) V12 MODELS F17
COACHLINES(TAPE)-3.6 MODELS F16
CRUCIFORM AND HEATSHIELD-CABRIOLET F02
DOOR GLASS-SEALS - NOT CONVERTIBLE F06
DOOR HANDLES-LOCKS-SOLENOIDS E07
DOOR HINGES D18
DOOR SEALS AND TREAD PLATES E02
FRONT BUMPER G07
FRONT DOOR GLASS-SEALS-CONVERTIBLE F07
FRONT WINDOW REGULATORS - CONVERTIBLE E05
FUEL FILLER LID AND LOCK-COUPE C09
GEARBOX COVER PLATES E13
GEARBOX TUNNEL COVER-AUTO G/BOX MODELS E11
GEARBOX TUNNEL COVER E10
HEADLAMP SURROUNDS F12
HEATED BACKLIGHT-SEAL-COUPE F11
MIRRORS G02
REAR BUMPER G09
REAR QTR GLASS AND SEALS-CONVERTIBLE F10
REAR QTR WINDOW REGULATORS - CONVERTIBLE E06
REAR QUARTER GLASS-SEALS-CABRIOLET F09
BADGE RH-B POST F09
REAR QUARTER GLASS-SEALS-COUPE F08
SPOILER-UNDERTRAY G06
STONE GUARDS D17
TRUNK REINFORCEMENT - CONVERTIBLE F03
WEATHERSTRIPS AND TREADPLATES E03
WINDOW REGULATORS-NOT CONVERTIBLE E04
WINDSCREEN - SEALS - CONVERTIBLE F05
WINDSCREEN-SEALS - NOT CONVERTIBLE F04

3G 04

JAGUAR XJS RANGE (JAN 1987 ON) | G04 fiche 3 | CHASSIS, SUBFRAMES, BODYSHELL, FITTINGS

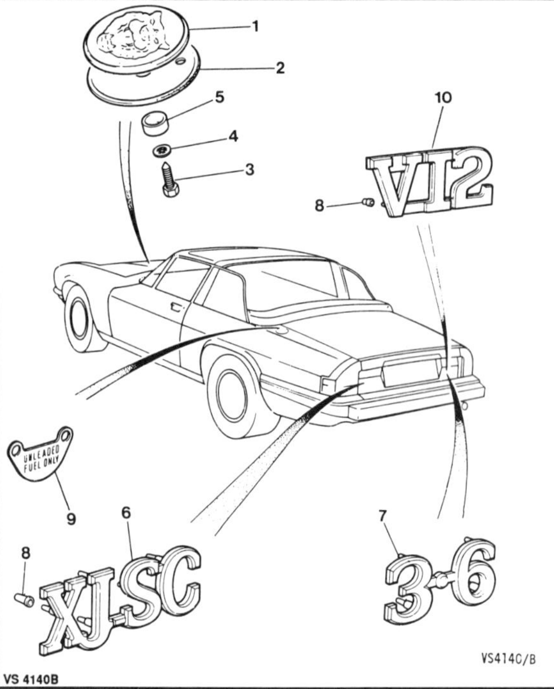

VS 4140B

BADGES-EMBLEMS-CABRIOLET

Illus	Part Number	123456	Description	Quantity	Change Point	Remarks
1	BAC 3838		Badge-bonnet	1		
2	BAC 3839		Mat-badge	1		
3	UCS 113 3R		Screw-set	1		
4	WF 702101 J		Washer-shakeproof	1		
5	BAC 3840		Washer-cup	1		
6	BCC 3667		Emblem-XJ-SC	1		
7	BCC 3668		Emblem-3.6	1		
8	BAC 1681		Pushnut-emblems	10		
9	BAC 6480		Badge-unleaded fuel	NLA		Use BDC 7261
9	BDC 7261		Badge-unleaded fuel	1		USA/CDN/AUS/J
10	BCC 3671		Emblem-V-12	1		

L66.02JP31/8/88

3G 05

JAGUAR XJS RANGE (JAN 1987 ON) | G05 fiche 3 | CHASSIS, SUBFRAMES, BODYSHELL, FITTINGS

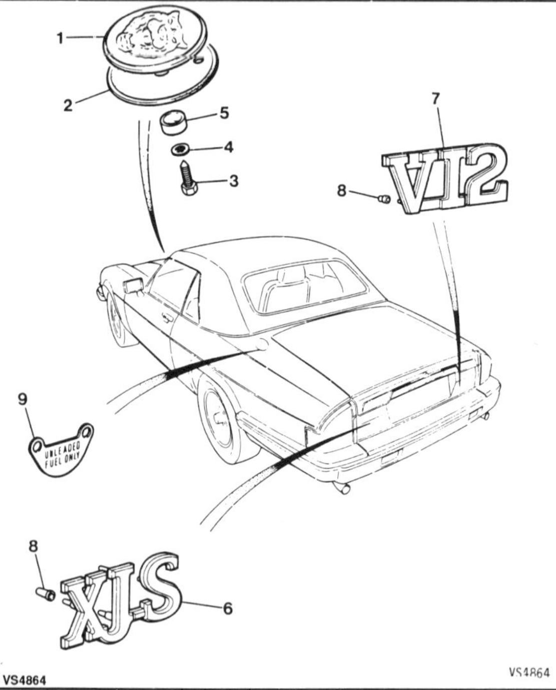

VS4864

BADGES - EMBLEMS - CONVERTIBLE

Illus	Part Number	123456	Description	Quantity	Change Point	Remarks
1	BAC 3838		Badge-bonnet	1		
2	BAC 3839		Mat-badge	1		
3	UCS 113 3R		Screw-set	1		
4	WF 702101 J		Washer-shakeproof	1		
5	BAC 3840		Washer-cup	1		
6	BCC 3665		Badge-XJ-S	1		
7	BCC 3671		Badge-V-12	1		
8	BAC 1681		Pushnut-emblems	8		
9	BAC 6480		Badge-unleaded fuel	NLA		Use BDC 7261
9	BDC 7261		Badge-unleaded fuel	1		

L66.04JP1/9/88

MASTER INDEX

ENGINE 1.C
FLYWHEEL AND CLUTCH 1.G
GEARBOX AND PROPSHAFT 1.H
AXLES, SUSPENSION, DRIVE SHAFTS, WHEELS 1.K
STEERING 1.M
BRAKES AND BRAKE CONTROLS 1.N
FUEL, EXHAUST AND EMISSION SYSTEMS 2.C
COOLING, HEATING AND AIR CONDITIONING 2.G
ELECTRICAL, WASHERS-WIPERS, INSTRUMENTS 2.J
CHASSIS, SUBFRAMES, BODYSHELL, FITTINGS 3.C
FASCIA, TRIM, SEATS AND FIXINGS 3.H
ACCESSORIES AND PAINT 4.C
NUMERICAL INDEX 4.D

GROUP INDEX

BADGES - EMBLEMS - CONVERTIBLE G05
BADGES-EMBLEMS-CABRIOLET G04
BADGES-EMBLEMS-COUPE. G03
BODY SHELL - CONVERTIBLE D03
BODY SHELL-CABRIOLET C10
BODY SHELL-COUPE. C02
REAR WINGS (TONNEAU) C08
BONNET LOCK MECHANISM D15
BONNET SAFETY CATCH AND HINGES D14
BONNET-WINGS-DOORS-BOOT LID. D12
BOOT HINGES-SEAL-LAMP HOUSING. E14
BRIGHT FINISHERS-NOT CONVERTIBLE F14
COACHLINES(TAPE) V12 MODELS. F17
COACHLINES(TAPE)-3.6 MODELS. F16
CRUCIFORM AND HEATSHIELD-CABRIOLET F02
DOOR GLASS-SEALS - NOT CONVERTIBLE F06
DOOR HANDLES-LOCKS-SOLENOIDS E07
DOOR HINGES D18
DOOR SEALS AND TREAD PLATES. E02
FRONT BUMPER. G07
FRONT DOOR GLASS-SEALS-CONVERTIBLE F07
FRONT WINDOW REGULATORS - CONVERTIBLE E05
FUEL FILLER LID AND LOCK-COUPE C09
GEARBOX COVER PLATES. E13
GEARBOX TUNNEL COVER-AUTO G/BOX MODELS E11
GEARBOX TUNNEL COVER. E10
HEADLAMP SURROUNDS. F12
HEATED BACKLIGHT-SEAL-COUPE. F11
MIRRORS. G02
REAR BUMPER. G09
REAR QTR GLASS AND SEALS-CONVERTIBLE F10
REAR QTR WINDOW REGULATORS - CONVERTIBLE E06
REAR QUARTER GLASS-SEALS-CABRIOLET F09
BADGE RH-B POST F09
REAR QUARTER GLASS-SEALS-COUPE. F08
SPOILER-UNDERTRAY. G06
STONE GUARDS. D17
TRUNK REINFORCEMENT - CONVERTIBLE F03
WEATHERSTRIPS AND TREADPLATES E03
WINDOW REGULATORS-NOT CONVERTIBLE E04
WINDSCREEN - SEALS - CONVERTIBLE F05
WINDSCREEN-SEALS - NOT CONVERTIBLE F04

SPOILER-UNDERTRAY.

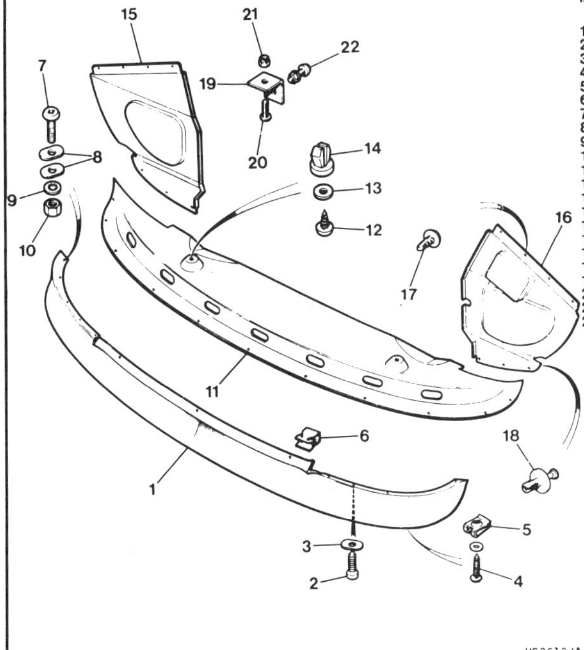

VS 3613A

Illus	Part Number	123456 Description	Quantity	Change Point	Remarks
1	BD 45624	Spoiler-front	1		
2	AR 610051 J	Screw-self-tap	8		
3	BD 542 9	Washer-oval	8		
4	AGU 1413 J	Screw-self tap	11		
5	AGU 2380 J	Nut-spire	11		
6	BD 21962	Clip	2		
7	UFS 819 4H	Screw-set	2		
8	BD 542 9	Washer-oval	4		
9	WF 702101 J	Washer-shakeproof	2		
10	UFN 119 L	Nut-plain	2		
11	BCC 4462	Undertray	1		
12	JLM 9577	Screw-self-tap	2		
13	WC 702101 J	Washer-plain	2		
14	C 31973 3	Nut-plastic	2		
15	BAC 4350	Baffle-RH	1		
16	BAC 4352	Baffle-LH	1		
17	JLM 9579	Fastener-drive	8		
18	BD 47854	Rivet-plastic	6		
19	BAC 4450	Bracket	4		
20	SE 604041 J	Screw-set	4		
21	C 8667 7	Nut-self-lock	4		
22	BAC 4451	Screw and nut-plastic	4		

FRONT BUMPER.

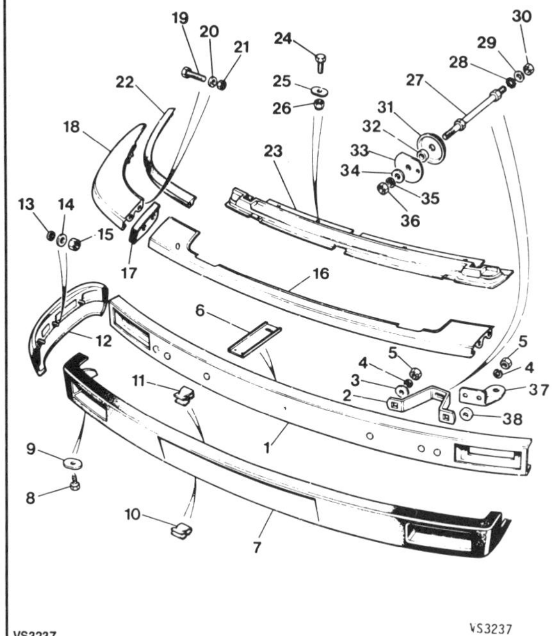

VS3237

Illus	Part Number	123456 Description	Quantity	Change Point	Remarks
1	BCC 5120	Beam-bumper	1		
2	BAC 3426	Bracket-sides	2		
3	BD 541 43	Washer-plain	4		
4	WM 600061 J	Washer-spring	4		
5	JLM 9689	Nut-plain	4		
6	BAC 3505	Bracket-centre	1		
7	BAC 1331	Cover-beam	1		
8	AZ 614061 J	Screw-self-tap	2		
9	BD 542 8	Washer-oval	2		
10	BD 21962	Clip-retainer	9		
11	BD 48049	Clip-retainer	8		
12	BAC 2934	Finisher-RH-side	1		
	BAC 2935	Finisher-LH-side	1		
13	WM 600041 J	Washer-spring	6		
14	BD 542 8	Washer-plain	6		
15	JLM 9683	Nut-plain	6		
16	BAC 3347	Blade-bumper	1	....	Use BCC3435
16	BCC 3435	Blade-bumper	1		
17	BAC 3480	Joint-RH	1		
	BAC 3481	Joint-LH	1		
18	BAC 3348	Blade-RH-side	1	....	Use BCC3436
18	BCC 3436	Blade-RH-side	1		
	BAC 3349	Blade-LH-side	1	....	Use BCC3437
	BCC 3437	Blade-LH-side	1		
19	UFS 131 6V	Screw-set	4		
20	WM 600041 J	Washer-spring	4		
21	JLM 9684	Nut-plain	4		
22	BAC 4383	Finisher-edge	2		
23	BAC 3504	Plinth-bumper	1		
24	SH 604051 J	Screw-set	7		
25	BD 542 8	Washer-oval	18		
26	JLM 9690	Nut-self-lock	13		
27	BAC 3074	Bolt-mounting	2		
28	WE 600062 J	Washer-shakeproof	2		
29	FW 206 T	Washer-plain	2		
30	AFU 1082 J	Nut-locking	2		
31	BD 44645	Locator	2		
32	C 37900	Spacer	2		
33	BAC 1357	Plate-support	2		
34	BD 541 43	Washer-plain	2		
35	WL 110001 J	Washer-spring	2		
36	JLM 9683	Nut-plain	2		
	BCC 5116	Bracket-fog lamp-RH	1		
37	BCC 5117	Bracket-fog lamp-LH	1		
38	BCC 5115	Spacer	2		

MASTER INDEX

ENGINE 1.C
FLYWHEEL AND CLUTCH 1.G
GEARBOX AND PROPSHAFT 1.H
AXLES, SUSPENSION, DRIVE SHAFTS, WHEELS 1.K
STEERING 1.M
BRAKES AND BRAKE CONTROLS 1.N
FUEL, EXHAUST AND EMISSION SYSTEMS 2.C
COOLING, HEATING AND AIR CONDITIONING 2.G
ELECTRICAL, WASHERS-WIPERS, INSTRUMENTS 2.J
CHASSIS, SUBFRAMES, BODYSHELL, FITTINGS 3.C
FASCIA, TRIM, SEATS AND FIXINGS 3.H
ACCESSORIES AND PAINT 4.C
NUMERICAL INDEX 4.D

GROUP INDEX

BADGES - EMBLEMS - CONVERTIBLE G05
BADGES-EMBLEMS-CABRIOLET G04
BADGES-EMBLEMS-COUPE G03
BODY SHELL - CONVERTIBLE D03
BODY SHELL-CABRIOLET C10
BODY SHELL-COUPE C02
REAR WINGS (TONNEAU) C08
BONNET LOCK MECHANISM D15
BONNET SAFETY CATCH AND HINGES D14
BONNET-WINGS-DOORS-BOOT LID D12
BOOT HINGES-SEAL-LAMP HOUSING E14
BRIGHT FINISHERS-NOT CONVERTIBLE F14
COACHLINES(TAPE) V12 MODELS F17
COACHLINES(TAPE)-3.6 MODELS F16
CRUCIFORM AND HEATSHIELD-CABRIOLET F02
DOOR GLASS-SEALS - NOT CONVERTIBLE F06
DOOR HANDLES-LOCKS-SOLENOIDS E07
DOOR HINGES D18
DOOR SEALS AND TREAD PLATES E02
FRONT BUMPER G07
FRONT DOOR GLASS-SEALS-CONVERTIBLE F07
FRONT WINDOW REGULATORS - CONVERTIBLE E05
FUEL FILLER LID AND LOCK-COUPE C09
GEARBOX COVER PLATES E13
GEARBOX TUNNEL COVER-AUTO G/BOX MODELS E11
GEARBOX TUNNEL COVER E10
HEADLAMP SURROUNDS F12
HEATED BACKLIGHT-SEAL-COUPE F11
MIRRORS G02
REAR BUMPER G09
REAR QTR GLASS AND SEALS-CONVERTIBLE F10
REAR QTR WINDOW REGULATORS - CONVERTIBLE E06
REAR QUARTER GLASS-SEALS-CABRIOLET F09
BADGE RH-B POST F09
REAR QUARTER GLASS-SEALS-COUPE F08
SPOILER-UNDERTRAY G06
STONE GUARDS D17
TRUNK REINFORCEMENT - CONVERTIBLE F03
WEATHERSTRIPS AND TREADPLATES E03
WINDOW REGULATORS-NOT CONVERTIBLE E04
WINDSCREEN - SEALS - CONVERTIBLE F05
WINDSCREEN-SEALS - NOT CONVERTIBLE F04

3G 08

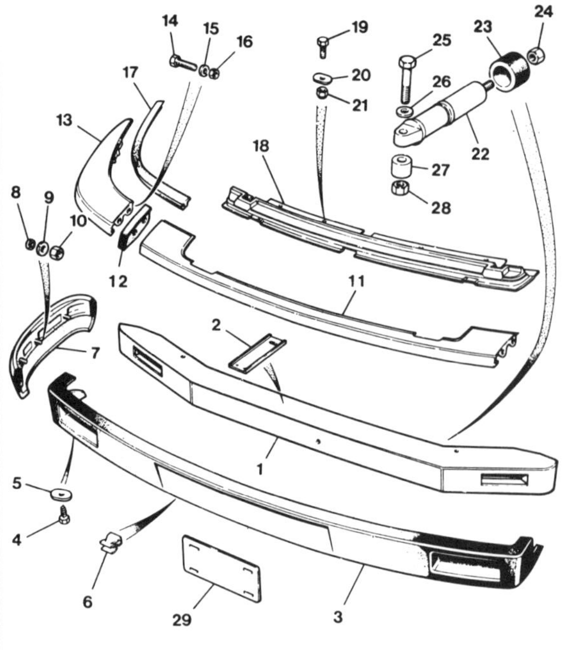

VS3319

VS3319

FRONT BUMPER-WITH ABSORBERS.

Illus	Part Number	Description	Quantity	Change Point	Remarks
1	BCC 5121	Beam-bumper	1		
2	BAC 3505	Bracket-centre	1		
3	BAC 1170	Cover-beam	1		
4	AZ 614061 J	Screw-self-tap	9		
5	BD 592 8	Washer-oval	9		
6	BD 21962	Clip-retainer	9		
7	BAC 2934	Finisher-RH-side	1		
	BAC 2935	Finisher-LH-side	1		
8	WM 604041 J	Washer-spring	6		
9	BD 542 8	Washer-plain	6		
10	JLM 9683	Nut-plain	6		
11	BAC 3347	Blade-bumper	1		Use BCC3435
11	BCC 3435	Blade-bumper	1		
12	BAC 3480	Joint-RH	1		
	BAC 3481	Joint-LH	1		
13	BAC 3348	Blade-RH-side	1		Use BCC3436
13	BCC 3436	Blade-RH-side	1		
	BAC 3349	Blade-LH-side	1		Use BCC3437
	BCC 3437	Blade-LH-side	1		
14	UFS 131 6V	Screw-set	4		
15	WM 600041 J	Washer-spring	4		
16	JLM 9683	Nut-plain	4		
17	BAC 4383	Finisher-edge	2		
18	BAC 3504	Plinth-bumper	1		
19	SH 604051 J	Screw-set	7		
20	BD 542 8	Washer-oval	18		
21	JLM 9690	Nut-self-lock	13		
22	BD 48649	Absorber-impact	2		
23	BRC 3242	Mounting-rubber	2		
24	AFU 1082 J	Nut-locking	2		
25	BH 608271 J	Bolt	2		
26	FW 206 T	Washer-plain	2		
27	BAC 1166	Bush	2		
28	NY 608041 J	Nut-plain	2		
29	BCC 4102	Reinforcement-licence plate	1		USA/CDN only

MASTER INDEX

GROUP INDEX

3G 09

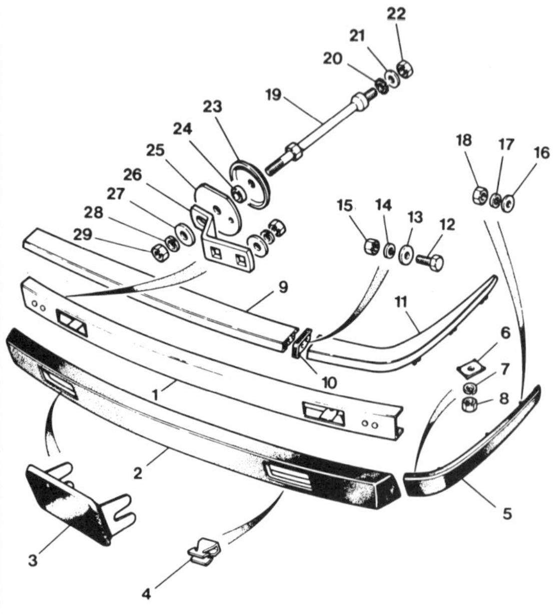

VS 3238

VS3238

REAR BUMPER.

Illus	Part Number	Description	Quantity	Change Point	Remarks
1	BAC 1338	Beam-bumper	1		
2	BAC 1337	Cover-beam	1		
3	BAC 1401	Filler-foglamp aperture	A/R		
4	BD 21962	Clip-retainer	9		
5	BAC 2928	Finisher-RH-side	1		
	BAC 2929	Finisher-LH-side	1		
6	BAC 2446	Washer-plate	10		
7	JLM 302	Washer-spring	10		
8	JLM 9690	Nut-self-lock	10		
9	BAC 3361	Blade-bumper	1		Use BCC3434
9	BCC 3434	Blade-bumper	1		
10	BAC 3479	Joint	2		
11	BAC 3362	Blade-RH-side	1		Use BCC3432
11	BCC 3432	Blade-RH-side	1		
	BAC 3363	Blade-LH-side	1		Use BCC3433
	BCC 3433	Blade-LH-side	1		
12	UFS 131 6V	Screw-set	4		
13	WC 108051 J	Washer-plain	8		
14	JLM 303	Washer-spring	4		
15	JLM 9684	Nut-plain	4		
16	BD 542 8	Washer-plain	6		
17	JLM 302	Washer-spring	6		
18	JLM 9690	Nut-self-lock	6		
19	BAC 3074	Bolt-mounting	2		
20	WE 600061 J	Washer-shakeproof	2		
21	FW 206 T	Washer-plain	2		
22	AFU 1082 J	Nut-locking	2		
23	BD 44645	Locator	2		
24	C 37900	Spacer	2		
25	BAC 1357	Plate-support	2		
26	BAC 2936	Bracket-RH-support	1		
	BAC 2937	Bracket-LH-support	1		
27	BD 541 43	Washer-plain	6		
28	WM 600061 J	Washer-spring	6		
29	JLM 9683	Nut-plain	6		

MASTER INDEX
ENGINE 1.C
FLYWHEEL AND CLUTCH 1.G
GEARBOX AND PROPSHAFT 1.H
AXLES, SUSPENSION, DRIVE SHAFTS, WHEELS 1.K
STEERING 1.M
BRAKES AND BRAKE CONTROLS 1.N
FUEL, EXHAUST AND EMISSION SYSTEMS 2.C
COOLING, HEATING AND AIR CONDITIONING 2.G
ELECTRICAL, WASHERS-WIPERS, INSTRUMENTS 2.J
CHASSIS, SUBFRAMES, BODYSHELL, FITTINGS 3.C
FASCIA, TRIM, SEATS AND FIXINGS 3.H
ACCESSORIES AND PAINT 4.C
NUMERICAL INDEX 4.D
GROUP INDEX
BADGES - EMBLEMS - CONVERTIBLE G05
BADGES-EMBLEMS-CABRIOLET G04
BADGES-EMBLEMS-COUPE. G03
BODY SHELL - CONVERTIBLE D03
BODY SHELL-CABRIOLET C10
BODY SHELL-COUPE. C02
REAR WINGS (TONNEAU) C08
BONNET LOCK MECHANISM D15
BONNET SAFETY CATCH AND HINGES D14
BONNET-WINGS-DOORS-BOOT LID. D12
BOOT HINGES-SEAL-LAMP HOUSING. E14
BRIGHT FINISHERS-NOT CONVERTIBLE F14
COACHLINES(TAPE) V12 MODELS. F17
COACHLINES(TAPE)-3.6 MODELS. F16
CRUCIFORM AND HEATSHIELD-CABRIOLET F02
DOOR GLASS-SEALS - NOT CONVERTIBLE F06
DOOR HANDLES-LOCKS-SOLENOIDS E07
DOOR HINGES D18
DOOR SEALS AND TREAD PLATES. E02
FRONT BUMPER. G07
FRONT DOOR GLASS-SEALS-CONVERTIBLE F07
FRONT WINDOW REGULATORS - CONVERTIBLE E05
FUEL FILLER LID AND LOCK-COUPE C09
GEARBOX COVER PLATES. E13
GEARBOX TUNNEL COVER-AUTO G/BOX MODELS E11
GEARBOX TUNNEL COVER. E10
HEADLAMP SURROUNDS. F12
HEATED BACKLIGHT-SEAL-COUPE. F11
MIRRORS. G02
REAR BUMPER. G09
REAR QTR GLASS AND SEALS-CONVERTIBLE F10
REAR QTR WINDOW REGULATORS - CONVERTIBLE E06
REAR QUARTER GLASS-SEALS-CABRIOLET F09
BADGE RH-B POST F09
REAR QUARTER GLASS-SEALS-COUPE. F08
SPOILER-UNDERTRAY. G06
STONE GUARDS. D17
TRUNK REINFORCEMENT - CONVERTIBLE F03
WEATHERSTRIPS AND TREADPLATES E03
WINDOW REGULATORS-NOT CONVERTIBLE E04
WINDSCREEN - SEALS - CONVERTIBLE F05
WINDSCREEN-SEALS - NOT CONVERTIBLE F04

REAR BUMPER-WITH ABSORBERS

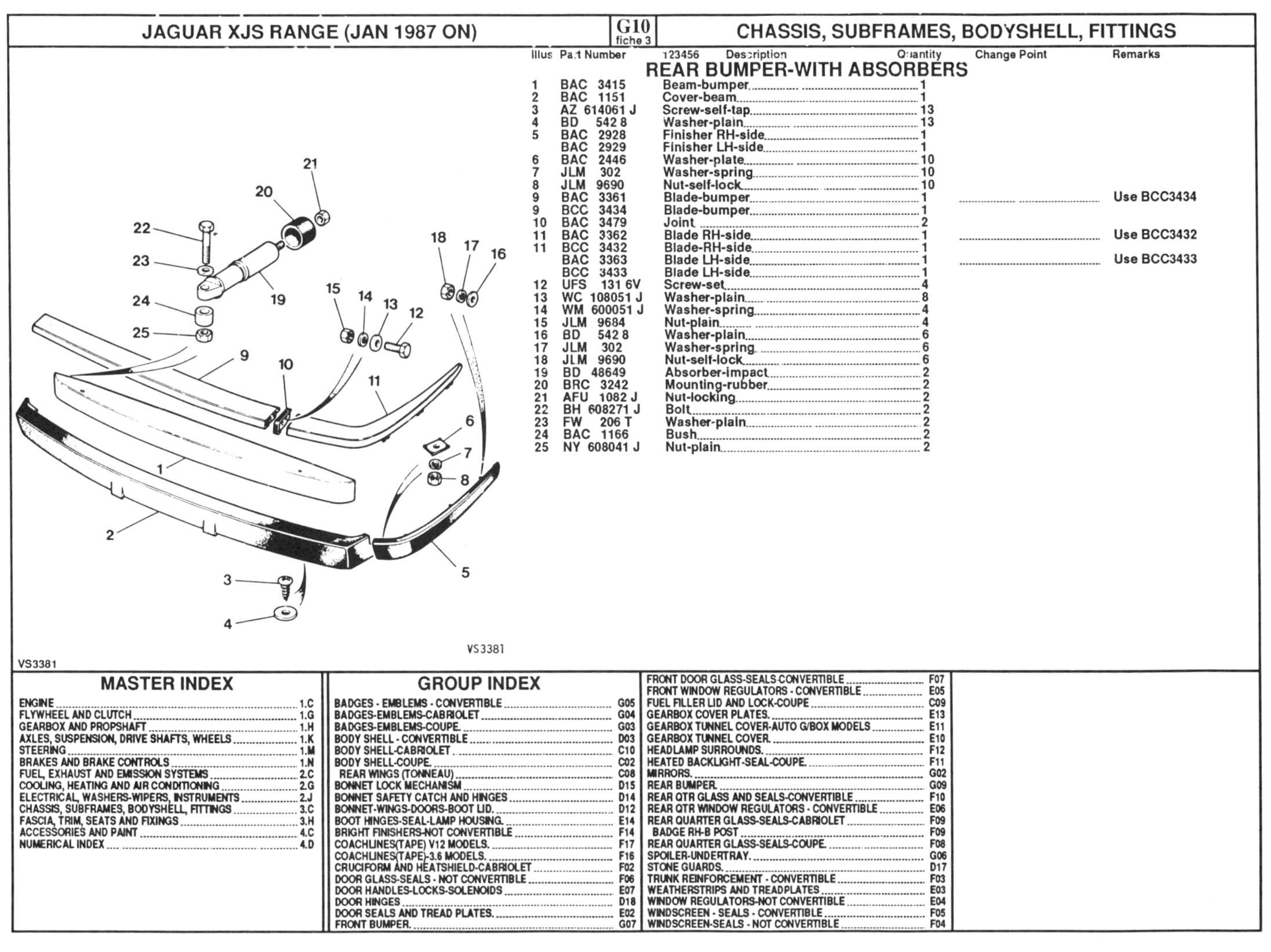

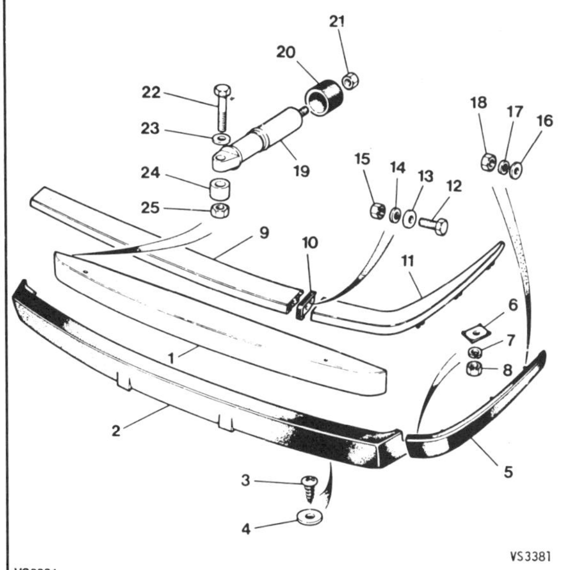

VS3381

Illus	Part Number	123456	Description	Quantity	Change Point	Remarks
1	BAC 3415		Beam-bumper	1		
2	BAC 1151		Cover-beam	1		
3	AZ 614061 J		Screw-self-tap	13		
4	BD 542 8		Washer-plain	13		
5	BAC 2928		Finisher RH-side	1		
	BAC 2929		Finisher LH-side	1		
6	BAC 2446		Washer-plate	10		
7	JLM 302		Washer-spring	10		
8	JLM 9690		Nut-self-lock	10		
9	BAC 3361		Blade-bumper	1		Use BCC3434
9	BCC 3434		Blade-bumper	1		
10	BAC 3479		Joint	2		
11	BAC 3362		Blade RH-side	1		Use BCC3432
11	BCC 3432		Blade-RH-side	1		
	BAC 3363		Blade LH-side	1		Use BCC3433
	BCC 3433		Blade LH-side	1		
12	UFS 131 6V		Screw-set	4		
13	WC 108051 J		Washer-plain	8		
14	WM 600051 J		Washer-spring	4		
15	JLM 9684		Nut-plain	4		
16	BD 542 8		Washer-plain	6		
17	JLM 302		Washer-spring	6		
18	JLM 9690		Nut-self-lock	6		
19	BD 48649		Absorber-impact	2		
20	BRC 3242		Mounting-rubber	2		
21	AFU 1082 J		Nut-locking	2		
22	BH 608271 J		Bolt	2		
23	FW 206 T		Washer-plain	2		
24	BAC 1166		Bush	2		
25	NY 608041 J		Nut-plain	2		

MASTER INDEX

GROUP INDEX

MASTER INDEX

GROUP INDEX

JAGUAR XJS RANGE (JAN 1987 ON)

H02 fiche 3

FASCIA, TRIM, SEATS AND FIXINGS

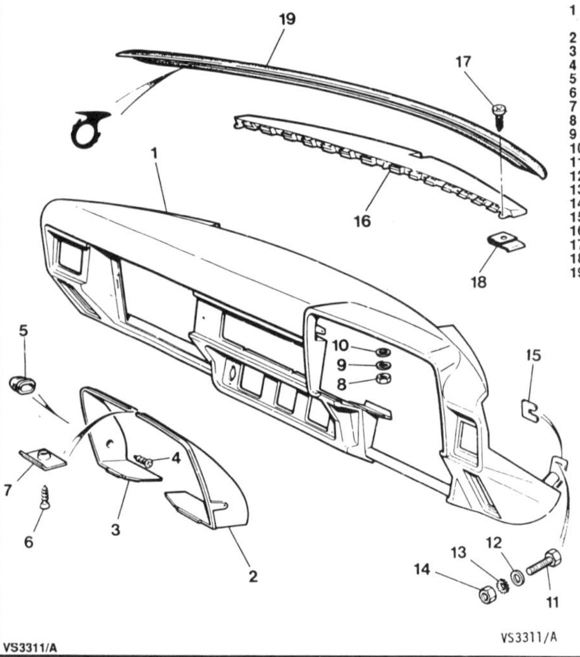

VS3311/A

FACIA PANEL.

Illus	Part Number	123456	Description	Quantity	Change Point	Remarks
1	BBC 5546		Facia panel-RHD	1		
	BBC 5547		Facia panel-LHD	1		
2	BBC 5486		Surround-instrument RH	1		
3	BBC 5487		Surround-instrument LH	1		
4	DAB 6 8C		Screw-self-tap	2		
5	BBC 5498		Nut-lokut	2		
6	DBB 6 6C		Screw-self-tap	1		
7	BD 47814		Retainer-surround	1		
8	JLM 9683		Nut-plain	2		
9	JLM 302		Washer-spring	2		
10	JLM 296		Washer-plain	2		
11	SH 605101 J		Screw-set	2		
12	C 15183		Washer-plain	4		
13	C 725		Washer-shakeproof	2		
14	JLM 9684		Nut-plain	2		
15	BD 24295		Packer	A/R		
16	BD 46439		Duct-demister	1		
17	DBB 808 8C		Screw-self-tap	2		
18	AJ 608025 J		Nut-spire	2		
19	BDC 8054		Seal-facia to windscreen	1		Convertible only

MASTER INDEX

GROUP INDEX

JAGUAR XJS RANGE (JAN 1987 ON)

H03 fiche 3

FASCIA, TRIM, SEATS AND FIXINGS

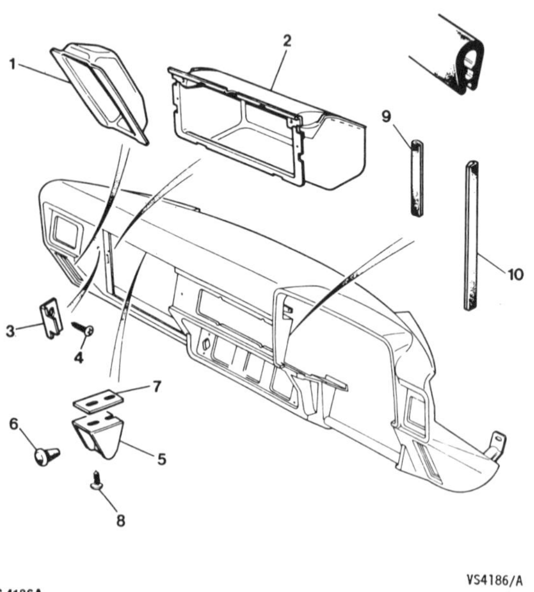

VS 4186A

GLOVEBOX-LIGHT SHIELDS.

Illus	Part Number	123456	Description	Quantity	Change Point	Remarks
1	BAC 4474		Shield-light	2		
2	BCC 5090		Box-glove	1		Use BDC 6496
2	BDC 6494		Box-glove	1		
3	BD 47898		Bracket-impact	2		
4	DAZ 406 8C		Screw-self-tap	4		
5	BCC 3441		Striker	1		
6	BCC 5329		Buffer-rubber	1		
7	BD 46831		Packing	A/R		
8	DAB 706 6C		Screw-self-tap	2		
9	BD 45909 2		Finisher-flange	1		
10	BD 44643 1		Finisher-flange	1		

MASTER INDEX

ENGINE 1.C
FLYWHEEL AND CLUTCH 1.G
GEARBOX AND PROPSHAFT 1.H
AXLES, SUSPENSION, DRIVE SHAFTS, WHEELS 1.K
STEERING 1.M
BRAKES AND BRAKE CONTROLS 1.N
FUEL, EXHAUST AND EMISSION SYSTEMS 2.C
COOLING, HEATING AND AIR CONDITIONING 2.G
ELECTRICAL, WASHERS-WIPERS, INSTRUMENTS 2.J
CHASSIS, SUBFRAMES, BODYSHELL, FITTINGS 3.C
FASCIA, TRIM, SEATS AND FIXINGS 3.H
ACCESSORIES AND PAINT 4.C
NUMERICAL INDEX 4.D

GROUP INDEX

BATTERY AND SPARE WHEEL COVERS. N05
BONNET INSULATION-3.6 LITRE N06
BONNET INSULATION-5.3 MODELS N08
BONNET INSULATION-5.3 MODELS. N07
BOOT CARPETS - NOT CONVERTIBLE N03
BOOT CARPETS-CONVERTIBLE N04
CARPETS-'A' POST FINISHERS-. L12
CONSOLE-AMBLA. H10
CONSOLE-LEATHER. H08
DOOR CASINGS-ARMRESTS AND POCKETS-AMBLA. H16
DOOR CASINGS-ARMRESTS-CONVERTIBLE I02
HEADLINING I15
DOOR CASINGS-ARMRESTS-LEATHER. H14
EXTERIOR TRIM COVERS N15
FACIA PANEL. H02
FRONT SEAT BELTS-CONVERTIBLE K15
FRONT SEAT SLIDES K16
MOTOR AND ANCHORAGE ASSY-RH L02
FRONT SEATS-AMBLA/CLOTH.-3.6 LITRE K04
FRONT SEATS-LEATHER. J03
FRONT SEATS-LEATHER/CLOTH-5.3 LITRE K07
FRT SEATS-LEATHER/CLOTH(SPORTSPACK) K10
FRONT BELT KIT-COUPE K14
FRONT SEAT SLIDES K13
FUEL APERATURE SEAL AND HOSES-COUPE. N13
HARDTOP & HOODBAG-CABRIOLET. N17
HOOD LIFT MECHANISM - CONVERTIBLE O04
HOOD SEALS-CONVERTIBLE O03
HOOD-CABRIOLET N16
HOOD-CONVERTIBLE O02
REAR CUBBY BOX-CONVERTIBLE L09
REAR LUGGAGE LOCKER-CABRIOLET L07
REAR SEAT BELTS. L11
REAR SEATS-LEATHER-COUPE. L03
REAR SEATS-LEATHER-COUPE. L04
REAR SEATS-LEATHER/CLOTH-COUPE. L06
TARGA PANELS-STOWAGEBAG-CABRIOLET N14
UNDER SCUTTLE CASINGS. H05
VENEERED FACIA PANELS-3.6 LITRE. H06
VENEERED FACIA PANELS-5.3 LITRE. H07

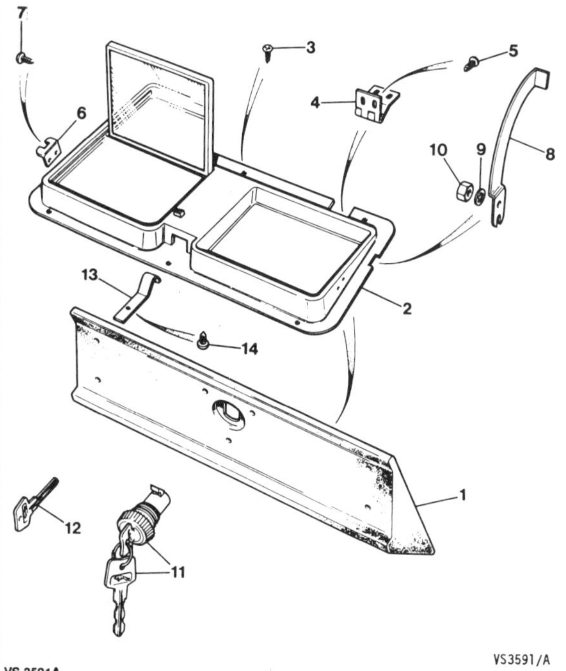

VS3591/A

VS 3591A

GLOVEBOX LID

Illus	Part Number	123456 Description	Quantity	Change Point	Remarks
1	BCC 2537	Lid-glovebox	1		
2	BCC 5129	Tray-vanity	1		Use BCC8633
2	BCC 8633	Tray-vanity	1		
3	DBB 6 4C	Screw-self-tap	6		
4	BD 45006	Hinge	2		
5	DBB 806 4C	Screw-self-tap	8		
6	BD 47897	Bracket-impact	2		
7	DBB 6 4C	Screw-self-tap	4		
8	BD 45555	Checkarm RH	1		
	BD 45556	Checkarm LH	1		
9	WF 703081 J	Washer-shakeproof	2		
10	UCN 116 L	Nut-plain	2		
11	BCC 2512	Lock and knob	1		
12		Key-blank	A/R		See locksets,Grp N
13	BD 45612	Spring-leaf	1		
14	DBZ 804 4C	Screw-self-tap	1		

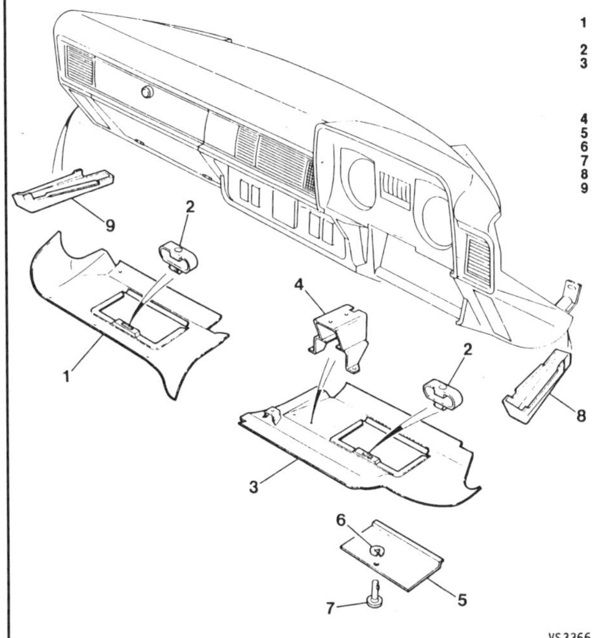

VS3366

VS 3366

UNDER SCUTTLE CASINGS.

Illus	Part Number	123456 Description	Quantity	Change Point	Remarks
1	BCC 6459 LZ	Casing-under scuttle-LH-RHD	1		Not passive restraint
	BCC 6460 LZ	Casing-under scuttle-RH-LHD	1		
2	BD 42805	Clip-casing	1		
3	BCC 6458 LZ	Casing-under scuttle-RH-RHD	1		
	BCC 6461 LZ	Casing-under scuttle-LH-LHD	1		
	BCC 6430 LZ	Casing-knee panel RH-LHD	1		With passive restraint
	BCC 6433 LZ	Casing-knee panel LH-LHD	1		
4	C 46205	Bracket-reostat	1		
5	BCC 8454	Cover-fuse aperture	2		
6	AAU 2364	Retainer-cover	1		
7	AAU 2365	Stud-cover	1		
8	BCC 4550	Finisher RH	1		
9	BCC 4551	Finisher LH	1		

MASTER INDEX

ENGINE 1.C
FLYWHEEL AND CLUTCH 1.G
GEARBOX AND PROPSHAFT 1.H
AXLES, SUSPENSION, DRIVE SHAFTS, WHEELS 1.K
STEERING 1.M
BRAKES AND BRAKE CONTROLS 1.N
FUEL, EXHAUST AND EMISSION SYSTEMS 2.C
COOLING, HEATING AND AIR CONDITIONING 2.G
ELECTRICAL, WASHERS-WIPERS, INSTRUMENTS 2.J
CHASSIS, SUBFRAMES, BODYSHELL, FITTINGS 3.C
FASCIA, TRIM, SEATS AND FIXINGS 3.H
ACCESSORIES AND PAINT 4.C
NUMERICAL INDEX 4.D

GROUP INDEX

BATTERY AND SPARE WHEEL COVERS. N05
BONNET INSULATION-3.6 LITRE N06
BONNET INSULATION-5.3 MODELS N08
BONNET INSULATION-5.3 MODELS. N07
BOOT CARPETS - NOT CONVERTIBLE N03
BOOT CARPETS-CONVERTIBLE N04
CARPETS-'A' POST FINISHERS-. L12
CONSOLE-AMBLA. H10
CONSOLE-LEATHER. H08
DOOR CASINGS-ARMRESTS AND POCKETS-AMBLA. H16
DOOR CASINGS-ARMRESTS-CONVERTIBLE I02
HEADLINING I15
DOOR CASINGS-ARMRESTS-LEATHER. H14
EXTERIOR TRIM COVERS N15
FACIA PANEL. H02
FRONT SEAT BELTS-CONVERTIBLE K15
FRONT SEAT SLIDES K16
MOTOR AND ANCHORAGE ASSY-RH L02
FRONT SEATS-AMBLA/CLOTH.-3.6 LITRE K04
FRONT SEATS-LEATHER. J03
FRONT SEATS-LEATHER/CLOTH-5.3 LITRE K07
FRT SEATS-LEATHER/CLOTH(SPORTSPACK) K10
FRONT BELT KIT-COUPE K14
FRONT SEAT SLIDES K13
FUEL APERATURE SEAL AND HOSES-COUPE. N13
HARDTOP & HOODBAG-CABRIOLET. N17
HOOD LIFT MECHANISM - CONVERTIBLE O04
HOOD SEALS-CONVERTIBLE O03
HOOD-CABRIOLET N16
HOOD-CONVERTIBLE O02
REAR CUBBY BOX-CONVERTIBLE L09
REAR LUGGAGE LOCKER-CABRIOLET L07
REAR SEAT BELTS. L11
REAR SEATS-LEATHER-COUPE. L03
REAR SEATS-LEATHER-COUPE. L04
REAR SEATS-LEATHER/CLOTH-COUPE. L06
TARGA PANELS-STOWAGEBAG-CABRIOLET N14
UNDER SCUTTLE CASINGS. H05
VENEERED FACIA PANELS-3.6 LITRE. H06
VENEERED FACIA PANELS-5.3 LITRE. H07

JAGUAR XJS RANGE (JAN 1987 ON) | H06 fiche 3 | FASCIA, TRIM, SEATS AND FIXINGS

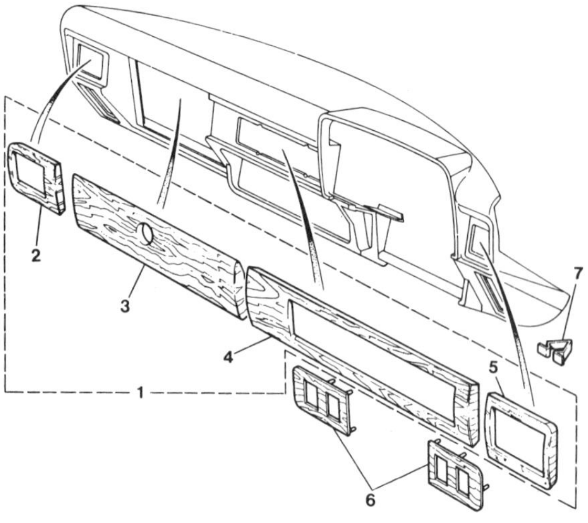

VS 3699A

VENEERED FACIA PANELS-3.6 LITRE.

Illus	Part Number	Description	Quantity	Change Point	Remarks
1	RTC 2769	Kit-veneered facia panels-RHD	1		
	RTC 2770	Kit-veneered facia panels-LHD	1		
2	BAC 2983	Panel-outer	1		Nearside
3	BAC 2986	Panel-glovebox lid	1		
4	BAC 2984	Panel-centre	1		
5	BAC 2980	Panel-outer	1		Offside
6	BAC 5655	Panel-switch finisher	2		With trip computer
	BBC 9299	Panel-switch finisher	1		With clock
7	BD 11309	Clip-panel fixing	8		

123456

MASTER INDEX

GROUP INDEX

JAGUAR XJS RANGE (JAN 1987 ON) | H07 fiche 3 | FASCIA, TRIM, SEATS AND FIXINGS

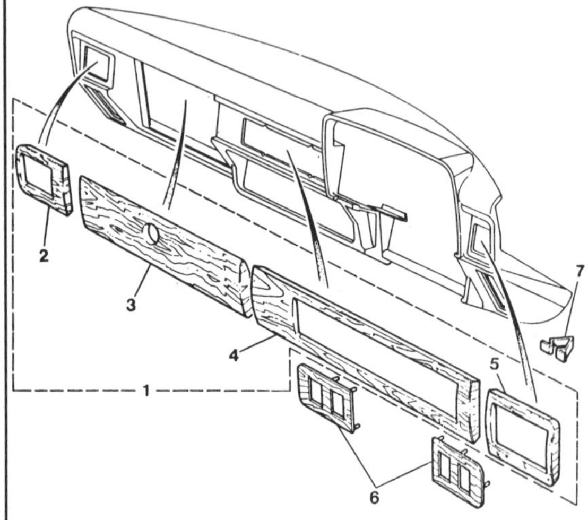

VS 3699A

VENEERED FACIA PANELS-5.3 LITRE.

Illus	Part Number	Description	Quantity	Change Point	Remarks
1	BAC 3026	KIT-VENEERED FACIA PANELS-RHD	1		
	BAC 3027	KIT-VENEERED FACIA PANELS-LHD	1		
2	BAC 2975	Panel-outer	1		Nearside
3	BAC 2978	Panel-glovebox lid	1		
4	BAC 2976	Panel-centre	1		
5	BAC 2972	Panel-outer	1		Offside
6	BBC 7417	Panel-switch finisher	2		
7	BD 11309	Clip-panel fixing	8		

123456

MASTER INDEX

ENGINE 1.C
FLYWHEEL AND CLUTCH 1.G
GEARBOX AND PROPSHAFT 1.H
AXLES, SUSPENSION, DRIVE SHAFTS, WHEELS 1.K
STEERING 1.M
BRAKES AND BRAKE CONTROLS 1.N
FUEL, EXHAUST AND EMISSION SYSTEMS 2.C
COOLING, HEATING AND AIR CONDITIONING 2.G
ELECTRICAL, WASHERS-WIPERS, INSTRUMENTS 2.J
CHASSIS, SUBFRAMES, BODYSHELL, FITTINGS 3.C
FASCIA, TRIM, SEATS AND FIXINGS 3.H
ACCESSORIES AND PAINT 4.C
NUMERICAL INDEX 4.D

GROUP INDEX

BATTERY AND SPARE WHEEL COVERS. N05
BONNET INSULATION-3.6 LITRE N06
BONNET INSULATION-5.3 MODELS N08
BONNET INSULATION-5.3 MODELS. N07
BOOT CARPETS - NOT CONVERTIBLE N03
BOOT CARPETS-CONVERTIBLE N04
CARPETS-'A' POST FINISHERS-. L12
CONSOLE-AMBLA. H10
CONSOLE-LEATHER. H08
DOOR CASINGS-ARMRESTS AND POCKETS-AMBLA. H16
DOOR CASINGS-ARMRESTS-CONVERTIBLE I02
HEADLINING I15
DOOR CASINGS-ARMRESTS-LEATHER. H14
EXTERIOR TRIM COVERS N15
FACIA PANEL. H02
FRONT SEAT BELTS-CONVERTIBLE K15
FRONT SEAT SLIDES K16
MOTOR AND ANCHORAGE ASSY-RH L02
FRONT SEATS-AMBLA/CLOTH.-3.6 LITRE K04
FRONT SEATS-LEATHER. J03
FRONT SEATS-LEATHER/CLOTH-5.3 LITRE K07
FRT SEATS-LEATHER/CLOTH(SPORTSPACK) K10
FRONT BELT KIT-COUPE K14
FRONT SEAT SLIDES K13
FUEL APERATURE SEAL AND HOSES-COUPE. N13
HARDTOP & HOODBAG-CABRIOLET. N17
HOOD LIFT MECHANISM - CONVERTIBLE O04
HOOD SEALS-CONVERTIBLE O03
HOOD-CABRIOLET N16
HOOD-CONVERTIBLE O02
REAR CUBBY BOX-CONVERTIBLE L09
REAR LUGGAGE LOCKER-CABRIOLET L07
REAR SEAT BELTS. L11
REAR SEATS-LEATHER-COUPE. L03
REAR SEATS-LEATHER-COUPE. L04
REAR SEATS-LEATHER/CLOTH-COUPE. L06
TARGA PANELS-STOWAGEBAG-CABRIOLET N14
UNDER SCUTTLE CASINGS. H05
VENEERED FACIA PANELS-3.6 LITRE. H06
VENEERED FACIA PANELS-5.3 LITRE. H07

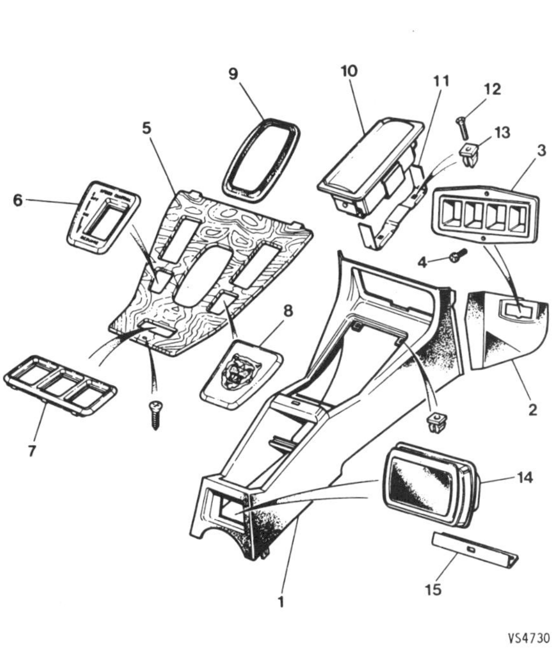

CONSOLE-LEATHER.

Illus	Part Number	123456 Description	Quantity	Change Point	Remarks
		CONSOLE ASSEMBLY			
1	BCC 7089 AE	Biscuit	1		
1	BCC 7089 LZ	Warm charcoal	1		
1	BCC 7089 AR	Buckskin	1		
1	BCC 7089 XE	Doeskin	1		
1	BCC 7089 JF	Isis blue	1		
1	BCC 7089 CM	Mulberry	1		
1	BCC 7089 LY	Savile grey	1		
1	BCC 7089 AW	Barley	1		
1	BCC 7089 ZM	Magnolia	1		
		PANEL-RH-CONSOLE			
2	BAC 3968 AE	Biscuit	1		
2	BAC 3968 LZ	Warm charcoal	1		
2	BAC 3968 AR	Buckskin	1		
2	BAC 3968 XE	Doeskin	1		
2	BAC 3968 JF	Isis blue	1		
2	BAC 3968 CM	Mulberry	1		
2	BAC 3968 LY	Savile grey	1		
2	BAC 3968 AW	Barley	1		
2	BAC 3968 ZM	Magnolia	1		
		PANEL-LH-CONSOLE			
	BAC 3969 AE	Biscuit	1		
	BAC 3969 LZ	Warm charcoal	1		
	BAC 3969 AR	Buckskin	1		
	BAC 3969 XE	Doeskin	1		
	BAC 3969 JF	Isis blue	1		
	BAC 3969 CM	Mulberry	1		
	BAC 3969 LY	Savile grey	1		
	BAC 3969 AW	Barley	1		
	BAC 3969 ZM	Magnolia	1		

MASTER INDEX

GROUP INDEX

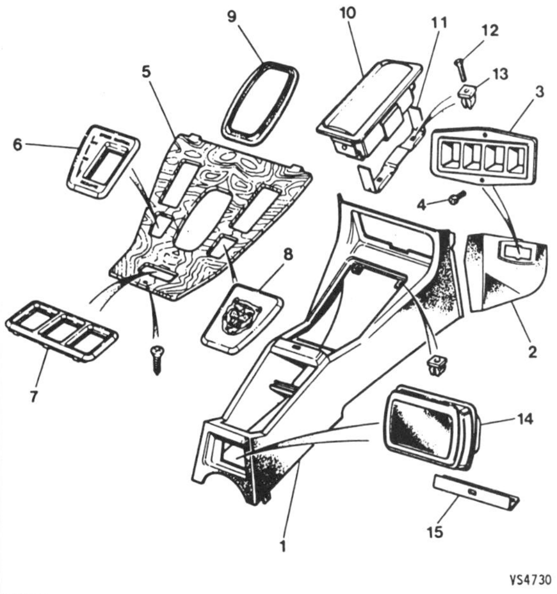

CONSOLE LEATHER-continued

Illus	Part Number	123456 Description	Quantity	Change Point	Remarks
3	BD 48244	Cover-RH-ventilation	1		
	BD 48245	Cover-LH-ventilation	1		
4	JLM 1460	Screw-self tap	4		
		PANEL-TOP FINISHER			
5	BCC 8252	Manual-elm burr	1		Use BCC 8273
5	BCC 8273	Manual-elm burr	1		
	BCC 8250	Auto-walnut burr	1		Use BCC 8274
	BCC 8274	Auto-walnut burr	1		
	BCC 8251	Auto-elm burr	1		Use BCC 8275
	BCC 8275	Auto-elm burr	1		
		BEZEL-CONSOLE SWITCH			
	BCC 2292	Speed control-RH	1		
6	BCC 2293	Speed control-LH	1		
	BCC 9056	Cigar lighter-RH	1		
	BCC 9057	Cigar lighter-LH	1		
7	BCC 2303	Window lift and roof	1		
	BCC 2302	Window lift	1		
8	BCC 2296	Badged insert-RH	1		
	BCC 2297	Badged insert-LH	1		
9	BBC 7581	Seal-selector cover	1		Auto only
10	BAC 1525	Ashtray	2		
11	BD 33031	Strap-ashtray	2		
12	DAZ 406 8C	Screw-self tap	4		
13	79013 J	Nut-plastic	6		
14	BAC 1537 LZ	Ashtray	1		
15	BAC 3661	Plate-ashtray	1		

MASTER INDEX
ENGINE 1.C
FLYWHEEL AND CLUTCH 1.G
GEARBOX AND PROPSHAFT 1.H
AXLES, SUSPENSION, DRIVE SHAFTS, WHEELS 1.K
STEERING 1.M
BRAKES AND BRAKE CONTROLS 1.N
FUEL, EXHAUST AND EMISSION SYSTEMS 2.C
COOLING, HEATING AND AIR CONDITIONING 2.G
ELECTRICAL, WASHERS-WIPERS, INSTRUMENTS 2.J
CHASSIS, SUBFRAMES, BODYSHELL, FITTINGS 3.C
FASCIA, TRIM, SEATS AND FIXINGS 3.H
ACCESSORIES AND PAINT 4.C
NUMERICAL INDEX 4.D
GROUP INDEX
BATTERY AND SPARE WHEEL COVERS. N05
BONNET INSULATION-3.6 LITRE N06
BONNET INSULATION-5.3 MODELS N08
BONNET INSULATION-5.3 MODELS. N07
BOOT CARPETS - NOT CONVERTIBLE N03
BOOT CARPETS-CONVERTIBLE N04
CARPETS-'A' POST FINISHERS-. L12
CONSOLE-AMBLA. H10
CONSOLE-LEATHER. H08
DOOR CASINGS-ARMRESTS AND POCKETS-AMBLA. H16
DOOR CASINGS-ARMRESTS-CONVERTIBLE I02
HEADLINING I15
DOOR CASINGS-ARMRESTS-LEATHER. H14
EXTERIOR TRIM COVERS N15
FACIA PANEL. H02
FRONT SEAT BELTS-CONVERTIBLE K15
FRONT SEAT SLIDES K16
MOTOR AND ANCHORAGE ASSY-RH L02
FRONT SEATS-AMBLA/CLOTH.-3.6 LITRE K04
FRONT SEATS-LEATHER. J03
FRONT SEATS-LEATHER/CLOTH-5.3 LITRE K07
FRT SEATS-LEATHER/CLOTH(SPORTSPACK) K10
FRONT BELT KIT-COUPE K14
FRONT SEAT SLIDES K13
FUEL APERATURE SEAL AND HOSES-COUPE. N13
HARDTOP & HOODBAG-CABRIOLET. N17
HOOD LIFT MECHANISM - CONVERTIBLE O04
HOOD SEALS-CONVERTIBLE O03
HOOD-CABRIOLET N16
HOOD-CONVERTIBLE O02
REAR CUBBY BOX-CONVERTIBLE L09
REAR LUGGAGE LOCKER-CABRIOLET L07
REAR SEAT BELTS. L11
REAR SEATS-LEATHER-COUPE. L03
REAR SEATS-LEATHER-COUPE. L04
REAR SEATS-LEATHER/CLOTH-COUPE. L06
TARGA PANELS-STOWAGEBAG-CABRIOLET N14
UNDER SCUTTLE CASINGS. H05
VENEERED FACIA PANELS-3.6 LITRE. H06
VENEERED FACIA PANELS-5.3 LITRE. H07

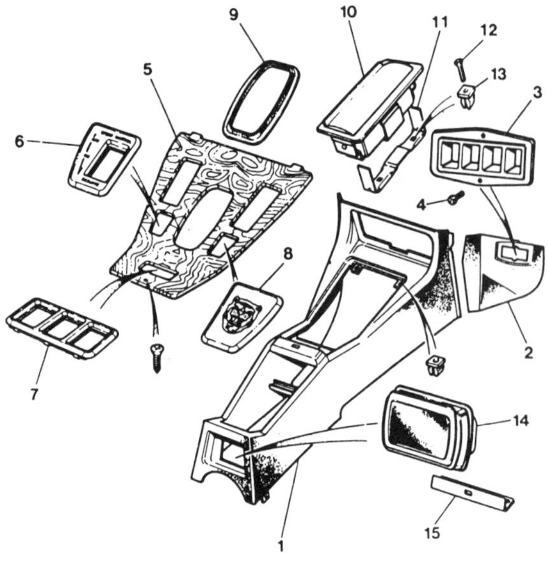

VS4730

CONSOLE-AMBLA.

Illus	Part Number	123456 Description	Quantity	Change Point	Remarks
		CONSOLE ASSEMBLY			
1	BCC 7091 AE	Biscuit	1		
1	BCC 7091 LZ	Warm Charcoal	1		
1	BCC 7091 AR	Buckskin	1		
1	BCC 7091 XE	Doeskin	1		
1	BCC 7091 JF	Isis Blue	1		
1	BCC 7091 CM	Mulberry	1		
1	BCC 7091 LY	Savile Grey	1		
1	BCC 7091 AW	Barley	1		
1	BCC 7091 ZM	Magnolia	1		
		PANEL-RH-CONSOLE			
2	BD 48228 AE	Biscuit	1		
2	BD 48228 LZ	Warm Charcoal	1		
2	BD 48228 AR	Buckskin	1		
2	BD 48228 XE	Doeskin	1		
2	BD 48228 JF	Isis Blue	1		
2	BD 48228 CM	Mulberry	1		
2	BD 48228 LY	Savile Grey	1		
2	BD 48228 AW	Barley	1		
2	BD 48228 ZM	Magnolia	1		
		PANEL-LH-CONSOLE			
	BD 48229 AE	Biscuit	1		
	BD 48229 LZ	Warm Charcoal	1		
	BD 48229 AR	Buckskin	1		
	BD 48229 XE	Doeskin	1		
	BD 48229 JF	Isis Blue	1		
	BD 48229 CM	Mulberry	1		
	BD 48229 LY	Savile Grey	1		
	BD 48229 AW	Barley	1		
	BD 48229 ZM	Magnolia	1		

3H 10

MASTER INDEX

GROUP INDEX

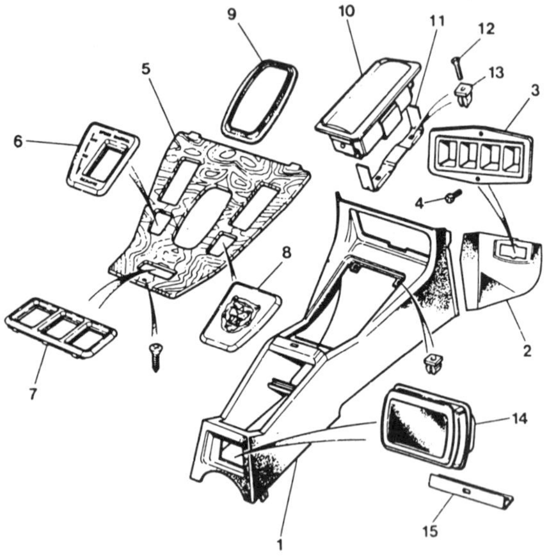

VS4730

CONSOLE-AMBLA-continued

Illus	Part Number	123456 Description	Quantity	Change Point	Remarks
3	BD 48244	Cover-RH-ventilation	1		
	BD 48245	Cover-LH-ventilation	1		
4	JLM 1460	Screw-self tap	4		
		PANEL-TOP FINISHER			
5	BCC 8252	Manual-elm burr	1		Use BCC 8273
5	BCC 8273	Manual-elm burr	1		
	BCC 8251	Auto-elm burr	1		Use BCC 8275
	BCC 8275	Auto-elm burr	1		
		BEZEL-CONSOLE SWITCH			
	BCC 2292	Speed control-RH	1		
6	BCC 2293	Speed control-LH	1		
	BCC 9056	Cigar lighter-RH	1		
	BCC 9057	Ciger lighter-LH	1		
7	BCC 2303	Window lift and roof	1		
	BCC 2302	Window lift	1		
8	BCC 2296	Badged insert RH	1		
	BCC 2297	Badged insert LH	1		
9	BBC 7581	Seal-selector cover	1		Auto only
10	BAC 1525	Ashtray	2		
11	BD 33031	Strap-ashtray	2		
12	DAZ 406 8C	Screw-self tap	4		
13	79013 J	Nut-plastic	6		
14	BAC 1537 LZ	Ashtray	1		
15	BAC 3661	Plate-ashtray	1		

3H 11

MASTER INDEX
ENGINE ... 1.C
FLYWHEEL AND CLUTCH ... 1.G
GEARBOX AND PROPSHAFT ... 1.H
AXLES, SUSPENSION, DRIVE SHAFTS, WHEELS ... 1.K
STEERING ... 1.M
BRAKES AND BRAKE CONTROLS ... 1.N
FUEL, EXHAUST AND EMISSION SYSTEMS ... 2.C
COOLING, HEATING AND AIR CONDITIONING ... 2.G
ELECTRICAL, WASHERS-WIPERS, INSTRUMENTS ... 2.J
CHASSIS, SUBFRAMES, BODYSHELL, FITTINGS ... 3.C
FASCIA, TRIM, SEATS AND FIXINGS ... 3.H
ACCESSORIES AND PAINT ... 4.C
NUMERICAL INDEX ... 4.D
GROUP INDEX
BATTERY AND SPARE WHEEL COVERS ... N05
BONNET INSULATION-3.6 LITRE ... N06
BONNET INSULATION-5.3 MODELS ... N08
BONNET INSULATION-5.3 MODELS ... N07
BOOT CARPETS - NOT CONVERTIBLE ... N03
BOOT CARPETS-CONVERTIBLE ... N04
CARPETS-'A' POST FINISHERS- ... L12
CONSOLE-AMBLA ... H10
CONSOLE-LEATHER ... H08
DOOR CASINGS-ARMRESTS AND POCKETS-AMBLA ... H16
DOOR CASINGS-ARMRESTS-CONVERTIBLE ... I02
HEADLINING ... I15
DOOR CASINGS-ARMRESTS-LEATHER ... H14
EXTERIOR TRIM COVERS ... N15
FACIA PANEL ... H02
FRONT SEAT BELTS-CONVERTIBLE ... K15
FRONT SEAT SLIDES ... K16
MOTOR AND ANCHORAGE ASSY-RH ... L02
FRONT SEATS-AMBLA/CLOTH.-3.6 LITRE ... K04
FRONT SEATS-LEATHER ... J03
FRONT SEATS-LEATHER/CLOTH-5.3 LITRE ... K07
FRT SEATS-LEATHER/CLOTH(SPORTSPACK) ... K10
FRONT BELT KIT-COUPE ... K14
FRONT SEAT SLIDES ... K13
FUEL APERATURE SEAL AND HOSES-COUPE ... N13
HARDTOP & HOODBAG-CABRIOLET ... N17
HOOD LIFT MECHANISM - CONVERTIBLE ... O04
HOOD SEALS-CONVERTIBLE ... O03
HOOD-CABRIOLET ... N16
HOOD-CONVERTIBLE ... O02
REAR CUBBY BOX-CONVERTIBLE ... L09
REAR LUGGAGE LOCKER-CABRIOLET ... L07
REAR SEAT BELTS ... L11
REAR SEATS-LEATHER-COUPE ... L03
REAR SEATS-LEATHER-COUPE ... L04
REAR SEATS-LEATHER/CLOTH-COUPE ... L06
TARGA PANELS-STOWAGEBAG-CABRIOLET ... N14
UNDER SCUTTLE CASINGS ... H05
VENEERED FACIA PANELS-3.6 LITRE ... H06
VENEERED FACIA PANELS-5.3 LITRE ... H07

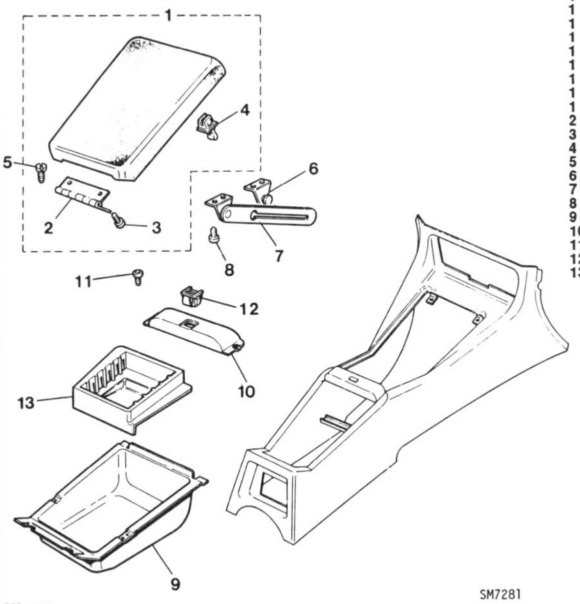

SM 7281

CONSOLE ARMREST/LID-LEATHER.

Illus	Part Number	123456 Description	Quantity	Change Point	Remarks
		ARMREST/LID-CONSOLE-NOT RADIO TELEPHONE			
1	BCC 5636 AE	Biscuit	1		
1	BCC 5636 LZ	Warm Charcoal	1		
1	BCC 5636 AR	Buckskin	1		
1	BCC 5636 XE	Doeskin	1		
1	BCC 5636 JF	Isis Blue	1		
1	BCC 5636 CM	Mulberry	1		
1	BCC 5636 LY	Savile Grey	1		
1	BCC 5636 AW	Barley	1		
1	BCC 5636 ZM	Magnolia	1		
2	BD 47321	Hinge-armrest/lid	1		
3	BD 13713 1	Screw-set	2		
4	BD 34716	Striker	1		
5	DAC 41012 C	Screw-self-tap	2		
6	BD 47837	Bracket	1		
7	BD 34753	Link stop	1		
8	AB 606032 J	Screw-self-tap	4		
9	BCC 7405	Pocket-armrest/lid	1		
10	BCC 7105	Cover-pocket armrest	1		
11	DAZ 406 8C	Screw-self-tap	3		
12	BD 34715	Latch-pocket	1		
13	BCC 4704	Storage rack-cassette	1		

MASTER INDEX

GROUP INDEX

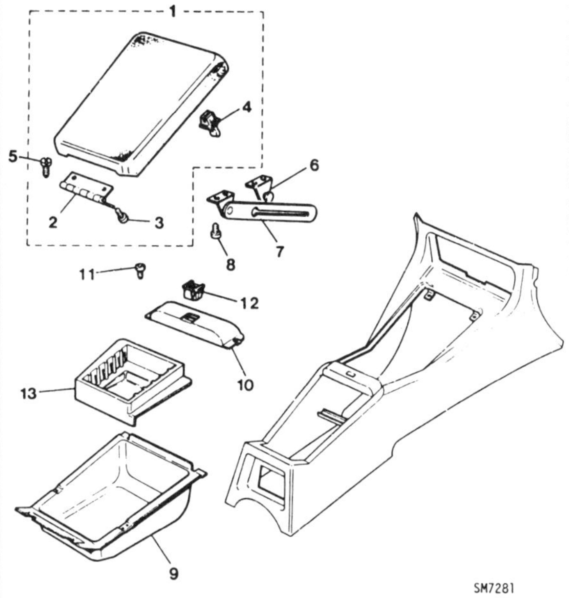

SM 7281

CONSOLE ARMREST/LID-AMBLA.

Illus	Part Number	123456 Description	Quantity	Change Point	Remarks
		ARMREST/LID-CONSOLE-NOT RADIO TELEPHONE			
1	BCC 5638 AE	Biscuit	1		
1	BCC 5638 LZ	Warm Charcoal	1		
1	BCC 5638 AR	Buckskin	1		
1	BCC 5638 XE	Doeskin	1		
1	BCC 5638 JF	Isis Blue	1		
1	BCC 5638 CM	Mulberry	1		
1	BCC 5638 LY	Savile Grey	1		
1	BCC 5638 AW	Barley	1		
1	BCC 5638 ZM	Magnolia	1		
2	BD 47321	Hinge-armrest/lid	1		
3	BD 13713 1	Screw-set	2		
4	BD 34716	Striker	1		
5	DAC 41012 C	Screw-self-tap	2		
6	BD 47837	Bracket	1		
7	BD 34753	Link stop	1		
8	AB 60632	Screw-self-tap	4		
9	BCC 7405	Pocket-armrest/lid	1		
10	BCC 7105	Cover-pocket armrest	1		
11	DAZ 406 8C	Screw-self-tap	3		
12	BD 34715	Latch-pocket	1		
13	BCC 4704	Storage rack-cassette	1		

MASTER INDEX
ENGINE 1.C
FLYWHEEL AND CLUTCH 1.G
GEARBOX AND PROPSHAFT 1.H
AXLES, SUSPENSION, DRIVE SHAFTS, WHEELS 1.K
STEERING 1.M
BRAKES AND BRAKE CONTROLS 1.N
FUEL, EXHAUST AND EMISSION SYSTEMS 2.C
COOLING, HEATING AND AIR CONDITIONING 2.G
ELECTRICAL, WASHERS-WIPERS, INSTRUMENTS 2.J
CHASSIS, SUBFRAMES, BODYSHELL, FITTINGS 3.C
FASCIA, TRIM, SEATS AND FIXINGS 3.H
ACCESSORIES AND PAINT 4.C
NUMERICAL INDEX 4.D
GROUP INDEX
BATTERY AND SPARE WHEEL COVERS. N05
BONNET INSULATION-3.6 LITRE N06
BONNET INSULATION-5.3 MODELS N08
BONNET INSULATION-5.3 MODELS. N07
BOOT CARPETS - NOT CONVERTIBLE N03
BOOT CARPETS-CONVERTIBLE N04
CARPETS-'A' POST FINISHERS-. L12
CONSOLE-AMBLA. H10
CONSOLE-LEATHER. H08
DOOR CASINGS-ARMRESTS AND POCKETS-AMBLA. H16
DOOR CASINGS-ARMRESTS-CONVERTIBLE I02
HEADLINING I15
DOOR CASINGS-ARMRESTS-LEATHER. H14
EXTERIOR TRIM COVERS N15
FACIA PANEL H02
FRONT SEAT BELTS-CONVERTIBLE K15
FRONT SEAT SLIDES K16
MOTOR AND ANCHORAGE ASSY-RH L02
FRONT SEATS-AMBLA/CLOTH.-3.6 LITRE K04
FRONT SEATS-LEATHER. J03
FRONT SEATS-LEATHER/CLOTH-5.3 LITRE K07
FRT SEATS-LEATHER/CLOTH(SPORTSPACK) K10
FRONT BELT KIT-COUPE K14
FRONT SEAT SLIDES K13
FUEL APERATURE SEAL AND HOSES-COUPE. N13
HARDTOP & HOODBAG-CABRIOLET. N17
HOOD LIFT MECHANISM - CONVERTIBLE O04
HOOD SEALS-CONVERTIBLE O03
HOOD-CABRIOLET N16
HOOD-CONVERTIBLE O02
REAR CUBBY BOX-CONVERTIBLE L09
REAR LUGGAGE LOCKER-CABRIOLET L07
REAR SEAT BELTS. L11
REAR SEATS-LEATHER-COUPE. L03
REAR SEATS-LEATHER-COUPE. L04
REAR SEATS-LEATHER/CLOTH-COUPE. L06
TARGA PANELS-STOWAGEBAG-CABRIOLET N14
UNDER SCUTTLE CASINGS. H05
VENEERED FACIA PANELS-3.6 LITRE. H06
VENEERED FACIA PANELS-5.3 LITRE. H07

3H 14

DOOR CASINGS-ARMRESTS-LEATHER.
NOT CONVERTIBLE

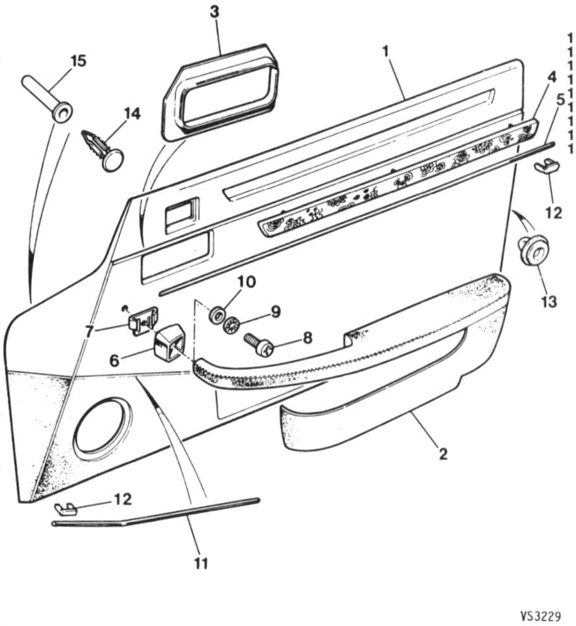

VS 3229

Illus	Part Number	123456 Description	Quantity	Change Point	Remarks
		CASING RH-RHD-DOOR			
1	BAC 5606 AE	Biscuit	1		
1	BAC 5606 LZ	Warm Charcoal	1		
1	BAC 5606 AR	Buckskin	1		
1	BAC 5606 XE	Doeskin	1		
1	BAC 5606 JF	Isis Blue	1		
1	BAC 5606 CM	Mulberry	1		
1	BAC 5606 LY	Savile Grey	1		
1	BAC 5606 AW	Barley	1		
1	BAC 5606 ZM	Magnolia	1		
		CASING-LH-RHD-DOOR			
	BAC 5615 AE	Biscuit	1		
	BAC 5615 LZ	Warm Charcoal	1		
	BAC 5615 AR	Buckskin	1		
	BAC 5615 XE	Doeskin	1		
	BAC 5615 JF	Isis Blue	1		
	BAC 5615 CM	Mulberry	1		
	BAC 5615 LY	Savile Grey	1		
	BAC 5615 AW	Barley	1		
	BAC 5615 ZM	Magnolia	1		
		CASING-RH-LHD-DOOR			
	BAC 5614 AE	Biscuit	1		
	BAC 5614 LZ	Warm Charcoal	1		
	BAC 5614 AR	Buckskin	1		
	BAC 5614 XE	Doeskin	1		
	BAC 5614 JF	Isis Blue	1		
	BAC 5614 CM	Mulberry	1		
	BAC 5614 LY	Savile Grey	1		
	BAC 5614 AW	Barley	1		
	BAC 5614 ZM	Magnolia	1		
		CASING LH-LHD-DOOR			
	BAC 5607 AE	Biscuit	1		
	BAC 5607 LZ	Warm Charcoal	1		
	BAC 5607 AR	Buckskin	1		
	BAC 5607 XE	Doeskin	1		
	BAC 5607 JF	Isis blue	1		
	BAC 5607 CM	Mulberry	1		
	BAC 5607 LY	Savile Grey	1		
	BAC 5607 AW	Barley	1		
	BAC 5607 ZM	Magnolia	1		

MASTER INDEX

GROUP INDEX

3H 15

DOOR CASINGS-ARMRESTS AND POCKETS-LEATHER.
NOT CONVERTIBLE

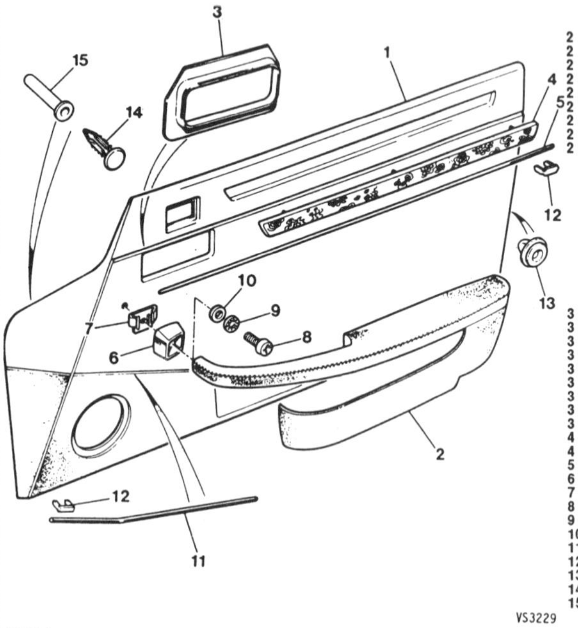

VS 3229

Illus	Part Number	123456 Description	Quantity	Change Point	Remarks
		ARMREST AND POCKET-RH			
2	BAC 4110 AE	Biscuit	1		
2	BAC 4110 LZ	Warm Charcoal	1		
2	BAC 4110 AR	Buckskin	1		
2	BAC 4110 XE	Doeskin	1		
2	BAC 4110 JF	Isis Blue	1		
2	BAC 4110 CM	Mulberry	1		
2	BAC 4110 LY	Savile Grey	1		
2	BAC 4110 AW	Barley	1		
2	BAC 4110 ZM	Magnolia	1		
		ARMREST AND POCKET-LH			
	BAC 4111 AE	Biscuit	1		
	BAC 4111 LZ	Warm Charcoal	1		
	BAC 4111 AR	Buckskin	1		
	BAC 4111 XE	Doeskin	1		
	BAC 4111 JF	Isis Blue	1		
	BAC 4111 CM	Mulberry	1		
	BAC 4111 LY	Savile Grey	1		
	BAC 4111 AW	Barley	1		
	BAC 4111 ZM	Magnolia	1		
		REINFORCEMENT-HANDLE			
3	BCC 2520 AE	Biscuit	2		
3	BCC 2520 LZ	Warm Charcoal	2		
3	BCC 2520 AR	Buckskin	2		
3	BCC 2520 XE	Doeskin	2		
3	BCC 2520 JF	Isis Blue	2		
3	BCC 2520 CM	Mulberry	2		
3	BCC 2520 LY	Savile Grey	2		
3	BCC 2520 AW	Barley	2		
3	BCC 2520 ZM	Magnolia	2		
4	BAC 4131	Capping-waist (elm)	2		
4	BBC 7422	Capping-waist(walnut burr)	2		
5	BD 46571	Moulding-upper	2		
6	BAC 1462	Capping-armrest	2		
7	BD 42434	Plate-base	2		
8	SE 604081 J	Screw-set	2		
9	C 724	Washer-shakeproof	2		
10	JLM 296	Washer-plain	2		
11	BAC 5616	Moulding-lower	2		
12	BAC 5298	Clip-mouldings	18		
13	AGU 1018	Clip-casing	18		
14	BD 13102	Button-pronged	8		
15	BD 47644	Snap sac	8		

MASTER INDEX

ENGINE 1.C
FLYWHEEL AND CLUTCH 1.G
GEARBOX AND PROPSHAFT 1.H
AXLES, SUSPENSION, DRIVE SHAFTS, WHEELS 1.K
STEERING 1.M
BRAKES AND BRAKE CONTROLS 1.N
FUEL, EXHAUST AND EMISSION SYSTEMS 2.C
COOLING, HEATING AND AIR CONDITIONING 2.G
ELECTRICAL, WASHERS-WIPERS, INSTRUMENTS 2.J
CHASSIS, SUBFRAMES, BODYSHELL, FITTINGS 3.C
FASCIA, TRIM, SEATS AND FIXINGS 3.H
ACCESSORIES AND PAINT 4.C
NUMERICAL INDEX 4.D

GROUP INDEX

BATTERY AND SPARE WHEEL COVERS N05
BONNET INSULATION-3.6 LITRE N06
BONNET INSULATION-5.3 MODELS N08
BONNET INSULATION-5.3 MODELS N07
BOOT CARPETS - NOT CONVERTIBLE N03
BOOT CARPETS-CONVERTIBLE N04
CARPETS-'A' POST FINISHERS- L12
CONSOLE-AMBLA H10
CONSOLE-LEATHER H08
DOOR CASINGS-ARMRESTS AND POCKETS-AMBLA H16
DOOR CASINGS-ARMRESTS-CONVERTIBLE I02
HEADLINING I15
DOOR CASINGS-ARMRESTS-LEATHER H14
EXTERIOR TRIM COVERS N15
FACIA PANEL H02
FRONT SEAT BELTS-CONVERTIBLE K15
FRONT SEAT SLIDES K16
MOTOR AND ANCHORAGE ASSY-RH L02
FRONT SEATS-AMBLA/CLOTH.-3.6 LITRE K04
FRONT SEATS-LEATHER J03
FRONT SEATS-LEATHER/CLOTH-5.3 LITRE K07
FRT SEATS-LEATHER/CLOTH(SPORTSPACK) K10
FRONT BELT KIT-COUPE K14
FRONT SEAT SLIDES K13
FUEL APERATURE SEAL AND HOSES-COUPE N13
HARDTOP & HOODBAG-CABRIOLET N17
HOOD LIFT MECHANISM - CONVERTIBLE O04
HOOD SEALS-CONVERTIBLE O03
HOOD-CABRIOLET N16
HOOD-CONVERTIBLE O02
REAR CUBBY BOX-CONVERTIBLE L09
REAR LUGGAGE LOCKER-CABRIOLET L07
REAR SEAT BELTS L11
REAR SEATS-LEATHER-COUPE L03
REAR SEATS-LEATHER-COUPE L04
REAR SEATS-LEATHER/CLOTH-COUPE L06
TARGA PANELS-STOWAGEBAG-CABRIOLET N14
UNDER SCUTTLE CASINGS H05
VENEERED FACIA PANELS-3.6 LITRE H06
VENEERED FACIA PANELS-5.3 LITRE H07

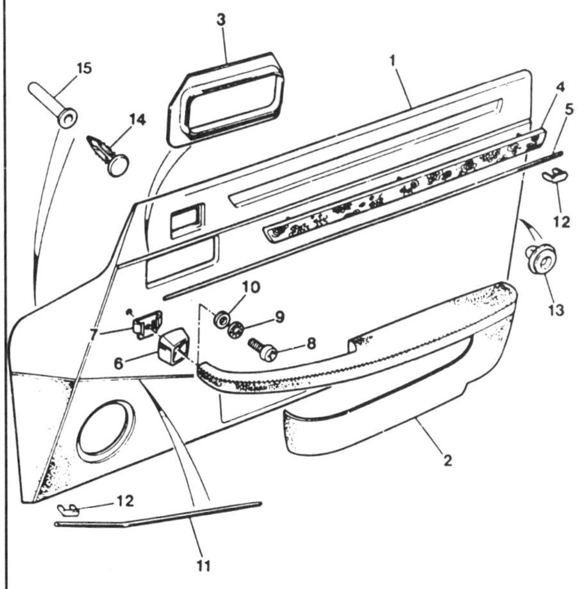

VS3229

VS 3229

DOOR CASINGS-ARMRESTS AND POCKETS-AMBLA.

Illus	Part Number	123456	Description	Quantity	Change Point	Remarks
			CASING-RH-RHD-DOOR			
1	BBC 1996 AE		Biscuit	1		
1	BBC 1996 LZ		Warm Charcoal	1		
1	BBC 1996 AR		Buckskin	1		
1	BBC 1996 XE		Doeskin	1		
1	BBC 1996 JF		Isis Blue	1		
1	BBC 1996 CM		Mulberry	1		
1	BBC 1996 LY		Savile Grey	1		
1	BBC 1996 AW		Barley	1		
1	BBC 1996 ZM		Magnolia	1		
			CASING-LH-RHD-DOOR			
	BBC 1997 AE		Biscuit	1		
	BBC 1997 LZ		Warm Charcoal	1		
	BBC 1997 AR		Buckskin	1		
	BBC 1997 XE		Doeskin	1		
	BBC 1997 JF		Isis Blue	1		
	BBC 1997 CM		Mulberry	1		
	BBC 1997 LY		Savile Grey	1		
	BBC 1997 AW		Barley	1		
	BBC 1997 ZM		Magnolia	1		
			CASING-RH-LHD-DOOR			
	BBC 1998 AE		Biscuit	1		
	BBC 1998 LZ		Warm Charcoal	1		
	BBC 1998 AR		Buckskin	1		
	BBC 1998 XE		Doeskin	1		
	BBC 1998 JF		Isis Blue	1		
	BBC 1998 CM		Mulberry	1		
	BBC 1998 LY		Savile Grey	1		
	BBC 1998 AW		Barley	1		
	BBC 1998 ZM		Magnolia	1		
			CASING-LH-LHD-DOOR			
	BBC 1999 AE		Biscuit	1		
	BBC 1999 LZ		Warm Charcoal	1		
	BBC 1999 AR		Buckskin	1		
	BBC 1999 XE		Doeskin	1		
	BBC 1999 JF		Isis Blue	1		
	BBC 1999 CM		Mulberry	1		
	BBC 1999 LY		Savile Grey	1		
	BBC 1999 AW		Barley	1		
	BBC 1999 ZM		Magnolia	1		

MASTER INDEX

GROUP INDEX

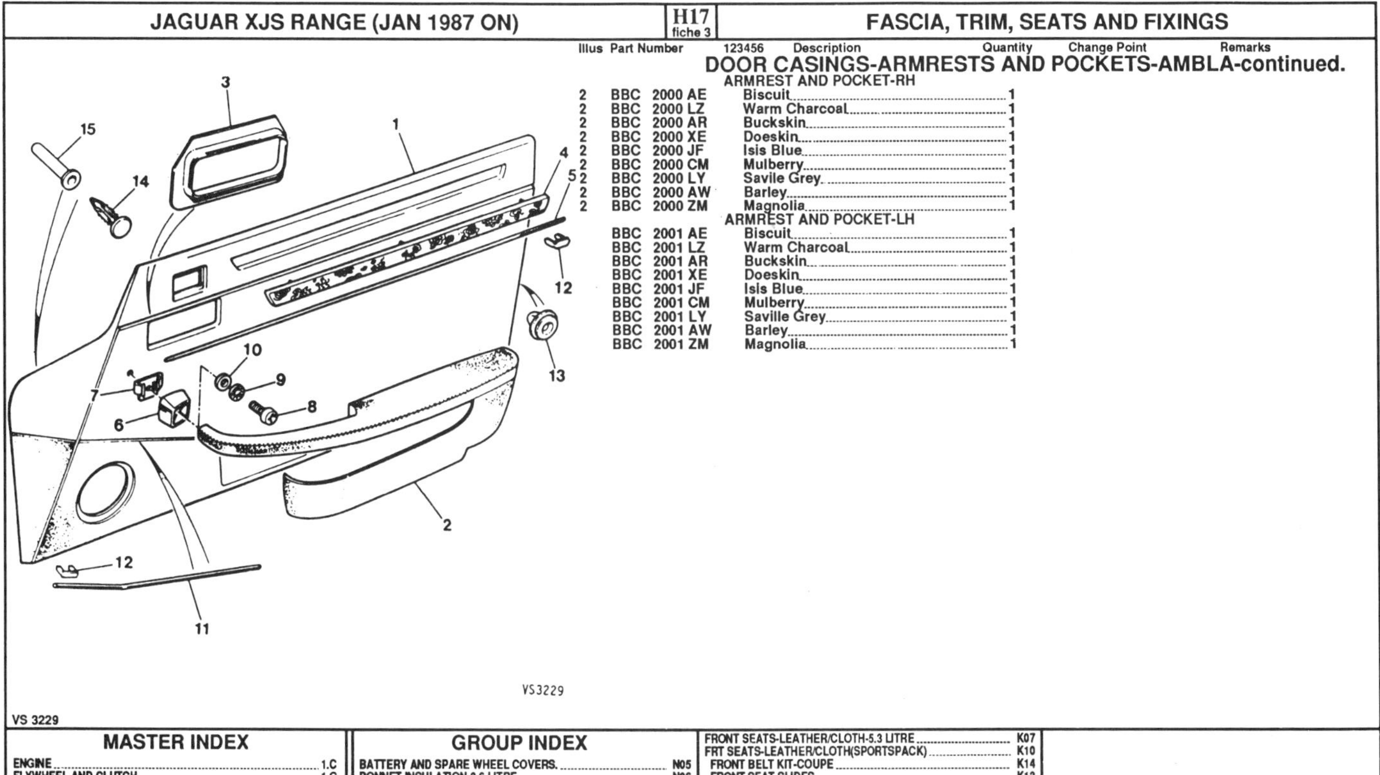

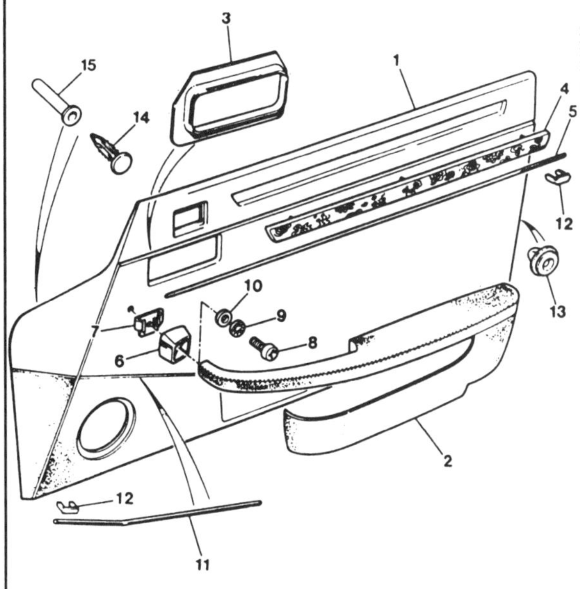

VS3229

VS 3229

DOOR CASINGS-ARMRESTS AND POCKETS-AMBLA-continued.

Illus	Part Number	123456	Description	Quantity	Change Point	Remarks
			ARMREST AND POCKET-RH			
2	BBC 2000 AE		Biscuit	1		
2	BBC 2000 LZ		Warm Charcoal	1		
2	BBC 2000 AR		Buckskin	1		
2	BBC 2000 XE		Doeskin	1		
2	BBC 2000 JF		Isis Blue	1		
2	BBC 2000 CM		Mulberry	1		
2	BBC 2000 LY		Savile Grey.	1		
2	BBC 2000 AW		Barley	1		
2	BBC 2000 ZM		Magnolia	1		
			ARMREST AND POCKET-LH			
	BBC 2001 AE		Biscuit	1		
	BBC 2001 LZ		Warm Charcoal	1		
	BBC 2001 AR		Buckskin	1		
	BBC 2001 XE		Doeskin	1		
	BBC 2001 JF		Isis Blue	1		
	BBC 2001 CM		Mulberry	1		
	BBC 2001 LY		Saville Grey	1		
	BBC 2001 AW		Barley	1		
	BBC 2001 ZM		Magnolia	1		

MASTER INDEX
ENGINE 1.C
FLYWHEEL AND CLUTCH 1.G
GEARBOX AND PROPSHAFT 1.H
AXLES, SUSPENSION, DRIVE SHAFTS, WHEELS 1.K
STEERING 1.M
BRAKES AND BRAKE CONTROLS 1.N
FUEL, EXHAUST AND EMISSION SYSTEMS 2.C
COOLING, HEATING AND AIR CONDITIONING 2.G
ELECTRICAL, WASHERS-WIPERS, INSTRUMENTS 2.J
CHASSIS, SUBFRAMES, BODYSHELL, FITTINGS 3.C
FASCIA, TRIM, SEATS AND FIXINGS 3.H
ACCESSORIES AND PAINT 4.C
NUMERICAL INDEX 4.D
GROUP INDEX
BATTERY AND SPARE WHEEL COVERS. N05
BONNET INSULATION-3.6 LITRE N06
BONNET INSULATION-5.3 MODELS N08
BONNET INSULATION-5.3 MODELS. N07
BOOT CARPETS - NOT CONVERTIBLE N03
BOOT CARPETS-CONVERTIBLE N04
CARPETS-'A' POST FINISHERS-. L12
CONSOLE-AMBLA. H10
CONSOLE-LEATHER. H08
DOOR CASINGS-ARMRESTS AND POCKETS-AMBLA. H16
DOOR CASINGS-ARMRESTS-CONVERTIBLE I02
HEADLINING I15
DOOR CASINGS-ARMRESTS-LEATHER. H14
EXTERIOR TRIM COVERS N15
FACIA PANEL H02
FRONT SEAT BELTS-CONVERTIBLE K15
FRONT SEAT SLIDES K16
MOTOR AND ANCHORAGE ASSY-RH L02
FRONT SEATS-AMBLA/CLOTH.-3.6 LITRE K04
FRONT SEATS-LEATHER. J03
FRONT SEATS-LEATHER/CLOTH-5.3 LITRE K07
FRT SEATS-LEATHER/CLOTH(SPORTSPACK) K10
FRONT BELT KIT-COUPE K14
FRONT SEAT SLIDES K13
FUEL APERATURE SEAL AND HOSES-COUPE. N13
HARDTOP & HOODBAG-CABRIOLET. N17
HOOD LIFT MECHANISM - CONVERTIBLE O04
HOOD SEALS-CONVERTIBLE O03
HOOD-CABRIOLET N16
HOOD-CONVERTIBLE O02
REAR CUBBY BOX-CONVERTIBLE L09
REAR LUGGAGE LOCKER-CABRIOLET L07
REAR SEAT BELTS. L11
REAR SEATS-LEATHER-COUPE. L03
REAR SEATS-LEATHER-COUPE. L04
REAR SEATS-LEATHER/CLOTH-COUPE. L06
TARGA PANELS-STOWAGEBAG-CABRIOLET N14
UNDER SCUTTLE CASINGS. H05
VENEERED FACIA PANELS-3.6 LITRE. H06
VENEERED FACIA PANELS-5.3 LITRE. H07

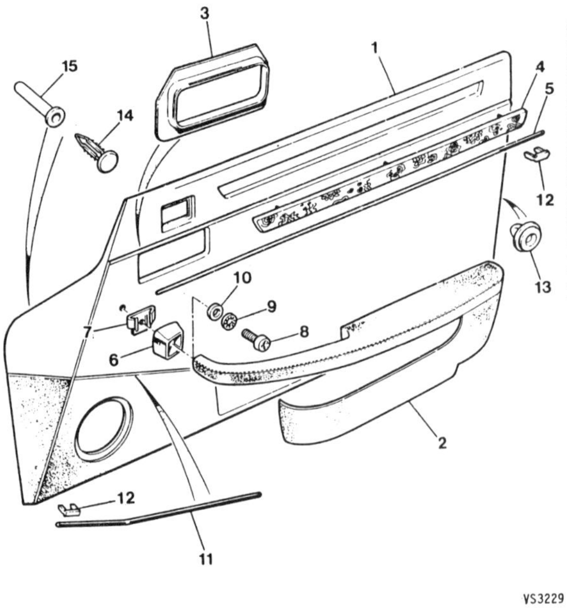

VS 3229

DOOR CASING-ARMREST AND POCKETS-AMBLA-continued.

Illus	Part Number	123456 Description	Quantity	Change Point	Remarks
		REINFORCEMENT-HANDLE			
3	BD 45521 AE	Biscuit	2		
3	BD 45521 PA	Black	2		
3	BD 45521 AR	Buckskin	2		
3	BD 45521 XE	Doeskin	2		
3	BD 45521 JF	Isis Blue	2		
3	BD 45521 CM	Mulberry	2		
3	BD 45521 LY	Savile Grey	2		
3	BD 45521 AW	Barley	2		
3	BD 45521 ZM	Magnolia	1		
4	BAC 4131	Capping-waist (elm)	2		
5	BD 46571	Moulding-upper	2		
6	BAC 1462	Capping-armrest	2		
7	BD 42434	Plate-base	2		
8	SE 604081 J	Screw-set	2		
9	C 724	Washer-shakeproof	2		
10	JLM 296	Washer-plain	2		
11	BAC 5616	Moulding-lower	2		
12	BAC 5298	Clip-mouldings	18		
13	AGU 1018	Clip-casing	18		
14	BD 13102	Button-pronged	8		
15	BD 47644	Snap sac	8		

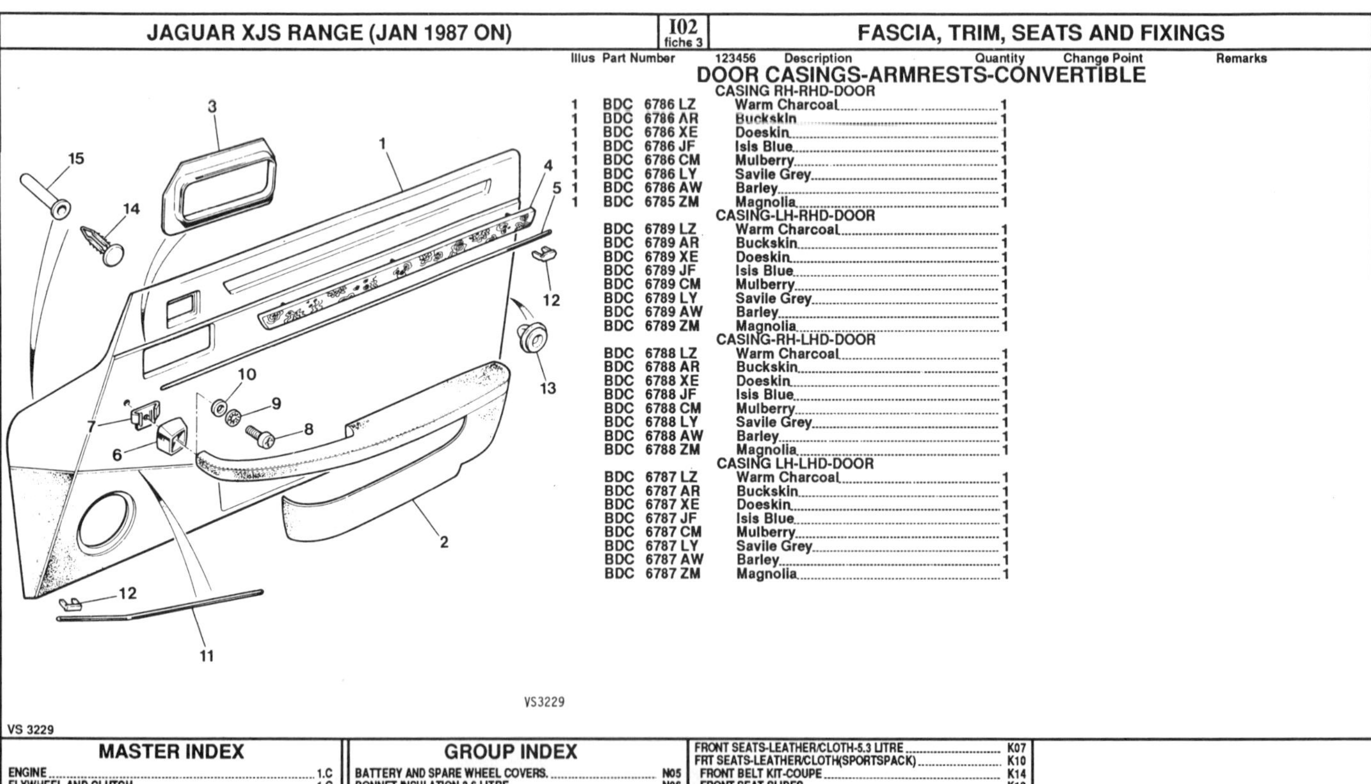

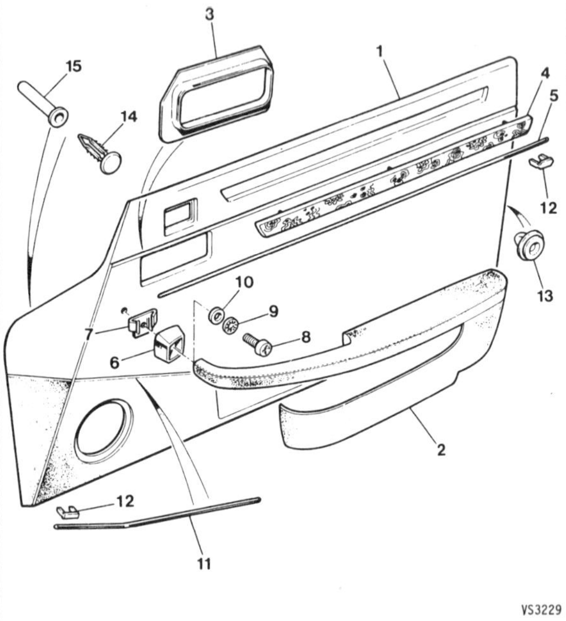

VS 3229

DOOR CASINGS-ARMRESTS-CONVERTIBLE

Illus	Part Number	123456 Description	Quantity	Change Point	Remarks
		CASING RH-RHD-DOOR			
1	BDC 6786 LZ	Warm Charcoal	1		
1	BDC 6786 AR	Buckskin	1		
1	BDC 6786 XE	Doeskin	1		
1	BDC 6786 JF	Isis Blue	1		
1	BDC 6786 CM	Mulberry	1		
1	BDC 6786 LY	Savile Grey	1		
1	BDC 6786 AW	Barley	1		
1	BDC 6785 ZM	Magnolia	1		
		CASING-LH-RHD-DOOR			
	BDC 6789 LZ	Warm Charcoal	1		
	BDC 6789 AR	Buckskin	1		
	BDC 6789 XE	Doeskin	1		
	BDC 6789 JF	Isis Blue	1		
	BDC 6789 CM	Mulberry	1		
	BDC 6789 LY	Savile Grey	1		
	BDC 6789 AW	Barley	1		
	BDC 6789 ZM	Magnolia	1		
		CASING-RH-LHD-DOOR			
	BDC 6788 LZ	Warm Charcoal	1		
	BDC 6788 AR	Buckskin	1		
	BDC 6788 XE	Doeskin	1		
	BDC 6788 JF	Isis Blue	1		
	BDC 6788 CM	Mulberry	1		
	BDC 6788 LY	Savile Grey	1		
	BDC 6788 AW	Barley	1		
	BDC 6788 ZM	Magnolia	1		
		CASING LH-LHD-DOOR			
	BDC 6787 LZ	Warm Charcoal	1		
	BDC 6787 AR	Buckskin	1		
	BDC 6787 XE	Doeskin	1		
	BDC 6787 JF	Isis Blue	1		
	BDC 6787 CM	Mulberry	1		
	BDC 6787 LY	Savile Grey	1		
	BDC 6787 AW	Barley	1		
	BDC 6787 ZM	Magnolia	1		

MASTER INDEX

ENGINE ... 1.C
FLYWHEEL AND CLUTCH ... 1.G
GEARBOX AND PROPSHAFT ... 1.H
AXLES, SUSPENSION, DRIVE SHAFTS, WHEELS ... 1.K
STEERING ... 1.M
BRAKES AND BRAKE CONTROLS ... 1.N
FUEL, EXHAUST AND EMISSION SYSTEMS ... 2.C
COOLING, HEATING AND AIR CONDITIONING ... 2.G
ELECTRICAL, WASHERS-WIPERS, INSTRUMENTS ... 2.J
CHASSIS, SUBFRAMES, BODYSHELL, FITTINGS ... 3.C
FASCIA, TRIM, SEATS AND FIXINGS ... 3.H
ACCESSORIES AND PAINT ... 4.C
NUMERICAL INDEX ... 4.D

GROUP INDEX

BATTERY AND SPARE WHEEL COVERS. ... N05
BONNET INSULATION-3.6 LITRE ... N06
BONNET INSULATION-5.3 MODELS ... N08
BONNET INSULATION-5.3 MODELS. ... N07
BOOT CARPETS - NOT CONVERTIBLE ... N03
BOOT CARPETS-CONVERTIBLE ... N04
CARPETS-'A' POST FINISHERS-. ... L12
CONSOLE-AMBLA. ... H10
CONSOLE-LEATHER. ... H08
DOOR CASINGS-ARMRESTS AND POCKETS-AMBLA. ... H16
DOOR CASINGS-ARMRESTS-CONVERTIBLE ... I02
HEADLINING ... I15
DOOR CASINGS-ARMRESTS-LEATHER. ... H14
EXTERIOR TRIM COVERS ... N15
FACIA PANEL. ... H02
FRONT SEAT BELTS-CONVERTIBLE ... K15
FRONT SEAT SLIDES ... K16
MOTOR AND ANCHORAGE ASSY-RH ... L02
FRONT SEATS-AMBLA/CLOTH.-3.6 LITRE ... K04
FRONT SEATS-LEATHER. ... J03
FRONT SEATS-LEATHER/CLOTH-5.3 LITRE ... K07
FRT SEATS-LEATHER/CLOTH(SPORTSPACK) ... K10
FRONT BELT KIT-COUPE ... K14
FRONT SEAT SLIDES ... K13
FUEL APERATURE SEAL AND HOSES-COUPE. ... N13
HARDTOP & HOODBAG-CABRIOLET. ... N17
HOOD LIFT MECHANISM - CONVERTIBLE ... O04
HOOD SEALS-CONVERTIBLE ... O03
HOOD-CABRIOLET ... N16
HOOD-CONVERTIBLE ... O02
REAR CUBBY BOX-CONVERTIBLE ... L09
REAR LUGGAGE LOCKER-CABRIOLET ... L07
REAR SEAT BELTS. ... L11
REAR SEATS-LEATHER-COUPE. ... L03
REAR SEATS-LEATHER-COUPE. ... L04
REAR SEATS-LEATHER/CLOTH-COUPE. ... L06
TARGA PANELS-STOWAGEBAG-CABRIOLET ... N14
UNDER SCUTTLE CASINGS. ... H05
VENEERED FACIA PANELS-3.6 LITRE. ... H06
VENEERED FACIA PANELS-5.3 LITRE. ... H07

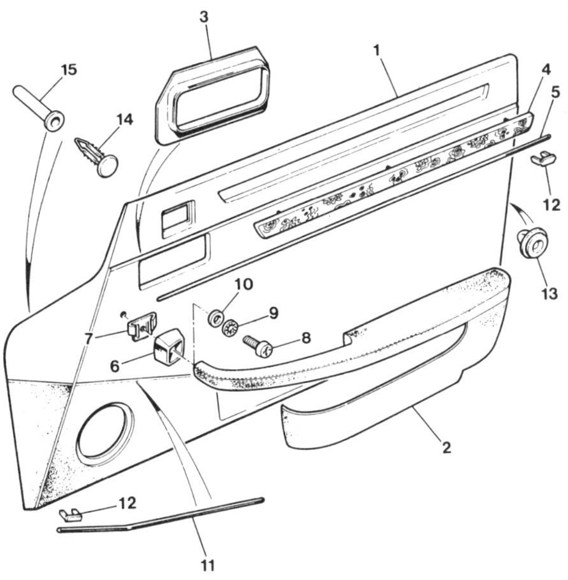

VS 3229

DOOR CASINGS-ARMRESTS AND POCKETS-CONVERTIBLE-continued

Illus	Part Number	123456 Description	Quantity	Change Point	Remarks
		ARMREST AND POCKET-RH			
2	BAC 4110 LZ	Warm Charcoal	1		
2	BAC 4110 AR	Buckskin	1		
2	BAC 4110 XE	Doeskin	1		
2	BAC 4110 JF	Isis Blue	1		
2	BAC 4110 CM	Mulberry	1		
2	BAC 4110 LY	Savile Grey	1		
2	BAC 4110 AW	Barley	1		
2	BAC 4110 ZM	Magnolia	1		
		ARMREST AND POCKET-LH			
	BAC 4111 LZ	Warm Charcoal	1		
	BAC 4111 AR	Buckskin	1		
	BAC 4111 XE	Doeskin	1		
	BAC 4111 JF	Isis Blue	1		
	BAC 4111 CM	Mulberry	1		
	BAC 4111 LY	Savile Grey	1		
	BAC 4111 AW	Barley	1		
	BAC 4111 ZM	Magnolia	1		
		REINFORCEMENT-HANDLE			
3	BCC 2520 LZ	Warm Charcoal	2		
3	BCC 2520 AR	Buckskin	2		
3	BCC 2520 XE	Doeskin	2		
3	BCC 2520 JF	Isis Blue	2		
3	BCC 2520 CM	Mulberry	2		
3	BCC 2520 LY	Savile Grey	2		
3	BCC 2520 AW	Barley	2		
3	BCC 2520 ZM	Magnolia	2		
4	BDC 3038	Capping-waist(walnut burr)	2		
5	BD 46571	Moulding-upper	2		
6	BAC 1462	Capping-armrest	2		
7	BD 42434	Plate-base	2		
8	SE 604081 J	Screw-set	2		
9	C 724	Washer-shakeproof	2		
10	JLM 296	Washer-plain	2		
11	BAC 5616	Moulding-lower	2		
12	BAC 5298	Clip-mouldings	18		
13	AGU 1018	Clip-casing	18		
14	BD 13102	Button-pronged	8		
15	BD 47644	Snap sac	8		

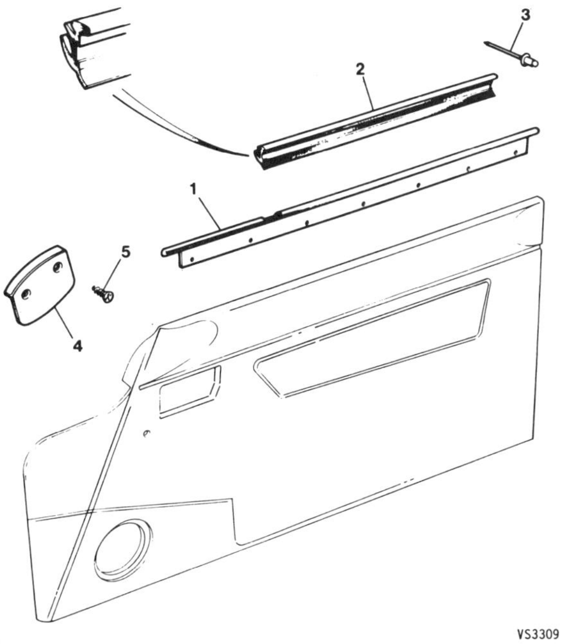

VS3309

DOOR TRIM

Illus	Part Number	123456 Description	Quantity	Change Point	Remarks
1	BD 47401	Moulding-waist	2		Not Convertible
2	BD 46458 1	Seal-waist	2		Not Convertible
1	BDC 6636	Moulding-waist-RH	1		Convertible
	BDC 6637	Moulding-waist-LH	1		Convertible
2	BDC 5800	Seal-waist	2		Convertible
3	BD 1818 1	Rivet	14		
4	BD 48211	Plate-striker	2		
5	AC 606042 J	Screw-self tap	4		

MASTER INDEX

ENGINE ... 1.C
FLYWHEEL AND CLUTCH ... 1.G
GEARBOX AND PROPSHAFT ... 1.H
AXLES, SUSPENSION, DRIVE SHAFTS, WHEELS ... 1.K
STEERING ... 1.M
BRAKES AND BRAKE CONTROLS ... 1.N
FUEL, EXHAUST AND EMISSION SYSTEMS ... 2.C
COOLING, HEATING AND AIR CONDITIONING ... 2.G
ELECTRICAL, WASHERS-WIPERS, INSTRUMENTS ... 2.J
CHASSIS, SUBFRAMES, BODYSHELL, FITTINGS ... 3.C
FASCIA, TRIM, SEATS AND FIXINGS ... 3.H
ACCESSORIES AND PAINT ... 4.C
NUMERICAL INDEX ... 4.D

GROUP INDEX

BATTERY AND SPARE WHEEL COVERS. ... N05
BONNET INSULATION-3.6 LITRE ... N06
BONNET INSULATION-5.3 MODELS ... N08
BONNET INSULATION-5.3 MODELS. ... N07
BOOT CARPETS - NOT CONVERTIBLE ... N03
BOOT CARPETS-CONVERTIBLE ... N04
CARPETS-'A' POST FINISHERS-. ... L12
CONSOLE-AMBLA. ... H10
CONSOLE-LEATHER. ... H08
DOOR CASINGS-ARMRESTS AND POCKETS-AMBLA. ... H16
DOOR CASINGS-ARMRESTS-CONVERTIBLE ... I02
HEADLINING ... I15
DOOR CASINGS-ARMRESTS-LEATHER. ... H14
EXTERIOR TRIM COVERS ... N15
FACIA PANEL. ... H02
FRONT SEAT BELTS-CONVERTIBLE ... K15
FRONT SEAT SLIDES ... K16
MOTOR AND ANCHORAGE ASSY-RH ... L02
FRONT SEATS-AMBLA/CLOTH.-3.6 LITRE ... K04
FRONT SEATS-LEATHER. ... J03
FRONT SEATS-LEATHER/CLOTH-5.3 LITRE ... K07
FRT SEATS-LEATHER/CLOTH(SPORTSPACK) ... K10
FRONT BELT KIT-COUPE ... K14
FRONT SEAT SLIDES ... K13
FUEL APERATURE SEAL AND HOSES-COUPE. ... N13
HARDTOP & HOODBAG-CABRIOLET. ... N17
HOOD LIFT MECHANISM - CONVERTIBLE ... O04
HOOD SEALS-CONVERTIBLE ... O03
HOOD-CABRIOLET ... N16
HOOD-CONVERTIBLE ... O02
REAR CUBBY BOX-CONVERTIBLE ... L09
REAR LUGGAGE LOCKER-CABRIOLET ... L07
REAR SEAT BELTS. ... L11
REAR SEATS-LEATHER-COUPE. ... L03
REAR SEATS-LEATHER-COUPE. ... L04
REAR SEATS-LEATHER/CLOTH-COUPE. ... L06
TARGA PANELS-STOWAGEBAG-CABRIOLET ... N14
UNDER SCUTTLE CASINGS. ... H05
VENEERED FACIA PANELS-3.6 LITRE. ... H06
VENEERED FACIA PANELS-5.3 LITRE. ... H07

3I 05

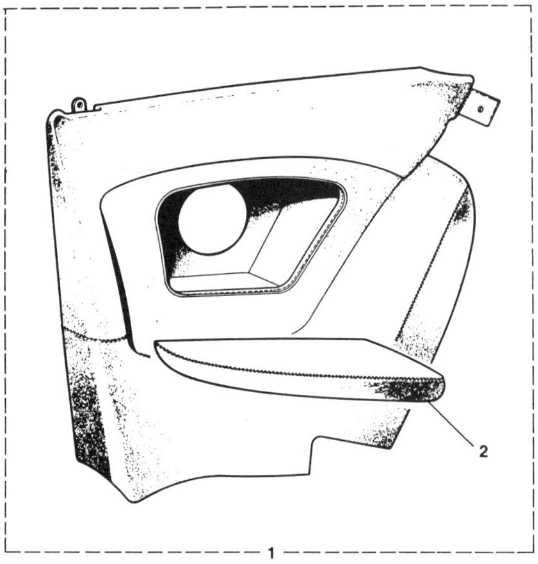

VS4732

REAR QUARTER CASINGS-LEATHER-COUPE NOT PASS.REST.

UP TO (V)149071

Illus	Part Number	Description	Quantity	Change Point	Remarks
		CASING RH-REAR QUARTER			
1	BCC 5804 AE	Biscuit	1		
1	BCC 5804 LZ	Warm charcoal	1		
1	BCC 5804 AR	Buckskin	1		
1	BCC 5804 XE	Doeskin	1		
1	BCC 5804 JF	Isis blue	1		
1	BCC 5804 CM	Mulberry	1		
1	BCC 5804 LY	Savile grey	1		
1	BCC 5804 AW	Barley	1		
1	BCC 5804 ZM	Magnolia	1		
		ARMREST RH-REAR QUARTER			
2	BAC 3986 AE	Biscuit	1		
2	BAC 3986 LZ	Warm charcoal	1		
2	BAC 3986 AR	Buckskin	1		
2	BAC 3986 XE	Doeskin	1		
2	BAC 3986 JF	Isis blue	1		
2	BAC 3986 CM	Mulberry	1		
2	BAC 3986 LY	Savile grey	1		
2	BAC 3986 AW	Barley	1		
2	BAC 3986 ZM	Magnolia	1		
		CASING LH-REAR QUARTER			
	BCC 5805 AE	Biscuit	1		
	BCC 5805 LZ	Warm charcoal	1		
	BCC 5805 AR	Buckskin	1		
	BCC 5805 XE	Doeskin	1		
	BCC 5805 JF	Isis blue	1		
	BCC 5805 CM	Mulberry	1		
	BCC 5805 LY	Savile grey	1		
	BCC 5805 AW	Barley	1		
	BCC 5895 ZM	Magnolia	1		
		ARMREST LH-REAR QUARTER			
	BAC 3987 AE	Biscuit	1		
	BAC 3987 LZ	Warm charcoal	1		
	BAC 3987 AR	Buckskin	1		
	BAC 3987 XE	Doeskin	1		
	BAC 3987 JF	Isis blue	1		
	BAC 3987 CM	Mulberry	1		
	BAC 3987 LY	Savile grey	1		
	BAC 3987 AW	Barley	1		
	BAC 3987 ZM	Magnolia	1		

M20.02JP24/6/88

VS4732

MASTER INDEX

GROUP INDEX

3I 06

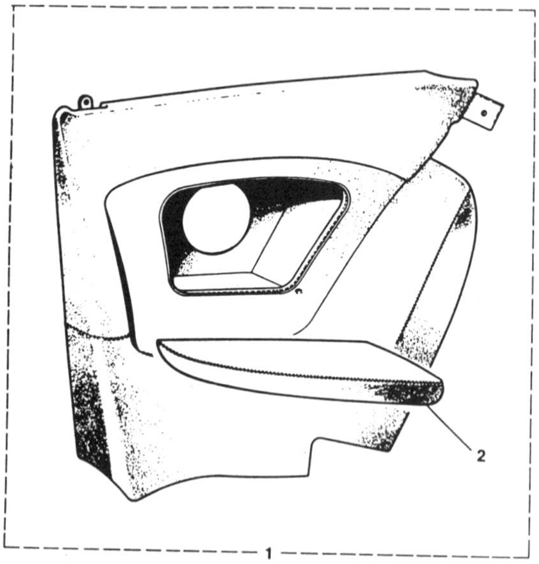

VS4732

REAR QUARTER CASINGS-LEATHER/AMBLA-COUPE NOT PASS.REST.

FROM (V)149072

Illus	Part Number	Description	Quantity	Change Point	Remarks
		CASING RH-REAR QUARTER			
1	BDC 7630 LZ	Warm Charcoal	1		
1	BDC 7630 AR	Buckskin	1		
1	BDC 7630 XE	Doeskin	1		
1	BDC 7630 JF	Isis Blue	1		
1	BDC 7630 CM	Mulberry	1		
1	BDC 7630 LY	Savile Grey	1		
1	BDC 7630 AW	Barley	1		
1	BDC 7630 ZM	Magnolia	1		
		ARMREST RH-REAR QUARTER			
2	BAC 3986 LZ	Warm Charcoal	1		
2	BAC 3986 AR	Buckskin	1		
2	BAC 3986 XE	Doeskin	1		
2	BAC 3986 JF	Isis Blue	1		
2	BAC 3986 CM	Mulberry	1		
2	BAC 3986 LY	Savile Grey	1		
2	BAC 3986 AW	Barley	1		
2	BAC 3986 ZM	Magnolia	1		
		CASING LH-REAR QUARTER			
	BDC 7631 LZ	Warm Charcoal	1		
	BDC 7631 AR	Buckskin	1		
	BDC 7631 XE	Doeskin	1		
	BDC 7631 JF	Isis Blue	1		
	BDC 7631 CM	Mulberry	1		
	BDC 76315 LY	Savile Grey	1		
	BDC 7631 AW	Barley	1		
	BDC 7631 ZM	Magnolia	1		
		ARMREST LH-REAR QUARTER			
	BAC 3987 LZ	Warm Charcoal	1		
	BAC 3987 AR	Buckskin	1		
	BAC 3987 XE	Doeskin	1		
	BAC 3987 JF	Isis Blue	1		
	BAC 3987 CM	Mulberry	1		
	BAC 3987 LY	Savile Grey	1		
	BAC 3987 AW	Barley	1		
	BAC 3987 ZM	Magnolia	1		

M20.04JP24/6/88

VS4732

MASTER INDEX

ENGINE 1.C
FLYWHEEL AND CLUTCH 1.G
GEARBOX AND PROPSHAFT 1.H
AXLES, SUSPENSION, DRIVE SHAFTS, WHEELS 1.K
STEERING 1.M
BRAKES AND BRAKE CONTROLS 1.N
FUEL, EXHAUST AND EMISSION SYSTEMS 2.C
COOLING, HEATING AND AIR CONDITIONING 2.G
ELECTRICAL, WASHERS-WIPERS, INSTRUMENTS 2.J
CHASSIS, SUBFRAMES, BODYSHELL, FITTINGS 3.C
FASCIA, TRIM, SEATS AND FIXINGS 3.H
ACCESSORIES AND PAINT 4.C
NUMERICAL INDEX 4.D

GROUP INDEX

BATTERY AND SPARE WHEEL COVERS N05
BONNET INSULATION-3.6 LITRE N06
BONNET INSULATION-5.3 MODELS N08
BONNET INSULATION-5.3 MODELS N07
BOOT CARPETS - NOT CONVERTIBLE N03
BOOT CARPETS-CONVERTIBLE N04
CARPETS-'A' POST FINISHERS- L12
CONSOLE-AMBLA H10
CONSOLE-LEATHER H08
DOOR CASINGS-ARMRESTS AND POCKETS-AMBLA H16
DOOR CASINGS-ARMRESTS-CONVERTIBLE I02
HEADLINING I15
DOOR CASINGS-ARMRESTS-LEATHER H14
EXTERIOR TRIM COVERS N15
FACIA PANEL H02
FRONT SEAT BELTS-CONVERTIBLE K15
FRONT SEAT SLIDES K16
MOTOR AND ANCHORAGE ASSY-RH L02
FRONT SEATS-AMBLA/CLOTH.-3.6 LITRE K04
FRONT SEATS-LEATHER J03
FRONT SEATS-LEATHER/CLOTH-5.3 LITRE K07
FRT SEATS-LEATHER/CLOTH(SPORTSPACK) K10
FRONT BELT KIT-COUPE K14
FRONT SEAT SLIDES K13
FUEL APERATURE SEAL AND HOSES-COUPE N13
HARDTOP & HOODBAG-CABRIOLET N17
HOOD LIFT MECHANISM - CONVERTIBLE O04
HOOD SEALS-CONVERTIBLE O03
HOOD-CABRIOLET N16
HOOD-CONVERTIBLE O02
REAR CUBBY BOX-CONVERTIBLE L09
REAR LUGGAGE LOCKER-CABRIOLET L07
REAR SEAT BELTS L11
REAR SEATS-LEATHER-COUPE L03
REAR SEATS-LEATHER-COUPE L04
REAR SEATS-LEATHER/CLOTH-COUPE L06
TARGA PANELS-STOWAGEBAG-CABRIOLET N14
UNDER SCUTTLE CASINGS H05
VENEERED FACIA PANELS-3.6 LITRE H06
VENEERED FACIA PANELS-5.3 LITRE H07

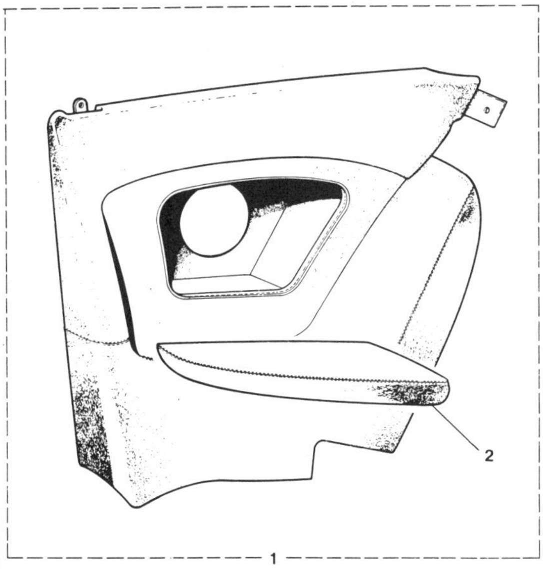

REAR QUARTER CASINGS-LEATHER-COUPE WITH PASS.REST.

Illus	Part Number	123456	Description	Quantity	Change Point	Remarks
			UP TO (V)149071			
			CASING RH-REAR QUARTER			
1	BCC 2030 AW		Barley	1		
1	BCC 2030 AE		Biscuit	1		
1	BCC 2030 AR		Buckskin	1		
1	BCC 2030 XE		Doeskin	1		
1	BCC 2030 JF		Isis blue	[illegible]		
1	BCC 2030 ZM		Magnolia	1		
1	BCC 2030 CM		Mulberry	1		
1	BCC 2030 LY		Savile grey	1		
1	BCC 2030 LZ		Warm charcoal	1		
			ARMREST RH-REAR QUARTER			
2	BAC 3986 AW		Barley	1		
2	BAC 3986 AE		Biscuit	1		
2	BAC 3986 AR		Buckskin	1		
2	BAC 3986 XE		Doeskin	1		
2	BAC 3986 JF		Isis blue	1		
2	BAC 3986 ZM		Magnolia	1		
2	BAC 3986 CM		Mulberry	1		
2	BAC 3986 LY		Savile grey	1		
2	BAC 3986 LZ		Warm charcoal	1		
			CASING LH-REAR QUARTER			
	BCC 2031 AW		Barley	1		
	BCC 2031 AE		Biscuit	1		
	BCC 2031 AR		Buckskin	1		
	BCC 2031 XE		Doeskin	1		
	BCC 2031 JF		Isis blue	1		
	BCC 2031 ZM		Magnolia	1		
	BCC 2031 CM		Mulberry	1		
	BCC 2031 LY		Savile grey	1		
	BCC 2031 LZ		Warm charcoal	1		
			ARMREST LH-REAR QUARTER			
	BAC 3987 AW		Barley	1		
	BAC 3987 AE		Biscuit	1		
	BAC 3987 AR		Buckskin	1		
	BAC 3987 XE		Doeskin	1		
	BAC 3987 JF		Isis blue	1		
	BAC 3987 ZM		Magnolia	1		
	BAC 3987 CM		Mulberry	1		
	BAC 3987 LY		Savile grey	1		
	BAC 3987 LZ		Warm charcoal	1		

M22JP20/7/88

VS4732

VS4732

MASTER INDEX

GROUP INDEX

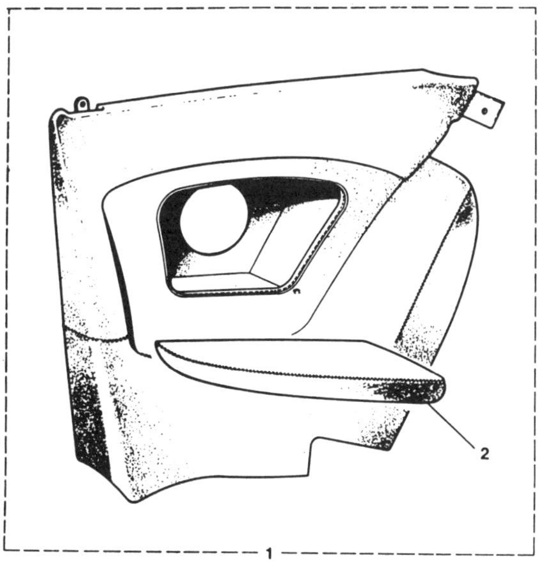

REAR QUARTER CASINGS-LEATHER/AMBLA COUPE WITH PASS.REST.

Illus	Part Number	123456	Description	Quantity	Change Point	Remarks
			FROM (V)149072			
			CASING RH-REAR QUARTER			
1	BDC 7636 AW		Barley	1		
1	BDC 7636 AR		Buckskin	1		
1	BDC 7636 XE		Doeskin	1		
1	BDC 7636 JF		Isis blue	1		
1	BDC 7636 ZM		Magnolia	1		
1	BDC 7636 CM		Mulberry	1		
1	BDC 7636 LY		Savile grey	1		
1	BDC 7636 LZ		Warm charcoal	1		
			ARMREST RH-REAR QUARTER			
2	BAC 3986 AW		Barley	1		
2	BAC 3986 AR		Buckskin	1		
2	BAC 3986 XE		Doeskin	1		
2	BAC 3986 JF		Isis blue	1		
2	BAC 3986 ZM		Magnolia	1		
2	BAC 3986 CM		Mulberry	1		
2	BAC 3986 LY		Savile grey	1		
2	BAC 3986 LZ		Warm charcoal	1		
			CASING LH-REAR QUARTER			
	BDC 7637 AW		Barley	1		
	BDC 7637 AR		Buckskin	1		
	BDC 7637 XE		Doeskin	1		
	BDC 7637 JF		Isis blue	1		
	BDC 7637 ZM		Magnolia	1		
	BDC 7637 CM		Mulberry	1		
	BDC 7637 LY		Savile grey	1		
	BDC 7637 LZ		Warm charcoal	1		
			ARMREST LH-REAR QUARTER			
	BAC 3987 AW		Barley	1		
	BAC 3987 AR		Buckskin	1		
	BAC 3987 XE		Doeskin	1		
	BAC 3987 JF		Isis blue	1		
	BAC 3987 ZM		Magnolia	1		
	BAC 3987 CM		Mulberry	1		
	BAC 3987 LY		Savile grey	1		
	BAC 3987 LZ		Warm charcoal	1		

M22.02JP20/7/88

VS4732

VS4732

MASTER INDEX
ENGINE 1.C
FLYWHEEL AND CLUTCH 1.G
GEARBOX AND PROPSHAFT 1.H
AXLES, SUSPENSION, DRIVE SHAFTS, WHEELS 1.K
STEERING 1.M
BRAKES AND BRAKE CONTROLS 1.N
FUEL, EXHAUST AND EMISSION SYSTEMS 2.C
COOLING, HEATING AND AIR CONDITIONING 2.G
ELECTRICAL, WASHERS-WIPERS, INSTRUMENTS 2.J
CHASSIS, SUBFRAMES, BODYSHELL, FITTINGS 3.C
FASCIA, TRIM, SEATS AND FIXINGS 3.H
ACCESSORIES AND PAINT 4.C
NUMERICAL INDEX 4.D
GROUP INDEX
BATTERY AND SPARE WHEEL COVERS. N05
BONNET INSULATION-3.6 LITRE N06
BONNET INSULATION-5.3 MODELS N08
BONNET INSULATION-5.3 MODELS. N07
BOOT CARPETS - NOT CONVERTIBLE N03
BOOT CARPETS-CONVERTIBLE N04
CARPETS-'A' POST FINISHERS-. L12
CONSOLE-AMBLA. H10
CONSOLE-LEATHER. H08
DOOR CASINGS-ARMRESTS AND POCKETS-AMBLA. H16
DOOR CASINGS-ARMRESTS-CONVERTIBLE I02
HEADLINING I15
DOOR CASINGS-ARMRESTS-LEATHER. H14
EXTERIOR TRIM COVERS N15
FACIA PANEL. H02
FRONT SEAT BELTS-CONVERTIBLE K15
FRONT SEAT SLIDES K16
MOTOR AND ANCHORAGE ASSY-RH L02
FRONT SEATS-AMBLA/CLOTH.-3.6 LITRE K04
FRONT SEATS-LEATHER. J03
FRONT SEATS-LEATHER/CLOTH-5.3 LITRE K07
FRT SEATS-LEATHER/CLOTH(SPORTSPACK) K10
FRONT BELT KIT-COUPE K14
FRONT SEAT SLIDES K13
FUEL APERATURE SEAL AND HOSES-COUPE. N13
HARDTOP & HOODBAG-CABRIOLET. N17
HOOD LIFT MECHANISM - CONVERTIBLE O04
HOOD SEALS-CONVERTIBLE O03
HOOD-CABRIOLET N16
HOOD-CONVERTIBLE O02
REAR CUBBY BOX-CONVERTIBLE L09
REAR LUGGAGE LOCKER-CABRIOLET L07
REAR SEAT BELTS. L11
REAR SEATS-LEATHER-COUPE. L03
REAR SEATS-LEATHER-COUPE. L04
REAR SEATS-LEATHER/CLOTH-COUPE. L06
TARGA PANELS-STOWAGEBAG-CABRIOLET N14
UNDER SCUTTLE CASINGS. H05
VENEERED FACIA PANELS-3.6 LITRE. H06
VENEERED FACIA PANELS-5.3 LITRE. H07

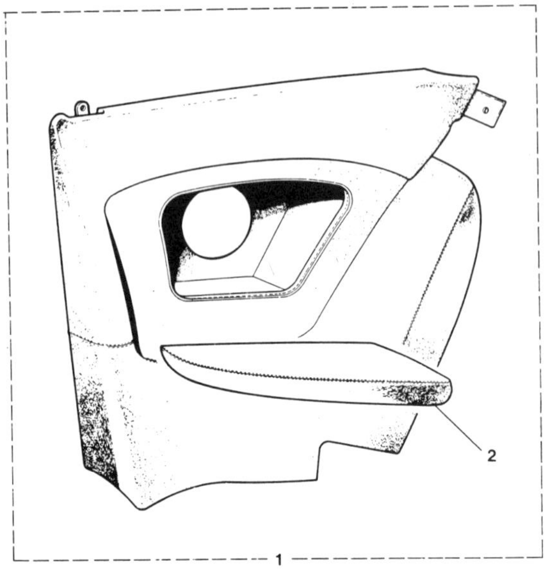

REAR QUARTER CASINGS-ARMRESTS-AMBLA-COUPE

UP TO (V)151099

Illus	Part Number	Description	Quantity	Change Point	Remarks
		CASING RH-REAR QUARTER			
1	BCC 5798 AE	Biscuit	1		
1	BCC 5798 LZ	Warm Charcoal	1		
1	BCC 5798 AR	Buckskin	1		
1	BCC 5798 XE	Doeskin	1		
1	BCC 5798 JF	Isis Blue	1		
1	BCC 5798 CM	Mulberry	1		
1	BCC 5798 LY	Savile Grey	1		
1	BCC 5798 AW	Barley	1		
1	BCC 5798 ZM	Magnolia	1		
		ARMREST RH-REAR QUARTER			
2	BD 48433 AE	Biscuit	1		
2	BD 48433 LZ	Warm Charcoal	1		
2	BD 48433 AR	Buckskin	1		
2	BD 48433 XE	Doeskin	1		
2	BD 48433 JF	Isis Blue	1		
2	BD 48433 CM	Mulberry	1		
2	BD 48433 LY	Savile Grey	1		
2	BD 48433 AW	Barley	1		
2	BD 48433 ZM	Magnolia	1		
		CASING LH-REAR QUARTER			
	BCC 5799 AE	Biscuit	1		
	BCC 5799 LZ	Warm Charcoal	1		
	BCC 5799 AR	Buckskin	1		
	BCC 5799 XE	Doeskin	1		
	BCC 5799 JF	Isis Blue	1		
	BCC 5799 CM	Mulberry	1		
	BCC 5799 LY	Savile Grey	1		
	BCC 5799 AW	Barley	1		
	BCC 5799 ZM	Magnolia	1		
		ARMREST LH-REAR QUARTER			
	BD 48434 AE	Biscuit	1		
	BD 48434 LZ	Warm Charcoal	1		
	BD 48434 AR	Buckskin	1		
	BD 48434 XE	Doeskin	1		
	BD 48434 JF	Isis Blue	1		
	BD 48434 CM	Mulberry	1		
	BD 48434 LY	Savile Grey	1		
	BD 48434 AW	Barley	1		
	BD 48434 ZM	Magnolia	1		

M22.04JP20/7/88

VS4732

3I 09

MASTER INDEX

GROUP INDEX

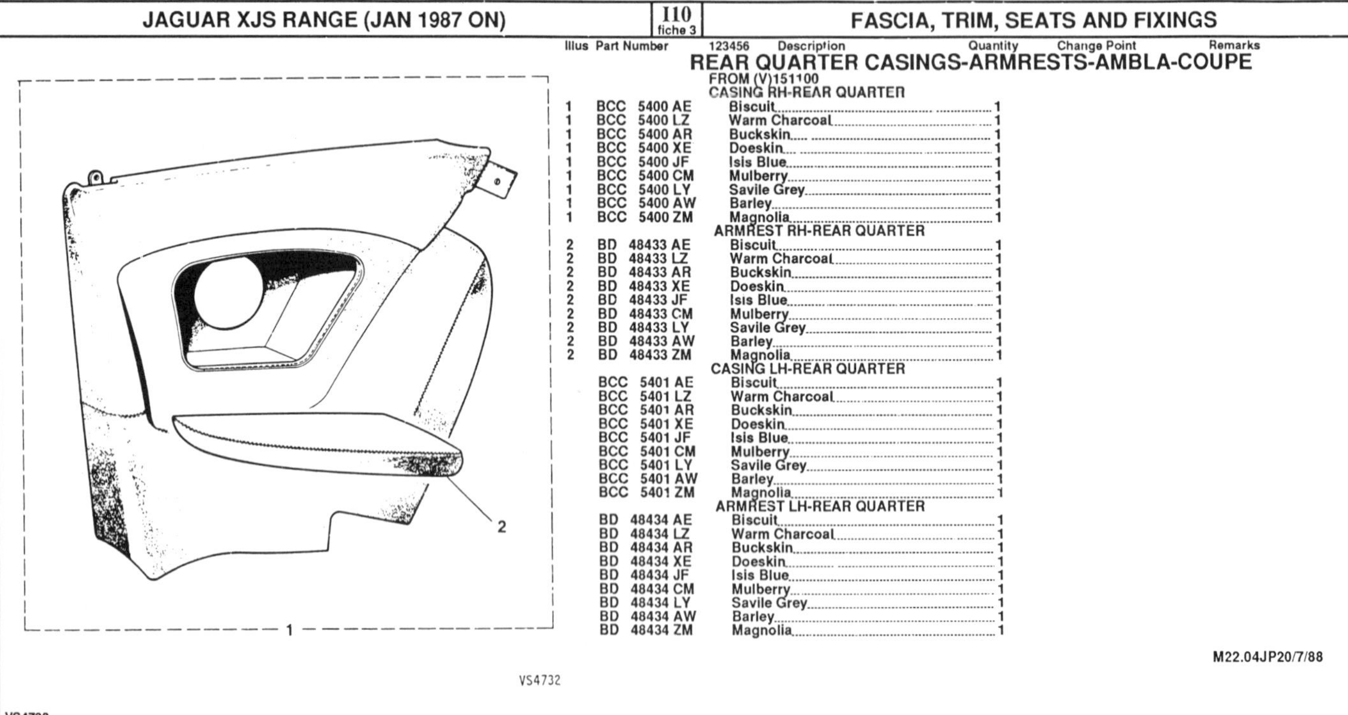

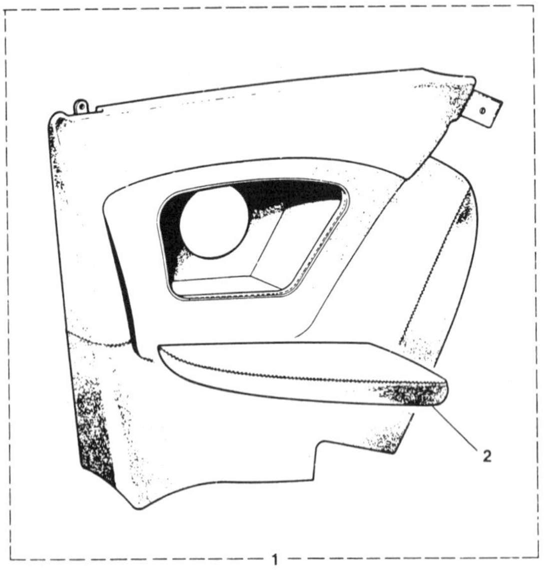

REAR QUARTER CASINGS-ARMRESTS-AMBLA-COUPE

FROM (V)151100

Illus	Part Number	Description	Quantity	Change Point	Remarks
		CASING RH-REAR QUARTER			
1	BCC 5400 AE	Biscuit	1		
1	BCC 5400 LZ	Warm Charcoal	1		
1	BCC 5400 AR	Buckskin	1		
1	BCC 5400 XE	Doeskin	1		
1	BCC 5400 JF	Isis Blue	1		
1	BCC 5400 CM	Mulberry	1		
1	BCC 5400 LY	Savile Grey	1		
1	BCC 5400 AW	Barley	1		
1	BCC 5400 ZM	Magnolia	1		
		ARMREST RH-REAR QUARTER			
2	BD 48433 AE	Biscuit	1		
2	BD 48433 LZ	Warm Charcoal	1		
2	BD 48433 AR	Buckskin	1		
2	BD 48433 XE	Doeskin	1		
2	BD 48433 JF	Isis Blue	1		
2	BD 48433 CM	Mulberry	1		
2	BD 48433 LY	Savile Grey	1		
2	BD 48433 AW	Barley	1		
2	BD 48433 ZM	Magnolia	1		
		CASING LH-REAR QUARTER			
	BCC 5401 AE	Biscuit	1		
	BCC 5401 LZ	Warm Charcoal	1		
	BCC 5401 AR	Buckskin	1		
	BCC 5401 XE	Doeskin	1		
	BCC 5401 JF	Isis Blue	1		
	BCC 5401 CM	Mulberry	1		
	BCC 5401 LY	Savile Grey	1		
	BCC 5401 AW	Barley	1		
	BCC 5401 ZM	Magnolia	1		
		ARMREST LH-REAR QUARTER			
	BD 48434 AE	Biscuit	1		
	BD 48434 LZ	Warm Charcoal	1		
	BD 48434 AR	Buckskin	1		
	BD 48434 XE	Doeskin	1		
	BD 48434 JF	Isis Blue	1		
	BD 48434 CM	Mulberry	1		
	BD 48434 LY	Savile Grey	1		
	BD 48434 AW	Barley	1		
	BD 48434 ZM	Magnolia	1		

M22.04JP20/7/88

VS4732

3I 10

MASTER INDEX
ENGINE 1.C
FLYWHEEL AND CLUTCH 1.G
GEARBOX AND PROPSHAFT 1.H
AXLES, SUSPENSION, DRIVE SHAFTS, WHEELS 1.K
STEERING 1.M
BRAKES AND BRAKE CONTROLS 1.N
FUEL, EXHAUST AND EMISSION SYSTEMS 2.C
COOLING, HEATING AND AIR CONDITIONING 2.G
ELECTRICAL, WASHERS-WIPERS, INSTRUMENTS 2.J
CHASSIS, SUBFRAMES, BODYSHELL, FITTINGS 3.C
FASCIA, TRIM, SEATS AND FIXINGS 3.H
ACCESSORIES AND PAINT 4.C
NUMERICAL INDEX 4.D
GROUP INDEX
BATTERY AND SPARE WHEEL COVERS. N05
BONNET INSULATION-3.6 LITRE N06
BONNET INSULATION-5.3 MODELS N08
BONNET INSULATION-5.3 MODELS. N07
BOOT CARPETS - NOT CONVERTIBLE N03
BOOT CARPETS-CONVERTIBLE N04
CARPETS-'A' POST FINISHERS-. L12
CONSOLE-AMBLA. H10
CONSOLE-LEATHER. H08
DOOR CASINGS-ARMRESTS AND POCKETS-AMBLA. H16
DOOR CASINGS-ARMRESTS-CONVERTIBLE I02
HEADLINING I15
DOOR CASINGS-ARMRESTS-LEATHER. H14
EXTERIOR TRIM COVERS N15
FACIA PANEL. H02
FRONT SEAT BELTS-CONVERTIBLE K15
FRONT SEAT SLIDES K16
MOTOR AND ANCHORAGE ASSY-RH L02
FRONT SEATS-AMBLA/CLOTH.-3.6 LITRE K04
FRONT SEATS-LEATHER. J03
FRONT SEATS-LEATHER/CLOTH-5.3 LITRE K07
FRT SEATS-LEATHER/CLOTH(SPORTSPACK) K10
FRONT BELT KIT-COUPE K14
FRONT SEAT SLIDES K13
FUEL APERATURE SEAL AND HOSES-COUPE. N13
HARDTOP & HOODBAG-CABRIOLET. N17
HOOD LIFT MECHANISM - CONVERTIBLE O04
HOOD SEALS-CONVERTIBLE O03
HOOD-CABRIOLET N16
HOOD-CONVERTIBLE O02
REAR CUBBY BOX-CONVERTIBLE L09
REAR LUGGAGE LOCKER-CABRIOLET L07
REAR SEAT BELTS. L11
REAR SEATS-LEATHER-COUPE. L03
REAR SEATS-LEATHER-COUPE. L04
REAR SEATS-LEATHER/CLOTH-COUPE. L06
TARGA PANELS-STOWAGEBAG-CABRIOLET N14
UNDER SCUTTLE CASINGS. H05
VENEERED FACIA PANELS-3.6 LITRE. H06
VENEERED FACIA PANELS-5.3 LITRE. H07

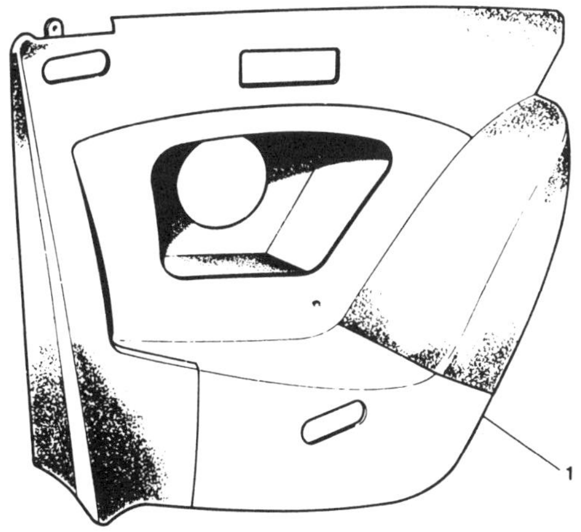

VS 4320

REAR QUARTER CASINGS-CABRIOLET-NOT PASS.

Illus	Part Number	123456	Description	Quantity	Change Point	Remarks
			CASING RH-REAR QUARTER			
1	BCC 6342 AE		Biscuit	1		
1	BCC 6342 LZ		Warm Charcoal	1		
1	BCC 6342 AR		Buckskin	1		
1	BCC 6342 XE		Doeskin	1		
1	BCC 6342 JF		Isis Blue	1		
1	BCC 6342 CM		Mulberry	1		
1	BCC 6342 LY		Savile Grey	1		
1	BCC 6342 AW		Barley	1		
1	BCC 6342 ZM		Magnolia	1		
			CASING LH-REAR QUARTER			
	BCC 6343 AE		Biscuit	1		
	BCC 6343 LZ		Warm Charcoal	1		
	BCC 6343 AR		Buckskin	1		
	BCC 6343 XE		Doeskin	1		
	BCC 6343 JF		Isis Blue	1		
	BCC 6343 CM		Mulberry	1		
	BCC 6343 LY		Savile Grey	1		
	BCC 6343 AW		Barley	1		
	BCC 6343 ZM		Magnolia	1		

MASTER INDEX

GROUP INDEX

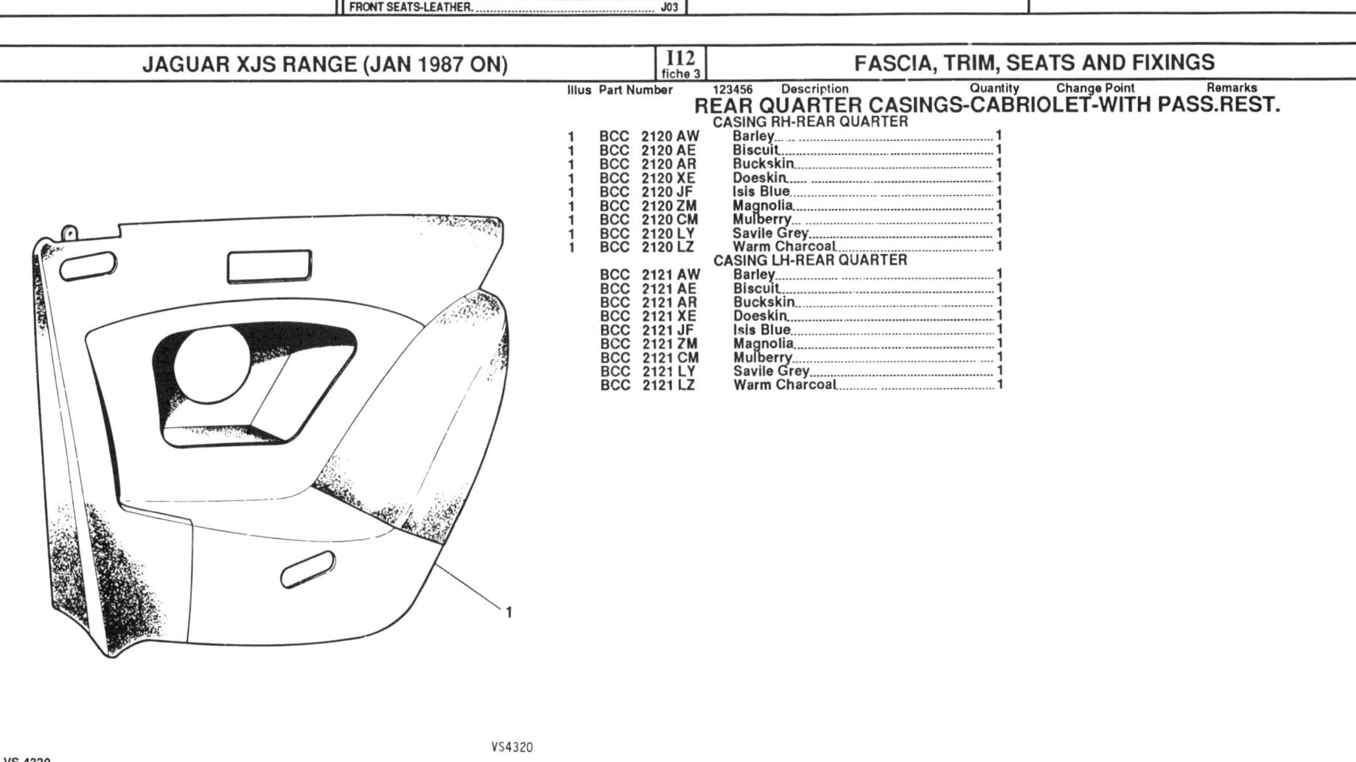

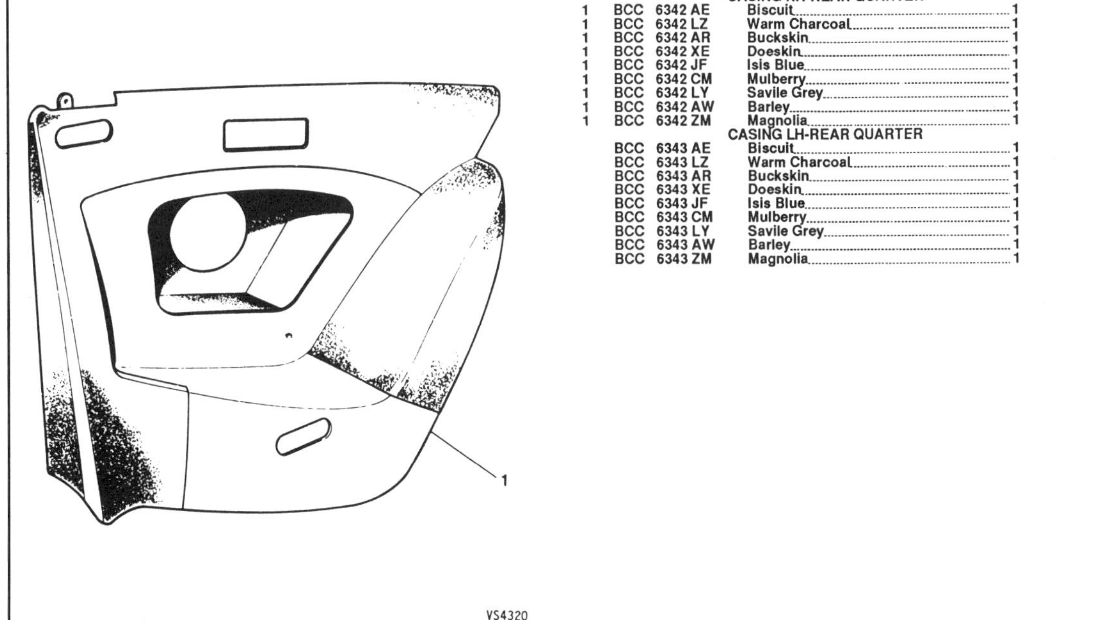

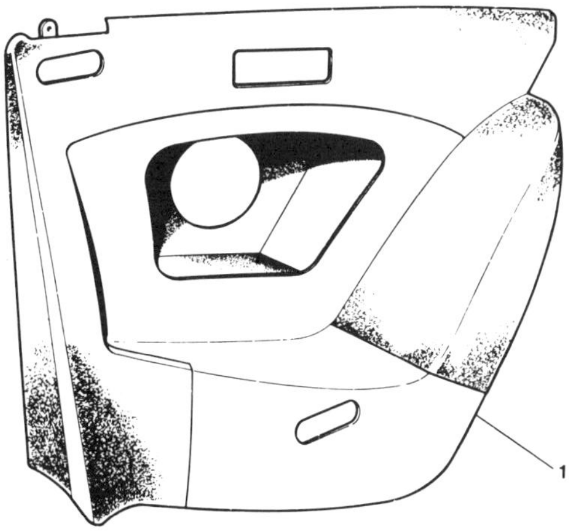

VS 4320

REAR QUARTER CASINGS-CABRIOLET-WITH PASS.REST.

Illus	Part Number	123456	Description	Quantity	Change Point	Remarks
			CASING RH-REAR QUARTER			
1	BCC 2120 AW		Barley	1		
1	BCC 2120 AE		Biscuit	1		
1	BCC 2120 AR		Buckskin	1		
1	BCC 2120 XE		Doeskin	1		
1	BCC 2120 JF		Isis Blue	1		
1	BCC 2120 ZM		Magnolia	1		
1	BCC 2120 CM		Mulberry	1		
1	BCC 2120 LY		Savile Grey	1		
1	BCC 2120 LZ		Warm Charcoal	1		
			CASING LH-REAR QUARTER			
	BCC 2121 AW		Barley	1		
	BCC 2121 AE		Biscuit	1		
	BCC 2121 AR		Buckskin	1		
	BCC 2121 XE		Doeskin	1		
	BCC 2121 JF		Isis Blue	1		
	BCC 2121 ZM		Magnolia	1		
	BCC 2121 CM		Mulberry	1		
	BCC 2121 LY		Savile Grey	1		
	BCC 2121 LZ		Warm Charcoal	1		

MASTER INDEX
ENGINE 1.C
FLYWHEEL AND CLUTCH 1.G
GEARBOX AND PROPSHAFT 1.H
AXLES, SUSPENSION, DRIVE SHAFTS, WHEELS 1.K
STEERING 1.M
BRAKES AND BRAKE CONTROLS 1.N
FUEL, EXHAUST AND EMISSION SYSTEMS 2.C
COOLING, HEATING AND AIR CONDITIONING 2.G
ELECTRICAL, WASHERS-WIPERS, INSTRUMENTS 2.J
CHASSIS, SUBFRAMES, BODYSHELL, FITTINGS 3.C
FASCIA, TRIM, SEATS AND FIXINGS 3.H
ACCESSORIES AND PAINT 4.C
NUMERICAL INDEX 4.D
GROUP INDEX
BATTERY AND SPARE WHEEL COVERS. N05
BONNET INSULATION-3.6 LITRE N06
BONNET INSULATION-5.3 MODELS N08
BONNET INSULATION-5.3 MODELS. N07
BOOT CARPETS - NOT CONVERTIBLE N03
BOOT CARPETS-CONVERTIBLE N04
CARPETS-'A' POST FINISHERS-. L12
CONSOLE-AMBLA. H10
CONSOLE-LEATHER. H08
DOOR CASINGS-ARMRESTS AND POCKETS-AMBLA. H16
DOOR CASINGS-ARMRESTS-CONVERTIBLE I02
HEADLINING I15
DOOR CASINGS-ARMRESTS-LEATHER. H14
EXTERIOR TRIM COVERS N15
FACIA PANEL. H02
FRONT SEAT BELTS-CONVERTIBLE K15
FRONT SEAT SLIDES K16
MOTOR AND ANCHORAGE ASSY-RH L02
FRONT SEATS-AMBLA/CLOTH.-3.6 LITRE K04
FRONT SEATS-LEATHER. J03
FRONT SEATS-LEATHER/CLOTH-5.3 LITRE K07
FRT SEATS-LEATHER/CLOTH(SPORTSPACK) K10
FRONT BELT KIT-COUPE K14
FRONT SEAT SLIDES K13
FUEL APERATURE SEAL AND HOSES-COUPE. N13
HARDTOP & HOODBAG-CABRIOLET. N17
HOOD LIFT MECHANISM - CONVERTIBLE O04
HOOD SEALS-CONVERTIBLE O03
HOOD-CABRIOLET N16
HOOD-CONVERTIBLE O02
REAR CUBBY BOX-CONVERTIBLE L09
REAR LUGGAGE LOCKER-CABRIOLET L07
REAR SEAT BELTS. L11
REAR SEATS-LEATHER-COUPE. L03
REAR SEATS-LEATHER-COUPE. L04
REAR SEATS-LEATHER/CLOTH-COUPE. L06
TARGA PANELS-STOWAGEBAG-CABRIOLET N14
UNDER SCUTTLE CASINGS. H05
VENEERED FACIA PANELS-3.6 LITRE. H06
VENEERED FACIA PANELS-5.3 LITRE. H07

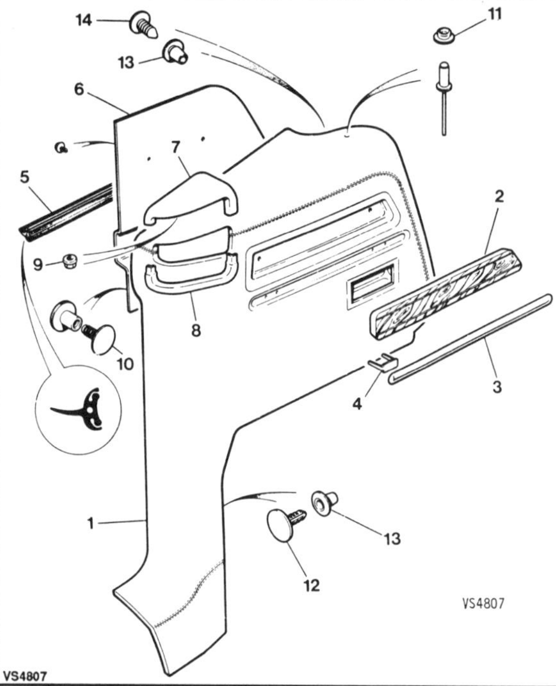

VS4807

REAR QTR CASINGS COVERTIBLE

Illus	Part Number	123456	Description	Quantity	Change Point	Remarks
			REAR QTR CASING RH			
1	JLM 1480 AW		Barley	1		
1	JLM 1480 AR		Buckskin	1		
1	JLM 1480 XF		Doeskin	1		
1	JLM 1480 JF		Isis Blue	1		
1	JLM 1480 ZM		Magnolia	1		
1	JLM 1480 CM		Mulberry	1		
1	JLM 1480 LY		Savile Grey	1		
1	JLM 1480 LZ		Warm Charcoal	1		
			REAR QTR CASING LH			
	JLM 1481 AW		Barley	1		
	JLM 1481 AR		Buckskin	1		
	JLM 1481 XE		Doeskin	1		
	JLM 1481 JF		Isis Blue	1		
	JLM 1481 ZM		Magnolia	1		
	JLM 1481 CM		Mulberry	1		
	JLM 1481 LY		Savile Grey	1		
	JLM 1481 LZ		Warm Charcoal	1		
2	BDC 7742		Capping (Walnutt Burr) RH	1		
	BDC 7743		Capping (Walnutt Burr) LH	1		
3	BDC 4937		Moulding - chrome	2		
4	BAC 5290		Clip - moulding	6		
5	BBC 7655		Seal - Inner	2		
6	BDC 6452		Cover - inner RH	1		
	BDC 6453		Cover - inner LH	1		
7	BDC 7050		Capping - upper RH	1		
	BDC 7051		Capping - upper LH	1		
8	BDC 7052		Capping - lower RH	1		
	BDC 7053		Capping - lower LH	1		
9	AGU 2517		Nut - push on	4		
10	CBC 4861		Rivet - ratchet	8		
11	AGU 2539 J		Stud - hood cover fastening	2		
12	AGU 2533 J		Fixing - fir tree	4		
13	AGU 2388 J		Grommet - panel	6		
			FIXING - BUTTON			
14	BBC 7525 AW		Barley	2		
14	BBC 7525 AR		Buckskin	2		
14	BBC 7525 XF		Doeskin	2		
14	BBC 7525 JF		Isis blue	2		
14	BBC 7525 ZM		Magnolia	2		
14	BBC 7525 CM		Mulberry	2		
14	BBC 7525 LY		Savile grey	2		
14	BBC 7525 LZ		Warm charcoal	2		

3I 13

MASTER INDEX

GROUP INDEX

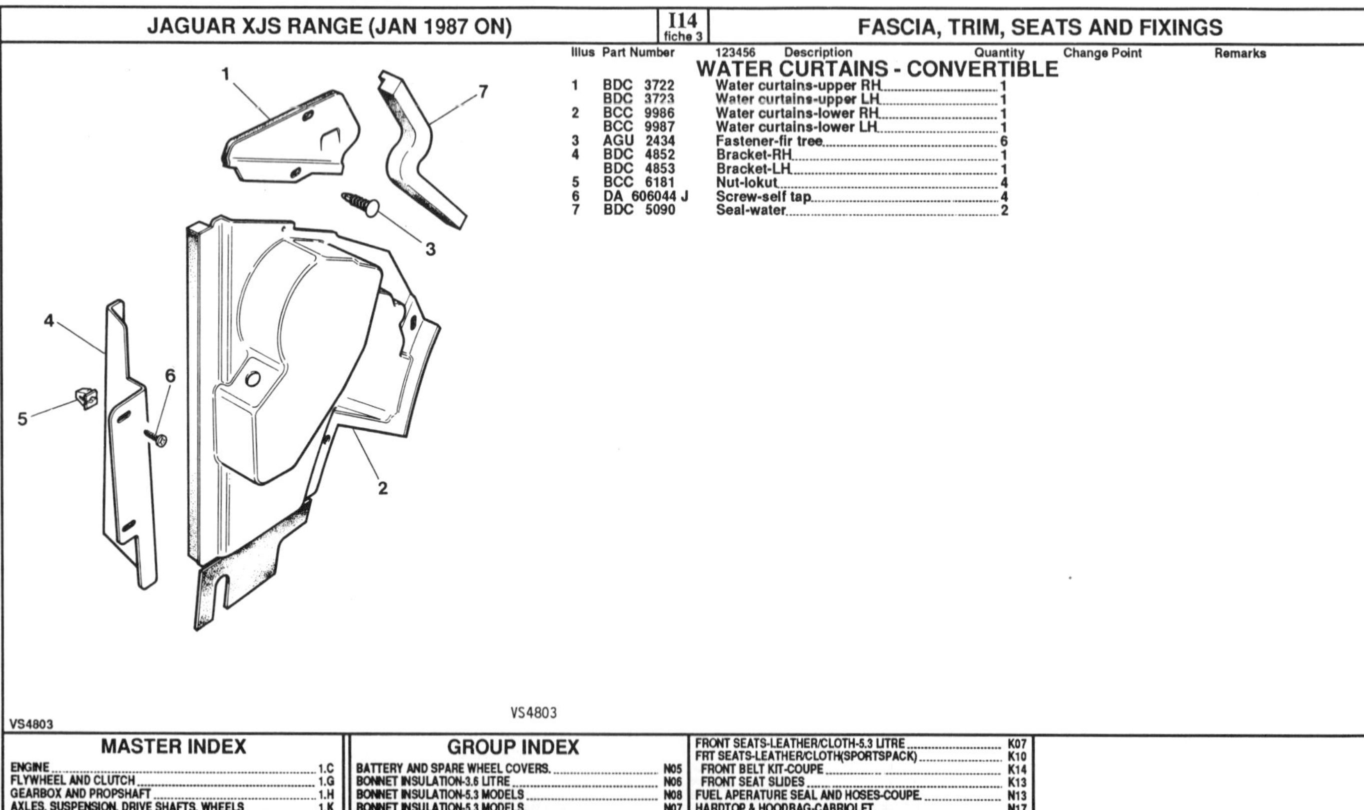

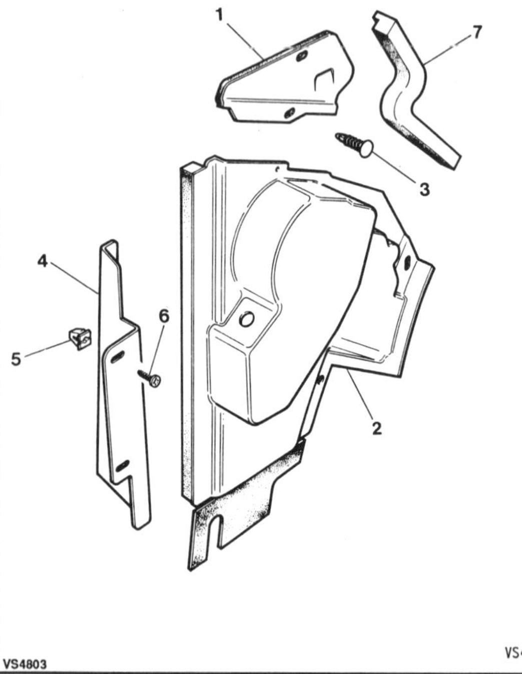

VS4803

WATER CURTAINS - CONVERTIBLE

Illus	Part Number	123456	Description	Quantity	Change Point	Remarks
1	BDC 3722		Water curtains-upper RH	1		
	BDC 3723		Water curtains-upper LH	1		
2	BCC 9986		Water curtains-lower RH	1		
	BCC 9987		Water curtains-lower LH	1		
3	AGU 2434		Fastener-fir tree	6		
4	BDC 4852		Bracket-RH	1		
	BDC 4853		Bracket-LH	1		
5	BCC 6181		Nut-lokut	4		
6	DA 606044 J		Screw-self tap	4		
7	BDC 5090		Seal-water	2		

3I 14

MASTER INDEX
ENGINE 1.C
FLYWHEEL AND CLUTCH 1.G
GEARBOX AND PROPSHAFT 1.H
AXLES, SUSPENSION, DRIVE SHAFTS, WHEELS 1.K
STEERING 1.M
BRAKES AND BRAKE CONTROLS 1.N
FUEL, EXHAUST AND EMISSION SYSTEMS 2.C
COOLING, HEATING AND AIR CONDITIONING 2.G
ELECTRICAL, WASHERS-WIPERS, INSTRUMENTS 2.J
CHASSIS, SUBFRAMES, BODYSHELL, FITTINGS 3.C
FASCIA, TRIM, SEATS AND FIXINGS 3.H
ACCESSORIES AND PAINT 4.C
NUMERICAL INDEX 4.D
GROUP INDEX
BATTERY AND SPARE WHEEL COVERS. N05
BONNET INSULATION-3.6 LITRE N06
BONNET INSULATION-5.3 MODELS N08
BONNET INSULATION-5.3 MODELS. N07
BOOT CARPETS - NOT CONVERTIBLE N03
BOOT CARPETS-CONVERTIBLE N04
CARPETS-'A' POST FINISHERS-. L12
CONSOLE-AMBLA. H10
CONSOLE-LEATHER. H08
DOOR CASINGS-ARMRESTS AND POCKETS-AMBLA. H16
DOOR CASINGS-ARMRESTS-CONVERTIBLE I02
HEADLINING I15
DOOR CASINGS-ARMRESTS-LEATHER. H14
EXTERIOR TRIM COVERS N15
FACIA PANEL. H02
FRONT SEAT BELTS-CONVERTIBLE K15
FRONT SEAT SLIDES K16
MOTOR AND ANCHORAGE ASSY-RH L02
FRONT SEATS-AMBLA/CLOTH.-3.6 LITRE K04
FRONT SEATS-LEATHER. J03
FRONT SEATS-LEATHER/CLOTH-5.3 LITRE K07
FRT SEATS-LEATHER/CLOTH(SPORTSPACK) K10
FRONT BELT KIT-COUPE K14
FRONT SEAT SLIDES K13
FUEL APERATURE SEAL AND HOSES-COUPE. N13
HARDTOP & HOODBAG-CABRIOLET. N17
HOOD LIFT MECHANISM - CONVERTIBLE O04
HOOD SEALS-CONVERTIBLE O03
HOOD-CABRIOLET N16
HOOD-CONVERTIBLE O02
REAR CUBBY BOX-CONVERTIBLE L09
REAR LUGGAGE LOCKER-CABRIOLET L07
REAR SEAT BELTS. L11
REAR SEATS-LEATHER-COUPE. L03
REAR SEATS-LEATHER-COUPE. L04
REAR SEATS-LEATHER/CLOTH-COUPE. L06
TARGA PANELS-STOWAGEBAG-CABRIOLET N14
UNDER SCUTTLE CASINGS. H05
VENEERED FACIA PANELS-3.6 LITRE. H06
VENEERED FACIA PANELS-5.3 LITRE. H07

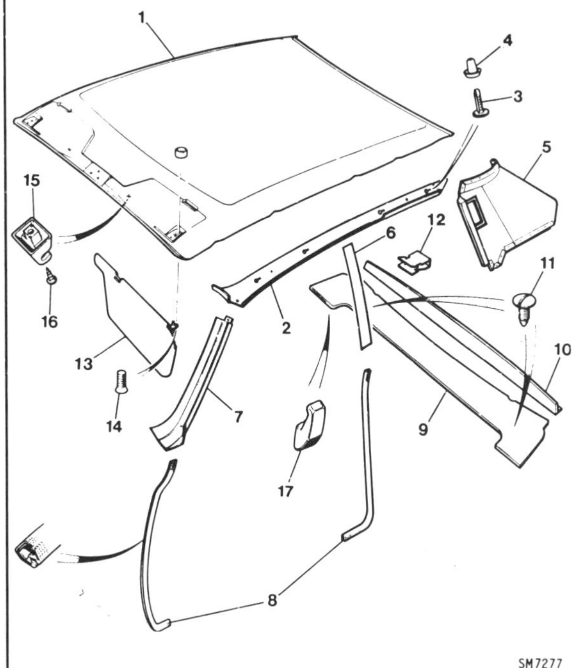

SM 7277

INTERIOR TRIM-COUPE NOT PASSIVE RESTRAINT.

Illus	Part Number	Description	Quantity	Change Point	Remarks
1	BD 47656 ND	HEADLINING	1		Use BCC 5292 ND
1	BCC 5292 ND	Headlining	NLA		Use BDC 4361 ND
1	BDC 4361 ND	Headlining	1		Limestone
	BD 45056 ND	Cantrail RH	1		Limestone
2	BD 45057 ND	Cantrail LH	1		Limestone
3	BD 13102	Button-pronged	8		
4	BD 47802	Socket-retainer	8		
	BBC 7862 ND	Trim-RH-rear	1		Limestone
5	BBC 7863 ND	Trim LH-rear	1		Limestone
	BBC 7544 XE	Finisher RH-'B' post	1		Doeskin
6	BBC 7545 XE	Finisher LH-'B' post	1		Doeskin
	BBC 7494 XE	Finisher RH-'A' post	1		Doeskin
7	BBC 7495 XE	Finisher LH-'A' post	1		Doeskin
8	BAC 2168 PA	Finisher-flange	4		
9	BBC 7870 PA	Shelf-parcel	1		
10	BAC 5566 PA	Strip-parcel shelf	1	Up to 142986	
10	BDC 4024 PA	Strip-parcel shelf	1	From 142987	
11	BAC 2904	Plug-parcel shelf	2		
12	BD 21962	Clip-retainer	7		
	BBC 7538 XE	Sunvisor RH	1		Doeskin
13	BBC 7539 XE	Sunvisor LH	1		Doeskin
14	BD 43278 4	Screw-set	4		
15	BBC 1727 XE	Stowage-sunvisors	2		Use BCC 9494 XE
15	BCC 9494 XE	Stowage - sunvisors	2		
16	AD 606051 J	Screw-self-tap	2		
17	BBC 9257	Hook-coat	2		

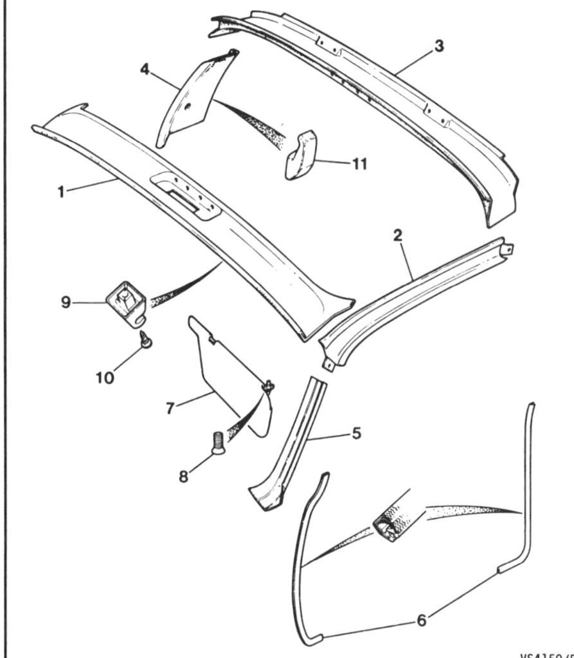

VS 4150B

INTERIOR TRIM-CABRIOLET-NOT PASSIVE RESTRAINT

Illus	Part Number	Description	Quantity	Change Point	Remarks
1	BCC 3624 ND	Trim-header	1		
	BCC 3622 ND	Trim-cantrail RH	1		
2	BCC 3623 ND	Trim-cantrail LH	1		
3	BCC 3626 XE	Trim-roll bar	1		
4	BCC 3620 XE	Trim-B post RH	1		
	BCC 3621 XE	Trim-B post LH	1		
	BBC 7494 XE	Finisher RH-'A' post	1		Doeskin
5	BBC 7495 XE	Finisher LH-'A' post	1		Doeskin
6	BAC 2168 PA	Finisher-flange	4		
	BBC 7538 XE	Sunvisor RH	1		Doeskin
7	BBC 7539 XE	Sunvisor LH	1		Doeskin
8	BD 43278 4	Screw-set	4		
9	BBC 1727 XE	Stowage sunvisors	2		Use BCC 9494 XE
9	BCC 9494 XE	Stowage sunvisors	2		
10	AD 606051 J	Screw-self-tap	2		
11	BBC 9257	Hook-coat	2		

MASTER INDEX

ENGINE 1.C
FLYWHEEL AND CLUTCH 1.G
GEARBOX AND PROPSHAFT 1.H
AXLES, SUSPENSION, DRIVE SHAFTS, WHEELS 1.K
STEERING 1.M
BRAKES AND BRAKE CONTROLS 1.N
FUEL, EXHAUST AND EMISSION SYSTEMS 2.C
COOLING, HEATING AND AIR CONDITIONING 2.G
ELECTRICAL, WASHERS-WIPERS, INSTRUMENTS 2.J
CHASSIS, SUBFRAMES, BODYSHELL, FITTINGS 3.C
FASCIA, TRIM, SEATS AND FIXINGS 3.H
ACCESSORIES AND PAINT 4.C
NUMERICAL INDEX 4.D

GROUP INDEX

BATTERY AND SPARE WHEEL COVERS N05
BONNET INSULATION-3.6 LITRE N06
BONNET INSULATION-5.3 MODELS N08
BONNET INSULATION-5.3 MODELS N07
BOOT CARPETS - NOT CONVERTIBLE N03
BOOT CARPETS-CONVERTIBLE N04
CARPETS-'A' POST FINISHERS- L12
CONSOLE-AMBLA H10
CONSOLE-LEATHER H08
DOOR CASINGS-ARMRESTS AND POCKETS-AMBLA H16
DOOR CASINGS-ARMRESTS-CONVERTIBLE I02
HEADLINING I15
DOOR CASINGS-ARMRESTS-LEATHER H14
EXTERIOR TRIM COVERS N15
FACIA PANEL H02
FRONT SEAT BELTS-CONVERTIBLE K15
FRONT SEAT SLIDES K16
MOTOR AND ANCHORAGE ASSY-RH L02
FRONT SEATS-AMBLA/CLOTH.-3.6 LITRE K04
FRONT SEATS-LEATHER J03
FRONT SEATS-LEATHER/CLOTH-5.3 LITRE K07
FRT SEATS-LEATHER/CLOTH(SPORTSPACK) K10
FRONT BELT KIT-COUPE K14
FRONT SEAT SLIDES K13
FUEL APERATURE SEAL AND HOSES-COUPE N13
HARDTOP & HOODBAG-CABRIOLET N17
HOOD LIFT MECHANISM - CONVERTIBLE O04
HOOD SEALS-CONVERTIBLE O03
HOOD-CABRIOLET N16
HOOD-CONVERTIBLE O02
REAR CUBBY BOX-CONVERTIBLE L09
REAR LUGGAGE LOCKER-CABRIOLET L07
REAR SEAT BELTS L11
REAR SEATS-LEATHER-COUPE L03
REAR SEATS-LEATHER-COUPE L04
REAR SEATS-LEATHER/CLOTH-COUPE L06
TARGA PANELS-STOWAGEBAG-CABRIOLET N14
UNDER SCUTTLE CASINGS H05
VENEERED FACIA PANELS-3.6 LITRE H06
VENEERED FACIA PANELS-5.3 LITRE H07

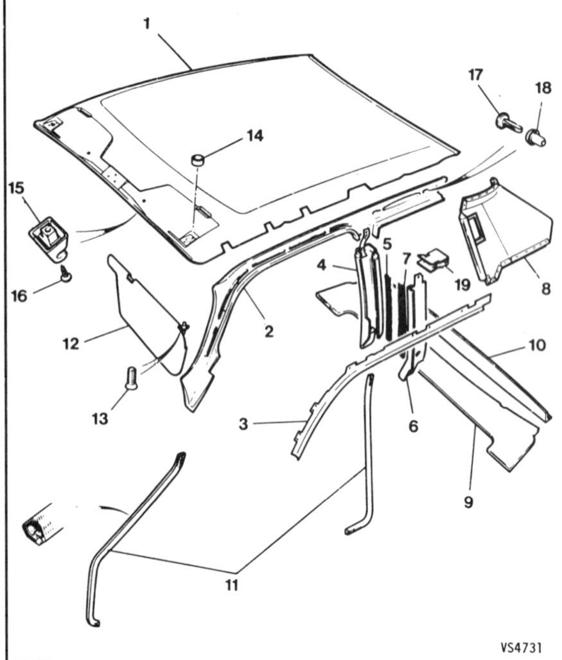

VS4731

INTERIOR TRIM-COUPE-WITH PASSIVE RESTRAINT

Illus	Part Number	123456 Description	Quantity	Change Point	Remarks
1	BCC 5292 ND	Lining-roof	NLA		Use BDC 4361ND
1	BDC 4361 ND	Lining-roof	1		
	BCC 3042 XE	Finisher-trim rail-RH	1		
2	BCC 3043 XE	Finisher-trim rail-LH	1		
	BBC 2520	Cover-'A' post & cantrail-RH	1		Use BDC 3018
	BDC 3018	Cover-'A' post & cantrail-RH	1		
3	BBC 2521	Cover-'A' post & cantrail-LH	1		Use BDC 3017
3	BDC 3017	Cover-'A' post & cantrail-LH	1		
	BCC 7306 XE	Trim-'B' post outer-RH	1	Up to (V) 149004	
4	BCC 7307 XE	Trim-'B' post outer-LH	1	Up to (V) 149004	
5	BCC 3364	Brush strip-outer	2	Up to (V) 149004	
	BCC 7364 XE	Trim-'B' post inner-RH	1	Up to (V) 149004	
6	BCC 7365 XE	Trim-'B' post inner-LH	1	Up to (V) 149004	
	BDC 4014 XE	Trim-'B' post outer RH	1	From (V) 149005	
4	BDC 4015 XE	Trim-'B' post outer LH	1	From (V) 149005	
	BDC 4012	Brush strip-outer RH	1	From (V) 149005	
5	BDC 4013	Brush strip-outer LH	1	From (V) 149005	
	BDC 3796 XE	Trim-'B' post inner RH	1	From (V) 149005	
6	BDC 3797 XE	Trim-'B' post inner LH	1	From (V) 149005	
7	BCC 3382	Brush strip-inner	2		
	BBC 7862 ND	Trim-RH rear	1		
8	BBC 7863 ND	Trim-LH rear	1		
9	BBC 7870 PA	Shelf-parcel	1		
10	BAC 5566 PA	Strip-parcel shelf	1	Up to 142986	
10	BDC 4024 PA	Strip-parcel shelf	1	From 142987	
11	BAC 2168 PA	Finisher-flange	4		
	BDC 4308 XE	Sunvisor-RH	1		
12	BDC 4309 XE	Sunvisor-LH	1		
13	BD 43278 4	Screw-set	4		
14	BD 18957 6	Spacer	4		
15	BBC 1727 XE	Storage-sunvisors	2		Use BCC 9494 XE
15	BCC 9494 XE	Storage-sunvisors	2		
16	AD 606051 J	Screw-self tap	2		
17	BD 13102	Button-pronged	4		
18	BD 47802	Socket-retainer	4		
19	BD 21962	Clip-retainer	7		

3I 17

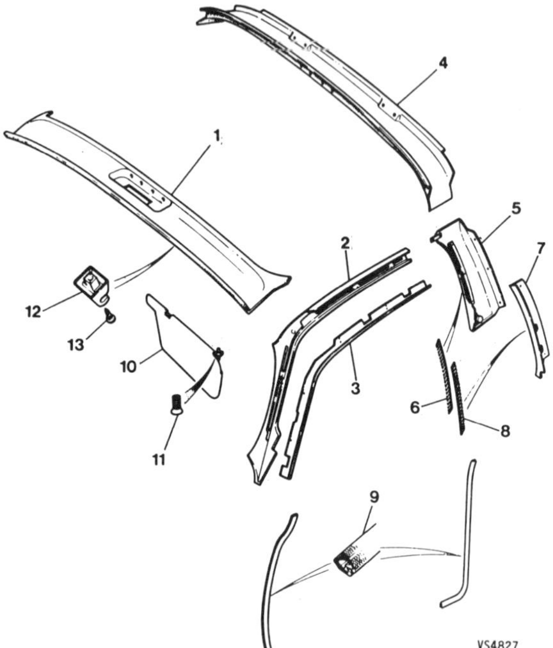

VS 4827

INTERIOR TRIM-CABRIOLET-WITH PASSIVE RESTRAINT

Illus	Part Number	123456 Description	Quantity	Change Point	Remarks
1	BCC 3624 ND	Trim header	1		
	BBC 6722 XE	Trim-rail finisher	1		
2	BBC 6723 XE	Trim-rail finisher	1		
	BCC 8078	Finisher-'A' post and cantrails-RH	1		
3	BCC 8079	Finisher-'A' post and cantrails-LH	1		
4	BCC 3626 XE	Trim-roll bar	1		
	BBC 8682 XE	Armature-'B-C' post-RH	1	Up to (V)142242	
	BDC 3214 XE	Armature-'B-C' post RH	1	From (V)142243	
5	BBC 8683 XE	Armature-'B-C' post LH	1	Up to (V)142242	
5	BDC 3215 XE	Armature-'B-C' post LH	1	From (V)142243	
	BCC 7476	Brush assembly inner-RH	1		
6	BCC 7477	Brush assembly inner-LH	1		
	BBC 8686 XE	Cover-'B' post-RH	1	Up to (V)142242	
	BCC 9968 XE	Cover-'B' post RH	1	From (V)142243	
7	BBC 8687 XE	Cover-'B' post-LH	1	Up to (V)142242	
7	BCC 9969 XE	Cover-'B' post LH	1	From (V)142243	
	BCC 7478	Brush strip outer-RH	1		
8	BCC 7479	Brush strip outer-LH	1		
9	BAC 2168 PA	Finisher-flange	4		
	BDC 4308 XE	Sunvisor-RH	1		
10	BDC 4309 XE	Sunvisor-LH	1		
11	BD 43278 4	Screw-set	4		
12	BBC 1727 XE	Stowage-sunvisors	2		Use BCC 9494 XE
12	BCC 9494 XE	Stowage-sunvisors	2		
13	AD 606051 J	Screw-self tap	2		

3I 18

MASTER INDEX

ENGINE 1.C
FLYWHEEL AND CLUTCH 1.G
GEARBOX AND PROPSHAFT 1.H
AXLES, SUSPENSION, DRIVE SHAFTS, WHEELS 1.K
STEERING 1.M
BRAKES AND BRAKE CONTROLS 1.N
FUEL, EXHAUST AND EMISSION SYSTEMS 2.C
COOLING, HEATING AND AIR CONDITIONING 2.G
ELECTRICAL, WASHERS-WIPERS, INSTRUMENTS 2.J
CHASSIS, SUBFRAMES, BODYSHELL, FITTINGS 3.C
FASCIA, TRIM, SEATS AND FIXINGS 3.H
ACCESSORIES AND PAINT 4.C
NUMERICAL INDEX 4.D

GROUP INDEX

BATTERY AND SPARE WHEEL COVERS N05
BONNET INSULATION-3.6 LITRE N06
BONNET INSULATION-5.3 MODELS N08
BONNET INSULATION-5.3 MODELS N07
BOOT CARPETS - NOT CONVERTIBLE N03
BOOT CARPETS-CONVERTIBLE N04
CARPETS-'A' POST FINISHERS- L12
CONSOLE-AMBLA H10
CONSOLE-LEATHER H08
DOOR CASINGS-ARMRESTS AND POCKETS-AMBLA H16
DOOR CASINGS-ARMRESTS-CONVERTIBLE I02
HEADLINING I15
DOOR CASINGS-ARMRESTS-LEATHER H14
EXTERIOR TRIM COVERS N15
FACIA PANEL H02
FRONT SEAT BELTS-CONVERTIBLE K15
FRONT SEAT SLIDES K16
MOTOR AND ANCHORAGE ASSY-RH L02
FRONT SEATS-AMBLA/CLOTH.-3.6 LITRE K04
FRONT SEATS-LEATHER J03
FRONT SEATS-LEATHER/CLOTH-5.3 LITRE K07
FRT SEATS-LEATHER/CLOTH(SPORTSPACK) K10
FRONT BELT KIT-COUPE K14
FRONT SEAT SLIDES K13
FUEL APERATURE SEAL AND HOSES-COUPE N13
HARDTOP & HOODBAG-CABRIOLET N17
HOOD LIFT MECHANISM - CONVERTIBLE O04
HOOD SEALS-CONVERTIBLE O03
HOOD-CABRIOLET N16
HOOD-CONVERTIBLE O02
REAR CUBBY BOX-CONVERTIBLE L09
REAR LUGGAGE LOCKER-CABRIOLET L07
REAR SEAT BELTS L11
REAR SEATS-LEATHER-COUPE L03
REAR SEATS-LEATHER-COUPE L04
REAR SEATS-LEATHER/CLOTH-COUPE L06
TARGA PANELS-STOWAGEBAG-CABRIOLET N14
UNDER SCUTTLE CASINGS H05
VENEERED FACIA PANELS-3.6 LITRE H06
VENEERED FACIA PANELS-5.3 LITRE H07

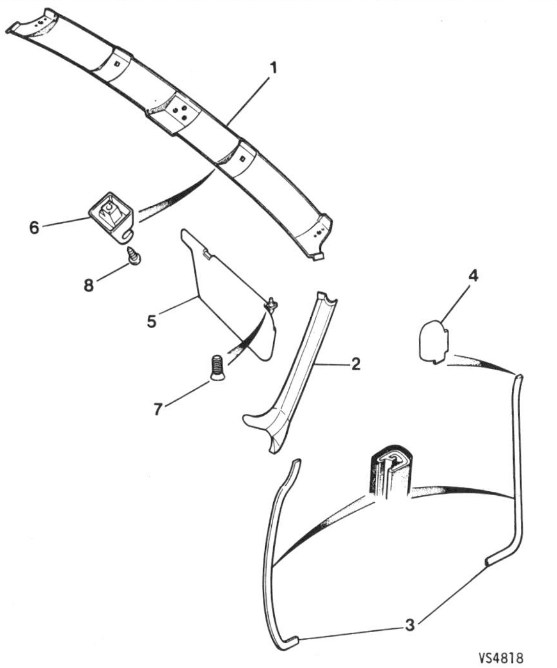

VS4818

Illus	Part Number	123456 Description	Quantity	Change Point	Remarks
		INTERIOR TRIM - CONVERTIBLE			
1	BDC 3595 XE	Finisher-upper windshield	1		
	BDC 3596 XE	Finisher-"A" post RH	1		
2	BDC 3597 XE	Finisher-"A" post LH	1		
3	BDC 5081	Finisher-flange	4		
5	BDC 5336	Cap-finisher-"B" post	2		
	BDC 4308 XE	Sunvisor RH	1		
5	BDC 4309 XE	Sunvisor LH	1		
6	BCC 9494 XE	Stowage-sunvisors	2		
7	BD 43278 4	Screw-set	4		
8	AD 606051 J	Screw-self tap	2		

MASTER INDEX

GROUP INDEX

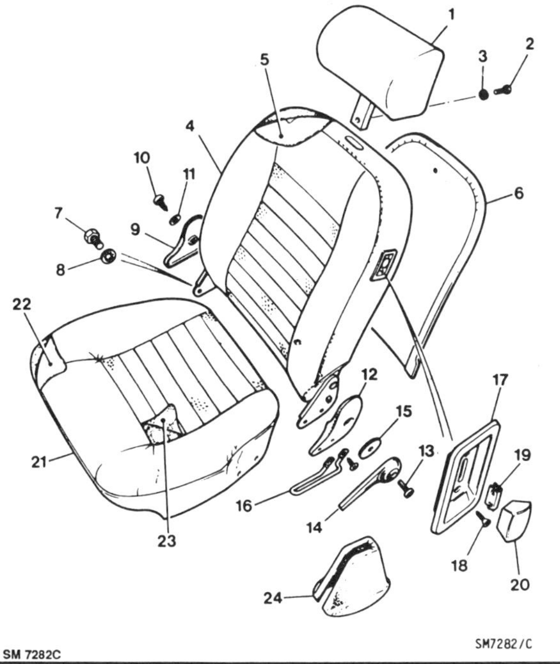

SM 7282C

Illus	Part Number	123456 Description	Quantity	Change Point	Remarks
		FRONT SEATS-LEATHER.			
		HEADRESTRAINT-FRONT			
1	BD 46770 AE	Biscuit	2		
1	BD 46770 LZ	Warm Charcoal	2		
1	BD 46770 AR	Buckskin	2		
1	BD 46770 XE	Doeskin	2		
1	BD 46770 JF	Isis Blue	2		
1	BD 46770 CM	Mulberry	2		
1	BD 46770 LY	Savile Grey	2		
1	BD 46770 AW	Barley	2		
1	BD 46770 ZM	Magnolia	2		
2	SH 605101 J	Screw-set	2		
3	JLM 303	Washer-spring	2		
		SQUAB RH-WITH LUMBAR ADJUST			
	BCC 3016 AE	Biscuit	1		
	BCC 3016 LZ	Warm Charcoal	1		
	BCC 3016 AR	Buckskin	1		
	BCC 3016 XE	Doeskin	1		
	BCC 3016 JF	Isis Blue	1		
	BCC 3016 CM	Mulberry	1		
	BCC 3016 LY	Savile Grey	1		
	BCC 3016 AW	Barley	1		
	BCC 3016 ZM	Magnolia	1		
	BCC 8504	Diaphragm-lumbar RH	1		
		SQUAB LH WITH LUMBAR ADJUST			
4	BCC 3017 AE	Biscuit	1		
4	BCC 3017 LZ	Warm Charcoal	1		
4	BCC 3017 AR	Buckskin	1		
4	BCC 3017 XE	Doeskin	1		
4	BCC 3017 JF	Isis Blue	1		
4	BCC 3017 CM	Mulberry	1		
4	BCC 3017 LY	Savile Grey	1		
4	BCC 3017 AW	Barley	1		
4	BCC 3017 ZM	Magnolia	1		
	BCC 8505	Diaphragm-lumbar LH	1		

MASTER INDEX
ENGINE 1.C
FLYWHEEL AND CLUTCH 1.G
GEARBOX AND PROPSHAFT 1.H
AXLES, SUSPENSION, DRIVE SHAFTS, WHEELS 1.K
STEERING 1.M
BRAKES AND BRAKE CONTROLS 1.N
FUEL, EXHAUST AND EMISSION SYSTEMS 2.C
COOLING, HEATING AND AIR CONDITIONING 2.G
ELECTRICAL, WASHERS-WIPERS, INSTRUMENTS 2.J
CHASSIS, SUBFRAMES, BODYSHELL, FITTINGS 3.C
FASCIA, TRIM, SEATS AND FIXINGS 3.H
ACCESSORIES AND PAINT 4.C
NUMERICAL INDEX 4.D
GROUP INDEX
BATTERY AND SPARE WHEEL COVERS. N05
BONNET INSULATION-3.6 LITRE N06
BONNET INSULATION-5.3 MODELS N08
BONNET INSULATION-5.3 MODELS. N07
BOOT CARPETS - NOT CONVERTIBLE N03
BOOT CARPETS-CONVERTIBLE N04
CARPETS-'A' POST FINISHERS-. L12
CONSOLE-AMBLA. H10
CONSOLE-LEATHER. H08
DOOR CASINGS-ARMRESTS AND POCKETS-AMBLA. H16
DOOR CASINGS-ARMRESTS-CONVERTIBLE I02
HEADLINING I15
DOOR CASINGS-ARMRESTS-LEATHER. H14
EXTERIOR TRIM COVERS N15
FACIA PANEL H02
FRONT SEAT BELTS-CONVERTIBLE K15
FRONT SEAT SLIDES K16
MOTOR AND ANCHORAGE ASSY-RH L02
FRONT SEATS-AMBLA/CLOTH.-3.6 LITRE K04
FRONT SEATS-LEATHER. J03
FRONT SEATS-LEATHER/CLOTH-5.3 LITRE K07
FRT SEATS-LEATHER/CLOTH(SPORTSPACK) K10
FRONT BELT KIT-COUPE K14
FRONT SEAT SLIDES K13
FUEL APERATURE SEAL AND HOSES-COUPE. N13
HARDTOP & HOODBAG-CABRIOLET. N17
HOOD LIFT MECHANISM - CONVERTIBLE O04
HOOD SEALS-CONVERTIBLE O03
HOOD-CABRIOLET N16
HOOD-CONVERTIBLE O02
REAR CUBBY BOX-CONVERTIBLE L09
REAR LUGGAGE LOCKER-CABRIOLET L07
REAR SEAT BELTS. L11
REAR SEATS-LEATHER-COUPE. L03
REAR SEATS-LEATHER-COUPE. L04
REAR SEATS-LEATHER/CLOTH-COUPE. L06
TARGA PANELS-STOWAGEBAG-CABRIOLET N14
UNDER SCUTTLE CASINGS. H05
VENEERED FACIA PANELS-3.6 LITRE. H06
VENEERED FACIA PANELS-5.3 LITRE. H07

3J 04

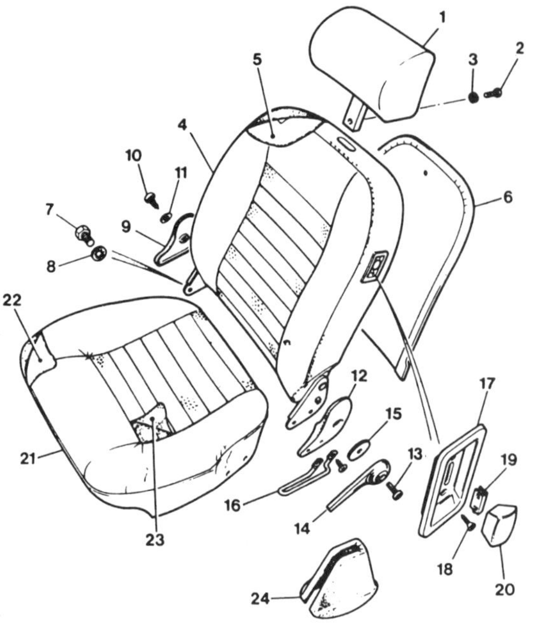

SM7282C SM7282/C

Illus	Part Number	123456	Description	Quantity	Change Point	Remarks
			FRONT SEATS-LEATHER-continued			
			SQUAB RH-NOT LUMBAR ADJUST			
	BCC 3014 AW		Barley	1		
	BCC 3014 AE		Biscuit	1		
	BCC 3014 AR		Buckskin	1		
	BCC 3014 XE		Doeskin	1		
	BCC 3014 JF		Isis Blue	1		
	BCC 3014 ZM		Magnolia	1		
	BCC 3014 CM		Mulberry	1		
	BCC 3014 LY		Savile Grey	1		
	BCC 3014 LZ		Warm charcoal	1		
			SQUAB LH-NOT LUMBAR ADJUST			
4	BCC 3015 AW		Barley	1		
4	BCC 3015 AE		Biscuit	1		
4	BCC 3015 AR		Buckskin	1		
4	BCC 3015 XE		Doeskin	1		
4	BCC 3015 JF		Isis Blue	1		
4	BCC 3015 ZM		Magnolia	1		
4	BCC 3015 CM		Mulberry	1		
4	BCC 3015 LY		Savile Grey	1		
4	BCC 3015 LZ		Warm charcoal	1		

3J 05

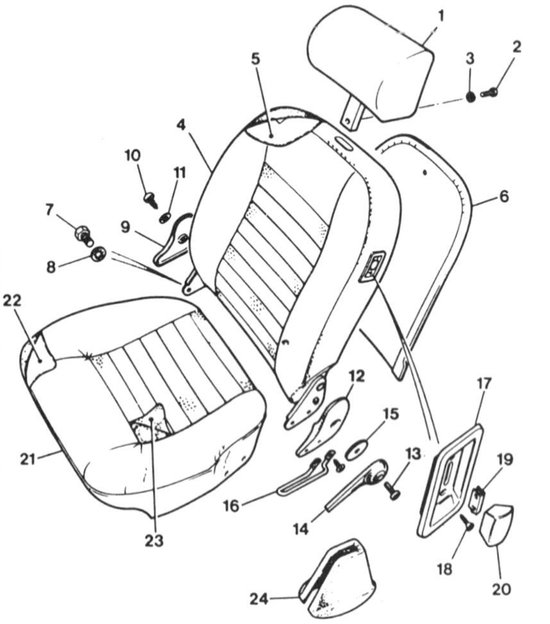

SM 7282C SM7282/C

Illus	Part Number	123456	Description	Quantity	Change Point	Remarks
			FRONT SEATS-LEATHER-continued.			
			COVER-FRONT SQUABS			
5	BAC 4040 AE		Biscuit	2		
5	BAC 4040 LZ		Warm Charcoal	2		
5	BAC 4040 AR		Buckskin	2		
5	BAC 4040 XE		Doeskin	2		
5	BAC 4040 JF		Isis Blue	2		
5	BAC 4040 CM		Mulberry	2		
5	BAC 4040 LY		Savile Grey	2		
5	BAC 4040 AW		Barley	1		
5	BAC 4040 ZM		Magnolia	1		
			PANEL RH-SQUAB BACK			
	BAC 4038 AE		Biscuit	1		
	BAC 4038 LZ		Warm Charcoal	1		
	BAC 4038 AR		Buckskin	1		
	BAC 4038 XE		Doeskin	1		
	BAC 4038 JF		Isis Blue	1		
	BAC 4038 CM		Mulberry	1		
	BAC 4038 LY		Savile Grey	1		
	BAC 4038 AW		Barley	1		
	BAC 4038 ZM		Magnolia	1		

MASTER INDEX
ENGINE 1.C
FLYWHEEL AND CLUTCH 1.G
GEARBOX AND PROPSHAFT 1.H
AXLES, SUSPENSION, DRIVE SHAFTS, WHEELS 1.K
STEERING 1.M
BRAKES AND BRAKE CONTROLS 1.N
FUEL, EXHAUST AND EMISSION SYSTEMS 2.C
COOLING, HEATING AND AIR CONDITIONING 2.G
ELECTRICAL, WASHERS-WIPERS, INSTRUMENTS 2.J
CHASSIS, SUBFRAMES, BODYSHELL, FITTINGS 3.C
FASCIA, TRIM, SEATS AND FIXINGS 3.H
ACCESSORIES AND PAINT 4.C
NUMERICAL INDEX 4.D

GROUP INDEX
BATTERY AND SPARE WHEEL COVERS N05
BONNET INSULATION-3.6 LITRE N06
BONNET INSULATION-5.3 MODELS N08
BONNET INSULATION-5.3 MODELS N07
BOOT CARPETS - NOT CONVERTIBLE N03
BOOT CARPETS-CONVERTIBLE N04
CARPETS-'A' POST FINISHERS- L12
CONSOLE-AMBLA H10
CONSOLE-LEATHER H08
DOOR CASINGS-ARMRESTS AND POCKETS-AMBLA H16
DOOR CASINGS-ARMRESTS-CONVERTIBLE I02
HEADLINING I15
DOOR CASINGS-ARMRESTS-LEATHER H14
EXTERIOR TRIM COVERS N15
FACIA PANEL H02
FRONT SEAT BELTS-CONVERTIBLE K15
FRONT SEAT SLIDES K16
MOTOR AND ANCHORAGE ASSY-RH L02
FRONT SEATS-AMBLA/CLOTH.-3.6 LITRE K04
FRONT SEATS-LEATHER J03
FRONT SEATS-LEATHER/CLOTH-5.3 LITRE K07
FRT SEATS-LEATHER/CLOTH(SPORTSPACK) K10
FRONT BELT KIT-COUPE K14
FRONT SEAT SLIDES K13
FUEL APERATURE SEAL AND HOSES-COUPE N13
HARDTOP & HOODBAG-CABRIOLET N17
HOOD LIFT MECHANISM - CONVERTIBLE O04
HOOD SEALS-CONVERTIBLE O03
HOOD-CABRIOLET N16
HOOD-CONVERTIBLE O02
REAR CUBBY BOX-CONVERTIBLE L09
REAR LUGGAGE LOCKER-CABRIOLET L07
REAR SEAT BELTS L11
REAR SEATS-LEATHER-COUPE L03
REAR SEATS-LEATHER-COUPE L04
REAR SEATS-LEATHER/CLOTH-COUPE L06
TARGA PANELS-STOWAGEBAG-CABRIOLET N14
UNDER SCUTTLE CASINGS H05
VENEERED FACIA PANELS-3.6 LITRE H06
VENEERED FACIA PANELS-5.3 LITRE H07

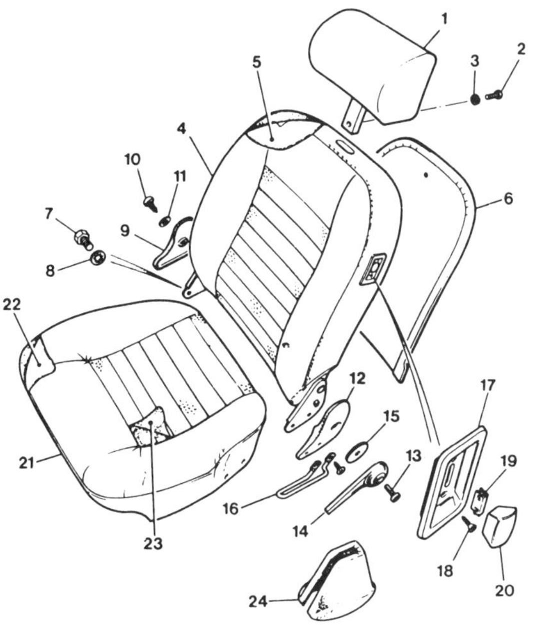

SM7282C

FRONT SEATS-LEATHER-continued

Illus	Part Number	123456	Description	Quantity	Change Point	Remarks
			PANEL LH-SQUAB BACK			
6	BAC 4039 AE		Biscuit	1		
6	BAC 4039 LZ		Warm Charcoal	1		
6	BAC 4039 AR		Buckskin	1		
6	BAC 4039 XE		Doeskin	1		
6	BAC 4039 JF		Isis Blue	1		
6	BAC 4039 CM		Mulberry	1		
6	BAC 4039 LY		Savile Grey	1		
6	BAC 4039 AW		Barley	1		
6	BAC 4039 ZM		Magnolia	1		
7	SH 604041 J		Screw-Set	8		
8	C 724		Washer-Shakeproof	8		
			COVER RH-INNER			
	BCC 3036 AE		Biscuit	1		
	BCC 3036 LZ		Warm Charcoal	1		
	BCC 3036 AR		Buckskin	1		
	BCC 3036 XE		Doeskin	1		
	BCC 3036 JF		Isis Blue	1		
	BCC 3036 CM		Mulberry	1		
	BCC 3036 LY		Savile Grey	1		
	BCC 3036 AW		Barley	1		
	BCC 3036 ZM		Magnolia	1		
			COVER LH-INNER			
9	BCC 3037 AE		Biscuit	1		
9	BCC 3037 LZ		Warm Charcoal	1		
9	BCC 3037 AR		Buckskin	1		
9	BCC 3037 XE		Doeskin	1		
9	BCC 3037 JF		Isis Blue	1		
9	BCC 3037 CM		Mulberry	1		
9	BCC 3037 LY		Savile Grey	1		
9	BCC 3037 AW		Barley	1		
9	BCC 3037 ZM		Magnolia	1		
10	JLM 307		Screw-Self Tap	2		
11	WC 702102 J		Washer-Plain	2		

MASTER INDEX

GROUP INDEX

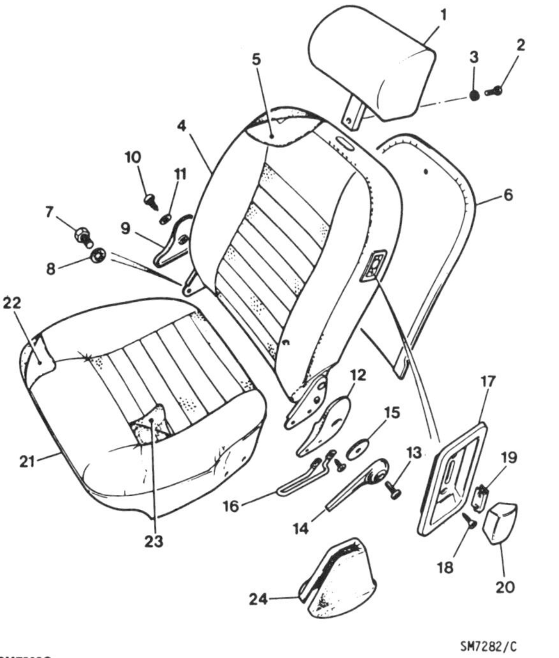

SM7282C

FRONT SEATS-LEATHER-continued

Illus	Part Number	123456	Description	Quantity	Change Point	Remarks
			COVER RH-OUTER			
	BCC 3034 AE		Biscuit	1		
	BCC 3034 LZ		Warm Charcoal	1		
	BCC 3034 AR		Buckskin	1		
	BCC 3034 XE		Doeskin	1		
	BCC 3034 JF		Isis Blue	1		
	BCC 3034 CM		Mulberry	1		
	BCC 3034 LY		Savile Grey	1		
	BCC 3034 AW		Barley	1		
	BCC 3034 ZM		Magnolia	1		
			COVER LH-OUTER			
12	BCC 3035 AE		Biscuit	1		
12	BCC 3035 LZ		Warm Charcoal	1		
12	BCC 3035 AR		Buckskin	1		
12	BCC 3035 XE		Doeskin	1		
12	BCC 3035 JF		Isis Blue	1		
12	BCC 3035 CM		Mulberry	1		
12	BCC 3035 LY		Savile Grey	1		
	BCC 3035 AW		Barley	1		
	BCC 3035 ZM		Magnolia	1		

MASTER INDEX
ENGINE 1.C
FLYWHEEL AND CLUTCH 1.G
GEARBOX AND PROPSHAFT 1.H
AXLES, SUSPENSION, DRIVE SHAFTS, WHEELS 1.K
STEERING 1.M
BRAKES AND BRAKE CONTROLS 1.N
FUEL, EXHAUST AND EMISSION SYSTEMS 2.C
COOLING, HEATING AND AIR CONDITIONING 2.G
ELECTRICAL, WASHERS-WIPERS, INSTRUMENTS 2.J
CHASSIS, SUBFRAMES, BODYSHELL, FITTINGS 3.C
FASCIA, TRIM, SEATS AND FIXINGS 3.H
ACCESSORIES AND PAINT 4.C
NUMERICAL INDEX 4.D
GROUP INDEX
BATTERY AND SPARE WHEEL COVERS N05
BONNET INSULATION-3.6 LITRE N06
BONNET INSULATION-5.3 MODELS N08
BONNET INSULATION-5.3 MODELS N07
BOOT CARPETS - NOT CONVERTIBLE N03
BOOT CARPETS-CONVERTIBLE N04
CARPETS-'A' POST FINISHERS- L12
CONSOLE-AMBLA H10
CONSOLE-LEATHER H08
DOOR CASINGS-ARMRESTS AND POCKETS-AMBLA H16
DOOR CASINGS-ARMRESTS-CONVERTIBLE I02
HEADLINING I15
DOOR CASINGS-ARMRESTS-LEATHER H14
EXTERIOR TRIM COVERS N15
FACIA PANEL H02
FRONT SEAT BELTS-CONVERTIBLE K15
FRONT SEAT SLIDES K16
MOTOR AND ANCHORAGE ASSY-RH L02
FRONT SEATS-AMBLA/CLOTH.-3.6 LITRE K04
FRONT SEATS-LEATHER J03
FRONT SEATS-LEATHER/CLOTH-5.3 LITRE K07
FRT SEATS-LEATHER/CLOTH(SPORTSPACK) K10
FRONT BELT KIT-COUPE K14
FRONT SEAT SLIDES K13
FUEL APERATURE SEAL AND HOSES-COUPE N13
HARDTOP & HOODBAG-CABRIOLET N17
HOOD LIFT MECHANISM - CONVERTIBLE O04
HOOD SEALS-CONVERTIBLE O03
HOOD-CABRIOLET N16
HOOD-CONVERTIBLE O02
REAR CUBBY BOX-CONVERTIBLE L09
REAR LUGGAGE LOCKER-CABRIOLET L07
REAR SEAT BELTS L11
REAR SEATS-LEATHER-COUPE L03
REAR SEATS-LEATHER-COUPE L04
REAR SEATS-LEATHER/CLOTH-COUPE L06
TARGA PANELS-STOWAGEBAG-CABRIOLET N14
UNDER SCUTTLE CASINGS H05
VENEERED FACIA PANELS-3.6 LITRE H06
VENEERED FACIA PANELS-5.3 LITRE H07

3J 08

JAGUAR XJS RANGE (JAN 1987 ON) | J08 fiche 3 | FASCIA, TRIM, SEATS AND FIXINGS

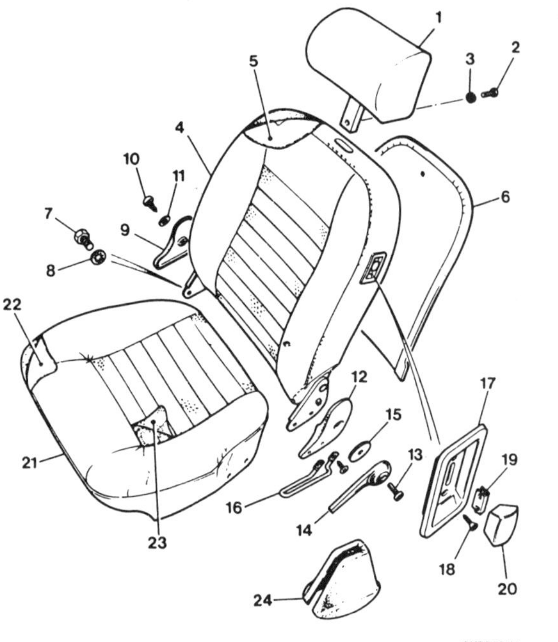

SM7282C

FRONT SEATS-LEATHER-continued

Illus	Part Number	123456	Description	Quantity	Change Point	Remarks
13	BD 34524 1		Screw-taptite	2		
	BCC 7508		Lever-recline-RH	1		
14	BCC 7509		Lever-recline-LH	1		
15	BD 46805		Washer-plastic	2		
16	BBC 8968		Guide-seat belt-RH	1		With passive rest.
16	BBC 8969		Guide-seat belt-LH	1		
17	715721		Escutcheon	2		
18	DAB 6 6C		Screw-self tap	4		
19	BD 42638		Clip-retainer	2		
20	624571		Knob	2		
			CUSHION-ASSEMBLY FRONT			
21	BCC 2975 AE		Biscuit	2		
21	BCC 2975 LZ		Warm Charcoal	2		
21	BCC 2975 AR		Buckskin	2		
21	BCC 2975 XE		Doeskin	2		
21	BCC 2975 JF		Isis Blue	2		
21	BCC 2975 CM		Mulberry	2		
21	BCC 2975 LY		Savile Grey	2		
21	BCC 2975 AW		Barley	2		
21	BCC 2975 ZM		Magnolia	2		

MASTER INDEX

GROUP INDEX

3J 09

JAGUAR XJS RANGE (JAN 1987 ON) | J09 fiche 3 | FASCIA, TRIM, SEATS AND FIXINGS

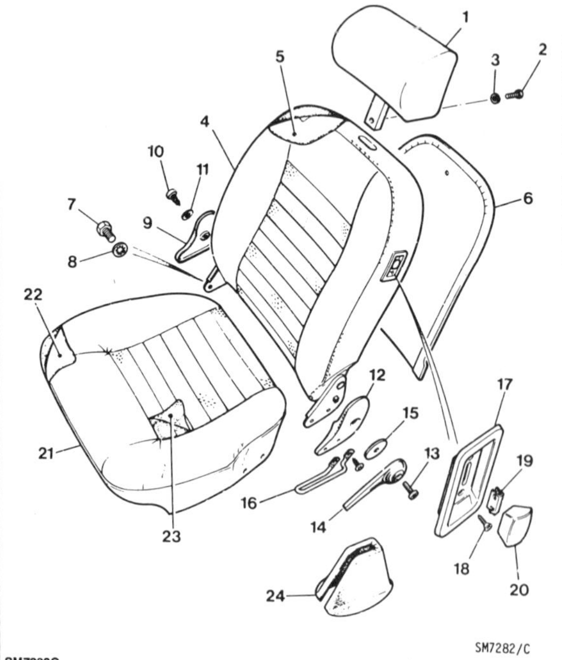

SM7282C

FRONT SEATS LEATHER continued

Illus	Part Number	123456	Description	Quantity	Change Point	Remarks
			COVER-OUTER-CUSHION			
22	BAC 4060 AE		Biscuit	2		
22	BAC 4060 LZ		Warm Charcoal	2		
22	BAC 4060 AR		Buckskin	2		
22	BAC 4060 XE		Doeskin	2		
22	BAC 4060 JF		Isis Blue	2		
22	BAC 4060 CM		Mulberry	2		
22	BAC 4060 LY		Savile Grey	2		
22	BAC 4060 AW		Barley	2		
22	BAC 4060 ZM		Magnolia	2		
			COVER-INNER-CUSHION			
23	BAC 4064 AE		Biscuit	2		
23	BAC 4064 LZ		Warm Charcoal	2		
23	BAC 4064 AR		Buckskin	2		
23	BAC 4064 XE		Doeskin	2		
23	BAC 4064 JF		Isis Blue	2		
23	BAC 4064 CM		Mulberry	2		
23	BAC 4064 LY		Savile Grey	2		
23	BAC 4064 AW		Barley	2		
23	BAC 4064 ZM		Magnolia	2		

MASTER INDEX
ENGINE 1.C
FLYWHEEL AND CLUTCH 1.G
GEARBOX AND PROPSHAFT 1.H
AXLES, SUSPENSION, DRIVE SHAFTS, WHEELS 1.K
STEERING 1.M
BRAKES AND BRAKE CONTROLS 1.N
FUEL, EXHAUST AND EMISSION SYSTEMS 2.C
COOLING, HEATING AND AIR CONDITIONING 2.G
ELECTRICAL, WASHERS-WIPERS, INSTRUMENTS 2.J
CHASSIS, SUBFRAMES, BODYSHELL, FITTINGS 3.C
FASCIA, TRIM, SEATS AND FIXINGS 3.H
ACCESSORIES AND PAINT 4.C
NUMERICAL INDEX 4.D
GROUP INDEX
BATTERY AND SPARE WHEEL COVERS N05
BONNET INSULATION-3.6 LITRE N06
BONNET INSULATION-5.3 MODELS N08
BONNET INSULATION-5.3 MODELS N07
BOOT CARPETS - NOT CONVERTIBLE N03
BOOT CARPETS-CONVERTIBLE N04
CARPETS-'A' POST FINISHERS- L12
CONSOLE-AMBLA H10
CONSOLE-LEATHER H08
DOOR CASINGS-ARMRESTS AND POCKETS-AMBLA H16
DOOR CASINGS-ARMRESTS-CONVERTIBLE I02
HEADLINING I15
DOOR CASINGS-ARMRESTS-LEATHER H14
EXTERIOR TRIM COVERS N15
FACIA PANEL H02
FRONT SEAT BELTS-CONVERTIBLE K15
FRONT SEAT SLIDES K16
MOTOR AND ANCHORAGE ASSY-RH L02
FRONT SEATS-AMBLA/CLOTH.-3.6 LITRE K04
FRONT SEATS-LEATHER J03
FRONT SEATS-LEATHER/CLOTH-5.3 LITRE K07
FRT SEATS-LEATHER/CLOTH(SPORTSPACK) K10
FRONT BELT KIT-COUPE K14
FRONT SEAT SLIDES K13
FUEL APERATURE SEAL AND HOSES-COUPE N13
HARDTOP & HOODBAG-CABRIOLET N17
HOOD LIFT MECHANISM - CONVERTIBLE O04
HOOD SEALS-CONVERTIBLE O03
HOOD-CABRIOLET N16
HOOD-CONVERTIBLE O02
REAR CUBBY BOX-CONVERTIBLE L09
REAR LUGGAGE LOCKER-CABRIOLET L07
REAR SEAT BELTS L11
REAR SEATS-LEATHER-COUPE L03
REAR SEATS-LEATHER-COUPE L04
REAR SEATS-LEATHER/CLOTH-COUPE L06
TARGA PANELS-STOWAGEBAG-CABRIOLET N14
UNDER SCUTTLE CASINGS H05
VENEERED FACIA PANELS-3.6 LITRE H06
VENEERED FACIA PANELS-5.3 LITRE H07

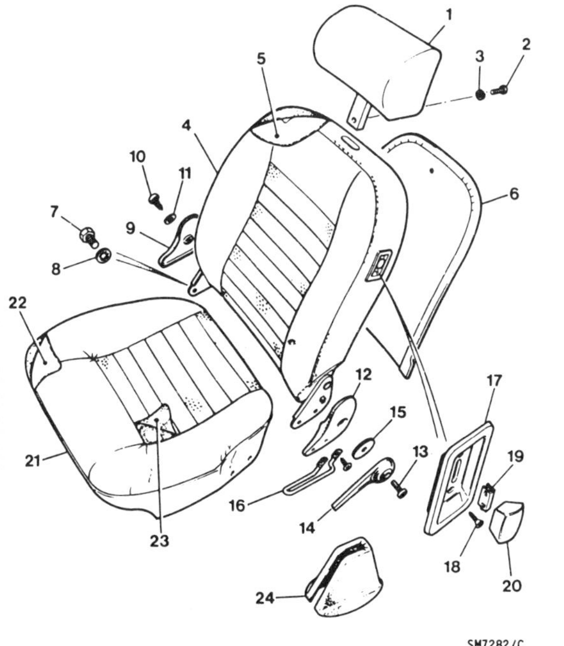

FRONT SEATS - LEATHER - continued

Illus	Part Number	Description	Quantity	Change Point	Remarks
		TRIM - OUTER SEAT MOUNTING			
24	BCC 7828 AE	Biscuit	2		
24	BCC 7828 LZ	Warm Charcoal	2		
24	BCC 7828 AR	Buckskin	2		
24	BCC 7828 XE	Doeskin	2		
24	BCC 7828 JF	Isis Blue	2		
24	BCC 7828 CM	Mulberry	2		
24	BCC 7828 LY	Savile Grey	2		
24	BCC 7828 AW	Barley	2		
24	BCC 7828 ZM	Magnolia	2		
		TRIM-INNER SEAT MOUNTING			
	BCC 7829 AE	Biscuit	2		
	BCC 7829 LZ	Warm Charcoal	2		
	BCC 7829 AR	Buckskin	2		
	BCC 7829 XE	Doeskin	2		
	BCC 7829 JF	Isis Blue	2		
	BCC 7829 CM	Mulberry	2		
	BCC 7829 LY	Savile Grey	2		
	BCC 7829 AW	Barley	2		
	BCC 7829 ZM	Magnolia	2		

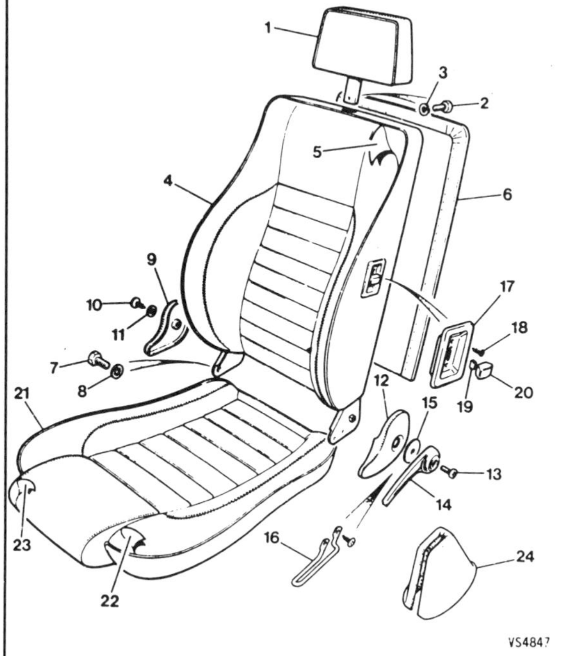

FRT SEATS-LEATHER(SPORTSPACK)

Illus	Part Number	Description	Quantity	Change Point	Remarks
		HEAD RESTRAINT-FRONT			
1	BCC 9320 AW	Barley	2		
1	BCC 9320 AR	Buckskin	2		
1	BCC 9320 XE	Doeskin	2		
1	BCC 9320 ZXE	Doeskin/blue piping	2		
1	BCC 9320 YXE	Doeskin/red piping	2		
1	BCC 9320 JF	Isis blue	2		
1	BCC 9320 ZM	Magnolia	2		
1	BCC 9320 ZZM	Magnolia/blue piping	2		
1	BCC 9320 YZM	Magnolia/red piping	2		
1	BCC 9320 CM	Mulberry	2		
1	BCC 9320 LY	Savile grey	2		
1	BCC 9320 ZLY	Savile grey/blue piping	2		
1	BCC 9320 YLY	Savile grey/red piping	2		
1	BCC 9320 LZ	Warm charcoal	2		
2	SH 605101 J	Screw-set	2		
3	JLM 303	Washer-spring	2		
		SQUAB RH-WITH LUMBAR ADJUST			
	BCC 9278 AW	Barley	1		
	BCC 9278 AR	Buckskin	1		
	BCC 9278 XE	Doeskin	1		
	BCC 9278 ZXE	Doeskin/blue piping	1		
	BCC 9278 YXE	Doeskin/red piping	1		
	BCC 9278 JF	Isis Blue	1		
	BCC 9278 ZM	Magnolia	1		
	BCC 9278 ZZM	Magnolia/blue piping	1		
	BCC 9278 YZM	Magnolia/red piping	1		
	BCC 9278 CM	Mulberry	1		
	BCC 9278 LY	Savile grey	1		
	BCC 9278 ZLY	Savile grey/blue piping	1		
	BCC 9278 YLY	Savile grey/red piping	1		
	BCC 9278 LZ	Warm charcoal	1		
	BCC 8504	Diaphragm - lumbar RH	1		

M32.08JP22/9/88

MASTER INDEX

ENGINE 1.C
FLYWHEEL AND CLUTCH 1.G
GEARBOX AND PROPSHAFT 1.H
AXLES, SUSPENSION, DRIVE SHAFTS, WHEELS 1.K
STEERING 1.M
BRAKES AND BRAKE CONTROLS 1.N
FUEL, EXHAUST AND EMISSION SYSTEMS 2.C
COOLING, HEATING AND AIR CONDITIONING 2.G
ELECTRICAL, WASHERS-WIPERS, INSTRUMENTS 2.J
CHASSIS, SUBFRAMES, BODYSHELL, FITTINGS 3.C
FASCIA, TRIM, SEATS AND FIXINGS 3.H
ACCESSORIES AND PAINT 4.C
NUMERICAL INDEX 4.D

GROUP INDEX

BATTERY AND SPARE WHEEL COVERS N05
BONNET INSULATION-3.6 LITRE N06
BONNET INSULATION-5.3 MODELS N08
BONNET INSULATION-5.3 MODELS N07
BOOT CARPETS - NOT CONVERTIBLE N03
BOOT CARPETS-CONVERTIBLE N04
CARPETS-'A' POST FINISHERS- L12
CONSOLE-AMBLA H10
CONSOLE-LEATHER H08
DOOR CASINGS-ARMRESTS AND POCKETS-AMBLA H16
DOOR CASINGS-ARMRESTS-CONVERTIBLE I02
HEADLINING I15
DOOR CASINGS-ARMRESTS-LEATHER H14
EXTERIOR TRIM COVERS N15
FACIA PANEL H02
FRONT SEAT BELTS-CONVERTIBLE K15
FRONT SEAT SLIDES K16
MOTOR AND ANCHORAGE ASSY-RH L02
FRONT SEATS-AMBLA/CLOTH-3.6 LITRE K04
FRONT SEATS-LEATHER J03
FRONT SEATS-LEATHER/CLOTH-5.3 LITRE K07
FRT SEATS-LEATHER/CLOTH(SPORTSPACK) K10
FRONT BELT KIT-COUPE K14
FRONT SEAT SLIDES K13
FUEL APERATURE SEAL AND HOSES-COUPE N13
HARDTOP & HOODBAG-CABRIOLET N17
HOOD LIFT MECHANISM - CONVERTIBLE O04
HOOD SEALS-CONVERTIBLE O03
HOOD-CABRIOLET N16
HOOD-CONVERTIBLE O02
REAR CUBBY BOX-CONVERTIBLE L09
REAR LUGGAGE LOCKER-CABRIOLET L07
REAR SEAT BELTS L11
REAR SEATS-LEATHER-COUPE L03
REAR SEATS-LEATHER-COUPE L04
REAR SEATS-LEATHER/CLOTH-COUPE L06
TARGA PANELS-STOWAGEBAG-CABRIOLET N14
UNDER SCUTTLE CASINGS H05
VENEERED FACIA PANELS-3.6 LITRE H06
VENEERED FACIA PANELS-5.3 LITRE H07

3J 12

JAGUAR XJS RANGE (JAN 1987 ON) | J12 fiche 3 | FASCIA, TRIM, SEATS AND FIXINGS

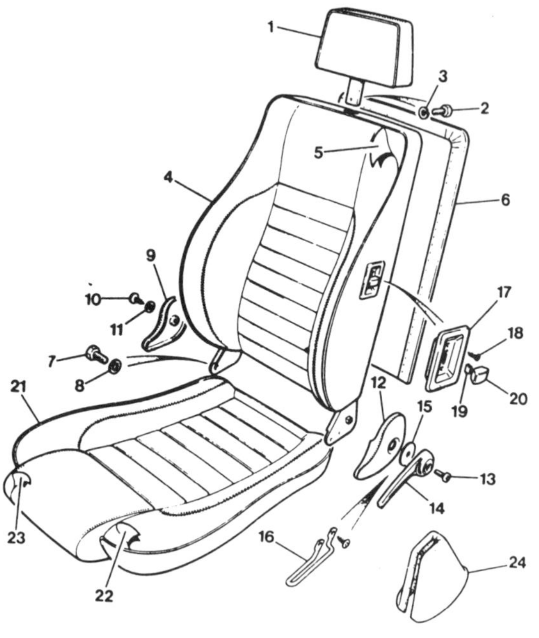

VS4847

FRT SEATS-LEATHER(SPORTSPACK)

Illus	Part Number	123456	Description	Quantity	Change Point	Remarks
			SQUAB LH-WITH LUMBAR ADJUST			
4	BCC 9279 AW		Barley	1		
4	BCC 9279 AR		Buckskin	1		
4	BCC 9279 XE		Doeskin	1		
4	BCC 9279 ZXE		Doeskin/blue piping	1		
4	BCC 9279 YXE		Doeskin/red piping	1		
4	BCC 9279 JF		Isis blue	1		
4	BCC 9279 ZM		Magnolia	1		
4	BCC 9279 ZZM		Magnolia/blue piping	1		
4	BCC 9279 YZM		Magnolia/red piping	1		
4	BCC 9279 CM		Mulberry	1		
4	BCC 9279 LY		Savile grey	1		
4	BCC 9279 ZLY		Savile grey/blue piping	1		
4	BCC 9279 YLY		Savile grey/red piping	1		
4	BCC 9279 LZ		Warm Charcoal	1		
4	BCC 8505		Diaphragm-lumbar LH	1		

M34JP22/9/88

3J 13

JAGUAR XJS RANGE (JAN 1987 ON) | J13 fiche 3 | FASCIA, TRIM, SEATS AND FIXINGS

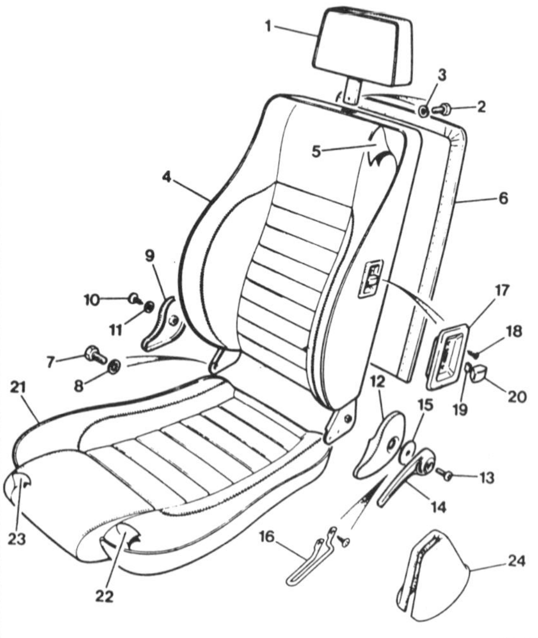

VS4847

FRONT SEATS-LEATHER(SPORTSPACK)-Cont'd

Illus	Part Number	123456	Description	Quantity	Change Point	Remarks
			SQUAB RH-NOT LUMBAR ADJUST			
	BCC 9284 AW		Barley	NLA		Use BDC 7456 AW
	BCC 9284 AR		Buckskin	NLA		Use BDC 7456 AR
	BCC 9284 XE		Doeskin	NLA		Use BDC 7456 XE
	BCC 9284 JF		Isis blue	NLA		Use BDC 7456 JF
	BCC 9284 ZM		Magnolia	NLA		Use BDC 7456 ZM
	BCC 9284 CM		Mulberry	NLA		Use BDC 7456 CM
	BCC 9284 LY		Savile grey	NLA		Use BDC 7456 LY
	BCC 9284 LZ		Warm charcoal	NLA		Use BDC 7456 LZ
			SQUAB RH-NOT LUMBAR ADJUST			
	BDC 7456 AW		Barley	1		
	BDC 7456 AR		Buckskin	1		
	BDC 7456 XE		Doeskin	1		
	BDC 7456 ZXE		Doeskin/blue piping	1		
	BDC 7456 YXE		Doeskin/red piping	1		
	BDC 7456 JF		Isis blue	1		
	BDC 7456 ZM		Magnolia	1		
	BDC 7456 ZZM		Magnolia/blue piping	1		
	BDC 7456 YZM		Magnolia/red piping	1		
	BDC 7456 CM		Mulberry	1		
	BDC 7456 LY		Savile grey	1		
	BDC 7456 ZLY		Savile grey/blue piping	1		
	BDC 7456 YLY		Savile grey/red piping	1		
	BDC 7456 LZ		Warm charcoal	1		

M34JP23/9/88

MASTER INDEX

ENGINE 1.C
FLYWHEEL AND CLUTCH 1.G
GEARBOX AND PROPSHAFT 1.H
AXLES, SUSPENSION, DRIVE SHAFTS, WHEELS 1.K
STEERING 1.M
BRAKES AND BRAKE CONTROLS 1.N
FUEL, EXHAUST AND EMISSION SYSTEMS 2.C
COOLING, HEATING AND AIR CONDITIONING 2.G
ELECTRICAL, WASHERS-WIPERS, INSTRUMENTS 2.J
CHASSIS, SUBFRAMES, BODYSHELL, FITTINGS 3.C
FASCIA, TRIM, SEATS AND FIXINGS 3.H
ACCESSORIES AND PAINT 4.C
NUMERICAL INDEX 4.D

GROUP INDEX

BATTERY AND SPARE WHEEL COVERS. N05
BONNET INSULATION-3.6 LITRE N06
BONNET INSULATION-5.3 MODELS N08
BONNET INSULATION-5.3 MODELS. N07
BOOT CARPETS - NOT CONVERTIBLE N03
BOOT CARPETS-CONVERTIBLE N04
CARPETS-'A' POST FINISHERS-. L12
CONSOLE-AMBLA. H10
CONSOLE-LEATHER. H08
DOOR CASINGS-ARMRESTS AND POCKETS-AMBLA. H16
DOOR CASINGS-ARMRESTS-CONVERTIBLE I02
HEADLINING I15
DOOR CASINGS-ARMRESTS-LEATHER. H14
EXTERIOR TRIM COVERS N15
FACIA PANEL. H02
FRONT SEAT BELTS-CONVERTIBLE K15
FRONT SEAT SLIDES K16
MOTOR AND ANCHORAGE ASSY-RH L02
FRONT SEATS-AMBLA/CLOTH.-3.6 LITRE K04
FRONT SEATS-LEATHER. J03
FRONT SEATS-LEATHER/CLOTH-5.3 LITRE K07
FRT SEATS-LEATHER/CLOTH(SPORTSPACK) K10
FRONT BELT KIT-COUPE K14
FRONT SEAT SLIDES K13
FUEL APERATURE SEAL AND HOSES-COUPE. N13
HARDTOP & HOODBAG-CABRIOLET. N17
HOOD LIFT MECHANISM - CONVERTIBLE O04
HOOD SEALS-CONVERTIBLE O03
HOOD-CABRIOLET N16
HOOD-CONVERTIBLE O02
REAR CUBBY BOX-CONVERTIBLE L09
REAR LUGGAGE LOCKER-CABRIOLET L07
REAR SEAT BELTS. L11
REAR SEATS-LEATHER-COUPE. L03
REAR SEATS-LEATHER-COUPE. L04
REAR SEATS-LEATHER/CLOTH-COUPE. L06
TARGA PANELS-STOWAGEBAG-CABRIOLET N14
UNDER SCUTTLE CASINGS. H05
VENEERED FACIA PANELS-3.6 LITRE. H06
VENEERED FACIA PANELS-5.3 LITRE. H07

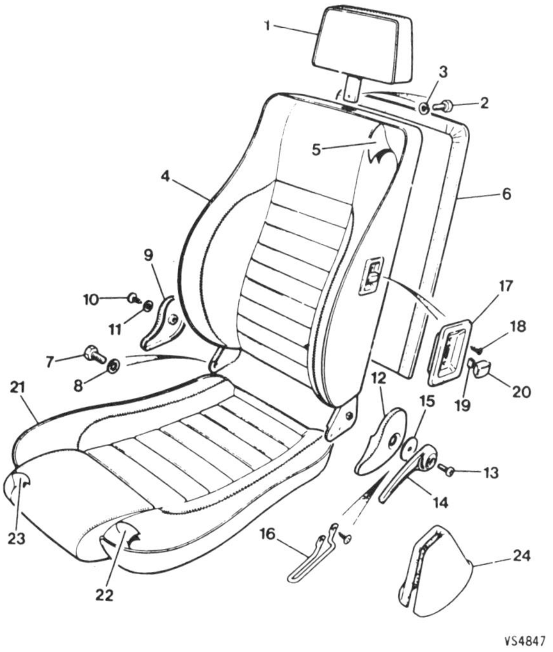

VS4847

FRONT SEATS-LEATHER(SPORTSPACK)-continued.

Illus	Part Number	123456 Description	Quantity	Change Point	Remarks
		SQUAB LH-NOT LUMBAR ADJUST			
4	BCC 9285 AW	Barley	NLA		Use BDC 7457 AW
4	BCC 9285 AR	Buckskin	NLA		Use BDC 7457 AR
4	BCC 9285 XE	Doeskin	NLA		Use BDC 7457 XE
4	BCC 9285 JF	Isis Blue	NLA		Use BDC 7457 JF
4	BCC 9285 ZM	Magnolia	NLA		Use BDC 7457 ZM
4	BCC 9285 CM	Mulberry	NLA		Use BDC 7457 CM
4	BCC 9285 LY	Savile Grey	NLA		Use BDC 7457 LY
4	BCC 9285 LZ	Warm Charcoal	NLA		Use BDC 7457 LZ
		SQUAB LH-NOT LUMBAR ADJUST			
4	BDC 7457 AW	Barley	1		
4	BDC 7457 AR	Buckskin	1		
4	BDC 7457 XE	Doeskin	1		
4	BDC 7457 ZXE	Doeskin/blue piping	1		
4	BDC 7457 YXE	Doeskin/red piping	1		
4	BDC 7457 JF	Isis blue	1		
4	BDC 7457 ZM	Magnolia	1		
4	BDC 7457 ZZM	Magnolia/blue piping	1		
4	BDC 7457 YZM	Magnolia/red piping	1		
4	BDC 7457 CM	Mulberry	1		
4	BDC 7457 LY	Savile grey	1		
4	BDC 7457 ZLY	Savile grey/blue piping	1		
4	BDC 7457 YLY	Savile grey/red piping	1		
4	BDC 7457 LZ	Warm charcoal	1		

M34.02JP23/9/88

MASTER INDEX

GROUP INDEX

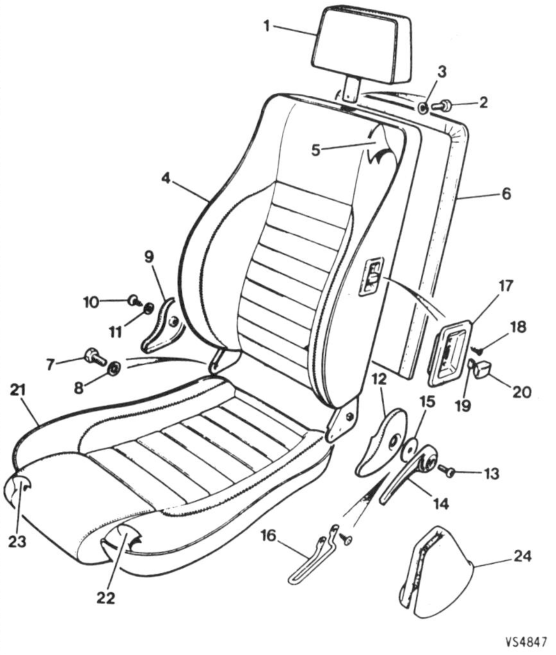

VS4847

FRONT SEATS-LEATHER(SPORTSPACK)-continued.

Illus	Part Number	123456 Description	Quantity	Change Point	Remarks
		COVER-FRONT SQUABS			
5	BCC 9324 AW	Barley	2		
5	BCC 9324 AR	Buckskin	2		
5	BCC 9324 XE	Doeskin	2		
5	BCC 9324 ZXE	Doeskin/blue piping	2		
5	BCC 9324 YXE	Doeskin/red piping	2		
5	BCC 9324 JF	Isis blue	2		
5	BCC 9324 ZM	Magnolia	2		
5	BCC 9324 ZZM	Magnolia/blue piping	2		
5	BCC 9324 YZM	Magnoia/red piping	2		
5	BCC 9324 CM	Mulberry	2		
5	BCC 9324 LY	Savile grey	2		
5	BCC 9324 ZLY	Savile grey/blue piping	2		
5	BCC 9324 YLY	Savile grey/red piping	2		
5	BCC 9324 LZ	Warm charcoal	2		
		PANEL RH-SQUAB BACK			
	BCC 9306 AW	Barley	1		
	BCC 9306 AR	Buckskin	1		
	BCC 9306 XE	Doeskin	1		
	BCC 9306 JF	Isis blue	1		
	BCC 9306 ZM	Magnolia	1		
	BCC 9306 CM	Mulberry	1		
	BCC 9306 LY	Savile grey	1		
	BCC 9306 LZ	Warm charcoal	1		

M34.04JP22/9/88

MASTER INDEX
ENGINE 1.C
FLYWHEEL AND CLUTCH 1.G
GEARBOX AND PROPSHAFT 1.H
AXLES, SUSPENSION, DRIVE SHAFTS, WHEELS 1.K
STEERING 1.M
BRAKES AND BRAKE CONTROLS 1.N
FUEL, EXHAUST AND EMISSION SYSTEMS 2.C
COOLING, HEATING AND AIR CONDITIONING 2.G
ELECTRICAL, WASHERS-WIPERS, INSTRUMENTS 2.J
CHASSIS, SUBFRAMES, BODYSHELL, FITTINGS 3.C
FASCIA, TRIM, SEATS AND FIXINGS 3.H
ACCESSORIES AND PAINT 4.C
NUMERICAL INDEX 4.D
GROUP INDEX
BATTERY AND SPARE WHEEL COVERS. N05
BONNET INSULATION-3.6 LITRE N06
BONNET INSULATION-5.3 MODELS N08
BONNET INSULATION-5.3 MODELS. N07
BOOT CARPETS - NOT CONVERTIBLE N03
BOOT CARPETS-CONVERTIBLE N04
CARPETS-'A' POST FINISHERS-. L12
CONSOLE-AMBLA. H10
CONSOLE-LEATHER. H08
DOOR CASINGS-ARMRESTS AND POCKETS-AMBLA. H16
DOOR CASINGS-ARMRESTS-CONVERTIBLE I02
HEADLINING I15
DOOR CASINGS-ARMRESTS-LEATHER. H14
EXTERIOR TRIM COVERS N15
FACIA PANEL. H02
FRONT SEAT BELTS-CONVERTIBLE K15
FRONT SEAT SLIDES K16
MOTOR AND ANCHORAGE ASSY-RH L02
FRONT SEATS-AMBLA/CLOTH.-3.6 LITRE K04
FRONT SEATS-LEATHER. J03
FRONT SEATS-LEATHER/CLOTH-5.3 LITRE K07
FRT SEATS-LEATHER/CLOTH(SPORTSPACK) K10
FRONT BELT KIT-COUPE K14
FRONT SEAT SLIDES K13
FUEL APERATURE SEAL AND HOSES-COUPE. N13
HARDTOP & HOODBAG-CABRIOLET. N17
HOOD LIFT MECHANISM - CONVERTIBLE O04
HOOD SEALS-CONVERTIBLE O03
HOOD-CABRIOLET N16
HOOD-CONVERTIBLE O02
REAR CUBBY BOX-CONVERTIBLE L09
REAR LUGGAGE LOCKER-CABRIOLET L07
REAR SEAT BELTS. L11
REAR SEATS-LEATHER-COUPE. L03
REAR SEATS-LEATHER-COUPE. L04
REAR SEATS-LEATHER/CLOTH-COUPE. L06
TARGA PANELS-STOWAGEBAG-CABRIOLET N14
UNDER SCUTTLE CASINGS. H05
VENEERED FACIA PANELS-3.6 LITRE. H06
VENEERED FACIA PANELS-5.3 LITRE. H07

3J 16

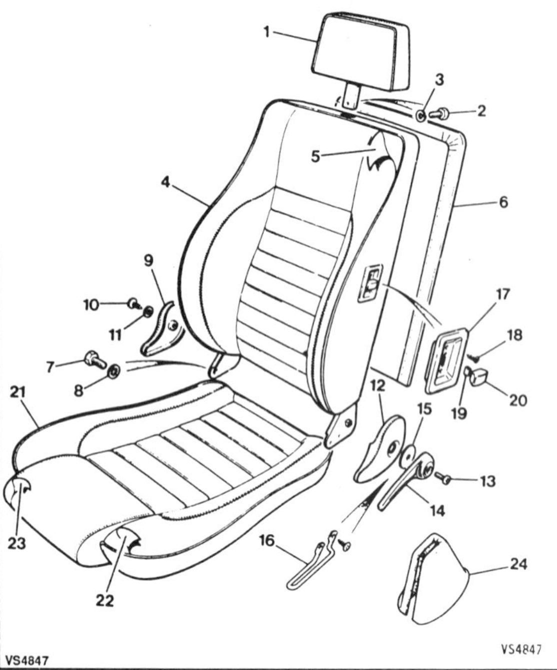

FRONT SEATS-LEATHER(SPORTSPACK)-continued.

Illus	Part Number	123456 Description	Quantity	Change Point	Remarks
		PANEL LH-SQUAB BACK			
6	BCC 9307 AW	Barley	1		
6	BCC 9307 AR	Buckskin	1		
6	BCC 9307 XE	Doeskin	1		
6	BCC 9307 JF	Isis Blue	1		
6	BCC 9307 ZM	Magnolia	1		
6	BCC 9307 CM	Mulberry	1		
6	BCC 9307 LY	Savile Grey	1		
6	BCC 9307 LZ	Warm Charcoal	1		
7	SH 604041 J	Screw-set	8		
8	C 724	Washer-shakeproof	8		
		COVER RH-INNER			
	BCC 3036 AW	Barley	1		
	BCC 3036 AR	Buckskin	1		
	BCC 3036 XE	Doeskin	1		
	BCC 3036 JF	Isis Blue	1		
	BCC 3036 ZM	Magnolia	1		
	BCC 3036 CM	Mulberry	1		
	BCC 3036 LY	Savile Grey	1		
	BCC 3036 LZ	Warm Charcoal	1		
		COVER LH-INNER			
9	BCC 3037 AW	Barley	1		
9	BCC 3037 AR	Buckskin	1		
9	BCC 3037 XE	Doeskin	1		
9	BCC 3037 JF	Isis Blue	1		
9	BCC 3037 ZM	Magnolia	1		
9	BCC 3037 CM	Mulberry	1		
9	BCC 3037 LY	Savile Grey	1		
9	BCC 3037 LZ	Warm Charcoal	1		
10	JLM 307	Screw-self tap	2		
11	WC 702102 J	Washer-plain	2		

MASTER INDEX

GROUP INDEX

3J 17

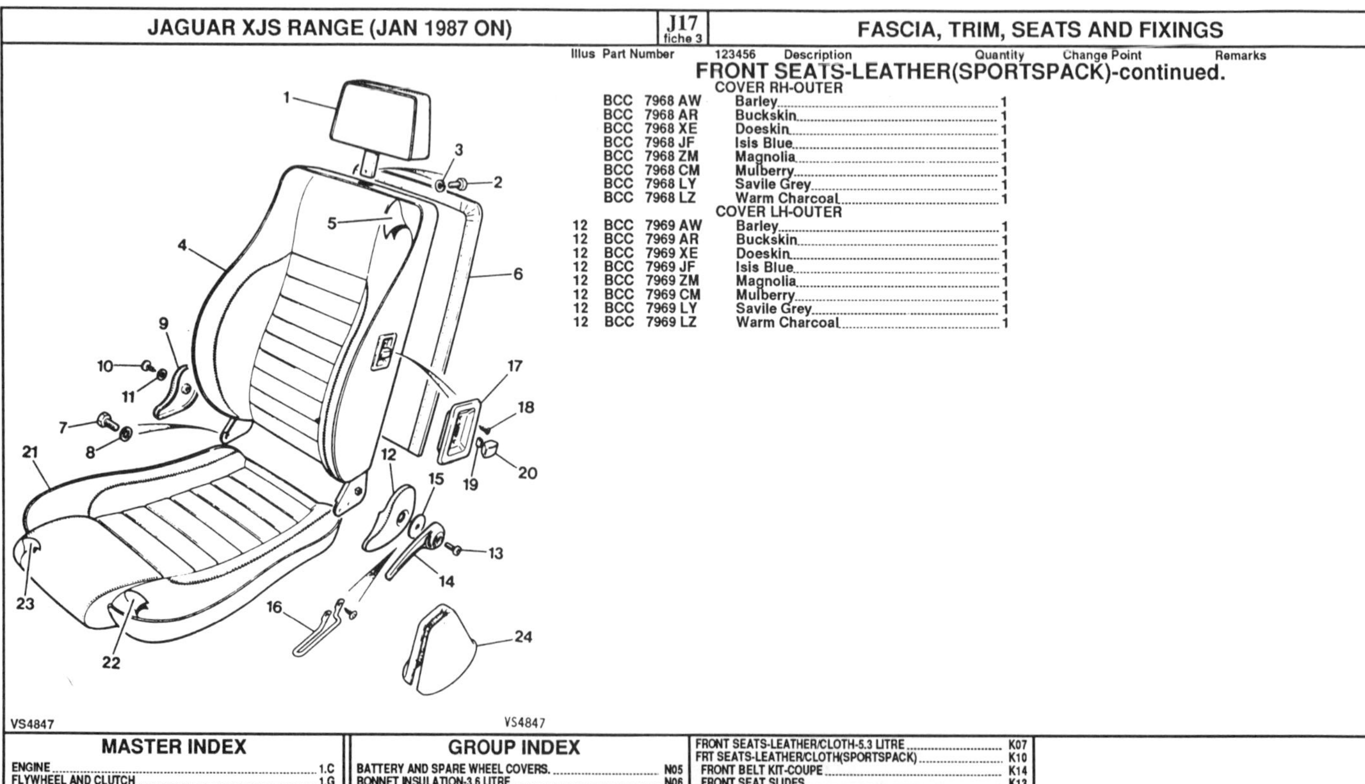

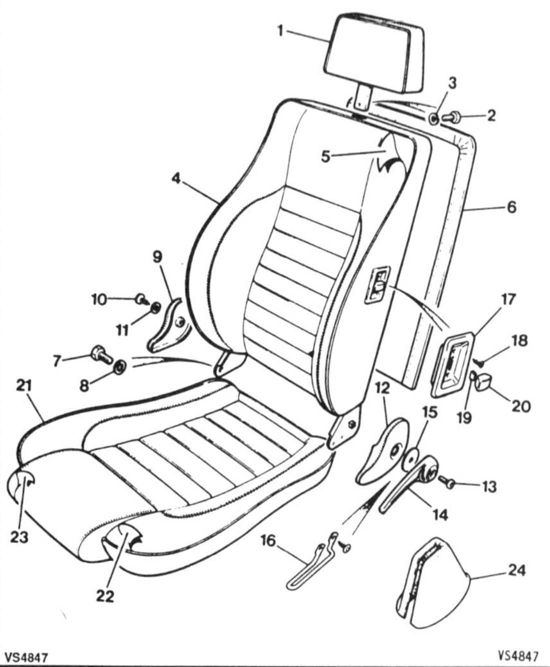

FRONT SEATS-LEATHER(SPORTSPACK)-continued.

Illus	Part Number	123456 Description	Quantity	Change Point	Remarks
		COVER RH-OUTER			
	BCC 7968 AW	Barley	1		
	BCC 7968 AR	Buckskin	1		
	BCC 7968 XE	Doeskin	1		
	BCC 7968 JF	Isis Blue	1		
	BCC 7968 ZM	Magnolia	1		
	BCC 7968 CM	Mulberry	1		
	BCC 7968 LY	Savile Grey	1		
	BCC 7968 LZ	Warm Charcoal	1		
		COVER LH-OUTER			
12	BCC 7969 AW	Barley	1		
12	BCC 7969 AR	Buckskin	1		
12	BCC 7969 XE	Doeskin	1		
12	BCC 7969 JF	Isis Blue	1		
12	BCC 7969 ZM	Magnolia	1		
12	BCC 7969 CM	Mulberry	1		
12	BCC 7969 LY	Savile Grey	1		
12	BCC 7969 LZ	Warm Charcoal	1		

MASTER INDEX
ENGINE 1.C
FLYWHEEL AND CLUTCH 1.G
GEARBOX AND PROPSHAFT 1.H
AXLES, SUSPENSION, DRIVE SHAFTS, WHEELS 1.K
STEERING 1.M
BRAKES AND BRAKE CONTROLS 1.N
FUEL, EXHAUST AND EMISSION SYSTEMS 2.C
COOLING, HEATING AND AIR CONDITIONING 2.G
ELECTRICAL, WASHERS-WIPERS, INSTRUMENTS 2.J
CHASSIS, SUBFRAMES, BODYSHELL, FITTINGS 3.C
FASCIA, TRIM, SEATS AND FIXINGS 3.H
ACCESSORIES AND PAINT 4.C
NUMERICAL INDEX 4.D
GROUP INDEX
BATTERY AND SPARE WHEEL COVERS N05
BONNET INSULATION-3.6 LITRE N06
BONNET INSULATION-5.3 MODELS N08
BONNET INSULATION-5.3 MODELS N07
BOOT CARPETS - NOT CONVERTIBLE N03
BOOT CARPETS-CONVERTIBLE N04
CARPETS-'A' POST FINISHERS- L12
CONSOLE-AMBLA H10
CONSOLE-LEATHER H08
DOOR CASINGS-ARMRESTS AND POCKETS-AMBLA H16
DOOR CASINGS-ARMRESTS-CONVERTIBLE I02
HEADLINING I15
DOOR CASINGS-ARMRESTS-LEATHER H14
EXTERIOR TRIM COVERS N15
FACIA PANEL H02
FRONT SEAT BELTS-CONVERTIBLE K15
FRONT SEAT SLIDES K16
MOTOR AND ANCHORAGE ASSY-RH L02
FRONT SEATS-AMBLA/CLOTH.-3.6 LITRE K04
FRONT SEATS-LEATHER J03
FRONT SEATS-LEATHER/CLOTH-5.3 LITRE K07
FRT SEATS-LEATHER/CLOTH(SPORTSPACK) K10
FRONT BELT KIT-COUPE K14
FRONT SEAT SLIDES K13
FUEL APERATURE SEAL AND HOSES-COUPE N13
HARDTOP & HOODBAG-CABRIOLET N17
HOOD LIFT MECHANISM - CONVERTIBLE O04
HOOD SEALS-CONVERTIBLE O03
HOOD-CABRIOLET N16
HOOD-CONVERTIBLE O02
REAR CUBBY BOX-CONVERTIBLE L09
REAR LUGGAGE LOCKER-CABRIOLET L07
REAR SEAT BELTS L11
REAR SEATS-LEATHER-COUPE L03
REAR SEATS-LEATHER-COUPE L04
REAR SEATS-LEATHER/CLOTH-COUPE L06
TARGA PANELS-STOWAGEBAG-CABRIOLET N14
UNDER SCUTTLE CASINGS H05
VENEERED FACIA PANELS-3.6 LITRE H06
VENEERED FACIA PANELS-5.3 LITRE H07

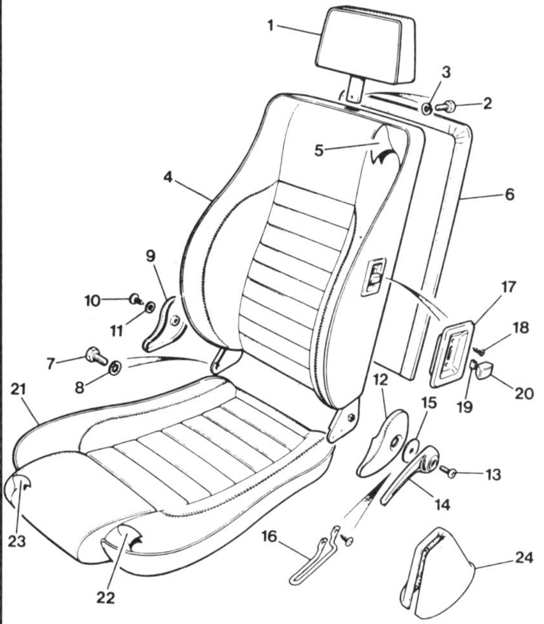

FRONT SEATS-LEATHER(SPORTSPACK)-continued.

Illus	Part Number	123456 Description	Quantity	Change Point	Remarks
13	BD 34524 1	Screw-taptite	2		
	BCC 7508	Lever-recline-RH	1		
14	BCC 7509	Lever-recline-LH	1		
15	BD 46805	Washer-plastic	2		
16	BBC 8968	Guide-seat belt-RH	1		With passive restraint
16	BBC 8969	Guide-seat belt-LH	1		With passive restraint
17	715721	Escutcheon	2		
18	DAB 006 6C	Screw-self tap	4		
19	BD 42638	Clip-retainer	2		
20	624571	Knob	2		
		CUSHION ASSEMBLY-FRONT			
21	BCC 9310 AW	Barley	2		
21	BCC 9310 AR	Buckskin	2		
21	BCC 9310 XE	Doeskin	2		
21	BCC 9310 ZXE	Doeskin/blue piping	2		
21	BCC 9310 YXE	Doeskin/red piping	2		
21	BCC 9310 JF	Isis blue	2		
21	BCC 9310 ZM	Magnolia	2		
21	BCC 9310 ZZM	Magnolia/blue piping	2		
21	BCC 9310 YZM	Magnolia/red piping	2		
21	BCC 9310 CM	Mulberry	2		
21	BCC 9310 LY	Savile Grey	2		
21	BCC 9310 ZLY	Savile grey/blue piping	2		
21	BCC 9310 YLY	Savile grey/red piping	2		
21	BCC 9310 LZ	Warm charcoal	2		
		COVER-OUTER CUSHION			
22	BCC 9345 AW	Barley	2		
22	BCC 9345 AR	Buckskin	2		
22	BCC 9345 XE	Doeskin	2		
22	BCC 9345 ZXE	Doeskin/blue piping	2		
22	BCC 9345 YXE	Doeskin/red piping	2		
22	BCC 9345 JF	Isis blue	2		
22	BCC 9345 ZM	Magnolia	2		
22	BCC 9345 ZZM	Magnolia/blue piping	2		
22	BCC 9345 YZM	Magnolia/red piping	2		
22	BCC 9345 CM	Mulberry	2		
22	BCC 9345 LY	Savile grey	2		
22	BCC 9345 ZLY	Savile grey/blue piping	2		
22	BCC 9345 YLY	Savile grey/red piping	2		
22	BCC 9345 LZ	Warm charcoal	2		

M34.10JP22/9/88

VS4847

MASTER INDEX

GROUP INDEX

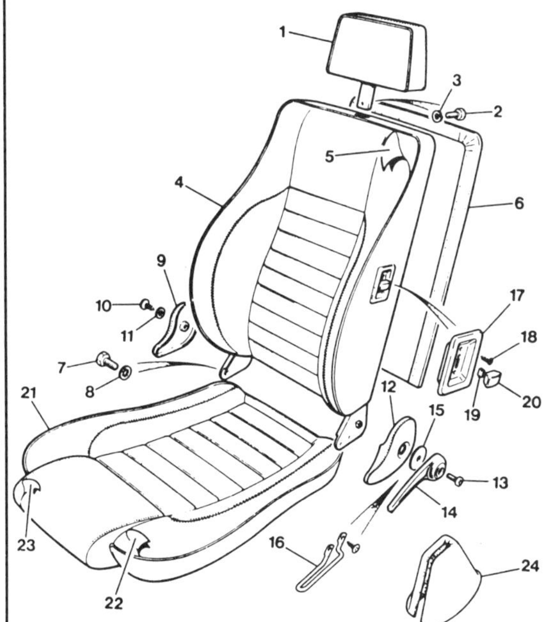

FRONT SEATS-LEATHER(SPORTSPACK)-continued.

Illus	Part Number	123456 Description	Quantity	Change Point	Remarks
		COVER-INNER CUSHION			
23	BCC 9355 AW	Barley	2		
23	BCC 9355 AR	Buckskin	2		
23	BCC 9355 XE	Doeskin	2		
23	BCC 9355 JF	Isis Blue	2		
23	BCC 9355 ZM	Magnolia	2		
23	BCC 9355 CM	Mulberry	2		
23	BCC 9355 LY	Savile Grey	2		
23	BCC 9355 LZ	Warm Charcoal	2		

VS4847

MASTER INDEX
ENGINE 1.C
FLYWHEEL AND CLUTCH 1.G
GEARBOX AND PROPSHAFT 1.H
AXLES, SUSPENSION, DRIVE SHAFTS, WHEELS 1.K
STEERING 1.M
BRAKES AND BRAKE CONTROLS 1.N
FUEL, EXHAUST AND EMISSION SYSTEMS 2.C
COOLING, HEATING AND AIR CONDITIONING 2.G
ELECTRICAL, WASHERS-WIPERS, INSTRUMENTS 2.J
CHASSIS, SUBFRAMES, BODYSHELL, FITTINGS 3.C
FASCIA, TRIM, SEATS AND FIXINGS 3.H
ACCESSORIES AND PAINT 4.C
NUMERICAL INDEX 4.D
GROUP INDEX
BATTERY AND SPARE WHEEL COVERS. N05
BONNET INSULATION-3.6 LITRE N06
BONNET INSULATION-5.3 MODELS N08
BONNET INSULATION-5.3 MODELS. N07
BOOT CARPETS - NOT CONVERTIBLE N03
BOOT CARPETS-CONVERTIBLE N04
CARPETS-'A' POST FINISHERS-. L12
CONSOLE-AMBLA. H10
CONSOLE-LEATHER. H08
DOOR CASINGS-ARMRESTS AND POCKETS-AMBLA. H16
DOOR CASINGS-ARMRESTS-CONVERTIBLE I02
HEADLINING I15
DOOR CASINGS-ARMRESTS-LEATHER. H14
EXTERIOR TRIM COVERS N15
FACIA PANEL. H02
FRONT SEAT BELTS-CONVERTIBLE K15
FRONT SEAT SLIDES K16
MOTOR AND ANCHORAGE ASSY-RH L02
FRONT SEATS-AMBLA/CLOTH.-3.6 LITRE K04
FRONT SEATS-LEATHER. J03
FRONT SEATS-LEATHER/CLOTH-5.3 LITRE K07
FRT SEATS-LEATHER/CLOTH(SPORTSPACK) K10
FRONT BELT KIT-COUPE K14
FRONT SEAT SLIDES K13
FUEL APERATURE SEAL AND HOSES-COUPE. N13
HARDTOP & HOODBAG-CABRIOLET. N17
HOOD LIFT MECHANISM - CONVERTIBLE O04
HOOD SEALS-CONVERTIBLE O03
HOOD-CABRIOLET N16
HOOD-CONVERTIBLE O02
REAR CUBBY BOX-CONVERTIBLE L09
REAR LUGGAGE LOCKER-CABRIOLET L07
REAR SEAT BELTS. L11
REAR SEATS-LEATHER-COUPE. L03
REAR SEATS-LEATHER-COUPE. L04
REAR SEATS-LEATHER/CLOTH-COUPE. L06
TARGA PANELS-STOWAGEBAG-CABRIOLET N14
UNDER SCUTTLE CASINGS. H05
VENEERED FACIA PANELS-3.6 LITRE. H06
VENEERED FACIA PANELS-5.3 LITRE. H07

3K 03

JAGUAR XJS RANGE (JAN 1987 ON) | K03 fiche 3 | FASCIA, TRIM, SEATS AND FIXINGS

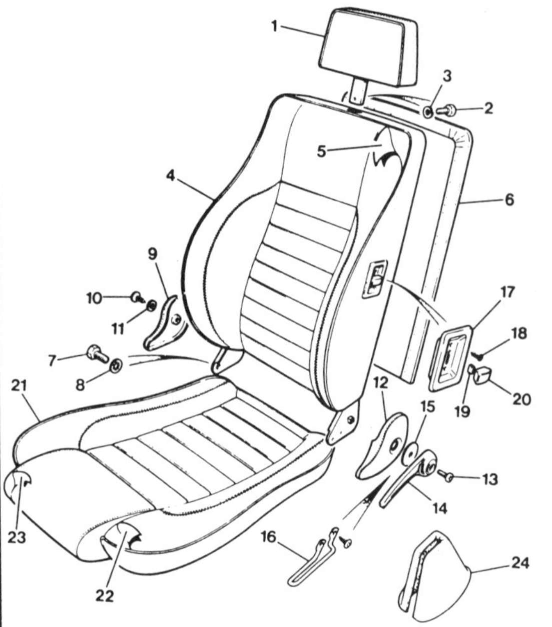

VS 4847

FRONT SEATS-LEATHER(SPORTSPACK)-continued.

Illus	Part Number	123456 Description	Quantity	Change Point	Remarks
		TRIM-OUTER SEAT MOUNTING			
24	BCC 7828 AW	Barley	2		
24	BCC 7828 AR	Buckskin	2		
24	BCC 7828 XE	Doeskin	2		
24	BCC 7828 JF	Isis Blue	2		
24	BCC 7828 ZM	Magnolia	2		
24	BCC 7828 CM	Mulberry	2		
24	BCC 7828 LY	Savile Grey	2		
24	BCC 7828 LZ	Warm Charcoal	2		
		TRIM-INNER SEAT MOUNTING			
	BCC 7829 AW	Barley	2		
	BCC 7829 AR	Buckskin	2		
	BCC 7829 XE	Doeskin	2		
	BCC 7829 JF	Isis Blue	2		
	BCC 7829 ZM	Magnolia	2		
	BCC 7829 CM	Mulberry	2		
	BCC 7829 LY	Savile Grey	2		
	BCC 7829 LZ	Warm Charcoal	2		

MASTER INDEX

GROUP INDEX

3K 04

JAGUAR XJS RANGE (JAN 1987 ON) | K04 fiche 3 | FASCIA, TRIM, SEATS AND FIXINGS

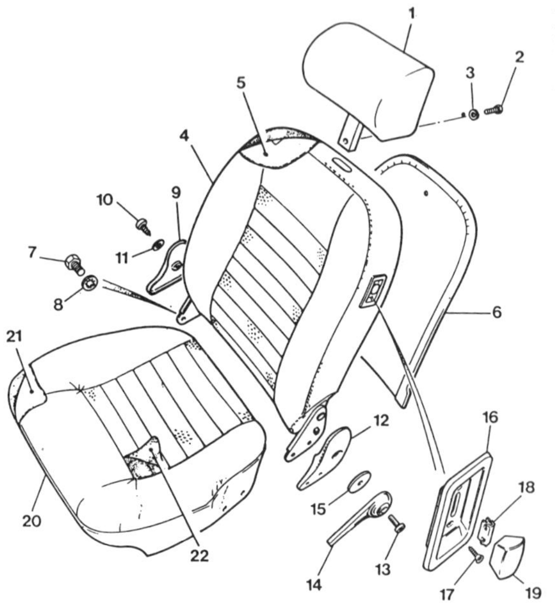

VS4739A

FRONT SEATS-AMBLA/CLOTH.-3.6 LITRE

Illus	Part Number	123456 Description	Quantity	Change Point	Remarks
		HEAD RESTRAINT-FRONT			
1	BBC 8325 YF	Cheviot tweed	2		
1	BBC 8325 AY	Chiltern tweed	2		
1	BBC 8325 AX	Cotswold tweed	2		
1	BBC 878325 YE	Pennine tweed	2		
2	SH 605101 J	Screw-set	2		
3	JLM 303	Washer-spring	2		
		SQUAB RH-WITH LUMBAR ADJUST			
	BCC 6720 YF	Cheviot tweed			
	BCC 6720 AY	Chiltern tweed	1		
	BCC 6720 AX	Cotswold tweed	1		
	BCC 6720 YE	Pennine tweed	1		
	BCC 8504	Diaphragm-lumbar-RH	1		
		SQUAB LH-WITH LUMBAR ADJUST			
4	BCC 6721 YF	Cheviot tweed	1		
4	BCC 6721 AY	Chiltern tweed	1		
4	BCC 6721 AX	Cotswold tweed	1		
4	BCC 6721 YE	Pennine tweed	1		
	BCC 8505	Diaphragm-lumbar-LH	1		
		SQUAB RH-NOT LUMBAR ADJUST			
	BCC 6716 YF	Cheviot tweed	1		
	BCC 6716 AY	Chiltern tweed	1		
	BCC 6716 AX	Cotswold tweed	1		
	BCC 6716 YE	Pennine tweed	1		
		SQUAB LH-NOT LUMBAR ADJUST			
4	BCC 6717 YF	Cheviot tweed	1		
4	BCC 6717 AY	Chiltern tweed	1		
4	BCC 6717 AX	Cotswold tweed	1		
4	BCC 6721 YE	Pennine tweed	1		
		COVER-FRONT SQUABS			
5	BBC 8306 YF	Cheviot tweed	2		
5	BBC 8306 AY	Chiltern tweed	2		
5	BBC 8306 AX	Cotswold tweed	2		
5	BBC 878306 YE	Pennine tweed	2		

MASTER INDEX

ENGINE 1.C
FLYWHEEL AND CLUTCH 1.G
GEARBOX AND PROPSHAFT 1.H
AXLES, SUSPENSION, DRIVE SHAFTS, WHEELS 1.K
STEERING 1.M
BRAKES AND BRAKE CONTROLS 1.N
FUEL, EXHAUST AND EMISSION SYSTEMS 2.C
COOLING, HEATING AND AIR CONDITIONING 2.G
ELECTRICAL, WASHERS-WIPERS, INSTRUMENTS 2.J
CHASSIS, SUBFRAMES, BODYSHELL, FITTINGS 3.C
FASCIA, TRIM, SEATS AND FIXINGS 3.H
ACCESSORIES AND PAINT 4.C
NUMERICAL INDEX 4.D

GROUP INDEX

BATTERY AND SPARE WHEEL COVERS N05
BONNET INSULATION-3.6 LITRE N06
BONNET INSULATION-5.3 MODELS N08
BONNET INSULATION-5.3 MODELS N07
BOOT CARPETS - NOT CONVERTIBLE N03
BOOT CARPETS-CONVERTIBLE N04
CARPETS-'A' POST FINISHERS- L12
CONSOLE-AMBLA H10
CONSOLE-LEATHER H08
DOOR CASINGS-ARMRESTS AND POCKETS-AMBLA H16
DOOR CASINGS-ARMRESTS-CONVERTIBLE I02
HEADLINING I15
DOOR CASINGS-ARMRESTS-LEATHER H14
EXTERIOR TRIM COVERS N15
FACIA PANEL H02
FRONT SEAT BELTS-CONVERTIBLE K15
FRONT SEAT SLIDES K16
MOTOR AND ANCHORAGE ASSY-RH L02
FRONT SEATS-AMBLA/CLOTH.-3.6 LITRE K04
FRONT SEATS-LEATHER J03
FRONT SEATS-LEATHER/CLOTH-5.3 LITRE K07
FRT SEATS-LEATHER/CLOTH(SPORTSPACK) K10
FRONT BELT KIT-COUPE K14
FRONT SEAT SLIDES K13
FUEL APERATURE SEAL AND HOSES-COUPE N13
HARDTOP & HOODBAG-CABRIOLET N17
HOOD LIFT MECHANISM - CONVERTIBLE O04
HOOD SEALS-CONVERTIBLE O03
HOOD-CABRIOLET N16
HOOD-CONVERTIBLE O02
REAR CUBBY BOX-CONVERTIBLE L09
REAR LUGGAGE LOCKER-CABRIOLET L07
REAR SEAT BELTS L11
REAR SEATS-LEATHER-COUPE L03
REAR SEATS-LEATHER-COUPE L04
REAR SEATS-LEATHER/CLOTH-COUPE L06
TARGA PANELS-STOWAGEBAG-CABRIOLET N14
UNDER SCUTTLE CASINGS H05
VENEERED FACIA PANELS-3.6 LITRE H06
VENEERED FACIA PANELS-5.3 LITRE H07

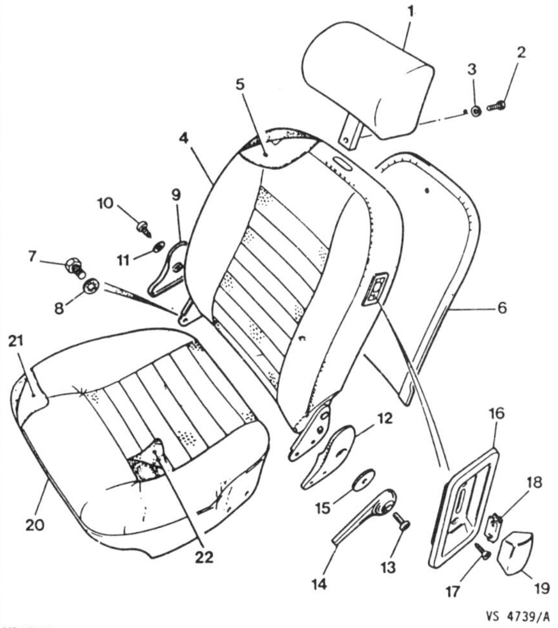

VS4739A

FRONT SEATS-AMBLA/CLOTH.-3.6 LITRE

Illus	Part Number	123456	Description	Quantity	Change Point	Remarks
			PANEL RH-SQUAB BACK			
	BD 47029 AW		Barley	1		
	BD 47029 PA		Black	1		
	BD 47029 XE		Doeskin	1		
	BD 47029 LY		Savile Grey	1		
			PANEL LH-SQUAB BACK			
6	BD 47030 AW		Barley	1		
6	BD 47030 PA		Black	1		
6	BD 47030 XE		Doeskin	1		
6	BD 47030 LY		Savile Grey	1		
7	SH 604041 J		Screw-set	8		
8	C 724		Washer-shakeproof	8		
			COVER RH-INNER			
	BCC 3040 AW		Barley	1		
	BCC 3040 PA		Black	1		
	BCC 3040 XE		Doeskin	1		
	BCC 3040 LY		Savile Grey	1		
			COVER LH-INNER			
9	BCC 3041 AW		Barley	1		
9	BCC 3041 PA		Black	1		
9	BCC 3041 XE		Doeskin	1		
9	BCC 3041 LY		Savile Grey	1		

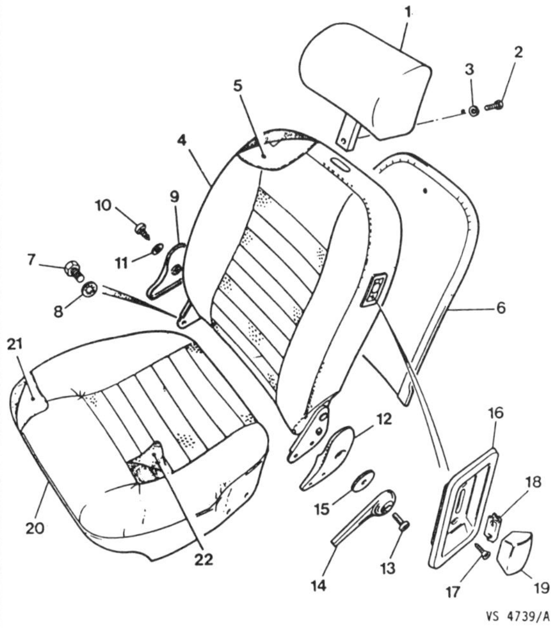

VS4739A

FRONT SEATS-AMBLA/CLOTH-3.6 LITRE-continued.

Illus	Part Number	123456	Description	Quantity	Change Point	Remarks
10	JLM 307		Screw-self tap	2		
11	WC 702102 J		Washer-plain	2		
			COVER RH-OUTER			
	BCC 3038 AW		Barley	1		
	BCC 3038 LZ		Warm Charcoal	1		
	BCC 3038 XE		Doeskin	1		
	BCC 3038 LY		Savile Grey	1		
			COVER LH-OUTER			
12	BCC 3039 AW		Barley	1		
12	BCC 3039 LZ		Warm charcoal	1		
12	BCC 3039 XE		Doeskin	1		
12	BCC 3039 LY		Savile grey	1		
13	BD 34524 1		Screw-taptite	2		
	BCC 7508		Lever-recline-RH	1		
14	BCC 7509		Lever-recline-LH	1		
15	BD 46805		Washer-plastic	2		
16	715721		Escutcheon	2		
17	DAB 6 6C		Screw-self tap	4		
18	BD 42638		Clip-retainer	2		
19	624571		Knob	2		
			CUSHION ASSEMBLY-FRONT			
20	BCC 2993 YF		Cheviot tweed	2		
20	BCC 2993 AY		Chiltern tweed	2		
20	BCC 2993 AX		Cotswold tweed	2		
20	BCC 2993 YE		Pennine tweed	2		
			COVER-OUTER-CUSHION			
21	BAC 4060 AW		Barley	2		
21	BAC 4060 LZ		Warm Charcoal	2		
21	BAC 4060 XE		Doeskin	2		
21	BAC 4060 LY		Savile Grey	2		
			COVER-INNER-CUSHION			
22	BBC 8320 YF		Cheviot tweed	2		
22	BBC 8320 AY		Chiltern tweed	2		
22	BBC 8320 AX		Cotswold tweed	2		
22	BBC 8320 YE		Pennine tweed	2		

MASTER INDEX

ENGINE ... 1.C
FLYWHEEL AND CLUTCH ... 1.G
GEARBOX AND PROPSHAFT ... 1.H
AXLES, SUSPENSION, DRIVE SHAFTS, WHEELS ... 1.K
STEERING ... 1.M
BRAKES AND BRAKE CONTROLS ... 1.N
FUEL, EXHAUST AND EMISSION SYSTEMS ... 2.C
COOLING, HEATING AND AIR CONDITIONING ... 2.G
ELECTRICAL, WASHERS-WIPERS, INSTRUMENTS ... 2.J
CHASSIS, SUBFRAMES, BODYSHELL, FITTINGS ... 3.C
FASCIA, TRIM, SEATS AND FIXINGS ... 3.H
ACCESSORIES AND PAINT ... 4.C
NUMERICAL INDEX ... 4.D

GROUP INDEX

BATTERY AND SPARE WHEEL COVERS ... N05
BONNET INSULATION-3.6 LITRE ... N06
BONNET INSULATION-5.3 MODELS ... N08
BONNET INSULATION-5.3 MODELS ... N07
BOOT CARPETS - NOT CONVERTIBLE ... N03
BOOT CARPETS-CONVERTIBLE ... N04
CARPETS-'A' POST FINISHERS- ... L12
CONSOLE-AMBLA ... H10
CONSOLE-LEATHER ... H08
DOOR CASINGS-ARMRESTS AND POCKETS-AMBLA ... H16
DOOR CASINGS-ARMRESTS-CONVERTIBLE ... I02
HEADLINING ... I15
DOOR CASINGS-ARMRESTS-LEATHER ... H14
EXTERIOR TRIM COVERS ... N15
FACIA PANEL ... H02
FRONT SEAT BELTS-CONVERTIBLE ... K15
FRONT SEAT SLIDES ... K16
MOTOR AND ANCHORAGE ASSY-RH ... L02
FRONT SEATS-AMBLA/CLOTH.-3.6 LITRE ... K04
FRONT SEATS-LEATHER ... J03
FRONT SEATS-LEATHER/CLOTH-5.3 LITRE ... K07
FRT SEATS-LEATHER/CLOTH(SPORTSPACK) ... K10
FRONT BELT KIT-COUPE ... K14
FRONT SEAT SLIDES ... K13
FUEL APERATURE SEAL AND HOSES-COUPE ... N13
HARDTOP & HOODBAG-CABRIOLET ... N17
HOOD LIFT MECHANISM - CONVERTIBLE ... O04
HOOD SEALS-CONVERTIBLE ... O03
HOOD-CABRIOLET ... N16
HOOD-CONVERTIBLE ... O02
REAR CUBBY BOX-CONVERTIBLE ... L09
REAR LUGGAGE LOCKER-CABRIOLET ... L07
REAR SEAT BELTS ... L11
REAR SEATS-LEATHER-COUPE ... L03
REAR SEATS-LEATHER-COUPE ... L04
REAR SEATS-LEATHER/CLOTH-COUPE ... L06
TARGA PANELS-STOWAGEBAG-CABRIOLET ... N14
UNDER SCUTTLE CASINGS ... H05
VENEERED FACIA PANELS-3.6 LITRE ... H06
VENEERED FACIA PANELS-5.3 LITRE ... H07

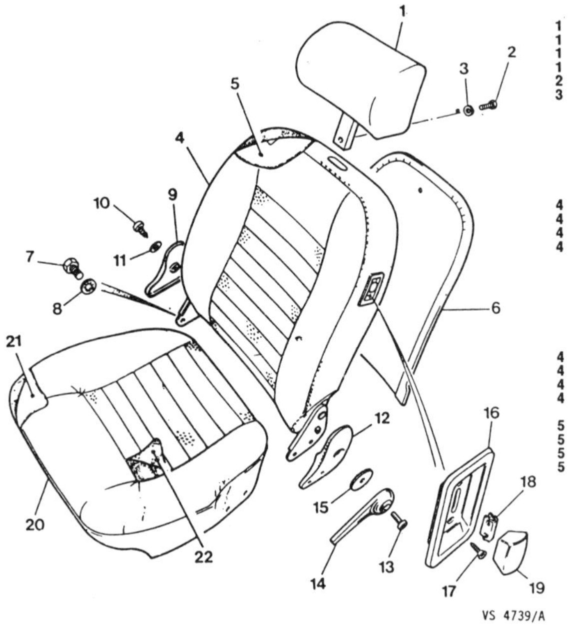

VS4739A

FRONT SEATS-LEATHER/CLOTH-5.3 LITRE

Illus	Part Number	123456 Description	Quantity	Change Point	Remarks
		HEADRESTRAINT-FRONT			
1	BBC 8325 YF	Cheviot tweed	2		
1	BBC 8325 AY	Chiltern tweed	2		
1	BBC 8325 AX	Cotswold tweed	2		
1	BBC 878325 YE	Pennine tweed	2		
2	SH 605101 J	Screw-set	2		
3	JLM 303	Washer-spring	2		
		SQUAB RH-WITH LUMBAR ADJUST			
	BCC 3024 YF	Cheviot tweed	1		
	BCC 3024 AY	Chiltern tweed	1		
	BCC 3024 AX	Cotswold tweed	1		
	BCC 3024 YE	Pennine tweed	1		
	BCC 8504	Diaphragm-lumbar-RH	1		
		SQUAB LH-WITH LUMBAR ADJUST			
4	BCC 3025 YF	Cheviot tweed	1		
4	BCC 3025 AY	Chiltern tweed	1		
4	BCC 3025 AX	Cotswold tweed	1		
4	BCC 3025 YE	Pennine tweed	1		
	BCC 8505	Diaphragm-lumbar LH	1		
		SQUAB RH-NOT LUMBAR ADJUST			
	BCC 3022 YF	Cheviot tweed	1		
	BCC 3022 AY	Chiltern tweed	1		
	BCC 3022 AX	Cotswold tweed	1		
	BBC 3022 YE	Pennine tweed	1		
		SQUAB LH-NOT LUMBAR ADJUST			
4	BCC 3023 YF	Cheviot tweed	1		
4	BCC 3023 AY	Chiltern tweed	1		
4	BCC 3023 AX	Cotswold tweed	1		
4	BCC 3023 YE	Pennine tweed	1		
		COVER-FRONT SQUABS			
5	BBC 8314 YF	Cheviot tweed	1		
5	BBC 8314 AY	Chiltern tweed	1		
5	BBC 8314 AX	Cotswold tweed	1		
5	BBC 878325 YE	Pennine tweed	1		

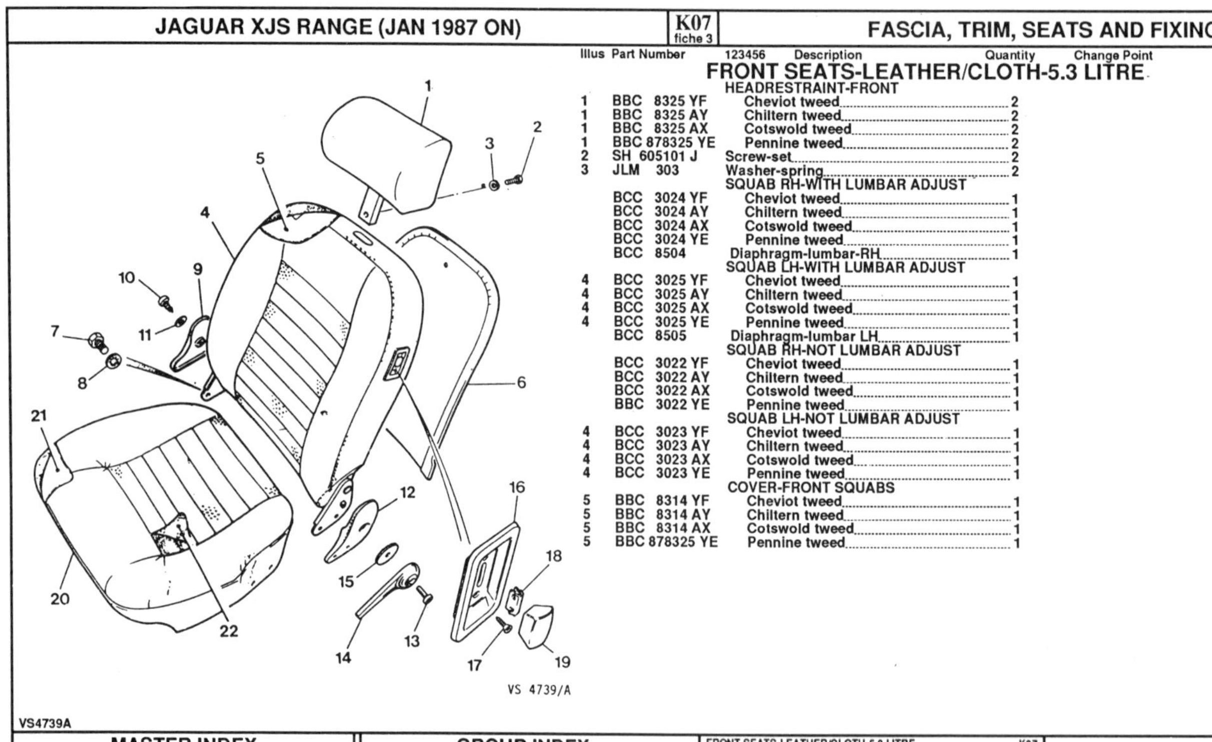

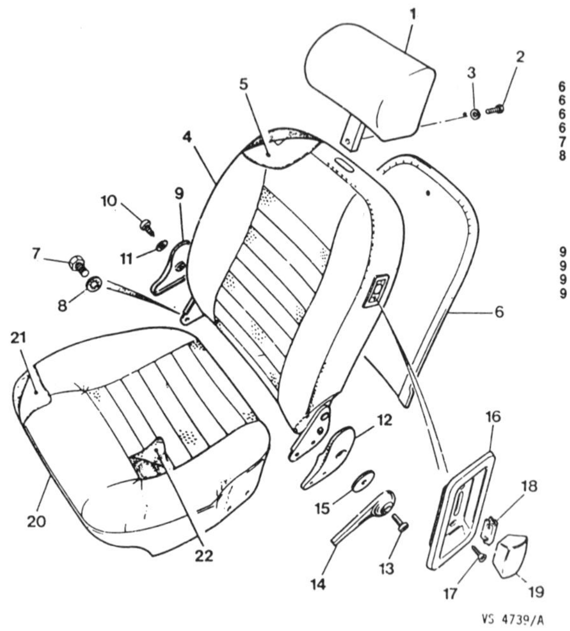

VS4739A

FRONT SEATS-LEATHER/CLOTH-5.3 LITRE

Illus	Part Number	123456 Description	Quantity	Change Point	Remarks
		PANEL RH-SQUAB BACK			
	BAC 4038 AW	Barley	1		
	BAC 4038 LZ	Warm Charcoal	1		
	BAC 4038 XE	Doeskin	1		
	BAC 4038 LY	Savile Grey	1		
		PANEL LH-SQUAB BACK			
6	BAC 4039 AW	Barley	1		
6	BAC 4039 LZ	Warm Charcoal	1		
6	BAC 4039 XE	Doeskin	1		
6	BAC 4039 LY	Savile Grey	1		
7	SH 604041 J	Screw-set	8		
8	C 724	Washer-shakeproof	8		
		COVER RH-INNER			
	BCC 3036 AW	Barley	1		
	BCC 3036 LZ	Warm Charcoal	1		
	BCC 3036 XE	Doeskin	1		
	BCC 3036 LY	Savile Grey	1		
		COVER LH-INNER			
9	BCC 3037 AW	Barley	1		
9	BCC 3037 LZ	Warm Charcoal	1		
9	BCC 3037 XE	Doeskin	1		
9	BCC 3037 LY	Savile Grey	1		

MASTER INDEX

ENGINE ... 1.C
FLYWHEEL AND CLUTCH ... 1.G
GEARBOX AND PROPSHAFT ... 1.H
AXLES, SUSPENSION, DRIVE SHAFTS, WHEELS ... 1.K
STEERING ... 1.M
BRAKES AND BRAKE CONTROLS ... 1.N
FUEL, EXHAUST AND EMISSION SYSTEMS ... 2.C
COOLING, HEATING AND AIR CONDITIONING ... 2.G
ELECTRICAL, WASHERS-WIPERS, INSTRUMENTS ... 2.J
CHASSIS, SUBFRAMES, BODYSHELL, FITTINGS ... 3.C
FASCIA, TRIM, SEATS AND FIXINGS ... 3.H
ACCESSORIES AND PAINT ... 4.C
NUMERICAL INDEX ... 4.D

GROUP INDEX

BATTERY AND SPARE WHEEL COVERS ... N05
BONNET INSULATION-3.6 LITRE ... N06
BONNET INSULATION-5.3 MODELS ... N08
BONNET INSULATION-5.3 MODELS ... N07
BOOT CARPETS - NOT CONVERTIBLE ... N03
BOOT CARPETS-CONVERTIBLE ... N04
CARPETS-'A' POST FINISHERS- ... L12
CONSOLE-AMBLA ... H10
CONSOLE-LEATHER ... H08
DOOR CASINGS-ARMRESTS AND POCKETS-AMBLA ... H16
DOOR CASINGS-ARMRESTS-CONVERTIBLE ... I02
HEADLINING ... I15
DOOR CASINGS-ARMRESTS-LEATHER ... H14
EXTERIOR TRIM COVERS ... N15
FACIA PANEL ... H02
FRONT SEAT BELTS-CONVERTIBLE ... K15
FRONT SEAT SLIDES ... K16
MOTOR AND ANCHORAGE ASSY-RH ... L02
FRONT SEATS-AMBLA/CLOTH.-3.6 LITRE ... K04
FRONT SEATS-LEATHER ... J03
FRONT SEATS-LEATHER/CLOTH-5.3 LITRE ... K07
FRT SEATS-LEATHER/CLOTH(SPORTSPACK) ... K10
FRONT BELT KIT-COUPE ... K14
FRONT SEAT SLIDES ... K13
FUEL APERATURE SEAL AND HOSES-COUPE ... N13
HARDTOP & HOODBAG-CABRIOLET ... N17
HOOD LIFT MECHANISM - CONVERTIBLE ... O04
HOOD SEALS-CONVERTIBLE ... O03
HOOD-CABRIOLET ... N16
HOOD-CONVERTIBLE ... O02
REAR CUBBY BOX-CONVERTIBLE ... L09
REAR LUGGAGE LOCKER-CABRIOLET ... L07
REAR SEAT BELTS ... L11
REAR SEATS-LEATHER-COUPE ... L03
REAR SEATS-LEATHER-COUPE ... L04
REAR SEATS-LEATHER/CLOTH-COUPE ... L06
TARGA PANELS-STOWAGEBAG-CABRIOLET ... N14
UNDER SCUTTLE CASINGS ... H05
VENEERED FACIA PANELS-3.6 LITRE ... H06
VENEERED FACIA PANELS-5.3 LITRE ... H07

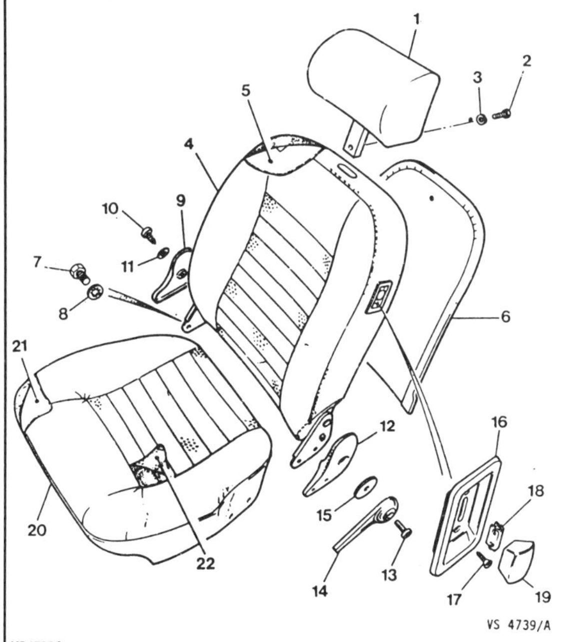

VS4739A

FRONT SEATS-LEATHER/CLOTH-5.3 LITRE-continued.

Illus	Part Number	123456	Description	Quantity	Change Point	Remarks
10	JLM 307		Screw-self tap	2		
11	WC 702102 J		Washer-plain	2		
			COVER RH-OUTER			
	BCC 3034 AW		Barley	1		
	BCC 3034 LZ		Warm Charcoal	1		
	BCC 3034 XE		Doeskin	1		
	BCC 3034 LY		Savile Grey	1		
			COVER LH-OUTER			
12	BCC 3035 AW		Barley	1		
12	BCC 3035 LZ		Warm Charcoal	1		
12	BCC 3035 XE		Doeskin	1		
12	BCC 3035 LY		Savile Grey	1		
13	BD 34541 1		Screw-taptite	2		
	BCC 7508		Lever-recline-RH	1		
14	BCC 7509		Lever-recline-LH	1		
15	BD 46805		Washer-plastic	2		
16	715721		Escutcheon	2		
17	DAB 006 6C		Screw-self tap	4		
18	BD 42638		Clip-retainer	2		
19	624571		Knob	2		
			CUSHION ASSEMBLY-FRONT			
20	BCC 2993 YF		Cheviot tweed	2		
20	BCC 2993 AY		Chiltern tweed	2		
20	BCC 2993 AX		Cotswold tweed	2		
20	BCC 2993 YE		Penine tweed	2		
			COVER-OUTER CUSHION			
21	BAC 4060 AW		Barley	2		
21	BAC 4060 LZ		Warm Charcoal	2		
21	BAC 4060 XE		Doeskin	2		
21	BAC 4060 LY		Savile Grey	2		
			COVER-INNER CUSHION			
22	BBC 8320 YF		Cheviot tweed	2		
22	BBC 8320 AY		Chiltern tweed	2		
22	BBC 8320 AX		Cotswold tweed	2		
22	BBC 8320 YE		Pennine tweed	2		

MASTER INDEX

GROUP INDEX

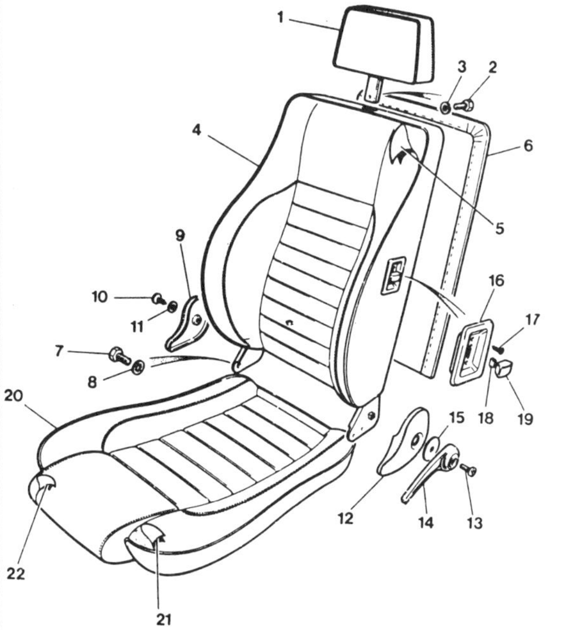

VS4797

FRT SEATS-LEATHER/CLOTH(SPORTSPACK)

Illus	Part Number	123456	Description	Quantity	Change Point	Remarks
			HEADRESTRAINT-FRONT			
1	BCC 9322 YF		Cheviot tweed	2		
1	BCC 9322 AY		Chiltern tweed	2		
1	BCC 9322 AX		Cotswold tweed	2		
1	BCC 9322 YE		Pennine tweed	2		
2	SH 605101 J		Screw-set	2		
3	JLM 303		Washer-spring	2		
			SQUAB RH-WITH LUMBAR ADJUST			
	BCC 9286 YF		Cheviot tweed	1		
	BCC 9286 AY		Chiltern tweed	1		
	BCC 9286 AX		Cotswold tweed	1		
	BCC 9286 YE		Pennine tweed	1		
	BCC 8504		Lumbar diaphragm RH	1		
			SQUAB LH-WITH LUMBAR ADJUST			
4	BCC 9287 YF		Cheviot tweed	1		
4	BCC 9287 AY		Chiltern tweed	1		
4	BCC 9287 AX		Cotswold tweed	1		
4	BCC 9287 YE		Pennine tweed	1		
	BCC 8505		Lumbar diaphragm LH	1		
			SQUAB RH-NOT LUMBAR ADJUST			
	BCC 9292 YF		Cheviot tweed	1		
	BCC 9292 AY		Chiltern tweed	1		
	BCC 9292 AX		Cotswold tweed	1		
	BCC 9292 YE		Pennine tweed	1		
			SQUAB LH-NOT LUMBAR ADJUST			
4	BCC 9293 YF		Cheviot tweed	1		
4	BCC 9293 AY		Chiltern tweed	1		
4	BCC 9293 AX		Cotswold tweed	1		
4	BCC 9293 YE		Pennine tweed	1		
			COVER-FRONT SQUABS			
5	BCC 9370 YF		Cheviot tweed	1		
5	BCC 9370 AY		Chiltern tweed	1		
5	BCC 9370 AX		Cotswold tweed	1		
5	BCC 9370 YE		Pennine tweed	1		

MASTER INDEX

ENGINE 1.C
FLYWHEEL AND CLUTCH 1.G
GEARBOX AND PROPSHAFT 1.H
AXLES, SUSPENSION, DRIVE SHAFTS, WHEELS 1.K
STEERING 1.M
BRAKES AND BRAKE CONTROLS 1.N
FUEL, EXHAUST AND EMISSION SYSTEMS 2.C
COOLING, HEATING AND AIR CONDITIONING 2.G
ELECTRICAL, WASHERS-WIPERS, INSTRUMENTS 2.J
CHASSIS, SUBFRAMES, BODYSHELL, FITTINGS 3.C
FASCIA, TRIM, SEATS AND FIXINGS 3.H
ACCESSORIES AND PAINT 4.C
NUMERICAL INDEX 4.D

GROUP INDEX

BATTERY AND SPARE WHEEL COVERS N05
BONNET INSULATION-3.6 LITRE N06
BONNET INSULATION-5.3 MODELS N08
BONNET INSULATION-5.3 MODELS N07
BOOT CARPETS - NOT CONVERTIBLE N03
BOOT CARPETS-CONVERTIBLE N04
CARPETS-'A' POST FINISHERS- L12
CONSOLE-AMBLA H10
CONSOLE-LEATHER H08
DOOR CASINGS-ARMRESTS AND POCKETS-AMBLA H16
DOOR CASINGS-ARMRESTS-CONVERTIBLE I02
HEADLINING I15
DOOR CASINGS-ARMRESTS-LEATHER H14
EXTERIOR TRIM COVERS N15
FACIA PANEL H02
FRONT SEAT BELTS-CONVERTIBLE K15
FRONT SEAT SLIDES K16
MOTOR AND ANCHORAGE ASSY-RH L02
FRONT SEATS-AMBLA/CLOTH.-3.6 LITRE K04
FRONT SEATS-LEATHER J03
FRONT SEATS-LEATHER/CLOTH-5.3 LITRE K07
FRT SEATS-LEATHER/CLOTH(SPORTSPACK) K10
FRONT BELT KIT-COUPE K14
FRONT SEAT SLIDES K13
FUEL APERATURE SEAL AND HOSES-COUPE N13
HARDTOP & HOODBAG-CABRIOLET N17
HOOD LIFT MECHANISM - CONVERTIBLE O04
HOOD SEALS-CONVERTIBLE O03
HOOD-CABRIOLET N16
HOOD-CONVERTIBLE O02
REAR CUBBY BOX-CONVERTIBLE L09
REAR LUGGAGE LOCKER-CABRIOLET L07
REAR SEAT BELTS L11
REAR SEATS-LEATHER-COUPE L03
REAR SEATS-LEATHER-COUPE L04
REAR SEATS-LEATHER/CLOTH-COUPE L06
TARGA PANELS-STOWAGEBAG-CABRIOLET N14
UNDER SCUTTLE CASINGS H05
VENEERED FACIA PANELS-3.6 LITRE H06
VENEERED FACIA PANELS-5.3 LITRE H07

3K 11

JAGUAR XJS RANGE (JAN 1987 ON)	K11 fiche 3	FASCIA, TRIM, SEATS AND FIXINGS

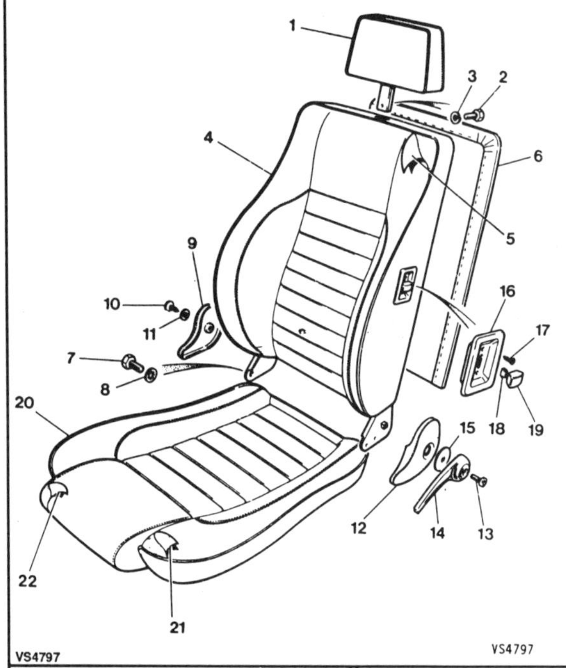

VS4797

FRONT SEATS-LEATHER/CLOTH(SPORTSPACK)-continued.

Illus	Part Number	123456 Description	Quantity	Change Point	Remarks
		PANEL RH-SQUAB BACK			
	BCC 9306 AW	Barley	1		
	BCC 9306 LZ	Warm Charcoal	1		
	BCC 9306 XE	Doeskin	1		
	BCC 9306 LY	Savile Grey	1		
		PANEL LH-SQUAB BACK			
6	BCC 9307 AW	Barley	1		
6	BCC 9307 LZ	Warm Charcoal	1		
6	BCC 9307 XE	Doeskin	1		
6	BCC 9307 LY	Savile Grey	1		
7	SH 604041 J	Screw-set	8		
8	C 724	Washer-shakeproof	8		
		COVER RH-INNER			
	BCC 3036 AW	Barley	1		
	BCC 3036 LZ	Warm Charcoal	1		
	BCC 3036 XE	Doeskin	1		
	BCC 3036 LY	Savile Grey	1		
		COVER LH-INNER			
9	BCC 3037 AW	Barley	1		
9	BCC 3037 LZ	Warm Charcoal	1		
9	BCC 3037 XE	Doeskin	1		
9	BCC 3037 LY	Savile Grey	1		

3K 12

JAGUAR XJS RANGE (JAN 1987 ON)	K12 fiche 3	FASCIA, TRIM, SEATS AND FIXINGS

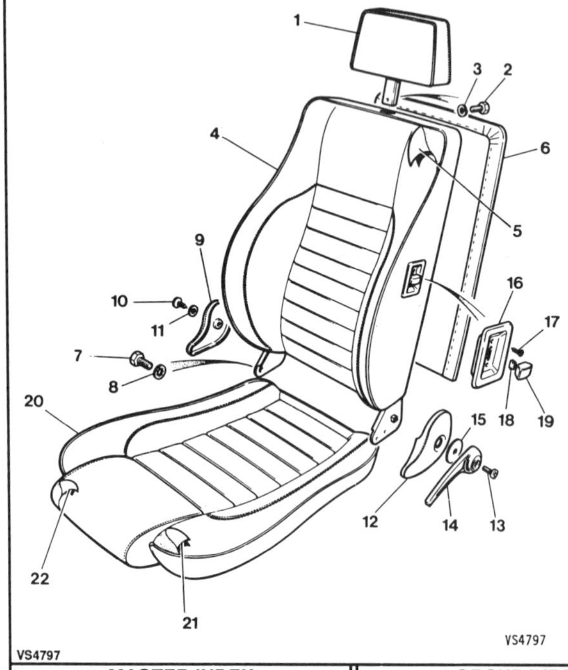

VS4797

FRONT SEATS-LEATHER/CLOTH(SPORTSPACK)-continued.

Illus	Part Number	123456 Description	Quantity	Change Point	Remarks
10	JLM 307	Screw-self tap	2		
11	WC 702101 J	Washer-plain	2		
		COVER RH-OUTER			
	BCC 7968 AW	Barley	1		
	BCC 7968 LZ	Warm Charcoal	1		
	BCC 7968 XE	Doeskin	1		
	BCC 7968 LY	Savile Grey	1		
		COVER LH-OUTER			
12	BCC 7969 AW	Barley	1		
12	BCC 7969 LZ	Warm Charcoal	1		
12	BCC 7969 XE	Doeskin	1		
12	BCC 7969 LY	Savile Grey	1		
13	BD 34541 1	Screw-taptite	2		
	BCC 7508	Lever-recline-RH	1		
14	BCC 7509	Lever-recline-LH	1		
15	BD 46805	Washer-plastic	2		
16	715721	Escutcheon	2		
17	DAB 006 6C	Screw-self tap	4		
18	BD 42638	Clip-retainer	2		
19	624571	Knob	2		
		CUSHION ASSEMBLY-FRONT			
20	BCC 9315 YF	Cheviot tweed	2		
20	BCC 9315 AY	Chiltern tweed	2		
20	BCC 9315 AX	Cotswold tweed	2		
20	BCC 9315 YE	Pennine tweed	2		
		COVER OUTER-CUSHION			
21	BCC 9379 AW	Barley	2		
21	BCC 9379 LZ	Warm Charcoal	2		
21	BCC 9379 XE	Doeskin	2		
21	BCC 9379 LY	Savile Grey	2		
		COVER-INNER CUSHION			
22	BCC 9384 YF	Cheviot tweed	2		
22	BCC 9384 AY	Chiltern tweed	2		
22	BCC 9384 AX	Cotswold tweed	2		
22	BCC 9384 YE	Pennine tweed	2		

MASTER INDEX

ENGINE	1.C
FLYWHEEL AND CLUTCH	1.G
GEARBOX AND PROPSHAFT	1.H
AXLES, SUSPENSION, DRIVE SHAFTS, WHEELS	1.K
STEERING	1.M
BRAKES AND BRAKE CONTROLS	1.N
FUEL, EXHAUST AND EMISSION SYSTEMS	2.C
COOLING, HEATING AND AIR CONDITIONING	2.G
ELECTRICAL, WASHERS-WIPERS, INSTRUMENTS	2.J
CHASSIS, SUBFRAMES, BODYSHELL, FITTINGS	3.C
FASCIA, TRIM, SEATS AND FIXINGS	3.H
ACCESSORIES AND PAINT	4.C
NUMERICAL INDEX	4.D

GROUP INDEX

BATTERY AND SPARE WHEEL COVERS	N05
BONNET INSULATION-3.6 LITRE	N06
BONNET INSULATION-5.3 MODELS	N08
BONNET INSULATION-5.3 MODELS	N07
BOOT CARPETS - NOT CONVERTIBLE	N03
BOOT CARPETS-CONVERTIBLE	N04
CARPETS-'A' POST FINISHERS-	L12
CONSOLE-AMBLA	H10
CONSOLE-LEATHER	H08
DOOR CASINGS-ARMRESTS AND POCKETS-AMBLA	H16
DOOR CASINGS-ARMRESTS-CONVERTIBLE	I02
HEADLINING	I15
DOOR CASINGS-ARMRESTS-LEATHER	H14
EXTERIOR TRIM COVERS	N15
FACIA PANEL	H02
FRONT SEAT BELTS-CONVERTIBLE	K15
FRONT SEAT SLIDES	K16
MOTOR AND ANCHORAGE ASSY-RH	L02
FRONT SEATS-AMBLA/CLOTH-3.6 LITRE	K04
FRONT SEATS-LEATHER	J03
FRONT SEATS-LEATHER/CLOTH-5.3 LITRE	K07
FRT SEATS-LEATHER/CLOTH(SPORTSPACK)	K10
FRONT BELT KIT-COUPE	K14
FRONT SEAT SLIDES	K13
FUEL APERATURE SEAL AND HOSES-COUPE	N13
HARDTOP & HOODBAG-CABRIOLET	N17
HOOD LIFT MECHANISM - CONVERTIBLE	O04
HOOD SEALS-CONVERTIBLE	O03
HOOD-CABRIOLET	N16
HOOD-CONVERTIBLE	O02
REAR CUBBY BOX-CONVERTIBLE	L09
REAR LUGGAGE LOCKER-CABRIOLET	L07
REAR SEAT BELTS	L11
REAR SEATS-LEATHER-COUPE	L03
REAR SEATS-LEATHER-COUPE	L04
REAR SEATS-LEATHER/CLOTH-COUPE	L06
TARGA PANELS-STOWAGEBAG-CABRIOLET	N14
UNDER SCUTTLE CASINGS	H05
VENEERED FACIA PANELS-3.6 LITRE	H06
VENEERED FACIA PANELS-5.3 LITRE	H07

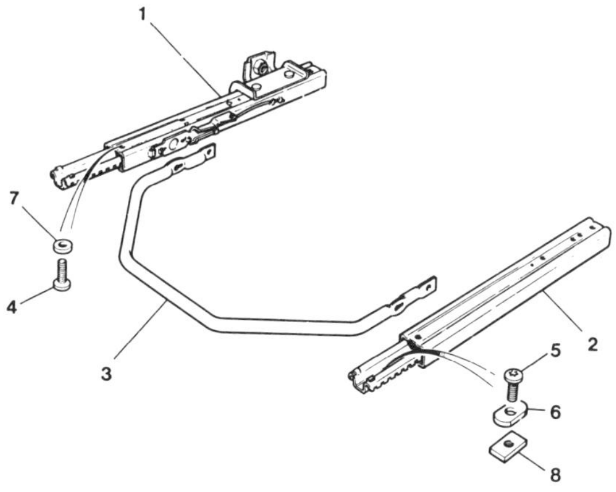

VS4734

VS4734

FRONT SEAT SLIDES-NOT PASSIVE RESTRAINT

Illus	Part Number	123456	Description	Quantity	Change Point	Remarks
			FRONT SEAT SLIDES			
1	BCC 6316		RH-LH seat	1		
	BCC 6317		LH-RH seat	1		
2	BCC 6311		LH-LH seat	1		
	BCC 6310		RH-RH seat	1		
3	BCC 6305		Handle-operating	2		
4	BBC 7759		Screw-torx	8		
5	BCC 4522		Screw-torx-pan head	8		
6	BCC 2522		Washer	16		
7	BCC 4518		Spacer	8		
8	BCC 8038		Plate-tapping	4		

MASTER INDEX

GROUP INDEX

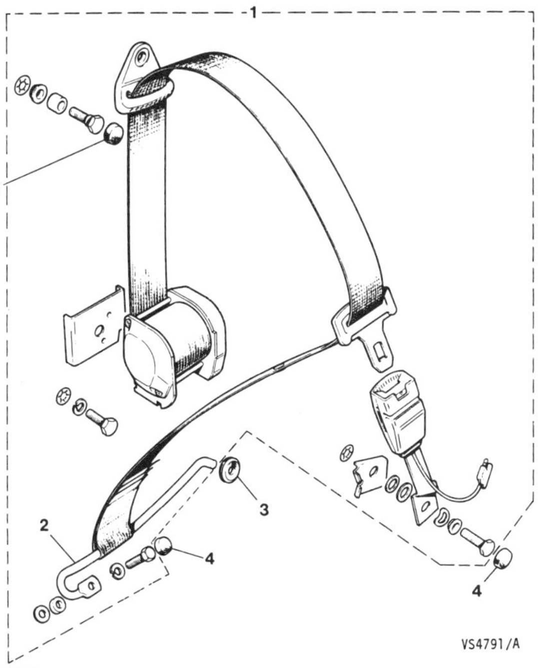

VS4791/A

VS4791/A

FRONT SEAT BELTS-NOT PASSIVE RESTRAINT-NOT CONVERTIBLE

Illus	Part Number	123456	Description	Quantity	Change Point	Remarks
			FRONT BELT KIT-COUPE			
1	JLM 1316		RH	1		
	JLM 1317		LH	1		
1	JLM 1320		RH	1		Aus only
	JLM 1321		LH	1		Aus only
			FRONT BELT KIT-CABRIOLET			
1	JLM 1318		RH	1		
	JLM 1319		LH	1		
1	JLM 1322		RH	1		Aus only
	JLM 1323		LH	1		Aus only
2	BBC 7800		Assembly-slider bar RH	1		Use BCC 8524
2	BCC 8524		Assembly-slider bar RH	1		
	BBC 7801		Assembly-slider bar LH	1		Use BCC 8525
	BCC 8525		Assembly-slider bar LH	1		
3	CAC 6426		Grommet	2		
4	BCC 9037		Cap-bolt	4		

MASTER INDEX
ENGINE ... 1.C
FLYWHEEL AND CLUTCH ... 1.G
GEARBOX AND PROPSHAFT ... 1.H
AXLES, SUSPENSION, DRIVE SHAFTS, WHEELS ... 1.K
STEERING ... 1.M
BRAKES AND BRAKE CONTROLS ... 1.N
FUEL, EXHAUST AND EMISSION SYSTEMS ... 2.C
COOLING, HEATING AND AIR CONDITIONING ... 2.G
ELECTRICAL, WASHERS-WIPERS, INSTRUMENTS ... 2.J
CHASSIS, SUBFRAMES, BODYSHELL, FITTINGS ... 3.C
FASCIA, TRIM, SEATS AND FIXINGS ... 3.H
ACCESSORIES AND PAINT ... 4.C
NUMERICAL INDEX ... 4.D
GROUP INDEX
BATTERY AND SPARE WHEEL COVERS ... N05
BONNET INSULATION-3.6 LITRE ... N06
BONNET INSULATION-5.3 MODELS ... N08
BONNET INSULATION-5.3 MODELS ... N07
BOOT CARPETS - NOT CONVERTIBLE ... N03
BOOT CARPETS-CONVERTIBLE ... N04
CARPETS-'A' POST FINISHERS- ... L12
CONSOLE-AMBLA ... H10
CONSOLE-LEATHER ... H08
DOOR CASINGS-ARMRESTS AND POCKETS-AMBLA ... H16
DOOR CASINGS-ARMRESTS-CONVERTIBLE ... I02
HEADLINING ... I15
DOOR CASINGS-ARMRESTS-LEATHER ... H14
EXTERIOR TRIM COVERS ... N15
FACIA PANEL ... H02
FRONT SEAT BELTS-CONVERTIBLE ... K15
FRONT SEAT SLIDES ... K16
MOTOR AND ANCHORAGE ASSY-RH ... L02
FRONT SEATS-AMBLA/CLOTH.-3.6 LITRE ... K04
FRONT SEATS-LEATHER ... J03
FRONT SEATS-LEATHER/CLOTH-5.3 LITRE ... K07
FRT SEATS-LEATHER/CLOTH(SPORTSPACK) ... K10
FRONT BELT KIT-COUPE ... K14
FRONT SEAT SLIDES ... K13
FUEL APERATURE SEAL AND HOSES-COUPE ... N13
HARDTOP & HOODBAG-CABRIOLET ... N17
HOOD LIFT MECHANISM - CONVERTIBLE ... O04
HOOD SEALS-CONVERTIBLE ... O03
HOOD-CABRIOLET ... N16
HOOD-CONVERTIBLE ... O02
REAR CUBBY BOX-CONVERTIBLE ... L09
REAR LUGGAGE LOCKER-CABRIOLET ... L07
REAR SEAT BELTS ... L11
REAR SEATS-LEATHER-COUPE ... L03
REAR SEATS-LEATHER-COUPE ... L04
REAR SEATS-LEATHER/CLOTH-COUPE ... L06
TARGA PANELS-STOWAGEBAG-CABRIOLET ... N14
UNDER SCUTTLE CASINGS ... H05
VENEERED FACIA PANELS-3.6 LITRE ... H06
VENEERED FACIA PANELS-5.3 LITRE ... H07

3K 15

JAGUAR XJS RANGE (JAN 1987 ON)	K15 fiche 3	FASCIA, TRIM, SEATS AND FIXINGS

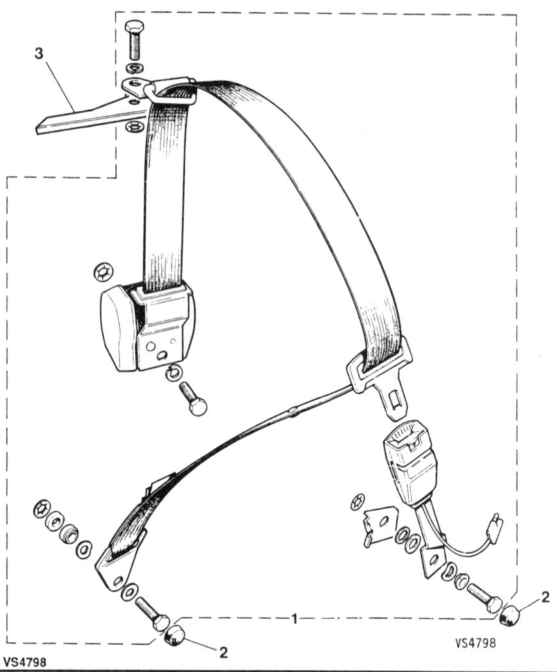

VS4798

FRONT SEAT BELTS-CONVERTIBLE

Illus	Part Number	123456 Description	Quantity	Change Point	Remarks
1	JLM 1447	Front seat belt kit RH	1		Not AUS
	JLM 1448	Front seat belt kit LH	1		Not AUS
	JLM 1614	Front seat belt kit RH	1		AUS only
	JLM 1615	Front seat belt kit LH	1		AUS only
2	BCC 9037	Cap-bolt	4		
3	BDC 6574	Plate-running loop RH	1		
	BDC 6575	Plate-running loop LH	1		

3K 16

JAGUAR XJS RANGE (JAN 1987 ON)	K16 fiche 3	FASCIA, TRIM, SEATS AND FIXINGS

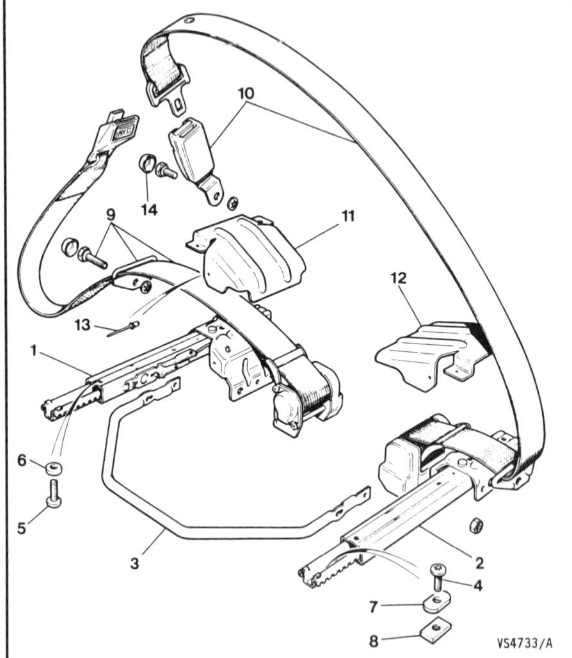

VS 4733A

FRONT SEAT SLIDES AND BELTS-PASSIVE RESTRAINT

UP TO (V)147543

Illus	Part Number	123456 Description	Quantity	Change Point	Remarks
		FRONT SEAT SLIDES			
1	BCC 6412	RH-LH seat	1		
2	BCC 6415	LH-LH seat	1		
	BCC 6414	RH-RH seat	1		
	BCC 6413	LH-RH seat	1		
3	BCC 6305	Handle operating	2		
4	BBC 7759	Screw-torx pan head	8		Slide to body
5	BCC 4522	Screw-torx pan head	8		Slide to seat
6	BCC 4518	Spacer	8		
7	BCC 2522	Washer	16		
8	BCC 8038	Plate-tapping	4		
	BCC 9150	Seat belt assembly-passive-RH	1		
9	BCC 9151	Seat belt assembly-passive-LH	1		
	JLM 1265	Seat belt kit-lap-RH	1		
10	JLM 1266	Seat belt kit-lap-LH	1		
11	BCC 4142	Shield-seat belt reel-RH	2		
12	BCC 4143	Shield-seat belt reel-LH	2		
13	RA 608255 J	Rivet-pop	6		
14	BCC 9037	Cap-bolt	4		

MASTER INDEX

ENGINE	1.C
FLYWHEEL AND CLUTCH	1.G
GEARBOX AND PROPSHAFT	1.H
AXLES, SUSPENSION, DRIVE SHAFTS, WHEELS	1.K
STEERING	1.M
BRAKES AND BRAKE CONTROLS	1.N
FUEL, EXHAUST AND EMISSION SYSTEMS	2.C
COOLING, HEATING AND AIR CONDITIONING	2.G
ELECTRICAL, WASHERS-WIPERS, INSTRUMENTS	2.J
CHASSIS, SUBFRAMES, BODYSHELL, FITTINGS	3.C
FASCIA, TRIM, SEATS AND FIXINGS	3.H
ACCESSORIES AND PAINT	4.C
NUMERICAL INDEX	4.D

GROUP INDEX

BATTERY AND SPARE WHEEL COVERS	N05
BONNET INSULATION-3.6 LITRE	N06
BONNET INSULATION-5.3 MODELS	N08
BONNET INSULATION-5.3 MODELS	N07
BOOT CARPETS - NOT CONVERTIBLE	N03
BOOT CARPETS-CONVERTIBLE	N04
CARPETS-'A' POST FINISHERS-	L12
CONSOLE-AMBLA	H10
CONSOLE-LEATHER	H08
DOOR CASINGS-ARMRESTS AND POCKETS-AMBLA	H16
DOOR CASINGS-ARMRESTS-CONVERTIBLE	I02
HEADLINING	I15
DOOR CASINGS-ARMRESTS-LEATHER	H14
EXTERIOR TRIM COVERS	N15
FACIA PANEL	H02
FRONT SEAT BELTS-CONVERTIBLE	K15
FRONT SEAT SLIDES	K16
MOTOR AND ANCHORAGE ASSY-RH	L02
FRONT SEATS-AMBLA/CLOTH.-3.6 LITRE	K04
FRONT SEATS-LEATHER	J03
FRONT SEATS-LEATHER/CLOTH-5.3 LITRE	K07
FRT SEATS-LEATHER/CLOTH(SPORTSPACK)	K10
FRONT BELT KIT-COUPE	K14
FRONT SEAT SLIDES	K13
FUEL APERATURE SEAL AND HOSES-COUPE	N13
HARDTOP & HOODBAG-CABRIOLET	N17
HOOD LIFT MECHANISM - CONVERTIBLE	O04
HOOD SEALS-CONVERTIBLE	O03
HOOD-CABRIOLET	N16
HOOD-CONVERTIBLE	O02
REAR CUBBY BOX-CONVERTIBLE	L09
REAR LUGGAGE LOCKER-CABRIOLET	L07
REAR SEAT BELTS	L11
REAR SEATS-LEATHER-COUPE	L03
REAR SEATS-LEATHER-COUPE	L04
REAR SEATS-LEATHER/CLOTH-COUPE	L06
TARGA PANELS-STOWAGEBAG-CABRIOLET	N14
UNDER SCUTTLE CASINGS	H05
VENEERED FACIA PANELS-3.6 LITRE	H06
VENEERED FACIA PANELS-5.3 LITRE	H07

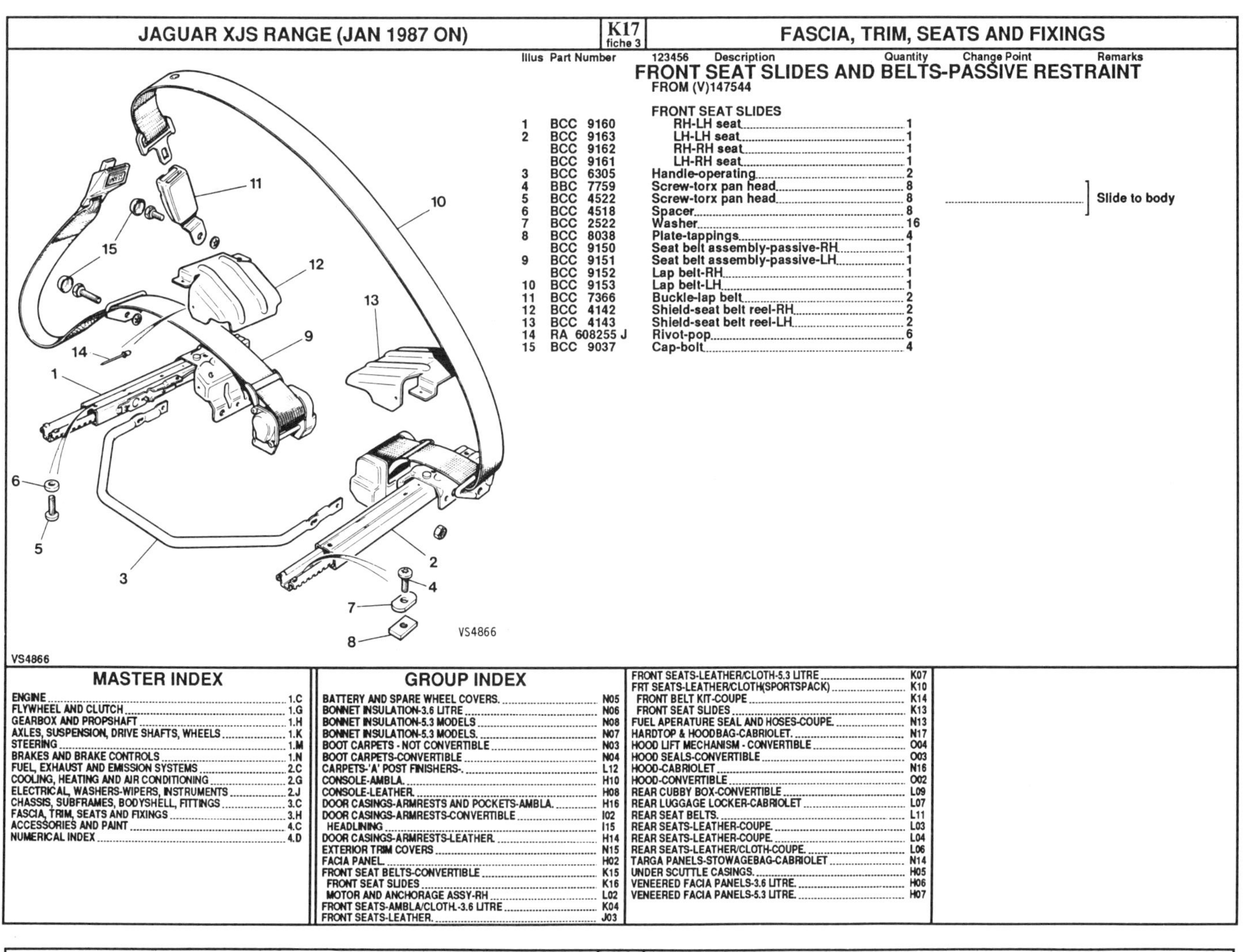

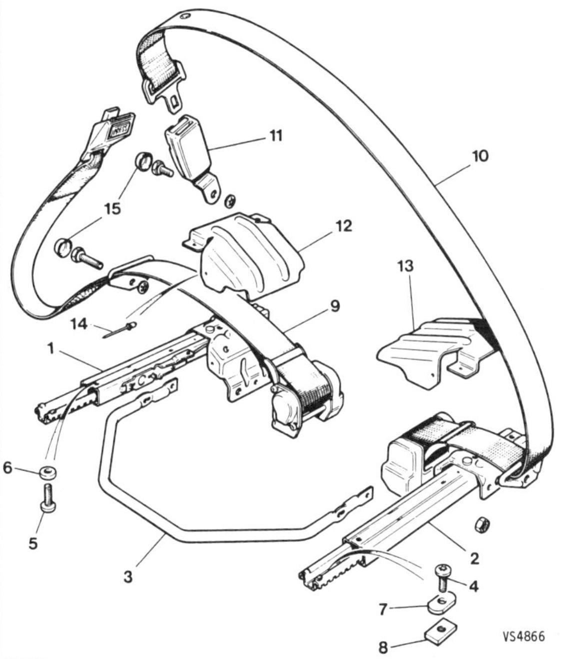

VS4866

FRONT SEAT SLIDES AND BELTS-PASSIVE RESTRAINT

FROM (V)147544

Illus	Part Number	123456 Description	Quantity	Change Point	Remarks
		FRONT SEAT SLIDES			
1	BCC 9160	RH-LH seat	1		
2	BCC 9163	LH-LH seat	1		
	BCC 9162	RH-RH seat	1		
	BCC 9161	LH-RH seat	1		
3	BCC 6305	Handle-operating	2		
4	BBC 7759	Screw-torx pan head	8		Slide to body
5	BCC 4522	Screw-torx pan head	8		Slide to body
6	BCC 4518	Spacer	8		Slide to body
7	BCC 2522	Washer	16		Slide to body
8	BCC 8038	Plate-tappings	4		
	BCC 9150	Seat belt assembly-passive-RH	1		
9	BCC 9151	Seat belt assembly-passive-LH	1		
	BCC 9152	Lap belt-RH	1		
10	BCC 9153	Lap belt-LH	1		
11	BCC 7366	Buckle-lap belt	2		
12	BCC 4142	Shield-seat belt reel-RH	2		
13	BCC 4143	Shield-seat belt reel-LH	2		
14	RA 608255 J	Rivot-pop	6		
15	BCC 9037	Cap-bolt	4		

MASTER INDEX

GROUP INDEX

EMERGENCY TONGUE KIT

Illus	Part Number	123456 Description	Quantity	Change Point	Remarks
	BCC 6475	Tongue kit-emergency	NLA		Use BCC 7959
	BCC 7959	Tongue Kit-emergency	1		

ILLUSTRATION NOT USED

ND

MASTER INDEX
ENGINE 1.C
FLYWHEEL AND CLUTCH 1.G
GEARBOX AND PROPSHAFT 1.H
AXLES, SUSPENSION, DRIVE SHAFTS, WHEELS 1.K
STEERING 1.M
BRAKES AND BRAKE CONTROLS 1.N
FUEL, EXHAUST AND EMISSION SYSTEMS 2.C
COOLING, HEATING AND AIR CONDITIONING 2.G
ELECTRICAL, WASHERS-WIPERS, INSTRUMENTS 2.J
CHASSIS, SUBFRAMES, BODYSHELL, FITTINGS 3.C
FASCIA, TRIM, SEATS AND FIXINGS 3.H
ACCESSORIES AND PAINT 4.C
NUMERICAL INDEX 4.D
GROUP INDEX
BATTERY AND SPARE WHEEL COVERS. N05
BONNET INSULATION-3.6 LITRE N06
BONNET INSULATION-5.3 MODELS N08
BONNET INSULATION-5.3 MODELS. N07
BOOT CARPETS - NOT CONVERTIBLE N03
BOOT CARPETS-CONVERTIBLE N04
CARPETS-'A' POST FINISHERS-. L12
CONSOLE-AMBLA. H10
CONSOLE-LEATHER. H08
DOOR CASINGS-ARMRESTS AND POCKETS-AMBLA. H16
DOOR CASINGS-ARMRESTS-CONVERTIBLE I02
HEADLINING I15
DOOR CASINGS-ARMRESTS-LEATHER. H14
EXTERIOR TRIM COVERS N15
FACIA PANEL. H02
FRONT SEAT BELTS-CONVERTIBLE K15
FRONT SEAT SLIDES K16
MOTOR AND ANCHORAGE ASSY-RH L02
FRONT SEATS-AMBLA/CLOTH.-3.6 LITRE K04
FRONT SEATS-LEATHER. J03
FRONT SEATS-LEATHER/CLOTH-5.3 LITRE K07
FRT SEATS-LEATHER/CLOTH(SPORTSPACK) K10
FRONT BELT KIT-COUPE K14
FRONT SEAT SLIDES K13
FUEL APERATURE SEAL AND HOSES-COUPE. N13
HARDTOP & HOODBAG-CABRIOLET. N17
HOOD LIFT MECHANISM - CONVERTIBLE O04
HOOD SEALS-CONVERTIBLE O03
HOOD-CABRIOLET N16
HOOD-CONVERTIBLE O02
REAR CUBBY BOX-CONVERTIBLE L09
REAR LUGGAGE LOCKER-CABRIOLET L07
REAR SEAT BELTS. L11
REAR SEATS-LEATHER-COUPE. L03
REAR SEATS-LEATHER-COUPE. L04
REAR SEATS-LEATHER/CLOTH-COUPE. L06
TARGA PANELS-STOWAGEBAG-CABRIOLET N14
UNDER SCUTTLE CASINGS. H05
VENEERED FACIA PANELS-3.6 LITRE. H06
VENEERED FACIA PANELS-5.3 LITRE. H07

3L 02

JAGUAR XJS RANGE (JAN 1987 ON) | L02 fiche 3 | FASCIA, TRIM, SEATS AND FIXINGS

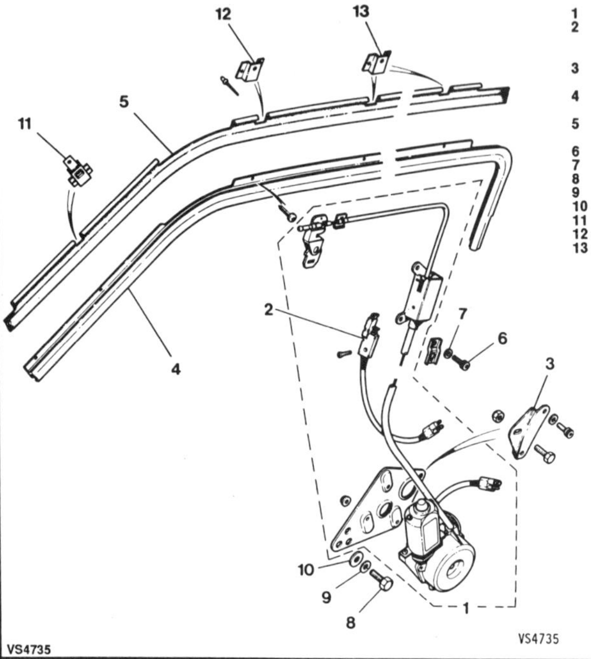

VS4735

MOTOR-ANCHORAGE & RAIL-PASSIVE RESTRAINT

Illus	Part Number	123456	Description	Quantity	Change Point	Remarks
1	BBC 2186		MOTOR AND ANCHORAGE ASSY-RH	1		
2	DAC 4592		Switch-docking & proximity-RH	1		
	BBC 2187		MOTOR AND ANCHORAGE ASSY-LH	1		
	DAC 4593		Switch-docking & proxmity-LH	1		
3	BBC 7158		Bracket-motor mounting-RH	1		
	BBC 7159		Bracket-motor mounting-LH	1		
4	BAC 4722		Rail-RH	1		
	BAC 4723		Rail-LH	1		
5	BCC 6422		Seal-rail-RH	1		
	BCC 6423		Seal-rail-LH	1		
6	BD 48947 2		Screw torx head	8		
7	FW 105 E		Washer-plain	8		
8	SH 607061 J		Screw-set	2		
9	WC 600071 J		Washer-spring	2		
10	WA 600071 J		Washer-plain	2		
11	BCC 7070		Bracket-'A' post	4		
12	BBC 2035		Bracket-cantrail-front	2		
13	BBC 2039		Bracket-cantrail	4		

3L 03

JAGUAR XJS RANGE (JAN 1987 ON) | L03 fiche 3 | FASCIA, TRIM, SEATS AND FIXINGS

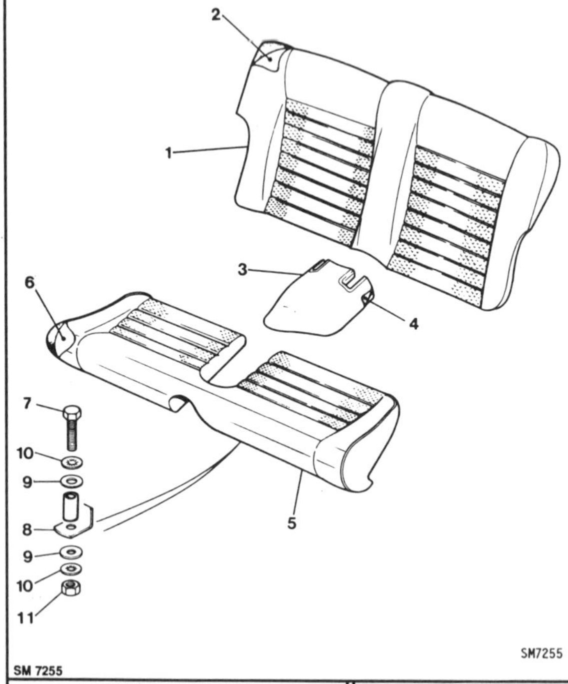

SM7255

REAR SEATS-LEATHER-COUPE.

Illus	Part Number	123456	Description	Quantity	Change Point	Remarks
			SQUAB-REAR SEAT			
1	BCC 7000 AE		Biscuit	1		
1	BCC 7000 LZ		Warm Charcoal	NLA		Use BCC9996LZ
1	BCC 7000 AR		Buckskin	NLA		Use BCC9996AR
1	BCC 7000 XE		Doeskin	NLA		Use BCC9996XE
1	BCC 7000 JF		Isis Blue	NLA		Use BCC9996JF
1	BCC 7000 CM		Mulberry	NLA		Use BCC9996CM
1	BCC 7000 LY		Savile Grey	NLA		Use BCC9996LY
1	BCC 7000 AW		Barley	NLA		Use BCC9996AW
1	BCC 7000 ZM		Magnolia	NLA		Use BCC9996ZM
			SQUAB-REAR SEAT			
1	BCC 9996 LZ		Warm charcoal	1		
1	BCC 9996 AR		Buckskin	1		
1	BCC 9996 XE		Doeskin	1		
1	BCC 9996 JF		Isis blue	1		
1	BCC 9996 CM		Mulberry	1		
1	BCC 9996 LY		Savile grey	1		
1	BCC 9996 AW		Barley	1		
1	BCC 9996 ZM		Magnolia	1		
			COVER-REAR-SQUAB			
2	BCC 7001 AE		Biscuit	1		
2	BCC 7001 LZ		Warm charcoal	1		
2	BCC 7001 AR		Buckskin	1		
2	BCC 7001 XE		Doeskin	1		
2	BCC 7001 JF		Isis blue	1		
2	BCC 7001 CM		Mulberry	1		
2	BCC 7001 LY		Savile Grey	1		
2	BCC 7001 AW		Barley	1		
2	BCC 7001 ZM		Magnolia	1		

M40JP24/6/88

MASTER INDEX

ENGINE ... 1.C
FLYWHEEL AND CLUTCH ... 1.G
GEARBOX AND PROPSHAFT ... 1.H
AXLES, SUSPENSION, DRIVE SHAFTS, WHEELS ... 1.K
STEERING ... 1.M
BRAKES AND BRAKE CONTROLS ... 1.N
FUEL, EXHAUST AND EMISSION SYSTEMS ... 2.C
COOLING, HEATING AND AIR CONDITIONING ... 2.G
ELECTRICAL, WASHERS-WIPERS, INSTRUMENTS ... 2.J
CHASSIS, SUBFRAMES, BODYSHELL, FITTINGS ... 3.C
FASCIA, TRIM, SEATS AND FIXINGS ... 3.H
ACCESSORIES AND PAINT ... 4.C
NUMERICAL INDEX ... 4.D

GROUP INDEX

BATTERY AND SPARE WHEEL COVERS ... N05
BONNET INSULATION-3.6 LITRE ... N06
BONNET INSULATION-5.3 MODELS ... N08
BONNET INSULATION-5.3 MODELS ... N07
BOOT CARPETS - NOT CONVERTIBLE ... N03
BOOT CARPETS-CONVERTIBLE ... N04
CARPETS-'A' POST FINISHERS- ... L12
CONSOLE-AMBLA ... H10
CONSOLE-LEATHER ... H08
DOOR CASINGS-ARMRESTS AND POCKETS-AMBLA ... H16
DOOR CASINGS-ARMRESTS-CONVERTIBLE ... I02
HEADLINING ... I15
DOOR CASINGS-ARMRESTS-LEATHER ... H14
EXTERIOR TRIM COVERS ... N15
FACIA PANEL ... H02
FRONT SEAT BELTS-CONVERTIBLE ... K15
FRONT SEAT SLIDES ... K16
MOTOR AND ANCHORAGE ASSY-RH ... L02
FRONT SEATS-AMBLA/CLOTH.-3.6 LITRE ... K04
FRONT SEATS-LEATHER ... J03
FRONT SEATS-LEATHER/CLOTH-5.3 LITRE ... K07
FRT SEATS-LEATHER/CLOTH(SPORTSPACK) ... K10
FRONT BELT KIT-COUPE ... K14
FRONT SEAT SLIDES ... K13
FUEL APERATURE SEAL AND HOSES-COUPE ... N13
HARDTOP & HOODBAG-CABRIOLET ... N17
HOOD LIFT MECHANISM - CONVERTIBLE ... O04
HOOD SEALS-CONVERTIBLE ... O03
HOOD-CABRIOLET ... N16
HOOD-CONVERTIBLE ... O02
REAR CUBBY BOX-CONVERTIBLE ... L09
REAR LUGGAGE LOCKER-CABRIOLET ... L07
REAR SEAT BELTS ... L11
REAR SEATS-LEATHER-COUPE ... L03
REAR SEATS-LEATHER-COUPE ... L04
REAR SEATS-LEATHER/CLOTH-COUPE ... L06
TARGA PANELS-STOWAGEBAG-CABRIOLET ... N14
UNDER SCUTTLE CASINGS ... H05
VENEERED FACIA PANELS-3.6 LITRE ... H06
VENEERED FACIA PANELS-5.3 LITRE ... H07

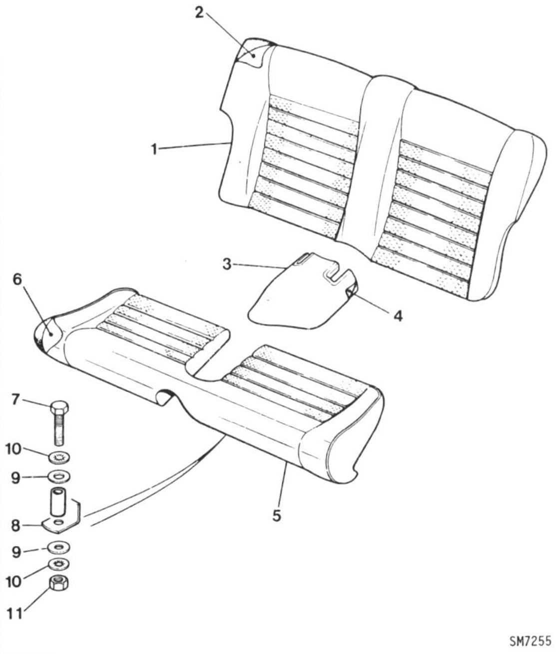

SM 7255

REAR SEATS-LEATHER-COUPE.

Illus	Part Number	123456 Description	Quantity	Change Point	Remarks
		PANEL-TUNNEL			
3	BCC 2474 AE	Biscuit	1		
3	BCC 2474 LZ	Warm Charcoal	1		
3	BCC 2474 AR	Buckskin	1		
3	BCC 2474 XE	Doeskin	1		
3	BCC 2474 JF	Isis Blue	1		
3	BCC 2474 CM	Mulberry	1		
3	BCC 2474 LY	Savile Grey	1		
3	BCC 2474 AW	Barley	1		
3	BCC 2474 ZM	Magnolia	1		
		COVER-TUNNEL PANEL			
4	BCC 2472 AE	Biscuit	1		
4	BCC 2472 LZ	Warm Charcoal	1		
4	BCC 2472 AR	Buckskin	1		
4	BCC 2472 XE	Doeskin	1		
4	BCC 2472 JF	Isis Blue	1		
4	BCC 2472 CM	Mulberry	1		
4	BCC 2472 LY	Savile Grey	1		
4	BCC 2472 AW	Barley	1		
4	BCC 2472 ZM	Magnolia	1		M40.02JP24/6/88

MASTER INDEX

GROUP INDEX

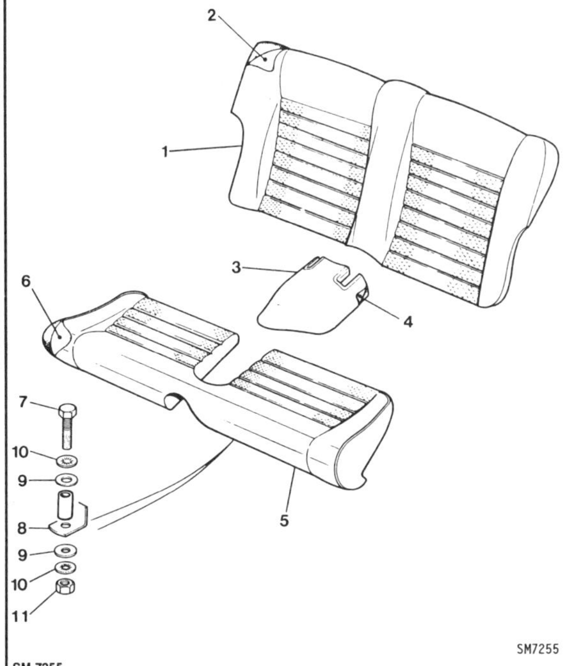

SM 7255

REAR SEATS-LEATHER-COUPE-continued.

Illus	Part Number	123456 Description	Quantity	Change Point	Remarks
		CUSHION-REAR SEAT			
5	BAC 4068 AE	Biscuit	1		
5	BAC 4068 LZ	Warm Charcoal	1		
5	BAC 4068 AR	Buckskin	1		
5	BAC 4068 XE	Doeskin	1		
5	BAC 4068 JF	Isis Blue	1		
5	BAC 4068 CM	Mulberry	1		
5	BAC 4068 LY	Savile Grey	1		
5	BAC 4068 AW	Barley	1		
5	BAC 4068 ZM	Magnolia	1		
		COVER-REAR CUSHION			
6	BAC 4069 AE	Biscuit	1		
6	BAC 4069 LZ	Warm Charcoal	1		
6	BAC 4069 AR	Buckskin	1		
6	BAC 4069 XE	Doeskin	1		
6	BAC 4069 JF	Isis Blue	1		
6	BAC 4069 CM	Mulberry	1		
6	BAC 4069 LY	Savile Grey	1		
6	BAC 4069 AW	Barley	1		
6	BAC 4069 ZM	Magnolia	1		
7	UF 119 6R	Screw-set	2		
8	BD 48903	Clip-fixing	2		
9	WC 702101 J	Washer-plain	4		
10	WF 702101 J	Washer-shakeproof	4		
11	UFN 119 L	Nut-plain	2		

MASTER INDEX
ENGINE 1.C
FLYWHEEL AND CLUTCH 1.G
GEARBOX AND PROPSHAFT 1.H
AXLES, SUSPENSION, DRIVE SHAFTS, WHEELS 1.K
STEERING 1.M
BRAKES AND BRAKE CONTROLS 1.N
FUEL, EXHAUST AND EMISSION SYSTEMS 2.C
COOLING, HEATING AND AIR CONDITIONING 2.G
ELECTRICAL, WASHERS-WIPERS, INSTRUMENTS 2.J
CHASSIS, SUBFRAMES, BODYSHELL, FITTINGS 3.C
FASCIA, TRIM, SEATS AND FIXINGS 3.H
ACCESSORIES AND PAINT 4.C
NUMERICAL INDEX 4.D
GROUP INDEX
BATTERY AND SPARE WHEEL COVERS. N05
BONNET INSULATION-3.6 LITRE N06
BONNET INSULATION-5.3 MODELS N08
BONNET INSULATION-5.3 MODELS. N07
BOOT CARPETS - NOT CONVERTIBLE N03
BOOT CARPETS-CONVERTIBLE N04
CARPETS-'A' POST FINISHERS-. L12
CONSOLE-AMBLA. H10
CONSOLE-LEATHER. H08
DOOR CASINGS-ARMRESTS AND POCKETS-AMBLA. H16
DOOR CASINGS-ARMRESTS-CONVERTIBLE I02
HEADLINING I15
DOOR CASINGS-ARMRESTS-LEATHER. H14
EXTERIOR TRIM COVERS N15
FACIA PANEL. H02
FRONT SEAT BELTS-CONVERTIBLE K15
FRONT SEAT SLIDES K16
MOTOR AND ANCHORAGE ASSY-RH L02
FRONT SEATS-AMBLA/CLOTH.-3.6 LITRE K04
FRONT SEATS-LEATHER. J03
FRONT SEATS-LEATHER/CLOTH-5.3 LITRE K07
FRT SEATS-LEATHER/CLOTH(SPORTSPACK) K10
FRONT BELT KIT-COUPE K14
FRONT SEAT SLIDES K13
FUEL APERATURE SEAL AND HOSES-COUPE. N13
HARDTOP & HOODBAG-CABRIOLET. N17
HOOD LIFT MECHANISM - CONVERTIBLE O04
HOOD SEALS-CONVERTIBLE O03
HOOD-CABRIOLET N16
HOOD-CONVERTIBLE O02
REAR CUBBY BOX-CONVERTIBLE L09
REAR LUGGAGE LOCKER-CABRIOLET L07
REAR SEAT BELTS. L11
REAR SEATS-LEATHER-COUPE. L03
REAR SEATS-LEATHER-COUPE. L04
REAR SEATS-LEATHER/CLOTH-COUPE. L06
TARGA PANELS-STOWAGEBAG-CABRIOLET N14
UNDER SCUTTLE CASINGS. H05
VENEERED FACIA PANELS-3.6 LITRE. H06
VENEERED FACIA PANELS-5.3 LITRE. H07

3L 06

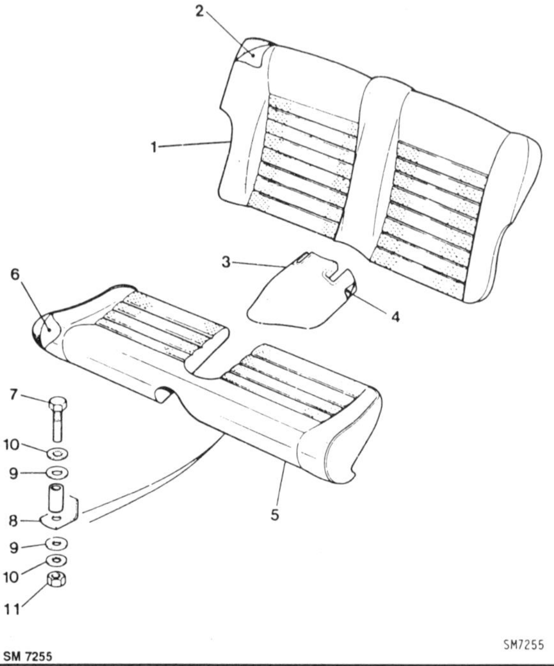

SM 7255

REAR SEATS-LEATHER/CLOTH-COUPE.

Illus	Part Number	123456 Description	Quantity	Change Point	Remarks
		SQUAB-REAR SEAT			
1	BCC 7003 YF	Cheviot tweed	NLA		Use BCC9997YF
1	BCC 7003 AY	Chiltern tweed	NLA		Use BCC9997AY
1	BCC 7003 AX	Cotswold tweed	NLA		Use BCC9997AX
1	BCC 7003 YE	Pennine tweed	NLA		Use BCC9997YE
		SQUAB-REAR SEAT			
1	BCC 9997 YF	Cheviot tweed	1		
1	BCC 9997 AY	Chiltern tweed	1		
1	BCC 9997 AX	Cotswold tweed	1		
1	BCC 9997 YE	Pennine tweed	1		
		COVER-REAR-SQUAB			
2	BCC 7004 YF	Cheviot tweed	1		
2	BCC 7004 AY	Chiltern tweed	1		
2	BCC 7004 AX	Cotswold tweed	1		
2	BCC 7004 YE	Pennine tweed	1		
		PANEL-TUNNEL			
3	BCC 2474 AW	Barley	1		
3	BCC 2474 LZ	Warm Charcoal	1		
3	BCC 2474 XE	Doeskin	1		
3	BCC 2474 LY	Savile Grey	1		
		COVER-TUNNEL PANEL			
4	BCC 2472 AW	Barley	1		
4	BCC 2472 LZ	Warm Charcoal	1		
4	BCC 2472 XE	Doeskin	1		
4	BCC 2472 LY	Savile Grey	1		
		CUSHION-REAR SEAT			
5	BBC 8332 YF	Cheviot tweed	1		
5	BBC 8332 AY	Chiltern tweed	1		
5	BBC 8332 AX	Cotswold tweed	1		
5	BBC 578332 YE	Pennine tweed	1		
		COVER-REAR CUSHION			
6	BBC 8331 YF	Cheviot tweed	1		
6	BBC 8331 AY	Chiltern tweed	1		
6	BBC 8331 AX	Cotswold tweed	1		
6	BBC 878331 YE	Pennine tweed	1		
7	UFS 119 6R	Screw-set	2		
8	BD 48903	Clip-fixing	2		
9	WC 702101 J	Washer-plain	4		
10	WF 702101 J	Washer-shakeproof	4		
11	UFN 119 L	Nut-plain	2		

M42.02JP24/6/88

MASTER INDEX

GROUP INDEX

3L 07

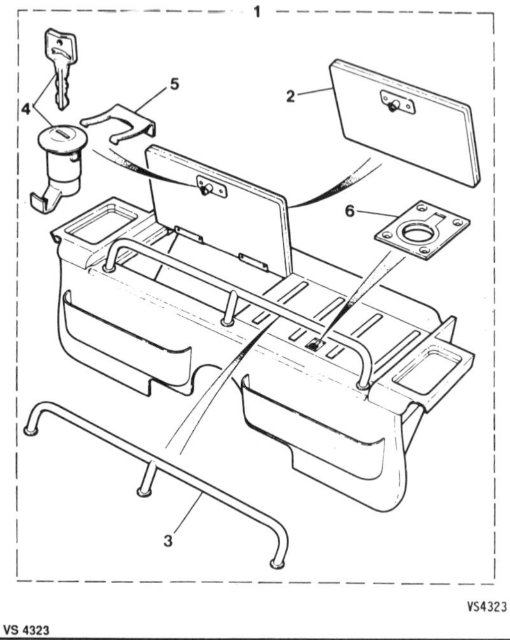

VS 4323

REAR LUGGAGE LOCKER-CABRIOLET

Illus	Part Number	123456 Description	Quantity	Change Point	Remarks
		LOCKER-LUGGAGE			
1	BCC 6093 AE	Biscuit	1		
1	BCC 6093 LZ	Warm Charcoal	1		
1	BCC 6093 AR	Buckskin	1		
1	BCC 6093 XE	Doeskin	1		
1	BCC 6093 JF	Isis Blue	1		
1	BCC 6093 CM	Mulberry	1		
1	BCC 6093 LY	Savile Grey	1		
1	BCC 6093 AW	Barley	1		
1	BCC 6093 ZM	Magnolia	1		
		LID-LOCKER			
2	BBC 9980 AP	Mink	2		
2	BBC 9980 XF	Rattan	2		
2	BBC 9980 LE	Slate	2		
2	BBC 9980 XA	Smoke Grey	2		
2	BBC 9980 CN	Wine Red	2		
3	BBC 1856	Rail-luggage	1		
4	BBC 9779	Assembly-lock	2		
5	BBC 9773	Clip-retaining	2		
6	BBC 9774	Finger pull	2		

MASTER INDEX

ENGINE 1.C
FLYWHEEL AND CLUTCH 1.G
GEARBOX AND PROPSHAFT 1.H
AXLES, SUSPENSION, DRIVE SHAFTS, WHEELS 1.K
STEERING 1.M
BRAKES AND BRAKE CONTROLS 1.N
FUEL, EXHAUST AND EMISSION SYSTEMS 2.C
COOLING, HEATING AND AIR CONDITIONING 2.G
ELECTRICAL, WASHERS-WIPERS, INSTRUMENTS 2.J
CHASSIS, SUBFRAMES, BODYSHELL, FITTINGS 3.C
FASCIA, TRIM, SEATS AND FIXINGS 3.H
ACCESSORIES AND PAINT 4.C
NUMERICAL INDEX 4.D

GROUP INDEX

BATTERY AND SPARE WHEEL COVERS N05
BONNET INSULATION-3.6 LITRE N06
BONNET INSULATION-5.3 MODELS N08
BONNET INSULATION-5.3 MODELS N07
BOOT CARPETS - NOT CONVERTIBLE N03
BOOT CARPETS-CONVERTIBLE N04
CARPETS-'A' POST FINISHERS- L12
CONSOLE-AMBLA H10
CONSOLE-LEATHER H08
DOOR CASINGS-ARMRESTS AND POCKETS-AMBLA H16
DOOR CASINGS-ARMRESTS-CONVERTIBLE I02
HEADLINING I15
DOOR CASINGS-ARMRESTS-LEATHER H14
EXTERIOR TRIM COVERS N15
FACIA PANEL H02
FRONT SEAT BELTS-CONVERTIBLE K15
FRONT SEAT SLIDES K16
MOTOR AND ANCHORAGE ASSY-RH L02
FRONT SEATS-AMBLA/CLOTH.-3.6 LITRE K04
FRONT SEATS-LEATHER J03
FRONT SEATS-LEATHER/CLOTH-5.3 LITRE K07
FRT SEATS-LEATHER/CLOTH(SPORTSPACK) K10
FRONT BELT KIT-COUPE K14
FRONT SEAT SLIDES K13
FUEL APERATURE SEAL AND HOSES-COUPE N13
HARDTOP & HOODBAG-CABRIOLET N17
HOOD LIFT MECHANISM - CONVERTIBLE O04
HOOD SEALS-CONVERTIBLE O03
HOOD-CABRIOLET N16
HOOD-CONVERTIBLE O02
REAR CUBBY BOX-CONVERTIBLE L09
REAR LUGGAGE LOCKER-CABRIOLET L07
REAR SEAT BELTS L11
REAR SEATS-LEATHER-COUPE L03
REAR SEATS-LEATHER-COUPE L04
REAR SEATS-LEATHER/CLOTH-COUPE L06
TARGA PANELS-STOWAGEBAG-CABRIOLET N14
UNDER SCUTTLE CASINGS H05
VENEERED FACIA PANELS-3.6 LITRE H06
VENEERED FACIA PANELS-5.3 LITRE H07

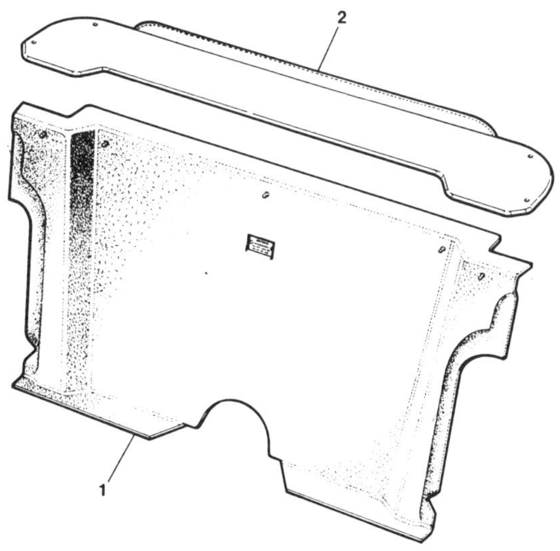

VS 4322

SQUAB PANEL-PARCEL SHELF-CABRIOLET

Illus	Part Number	123456 Description	Quantity	Change Point	Remarks
		PANEL-REAR SQUAB			
1	BBC 2260 AP	Mink	1		
1	BBC 2260 XF	Rattan	1		
1	BBC 2260 LE	Slate	1		
1	BBC 2260 XA	Smoke Grey	1		
1	BBC 2260 CN	Wine Red	1		
		SHELF-REAR PARCEL			
2	BBC 2276 AE	Biscuit	1		
2	BBC 2276 LZ	Warm Charcoal	1		
2	BBC 2276 AR	Buckskin	1		
2	BBC 2276 XE	Doeskin	1		
2	BBC 2276 JF	Isis Blue	1		
2	BBC 2276 CM	Mulberry	1		
2	BBC 2276 LY	Savile Grey	1		
2	BBC 2276 AW	Barley	1		
2	BBC 2276 ZM	Magnolia	1		

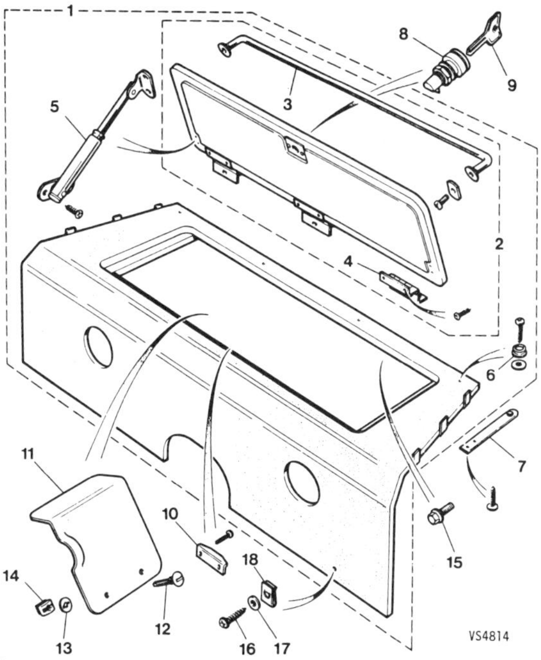

VS4814

REAR CUBBY BOX-CONVERTIBLE

Illus	Part Number	123456 Description	Quantity	Change Point	Remarks
		BOX-CUBBY			
1	JLM 1555 AP	Mink	1		
1	JLM 1555 XF	Rattan	1		
1	JLM 1555 LE	Slate	1		
1	JLM 1555 XA	Smoke Grey	1		
1	JLM 1555 CN	Wine Red	1		
		LID CUBBY BOX			
2	BDC 8523 AP	Mink	1		
2	BCC 8523 XF	Rattan	1		
2	BCC 8523 LE	Slate	1		
2	BCC 8523 XA	Smoke Grey	1		
2	BCC 8523 CN	Wine Red	1		
3	BCC 2761	RAIL-LUGGAGE	1		
4	BCC 2798	Hinge	3		
5	BCC 5520	Strut	2		
6	BBC 1342	Fastener-male	2		
		STRAP-HOOD COVER FIXING			
7	BDC 6563 AW	Barley	2		
7	BDC 6563 AR	Buckskin	2		
7	BDC 6563 XE	Doeskin	2		
7	BDC 6563 JF	Isis Blue	2		
7	BDC 6563 ZM	Magnolia	2		
7	BDC 6563 CM	Mulberry	2		
7	BDC 6563 LY	Savile Grey	2		
7	BDC 6563 LZ	Warm Charcoal	2		
8	BCC 7892	Lock-cubby box	1		Complete with key
		Key-blank	1		See locksets Group N
10	BDC 5076	Striker-cubby lock	1		

M44.04JP13/10/88

MASTER INDEX
ENGINE 1.C
FLYWHEEL AND CLUTCH 1.G
GEARBOX AND PROPSHAFT 1.H
AXLES, SUSPENSION, DRIVE SHAFTS, WHEELS 1.K
STEERING 1.M
BRAKES AND BRAKE CONTROLS 1.N
FUEL, EXHAUST AND EMISSION SYSTEMS 2.C
COOLING, HEATING AND AIR CONDITIONING 2.G
ELECTRICAL, WASHERS-WIPERS, INSTRUMENTS 2.J
CHASSIS, SUBFRAMES, BODYSHELL, FITTINGS 3.C
FASCIA, TRIM, SEATS AND FIXINGS 3.H
ACCESSORIES AND PAINT 4.C
NUMERICAL INDEX 4.D

GROUP INDEX
BATTERY AND SPARE WHEEL COVERS N05
BONNET INSULATION-3.6 LITRE N06
BONNET INSULATION-5.3 MODELS N08
BONNET INSULATION-5.3 MODELS N07
BOOT CARPETS - NOT CONVERTIBLE N03
BOOT CARPETS-CONVERTIBLE N04
CARPETS-'A' POST FINISHERS- L12
CONSOLE-AMBLA H10
CONSOLE-LEATHER H08
DOOR CASINGS-ARMRESTS AND POCKETS-AMBLA H16
DOOR CASINGS-ARMRESTS-CONVERTIBLE I02
HEADLINING I15
DOOR CASINGS-ARMRESTS-LEATHER H14
EXTERIOR TRIM COVERS N15
FACIA PANEL H02
FRONT SEAT BELTS-CONVERTIBLE K15
FRONT SEAT SLIDES K16
MOTOR AND ANCHORAGE ASSY-RH L02
FRONT SEATS-AMBLA/CLOTH.-3.6 LITRE K04
FRONT SEATS-LEATHER J03
FRONT SEATS-LEATHER/CLOTH-5.3 LITRE K07
FRT SEATS-LEATHER/CLOTH(SPORTSPACK) K10
FRONT BELT KIT-COUPE K14
FRONT SEAT SLIDES K13
FUEL APERATURE SEAL AND HOSES-COUPE N13
HARDTOP & HOODBAG-CABRIOLET N17
HOOD LIFT MECHANISM - CONVERTIBLE O04
HOOD SEALS-CONVERTIBLE O03
HOOD-CABRIOLET N16
HOOD-CONVERTIBLE O02
REAR CUBBY BOX-CONVERTIBLE L09
REAR LUGGAGE LOCKER-CABRIOLET L07
REAR SEAT BELTS L11
REAR SEATS-LEATHER-COUPE L03
REAR SEATS-LEATHER-COUPE L04
REAR SEATS-LEATHER/CLOTH-COUPE L06
TARGA PANELS-STOWAGEBAG-CABRIOLET N14
UNDER SCUTTLE CASINGS H05
VENEERED FACIA PANELS-3.6 LITRE H06
VENEERED FACIA PANELS-5.3 LITRE H07

3L 10

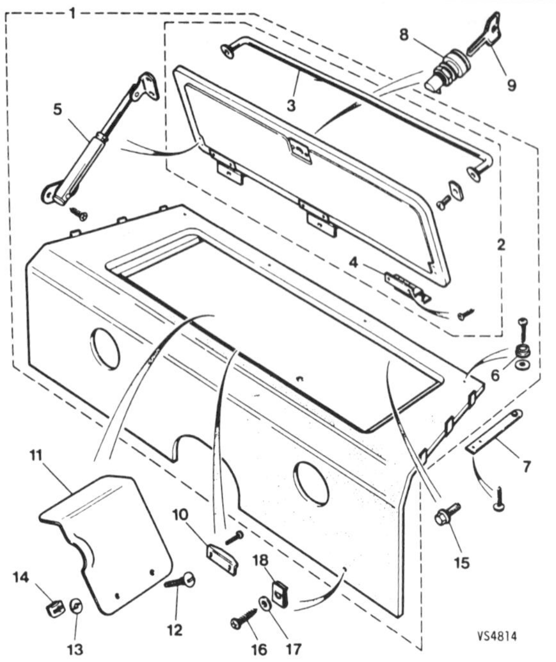

VS4814

REAR CUBBY BOX-CONVERTIBLE-continued

Illus	Part Number	123456 Description	Quantity	Change Point	Remarks
		ACCESS COVER-HOOD PUMP			
11	BCC 7786 AW	Barley	1		
11	BCC 7786 AR	Buckskin	1		
11	BCC 7786 XE	Doeskin	1		
11	BCC 7786 JF	Isis Blue	1		
11	BCC 7786 ZM	Magnolia	1		
11	BCC 7786 CM	Mulberry	1		
11	BCC 7786 LY	Savile Grey	1		
11	BCC 7786 LZ	Warm Charcoal	1		
		STUD-TURN			
12	AGU 2229 AW	Barley	2		
12	AGU 2229 AR	Buckskin	2		
12	AGU 2229 XE	Doeskin	2		
12	AGU 2229 JF	Isis Blue	2		
12	AGU 2229 ZM	Magnolia	2		
12	AGU 2229 CM	Mulberry	2		
12	AGU 2229 LY	Savile Grey	2		
12	AGU 2229 LZ	Warm Charcoal	2		
13	AGU 2390	Washer	2		
14	AGU 1863	Camm	2		
15	FS 106121 J	Bolt-flanged	3		Cubby to bulkhead
16	AB 608051 J	Screw-self tap	3		
17	WC 105001 J	Washer	3		
18	AK 608081 J	Nut-spire	2		
	AK 608051 J	Nut-spire	1		Cubby to tunnel

3L 11

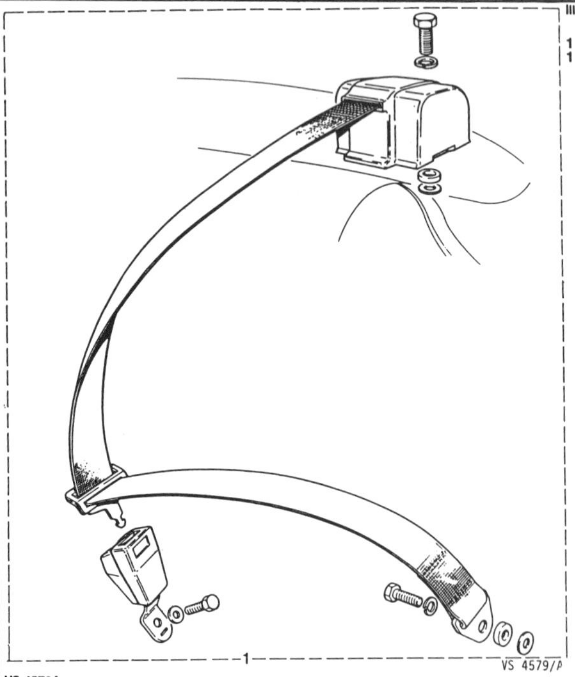

VS 4579A

REAR SEAT BELTS.

Illus	Part Number	123456 Description	Quantity	Change Point	Remarks
1	JLM 830	Belt kit-rear seat	2		Not Aus.
1	JLM 831	Belt kit-rear seat	2		Aus. only

MASTER INDEX

ENGINE 1.C
FLYWHEEL AND CLUTCH 1.G
GEARBOX AND PROPSHAFT 1.H
AXLES, SUSPENSION, DRIVE SHAFTS, WHEELS 1.K
STEERING 1.M
BRAKES AND BRAKE CONTROLS 1.N
FUEL, EXHAUST AND EMISSION SYSTEMS 2.C
COOLING, HEATING AND AIR CONDITIONING 2.G
ELECTRICAL, WASHERS-WIPERS, INSTRUMENTS 2.J
CHASSIS, SUBFRAMES, BODYSHELL, FITTINGS 3.C
FASCIA, TRIM, SEATS AND FIXINGS 3.H
ACCESSORIES AND PAINT 4.C
NUMERICAL INDEX 4.D

GROUP INDEX

BATTERY AND SPARE WHEEL COVERS. N05
BONNET INSULATION-3.6 LITRE N06
BONNET INSULATION-5.3 MODELS N08
BONNET INSULATION-5.3 MODELS. N07
BOOT CARPETS - NOT CONVERTIBLE N03
BOOT CARPETS-CONVERTIBLE N04
CARPETS-'A' POST FINISHERS-. L12
CONSOLE-AMBLA. H10
CONSOLE-LEATHER. H08
DOOR CASINGS-ARMRESTS AND POCKETS-AMBLA. H16
DOOR CASINGS-ARMRESTS-CONVERTIBLE I02
HEADLINING I15
DOOR CASINGS-ARMRESTS-LEATHER. H14
EXTERIOR TRIM COVERS N15
FACIA PANEL. H02
FRONT SEAT BELTS-CONVERTIBLE K15
FRONT SEAT SLIDES K16
MOTOR AND ANCHORAGE ASSY-RH L02
FRONT SEATS-AMBLA/CLOTH.-3.6 LITRE K04
FRONT SEATS-LEATHER. J03
FRONT SEATS-LEATHER/CLOTH-5.3 LITRE K07
FRT SEATS-LEATHER/CLOTH(SPORTSPACK) K10
FRONT BELT KIT-COUPE K14
FRONT SEAT SLIDES K13
FUEL APERATURE SEAL AND HOSES-COUPE. N13
HARDTOP & HOODBAG-CABRIOLET. N17
HOOD LIFT MECHANISM - CONVERTIBLE O04
HOOD SEALS-CONVERTIBLE O03
HOOD-CABRIOLET N16
HOOD-CONVERTIBLE O02
REAR CUBBY BOX-CONVERTIBLE L09
REAR LUGGAGE LOCKER-CABRIOLET L07
REAR SEAT BELTS. L11
REAR SEATS-LEATHER-COUPE. L03
REAR SEATS-LEATHER-COUPE. L04
REAR SEATS-LEATHER/CLOTH-COUPE. L06
TARGA PANELS-STOWAGEBAG-CABRIOLET N14
UNDER SCUTTLE CASINGS. H05
VENEERED FACIA PANELS-3.6 LITRE. H06
VENEERED FACIA PANELS-5.3 LITRE. H07

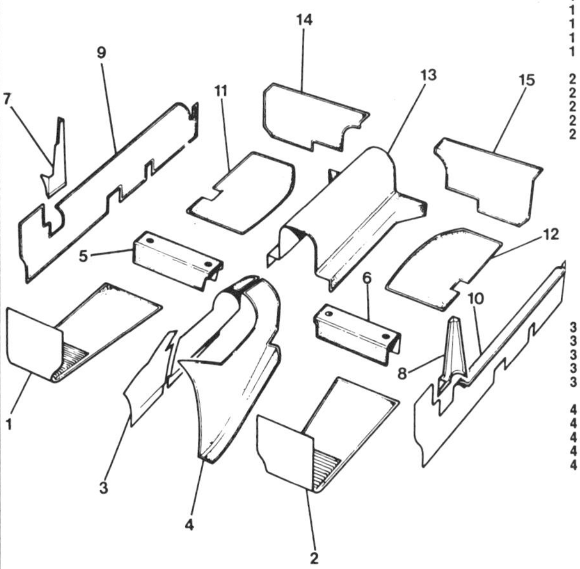

VS 3700

CARPETS-'A' POST FINISHERS-.

3.6 LITRE COUPE

Illus	Part Number	123456	Description	Quantity	Change Point	Remarks
			CARPET RH-RHD-FRONT			
1	BBC 3122 AP		Mink	1		
1	BBC 3122 XF		Rattan	1		
1	BBC 3122 LE		Slate	1		
1	BBC 3122 XA		Smoke Grey	1		
1	BBC 3122 CN		Wine Red	1		
			CARPET LH-RHD-FRONT			
2	BCC 2067 AP		Mink	1		
2	BCC 2067 XF		Rattan	1		
2	BCC 2067 LE		Slate	1		
2	BCC 2067 XA		Smoke Grey	1		
2	BCC 2067 CN		Wine Red	1		
			CARPET RH-LHD-FRONT			
	BCC 2066 AP		Mink	1		
	BCC 2066 XF		Rattan	1		
	BCC 2066 LE		Slate	1		
	BCC 2066 XA		Smoke Grey	1		
	BCC 2066 CN		Wine Red	1		
			CARPET LH-LHD-FRONT			
	BBC 3123 AP		Mink	1		
	BBC 3123 XF		Rattan	1		
	BBC 3123 LE		Slate	1		
	BBC 3123 XA		Smoke Grey	1		
	BBC 3123 CN		Wine Red	1		
			CARPET-RH-GEARBOX			
3	BAC 5678 AP		Mink	1		
3	BAC 5678 XF		Rattan	1		
3	BAC 5678 LE		Slate	1		
3	BAC 5678 XA		Smoke Grey	1		
3	BAC 5678 CN		Wine Red	1		
			CARPET LH-RHD-GEARBOX			
4	BAC 2564 AP		Mink	1		
4	BAC 2564 XF		Rattan	1		
4	BAC 2564 LE		Slate	1		
4	BAC 2564 XA		Smoke Grey	1		
4	BAC 2564 CN		Wine Red	1		
			CARPET LH-LHD-GEARBOX			
	BAC 2565 AP		Mink	1		
	BAC 2565 XF		Rattan	1		
	BAC 2565 LE		Slate	1		
	BAC 2565 XA		Smoke Grey	1		
	BAC 2565 CN		Wine Red	1		

MASTER INDEX

GROUP INDEX

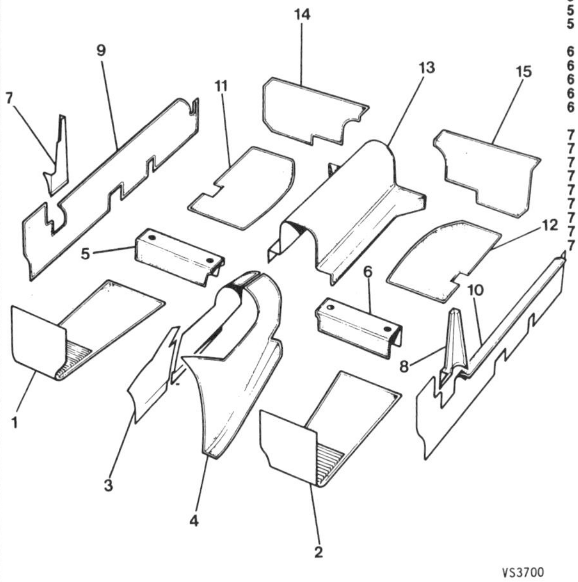

VS 3700

CARPETS-'A'POST FINISHERS-3.6 LITRE COUPE-continued

Illus	Part Number	123456	Description	Quantity	Change Point	Remarks
			CARPET RH-CROSSMEMBER			
5	BCC 2314 AP		Mink	1		
5	BCC 2314 XF		Rattan	1		
5	BCC 2314 LE		Slate	1		
5	BCC 2314 XA		Smoke Grey	1		
5	BCC 2314 CN		Wine Red	1		
			CARPET LH-CROSSMEMBER			
6	BCC 2315 AP		Mink	1		
6	BCC 2315 XF		Rattan	1		
6	BCC 2315 LE		Slate	1		
6	BCC 2315 XA		Smoke Grey	1		
6	BCC 2315 CN		Wine Red	1		
			FINISHER RH-'A' POST			
7	BBC 7998 AE		Biscuit	1		
7	BBC 7998 LZ		Warm Charcoal	1		
7	BBC 7998 AR		Buckskin	1		
7	BBC 7998 XE		Doeskin	1		
7	BBC 7998 JF		Isis Blue	1		
7	BBC 7998 CM		Mulberry	1		
7	BBC 7998 LY		Savile Grey	1		
7	BBC 7998 AW		Barley	1		
7	BBC 7998 ZM		Magnolia	1		

MASTER INDEX
ENGINE 1.C
FLYWHEEL AND CLUTCH 1.G
GEARBOX AND PROPSHAFT 1.H
AXLES, SUSPENSION, DRIVE SHAFTS, WHEELS 1.K
STEERING 1.M
BRAKES AND BRAKE CONTROLS 1.N
FUEL, EXHAUST AND EMISSION SYSTEMS 2.C
COOLING, HEATING AND AIR CONDITIONING 2.G
ELECTRICAL, WASHERS-WIPERS, INSTRUMENTS 2.J
CHASSIS, SUBFRAMES, BODYSHELL, FITTINGS 3.C
FASCIA, TRIM, SEATS AND FIXINGS 3.H
ACCESSORIES AND PAINT 4.C
NUMERICAL INDEX 4.D
GROUP INDEX
BATTERY AND SPARE WHEEL COVERS. N05
BONNET INSULATION-3.6 LITRE N06
BONNET INSULATION-5.3 MODELS N08
BONNET INSULATION-5.3 MODELS. N07
BOOT CARPETS - NOT CONVERTIBLE N03
BOOT CARPETS-CONVERTIBLE N04
CARPETS-'A' POST FINISHERS-. L12
CONSOLE-AMBLA. H10
CONSOLE-LEATHER. H08
DOOR CASINGS-ARMRESTS AND POCKETS-AMBLA. H16
DOOR CASINGS-ARMRESTS-CONVERTIBLE I02
HEADLINING I15
DOOR CASINGS-ARMRESTS-LEATHER. H14
EXTERIOR TRIM COVERS N15
FACIA PANEL. H02
FRONT SEAT BELTS-CONVERTIBLE K15
FRONT SEAT SLIDES K16
MOTOR AND ANCHORAGE ASSY-RH L02
FRONT SEATS-AMBLA/CLOTH.-3.6 LITRE K04
FRONT SEATS-LEATHER. J03
FRONT SEATS-LEATHER/CLOTH-5.3 LITRE K07
FRT SEATS-LEATHER/CLOTH(SPORTSPACK) K10
FRONT BELT KIT-COUPE K14
FRONT SEAT SLIDES K13
FUEL APERATURE SEAL AND HOSES-COUPE. N13
HARDTOP & HOODBAG-CABRIOLET. N17
HOOD LIFT MECHANISM - CONVERTIBLE O04
HOOD SEALS-CONVERTIBLE O03
HOOD-CABRIOLET N16
HOOD-CONVERTIBLE O02
REAR CUBBY BOX-CONVERTIBLE L09
REAR LUGGAGE LOCKER-CABRIOLET L07
REAR SEAT BELTS. L11
REAR SEATS-LEATHER-COUPE. L03
REAR SEATS-LEATHER-COUPE. L04
REAR SEATS-LEATHER/CLOTH-COUPE. L06
TARGA PANELS-STOWAGEBAG-CABRIOLET N14
UNDER SCUTTLE CASINGS. H05
VENEERED FACIA PANELS-3.6 LITRE. H06
VENEERED FACIA PANELS-5.3 LITRE. H07

3L 14

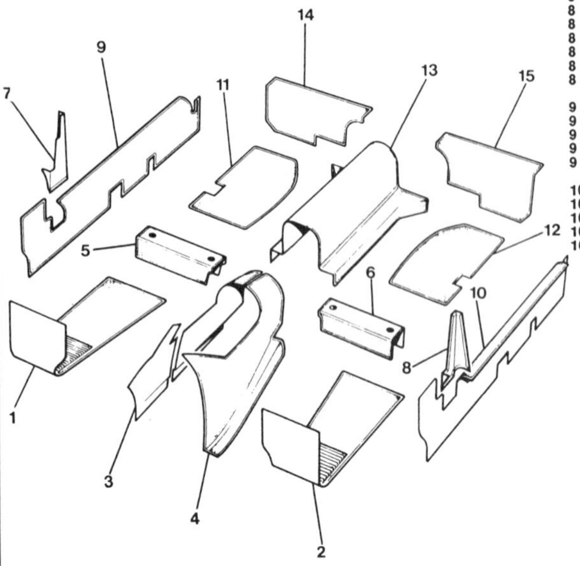

VS 3700

CARPETS-'A'POST FINISHERS-3.6 LITRE COUPE-continued

Illus	Part Number	123456	Description	Quantity	Change Point	Remarks
			FINISHER LH-'A' POST			
8	BBC 7999 AE		Biscuit	1		
8	BBC 7999 LZ		Warm Charcoal	1		
8	BBC 7999 AR		Buckskin	1		
8	BBC 7999 XE		Doeskin	1		
8	BBC 7999 JF		Isis Blue	1		
8	BBC 7999 CM		Mulberry	1		
8	BBC 7999 LY		Savile Grey	1		
8	BBC 7999 AW		Barley	1		
8	BBC 7999 ZM		Magnolia	1		
			CARPET RH-SILL			
9	BCC 8152 AP		Mink	1		
9	BCC 8152 XF		Rattan	1		
9	BCC 8152 LE		Slate	1		
9	BCC 8152 XA		Smoke Grey	1		
9	BCC 8152 CN		Wine Red	1		
			CARPET LH-SILL			
10	BCC 8153 AP		Mink	1		
10	BCC 8153 XF		Rattan	1		
10	BCC 8153 LE		Slate	1		
10	BCC 8153 XA		Smoke Grey	1		
10	BCC 8153 CN		Wine Red	1		

3L 15

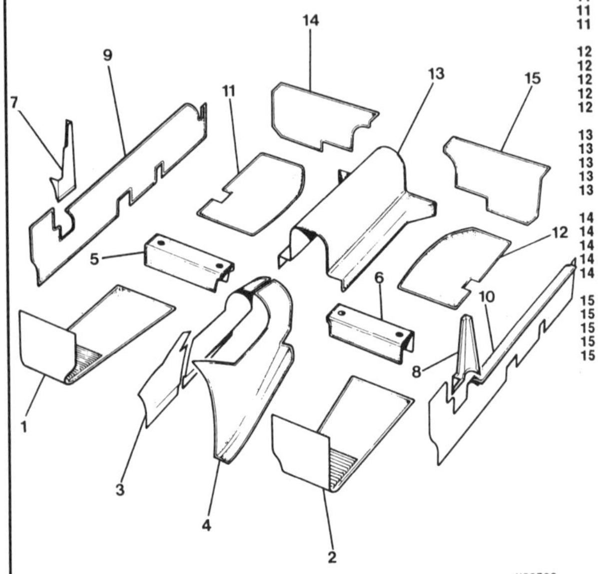

VS 3700

CARPETS-'A' POST FINISHERS-3.6 LITRE COUPE-continued

Illus	Part Number	123456	Description	Quantity	Change Point	Remarks
			CARPET RH-REAR			
11	BCC 2580 AP		Mink	1		
11	BCC 2580 XF		Rattan	1		
11	BCC 2580 LE		Slate	1		
11	BCC 2580 XA		Smoke Grey	1		
11	BCC 2580 CN		Wine Red	1		
			CARPET LH-REAR			
12	BCC 2581 AP		Mink	1		
12	BCC 2581 XF		Rattan	1		
12	BCC 2581 LE		Slate	1		
12	BCC 2581 XA		Smoke Grey	1		
12	BCC 2581 CN		Wine Red	1		
			CARPET-TUNNEL-REAR			
13	BCC 2312 AP		Mink	1		
13	BCC 2312 XF		Rattan	1		
13	BCC 2312 LE		Slate	1		
13	BCC 2312 XA		Smoke Grey	1		
13	BCC 2312 CN		Wine Red	1		
			CARPET RH-CROSSMEMBER			
14	BCC 6212 AP		Mink	1		
14	BCC 6212 XF		Rattan	1		
14	BCC 6212 LE		Slate	1		
14	BCC 6212 XA		Smoke Grey	1		
14	BCC 6212 CN		Wine Red	1		
			CARPET LH-CROSSMEMBER			
15	BCC 6213 AP		Mink	1		
15	BCC 6213 XF		Rattan	1		
15	BCC 6213 LE		Slate	1		
15	BCC 6213 XA		Smoke Grey	1		
15	BCC 6213 CN		Wine Red	1		

MASTER INDEX

ENGINE ... 1.C
FLYWHEEL AND CLUTCH ... 1.G
GEARBOX AND PROPSHAFT ... 1.H
AXLES, SUSPENSION, DRIVE SHAFTS, WHEELS ... 1.K
STEERING ... 1.M
BRAKES AND BRAKE CONTROLS ... 1.N
FUEL, EXHAUST AND EMISSION SYSTEMS ... 2.C
COOLING, HEATING AND AIR CONDITIONING ... 2.G
ELECTRICAL, WASHERS-WIPERS, INSTRUMENTS ... 2.J
CHASSIS, SUBFRAMES, BODYSHELL, FITTINGS ... 3.C
FASCIA, TRIM, SEATS AND FIXINGS ... 3.H
ACCESSORIES AND PAINT ... 4.C
NUMERICAL INDEX ... 4.D

GROUP INDEX

BATTERY AND SPARE WHEEL COVERS ... N05
BONNET INSULATION-3.6 LITRE ... N06
BONNET INSULATION-5.3 MODELS ... N08
BONNET INSULATION-5.3 MODELS ... N07
BOOT CARPETS - NOT CONVERTIBLE ... N03
BOOT CARPETS-CONVERTIBLE ... N04
CARPETS-'A' POST FINISHERS- ... L12
CONSOLE-AMBLA ... H10
CONSOLE-LEATHER ... H08
DOOR CASINGS-ARMRESTS AND POCKETS-AMBLA ... H16
DOOR CASINGS-ARMRESTS-CONVERTIBLE ... I02
HEADLINING ... I15
DOOR CASINGS-ARMRESTS-LEATHER ... H14
EXTERIOR TRIM COVERS ... N15
FACIA PANEL ... H02
FRONT SEAT BELTS-CONVERTIBLE ... K15
FRONT SEAT SLIDES ... K16
MOTOR AND ANCHORAGE ASSY-RH ... L02
FRONT SEATS-AMBLA/CLOTH.-3.6 LITRE ... K04
FRONT SEATS-LEATHER ... J03
FRONT SEATS-LEATHER/CLOTH-5.3 LITRE ... K07
FRT SEATS-LEATHER/CLOTH(SPORTSPACK) ... K10
FRONT BELT KIT-COUPE ... K14
FRONT SEAT SLIDES ... K13
FUEL APERATURE SEAL AND HOSES-COUPE ... N13
HARDTOP & HOODBAG-CABRIOLET ... N17
HOOD LIFT MECHANISM - CONVERTIBLE ... O04
HOOD SEALS-CONVERTIBLE ... O03
HOOD-CABRIOLET ... N16
HOOD-CONVERTIBLE ... O02
REAR CUBBY BOX-CONVERTIBLE ... L09
REAR LUGGAGE LOCKER-CABRIOLET ... L07
REAR SEAT BELTS ... L11
REAR SEATS-LEATHER-COUPE ... L03
REAR SEATS-LEATHER-COUPE ... L04
REAR SEATS-LEATHER/CLOTH-COUPE ... L06
TARGA PANELS-STOWAGEBAG-CABRIOLET ... N14
UNDER SCUTTLE CASINGS ... H05
VENEERED FACIA PANELS-3.6 LITRE ... H06
VENEERED FACIA PANELS-5.3 LITRE ... H07

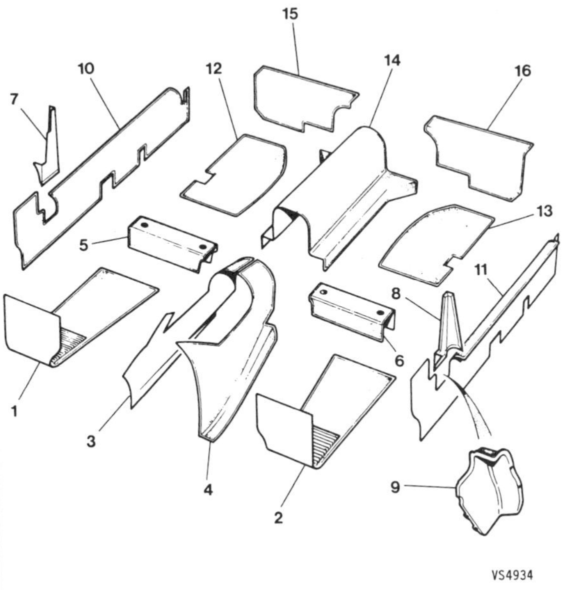

VS4934

CARPETS-'A' POST FINISHERS-5.3 LITRE COUPE

Illus	Part Number	123456	Description	Quantity	Change Point	Remarks
			CARPET RH-RHD-FRONT			
1	BBC 3122 AP		Mink	1		
1	BBC 3122 XF		Rattan	1		
1	BBC 3122 LE		Slate	1		
1	BBC 3122 XA		Smoke Grey	1		
1	BBC 3122 CN		Wine Red	1		
			CARPET LH-RHD-FRONT			
2	BAC 4259 AP		Mink	1	Up to 89 MY	
2	BAC 4259 XF		Rattan	1	Up to 89 MY	
2	BAC 4259 LE		Slate	1	Up to 89 MY	
2	BAC 4259 XA		Smoke Grey	1	Up to 89 MY	
2	BAC 4259 CN		Wine Red	1	Up to 89 MY	
			CARPET LH-RHD-FRONT			
	BDC 3181 AP		Mink	1	From 89 MY	
	BDC 3181 XF		Rattan	1	From 89 MY	
	BDC 3181 LE		Slate	1	From 89 MY	
	BDC 3181 XA		Smoke grey	1	From 89 MY	
	BDC 3181 CN		Wine red	1	From 89 MY	
			CARPET RH-LHD-FRONT			
	BAC 4258 AP		Mink	1	Up to 89 MY	
	BAC 4258 XF		Rattan	1	Up to 89 MY	
	BAC 4258 LE		Slate	1	Up to 89 MY	
	BAC 4258 XA		Smoke Grey	1	Up to 89 MY	
	BAC 4258 CN		Wine Red	1	Up to 89 MY	
			CARPET RH-LHD-FRONT			
	BDC 3180 AP		Mink	1	From 89 MY	
	BDC 3180 XF		Rattan	1	From 89 MY	
	BDC 3180 LE		Slate	1	From 89 MY	
	BDC 3180 XA		Smoke grey	1	From 89 MY	
	BDC 3180 CN		Wine red	1	From 89 MY	
			CARPET LH-LHD-FRONT			
	BBC 3123 AP		Mink	1		
	BBC 3123 XF		Rattan	1		
	BBC 3123 LE		Slate	1		
	BBC 3123 XA		Smoke Grey	1		
	BBC 3123 CN		Wine Red	1		

M50.06JP13/10/88

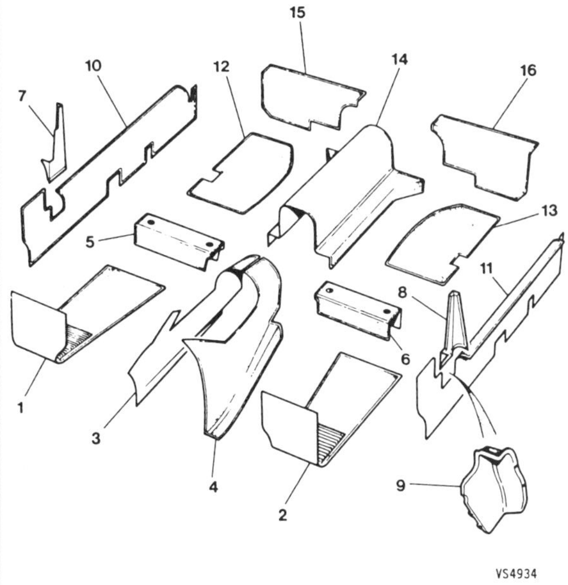

VS4934

CARPETS-'A' POST FINISHERS-5.3 LITRE COUPE

Illus	Part Number	123456	Description	Quantity	Change Point	Remarks
			CARPET-RH-GEARBOX			
3	BAC 5678 AP		Mink	1		
3	BAC 5678 XF		Rattan	1		
3	BAC 5678 LE		Slate	1		
3	BAC 5678 XA		Smoke grey	1		
3	BAC 5678 CN		Wine red	1		
			CARPET-LH-RHD-GEARBOX			
4	BAC 2564 AP		Mink	1		
4	BAC 2564 XF		Rattan	1		
4	BAC 2564 LE		Slate	1		
4	BAC 2564 XA		Smoke grey	1		
4	BAC 2564 CN		Wine red	1		
			CARPET LH-LHD-GEARBOX			
	BAC 2565 AP		Mink	1		
	BAC 2565 XF		Rattan	1		
	BAC 2565 LE		Slate	1		
	BAC 2565 XA		Smoke grey	1		
	BAC 2565 CN		Wine red	1		

M50.06JP13/10/88

MASTER INDEX

ENGINE 1.C
FLYWHEEL AND CLUTCH 1.G
GEARBOX AND PROPSHAFT 1.H
AXLES, SUSPENSION, DRIVE SHAFTS, WHEELS 1.K
STEERING 1.M
BRAKES AND BRAKE CONTROLS 1.N
FUEL, EXHAUST AND EMISSION SYSTEMS 2.C
COOLING, HEATING AND AIR CONDITIONING 2.G
ELECTRICAL, WASHERS-WIPERS, INSTRUMENTS 2.J
CHASSIS, SUBFRAMES, BODYSHELL, FITTINGS 3.C
FASCIA, TRIM, SEATS AND FIXINGS 3.H
ACCESSORIES AND PAINT 4.C
NUMERICAL INDEX 4.D

GROUP INDEX

BATTERY AND SPARE WHEEL COVERS N05
BONNET INSULATION-3.6 LITRE N06
BONNET INSULATION-5.3 MODELS N08
BONNET INSULATION-5.3 MODELS N07
BOOT CARPETS - NOT CONVERTIBLE N03
BOOT CARPETS-CONVERTIBLE N04
CARPETS-'A' POST FINISHERS- L12
CONSOLE-AMBLA H10
CONSOLE-LEATHER H08
DOOR CASINGS-ARMRESTS AND POCKETS-AMBLA H16
DOOR CASINGS-ARMRESTS-CONVERTIBLE I02
HEADLINING I15
DOOR CASINGS-ARMRESTS-LEATHER H14
EXTERIOR TRIM COVERS N15
FACIA PANEL H02
FRONT SEAT BELTS-CONVERTIBLE K15
FRONT SEAT SLIDES K16
MOTOR AND ANCHORAGE ASSY-RH L02
FRONT SEATS-AMBLA/CLOTH-3.6 LITRE K04
FRONT SEATS-LEATHER J03
FRONT SEATS-LEATHER/CLOTH-5.3 LITRE K07
FRT SEATS-LEATHER/CLOTH(SPORTSPACK) K10
FRONT BELT KIT-COUPE K14
FRONT SEAT SLIDES K13
FUEL APERATURE SEAL AND HOSES-COUPE N13
HARDTOP & HOODBAG-CABRIOLET N17
HOOD LIFT MECHANISM - CONVERTIBLE O04
HOOD SEALS-CONVERTIBLE O03
HOOD-CABRIOLET N16
HOOD-CONVERTIBLE O02
REAR CUBBY BOX-CONVERTIBLE L09
REAR LUGGAGE LOCKER-CABRIOLET L07
REAR SEAT BELTS L11
REAR SEATS-LEATHER-COUPE L03
REAR SEATS-LEATHER-COUPE L04
REAR SEATS-LEATHER/CLOTH-COUPE L06
TARGA PANELS-STOWAGEBAG-CABRIOLET N14
UNDER SCUTTLE CASINGS H05
VENEERED FACIA PANELS-3.6 LITRE H06
VENEERED FACIA PANELS-5.3 LITRE H07

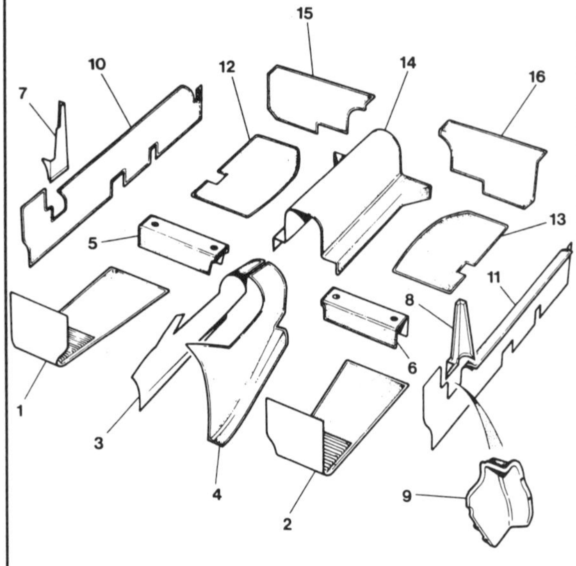

VS4934

CARPETS-'A'POST FINISHERS-5.3 LITRE COUPE-continued

Illus	Part Number	Description	Quantity	Change Point	Remarks
		CARPET RH-CROSSMEMBER			
5	BCC 2314 AP	Mink	1		
5	BCC 2314 XF	Rattan	1		
5	BCC 2314 LE	Slate	1		
5	BCC 2314 XA	Smoke Grey	1		
5	BCC 2314 CN	Wine Red	1		
		CARPET LH-CROSSMEMBER			
6	BCC 2315 AP	Mink	1		
6	BCC 2315 XF	Rattan	1		
6	BCC 2315 LE	Slate	1		
6	BCC 2315 XA	Smoke Grey	1		
6	BCC 2315 CN	Wine Red	1		
		FINISHER RH-'A' POST			
7	BBC 7998 AE	Biscuit	1		
7	BBC 7998 LZ	Warm Charcoal	1		
7	BBC 7998 AR	Buckskin	1		
7	BBC 7998 XE	Doeskin	1		
7	BBC 7998 JF	Isis Blue	1		
7	BBC 7998 CM	Mulberry	1		
7	BBC 7998 LY	Savile Grey	1		
7	BBC 7998 AW	Barley	1		
7	BBC 7998 ZM	Magnolia	1		
		FINISHER RH-LHD-'A' POST			
7	BEC 2040 AW	Barley	1		From 89 MY
7	BEC 2040 AR	Buckskin	1		From 89 MY
7	BEC 2040 XE	Doeskin	1		From 89 MY
7	BEC 2040 JF	Isis blue	1		From 89 MY
7	BEC 2040 ZM	Magnolia	1		From 89 MY
7	BEC 2040 CM	Mulberry	1		From 89 MY
7	BEC 2040 LY	Savile grey	1		From 89 MY
7	BEC 2040 LZ	Warm charcoal	1		From 89 MY

M50.10JP13/10/88

MASTER INDEX

GROUP INDEX

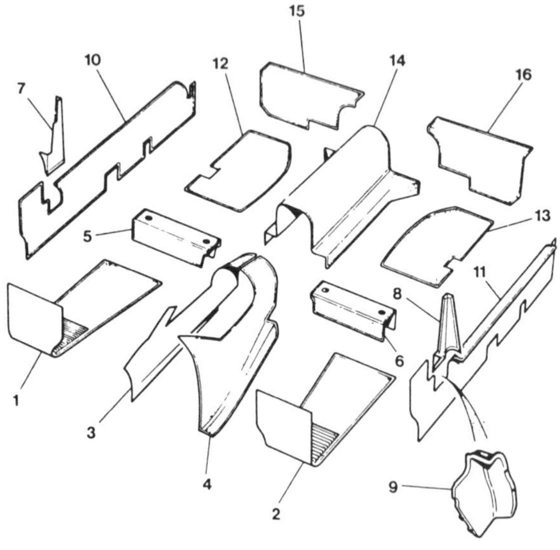

VS4934

CARPETS-'A'POST FINISHERS-5.3 LITRE COUPE-continued

Illus	Part Number	Description	Quantity	Change Point	Remarks
		FINISHER LH-'A' POST			
8	BBC 7999 AE	Biscuit	1		LHD only from 89 MY
8	BBC 7999 LZ	Warm Charcoal	1		LHD only from 89 MY
8	BBC 7999 AR	Buckskin	1		LHD only from 89 MY
8	BBC 7999 XE	Doeskin	1		LHD only from 89 MY
8	BBC 7999 JF	Isis Blue	1		LHD only from 89 MY
8	BBC 7999 CM	Mulberry	1		LHD only from 89 MY
8	BBC 7999 LY	Savile Grey	1		LHD only from 89 MY
8	BBC 7999 AW	Barley	1		LHD only from 89 MY
8	BBC 7999 ZM	Magnolia	1		LHD only from 89 MY
		FINISHER LH-RHD-'A' POST			
8	BEC 2041 AW	Barley	1		From 89 MY
8	BEC 2041 AR	Buckskin	1		From 89 MY
8	BEC 2041 XE	Doeskin	1		From 89 MY
8	BEC 2041 JF	Isis blue	1		From 89 MY
8	BEC 2041 ZM	Magnolia	1		From 89 MY
8	BEC 2041 CM	Mulberry	1		From 89 MY
8	BEC 2041 LY	Savile grey	1		From 89 MY
8	BEC 2041 LZ	Warm charcoal	1		From 89 MY
		MOUNTING COVER-ECU-LH-RHD			
9	BEC 1693 AP	Mink	1		From 89 MY
9	BEC 1693 XF	Rattan	1		From 89 MY
9	BEC 1693 LE	Slate	1		From 89 MY
9	BEC 1693 XA	Smoke grey	1		From 89 MY
9	BEC 1693 CN	Wine red	1		From 89 MY
		MOUNTING COVER-ECU-RH-LHD			
	BEC 1692 AP	Mink	1		
	BEC 1692 XF	Rattan	1		
	BEC 1692 LE	Slate	1		
	BEC 1692 XA	Smoke grey	1		
	BEC 1692 CN	Wine red	1		

VS4934

MASTER INDEX

ENGINE 1.C
FLYWHEEL AND CLUTCH 1.G
GEARBOX AND PROPSHAFT 1.H
AXLES, SUSPENSION, DRIVE SHAFTS, WHEELS 1.K
STEERING 1.M
BRAKES AND BRAKE CONTROLS 1.N
FUEL, EXHAUST AND EMISSION SYSTEMS 2.C
COOLING, HEATING AND AIR CONDITIONING 2.G
ELECTRICAL, WASHERS-WIPERS, INSTRUMENTS 2.J
CHASSIS, SUBFRAMES, BODYSHELL, FITTINGS 3.C
FASCIA, TRIM, SEATS AND FIXINGS 3.H
ACCESSORIES AND PAINT 4.C
NUMERICAL INDEX 4.D

GROUP INDEX

BATTERY AND SPARE WHEEL COVERS N05
BONNET INSULATION-3.6 LITRE N06
BONNET INSULATION-5.3 MODELS N08
BONNET INSULATION-5.3 MODELS N07
BOOT CARPETS - NOT CONVERTIBLE N03
BOOT CARPETS-CONVERTIBLE N04
CARPETS-'A' POST FINISHERS- L12
CONSOLE-AMBLA H10
CONSOLE-LEATHER H08
DOOR CASINGS-ARMRESTS AND POCKETS-AMBLA H16
DOOR CASINGS-ARMRESTS-CONVERTIBLE I02
HEADLINING I15
DOOR CASINGS-ARMRESTS-LEATHER H14
EXTERIOR TRIM COVERS N15
FACIA PANEL H02
FRONT SEAT BELTS-CONVERTIBLE K15
FRONT SEAT SLIDES K16
MOTOR AND ANCHORAGE ASSY-RH L02
FRONT SEATS-AMBLA/CLOTH.-3.6 LITRE K04
FRONT SEATS-LEATHER J03
FRONT SEATS-LEATHER/CLOTH-5.3 LITRE K07
FRT SEATS-LEATHER/CLOTH(SPORTSPACK) K10
FRONT BELT KIT-COUPE K14
FRONT SEAT SLIDES K13
FUEL APERATURE SEAL AND HOSES-COUPE N13
HARDTOP & HOODBAG-CABRIOLET N17
HOOD LIFT MECHANISM - CONVERTIBLE O04
HOOD SEALS-CONVERTIBLE O03
HOOD-CABRIOLET N16
HOOD-CONVERTIBLE O02
REAR CUBBY BOX-CONVERTIBLE L09
REAR LUGGAGE LOCKER-CABRIOLET L07
REAR SEAT BELTS L11
REAR SEATS-LEATHER-COUPE L03
REAR SEATS-LEATHER-COUPE L04
REAR SEATS-LEATHER/CLOTH-COUPE L06
TARGA PANELS-STOWAGEBAG-CABRIOLET N14
UNDER SCUTTLE CASINGS H05
VENEERED FACIA PANELS-3.6 LITRE H06
VENEERED FACIA PANELS-5.3 LITRE H07

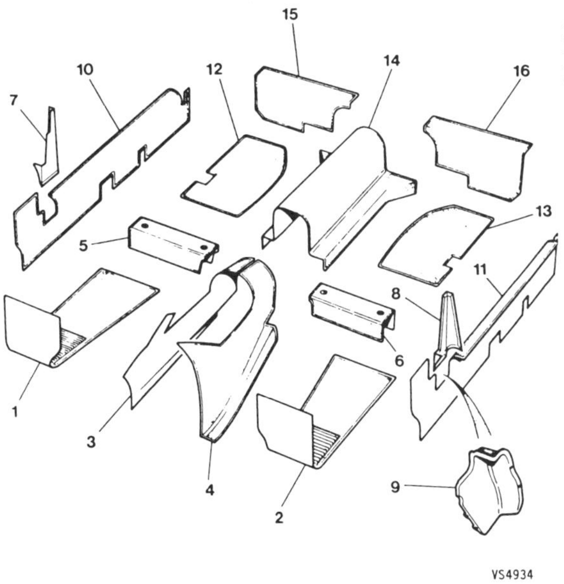

VS4934

VS4934

CARPETS-'A' POST FINISHERS-5.3 LITRE COUPE(Cont'd)

Illus	Part Number	123456	Description	Quantity	Change Point	Remarks
			CARPET RH-SILL			
10	BCC 8152 AP		Mink	1		Up to 89 MY
10	BCC 8152 XF		Rattan	1		
10	BCC 8152 LE		Slate	1		
10	BCC 8152 XA		Smoke grey	1		
10	BCC 8152 CN		Wine red	1		
			CARPET RH-SILL			
	BDC 3195 AP		Mink	1		From 89 MY
	BDC 3195 XF		Rattan	1		
	BDC 3195 LE		Slate	1		
	BDC 3195 XA		Smoke grey	1		
	BDC 3195 CN		Wine red	1		
			CARPET LH-SILL			
11	BCC 8153 AP		Mink	1		Up to 89 MY
11	BCC 8153 XF		Rattan	1		
11	BCC 8153 LE		Slate	1		
11	BCC 8153 XA		Smoke grey	1		
11	BCC 8153 CN		Wine red	1		
			CARPET LH-SILL			
	BDC 3194 AP		Mink	1		From 89 MY
	BDC 3194 XF		Rattan	1		
	BDC 3194 LE		Slate	1		
	BDC 3194 XA		Smoke grey	1		
	BDC 3194 CN		Wine red	1		

M50.14JP12/10/88

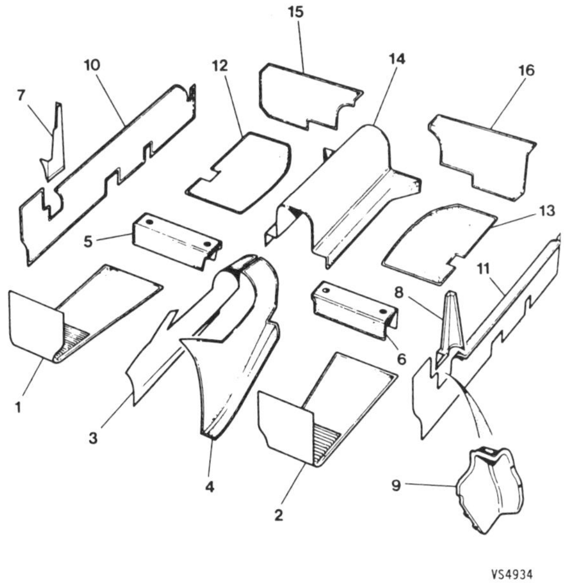

VS4934

VS4934

CARPETS-'A' POST FINISHERS-5.3 LITRE COUPE-continued

Illus	Part Number	123456	Description	Quantity	Change Point	Remarks
			CARPET RH-REAR			
12	BCC 2580 AP		Mink	1		
12	BCC 2580 XF		Rattan	1		
12	BCC 2580 LE		Slate	1		
12	BCC 2580 XA		Smoke Grey	1		
12	BCC 2580 CN		Wine Red	1		
			CARPET LH-REAR			
13	BCC 2581 AP		Mink	1		
13	BCC 2581 XF		Rattan	1		
13	BCC 2581 LE		Slate	1		
13	BCC 2581 XA		Smoke Grey	1		
13	BCC 2581 CN		Wine Red	1		
			CARPET-TUNNEL-REAR			
14	BCC 2312 AP		Mink	1		
14	BCC 2312 XF		Rattan	1		
14	BCC 2312 LE		Slate	1		
14	BCC 2312 XA		Smoke Grey	1		
14	BCC 2312 CN		Wine Red	1		
			CARPET RH-CROSSMEMBER			
15	BCC 6212 AP		Mink	1		
15	BCC 6212 XF		Rattan	1		
15	BCC 6212 LE		Slate	1		
15	BCC 6212 XA		Smoke Grey	1		
15	BCC 6212 CN		Wine Red	1		
			CARPET LH-CROSSMEMBER			
16	BCC 6213 AP		Mink	1		
16	BCC 6213 XF		Rattan	1		
16	BCC 6213 LE		Slate	1		
16	BCC 6213 XA		Smoke Grey	1		
16	BCC 6213 CN		Wine Red	1		

M52JP13/10/88

MASTER INDEX

ENGINE 1.C
FLYWHEEL AND CLUTCH 1.G
GEARBOX AND PROPSHAFT 1.H
AXLES, SUSPENSION, DRIVE SHAFTS, WHEELS 1.K
STEERING 1.M
BRAKES AND BRAKE CONTROLS 1.N
FUEL, EXHAUST AND EMISSION SYSTEMS 2.C
COOLING, HEATING AND AIR CONDITIONING 2.G
ELECTRICAL, WASHERS-WIPERS, INSTRUMENTS 2.J
CHASSIS, SUBFRAMES, BODYSHELL, FITTINGS 3.C
FASCIA, TRIM, SEATS AND FIXINGS 3.H
ACCESSORIES AND PAINT 4.C
NUMERICAL INDEX 4.D

GROUP INDEX

BATTERY AND SPARE WHEEL COVERS. N05
BONNET INSULATION-3.6 LITRE N06
BONNET INSULATION-5.3 MODELS N08
BONNET INSULATION-5.3 MODELS. N07
BOOT CARPETS - NOT CONVERTIBLE N03
BOOT CARPETS-CONVERTIBLE N04
CARPETS-'A' POST FINISHERS-. L12
CONSOLE-AMBLA. H10
CONSOLE-LEATHER. H08
DOOR CASINGS-ARMRESTS AND POCKETS-AMBLA. H16
DOOR CASINGS-ARMRESTS-CONVERTIBLE I02
HEADLINING I15
DOOR CASINGS-ARMRESTS-LEATHER. H14
EXTERIOR TRIM COVERS N15
FACIA PANEL. H02
FRONT SEAT BELTS-CONVERTIBLE K15
FRONT SEAT SLIDES K16
MOTOR AND ANCHORAGE ASSY-RH L02
FRONT SEATS-AMBLA/CLOTH.-3.6 LITRE K04
FRONT SEATS-LEATHER. J03
FRONT SEATS-LEATHER/CLOTH-5.3 LITRE K07
FRT SEATS-LEATHER/CLOTH(SPORTSPACK) K10
FRONT BELT KIT-COUPE K14
FRONT SEAT SLIDES K13
FUEL APERATURE SEAL AND HOSES-COUPE. N13
HARDTOP & HOODBAG-CABRIOLET. N17
HOOD LIFT MECHANISM - CONVERTIBLE O04
HOOD SEALS-CONVERTIBLE O03
HOOD-CABRIOLET N16
HOOD-CONVERTIBLE O02
REAR CUBBY BOX-CONVERTIBLE L09
REAR LUGGAGE LOCKER-CABRIOLET L07
REAR SEAT BELTS. L11
REAR SEATS-LEATHER-COUPE. L03
REAR SEATS-LEATHER-COUPE. L04
REAR SEATS-LEATHER/CLOTH-COUPE. L06
TARGA PANELS-STOWAGEBAG-CABRIOLET N14
UNDER SCUTTLE CASINGS. H05
VENEERED FACIA PANELS-3.6 LITRE. H06
VENEERED FACIA PANELS-5.3 LITRE. H07

3M 05

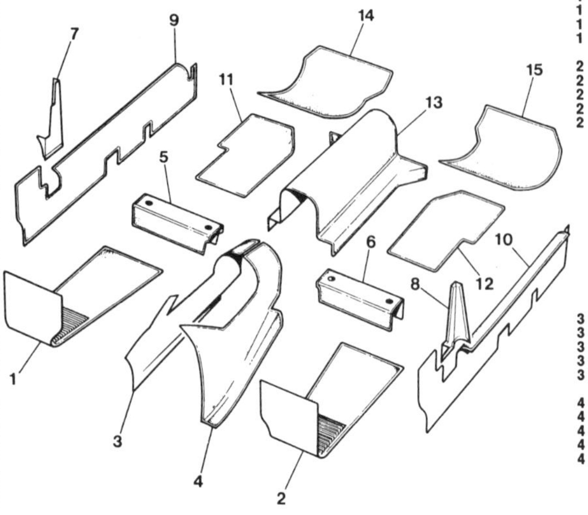

VS 4329

CARPETS-'A' POST FINISHERS-3.6 LITRE CABRIOLET

Illus	Part Number	Description	Quantity	Change Point	Remarks
		CARPET RH-RHD-FRONT			
1	BBC 3122 AP	Mink	1		
1	BBC 3122 XF	Rattan	1		
1	BBC 3122 LE	Slate	1		
1	BBC 3122 XA	Smoke Grey	1		
1	BBC 3122 CN	Wine Red	1		
		CARPET LH-RHD-FRONT			
2	BCC 2067 AP	Mink	1		
2	BCC 2067 XF	Rattan	1		
2	BCC 2067 LE	Slate	1		
2	BCC 2067 XA	Smoke Grey	1		
2	BCC 2067 CN	Wine Red	1		
		CARPET RH-LHD FRONT			
	BCC 2066 AP	Mink	1		
	BCC 2066 XF	Rattan	1		
	BCC 2066 LE	Slate	1		
	BCC 2066 XA	Smoke Grey	1		
	BCC 2066 CN	Wine Red	1		
		CARPET LH-LHD-FRONT			
	BBC 3123 AP	Mink	1		
	BBC 3123 XF	Rattan	1		
	BBC 3123 LE	Slate	1		
	BBC 3123 XA	Smoke Grey	1		
	BBC 3123 CN	Wine Red	1		
		CARPET-RH-GEARBOX			
3	BAC 5678 AP	Mink	1		
3	BAC 5678 XF	Rattan	1		
3	BAC 5678 LE	Slate	1		
3	BAC 5678 XA	Smoke Grey	1		
3	BAC 5678 CN	Wine Red	1		
		CARPET LH-RHD-GEARBOX			
4	BAC 2564 AP	Mink	1		
4	BAC 2564 XF	Rattan	1		
4	BAC 2564 LE	Slate	1		
4	BAC 2564 XA	Smoke Grey	1		
4	BAC 2564 CN	Wine Red	1		
		CARPET LH-LHD-GEARBOX			
	BAC 2565 AP	Mink	1		
	BAC 2565 XF	Rattan	1		
	BAC 2565 LE	Slate	1		
	BAC 2565 XA	Smoke Grey	1		
	BAC 2565 CN	Wine Red	1		

3M 06

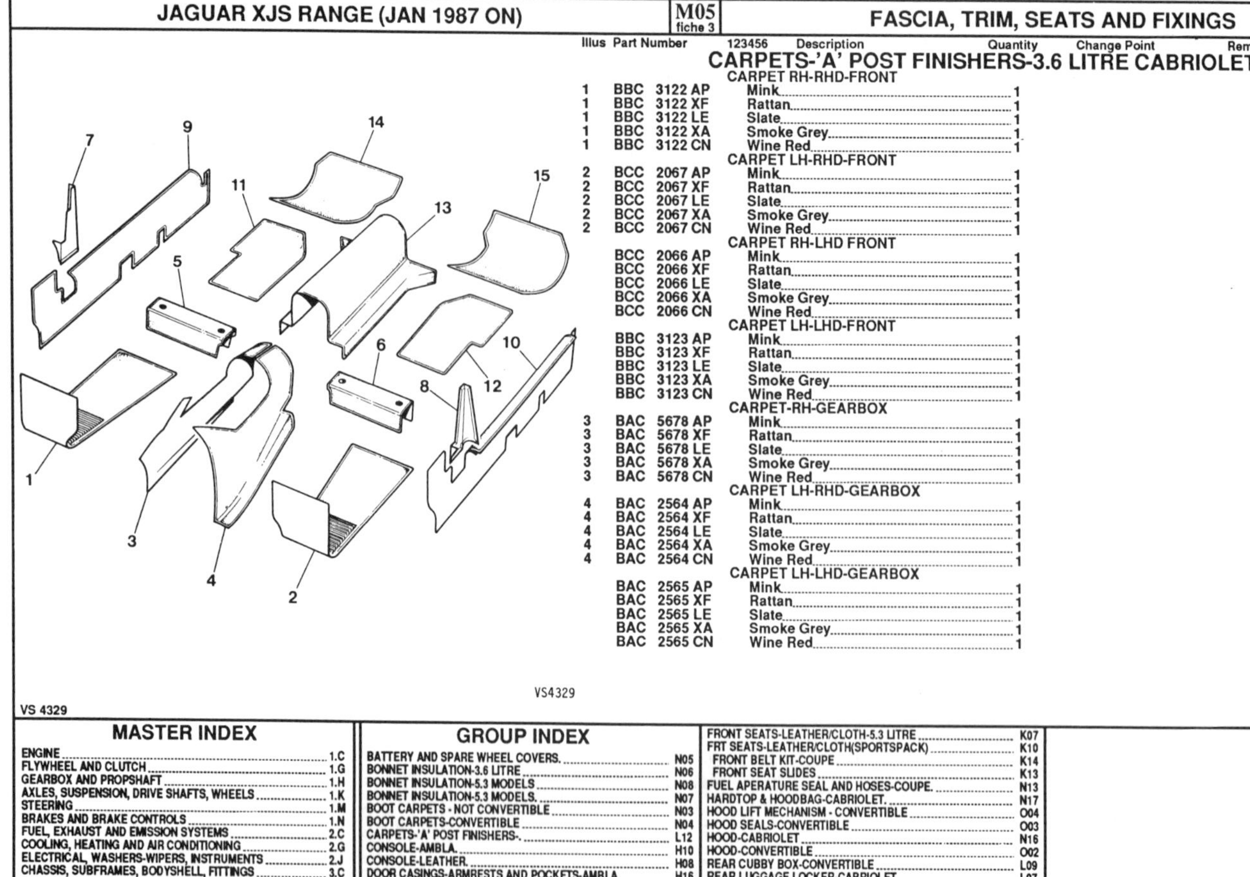

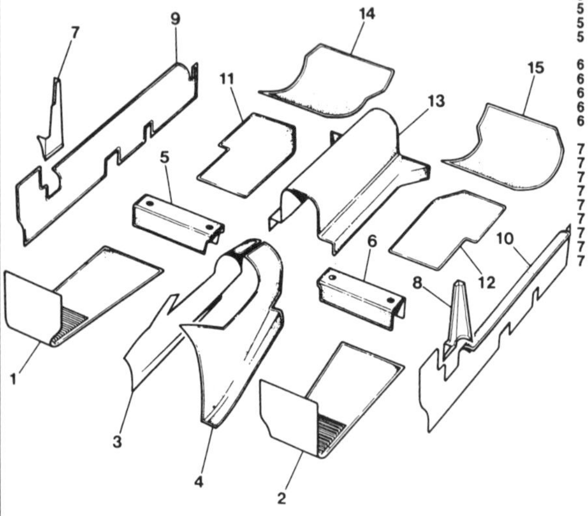

VS 4329

CARPETS-'A' POST FINISHERS-3.6 LITRE CABRIOLET-continued

Illus	Part Number	Description	Quantity	Change Point	Remarks
		CARPET RH-CROSSMEMBER			
5	BCC 2314 AP	Mink	1		
5	BCC 2314 XF	Rattan	1		
5	BCC 2314 LE	Slate	1		
5	BCC 2314 XA	Smoke Grey	1		
5	BCC 2314 CN	Wine Red	1		
		CARPET LH-CROSSMEMBER			
6	BCC 2315 AP	Mink	1		
6	BCC 2315 XF	Rattan	1		
6	BCC 2315 LE	Slate	1		
6	BCC 2315 XA	Smoke Grey	1		
6	BCC 2315 CN	Wine Red	1		
		FINISHER RH-'A' POST			
7	BBC 7998 AE	Biscuit	1		
7	BBC 7998 LZ	Warm Charcoal	1		
7	BBC 7998 AR	Buckskin	1		
7	BBC 7998 XE	Doeskin	1		
7	BBC 7998 JF	Isis Blue	1		
7	BBC 7998 CM	Mulberry	1		
7	BBC 7998 LY	Savile Grey	1		
7	BBC 7998 AW	Barley	1		
7	BBC 7998 ZM	Magnolia	1		

MASTER INDEX

ENGINE 1.C
FLYWHEEL AND CLUTCH 1.G
GEARBOX AND PROPSHAFT 1.H
AXLES, SUSPENSION, DRIVE SHAFTS, WHEELS 1.K
STEERING 1.M
BRAKES AND BRAKE CONTROLS 1.N
FUEL, EXHAUST AND EMISSION SYSTEMS 2.C
COOLING, HEATING AND AIR CONDITIONING 2.G
ELECTRICAL, WASHERS-WIPERS, INSTRUMENTS 2.J
CHASSIS, SUBFRAMES, BODYSHELL, FITTINGS 3.C
FASCIA, TRIM, SEATS AND FIXINGS 3.H
ACCESSORIES AND PAINT 4.C
NUMERICAL INDEX 4.D

GROUP INDEX

BATTERY AND SPARE WHEEL COVERS N05
BONNET INSULATION-3.6 LITRE N06
BONNET INSULATION-5.3 MODELS N08
BONNET INSULATION-5.3 MODELS N07
BOOT CARPETS - NOT CONVERTIBLE N03
BOOT CARPETS-CONVERTIBLE N04
CARPETS-'A' POST FINISHERS- L12
CONSOLE-AMBLA H10
CONSOLE-LEATHER H08
DOOR CASINGS-ARMRESTS AND POCKETS-AMBLA H16
DOOR CASINGS-ARMRESTS-CONVERTIBLE I02
HEADLINING I15
DOOR CASINGS-ARMRESTS-LEATHER H14
EXTERIOR TRIM COVERS N15
FACIA PANEL H02
FRONT SEAT BELTS-CONVERTIBLE K15
FRONT SEAT SLIDES K16
MOTOR AND ANCHORAGE ASSY-RH L02
FRONT SEATS-AMBLA/CLOTH.-3.6 LITRE K04
FRONT SEATS-LEATHER J03
FRONT SEATS-LEATHER/CLOTH-5.3 LITRE K07
FRT SEATS-LEATHER/CLOTH(SPORTSPACK) K10
FRONT BELT KIT-COUPE K14
FRONT SEAT SLIDES K13
FUEL APERATURE SEAL AND HOSES-COUPE N13
HARDTOP & HOODBAG-CABRIOLET N17
HOOD LIFT MECHANISM - CONVERTIBLE O04
HOOD SEALS-CONVERTIBLE O03
HOOD-CABRIOLET N16
HOOD-CONVERTIBLE O02
REAR CUBBY BOX-CONVERTIBLE L09
REAR LUGGAGE LOCKER-CABRIOLET L07
REAR SEAT BELTS L11
REAR SEATS-LEATHER-COUPE L03
REAR SEATS-LEATHER-COUPE L04
REAR SEATS-LEATHER/CLOTH-COUPE L06
TARGA PANELS-STOWAGEBAG-CABRIOLET N14
UNDER SCUTTLE CASINGS H05
VENEERED FACIA PANELS-3.6 LITRE H06
VENEERED FACIA PANELS-5.3 LITRE H07

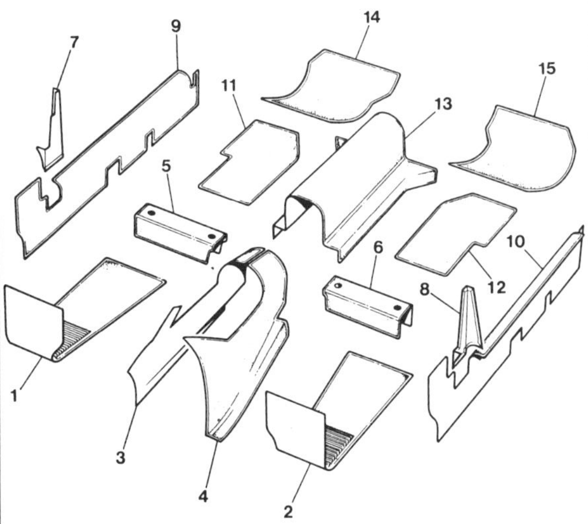

Illus	Part Number	123456	Description	Quantity	Change Point	Remarks
			CARPETS-'A' POST FINISHERS-3.6 LITRE CABRIOLET-continued			
			FINISHER LH-'A' POST			
8	BBC 7999 AE		Biscuit	1		
8	BBC 7999 LZ		Warm Charcoal	1		
8	BBC 7999 AR		Buckskin	1		
8	BBC 7999 XE		Doeskin	1		
8	BBC 7999 JF		Isis Blue	1		
8	BBC 7999 CM		Mulberry	1		
8	BBC 7999 LY		Savile Grey	1		
8	BBC 7999 AW		Barley	1		
8	BBC 7999 ZM		Magnolia	1		
			CARPET RH-SILL			
9	BCC 8152 AP		Mink	1		
9	BCC 8152 XF		Rattan	1		
9	BCC 8152 LE		Slate	1		
9	BCC 8152 XA		Smoke Grey	1		
9	BCC 8512 CN		Wine Red	1		
			CARPET LH-SILL			
10	BCC 8153 AP		Mink	1		
10	BCC 8153 XF		Rattan	1		
10	BCC 8153 LE		Slate	1		
10	BCC 8153 XA		Smoke Grey	1		
10	BCC 8153 CN		Wine Red	1		

VS 4329

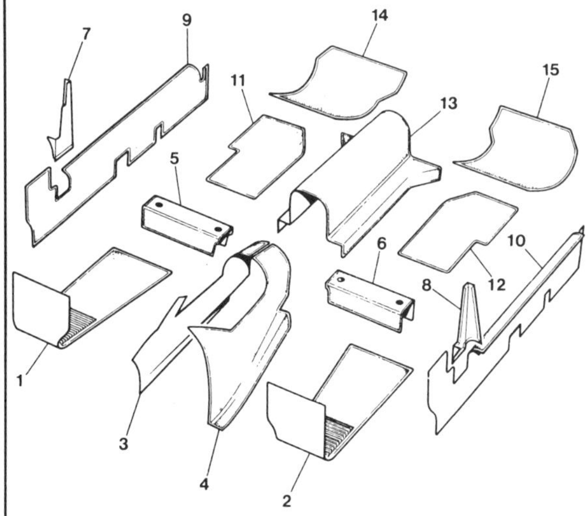

Illus	Part Number	123456	Description	Quantity	Change Point	Remarks
			CARPETS-'A' POST FINISHERS-3.6 LITRE CABRIOLET-continued			
			CARPET RH-REAR			
11	BCC 7992 AP		Mink	1		
11	BCC 7992 XF		Rattan	1		
11	BCC 7992 LE		Slate	1		
11	BCC 7992 XA		Smoke Grey	1		
11	BCC 7992 CN		Wine Red	1		
			CARPET LH-REAR			
12	BCC 7993 AP		Mink	1		
12	BCC 7993 XF		Rattan	1		
12	BCC 7993 LE		Slate	1		
12	BCC 7993 XA		Smoke Grey	1		
12	BCC 7993 CN		Wine Red	1		
			CARPET-TUNNEL-REAR			
13	BBC 2182 AP		Mink	1		
13	BBC 2182 XF		Rattan	1		
13	BBC 2182 LE		Slate	1		
13	BBC 2182 XA		Smoke Grey	1		
13	BBC 2182 CN		Wine Red	1		
			CARPET RH-LOCKER FLOOR			
14	BBC 2620 AP		Mink	1		
14	BBC 2620 XF		Rattan	1		
14	BBC 2620 LE		Slate	1		
14	BBC 2620 XA		Smoke Grey	1		
14	BBC 2620 CN		Wine Red	1		
			CARPET LH-LOCKER FLOOR			
15	BBC 2621 AP		Mink	1		
15	BBC 2621 XF		Rattan	1		
15	BBC 2621 LE		Slate	1		
15	BBC 2621 XA		Smoke Grey	1		
15	BBC 2621 CN		Wine Red	1		

VS 4329

MASTER INDEX

ENGINE ... 1.C
FLYWHEEL AND CLUTCH ... 1.G
GEARBOX AND PROPSHAFT ... 1.H
AXLES, SUSPENSION, DRIVE SHAFTS, WHEELS ... 1.K
STEERING ... 1.M
BRAKES AND BRAKE CONTROLS ... 1.N
FUEL, EXHAUST AND EMISSION SYSTEMS ... 2.C
COOLING, HEATING AND AIR CONDITIONING ... 2.G
ELECTRICAL, WASHERS-WIPERS, INSTRUMENTS ... 2.J
CHASSIS, SUBFRAMES, BODYSHELL, FITTINGS ... 3.C
FASCIA, TRIM, SEATS AND FIXINGS ... 3.H
ACCESSORIES AND PAINT ... 4.C
NUMERICAL INDEX ... 4.D

GROUP INDEX

BATTERY AND SPARE WHEEL COVERS. ... N05
BONNET INSULATION-3.6 LITRE ... N06
BONNET INSULATION-5.3 MODELS ... N08
BONNET INSULATION-5.3 MODELS. ... N07
BOOT CARPETS - NOT CONVERTIBLE ... N03
BOOT CARPETS-CONVERTIBLE ... N04
CARPETS-'A' POST FINISHERS-. ... L12
CONSOLE-AMBLA. ... H10
CONSOLE-LEATHER. ... H08
DOOR CASINGS-ARMRESTS AND POCKETS-AMBLA. ... H16
DOOR CASINGS-ARMRESTS-CONVERTIBLE ... I02
HEADLINING ... I15
DOOR CASINGS-ARMRESTS-LEATHER. ... H14
EXTERIOR TRIM COVERS ... N15
FACIA PANEL. ... H02
FRONT SEAT BELTS-CONVERTIBLE ... K15
FRONT SEAT SLIDES ... K16
MOTOR AND ANCHORAGE ASSY-RH ... L02
FRONT SEATS-AMBLA/CLOTH.-3.6 LITRE ... K04
FRONT SEATS-LEATHER. ... J03
FRONT SEATS-LEATHER/CLOTH-5.3 LITRE ... K07
FRT SEATS-LEATHER/CLOTH(SPORTSPACK) ... K10
FRONT BELT KIT-COUPE ... K14
FRONT SEAT SLIDES ... K13
FUEL APERATURE SEAL AND HOSES-COUPE. ... N13
HARDTOP & HOODBAG-CABRIOLET. ... N17
HOOD LIFT MECHANISM - CONVERTIBLE ... O04
HOOD SEALS-CONVERTIBLE ... O03
HOOD-CABRIOLET ... N16
HOOD-CONVERTIBLE ... O02
REAR CUBBY BOX-CONVERTIBLE ... L09
REAR LUGGAGE LOCKER-CABRIOLET ... L07
REAR SEAT BELTS. ... L11
REAR SEATS-LEATHER-COUPE. ... L03
REAR SEATS-LEATHER-COUPE. ... L04
REAR SEATS-LEATHER/CLOTH-COUPE. ... L06
TARGA PANELS-STOWAGEBAG-CABRIOLET ... N14
UNDER SCUTTLE CASINGS. ... H05
VENEERED FACIA PANELS-3.6 LITRE. ... H06
VENEERED FACIA PANELS-5.3 LITRE. ... H07

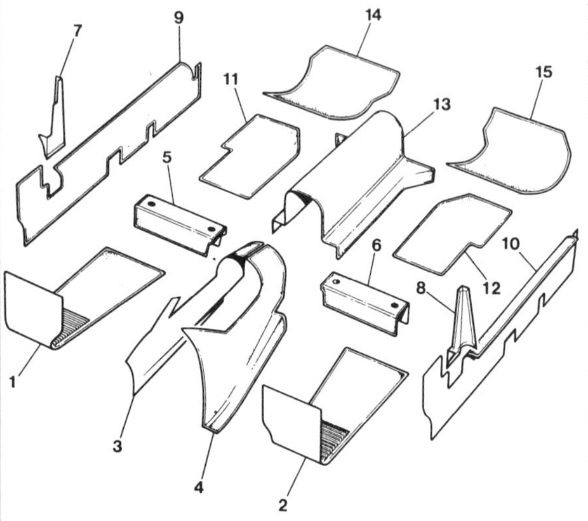

VS4329

VS 4329

CARPETS-'A' POST FINISHERS-5.3 LITRE CABRIOLET

Illus	Part Number	Description	Quantity	Change Point	Remarks
		CARPET RH-RHD-FRONT			
1	BBC 3122 AP	Mink	1		
1	BBC 3122 XF	Rattan	1		
1	BBC 3122 LE	Slate	1		
1	BBC 3122 XA	Smoke Grey	1		
1	BBC 3122 CN	Wine Red	1		
		CARPET LH-RHD-FRONT			
2	BAC 4259 AP	Mink	1		
2	BAC 4259 XF	Rattan	1		
2	BAC 4259 LE	Slate	1		
2	BAC 4259 XA	Smoke Grey	1		
2	BAC 4259 CN	Wine Red	1		
		CARPET RH-LHD FRONT			
	BAC 4258 AP	Mink	1		
	BAC 4258 XF	Rattan	1		
	BAC 4258 LE	Slate	1		
	BAC 4258 XA	Smoke Grey	1		
	BAC 4258 CN	Wine Red	1		
		CARPET LH-LHD-FRONT			
	BBC 3123 AP	Mink	1		
	BBC 3123 XF	Rattan	1		
	BBC 3123 LE	Slate	1		
	BBC 3123 XA	Smoke Grey	1		
	BBC 3123 CN	Wine Red	1		
		CARPET-RH-GEARBOX			
3	BAC 5678 AP	Mink	1		
3	BAC 5678 XF	Rattan	1		
3	BAC 5678 LE	Slate	1		
3	BAC 5678 XA	Smoke Grey	1		
3	BAC 5678 CN	Wine Red	1		
		CARPET LH-RHD-GEARBOX			
4	BAC 2564 AP	Mink	1		
4	BAC 2564 XF	Rattan	1		
4	BAC 2564 LE	Slate	1		
4	BAC 2564 XA	Smoke Grey	1		
4	BAC 2564 CN	Wine Red	1		
		CARPET LH-LHD-GEARBOX			
	BAC 2565 AP	Mink	1		
	BAC 2565 XF	Rattan	1		
	BAC 2565 LE	Slate	1		
	BAC 2565 XA	Smoke Grey	1		
	BAC 2565 CN	Wine Red	1		

3M 09

MASTER INDEX

GROUP INDEX

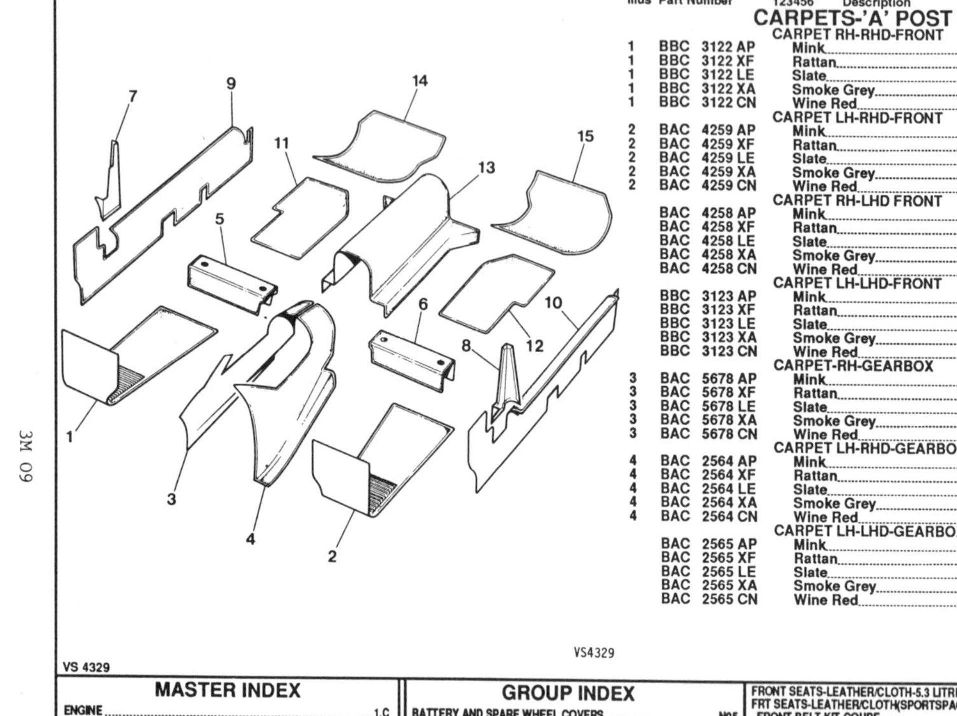

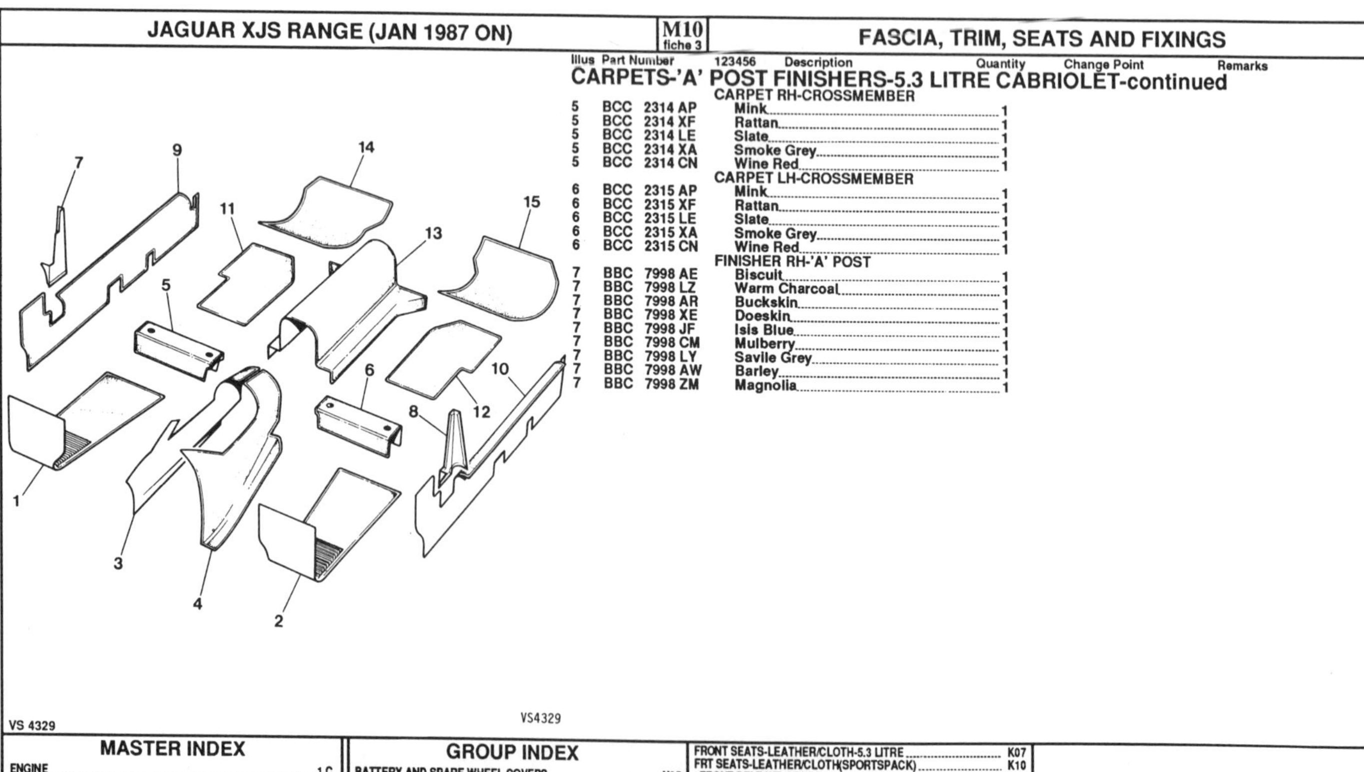

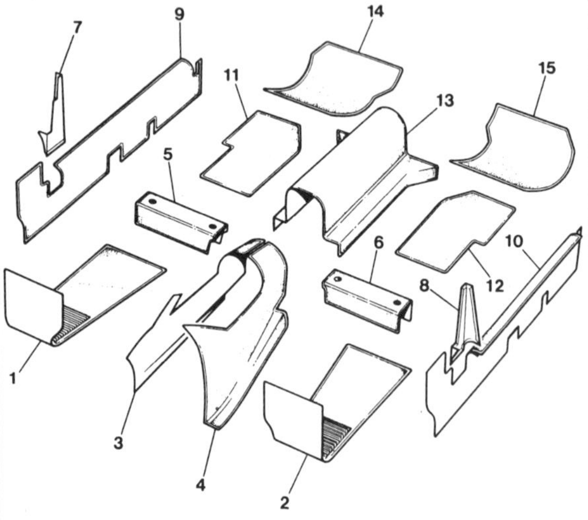

VS4329

VS 4329

CARPETS-'A' POST FINISHERS-5.3 LITRE CABRIOLET-continued

Illus	Part Number	Description	Quantity	Change Point	Remarks
		CARPET RH-CROSSMEMBER			
5	BCC 2314 AP	Mink	1		
5	BCC 2314 XF	Rattan	1		
5	BCC 2314 LE	Slate	1		
5	BCC 2314 XA	Smoke Grey	1		
5	BCC 2314 CN	Wine Red	1		
		CARPET LH-CROSSMEMBER			
6	BCC 2315 AP	Mink	1		
6	BCC 2315 XF	Rattan	1		
6	BCC 2315 LE	Slate	1		
6	BCC 2315 XA	Smoke Grey	1		
6	BCC 2315 CN	Wine Red	1		
		FINISHER RH-'A' POST			
7	BBC 7998 AE	Biscuit	1		
7	BBC 7998 LZ	Warm Charcoal	1		
7	BBC 7998 AR	Buckskin	1		
7	BBC 7998 XE	Doeskin	1		
7	BBC 7998 JF	Isis Blue	1		
7	BBC 7998 CM	Mulberry	1		
7	BBC 7998 LY	Savile Grey	1		
7	BBC 7998 AW	Barley	1		
7	BBC 7998 ZM	Magnolia	1		

3M 10

MASTER INDEX
ENGINE 1.C
FLYWHEEL AND CLUTCH 1.G
GEARBOX AND PROPSHAFT 1.H
AXLES, SUSPENSION, DRIVE SHAFTS, WHEELS 1.K
STEERING 1.M
BRAKES AND BRAKE CONTROLS 1.N
FUEL, EXHAUST AND EMISSION SYSTEMS 2.C
COOLING, HEATING AND AIR CONDITIONING 2.G
ELECTRICAL, WASHERS-WIPERS, INSTRUMENTS 2.J
CHASSIS, SUBFRAMES, BODYSHELL, FITTINGS 3.C
FASCIA, TRIM, SEATS AND FIXINGS 3.H
ACCESSORIES AND PAINT 4.C
NUMERICAL INDEX 4.D
GROUP INDEX
BATTERY AND SPARE WHEEL COVERS. N05
BONNET INSULATION-3.6 LITRE N06
BONNET INSULATION-5.3 MODELS N08
BONNET INSULATION-5.3 MODELS. N07
BOOT CARPETS - NOT CONVERTIBLE N03
BOOT CARPETS-CONVERTIBLE N04
CARPETS-'A' POST FINISHERS-. L12
CONSOLE-AMBLA. H10
CONSOLE-LEATHER. H08
DOOR CASINGS-ARMRESTS AND POCKETS-AMBLA. H16
DOOR CASINGS-ARMRESTS-CONVERTIBLE I02
HEADLINING I15
DOOR CASINGS-ARMRESTS-LEATHER. H14
EXTERIOR TRIM COVERS N15
FACIA PANEL. H02
FRONT SEAT BELTS-CONVERTIBLE K15
FRONT SEAT SLIDES K16
MOTOR AND ANCHORAGE ASSY-RH L02
FRONT SEATS-AMBLA/CLOTH.-3.6 LITRE K04
FRONT SEATS-LEATHER. J03
FRONT SEATS-LEATHER/CLOTH-5.3 LITRE K07
FRT SEATS-LEATHER/CLOTH(SPORTSPACK) K10
FRONT BELT KIT-COUPE K14
FRONT SEAT SLIDES K13
FUEL APERATURE SEAL AND HOSES-COUPE. N13
HARDTOP & HOODBAG-CABRIOLET. N17
HOOD LIFT MECHANISM - CONVERTIBLE O04
HOOD SEALS-CONVERTIBLE O03
HOOD-CABRIOLET N16
HOOD-CONVERTIBLE O02
REAR CUBBY BOX-CONVERTIBLE L09
REAR LUGGAGE LOCKER-CABRIOLET L07
REAR SEAT BELTS. L11
REAR SEATS-LEATHER-COUPE. L03
REAR SEATS-LEATHER-COUPE. L04
REAR SEATS-LEATHER/CLOTH-COUPE. L06
TARGA PANELS-STOWAGEBAG-CABRIOLET N14
UNDER SCUTTLE CASINGS. H05
VENEERED FACIA PANELS-3.6 LITRE. H06
VENEERED FACIA PANELS-5.3 LITRE. H07

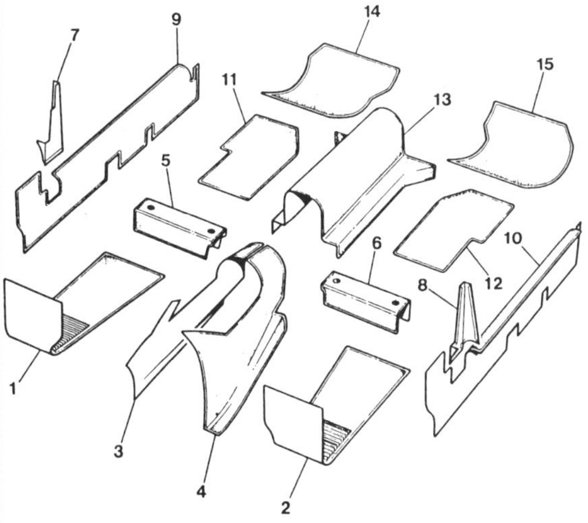

VS 4329

CARPETS-'A' POST FINISHERS-5.3 LITRE CABRIOLET-continued

Illus	Part Number	123456	Description	Quantity	Change Point	Remarks
			FINISHER LH-'A' POST			
8	BBC 7999 AE		Biscuit	1		
8	BBC 7999 LZ		Warm Charcoal	1		
8	BBC 7999 AR		Buckskin	1		
8	BBC 7999 XE		Doeskin	1		
8	BBC 7999 JF		Isis Blue	1		
8	BBC 7999 CM		Mulberry	1		
8	BBC 7999 LY		Savile Grey	1		
8	BBC 7999 AW		Barley	1		
8	BBC 7999 ZM		Magnolia	1		
			CARPET RH-SILL			
9	BCC 8152 AP		Mink	1		
9	BCC 8152 XF		Rattan	1		
9	BCC 8152 LE		Slate	1		
9	BCC 8152 XA		Smoke Grey	1		
9	BCC 8152 CN		Wine Red	1		
			CARPET LH-SILL			
10	BCC 8153 AP		Mink	1		
10	BCC 8153 XF		Rattan	1		
10	BCC 8153 LE		Slate	1		
10	BCC 8153 XA		Smoke Grey	1		
10	BCC 8153 CN		Wine Red	1		

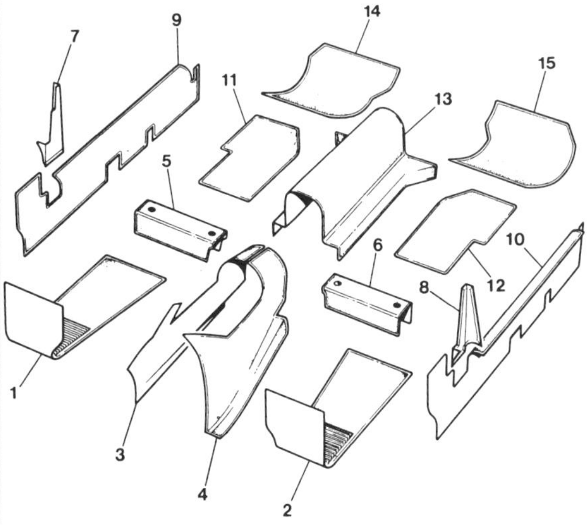

VS 4329

CARPETS-'A' POST FINISHERS-5.3 LITRE CABRIOLET-continued

Illus	Part Number	123456	Description	Quantity	Change Point	Remarks
			CARPET RH-REAR			
11	BCC 7992 AP		Mink	1		
11	BCC 7992 XF		Rattan	1		
11	BCC 7992 LE		Slate	1		
11	BCC 7992 XA		Smoke Grey	1		
11	BCC 7992 CN		Wine Red	1		
			CARPET LH-REAR			
12	BCC 7993 AP		Mink	1		
12	BCC 7993 XF		Rattan	1		
12	BCC 7993 LE		Slate	1		
12	BCC 7993 XA		Smoke Grey	1		
12	BCC 7993 CN		Wine Red	1		
			CARPET-TUNNEL-REAR			
13	BBC 2182 AP		Mink	1		
13	BBC 2182 XF		Rattan	1		
13	BBC 2182 LE		Slate	1		
13	BBC 2182 XA		Smoke Grey	1		
13	BBC 2182 CN		Wine Red	1		
			CARPET RH-LOCKER FLOOR			
14	BBC 2620 AP		Mink	1		
14	BBC 2620 XF		Rattan	1		
14	BBC 2620 LE		Slate	1		
14	BBC 2620 XA		Smoke Grey	1		
14	BBC 2620 CN		Wine Red	1		
			CARPET LH-LOCKER FLOOR			
15	BBC 2621 AP		Mink	1		
15	BBC 2621 XF		Rattan	1		
15	BBC 2621 LE		Slate	1		
15	BBC 2621 XA		Smoke Grey	1		
15	BBC 2621 CN		Wine Red	1		

MASTER INDEX

ENGINE 1.C
FLYWHEEL AND CLUTCH 1.G
GEARBOX AND PROPSHAFT 1.H
AXLES, SUSPENSION, DRIVE SHAFTS, WHEELS 1.K
STEERING 1.M
BRAKES AND BRAKE CONTROLS 1.N
FUEL, EXHAUST AND EMISSION SYSTEMS 2.C
COOLING, HEATING AND AIR CONDITIONING 2.G
ELECTRICAL, WASHERS-WIPERS, INSTRUMENTS 2.J
CHASSIS, SUBFRAMES, BODYSHELL, FITTINGS 3.C
FASCIA, TRIM, SEATS AND FIXINGS 3.H
ACCESSORIES AND PAINT 4.C
NUMERICAL INDEX 4.D

GROUP INDEX

BATTERY AND SPARE WHEEL COVERS N05
BONNET INSULATION-3.6 LITRE N06
BONNET INSULATION-5.3 MODELS N08
BONNET INSULATION-5.3 MODELS N07
BOOT CARPETS - NOT CONVERTIBLE N03
BOOT CARPETS-CONVERTIBLE N04
CARPETS-'A' POST FINISHERS- L12
CONSOLE-AMBLA H10
CONSOLE-LEATHER H08
DOOR CASINGS-ARMRESTS AND POCKETS-AMBLA H16
DOOR CASINGS-ARMRESTS-CONVERTIBLE I02
HEADLINING I15
DOOR CASINGS-ARMRESTS-LEATHER H14
EXTERIOR TRIM COVERS N15
FACIA PANEL H02
FRONT SEAT BELTS-CONVERTIBLE K15
FRONT SEAT SLIDES K16
MOTOR AND ANCHORAGE ASSY-RH L02
FRONT SEATS-AMBLA/CLOTH.-3.6 LITRE K04
FRONT SEATS-LEATHER J03
FRONT SEATS-LEATHER/CLOTH-5.3 LITRE K07
FRT SEATS-LEATHER/CLOTH(SPORTSPACK) K10
FRONT BELT KIT-COUPE K14
FRONT SEAT SLIDES K13
FUEL APERATURE SEAL AND HOSES-COUPE N13
HARDTOP & HOODBAG-CABRIOLET N17
HOOD LIFT MECHANISM - CONVERTIBLE O04
HOOD SEALS-CONVERTIBLE O03
HOOD-CABRIOLET N16
HOOD-CONVERTIBLE O02
REAR CUBBY BOX-CONVERTIBLE L09
REAR LUGGAGE LOCKER-CABRIOLET L07
REAR SEAT BELTS L11
REAR SEATS-LEATHER-COUPE L03
REAR SEATS-LEATHER-COUPE L04
REAR SEATS-LEATHER/CLOTH-COUPE L06
TARGA PANELS-STOWAGEBAG-CABRIOLET N14
UNDER SCUTTLE CASINGS H05
VENEERED FACIA PANELS-3.6 LITRE H06
VENEERED FACIA PANELS-5.3 LITRE H07

3M 13

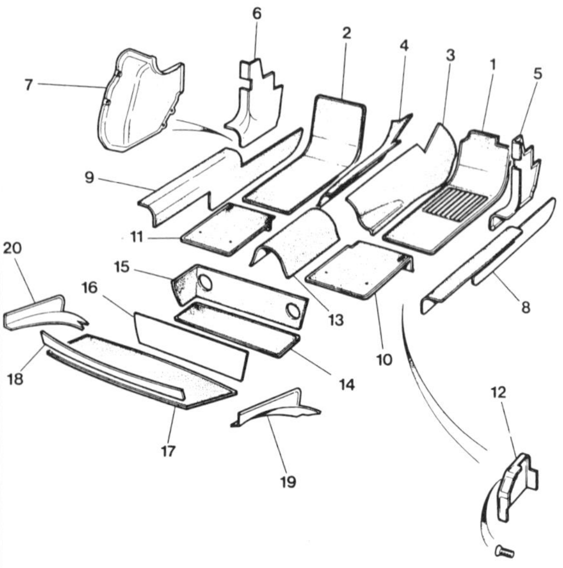

VS4846/B

CARPETS - "A" POST FINISHERS - CONVERTIBLE

Illus	Part Number	Description	Quantity	Change Point	Remarks
		CARPET-RH-RHD FRONT			
1	BDC 5584 AP	Mink	1		
1	BDC 5584 XF	Rattan	1		
1	BDC 5584 LE	Slate	1		
1	BDC 5584 XA	Smoke Grey	1		
1	BDC 5584 CN	Wine Red	1		
		CARPET-LH-RHD FRONT			
2	BDC 5595 AP	Mink	1		Up to 89 MY
2	BDC 5595 XF	Rattan	1		
2	BDC 5595 LE	Slate	1		
2	BDC 5595 XA	Smoke Grey	1		
2	BDC 5595 CN	Wine Red	1		
		CARPET - LH - RHD FRONT			
	BDC 3177 AP	Mink	1		From 89 MY
	BDC 3177 XF	Rattan	1		
	BDC 3177 LE	Slate	1		
	BDC 3177 XA	Smoke Grey	1		
	BDC 3177 CN	Wine Red	1		
		CARPET-RH-LHD FRONT			
	BDC 5592 AP	Mink	1		Up to 89 MY
	BDC 5592 XF	Rattan	1		
	BDC 5592 LE	Slate	1		
	BDC 5592 XA	Smoke Grey	1		
	BDC 5592 CN	Wine Red	1		
		CARPET - RH - LHD FRONT			
	BDC 3176 AP	Mink	1		From 89 MY
	BDC 3176 XF	Rattan	1		
	BDC 3176 LE	Slate	1		
	BDC 3176 XA	Smoke Grey	1		
	BDC 3176 CN	Wine Red	1		
		CARPET-LH-LHD FRONT			
	BDC 5599 AP	Mink	1		
	BDC 5599 XF	Rattan	1		
	BDC 5599 LE	Slate	1		
	BDC 5599 XA	Smoke Grey	1		
	BDC 5599 CN	Wine Red	1		
		CARPET-GEARBOX-RH-RHD			
3	BDC 5576 AP	Mink	1		
3	BDC 5576 XF	Rattan	1		
3	BDC 5576 LE	Slate	1		
3	BDC 5576 XA	Smoke Grey	1		
3	BDC 5576 CN	Wine Red	1		

M52.18JP12/10/88

VS4846/B

3M 14

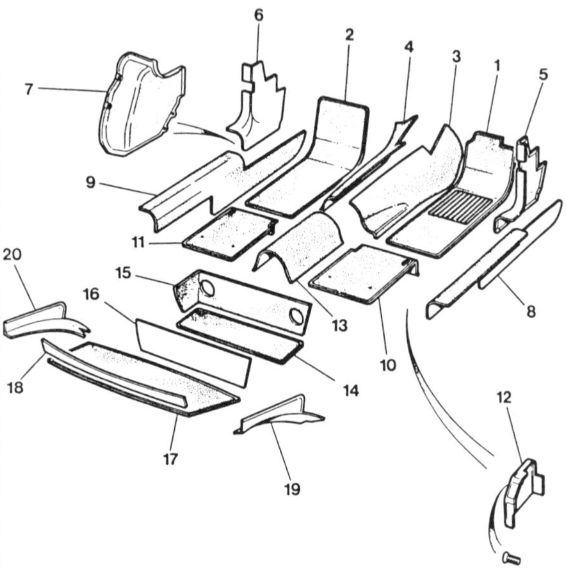

VS4846/B

CARPETS "A" POST FINISHERS-CONVERTIBLE-continued

Illus	Part Number	Description	Quantity	Change Point	Remarks
		CARPET-GEARBOX-LH-RHD			
4	BDC 5577 AP	Mink	1		
4	BDC 5577 XF	Rattan	1		
4	BDC 5577 LE	Slate	1		
4	BDC 5577 XA	Smoke Grey	1		
4	BDC 5577 CN	Wine Red	1		
		CARPET-GEARBOX-RH-LHD			
	BDC 5578 AP	Mink	1		
	BDC 5578 XF	Rattan	1		
	BDC 5578 LE	Slate	1		
	BDC 5578 XA	Smoke Grey	1		
	BDC 5578 CN	Wine Red	1		
		CARPET-GEARBOX-LH-LHD			
	BDC 5581 AP	Mink	1		
	BDC 5581 XF	Rattan	1		
	BDC 5581 LE	Slate	1		
	BDC 5581 XA	Smoke Grey	1		
	BDC 5581 CN	Wine Red	1		
		FINISHER-"A" POST RH			
5	BDC 6240 AW	Barley	1		Up to 89 MY
5	BDC 6240 AR	Buckskin	1		
5	BDC 6240 XE	Doeskin	1		
5	BDC 6240 JF	Isis Blue	1		
5	BDC 6240 ZM	Magnolia	1		
5	BDC 6240 CM	Mulberry	1		
5	BDC 6240 LY	Savile Grey	1		
5	BDC 6240 LZ	Warm Charcoal	1		
		FINISHER-"A" POST RH-RHD			
6	BEC 2004 AW	Barley	1		From 89 MY
6	BEC 2004 AR	Buckskin	1		
6	BEC 2004 XE	Doeskin	1		
6	BEC 2004 JF	Isis Blue	1		
6	BEC 2004 ZM	Magnolia	1		
6	BEC 2004 CM	Mulberry	1		
6	BEC 2004 LY	Savile Grey	1		
6	BEC 2004 LZ	Warm Charcoal	1		

M52.20JP12/10/88

VS4846/B

MASTER INDEX

ENGINE 1.C
FLYWHEEL AND CLUTCH 1.G
GEARBOX AND PROPSHAFT 1.H
AXLES, SUSPENSION, DRIVE SHAFTS, WHEELS 1.K
STEERING 1.M
BRAKES AND BRAKE CONTROLS 1.N
FUEL, EXHAUST AND EMISSION SYSTEMS 2.C
COOLING, HEATING AND AIR CONDITIONING 2.G
ELECTRICAL, WASHERS-WIPERS, INSTRUMENTS 2.J
CHASSIS, SUBFRAMES, BODYSHELL, FITTINGS 3.C
FASCIA, TRIM, SEATS AND FIXINGS 3.H
ACCESSORIES AND PAINT 4.C
NUMERICAL INDEX 4.D

GROUP INDEX

BATTERY AND SPARE WHEEL COVERS. N05
BONNET INSULATION-3.6 LITRE N06
BONNET INSULATION-5.3 MODELS N08
BONNET INSULATION-5.3 MODELS. N07
BOOT CARPETS - NOT CONVERTIBLE N03
BOOT CARPETS-CONVERTIBLE N04
CARPETS-'A' POST FINISHERS-. L12
CONSOLE-AMBLA. H10
CONSOLE-LEATHER. H08
DOOR CASINGS-ARMRESTS AND POCKETS-AMBLA. H16
DOOR CASINGS-ARMRESTS-CONVERTIBLE I02
HEADLINING I15
DOOR CASINGS-ARMRESTS-LEATHER. H14
EXTERIOR TRIM COVERS N15
FACIA PANEL. H02
FRONT SEAT BELTS-CONVERTIBLE K15
FRONT SEAT SLIDES K16
MOTOR AND ANCHORAGE ASSY-RH L02
FRONT SEATS-AMBLA/CLOTH.-3.6 LITRE K04
FRONT SEATS-LEATHER. J03
FRONT SEATS-LEATHER/CLOTH-5.3 LITRE K07
FRT SEATS-LEATHER/CLOTH(SPORTSPACK) K10
FRONT BELT KIT-COUPE K14
FRONT SEAT SLIDES K13
FUEL APERATURE SEAL AND HOSES-COUPE. N13
HARDTOP & HOODBAG-CABRIOLET. N17
HOOD LIFT MECHANISM - CONVERTIBLE O04
HOOD SEALS-CONVERTIBLE O03
HOOD-CABRIOLET N16
HOOD-CONVERTIBLE O02
REAR CUBBY BOX-CONVERTIBLE L09
REAR LUGGAGE LOCKER-CABRIOLET L07
REAR SEAT BELTS. L11
REAR SEATS-LEATHER-COUPE. L03
REAR SEATS-LEATHER-COUPE. L04
REAR SEATS-LEATHER/CLOTH-COUPE. L06
TARGA PANELS-STOWAGEBAG-CABRIOLET N14
UNDER SCUTTLE CASINGS. H05
VENEERED FACIA PANELS-3.6 LITRE. H06
VENEERED FACIA PANELS-5.3 LITRE. H07

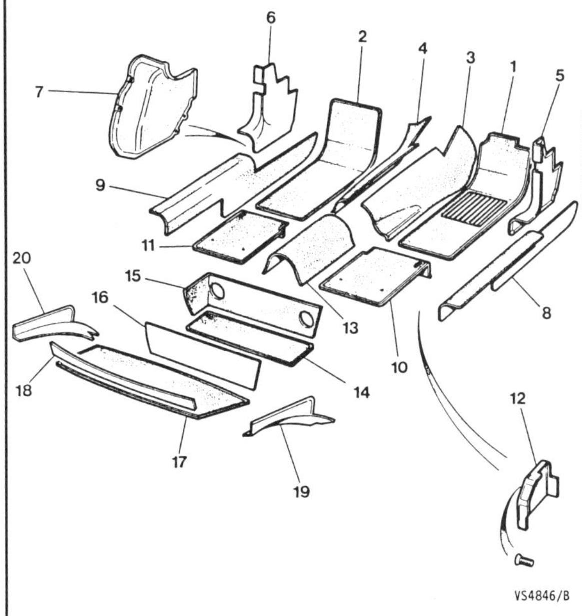

CARPETS 'A' POST FINISHERS-CONVERTIBLE-(Cont'd)

Illus	Part Number	123456 Description	Quantity	Change Point	Remarks
		FINISHER-'A' POST RH-LHD			
	BEC 2038 AW	Barley	1	From 89 MY	
	BEC 2038 AR	Buckskin	1		
	BEC 2038 XE	Doeskin	1		
	BEC 2038 JF	Isis blue	1		
	BEC 2038 ZM	Magnolia	1		
	BEC 2038 CM	Mulberry	1		
	BEC 2038 LY	Savile grey	1		
	BEC 2038 LZ	Warm charcoal	1		
		FINISHER-'A' POST LH			
6	BDC 6239 AW	Barley	1	Up to 89 MY	
6	BDC 6239 AR	Buckskin	1		
6	BDC 6239 XE	Doeskin	1		
6	BDC 6239 JF	Isis blue	1		
6	BDC 6239 ZM	Magnolia	1		
6	BDC 6239 CM	Mulberry	1		
6	BDC 6239 LY	Savile grey	1		
6	BDC 6239 LZ	Warm charcoal	1		
		FINSIHER-'A' POST LH-RHD			
	BEC 2039 AW	Barley	1	From 89 MY	
	BEC 2039 AR	Buckskin	1		
	BEC 2039 XE	Doeskin	1		
	BEC 2039 JF	Isis blue	1		
	BEC 2039 ZM	Magnolia	1		
	BEC 2039 CM	Mulberry	1		
	BEC 2039 LY	Savile grey	1		
	BEC 2039 LZ	Warm charcoal	1		
		FINSIHER-'A' POST LH-LHD			
	BEC 1431 AW	Barley	1	From 89 MY	
	BEC 1431 AR	Buckskin	1		
	BEC 1431 XE	Doeskin	1		
	BEC 1431 JF	Isis blue	1		
	BEC 1431 ZM	Magnolia	1		
	BEC 1431 CM	Mulberry	1		
	BEC 1431 LY	Savile grey	1		
	BEC 1431 LZ	Warm charcoal	1		

M52.22JP12/10/88

VS4846/B

MASTER INDEX

GROUP INDEX

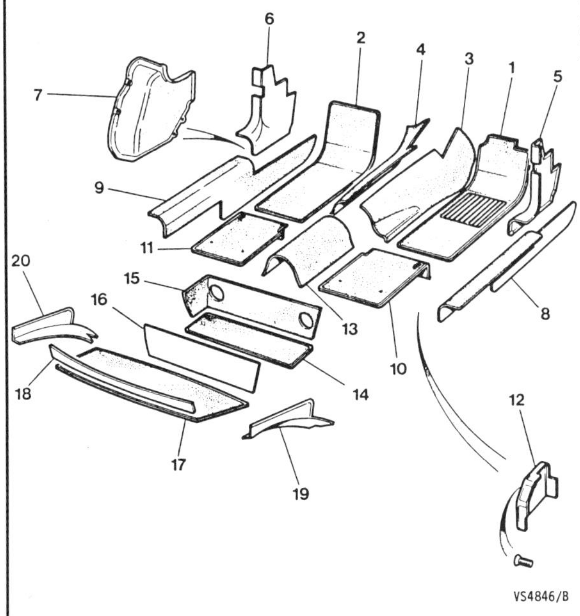

CARPETS-'A' POST FINISHERS-CONVERTIBLE-continued

Illus	Part Number	123456 Description	Quantity	Change Point	Remarks
		MOUNTING COVER-ECU-LH-RHD			
7	BEC 1693 AP	Mink	1	From 89 MY	
7	BEC 1693 XF	Rattan	1		
7	BEC 1693 LE	Slate	1		
7	BEC 1693 XA	Smoke grey	1		
7	BEC 1693 CN	Wine red	1		
		MOUNTING COVER-ECU-RH-LHD			
	BEC 1692 AP	Mink	1	From 89 MY	
	BEC 1692 XF	Rattan	1		
	BEC 1692 LE	Slate	1		
	BEC 1692 XA	Smoke grey	1		
	BEC 1692 CN	Wine red	1		
		CARPET-SILL-RH			
8	BDC 5702 AP	Mink	1	Up to 89 MY	
8	BDC 5702 XF	Rattan	1		
8	BDC 5702 LE	Slate	1		
8	BDC 5702 XA	Smoke Grey	1		
8	BDC 5702 CN	Wine Red	1		
		CARPET-SILL RH			
	BEC 1486 AP	Mink	1	From 89 MY	
	BEC 1486 XF	Rattan	1		
	BEC 1486 LE	Slate	1		
	BEC 1486 XA	Smoke grey	1		
	BEC 1486 CN	Wine red	1		
		CARPET-SILL-LH			
9	BDC 5701 AP	Mink	1	Up to 89 MY	
9	BDC 5701 XF	Rattan	1		
9	BDC 5701 LE	Slate	1		
9	BDC 5701 XA	Smoke Grey	1		
9	BDC 5701 CN	Wine Red	1		
		CARPET-SILL-LH			
	BEC 1487 AP	Mink	1	From 89 MY	
	BEC 1487 XF	Rattan	1		
	BEC 1487 LE	Slate	1		
	BEC 1487 XA	Smoke grey	1		
	BEC 1487 CN	Wine red	1		

M52.24JP12/10/88

VS4846/B

MASTER INDEX
ENGINE 1.C
FLYWHEEL AND CLUTCH 1.G
GEARBOX AND PROPSHAFT 1.H
AXLES, SUSPENSION, DRIVE SHAFTS, WHEELS 1.K
STEERING 1.M
BRAKES AND BRAKE CONTROLS 1.N
FUEL, EXHAUST AND EMISSION SYSTEMS 2.C
COOLING, HEATING AND AIR CONDITIONING 2.G
ELECTRICAL, WASHERS-WIPERS, INSTRUMENTS 2.J
CHASSIS, SUBFRAMES, BODYSHELL, FITTINGS 3.C
FASCIA, TRIM, SEATS AND FIXINGS 3.H
ACCESSORIES AND PAINT 4.C
NUMERICAL INDEX 4.D
GROUP INDEX
BATTERY AND SPARE WHEEL COVERS N05
BONNET INSULATION-3.6 LITRE N06
BONNET INSULATION-5.3 MODELS N08
BONNET INSULATION-5.3 MODELS N07
BOOT CARPETS - NOT CONVERTIBLE N03
BOOT CARPETS-CONVERTIBLE N04
CARPETS-'A' POST FINISHERS- L12
CONSOLE-AMBLA H10
CONSOLE-LEATHER H08
DOOR CASINGS-ARMRESTS AND POCKETS-AMBLA H16
DOOR CASINGS-ARMRESTS-CONVERTIBLE I02
HEADLINING I15
DOOR CASINGS-ARMRESTS-LEATHER H14
EXTERIOR TRIM COVERS N15
FACIA PANEL H02
FRONT SEAT BELTS-CONVERTIBLE K15
FRONT SEAT SLIDES K16
MOTOR AND ANCHORAGE ASSY-RH L02
FRONT SEATS-AMBLA/CLOTH.-3.6 LITRE K04
FRONT SEATS-LEATHER J03
FRONT SEATS-LEATHER/CLOTH-5.3 LITRE K07
FRT SEATS-LEATHER/CLOTH(SPORTSPACK) K10
FRONT BELT KIT-COUPE K14
FRONT SEAT SLIDES K13
FUEL APERATURE SEAL AND HOSES-COUPE N13
HARDTOP & HOODBAG-CABRIOLET N17
HOOD LIFT MECHANISM - CONVERTIBLE O04
HOOD SEALS-CONVERTIBLE O03
HOOD-CABRIOLET N16
HOOD-CONVERTIBLE O02
REAR CUBBY BOX-CONVERTIBLE L09
REAR LUGGAGE LOCKER-CABRIOLET L07
REAR SEAT BELTS L11
REAR SEATS-LEATHER-COUPE L03
REAR SEATS-LEATHER-COUPE L04
REAR SEATS-LEATHER/CLOTH-COUPE L06
TARGA PANELS-STOWAGEBAG-CABRIOLET N14
UNDER SCUTTLE CASINGS H05
VENEERED FACIA PANELS-3.6 LITRE H06
VENEERED FACIA PANELS-5.3 LITRE H07

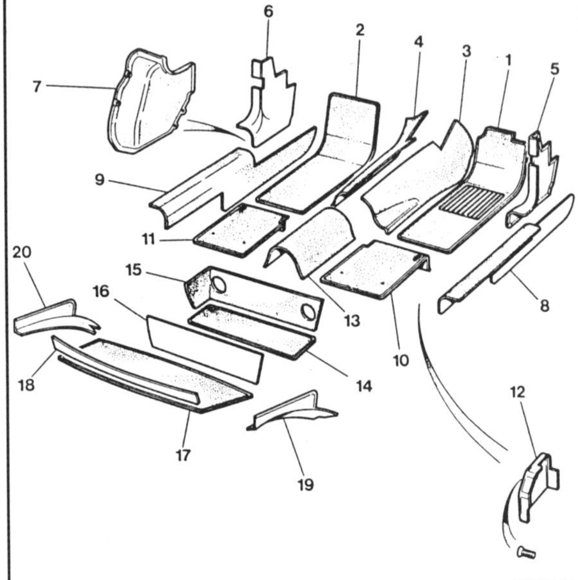

VS4846/B

3M 17

CARPETS-'A' POST FINISHERS-CONVERTIBLE-(Cont'd)

Illus	Part Number	Description	Quantity	Change Point	Remarks
		CARPET-REAR FLOOR RH			
10	BDC 8048 AP	Mink	1		
10	BDC 8048 XF	Rattan	1		
10	BDC 8048 LE	Slate	1		
10	BDC 8048 XA	Smoke grey	1		
10	BDC 8048 CN	Wine red	1		
		CARPET-REAR FLOOR LH			
11	BDC 8049 AP	Mink	1		
11	BDC 8049 XF	Rattan	1		
11	BDC 8049 LE	Slate	1		
11	BDC 8049 XA	Smoke grey	1		
11	BDC 8049 CN	Wine red	1		
12	BCC 5220	Cover-handbrake RHD	1		
	BCC 5219	Cover-handbrake LHD	1		

M52.26JP12/10/88

MASTER INDEX

GROUP INDEX

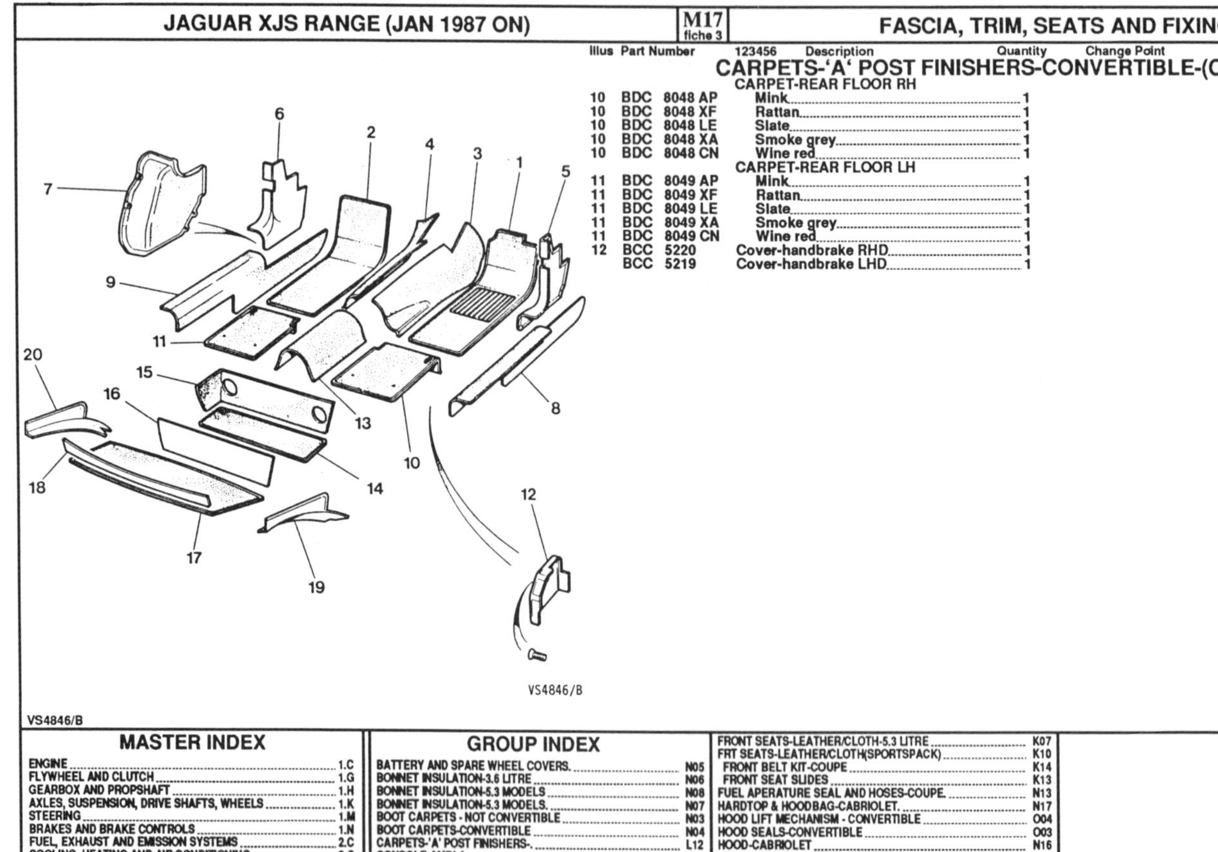

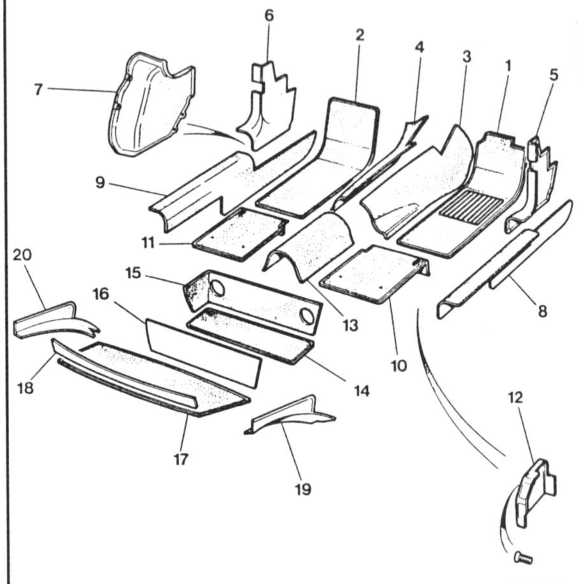

VS4846/B

3M 18

CARPETS-'A' POST FINISHERS-CONVERTIBLE-Continued

Illus	Part Number	Description	Quantity	Change Point	Remarks
		CARPET-TUNNEL			
13	BDC 8703 AP	Mink	1		
13	BDC 8703 XF	Rattan	1		
13	BDC 8703 LE	Slate	1		
13	BDC 8703 XA	Smoke Grey	1		
13	BDC 8703 CN	Wine Red	1		
		CARPET-CUBBY FLOOR			
14	BDC 5596 AP	Mink	1		
14	BDC 5596 XF	Rattan	1		
14	BDC 5596 LE	Slate	1		
14	BDC 5596 XA	Smoke Grey	1		
14	BDC 5596 CN	Wine Red	1		
		CARPET-CUBBY FRONT AND SIDE			
15	BDC 8428 AP	Mink	1		
15	BDC 8428 XF	Rattan	1		
15	BDC 8428 LE	Slate	1		
15	BDC 8428 XA	Smoke Grey	1		
15	BDC 8428 CM	Wine Red	1		
		CARPET-CUBBY REAR			
16	BDC 8417 AP	Mink	1		
16	BDC 8417 XF	Rattan	1		
16	BDC 8417 LE	Slate	1		
16	BDC 8417 XA	Smoke Grey	1		
16	BDC 8417 CN	Wine Red	1		
		CARPET-ROOF CONTAINER			
17	BDC 8424 AP	Mink	1		
17	BDC 8424 XF	Rattan	1		
17	BDC 8424 LE	Slate	1		
17	BDC 8424 XA	Smoke Grey	1		
17	BDC 8424 CN	Wine Red	1		
		CARPET-ROOF CONTAINER INFIL			
18	BDC 8426 AP	Mink	1		
18	BDC 8426 XF	Rattan	1		
18	BDC 8426 LE	Slate	1		
18	BDC 8426 XA	Smoke Grey	1		
18	BDC 8426 CN	Wine Red	1		

M52.28JP12/10/88

MASTER INDEX
ENGINE 1.C
FLYWHEEL AND CLUTCH 1.G
GEARBOX AND PROPSHAFT 1.H
AXLES, SUSPENSION, DRIVE SHAFTS, WHEELS 1.K
STEERING 1.M
BRAKES AND BRAKE CONTROLS 1.N
FUEL, EXHAUST AND EMISSION SYSTEMS 2.C
COOLING, HEATING AND AIR CONDITIONING 2.G
ELECTRICAL, WASHERS-WIPERS, INSTRUMENTS 2.J
CHASSIS, SUBFRAMES, BODYSHELL, FITTINGS 3.C
FASCIA, TRIM, SEATS AND FIXINGS 3.H
ACCESSORIES AND PAINT 4.C
NUMERICAL INDEX 4.D
GROUP INDEX
BATTERY AND SPARE WHEEL COVERS N05
BONNET INSULATION-3.6 LITRE N06
BONNET INSULATION-5.3 MODELS N08
BONNET INSULATION-5.3 MODELS N07
BOOT CARPETS - NOT CONVERTIBLE N03
BOOT CARPETS-CONVERTIBLE N04
CARPETS-'A' POST FINISHERS- L12
CONSOLE-AMBLA H10
CONSOLE-LEATHER H08
DOOR CASINGS-ARMRESTS AND POCKETS-AMBLA H16
DOOR CASINGS-ARMRESTS-CONVERTIBLE I02
HEADLINING I15
DOOR CASINGS-ARMRESTS-LEATHER H14
EXTERIOR TRIM COVERS N15
FACIA PANEL H02
FRONT SEAT BELTS-CONVERTIBLE K15
FRONT SEAT SLIDES K16
MOTOR AND ANCHORAGE ASSY-RH L02
FRONT SEATS-AMBLA/CLOTH.-3.6 LITRE K04
FRONT SEATS-LEATHER J03
FRONT SEATS-LEATHER/CLOTH-5.3 LITRE K07
FRT SEATS-LEATHER/CLOTH(SPORTSPACK) K10
FRONT BELT KIT-COUPE K14
FRONT SEAT SLIDES K13
FUEL APERATURE SEAL AND HOSES-COUPE N13
HARDTOP & HOODBAG-CABRIOLET N17
HOOD LIFT MECHANISM - CONVERTIBLE O04
HOOD SEALS-CONVERTIBLE O03
HOOD-CABRIOLET N16
HOOD-CONVERTIBLE O02
REAR CUBBY BOX-CONVERTIBLE L09
REAR LUGGAGE LOCKER-CABRIOLET L07
REAR SEAT BELTS L11
REAR SEATS-LEATHER-COUPE L03
REAR SEATS-LEATHER-COUPE L04
REAR SEATS-LEATHER/CLOTH-COUPE L06
TARGA PANELS-STOWAGEBAG-CABRIOLET N14
UNDER SCUTTLE CASINGS H05
VENEERED FACIA PANELS-3.6 LITRE H06
VENEERED FACIA PANELS-5.3 LITRE H07

JAGUAR XJS RANGE (JAN 1987 ON) | N02 fiche 3 | FASCIA, TRIM, SEATS AND FIXINGS

CARPETS-"A" POST FINISHERS-CONVERTIBLE-continued

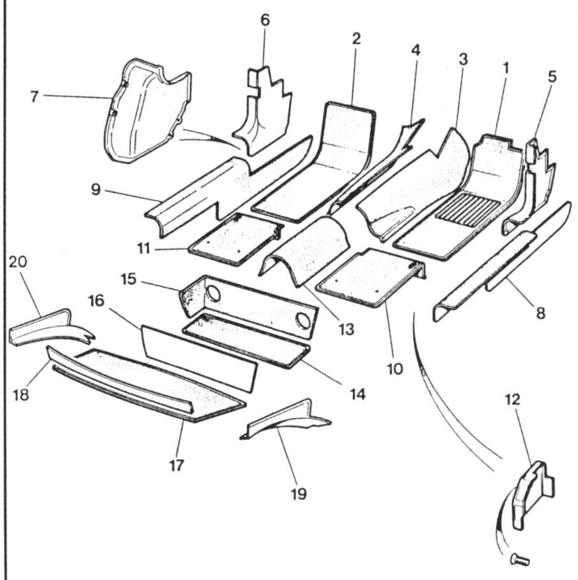

VS4846/B

VS4846/B

Illus	Part Number	123456	Description	Quantity	Change Point	Remarks
			CARPET-WHEELARCH-RH			
19	BDC 8418 AP		Mink	1		
19	BDC 8418 XF		Rattan	1		
19	BDC 8418 LE		Slate	1		
19	BDC 8418 XA		Smoke Grey	1		
19	BDC 8418 CN		Wine Red	1		
			CARPET-WHEELARCH-LH			
20	BDC 8419 AP		Mink	1		
20	BDC 8419 XF		Rattan	1		
20	BDC 8419 LE		Slate	1		
20	BDC 8419 XA		Smoke Grey	1		
20	BDC 8419 CN		Wine Red	1		

M52.30JP12/10/88

JAGUAR XJS RANGE (JAN 1987 ON) | N03 fiche 3 | FASCIA, TRIM, SEATS AND FIXINGS

BOOT CARPETS - NOT CONVERTIBLE

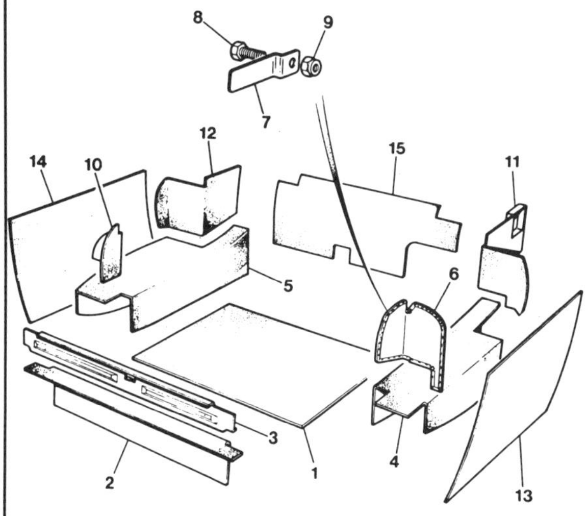

VS4491

VS 4491

Illus	Part Number	123456	Description	Quantity	Change Point	Remarks
1	BAC 4364		Carpet-boot floor	1		
2	BAC 4085		Carpet-rear panel	1		
3	BD 46684		Retainer-finisher	1		
4	BBC 7546		Carpet RH-boot beam	1		
5	BAC 4089		Carpet LH-boot beam	1		
6	BBC 2583		Carpet RH-aerial motor	1		
7	BBC 2584		Bracket	1		
8	UFS 819 4H		Screw-set	1		
9	C 8737 9		Nut-nyloc	1		
10	BAC 4087		Carpet LH-tail lamp	1		
11	BBC 7548		Carpet RH-wheelarch	1		
12	BAC 4187		Carpet LH-wheelarch	1		Coupe
12	BBC 2649		Carpet LH-wheelarch	1		Cabriolet
13	BBC 3286		Carpet RH-boot side	1		
14	BAC 5911		Carpet LH-boot side	1		
15	BAC 4084		Carpet-fuel tank	1	Up to (V)142986	Coupe
15	BCC 8515		Carpet-fuel tank	1	From (V)142987	Coupe
15	BBC 4460		Carpet-fuel tank	1		Cabriolet

MASTER INDEX

ENGINE 1.C
FLYWHEEL AND CLUTCH 1.G
GEARBOX AND PROPSHAFT 1.H
AXLES, SUSPENSION, DRIVE SHAFTS, WHEELS 1.K
STEERING 1.M
BRAKES AND BRAKE CONTROLS 1.N
FUEL, EXHAUST AND EMISSION SYSTEMS 2.C
COOLING, HEATING AND AIR CONDITIONING 2.G
ELECTRICAL, WASHERS-WIPERS, INSTRUMENTS 2.J
CHASSIS, SUBFRAMES, BODYSHELL, FITTINGS 3.C
FASCIA, TRIM, SEATS AND FIXINGS 3.H
ACCESSORIES AND PAINT 4.C
NUMERICAL INDEX 4.D

GROUP INDEX

BATTERY AND SPARE WHEEL COVERS N05
BONNET INSULATION-3.6 LITRE N06
BONNET INSULATION-5.3 MODELS N08
BONNET INSULATION-5.3 MODELS N07
BOOT CARPETS - NOT CONVERTIBLE N03
BOOT CARPETS-CONVERTIBLE N04
CARPETS-'A' POST FINISHERS- L12
CONSOLE-AMBLA H10
CONSOLE-LEATHER H08
DOOR CASINGS-ARMRESTS AND POCKETS-AMBLA H16
DOOR CASINGS-ARMRESTS-CONVERTIBLE I02
HEADLINING I15
DOOR CASINGS-ARMRESTS-LEATHER H14
EXTERIOR TRIM COVERS N15
FACIA PANEL H02
FRONT SEAT BELTS-CONVERTIBLE K15
FRONT SEAT SLIDES K16
MOTOR AND ANCHORAGE ASSY-RH L02
FRONT SEATS-AMBLA/CLOTH.-3.6 LITRE K04
FRONT SEATS-LEATHER J03
FRONT SEATS-LEATHER/CLOTH-5.3 LITRE K07
FRT SEATS-LEATHER/CLOTH(SPORTSPACK) K10
FRONT BELT KIT-COUPE K14
FRONT SEAT SLIDES K13
FUEL APERATURE SEAL AND HOSES-COUPE N13
HARDTOP & HOODBAG-CABRIOLET N17
HOOD LIFT MECHANISM - CONVERTIBLE O04
HOOD SEALS-CONVERTIBLE O03
HOOD-CABRIOLET N16
HOOD-CONVERTIBLE O02
REAR CUBBY BOX-CONVERTIBLE L09
REAR LUGGAGE LOCKER-CABRIOLET L07
REAR SEAT BELTS L11
REAR SEATS-LEATHER-COUPE L03
REAR SEATS-LEATHER-COUPE L04
REAR SEATS-LEATHER/CLOTH-COUPE L06
TARGA PANELS-STOWAGEBAG-CABRIOLET N14
UNDER SCUTTLE CASINGS H05
VENEERED FACIA PANELS-3.6 LITRE H06
VENEERED FACIA PANELS-5.3 LITRE H07

3N 04

BOOT CARPETS-CONVERTIBLE

Illus	Part Number	123456	Description	Quantity	Change Point	Remarks
1	BAC 4364		Carpet-boot floor	1		
2	BAC 4085		Carpet-rear panel	1		
3	BD 46684		Retainer-finisher	1		
4	BBC 7546		Carpet RH-boot beam	1		
5	BAC 4089		Carpet LH-boot beam	1		
6	BBC 2583		Carpet RH-aerial motor	1		
7	BAC 4087		Carpet LH-tail lamp	1		
8	BDC 4538		Carpet RH-wheelarch	1		
9	BDC 4559		Carpet LH-wheelarch	1		
10	BDC 3122		Cover-relay	1		
11	BDC 4994		Stud-fixing	1		
12	BBC 3286		Carpet RH-boot side	1		
13	BAC 5911		Carpet LH-boot side	1		
14	BDC 6358		Carpet-fuel tank	1		
15	BDC 6770		Cover-tank clamp	1		
16	AGU 1091		Button-'W'	2		

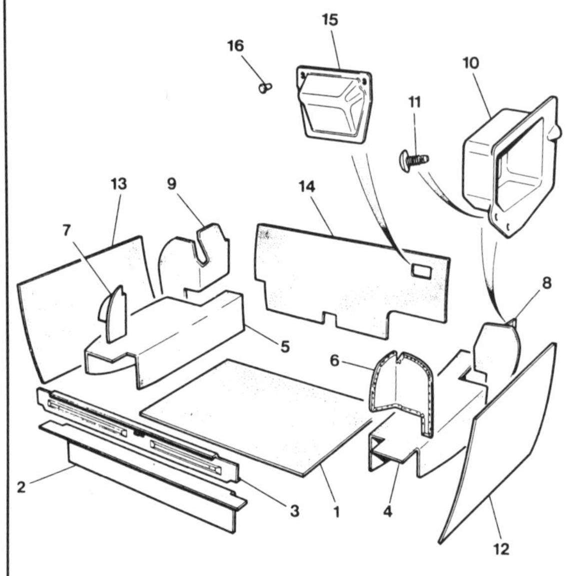

VS4857

VS4857

3N 05

BATTERY AND SPARE WHEEL COVERS.

Illus	Part Number	123456	Description	Quantity	Change Point	Remarks
1	BBC 5180		Cover-battery	1	Up to (V)145729	
1	BCC 9764		Cover-battery	1	From (V)145730	
2	C 46024		Panel-support	1	Up to (V)145729	
2	BDC 6091		Panel-support	1	From (V)145730	
3	BD 44643 1		Strip edging	1		
4	BD 46140 1		Stud-turnbuckle	2		
5	BD 46140 3		Washer-turnbuckle	2		
6	BD 46140 2		Cam-turnbuckle	2		
7	DBB 80810 C		Screw-self-tap	2		
8	BD 46141		Clip-edge	2		
9	CAC 5099		Bolt-hook	NLA		Use CBC4388
9	CBC 4388		Bolt-hook	1		
10	CAC 5097		Spacer	1		
11	C 44668		Screw-thumb	1		
12	BAC 5913		Cover-spare wheel	NLA		Use BCC9414
12	BCC 9414		Cover-spare wheel	1		

M56JP11/7/88

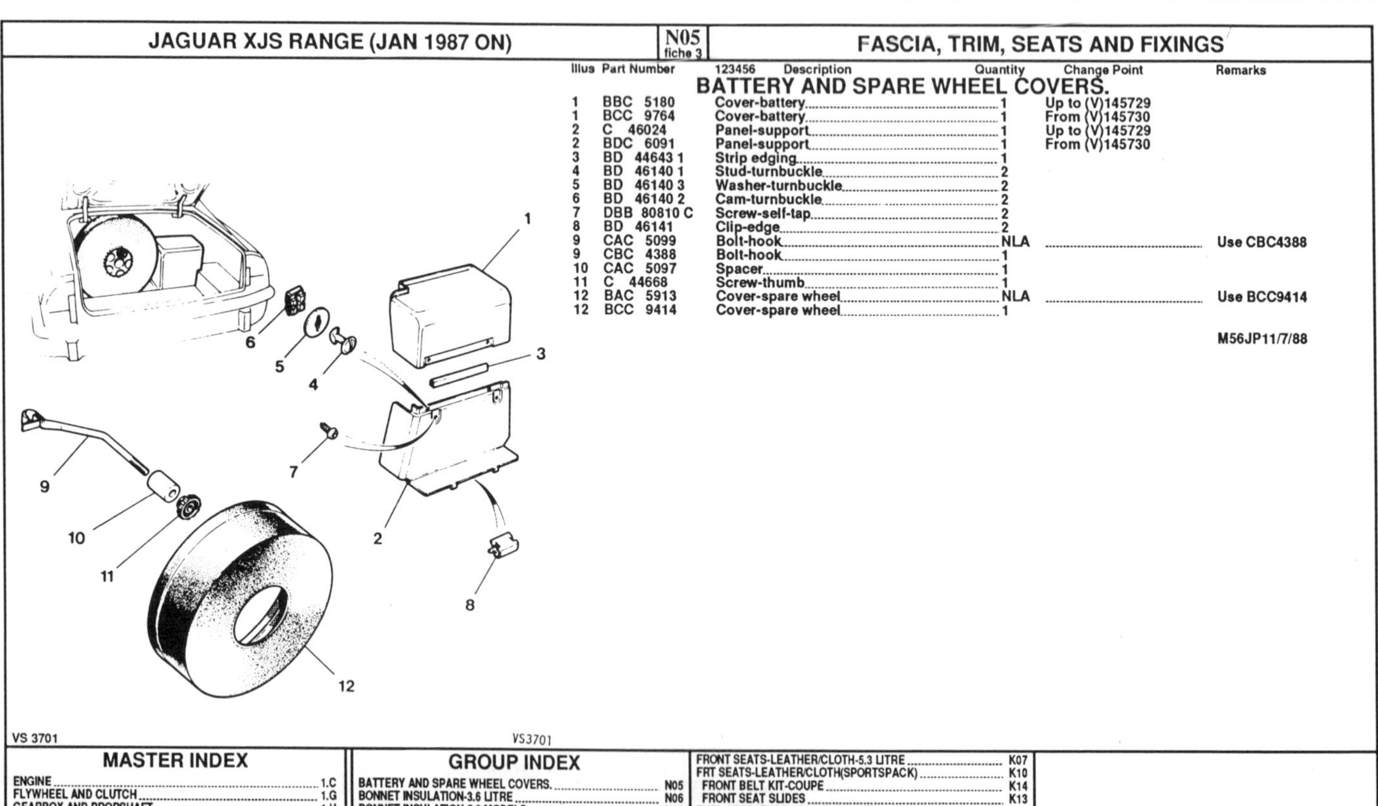

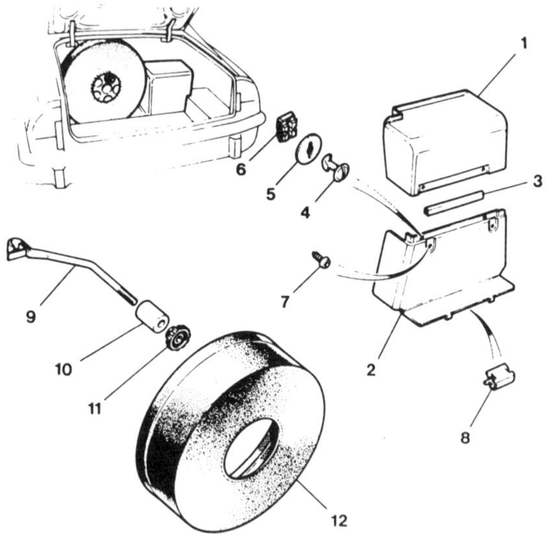

VS 3701

VS3701

MASTER INDEX

ENGINE 1.C
FLYWHEEL AND CLUTCH 1.G
GEARBOX AND PROPSHAFT 1.H
AXLES, SUSPENSION, DRIVE SHAFTS, WHEELS 1.K
STEERING 1.M
BRAKES AND BRAKE CONTROLS 1.N
FUEL, EXHAUST AND EMISSION SYSTEMS 2.C
COOLING, HEATING AND AIR CONDITIONING 2.G
ELECTRICAL, WASHERS-WIPERS, INSTRUMENTS 2.J
CHASSIS, SUBFRAMES, BODYSHELL, FITTINGS 3.C
FASCIA, TRIM, SEATS AND FIXINGS 3.H
ACCESSORIES AND PAINT 4.C
NUMERICAL INDEX 4.D

GROUP INDEX

BATTERY AND SPARE WHEEL COVERS N05
BONNET INSULATION-3.6 LITRE N06
BONNET INSULATION-5.3 MODELS N08
BONNET INSULATION-5.3 MODELS N07
BOOT CARPETS - NOT CONVERTIBLE N03
BOOT CARPETS-CONVERTIBLE N04
CARPETS-'A' POST FINISHERS- L12
CONSOLE-AMBLA H10
CONSOLE-LEATHER H08
DOOR CASINGS-ARMRESTS AND POCKETS-AMBLA H16
DOOR CASINGS-ARMRESTS-CONVERTIBLE I02
HEADLINING I15
DOOR CASINGS-ARMRESTS-LEATHER H14
EXTERIOR TRIM COVERS N15
FACIA PANEL H02
FRONT SEAT BELTS-CONVERTIBLE K15
FRONT SEAT SLIDES K16
MOTOR AND ANCHORAGE ASSY-RH L02
FRONT SEATS-AMBLA/CLOTH.-3.6 LITRE K04
FRONT SEATS-LEATHER J03
FRONT SEATS-LEATHER/CLOTH-5.3 LITRE K07
FRT SEATS-LEATHER/CLOTH(SPORTSPACK) K10
FRONT BELT KIT-COUPE K14
FRONT SEAT SLIDES K13
FUEL APERATURE SEAL AND HOSES-COUPE N13
HARDTOP & HOODBAG-CABRIOLET N17
HOOD LIFT MECHANISM - CONVERTIBLE O04
HOOD SEALS-CONVERTIBLE O03
HOOD-CABRIOLET N16
HOOD-CONVERTIBLE O02
REAR CUBBY BOX-CONVERTIBLE L09
REAR LUGGAGE LOCKER-CABRIOLET L07
REAR SEAT BELTS L11
REAR SEATS-LEATHER-COUPE L03
REAR SEATS-LEATHER-COUPE L04
REAR SEATS-LEATHER/CLOTH-COUPE L06
TARGA PANELS-STOWAGEBAG-CABRIOLET N14
UNDER SCUTTLE CASINGS H05
VENEERED FACIA PANELS-3.6 LITRE H06
VENEERED FACIA PANELS-5.3 LITRE H07

BONNET INSULATION-3.6 LITRE

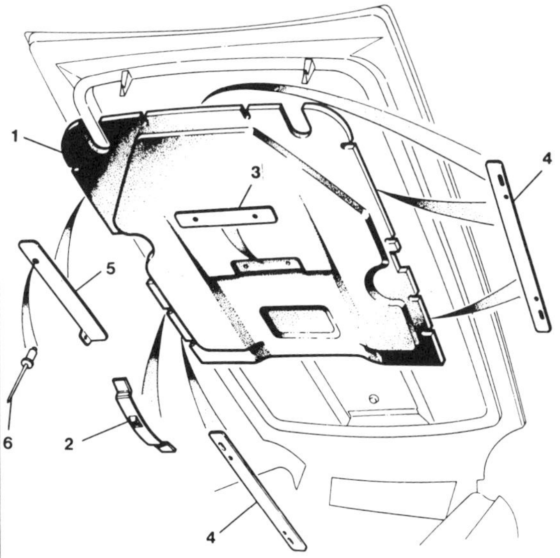

VS3714

Illus	Part Number	123456 Description	Quantity	Change Point	Remarks
1	BCC 9876	Pad-bonnet insulation	1		Use BDC 5796
1	BDC 5796	Pad-bonnet insulation	1		
2	BAC 3787	Clip-spring	11		
3	BBC 1454	Plate-centre	1		
4	BBC 1455	Plate-side and rear	4		
	BCC 2178	Plate-RH rear	1	Up to (V)148944	
5	BCC 2179	Plate-LH rear	1	Up to (V)148944	
	BDC 5797	Plate-retention	2	From (V)148945	
6	RA 608063 J	Rivet	1		

VS 3714

MASTER INDEX

GROUP INDEX

BONNET INSULATION-5.3 MODELS.

UPTO (V)148781

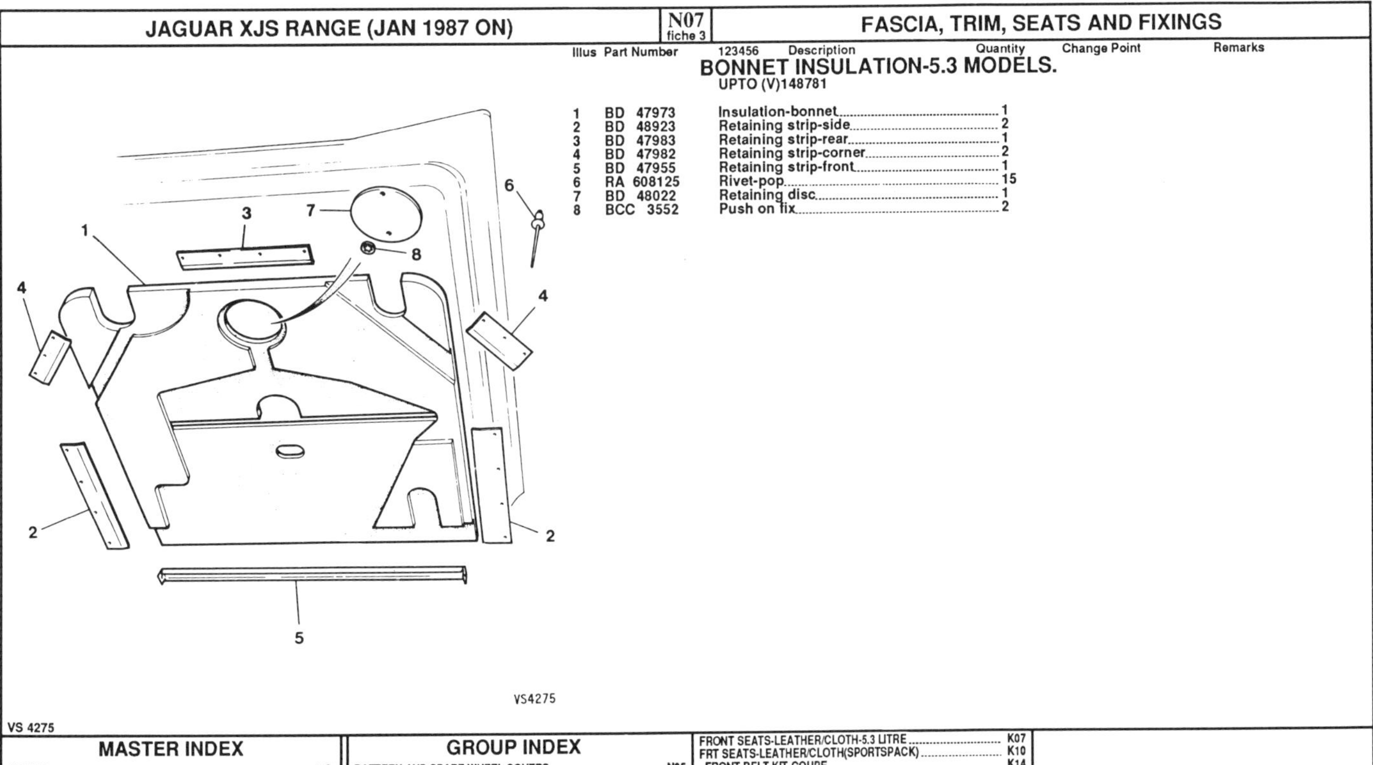

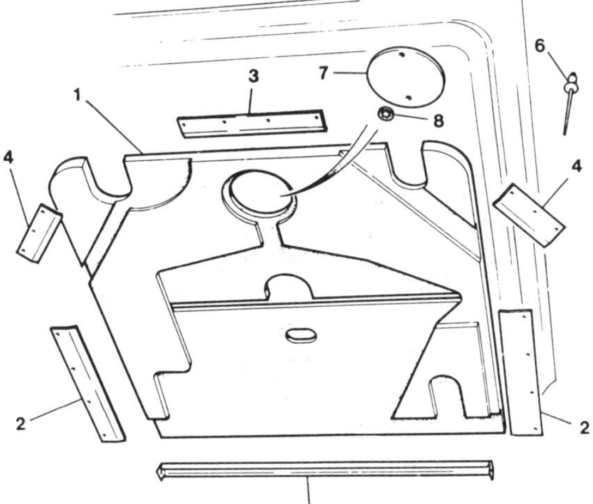

VS4275

Illus	Part Number	123456 Description	Quantity	Change Point	Remarks
1	BD 47973	Insulation-bonnet	1		
2	BD 48923	Retaining strip-side	2		
3	BD 47983	Retaining strip-rear	1		
4	BD 47982	Retaining strip-corner	2		
5	BD 47955	Retaining strip-front	1		
6	RA 608125	Rivet-pop	15		
7	BD 48022	Retaining disc	1		
8	BCC 3552	Push on fix	2		

VS 4275

MASTER INDEX
ENGINE 1.C
FLYWHEEL AND CLUTCH 1.G
GEARBOX AND PROPSHAFT 1.H
AXLES, SUSPENSION, DRIVE SHAFTS, WHEELS 1.K
STEERING 1.M
BRAKES AND BRAKE CONTROLS 1.N
FUEL, EXHAUST AND EMISSION SYSTEMS 2.C
COOLING, HEATING AND AIR CONDITIONING 2.G
ELECTRICAL, WASHERS-WIPERS, INSTRUMENTS 2.J
CHASSIS, SUBFRAMES, BODYSHELL, FITTINGS 3.C
FASCIA, TRIM, SEATS AND FIXINGS 3.H
ACCESSORIES AND PAINT 4.C
NUMERICAL INDEX 4.D
GROUP INDEX
BATTERY AND SPARE WHEEL COVERS. N05
BONNET INSULATION-3.6 LITRE N06
BONNET INSULATION-5.3 MODELS N08
BONNET INSULATION-5.3 MODELS. N07
BOOT CARPETS - NOT CONVERTIBLE N03
BOOT CARPETS-CONVERTIBLE N04
CARPETS-'A' POST FINISHERS-. L12
CONSOLE-AMBLA. H10
CONSOLE-LEATHER. H08
DOOR CASINGS-ARMRESTS AND POCKETS-AMBLA. H16
DOOR CASINGS-ARMRESTS-CONVERTIBLE I02
HEADLINING I15
DOOR CASINGS-ARMRESTS-LEATHER. H14
EXTERIOR TRIM COVERS N15
FACIA PANEL. H02
FRONT SEAT BELTS-CONVERTIBLE K15
FRONT SEAT SLIDES K16
MOTOR AND ANCHORAGE ASSY-RH L02
FRONT SEATS-AMBLA/CLOTH.-3.6 LITRE K04
FRONT SEATS-LEATHER. J03
FRONT SEATS-LEATHER/CLOTH-5.3 LITRE K07
FRT SEATS-LEATHER/CLOTH(SPORTSPACK) K10
FRONT BELT KIT-COUPE K14
FRONT SEAT SLIDES K13
FUEL APERATURE SEAL AND HOSES-COUPE. N13
HARDTOP & HOODBAG-CABRIOLET. N17
HOOD LIFT MECHANISM - CONVERTIBLE O04
HOOD SEALS-CONVERTIBLE O03
HOOD-CABRIOLET N16
HOOD-CONVERTIBLE O02
REAR CUBBY BOX-CONVERTIBLE L09
REAR LUGGAGE LOCKER-CABRIOLET L07
REAR SEAT BELTS. L11
REAR SEATS-LEATHER-COUPE. L03
REAR SEATS-LEATHER-COUPE. L04
REAR SEATS-LEATHER/CLOTH-COUPE. L06
TARGA PANELS-STOWAGEBAG-CABRIOLET N14
UNDER SCUTTLE CASINGS. H05
VENEERED FACIA PANELS-3.6 LITRE. H06
VENEERED FACIA PANELS-5.3 LITRE. H07

3N 08

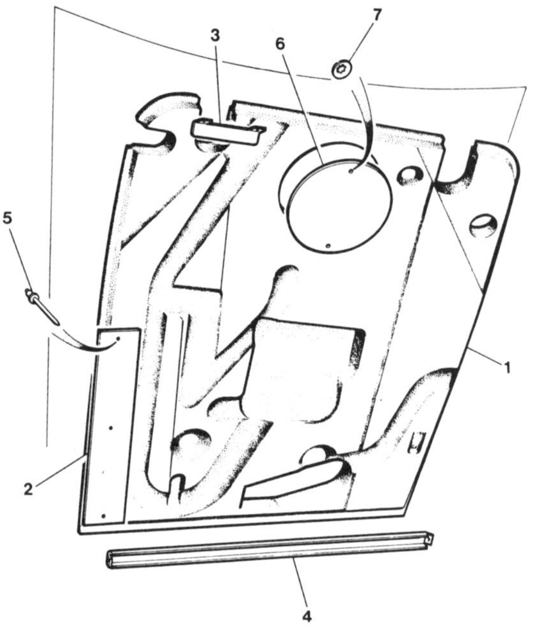

VS4845

BONNET INSULATION-5.3 MODELS

FROM (V)148782

Illus	Part Number	Description	Quantity	Change Point	Remarks
1	BDC 7973	Insulation-bonnet	1		
2	BDC 7951	Retaining strip-side	2		
3	BDC 5797	Retaining strip-rear	2		
4	BD 47955	Retaining strip-front	1		
5	RA 608125	Rivet-pop	12		
6	BD 48022	Retaining disc	1		
7	BCC 3552	Fix-push on	2		

M58.04JP1/9/88

MASTER INDEX

GROUP INDEX

3N 09

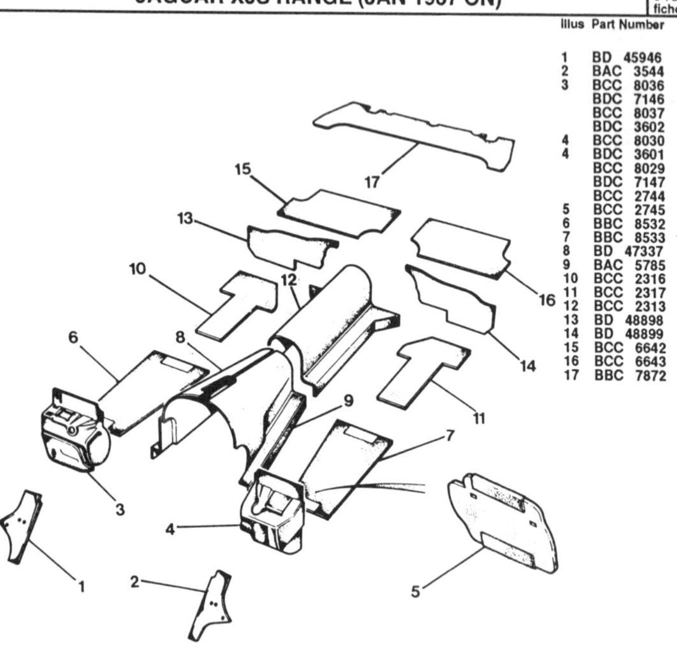

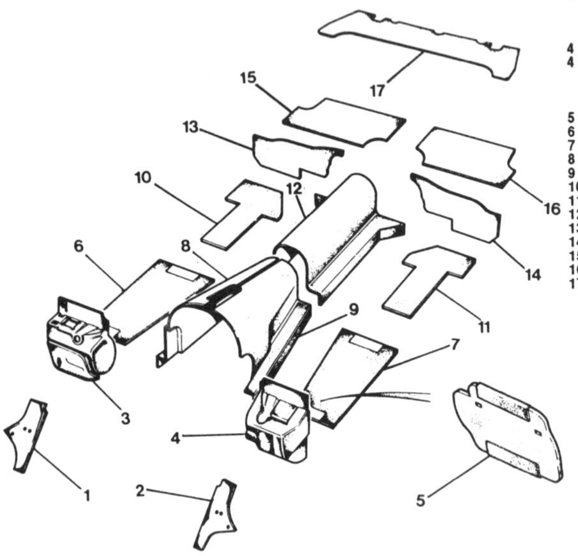

SM7287

SOUND INSULATION AND HEATSHIELDS-COUPE.

Illus	Part Number	Description	Quantity	Change Point	Remarks
1	BD 45946	Heatshield RH	1		
2	BAC 3544	Heatshield LH	1		
3	BCC 8036	Insulation RH-RHD-toeboard	1	....	Use BDC 7146
	BDC 7146	Insulation RH-RHD-toeboard	1		
	BCC 8037	Insulation RH-LHD-toeboard	1	....	Use BDC 3602
	BDC 3602	Insulation RH-LHD-toeboard	1		
4	BCC 8030	Insulation LH-RHD-toeboard	1	....	Use BDC 3601
4	BDC 3601	Insulation LH-RHD-toeboard	1		
	BCC 8029	Insulation LH-LHD-toeboard	1	....	Use BDC 7147
	BDC 7147	Insulation LH-LHD-toeboard	1		
	BCC 2744	Sound deadening RH-LHD-toeboard	1	....	3.6 litre only
5	BCC 2745	Sound deadening LH-RHD-toeboard	1	....	
6	BBC 8532	Insulation RH-front floor	1		
7	BBC 8533	Insulation LH-front floor	1		
8	BD 47337	Insulation RH-gearbox	1		
9	BAC 5785	Insulation LH-gearbox	1		
10	BCC 2316	Insulation RH-rear floor	1		
11	BCC 2317	Insulation LH-rear floor	1		
12	BCC 2313	Insulation-rear tunnel	1		
13	BD 48898	Insulation RH-crossmember	1		
14	BD 48899	Insulation LH-crossmember	1		
15	BCC 6642	Insulation RH-seat pan	1		
16	BCC 6643	Insulation LH-seat pan	1		
17	BBC 7872	Insulation-parcel shelf	1		

MASTER INDEX
ENGINE 1.C
FLYWHEEL AND CLUTCH 1.G
GEARBOX AND PROPSHAFT 1.H
AXLES, SUSPENSION, DRIVE SHAFTS, WHEELS 1.K
STEERING 1.M
BRAKES AND BRAKE CONTROLS 1.N
FUEL, EXHAUST AND EMISSION SYSTEMS 2.C
COOLING, HEATING AND AIR CONDITIONING 2.G
ELECTRICAL, WASHERS-WIPERS, INSTRUMENTS 2.J
CHASSIS, SUBFRAMES, BODYSHELL, FITTINGS 3.C
FASCIA, TRIM, SEATS AND FIXINGS 3.H
ACCESSORIES AND PAINT 4.C
NUMERICAL INDEX 4.D
GROUP INDEX
BATTERY AND SPARE WHEEL COVERS N05
BONNET INSULATION-3.6 LITRE N06
BONNET INSULATION-5.3 MODELS N08
BONNET INSULATION-5.3 MODELS N07
BOOT CARPETS - NOT CONVERTIBLE N03
BOOT CARPETS-CONVERTIBLE N04
CARPETS-'A' POST FINISHERS- L12
CONSOLE-AMBLA H10
CONSOLE-LEATHER H08
DOOR CASINGS-ARMRESTS AND POCKETS-AMBLA H16
DOOR CASINGS-ARMRESTS-CONVERTIBLE I02
HEADLINING I15
DOOR CASINGS-ARMRESTS-LEATHER H14
EXTERIOR TRIM COVERS N15
FACIA PANEL H02
FRONT SEAT BELTS-CONVERTIBLE K15
FRONT SEAT SLIDES K16
MOTOR AND ANCHORAGE ASSY-RH L02
FRONT SEATS-AMBLA/CLOTH.-3.6 LITRE K04
FRONT SEATS-LEATHER J03
FRONT SEATS-LEATHER/CLOTH-5.3 LITRE K07
FRT SEATS-LEATHER/CLOTH(SPORTSPACK) K10
FRONT BELT KIT-COUPE K14
FRONT SEAT SLIDES K13
FUEL APERATURE SEAL AND HOSES-COUPE N13
HARDTOP & HOODBAG-CABRIOLET N17
HOOD LIFT MECHANISM - CONVERTIBLE O04
HOOD SEALS-CONVERTIBLE O03
HOOD-CABRIOLET N16
HOOD-CONVERTIBLE O02
REAR CUBBY BOX-CONVERTIBLE L09
REAR LUGGAGE LOCKER-CABRIOLET L07
REAR SEAT BELTS L11
REAR SEATS-LEATHER-COUPE L03
REAR SEATS-LEATHER-COUPE L04
REAR SEATS-LEATHER/CLOTH-COUPE L06
TARGA PANELS-STOWAGEBAG-CABRIOLET N14
UNDER SCUTTLE CASINGS H05
VENEERED FACIA PANELS-3.6 LITRE H06
VENEERED FACIA PANELS-5.3 LITRE H07

JAGUAR XJS RANGE (JAN 1987 ON)	N10 fiche 3	FASCIA, TRIM, SEATS AND FIXINGS

SOUND INSULATION AND HEATSHIELD-CABRIOLET

Illus	Part Number	123456	Description	Quantity	Change Point	Remarks
1	BD 45946		Heatshield RH	1		
2	BAC 3544		Heatshield LH	1		
3	BCC 8036		Insulation RH-RHD-toeboard	1		Use BDC 7146
3	BDC 7146		Insulation RH-RHD-toeboard	1		
	BCC 8037		Insulation RH-LHD-toeboard	1		Use BDC 3602
	BDC 3602		Insulation RH-LHD-toeboard	1		
4	BCC 8030		Insulation-LH-RHD-toeboard	1		Use BDC 3601
4	BDC 3601		Insulation LH-RHD-toeboard	1		
	BCC 8029		Insulation LH-LHD-toeboard	1		Use BDC 7147
	BDC 7147		Insulation LH-LHD-toeboard	1		
	BCC 2744		Sound deadening RH-LHD-toeboard	1		3.6 litre only
5	BCC 2745		Sound deadening LH-RHD-toeboard	1		3.6 litre only
6	BBC 8532		Insulation RH-front floor	1		
7	BBC 8533		Insulation LH-front floor	1		
8	BD 47337		Insulation RH-gearbox	1		
9	BAC 5785		Insulation LH-gearbox	1		
10	BCC 2316		Insulation RH-rear floor	1		
11	BCC 2317		Insulation LH-rear floor	1		
12	BCC 2313		Insulation-rear tunnel	1		
13	BD 48898		Insulation RH-crossmember	1		
14	BD 48899		Insulation LH-crossmember	1		
15	BBC 2114		Insulation RH-luggage locker	1		
16	BBC 2115		Insulation LH-luggage locker	1		
17	BBC 2112		Insulation forward tunnel	1		
18	BBC 2113		Insulation-rear tunnel	1		

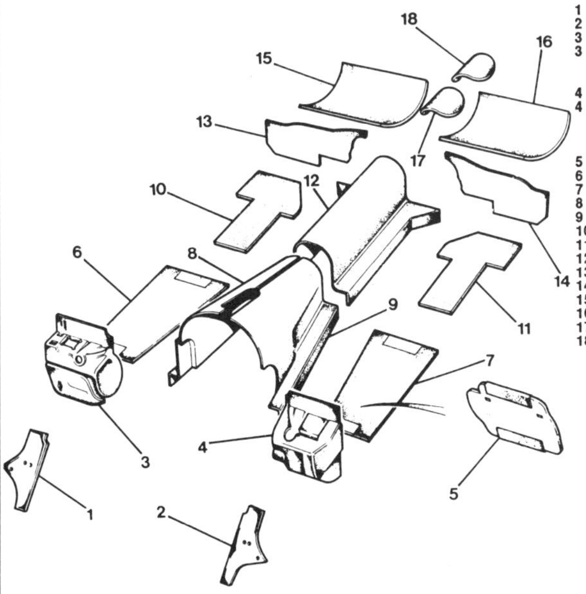

VS4328

MASTER INDEX

GROUP INDEX

JAGUAR XJS RANGE (JAN 1987 ON)	N11 fiche 3	FASCIA, TRIM, SEATS AND FIXINGS

SOUND INSULATION AND HEAT SHEILDS - CONVERTIBLE

Illus	Part Number	123456	Description	Quantity	Change Point	Remarks
1	BD 45946		Heatshield RH	1		
2	BAC 3544		Heatshield LH	1		
3	BCC 6870		Insulation RH - RHD toeboard	1		
	BCC 6872		Insulation RH - LHD toeboard	1		
4	BCC 6871		Insulation LH - RHD toeboard	1		
	BCC 6873		Insulation LH - LHD toeboard	1		
5	BDC 4006		Insulation RH - front floor	1		
6	BDC 4007		Insulation LH - front floor	1		
7	BDC 6332		Insulation RH - gearbox	1		
8	BDC 4027		Insulation LH - gearbox	1		
9	BDC 4004		Insulation - rear floor	2		
10	BDC 4874		Insulation - rear tunnel	1		
11	BDC 4363		Insulation box - hood pump	1		
12	BDC 4001		Insulation - rear bulkhead	1		
13	BDC 4000		Insulation - roof container	1		

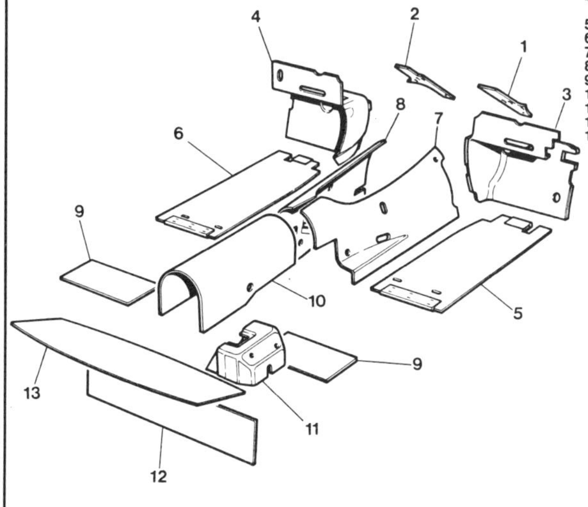

VS4795

MASTER INDEX

ENGINE 1.C
FLYWHEEL AND CLUTCH 1.G
GEARBOX AND PROPSHAFT 1.H
AXLES, SUSPENSION, DRIVE SHAFTS, WHEELS 1.K
STEERING 1.M
BRAKES AND BRAKE CONTROLS 1.N
FUEL, EXHAUST AND EMISSION SYSTEMS 2.C
COOLING, HEATING AND AIR CONDITIONING 2.G
ELECTRICAL, WASHERS-WIPERS, INSTRUMENTS 2.J
CHASSIS, SUBFRAMES, BODYSHELL, FITTINGS 3.C
FASCIA, TRIM, SEATS AND FIXINGS 3.H
ACCESSORIES AND PAINT 4.C
NUMERICAL INDEX 4.D

GROUP INDEX

BATTERY AND SPARE WHEEL COVERS N05
BONNET INSULATION-3.6 LITRE N06
BONNET INSULATION-5.3 MODELS N08
BONNET INSULATION-5.3 MODELS N07
BOOT CARPETS - NOT CONVERTIBLE N03
BOOT CARPETS-CONVERTIBLE N04
CARPETS-'A' POST FINISHERS- L12
CONSOLE-AMBLA H10
CONSOLE-LEATHER H08
DOOR CASINGS-ARMRESTS AND POCKETS-AMBLA H16
DOOR CASINGS-ARMRESTS-CONVERTIBLE I02
HEADLINING I15
DOOR CASINGS-ARMRESTS-LEATHER H14
EXTERIOR TRIM COVERS N15
FACIA PANEL H02
FRONT SEAT BELTS-CONVERTIBLE K15
FRONT SEAT SLIDES K16
MOTOR AND ANCHORAGE ASSY-RH L02
FRONT SEATS-AMBLA/CLOTH.-3.6 LITRE K04
FRONT SEATS-LEATHER J03
FRONT SEATS-LEATHER/CLOTH-5.3 LITRE K07
FRT SEATS-LEATHER/CLOTH(SPORTSPACK) K10
FRONT BELT KIT-COUPE K14
FRONT SEAT SLIDES K13
FUEL APERATURE SEAL AND HOSES-COUPE N13
HARDTOP & HOODBAG-CABRIOLET N17
HOOD LIFT MECHANISM - CONVERTIBLE O04
HOOD SEALS-CONVERTIBLE O03
HOOD-CABRIOLET N16
HOOD-CONVERTIBLE O02
REAR CUBBY BOX-CONVERTIBLE L09
REAR LUGGAGE LOCKER-CABRIOLET L07
REAR SEAT BELTS L11
REAR SEATS-LEATHER-COUPE L03
REAR SEATS-LEATHER-COUPE L04
REAR SEATS-LEATHER/CLOTH-COUPE L06
TARGA PANELS-STOWAGEBAG-CABRIOLET N14
UNDER SCUTTLE CASINGS H05
VENEERED FACIA PANELS-3.6 LITRE H06
VENEERED FACIA PANELS-5.3 LITRE H07

3N 12

BOOT INSULATION

Illus	Part Number	Description	Quantity	Change Point	Remarks
1	BAC 3375	Insulation-fuel tank	1		Not Convertible
	BDC 5068	Insulation-fuel tank	1		Convertible
2	BD 48890	Insulation RH-wheelarch	1		
3	BD 44971	Insulation LH-wheelarch	1		

M62JP13/7/88

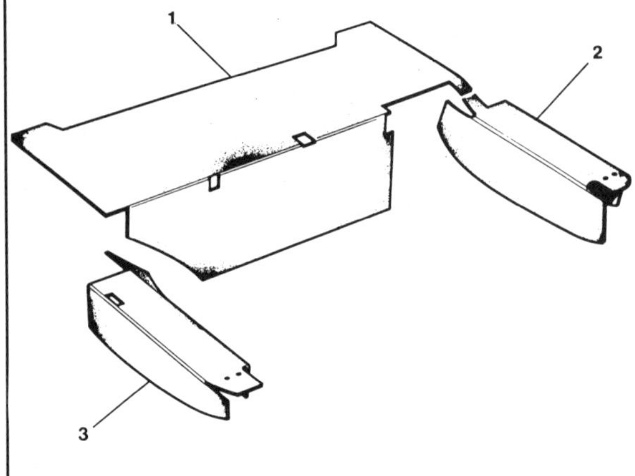

SM 7288

MASTER INDEX

GROUP INDEX

3N 13

FUEL APERATURE SEAL AND HOSES-COUPE.

Illus	Part Number	Description	Quantity	Change Point	Remarks
1	BD 48737	Seal-fuel aperature	1		
2	C 15150 8	Hose-140ins-overflow	4		
3	BAC 2864	'Y' piece	1		
4	CAC 8774	Clip-hose	1		

M64JP27/7/88

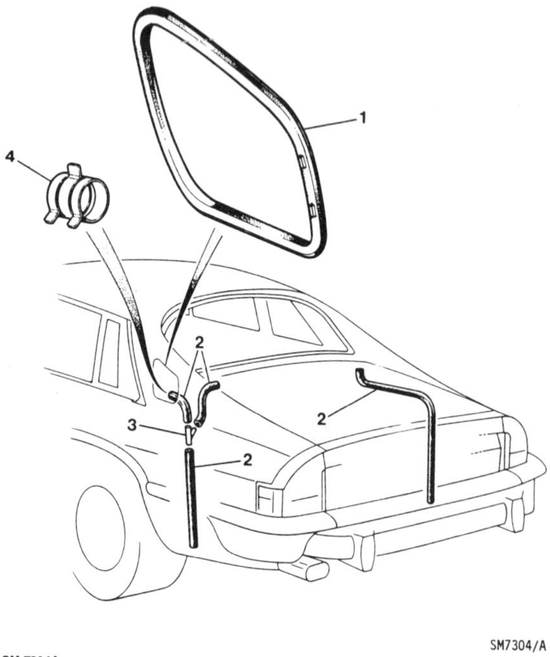

SM 7304A

MASTER INDEX
ENGINE 1.C
FLYWHEEL AND CLUTCH 1.G
GEARBOX AND PROPSHAFT 1.H
AXLES, SUSPENSION, DRIVE SHAFTS, WHEELS 1.K
STEERING 1.M
BRAKES AND BRAKE CONTROLS 1.N
FUEL, EXHAUST AND EMISSION SYSTEMS 2.C
COOLING, HEATING AND AIR CONDITIONING 2.G
ELECTRICAL, WASHERS-WIPERS, INSTRUMENTS 2.J
CHASSIS, SUBFRAMES, BODYSHELL, FITTINGS 3.C
FASCIA, TRIM, SEATS AND FIXINGS 3.H
ACCESSORIES AND PAINT 4.C
NUMERICAL INDEX 4.D
GROUP INDEX
BATTERY AND SPARE WHEEL COVERS N05
BONNET INSULATION-3.6 LITRE N06
BONNET INSULATION-5.3 MODELS N08
BONNET INSULATION-5.3 MODELS N07
BOOT CARPETS - NOT CONVERTIBLE N03
BOOT CARPETS-CONVERTIBLE N04
CARPETS-'A' POST FINISHERS- L12
CONSOLE-AMBLA H10
CONSOLE-LEATHER H08
DOOR CASINGS-ARMRESTS AND POCKETS-AMBLA H16
DOOR CASINGS-ARMRESTS-CONVERTIBLE I02
HEADLINING I15
DOOR CASINGS-ARMRESTS-LEATHER H14
EXTERIOR TRIM COVERS N15
FACIA PANEL H02
FRONT SEAT BELTS-CONVERTIBLE K15
FRONT SEAT SLIDES K16
MOTOR AND ANCHORAGE ASSY-RH L02
FRONT SEATS-AMBLA/CLOTH.-3.6 LITRE K04
FRONT SEATS-LEATHER J03
FRONT SEATS-LEATHER/CLOTH-5.3 LITRE K07
FRT SEATS-LEATHER/CLOTH(SPORTSPACK) K10
FRONT BELT KIT-COUPE K14
FRONT SEAT SLIDES K13
FUEL APERATURE SEAL AND HOSES-COUPE N13
HARDTOP & HOODBAG-CABRIOLET N17
HOOD LIFT MECHANISM - CONVERTIBLE O04
HOOD SEALS-CONVERTIBLE O03
HOOD-CABRIOLET N16
HOOD-CONVERTIBLE O02
REAR CUBBY BOX-CONVERTIBLE L09
REAR LUGGAGE LOCKER-CABRIOLET L07
REAR SEAT BELTS L11
REAR SEATS-LEATHER-COUPE L03
REAR SEATS-LEATHER-COUPE L04
REAR SEATS-LEATHER/CLOTH-COUPE L06
TARGA PANELS-STOWAGEBAG-CABRIOLET N14
UNDER SCUTTLE CASINGS H05
VENEERED FACIA PANELS-3.6 LITRE H06
VENEERED FACIA PANELS-5.3 LITRE H07

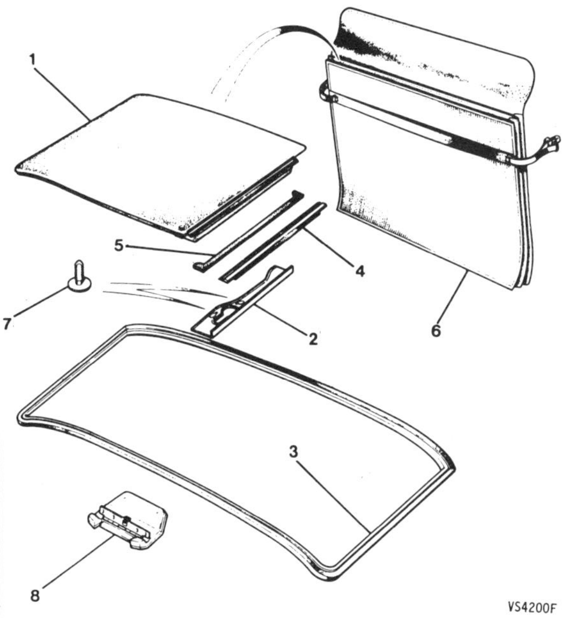

VS 4200F

JAGUAR XJS RANGE (JAN 1987 ON) — N14 fiche 3 — FASCIA, TRIM, SEATS AND FIXINGS

TARGA PANELS-STOWAGEBAG-CABRIOLET

Illus	Part Number	123456 Description	Quantity	Change Point	Remarks
		TARGA PANEL RHD-RH			
1	BBC 9750 PM	Black	1		Fosters
1	BBC 9750 JX	Blue	1		
1	BBC 9750 NF	Brown	1		
1	JLM 1645 PM	Black	1		Happich
1	JLM 1645 JX	Blue	1		
1	JLM 1645 NF	Brown	1		
		TARGA PANEL RHD-LH			
	BBC 9751 PM	Black	1		Fosters
	BBC 9751 JX	Blue	1		
	BBC 9751 NF	Brown	1		
	JLM 1646 PM	Black	1		Happich
	JLM 1646 JX	Brown	1		
	JLM 1646 NF	Blue	1		
		TARGA PANEL LHD-RH			
	BBC 9752 PM	Black	1		Fosters
	BBC 9752 JX	Blue	1		
	BBC 9752 NF	Brown	1		
	JLM 1647 PM	Black	1		Happich
	JLM 1647 JX	Blue	1		
	JLM 1647 NF	Brown	1		
		TARGA PANEL LHD-LH			
	BBC 9753 PM	Black	1		Fosters
	BBC 9753 JX	Blue	1		
	BBC 9753 NF	Brown	1		
	JLM 1648 PM	Black	1		Happich
	JLM 1648 JX	Blue	1		
	JLM 1648 NF	Brown	1		
	BAC 7795 ND	Cover-inner-targa panel	1		
2	BBC 5748	Cover-RHD-RH	1		
	BBC 5749	Cover-RHD-LH	1		
	BBC 5750	Cover-LHD-RH	1		
	BBC 5751	Cover-LHD-LH	1		
3	BBC 6664	Seal-targa aperture	1		
4	BBC 6691	Seal-centre-lower	1		
5	BBC 9565	Seal-centre-upper	1		
6	BAC 7936	Bag-stowage-in boot	1		
7	BAC 6074	Stud-trim fixing	4		
8	BAC 7864	Deflector-wind	1		
	JLM 1434	Fastener kit - Tenax	1		Part of wind deflector

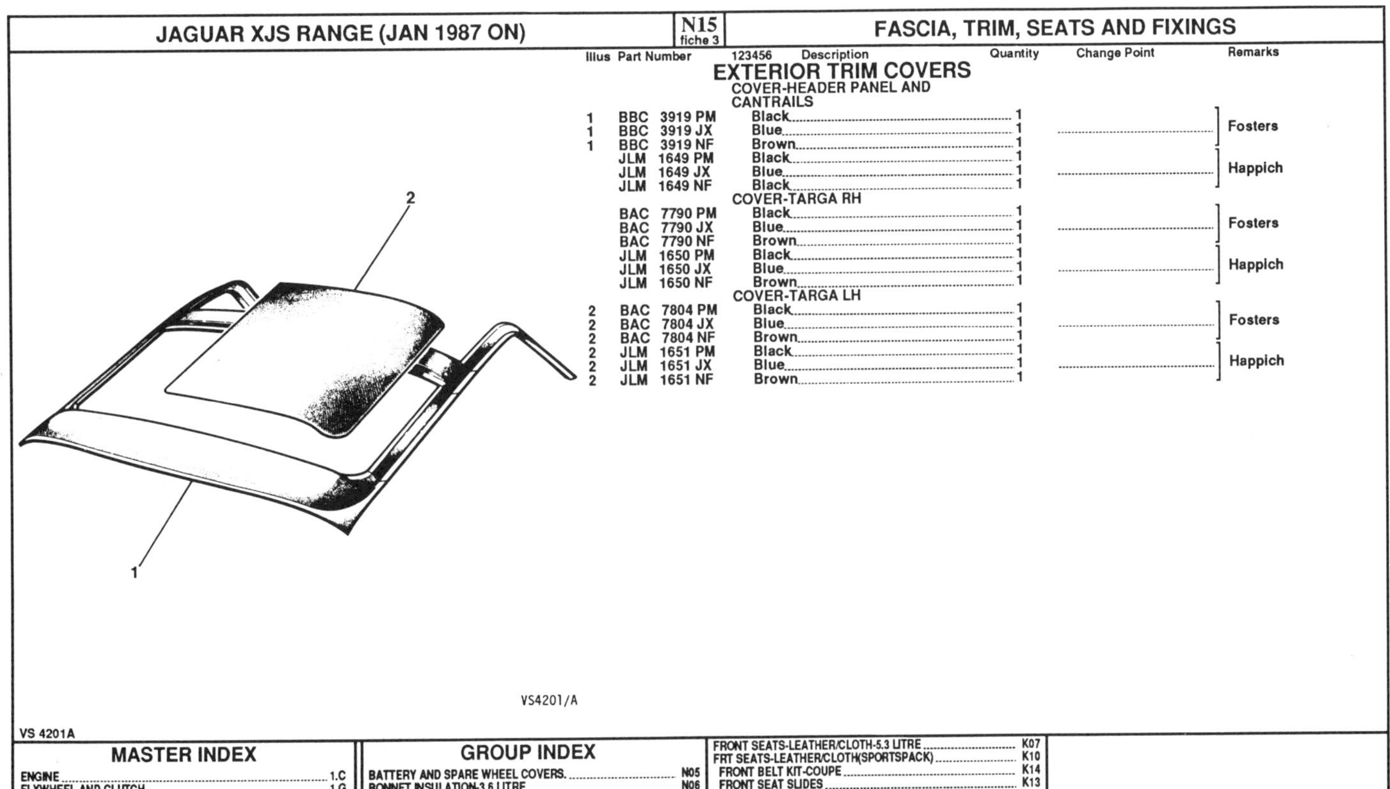

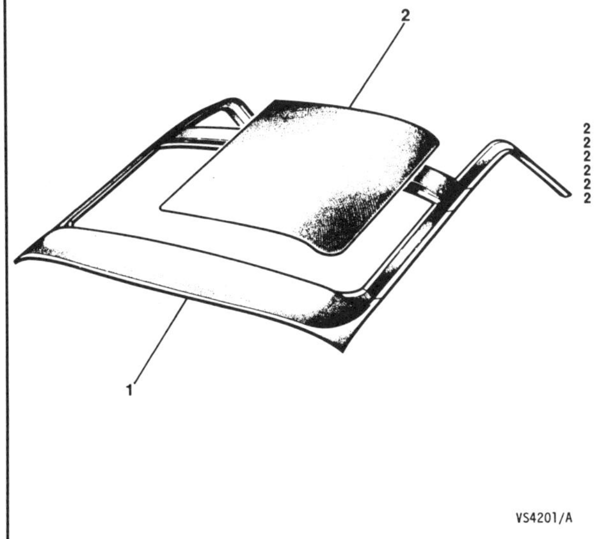

VS 4201A

EXTERIOR TRIM COVERS

Illus	Part Number	123456 Description	Quantity	Change Point	Remarks
		COVER-HEADER PANEL AND CANTRAILS			
1	BBC 3919 PM	Black	1		Fosters
1	BBC 3919 JX	Blue	1		
1	BBC 3919 NF	Brown	1		
	JLM 1649 PM	Black	1		Happich
	JLM 1649 JX	Blue	1		
	JLM 1649 NF	Black	1		
		COVER-TARGA RH			
	BAC 7790 PM	Black	1		Fosters
	BAC 7790 JX	Blue	1		
	BAC 7790 NF	Brown	1		
	JLM 1650 PM	Black	1		Happich
	JLM 1650 JX	Blue	1		
	JLM 1650 NF	Brown	1		
		COVER-TARGA LH			
2	BAC 7804 PM	Black	1		Fosters
2	BAC 7804 JX	Blue	1		
2	BAC 7804 NF	Brown	1		
2	JLM 1651 PM	Black	1		Happich
2	JLM 1651 JX	Blue	1		
2	JLM 1651 NF	Brown	1		

MASTER INDEX

ENGINE ... 1.C
FLYWHEEL AND CLUTCH ... 1.G
GEARBOX AND PROPSHAFT ... 1.H
AXLES, SUSPENSION, DRIVE SHAFTS, WHEELS ... 1.K
STEERING ... 1.M
BRAKES AND BRAKE CONTROLS ... 1.N
FUEL, EXHAUST AND EMISSION SYSTEMS ... 2.C
COOLING, HEATING AND AIR CONDITIONING ... 2.G
ELECTRICAL, WASHERS-WIPERS, INSTRUMENTS ... 2.J
CHASSIS, SUBFRAMES, BODYSHELL, FITTINGS ... 3.C
FASCIA, TRIM, SEATS AND FIXINGS ... 3.H
ACCESSORIES AND PAINT ... 4.C
NUMERICAL INDEX ... 4.D

GROUP INDEX

BATTERY AND SPARE WHEEL COVERS ... N05
BONNET INSULATION-3.6 LITRE ... N06
BONNET INSULATION-5.3 MODELS ... N08
BONNET INSULATION-5.3 MODELS ... N07
BOOT CARPETS - NOT CONVERTIBLE ... N03
BOOT CARPETS-CONVERTIBLE ... N04
CARPETS-'A' POST FINISHERS- ... L12
CONSOLE-AMBLA ... H10
CONSOLE-LEATHER ... H08
DOOR CASINGS-ARMRESTS AND POCKETS-AMBLA ... H16
DOOR CASINGS-ARMRESTS-CONVERTIBLE ... I02
HEADLINING ... I15
DOOR CASINGS-ARMRESTS-LEATHER ... H14
EXTERIOR TRIM COVERS ... N15
FACIA PANEL ... H02
FRONT SEAT BELTS-CONVERTIBLE ... K15
FRONT SEAT SLIDES ... K16
MOTOR AND ANCHORAGE ASSY-RH ... L02
FRONT SEATS-AMBLA/CLOTH.-3.6 LITRE ... K04
FRONT SEATS-LEATHER ... J03
FRONT SEATS-LEATHER/CLOTH-5.3 LITRE ... K07
FRT SEATS-LEATHER/CLOTH(SPORTSPACK) ... K10
FRONT BELT KIT-COUPE ... K14
FRONT SEAT SLIDES ... K13
FUEL APERATURE SEAL AND HOSES-COUPE ... N13
HARDTOP & HOODBAG-CABRIOLET ... N17
HOOD LIFT MECHANISM - CONVERTIBLE ... O04
HOOD SEALS-CONVERTIBLE ... O03
HOOD-CABRIOLET ... N16
HOOD-CONVERTIBLE ... O02
REAR CUBBY BOX-CONVERTIBLE ... L09
REAR LUGGAGE LOCKER-CABRIOLET ... L07
REAR SEAT BELTS ... L11
REAR SEATS-LEATHER-COUPE ... L03
REAR SEATS-LEATHER-COUPE ... L04
REAR SEATS-LEATHER/CLOTH-COUPE ... L06
TARGA PANELS-STOWAGEBAG-CABRIOLET ... N14
UNDER SCUTTLE CASINGS ... H05
VENEERED FACIA PANELS-3.6 LITRE ... H06
VENEERED FACIA PANELS-5.3 LITRE ... H07

JAGUAR XJS RANGE (JAN 1987 ON) — N16 fiche 3 — FASCIA, TRIM, SEATS AND FIXINGS

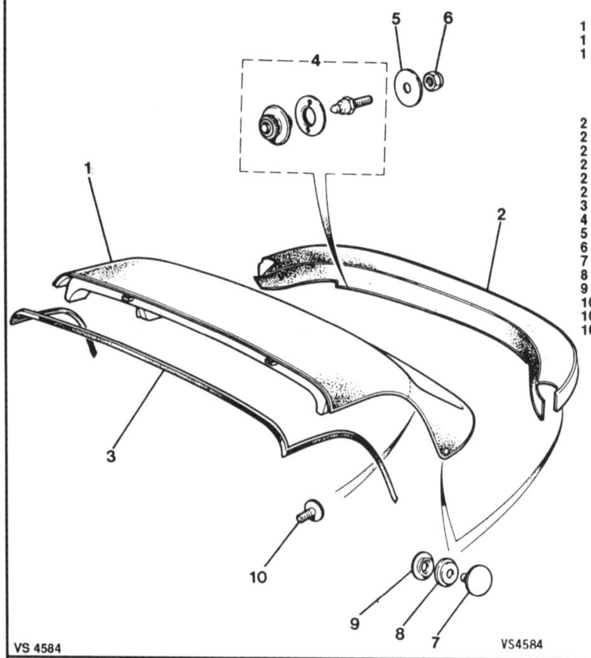

HOOD-CABRIOLET

Illus	Part Number	123456 Description	Quantity	Change Point	Remarks
		HOOD			
1	BAC 7816 PM	Black	1		Fosters
1	BAC 7816 JX	Blue	1		
1	BAC 7816 NF	Brown	1		
	JLM 1652 PM	Black	1		Happich
	JLM 1652 JX	Blue	1		
	JLM 1652 NF	Brown	1		
		COVER-HOOD			
2	BAC 7955 PM	Black	1		Fosters
2	BAC 7955 JX	Blue	1		
2	BAC 7955 NF	Brown	1		
2	JLM 1653 PM	Black	1		Happich
2	JLM 1653 JX	Blue	1		
2	JLM 1653 NF	Brown	1		
3	BCC 1001	Seal-aperture	1		
4	JLM 861	Kit-tenax fastener	5		
5	BD 32692	Washer-flat	5		
6	C 8737 9	Nut-nyloc	5		
7	BAC 7424	Button-fastener-female	6		
8	BBC 1343	Socket-fastener-female	6		
9	BBC 1342	Stud-fastener-male	6		
10	BAC 7420	Screw-c/sunk hex	11		Fixing hood to body
10	BAC 7952	Screw-c/sunk hex-chromed	4		
10	BAC 7808	Bolt-special	4		

JAGUAR XJS RANGE (JAN 1987 ON) — N17 fiche 3 — FASCIA, TRIM, SEATS AND FIXINGS

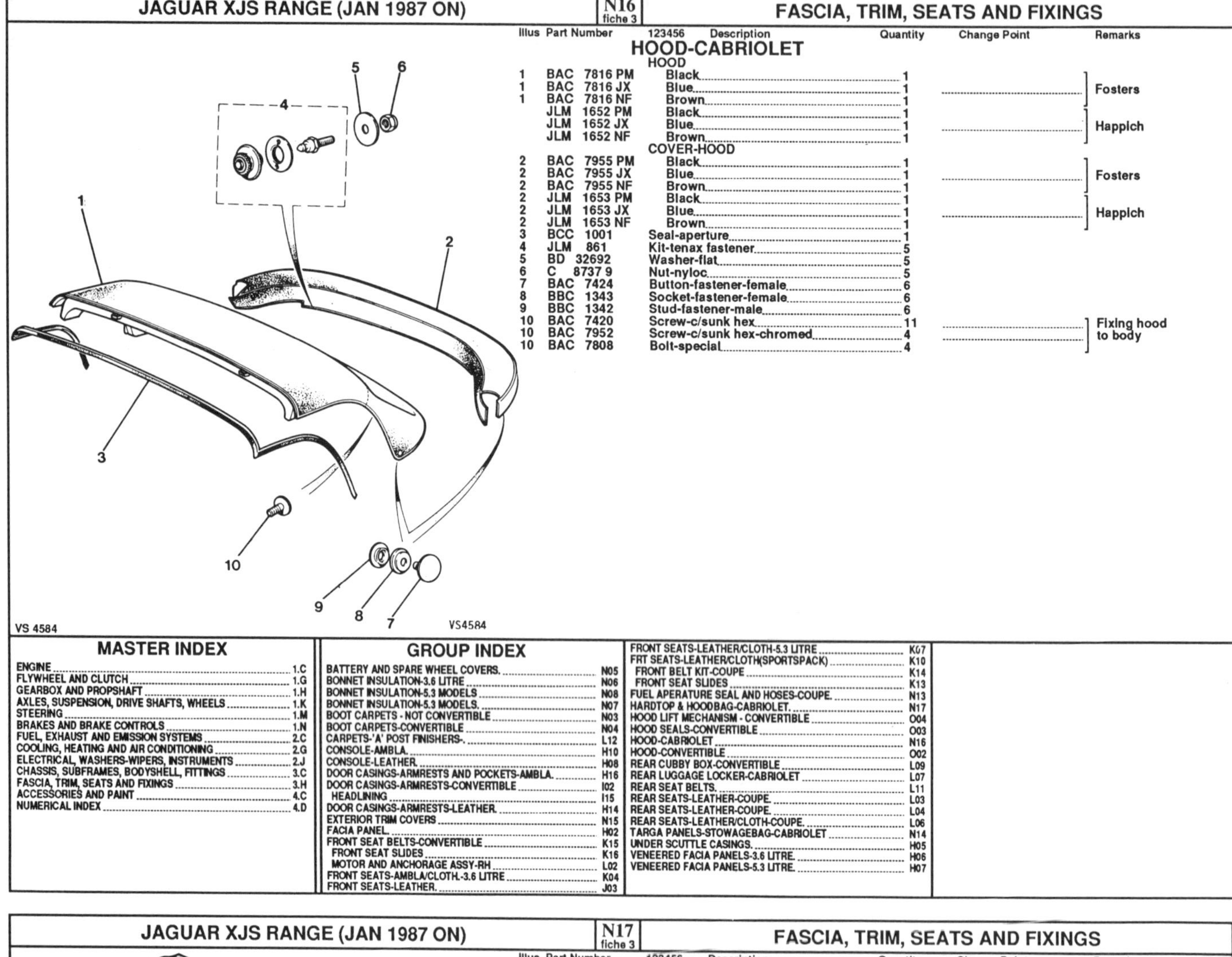

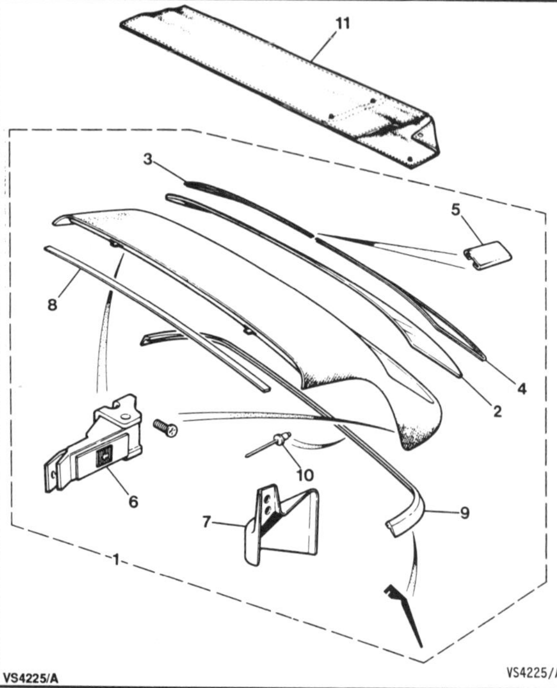

HARDTOP & HOODBAG-CABRIOLET.

Illus	Part Number	123456 Description	Quantity	Change Point	Remarks
		HARDTOP ASSEMBLY			
1	BCC 6247 PM	Black	1		Use BDC 5767 PM
1	BCC 6247 JX	Blue	1		Use BDC 5767 JX
1	BCC 6247 NF	Brown	1		Use BDC 5767 NF
		HARDTOP ASSEMBLY			
1	BDC 5767 PM	Black	1		Fosters
1	BDC 5767 JX	Blue	1		
1	BDC 5767 NF	Brown	1		
1	JLM 1654 PM	Black	1		Happich
1	JLM 1654 JX	Blue	1		
1	JLM 1654 NF	Brown	1		
		OUTER COVER			
	BBC 1317 PM	Black	1		Use BDC 5787 PM
	BBC 1317 JX	Blue	1		Use BDC 5787 JX
	BBC 1317 NF	Brown	1		Use BDC 5787 NF
		OUTER COVER			
	BDC 5787 PM	Black	1		Fosters
	BDC 5787 JX	Blue	1		
	BDC 5787 NF	Brown	1		
	JLM 1655 PM	Black	1		Happich
	JLM 1654 JX	Brown	1		
	JLM 1654 NF	Blue	1		
2	BAC 7426	Glass-heated backlight	1		
	RTC 2668	Betaseal kit-glass fixing	1		
3	BAC 6104	Finisher-glass-RH	1		
4	BAC 6105	Finisher-glass-LH	1		
5	BBC 1327	Finisher-joint	2		
6	BCC 5587	Catch-hardtop	2		
	BAC 6086 ND	Cover-hardtop catch RH	NLA		Use BAC 6084 ND
	BAC 6084 ND	Cover-hardtop catch RH	1		
7	BAC 6087 ND	Cover-hardtop catch LH	NLA		Use BAC 6085 ND
7	BAC 6085 ND	Cover-hardtop catch LH	1		
8	BCC 6597	Seal-header	1		
9	BBC 1324	Seal-rear flange	1		
10	AGU 2392	Rivet-special	18		Fixing Rear flange seal
11	BAC 7832	Bag-hood stowage	1		

MASTER INDEX

ENGINE 1.C
FLYWHEEL AND CLUTCH 1.G
GEARBOX AND PROPSHAFT 1.H
AXLES, SUSPENSION, DRIVE SHAFTS, WHEELS 1.K
STEERING 1.M
BRAKES AND BRAKE CONTROLS 1.N
FUEL, EXHAUST AND EMISSION SYSTEMS 2.C
COOLING, HEATING AND AIR CONDITIONING 2.G
ELECTRICAL, WASHERS-WIPERS, INSTRUMENTS 2.J
CHASSIS, SUBFRAMES, BODYSHELL, FITTINGS 3.C
FASCIA, TRIM, SEATS AND FIXINGS 3.H
ACCESSORIES AND PAINT 4.C
NUMERICAL INDEX 4.D

GROUP INDEX

BATTERY AND SPARE WHEEL COVERS. N05
BONNET INSULATION-3.6 LITRE N06
BONNET INSULATION-5.3 MODELS N08
BONNET INSULATION-5.3 MODELS. N07
BOOT CARPETS - NOT CONVERTIBLE N03
BOOT CARPETS-CONVERTIBLE N04
CARPETS-'A' POST FINISHERS-. L12
CONSOLE-AMBLA. H10
CONSOLE-LEATHER. H08
DOOR CASINGS-ARMRESTS AND POCKETS-AMBLA. H16
DOOR CASINGS-ARMRESTS-CONVERTIBLE I02
HEADLINING I15
DOOR CASINGS-ARMRESTS-LEATHER. H14
EXTERIOR TRIM COVERS N15
FACIA PANEL. H02
FRONT SEAT BELTS-CONVERTIBLE K15
FRONT SEAT SLIDES K16
MOTOR AND ANCHORAGE ASSY-RH L02
FRONT SEATS-AMBLA/CLOTH.-3.6 LITRE K04
FRONT SEATS-LEATHER. J03
FRONT SEATS-LEATHER/CLOTH-5.3 LITRE K07
FRT SEATS-LEATHER/CLOTH(SPORTSPACK) K10
FRONT BELT KIT-COUPE K14
FRONT SEAT SLIDES K13
FUEL APERATURE SEAL AND HOSES-COUPE. N13
HARDTOP & HOODBAG-CABRIOLET. N17
HOOD LIFT MECHANISM - CONVERTIBLE O04
HOOD SEALS-CONVERTIBLE O03
HOOD-CABRIOLET N16
HOOD-CONVERTIBLE O02
REAR CUBBY BOX-CONVERTIBLE L09
REAR LUGGAGE LOCKER-CABRIOLET L07
REAR SEAT BELTS. L11
REAR SEATS-LEATHER-COUPE. L03
REAR SEATS-LEATHER-COUPE. L04
REAR SEATS-LEATHER/CLOTH-COUPE. L06
TARGA PANELS-STOWAGEBAG-CABRIOLET N14
UNDER SCUTTLE CASINGS. H05
VENEERED FACIA PANELS-3.6 LITRE. H06
VENEERED FACIA PANELS-5.3 LITRE. H07

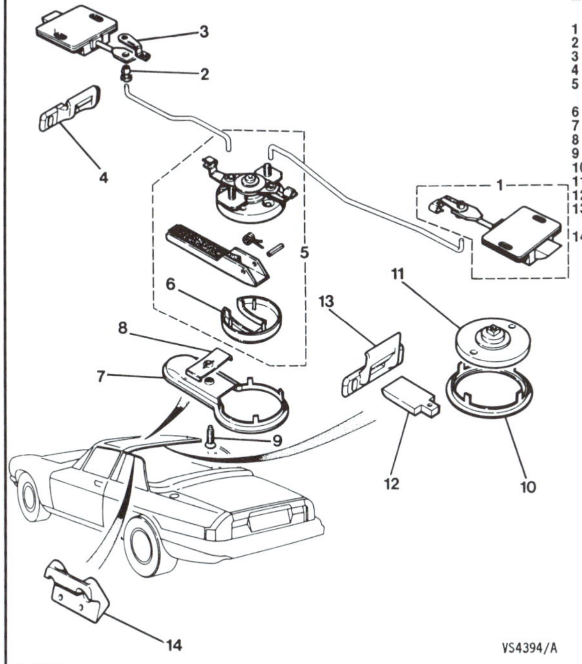

VS 4394A

TARGA/HOOD FIXINGS ETC.

Illus	Part Number	123456	Description	Quantity	Change Point	Remarks
1	BBC 7072		Latch-targa panels	4		
2	BBC 5732		Bush-plastic	8		
3	BBC 5733		Clip	8		
4	BBC 9132		Catchplate-latch	4		
5	BBC 9612		HANDLE ASSY-RH-TARGA PANEL	1		
	BBC 9613		HANDLE ASSY-LH-TARGA PANEL	1		
6	BBC 5757		Cover-pivot	2		
7	BBC 5716		Bezel-handle-targa	2		
8	BBC 5734		Clip-targa handle bezel	2		
9	AC 610041 J		Screw-self-tap	2		
10	BAC 7975		Bezel-handles-hood	1		
11	BAC 7907		Handle-hood	1		
12	BAC 7997		Bolt-sliding-hood	2		
13	BBC 9133		Catchplate-bolt	2		
	BAC 7412		Guide-RH-header side	1		
14	BAC 7413		Guide-LH-header side	1		

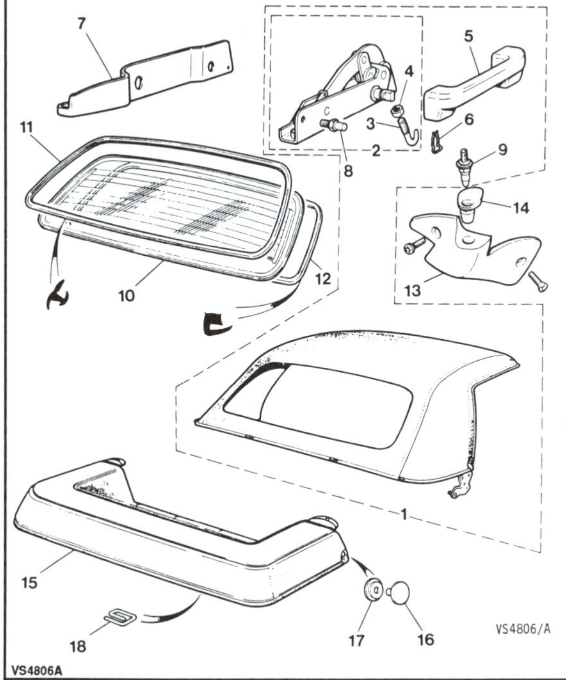

VS4806A

HOOD-CONVERTIBLE

Illus	Part Number	123456	Description	Quantity	Change Point	Remarks
			HOOD ASSEMBLY			
1	JLM 1570 PM		Black	1		
1	JLM 1570 JX		Blue	1		
1	JLM 1570 NF		Brown	1		
	BBC 8200		Latch-closing-RH	1		
2	BBC 8201		Latch-closing-LH	1		
3	BBC 8209		Hook	2		
4	BBC 8211		Nut hook	2		
	BBC 8216		Handle-latch RH	1		
5	BBC 8217		Handle-latch-LH	1		
6	BBC 8224		Circlip	2		
	BCC 9998		Cover-handle-RH	1		
7	BCC 9999		Cover-handle-LH	1		
8	BCC 9791		Pin-latching	2		
9	BBC 8268		Pin-bearing	2		
10	BBC 8243		Glass-backlight	1		
11	BDC 4936		Rubber-retaining	1		
12	BCC 7341		Dam-mastic retention	1		
	BDC 5092		Locking latch-lower-RH	1		
13	BDC 5093		Locking latch-lower-LH	1		
	BCC 5578		Cup-bearing pin-RH	1		
14	BCC 5579		Cup-bearing pin-LH	1		
			COVER-HOOD			
15	BDC 8958 PM		Black	1		
15	BDC 8958 JX		Blue	1		
15	BDC 8958 NF		Brown	1		
16	BAC 7424		Button	6		
17	BBC 1343		Fastener-female	6		
18	BCC 5560		Hook	6		
	BDC 6712		Bag-hood cover stowage	1		M74JP11/7/88

MASTER INDEX

ENGINE 1.C
FLYWHEEL AND CLUTCH 1.G
GEARBOX AND PROPSHAFT 1.H
AXLES, SUSPENSION, DRIVE SHAFTS, WHEELS 1.K
STEERING 1.M
BRAKES AND BRAKE CONTROLS 1.N
FUEL, EXHAUST AND EMISSION SYSTEMS 2.C
COOLING, HEATING AND AIR CONDITIONING 2.G
ELECTRICAL, WASHERS-WIPERS, INSTRUMENTS 2.J
CHASSIS, SUBFRAMES, BODYSHELL, FITTINGS 3.C
FASCIA, TRIM, SEATS AND FIXINGS 3.H
ACCESSORIES AND PAINT 4.C
NUMERICAL INDEX 4.D

GROUP INDEX

BATTERY AND SPARE WHEEL COVERS N05
BONNET INSULATION-3.6 LITRE N06
BONNET INSULATION-5.3 MODELS N08
BONNET INSULATION-5.3 MODELS N07
BOOT CARPETS - NOT CONVERTIBLE N03
BOOT CARPETS-CONVERTIBLE N04
CARPETS-'A' POST FINISHERS- L12
CONSOLE-AMBLA H10
CONSOLE-LEATHER H08
DOOR CASINGS-ARMRESTS AND POCKETS-AMBLA H16
DOOR CASINGS-ARMRESTS-CONVERTIBLE I02
HEADLINING I15
DOOR CASINGS-ARMRESTS-LEATHER H14
EXTERIOR TRIM COVERS N15
FACIA PANEL H02
FRONT SEAT BELTS-CONVERTIBLE K15
FRONT SEAT SLIDES K16
MOTOR AND ANCHORAGE ASSY-RH L02
FRONT SEATS-AMBLA/CLOTH.-3.6 LITRE K04
FRONT SEATS-LEATHER J03
FRONT SEATS-LEATHER/CLOTH-5.3 LITRE K07
FRT SEATS-LEATHER/CLOTH(SPORTSPACK) K10
FRONT BELT KIT-COUPE K14
FRONT SEAT SLIDES K13
FUEL APERATURE SEAL AND HOSES-COUPE N13
HARDTOP & HOODBAG-CABRIOLET N17
HOOD LIFT MECHANISM - CONVERTIBLE O04
HOOD SEALS-CONVERTIBLE O03
HOOD-CABRIOLET N16
HOOD-CONVERTIBLE O02
REAR CUBBY BOX-CONVERTIBLE L09
REAR LUGGAGE LOCKER-CABRIOLET L07
REAR SEAT BELTS L11
REAR SEATS-LEATHER-COUPE L03
REAR SEATS-LEATHER-COUPE L04
REAR SEATS-LEATHER/CLOTH-COUPE L06
TARGA PANELS-STOWAGEBAG-CABRIOLET N14
UNDER SCUTTLE CASINGS H05
VENEERED FACIA PANELS-3.6 LITRE H06
VENEERED FACIA PANELS-5.3 LITRE H07

30 03

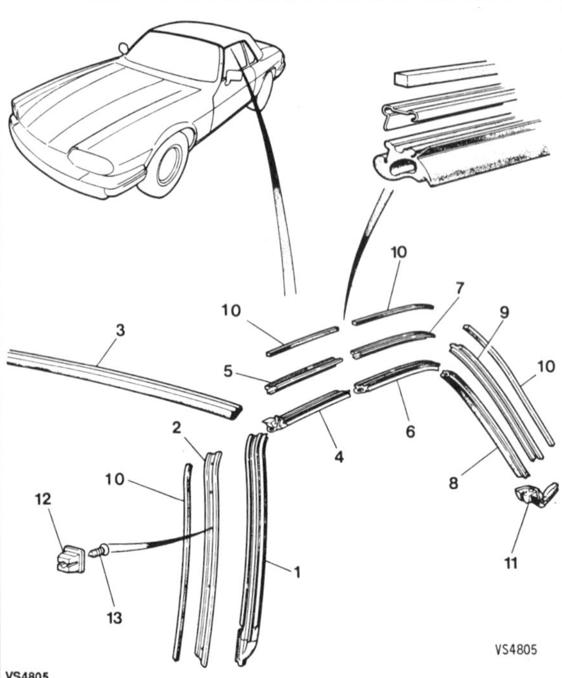

VS4805

HOOD SEALS-CONVERTIBLE

Illus	Part Number	123456	Description	Quantity	Change Point	Remarks
	BDC 8500		Seal-"A" post-RH	1		
1	BDC 8501		Seal-"A" post-LH	1		
	BDC 8498		Rail-seal mounting-RH	1		
2	BDC 8499		Rail-seal mounting-LH	1		
3	BBC 8194		Seal-header	1		
	BBC 8186		Seal-front frame-RH	1		
4	BBC 8187		Seal-front frame-LH	1		
	BBC 8188		Rail-seal mounting-RH	1		
5	BBC 8189		Rail-seal mounting LH	1		
	BBC 8178		Seal-rear frame RH	1		
6	BBC 8179		Seal-rear frame LH	1		
	BBC 8180		Rail-seal mounting RH	1		
7	BBC 8181		Rail-seal mounting LH	1		
	BBC 8190		Seal-main column-RH	1		
8	BBC 8191		Seal-main column-LH	1		
	BBC 8192		Rail-seal mounting-RH	1		
9	BBC 8193		Rail-seal mounting-LH	1		
10	BBC 8265		Seal strip-rails to frame	A/R		Cut to length
	BDC 6242		Seal-'C' post-RH	1		
11	BDC 6243		Seal-'C' post-LH	1		
12	BBC 8184		Nut-lokut	46		
13	AB 606031 J		Screw	46		

30 04

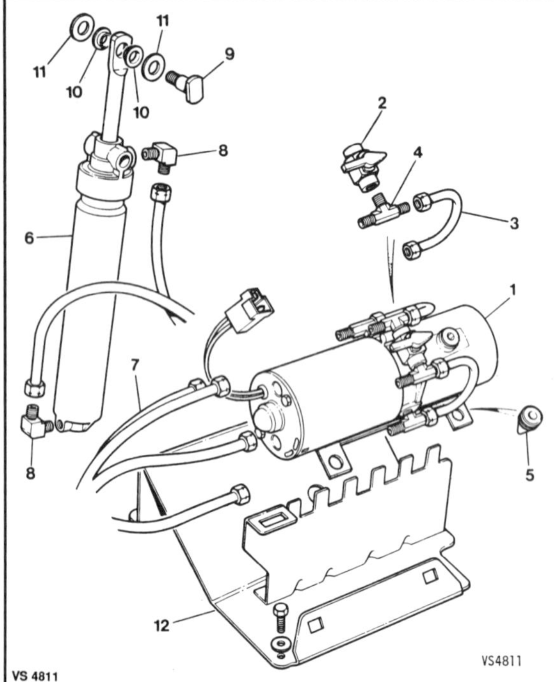

VS 4811

HOOD LIFT MECHANISM - CONVERTIBLE

Illus	Part Number	123456	Description	Quantity	Change Point	Remarks
1	BCC 2821		Pump Assy	1		
2	JLM 1547		Valve and handle assy	1		
3	BCC 2822		Tube	2		
4	BCC 2826		Fitting	4		
5	BCC 2824		Foot	4		
6	BCC 2829		Cylinder - hood lift	2		
7	BCC 2828		Hose	4		C/with nut and olive
8	BCC 2830		Elbow	4		
9	BBC 8042		Pin - pivot	2		
10	BCC 7642		Bush - pivot pin	4		
11	BBC 8049		Washer - pivot pin	4		
12	BDC 4362		Mtg plate-pump and relays	1		

MASTER INDEX

ENGINE 1.C
FLYWHEEL AND CLUTCH 1.G
GEARBOX AND PROPSHAFT 1.H
AXLES, SUSPENSION, DRIVE SHAFTS, WHEELS 1.K
STEERING 1.M
BRAKES AND BRAKE CONTROLS 1.N
FUEL, EXHAUST AND EMISSION SYSTEMS 2.C
COOLING, HEATING AND AIR CONDITIONING 2.G
ELECTRICAL, WASHERS-WIPERS, INSTRUMENTS 2.J
CHASSIS, SUBFRAMES, BODYSHELL, FITTINGS 3.C
FASCIA, TRIM, SEATS AND FIXINGS 3.H
ACCESSORIES AND PAINT 4.C
NUMERICAL INDEX 4.D

GROUP INDEX

BATTERY AND SPARE WHEEL COVERS. N05
BONNET INSULATION-3.6 LITRE N06
BONNET INSULATION-5.3 MODELS N08
BONNET INSULATION-5.3 MODELS. N07
BOOT CARPETS - NOT CONVERTIBLE N03
BOOT CARPETS-CONVERTIBLE N04
CARPETS-'A' POST FINISHERS-. L12
CONSOLE-AMBLA. H10
CONSOLE-LEATHER. H08
DOOR CASINGS-ARMRESTS AND POCKETS-AMBLA. H16
DOOR CASINGS-ARMRESTS-CONVERTIBLE I02
HEADLINING I15
DOOR CASINGS-ARMRESTS-LEATHER. H14
EXTERIOR TRIM COVERS N15
FACIA PANEL. H02
FRONT SEAT BELTS-CONVERTIBLE K15
FRONT SEAT SLIDES K16
MOTOR AND ANCHORAGE ASSY-RH L02
FRONT SEATS-AMBLA/CLOTH.-3.6 LITRE K04
FRONT SEATS-LEATHER. J03
FRONT SEATS-LEATHER/CLOTH-5.3 LITRE K07
FRT SEATS-LEATHER/CLOTH(SPORTSPACK) K10
FRONT BELT KIT-COUPE K14
FRONT SEAT SLIDES K13
FUEL APERATURE SEAL AND HOSES-COUPE. N13
HARDTOP & HOODBAG-CABRIOLET. N17
HOOD LIFT MECHANISM - CONVERTIBLE O04
HOOD SEALS-CONVERTIBLE O03
HOOD-CABRIOLET N16
HOOD-CONVERTIBLE O02
REAR CUBBY BOX-CONVERTIBLE L09
REAR LUGGAGE LOCKER-CABRIOLET L07
REAR SEAT BELTS. L11
REAR SEATS-LEATHER-COUPE. L03
REAR SEATS-LEATHER-COUPE. L04
REAR SEATS-LEATHER/CLOTH-COUPE. L06
TARGA PANELS-STOWAGEBAG-CABRIOLET N14
UNDER SCUTTLE CASINGS. H05
VENEERED FACIA PANELS-3.6 LITRE. H06
VENEERED FACIA PANELS-5.3 LITRE. H07

MASTER INDEX

GROUP INDEX

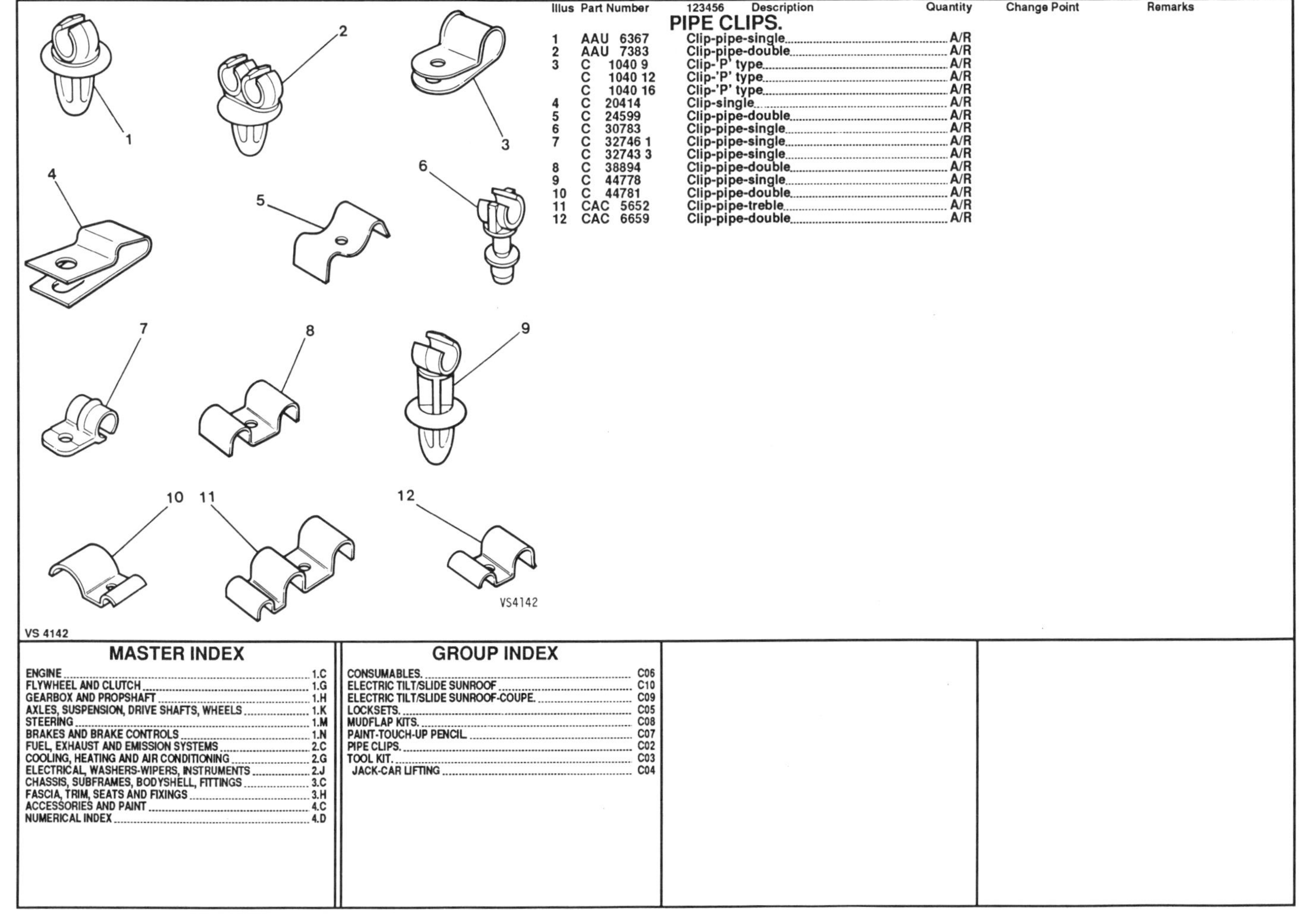

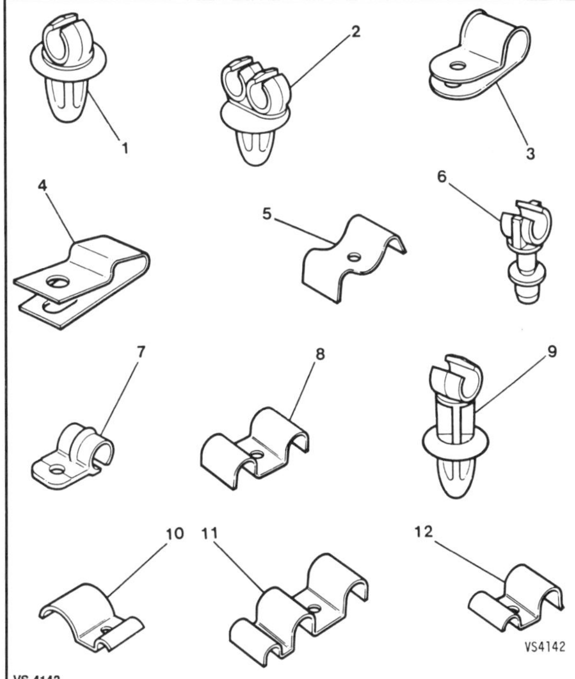

VS 4142

PIPE CLIPS.

Illus	Part Number	123456	Description	Quantity	Change Point	Remarks
1	AAU 6367		Clip-pipe-single	A/R		
2	AAU 7383		Clip-pipe-double	A/R		
3	C 1040 9		Clip-'P' type	A/R		
	C 1040 12		Clip-'P' type	A/R		
	C 1040 16		Clip-'P' type	A/R		
4	C 20414		Clip-single	A/R		
5	C 24599		Clip-pipe-double	A/R		
6	C 30783		Clip-pipe-single	A/R		
7	C 32746 1		Clip-pipe-single	A/R		
	C 32743 3		Clip-pipe-single	A/R		
8	C 38894		Clip-pipe-double	A/R		
9	C 44778		Clip-pipe-single	A/R		
10	C 44781		Clip-pipe-double	A/R		
11	CAC 5652		Clip-pipe-treble	A/R		
12	CAC 6659		Clip-pipe-double	A/R		

MASTER INDEX

GROUP INDEX

CONSUMABLES. C06
ELECTRIC TILT/SLIDE SUNROOF C10
ELECTRIC TILT/SLIDE SUNROOF-COUPE. C09
LOCKSETS. C05
MUDFLAP KITS. C08
PAINT-TOUCH-UP PENCIL. C07
PIPE CLIPS. C02
TOOL KIT. C03
JACK-CAR LIFTING C04

JAGUAR XJS RANGE (JAN 1987 ON) | C03 fiche 4 | ACCESSORIES AND PAINT

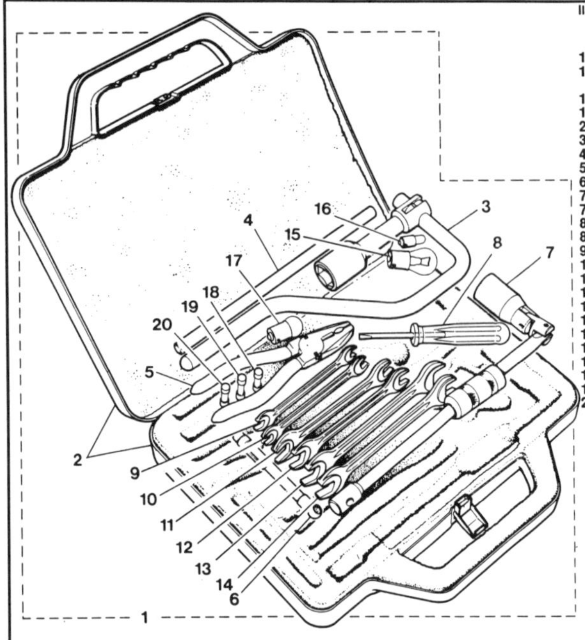

VS1572

VS 1572

TOOL KIT.

Illus	Part Number	123456 Description	Quantity	Change Point	Remarks
		KIT-TOOL(COMPLETE)-3.6 LITRE			
1	CAC 6998	Coupe	1		
1	CAC 9466	Cabriolet	1		
		KIT-TOOL(COMPLETE)-5.3 LITRE			
1	CAC 3437	Coupe	1		
1	CBC 2158	Cabriolet & Convertible	1		
2	CAC 3422	Box-tool kit	1		
3	CAC 2740	Brace-wheel	1		
4	C 36613 2	Bar-tommy	1		
5	AFU 1024	Pliers	1		
6	562019 J	Gauge-tyre pressure	1		
7	CAC 7000	Spanner-spark plug	1		3.6 litre
7	CAC 5368	Spanner-spark plug	1		5.3 litre
8	565770 J	Screwdriver	1		Coupe
8	CAC 8943	Screwdriver	1		Cabriolet
9	CBC 2159	Spanner-9x11mm	1		
10	CBC 2160	Spanner-0.375x0.4375AF	1		
11	CBC 2162	Spanner-13x15mm	1		
12	CBC 2161	Spanner-0.5x0.5625AF	1		
13	CBC 2164	Spanner-17x19mm	1		
14	CBC 2163	Spanner-0.25x0.75AF	1		
15	C 9126	Bulb-21w	2		
16	JLM 9589	Bulb-4w	1		
17	JLM 9587	Bulb-5w	1		
18	541572	Fuse-20A	1		
19	DAC 1480	Fuse-3A	1		
20	C 39572	Fuse-35A	1		

MASTER INDEX

GROUP INDEX

JAGUAR XJS RANGE (JAN 1987 ON) | C04 fiche 4 | ACCESSORIES AND PAINT

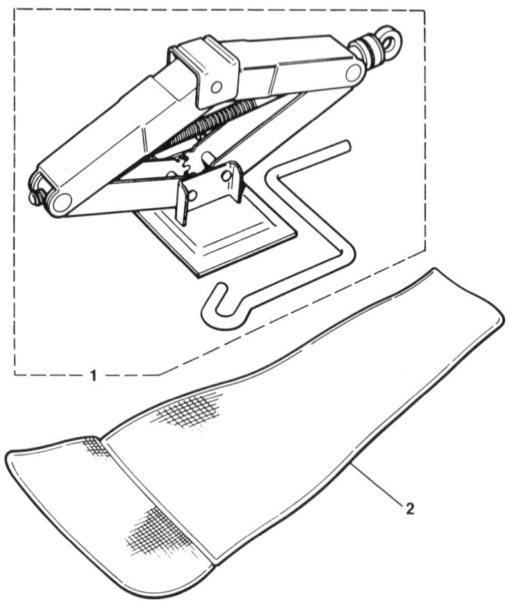

VS4690

VS 4690

JACK AND ENVELOPE

Illus	Part Number	123456 Description	Quantity	Change Point	Remarks
1	JLM 1020	JACK-CAR LIFTING	1		
2	CBC 1722	Envelope-jack	1		

MASTER INDEX

GROUP INDEX

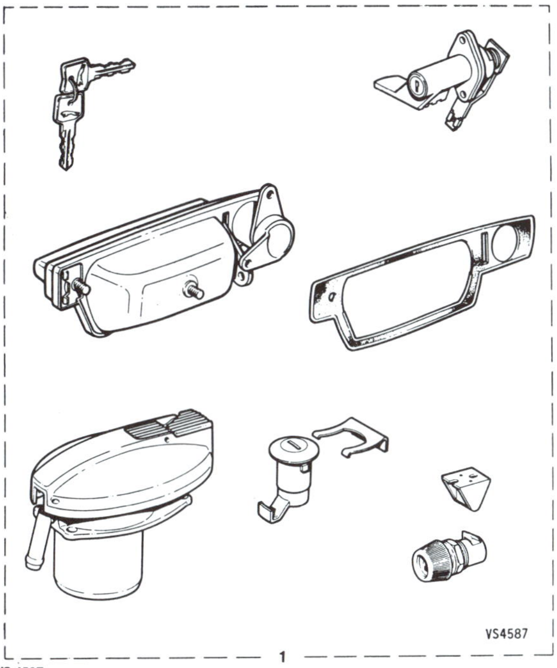

VS 4587

LOCKSETS.

Illus	Part Number	123456	Description	Quantity	Change Point	Remarks
			LOCKSET(COMPLETE)			
1	BCC 4033		Coupe	1	Up to (V)148781	Not USA/CDN
1	BCC 4032		Coupe	1		USA/CDN only
1	BDC 8811		Coupe	1	From (V)148782	Not USA/CDN
1	BDC 8812		Coupe	1		USA/CDN only
1	BBC 9226		Cabriolet	1		Not USA/CDN/AUS
1	BBC 9227		Cabriolet	1		USA/CDN/AUS
1	BDC 8813		Convertible	1		Not USA/CDN/AUS
1	BDC 8814		Convertible	1		AUS only
1	BDC 8815		Convertible	1		USA/CDN only
			KEY BLANKS			
	JLM 476		No 1 Precut	A/R		
	JLM 476 2		No 2 Precut	A/R		
	JLM 476 3		No 3 Precut	A/R		
	JLM 744		Workshop No 1 Precut	A/R		USA only
	JLM 744 2		Workshop No 2 Precut	A/R		USA only
	JLM 744 3		Workshop No 3 Precut	A/R		USA only

MASTER INDEX

GROUP INDEX

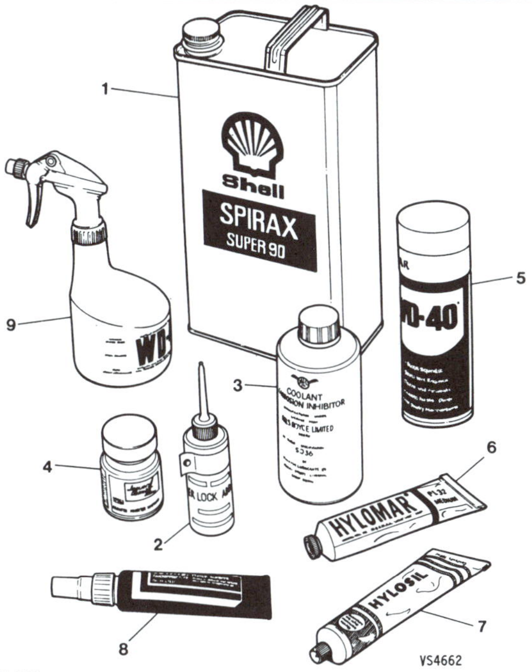

VS 4662

CONSUMABLES.

Illus	Part Number	123456	Description	Quantity	Change Point	Remarks
1	RTC 1479		Oil-final drive	A/R		
2	RTC 1001		Additive-power-lok	A/R		
3	605765 J		Inhibitor-coolant-corrosion	A/R		
4	12953		Inhibitor-radiator leak	A/R		
	AAU 1582		Fluid-brake cleaning 0.50 pint	A/R		
	37H 6082 J		Fluid-brake cleaning 1 gal	A/R		
5	JLM 9704		WD-40 Spray Can-Large	A/R		
	JLM 9705		WD-40 Spray Can-Small	A/R		
	JLM 9708		WD-40 Can-5 Litre	A/R		Workshop Pack
6	JLM 9709		Sealant-Hylomar-110g Tube	A/R		
7	JLM 9710		Sealant-Hylosil(RTV)-110g Tube	A/R		
8	JLM 9822		Sealant-Flange-50ml Tube	A/R		
9	JLM 9734		Dispenser-For 5 Litre WD-40	A/R		
	JLM 10153		Fluid-hood lift - 5 litre	A/R		Convertible
	JLM 1227		Loctite 510 - 50ml tube	A/R		
	JLM 1247		Loctite 574 - 50ml tube	A/R		
	JLM 1746		Loctite 542 - 50ml tube	A/R		
	JLM 1747		Loctite 638 - 50ml tube	A/R		
	JLM 1748		Loctite studlock 270 - 50ml tube	A/R		

N8JP18/7/88

MASTER INDEX

ENGINE 1.C
FLYWHEEL AND CLUTCH 1.G
GEARBOX AND PROPSHAFT 1.H
AXLES, SUSPENSION, DRIVE SHAFTS, WHEELS 1.K
STEERING 1.M
BRAKES AND BRAKE CONTROLS 1.N
FUEL, EXHAUST AND EMISSION SYSTEMS 2.C
COOLING, HEATING AND AIR CONDITIONING 2.G
ELECTRICAL, WASHERS-WIPERS, INSTRUMENTS 2.J
CHASSIS, SUBFRAMES, BODYSHELL, FITTINGS 3.C
FASCIA, TRIM, SEATS AND FIXINGS 3.H
ACCESSORIES AND PAINT 4.C
NUMERICAL INDEX 4.D

GROUP INDEX

CONSUMABLES. C06
ELECTRIC TILT/SLIDE SUNROOF C10
ELECTRIC TILT/SLIDE SUNROOF-COUPE. C09
LOCKSETS. C05
MUDFLAP KITS. C08
PAINT-TOUCH-UP PENCIL C07
PIPE CLIPS. C02
TOOL KIT. C03
JACK-CAR LIFTING C04

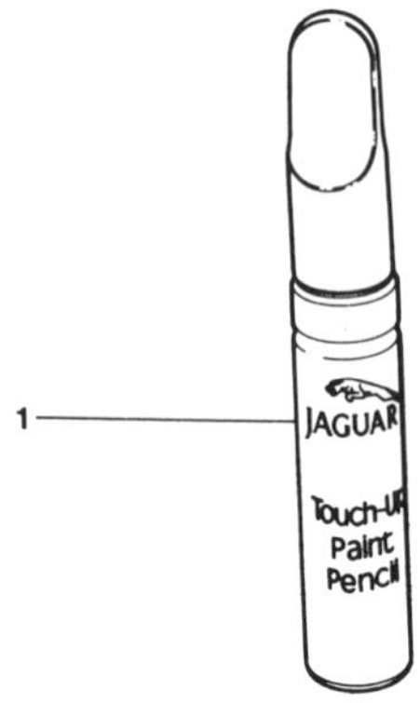

VS4495

VS 4495

PAINT-TOUCH-UP PENCIL.

Illus	Part Number	123456 Description	Quantity	Change Point	Remarks
		PENCIL-COLOUR TOUCH			
1	JLM 9070	Alpine green	A/R		JBC 709
1	JLM 9063	Arctic blue	A/R		JBC 337
1	JLM 9060	Bordeaux red	A/R		JBC 340
1	JLM 9061	Crimson	A/R		JBC 714
1	JLM 9066	Dorchester grey	A/R		JBC 342
1	JLM 9074	Glacier white	A/R		JBC 721
1	JLM 9056	Grenadier red	A/R		JBC 332
1	JLM 9057	Jaguar racing green	A/R		JBC 701
1	JLM 9059	Jet black	A/R		JBC 333
1	JLM 9071	Moorland green	A/R		JBC 717
1	JLM 9055	Nimbus white	A/R		JBC 700
1	JLM 9068	Satin beige	A/R		JBC 711
1	JLM 9073	Signal red	A/R		JBC 721
1	JLM 9069	Silver birch	A/R		JBC 716
1	JLM 9062	Solent blue	A/R		JBC 715
1	JLM 9067	Sovereign gold	A/R		JBC 341
1	JLM 9065	Talisman silver	A/R		JBC 336
1	JLM 9064	Tungsten	A/R		JBC 718
1	JLM 9058	Westminster blue	A/R		JBC 712
1	JLM 9072	Clear lacquer	A/R		

4C 07

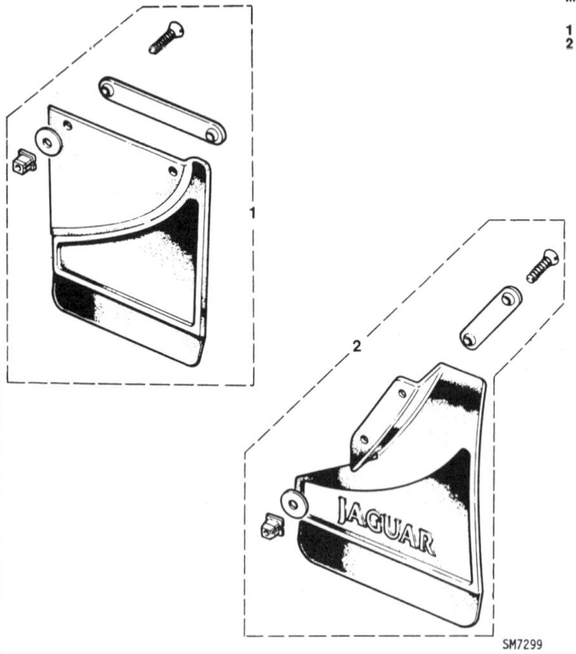

SM7299

SM 7299

MUDFLAP KITS.

Illus	Part Number	123456 Description	Quantity	Change Point	Remarks
1	JLM 9773	Mudflap kit-front	1		Not USA/CDN
2	JLM 341	Mudflap kit-rear	1		

4C 08

MASTER INDEX

ENGINE ... 1.C
FLYWHEEL AND CLUTCH ... 1.G
GEARBOX AND PROPSHAFT ... 1.H
AXLES, SUSPENSION, DRIVE SHAFTS, WHEELS ... 1.K
STEERING ... 1.M
BRAKES AND BRAKE CONTROLS ... 1.N
FUEL, EXHAUST AND EMISSION SYSTEMS ... 2.C
COOLING, HEATING AND AIR CONDITIONING ... 2.G
ELECTRICAL, WASHERS-WIPERS, INSTRUMENTS ... 2.J
CHASSIS, SUBFRAMES, BODYSHELL, FITTINGS ... 3.C
FASCIA, TRIM, SEATS AND FIXINGS ... 3.H
ACCESSORIES AND PAINT ... 4.C
NUMERICAL INDEX ... 4.D

GROUP INDEX

CONSUMABLES. ... C06
ELECTRIC TILT/SLIDE SUNROOF ... C10
ELECTRIC TILT/SLIDE SUNROOF-COUPE. ... C09
LOCKSETS. ... C05
MUDFLAP KITS. ... C08
PAINT-TOUCH-UP PENCIL. ... C07
PIPE CLIPS. ... C02
TOOL KIT. ... C03
JACK-CAR LIFTING ... C04

ELECTRIC TILT/SLIDE SUNROOF-COUPE.

Illus	Part Number	123456 Description	Quantity	Change Point	Remarks
1	BBC 2682	Kit-sunroof complete	1		
	JLM 440 ND	Headlining cloth	A/R		Limestone

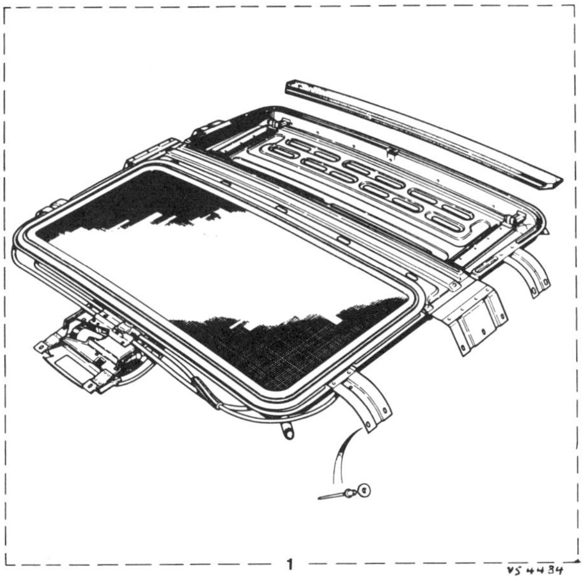

VS 4484

MASTER INDEX

GROUP INDEX

ELECTRIC TILT/SLIDE SUNROOF

COUPE-continued

Illus	Part Number	123456 Description	Quantity	Change Point	Remarks
1	JLM 411	GLASS PANEL ASSEMBLY-SUNROOF	1		
2	JLM 412	Panel seal-front	1		
3	JLM 413	Panel seal-rear	1		
4	JLM 431	Setscrew and washer	6		Fixing glass to Cullisse
5	JLM 426	Trim ring-roof aperture	1		
	JLM 429	Locking strip-trim ring	1		
6	JLM 430	Rivet	14		
7	JLM 414	Guide roller	2		
	JLM 415	Cam-panel lifting-RH	1		
8	JLM 416	Cam-panel lifting-LH	1		
9	JLM 417	Screw-set	4		
10	JLM 418	Washer-plain	4		
	JLM 419	Guide and cable assembly-RH	1		
11	JLM 420	Guide and cable assembly-LH	1		
12	JLM 427	Drain tube-sunroof tray	4		
13	EAC 3215 5	Clip-drain tube	4		
14	JLM 421	Sliding blind assembly	1		
15	JLM 432	Rear trim panel assembly	1		Fitted above Backlight
16	JLM 433	Stud-panel fixing	5		
17	JLM 434	Insert-panel fixing	5		
18	JLM 423	Stud	2		Fixing motor cover
19	BD 46140 3	Washer	2		Fixing motor cover
20	BD 46140 2	Cam-fastener	2		
21	JLM 1381	Deflector	1		
22	JLM 1662	Shoe - slide	2		

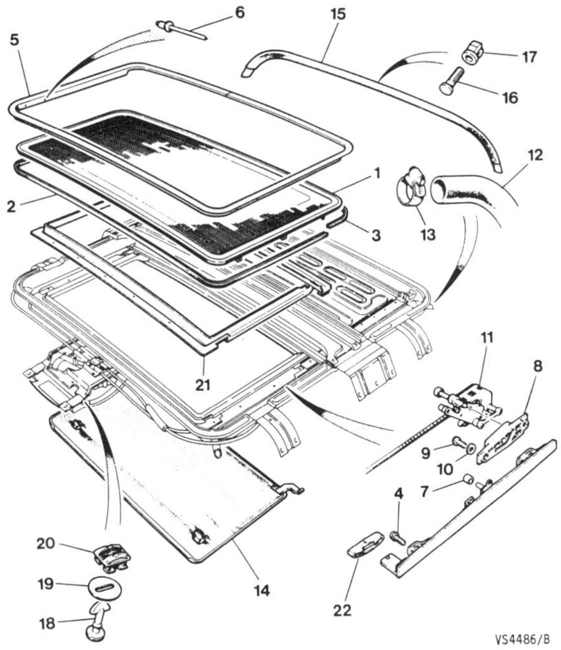

VS4486/B

MASTER INDEX

ENGINE 1.C
FLYWHEEL AND CLUTCH 1.G
GEARBOX AND PROPSHAFT 1.H
AXLES, SUSPENSION, DRIVE SHAFTS, WHEELS 1.K
STEERING 1.M
BRAKES AND BRAKE CONTROLS 1.N
FUEL, EXHAUST AND EMISSION SYSTEMS 2.C
COOLING, HEATING AND AIR CONDITIONING 2.G
ELECTRICAL, WASHERS-WIPERS, INSTRUMENTS 2.J
CHASSIS, SUBFRAMES, BODYSHELL, FITTINGS 3.C
FASCIA, TRIM, SEATS AND FIXINGS 3.H
ACCESSORIES AND PAINT 4.C
NUMERICAL INDEX 4.D

GROUP INDEX

CONSUMABLES C06
ELECTRIC TILT/SLIDE SUNROOF C10
ELECTRIC TILT/SLIDE SUNROOF-COUPE C09
LOCKSETS C05
MUDFLAP KITS C08
PAINT-TOUCH-UP PENCIL C07
PIPE CLIPS C02
TOOL KIT C03
JACK-CAR LIFTING C04

JAGUAR XJS RANGE (JAN 1987 ON) | C11 fiche 4 | ACCESSORIES AND PAINT

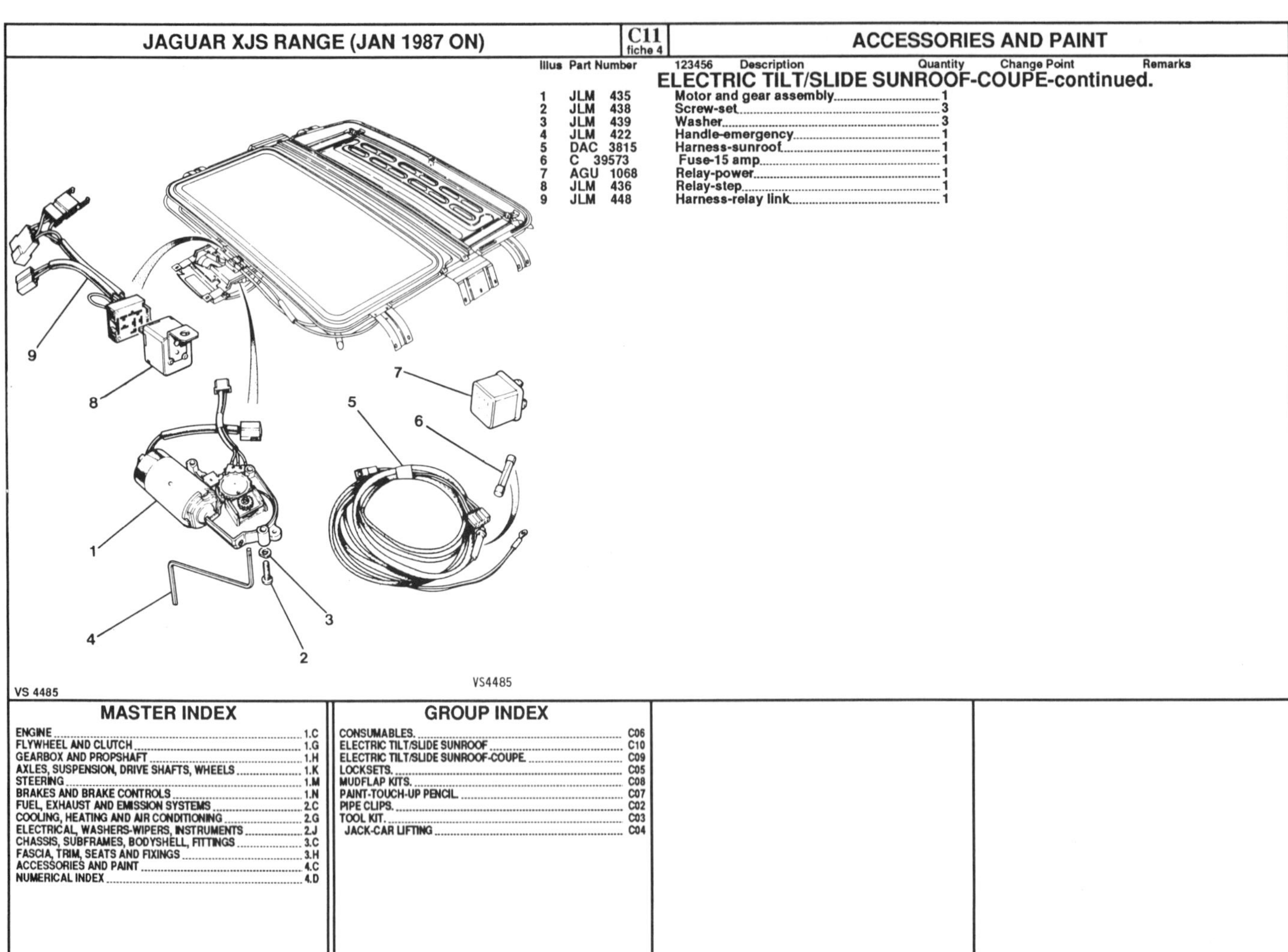

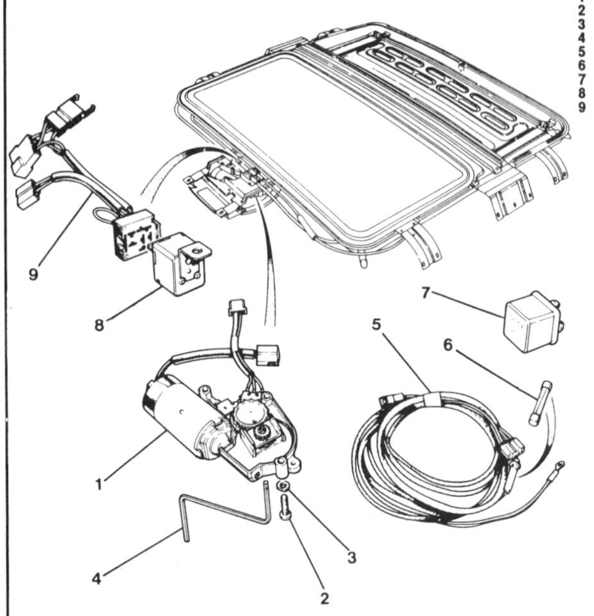

VS 4485

ELECTRIC TILT/SLIDE SUNROOF-COUPE-continued.

Illus	Part Number	123456	Description	Quantity	Change Point	Remarks
1	JLM 435		Motor and gear assembly	1		
2	JLM 438		Screw-set	3		
3	JLM 439		Washer	3		
4	JLM 422		Handle-emergency	1		
5	DAC 3815		Harness-sunroof	1		
6	C 39573		Fuse-15 amp	1		
7	AGU 1068		Relay-power	1		
8	JLM 436		Relay-step	1		
9	JLM 448		Harness-relay link	1		

MASTER INDEX

GROUP INDEX

JAGUAR XJS RANGE (JAN 1987 ON) | D01 Fiche 4 | NUMERICAL INDEX 1675 TO AGU 2436

1675	1.N03	12953	4.C06	AAU 2102	1.N02	AAU 6639	1.I18	AAU 6840	1.N10	AB 612061 J	1.N16	AEU 1617	2.D10	AFU 1024	4.C03
2390	1.N10	78319 J	2.K04	AAU 2103	1.N02	AAU 6639	1.J02	AAU 6841	1.I16	AC 604042 J	3.F06	AEU 1658	1.I17	AFU 1082 J	3.G07
2390	2.C16	78319 J	3.F12	AAU 2364	3.H05	AAU 6640	1.I10	AAU 7383	1.O02	AC 606042 J	3.I04	AEU 1659	1.I16	AFU 1082 J	3.G08
2390	2.E09	70723	1.M04	AAU 2365	3.H05	AAU 6641	1.I10	AAU 7383	1.O04	AC 610041 J	3.N18	AEU 1660	1.I09	AFU 1082 J	3.G09
2554	2.C12	79013 J	3.H09	AAU 3363	2.L17	AAU 6641	1.I12	AAU 7383	4.C02	ACU 2415	3.G02	AEU 1672	2.J02	AFU 1082 J	3.G10
3845	1.K16	79013 J	3.H11	AAU 3363	2.L18	AAU 6641	1.J02	AAU 7574	2.L17	ACU 3010	3.F05	AEU 1672	2.J04	AGU 1004	2.F07
3845	1.L03	108756	1.N03	AAU 3364	2.L17	AAU 6642	1.I11	AAU 7574	2.L18	ACU 6289	2.F06	AEU 1687	2.G18	AGU 1004	2.F15
4900	1.H05	12456	1.L03	AAU 3364	2.L18	AAU 6642	1.I12	AAU 8008	1.N06	AD 604031 J	1.H06	AEU 1687	2.I03	AGU 1018	3.H15
4900	1.H06	150653	2.K15	AAU 3377	1.N03	AAU 6642	1.I14	AAU 8271	1.K17	AD 604031 J	1.J05	AEU 1689 J	2.G18	AGU 1018	3.H18
4900	1.J04	150654	2.K15	AAU 3378	1.N03	AAU 6642	1.J02	AAU 8271	1.L04	AD 606042 J	3.G02	AEU 1689 J	2.I03	AGU 1018	3.I03
4900	1.J05	151857	1.K07	AAU 3379	1.N03	AAU 6643	1.I12	AAU 8639	1.I16	AD 606051 J	3.I15	AEU 1690 J	2.G18	AGU 1068	2.L09
4900	1.O10	234532 J	1.K04	AAU 3380	1.N03	AAU 6644	1.I12	AAU 8641	1.I16	AD 606051 J	3.I16	AEU 1690 J	2.I03	AGU 1068	2.L10
6114	1.K15	234532 J	1.K07	AAU 6066	1.I06	AAU 6645	1.I12	AAU 9520	1.I12	AD 606051 J	3.I17	AEU 1692	2.G18	AGU 1068	2.L15
6114	1.L02	234532 J	1.K09	AAU 6367	1.N18	AAU 6646	1.I13	AAU 9521	1.I12	AD 606051 J	3.I18	AEU 1692	2.I03	AGU 1068	2.M08
7680	1.K17	252167 J	1.K02	AAU 6367	1.O02	AAU 6647	1.I13	AAU 9522	1.I17	AD 606051 J	3.J02	AEU 1693	2.G18	AGU 1068	2.M16
7680	1.L04	338023 J	1.K14	AAU 6367	1.O03	AAU 6648	1.I16	AAU 9523	1.I14	AD 610052 J	3.F12	AEU 1693	2.I03	AGU 1068	4.C11
7685	1.K17	511598	2.K03	AAU 6367	1.O04	AAU 6649	1.I11	AAU 9524	1.I11	ADU 9028	1.D18	AEU 1694 J	2.G18	AGU 1070	2.L10
7685	1.L04	513682 J	2.J11	AAU 6367	4.C02	AAU 6650	1.I12	AAU 9525	1.I10	AEU 1038	1.I12	AEU 1694 J	2.I03	AGU 1070	2.L15
7686	1.K17	518564 J	1.M13	AAU 6600	1.I14	AAU 6651	1.I12	AAU 9526	1.I11	AEU 1039	1.I11	AEU 1695	2.G18	AGU 1079	3.D14
7686	1.L04	519542	2.K15	AAU 6601	1.I14	AAU 6652	1.I13	AAU 9527	1.I14	AEU 1040	1.I11	AEU 1695	2.I03	AGU 1091	3.N04
8016	1.O10	519544	2.K15	AAU 6603	1.I09	AAU 6653	1.I14	AAU 9528	1.I10	AEU 1048	1.I11	AEU 1696	2.G18	AGU 1097	2.N07
8337	1.O10	541572	4.C03	AAU 6605	1.I13	AAU 6654	1.I14	AAU 9529	1.I13	AEU 1049	1.I15	AEU 1696	2.I03	AGU 1108	2.L16
8489	1.K15	562019 J	4.C03	AAU 6606	1.I13	AAU 6655	1.I14	AAU 9530	1.I13	AEU 1050	1.I18	AEU 1697	2.G18	AGU 1109	2.L09
8489	1.L02	565770 J	4.C03	AAU 6607	1.I12	AAU 6656	1.I12	AAU 9531	1.I13	AEU 1057	3.D14	AEU 1697	2.I03	AGU 1117	2.J10
8836	1.O10	575047 J	2.K18	AAU 6608	1.I13	AAU 6657	1.I12	AAU 9532	1.I13	AEU 1065	1.I10	AEU 1699	2.G18	AGU 1135	2.H09
8840	1.O10	605410 J	2.J02	AAU 6610	1.I12	AAU 6658	1.I12	AAU 9533	1.I13	AEU 1068	1.N03	AEU 1699	2.I03	AGU 1168	1.C08
8841	1.O10	605410 J	2.J04	AAU 6610	1.I13	AAU 6659	1.I12	AAU 9534	1.I13	AEU 1103	1.I09	AEU 1702	1.I14	AGU 1169	2.J06
8845	1.O10	605765 J	4.C06	AAU 6611	1.I12	AAU 6660	1.I14	AAU 9595	1.I09	AEU 1183	1.M06	AEU 1721	2.J12	AGU 1198	1.N06
8938	1.O10	608094 J	2.J11	AAU 6612	1.I12	AAU 6661	1.I14	AB 60632	3.H13	AEU 1183	1.M07	AEU 1722	2.J12	AGU 1287 J	2.E04
8940	1.O10	608311 J	2.K07	AAU 6613	1.I12	AAU 6662	1.I11	AB 100031	2.H08	AEU 1197	1.I13	AEU 1723	2.J12	AGU 1299	2.N12
8980	1.O10	608312 J	2.K07	AAU 6614	1.I12	AAU 6663	1.I11	AB 604021 J	2.M13	AEU 1236	1.I09	AEU 1736	2.K17	AGU 1317	1.O07
9335	1.O10	624571	3.J08	AAU 6616	1.I12	AAU 6664	1.I11	AB 604044 J	2.M11	AEU 1248 J	1.M06	AEU 1737	2.K17	AGU 1329 J	2.C05
9336	1.O10	624571	3.J18	AAU 6617	1.I12	AAU 6666	1.I11	AB 606021	2.H14	AEU 1248 J	1.M07	AEU 1738	2.K17	AGU 1329 J	2.C09
9411	1.I07	624571	3.K06	AAU 6618	1.I12	AAU 6667	1.I11	AB 606021 J	2.K03	AEU 1360	1.I09	AEU 1763	2.K18	AGU 1329 J	2.E04
9412	1.I07	624571	3.K09	AAU 6619	1.I12	AAU 6668	1.I11	AB 606021 J	2.M08	AEU 1364	2.K03	AEU 1768	2.J09	AGU 1381	1.H03
9413	1.I07	624571	3.K12	AAU 6620	1.I14	AAU 6669	1.J03	AB 606031 J	2.O16	AEU 1365	2.K03	AEU 1773	2.K17	AGU 1413 J	3.G06
9750	1.O10	715721	3.J08	AAU 6621	1.I14	AAU 6670	1.I09	AB 606031 J	3.E14	AEU 1366	2.K03	AEU 1774	2.K17	AGU 1533	1.H03
9751	1.O10	715721	3.J18	AAU 6622	1.I14	AAU 6671	1.I16	AB 606031 J	3.E15	AEU 1385	1.K17	AEU 1858	2.G18	AGU 1863	3.L10
11099	1.K16	715721	3.K06	AAU 6623	1.I11	AAU 6672	1.I16	AB 606031 J	3.O03	AEU 1385	1.L04	AEU 1858	2.I03	AGU 1870	2.L16
11099	1.L03	715721	3.K09	AAU 6624	1.I11	AAU 6673	1.I16	AB 606032 J	3.H12	AEU 1386	1.K17	AEU 1930 N	2.J02	AGU 2074	2.N17
11290	1.K17	715721	3.K12	AAU 6625	1.I11	AAU 6674	1.I16	AB 606041 J	1.H06	AEU 1386	1.L04	AEU 1930 N	2.J04	AGU 2093	1.I05
11290	1.L04	12840	1.F08	AAU 6626	1.I11	AAU 6676	1.I15	AB 606041 J	1.J05	AEU 1387	1.K17	AEU 2519	1.M06	AGU 2098 J	2.C18
11291	1.K17	13H 5270 J	2.L17	AAU 6627	1.I11	AAU 6677	1.I09	AB 608031 J	2.K17	AEU 1387	1.L04	AEU 2519	1.M07	AGU 2229 AR	3.L10
11291	1.L04	13H 5270 J	2.L18	AAU 6628	1.I11	AAU 6679	1.I17	AB 608031 J	2.M04	AEU 1412	2.L16	AEU 2525	2.M04	AGU 2229 AW	3.L10
11356	2.K07	13H 9559	1.I09	AAU 6629	1.I11	AAU 6680	1.I17	AB 608051 J	1.M04	AEU 1427	2.K17	AEU 2526	2.M04	AGU 2229 CM	3.L10
11368	1.N03	27H 2403 J	2.K07	AAU 6630	1.I11	AAU 6682	1.I17	AB 608051 J	3.L10	AEU 1434	1.M06	AEU 2527	2.M04	AGU 2229 JF	3.L10
11369	1.N02	37H 6082 J	4.C06	AAU 6631	1.I10	AAU 6683	1.I17	AB 610031 J	1.N13	AEU 1434	1.M07	AEU 2553	2.K11	AGU 2229 LY	3.L10
11369	1.N03	37H 7193 J	1.M04	AAU 6632	1.I10	AAU 6685	1.I17	AB 610031 J	1.N15	AEU 1445	1.M06	AEU 2554	2.K11	AGU 2229 LZ	3.L10
11727	2.K07	37H 7627 J	1.M13	AAU 6633	1.I10	AAU 6686	1.I17	AB 610031 J	1.O06	AEU 1445	1.M07	AEU 2555	2.K11	AGU 2229 XE	3.L10
11728	2.K07	608 094 J	2.J12	AAU 6634	1.I10	AAU 6687	1.I17	AB 610031 J	1.O07	AEU 1465	2.D10	AEU 2556	2.K11	AGU 2229 ZM	3.L10
11730	2.K07	AAU 1312	2.K05	AAU 6635	1.I10	AAU 6688	1.I17	AB 610031 J	2.D04	AEU 1466	2.D10	AEU 2593	2.K04	AGU 2321	3.F07
11987	1.M06	AAU 1313	2.K05	AAU 6636	1.I10	AAU 6689	1.I18	AB 610031 J	2.F08	AEU 1467	2.D10	AEU 2594	2.K04	AGU 2380 J	3.G06
11987	1.M07	AAU 1497	2.K18	AAU 6637	1.I10	AAU 6690	1.I18	AB 610031 J	2.G12	AEU 1468	2.D10	AEU 2786	1.M11	AGU 2385 J	2.C13
12252	1.K16	AAU 1498	2.K18	AAU 6638	1.I09	AAU 6693	1.I18	AB 610051 J	2.D04	AEU 1469	2.D10	AEU 2787	1.M11	AGU 2388 J	3.I13
12252	1.L03	AAU 1582	4.C06	AAU 6639	1.I09	AAU 6694	1.I13	AB 610051 J	3.F12	AEU 1470	2.D10	AEU 2997	1.M06	AGU 2390	3.L10
12446	1.N10	AAU 1909 J	2.K18	AAU 6639	1.I10	AAU 6696	1.I14	AB 610052 J	3.F12	AEU 1471	2.D10	AEU 2997	1.M07	AGU 2392	3.N17
12456	1.K16	AAU 2019	2.G11	AAU 6639	1.I14	AAU 6697	1.I14	AB 610061	1.N18	AEU 1472	2.D10	AEU 4004	2.K14	AGU 2401 1	1.C09
12458 1	1.K16	AAU 2043	2.K16	AAU 6639	1.I15	AAU 6698	1.I10	AB 610081 J	2.J14	AEU 1590	2.K14	AEU 4005	1.H02	AGU 2414	1.N08
12798	1.N02	AAU 2044	2.K16	AAU 6639	1.I16	AAU 6699	1.I10	AB 612051 J	2.C05	AEU 1616 J	2.J02	AEU 4006	1.H02	AGU 2434	3.I14
12803	1.K04	AAU 2045	2.K16	AAU 6639	1.I17	AAU 6839	1.N10	AB 612061 J	1.N14	AEU 1616 J	2.J04	AEU 4174	2.D10	AGU 2436	2.D04

JAGUAR XJS RANGE (JAN 1987 ON) — D02 Fiche 4 — NUMERICAL INDEX AGU 2484 TO BAC 5678 CN

AGU 2484	1.K05
AGU 2517	3.I13
AGU 2527 J	1.C15
AGU 2533 J	3.I13
AGU 2536 J	2.C11
AGU 2539 J	3.I13
AH 610011	1.J03
AJ 608025 J	3.H02
AJ 612011	3.E11
AK 608051 J	3.L10
AK 608081 J	3.L10
AK 610021 J	2.M10
AR 10051 J	3.E11
AR 606031	2.N06
AR 608051 J	2.F08
AR 610051 J	3.G06
AR 610061 J	2.D12
AR 610071	2.F08
ARA 1501 J	2.D05
ARA 1501 J	2.D06
ARA 1501 J	2.D07
ARA 1501 J	2.D09
ARA 1502 J	2.D05
ARA 1502 J	2.D06
ARA 1502 J	2.D07
ARA 1502 J	2.D09
ATA 7166	1.K16
ATA 7166	1.L03
AUC 1358 J	2.C06
AUC 1358 J	2.E05
AZ 610051	1.J03
AZ 614061 J	3.G07
AZ 614061 J	3.G08
AZ 614061 J	3.G10
BAC 1151	3.G10
BAC 1166	3.G08
BAC 1166	3.G10
BAC 1170	3.G08
BAC 1331	3.G07
BAC 1337	3.G09
BAC 1338	3.G09
BAC 1357	3.G07
BAC 1357	3.G09
BAC 1461	3.G09
BAC 1462	3.H15
BAC 1462	3.H18
BAC 1462	3.I03
BAC 1488	3.C09
BAC 1525	3.H09
BAC 1525	3.H11
BAC 1537 LZ	3.H09
BAC 1537 LZ	3.H11
BAC 1681	3.G03
BAC 1681	3.G04
BAC 1681	3.G05
BAC 2168 PA	3.I15
BAC 2168 PA	3.I16
BAC 2168 PA	3.I17
BAC 2168 PA	3.I18
BAC 2446	3.G09
BAC 2446	3.G10
BAC 2564 AP	3.L12
BAC 2564 AP	3.L17
BAC 2564 AP	3.M05
BAC 2564 AP	3.M09
BAC 2564 CN	3.L12
BAC 2564 CN	3.L17
BAC 2564 CN	3.M05
BAC 2564 CN	3.M09
BAC 2564 LE	3.L12
BAC 2564 LE	3.L17
BAC 2564 LE	3.M05
BAC 2564 LE	3.M09
BAC 2564 XA	3.L12
BAC 2564 XA	3.L17
BAC 2564 XA	3.M05
BAC 2564 XA	3.M09
BAC 2564 XF	3.L12
BAC 2564 XF	3.L17
BAC 2564 XF	3.M05
BAC 2564 XF	3.M09
BAC 2565 AP	3.L12
BAC 2565 AP	3.L17
BAC 2565 AP	3.M05
BAC 2565 AP	3.M09
BAC 2565 CN	3.L12
BAC 2565 CN	3.L17
BAC 2565 CN	3.M05
BAC 2565 CN	3.M09
BAC 2565 LE	3.L12
BAC 2565 LE	3.L17
BAC 2565 LE	3.M05
BAC 2565 LE	3.M09
BAC 2565 XA	3.L12
BAC 2565 XA	3.L17
BAC 2565 XA	3.M05
BAC 2565 XA	3.M09
BAC 2565 XF	3.L12
BAC 2565 XF	3.L17
BAC 2565 XF	3.M05
BAC 2565 XF	3.M09
BAC 2864	3.N13
BAC 2904	3.I15
BAC 2928	3.G09
BAC 2928	3.G10
BAC 2929	3.G09
BAC 2929	3.G10
BAC 2934	3.G07
BAC 2934	3.G08
BAC 2935	3.G07
BAC 2935	3.G08
BAC 2936	3.G09
BAC 2937	3.G09
BAC 2953	2.H14
BAC 2972	3.H07
BAC 2975	3.H07
BAC 2976	3.H07
BAC 2978	3.H07
BAC 2980	3.H06
BAC 2983	3.H06
BAC 2984	3.H06
BAC 2986	3.H06
BAC 3026	3.H07
BAC 3027	3.H07
BAC 3074	3.G07
BAC 3074	3.G09
BAC 3075	3.G02
BAC 3347	3.G07
BAC 3347	3.G08
BAC 3348	3.G07
BAC 3348	3.G08
BAC 3349	3.G07
BAC 3349	3.G08
BAC 3361	3.G09
BAC 3361	3.G10
BAC 3362	3.G09
BAC 3362	3.G10
BAC 3363	3.G09
BAC 3363	3.G10
BAC 3375	3.N12
BAC 3411	2.H12
BAC 3415	3.G10
BAC 3426	3.G07
BAC 3477	3.D13
BAC 3479	3.G09
BAC 3479	3.G10
BAC 3480	3.G07
BAC 3480	3.G08
BAC 3481	3.G07
BAC 3481	3.G08
BAC 3504	3.G07
BAC 3504	3.G08
BAC 3505	3.G07
BAC 3505	3.G08
BAC 3528	1.J08
BAC 3544	3.N09
BAC 3544	3.N10
BAC 3544	3.N11
BAC 3596	3.E07
BAC 3597	3.E07
BAC 3661	3.H09
BAC 3661	3.H11
BAC 3748	2.H14
BAC 3787	3.N06
BAC 3838	3.G03
BAC 3838	3.G04
BAC 3838	3.G05
BAC 3839	3.G03
BAC 3839	3.G04
BAC 3839	3.G05
BAC 3840	3.G03
BAC 3840	3.G04
BAC 3840	3.G05
BAC 3968 AE	3.H08
BAC 3968 AR	3.H08
BAC 3968 AW	3.H08
BAC 3968 CM	3.H08
BAC 3968 JF	3.H08
BAC 3968 LY	3.H08
BAC 3968 LZ	3.H08
BAC 3968 XE	3.H08
BAC 3968 ZM	3.H08
BAC 3969 AE	3.H08
BAC 3969 AR	3.H08
BAC 3969 AW	3.H08
BAC 3969 CM	3.H08
BAC 3969 JF	3.H08
BAC 3969 LY	3.H08
BAC 3969 LZ	3.H08
BAC 3969 XE	3.H08
BAC 3969 ZM	3.H08
BAC 3986 AE	3.I05
BAC 3986 AE	3.I07
BAC 3986 AR	3.I05
BAC 3986 AR	3.I06
BAC 3986 AR	3.I07
BAC 3986 AR	3.I08
BAC 3986 AW	3.I05
BAC 3986 AW	3.I06
BAC 3986 AW	3.I07
BAC 3986 AW	3.I08
BAC 3986 CM	3.I05
BAC 3986 CM	3.I06
BAC 3986 CM	3.I07
BAC 3986 CM	3.I08
BAC 3986 JF	3.I05
BAC 3986 JF	3.I06
BAC 3986 JF	3.I07
BAC 3986 JF	3.I08
BAC 3986 LY	3.I05
BAC 3986 LY	3.I06
BAC 3986 LY	3.I07
BAC 3986 LY	3.I08
BAC 3986 LZ	3.I05
BAC 3986 LZ	3.I06
BAC 3986 LZ	3.I07
BAC 3986 LZ	3.I08
BAC 3986 XE	3.I05
BAC 3986 XE	3.I06
BAC 3986 XE	3.I07
BAC 3986 XE	3.I08
BAC 3986 ZM	3.I05
BAC 3986 ZM	3.I06
BAC 3986 ZM	3.I07
BAC 3986 ZM	3.I08
BAC 3987 AE	3.I05
BAC 3987 AE	3.I07
BAC 3987 AR	3.I05
BAC 3987 AR	3.I06
BAC 3987 AR	3.I07
BAC 3987 AR	3.I08
BAC 3987 AW	3.I05
BAC 3987 AW	3.I06
BAC 3987 AW	3.I07
BAC 3987 AW	3.I08
BAC 3987 CM	3.I05
BAC 3987 CM	3.I06
BAC 3987 CM	3.I07
BAC 3987 CM	3.I08
BAC 3987 JF	3.I05
BAC 3987 JF	3.I06
BAC 3987 JF	3.I07
BAC 3987 JF	3.I08
BAC 3987 LY	3.I05
BAC 3987 LY	3.I06
BAC 3987 LY	3.I07
BAC 3987 LY	3.I08
BAC 3987 LZ	3.I05
BAC 3987 LZ	3.I06
BAC 3987 LZ	3.I07
BAC 3987 LZ	3.I08
BAC 3987 XE	3.I05
BAC 3987 XE	3.I06
BAC 3987 XE	3.I07
BAC 3987 XE	3.I08
BAC 3987 ZM	3.I05
BAC 3987 ZM	3.I06
BAC 3987 ZM	3.I07
BAC 3987 ZM	3.I08
BAC 4038 AE	3.J05
BAC 4038 AR	3.J05
BAC 4038 AW	3.J05
BAC 4038 AW	3.K08
BAC 4038 CM	3.J05
BAC 4038 JF	3.J05
BAC 4038 LY	3.J05
BAC 4038 LY	3.K08
BAC 4038 LZ	3.J05
BAC 4038 LZ	3.K08
BAC 4038 XE	3.J05
BAC 4038 XE	3.K08
BAC 4038 ZM	3.J05
BAC 4039 AE	3.J06
BAC 4039 AR	3.J06
BAC 4039 AW	3.J06
BAC 4039 AW	3.K08
BAC 4039 CM	3.J06
BAC 4039 JF	3.J06
BAC 4039 LY	3.J06
BAC 4039 LY	3.K08
BAC 4039 LZ	3.J06
BAC 4039 LZ	3.K08
BAC 4039 XE	3.J06
BAC 4039 XE	3.K08
BAC 4039 ZM	3.J06
BAC 4040 AE	3.J05
BAC 4040 AR	3.J05
BAC 4040 AW	3.J05
BAC 4040 CM	3.J05
BAC 4040 JF	3.J05
BAC 4040 LY	3.J05
BAC 4040 LZ	3.J05
BAC 4040 XE	3.J05
BAC 4040 ZM	3.J05
BAC 4060 AE	3.J09
BAC 4060 AR	3.J09
BAC 4060 AW	3.J09
BAC 4060 AW	3.K06
BAC 4060 AW	3.K09
BAC 4060 CM	3.J09
BAC 4060 JF	3.J09
BAC 4060 LY	3.J09
BAC 4060 LY	3.K06
BAC 4060 LY	3.K09
BAC 4060 LZ	3.J09
BAC 4060 LZ	3.K06
BAC 4060 LZ	3.K09
BAC 4060 XE	3.J09
BAC 4060 XE	3.K06
BAC 4060 XE	3.K09
BAC 4060 ZM	3.J09
BAC 4064 AE	3.J09
BAC 4064 AR	3.J09
BAC 4064 AW	3.J09
BAC 4064 CM	3.J09
BAC 4064 JF	3.J09
BAC 4064 LY	3.J09
BAC 4064 LZ	3.J09
BAC 4064 XE	3.J09
BAC 4064 ZM	3.J09
BAC 4068 AE	3.L05
BAC 4068 AR	3.L05
BAC 4068 AW	3.L05
BAC 4068 CM	3.L05
BAC 4068 JF	3.L05
BAC 4068 LY	3.L05
BAC 4068 LZ	3.L05
BAC 4068 XE	3.L05
BAC 4068 ZM	3.L05
BAC 4069 AE	3.L05
BAC 4069 AR	3.L05
BAC 4069 AW	3.L05
BAC 4069 CM	3.L05
BAC 4069 JF	3.L05
BAC 4069 LY	3.L05
BAC 4069 LZ	3.L05
BAC 4069 XE	3.L05
BAC 4069 ZM	3.L05
BAC 4084	3.N03
BAC 4085	3.N03
BAC 4085	3.N04
BAC 4087	3.N03
BAC 4087	3.N04
BAC 4089	3.N03
BAC 4089	3.N04
BAC 4110 AE	3.H15
BAC 4110 AR	3.H15
BAC 4110 AR	3.I03
BAC 4110 AW	3.H15
BAC 4110 AW	3.I03
BAC 4110 CM	3.H15
BAC 4110 CM	3.I03
BAC 4110 JF	3.H15
BAC 4110 JF	3.I03
BAC 4110 LY	3.H15
BAC 4110 LY	3.I03
BAC 4110 LZ	3.H15
BAC 4110 LZ	3.I03
BAC 4110 XE	3.H15
BAC 4110 XE	3.I03
BAC 4110 ZM	3.H15
BAC 4110 ZM	3.I03
BAC 4111 AE	3.H15
BAC 4111 AR	3.H15
BAC 4111 AR	3.I03
BAC 4111 AW	3.H15
BAC 4111 AW	3.I03
BAC 4111 CM	3.H15
BAC 4111 CM	3.I03
BAC 4111 JF	3.H15
BAC 4111 JF	3.I03
BAC 4111 LY	3.H15
BAC 4111 LY	3.I03
BAC 4111 LZ	3.H15
BAC 4111 LZ	3.I03
BAC 4111 XE	3.H15
BAC 4111 XE	3.I03
BAC 4111 ZM	3.H15
BAC 4111 ZM	3.I03
BAC 4131	3.H15
BAC 4131	3.H18
BAC 4187	3.N03
BAC 4252	2.L06
BAC 4253	2.L06
BAC 4255	2.H14
BAC 4256	3.F14
BAC 4257	3.F14
BAC 4258 AP	3.L16
BAC 4258 AP	3.M09
BAC 4258 CN	3.L16
BAC 4258 CN	3.M09
BAC 4258 LE	3.L16
BAC 4258 LE	3.M09
BAC 4258 XA	3.L16
BAC 4258 XA	3.M09
BAC 4258 XF	3.L16
BAC 4258 XF	3.M09
BAC 4259 AP	3.L16
BAC 4259 AP	3.M09
BAC 4259 CN	3.L16
BAC 4259 CN	3.M09
BAC 4259 LE	3.L16
BAC 4259 LE	3.M09
BAC 4259 XA	3.L16
BAC 4259 XA	3.M09
BAC 4259 XF	3.L16
BAC 4259 XF	3.M09
BAC 4300	3.D14
BAC 4333	2.L06
BAC 4343	3.D13
BAC 4350	3.G06
BAC 4352	3.G06
BAC 4364	3.N03
BAC 4364	3.N04
BAC 4383	3.G07
BAC 4383	3.G08
BAC 4450	3.G06
BAC 4451	3.G06
BAC 4474	3.H03
BAC 4486	3.F12
BAC 4487	3.F12
BAC 4722	3.L02
BAC 4723	3.L02
BAC 5290	3.I13
BAC 5298	3.H15
BAC 5298	3.H18
BAC 5298	3.I03
BAC 5416	3.F12
BAC 5417	3.F12
BAC 5427	3.E14
BAC 5427	3.E15
BAC 5566 PA	3.I15
BAC 5566 PA	3.I17
BAC 5606 AE	3.H14
BAC 5606 AR	3.H14
BAC 5606 AW	3.H14
BAC 5606 CM	3.H14
BAC 5606 JF	3.H14
BAC 5606 LY	3.H14
BAC 5606 LZ	3.H14
BAC 5606 XE	3.H14
BAC 5606 ZM	3.H14
BAC 5607 AE	3.H14
BAC 5607 AR	3.H14
BAC 5607 AW	3.H14
BAC 5607 CM	3.H14
BAC 5607 JF	3.H14
BAC 5607 LY	3.H14
BAC 5607 LZ	3.H14
BAC 5607 XE	3.H14
BAC 5607 ZM	3.H14
BAC 5614 AE	3.H14
BAC 5614 AR	3.H14
BAC 5614 AW	3.H14
BAC 5614 CM	3.H14
BAC 5614 JF	3.H14
BAC 5614 LY	3.H14
BAC 5614 LZ	3.H14
BAC 5614 XE	3.H14
BAC 5614 ZM	3.H14
BAC 5615 AE	3.H14
BAC 5615 AR	3.H14
BAC 5615 AW	3.H14
BAC 5615 CM	3.H14
BAC 5615 JF	3.H14
BAC 5615 LY	3.H14
BAC 5615 LZ	3.H14
BAC 5615 XE	3.H14
BAC 5615 ZM	3.H14
BAC 5616	3.H15
BAC 5616	3.H18
BAC 5616	3.I03
BAC 5655	3.H06
BAC 5678 AP	3.L12
BAC 5678 AP	3.L17
BAC 5678 AP	3.M05
BAC 5678 AP	3.M09
BAC 5678 CN	3.L12
BAC 5678 CN	3.L17
BAC 5678 CN	3.M05

JAGUAR XJS RANGE (JAN 1987 ON) — D03 Fiche 4 — NUMERICAL INDEX BAC 5678 CN TO BBC 8209

BAC 5678 CN	3.M09
BAC 5678 LE	3.L12
BAC 5678 LE	3.L17
BAC 5678 LE	3.M05
BAC 5678 LE	3.M09
BAC 5678 XA	3.L12
BAC 5678 XA	3.L17
BAC 5678 XA	3.M05
BAC 5678 XA	3.M09
BAC 5678 XF	3.L12
BAC 5678 XF	3.L17
BAC 5678 XF	3.M05
BAC 5678 XF	3.M09
BAC 5764	3.F13
BAC 5785	3.N09
BAC 5785	3.N10
BAC 5911	3.N03
BAC 5911	3.N04
BAC 5913	3.N05
BAC 6019	3.E02
BAC 6074	3.N14
BAC 6084 ND	3.N17
BAC 6085 ND	3.N17
BAC 6086 ND	3.N17
BAC 6087 ND	3.N17
BAC 6104	3.N17
BAC 6105	3.N17
BAC 6236	3.F14
BAC 6306	3.D13
BAC 6449	1.J08
BAC 6480	3.G04
BAC 6480	3.G05
BAC 6481	3.G03
BAC 7412	3.N18
BAC 7413	3.N18
BAC 7420	3.N16
BAC 7424	3.N16
BAC 7424	3.O02
BAC 7426	3.N17
BAC 7700	3.C16
BAC 7701	3.C16
BAC 7728	3.C16
BAC 7735	2.L12
BAC 7736	3.C15
BAC 7738	3.F09
BAC 7739	3.F09
BAC 7740	3.F09
BAC 7741	3.F09
BAC 7742	3.F09
BAC 7743	3.F09
BAC 7788	3.F09
BAC 7789	3.F09
BAC 7790 JX	3.N15
BAC 7790 NF	3.N15
BAC 7790 PM	3.N15
BAC 7795 ND	3.N14
BAC 7802	3.F09
BAC 7803	3.F09
BAC 7804 JX	3.N15
BAC 7804 NF	3.N15
BAC 7804 PM	3.N15
BAC 7808	3.N16
BAC 7816 JX	3.N16
BAC 7816 NF	3.N16
BAC 7816 PM	3.N16
BAC 7820	3.C15
BAC 7821	3.C15
BAC 7832	3.N17
BAC 7853	3.C15
BAC 7854	3.C15
BAC 7864	3.N14
BAC 7879	3.C18
BAC 7897	3.C15
BAC 7907	3.N18
BAC 7936	3.N14
BAC 7952	3.N16
BAC 7955 JX	3.N16
BAC 7955 NF	3.N16
BAC 7955 PM	3.N16
BAC 7975	3.N18
BAC 7997	3.N18
BAC 8272	3.F09
BAC 8273	3.F09
BAC 8274	3.C18
BAC 8276	3.C18
BAC 8277	3.C18
BAC 8279	3.C18
BAC 8285	3.C15
BAC 8287	3.C15
BAC 8288	3.C15
BAC 8294	3.C16
BAC 8295	3.C16
BAC 8976	3.C18
BAC 9496	3.C15
BAU 1124	2.L03
BAU 2043	2.L04
BAU 2043	2.L07
BAU 2043	2.L08
BAU 2135 J	1.M11
BAU 2136 J	1.M11
BAU 2137 J	1.M11
BAU 4604 J	1.K16
BAU 4604 J	1.L03
BAU 4781 A	1.G06
BBC 1305	3.E10
BBC 1317 JX	3.N17
BBC 1317 NF	3.N17
BBC 1317 PM	3.N17
BBC 1324	3.N17
BBC 1327	3.N17
BBC 1342	3.L09
BBC 1342	3.N16
BBC 1343	3.N16
BBC 1343	3.O02
BBC 1454	3.N06
BBC 1455	3.N06
BBC 1727 XE	3.I15
BBC 1727 XE	3.I16
BBC 1727 XE	3.I17
BBC 1727 XE	3.I18
BBC 1856	3.L07
BBC 1996 AE	3.H16
BBC 1996 AR	3.H16
BBC 1996 AW	3.H16
BBC 1996 CM	3.H16
BBC 1996 JF	3.H16
BBC 1996 LY	3.H16
BBC 1996 LZ	3.H16
BBC 1996 XE	3.H16
BBC 1996 ZM	3.H16
BBC 1997 AE	3.H16
BBC 1997 AR	3.H16
BBC 1997 AW	3.H16
BBC 1997 CM	3.H16
BBC 1997 JF	3.H16
BBC 1997 LY	3.H16
BBC 1997 LZ	3.H16
BBC 1997 XE	3.H16
BBC 1997 ZM	3.H16
BBC 1998 AE	3.H16
BBC 1998 AR	3.H16
BBC 1998 AW	3.H16
BBC 1998 CM	3.H16
BBC 1998 JF	3.H16
BBC 1998 LY	3.H16
BBC 1998 LZ	3.H16
BBC 1998 XE	3.H16
BBC 1998 ZM	3.H16
BBC 1999 AE	3.H16
BBC 1999 AR	3.H16
BBC 1999 AW	3.H16
BBC 1999 CM	3.H16
BBC 1999 JF	3.H16
BBC 1999 LY	3.H16
BBC 1999 LZ	3.H16
BBC 1999 XE	3.H16
BBC 1999 ZM	3.H16
BBC 2000 AE	3.H17
BBC 2000 AR	3.H17
BBC 2000 AW	3.H17
BBC 2000 CM	3.H17
BBC 2000 JF	3.H17
BBC 2000 LY	3.H17
BBC 2000 LZ	3.H17
BBC 2000 XE	3.H17
BBC 2000 ZM	3.H17
BBC 2001 AE	3.H17
BBC 2001 AR	3.H17
BBC 2001 AW	3.H17
BBC 2001 CM	3.H17
BBC 2001 JF	3.H17
BBC 2001 LY	3.H17
BBC 2001 LZ	3.H17
BBC 2001 XE	3.H17
BBC 2001 ZM	3.H17
BBC 2035	3.L02
BBC 2039	3.L02
BBC 2112	3.N10
BBC 2113	3.N10
BBC 2114	3.N10
BBC 2115	3.N10
BBC 2182 AP	3.M08
BBC 2182 AP	3.M12
BBC 2182 CN	3.M08
BBC 2182 CN	3.M12
BBC 2182 LE	3.M08
BBC 2182 LE	3.M12
BBC 2182 XA	3.M08
BBC 2182 XA	3.M12
BBC 2182 XF	3.M08
BBC 2182 XF	3.M12
BBC 2186	3.L02
BBC 2187	3.L02
BBC 2260 AP	3.L08
BBC 2260 CN	3.L08
BBC 2260 LE	3.L08
BBC 2260 XA	3.L08
BBC 2260 XF	3.L08
BBC 2276 AE	3.L08
BBC 2276 AR	3.L08
BBC 2276 AW	3.L08
BBC 2276 CM	3.L08
BBC 2276 JF	3.L08
BBC 2276 LY	3.L08
BBC 2276 LZ	3.L08
BBC 2276 XE	3.L08
BBC 2276 ZM	3.L08
BBC 2483	2.N02
BBC 2520	3.I17
BBC 2521	3.I17
BBC 2583	3.N03
BBC 2583	3.N04
BBC 2584	3.N03
BBC 2620 AP	3.M08
BBC 2620 AP	3.M12
BBC 2620 CN	3.M08
BBC 2620 CN	3.M12
BBC 2620 LE	3.M08
BBC 2620 LE	3.M12
BBC 2620 XA	3.M08
BBC 2620 XA	3.M12
BBC 2620 XF	3.M08
BBC 2620 XF	3.M12
BBC 2621 AP	3.M08
BBC 2621 AP	3.M12
BBC 2621 CN	3.M08
BBC 2621 CN	3.M12
BBC 2621 LE	3.M08
BBC 2621 LE	3.M12
BBC 2621 XA	3.M08
BBC 2621 XA	3.M12
BBC 2621 XF	3.M08
BBC 2621 XF	3.M12
BBC 2649	3.N03
BBC 2682	4.C09
BBC 2996	3.F06
BBC 2997	3.F06
BBC 3022 YE	3.K07
BBC 3122 AP	3.L12
BBC 3122 AP	3.L16
BBC 3122 AP	3.M05
BBC 3122 AP	3.M09
BBC 3122 CN	3.L12
BBC 3122 CN	3.L16
BBC 3122 CN	3.M05
BBC 3122 CN	3.M09
BBC 3122 LE	3.L12
BBC 3122 LE	3.L16
BBC 3122 LE	3.M05
BBC 3122 LE	3.M09
BBC 3122 XA	3.L12
BBC 3122 XA	3.L16
BBC 3122 XA	3.M05
BBC 3122 XA	3.M09
BBC 3122 XF	3.L12
BBC 3122 XF	3.L16
BBC 3122 XF	3.M05
BBC 3122 XF	3.M09
BBC 3123 AP	3.L12
BBC 3123 AP	3.L16
BBC 3123 AP	3.M05
BBC 3123 AP	3.M09
BBC 3123 CN	3.L12
BBC 3123 CN	3.L16
BBC 3123 CN	3.M05
BBC 3123 CN	3.M09
BBC 3123 LE	3.L12
BBC 3123 LE	3.L16
BBC 3123 LE	3.M05
BBC 3123 LE	3.M09
BBC 3123 XA	3.L12
BBC 3123 XA	3.L16
BBC 3123 XA	3.M05
BBC 3123 XA	3.M09
BBC 3123 XF	3.L12
BBC 3123 XF	3.L16
BBC 3123 XF	3.M05
BBC 3123 XF	3.M09
BBC 3286	3.N03
BBC 3286	3.N04
BBC 3693	2.D10
BBC 3693	2.D11
BBC 3916	3.C16
BBC 3917	3.C16
BBC 3919 JX	3.N15
BBC 3919 NF	3.N15
BBC 3919 PM	3.N15
BBC 3922	3.C18
BBC 3923	3.C18
BBC 4411	3.D14
BBC 4441	3.E14
BBC 4441	3.E15
BBC 4444	3.D17
BBC 4454	3.D14
BBC 4460	3.N03
BBC 5180	3.N05
BBC 5486	3.H02
BBC 5487	3.H02
BBC 5498	3.H02
BBC 5534	3.E02
BBC 5535	3.E02
BBC 5546	3.H02
BBC 5547	3.H02
BBC 5636	2.D10
BBC 5716	3.N18
BBC 5732	3.N18
BBC 5733	3.N18
BBC 5734	3.N18
BBC 5748	3.N14
BBC 5749	3.N14
BBC 5750	3.N14
BBC 5751	3.N14
BBC 5757	3.N18
BBC 5932	3.E07
BBC 5933	3.E07
BBC 6292	3.F06
BBC 6619	3.F11
BBC 6664	3.N14
BBC 6691	3.N14
BBC 6722 XE	3.I18
BBC 6723 XE	3.I18
BBC 6907	3.F05
BBC 6908	3.F05
BBC 6918	3.F07
BBC 6919	3.F07
BBC 6934	3.F07
BBC 6935	3.F07
BBC 6942	3.E09
BBC 6966	3.D06
BBC 6991	3.E15
BBC 6996	3.F05
BBC 6997	3.F05
BBC 6998	3.F05
BBC 6999	3.F05
BBC 7000	3.E18
BBC 7072	3.N18
BBC 7158	3.L02
BBC 7159	3.L02
BBC 7221	3.E09
BBC 7224	3.E09
BBC 7225	3.E09
BBC 7395	3.E09
BBC 7417	3.H07
BBC 7422	3.H15
BBC 7429	2.N15
BBC 7494 XE	3.I15
BBC 7494 XE	3.I16
BBC 7495 XE	3.I15
BBC 7495 XE	3.I16
BBC 7525 AR	3.I13
BBC 7525 AW	3.I13
BBC 7525 CM	3.I13
BBC 7525 JF	3.I13
BBC 7525 LY	3.I13
BBC 7525 LZ	3.I13
BBC 7525 XF	3.I13
BBC 7525 ZM	3.I13
BBC 7538 XE	3.I15
BBC 7538 XE	3.I16
BBC 7539 XE	3.I15
BBC 7539 XE	3.I16
BBC 7544 XE	3.I15
BBC 7545 XE	3.I15
BBC 7546	3.N03
BBC 7546	3.N04
BBC 7548	3.N03
BBC 7581	3.H09
BBC 7581	3.H11
BBC 7654	3.E15
BBC 7655	3.I13
BBC 7680	3.F07
BBC 7681	3.F07
BBC 7684	3.D17
BBC 7685	3.D17
BBC 7730	3.F07
BBC 7731	3.F07
BBC 7743	2.H18
BBC 7759	3.K13
BBC 7759	3.K16
BBC 7759	3.K17
BBC 7800	3.K14
BBC 7801	3.K14
BBC 7862 ND	3.I15
BBC 7862 ND	3.I17
BBC 7863 ND	3.I15
BBC 7863 ND	3.I17
BBC 7870 PA	3.I15
BBC 7870 PA	3.I17
BBC 7872	3.N09
BBC 7998 AE	3.L13
BBC 7998 AE	3.L18
BBC 7998 AE	3.M06
BBC 7998 AE	3.M10
BBC 7998 AR	3.L13
BBC 7998 AR	3.L18
BBC 7998 AR	3.M06
BBC 7998 AR	3.M10
BBC 7998 AW	3.L13
BBC 7998 AW	3.L18
BBC 7998 AW	3.M06
BBC 7998 AW	3.M10
BBC 7998 CM	3.L13
BBC 7998 CM	3.L18
BBC 7998 CM	3.M06
BBC 7998 CM	3.M10
BBC 7998 JF	3.L13
BBC 7998 JF	3.L18
BBC 7998 JF	3.M06
BBC 7998 JF	3.M10
BBC 7998 LY	3.L13
BBC 7998 LY	3.L18
BBC 7998 LY	3.M06
BBC 7998 LY	3.M10
BBC 7998 LZ	3.L13
BBC 7998 LZ	3.L18
BBC 7998 LZ	3.M06
BBC 7998 LZ	3.M10
BBC 7998 XE	3.L13
BBC 7998 XE	3.L18
BBC 7998 XE	3.M06
BBC 7998 XE	3.M10
BBC 7998 ZM	3.L13
BBC 7998 ZM	3.L18
BBC 7998 ZM	3.M06
BBC 7998 ZM	3.M10
BBC 7999 AE	3.L14
BBC 7999 AE	3.M02
BBC 7999 AE	3.M07
BBC 7999 AE	3.M11
BBC 7999 AR	3.L14
BBC 7999 AR	3.M02
BBC 7999 AR	3.M07
BBC 7999 AR	3.M11
BBC 7999 AW	3.L14
BBC 7999 AW	3.M02
BBC 7999 AW	3.M07
BBC 7999 AW	3.M11
BBC 7999 CM	3.L14
BBC 7999 CM	3.M02
BBC 7999 CM	3.M07
BBC 7999 CM	3.M11
BBC 7999 JF	3.L14
BBC 7999 JF	3.M02
BBC 7999 JF	3.M07
BBC 7999 JF	3.M11
BBC 7999 LY	3.L14
BBC 7999 LY	3.M02
BBC 7999 LY	3.M07
BBC 7999 LY	3.M11
BBC 7999 LZ	3.L14
BBC 7999 LZ	3.M02
BBC 7999 LZ	3.M07
BBC 7999 LZ	3.M11
BBC 7999 XE	3.L14
BBC 7999 XE	3.M02
BBC 7999 XE	3.M07
BBC 7999 XE	3.M11
BBC 7999 ZM	3.L14
BBC 7999 ZM	3.M02
BBC 7999 ZM	3.M07
BBC 7999 ZM	3.M11
BBC 8042	3.O04
BBC 8049	3.O04
BBC 8178	3.O03
BBC 8179	3.O03
BBC 8180	3.O03
BBC 8181	3.O03
BBC 8184	3.O03
BBC 8186	3.O03
BBC 8187	3.O03
BBC 8188	3.O03
BBC 8189	3.O03
BBC 8190	3.O03
BBC 8191	3.O03
BBC 8192	3.O03
BBC 8193	3.O03
BBC 8194	3.O03
BBC 8200	3.O02
BBC 8201	3.O02
BBC 8209	3.O02

JAGUAR XJS RANGE (JAN 1987 ON) — D04 Fiche 4 — NUMERICAL INDEX BBC 8211 TO BCC 3038 XE

BBC 8211	3.O02
BBC 8216	3.O02
BBC 8217	3.O02
BBC 8224	3.O02
BBC 8243	3.O02
BBC 8265	3.O03
BBC 8268	3.O02
BBC 8306 AX	3.K04
BBC 8306 AY	3.K04
BBC 8306 YF	3.K04
BBC 8314 AX	3.K07
BBC 8314 AY	3.K07
BBC 8314 YF	3.K07
BBC 8320 AX	3.K06
BBC 8320 AX	3.K09
BBC 8320 AY	3.K06
BBC 8320 AY	3.K09
BBC 8320 YE	3.K06
BBC 8320 YE	3.K09
BBC 8320 YF	3.K06
BBC 8320 YF	3.K09
BBC 8325 AX	3.K04
BBC 8325 AX	3.K07
BBC 8325 AY	3.K04
BBC 8325 AY	3.K07
BBC 8325 YF	3.K04
BBC 8325 YF	3.K07
BBC 8331 AX	3.L06
BBC 8331 AY	3.L06
BBC 8331 YF	3.L06
BBC 8332 AX	3.L06
BBC 8332 AY	3.L06
BBC 8332 YF	3.L06
BBC 8532	3.N09
BBC 8532	3.N10
BBC 8533	3.N09
BBC 8533	3.N10
BBC 8554	3.E08
BBC 8606	3.G02
BBC 8607	3.G02
BBC 8608	3.G02
BBC 8609	3.G02
BBC 8610	3.G02
BBC 8611	3.G02
BBC 8679	3.D17
BBC 8682 XE	3.I18
BBC 8683 XE	3.I18
BBC 8686 XE	3.I18
BBC 8687 XE	3.I18
BBC 8710	3.D17
BBC 8968	3.J08
BBC 8968	3.J18
BBC 8969	3.J08
BBC 8969	3.J18
BBC 9132	3.N18
BBC 9133	3.N18
BBC 9226	4.C05
BBC 9227	4.C05
BBC 9257	3.I15
BBC 9257	3.I16
BBC 9299	3.H06
BBC 9384	3.C09
BBC 9565	3.N14
BBC 9612	3.N18
BBC 9613	3.N18
BBC 9616	3.C09
BBC 9716	3.F16
BBC 9717	3.F16
BBC 9718	3.F16
BBC 9719	3.F16
BBC 9720	3.F16
BBC 9721	3.F16
BBC 9722	3.F16
BBC 9723	3.F16
BBC 9724	3.F16
BBC 9726	3.F16
BBC 9727	3.F16
BBC 9728	3.F16
BBC 9729	3.F16
BBC 9730	3.F16
BBC 9731	3.F16
BBC 9750 JX	3.N14
BBC 9750 NF	3.N14
BBC 9750 PM	3.N14
BBC 9751 JX	3.N14
BBC 9751 NF	3.N14
BBC 9751 PM	3.N14
BBC 9752 JX	3.N14
BBC 9752 NF	3.N14
BBC 9752 PM	3.N14
BBC 9753 JX	3.N14
BBC 9753 NF	3.N14
BBC 9753 PM	3.N14
BBC 9773	3.L07
BBC 9774	3.L07
BBC 9779	3.L07
BBC 9853	3.E02
BBC 9980 AP	3.L07
BBC 9980 CN	3.L07
BBC 9980 LE	3.L07
BBC 9980 XA	3.L07
BBC 9980 XF	3.L07
BBC 578332 YE	3.L06
BBC 878306 YE	3.K04
BBC 878325 YE	3.K04
BBC 878325 YE	3.K07
BBC 878331 YE	3.L06
BCC 1001	3.N16
BCC 2030 AE	3.I07
BCC 2030 AR	3.I07
BCC 2030 AW	3.I07
BCC 2030 CM	3.I07
BCC 2030 JF	3.I07
BCC 2030 LY	3.I07
BCC 2030 LZ	3.I07
BCC 2030 XE	3.I07
BCC 2030 ZM	3.I07
BCC 2031 AE	3.I07
BCC 2031 AR	3.I07
BCC 2031 AW	3.I07
BCC 2031 CM	3.I07
BCC 2031 JF	3.I07
BCC 2031 LY	3.I07
BCC 2031 LZ	3.I07
BCC 2031 XE	3.I07
BCC 2031 ZM	3.I07
BCC 2066 AP	3.L12
BCC 2066 AP	3.M05
BCC 2066 CN	3.L12
BCC 2066 CN	3.M05
BCC 2066 LE	3.L12
BCC 2066 LE	3.M05
BCC 2066 XA	3.L12
BCC 2066 XA	3.M05
BCC 2066 XF	3.L12
BCC 2066 XF	3.M05
BCC 2067 AP	3.L12
BCC 2067 AP	3.M05
BCC 2067 CN	3.L12
BCC 2067 CN	3.M05
BCC 2067 LE	3.L12
BCC 2067 LE	3.M05
BCC 2067 XA	3.L12
BCC 2067 XA	3.M05
BCC 2067 XF	3.L12
BCC 2067 XF	3.M05
BCC 2120 AE	3.I12
BCC 2120 AR	3.I12
BCC 2120 AW	3.I12
BCC 2120 CM	3.I12
BCC 2120 JF	3.I12
BCC 2120 LY	3.I12
BCC 2120 LZ	3.I12
BCC 2120 XE	3.I12
BCC 2120 ZM	3.I12
BCC 2121 AE	3.I12
BCC 2121 AR	3.I12
BCC 2121 AW	3.I12
BCC 2121 CM	3.I12
BCC 2121 JF	3.I12
BCC 2121 LY	3.I12
BCC 2121 LZ	3.I12
BCC 2121 XE	3.I12
BCC 2121 ZM	3.I12
BCC 2178	3.N06
BCC 2179	3.N06
BCC 2287	3.E10
BCC 2292	3.H09
BCC 2292	3.H11
BCC 2293	3.H09
BCC 2293	3.H11
BCC 2296	3.H09
BCC 2296	3.H11
BCC 2297	3.H09
BCC 2297	3.H11
BCC 2302	3.H09
BCC 2302	3.H11
BCC 2303	3.H09
BCC 2303	3.H11
BCC 2312 AP	3.L15
BCC 2312 AP	3.M04
BCC 2312 CM	3.L15
BCC 2312 CN	3.M04
BCC 2312 LE	3.L15
BCC 2312 LE	3.M04
BCC 2312 XA	3.L15
BCC 2312 XA	3.M04
BCC 2312 XF	3.L15
BCC 2312 XF	3.M04
BCC 2313	3.N09
BCC 2313	3.N10
BCC 2314 AP	3.L13
BCC 2314 AP	3.L18
BCC 2314 AP	3.M06
BCC 2314 AP	3.M10
BCC 2314 CN	3.L13
BCC 2314 CN	3.L18
BCC 2314 CN	3.M06
BCC 2314 CN	3.M10
BCC 2314 LE	3.L13
BCC 2314 LE	3.L18
BCC 2314 LE	3.M06
BCC 2314 LE	3.M10
BCC 2314 XA	3.L13
BCC 2314 XA	3.L18
BCC 2314 XA	3.M06
BCC 2314 XA	3.M10
BCC 2314 XF	3.L13
BCC 2314 XF	3.L18
BCC 2314 XF	3.M06
BCC 2314 XF	3.M10
BCC 2315 AP	3.L13
BCC 2315 AP	3.L18
BCC 2315 AP	3.M06
BCC 2315 AP	3.M10
BCC 2315 CN	3.L13
BCC 2315 CN	3.L18
BCC 2315 CN	3.M06
BCC 2315 CN	3.M10
BCC 2315 LE	3.L13
BCC 2315 LE	3.L18
BCC 2315 LE	3.M06
BCC 2315 LE	3.M10
BCC 2315 XA	3.L13
BCC 2315 XA	3.L18
BCC 2315 XA	3.M06
BCC 2315 XA	3.M10
BCC 2315 XF	3.L13
BCC 2315 XF	3.L18
BCC 2315 XF	3.M06
BCC 2315 XF	3.M10
BCC 2316	3.N09
BCC 2316	3.N10
BCC 2317	3.N09
BCC 2317	3.N10
BCC 2392	3.E09
BCC 2393	3.E09
BCC 2472 AE	3.L04
BCC 2472 AR	3.L04
BCC 2472 AW	3.L04
BCC 2472 AW	3.L06
BCC 2472 CM	3.L04
BCC 2472 JF	3.L04
BCC 2472 LY	3.L04
BCC 2472 LY	3.L06
BCC 2472 LZ	3.L04
BCC 2472 LZ	3.L06
BCC 2472 XE	3.L04
BCC 2472 XE	3.L06
BCC 2472 ZM	3.L04
BCC 2474 AE	3.L04
BCC 2474 AR	3.L04
BCC 2474 AW	3.L04
BCC 2474 AW	3.L06
BCC 2474 CM	3.L04
BCC 2474 JF	3.L04
BCC 2474 LY	3.L04
BCC 2474 LY	3.L06
BCC 2474 LZ	3.L04
BCC 2474 LZ	3.L06
BCC 2474 XE	3.L04
BCC 2474 XE	3.L06
BCC 2474 ZM	3.L04
BCC 2512	3.H04
BCC 2520 AE	3.H15
BCC 2520 AR	3.H15
BCC 2520 AR	3.I03
BCC 2520 AW	3.H15
BCC 2520 AW	3.I03
BCC 2520 CM	3.H15
BCC 2520 CM	3.I03
BCC 2520 JF	3.H15
BCC 2520 JF	3.I03
BCC 2520 LY	3.H15
BCC 2520 LY	3.I03
BCC 2520 LZ	3.H15
BCC 2520 LZ	3.I03
BCC 2520 XE	3.H15
BCC 2520 XE	3.I03
BCC 2520 ZM	3.H15
BCC 2520 ZM	3.I03
BCC 2522	3.K13
BCC 2522	3.K16
BCC 2522	3.K17
BCC 2537	3.H04
BCC 2541	3.D15
BCC 2580 AP	3.L15
BCC 2580 AP	3.M04
BCC 2580 CN	3.L15
BCC 2580 CN	3.M04
BCC 2580 LE	3.L15
BCC 2580 LE	3.M04
BCC 2580 XA	3.L15
BCC 2580 XA	3.M04
BCC 2580 XF	3.L15
BCC 2580 XF	3.M04
BCC 2581 AP	3.L15
BCC 2581 AP	3.M04
BCC 2581 CN	3.L15
BCC 2581 CN	3.M04
BCC 2581 LE	3.L15
BCC 2581 LE	3.M04
BCC 2581 XA	3.L15
BCC 2581 XA	3.M04
BCC 2581 XF	3.L15
BCC 2581 XF	3.M04
BCC 2614	3.D17
BCC 2695	3.F13
BCC 2696	3.F13
BCC 2712 1	3.D17
BCC 2712 3	2.L10
BCC 2744	3.N09
BCC 2744	3.N10
BCC 2745	3.N09
BCC 2745	3.N10
BCC 2761	3.L09
BCC 2798	3.L09
BCC 2816	3.F07
BCC 2817	3.F07
BCC 2821	3.O04
BCC 2822	3.O04
BCC 2824	3.O04
BCC 2826	3.O04
BCC 2828	3.O04
BCC 2829	3.O04
BCC 2830	3.O04
BCC 2838	3.F07
BCC 2839	3.F07
BCC 2841	3.E15
BCC 2844	3.F07
BCC 2845	3.F07
BCC 2934	3.F13
BCC 2975 AE	3.J08
BCC 2975 AR	3.J08
BCC 2975 AW	3.J08
BCC 2975 CM	3.J08
BCC 2975 JF	3.J08
BCC 2975 LY	3.J08
BCC 2975 LZ	3.J08
BCC 2975 XE	3.J08
BCC 2975 ZM	3.J08
BCC 2993 AX	3.K06
BCC 2993 AX	3.K09
BCC 2993 AY	3.K06
BCC 2993 AY	3.K09
BCC 2993 YE	3.K06
BCC 2993 YE	3.K09
BCC 2993 YF	3.K06
BCC 2993 YF	3.K09
BCC 3014 AE	3.J04
BCC 3014 AR	3.J04
BCC 3014 AW	3.J04
BCC 3014 CM	3.J04
BCC 3014 JF	3.J04
BCC 3014 LY	3.J04
BCC 3014 LZ	3.J04
BCC 3014 XE	3.J04
BCC 3014 ZM	3.J04
BCC 3015 AE	3.J04
BCC 3015 AR	3.J04
BCC 3015 AW	3.J04
BCC 3015 CM	3.J04
BCC 3015 JF	3.J04
BCC 3015 LY	3.J04
BCC 3015 LZ	3.J04
BCC 3015 XE	3.J04
BCC 3015 ZM	3.J04
BCC 3016 AE	3.J03
BCC 3016 AR	3.J03
BCC 3016 AW	3.J03
BCC 3016 CM	3.J03
BCC 3016 JF	3.J03
BCC 3016 LY	3.J03
BCC 3016 LZ	3.J03
BCC 3016 XE	3.J03
BCC 3016 ZM	3.J03
BCC 3017 AE	3.J03
BCC 3017 AR	3.J03
BCC 3017 AW	3.J03
BCC 3017 CM	3.J03
BCC 3017 JF	3.J03
BCC 3017 LY	3.J03
BCC 3017 LZ	3.J03
BCC 3017 XE	3.J03
BCC 3017 ZM	3.J03
BCC 3022 AX	3.K07
BCC 3022 AY	3.K07
BCC 3022 YF	3.K07
BCC 3023 AX	3.K07
BCC 3023 AY	3.K07
BCC 3023 YE	3.K07
BCC 3023 YF	3.K07
BCC 3024 AX	3.K07
BCC 3024 AY	3.K07
BCC 3024 YE	3.K07
BCC 3024 YF	3.K07
BCC 3025 AX	3.K07
BCC 3025 AY	3.K07
BCC 3025 YE	3.K07
BCC 3025 YF	3.K07
BCC 3034 AE	3.J07
BCC 3034 AR	3.J07
BCC 3034 AW	3.J07
BCC 3034 AW	3.K09
BCC 3034 CM	3.J07
BCC 3034 JF	3.J07
BCC 3034 LY	3.J07
BCC 3034 LY	3.K09
BCC 3034 LZ	3.J07
BCC 3034 LZ	3.K09
BCC 3034 XE	3.J07
BCC 3034 XE	3.K09
BCC 3034 ZM	3.J07
BCC 3035 AE	3.J07
BCC 3035 AR	3.J07
BCC 3035 AW	3.J07
BCC 3035 AW	3.K09
BCC 3035 CM	3.J07
BCC 3035 JF	3.J07
BCC 3035 LY	3.J07
BCC 3035 LY	3.K09
BCC 3035 LZ	3.J07
BCC 3035 LZ	3.K09
BCC 3035 XE	3.J07
BCC 3035 XE	3.K09
BCC 3035 ZM	3.J07
BCC 3036 AE	3.J06
BCC 3036 AR	3.J06
BCC 3036 AR	3.J16
BCC 3036 AW	3.J06
BCC 3036 AW	3.J16
BCC 3036 AW	3.K08
BCC 3036 AW	3.K11
BCC 3036 CM	3.J06
BCC 3036 CM	3.J16
BCC 3036 JF	3.J06
BCC 3036 JF	3.J16
BCC 3036 LY	3.J06
BCC 3036 LY	3.J16
BCC 3036 LY	3.K08
BCC 3036 LY	3.K11
BCC 3036 LZ	3.J06
BCC 3036 LZ	3.J16
BCC 3036 LZ	3.K08
BCC 3036 LZ	3.K11
BCC 3036 XE	3.J06
BCC 3036 XE	3.J16
BCC 3036 XE	3.K08
BCC 3036 XE	3.K11
BCC 3036 ZM	3.J06
BCC 3036 ZM	3.J16
BCC 3037 AE	3.J06
BCC 3037 AR	3.J06
BCC 3037 AR	3.J16
BCC 3037 AW	3.J06
BCC 3037 AW	3.J16
BCC 3037 AW	3.K08
BCC 3037 AW	3.K11
BCC 3037 CM	3.J06
BCC 3037 CM	3.J16
BCC 3037 JF	3.J06
BCC 3037 JF	3.J16
BCC 3037 LY	3.J06
BCC 3037 LY	3.J16
BCC 3037 LY	3.K08
BCC 3037 LY	3.K11
BCC 3037 LZ	3.J06
BCC 3037 LZ	3.J16
BCC 3037 LZ	3.K08
BCC 3037 LZ	3.K11
BCC 3037 XE	3.J06
BCC 3037 XE	3.J16
BCC 3037 XE	3.K08
BCC 3037 XE	3.K11
BCC 3037 ZM	3.J06
BCC 3037 ZM	3.J16
BCC 3038 AW	3.K06
BCC 3038 LY	3.K06
BCC 3038 LZ	3.K06
BCC 3038 XE	3.K06

JAGUAR XJS RANGE (JAN 1987 ON) — D05 Fiche 4 — NUMERICAL INDEX BCC 3039 AW TO BCC 7969 AW

BCC 3039 AW	3.K06
BCC 3039 LY	3.K06
BCC 3039 LZ	3.K06
BCC 3039 XE	3.K06
BCC 3040 AW	3.K05
BCC 3040 LY	3.K05
BCC 3040 PA	3.K05
BCC 3040 XE	3.K05
BCC 3041 AW	3.K05
BCC 3041 LY	3.K05
BCC 3041 PA	3.K05
BCC 3041 XE	3.K05
BCC 3042 XE	3.I17
BCC 3043 XE	3.I17
BCC 3364	3.I17
BCC 3382	3.I17
BCC 3416	3.C08
BCC 3417	3.C08
BCC 3432	3.G09
BCC 3432	3.G10
BCC 3433	3.G09
BCC 3433	3.G10
BCC 3434	3.G09
BCC 3434	3.G10
BCC 3435	3.G07
BCC 3435	3.G08
BCC 3436	3.G07
BCC 3436	3.G08
BCC 3437	3.G07
BCC 3437	3.G08
BCC 3441	3.H03
BCC 3552	3.N07
BCC 3552	3.N08
BCC 3594	3.D17
BCC 3620 XE	3.I16
BCC 3621 XE	3.I16
BCC 3622 ND	3.I16
BCC 3623 ND	3.I16
BCC 3624 ND	3.I16
BCC 3624 ND	3.I18
BCC 3626 XE	3.I16
BCC 3626 XE	3.I18
BCC 3664	3.E18
BCC 3665	3.G03
BCC 3665	3.G05
BCC 3667	3.G04
BCC 3668	3.G03
BCC 3668	3.G04
BCC 3671	3.G03
BCC 3671	3.G04
BCC 3671	3.G05
BCC 3756	3.E09
BCC 4032	4.C05
BCC 4033	4.C05
BCC 4048	3.D17
BCC 4049	3.D17
BCC 4091	3.G02
BCC 4102	3.G08
BCC 4142	3.K16
BCC 4142	3.K17
BCC 4143	3.K16
BCC 4143	3.K17
BCC 4462	3.G06
BCC 4478	3.E02
BCC 4479	3.E02
BCC 4518	3.K13
BCC 4518	3.K16
BCC 4518	3.K17
BCC 4522	3.K13
BCC 4522	3.K16
BCC 4522	3.K17
BCC 4550	3.H05
BCC 4551	3.H05
BCC 4578	3.F16
BCC 4579	3.F16
BCC 4580	3.F16
BCC 4582	3.F16
BCC 4583	3.F16
BCC 4604	3.F17
BCC 4605	3.F17
BCC 4606	3.F17
BCC 4607	3.F17
BCC 4608	3.F17
BCC 4609	3.F17
BCC 4610	3.F17
BCC 4611	3.F17
BCC 4612	3.F17
BCC 4613	3.F17
BCC 4614	3.F17
BCC 4615	3.F17
BCC 4616	3.F17
BCC 4617	3.F17
BCC 4618	3.F17
BCC 4619	3.F17
BCC 4620	3.F17
BCC 4621	3.F17
BCC 4622	3.F17
BCC 4623	3.F17
BCC 4624	3.F17
BCC 4625	3.F17
BCC 4626	3.F17
BCC 4627	3.F17
BCC 4628	3.F17
BCC 4629	3.F17
BCC 4630	3.F17
BCC 4631	3.F17
BCC 4632	3.F18
BCC 4633	3.F18
BCC 4634	3.F18
BCC 4635	3.F18
BCC 4636	3.F18
BCC 4637	3.F18
BCC 4638	3.F18
BCC 4639	3.F18
BCC 4640	3.F18
BCC 4641	3.F18
BCC 4642	3.F18
BCC 4643	3.F18
BCC 4644	3.F18
BCC 4645	3.F18
BCC 4704	3.H12
BCC 4704	3.H13
BCC 4808	3.C16
BCC 4809	3.C16
BCC 5090	3.H03
BCC 5115	3.G07
BCC 5116	3.G07
BCC 5117	3.G07
BCC 5120	3.G07
BCC 5121	3.G08
BCC 5129	3.H04
BCC 5219	3.M17
BCC 5220	3.M17
BCC 5291	2.D10
BCC 5292 ND	3.I15
BCC 5292 ND	3.I17
BCC 5329	3.H03
BCC 5400 AE	3.I10
BCC 5400 AR	3.I10
BCC 5400 AW	3.I10
BCC 5400 CM	3.I10
BCC 5400 JF	3.I10
BCC 5400 LY	3.I10
BCC 5400 LZ	3.I10
BCC 5400 XE	3.I10
BCC 5400 ZM	3.I10
BCC 5401 AE	3.I10
BCC 5401 AR	3.I10
BCC 5401 AW	3.I10
BCC 5401 CM	3.I10
BCC 5401 JF	3.I10
BCC 5401 LY	3.I10
BCC 5401 LZ	3.I10
BCC 5401 XE	3.I10
BCC 5401 ZM	3.I10
BCC 5430	2.H15
BCC 5431	2.H15
BCC 5518	3.F10
BCC 5520	3.L09
BCC 5560	3.O02
BCC 5578	3.O02
BCC 5579	3.O02
BCC 5587	3.N17
BCC 5636 AE	3.H12
BCC 5636 AR	3.H12
BCC 5636 AW	3.H12
BCC 5636 CM	3.H12
BCC 5636 JF	3.H12
BCC 5636 LY	3.H12
BCC 5636 LZ	3.H12
BCC 5636 XE	3.H12
BCC 5636 ZM	3.H12
BCC 5638 AE	3.H13
BCC 5638 AR	3.H13
BCC 5638 AW	3.H13
BCC 5638 CM	3.H13
BCC 5638 JF	3.H13
BCC 5638 LY	3.H13
BCC 5638 LZ	3.H13
BCC 5638 XE	3.H13
BCC 5638 ZM	3.H13
BCC 5642	2.L10
BCC 5643	2.L10
BCC 5656	3.F16
BCC 5657	3.F16
BCC 5658	3.F16
BCC 5659	3.F16
BCC 5660	3.F16
BCC 5661	3.F16
BCC 5662	3.F16
BCC 5663	3.F16
BCC 5774	3.E04
BCC 5775	3.E04
BCC 5798 AE	3.I09
BCC 5798 AR	3.I09
BCC 5798 AW	3.I09
BCC 5798 CM	3.I09
BCC 5798 JF	3.I09
BCC 5798 LY	3.I09
BCC 5798 LZ	3.I09
BCC 5798 XE	3.I09
BCC 5798 ZM	3.I09
BCC 5799 AE	3.I09
BCC 5799 AR	3.I09
BCC 5799 AW	3.I09
BCC 5799 CM	3.I09
BCC 5799 JF	3.I09
BCC 5799 LY	3.I09
BCC 5799 LZ	3.I09
BCC 5799 XE	3.I09
BCC 5799 ZM	3.I09
BCC 5804 AE	3.I05
BCC 5804 AR	3.I05
BCC 5804 AW	3.I05
BCC 5804 CM	3.I05
BCC 5804 JF	3.I05
BCC 5804 LY	3.I05
BCC 5804 LZ	3.I05
BCC 5804 XE	3.I05
BCC 5804 ZM	3.I05
BCC 5805 AE	3.I05
BCC 5805 AR	3.I05
BCC 5805 AW	3.I05
BCC 5805 CM	3.I05
BCC 5805 JF	3.I05
BCC 5805 LY	3.I05
BCC 5805 LZ	3.I05
BCC 5805 XE	3.I05
BCC 5895 ZM	3.I05
BCC 6070	3.E03
BCC 6071	3.E03
BCC 6082	3.C16
BCC 6083	3.C16
BCC 6084	3.C16
BCC 6085	3.C16
BCC 6093 AE	3.L07
BCC 6093 AR	3.L07
BCC 6093 AW	3.L07
BCC 6093 CM	3.L07
BCC 6093 JF	3.L07
BCC 6093 LY	3.L07
BCC 6093 LZ	3.L07
BCC 6093 XE	3.L07
BCC 6093 ZM	3.L07
BCC 6181	3.I14
BCC 6212 AP	3.L15
BCC 6212 AP	3.M04
BCC 6212 CN	3.L15
BCC 6212 CN	3.M04
BCC 6212 LE	3.L15
BCC 6212 LE	3.M04
BCC 6212 XA	3.L15
BCC 6212 XA	3.M04
BCC 6212 XF	3.L15
BCC 6212 XF	3.M04
BCC 6213 AP	3.L15
BCC 6213 AP	3.M04
BCC 6213 CN	3.L15
BCC 6213 CN	3.M04
BCC 6213 LE	3.L15
BCC 6213 LE	3.M04
BCC 6213 XA	3.L15
BCC 6213 XA	3.M04
BCC 6213 XF	3.L15
BCC 6213 XF	3.M04
BCC 6247 JX	3.N17
BCC 6247 NF	3.N17
BCC 6247 PM	3.N17
BCC 6305	3.K13
BCC 6305	3.K16
BCC 6305	3.K17
BCC 6310	3.K13
BCC 6311	3.K13
BCC 6316	3.K13
BCC 6317	3.K13
BCC 6342 AE	3.I11
BCC 6342 AR	3.I11
BCC 6342 AW	3.I11
BCC 6342 CM	3.I11
BCC 6342 JF	3.I11
BCC 6342 LY	3.I11
BCC 6342 LZ	3.I11
BCC 6342 XE	3.I11
BCC 6342 ZM	3.I11
BCC 6343 AE	3.I11
BCC 6343 AR	3.I11
BCC 6343 AW	3.I11
BCC 6343 CM	3.I11
BCC 6343 JF	3.I11
BCC 6343 LY	3.I11
BCC 6343 LZ	3.I11
BCC 6343 XE	3.I11
BCC 6343 ZM	3.I11
BCC 6412	3.K16
BCC 6413	3.K16
BCC 6414	3.K16
BCC 6415	3.K16
BCC 6422	3.L02
BCC 6423	3.L02
BCC 6430 LZ	3.H05
BCC 6433 LZ	3.H05
BCC 6458 LZ	3.H05
BCC 6459 LZ	3.H05
BCC 6460 LZ	3.H05
BCC 6461 LZ	3.H05
BCC 6475	3.K18
BCC 6597	3.N17
BCC 6642	3.N09
BCC 6643	3.N09
BCC 6659	3.C18
BCC 6690	3.E02
BCC 6691	3.E02
BCC 6716 AX	3.K04
BCC 6716 AY	3.K04
BCC 6716 YE	3.K04
BCC 6716 YF	3.K04
BCC 6717 AX	3.K04
BCC 6717 AY	3.K04
BCC 6717 YF	3.K04
BCC 6720 AX	3.K04
BCC 6720 AY	3.K04
BCC 6720 YE	3.K04
BCC 6720 YF	3.K04
BCC 6721 AX	3.K04
BCC 6721 AY	3.K04
BCC 6721 YE	3.K04
BCC 6721 YF	3.K04
BCC 6778	3.E18
BCC 6804	3.D11
BCC 6805	3.D11
BCC 6831	3.F03
BCC 6870	3.N11
BCC 6871	3.N11
BCC 6872	3.N11
BCC 6873	3.N11
BCC 6886	3.E03
BCC 6887	3.E03
BCC 7000 AE	3.L03
BCC 7000 AR	3.L03
BCC 7000 AW	3.L03
BCC 7000 CM	3.L03
BCC 7000 JF	3.L03
BCC 7000 LY	3.L03
BCC 7000 LZ	3.L03
BCC 7000 XE	3.L03
BCC 7000 ZM	3.L03
BCC 7001 AE	3.L03
BCC 7001 AR	3.L03
BCC 7001 AW	3.L03
BCC 7001 CM	3.L03
BCC 7001 JF	3.L03
BCC 7001 LY	3.L03
BCC 7001 LZ	3.L03
BCC 7001 XE	3.L03
BCC 7001 ZM	3.L03
BCC 7003 AX	3.L06
BCC 7003 AY	3.L06
BCC 7003 YE	3.L06
BCC 7003 YF	3.L06
BCC 7004 AX	3.L06
BCC 7004 AY	3.L06
BCC 7004 YE	3.L06
BCC 7004 YF	3.L06
BCC 7070	3.L02
BCC 7073	3.D17
BCC 7089 AE	3.H08
BCC 7089 AR	3.H08
BCC 7089 AW	3.H08
BCC 7089 CM	3.H08
BCC 7089 JF	3.H08
BCC 7089 LY	3.H08
BCC 7089 LZ	3.H08
BCC 7089 XE	3.H08
BCC 7089 ZM	3.H08
BCC 7091 AE	3.H10
BCC 7091 AR	3.H10
BCC 7091 AW	3.H10
BCC 7091 CM	3.H10
BCC 7091 JF	3.H10
BCC 7091 LY	3.H10
BCC 7091 LZ	3.H10
BCC 7091 XE	3.H10
BCC 7091 ZM	3.H10
BCC 7105	3.H12
BCC 7105	3.H13
BCC 7196	2.K08
BCC 7196	2.K09
BCC 7196	2.K10
BCC 7306 XE	3.I17
BCC 7307 XE	3.I17
BCC 7335	3.F05
BCC 7341	3.O02
BCC 7364 XE	3.I17
BCC 7365 XE	3.I17
BCC 7366	3.K17
BCC 7405	3.H12
BCC 7405	3.H13
BCC 7408	3.E03
BCC 7476	3.I18
BCC 7477	3.I18
BCC 7478	3.I18
BCC 7479	3.I18
BCC 7508	3.J08
BCC 7508	3.J18
BCC 7508	3.K06
BCC 7508	3.K09
BCC 7508	3.K12
BCC 7509	3.J08
BCC 7509	3.J18
BCC 7509	3.K06
BCC 7509	3.K09
BCC 7509	3.K12
BCC 7626	3.C16
BCC 7627	3.C16
BCC 7642	3.O04
BCC 7685	3.E08
BCC 7689	2.D04
BCC 7784	3.G02
BCC 7785	3.G02
BCC 7786 AR	3.L10
BCC 7786 AW	3.L10
BCC 7786 CM	3.L10
BCC 7786 JF	3.L10
BCC 7786 LY	3.L10
BCC 7786 LZ	3.L10
BCC 7786 XE	3.L10
BCC 7786 ZM	3.L10
BCC 7828 AE	3.J10
BCC 7828 AR	3.J10
BCC 7828 AR	3.K03
BCC 7828 AW	3.J10
BCC 7828 AW	3.K03
BCC 7828 CM	3.J10
BCC 7828 CM	3.K03
BCC 7828 JF	3.J10
BCC 7828 JF	3.K03
BCC 7828 LY	3.J10
BCC 7828 LY	3.K03
BCC 7828 LZ	3.J10
BCC 7828 LZ	3.K03
BCC 7828 XE	3.J10
BCC 7828 XE	3.K03
BCC 7828 ZM	3.J10
BCC 7828 ZM	3.K03
BCC 7829 AE	3.J10
BCC 7829 AR	3.J10
BCC 7829 AR	3.K03
BCC 7829 AW	3.J10
BCC 7829 AW	3.K03
BCC 7829 CM	3.J10
BCC 7829 CM	3.K03
BCC 7829 JF	3.J10
BCC 7829 JF	3.K03
BCC 7829 LY	3.J10
BCC 7829 LY	3.K03
BCC 7829 LZ	3.J10
BCC 7829 LZ	3.K03
BCC 7829 XE	3.J10
BCC 7829 XE	3.K03
BCC 7829 ZM	3.J10
BCC 7829 ZM	3.K03
BCC 7892	3.L09
BCC 7900	2.L10
BCC 7901	2.L10
BCC 7959	3.K18
BCC 7968 AR	3.J17
BCC 7968 AW	3.J17
BCC 7968 AW	3.K12
BCC 7968 CM	3.J17
BCC 7968 JF	3.J17
BCC 7968 LY	3.J17
BCC 7968 LY	3.K12
BCC 7968 LZ	3.J17
BCC 7968 LZ	3.K12
BCC 7968 XE	3.J17
BCC 7968 XE	3.K12
BCC 7968 ZM	3.J17
BCC 7969 AR	3.J17
BCC 7969 AW	3.J17
BCC 7969 AW	3.K12

JAGUAR XJS RANGE (JAN 1987 ON) — D06 Fiche 4 — NUMERICAL INDEX BCC 7969 CM TO BD 36436

BCC 7969 CM	3.J17
BCC 7969 JF	3.J17
BCC 7969 LY	3.J17
BCC 7969 LY	3.K12
BCC 7969 LZ	3.J17
BCC 7969 LZ	3.K12
BCC 7969 XE	3.J17
BCC 7969 XE	3.K12
BCC 7969 ZM	3.J17
BCC 7992 AP	3.M08
BCC 7992 AP	3.M12
BCC 7992 CN	3.M08
BCC 7992 CN	3.M12
BCC 7992 LE	3.M08
BCC 7992 LE	3.M12
BCC 7992 XA	3.M08
BCC 7992 XA	3.M12
BCC 7992 XF	3.M08
BCC 7992 XF	3.M12
BCC 7993 AP	3.M08
BCC 7993 AP	3.M12
BCC 7993 CN	3.M08
BCC 7993 CN	3.M12
BCC 7993 LE	3.M08
BCC 7993 LE	3.M12
BCC 7993 XA	3.M08
BCC 7993 XA	3.M12
BCC 7993 XF	3.M08
BCC 7993 XF	3.M12
BCC 8003	2.K10
BCC 8029	3.N09
BCC 8029	3.N10
BCC 8030	3.N09
BCC 8030	3.N10
BCC 8032	3.E09
BCC 8033	3.E09
BCC 8036	3.N09
BCC 8036	3.N10
BCC 8037	3.N09
BCC 8037	3.N10
BCC 8038	3.K13
BCC 8038	3.K16
BCC 8038	3.K17
BCC 8078	3.I18
BCC 8079	3.I18
BCC 8098	2.L10
BCC 8099	2.L10
BCC 8152 AP	3.L14
BCC 8152 AP	3.M03
BCC 8152 AP	3.M07
BCC 8152 AP	3.M11
BCC 8152 CN	3.L14
BCC 8152 CN	3.M03
BCC 8152 CN	3.M11
BCC 8152 LE	3.L14
BCC 8152 LE	3.M03
BCC 8152 LE	3.M07
BCC 8152 LE	3.M11
BCC 8152 XA	3.L14
BCC 8152 XA	3.M03
BCC 8152 XA	3.M07
BCC 8152 XA	3.M11
BCC 8152 XF	3.L14
BCC 8152 XF	3.M03
BCC 8152 XF	3.M07
BCC 8152 XF	3.M11
BCC 8153 AP	3.L14
BCC 8153 AP	3.M03
BCC 8153 AP	3.M07
BCC 8153 AP	3.M11
BCC 8153 CN	3.L14
BCC 8153 CN	3.M03
BCC 8153 CN	3.M07
BCC 8153 CN	3.M11
BCC 8153 LE	3.L14
BCC 8153 LE	3.M03
BCC 8153 LE	3.M07
BCC 8153 LE	3.M11
BCC 8153 XA	3.L14
BCC 8153 XA	3.M03
BCC 8153 XA	3.M07
BCC 8153 XA	3.M11
BCC 8153 XF	3.L14
BCC 8153 XF	3.M03
BCC 8153 XF	3.M07
BCC 8153 XF	3.M11
BCC 8250	3.H09
BCC 8251	3.H09
BCC 8251	3.H11
BCC 8252	3.H09
BCC 8252	3.H11
BCC 8273	3.H09
BCC 8273	3.H11
BCC 8274	3.H09
BCC 8275	3.H09
BCC 8275	3.H11
BCC 8437	3.F10
BCC 8438	3.F10
BCC 8439	3.F10
BCC 8440	3.F10
BCC 8454	3.H05
BCC 8504	3.J03
BCC 8504	3.J11
BCC 8504	3.K04
BCC 8504	3.K07
BCC 8504	3.K10
BCC 8505	3.J03
BCC 8505	3.J12
BCC 8505	3.K04
BCC 8505	3.K07
BCC 8505	3.K10
BCC 8509	2.H17
BCC 8512 CN	3.M07
BCC 8515	3.N03
BCC 8523 CN	3.L09
BCC 8523 LE	3.L09
BCC 8523 XA	3.L09
BCC 8523 XF	3.L09
BCC 8524	3.K14
BCC 8525	3.K14
BCC 8633	3.H04
BCC 9037	3.K14
BCC 9037	3.K15
BCC 9037	3.K16
BCC 9037	3.K17
BCC 9056	3.H09
BCC 9056	3.H11
BCC 9057	3.H09
BCC 9057	3.H11
BCC 9150	3.K16
BCC 9150	3.K17
BCC 9151	3.K16
BCC 9151	3.K17
BCC 9152	3.K17
BCC 9153	3.K17
BCC 9160	3.K17
BCC 9161	3.K17
BCC 9162	3.K17
BCC 9163	3.K17
BCC 9278 AR	3.J11
BCC 9278 AW	3.J11
BCC 9278 CM	3.J11
BCC 9278 JF	3.J11
BCC 9278 LY	3.J11
BCC 9278 LZ	3.J11
BCC 9278 XE	3.J11
BCC 9278 YLY	3.J11
BCC 9278 YXE	3.J11
BCC 9278 YZM	3.J11
BCC 9278 ZLY	3.J11
BCC 9278 ZM	3.J11
BCC 9278 ZXE	3.J11
BCC 9278 ZZM	3.J11
BCC 9279 AR	3.J12
BCC 9279 AW	3.J12
BCC 9279 CM	3.J12
BCC 9279 JF	3.J12
BCC 9279 LY	3.J12
BCC 9279 LZ	3.J12
BCC 9279 XE	3.J12
BCC 9279 YLY	3.J12
BCC 9279 YXE	3.J12
BCC 9279 YZM	3.J12
BCC 9279 ZLY	3.J12
BCC 9279 ZM	3.J12
BCC 9279 ZXE	3.J12
BCC 9279 ZZM	3.J12
BCC 9284 AR	3.J13
BCC 9284 AW	3.J13
BCC 9284 CM	3.J13
BCC 9284 JF	3.J13
BCC 9284 LY	3.J13
BCC 9284 LZ	3.J13
BCC 9284 XE	3.J13
BCC 9284 ZM	3.J13
BCC 9285 AR	3.J14
BCC 9285 AW	3.J14
BCC 9285 CM	3.J14
BCC 9285 JF	3.J14
BCC 9285 LY	3.J14
BCC 9285 LZ	3.J14
BCC 9285 XE	3.J14
BCC 9285 ZM	3.J14
BCC 9286 AX	3.K10
BCC 9286 AY	3.K10
BCC 9286 YE	3.K10
BCC 9286 YF	3.K10
BCC 9287 AX	3.K10
BCC 9287 AY	3.K10
BCC 9287 YE	3.K10
BCC 9287 YF	3.K10
BCC 9292 AX	3.K10
BCC 9292 AY	3.K10
BCC 9292 YE	3.K10
BCC 9292 YF	3.K10
BCC 9293 AX	3.K10
BCC 9293 AY	3.K10
BCC 9293 YE	3.K10
BCC 9293 YF	3.K10
BCC 9306 AR	3.J15
BCC 9306 AW	3.J15
BCC 9306 AW	3.K11
BCC 9306 CM	3.J15
BCC 9306 JF	3.J15
BCC 9306 LY	3.J15
BCC 9306 LY	3.K11
BCC 9306 LZ	3.J15
BCC 9306 LZ	3.K11
BCC 9306 XE	3.J15
BCC 9306 XE	3.K11
BCC 9306 ZM	3.J15
BCC 9307 AR	3.J16
BCC 9307 AW	3.J16
BCC 9307 AW	3.K11
BCC 9307 CM	3.J16
BCC 9307 JF	3.J16
BCC 9307 LY	3.J16
BCC 9307 LY	3.K11
BCC 9307 LZ	3.J16
BCC 9307 LZ	3.K11
BCC 9307 XE	3.J16
BCC 9307 XE	3.K11
BCC 9307 ZM	3.J16
BCC 9310 AR	3.J18
BCC 9310 AW	3.J18
BCC 9310 CM	3.J18
BCC 9310 JF	3.J18
BCC 9310 LY	3.J18
BCC 9310 LZ	3.J18
BCC 9310 XE	3.J18
BCC 9310 YLY	3.J18
BCC 9310 YXE	3.J18
BCC 9310 YZM	3.J18
BCC 9310 ZLY	3.J18
BCC 9310 ZM	3.J18
BCC 9310 ZXE	3.J18
BCC 9310 ZZM	3.J18
BCC 9315 AX	3.K12
BCC 9315 AY	3.K12
BCC 9315 YE	3.K12
BCC 9315 YF	3.K12
BCC 9320 AR	3.J11
BCC 9320 AW	3.J11
BCC 9320 CM	3.J11
BCC 9320 JF	3.J11
BCC 9320 LY	3.J11
BCC 9320 LZ	3.J11
BCC 9320 XE	3.J11
BCC 9320 YLY	3.J11
BCC 9320 YXE	3.J11
BCC 9320 YZM	3.J11
BCC 9320 ZLY	3.J11
BCC 9320 ZM	3.J11
BCC 9320 ZXE	3.J11
BCC 9320 ZZM	3.J11
BCC 9322 AX	3.K10
BCC 9322 AY	3.K10
BCC 9322 YE	3.K10
BCC 9322 YF	3.K10
BCC 9324 AR	3.J15
BCC 9324 AW	3.J15
BCC 9324 CM	3.J15
BCC 9324 JF	3.J15
BCC 9324 LY	3.J15
BCC 9324 LZ	3.J15
BCC 9324 XE	3.J15
BCC 9324 YLY	3.J15
BCC 9324 YXE	3.J15
BCC 9324 YZM	3.J15
BCC 9324 ZLY	3.J15
BCC 9324 ZM	3.J15
BCC 9324 ZXE	3.J15
BCC 9324 ZZM	3.J15
BCC 9345 AR	3.J18
BCC 9345 AW	3.J18
BCC 9345 CM	3.J18
BCC 9345 JF	3.J18
BCC 9345 LY	3.J18
BCC 9345 LZ	3.J18
BCC 9345 XE	3.J18
BCC 9345 YLY	3.J18
BCC 9345 YXE	3.J18
BCC 9345 YZM	3.J18
BCC 9345 ZLY	3.J18
BCC 9345 ZM	3.J18
BCC 9345 ZXE	3.J18
BCC 9345 ZZM	3.J18
BCC 9355 AR	3.K02
BCC 9355 AW	3.K02
BCC 9355 CM	3.K02
BCC 9355 JF	3.K02
BCC 9355 LY	3.K02
BCC 9355 LZ	3.K02
BCC 9355 XE	3.K02
BCC 9355 ZM	3.K02
BCC 9370 AX	3.K10
BCC 9370 AY	3.K10
BCC 9370 YE	3.K10
BCC 9370 YF	3.K10
BCC 9379 AW	3.K12
BCC 9379 LY	3.K12
BCC 9379 LZ	3.K12
BCC 9379 XE	3.K12
BCC 9384 AX	3.K12
BCC 9384 AY	3.K12
BCC 9384 YE	3.K12
BCC 9384 YF	3.K12
BCC 9414	3.N05
BCC 9494 XE	3.I15
BCC 9494 XE	3.I16
BCC 9494 XE	3.I17
BCC 9494 XE	3.I18
BCC 9494 XE	3.J02
BCC 9550	3.E18
BCC 9551	3.E18
BCC 9561	2.O02
BCC 9742	3.E02
BCC 9743	3.E02
BCC 9764	3.N05
BCC 9791	3.O02
BCC 9876	3.N06
BCC 9968 XE	3.I18
BCC 9969 XE	3.I18
BCC 9986	3.I14
BCC 9987	3.I14
BCC 9996 AR	3.L03
BCC 9996 AW	3.L03
BCC 9996 CM	3.L03
BCC 9996 JF	3.L03
BCC 9996 LY	3.L03
BCC 9996 LZ	3.L03
BCC 9996 XE	3.L03
BCC 9996 ZM	3.L03
BCC 9997 AX	3.L06
BCC 9997 AY	3.L06
BCC 9997 YE	3.L06
BCC 9997 YF	3.L06
BCC 9998	3.O02
BCC 9999	3.O02
BD 541 20	2.G10
BD 541 20	2.H08
BD 541 20	2.L02
BD 541 20	2.N18
BD 541 20	2.O02
BD 541 20	3.F13
BD 541 27	3.F13
BD 541 28	2.C15
BD 541 28	2.C16
BD 541 28	2.C17
BD 541 28	2.E09
BD 541 32	2.E09
BD 541 34	2.F08
BD 541 34	2.L16
BD 541 34	3.E11
BD 541 34	3.F12
BD 541 37	3.D14
BD 541 41	2.F07
BD 541 41	2.F15
BD 541 43	2.G02
BD 541 43	3.G07
BD 541 43	3.G09
BD 541 47	2.J15
BD 541 48	1.H04
BD 541 48	1.I04
BD 541 5	3.D14
BD 541 51	2.F09
BD 541 9	1.K13
BD 541 9	2.J14
BD 541 9	2.J15
BD 541 9	2.K04
BD 541 9	3.E04
BD 541 9	3.E16
BD 541 9	3.E17
BD 542 10	1.J08
BD 542 12	2.F09
BD 542 6	1.J07
BD 542 7	1.J07
BD 542 7	1.M03
BD 542 8	2.F17
BD 542 8	3.E11
BD 542 8	3.G07
BD 542 8	3.G08
BD 542 8	3.G09
BD 542 8	3.G10
BD 542 9	2.F08
BD 542 9	2.F16
BD 542 9	3.G06
BD 592 8	3.G08
BD 1814 4	1.K14
BD 1818 1	3.F15
BD 1818 1	3.I04
BD 2906 3	1.O10
BD 8633 10	2.F08
BD 8633 12	2.J15
BD 8633 15	2.F08
BD 8633 2	2.O14
BD 8633 3	1.M04
BD 8633 3	2.F09
BD 8633 6	3.D17
BD 10224	3.D14
BD 11309	3.H06
BD 11309	3.H07
BD 11712 4	2.H16
BD 13102	3.H15
BD 13102	3.H18
BD 13102	3.I03
BD 13102	3.I15
BD 13102	3.I17
BD 13565 3	3.D14
BD 13713 1	3.H12
BD 13713 1	3.H13
BD 15466 2	3.F04
BD 15588 8	2.D05
BD 18957 6	3.I17
BD 19203	1.D18
BD 20013 5	2.C12
BD 20013 7	2.C16
BD 20013 7	2.C17
BD 20013 9	2.C16
BD 20013 9	2.C17
BD 21181	3.F04
BD 21468	3.F06
BD 21899	2.D02
BD 21962	3.G06
BD 21962	3.G07
BD 21962	3.G08
BD 21962	3.G09
BD 21962	3.I15
BD 21962	3.I17
BD 23149	2.J10
BD 24295	3.H02
BD 24978	1.O06
BD 24978 1	2.D02
BD 24978 1	2.D13
BD 24978 1	2.D14
BD 24978 1	2.D16
BD 24978 2	2.C18
BD 26562	2.D10
BD 26706 1	2.H16
BD 26706 1	3.F14
BD 27546	2.C15
BD 27546 2	2.C12
BD 28658	3.C04
BD 28658	3.C12
BD 28658	3.D05
BD 30541 1	3.D17
BD 31067	1.H06
BD 31067	1.J05
BD 31101	3.E16
BD 31101	3.E17
BD 31104	3.E16
BD 31104	3.E17
BD 31316	3.D14
BD 32396	2.K06
BD 32396	3.E09
BD 32692	3.N16
BD 33031	3.H09
BD 33031	3.H11
BD 33506 1	2.F08
BD 33506 1	2.F09
BD 33506 1	2.F16
BD 34524 1	1.J03
BD 34524 1	3.J08
BD 34524 1	3.J18
BD 34524 1	3.K06
BD 34540	3.E09
BD 34541 1	3.K09
BD 34541 1	3.K12
BD 34709 1	3.E13
BD 34715	3.H12
BD 34715	3.H13
BD 34716	3.H12
BD 34716	3.H13
BD 34753	3.H12
BD 34753	3.H13
BD 34982	3.D15
BD 35306 1	2.F08
BD 36002	3.E16
BD 36002	3.E17
BD 36436	2.G12

JAGUAR XJS RANGE (JAN 1987 ON) — D07 Fiche 4 — NUMERICAL INDEX BD 36746 2 TO BDC 5592 CN

BD 36746 2	1.M05
BD 37522	2.C18
BD 37522	2.D02
BD 37824 1	1.F06
BD 37943	3.E14
BD 37943	3.E15
BD 38393	1.J05
BD 38427 1	1.H06
BD 38427 1	1.J05
BD 38852 3	1.M02
BD 39893	3.D15
BD 40748 3	3.C09
BD 40755	3.D15
BD 42123	3.E13
BD 42434	3.H15
BD 42434	3.H18
BD 42434	3.I03
BD 42638	3.J08
BD 42638	3.J18
BD 42638	3.K06
BD 42638	3.K09
BD 42638	3.K12
BD 42805	3.H05
BD 43246	3.D15
BD 43247	3.D15
BD 43266	3.E07
BD 43266	3.E08
BD 43267	3.E07
BD 43267	3.E08
BD 43278 3	3.G02
BD 43278 4	3.G02
BD 43278 4	3.I15
BD 43278 4	3.I16
BD 43278 4	3.I17
BD 43278 4	3.I18
BD 43278 4	3.J02
BD 44093 1	3.F06
BD 44094 1	3.F06
BD 44130	2.H14
BD 44162	3.E16
BD 44163	3.E16
BD 44164	3.E16
BD 44164	3.E17
BD 44231	3.F04
BD 44232	3.F04
BD 44236	3.F04
BD 44237	3.F14
BD 44240	3.F14
BD 44241	3.F14
BD 44243	3.F14
BD 44315	3.E14
BD 44315	3.E15
BD 44385	3.F15
BD 44391	3.F11
BD 44393	3.F11
BD 44394	3.F15
BD 44487 1	3.F08
BD 44488 1	3.F08
BD 44533	3.F13
BD 44643 1	1.N12
BD 44643 1	2.N12
BD 44643 1	3.H03
BD 44643 1	3.N05
BD 44643 12	1.N12
BD 44645	3.G07
BD 44645	3.G09
BD 44650	3.E07
BD 44650	3.E08
BD 44650	3.E09
BD 44773	3.F06
BD 44956	3.C07
BD 44956	3.C17
BD 44956	3.D09
BD 44957	3.C07
BD 44957	3.C17
BD 44957	3.D09
BD 44971	3.N12
BD 44991	3.C05
BD 44991	3.C13
BD 44991	3.D06
BD 44992	3.C05
BD 44992	3.C13
BD 45006	3.H04
BD 45056 ND	3.I15
BD 45057 ND	3.I15
BD 45088	3.F06
BD 45089	3.F06
BD 45170	3.F14
BD 45171	3.F14
BD 45172	3.F14
BD 45173	3.F14
BD 45175	3.F14
BD 45176	3.F14
BD 45521 AE	3.H18
BD 45521 AR	3.H18
BD 45521 AW	3.H18
BD 45521 CM	3.H18
BD 45521 JF	3.H18
BD 45521 LY	3.H18
BD 45521 PA	3.H18
BD 45521 XE	3.H18
BD 45521 ZM	3.H18
BD 45555	3.H04
BD 45556	3.H04
BD 45612	3.H04
BD 45619	3.E13
BD 45620	3.E13
BD 45624	3.G06
BD 45628	3.F14
BD 45629	3.F14
BD 45640	3.F15
BD 45663	3.E07
BD 45664	3.E07
BD 45665	3.E07
BD 45672	3.F14
BD 45673	3.F14
BD 45850	3.E07
BD 45850	3.E08
BD 45851	3.E07
BD 45851	3.E08
BD 45909 2	3.H03
BD 45919	3.F15
BD 45920	3.F15
BD 45934	3.F06
BD 45935	3.F06
BD 45946	3.N09
BD 45946	3.N10
BD 45946	3.N11
BD 45974	3.E09
BD 46085	3.D13
BD 46139	3.D15
BD 46140 1	3.N05
BD 46140 2	3.N05
BD 46140 2	4.C10
BD 46140 3	3.D17
BD 46140 3	3.N05
BD 46140 3	4.C10
BD 46141	3.N05
BD 46143	2.H14
BD 46151	2.N18
BD 46404	3.D17
BD 46405	3.E04
BD 46406	3.E04
BD 46407	3.D17
BD 46408	3.D17
BD 46439	3.H02
BD 46440 3	3.F06
BD 46458 1	3.I04
BD 46571	3.H15
BD 46571	3.H18
BD 46571	3.I03
BD 46664 1	2.H12
BD 46684	3.N03
BD 46684	3.N04
BD 46717	2.K07
BD 46718	2.K07
BD 46763	3.F08
BD 46764	3.F08
BD 46767	3.D15
BD 46770 AE	3.J03
BD 46770 AR	3.J03
BD 46770 AW	3.J03
BD 46770 CM	3.J03
BD 46770 JF	3.J03
BD 46770 LY	3.J03
BD 46770 LZ	3.J03
BD 46770 XE	3.J03
BD 46770 ZM	3.J03
BD 46805	3.J08
BD 46805	3.J18
BD 46805	3.K06
BD 46805	3.K12
BD 46831	3.H03
BD 46908 1	3.F15
BD 46908 3	3.F14
BD 46958	3.E10
BD 46958	3.E11
BD 46985	3.E16
BD 46985	3.E17
BD 47029 AW	3.K05
BD 47029 LY	3.K05
BD 47029 PA	3.K05
BD 47029 XE	3.K05
BD 47030 AW	3.K05
BD 47030 LY	3.K05
BD 47030 PA	3.K05
BD 47030 XE	3.K05
BD 47321	3.H12
BD 47321	3.H13
BD 47322	3.E16
BD 47322	3.E17
BD 47337	3.N09
BD 47337	3.N10
BD 47401	3.I04
BD 47644	3.H15
BD 47644	3.H18
BD 47644	3.I03
BD 47656 ND	3.I15
BD 47679	3.E09
BD 47680	3.E09
BD 47684	3.E09
BD 47755	3.E09
BD 47756	3.E09
BD 47775 1	2.L17
BD 47775 1	2.L18
BD 47779	3.F06
BD 47802	3.I15
BD 47802	3.I17
BD 47814	3.H02
BD 47837	3.H12
BD 47837	3.H13
BD 47854	3.G06
BD 47897	3.H04
BD 47898	3.H03
BD 47899	3.E07
BD 47899	3.E08
BD 47923	3.D15
BD 47931	3.F15
BD 47948	3.C07
BD 47948	3.C17
BD 47948	3.D09
BD 47955	3.N07
BD 47955	3.N08
BD 47973	3.N07
BD 47980	3.D15
BD 47982	3.N07
BD 47983	3.N07
BD 48015	3.F14
BD 48016	3.F14
BD 48022	3.N07
BD 48022	3.N08
BD 48027	3.D16
BD 48028	3.D16
BD 48029	3.D15
BD 48038	1.J08
BD 48042	3.E11
BD 48049	3.E09
BD 48049	3.G07
BD 48128	3.E09
BD 48203	3.D14
BD 48211	3.I04
BD 48228 AE	3.H10
BD 48228 AR	3.H10
BD 48228 AW	3.H10
BD 48228 CM	3.H10
BD 48228 JF	3.H10
BD 48228 LY	3.H10
BD 48228 LZ	3.H10
BD 48228 XE	3.H10
BD 48228 ZM	3.H10
BD 48229 AE	3.H10
BD 48229 AR	3.H10
BD 48229 AW	3.H10
BD 48229 CM	3.H10
BD 48229 JF	3.H10
BD 48229 LY	3.H10
BD 48229 LZ	3.H10
BD 48229 XE	3.H10
BD 48229 ZM	3.H10
BD 48244	3.H09
BD 48244	3.H11
BD 48245	3.H09
BD 48245	3.H11
BD 48281	2.H15
BD 48286	3.E04
BD 48322	3.D15
BD 48323	3.D15
BD 48339	3.F13
BD 48375	2.K04
BD 48393	3.D14
BD 48396	3.D14
BD 48433 AE	3.I09
BD 48433 AE	3.I10
BD 48433 AR	3.I09
BD 48433 AR	3.I10
BD 48433 AW	3.I09
BD 48433 AW	3.I10
BD 48433 CM	3.I09
BD 48433 CM	3.I10
BD 48433 JF	3.I09
BD 48433 JF	3.I10
BD 48433 LY	3.I09
BD 48433 LY	3.I10
BD 48433 LZ	3.I09
BD 48433 LZ	3.I10
BD 48433 XE	3.I09
BD 48433 XE	3.I10
BD 48433 ZM	3.I09
BD 48433 ZM	3.I10
BD 48434 AE	3.I09
BD 48434 AE	3.I10
BD 48434 AR	3.I09
BD 48434 AR	3.I10
BD 48434 AW	3.I09
BD 48434 AW	3.I10
BD 48434 CM	3.I09
BD 48434 CM	3.I10
BD 48434 JF	3.I09
BD 48434 JF	3.I10
BD 48434 LY	3.I09
BD 48434 LY	3.I10
BD 48434 LZ	3.I09
BD 48434 LZ	3.I10
BD 48434 XE	3.I09
BD 48434 XE	3.I10
BD 48434 ZM	3.I09
BD 48434 ZM	3.I10
BD 48649	3.G08
BD 48649	3.G10
BD 48684 1	3.G02
BD 48737	3.N13
BD 48752	3.F14
BD 48753	3.F14
BD 48757	3.G02
BD 48890	3.N12
BD 48898	3.N09
BD 48898	3.N10
BD 48899	3.N09
BD 48899	3.N10
BD 48903	3.L05
BD 48903	3.L06
BD 48923	3.N07
BD 48947 2	3.L02
BD 48986	3.E08
BD 46805	3.K09
BDC 3014	3.G02
BDC 3017	3.I17
BDC 3018	3.I17
BDC 3035	3.F05
BDC 3036	3.F05
BDC 3037	3.F05
BDC 3038	3.I03
BDC 3122	3.N04
BDC 3174	3.E12
BDC 3176 AP	3.M13
BDC 3176 CN	3.M13
BDC 3176 LE	3.M13
BDC 3176 XA	3.M13
BDC 3176 XF	3.M13
BDC 3177 AP	3.M13
BDC 3177 CN	3.M13
BDC 3177 LE	3.M13
BDC 3177 XA	3.M13
BDC 3177 XF	3.M13
BDC 3180 AP	3.L16
BDC 3180 CN	3.L16
BDC 3180 LE	3.L16
BDC 3180 XA	3.L16
BDC 3180 XF	3.L16
BDC 3181 AP	3.L16
BDC 3181 CN	3.L16
BDC 3181 LE	3.L16
BDC 3181 XA	3.L16
BDC 3181 XF	3.L16
BDC 3194 AP	3.M03
BDC 3194 CN	3.M03
BDC 3194 LE	3.M03
BDC 3194 XA	3.M03
BDC 3194 XF	3.M03
BDC 3195 AP	3.M03
BDC 3195 CN	3.M03
BDC 3195 LE	3.M03
BDC 3195 XA	3.M03
BDC 3195 XF	3.M03
BDC 3214 XE	3.I18
BDC 3215 XE	3.I18
BDC 3534	3.E05
BDC 3535	3.E05
BDC 3536	2.O02
BDC 3556	3.F07
BDC 3557	3.F07
BDC 3562	2.H16
BDC 3563	2.H16
BDC 3572	3.E15
BDC 3573	3.E15
BDC 3574	3.E15
BDC 3575	3.E15
BDC 3595 XE	3.J02
BDC 3596 XE	3.J02
BDC 3597 XE	3.J02
BDC 3601	3.N09
BDC 3601	3.N10
BDC 3602	3.N09
BDC 3602	3.N10
BDC 3603	3.G02
BDC 3722	3.I14
BDC 3723	3.I14
BDC 3725	3.F03
BDC 3796 XE	3.I17
BDC 3797 XE	3.I17
BDC 4000	3.N11
BDC 4001	3.N11
BDC 4004	3.N11
BDC 4006	3.N11
BDC 4007	3.N11
BDC 4012	3.I17
BDC 4013	3.I17
BDC 4014 XE	3.I17
BDC 4015 XE	3.I17
BDC 4021	3.E08
BDC 4024 PA	3.I15
BDC 4024 PA	3.I17
BDC 4027	3.N11
BDC 4060	3.F03
BDC 4096	3.E06
BDC 4097	3.E06
BDC 4308 XE	3.I17
BDC 4308 XE	3.I18
BDC 4308 XE	3.J02
BDC 4309 XE	3.I17
BDC 4309 XE	3.I18
BDC 4309 XE	3.J02
BDC 4361 ND	3.I15
BDC 4361 ND	3.I17
BDC 4362	3.O04
BDC 4363	3.N11
BDC 4526	3.E09
BDC 4527	3.E09
BDC 4538	3.N04
BDC 4554	3.F10
BDC 4555	3.F10
BDC 4556	3.F10
BDC 4557	3.F10
BDC 4559	3.N04
BDC 4642	3.E02
BDC 4643	3.E02
BDC 4852	3.I14
BDC 4853	3.I14
BDC 4874	3.N11
BDC 4878	3.F07
BDC 4879	3.F07
BDC 4912	3.F10
BDC 4936	3.O02
BDC 4937	3.I13
BDC 4981	3.E08
BDC 4982	3.E08
BDC 4983	3.E08
BDC 4992	3.E08
BDC 4993	3.E08
BDC 4994	3.N04
BDC 5068	3.N12
BDC 5076	3.L09
BDC 5078	3.E05
BDC 5079	3.E05
BDC 5081	3.J02
BDC 5090	3.I14
BDC 5092	3.O02
BDC 5093	3.O02
BDC 5336	3.J02
BDC 5360	3.F10
BDC 5361	3.F10
BDC 5566	3.F07
BDC 5567	3.F07
BDC 5576 AP	3.M13
BDC 5576 CN	3.M13
BDC 5576 LE	3.M13
BDC 5576 XA	3.M13
BDC 5576 XF	3.M13
BDC 5577 AP	3.M14
BDC 5577 CN	3.M14
BDC 5577 LE	3.M14
BDC 5577 XA	3.M14
BDC 5577 XF	3.M14
BDC 5578 AP	3.M14
BDC 5578 CN	3.M14
BDC 5578 LE	3.M14
BDC 5578 XA	3.M14
BDC 5578 XF	3.M14
BDC 5581 AP	3.M14
BDC 5581 CN	3.M14
BDC 5581 LE	3.M14
BDC 5581 XA	3.M14
BDC 5581 XF	3.M14
BDC 5584 AP	3.M13
BDC 5584 CN	3.M13
BDC 5584 LE	3.M13
BDC 5584 XA	3.M13
BDC 5584 XF	3.M13
BDC 5592 AP	3.M13
BDC 5592 CN	3.M13

JAGUAR XJS RANGE (JAN 1987 ON) — D08 Fiche 4 — NUMERICAL INDEX BDC 5592 LE TO BJC 1517

JAGUAR XJS RANGE (JAN 1987 ON) — D09 Fiche 4 — NUMERICAL INDEX BJC 1518 TO C 29283

C 29302	2.E09	C 30075 2	1.D12	C 30783	4.C02	C 32533	1.N02	C 33835	2.E03	C 35515	1.M02	C 37021	2.C06	C 38421	1.E02
C 29302	2.H12	C 30075 2	1.D13	C 30952	1.K04	C 32534	1.D11	C 33835	2.E13	C 35521	1.D16	C 37133 09	1.F06	C 38494 3	2.D02
C 29308	1.D14	C 30075 2	1.D15	C 30953	1.K04	C 32535	1.D11	C 33835	2.E14	C 35521	1.D17	C 37175 7	1.J03	C 38494 3	2.D04
C 29310	1.K12	C 30075 2	1.D16	C 30954	1.K04	C 32544 A	1.D11	C 33835	2.E18	C 35557	2.F03	C 37273	1.K10	C 38494 3	2.D05
C 29330	1.D13	C 30075 2	1.D17	C 30955	1.K04	C 32544 B	1.D11	C 33835	2.H13	C 35568	1.D16	C 37278	2.F03	C 38494 3	2.H10
C 29335	1.E02	C 30075 2	1.D18	C 30993	1.K03	C 32679	1.D11	C 33835	2.L06	C 35568	1.D17	C 37278	2.I05	C 38494 5	2.D16
C 29337	1.D18	C 30075 2	1.E02	C 30993	1.N08	C 32680	1.D11	C 33851	1.D16	C 35570	1.D16	C 37330 1	2.E13	C 38494 6	1.H06
C 29338	1.D18	C 30075 2	1.E03	C 31063	1.D16	C 32726	1.D10	C 33855	1.D16	C 35570	1.D17	C 37330 1	2.E15	C 38494 6	1.J05
C 29339	1.D18	C 30075 2	1.E04	C 31063	1.D17	C 32743 3	4.C02	C 33869	1.D16	C 35571	1.D16	C 37330 1	2.E16	C 38518	2.H10
C 29340	1.D18	C 30075 2	1.E09	C 31064	1.D16	C 32746	1.O06	C 33884	2.D18	C 35571	1.D17	C 37330 1	2.F02	C 38522	1.F06
C 29349	1.E02	C 30075 2	1.E10	C 31064	1.D17	C 32746 1	1.N18	C 33889	1.C09	C 35572	1.D12	C 37346	1.J06	C 38523	1.M02
C 29373	1.F09	C 30075 2	1.E11	C 31068	1.E02	C 32746 1	1.O07	C 33889	1.E02	C 35657	1.D16	C 37355	1.D12	C 38524	1.M02
C 29374	1.K10	C 30075 2	1.E12	C 31075	2.F07	C 32746 1	2.C18	C 33917	1.C09	C 35732	1.C11	C 37435	2.E08	C 38526	1.M02
C 29390	1.C11	C 30075 2	1.E13	C 31075	2.F15	C 32746 1	2.D13	C 33917	1.E02	C 35732	1.E05	C 37523	1.E03	C 38527	1.M02
C 29390	1.E03	C 30075 2	1.E14	C 31076	1.E02	C 32746 1	4.C02	C 33918	1.C09	C 35770	2.F04	C 37566	2.I04	C 38528	1.M02
C 29391	1.C11	C 30075 2	1.E15	C 31106	2.K08	C 32746 2	2.D17	C 33918	1.E02	C 35779	2.F04	C 37576	2.I04	C 38549	1.D13
C 29391	1.E03	C 30075 2	1.E16	C 31106	2.K09	C 32746 3	2.D02	C 33921	1.F02	C 35800	1.E05	C 37627	1.O06	C 38559	1.O06
C 29406	1.M09	C 30075 2	1.E17	C 31131	2.D06	C 32746 3	2.D14	C 34006	1.M13	C 35801	1.E05	C 37817	1.E10	C 38559	1.O07
C 29407 1	1.M08	C 30075 2	1.E18	C 31131	2.D09	C 32746 3	2.D16	C 34006	2.I02	C 35840	1.H05	C 37873	1.D12	C 38559	2.F08
C 29407 1	1.M09	C 30075 2	2.E04	C 31193	1.D15	C 33050	1.H05	C 34016	1.D11	C 35840	1.J04	C 37874	1.E08	C 38616	2.N02
C 29419	1.E03	C 30075 2	2.F03	C 31201	1.E02	C 33050	1.J04	C 34017	2.F04	C 35840	1.O08	C 37874	1.E09	C 38616	2.N04
C 29420	1.E03	C 30075 2	2.I02	C 31228	1.H05	C 33081	1.N17	C 34059	3.E09	C 35840	2.C14	C 37875	2.I05	C 38616	2.N05
C 29428	1.E05	C 30075 2	2.J15	C 31228	1.J04	C 33139 1	1.M14	C 34230	2.I02	C 35840	2.C17	C 37880	2.G11	C 38616	2.N06
C 29429	1.E05	C 30075 3	2.L06	C 31232	1.H05	C 33139 4	1.M14	C 34241	1.M13	C 35840	2.E06	C 37882	1.D16	C 38720	1.E12
C 29482	1.G04	C 30075 5	2.I02	C 31232	1.J04	C 33244	1.C12	C 34241	2.H02	C 35840	2.E07	C 37886	2.I05	C 38720	1.E13
C 29485	1.D11	C 30114	1.K05	C 31301	3.C05	C 33244	1.E02	C 34241	2.J06	C 35840	2.O10	C 37888	1.D12	C 38720	1.E14
C 29501	2.G02	C 30114	1.K06	C 31302	3.C13	C 33244	1.E04	C 34283	1.E03	C 35840	2.O13	C 37888	1.E08	C 38720	1.E15
C 29501	2.G05	C 30115	1.K05	C 31302	3.D06	C 33280	2.D18	C 34388	2.E08	C 35864	2.I04	C 37892	1.D13	C 38720	1.E16
C 29590	1.E02	C 30115	1.K06	C 31304 4	1.M12	C 33280	2.E02	C 34403	2.G03	C 36013	1.D13	C 37900	3.G07	C 38791	1.M02
C 29626	1.E10	C 30117	1.K06	C 31304 4	1.M14	C 33280	2.E05	C 34608	1.F06	C 36020	1.E11	C 37900	3.G09	C 38801	1.F06
C 29635	1.E02	C 30213	1.M03	C 31343	2.F17	C 33440	1.N03	C 34608	2.J10	C 36119	2.F03	C 37921	2.G11	C 38802	1.E09
C 29636	1.E02	C 30251	1.N03	C 31352	2.N18	C 33440 1	1.N03	C 34632	2.F04	C 36119	2.I05	C 37925	1.D18	C 38844 30	1.I09
C 29652	1.E07	C 30314	1.K02	C 31352	2.O02	C 33440 2	1.N03	C 34646 23	2.D18	C 36120	2.J06	C 37939	2.C06	C 38844 30	1.I15
C 29699	1.C06	C 30316	1.M03	C 31422	2.D18	C 33509	1.H05	C 34646 23	2.E02	C 36120	2.J07	C 37941 1	2.G04	C 38850	2.M16
C 29699	1.D14	C 30344	1.E03	C 31422	2.E02	C 33509	1.J04	C 34697	1.C05	C 36137	1.K09	C 37941 1	2.G07	C 38887	1.N18
C 29828	1.D16	C 30344	1.E12	C 31657	1.K11	C 33509	1.J06	C 34697	1.D13	C 36138	1.K09	C 37941 7	2.G04	C 38887	2.D02
C 29829	1.D17	C 30344	1.E13	C 31973 11	2.H15	C 33509	2.C14	C 34697 1	1.C05	C 36143	2.F03	C 37990	1.E12	C 38894	1.I05
C 29874	2.G10	C 30344	1.E14	C 31973 11	2.J14	C 33509	2.C17	C 34697 1	1.D13	C 36151	2.J07	C 37990	1.E13	C 38894	1.M14
C 29875	2.L02	C 30344	1.E15	C 31973 11	2.L09	C 33600 1	2.C15	C 34734	1.D10	C 36175 11	2.G12	C 37990	1.E14	C 38894	1.N13
C 29875	3.F13	C 30344	1.E16	C 31973 3	3.G06	C 33620	1.N13	C 34734	1.E03	C 36175 11	2.G15	C 37990	1.E15	C 38894	1.N14
C 29888	1.O06	C 30370	1.K02	C 31973 5	3.F13	C 33620	1.N14	C 34757	1.E05	C 36175 20	2.G12	C 37990	1.E16	C 38894	1.N15
C 29888	1.O07	C 30371	1.K02	C 31973 6	2.O14	C 33620	1.N15	C 34920	1.D18	C 36189	1.M13	C 37990	2.E10	C 38894	1.N16
C 29890	1.K03	C 30381	2.G03	C 31973 6	3.F12	C 33620	1.N16	C 34937	1.I15	C 36196	1.E10	C 37990	2.F03	C 38894	4.C02
C 29975	1.K03	C 30427	1.E02	C 31973 6	3.F13	C 33620	2.D12	C 34946	2.E11	C 36217	1.M13	C 37990	2.F04	C 38955	2.M12
C 29979	1.K03	C 30431	1.C11	C 32185	1.D13	C 33620	2.G14	C 35093 1	1.C04	C 36298	1.E10	C 37999	1.D12	C 38966	2.L17
C 30002	1.D18	C 30431	1.E05	C 32197	1.D13	C 33622	1.N06	C 35093 1	1.D12	C 36389	2.G07	C 38005	2.N15	C 38966	2.L18
C 30064	1.N02	C 30484	1.E02	C 32202	1.N04	C 33682	1.K02	C 35093 1	1.G04	C 36471	1.N13	C 38005	2.N16	C 39032	2.M07
C 30075	1.E08	C 30485	1.E02	C 32202	1.N06	C 33745 1	1.K03	C 35093 2	2.J10	C 36471	1.N14	C 38009	2.G11	C 39032 J	2.M06
C 30075 1	1.D12	C 30486	1.K03	C 32202	1.N07	C 33784	1.H05	C 35099	2.F04	C 36471	1.N15	C 38037	1.M04	C 39213	2.L06
C 30075 1	1.D15	C 30493	1.E02	C 32227	1.E02	C 33784	1.J04	C 35125	2.E15	C 36471	1.N16	C 38074	1.F06	C 39246	1.H08
C 30075 1	1.D16	C 30611	1.K03	C 32272	1.D10	C 33784	1.J06	C 35125	2.E16	C 36520	1.D12	C 38075	1.F06	C 39314	2.N07
C 30075 1	1.E07	C 30612	1.K03	C 32273	1.D10	C 33784	1.O08	C 35125	2.E17	C 36542	1.E10	C 38112	1.K06	C 39315	2.N07
C 30075 1	1.E08	C 30613	1.K03	C 32312 3	3.G02	C 33784	2.C14	C 35125 1	2.H12	C 36546	2.I02	C 38113	1.K06	C 39341	1.M06
C 30075 1	1.E09	C 30614	1.K03	C 32331	1.D11	C 33784	2.C17	C 35401	2.F10	C 36611	2.N07	C 38189	1.F06	C 39341	1.M07
C 30075 1	2.E06	C 30721	1.D08	C 32332	1.D11	C 33784	2.E06	C 35401	2.F13	C 36613 2	4.C03	C 38195	1.F09	C 39413	2.G16
C 30075 1	2.E07	C 30721	1.F09	C 32341	2.O17	C 33784	2.E07	C 35401	2.F14	C 36734	2.L07	C 38302	2.G10	C 39456	1.M06
C 30075 1	2.E11	C 30722	1.K04	C 32366	1.E06	C 33784	2.O10	C 35468	1.J04	C 36881	1.K12	C 38302	2.G11	C 39456	1.M07
C 30075 1	2.I05	C 30741	1.M05	C 32367 1	1.C05	C 33784	2.O13	C 35509	2.E06	C 36887 1	1.K02	C 38302	2.H08	C 39457	1.M06
C 30075 2	1.D02	C 30767	1.C09	C 32367 1	1.D13	C 33835	1.I09	C 35509	2.E07	C 36983	2.I04	C 38333	2.G02	C 39457	1.M07
C 30075 2	1.D11	C 30767	1.E02	C 32458 7	1.J07	C 33835	1.I15	C 35512	1.D16	C 36996	2.G06	C 38333	2.G05	C 39462	2.N12

C 39515	1.N05	C 41769	1.O08	C 42301 11	2.E08	C 43206 1	1.M14	C 43599 12	1.N15	C 44014	1.O06	C 44630	2.C17	C 45307	1.M06
C 39515	1.N08	C 41770	2.C18	C 42301 2	1.E18	C 43206 1	2.C18	C 43599 12	1.N16	C 44014	1.O07	C 44649	1.D07	C 45307	1.M07
C 39520	2.H13	C 41770	2.D02	C 42301 5	1.E17	C 43206 1	2.G14	C 43599 14	2.D05	C 44015	1.O06	C 44649	2.L02	C 45360	1.K07
C 39559	1.M02	C 41794	2.D05	C 42301 5	1.E18	C 43206 10	2.D04	C 43599 3	2.C18	C 44015	1.O07	C 44649	3.F13	C 45361	1.K07
C 39572	4.C03	C 41801	2.D05	C 42301 6	1.E17	C 43206 2	1.M12	C 43599 6	2.D06	C 44067	2.E08	C 44668	3.N05	C 45367	1.K02
C 39573	4.C11	C 41802	2.D05	C 42301 6	1.E18	C 43206 3	1.K14	C 43599 6	2.D13	C 44069	2.E05	C 44673	1.M03	C 45368	1.K02
C 39597	1.N04	C 41803	2.D05	C 42308	2.H13	C 43206 3	2.G06	C 43599 7	2.D14	C 44083	2.F07	C 44674	1.M03	C 45467	1.D12
C 39597	1.N09	C 41831	1.K12	C 42325 12	2.E05	C 43206 4	2.G03	C 43599 7	2.D16	C 44083	2.F15	C 44676	1.M03	C 45512	2.K08
C 39619	1.N04	C 41856	2.D05	C 42325 4	2.E11	C 43206 5	2.G03	C 43599 8	2.D13	C 44085 2	2.G05	C 44700	2.L17	C 45512	2.K09
C 39619	1.N05	C 41884	2.E08	C 42372	2.K17	C 43206 5	2.G06	C 43599 9	2.D02	C 44097	2.E05	C 44700	2.L18	C 45551	1.M13
C 39619	1.N08	C 41887	1.E17	C 42373	2.K17	C 43206 7	2.D06	C 43599 9	2.D03	C 44117	2.H14	C 44703	1.M10	C 45625	2.K08
C 39619	1.N09	C 41887	1.E18	C 42381	2.M10	C 43206 8	2.G13	C 43599 9	2.D04	C 44142	2.H14	C 44718	2.N17	C 45625	2.K09
C 39624	1.N05	C 41888	1.E18	C 42421	1.N17	C 43206 8	2.G14	C 43599 9	2.D15	C 44143	2.H14	C 44719	2.N17	C 45625	2.K10
C 39624	1.N08	C 41902	2.E08	C 42452	2.G14	C 43208	2.E06	C 43604 1	1.M14	C 44144	2.D06	C 44760	1.M04	C 45666	1.K02
C 39655	1.D11	C 41920	2.H04	C 42511	1.C15	C 43208	2.E07	C 43608	2.F04	C 44146 1	1.K07	C 44763	1.M04	C 45700	1.O09
C 39686	1.F02	C 41920	2.H12	C 42511	1.E05	C 43216	1.K03	C 43686 1	2.F08	C 44151	1.D12	C 44778	1.N18	C 45709	1.K07
C 39687	1.F02	C 41920	2.H14	C 42525	2.O15	C 43216	1.K04	C 43686 1	2.F16	C 44152	1.D12	C 44778	4.C02	C 45710	1.K07
C 39692	1.K10	C 41921	2.H04	C 42566	2.F04	C 43221	2.G05	C 43687	1.N14	C 44163	2.N17	C 44781	1.O06	C 45711	1.K07
C 39929	2.F04	C 41943	1.E06	C 42595	1.E12	C 43221	2.G12	C 43687	1.N16	C 44183	2.N17	C 44781	2.D02	C 45724	1.K07
C 39930	2.F04	C 41988 5	1.O08	C 42595	1.E13	C 43221	2.G15	C 43689	1.I05	C 44183	2.O05	C 44781	4.C02	C 45726	1.K07
C 40091	2.K07	C 41993	2.D09	C 42595	1.E14	C 43222	2.G05	C 43689	1.N13	C 44183	2.O06	C 44896	1.N04	C 45753 4	1.N17
C 40092	2.K07	C 41994	2.D17	C 42595	1.E15	C 43222	2.G12	C 43689	1.N14	C 44184	2.O03	C 44896	1.N07	C 45759 3	1.E11
C 40158	1.K11	C 41995	2.D17	C 42595	1.E16	C 43222	2.G15	C 43708	2.E08	C 44184	2.O04	C 44896	1.N09	C 45759 3	2.C11
C 40177	1.D15	C 42001	1.M05	C 42599	1.E16	C 43231	2.E08	C 43719	1.M14	C 44184	2.O05	C 44896	2.L04	C 45759 3	2.G02
C 40184	2.M05	C 42013	2.D13	C 42606	1.E12	C 43232	2.E08	C 43722	1.M14	C 44184	2.O06	C 44896	2.L07	C 45801	1.D07
C 40187	2.G07	C 42072	1.E07	C 42606	1.E13	C 43253	1.O06	C 43733	1.M05	C 44190	1.E17	C 44969	2.H14	C 45801	3.F13
C 41013	2.H08	C 42077	2.E03	C 42606	1.E14	C 43253	1.O07	C 43748	1.N13	C 44194	1.N02	C 44979	1.F02	C 45804	1.M03
C 41014	2.H08	C 42100	1.H06	C 42618	2.E08	C 43254	1.O06	C 43748	1.N15	C 44195	1.N02	C 44979 1	1.F02	C 45906	1.M14
C 41028	1.D16	C 42100	1.J05	C 42701	2.O07	C 43265 1	2.E05	C 43760	2.N18	C 44198	1.N02	C 44998	1.M10	C 45907	1.M14
C 41028	1.D17	C 42106	1.O08	C 42701	2.O08	C 43266	2.G06	C 43761	2.N18	C 44199	1.N02	C 44999	1.M10	C 45917	2.N15
C 41051	2.G13	C 42123	2.D06	C 42701	2.O09	C 43267	2.G06	C 43762	2.N18	C 44219	1.E06	C 45016	2.G13	C 45950	2.N18
C 41051	2.G14	C 42123	2.D09	C 42717	1.E12	C 43280	2.E05	C 43762	2.O02	C 44221	2.H12	C 45019	2.G07	C 45981	2.E08
C 41099	2.G06	C 42150	1.N18	C 42717	1.E13	C 43317	2.J10	C 43773	2.M16	C 44221	2.H14	C 45021	2.N17	C 46014	1.O08
C 41102	2.G06	C 42150	1.O02	C 42717	1.E14	C 43318	2.J10	C 43790	1.K05	C 44230	1.C04	C 45047	1.M10	C 46015	1.O08
C 41103	2.G06	C 42150	1.O03	C 42717	1.E15	C 43319	1.J03	C 43790	1.K06	C 44230	1.D12	C 45055	1.N17	C 46021	1.D04
C 41115 1	2.G11	C 42150	1.O04	C 42736	2.H14	C 43340	2.J10	C 43791	1.K05	C 44231	2.E10	C 45066	1.M14	C 46021	1.I08
C 41116 29	1.E06	C 42150	1.O05	C 42737	2.H14	C 43354	1.E17	C 43791	1.K06	C 44250	1.D15	C 45068	1.M06	C 46021	1.I09
C 41116 6	1.M05	C 42165	2.E11	C 42794	1.E13	C 43354	1.E18	C 43807	1.K13	C 44251	1.D15	C 45068	1.M07	C 46024	3.N05
C 41271	1.K05	C 42176	1.E04	C 42795	1.E16	C 43375	1.N13	C 43809	1.D14	C 44287	1.E17	C 45068	1.M14	C 46063	2.C15
C 41271	1.K06	C 42177	1.E04	C 42797	1.C15	C 43375	1.N14	C 43829	1.N05	C 44353	1.N09	C 45083	1.O09	C 46064	1.D10
C 41304	1.K07	C 42178 3	1.K13	C 42797	1.E08	C 43375	1.N15	C 43829	1.N09	C 44386	1.E06	C 45084	2.F07	C 46070	1.I15
C 41305	1.K07	C 42191	1.L05	C 42797	1.E09	C 43375	1.N16	C 43831	2.F07	C 44392	2.E10	C 45084	2.F15	C 46109 1	2.D05
C 41314	1.K04	C 42200	1.C15	C 42799 1	2.H14	C 43378	1.N13	C 43831	2.F15	C 44410	1.E06	C 45088	1.N17	C 46145	2.L17
C 41315	1.K04	C 42200	1.E07	C 42799 17	2.L13	C 43378	1.N14	C 43861	2.N17	C 44412	2.E10	C 45089	1.N17	C 46145	2.L18
C 41351 9	1.E17	C 42220	1.K12	C 42803	2.D06	C 43378	1.N15	C 43863	2.K03	C 44468 2	2.H12	C 45090	1.N17	C 46157	2.D06
C 41373	2.E11	C 42238	2.L17	C 42804	2.D06	C 43378	1.N16	C 43865	2.N17	C 44477 1	2.E06	C 45125	1.H05	C 46186	1.K02
C 41373 1	1.E11	C 42238	2.L18	C 42805	2.D06	C 43378	2.D12	C 43898	2.L17	C 44477 1	2.E07	C 45125	1.J04	C 46194	2.M12
C 41374	1.E11	C 42239	2.L17	C 42806	2.D06	C 43378	2.D17	C 43898	2.L18	C 44478	2.E08	C 45125	1.J06	C 46205	3.H05
C 41506	1.O08	C 42239	2.L18	C 42817	2.D06	C 43424	1.O08	C 43923	1.F06	C 44479	2.D02	C 45125	2.O10	C 46221	2.M12
C 41508	1.O08	C 42240	2.L17	C 42853	2.N18	C 43502	2.N17	C 43943	2.N18	C 44479	2.G08	C 45125	2.O13	C 46269	2.J09
C 41606	1.N02	C 42240	2.L18	C 42853	2.O02	C 43578 4	2.G02	C 43958	2.D06	C 44479	2.G09	C 45173	2.F13	C 46272	1.C15
C 41607	1.N02	C 42289	2.L17	C 42879	1.M05	C 43578 5	2.G02	C 43994 16	2.G14	C 44511	1.M04	C 45173	2.F14	C 46272	1.E07
C 41618	1.K07	C 42289	2.L18	C 42907	1.K13	C 43599 10	1.N13	C 43999	2.G14	C 44534	1.N09	C 45230	1.I13	C 46337	2.F04
C 41712	1.N18	C 42291	2.M08	C 42967	2.G14	C 43599 10	1.N14	C 44001	2.M12	C 44543	2.F04	C 45231	1.I13	C 87371	2.O17
C 41712	1.O02	C 42301 1	1.E17	C 43058	1.K05	C 43599 10	1.N15	C 44001	2.N10	C 44551	1.M03	C 45232	1.G04	C 273286 6Z	2.D06
C 41712	1.O03	C 42301 1	2.E17	C 43058	1.K06	C 43599 10	1.N16	C 44006	1.F02	C 44553	1.M03	C 45247	1.I16	CAC 1018	2.F17
C 41712	1.O04	C 42301 10	1.E17	C 43070	1.F03	C 43599 10	2.D02	C 44007	1.F02	C 44602	2.C16	C 45248	1.H06	CAC 1040 3	2.D12
C 41712	1.O05	C 42301 10	1.E18	C 43196	2.D05	C 43599 10	2.E10	C 44008	1.F02	C 44602	2.C17	C 45248	1.J05	CAC 1099	1.K07
C 41712	1.O06	C 42301 11	1.E17	C 43200	2.D02	C 43599 12	1.N13	C 44009	1.F02	C 44616	2.C12	C 45287	2.C12	CAC 1100	1.K07
C 41712	1.O07	C 42301 11	1.E18	C 43206 1	1.M12	C 43599 12	1.N14	C 44010	1.E04	C 44630	2.C16	C 45287	2.C15	CAC 1101	1.K07

NUMERICAL INDEX CAC 1102 TO CAC 6966

CAC 1102	1.K07	CAC 2398	2.D08	CAC 3132	2.D04	CAC 3801 3	2.F07	CAC 3917	2.D03	CAC 4763	1 L06	CAC 5439	1.N18	CAC 5866	2.D06
CAC 1162	1.K02	CAC 2403	2.D08	CAC 3144	2.F17	CAC 3801 3	2.F15	CAC 3923	1.N13	CAC 4797	2.G04	CAC 5439	1.O02	CAC 5866	2.D09
CAC 1227	1.N10	CAC 2404	2.D08	CAC 3145	2.F17	CAC 3801 4	2.F06	CAC 4029	2.E09	CAC 4845	1.H05	CAC 5440	1.O03	CAC 5866	2.D13
CAC 1384	1.N15	CAC 2405	2.D08	CAC 3227	1.J07	CAC 3802	2.F06	CAC 4033 22	2.D14	CAC 4845	1.J04	CAC 5440	1.O04	CAC 5866 4	2.D17
CAC 1402	2.F10	CAC 2416	1.K09	CAC 3231 4	2.D14	CAC 3802	2.F07	CAC 4033 22	2.D16	CAC 4968	2.D14	CAC 5441	1.O03	CAC 5867	2.D02
CAC 1403	2.F10	CAC 2418	2.M05	CAC 3231 4	2.D16	CAC 3802	2.F10	CAC 4040	2.D17	CAC 4968	2.D16	CAC 5441	1.O04	CAC 5867	2.D09
CAC 1410	2.F14	CAC 2430	1.J07	CAC 3231 4	2.O15	CAC 3802	2.F11	CAC 4067	1.O08	CAC 5009	2.H13	CAC 5442	1.O03	CAC 5867	2.D13
CAC 1429	2.C12	CAC 2462	1.J07	CAC 3242	1.J07	CAC 3802	2.F12	CAC 4068	1.O08	CAC 5058	2.O10	CAC 5442	1.O04	CAC 5867 22	2.C18
CAC 1582	1.N10	CAC 2469	1.J07	CAC 3293	1.O08	CAC 3802	2.F13	CAC 4073	2.F05	CAC 5058	2.O13	CAC 5443	1.O03	CAC 5867 22	2.E10
CAC 1583	1.N10	CAC 2481	2.D04	CAC 3301	1.K08	CAC 3802	2.F14	CAC 4073	2.F06	CAC 5073	1.L06	CAC 5443	1.O04	CAC 5868	2.D02
CAC 1610	2.C15	CAC 2494	2.F07	CAC 3321	1.M02	CAC 3802	2.F15	CAC 4088	1.O08	CAC 5081	2.F05	CAC 5446	1.O03	CAC 5868	2.D03
CAC 1635	1.M08	CAC 2494	2.F15	CAC 3398	2.H04	CAC 3803	2.F07	CAC 4089	1.O08	CAC 5092 1	2.G05	CAC 5446	1.O04	CAC 5869 19	2.D15
CAC 1635	1.M09	CAC 2522	1.M12	CAC 3416	1.K02	CAC 3803	2.F10	CAC 4123	1.L05	CAC 5093 1	2.G05	CAC 5451	1.O06	CAC 5885	1.M02
CAC 1646	2.C18	CAC 2523	1.J07	CAC 3417	1.K02	CAC 3803	2.F11	CAC 4123	1.L06	CAC 5095	2.G12	CAC 5451	1.O07	CAC 5893	1.H06
CAC 1646	2.D02	CAC 2546	1.M10	CAC 3422	4.C03	CAC 3803	2.F12	CAC 4188	1.N14	CAC 5095	2.G14	CAC 5452	1.O06	CAC 5893	1.J05
CAC 1648	2.D05	CAC 2549	1.M10	CAC 3437	4.C03	CAC 3803	2.F13	CAC 4188	1.N15	CAC 5097	3.N05	CAC 5452	1.O07	CAC 5951	1.H06
CAC 1660	2.F15	CAC 2552	2.E09	CAC 3441	1.F09	CAC 3803	2.F14	CAC 4188	1.N16	CAC 5099	3.N05	CAC 5453	1.N18	CAC 5952	1.H06
CAC 1661	2.F15	CAC 2571	2.D12	CAC 3442	1.F09	CAC 3803	2.F15	CAC 4188	2.H13	CAC 5100	2.G03	CAC 5453	1.O02	CAC 5957	1.H05
CAC 1706	2.D04	CAC 2582	2.F09	CAC 3470	1.M05	CAC 3814	2.D12	CAC 4267	2.F11	CAC 5109 1	2.F16	CAC 5454	1.O06	CAC 5970	2.F08
CAC 1742	1.N07	CAC 2583	2.F09	CAC 3477	2.D04	CAC 3814	2.D17	CAC 4267	2.F12	CAC 5118	1.D07	CAC 5455	1.O03	CAC 6012	1.H05
CAC 1765	2.F05	CAC 2588 4	2.F09	CAC 3565	2.D08	CAC 3818 10	1.K11	CAC 4268	2.E09	CAC 5125	2.G14	CAC 5455	1.O04	CAC 6068	2.F07
CAC 1786 8	2.G14	CAC 2588 5	2.F09	CAC 3589 2	1.M12	CAC 3818 12	1.K11	CAC 4269	2.D04	CAC 5146	2.C12	CAC 5456	1.O06	CAC 6068	2.F15
CAC 1786 8	2.H14	CAC 2589	2.F09	CAC 3589 2	1.M14	CAC 3818 14	1.K11	CAC 4307 2	2.D02	CAC 5180	2.C12	CAC 5457	1.O06	CAC 6069	2.F07
CAC 1792	2.F17	CAC 2590	2.F09	CAC 3590	1.O08	CAC 3818 16	1.K11	CAC 4316	2.C13	CAC 5181	2.C12	CAC 5457	1.O07	CAC 6069	2.F15
CAC 1793	2.F17	CAC 2591	2.O17	CAC 3591	1.O08	CAC 3818 18	1.K11	CAC 4316	2.C14	CAC 5201	2.G04	CAC 5458	1.O06	CAC 6136	1.H05
CAC 1823	2.G10	CAC 2602	2.O10	CAC 3603	2.D12	CAC 3818 20	1.K11	CAC 4319	2.F11	CAC 5215	2.G10	CAC 5458	1.O07	CAC 6140	1.N18
CAC 1823	3.E08	CAC 2602	2.O13	CAC 3603	2.D17	CAC 3818 22	1.K11	CAC 4319	2.F12	CAC 5219	2.G10	CAC 5466	1.N18	CAC 6140	1.O02
CAC 1825	2.H08	CAC 2631	2.F09	CAC 3654	1.M14	CAC 3818 24	1.K11	CAC 4328	1.M10	CAC 5225	2.G03	CAC 5466	1.O02	CAC 6140	1.O03
CAC 1853 1	2.D12	CAC 2716	1.O10	CAC 3655	1.M14	CAC 3818 26	1.K11	CAC 4330	1.M10	CAC 5240	2.G10	CAC 5468	2.F08	CAC 6140	1.O04
CAC 1881	2.H08	CAC 2717	1.O10	CAC 3656	1.M14	CAC 3818 28	1.K11	CAC 4333	2.F07	CAC 5242	2.G10	CAC 5468	2.F16	CAC 6143	1.O06
CAC 1922	2.F17	CAC 2740	4.C03	CAC 3672	2.G08	CAC 3818 30	1.K11	CAC 4333	2.F15	CAC 5261	1.D07	CAC 5469	2.F08	CAC 6200	2.C16
CAC 1923	2.F17	CAC 2750	2.F07	CAC 3672	2.G09	CAC 3818 32	1.K11	CAC 4336	2.F05	CAC 5304	1.J04	CAC 5469	2.F16	CAC 6241	2.C18
CAC 1928	2.F13	CAC 2750	2.F15	CAC 3673 9	2.G09	CAC 3818 34	1.K11	CAC 4336	2.F06	CAC 5328	2.C12	CAC 5477	1.D07	CAC 6254 1	2.C16
CAC 1929	2.F13	CAC 2751	2.F07	CAC 3674	2.G08	CAC 3818 36	1.K11	CAC 4347	1.N04	CAC 5340	1.H05	CAC 5483	1.J04	CAC 6272	3.E13
CAC 1930	2.F15	CAC 2751	2.F15	CAC 3674	2.G09	CAC 3818 38	1.K11	CAC 4347	1.N09	CAC 5340	1.J04	CAC 5483	1.J06	CAC 6327	1.J07
CAC 1931	2.F15	CAC 2753	2.F07	CAC 3677	2.D14	CAC 3818 40	1.K11	CAC 4373 1	2.F16	CAC 5343	2.C18	CAC 5491	1.J05	CAC 6377	1.L05
CAC 1974	2.F10	CAC 2753	2.F15	CAC 3677	2.D16	CAC 3818 42	1.K11	CAC 4374	2.F16	CAC 5345	2.D12	CAC 5509	1.M12	CAC 6380	1.L05
CAC 1975	2.F10	CAC 2765 3	1.M05	CAC 3678	2.G08	CAC 3818 44	1.K11	CAC 4375	2.F16	CAC 5347	1.O06	CAC 5537	1.D07	CAC 6426	3.K14
CAC 2015	2.D06	CAC 2823	2.F07	CAC 3678	2.G09	CAC 3818 46	1.K11	CAC 4379	1.L06	CAC 5355	1.N04	CAC 5574	1.N16	CAC 6442	1.L05
CAC 2036	1.M08	CAC 2823	2.F15	CAC 3682	2.G09	CAC 3818 48	1.K11	CAC 4407	2.E09	CAC 5355	1.N07	CAC 5596	2.H09	CAC 6442	1.L07
CAC 2036	1.M09	CAC 2878	2.D02	CAC 3684	2.D04	CAC 3818 50	1.K11	CAC 4408 1	2.F08	CAC 5366	1.H05	CAC 5596	2.H10	CAC 6502	1.L06
CAC 2037	1.M08	CAC 2879	2.D02	CAC 3685	2.D04	CAC 3821	2.G05	CAC 4410	2.F16	CAC 5366	1.J04	CAC 5608	1.N05	CAC 6502	1.L07
CAC 2037	1.M09	CAC 2884	2.M04	CAC 3696	2.H10	CAC 3823	2.C16	CAC 4411	2.F16	CAC 5368	4.C03	CAC 5609	1.N04	CAC 6593	2.G03
CAC 2040	1.H05	CAC 2885	2.M04	CAC 3697	2.H10	CAC 3823	2.C17	CAC 4452	1.J04	CAC 5406	2.C18	CAC 5612	1.N05	CAC 6594	2.G03
CAC 2040	1.J04	CAC 2886	2.M04	CAC 3713	1.N16	CAC 3826	2.C16	CAC 4474	2.G05	CAC 5408	1.N14	CAC 5612	1.N08	CAC 6612	1.D08
CAC 2040	1.J06	CAC 2917	2.G14	CAC 3714	1.N15	CAC 3826	2.C17	CAC 4504	1.L06	CAC 5409	1.N13	CAC 5613	1.N05	CAC 6613	1.D08
CAC 2040	2.O10	CAC 2924	2.F17	CAC 3722	2.D03	CAC 3827	2.C16	CAC 4508	2.G14	CAC 5423	2.F08	CAC 5613	1.N08	CAC 6617	1.D08
CAC 2040	2.O13	CAC 2961	2.D08	CAC 3723	2.D03	CAC 3827	2.C17	CAC 4549	1.L06	CAC 5434	1.N18	CAC 5620	2.F07	CAC 6618	1.D08
CAC 2047	2.G10	CAC 2967	1.M02	CAC 3730	2.G08	CAC 3832	2.E09	CAC 4559	2.G05	CAC 5434	1.O02	CAC 5620	2.F15	CAC 6647	1.D07
CAC 2191 1	1.I07	CAC 2970	1.M02	CAC 3730	2.G09	CAC 3835	2.C16	CAC 4560 2	1.I06	CAC 5434	1.O05	CAC 5621	2.F07	CAC 6659	2.D12
CAC 2236	2.F07	CAC 2995	2.F17	CAC 3762	2.D05	CAC 3835	2.C17	CAC 4568	2.G15	CAC 5435	1.N18	CAC 5621	2.F15	CAC 6659	4.C02
CAC 2237	2.F07	CAC 3014	1.M14	CAC 3785	1.F04	CAC 3842	2.C15	CAC 4572 1	2.G05	CAC 5435	1.O02	CAC 5632	1.J05	CAC 6724	1.H05
CAC 2299	2.H03	CAC 3039	2.E09	CAC 3785	1.F06	CAC 3858	2.C15	CAC 4572 1	2.G12	CAC 5435	1.O03	CAC 5636	1.J05	CAC 6746	2.H07
CAC 2299	2.H12	CAC 3067	1.K08	CAC 3794	2.G07	CAC 3891	1.N09	CAC 4572 1	2.G14	CAC 5435	1.O04	CAC 5652	2.C18	CAC 6747	2.H07
CAC 2316	1.K09	CAC 3101	2.D09	CAC 3797	2.H14	CAC 3897	1.N09	CAC 4578	1.N17	CAC 5435	1.O05	CAC 5652	4.C02	CAC 6853	1.M12
CAC 2321	2.G02	CAC 3101	2.D13	CAC 3801 1	2.F10	CAC 3899	1.N09	CAC 4579	1.N17	CAC 5436	1.O02	CAC 5667	1.L05	CAC 6873	2.D15
CAC 2321	2.G05	CAC 3101	2.D17	CAC 3801 1	2.F11	CAC 3907	1.M10	CAC 4602	1.N17	CAC 5437	1.N18	CAC 5804	1.D07	CAC 6874	2.D12
CAC 2327	1.J07	CAC 3103	2.F07	CAC 3801 1	2.F12	CAC 3915	2.F08	CAC 4651 7	1.K02	CAC 5437	1.O02	CAC 5805	1.D07	CAC 6874	2.D15
CAC 2341	2.H10	CAC 3103	2.F15	CAC 3801 1	2.F13	CAC 3915	2.F16	CAC 4723	2.G05	CAC 5438	1.N18	CAC 5866	1.E08	CAC 6941	1.M12
CAC 2395	2.D08	CAC 3129	2.G02	CAC 3801 1	2.F14	CAC 3916	2.D03	CAC 4748	1.J04	CAC 5438	1.O02	CAC 5866	1.E09	CAC 6966	1.D02

NUMERICAL INDEX CAC 6966 TO DAB 006 6C

CAC 6966	1.E17	CAC 8900 11	2.D09	CBC 1416	1.D07	CBC 2163	4.C03	CBC 4103	1.K06	CBC 4851	2.F12	CBC 5616	1.K06	CBC 6550	1.K05
CAC 6995	1.H06	CAC 8900 11	2.D12	CBC 1448	2.G11	CBC 2164	4.C03	CBC 4118 1	2.D13	CBC 4861	3.I13	CBC 5649	2.D05	CBC 6559	1.K06
CAC 6998	4.C03	CAC 8900 11	2.D15	CBC 1448	2.H08	CBC 2179	2.D02	CBC 4125	1.H04	CBC 4884	1.K07	CBC 5652	1.N02	CBC 6579 1	2.D07
CAC 7000	4.C03	CAC 8900 11	2.E10	CBC 1448	3.F13	CBC 2179	2.E04	CBC 4175	1.N05	CBC 4885	1.K07	CBC 5653	1.N02	CBC 6606	2.H03
CAO 7191	3.F02	CAC 8943	4.C03	CBC 1491	1.L05	CBC 2181	2.F08	CBC 4176	1.N05	CBC 4888	2.F05	CBC 5657	2.D04	CBC 6607	2.H03
CAC 7193	3.F02	CAC 8968 2	1.I05	CBC 1491	1.L06	CBC 2185	2.H08	CBC 4179	2.F05	CBC 4889	2.G08	CBC 5682 E	1.M07	CBC 6608	1.K05
CAC 7196	3.F02	CAC 9089	1.K05	CBC 1491	1.L07	CBC 2292	1.K02	CBC 4185	1.L07	CBC 4895	2.D04	CBC 5682 N	1.M07	CBC 6644 1	2.D07
CAC 7203	2.D09	CAC 9089	1.K06	CBC 1530	2.F12	CBC 2293	1.K02	CBC 4233	1.O07	CBC 4896	2.D02	CBC 5683	1.M07	CBC 6710	2.D02
CAC 7204	2.D09	CAC 9091 1	1.K10	CBC 1531	2.F12	CBC 2352	1.O05	CBC 4234	1.O07	CBC 4897	2.C18	CBC 5684	1.M07	CBC 6766	1.K07
CAC 7207	2.F16	CAC 9252	1.N18	CBC 1717 E	1.M07	CBC 2353	1.O05	CBC 4235	1.O07	CBC 4901	1.K13	CBC 5688	2.F08	CBC 6888	2.F16
CAC 7207	3.F02	CAC 9252	1.O03	CBC 1717 N	1.M07	CBC 2354	1.O07	CBC 4236	1.O07	CBC 4918	1.M12	CBC 5691	2.F08	CBC 6889	2.F16
CAC 7218	1.L05	CAC 9281 H	1.K18	CBC 1718	1.M07	CBC 2355	1.O07	CBC 4238	2.D07	CBC 4938	2.D02	CBC 5708 N	1.M07	CBC 7209	1.N07
CAC 7218	1.L06	CAC 9333	1.H04	CBC 1719 N	1.M07	CBC 2357	1.O05	CBC 4266	2.G12	CBC 4939	2.D02	CBC 5709	1.M07	CBC 7210	1.N07
CAC 7218	1.L07	CAC 9333	1.I04	CBC 1720 E	1.M06	CBC 2358	1.O05	CBC 4266	2.G15	CBC 4940	2.D02	CBC 5710	1.M07	CBC 7216	1.N08
CAC 7455	1.N06	CAC 9343	1.H04	CBC 1720 N	1.M06	CBC 2366	2.H09	CBC 4388	3.N05	CBC 4941	2.D02	CBC 5728	2.F06	CBC 7217	1.N08
CAC 7461	1.N06	CAC 9343	1.I04	CBC 1721	1.M06	CBC 2368	2.F17	CBC 4392	1.O07	CBC 4967	2.H11	CBC 5735	1.K02	CBC 7297	2.G04
CAC 7508	2.D09	CAC 9344	1.H04	CBC 1722	4.C04	CBC 2369	2.F17	CBC 4393	1.O05	CBC 5019	1.N11	CBC 5736	1.K02	CBC 7361	1.K03
CAC 7509	2.D07	CAC 9344	1.I04	CBC 1736	1.K07	CBC 2388	2.F13	CBC 4436	1.O08	CBC 5056	2.C14	CBC 5737	1.K08	CBC 7462	2.H03
CAC 7509	2.D09	CAC 9466	4.C03	CBC 1737	1.K07	CBC 2389	2.F13	CBC 4437	1.O08	CBC 5110	2.H09	CBC 5738	1.M03	CBC 7463	2.H03
CAC 7556	1.M05	CAC 9475	1.N06	CBC 1740	1.K11	CBC 2469	1.L07	CBC 4439	1.O08	CBC 5173	1.K05	CBC 5740	1.K05	CBC 7958	2.F11
CAC 7586	2.D09	CAC 9476	1.N06	CBC 1784	1.K11	CBC 2473	2.C12	CBC 4440	1.O08	CBC 5173	1.K06	CBC 5742	1.K10	CBC 7959	2.F11
CAC 7599	2.D09	CAC 9477	1.N06	CBC 1785	1.K07	CBC 2482	1.J05	CBC 4441	1.O08	CBC 5174	1.K05	CBC 5758	1.N07	CBC 7979	2.D02
CAC 7640	1.O08	CAC 9478	1.N06	CBC 1792 28	1.L02	CBC 2521	2.F05	CBC 4444	2.H03	CBC 5174	1.K06	CBC 5759	1.N07	CBC 8104	2.F06
CAC 7657	2.F08	CAC 9479	1.N06	CBC 1792 35	1.K15	CBC 2526	1.H06	CBC 4491	1.J03	CBC 5190	1.N02	CBC 5865	2.D15	CBC 8106	2.F06
CAC 7657	2.F16	CAC 9508	1.N06	CBC 1805	1.K04	CBC 2550	1.M10	CBC 4492	1.J03	CBC 5191	1.N02	CBC 5928	1.M08	CBC 8189	2.C18
CAC 7670	2.F06	CAC 9621 1	1.N06	CBC 1864	2.G03	CBC 2590 8	2.G13	CBC 4516	1.O08	CBC 5201	2.G12	CBC 5928	1.M09	CBC 8189	2.D02
CAC 7849	2.D14	CAC 9734	1.J03	CBC 1867	1.K07	CBC 2591 4	2.D14	CBC 4517	1.O08	CBC 5201	2.G15	CBC 5931	1.D08	CBC 8212	2.F11
CAC 7849	2.D16	CAC 9735	1.J03	CBC 1883	1.H04	CBC 2591 4	2.D16	CBC 4518	2.D05	CBC 5212	2.H14	CBC 5931	1.F09	CBC 8212	2.F12
CAC 7876	1.N06	CAC 9759	1.N08	CBC 1883	1.I04	CBC 2591 4	2.E03	CBC 4572 1	2.G15	CBC 5213	2.H14	CBC 5942	2.G09	CBC 8213	2.F11
CAC 7877	2.D02	CAC 9760	1.N08	CBC 1904 1	2.C13	CBC 2591 4	2.E04	CBC 4576 1	2.C13	CBC 5224	1.J05	CBC 5945	2.G09	CBC 8213	2.F12
CAC 7878	2.D02	CAC 9800	2.G13	CBC 1910 1	2.C17	CBC 2591 4	2.E12	CBC 4577 1	2.C16	CBC 5253	1.N07	CBC 5984	2.G09	CBC 8388	2.F15
CAC 7912	2.D03	CAC 9801	2.G13	CBC 1915 4	2.C14	CBC 2591 4	2.E13	CBC 4648 E	1.M06	CBC 5255	1.N07	CBC 6014	2.F06	CBP 221	3.C09
CAC 7912	2.H10	CAC 9816	1.D08	CBC 1925 1	2.C16	CBC 2591 4	2.E15	CBC 4648 N	1.M06	CBC 5305	1.L07	CBC 6027	2.D08	CBP 224	3.C03
CAC 7936	1.N10	CAC 9816	1.F09	CBC 1931	2.G08	CBC 2591 4	2.E16	CBC 4649	1.M06	CBC 5307	1.M12	CBC 6036	2.H10	CBP 224	3.C11
CAC 7996	1.H04	CAC 9937	1.K04	CBC 1948	2.G12	CBC 2591 4	2.E18	CBC 4652	1.M03	CBC 5320	2.F11	CBC 6037	2.H08	CBP 224	3.D04
CAC 8032	2.H06	CAC 9938	1.K03	CBC 1948	2.G15	CBC 2591 4	2.F02	CBC 4681	1.N07	CBC 5321	2.F11	CBC 6044	1.N05	CBP 303	3.D12
CAC 8050	2.E03	CAC 80881 1	2.H04	CBC 1955	2.H09	CBC 2600	1.N08	CBC 4704	1.K07	CBC 5324	2.F11	CBC 6072	2.D07	CBP 313	3.C03
CAC 8073	1.N17	CAM 1978 J	1.C05	CBC 1956	2.H09	CBC 2605	1.J05	CBC 4706	1.K11	CBC 5324	2.F12	CBC 6083	1.N04	CBP 313	3.C11
CAC 8091	1.N06	CBC 1262	1.L05	CBC 1958	2.G12	CBC 2616	1.M12	CBC 4721	1.H06	CBC 5325	2.F11	CBC 6083	1.N07	CBP 313	3.D04
CAC 8106	1.N06	CBC 1262	1.L07	CBC 1958	2.G15	CBC 2630	2.F06	CBC 4722	2.D05	CBC 5325	2.F12	CBC 6083	1.N09	CBP 314	3.C03
CAC 8127	1.I04	CBC 1318	1.H04	CBC 1976	2.C18	CBC 2650	1.O05	CBC 4743	2.C18	CBC 5328	2.F14	CBC 6140	1.N12	CBP 314	3.C11
CAC 8187 1	1.M05	CBC 1318	1.I04	CBC 1995	1.I06	CBC 2671	1.N18	CBC 4744	1.H04	CBC 5329	2.F14	CBC 6141	1.N12	CBP 314	3.D04
CAC 8285	2.D13	CBC 1319	1.H04	CBC 2046	1.O08	CBC 2690	1.F04	CBC 4744	1.I04	CBC 5336	1.K02	CBC 6142	1.N12	CBP 318	3.D12
CAC 8324	1.N06	CBC 1319	1.I04	CBC 2054	2.E09	CBC 2691	1.F04	CBC 4745	1.H04	CBC 5337	1.K02	CBC 6143	1.N12	CP 108101	1.I05
CAC 8341	1.H05	CBC 1320	1.H04	CBC 2072	2.F11	CBC 2692	1.F04	CBC 4745	1.I04	CBC 5355	2.C18	CBC 6174	1.O05	CP 108161 J	1.E17
CAC 8357	1.H06	CBC 1320	1.I04	CBC 2073	2.F11	CBC 2705	2.D12	CBC 4760 1	2.D06	CBC 5357	1.M12	CBC 6182	1.O05	CP 108161 J	2.E17
CAC 8471	1.N04	CBC 1321	1.H04	CBC 2074	2.F11	CBC 2731	1.M02	CBC 4762 1	2.D09	CBC 5361	2.F08	CBC 6183	1.O05	CR 120225 J	1.C08
CAC 8478	1.D07	CBC 1321	1.I04	CBC 2074	2.F12	CBC 2793	1.K10	CBC 4830	2.G02	CBC 5428	1.O05	CBC 6209	2.G04	CRC 1977	2.G08
CAC 8478	2.C03	CBC 1324	1.H04	CBC 2075	2.F11	CBC 2846	2.H08	CBC 4834	2.F06	CBC 5429	1.O05	CBC 6280	2.D13	CRC 1977	2.G09
CAC 8535	2.F06	CBC 1324	1.I04	CBC 2075	2.F12	CBC 2855	2.H04	CBC 4834	2.F10	CBC 5432	1.I06	CBC 6280	2.D17	CRC 2224	2.G10
CAC 8614	2.G14	CBC 1329	1.H04	CBC 2096	2.G12	CBC 2887	2.D07	CBC 4834	2.F11	CBC 5502	1.N04	CBC 6333	1.M02	CRC 4787 J	2.O11
CAC 8635	2.C18	CBC 1329	1.I04	CBC 2096	2.G15	CBC 2887	2.D09	CBC 4834	2.F13	CBC 5503	1.N04	CBC 6336	2.G09	CZG 4707 J	2.H14
CAC 8662	2.C18	CBC 1380	1.K04	CBC 2132	2.C12	CBC 4047	2.G05	CBC 4834	2.F14	CBC 5504	1.N04	CBC 6345	1.M02	DA 706 6S	3.F14
CAC 8666	1.O08	CBC 1382 1	2.E09	CBC 2141	1.N07	CBC 4048	2.G05	CBC 4838	2.G06	CBC 5505	1.N04	CBC 6345	2.M05	DA 81012 C	3.F13
CAC 8667	1.O08	CBC 1382 2	2.E09	CBC 2142	1.N07	CBC 4049	2.G02	CBC 4843	2.F05	CBC 5525 1	2.D06	CBC 6346	2.D05	DA 606044 J	3.I14
CAC 8767 25	2.M02	CBC 1390	1.N18	CBC 2158	4.C03	CBC 4060	2.F14	CBC 4843	2.F12	CBC 5542	2.F13	CBC 6416 1	1.K03	DA 606084	2.N16
CAC 8770	2.J12	CBC 1390	1.O05	CBC 2159	4.C03	CBC 4061	2.F14	CBC 4844	2.C12	CBC 5543	2.F13	CBC 6416 2	1.K03	DAB 6 6C	3.J08
CAC 8770	2.J13	CBC 1393	1.O06	CBC 2160	4.C03	CBC 4102	1.K05	CBC 4850	2.F11	CBC 5579	1.K02	CBC 6443	2.G16	DAB 6 6C	3.K06
CAC 8772	2.D07	CBC 1393	1.O07	CBC 2161	4.C03	CBC 4102	1.K06	CBC 4850	2.F12	CBC 5580	1.K02	CBC 6466	1.N02	DAB 6 8C	3.H02
CAC 8774	3.N13	CBC 1401	2.F06	CBC 2162	4.C03	CBC 4103	1.K05	CBC 4851	2.F11	CBC 5611	1.K06	CBC 6467	1.N02	DAB 006 6C	3.J18

JAGUAR XJS RANGE (JAN 1987 ON)

D14 Fiche 4

NUMERICAL INDEX DAB 006 6C TO DAC 5565

DAB 006 6C	3.K09	DAC 1996	2.M18	DAC 2650	2.O03	DAC 3244	2.K10	DAC 4118	2.L11	DAC 4388	2.M12	DAC 4611	2.K12	DAC 4839	2.N09
DAB 006 6C	3.K12	DAC 1996	2.N02	DAC 2650	2.O05	DAC 3247	2.J12	DAC 4118	2.L12	DAC 4389	2.M12	DAC 4613	2.M12	DAC 4840	2.N09
DAB 706 6C	3.H03	DAC 1996	2.N04	DAC 2654	2.O08	DAC 3255 3	2.N12	DAC 4131	2.O08	DAC 4403	2.M08	DAC 4613	2.M16	DAC 4841	2.N09
DAB 80810 C	3.F13	DAC 2005	2.L06	DAC 2660	2.K10	DAC 3269 N	2.J09	DAC 4151	1.M04	DAC 4404	2.N10	DAC 4619	2.O03	DAC 4842	2.N09
DAC 706 8C	3.F05	DAC 2007	2.L06	DAC 2662 1	2.M14	DAC 3351	1.O08	DAC 4154	1.M04	DAC 4405	2.D07	DAC 4619	2.O05	DAC 4849	2.O03
DAC 1018	2.O16	DAC 2035	2.O17	DAC 2662 2	2.M14	DAC 3386	2.L04	DAC 4168	2.J12	DAC 4408	2.N10	DAC 4628	2.O07	DAC 4849	2.O04
DAC 1027	2.L09	DAC 2044	2.L16	DAC 2662 8	2.M14	DAC 3407	2.L18	DAC 4179	2.J10	DAC 4418	2.N16	DAC 4628	2.O09	DAC 4859	2.L08
DAC 1027	2.M16	DAC 2050	2.O03	DAC 2662 9	2.M14	DAC 3408	2.L18	DAC 4192	1.H05	DAC 4423	2.M15	DAC 4629	2.O04	DAC 4862	2.M11
DAC 1027	2.M18	DAC 2050	2.O04	DAC 2663 2	2.M14	DAC 3416	2.N15	DAC 4192	1.J04	DAC 4434	2.G16	DAC 4629	2.O06	DAC 4864	2.L14
DAC 1027	2.N06	DAC 2050	2.O05	DAC 2667	1.D07	DAC 3416	2.N16	DAC 4197	2.J18	DAC 4435	2.M09	DAC 4630	2.O04	DAC 4870	2.M11
DAC 1028	2.M18	DAC 2050	2.O06	DAC 2667	2.L02	DAC 3436	2.L03	DAC 4214	2.O03	DAC 4436	2.M09	DAC 4630	2.O06	DAC 4876	2.O06
DAC 1028	2.N06	DAC 2081	2.O03	DAC 2667	3.F13	DAC 3437	2.L03	DAC 4214	2.O04	DAC 4438	2.M09	DAC 4639	2.O04	DAC 4878	2.N07
DAC 1043	2.O16	DAC 2081	2.O04	DAC 2689	2.O05	DAC 3438	2.L03	DAC 4241	2.K13	DAC 4452	2.M06	DAC 4639	2.O06	DAC 4886	2.O03
DAC 1054	1.M04	DAC 2081	2.O05	DAC 2689	2.O06	DAC 3439	2.L03	DAC 4242	2.K13	DAC 4453	2.O16	DAC 4651	2.I04	DAC 4886	2.O04
DAC 1185	2.K08	DAC 2081	2.O06	DAC 2691	2.J15	DAC 3475	2.G16	DAC 4244	2.L04	DAC 4460	2.O04	DAC 4653	2.N09	DAC 4888	2.L08
DAC 1185	2.K09	DAC 2084	2.O07	DAC 2693	2.J15	DAC 3489	2.O05	DAC 4244	2.L08	DAC 4478	2.L11	DAC 4660	2.M16	DAC 4889	2.M14
DAC 1185	2.K10	DAC 2084	2.O08	DAC 2695	2.K14	DAC 3499	2.O16	DAC 4246	2.O05	DAC 4478	2.L12	DAC 4661	2.M16	DAC 4890	2.M08
DAC 1226	2.O16	DAC 2090	2.K16	DAC 2701	2.M12	DAC 3516	2.N12	DAC 4246	2.O06	DAC 4480	2.O08	DAC 4664	2.M15	DAC 4891	2.M08
DAC 1234 3	2.L04	DAC 2095 1	2.M14	DAC 2711	2.M17	DAC 3518	2.N12	DAC 4254	2.N13	DAC 4480	2.O09	DAC 4665	2.M15	DAC 5051	2.M18
DAC 1234 3	2.L06	DAC 2096	2.L07	DAC 2712	2.O08	DAC 3541	2.O03	DAC 4259	2.G16	DAC 4481	2.O08	DAC 4693	2.L17	DAC 5053	2.N03
DAC 1234 3	2.L08	DAC 2097	2.O03	DAC 2760	2.M17	DAC 3541	2.O04	DAC 4270	2.O03	DAC 4481	2.O09	DAC 4693	2.L18	DAC 5061	2.E02
DAC 1235	2.O16	DAC 2097	2.O04	DAC 2804	2.G16	DAC 3541	2.O05	DAC 4271	2.O05	DAC 4484	2.L08	DAC 4694	2.L17	DAC 5064	2.M14
DAC 1242	2.O16	DAC 2097	2.O05	DAC 2846	2.N17	DAC 3541	2.O06	DAC 4272	2.O05	DAC 4485	2.L08	DAC 4694	2.L18	DAC 5065	2.M08
DAC 1308	1.N09	DAC 2097	2.O06	DAC 2851	2.L07	DAC 3551	2.O07	DAC 4273	2.O03	DAC 4486	2.O10	DAC 4695	2.L18	DAC 5073	2.M08
DAC 1328	2.M16	DAC 2106	2.M06	DAC 2852	1.M04	DAC 3551	2.O08	DAC 4274	2.O05	DAC 4506	2.N09	DAC 4696	2.L17	DAC 5076	2.M09
DAC 1329	2.M17	DAC 2106	2.M07	DAC 2854	2.O03	DAC 3567	2.O08	DAC 4277	2.O08	DAC 4519	2.M09	DAC 4696	2.L18	DAC 5077	2.M09
DAC 1345	2.O13	DAC 2128	2.M06	DAC 2854	2.O04	DAC 3567	2.O09	DAC 4278	2.O03	DAC 4522	2.M09	DAC 4697	2.L17	DAC 5092	2.N04
DAC 1355	2.M06	DAC 2128	2.M07	DAC 2854	2.O05	DAC 3575	2.O16	DAC 4279	2.O05	DAC 4523	2.M09	DAC 4697	2.L18	DAC 5096	2.N09
DAC 1355	2.M07	DAC 2129	2.M06	DAC 2854	2.O06	DAC 3579	2.N12	DAC 4280	2.O03	DAC 4524	2.L09	DAC 4699	2.L18	DAC 5205	2.N08
DAC 1373	2.K07	DAC 2129	2.M07	DAC 2866	2.M08	DAC 3588	2.N12	DAC 4281	2.O05	DAC 4528	2.K12	DAC 4727	2.O02	DAC 5224 N	2.J05
DAC 1391	2.O17	DAC 2136	3.F12	DAC 2876	2.L04	DAC 3636	2.N07	DAC 4282	2.O03	DAC 4532	2.L06	DAC 4729	2.J05	DAC 5295	2.O12
DAC 1394	2.O17	DAC 2137	3.F12	DAC 2876	2.L07	DAC 3650	2.O17	DAC 4285	2.O07	DAC 4533	2.L06	DAC 4731	2.J14	DAC 5316	2.M17
DAC 1480	4.C03	DAC 2163	1.H06	DAC 2876	2.L08	DAC 3673	3.E14	DAC 4285	2.O08	DAC 4536	2.M05	DAC 4732 1	2.L08	DAC 5317	2.M17
DAC 1562	2.M16	DAC 2163	1.J05	DAC 2877	2.L08	DAC 3674	2.O05	DAC 4286	2.O03	DAC 4537	2.M05	DAC 4732 2	2.L08	DAC 5320	2.M17
DAC 1571	2.M07	DAC 2173	2.J15	DAC 2882	1.M02	DAC 3674	2.O06	DAC 4286	2.O04	DAC 4541	2.O04	DAC 4732 3	2.L08	DAC 5322	2.L09
DAC 1594	2.O14	DAC 2181	2.L18	DAC 2898	2.L06	DAC 3718	2.O05	DAC 4287	2.O03	DAC 4543	2.L09	DAC 4734	2.L08	DAC 5482	2.O03
DAC 1633	2.K04	DAC 2182	2.L17	DAC 2899	2.L06	DAC 3718	2.O06	DAC 4287	2.O04	DAC 4545	2.N17	DAC 4737	1.E12	DAC 5482	2.O04
DAC 1673	2.O05	DAC 2182	2.L18	DAC 2903	2.L07	DAC 3719	2.O12	DAC 4288	2.O03	DAC 4550	2.O05	DAC 4737	1.E13	DAC 5492	2.O06
DAC 1673	2.O06	DAC 2237	2.N15	DAC 2926	2.L17	DAC 3726	2.N12	DAC 4288	2.O05	DAC 4550	2.O06	DAC 4737	1.E14	DAC 5493	2.O04
DAC 1676	2.O16	DAC 2432	2.L06	DAC 2926	2.L18	DAC 3796	2.M17	DAC 4289	2.O03	DAC 4552	2.O05	DAC 4750	2.L11	DAC 5500	2.D06
DAC 1723	2.O16	DAC 2488	2.L06	DAC 2937	2.L17	DAC 3797	2.M12	DAC 4289	2.O05	DAC 4552	2.O06	DAC 4754	2.O02	DAC 5504	2.L02
DAC 1731	2.M16	DAC 2500	2.K05	DAC 2940	2.L06	DAC 3800	2.K11	DAC 4292	2.N08	DAC 4559	2.M09	DAC 4758	2.J13	DAC 5505	2.L02
DAC 1750	2.L18	DAC 2501	2.K05	DAC 2941	2.L06	DAC 3801	2.K11	DAC 4293	2.O14	DAC 4569	2.M03	DAC 4772	2.L08	DAC 5511	2.G16
DAC 1762	2.M13	DAC 2507	2.K05	DAC 2945	2.J15	DAC 3802	2.K11	DAC 4295	2.M06	DAC 4579	2.L17	DAC 4773	2.L08	DAC 5517	2.N02
DAC 1820	2.N12	DAC 2508	2.K05	DAC 2963	2.L04	DAC 3803	2.K11	DAC 4296	2.M06	DAC 4579	2.L18	DAC 4775	1.M04	DAC 5525	2.D09
DAC 1821	2.M13	DAC 2583	1.C18	DAC 2963	2.L06	DAC 3815	4.C11	DAC 4302	2.K12	DAC 4585	2.L11	DAC 4777	2.K08	DAC 5531	2.L18
DAC 1836	2.M08	DAC 2583	1.E12	DAC 2963	2.L08	DAC 3844	1.H05	DAC 4303	2.K12	DAC 4585	2.L12	DAC 4777	2.K09	DAC 5550	2.O04
DAC 1847 5	2.L06	DAC 2583	1.E13	DAC 2993	2.L06	DAC 3844	1.J04	DAC 4307	2.M14	DAC 4586	2.L11	DAC 4777	2.K10	DAC 5550	2.O06
DAC 1855	2.K14	DAC 2583	1.E14	DAC 3013	2.N15	DAC 3962	2.K12	DAC 4310	2.G16	DAC 4586	2.L12	DAC 4781	1.M04	DAC 5551	2.O04
DAC 1862	2.K06	DAC 2583	1.E15	DAC 3046	3.E07	DAC 3966	2.K12	DAC 4311	2.G16	DAC 4591	2.L14	DAC 4792	2.L08	DAC 5551	2.O06
DAC 1908	2.M12	DAC 2589	2.L15	DAC 3047	3.E07	DAC 3978	2.N05	DAC 4329	2.J12	DAC 4592	3.L02	DAC 4793	2.L08	DAC 5552	2.O06
DAC 1910	2.M12	DAC 2592	2.N06	DAC 3113	2.L15	DAC 4062	2.J12	DAC 4334 N	2.J09	DAC 4593	3.L02	DAC 4795	2.L07	DAC 5553	2.O06
DAC 1911	2.N11	DAC 2593	2.N06	DAC 3119	2.L07	DAC 4062	2.J13	DAC 4368	2.L07	DAC 4594	2.O04	DAC 4798	2.L06	DAC 5556	2.L08
DAC 1915	2.O03	DAC 2621	2.M14	DAC 3120	2.O05	DAC 4063	2.J12	DAC 4372	2.O05	DAC 4595	2.O03	DAC 4799	2.L06	DAC 5559	2.O06
DAC 1915	2.O04	DAC 2641	2.K17	DAC 3120	2.O06	DAC 4066	2.N13	DAC 4377	2.M09	DAC 4595	2.O05	DAC 4808	2.L04	DAC 5560	2.O06
DAC 1915	2.O05	DAC 2644	2.K17	DAC 3152	2.L17	DAC 4087 2	2.J12	DAC 4379	2.J12	DAC 4595	2.O06	DAC 4816	2.K12	DAC 5561	2.O06
DAC 1915	2.O06	DAC 2646	2.K06	DAC 3152	2.L18	DAC 4087 2	2.J13	DAC 4380	2.M17	DAC 4606	1.D18	DAC 4820	2.K07	DAC 5562	2.O04
DAC 1994	2.E09	DAC 2647	2.K06	DAC 3170	1.M04	DAC 4093	2.O08	DAC 4381	2.M17	DAC 4606	1.G04	DAC 4834	3.E14	DAC 5563	2.O04
DAC 1995	2.O15	DAC 2648	3.G02	DAC 3244	2.K08	DAC 4094	2.O08	DAC 4384	2.O05	DAC 4607	2.J16	DAC 4834	3.E15	DAC 5564	2.O09
DAC 1996	2.L09	DAC 2649	3.G02	DAC 3244	2.K09	DAC 4104	2.J15	DAC 4385	2.O03	DAC 4608	2.J16	DAC 4838	2.L09	DAC 5565	2.O07

JAGUAR XJS RANGE (JAN 1987 ON)

D15 Fiche 4

NUMERICAL INDEX DAC 5565 TO EAC 4271

DAC 5565	2.O09	DAC 41012 C	3.H12	DBC 2972	2.N04	DGC 2494	2.M15	EAC 1726	1.I15	EAC 3032	2.E07	EAC 3215 7	2.G15	EAC 3954 3	2.E16
DAC 5566	2.O09	DAC 41012 C	3.H13	DBC 2973	2.M18	DGC 2494	2.M17	EAC 1729	1.D18	EAC 3038	2.I02	EAC 3215 8	1.C17	EAC 3954 3	2.E17
DAC 5567	2.O06	DAZ 406 8C	3.H03	DBC 2973	2.N02	DGC 2494	2.M18	EAC 1828	2.D18	EAC 3041	2.I05	EAC 3215 8	1.E11	EAC 3964	1.C03
DAC 5568	2.O06	DAZ 406 8C	3.H09	DBC 2973	2.N03	DGC 2494	2.N02	EAC 1828	2.E02	EAC 3042	2.I05	EAC 3215 8	2.C04	EAC 3964	1.C04
DAC 5569	2.O04	DAZ 406 8C	3.H11	DBC 2973	2.N04	DGC 2494	2.N03	EAC 1864	1.C03	EAC 3047	2.E06	EAC 3215 8	2.C09	EAC 3964	1.E08
DAC 5570	2.O04	DAZ 406 8C	3.H12	DBC 3047	2.O11	DGC 2494	2.N04	EAC 1884	1.J03	EAC 3047	2.E07	EAC 3215 8	2.C11	EAC 3968	1.M11
DAC 5574	2.N14	DAZ 406 8C	3.H13	DBC 3048	2.O11	DGC 2495	2.M15	EAC 1896	1.H03	EAC 3051	1.E17	EAC 3215 8	2.D12	EAC 3970	1.M11
DAC 5575	2.N14	DAZ 408 8C	1.M04	DBC 3056	2.O11	DGC 2495	2.M17	EAC 1942	2.E08	EAC 3051	1.E18	EAC 3215 8	2.E11	EAC 3972	1.C03
DAC 5580	2.O09	DAZ 408 8C	2.C16	DBC 3060	2.M17	DGC 2495	2.M18	EAC 1954	1.E03	EAC 3090	2.E11	EAC 3215 9	2.C04	EAC 3972	1.C04
DAC 5583	2.J17	DAZ 408 8C	2.C17	DBC 3060	2.M18	DGC 2495	2.N02	EAC 1955	1.E03	EAC 3094	1.I15	EAC 3215 9	2.L08	EAC 3977	1.C10
DAC 5591	2.O04	DAZ 408 8C	2.O14	DBC 3060	2.N02	DGC 2495	2.N03	EAC 2127	2.J07	EAC 3099	1.E17	EAC 3234	1.D12	EAC 4012	2.E13
DAC 5591	2.O06	DAZ 810 8C	2.F08	DBC 3060	2.N03	DGC 2495	2.N04	EAC 2206	2.C11	EAC 3099	1.E18	EAC 3265	2.I02	EAC 4013	2.E13
DAC 5592	2.O09	DAZ 810 8C	2.F16	DBC 3060	2.N04	DGC 2496	2.M15	EAC 2249 2	2.E13	EAC 3112	2.D04	EAC 3385	2.J06	EAC 4017	1.E12
DAC 5594	2.O09	DAZ 810 8C	2.G07	DBC 3085	2.M15	DGC 2496	2.M17	EAC 2258	1.C04	EAC 3164	2.E02	EAC 3389	2.E10	EAC 4018	1.E16
DAC 5608	2.N17	DAZ 810 8C	2.H15	DBC 3092	1.N09	DGC 2496	2.M18	EAC 2273	2.E11	EAC 3167	1.M13	EAC 3419	2.E12	EAC 4025	2.D15
DAC 5863	2.N09	DAZ 810 8C	3.D17	DBC 3102	2.J03	DGC 2496	2.N02	EAC 2414	2.E03	EAC 3170	1.M13	EAC 3437	2.I02	EAC 4025	2.E13
DAC 5869	2.L13	DAZ 810 8L	2.M10	DBC 3229	2.J17	DGC 2496	2.N03	EAC 2415	2.E03	EAC 3171	1.M13	EAC 3438	2.I02	EAC 4025	2.E14
DAC 5870	2.L13	DBB 6 4C	3.H04	DBC 3230	2.J03	DGC 2496	2.N04	EAC 2428	1.C15	EAC 3178	2.E06	EAC 3454	1.C17	EAC 4025	2.E15
DAC 5871	2.L13	DBB 6 6C	3.H02	DBC 3230	2.J05	DGC 2497	2.M15	EAC 2456	1.C05	EAC 3178	2.E07	EAC 3455	1.C08	EAC 4025	2.E16
DAC 5881	2.N10	DBB 806 4C	3.H04	DBC 3231	2.J03	DGC 2497	2.M17	EAC 2464	1.C05	EAC 3179	2.E10	EAC 3488	2.E13	EAC 4025	2.E17
DAC 5884	2.N07	DBB 808 8C	2.F09	DBC 3231	2.J05	DGC 2497	2.M18	EAC 2494	1.C03	EAC 3185	1.I15	EAC 3489	1.E07	EAC 4025	2.F02
DAC 5895	2.O06	DBB 808 8C	3.H02	DBC 3232	2.J03	DGC 2497	2.N02	EAC 2495	1.C03	EAC 3191	1.E03	EAC 3569	1.E07	EAC 4029	2.E05
DAC 5931	2.N14	DBB 80810 C	3.N05	DBC 3232	2.J05	DGC 2497	2.N03	EAC 2510	1.E11	EAC 3192	1.E03	EAC 3610	2.E06	EAC 4038	1.C08
DAC 5933	2.N14	DBC 1140	2.J14	DBC 3710	2.L10	DGC 2497	2.N04	EAC 2510	2.G02	EAC 3195	1.E11	EAC 3610	2.E07	EAC 4048	1.I17
DAC 5936	2.N14	DBC 1167	2.O11	DBC 3711	2.L10	DGC 2498	2.M15	EAC 2522	1.C03	EAC 3212	1.E10	EAC 3610 1	2.E06	EAC 4049	1.C11
DAC 5937	2.N14	DBC 1168 1	2.O11	DBC 3722	2.O11	DGC 2498	2.M17	EAC 2522	1.C08	EAC 3215 1	1.C15	EAC 3610 1	2.E07	EAC 4056	1.C05
DAC 5942	2.L08	DBC 1169	2.L10	DBC 3778	2.M18	DGC 2498	2.M18	EAC 2522	1.C10	EAC 3215 1	1.C17	EAC 3610 2	2.E06	EAC 4057	2.E02
DAC 5943	2.O16	DBC 1626	1.C11	DBC 3778	2.N02	DGC 2498	2.N02	EAC 2547	1.G05	EAC 3215 1	1.E08	EAC 3610 2	2.E07	EAC 4061	1.I13
DAC 5955	2.O16	DBC 2000	2.J14	DBC 3778	2.N03	DGC 2498	2.N03	EAC 2556	2.E05	EAC 3215 1	1.E09	EAC 3634	2.E11	EAC 4069	2.E13
DAC 5956	1.M04	DBC 2013	3.F13	DBC 3778	2.N04	DGC 2498	2.N04	EAC 2577	1.C07	EAC 3215 1	2.D09	EAC 3636	1.E03	EAC 4069	2.E15
DAC 5962	2.N16	DBC 2022	2.M13	DBC 4410	2.L10	DGC 2499	2.M15	EAC 2630	1.C18	EAC 3215 10	1.C18	EAC 3637	1.E03	EAC 4069	2.E16
DAC 5963	2.N16	DBC 2134	1.N05	DBC 4411	2.L10	DGC 2499	2.M17	EAC 2630	1.E16	EAC 3215 11	1.E11	EAC 3639	1.E03	EAC 4069	2.E18
DAC 5964	2.E09	DBC 2139	1.C10	DBZ 804 4C	3.H04	DGC 2499	2.M18	EAC 2630	2.E12	EAC 3215 13	1.C17	EAC 3641	1.E03	EAC 4069	2.F02
DAC 5966	2.N16	DBC 2145	2.G02	DBZ 812 10C	3.E11	DGC 2499	2.N02	EAC 2632	1.C15	EAC 3215 13	2.C02	EAC 3653	1.E04	EAC 4082	2.E15
DAC 5998	1.D18	DBC 2175	2.O03	DBZ 11010 C	3.D17	DGC 2499	2.N03	EAC 2644	2.E03	EAC 3215 13	2.D07	EAC 3671	1.C15	EAC 4082	2.E16
DAC 5998	1.G04	DBC 2182	2.O11	DBZ 80812 C	2.D02	DGC 2499	2.N10	EAC 2648	2.E08	EAC 3215 13	2.D09	EAC 3701	1.D14	EAC 4082	2.E17
DAC 6039	2.M12	DBC 2200	2.N12	DGC 2490	2.M15	DJZ 810 10C	2.O14	EAC 2650	1.E17	EAC 3215 14	2.D07	EAC 3714	2.F04	EAC 4083	1.E14
DAC 6040	2.J17	DBC 2483	2.M16	DGC 2490	2.M17	DRC 1449	2.G02	EAC 2650	1.E18	EAC 3215 14	2.D09	EAC 3762	1.E12	EAC 4083	1.E15
DAC 6049	2.E08	DBC 2483	2.N04	DGC 2490	2.M18	DRC 1449	2.G10	EAC 2654	2.E11	EAC 3215 15	2.C03	EAC 3765	2.J02	EAC 4083	2.E15
DAC 6053	2.M17	DBC 2483	2.N08	DGC 2490	2.N02	DRC 1449 J	2.C16	EAC 2655	2.E11	EAC 3215 2	2.C08	EAC 3765	2.J04	EAC 4083	2.E16
DAC 6072	2.O06	DBC 2484	2.M18	DGC 2490	2.N03	DRC 1449 J	3.E08	EAC 2658	2.E08	EAC 3215 2	2.D09	EAC 3793	1.C07	EAC 4083	2.E17
DAC 6074	2.O06	DBC 2484	2.N02	DGC 2490	2.N04	EAC 1027	1.C06	EAC 2670	2.E06	EAC 3215 2	2.E02	EAC 3803	1.C18	EAC 4084	1.E14
DAC 6103	2.D07	DBC 2484	2.N03	DGC 2491	2.M15	EAC 1133	2.D18	EAC 2732	1.C06	EAC 3215 2	2.E04	EAC 3810	1.C05	EAC 4096 1	2.E02
DAC 6134	1.N08	DBC 2484	2.N04	DGC 2491	2.M17	EAC 1133	2.E02	EAC 2757	1.D13	EAC 3215 2	2.E10	EAC 3823	1.D03	EAC 4100	2.E15
DAC 6148	2.L03	DBC 2484	2.N05	DGC 2491	2.M18	EAC 1301	2.C03	EAC 2815	2.E06	EAC 3215 2	2.G05	EAC 3823	1.F02	EAC 4100	2.E18
DAC 6149	2.L03	DBC 2484	2.N07	DGC 2491	2.N02	EAC 1323	1.F03	EAC 2863	2.C03	EAC 3215 2	2.G12	EAC 3827	2.E15	EAC 4119	1.C03
DAC 6150	2.L03	DBC 2632	2.M02	DGC 2491	2.N03	EAC 1380	1.F06	EAC 2863	2.D18	EAC 3215 2	2.G15	EAC 3827	2.E16	EAC 4123	2.E13
DAC 6204	2.L18	DBC 2640	2.L05	DGC 2491	2.N04	EAC 1381	1.F06	EAC 2899	1.E17	EAC 3215 20	2.C03	EAC 3827	2.E17	EAC 4123	2.E14
DAC 6221	2.M17	DBC 2641	2.C07	DGC 2492	2.M15	EAC 1424	1.I15	EAC 2909	1.C14	EAC 3215 5	2.D05	EAC 3833	2.F03	EAC 4123	2.E15
DAC 6222	2.M17	DBC 2690	2.J11	DGC 2492	2.M17	EAC 1425	1.I15	EAC 2911	1.C11	EAC 3215 5	4.C10	EAC 3850	1.D05	EAC 4123	2.E18
DAC 6337	2.L11	DBC 2735	2.L10	DGC 2492	2.M18	EAC 1426	1.I15	EAC 2912	1.C11	EAC 3215 6	2.C04	EAC 3864	2.H02	EAC 4126	1.C18
DAC 6337	2.L12	DBC 2910	2.L10	DGC 2492	2.N02	EAC 1427	1.I15	EAC 2928 4	2.G12	EAC 3215 6	2.C08	EAC 3866	2.H02	EAC 4126	1.E11
DAC 6338	2.L11	DBC 2911	2.L10	DGC 2492	2.N03	EAC 1428	1.I15	EAC 2928 4	2.G15	EAC 3215 6	2.E15	EAC 3867	2.H02	EAC 4161	2.E02
DAC 6338	2.L12	DBC 2937 N	2.J08	DGC 2492	2.N04	EAC 1449	1.F09	EAC 2928 7	2.G08	EAC 3215 6	2.E16	EAC 3878	1.D05	EAC 4175	1.C11
DAC 6347	2.L02	DBC 2960	1.N09	DGC 2493	2.M15	EAC 1450	1.F09	EAC 2928 7	2.G09	EAC 3215 6	2.E17	EAC 3909	2.F03	EAC 4181	2.J07
DAC 6348	2.L02	DBC 2967	2.M15	DGC 2493	2.M17	EAC 1605	1.F03	EAC 2929	2.E06	EAC 3215 6	2.F04	EAC 3921	1.C08	EAC 4185	2.I05
DAC 6556	2.L18	DBC 2972	2.M17	DGC 2493	2.M18	EAC 1616	1.F03	EAC 2929	2.E07	EAC 3215 7	1.C17	EAC 3926	1.C10	EAC 4191	1.E11
DAC 6557	2.L17	DBC 2972	2.M18	DGC 2493	2.N02	EAC 1620	1.C06	EAC 2946	2.D18	EAC 3215 7	2.C08	EAC 3927	1.C18	EAC 4192	1.E11
DAC 6557	2.L18	DBC 2972	2.N02	DGC 2493	2.N03	EAC 1621	2.F04	EAC 2978	1.J03	EAC 3215 7	2.E10	EAC 3927	1.E16	EAC 4262	2.D18
DAC 6558	2.L18	DBC 2972	2.N03	DGC 2493	2.N04	EAC 1675	1.J03	EAC 3032	2.E06	EAC 3215 7	2.G12	EAC 3954 3	2.E15	EAC 4271	2.E02

JAGUAR XJS RANGE (JAN 1987 ON) — D16 Fiche 4 — NUMERICAL INDEX EAC 4326 TO EAC 9286

EAC 4326	1.C07	EAC 4989	1.C11	EAC 5063	1.C18	EAC 5944	2.C11	EAC 6439	1.H07	EAC 7243	1.C16	EAC 7921	2.E08	EAC 8195	1.H02
EAC 4327	1.C07	EAC 5004	1.C15	EAC 5070	2.C03	EAC 5945	2.C11	EAC 6447	1.C16	EAC 7246	1.C16	EAC 7922	2.C04	EAC 8209	1.C18
EAC 4328	1.C07	EAC 5013 10	1.C12	EAC 5086	2.E03	EAC 5958	2.C10	EAC 6447	1.E10	EAC 7251	1.D12	EAC 7934	2.E13	EAC 8264	2.C03
EAC 4329 1	1.C07	EAC 5013 11	1.C12	EAC 5112	1.E03	EAC 5961	2.C10	EAC 6523	1.C15	EAC 7252	1.D12	EAC 7935	2.E15	EAC 8272	2.C09
EAC 4329 2	1.C07	EAC 5013 12	1.C12	EAC 5113	1.E03	EAC 6001	1.C10	EAC 6560	2.C04	EAC 7255	1.C14	EAC 7935	2.E18	EAC 8275	2.E03
EAC 4329 3	1.C07	EAC 5013 13	1.C12	EAC 5157	2.E15	EAC 6017	1.C07	EAC 6601	2.H02	EAC 7256	1.C14	EAC 7954	1.C10	EAC 8295	2.J17
EAC 4371	2.E13	EAC 5013 14	1.C12	EAC 5157	2.E16	EAC 6023	1.H02	EAC 6620	3.E10	EAC 7256	1.E03	EAC 7965	2.E04	EAC 8307	2.C04
EAC 4371	2.E14	EAC 5013 15	1.C12	EAC 5157	2.E18	EAC 6024	1.H03	EAC 6642	2.H02	EAC 7261	1.D03	EAC 7965	2.E13	EAC 8313	1.J03
EAC 4378	1.D04	EAC 5013 16	1.C12	EAC 5157	2.F02	EAC 6025	1.H03	EAC 6651	1.C16	EAC 7263	1.C08	EAC 7965	2.E14	EAC 8367	1.C08
EAC 4380	1.C07	EAC 5013 17	1.C13	EAC 5164	1.C04	EAC 6026	1.H03	EAC 6741	1.C08	EAC 7263	1.C10	EAC 7968	2.E11	EAC 8415	2.E06
EAC 4431	1.C06	EAC 5013 18	1.C13	EAC 5192	2.C09	EAC 6027	1.H03	EAC 6744	1.C09	EAC 7264	1.D03	EAC 7972	2.E04	EAC 8418	1.C12
EAC 4438	2.E11	EAC 5013 19	1.C13	EAC 5221	1.C04	EAC 6063	1.C18	EAC 6766	1.C08	EAC 7275	1.D03	EAC 7973	2.E03	EAC 8439 N	1.I08
EAC 4451	1.C03	EAC 5013 2	1.C12	EAC 5226	2.E05	EAC 6089	1.G06	EAC 6779	1.C08	EAC 7276	1.D03	EAC 7982	2.E04	EAC 8447	2.C02
EAC 4504	1.C07	EAC 5013 20	1.C13	EAC 5259	1.H03	EAC 6106	1.C17	EAC 6779	1.C09	EAC 7291	1.C05	EAC 7983	2.E04	EAC 8448	2.C02
EAC 4505	1.C07	EAC 5013 21	1.C13	EAC 5261	1.H03	EAC 6123	1.C17	EAC 6783	1.C09	EAC 7301	2.H02	EAC 7991	2.C08	EAC 8523	1.C12
EAC 4512	1.C15	EAC 5013 22	1.C13	EAC 5268	1.H03	EAC 6124	1.C17	EAC 6789	1.E08	EAC 7309	1.E03	EAC 8011	2.C06	EAC 8525	1.C12
EAC 4519	1.C08	EAC 5013 23	1.C13	EAC 5271	1.H03	EAC 6140	1.C18	EAC 6790	1.F05	EAC 7318	1.E03	EAC 8021 18	2.C11	EAC 8554	2.J18
EAC 4520	1.C09	EAC 5013 24	1.C13	EAC 5277	1.F02	EAC 6142	1.C18	EAC 6800	1.F05	EAC 7320	1.I10	EAC 8021 18	2.D15	EAC 8554	2.K02
EAC 4527	1.C08	EAC 5013 25	1.C13	EAC 5308	1.D03	EAC 6156	1.C08	EAC 6811	1.C10	EAC 7329	2.C06	EAC 8021 18	2.E13	EAC 8595	1.G05
EAC 4528	1.C08	EAC 5013 26	1.C13	EAC 5310	1.G06	EAC 6191	2.C11	EAC 6813	1.C18	EAC 7339	1.C17	EAC 8021 18	2.E14	EAC 8616	2.C07
EAC 4531	1.C08	EAC 5013 27	1.C13	EAC 5317	1.C08	EAC 6215	2.C03	EAC 6815	1.E17	EAC 7340	2.C04	EAC 8021 18	2.E18	EAC 8619	1.E17
EAC 4532	1.C08	EAC 5013 28	1.C13	EAC 5338	1.G06	EAC 6252 PA	1.D14	EAC 6815	1.E18	EAC 7341	2.C06	EAC 8021 30	2.E15	EAC 8730	1.H02
EAC 4533	1.C08	EAC 5013 29	1.C13	EAC 5371	1.H08	EAC 6252 PB	1.D14	EAC 6816	1.E17	EAC 7343	1.C17	EAC 8022 30	2.C05	EAC 8744 20	1.C06
EAC 4534	1.C09	EAC 5013 3	1.C12	EAC 5378	1.G06	EAC 6253 PA	1.D14	EAC 6816	1.E18	EAC 7353	1.C04	EAC 8022 30	2.C06	EAC 8744 S	1.C06
EAC 4535	1.C09	EAC 5013 30	1.C13	EAC 5405	1.G06	EAC 6253 PB	1.D14	EAC 6818	1.E17	EAC 7355	1.C07	EAC 8022 30	2.C11	EAC 8745 S	1.C06
EAC 4536	1.C09	EAC 5013 31	1.C13	EAC 5410	2.C06	EAC 6254	2.C04	EAC 6818	1.E18	EAC 7398	1.C07	EAC 8022 30	2.E03	EAC 8746 20	1.C06
EAC 4538	1.C09	EAC 5013 32	1.C13	EAC 5438	2.C10	EAC 6255	2.C04	EAC 6819	1.E17	EAC 7407	1.C15	EAC 8022 30	2.E12	EAC 8778	2.C07
EAC 4540	1.C09	EAC 5013 33	1.C13	EAC 5442	1.C16	EAC 6263	2.C04	EAC 6819	1.E18	EAC 7418	2.C08	EAC 8022 30	2.E13	EAC 8809	2.C09
EAC 4542	2.J06	EAC 5013 34	1.C13	EAC 5513	1.C17	EAC 6266	2.C03	EAC 6821	1.E12	EAC 7428	1.C04	EAC 8022 30	2.E14	EAC 8810	2.C09
EAC 4557	1.C09	EAC 5013 35	1.C13	EAC 5536	1.C07	EAC 6266	2.C06	EAC 6821	1.E13	EAC 7436	1.C04	EAC 8022 30	2.E15	EAC 8811	2.C09
EAC 4579	2.C05	EAC 5013 36	1.C13	EAC 5541	1.C06	EAC 6268	1.H03	EAC 6821	1.E14	EAC 7441	1.C15	EAC 8022 30	2.E16	EAC 8839	2.F03
EAC 4604	1.G03	EAC 5013 37	1.C13	EAC 5541	1.D14	EAC 6278	2.C05	EAC 6821	1.E15	EAC 7443	1.C15	EAC 8022 30	2.E17	EAC 8841	1.D17
EAC 4619	1.G02	EAC 5013 38	1.C13	EAC 5544	1.J03	EAC 6305	1.C16	EAC 6850	1.C05	EAC 7447	2.C08	EAC 8022 30	2.E18	EAC 8858	2.F03
EAC 4620	1.G02	EAC 5013 39	1.C13	EAC 5557	1.G06	EAC 6305	1.E10	EAC 6878	1.C08	EAC 7478	2.C08	EAC 8022 30	2.F02	EAC 8858	2.I05
EAC 4644	1.C17	EAC 5013 4	1.C12	EAC 5580	1.G02	EAC 6314	2.C05	EAC 6923	1.H08	EAC 7504	1.C11	EAC 8022 30	2.H13	EAC 8926	1.D12
EAC 4647	1.C08	EAC 5013 40	1.C14	EAC 5580	1.G03	EAC 6314	2.E04	EAC 6962	2.C03	EAC 7508	2.C10	EAC 8036 5	2.E11	EAC 8927	1.D13
EAC 4650	1.C09	EAC 5013 41	1.C14	EAC 5599	2.E05	EAC 6335	1.D05	EAC 6995	2.C05	EAC 7544	1.D02	EAC 8043	2.E04	EAC 8930	2.E03
EAC 4674	2.J11	EAC 5013 42	1.C14	EAC 5604	2.C04	EAC 6337	1.E08	EAC 6996	2.C05	EAC 7557	2.C08	EAC 8055	2.E04	EAC 8950	2.C09
EAC 4691	1.D02	EAC 5013 43	1.C14	EAC 5620	2.C02	EAC 6337	1.E09	EAC 7002	1.C03	EAC 7561	2.C03	EAC 8065	1.I05	EAC 8951	2.C09
EAC 4709	1.C14	EAC 5013 44	1.C14	EAC 5625	1.I05	EAC 6370	2.C03	EAC 7005	1.C05	EAC 7642	1.C09	EAC 8088	2.E04	EAC 8954	1.F05
EAC 4714	2.C02	EAC 5013 45	1.C14	EAC 5659	1.C10	EAC 6382	1.D02	EAC 7019	1.C17	EAC 7643	1.C08	EAC 8097	2.I02	EAC 8956	1.F05
EAC 4744	1.C11	EAC 5013 46	1.C14	EAC 5674	1.C07	EAC 6398	1.E08	EAC 7033	1.D05	EAC 7649	1.J03	EAC 8101	1.D12	EAC 8958	1.C04
EAC 4751	2.I02	EAC 5013 47	1.C14	EAC 5681	1.H02	EAC 6400	1.E08	EAC 7047	1.E16	EAC 7650	1.J03	EAC 8101	1.E08	EAC 8963	2.C02
EAC 4808	1.C04	EAC 5013 48	1.C14	EAC 5683	1.H02	EAC 6400	1.E09	EAC 7048	1.E12	EAC 7651	1.G06	EAC 8101	1.E09	EAC 8989	1.D02
EAC 4820	2.C06	EAC 5013 5	1.C12	EAC 5706	1.J03	EAC 6401	1.E08	EAC 7048	1.E13	EAC 7657	1.H03	EAC 8102	1.D16	EAC 8994	1.D03
EAC 4834	2.C06	EAC 5013 6	1.C12	EAC 5721	1.C10	EAC 6401	1.E09	EAC 7048	1.E14	EAC 7730	1.C05	EAC 8102	1.D17	EAC 8995	1.D03
EAC 4840	2.C02	EAC 5013 7	1.C12	EAC 5751	1.H03	EAC 6402	1.E08	EAC 7048	1.E15	EAC 7755	1.E09	EAC 8102	1.D18	EAC 8997	2.C08
EAC 4861	1.D02	EAC 5013 8	1.C12	EAC 5752	1.H03	EAC 6404	1.E08	EAC 7067	1.D04	EAC 7759	1.D02	EAC 8102	1.E08	EAC 8999	1.C18
EAC 4864	2.E04	EAC 5013 9	1.C12	EAC 5753	1.H03	EAC 6413	1.F05	EAC 7080	1.C08	EAC 7769	1.C03	EAC 8102	1.E09	EAC 9172	1.C16
EAC 4867	2.C05	EAC 5032	1.C14	EAC 5838	1.D02	EAC 6414	1.F05	EAC 7081	1.M11	EAC 7774	1.C11	EAC 8106	1.H18	EAC 9186	2.J18
EAC 4886	1.C15	EAC 5032	1.E03	EAC 5864	2.C02	EAC 6419	1.F05	EAC 7107	1.M11	EAC 7786	2.H02	EAC 8114	2.E06	EAC 9186	2.K02
EAC 4888	2.C07	EAC 5038	2.G03	EAC 5869	1.D02	EAC 6420	1.D12	EAC 7149	1.E08	EAC 7874	2.E03	EAC 8114	2.E07	EAC 9211	1.C16
EAC 4897	2.C10	EAC 5040	2.C11	EAC 5875	1.C14	EAC 6421	1.D12	EAC 7155	2.C08	EAC 7875	2.E03	EAC 8122	2.E04	EAC 9248	1.D13
EAC 4898	2.C10	EAC 5042	2.C11	EAC 5896	2.C11	EAC 6422	1.D17	EAC 7189	1.C11	EAC 7876	2.E03	EAC 8130	2.C02	EAC 9270	2.E12
EAC 4902	1.H03	EAC 5044	2.E06	EAC 5901	2.C05	EAC 6423	1.D17	EAC 7220	1.C16	EAC 7877	2.E04	EAC 8162	2.C07	EAC 9276	1.C04
EAC 4927	1.C15	EAC 5044	2.E07	EAC 5910	1.J03	EAC 6423	1.F05	EAC 7222	1.C17	EAC 7904	2.E13	EAC 8164	2.C07	EAC 9286	2.E03
EAC 4927	2.C11	EAC 5045	1.C11	EAC 5911	2.C09	EAC 6424	1.D17	EAC 7224	1.C17	EAC 7904	2.E14	EAC 8166	2.C07	EAC 9286	2.E13
EAC 4929	1.C15	EAC 5045	2.J17	EAC 5939	2.C11	EAC 6434	1.C15	EAC 7227	1.C14	EAC 7905	2.C11	EAC 8167	2.C07	EAC 9286	2.E15
EAC 4929	2.C11	EAC 5046	1.C11	EAC 5940	2.C11	EAC 6435	1.C11	EAC 7236	1.C16	EAC 7920	2.E06	EAC 8183	1.I05	EAC 9286	2.E16
EAC 4974	1.C03	EAC 5046	2.J17	EAC 5943	2.C11	EAC 6437	1.C11	EAC 7237	2.C10	EAC 7920	2.E07	EAC 8192	1.G05	EAC 9286	2.E18

JAGUAR XJS RANGE (JAN 1987 ON) — D17 Fiche 4 — NUMERICAL INDEX EAC 9286 TO JLM 322

EAC 9286	2.F02	EBC 1077	2.D18	EBC 2525	2.D18	FB 110171 J	2.J08	FS 108251 J	2.H02	FW 106 T	2.F13	JLM 296	1.O04	JLM 300	2.I02
EAC 9299	2.E12	EBC 1084	2.D18	EBC 2554	2.E05	FG 104 X	1.J03	FS 108251 J	2.J06	FW 106 T	2.F14	JLM 296	1.O06	JLM 300	2.I05
EAC 9319	2.J05	EBC 1084	2.E02	EBC 2556	2.E05	FG 104 X	1.N09	FS 108301 J	1.C03	FW 205 T	1.I06	JLM 296	1.O07	JLM 301	2.M05
EAC 9320	2.J07	EBC 1091	2.I05	EBC 2675	2.C04	FG 104 X	2.E03	FS 108301 J	1.C09	FW 206 T	2.F07	JLM 296	2.C12	JLM 301	2.N17
EAC 9369	1.F03	EBC 1131	1.C11	EBC 2735	1.D02	FG 104 X	2.E06	FS 108301 J	1.C16	FW 206 T	2.F15	JLM 296	2.C15	JLM 301	3.C05
EAC 9382	1.E16	EBC 1147	1.C05	EBC 2755	1.C12	FG 104 X	2.G07	FS 110251 J	1.D08	FW 206 T	3.F03	JLM 296	2.D12	JLM 301	3.C13
EAC 9440	2.C04	EBC 1178	1.H09	EBC 2756	1.C12	FG 104 X	3.D06	FS 110251 J	1.M11	FW 206 T	3.G07	JLM 296	2.F07	JLM 301	3.D13
EAC 9477	2.C07	EBC 1180	2.D18	EBC 2871	1.D13	FG 104 X	3.E05	FS 110251 J	2.H02	FW 206 T	3.G08	JLM 296	2.G10	JLM 302	1.D07
EAC 9508	1.C09	EBC 1181	2.E02	EBC 2925	1.C11	FG 105 X	1.H06	FS 110301 J	2.H02	FW 206 T	3.G09	JLM 296	2.J10	JLM 302	1.M10
EAC 9510	2.J06	EBC 1197	1.E13	EBC 2995	1.G04	FG 105 X	1.J05	FS 110351 J	1.D04	FW 206 T	3.G10	JLM 296	2.J15	JLM 302	1.N05
EAC 9534	1.C12	EBC 1197	1.E14	EBC 3003	2.E08	FG 105 X	1.N07	FS 110351 J	2.H02	FX 106041 J	3.E05	JLM 296	2.K04	JLM 302	2.J10
EAC 9535	1.C12	EBC 1197	1.E15	EBC 3375	2.C03	FG 105 X	1.O06	FT 108255 J	1.C08	GHF 117	2.J15	JLM 296	2.L04	JLM 302	2.L11
EAC 9552	2.C03	EBC 1197	2.E16	FB 106061 J	2.C18	FG 105 X	1.O07	FU 2545	2.D04	GL 604071	3.D06	JLM 296	2.L06	JLM 302	2.L12
EAC 9566	2.E07	EBC 1197	2.F02	FB 106071 M	2.C03	FG 105 X	2.C07	FU 2545 4	2.H10	GS 104161 J	2.E08	JLM 296	2.L11	JLM 302	3.D13
EAC 9567	2.E07	EBC 1198	2.E11	FB 106121 J	2.C04	FG 105 X	2.D17	FW 104 T	1.H06	GT 105161 J	3.C09	JLM 296	2.L12	JLM 302	3.D15
EAC 9568	2.E07	EBC 1200	1.E14	FB 108061 J	1.C07	FG 106 X	1.F09	FW 104 T	1.J05	GT 604071	2.K16	JLM 296	2.N18	JLM 302	3.E04
EAC 9587	2.G06	EBC 1200	1.E15	FB 108061 J	1.C08	FG 106 X	1.K02	FW 104 T	1.M12	J 103 9S	2.C15	JLM 296	2.O02	JLM 302	3.G09
EAC 9588	2.G06	EBC 1202	1.E13	FB 108061 J	1.C10	FN 105041	2.L10	FW 104 T	1.N09	J 205 8S	1.N05	JLM 296	3.D15	JLM 302	3.G10
EAC 9607	1.C09	EBC 1202	1.E15	FB 108061 J	1.D02	FN 105041	2.L13	FW 104 T	2.D05	J 205 8S	1.N06	JLM 296	3.E04	JLM 302	3.H02
EAC 9616	2.G18	EBC 1223	1.C08	FB 108061 J	2.H02	FN 106041 J	2.C03	FW 104 T	2.D06	J 205 8S	1.N08	JLM 296	3.E16	JLM 303	1.F06
EAC 9616	2.I03	EBC 1233	2.C08	FB 108061 J	2.J06	FN 106041 J	2.F06	FW 104 T	2.F06	J 10517 S	1.N10	JLM 296	3.E17	JLM 303	1.J07
EAC 9634	2.E07	EBC 1253	1.C10	FB 108071 J	1.C09	FN 108041 J	3.F10	FW 104 T	2.F12	J 20829 S	2.D06	JLM 296	3.F06	JLM 303	1.K08
EAC 9648	1.I05	EBC 1284	1.C11	FB 108071 J	1.C10	FS 11025 1	1.I05	FW 104 T	2.F13	JLM 113	1.D06	JLM 296	3.H02	JLM 303	1.O08
EAC 9649	1.I05	EBC 1288 20	1.C06	FB 108071 J	1.C11	FS 105121 J	2.L13	FW 104 T	2.F14	JLM 150	2.J11	JLM 296	3.H15	JLM 303	2.E13
EAC 9657	1.C10	EBC 1288 S	1.C06	FB 108071 J	2.H02	FS 105161 J	1.O07	FW 104 T	3.E05	JLM 193	2.L17	JLM 296	3.H18	JLM 303	2.E14
EAC 9675	1.D03	EBC 1292	1.D03	FB 108081 J	2.C10	FS 106101 J	1.C11	FW 105 E	1.N13	JLM 193	2.L18	JLM 296	3.I03	JLM 303	2.E15
EAC 9690	2.C05	EBC 1416	1.D18	FB 108091 J	1.C05	FS 106111 J	1.C17	FW 105 E	1.N15	JLM 215	2.K03	JLM 297	2.G14	JLM 303	2.E18
EAC 9692	1.D13	EBC 1488	2.F04	FB 108091 J	1.C08	FS 106121 J	1.C15	FW 105 E	1.N16	JLM 216	2.K03	JLM 298	1.D07	JLM 303	2.F07
EAC 9693	1.D13	EBC 1548	1.G04	FB 108091 J	1.E12	FS 106121 J	1.G04	FW 105 E	2.C14	JLM 238	2.L17	JLM 298	1.D08	JLM 303	2.F15
EAC 9745	1.C17	EBC 1556	1.G04	FB 108091 J	1.E13	FS 106121 J	3.L10	FW 105 E	2.D17	JLM 238	2.L18	JLM 298	1.F02	JLM 303	2.H14
EAC 9797	2.E07	EBC 1632	2.J08	FB 108091 J	1.E14	FS 106121 MJ	1.K07	FW 105 E	2.F10	JLM 239	2.L17	JLM 298	1.N05	JLM 303	2.I05
EAC 9799	1.G02	EBC 1647	1.D04	FB 108091 J	1.E15	FS 106141 J	2.C04	FW 105 E	3.L02	JLM 239	2.L18	JLM 298	1.N14	JLM 303	2.J07
EAC 9842	2.E07	EBC 1771	1.C11	FB 108091 J	1.E16	FS 106141 J	2.C05	FW 105 T	1.F04	JLM 267	1.I10	JLM 298	1.N17	JLM 303	3.D14
EAC 9844	1.E12	EBC 1789	2.C04	FB 108091 J	2.J06	FS 106141 J	2.C07	FW 105 T	1.F09	JLM 277	2.J02	JLM 298	2.D12	JLM 303	3.G09
EAC 9844	1.E13	EBC 1832	1.G05	FB 108101 J	1.C07	FS 106161 J	1.C04	FW 105 T	1.N04	JLM 277	2.J04	JLM 298	2.F05	JLM 303	3.J03
EAC 9844	1.E14	EBC 1908	1.G02	FB 108101 J	1.C09	FS 106161 J	1.C10	FW 105 T	1.N07	JLM 278	2.J02	JLM 298	2.F07	JLM 303	3.J11
EAC 9846	2.E02	EBC 1939	1.G02	FB 108101 J	1.C11	FS 106161 J	2.C07	FW 105 T	1.N13	JLM 278	2.J04	JLM 298	2.G12	JLM 303	3.K04
EAC 9848	1.D18	EBC 1974	2.E02	FB 108101 J	1.E12	FS 106161 MJ	1.K07	FW 105 T	1.N15	JLM 279	2.J09	JLM 298	2.L04	JLM 303	3.K07
EAC 9855	1.H03	EBC 1979	2.E07	FB 108101 J	1.E13	FS 106201 J	1.C07	FW 105 T	1.N16	JLM 280	2.J09	JLM 298	3.D14	JLM 303	3.K10
EAC 9866	2.E11	EBC 2054	1.D02	FB 108101 J	1.E14	FS 106201 J	1.D02	FW 105 T	2.C13	JLM 281	1.K18	JLM 298	3.D16	JLM 304	1.D08
EAC 9903	1.E12	EBC 2057	1.D02	FB 108101 J	1.E15	FS 106201 J	2.C07	FW 105 T	2.C14	JLM 282	1.K18	JLM 298	3.D18	JLM 304	1.J07
EAC 9903	1.E13	EBC 2084	1.C04	FB 108101 J	1.E16	FS 108121 J	2.H02	FW 105 T	2.D17	JLM 283	1.K18	JLM 298	3.E14	JLM 304	1.K05
EAC 9903	1.E14	EBC 2107	2.C06	FB 108111 J	1.C10	FS 108141 J	1.C17	FW 105 T	2.F05	JLM 284	1.K18	JLM 298	3.F02	JLM 304	1.K06
EAC 9903	1.E15	EBC 2108	2.C06	FB 108111 J	2.C10	FS 108141 J	1.D03	FW 105 T	2.F06	JLM 285	1.K18	JLM 298 T	2.D04	JLM 304	2.H02
EAC 9918	2.C07	EBC 2149	2.J07	FB 108121 J	1.C04	FS 108161 J	1.C04	FW 105 T	2.F10	JLM 286	1.K18	JLM 299	1.E02	JLM 304	2.H13
EAC 9921 A	1.C05	EBC 2195 N	1.H07	FB 108121 J	1.C07	FS 108161 J	1.C09	FW 105 T	2.F11	JLM 287	1.K18	JLM 299	1.F06	JLM 304	2.I02
EAC 9921 B	1.C05	EBC 2196	1.H15	FB 108121 J	2.F06	FS 108161 J	1.E06	FW 105 T	2.F13	JLM 288	1.K18	JLM 299	1.G06	JLM 304	2.I05
EAC 9921 C	1.C05	EBC 2197	1.C12	FB 108131 J	1.C10	FS 108201 J	1.C03	FW 105 T	2.F14	JLM 289	1.K18	JLM 299	1.J07	JLM 304	2.J07
EAC 9921 D	1.C05	EBC 2203	2.E08	FB 108141 J	1.C09	FS 108201 J	1.C10	FW 105 T	2.F15	JLM 290	1.K18	JLM 299	1.K08	JLM 304	3.D13
EAC 9921 E	1.C05	EBC 2209	2.E08	FB 108141 J	1.C15	FS 108201 J	1.C18	FW 105 T	2.O11	JLM 291	1.K18	JLM 299	1.M03	JLM 307	3.F06
EAC 9921 F	1.C05	EBC 2230	2.J17	FB 108181 J	1.C10	FS 108201 J	2.C10	FW 106 E	1.H05	JLM 292	1.K18	JLM 299	1.M05	JLM 307	3.J06
EAC 9921 G	1.C05	EBC 2241	2.D18	FB 108181 J	2.C10	FS 108201 J	2.H02	FW 106 E	1.J04	JLM 293	2.K08	JLM 299	1.M08	JLM 307	3.J16
EAC 9921 H	1.C05	EBC 2241	2.E02	FB 108201 J	1.C07	FS 108251 J	1.C04	FW 106 E	1.N07	JLM 293	2.K09	JLM 299	1.M09	JLM 307	3.K06
EAC 9921 I	1.C05	EBC 2255	1.D10	FB 108201 J	1.C18	FS 108251 J	1.C09	FW 106 E	1.N09	JLM 293	2.K10	JLM 299	1.N06	JLM 307	3.K09
EAC 9961	1.G02	EBC 2344	2.C06	FB 108201 J	1.E03	FS 108251 J	1.C11	FW 106 E	2.F06	JLM 293	2.K16	JLM 299	1.N10	JLM 307	3.K12
EAC 9985	1.G05	EBC 2351	1.D02	FB 108901 J	1.C08	FS 108251 J	1.C15	FW 106 T	2.D06	JLM 295	1.G06	JLM 299	1.N14	JLM 308	2.H10
EAC 9986	2.E04	EBC 2381	1.D02	FB 110071 J	1.G06	FS 108251 J	1.C17	FW 106 T	2.F05	JLM 295	1.K13	JLM 299	2.J07	JLM 308	3.F13
EAC 9987	2.E04	EBC 2479	1.E18	FB 110071 J	2.J08	FS 108251 J	1.G05	FW 106 T	2.F10	JLM 296	1.N18	JLM 299	2.L07	JLM 313	1.K18
EAW 3502	1.N03	EBC 2489	1.E18	FB 110125 J	1.G06	FS 108251 J	1.M11	FW 106 T	2.F11	JLM 296	1.O02	JLM 300	1.K08	JLM 314	1.E10
EBC 1052	2.J06	EBC 2524	2.E02	FB 110171 J	1.G06	FS 108251 J	2.C06	FW 106 T	2.F12	JLM 296	1.O03	JLM 300	1.N04	JLM 322	1.C18

JAGUAR XJS RANGE (JAN 1987 ON) — D18 Fiche 4 — NUMERICAL INDEX JLM 336 TO JLM 1348

JAGUAR XJS RANGE (JAN 1987 ON) — E01 Fiche 4 — NUMERICAL INDEX JLM 1351 TO RA 608255 J

JAGUAR XJS RANGE (JAN 1987 ON) — E02 Fiche 4 — NUMERICAL INDEX RKC 5259 J TO SH 605071 J

JAGUAR XJS RANGE (JAN 1987 ON) — E03 Fiche 4 — NUMERICAL INDEX SH 605071 J TO WW 108001 J

Jaguar Factory Manuals

Workshop Manuals

Model	Part no.
Jaguar 1.5, 2.5, 3.5 1946-48	
XK120/140/150/150S & Mk. 7/8/9 (SC)	
XK120/140/150/150S & Mk. 7/8/9 (HC)	
Mk. 2 (2.4, 3.4, 3.8, 240, 340)	E121/7
Mk. 10 (3.8/4.2 & 420G) (SC)	E136/2
Mk. 10 (3.8/4.2 & 420G) (HC)	E136/2
'S' Type 3.4 & 3.8	E133/3
Auto Transmission 1950-64	E113/6
E-Type 3.8/4.2 Ser. 1 & 2 (SC)	E123/8, E123
E-Type 3.8/4.2 Ser. 1 & 2 (HC)	B/3 & E156/1
E-Type V12 Ser. 3	E165/3

Model	Part no.
420 (SC)	E143/2
420 (HC)	E143/2
XJ6 2.8 & 4.2 Ser. 1	E155/3
XJ6 3.4 & 4.2 Ser. 2	E188/4
XJ12 Ser. 1	E172/1
XJ12 Ser. 2 & DD6 (SC)	E190/4
XJ12 Ser. 2 & DD6 (HC)	E190/4
XJ6 & XJ12 Ser. 3	AKM9006
XJ-S 6 Cylinder 3.6 & 4.0	AKM9063
XJ-S V12 5.3 & 6.0	AKM3455

Jaguar XJ6 (XJ40) 1986-94 Workshop Manual Owners Edition (SC)

Parts Catalogues

Model	Part no.
Mk. 2 (3.4, 3.8 & 340)	J34
'S' Type (3.4 & 3.8)	J35
E-Type 3.8 Ser. 1	J30
E-Type 4.2 Ser. 1	J37
E-Type Ser. 2 GT	IPL5/2
E-Type V12 Ser. 3 Open 2 Seater	RTC9014

Model	Part no.
XJ6 Ser. 1	RTC9016
XJ6 & Daimler Sovereign Ser. 2	RTC9883CA
XJ6 & Daimler Sovereign Ser. 3	RTC9885CF
XJ6 2.9 & 3.6 Litre Saloons 1986-1989	RTC9893CB
XJ-S 3.6 & 5.3 Jan 1987 on	RTC9900CA

Owners' Handbooks

Model	Part no.
XK120	
XK140	E101/2
XK150	E111/2
Mk. 2, 3.4	E116/10
Mk. 2, 3.8	E115/10
E-Type tuning & prep. for competition	
E-Type 3.8 Ser. 1	E122/7
E-Type 4.2, 2+2 Ser. 1	E131/6
E-Type 4.2 Ser. 2	E154/5
E-Type V12 Ser. 3	E160/2
E-Type V12 Ser. 3 (US)	A181/2

Model	Part no.
XJ 3.4 & 4.2 Ser. 2	E200/8
XJ6C Ser. 2	E184/1
XJ5.3 Ser. 2 (fuel injection)	E196/4
XJ12 Ser. 3	AKM4181
XJ12 Ser. 3 (US)	AKM4179
XJ-S (US) pub. '77	AKM3980

Note: SC - soft cover HC - hard cover

From Jaguar dealers or, in case of difficulty, direct from the distributors:

Brooklands Books Ltd., PO Box 146, Cobham, Surrey, KT11 1LG, England
Phone: 01932 865051 Fax 01932 868803
e-mail sales@brooklands-books.com www.brooklands-books.com
Brooklands Books Ltd., 1/81 Darley St., PO Box 199, Mona Vale, NSW 2103, Australia
Phone: 2 9997 8428 Fax: 2 9799 5799
Car Tech, 11605 Kost Dam Road, North Branch, MN 55056, USA
Phone: 800 551 4754 & 651 583 3471 Fax: 651 583 2023

Printed in England and distributed by Brooklands Books Ltd.,
PO Box 146, Cobham, Surrey, KT11 1LG, England
Phone: 01932 865051 Fax: 01932 868803
e-mail info@brooklands-books.com or visit our website www.brooklands-books.com

ISBN 1 85520 4002 B-J85PH Pub. No. RTC9900CA 10/7Z0

Copyright © Jaguar Cars Limited 1989 and Brooklands Books Limited 1994

This book is published by Brooklands Books Limited and based upon text and illustrations protected by copyright and first published by Jaguar Cars Limited in 1989 and may not be reproduced transmitted or copied by any means without the prior written permission of Jaguar Cars Limited and Brooklands Books Limited.

Printed in England and distributed by Brooklands Books Ltd.,
PO Box 146, Cobham, Surrey, KT11 1LG, England
Phone: 01932 865051 Fax: 01932 868803
e-mail info@brooklands-books.com or visit our website www.brooklands-books.com

ISBN 1 85520 4002 B-J85PH Pub. No. RTC9900CA 10/7Z0